Feinendegen · Shreeve · Eckelman · Bahk · Wagner (Eds.): **Molecular Nuclear Medicine**

Springer-Verlag Berlin Heidelberg GmbH

L. E. Feinendegen · W. W. Shreeve
W. C. Eckelman · Y.-W. Bahk · H. N. Wagner Jr.
(Eds.)

Molecular Nuclear Medicine

The Challenge of Genomics and Proteomics
to Clinical Practice

Contributors
C. Anderson · S. H. Audi · Y.-W. Bahk · J. B. Bassingthwaighte
S. R. Bergmann · V. Berlin · R. G. Blasberg · O. C. Boerman · C. Bremer
K. E. Britton · L. H. Bryant Jr. · J. W. M. Bulte · J.-K. Chung · S.-K. Chung
C. A. Combs · F. H. M. Corstens · C. A. Dawson · Y.-S. Ding
B. A. Dwamena · W. C. Eckelman · S. Egert · S. Epstein
L. E. Feinendegen · E. H. Fischer · J. S. Fowler · J. A. Frank · M. Fukunaga
A. H. Gjedde · M. D. Gross · J. Gyuris · T. Isawa · M. de Jong · N. Kley
J. Konishi · E. P. Krenning · G. S. Krenz · T. Kuwert · K.-J. Langen,
R. D. Lele · V. R. Lele · H. Loferer · J. Logan · G. D. Luker · R. R. MacGregor
L. S. Malmud · S. McCallum · T. Misaki · R. D. Neumann · H.-Y. Oei
W. J. G. Oyen · I. G. Panyutin · E. Passarge · R. N. Pierson Jr.
D. A. Piwnica-Worms · H. J. J. M. Rennen · D. L. Roerig · L. Rosenthall
S. Schmidt · M. Schwaiger · B. Shapiro · P. F. Sharp · W. W. Shreeve
J. C. Sisson · J. G. G. Tjuvajev · J.-L. C. Urbain · M.-C. M. Vekemans
N. D. Volkow · H. N. Wagner Jr. · R. L. Wahl · G.-J. Wang · R. Weissleder
A. Welch · T. A. Winters

With 292 Figures, Partly in Color, and 49 Tables

Springer

Ludwig E. Feinendegen, M.D.
Professor Emeritus Nuclear Medicine
Heinrich-Heine-University Düsseldorf
Wannental 45
88131 Lindau
Germany

Walton W. Shreeve, M.D., Ph.D.
Professor Emeritus
Radiology and Medicine
SUNY, Stonybrook
Medical Department
Brookhaven National Laboratory
Upton, NY 11973
USA

William C. Eckelman, Ph.D.
Chief PET Department
National Institutes of Health/
Clinical Center
Bldg. 10, Rm 1C401
10 Center Drive, MSC 1180
Bethesda, MD 20892-1180
USA

Yong-Whee Bahk, M.D., Ph.D.
Professor Emeritus (CUK) and Director
Department of Nuclear Medicine
Sung-Ae General Hospital
Shingil-1-Dong, Youngdungpo-Gu
Seoul 121-200
Korea

Henry N. Wagner Jr., M.D.
Nuclear Medicine
and Radiation Health Sciences
The Johns Hopkins University
School of Hygiene and Public Health
615 North Wolfe Street
Baltimore, MD 21205-2179
USA

Library of Congress Cataloging-in-Publication Data
Molecular nuclear medicine: the challenge of genomics and proteomics to clinical practice
/ L. E. Feinendegen ... [et al.], (eds.). p. cm.
  Includes bibliographical references.
  ISBN 978-3-642-62427-8   ISBN 978-3-642-55539-8 (eBook)
  DOI 10.1007/978-3-642-55539-8
  1. Radioisotope scanning. 2. Genomics. 3. Proteomics, 4. Nuclear
medicine – Technological innovations. I. Feinendegen, Ludwig E.
RC78.7.R4M65 2003   616.07'575–dc21   2003042530

http://www.springer.de

© Springer-Verlag Berlin Heidelberg 2003

Cover-Design: e studio calamar, Pau/Girona
Typesetting: K + V Fotosatz GmbH, Beerfelden

Printed on acid-free paper   21/3150 PF   5 4 3 2 1 0

Preface

This book was conceived in the 1990s when the human genome project began to advance rapidly and much more knowledge of genes and gene function was accumulating. It was evident that the many gene variants, such as multiple polymorphisms due to mutations, mixing of gene pools, environmental pressures, etc, needed to be matched to the present and future findings of laboratory and clinical medicine. Such matching would facilitate better diagnosis and prognosis and, hopefully, therapy. Particularly, a jump-start on early diagnosis would enhance the field of preventive medicine.

Nuclear medicine, with its ability not only to detect subtle biochemical changes in vivo but to quantify their degree of abnormality, ought to be in the forefront of this matching effort. While nuclear medicine had already helped with various overt genetic diseases, a more concerted effort would be needed in the future. To this end the intent of the editors was to gather the contributions of basic scientists in relevant fields of biochemistry, physiology, instrumentation and various nuclear medical subspecialties, all charged with relating to some extent the work in their own field to the significance of the emerging genome.

The editors believe that this goal has been achieved quite satisfactorily. Depending on the subject at hand, some authors place much emphasis on genomic relationships, whereas others make more casual or minor references. Even in the latter cases, readers will appreciate that the technological potentials provide fine opportunities for exploiting the revelations of the genome by the diagnostic abilities within the resources of nuclear medicine. Further goals of the book have been to recognize the large scope of techniques available to nuclear medicine, to offer the latest information on evaluation of risk/benefit ratios of ionizing radiation, and (last, but not least) to remind of the importance of placing pathology and clinical medicine on a solid foundation of understanding of biochemistry and physiology. Even as the book was being written it was becoming apparent that proteomics, the "sister" field to genomics, was going to be a systematic part of biology in its own right. Indeed, for several years, nuclear medicine has been dealing with functions and variations of proteins as membrane receptors, hormones, labeled targeting compounds (antibodies, peptides) and other entities. Such subjects appear throughout the book and, thus, both genomics and proteomics have been included in the sub-title.

The editors deeply appreciate the engagement, cooperation and enthusiasm by all authors of this book. Following the genesis of an initial concept by an international group of some of the editors, many authors regularly met with the editors and discussed the themes and thrust of the book at various societies of nuclear medicine around the world. Many authors also participated in two workshops, which took place in 1997 through efforts by the US Department of Energy, Washington DC, USA, also in conjunction with the Department of Nuclear Medicine at the Clinical Center of the National Institutes of Health in Bethesda, Maryland, USA. These workshops fostered the new challenges by debating on: "Functional Consequences of Gene Expression in Health and Disease" and "Isotope-Based Medical Research in the Post Genome Era". The initial stages of the book were in response to a request from ecomed-Publishers in Landsberg, Germany, whose head, Mr. Harald Heim, persistently and patiently supported the evolution of the book. The editors are indebted to Mr. Heim for his unswerving efforts on their behalf. Special gratitude with applause is due to the Springer Publishing House, which consented to continue the efforts by Mr. Heim. Dr. Ute Heilmann, Mrs. Wilma McHugh and Mrs. Ute Pfaff with their international staff of copyeditors were immensely open, efficient and patient in giving advice and producing the book in closest cooperation with the editors and authors. The editors, and surely authors alike, are very grateful for loyal secretarial help. May the concerted efforts by all participants help in the appreciation of the use of radionuclides and other tracers for in vivo observations of molecular interactions and their consequences and, thus, aid in bridging the gap between basic biomedical research and clinical application for the benefit of patients.

L. E. FEINENDEGEN
W. W. SHREEVE
W. C. ECKELMAN
Y.-W. BAHK
H. N. WAGNER Jr.

Foreword

"The essence of science lies not in discovering facts,
but in discovering new ways of thinking about them."
Lawrence Bragg, The History of Science, 1948

In nearly all parts of the world, in industrialized countries more than in predominantly rural living nations, health care has become so expensive that its maintenance at a socially broad level is an economic burden for individuals as well as health care providers, be they private or governmental. At present, in the United States alone, 1.7 trillion dollars, or 14% of the gross national product (GNP) is spent on health. Although the limit that will be accepted by the public remains uncertain, what is certain is that there will be a shift in emphasis from treatment of overt disease to the prevention of disease.

This foreseen re-orientation will bring new challenges to both basic research and clinical application. It will involve conceptual and technological adaptations especially of biomedical, analytical and diagnostic procedures, also with a new look at therapy. Nuclear medicine is expected to play a major role in clinically serving this development in an interdisciplinary approach. The key will be to better understand the origin, pathogenesis and progress of diseases and change their course in their earliest stages.

Genetics is likely to have an enormous impact on this re-orientation process in the decades ahead. As the individual genome determines the representative phenotype, which appears as a consequence of individual "genomics", identification of phenotypic differences characteristic of a given disease will bring the need of monitoring gene-linked biochemical reactions largely expressing protein functions, the sum of which is currently termed "proteomics."

It is now possible to detect not a few genetic mutations that occurred in the evolution of the human race or because of mutations occurring during the life of an individual. This information may provide estimates of the individual's risk of developing specific diseases. Mutations in human genetic material have occurred at a relatively steady rate over tens of thousands of years. Most have had a deleterious effect on

human life and reproduction, and so are lost. Others persist and are transmitted to one's children and grandchildren. Thus, a person's genome contains both the history of the human race, the specific population from which the person descended on both parental sides and mutations that have occurred over the course of the person's life. A person's genetic profile can characterize mutations reflected in SNPs (single nucleotide polymorphisms) that are responsible for phenotypic diversity, including the risk of subsequent specific diseases. A person's genome can be thought of as both a blueprint and subsequent templates upon which all the material of the specific human being is constructed throughout the person's life.

What is needed now is a new definition of disease with a richer and broader definition of what is normal in the light of advances in genomics, together with advances in molecular imaging, including positron emission tomography (PET), single photon emission computed tomography (SPECT) and magnification scintigraphy, such as described in Chapters 6 and 26. An important question is whether and when deviations from the "average" genome with their immediate and sometimes delayed occurring downstream subtle changes in molecular and cellular functions should be considered a possible disease or mere variant as in anatomy and physiology. If so, what would be the categorical criteria? Should persons without manifestations of disease but with a significant risk of disease be considered abnormal, with all the medical and social baggage, which this would entail? To what degree should the genomic profile lead to life style changes, or avoidance of environmental or occupational hazards, control of emotional stress, or dietary changes? Which potential manifestations of disease should be monitored? Which blood chemistry tests are preferable? Which should be imaging procedures? Which molecular targets, such as receptors, should be monitored? Many chapters in this book provide examples of the role of receptors in pathophysiology.

For example, early recognition of predisposition as reflected in regional brain chemistry can motivate an individual to alter harmful life styles in order to improve the quality and length of life and reduce societal burdens. Another example is the assessment of intestinal absorption of calcium to detect a predisposition to osteoporosis. Increasing calcium intake could have dramatic beneficial effects in persons at high risk.

The tracer technology of nuclear medicine will undoubtedly be widely demanded and play a major role. As in to-days practice of nuclear medicine and in facing future challenges, the projected benefits of nuclear medicine procedures need to be scaled against risks that are currently and widely portrayed as an exaggerated fear of low-dose radiation. To place the risk-benefit ratio into proper perspective, radiation biology at low doses should be understood in helping to ameliorate unwarranted alarm, as described in Chapter 33.

Huntington's Disease

An example of the increasing role of genetics in medicine is Huntington's disease (HD), which results from genetically programmed degeneration of neurons in specific regions of the brain. This disease passes from parent to child through a mutation of a normal gene. Each child who has this mutated gene has a 50/50 chance of subsequently developing the disease in later life. Pre-symptomatic testing is now commercially available for detection of this mutated gene.

How much do we really want to know about what lies ahead of us? Never before has this question been so important, as more and more information becomes available about our own genome, and what it entails in estimating the risks of disease that may lie ahead. The time of onset and relative progression of many diseases has been found to correlate with genetic mutations that vary from person to person. Genetic tests will increasingly be able to detect those persons who carry specifically mutated genes. Unfortunately, for many genetically caused diseases, such as Huntington's disease, there is currently no way to stop or reverse the disease, an important factor in making the decision of whether to take the test. In 1983, Huntington's disease was found to be related to a mutation of an autosomal dominant gene, localized on the fourth autosomal chromosome. This discovery

led to a pre-symptomatic test for Huntington's disease using DNA linkage analysis. Because it relied on tracing the inheritance of markers linked to the gene rather than the gene itself, the test required the analysis of DNA samples from multiple family members and was 95% accurate at best. As more and more markers for the Huntington's disease gene were identified, the test became more accurate. In 1993, the gene itself had been isolated. At one end of the gene, a pattern of three DNA bases (CAG), or nucleotides, repeats itself in all cases. In normal individuals, this tri-nucleotide, or triplet, repeat occurs between 11 and 29 times. In people with Huntington's disease, the repeat occurs over and over again, from 40 times to more than 80. The new test analyzes DNA directly for the presence of the Huntington's disease mutation, obviating the need for collection and analysis of samples from multiple family members. Many at risk individuals indicate a desire to know their gene carrier status, but far fewer actually undergo testing. When confronted with the opportunity for testing, the majority find that the emotional toll or risks to confidentiality outweigh the benefits of learning their gene profile. Of course, in the case where there is an effective treatment for a specific disease, especially if it is effective in the earliest stages of the disease, most people would agree to take the test.

Ron Cleveland, who described his experience on the Internet, was at great personal risk of developing Huntington's disease in the light of his family history. He wrote, "It has been nine days since I received my (genetic) test results for Huntington's disease. The result was 'Negative.' I should be thankful that I have been spared a life filled with the horror of Huntington's disease. Can I now walk away from my life in the shadow of this disease? I won the coin toss! How we act, think, react with others, we are defined by our wealth, our profession, our beauty – both physical and ethereal, and when our lives are affected by a disease as hideous as HD, then yes we are defined by HD."

He continued: "I have been told many times, never let HD define your life. I disagree. We need a definition for our life, no matter what that definition is; it dictates who we are."

Genes and Health Care

Genetic testing will be an important part of health care in the future, and people will be characterized as being at risk for a whole host of diseases. This raises the important question of the nosology of disease, that is, the system of classification of diseases. In genetic terms, this means that phenotypes can be related to (caused by) a specific genetic pattern. For some diseases, e.g. sickle cell anemia, a single gene mutation can have multiple phenotypes, affecting different tissues or organs, a phenomenon called "pleotropism." A second principle of genetics is that mutations of different genes can result in the same diseases, a phenomenon called "genetic heterogeneity". Thus, it can generally be said that the relationship between genotypes and phenotypes is complex and variable.

The challenge, addressed in this book, is to approach the understanding of genomics, proteomics and phenotyping by linking molecular processes to genes and related functions at the molecular, cellular and tissue level, as they occur at specific sites or in the whole body of patients. These molecular processes are found not only to involve the basic genetic material, such as nuclear and mitochondrial DNA and RNA, but also can be characterized in all body fluids, cells, tissues, and organs. It is now possible to do so with the new technologies of molecular imaging in health and disease. Regional biochemistry can be examined at progressively increasing spatial and temporal resolution. This book, which arguably could become a classic in view of its timeliness, bridges the gap between genotype and phenotype at the level of molecular processes. It is intended to be of value to basic scientists, clinicians, and clinical investigators.

A New Definition of Disease

The idea that a disease is a specific entity has been around since the time of Hippocrates. Faced with an individual patient with problems, the physician considers the patient both as a unique individual and as an example of other patients with similar "diseases". Making the "diagnosis" of one or more specific diseases provides the basis of prognosis and treatment of the patient's illness. Thus, the process of putting the patient in a diagnostic category ("making the diagnosis") is the first step in caring for the sick. The patient's illness is transformed from his problems as an individual to the diagnosis of a specific disease. Advances in nuclear medicine and, more recently in genetics, raises the question of whether the current approach to defining disease should be modified. Is it time for a change in how diseases should be defined, that is, in the nosology of disease? Today, health statistics are followed in every developed and most of developing countries. Conforming to World Health Organization practices, the National Centers for Health Statistics use an International Statistical Classification of Diseases.

A disease is characterized into one of three digit categories and four-digit codes, which define diseases on the basis of their location in the body and the cause of the disease. This statistics-oriented definition of disease began over three centuries ago. Fundamental to the system is designation of the part of the body affected. Next is the classification of the nature of the process, for example, is it infectious or is it due to a chemical toxin? If possible, the cause is stated, for example, trauma. The internationally accepted system being based on anatomical site, cause, and circumstances of onset is used primarily for statistical analysis, storage and retrieval of information and tabulation of data. However, the practicing physician is concerned primarily with the manifestations of disease that can be treated successfully to help solve the patient's problems. Will the time eventually come to revise the system based on the explosion of information regarding genotypes and phenotypes? Instead of primarily searching for the "ultimate cause" of a disease, abnormal molecular processes in networks of functional interactions can be identified as being the critical manifestations of a disease, and provide the basis for its diagnosis and therapy.

In 1878, Claude Bernard, the father of scientific medicine, wrote: "Vital phenomena possess their rigorously determined physico-chemical conditions. When one or more of these physico-chemical processes are abnormal in specific regions of the body, their elucidation provides "manifestations" of the patient's disease, which becomes the basis of prognosis and treatment." Claude Bernard's greatest contribution to scientific medicine is the idea that: "The circulating body fluids, the blood serum and the intraorganic fluids all constitute the internal environment, which preserves the necessary relations of exchange and equilibrium with the external cosmic environment." In the process of evolution, organic molecules separated themselves from their surrounding environment by forming membranes that contained these molecules, thus becoming the first cells. Groups of cells eventually became tissues and organs, as progressively more complex living organisms evolved. System failures came to be called "diseases", and a collection of laws defined healthy and diseased organisms in the practice of medicine. Bernard rejected the idea of looking for ultimate causes of disease, and searched for their "immediate causes, so as to be able to regulate their manifestations." This approach describes the role of nuclear medicine in the diagnostic process.

Molecular Nuclear Medicine

Observing and affecting molecular interactions as such in the living body by the radiotracer technique has become known as "molecular nuclear medicine." This approach is to view regional function in terms of biochemistry in the light of "normal" regional physiological and biochemical processes, defining these normal ranges by statistical criteria. The patient's individual problems are defined as deviations from "normal", that is, quantifiable deviations from the quantitative specific biochemical processes in persons without known abnormalities of these specific biochemical processes. These processes nearly always occur in one or several reaction circuits or loops. Their measurements can be used to characterize a specific person. The totality of these multiple and interrelated processes affecting the molecular, cellular and tissue levels varies from person to person. When a person becomes ill, the disease can be characterized by abnormalities in one or more of these regional biochemical processes. The nature of the disease is defined in terms of the nature of the individual sick person. People can often have similar manifestations of disease, but no two patients will be the same.

Until recently, radiologists being anatomically oriented in principle were often unimpressed by nuclear medicine images. Indeed, these images whether taken by PET, SPECT or magnification scintigraphy have even today limited spatial resolution. However, the visualization of biochemical processes in vivo such as in PET studies with ^{18}F-2-deoxyglucose (FDG) in many different organs and in cancerous tissues (described in many chapters in this book) brought radiologists and other physicians to take a new look at nuclear medicine as a whole new way to look at disease.

In 1968, I presented an equation that, using the principles of nuclear medicine, defined diseases as:

$$D_i = \Sigma m_i w_i$$

where D = disease; m = manifestation; w = weight. Subscript i means individual specific diseases, manifestations and weights.

The weight of each manifestation is determined chiefly by its specificity, that is, how unique its involvement is to the patient's problem. One cannot use every molecular process in the body as the basis of defining disease, or there would be millions of "diseases." It is necessary to categorize groups of molecular manifestations of disease based on the value of the categorization in prognosis and the planning and monitoring of treatment.

If the definition of disease encompasses a sum of specific groups of regional molecular manifestations, we face a new problem: We must begin to develop a new schema of disease classification on the basis of the regional molecular processes in patients with similar problems and similar manifestations of illness. We must develop a diagnostic classification schema based on grouping of abnormal regional biochemical manifestations. We must relate these groups of molecular manifestations to prognosis and treatment. This requires identification of the manifestations as they are expressed by loops of functional interactions of metabolic substrates amongst each other and with their respective molecular targets be they transporters, receptors, enzymes or other proteins.

Victor A. McKusick, a pioneer in clinical genetics has written: "Hardly more than a century ago such phenomena as jaundice, dropsy, and anemia were still classified and treated as 'diseases'". Subsequently, diseases were classified according to distinct histopathological findings. As outlined in this book, regional molecular manifestations are likely to play an important role in the future definitions of disease.

The authors of this book describe the application of nuclear medicine techniques to analyze cellular biochemistry and its relationship to disease processes expressed in tissue and organ dysfunction, for diagnostic and therapeutic purposes. They discuss both the analysis of the pathophysiology of various diseases and their potential link to individual genes and genomic factors. The book gives examples of how various tracer techniques now being used in nuclear medicine have advanced this process in the past, and how they are likely to be expanded in the future. Disease is increasingly defined as a deficiency or imbalance of regional molecular processes, brought about by the effects of genomic, proteomic, metabolic and environmental factors. Increasingly, a large number of polymorphisms present in critical genes are being shown to be linked to the expression of somatic differences at the molecular clinical levels.

The tools of nuclear medicine have shown their enormous potential to observe in vivo complex system functions at the various levels of biological organizations: the molecular, cellular, tissue and organ, as well as whole body level. Following the great success in qualitatively imaging of metabolism at various body sites in vivo with the help of refined tomographic and high-resolution techniques it is obviously courageous and demanding now to focus on quantifying manifestations of functional genomics and proteomics in situ. In this way, molecular nuclear medicine will play an increasing role in defining disease, in helping preventive medicine and in the planning and monitoring of treatment with a probably substantial cost saving effect.

HENRY N. WAGNER Jr.
(With contributions by all editors, gratefully acknowledged)

Table of Contents

Part IV Tumors

Part V Special Topics

List of Contributors

ANDERSON, CAROLYN, Ph.D.
Molecular Imaging Center
Mallinckrodt Institute of Radiology
and Department of Molecular Biology
and Pharmacology
Washington University Medical School
St. Louis, MO 63110
USA

AUDI, SAID H., Ph.D.
Department of Biomedical Engineering
Marquette University
P.O. Box 1881
Milwaukee, WI 53201-1881
USA

BAHK, YONG-WHEE, M.D., Ph.D.
Professor Emeritus (CUK) and Director
Department of Nuclear Medicine
Sung-Ae General Hospital
Shingil-1-Dong, Youngdungpo-Gu
Seoul 121-200
Korea

BASSINGTHWAIGHTE, JAMES B., M.D., Ph.D.
Bioengineering
University of Washington
Box 357962
Seattle, WA 98195-7962
USA

BERGMANN, STEVEN R., M.D., Ph.D.
Division of Cardiology, PH 10-405
College of Physicians and Surgeons
Columbia University
630 West 168 Street
New York, NY 10032
USA

BERLIN, VIVIAN, Ph.D.
GPC Biotech Inc.
610 Lincoln Street
Waltham, MA 02541
USA

BLASBERG, RONALD G., M.D.
Department of Neurology
Department of Radiology
Memorial Sloan-Kettering Cancer Center
1275 York Avenue, Box 52
New York, NY 10021
USA

BOERMAN, OTTO C., Ph.D.
Department of Nuclear Medicine
University Medical Center Nijmegen
P.O. Box 9101
6500 HB Nijmegen
The Netherlands

BREMER, CHRISTOPH, M.D.
Institute for Clinical Radiology
Universitätsklinikum Münster
Albert-Schweitzer-Straße 33
48129 Münster
Germany

BRITTON, KEITH E., M.D.
Department of Nuclear Medicine
The Royal Hospital of St. Bartholomew
West Smithfield
London EC1A, 7BE
UK

BRYANT, L. HENRY Jr., Ph.D.
Laboratory of Diagnostic Radiology Research
National Institutes of Health
Bldg. 10, Room B1N256
10 Center Drive, MSC 1074
Bethesda, MD 20892
USA

BULTE, JEFF W. M., Ph.D.
Department of Radiology
Johns Hopkins University School
of Medicine
217 Traylor Bldg., 720 Rutland Ave.
Baltimore, MD 21205-2195
USA

CHUNG, JUNE-KEY, M.D.
Department of Nuclear Medicine
Seoul National University Hospital
Seoul
Korea

CHUNG, SOO-KYO, M.D.
Department of Nuclear Medicine
St. Mary's Hospital
Catholic University Medical School
Seoul
Korea

COMBS, CHRISTIAN A., Ph.D.
The Light Microscopy Imaging Facility
National Heart Lung and Blood Institute
National Institutes of Health,
Bld. 10, Rm. B1D-416
Bethesda, Maryland 20892-1061
USA

CORSTENS, FRANS H. M., M.D. Ph.D.
Department of Nuclear Medicine
University Medical Center Nijmegen
P.O. Box 9101
6500 HB Nijmegen
The Netherlands

DAWSON, CHRISTOPHER A., Ph.D.
Research Service 151
Zablocki VA Medical Center
Medical College of Wisconsin
5000 W. National Ave.
Milwaukee, WI 53295-1000
USA

DING, YU-SHIN, Ph.D.
Chemistry Department
Brookhaven National Laboratory
Upton, NY 11973
USA

DWAMENA, BEN A., M.B., Ch.B.
Department of Radiology
Division of Nuclear Medicine
University of Michigan Medical School
Nuclear Medicine Service
B1 G412 University Hospital
Ann Arbor, Michigan 48109-0028
USA

ECKELMAN, WILLIAM C., Ph.D.
Chief PET Department
National Institutes of Health
Bldg. 10, Room 1C401
10 Center Drive, MSC 1180
Bethesda, MD 20892-1180
USA

EGERT, SILVIA, Ph.D.
Clinic of Nuclear Medicine and Policlinic
rechts der Isar
Technische Universität München
Ismaninger Straße 22
81657 München
Germany

EPSTEIN, SHILPI, M.D.
College of Physicians and Surgeons
Columbia University
Babies Hospital North 517
3959 Broadway
New York, NY 10032
USA

FEINENDEGEN, LUDWIG E., M.D.
Professor Emeritus Nuclear Medicine
Heinrich-Heine-University Düsseldorf
Wannental 45
88131 Lindau
Germany

FISCHER, EDMOND H., Ph.D.
Nobel Laureate
Department of Biochemistry
University of Washington, S.J.-70
Seattle, WA 98195-7350
USA

FOWLER, JOANNA S., Ph.D.
Chemistry Department
Brookhaven National Laboratory
Upton, NY 11973
USA

FRANK, JOSEPH A., M.D.
Laboratory of Diagnostic Radiology Research
National Institutes of Health
Bldg. 10, Room B1N256
10 Center Drive, MSC 1074
Bethesda, MD 20892
USA

FUKUNAGA, MASAO, M.D.
Department of Nuclear Medicine
Kawasaki Medical School
577 Matsushima
Kurashiki, Okayama 701-0192
Japan

GJEDDE, ALBERT H., M.D., Ph.D.
The Positron Emission
Tomography Center
Aarhus University Hospital
Norrebrogade 44
8000-Aarhus C
Denmark

GROSS, MILTON D., M.D.
Department of Radiology
Division of Nuclear Medicine
University of Michigan
Medical Center and Nuclear Medicine Service
Veterans Affairs Medical Center
1500 East Medical Center Drive
Ann Arbor, Michigan 48109-0028
USA

GYURIS, JENO, Ph.D.
GPC Biotech AG
610 Lincoln Street
Waltham MA 02541
USA

ISAWA, TOYOHARU, M.D., Ph.D.
Seirei Yokohama Hospital
215, Iwai-cho
Hodogaya-ku
Yokohama 240-8521
Japan

DE JONG, MARION, Ph.D.
Department of Nuclear Medicine
Erasmus University Medical Center
Rotterdam
P.O. Box 2040
3000 CA Rotterdam
The Netherlands

KLEY, NIKOLAI, Ph.D.
GPC Biotech Inc.
610 Lincoln Street
Waltham, MA 02541
USA

KONISHI, JUNJI, M.D.
Department of Nuclear Medicine
and Diagnostic Imaging
Graduate School of Medicine
Kyoto University
Sakyo-ku, Kyoto 606-8507
Japan

KRENNING, ERIC P., M.D., Ph.D.
Department of Nuclear Medicine
Erasmus University Medical Center
Rotterdam
P.O. Box 2040
3000 CA Rotterdam
The Netherlands

KRENZ, GARY S., Ph.D.
Department of Mathematics,
Statistics and Computer Science
Marquette University
P.O. Box 1881
Milwaukee, WI 53201-1881
USA

KUWERT, TORSTEN, M.D.
Department of Nuclear Medicine
Friedrich-Alexander-University
Erlangen-Nürnberg
Krankenhausstraße 12
91054 Erlangen
Germany

LANGEN, KARL-JOSEF, M.D.
Institute of Medicine
Research Center Jülich
Wilhelm-Johnen-Straße
52425 Jülich
Germany

LELE, RAMCHANDRA D., M.D.
Department of Nuclear Medicine
Jaslok Hospital and Research Centre
Lilavati Hospital and Research Centre
102 Buena-Vista, Gen. Bhosale Marg
Mumbai 400021
India

LELE, VIKRAM R., M.D.
Department of Nuclear Medicine
Jaslok Hospital and Research Center
15, G. Deshmukh Marg
Mumbai 400026
India

LOFERER, HANNES, Ph.D.
GPC Biotech AG
Fraunhoferstraße 20
82152 Martinsried
Germany

LOGAN, JANE, Ph.D.
Chemistry Department
Brookhaven National Laboratory
Upton, NY 11973
USA

LUKER, GARY D., M.D.
Molecular Imaging Center
Mallinckrodt Institute of Radiology
and Department of Molecular Biology
and Pharmacology
Washington University Medical School
St. Louis, MO 63110
USA

MACGREGOR, ROBERT R., B.S.
Chemistry Department
Brookhaven National Laboratory
Upton, New York, 11973
USA

MALMUD, LEON S., M.D.
Health Science Center
Temple University Hospital
3401 N Broad Street
Philadelphia, PA 19140
USA

MCCALLUM, STEPHEN, M.D.
Department of Bio-Medical Physics
and Bio-Engineering
Forsterhill
Aberdeen AB25 2ZD
Scotland

MISAKI, TAKASHI, M.D.
Tenri Hospital
Radioisotope Center
200 Mishimacho
Tenri city, Nara 632-8552
Japan

NEUMANN, RONALD D., M.D.
Department of Nuclear Medicine
National Institutes of Health
Bld. 10, Room 1C-401
10 Center Drive, MSC 1180
Bethesda, MD 20892-1180
USA

OEI, HONG-YOE, M.D., Ph.D.
Department of Nuclear Medicine
Erasmus University Medical Center
Rotterdam
P.O. Box 2040
3000 CA Rotterdam
The Netherlands

OYEN, WIM J. G., M.D.
Department of Nuclear Medicine
University Medical Center Nijmegen
P.O. Box 9101
6500 HB Nijmegen
The Netherlands

PANYUTIN, IGOR G., Ph.D.
Department of Nuclear Medicine
National Institutes of Health
Bld. 10, Room 1C-401
10 Center Drive, MSC 1180
Bethesda, MD 20892-1180
USA

PASSARGE, EBERHARD, M.D.
Institute of Human Genetics
University Hospital Essen
Hufelandstraße 52
45122 Essen
Germany

PIERSON, RICHARD N. JR., M.D., FACP
Nutrition Research Center
St. Luke's/Roosevelt Hospital Center
111 Amsterdam Avenue
New York, NY 10025
USA

PIWNICA-WORMS, DAVID A., M.D., Ph.D.
Molecular Imaging Center
Mallinckrodt Institute of Radiology
Washington University Medical School
510 S. Kingshighway Blvd., Box 8225
St. Louis, MO 63110
USA

RENNEN, HUUB J.J.M., MSc.
Department of Nuclear Medicine
University Medical Center Nijmegen
P.O. Box 9101
6500 HB Nijmegen
The Netherlands

ROERIG, DAVID L., Ph.D.
Research Service
Zablocki VA Medical Center
5000 W. National Ave.
Milwaukee, WI 53295-1000
USA

ROSENTHALL, LEONARD, M.D.
Department of Radiology
The Montreal General Hospital
1650 Cedar Ave.
Montreal, H3G 1A4
Canada

SCHMIDT, STEFAN, Ph.D.
GPC Biotech AG
Fraunhoferstraße 20
82152 Martinsried
Germany

SCHWAIGER, MARKUS, M.D.
Clinic of Nuclear Medicine and Policlinic
rechts der Isar
Technische Universität München
Ismaninger Straße 22
81657 München
Germany

SHAPIRO, BRAHM, M.B., Ch.B., Ph.D.
Department of Radiology
Division of Nuclear Medicine
University of Michigan
Medical Center and Nuclear Medicine Service
Veterans Affairs Medical Center
1500 East Medical Center Drive
Ann Arbor, Michigan 48109-0028
USA

SHARP, PETER F., Ph.D.
Department of Bio-Medical Physics
and Bio-Engineering
University of Aberdeen
Grampian Hospitals NHS Trust
Foresterhill
Aberdeen AB25 2ZD
Scotland

SHREEVE, WALTON W., M.D., Ph.D.
Professor Emeritus Radiology and Medicine
SUNY, Stonybrook
Medical Department
Brookhaven National Laboratory
Upton, NY 11973
USA

SISSON, J. C., M.D.
Department of Nuclear Medicine
BIG 412 University Hospital
University of Michigan Medical Center
1500 E. Medical Center Dr.
Ann Arbor, MI 48109-0028
USA

TJUVAJEV, JURI G. GELOVANI
Department of Neurology
Sloan-Kettering Institute
Memorial Sloan-Kettering Cancer Center
1275 York Avenue
New York, NY 10021-6007
USA

URBAIN, JEAN-LUC C., M.D., Ph.D.
Division of Molecular
and Functional Imaging
Cleveland Clinic Foundation
9500 Euclid Avenue
Cleveland, OH 44195
USA

VEKEMANS, MARIE-CHRISTIANE M., M.D.
Centre Hospitalier du Grand Hornu
Toute de Mons 63
7300 Hornu
Belgium

VOLKOW, NORA D., M.D.
National Institute of Drug Abuse
National Institutes of Health
6001 Executive Boulevard
Room 5274, MSC 9581
Bethesda, MD 20892-9581
USA

WAGNER, HENRY N. Jr., M.D.
Nuclear Medicine and Radiation Health Sciences
The Johns Hopkins University
School of Hygiene and Public Health
615 North Wolfe Street
Baltimore, MD 21205-2179
USA

WAHL, RICHARD L., M.D.
Nuclear Medicine Division
Johns Hopkins Medical Institute
Baltimore, MD 21305-2179
USA

WANG, GENE-JACK, M.D.
Medical Department
Brookhaven National Laboratory
Upton, New York 11973
USA

WEISSLEDER, RALPH, M.D., Ph.D.
731 Flagship Wharf
8th Street
Charleston, MA 02129
USA

WELCH, ANDY, M.D.
John Mallard Scottish PET Centre
Department of Bio-Medical Physics
and Bio-Engineering
Foresterhill
Aberdeen AB25 2ZD
Scotland

WINTERS, THOMAS A., Ph.D.
Department of Nuclear Medicine
National Institutes of Health
Bld. 10, Room 1C-401
10 Center Drive, MSC 1180
Bethesda, MD 20892-1180
USA

Part I   Primary Topics

CHAPTER 1

Measurements of Biochemical Reactions In Vivo

1

Ludwig E. Feinendegen, Walton W. Shreeve, Henry N. Wagner Jr.

Contents

1.1
Introduction

Immense progress has been made over the last century in medical diagnosis, especially in the techniques of radiological imaging. The evolution began in the mid-1890s by the epochal discoveries of Röntgen and Becquerel (Blaufox 1996). The initial limitation of radiology to observing externally radio-opaque structures such as bones in the living body rapidly evolved into functional imaging of organs such as kidney and liver and the circulatory system using radio-opaque contrast agents. With the advent of computed tomography soft tissues also became visible in high-resolution images, augmenting the spectrum of function studies of organ segments. Yet, the molecular-atomic level of biological organization in living tissues became observable only with the introduction of the radionuclide tracer technique by Hevesy (1913). Further studies with radioactive and stable isotopes of common elements in organic compounds (C, H, O, N, S, P) in the 1930s and 1940s revealed the surprising extent to which both "functional" and "structural" compounds in the body are ceaselessly formed and broken down, even with no change in overall amount, form, or function (Schoenheimer 1946; Hevesy 1948). Today, the specialty of nuclear medicine is indispensable in clinical diagnosis and therapy.

The current literature on nuclear medicine shows the wide usefulness of radionuclides as tracers of constantly occurring "life events" in the body, encompassing gross substrate turnover, blood perfusion, particular metabolic changes, and certain biochemical reactions at large or at defined sites. In fact, the tracer techniques have opened totally new insights into normal and pathologically altered cell functions in the living body, its organs and tissue segments. The technical advances that were needed to reach the present state of art required interdisciplinary efforts often on a relatively large scale. The powerful service to clinical medicine afforded by nuclear medicine is seen in: the introduction of useful, mainly gamma-emitting, radionuclides and of radiopharmaceutical chemistry for proper substrate labeling; the development of counting and imaging devices optimal for the task at hand; the burgeoning growth of data analysis with the help of models applicable to arrive at meaningful diagnostic information and the correlated radiation dosimetry and biology (Wagner et al. 1995).

Observing in vivo accumulation or kinetics of radionuclides in the whole body or certain regions of interest (ROI) demands that radiation is registered preferentially by dynamic modes. The instruments are single- or multiple-counting probe devices, planar gamma cameras, single-photon-emission tomographs (SPECT) or positron-emission tomographs (PET), as discussed in Chap. 6. The suitable radionuclides must emit gamma radiations or be positron emitters. In vitro analyses of radioactivity in samples of blood or breath and/or body excreta such as urine or saliva at times complement data obtained by imaging or counting, or are by themselves useful in evaluating biochemistry. Such in vitro assays use mainly beta-emitting radionuclides to be analyzed, for instance, by scintillation counting, or by autoradiography of isolated cells, tissue sections, or chromatograms. Special in vitro methods of great importance are the radioimmunoassays, particularly to clinical medicine. For both in vivo imaging and in vitro assays, stable isotopes also must be considered for use with magnetic resonance imaging (MRI) and spectroscopy (MRS) and for metabolic studies with labeled substrates (see Chaps. 12–14). Sometimes in conjunction with MRI, MRS can observe and quantify certain kinds and numbers of biochemical reactions in vivo. In vivo and in vitro techniques with stable isotopes may be limited by physical constraints imposed by the signal-to-target-atom ratio. In principle, the radionuclide tracer method allows the best overall observation of any substrate undisturbed in its system environment by imaging of defined sites in the living body, as well as by in vitro assays, as long as radiopharmaceutical chemistry allows the substrate to be labeled appropriately. Indeed, choice, production, testing, and supply of tracer-labeled pharmaceuticals suitable as indicators appear to be essential for the success of nuclear medicine in assessing biochemical reactions in the intact living body (see Chap. 5).

Over the past few decades, interest has intensified in observing biochemical reactions, not only qualitatively but also quantitatively, as they occur in vivo. The goal has become not only to localize the biochemical events to certain regions in the body but also to express them in a manner that is conventionally expected from in vitro laboratory methods. In vivo quantification considers the controlling networks of biochemical reactions as they respond to fluctuating inputs of nutrients and various environmental factors, which then also affect gene expression. The biochemical networks operate in patterns of mutual responses for the purpose of defense against, or adaptation to, challenges posed to the organism endogenously or from the environment. Defensive or adaptive responses of the organism occur at its molecular and cellular levels of organization and are principally triggered by cascades of molecular signals of different kinds not often easily understood from observations on isolated cells and tissue samples. A ready example is found in the operation of the immune system. The approach to quantitatively measure specific biochemical reactions as they occur in vivo brings new diagnostic information. It also opens new avenues to apply radionuclides as radiation sources for treating a variety of diseases at the level of single cells and molecules.

Quantification of single and/or multiple biochemical reactions is even more desirable in view of the nearly fully sequenced human genome; indeed, this development poses particular challenges to clinical medicine. Without doubt, gene expression patterns as they become increasingly available, for instance by using DNA array chips, are by themselves often insufficient for meaningful diagnostic revelation; they must be seen in the context of the whole biological system within its signaling network. Links between the genotype and phenotype need to be established in vivo at the level of molecular organization. This demands a close cooperation, even fusion, between molecular and cell biology, on the one hand, and the field of imaging techniques including radiopharmaceutical chemistry, on the other.

In the simplest case of one gene coding for one protein, the activity of one enzyme may depend upon the nature and integrity of one gene. However, most genes carry codes for various proteins with different functions. The diversities of genotypes and phenotypes are based on varying polymorphisms in individual genes, on a still largely elusive interplay between genes, and on regulatory mechanisms acting between gene products. Controls to be considered also are "epigenetic" phenomena, which involve changes in nuclear histones such as acetylation, methylation, phosphorylation, and appear, in addition to DNA-binding transcription factors, to govern the expression of particular genes. Post-translational alterations of proteins such as by glycation, methylation, and phosphorylation add to governing phenotypical diversity, as they may cause or be consequences of disease. Many clinical symptoms and findings require explanations at both the genome and proteome level.

Two examples of genetically determined or strongly influenced diseases are cystic fibrosis and diabetes mellitus. Cystic fibrosis will appear in about 1 of 2,400 newborn white children and in about 1 of 17,000 babies in the black population. This disease appears to be the consequence of changes in one known gene on chromosome 6 coding for the chloride channel protein. This gene has at least 350 known single nucleotide changes, or mutations (of which only a few, perhaps 7 or 8, seem to affect the clinical expression in terms of extent or location of pathology, be it in intestinal organs or lungs), mainly involving nearly all exocrine glands. Appropriate choices of nuclear medical tests may help to classify the disease (see Chaps. 2, 12). Type II diabetes has a much more complex genetic etiology, with at least 9 genetic allelic types of the insulin gene and 40 or more of the insulin receptor genes – and probably many more genes are also involved in causing this type of diabetes, yet to be discovered (see Chaps. 2, 14, 24). The ubiquity of insulin receptors in several organs suggests the possibility of genetically caused differences among them and the need for definition of multiple disease parameters, of which many may be evaluated by nuclear medicine procedures.

Most functional interactions between genes are little understood. To make these crucial reactions part of advanced diagnostic goals for eventual medical application, observations must include the intact signaling networks of living tissues in vivo. Any in vitro measurement on tissues, obtained for instance by biopsy, is a priori hampered by the fact that the physiological signaling within the complex system is practically excluded.

Thus, it appears timely to link in vivo biochemical reactions and reaction loops to genes. This may begin with a known pattern of expression of a defined set of genes in order to search for complementary downstream metabolic reactions. On the other hand, measured biochemical reactions, cell responses, and even tissue functions and local perfusion, may be the starting point, expressing a defined phenotype for the analysis of one or a set of known and unknown genes, an approach to what might be called "reverse genetics" (Wagner et al. 1995). In either approach, interrelationships between genes and metabolism are seen in the context of system cooperation under homeostatic control within physiologically functional networks or under pathologically operating conditions. Moreover, such systems may be studied under conditions of an intentional disturbance. The latter may then help in defining the tolerance of the system to stress.

The radionuclide and radiopharmaceutical application and measurement techniques have matured to an astonishing degree of precision and accuracy so that they will eventually offer in vivo assessment of what is generally called "functional genomics and proteomics" in clinical medicine. Here, even relatively minute changes in reaction kinetics, or phenotypically defined physiological responses, need to be monitored in vivo so that individual changes of a biochemical reaction rate or of the balance between functionally related reaction rates become informative. In order to reach that goal, though, a number of factors that complicate quantitative measurements require attention.

This chapter aims at giving a condensed overview and introduction to various potentially useful approaches to quantitatively measure biochemical reactions in vivo in such a way that the data may eventually be linked to gene expression, secondary control mechanisms, and individual physiological parameters. The emphasis is on summarizing relevant advances to assess biochemical reactions in vivo, so that an understanding may be gained of gene-orchestrated biochemical and physiological functions in complex adaptive systems. It is clear that the attempt is far from being comprehensive and the task of understanding larger complex system functions appears overwhelming. Yet, the time has come to try to do it.

1.2
Substrate/Ligand and the Corresponding Target Molecule

For dealing principally with the various technical approaches to in vivo measurements of biochemical reactions, it is necessary to dwell briefly on the fundamental relationship between substrate/ligand and the corresponding target molecule. Any biochemical reaction brings molecules into close contact with each other. The interaction of a given substrate with its target enzyme or receptor depends on specific molecular configurations, so that both comply with the demands for intermolecular bonds to initiate a biochemical reaction. The ubiquitous device of "lock-and-key" as the fundamental fine biochemical structure for interaction of compounds in metabolism, cell signaling, etc., was first articulated by Emil Fisher in 1890. Later, in 1952, this model was extended by Daniel Koshland to include the concept of a "molded fit" between compounds, e.g., a ligand or substrate and its target or receptor protein (Stryer 1995). Although the occurrence of either rigid "lock-and-key" or more flexible "molded fit" in any particular instance may be infrequently recognized, such concepts could be important considerations in nuclear medicine tactics and interpretation.

The molecular configuration of the target or receptor protein such as enzymes and cell surface receptors determines the accommodation of the labeled substrate or ligand. Thus, binding of substrate or ligand depends on the ultimate third-order structure of the target compound, which in turn may vary according to certain polymorphisms of the gene or genes that direct the structure. This may express different phenotypes that govern binding of the labeled compound. The concept applies importantly as well to drugs as ligands and to the individuality of response to drugs. Adaptive changes in the molecular configuration of the target protein come through various additional molecular interactions with the target and may, for instance, involve reaction products in feedback loops affecting enzyme-catalyzed reactions, or involve small molecular effectors, such Ca, Na, K, Mg, nitrous oxide (NO), and reactive oxygen species (ROS). These "signaling" molecules and ions may also be observed by applying the tracer technique. The ensuing controls affect the rate of a reaction with the primary substrate in various categories of biochemical reactions, be they receptor-ligand interactions, transport functions, or enzyme-catalyzed reactions. In addition, corresponding up- or down-regulation of a reaction may be brought about by altered numbers of available functional target molecules, such as receptors at cell membranes.

Moreover, slight changes in substrate configuration, which may occur for instance through radionuclide labeling, may significantly alter the rate of interaction between substrate and target, and subsequently influence the biochemical reaction. Intuitively, the "molded fit" model might permit binding of a tracer substrate to the appropriate target more variably different from the natural substrate than would be accommodated by a rigid "lock-and-key" case. This could allow for greater latitude of competition between the labeled substrate and its natural counterpart. The degree to which a labeled and often thereby unnatural analogue of a substrate fits the target mold could be approximately proportional to its binding affinity.

The interaction between substrate and target molecule may cause the substrate to remain bound to its target for a comparatively prolonged period of time; such is often the case in ligand-receptor interactions. The latter are essential again for activating biochemical reactions through changes of molecular target structures that are linked within the cell downstream from the receptor (see Chap. 4). A substrate may be bound to its target in order to be transported, for instance, across a cell membrane using a molecular channel system. Such active transport is common for most metabolic substrates. A substrate that binds to its target enzyme usually experiences a change in structure, either by molecular rearrangement, cleavage in some form, or by addition of a molecular subgroup. Some substrates undergo a sequence of such reactions. Enzyme-catalyzed reactions determine the vector of metabolic activity in the sense of substrate synthesis or catabolism.

For a substrate to meet its target at the reaction site, it may be synthesized on site, or diffuse to it, or be transported to it from the circulating blood through the interstitial space into the cell, or it may come from a neighbor cell or interstitial space. Within the cell, the substrate must pass through cellular compartments in order to reach its reaction site. These factors complicate in vivo measurements, as discussed later in this chapter.

1.3
Goal of Measurement

Enzyme-substrate interactions were recognized from early times as underlying a successive series of chemical changes in a substrate, for instance, from ingested food. This process called "metabolism" broadly encompasses either molecular breakdown, catabolism, or synthesis, anabolism. Metabolic changes can also be closely associated with transport across a cell membrane. Enzymes and transporting proteins appear to act in concert with other enzymes toward a physiological "goal," but as individual proteins may be separate from other related enzymes in a system and may have individual fates and integrity. Radionuclide tracers have helped to establish types and rates of biochemical reactions in functional circuits.

Those who measure biochemical reactions in the body may resort to selecting one or more specific labeled substrates for in vivo imaging or counting, or for in vitro analysis of tracer in exhaled breath, body fluids, or excreta. In the former approach, one explores at a selected ROI the rate and degree or amount of accumulation, or rate of disappearance of a labeled substrate either alone or in conjunction with an internal standard, as explained below. The latter approach determines the rate of appearance of one or more labeled reaction products in the breath, urine, or saliva of the patient consequent to metabolism of the labeled substrate through a known enzyme.

Ideally, the final outcome of a measurement of a biochemical reaction in vivo should be comparable to the data obtained by in vitro testing by conventional analytical procedures. In other words, the key parameters that conventionally describe a biochemical reaction are also the optimal goals for nuclear medicine approaches. A less ambitious task, but presently widely accepted in the practice of medicine, is simply the proof that a reaction occurs in the body or in a given ROI. This may result in a semi-quantitative statement regarding the degree of some increments by which a labeled substrate or ligand binds to its target molecule or undergoes a structural change perhaps associated with the release of a labeled reaction product. But for eventual assessment of consequences, for instance, of polymorphism-related functional changes of a target protein in a biochemical reaction, the measurement must be sensitive, specific, and quantitative.

In order to plan the measurement to be as quantitative as possible, the investigator should at the outset be familiar at least with first-order kinetics of biochemical reactions. These are well explained in biochemistry textbooks (Stryer 1995; Rodwell 1996). Thus, the Michaelis-Menten treatment of a biochemical reaction also appears to be a requirement for data measured in vivo. It leads to an understanding of reaction rates, their constants, and the velocity of a reaction as it depends on substrate concentration in relation to the available amount of enzyme (and similarly of receptor or transporter). In short, with:

E = total amount of enzyme available

S = substrate concentration

ES = amount of enzyme bound to substrate

k_1 = rate constant of substrate binding to enzyme

k_2 = rate constant of separation of substrate from enzyme

k_3 = rate constant of enzyme catalyzed reaction of substrate

then:

$$(E - [ES]) \cdot S \cdot k_1 = [ES] \cdot k_2 + [ES] \cdot k_3 \tag{1.1}$$

or by rearrangement:

$$\frac{E \cdot S - [ES] \cdot S}{[ES]} = \frac{k_2 + k_3}{k_1} \tag{1.2}$$

or

$$\frac{E \cdot S}{[ES]} - S = \frac{k_2 + k_3}{k_1} = K_m \tag{1.3}$$

K_m is the Michaelis-Menten constant.

If all E is involved in the reaction, the reaction has its maximal velocity V_{max}:

$$E \cdot k_1 = V_{max} \tag{1.4}$$

If only $(E-[ES])$ or $[ES]$ alone is involved in the reaction, the reaction velocity is v; let only $[ES]$ be involved:

$$[ES] \cdot k_1 = v \tag{1.5}$$

Inserting Eqs. 1.4 and 1.5 into Eq. 1.3:

$$\frac{V_{max} \cdot S}{k_1 \cdot v / k_1} - S = K_m \tag{1.6}$$

or

$$\frac{V_{max} \cdot S}{v} = K_m + S \tag{1.7}$$

and

$$v = \frac{V_{max} \cdot S}{K_m + S} \tag{1.8}$$

This then states that the velocity v in a first-order reaction is proportional to substrate concentration S and to $V_{max}/(K_m + S)$, i.e., with increasing S, the value of v asymptotically reaches a plateau at V_{max} when S becomes near equal to $(K_m + S)$. The expression $V_{max}/(K_m + S)$ in Eq. 1.8, for a given value of S, is a rate constant, which declines with increasing substrate concentration.

It is obvious that K_m relates to a value of S. In order to determine the K_m in terms of S, Eq. 1.8 is rearranged:

$$K_m = \frac{V_{max}}{v} \cdot S - S$$

It is clear that in the case of v being one half the value of V_{max}:

$$K_m = 2S' - S' = S' \tag{1.9}$$

Thus, Km is equal to the substrate concentration S' at v being one half of V_{max}.

The values of K_m and V_{max} can be easily obtained graphically by the reciprocal expression of Eq. 1.8:

$$\frac{1}{v} = \frac{K_m + S}{V_{max} \cdot S}$$

or:

$$\frac{1}{v} = \frac{1}{V_{max}} + \frac{K_m}{V_{max}} \cdot \frac{1}{S} \tag{1.10}$$

Equation 1.10 is the Lineweaver-Burke function. It gives V_{max} and K_m from plotting the reciprocal of the measured velocities v of a reaction against the reciprocal of different substrate concentrations S, at which the measurements were made. The resulting straight line intercepts the ordinate at $1/V_{max}$ and has the slope $K_{m/}V_{max}$.

It follows that the optimal goal of measurement of a biochemical reaction in vivo, if applicable, should be the determination of rate constants at different substrate concentrations, and subsequently K_m and V_{max}.

Rate constants may apply to different types of substrate transitions that comply with first-order reaction kinetics. Such transitions also pertain to compartmental analyses. The corresponding rate constants k individually conform to the differential equation $dC(t)/dt = k \cdot C(t)$, with $dC(t)$ giving the incremental change of radioactivity at time t during the time increment dt.

Indeed, as will be shown in more detail below, measurements on tracer movement in and out of a ROI can readily provide the inputs needed for compartmental analyses. Thus, in a simple one-tissue compartment model, the transit of an indicator from the circulating blood into the free substrate pool in tissue, and its return from there to the blood can be described by the clearance of tracer from the blood and the rate constant of tracer return to blood. An example of the application of a one-tissue compartment model describes the transport of glucose across the blood-brain barrier. The labeled glucose analogue 3-methyl-glucose is here the indicator. It is not metabolized but returns from the free glucose pool in the extravascular tissue space, mainly in glial cells, back into the circulating blood. The inflow of this indicator from the capillary flow system across the blood-brain barrier into the glial cells is given by K_1, which expresses the clearance rate. The indicator outflow back from the glial cells into the circulating blood adheres to the rate constant k_2. The corresponding expression is:

$$dC_T(t)/dt = K_1 \cdot C_p(t) - k_2 \cdot C_T(t)$$

with $C_p(t)$ being the radioactivity of indicator in the circulating blood plasma in counts per unit volume at time t and $C_T(t)$ the corresponding radioactivity in the extravascular tissue space at time t. It is to be noted that the ratio of amounts of indicator to glucose is different in blood and extravascular tissue space; in contrast to its indicator, glucose continues to be metabolized. This is important, because the value of k_2 here also depends on the amount of free glucose within the extravascular compartment, which varies with the extent of glucose metabolism.

A two-tissue compartment model applies when the indicator after entrance in to the tissue binds to a target molecule from where it again may separate intact or after a catabolic or anabolic alteration. In this instance, the clearance rate of the indicator from the blood into the tissue is conventionally designated k_1, the rate of return from tissue to blood is k_2, the acceptance rate of the labeled compound by its target molecule is k_3, and the rate of dissociation from this bond is k_4; in addition, the rate of biochemical generation of a labeled product that would transfer into the circulating blood may be k_5 (see Chaps. 7, 8).

Widely used in clinical practice today is the measurement of phosphorylation of glucose by using as indicator the labeled glucose analogue 5-deoxyglucose (Sokoloff et al. 1977). This glucose analogue when phosphorylated fails to be accepted into the next physiological metabolic reaction. It accumulates at the site of the primary phosphorylation. Thus, 5-deoxyglucose when labeled with ^{18}F allows in vivo quantification of glucose phosphorylation by measuring (with PET) the irreversible tracer binding, be it in brain, heart, or malignant tumor (Gallagher et al. 1977; Phelps et al. 1979; Reivich et al. 1979; Delbeke et al. 2002). In this instance, the rate constants k_4 and

k_5 in the two-tissue compartment model are practically irrelevant. The applicable equations relate to the two-tissue compartments: free tissue pool C_{TF} and the substrate binding site C_{TM}:

1) $dC_{TF}(t)/dt = K_1 \cdot C_P(t) - k_2 \cdot C_{TF}(t) - k_3 \cdot C_{TF}(t)$
2) $dC_{TM}(t)/dt = k_3 \cdot C_{TF}(t)$

with $C_{TF}(t)$ being the radioactivity of indicator in the free tissue pool in the ROI at time t; $C_{TM}(t)$ the radioactivity of indicator bonded at the reaction site in the ROI at time t, and $C_P(t)$ the radioactivity of indicator in arterial blood plasma at time t.

In a similar fashion, the practically irreversible binding of a labeled ligand to its receptor is readily accessible to compartmental analysis. More details appear later in this chapter and especially in Chap. 7.

From the above it is also clear that the measurement of certain reactions may well rely only on one or more rate constants without need to quantify substrate concentration. The advantage of restricting observation to the rate constants is the easy availability of measuring devices such as SPECT, that do not readily allow quantification of labeled substrate as PET does. Thus, biochemical measurements are open to anyone who has the conventional nuclear medical imaging devices that permit the dynamic mode of data registration.

1.4
Factors that Complicate Measurements of Biochemical Reactions

The quantitative measurement of a biochemical reaction in vivo using the tracer technique requires attention to various factors that potentially complicate the interpretation of measured data. Of course, the amount of indicator must be in the tracer range, so as not to preclude the observation of the wanted reaction. Moreover, external measurements of tracer only relay the fate of the tracer, such as a gamma-emitting radionuclide, and not necessarily that of the compound labeled as indicator to which the tracer is attached. Only if it is certain that the tracer remains bound to the labeled substance during the time of observation, does the measurement represent the indicator. On the other hand, separation of tracer from the indicator may testify to the occurrence of a biochemical reaction.

In the context of ROI imaging or counting, various factors need consideration that influence, for example, the path of the indicator in the body on its way to its molecular target within the ROI. The indicator is administered to the body usually by intravenous injection. Alternative routes are intraarterial, intracavitary, and interstitial injections, as well as ingestion and inhalation. Prior to reaching its target, the indicator is usually transported via the blood circulation into the target tissue, where local blood flow is an essential factor in indicator distribution. Within the organ, transport from the blood vessels into the interstitial space and diffusion through interstitial space, even if usually a fast process, may find barriers to the indicator's reaching the target cells for external binding. After having been bound, the indicator must be internalized to reach its intracellular target molecule. Finally, conditions within the target-containing cells may affect the final destination of the indicator. Similarly, measurements of tracer in body excreta demand attention to the kinetics of the labeled reaction product until it appears in the excreta. One needs to account for the various influences on measured indicator and tracer kinetics in order to quantitatively evaluate the kinetics of the wanted reaction.

An illustrative example is the imaging of tracer accumulation in a given ROI, under conditions of stability of tracer on the indicator. The kinetics of indicator inflow is supposed to signal indicator transport or binding to its receptor or enzyme, and the outflow rate should express either indicator release or a biochemical reaction with liberation of tracer. In any case, clearly, the chosen indicator must be appropriately labeled. Serial measurements must use an optimal image device such as a scintillation camera in the planar or tomographic mode and should begin right after the injection to register incremental tracer inflow, accumulation, and outflow in a chosen ROI. This dynamic imaging may be omitted in favor of a still image if it is properly timed after injection and if the indicator can be quantified on site, such as by PET (as explained in Chap. 7). This delayed imaging may help to minimize the influence of local blood supply on the data. The dynamic data collection may be transformed into a smooth time-activity curve. However, the rate of substrate or ligand binding, or of its release, or of its degradation in the ROI, can only be assessed when the measurements differentiate between indicator and its labeled degradation products that recirculate into the ROI. Moreover, local blood perfusion, transport, and diffusion of the labeled indicator and/or its labeled degradation product need to be accounted for so as to arrive at meaningful data.

It is obvious that three principal challenges must be solved: one pertains to the choice of tracer and to its positioning on the indicator; the second involves the data acquisition in such a form that supply of indicator to the reaction site is quantified and, if needed, indicator and its labeled reaction product are registered separately. The third challenge entails application of an appropriate model for data analysis – one that corresponds as closely as possible to the true life-events in the ROI, and if needed, considers confounding factors including those from local perfusion, substrate transport, and diffusion.

Any attempt at quantitative measurement of a biochemical reaction demands acknowledgement of the following:

- Imaging or counting of a labeled substrate in the stationary or dynamic mode only registers the relative amount or rate of uptake and/or release of the radionuclide tracer, and not necessarily of the labeled substrate within the chosen ROI in the body.
- For the labeled substrate to serve as indicator, the radionuclide must remain bound to the labeled substrate during the time of observation at least until the wanted reaction occurs.
- Local blood supply influences the rate of transport of the indicator to, and tracer removal from, the site of substrate reaction. A diminished local perfusion, for example, may cause a reduced local uptake of the indicator and/or concomitantly release of tracer and may mimic a reduced reaction rate.
- Local transport and diffusion from the capillary system into the extracellular space and cells determine the rate of supply of labeled substrate to, and removal of tracer from, the site of substrate reaction and, thus, may significantly influence the measurement of a reaction rate.
- The indicator within the cell may escape reaction and be transported or diffuse back into the extracellular space and capillary vessels. This loss of tracer from the reaction site may mimic a reaction rate.
- The tracer alone may recirculate into the ROI where the reaction occurs. Depending on the relative amount of recirculating tracer, it must be separately measured for final data assessment.

To overcome these principal challenges of directly observing biochemical reactions within a selected ROI, the following are required (Feinendegen et al. 1981):

1. Knowledge of the type of biochemical reaction to be studied in terms of substrate/molecular target interaction and biochemical fate of subsequent reaction products

2. Proper choice of radionuclide for labeling the substrate to be measured and for the type of imaging or counting instrument; for instance, a gamma-emitting radionuclide for conventional scintigraphic imaging (planar or tomographic mode), and a positron-emitting radionuclide for positron-emission-tomography (PET)

3. Optimal and usually stable positioning of the chosen radionuclide on the substrate in such a way that it becomes a useful indicator for tracing the wanted biochemical reaction

4. Measurement techniques (preferably in the dynamic mode) which generate time-activity curves for analyzing kinetics of labeled substrate, and, if needed, a second, separate indicator or tracer as internal standard

5. Data analysis with the help of models in compliance with local tissue physiology and biochemistry (compartment models usually); and

6. Proof of applicability of the method using collateral experiments

The above requirements 1, 2, and 3 are primary challenges for the expert in radiopharmaceutical chemistry. Requirements 4, 5, and 6 mainly address the nuclear medical specialist. Both specialists need to start by sharing attention to requirement 1. Usually more than one measurement is needed to describe the wanted reaction kinetics in vivo. Various applications of "dynamic multiple parameter analysis" have proven their effectiveness. The principal approaches will be briefly summarized in the sections below, and their modifications and applications will be described and discussed in various other chapters of this book.

1.5
Methods and Models
for Measuring Biochemical Reactions

This section gives various methods and models with some examples that aim at observing biochemical reactions in vivo both in a given ROI by imaging mainly in the dynamic mode, and, secondly, by whole body observation through dynamic in vitro analyses of tracer radionuclides in exhaled breath, body fluids, and excreta, with or without concomitant measurements in the peripheral blood. Of course, the measurements must be quantitative, highly sensitive, and specific in order to relate to fine-tuned control mechanisms in the signaling networks of living tissue. Two principle avenues answer to these demands. One of them focuses on observing a chosen single compartment, for instance, in a region of interest (ROI), and the other concentrates on observing a labeled substrate in two or more compartments, which usually consist of circulating arterial blood and the one or more tissue compartments it supplies with substrate where the reaction takes place. In the single compartment approach, a dual-parameter analysis is nearly always indispensable. The observation of a labeled substrate in two or more compartments simultaneously already provides the multiple-parameter analysis, without which a biochemical reaction in vivo can hardly be assessed properly.

1.5.1
Observations of Labeled Substrate(s) in a Compartment (ROI)

1.5.1.1
Single Parameter Analysis: Single Radionuclide on One Substrate

1.5.1.1.1
In Vivo Counting and ROI Imaging

In most instances, this approach does not allow one to measure a single biochemical reaction for reasons given above regarding the complicating factors. However, some measurements of tracer outflow from a given ROI may very well describe a reaction. This applies, for instance, to measuring lipid turnover in the myocardium using ^{11}C-palmitic acid or other suitably labeled fatty acids, provided that noise from recirculating non-specifically bound tracer in the ROI is minimal and that loss of indicator by simple back-transport/diffusion from the myocardial cells into the blood circulation has been excluded (Schelbert et al. 1986; Feinendegen 2000). Another example is the analysis of the effectiveness of membrane channel function in multi-drug resistance, for example, in breast cancer (see Chap. 27). The rate of washout of tracer from the cancer tissue here correlates directly with the channel function that is responsible for the degree of drug resistance. Because of easy pitfalls and misinterpretations, this mode of measurement re-

mains an exception yet an important one for imaging biochemical reactions.

1.5.1.1.2
In Vitro Counting Related to Regional or Whole Body Metabolism

In vitro counting of a tracer, for example in the patient's exhaled breath, at various times after administration of a labeled substrate may signal the reaction rate as it occurs at certain sites or anywhere in the body. A case in point is the study of intestinal absorption of labeled substrate such as ^{14}C-lactose given orally and measuring the appearance of ^{14}C-CO_2 in exhaled breath (Sasaki 1995). This serves to detect and define the common problem of "lactase deficiency" (see Chap. 14). Synthesis of hepatic proteins, e.g., albumin and/or fibrinogen, can be evaluated with carbon- or hydrogen-isotope-labeled amino acids in such conditions as cirrhosis or acute-phase response to inflammation (see Chap. 14). Total body changes in glucose utilization or production in relevant diseases, notably diabetes, have been studied with ^{14}C- or ^3H-labeled glucose under various conditions of glucose load with repetitive plasma and/or breath sampling for analysis of turnover, as discussed in Chap. 14. A relatively simple technique compares the amount of any ^{14}C-labeled precursor of glucose in gluconeogenesis that is converted to blood glucose against the amount converted to ^{14}C-CO_2 in breath and blood (De Meutter and Shreeve 1963). Such an approach might be adapted to clinical application for detection of early predisposition to diabetes mellitus. Many of such studies suggest increased glucose turnover in diabetes, which is consonant with other evidence using various tracers, such as ^{14}C, ^{13}C, ^3H, or ^2H, pointing to increased hepatic glucose production as a predominant effect of impaired insulin function. There is also evidence for an insulin effect on the genetic expression of key enzymes that switch intermediate 3-carbon substrates between oxidative and gluconeogenic fates (Taylor 1995).

With current techniques, some of these questions could perhaps be further explored in vivo with ^{11}C-glucose using PET, provided that the short half-life of ^{11}C allows for such observations. More futuristic is the possibility of analyzing the localization and biochemical transformation of administered ^{13}C-enriched glucose by MRI and MRS in vivo and/or in vitro over a certain period of time.

1.5.1.2
Dual Parameter Analysis:
Two Radionuclides in Different Positions
on One Substrate

1.5.1.2.1
In Vivo Counting and ROI Imaging

A rather simple approach to overcome many of the complicating factors in the in vivo measurement of biochemical reactions in a given ROI uses a double-labeled substrate. This technique is applicable in case a biochemical reaction leads to cleavage of the substrate into two differently labeled products. Double labeling with two different tracers that only separate upon substrate cleavage has been used in animal and human studies. If one of the labeled products remains at the reaction site and the other rapidly leaves the site, this approach also allows one to observe the site of the reaction. Thus, the enzyme ribonuclease was double-labeled with ^{131}I and ^{51}Cr and used in mice and rats to determine in tissue sections the rate and site of accumulation and degradation of the enzyme in vivo (Schultze et al. 1964). Autoradiography showed that the enzyme had rapidly accumulated in the proximal tubules of the kidneys where it was degraded within an hour. This was indicated by the loss of radioiodine whereas the chromium remained on site by entering new bonds. Thus, the change in isotopic ratio in the proximal tubular cells indicated that the enzyme had been degraded at that site.

This method was chosen in the early 1970s to image the degradation of insulin directly in vivo at various sites in the human body (Feinendegen and Ritzl 1971; Ritzl et al. 1974). In preparing this measurement, the insulin was double labeled with ^{131}I and ^{51}Cr in such a way that it retained its biological function. Both radionuclides become free upon insulin degradation in vitro and in vivo. Animal studies showed the free ^{131}I to quickly leave the site of degradation via the circulating blood and to enter the iodine pool of the body, supplying iodine to the thyroid gland. However, ^{51}Cr entered new bonds at the site of insulin degradation and thus remained behind (Ritzl and Feinendegen 1971). The special advantage of this type of measurement is the fact that the two different tracers remain bound to the substrate on its way through circulating blood, its transport and/or diffusion, until the reaction occurs; the externally measured quotient of the two tracers on the substrate remains constant until substrate degradation. This

eliminates a number of complicating factors, as previously discussed. The rate of change of the tracer quotient upon insulin degradation is the signal to be measured in the observed ROI. In fact, the tracer leaving the reaction site is the crucial signal and the other one remaining on site serves as an internal standard. A correction for recirculating radioactive iodine, if needed, is relatively easy; it requires measurement of peripheral blood and/or of quotient changes in an ROI where insulin degradation does not occur.

Dynamic imaging of the whole body begins immediately after intravenous injection of the double-labeled insulin. The change in tracer quotient shows the insulin to be degraded almost exclusively in the liver of patients with normal glucose metabolism, giving halftimes of about 40 min. The rates of degradation in the liver of patients with various forms of diabetes mellitus has been found to differ substantially from one another and from that of non-diabetic individuals. In a limited study, juveniles with type I diabetes had a significantly enhanced rate of insulin degradation, with half times ranging around 20–30 min, whereas in a second type of non-insulin dependent diabetes with a high insulin level a grossly reduced degradation rate was recorded, with half times over one hour (Ritzl et al. 1974). Such measurements may help in differentiating types and pathogenesis of diabetes mellitus. Moreover, calculation of the rate of synthesis of circulating insulin may result from measured turnover data, as long as substrate concentration at a non-degrading site is constant at a given homoeostatic equilibrium. This approach may also find application in quantitatively and locally assessing degradation and synthesis rate in vivo of other peptides and proteins.

A similar way of using two radionuclides on one substrate in different positions resulted in following the fate of isolated DNA for the first time in a living mammal (Friedrich et al. 1972; Meyers and Feinendegen 1975a, 1975b, 1976). DNA was simultaneously labeled with ^{3}H-thymidine and $5-^{125}$I-2'deoxyuridine. The latter is an analogue of thymidine but less readily incorporated into mammalian DNA in vivo and in many cells in vitro (Meyers and Feinendegen 1975). Following intravenous injection of the double-labeled DNA, the quotient of the two tracers is insensitive to local blood circulation, transport and/or diffusion. Upon DNA hydrolysis, both labeled pyrimidines are released and enter the re-incorporation pathway in DNA-synthesizing cells in the body. However, whereas

³H-thymidine is readily accepted into DNA synthesis, 5-¹²⁵I-2′deoxyuridine is discriminated against, so that after re-incorporation the quotient of the tracers in vivo changes in favor of ³H-thymidine in tissue and the reverse holds for the quotient in the circulating blood. Consequently, the change in the tracer quotient in favor of ³H-thymidine in any given tissue signals DNA hydrolysis. No change of quotient in the recipient cells, on the other hand, means that no hydrolysis has occurred and labeled DNA has been trapped as a whole (Feinendegen et al. 1973; Meyers and Feinendegen 1975, 1976). The data showed the probability of intercellular DNA transfer in lymph nodes in the examined mice.

In these studies, the labeled ³H-thymidine served as internal standard and the disappearance of 5-¹²⁵I-2′deoxyuridine against ³H-thymidine in tissue signaled reutilization of thymidine in favor of its analogue following DNA hydrolysis to the nucleoside level. By using thymidine or a metabolically equivalent analogue labeled, for instance, with positron emitting ¹¹C or ¹⁸F, together with 5-¹²³I-2′deoxyuridine for DNA double labeling, comparative studies on the fate of injected DNA may succeed in humans.

The dual-tracer-on-one-substrate technique in conjunction with dynamic imaging leads to the following mathematical representation. With:

$C[A/B]T(t) =$ the quotient of radioactivity from the two tracers, with A representing the signal and B the internal standard in the tissue region T at time t, and

$k =$ the rate constant of the reaction

$$dC[A/B]_\mathrm{T}(t)/dt = -k \cdot C[A/B]_\mathrm{T}(t)$$

and after integration

$$C[A/B]_\mathrm{T}(t) = -k \cdot \int C[A/B]_\mathrm{T} dt$$

The graphic display of this expression shows the slope, $-k$, to quantitatively describe the rate constant of the reaction in the observed tissue.

1.5.1.2.2
In Vitro Counting Related to Regional or Whole Body Metabolism

Following oral or intravascular administration of substrates labeled with two different tracers, for example ¹⁴C and ³H in two different positions of glucose, sev-

eral analyses are possible. After ingestion, the conversion of 1-³H-glucose to ³H-H₂O represents the rate of absorption and initial metabolism of glucose; this provides a control against which to measure the slower rate of ¹⁴C-CO₂ formation from 1-¹⁴C-glucose; the rate difference indicates the extent of glucose diversion to organic products such as fat in obese subjects (Shreeve et al. 1971). Another example is the use of double-labeled (²H and ¹⁸O) water injected intravenously. The difference in rates of decline of ²H in the peripheral blood signals H₂O turnover, whereas the decline of ¹⁸O indicates H₂O turnover plus CO₂ production and excretion in body water. The difference in the rates of decline allows calculation of energy expenditure in the total body (Ritz and Coward 1995), as discussed in more detail in Chap. (12). An equivalent approach sequentially measures labeled catabolites of substrates labeled with one tracer in two different positions, for example, ¹⁴C in position 1 or 6 of glucose (Kallie et al. 1968), or ¹⁴C in position 1 or 2 of glycine (Nyhan 1984; see Chap. 14). In these examples, the ratio of rates of appearance of the labeled catabolite CO₂ signals the relative rates of particular metabolic pathways involving particular reactions and can sensitively detect pathological deviations.

1.5.1.3
Dual Parameter Analysis: Two Different Labeled Substrates

1.5.1.3.1
In Vivo Counting and ROI Imaging

1.5.1.3.1.1
Subtraction Technique

In this category of methods, a labeled compound that as background noise disturbs the diagnostic signal is subtracted from the gross data in order to reveal the fate of the indicator in a biochemical reaction. A useful illustration of this approach gives the in vivo analysis of lipid metabolism in the myocardium with labeled fatty acids (Feinendegen 2000). The experience with double-labeled insulin, as referred to above, helped in the 1970s to develop techniques for measuring the metabolism of fatty acids in vivo in the myocardium (Freundlieb et al. 1978, 1980; Machulla et al. 1978; Feinendegen et al. 1981).

Fatty acids are essential sources of energy by way of beta-oxidation in the mitochondria of myocardium and muscle in general. Physiologically, fatty acids may be stored in lipids for retrieval into energy-delivering metabolism on demand. In attempting to relate the measured data obtained by dynamic planar scintigraphy to the in vivo fate of labeled fatty acids in various segments of the myocardium, a major problem has been the exclusion of non-specific signals from the scintigraphic image.

Fatty acids are long hydrocarbon chains with a carboxyl group on one end and a methyl group on the other. They can be labeled in different positions. Thus, radioactive carbon may be placed within the carboxyl group and radioactive iodine such as ^{123}I may be placed along the hydrocarbon chain wherever there is a double bond between two successive carbon atoms; or ^{123}I can replace the terminal methyl group, for instance, on the 16 or 17 C-position. The latter thus becomes ω-^{123}I-heptadecanoic acid. Labeled in various ways, fatty acids may retain the ability to at least partly participate in myocardial metabolism. No metabolic interference arises from labeling with radioactive carbon in the carboxyl group. When radioactive iodine replaces the terminal methyl-group, as in ω-^{123}I-heptadecanoic acid, the rate of indicator transfer into the myocardial lipid pool is reduced compared to physiological fatty acid. Most of the accumulated indicator in the myocardium rather rapidly transfers to beta-oxidation with quick liberation of free iodine tracer into the peripheral circulation. After having become incorporated into the lipid pool, ω-^{123}I-heptadecanoic acid has similar kinetics as the physiologically behaving carbon-labeled fatty acid and thus one can trace fatty acid transfer from the lipid pool into the beta-oxidation pathway (Feinendegen 1993). The advantage of terminally labeling with radioactive iodine is the use of conventional gamma cameras in planar or tomographic mode for analyzing in vivo fatty acid uptake and metabolism, especially the rate of lipid turnover. On the other hand, labeling with positron-emitting carbon requires the more expensive and labor-intensive PET, which allows observation of the unaltered fatty acid kinetics and quantification of substrate in tissue.

Depending on the type and labeling of fatty acid, only a certain fraction of the accumulating tracer becomes incorporated into the myocardial lipid pool (Feinendegen 1993, 2000). Degradation of fatty acids by beta-oxidation in the mitochondria is faster than the rate of fatty acid supply from the lipid pool for degradation. Upon full degradation, the free tracers, whether carbon or terminally bound iodine, quickly leave the degradation site via the circulating blood.

The rate of tracer disappearance from the ROI thus mirrors the rate of entrance of the indicator into the degradation pathway. However, contrary to the tracer carbon, the free iodine tracer rapidly reenters the ROI with the free iodine pool in the circulating blood. The catabolic iodine tracer thus adds to the signals from labeled fatty acid in the myocardium and obscures the fatty acid degradation signal in the dynamic mode of imaging.

The interfering signals from the pool of free iodine in blood and tissue have been separately counted following an intravenous injection of free radioactive iodine (Freundlieb et al. 1980; Feinendegen et al. 1981). The separate signals from the iodine pool were then used to compute the rate of release of the iodine tracer in the ROI as the signal of transfer of the labeled fatty acid from the lipid pool into rapid degradation in the mitochondria. In diseased heart muscle, tracer release may include fatty acid back diffusion or transport into the blood circulation. If needed, these two pathways can be separately observed, as described below in the discussion of the use of two labeled isomerically related substrates.

Thus with:

$C_{TM}(t)$ = radioactivity of labeled fatty acid in the myocardial lipids at time t, and

k = rate constant of loss of tracer from the myocardium, indicating physiological transfer of indicator from the lipid pool into mitochondrial beta-oxidation, followed by rapid washout of free iodine tracer into the circulating blood

$$dC_{TM}(t)/dt = -k \cdot C_{TM}$$

and after integration

$$C_{TM}(t) = -k \cdot \int C_{TM} dt$$

As stated above, one method of correcting for signal noise from iodine tracer in the recirculation pathway requires a second injection of free radioactive iodine such as ^{123}I-NaI and its measurement at time of equilibrium distribution, t', in tissue and blood. With

$Ir_{(T)}(t)$ = recirculating tracer after indicator degradation in tissue ROI at time t

$Ir_{(P)}(t)$ = recirculating tracer after indicator degradation per volume of peripheral blood at time t

$I^*_{(T)}(t')$ = injected standard tracer in tissue ROI at time t'

$I^*_{(P)}(t')$ = injected standard tracer per volume of peripheral blood at time t'

$C_{TM}(t) + Ir_{(T)}(t)$ = gross tracer in ROI image at time t

$C_{TM}(t)$ = $[C_{TM}(t) + Ir_{(T)}(t)] - Ir_{(P)}(t) \cdot [I^*_{(T)}(t')/I^*_{(P)}(t')]$

Following subtraction of the metabolically released and recirculating tracer in the ROI, given by the expression $\{Ir_{(P)}(t) \cdot [I^*_{(T)}(t')/I^*_{(P)}(t')]\}$ from the total radioactivity in the ROI, $[C_{TM}(t) + Ir_{(T)}(t)]$, the net signals from the myocardial lipids, $C_{TM}(t)$, in the ROI were obtained. Other correction techniques based on graphical analyses of radioactivity-time curves have been developed and successfully applied when needed (van Eenige et al. 1987).

Besides ^{11}C-labeled fatty acid, such as 1-^{11}C-palmitic acid, or a fatty acid with radioactive iodine replacing the terminal methyl group, such as ω-^{123}I-heptadecanoic acid, various other terminally phenylated fatty acids and those with side chains, mainly labeled with radioactive iodine, are being used today in clinical cardiology (Machulla et al. 1980; Knapp and Kropp 1995; Feinendegen 2000). These derivatives have the advantage of minimal recirculation of free tracer after indicator degradation and thus obviate the need for correction procedures. All these labeled fatty acids allow the measurement of not only the esterification of labeled fatty acids into the myocardial lipid pool, but also to various degrees the relatively slow rate of lipid turnover with fatty acid transfer into beta-oxidation. These diagnostic investigations, at times in conjunction with imaging of local blood perfusion, have added to understanding the pathology of ischemic heart disease and have helped in the non-invasive diagnosis of non-ischemic cardiomyopathies (Feinendegen 2000). Other applications of labeled fatty acids for metabolic studies in the liver are discussed below (Hoeck et al. 1986).

1.5.1.3.1.2
Internal Standard Technique

This particular method uses two differently labeled compounds, one the initial substrate that enters the reaction chain of interest, and the other a substrate downstream in that chain (see Chap. 16). Both labeled substrates are pulse-injected into the circulating blood and have similar kinetics en route to the reaction site. Dynamic imaging at a given ROI begins soon after injection of the indicators. Provided the labeled reaction products or liberated tracers rapidly

leave the reaction site without recirculation, the release rates of one tracer gives the rate of the upstream reaction and the other gives the rate of the downstream reaction in the chain. The release rate of tracer from the downstream substrate thus serves as an internal standard against which the release rate of tracer from the upstream substrate is scaled.

This technique has shown its success in linking some forms of non-ischemic cardiomyopathies to a genetically determined deficiency of one of the acyl-CoA-dehydrogenases that catalyze beta-oxidation of short-, medium-, and long-chain fatty acids. A deficiency in one of these enzymes may be accompanied by no clinical symptoms at all; or it may cause more or less grave cardiac insufficiency, differing in childhood and adolescence. In order to test for the type of the enzyme failure in fatty acid degradation, the release rates of tracer from the myocardium following intravenous injection of ^{11}C-labeled fatty acids of different chain length were measured. The downstream substrate was ^{11}C-acetate, the final substrate in the degradation chain, serving as an indicator of oxidative metabolism and, thus, as an internal standard (Brown et al. 1987; Armbrecht et al. 1990). Its intravenous injection shortly preceded or followed that of each one of the different ^{11}C-labeled fatty acids. The difference between the tracer release rate from the upstream fatty acid of a given chain length and the downstream internal standard release rate indicated the severity of inhibition of fatty acid degradation due to the deficiency of the appropriate acyl-CoA-dehydrogenase (see Chap. 16). To be diagnostically valid, the examination demands a steady state myocardial metabolism, i.e., exclusion of effects from substrate concentration in the circulating blood, hormones, workload, ischemia, and tracer back diffusion. This type of internal standard application is an excellent example of detecting in vivo the localized effect of a genetically caused disorder, with a relatively high sensitivity and without substantial risk to the patient (Kelly et al. 1993).

1.5.1.3.2
In Vitro Counting Related to Regional or Whole Body Metabolism

Dual-indicator techniques are potent also for in vitro measurements. For instance, a ^{14}C-labeled substrate proximal to an enzyme in a reaction chain and a ^{13}C-labeled substrate distal to it, are simultaneously administered, and the rates of formation of the differently labeled CO_2 are compared with each other. In this set, the second indicator serves as an internal standard, provided that the kinetics of both indicators is similar until they reach the reaction site. For instance, a crucial set of reactions of intermediary metabolism, concerning anabolic vs. catabolic fates of pyruvate, could be tested by simultaneously employing pyruvate and acetate, each carrying a different tracer such as ^{14}C and ^{13}C for measuring the rates of appearance of the differently labeled CO_2 in the exhaled breath. In this metabolic area, interpretations of altered ratios of formation of labeled CO_2 can be multiple and reaction identification is aided by analysis of tracers in relevant organic products such as glucose or ketone bodies. Possible applications could be in states of diabetes or excessive growth hormone (Shreeve et al. 1970). The breath CO_2 test for "lactase deficiency" using labeled lactose (see above) would be sharpened for better focus on intestinal lactase activity per se if both ^{14}C and ^{13}C should be used for dual labeling and simultaneous administration of labeled lactose and the downstream labeled product, glucose (see Chap. 14). In this case, there should be assurance of normal metabolism of the internal standard, glucose, e.g., no diabetes or obesity.

Regarding macromolecular synthesis, for instance of hepatic lipoprotein, the simultaneous use of ^{125}I-apo-A and ^{13}C-phenylalanine as a standard for general protein synthesis allows the quantification of the contribution to lipoprotein (LP) from the "good" apo-A (high-density LP, HDL) vs. general LP synthesis (Rader et al. 1993), as further discussed in Chap. 14. The kinetics of intestinal absorption of elemental calcium may be observed without contribution to data from subsequent post-absorptive metabolic fate, by the simultaneous use of two calcium isotopes, either of radioactive [45]Ca and [47]Ca (DeGrazia et al. 1965), or of stable [43]Ca and [46]Ca (Smith et al. 1996) (discussed in Chaps. 12 and 14). In this test, one of the two paired tracers is given orally and the other intravenously, with measurement of tracer quotients in urine or saliva. Such data has led to recognition of a genetic predisposition to impaired absorption of Ca^{++} as a prelude to osteoporosis. The kinetics of iron absorption can similarly be evaluated by the use of paired iron isotopes, be they radioactive or stable, one given orally, the other intravenously; subsequent measurements of tracer quotients in blood plasma and red cells may reveal, for example, the degree of Fe^{++} absorption in pregnant women (O'Brien et al. 1999) (see Chap. 14).

A more elaborate example of the double-labeling approach for in vitro counting is the study with long-term intravenous infusion of [3-^{14}C]-lactate together with ^{13}C-bicarbonate and oral administration of phenylacetate at various intervals over a period of six hours, in order to estimate Krebs cycle activity and hepatic gluconeogenesis (Landau et al. 1995). During indicator infusion, the ^{14}C and ^{13}C activities have been analyzed in breath CO_2, in blood glucose, and in urinary urea and glutamate from the excreted phenylacetylglutamine. This carefully designed investigation showed that in fasted normal subjects 80% of glucose production was via gluconeogenesis, whereas in fasting diabetic patients the metabolic state was more heterogeneous, gluconeogenesis being reduced to an average of about 45%. This lower contribution of gluconeogenesis to glucose production in fasting diabetic patients was explained as a consequence of remaining glycogen stores during the period of insulin withdrawal.

1.5.1.4
Dual Parameter Analysis:
Two Labeled Isomerically Related Substrates

Potentially powerful is the use of two isomeric substrates that are labeled by different tracers for simultaneous imaging, or with the same tracer for sequential imaging. Provided the two isomeric indicators share kinetics before reaching the site of the wanted reaction, measuring the two different rates of tracer accumulation or release at the chosen ROI gives the difference in the reaction rates of the two isomers.

The development and testing of two labeled isomerically related substrates, i.e., 15-(para-iodo-phenyl)-pentadecanoic acid (pPPA) and 15-(ortho-iodophenyl)-pentadecanoic acid (oPPA), stable-labeled with radioactive iodine (Machulla et al. 1980; Shreeve et al. 1984; Beckurts et al. 1985; Kaiser et al. 1990) provides an example of how one can observe the biochemical nature of the active site of an enzyme in vivo, while also exploiting this information for diagnostic purposes. The profound difference in metabolism of the para- vs. the ortho-iodinated species of phenylated fatty acid in humans suggests the involvement of a protein-binding site akin to that of enzymes such as tyrosine kinases and tyrosine phosphatases, which are widely involved in signal transduction, which is specially discussed in Chap. 4. The further curious differences in the vulnerabilities of reactions with pPPA and oPPA in hepatic lipid metabolism to noxious influences, such as chronic alcohol exposure (Shreeve et al. 1984) or adriamycin (Feinendegen et al. 1996), emphasize the value of employing other pairs of labeled isomeric substrates in parallel in order to reveal the possible occurrence or mechanisms of, or susceptibility to, disease. Genetically related changes in enzyme-substrate interactions should be observable by this technique. Thus far, myocardial and liver metabolism has been investigated by these two fatty acid isomers and may usefully serve to exemplify the paired isomers method.

1.5.1.4.1
Myocardium

In the context of labeling fatty acids with radioactive iodine for studies on myocardial lipid metabolism, the 15-(para-iodo-phenyl)-pentadecanoic acid, pPPA, proved metabolically very similar to natural fatty acids; moreover, its catabolites are rapidly released from the reaction site and excreted. This eliminates the need for time-consuming signal correction. Today pPPA is widely used in diagnostic cardiology with appropriate imaging devices (Feinendegen 2000).

Upon labeling phenyl-pentadecanoic acid with radioactive iodine, two thirds of the iodine binds in the para position, yielding pPPA and one third in the ortho position giving the isomeric 15-(ortho-iodo-phenyl)-pentadecanoic acid, oPPA (Machulla et al. 1980). This oPPA was found in studies with rodents to share with pPPA practically identical kinetics in blood circulation, transport and diffusion (Beckurts et al. 1985). Both isomers entered into the cells in parallel. However, both were significantly less readily bonded in vitro to CoA by the enzyme acyl-CoA-SH-thiolase than were palmitic acid and ω-iodo-heptadecanoic acid, probably by steric hindrance of the phenyl group in the ω position. Thus, after 30 min incubation the binding was about 30% for oPPA and about 45% for pPPA, but about 80% for ω-iodo-heptadecanoic acid, and about 90% for palmitic acid (Kaiser et al. 1990).

Greater differences between the two isomers were seen in rodents regarding esterification into complex lipids in the myocardium. Whereas pPPA readily entered the lipid pool, oPPA did so only to a minor degree (Beckurts et al. 1985). Moreover, in contrast to pPPA, oPPA in vivo hardly crossed into the mitochondria for beta oxidation, but left the cells nondegraded by back diffusion or transport into the circulating blood. When applied to humans in an initial attempt to label the free fatty acid pool as an internal standard in the measurement of myocardial lipid turn-

over, oPPA contrary to expectation was observed to have a much lower rate of release, with half times ranging to several hours (Antar et al. 1986). Indeed, only traces of labeled catabolites of oPPA appeared in the peripheral blood and urine. That the myocardial trapping of oPPA occurred outside mitochondria in the human myocardial cells was experimentally confirmed using a double labeling approach at diagnostic coronary angiography in patients. Both tracers, ^{123}I from the phenyl group of the oPPA and ^{14}C from the carboxyl group were retained in parallel and there was no change in the isotopic ratio in the coronary sinus. Loss of oPPA from the myocardium appeared mainly due to back diffusion or transport (Kaiser et al. 1990). Thus, after a practically parallel rate of uptake of oPPA and pPPA into the cells, pPPA may serve as an indicator of fatty acid metabolic degradation, while oPPA almost exclusively signals the rate of back diffusion or transport. It therefore appeared reasonable to use these two isomers to distinguish between losses of tracer from the myocardium by beta oxidation, with pPPA, versus back diffusion/transport, with oPPA (Feinendegen 1993).

The dual tracer technique with oPPA and pPPA both labeled with ^{123}I was applied diagnostically in patients with dilated cardiomyopathy (Feinendegen et al. 1995). The intravenous pulse injection of oPPA preceded the injection of pPPA by 50 min. After another 50 min, ^{123}I-NaI was intravenously given to correct for labeled catabolites in the tissue ROI by image subtraction, as referred to above. Regional uptakes of the two indicators did not differ between patients and control subjects. However, the release rates differed from control in at least 66% of the patients and the scatter was larger inter-individually than intra-individually. Since the release rate of tracer from oPPA indicated mainly back diffusion/transport, and that of pPPA both beta oxidation and back diffusion/transport, the difference between the two release rates are assumed to mainly give the rate of beta oxidation in the observed myocardial region. In this manner, three different patterns of lipid turnover appeared in these patients: (1) predominantly increased beta oxidation; (2) predominantly decreased beta oxidation, in part with increased back diffusion/transport; and (3) predominantly increased back diffusion/transport. These significantly distinct patterns of myocardial lipid turnover in different patients with non-ischemic cardiomyopathies express different pathologies at the molecular level and may involve genetically determined predispositions.

1.5.1.4.2
Liver

Following studies of in vivo liver metabolism with ω-^{123}I-heptadecanoic acid (Hoeck et al. 1986), the paired-isomers method with labeled fatty acids appeared even more revealing. As was shown for the myocardium, pPPA and oPPA in normal rats also had almost identical kinetics in the circulating blood before reaching the liver tissue cells. Both then had distinct patterns of esterification into the various lipid fractions, and this was independent of local blood perfusion at the anatomic site (Feinendegen et al. 1996). Adriamycin, which stimulates hepatic lipid synthesis, caused a highly significant change in rats in the ratios of liver uptake of pPPA and oPPA as into various lipid fractions compared to normal liver. The quotient oPPA/pPPA was 2.6 for whole liver in normal rats, and significantly lower at 1.5 after adriamycin treatment. The corresponding quotients oPPA/pPPA in liver triglycerides were significantly different by a factor of 5, with 2.0 in normal rats and 0.4 after adriamycin treatment (Feinendegen et al. 1996). By contrast, in mice exposed chronically to ethanol there was increased conversion of oPPA to liver triglycerides (TG) relative to nonexposed mice, with no effect of ethanol on incorporation of pPPA. In addition phasic differences occurred in the ethanolic mice in decline of activity in TG, as in total liver, from oPPA as well as from ω-^{123}I-heptadecanoic acid or ^{14}C-stearic acid, but not from ^{14}C-oleic or palmitic acids (Shreeve et al. 1984). The quotient change for paired fatty acid precursors converted to circulating triglycerides may be a sensitive diagnostic indicator of pathological hepatic lipid metabolism. Further study is also warranted in the context of genetically determined alterations in reaction rates of hepatic lipid synthesis.

In a preliminary clinical application, trace amounts of ^{123}I-labeled pPPA and oPPA were given intravenously one shortly after the other. The sequences of scintigrams displayed each indicator separately and as a quotient of the two. Indeed, the significant change of the normal pPPA/oPPA quotients in liver disease appeared as a clinically useful indicator of a disturbance of lipid metabolism (Ebert et al. 1993). Previously, local pathological alterations in liver metabolism could only be detected by microscopic examination of tissue specimens.

1.5.1.5
Dual Parameter Analysis:
Two-Isotope Effect on Labeled Substrate

For in vivo analyzing the function of substrate/ligand binding to its enzymes/receptor, the use of the isotopic effect of deuterium (^{2}H) in place of hydrogen (^{1}H) in hydrogen bonding to carbon was introduced by Fowler in a seminal study (Fowler et al. 1988; see Chap. 20). When ^{2}H replaces ^{1}H, where a carbon-hydrogen bond normally forms during substrate interaction with its specific target, the ensuing rate of binding is altered. The reason for this is the increased stability of a C-^{2}H bond compared with a C-^{1}H bond. By choosing this dual parameter technique for in vivo imaging of an appropriately labeled substrate, most confounding factors associated with in vivo imaging of biochemical reactions, as outlined above, can be excluded. If the concentration of receptor on site is to be assessed by the indicator ligand, the increased bond stability is an advantage in that it reduces the effect of an unfavorable relationship between a relatively low receptor concentration and a high rate of labeled substrate binding and loss.

This approach was employed for studying the function of monoamine oxidase, MAO, in the brain of humans and lower mammals. As given in more detail in Chap. 20, MAO type B binds L-deprenyl covalently and irreversibly to a co-factor that arises during MAO-B-catalyzed oxidation. In this way, L-deprenyl is a suicide inactivator of the enzyme. L-deprenyl labeled with ^{11}C therefore represents a potent indicator for measuring site and function of MAO-B at a given ROI in the brain by PET.

The hydrogen at various sites of ^{11}C-or ^{14}C-labeled L-deprenyl was replaced with deuterium. Deuterated and non-deuterated radioactively labeled L-deprenyl were then intravenously administered in experiments with baboons and mice and later in human studies. In the latter, PET imaging showed the two L-deprenyls to accumulate mostly in the basal ganglia and the thalamus, with lesser intensities found in the frontal cortex and the cingulate gyrus. Binding was lowest in the parietal and temporal cortices and the cerebellum. When the deuterium was incorporated in the methylene group of the propargyl group of L-deprenyl, an isotope effect appeared in that the rate of tracer uptake and binding was significantly reduced, as was the rate of its release compared to the control. The data thus established in vivo that the alpha-carbon-hydrogen bond on the propargyl group is a singular or major rate-limiting step in oxidation by MAO-B.

Obviously, this particular in vivo method appears powerful for investigating on the one hand the specific site with which a substrate/ligand binds to its specific enzyme/receptor, and on the other for observing the consequences of structural alterations on the binding due, for instance, to the action of molecular signaling, or to genetic polymorphism in the pathogenesis of disease. Moreover, this type of information may help to structurally tailor a substrate for developing drugs for effectively altering biochemical reactions in vivo.

1.5.2
Parallel Observation of Labeled Substrate in Two or More Compartments

Measuring the transfer of a given indicator between physiological compartments, such as between blood and tissue, body and urine, body water and tissue, etc., compartmental analyses have found wide application in nuclear medicine since its early days (see Chaps. 7, 8). Presently, with tracers and labeled indicators being available for imaging biochemical reactions, compartmental analyses are routinely used to assess a biochemical reaction by the kinetics of indicator transfer to, or loss from, or both, the site of a reaction – for instance within a chosen ROI. Two situations arise, one in which the indicator is practically irreversibly bound over the period of observation, and the other where the indicator after accumulation in the observed ROI leaves the site into the circulating blood or is degraded with release of tracer. The accuracy of compartmental analysis increasingly suffers with the number of interrelated compartments and complexity of the model. The observation of indicator binding alone, or together with subsequent indicator dissociation from its target site, in a chosen ROI makes the one- and two-tissue compartment analysis a near optimal tool. Chapters 7 and 8 extensively discuss modeling and tracer kinetics. This section will briefly introduce the compartmental analysis of reversible and irreversible indicator accumulation.

Usually, a single labeled substrate suffices for compartmental analysis of reversible or irreversible accumulation of indicator. The compartment model includes indicator supply from arterial blood to the tissue compartment where the reaction is to occur. The corresponding rate is expressed as clearance of indicator from the arterial blood plasma, and the rate of in-

dicator return into the blood is expressed by a rate constant. In this way, a dual parameter analysis results. A one-tissue compartment may suffice for describing reversible accumulation or indicator binding in tissue. A two-tissue compartment is composed of the extravascular tissue space of free indicator, i.e., the substrate pool in tissue, from where the indicator enters the irreversible bond to its target, which is denoted as the second tissue compartment. In either case, if the density of a defined receptor in terms of receptor concentration at the target site is the goal of measurement, the amount of indicator available and its binding affinity to the target are crucially important.

Obviously, the indicator must be structurally compatible with its specific target molecule, be it receptor, transporter, or enzyme in the interstitial space, and on or within a cell. In order to take advantage of the indicator's irreversible binding to its target, and thus its accumulation in the ROI as a function of the reaction rate, appropriate molecular tailoring of the indicator is a powerful approach, as discussed below in the context of glucose metabolism.

The actual measurement begins with the intravenous injection of the indicator and requires frequent scintigraphic recording of the quantities of the indicator in the ROI under study, and parallel analyses of indicator concentrations in samples of the arterial blood plasma, from where the indicator is supplied to the reaction site. The assumption, of course, is the rapid transfer of indicator from the blood to this site. These measurements result in a set of two time-radioactivity curves, one giving the temporal changes of indicator concentrations within the arterial blood plasma, and the other gives the corresponding changes of radioactivity within the ROI.

The analysis of the two curves describing the indicator concentration within arterial blood plasma and the ROI radioactivity may furnish the desired information regarding the biochemical reaction to be studied, as explained below. Nevertheless, various caveats need attention regarding the complex kinetics of indicator supply to the reaction site, the reaction itself and, if applicable, transfer of labeled reaction product away from the reaction site in the ROI with possible recirculation. The mathematical treatment of the curves may favor computerized fitting procedures with input of the model-given variables. Another choice is the graphic display of reaction rates and rate constants. In this section, the graphic analyses are briefly summarized and presented with examples of application. More details are in Chap. 7.

1.5.2.1
Labeled Substrate in One-Tissue Compartment Model

The one-tissue compartment model describes the indicator's arrival and accumulation at the reaction site via the circulating blood without having passed through a functionally relevant second compartment. The model also takes into consideration the release of indicator from its site of accumulation and removal by the circulating blood. This approach applies, for instance, to studying reversible transport across a biological barrier within a given ROI. Thus, with

$C_T(t)$ = radioactivity in extravascular tissue space at time t,

$C_P(t)$ = radioactivity per volume blood plasma at time t,

K_1 = clearance of indicator from blood plasma to the extravascular tissue space (radioactivity/min × volume)

k_2 = rate constant of indicator release from the extravascular tissue space to blood, (radioactivity/min × mass)

$$dC_T(t)/dt = K_1 C_P(t) - k_2 C_T(t) \qquad (1.11)$$

After equilibrium is reached between indicator concentration in blood and tissue,

$$K_1 C_P(t_{equ}) = k_2 C_T(t_{equ})$$

and the two measured time-radioactivity curves in blood plasma and tissue run in parallel. The ratio of the two measured values $C_T(t_{equ})/C_P(t_{equ})$ is equal to k_1/k_2 and expresses the tissue mass into which distributes an amount of indicator cleared per unit volume of blood plasma. This value is conventionally called the distribution volume.

By integration of Eq. 1.11,

$$C_T(t) = K_1 \int C_P dt - k_2 \int C_T dt \qquad (1.12)$$

This equation may be rearranged in various ways leading to linearized expressions. One (Gjedde and Diemer 1983) is:

$$C_T(t)/\int C_P dt = K_1 - k_2 \int C_T dt / \int C_P dt \qquad (1.13)$$

The graphical display of this equation results in a linear plot that extrapolates to the intercept at the value of K_1 and has the slope of $-k_2$.

Another way of linearization of Eq. 1.12 is (Logan et al. 1990):

$$\int C_T dt / C_T(t) = K_1/k_2 \int C_P dt / C_T(t) - 1/k_2 \qquad (1.14)$$

Here, the linear portion of the plot gives the intercept as $-1/k_2$ and the slope describes the distribution volume K_1/k_2.

The analysis thus results in the values of the clearance and the rate constant that governs back transport of the indicator from the tissue into the circulating blood. If such measurements are made at different substrate concentrations, which the indicator traces, the resulting rate constants may be interpreted in terms of first order reaction kinetics expressed by the Michaelis-Menten equation, described above.

1.5.2.1.1
Active Transport of Substrate

The one-tissue compartment model serves to quantify reversible transport of substrates across biological barriers such as the blood brain barrier (BBB). In analyzing glucose transport across the BBB at various blood glucose concentrations, ^{11}C labeled 3-O-methylglucose was employed as an indicator that is transported across the blood brain barrier similarly to d-glucose but is not metabolized (Feinendegen et al. 2001). It enters the free glucose pool in brain tissue and returns from there into the circulating blood. The potential of a very small second tissue compartment is neglected here. In the absence of a second tissue compartment, the indicator accumulation in the brain with time after indicator injection reaches equilibrium with the indicator concentration in blood plasma and the two time-radioactivity curves run in parallel. This indicates the reversible transport of this glucose analogue across the BBB between the capillary circulation and the extravascular free glucose pool in tissue. Diagnostic studies in humans showed that the rate constant k_2 indicating indicator outflow from the extravascular tissue space to blood differed among individuals to a considerably greater extent than the clearance rate of inflow, K_1. Measuring these values throughout the cerebral cortex at different plasma glucose concentrations revealed a highly significant inter-individual but not intra-individual variability more for k_2 than K_1 at normal and elevated blood glucose levels. The results were analyzed in the context of present day knowledge of glial cell biochemistry and indicate highly individualized glial cell function, possibly genetically determined (Feinendegen et al. 2001). Indeed, the data suggest the existence of individually regulated control mechanisms of glucose metabolism in the brain.

1.5.2.2
Labeled Substrate in Two-Tissue Compartment Model

The two-tissue compartment model contains, for instance, an extravascular tissue space into which the indicator enters from the circulating blood and from where it leaves, either by back transfer into the circulating blood or by irreversible binding to its target at the tissue site under observation; this irreversible binding site presents the second tissue compartment. The resulting two time-radioactivity curves, one from arterial blood plasma and the other from the tissue ROI, tend to converge, i.e., the tissue ROI curve rises where the one from blood plasma falls. Again, curve-fitting procedures as well as graphic displays are common for the determination of rate constants. Analogous to the one-tissue compartment analysis, here again the graphic approach is chosen to show the rate constants of indicator transfers in the ROI (Patlak et al. 1983; see Chap. 7). Thus, in a simplified manner with:

$C_T(t)$ = radioactivity in extravascular tisseu space at time t

$C_T(t)$ = $C_{TF}(t) + C_{TM}(t)$

$C_{TF}(t)$ = radioactivity in the extravascular pool of free indicator in tissue at time t (compart. 1)

$C_{TM}(t)$ = radioactivity irreversibly bound in extravascular tissue space at time t (compart. 2)

$C_P(t)$ = radioactivity per volume blood plasma at time t

K_1 = clearance of indicator from blood to extravascular pool of free indicator in tissue (radioactivity/min × volume)

k_2 = rate constant of indicator release from extravascular free indicator pool to blood (radioactivity/min × mass)

k_3 = rate constant of irreversible binding of radioactivity in extravascular tissue space radioactivity/min × mass)

the following differential equation arises for tissue compartment 1:

$$dC_{TF}(t)/dt = K_1 C_P(t) - k_2 C_{TF}(t) - k_2 C_{TF}(dt) \quad (1.15)$$

and for tissue compartment 2:

$$dC_{TM}(t)/dt = k_3 dC_{TF}(t) \quad (1.16)$$

Following integration of Eq. 1.16, tissue compartment 2 becomes:

$$C_{TM}(t) = k_3 \int C_{TF} dt \quad (1.17)$$

Also, at equilibrium between $C_P(t)$ and $C_{TF}(t)$:

$$K_1 C_P(t) = k_2 dC_{TF}(t) + k_3 C_{TF}(t) \quad (1.18)$$

and by rearranging Eq. 1.18 tissue compartment 1 becomes:

$$C_{TF}(t) = \frac{K_1}{k_2 + k_3} \cdot C_P(t) \quad (1.19)$$

Because the measured $C_T(t)$ covers the 2 compartments $C_{TF}(t) + C_{TM}(t)$, Eqs. 1.19 And 1.17 need be added:

$$C_T(t) = \frac{K_1}{k_2 + k_3} \cdot C_P(t) + k_3 \int C_{TF} dt \quad (1.20)$$

and by substituting Eq. 1.19 into Eq. 1.20:

$$C_t(t) = \frac{K_1}{k_2 + k_3} \cdot C_P(t) + k_3 \cdot \frac{K_1}{k_2 + k_3} \int C_P dt \quad (1.21)$$

Dividing both sides of Eq. 1.21 by the measured $C_P(t)$:

$$\frac{C_T(t)}{C_P(t)} = \frac{K_1}{k_2 + k_3} + k_3 \cdot \frac{K_1}{k_2 + k_3} \frac{\int C_P dt}{C_P(t)} \quad (1.22)$$

Thus, the parallel measurements of $C_T(t)$ and $C_P(t)$ at given time intervals allow one to plot Eq. 1.22 graphically, and the straight line in the plot extrapolates to the intercept value $[K_1/(k_2 + k_3)]$, the slope given by $[k_3 K_1/(k_2 + k_3)]$.

Therefore, dividing the slope $[k_3 K_1/(k_2 + k_3)]$ by the intercept $[K_1/(k_2 + k_3)]$ yields k_3, the rate constant of the irreversible binding.

1.5.2.2.1
Enzyme-Substrate Interaction

The two-tissue compartment model is most widely used for measuring glucose metabolism by labeled 2-deoxyglucose (Sokoloff et al. 1977). This glucose analogue is singularly suited to in vivo observation of the phosphorylation of d-glucose, the first step in glucose metabolism. D-glucose-phosphate is not further metabolized but is trapped at the reaction site. Thus, 2-deoxyglucose enters the free glucose pool in tissue in a manner similar to that of glucose, for instance, via the BBB into glial cells, where it either binds to its target molecule hexokinase to become phosphorylated and trapped, or from where it returns to the circulating blood. By labeling deoxyglucose with ^{14}C a superb indicator was introduced for the experimental in vivo observation of the rate of glucose phosphorylation throughout the brain, by autoradiographically measuring the accumulation of the indicator in tissue slices from the entire brain. The analysis demonstrated that local tracer accumulation, with full consideration of local flow, transport, and difference in binding affinities between d-glucose and its analogue, correlated with the local rate of d-glucose phosphorylation (Sokoloff et al. 1977). This seminal work opened a new field of studies in the living brain, which relies predominantly on d-glucose for energy. A normal human brain consumes on average about 120 g of d-glucose per day.

Following extensive experimental work with ^{14}C-labeled 2-deoxyglucose, labeling with the positron-emitting ^{18}F introduced the ^{18}F-2-deoxyglucose (FDG) for in vivo imaging in humans with PET (Gallagher et al. 1977). For quantifying the rate of d-glucose phosphorylation in the human brain with the help of the two-tissue compartmental model, accumulation of FDG in various brain regions is dynamically imaged over a given period of time in parallel with measuring the concentration changes of indicator in arterial blood plasma. Other tissues such as myocardium and malignant tumors are similarly examined with FDG using the dual parameter approach. Indeed, FDG is presently by far the most widely used indicator for metabolic studies in vivo using PET (Phelps et al. 1979; Reivich et al. 1979; Delbeke et al. 2002). Various applications are presented in this book (see Chaps. 17, 18, 20, 27).

Regarding glucose metabolism in the human brain, one may summarize that activated neurons in the brain trigger an increased amount of glucose into metabolism mainly in the glial cells adjacent to neurons. This appears intimately connected with an increase in local cerebral blood perfusion. Changes in both local glucose consumption and blood perfusion allow one to localize neuronal stimulation in the living brain in specified motor, sensory, and mental activities. Thus, brain function mapping has become a reality. On the other hand, the state of total relaxation as in deep yoga meditation has been seen to cause the entire brain and not just certain regions to be affected, and different individuals reacted either by a decrease or an increase in global glucose metabolism

(Herzog et al. 1990/1991). In clinical practice, brain PET imaging with FDG is now a routine procedure in patients suffering from circulatory disorders including cerebral ischemia, tumor, and neural degeneration such as Alzheimer's disease. In current brain research and clinical medicine functional MRI of local blood perfusion increasingly substitutes for or complements the classical FDG studies on cerebral energy metabolism. Regarding effects of genetic determinants in controlling local cerebral energy metabolism, it appears tempting to correlate measurements of glucose metabolism or local blood perfusion with those of neurotransmitter and receptor activity in the same and interconnected brain regions.

In the myocardium, accumulation of FDG in relation to labeled fatty acids may indicate a shift toward anaerobic metabolism for energy supply in areas of diminished blood flow with reduced oxygen supply. Thus, tissue vitality in ischemic myocardial regions can be correlated with relatively increased levels of glycolysis. This test in conjunction with coronary blood flow measurements may also help to distinguish between stunned and hibernating myocardium and sometimes appears crucial in deciding, for instance, on coronary bypass surgery. Myocardial imaging with FDG in cardiological research and practice has been extensively reviewed (Schroeder and Schelbert 2000; see Chap. 17).

The demand for FDG with PET currently also rises in the diagnosis of malignant tumors (Delbeke et al. 2002). Because of preferential glucose consumption for energy supply in many cancerous tissues, whole body imaging with FDG can reveal the site and often also the type of a tumor and its metastases several months earlier than their appearance in conventional structural images with x-ray or MRI. Depending on tumor type, tumor glucose consumption is highly sensitive to chemical and radiation therapy. By its quick response, a reduction in FDG uptake in a tumor soon after treatment may be an early sign of long-term therapeutic efficiency and allows for individual planning for optimal therapy. Here, also attempts at gene therapy and gene expression in vivo, as discussed below and in Chap. 30, need be included in the potential treatment modalities.

1.5.2.2.2
Ligand-Receptor Interactions

Probably the most frequently applied nuclear medical procedure in current molecular imaging in vivo addresses ligand-receptor interactions. Both, the one-tissue and two-tissue compartment models are indispensable in the quantification of reaction kinetics.

The concept of specific receptors on cell membrane surfaces being sensitive to specific circulating substrates was first advanced in 1900 by Paul Ehrlich (Rensberger 1996). Now it is estimated that about 40% of the genome codes for such receptors. They are sensitive to a variety of chemical messengers such as hormones, growth factors, neurotransmitters, and pharmacological agents, which usually circulate in the blood and interstitial space. Most messengers between cells function as ligands through binding to receptors on membranes. These receptors typically extend through the cell membrane, having external domains for receptivity and internal domains for attachment to specific cell proteins involved in signal transfers. Bonding of ligand triggers the receptor into phosphorylation or transphosphorylation in the region of the internal domain, as outlined in Chap. 4. Signal transductions through further protein-protein interactions inside the cell directly determine enzyme activities for metabolism or are transmitted to the nucleus, causing alterations in gene expression.

There is much evidence to indicate that polymorphisms of genes for receptors change the effectiveness of ligands and underlie various disease states. To cite an example: insulin resistance in Type II diabetes may be caused by heritable abnormalities of insulin receptors on the surface of peripheral target cells (Stern 2000; see Chap. 14). Another: genetically determined changes in serotonin receptor function in the mid-brain or even on platelets can be linked to predisposition to depression or alcoholism (see Chaps. 4, 14). Another: different phenotypes for estrogen receptor (ER) activity in breast carcinoma can be correlated to different patterns of gene expression (Gruvberger et al. 2001). And: mutations determining somatostatin receptor subtypes appear to account for differing responses of acromegalic patients to octreotide (Ballare et al. 2001). The challenge in nuclear medicine is to devise practical ways, by imaging or other means, to detect these gene-related abnormalities by differential binding of well-designed radioactively labeled ligands, which have physiological or pharmacological effects on the targeted receptors.

A particular advantage to nuclear medicine has been the appreciation that cell surface receptors are, in general, excessively or liberally expressed in tumor cells. One fruitful area has been the development of radioactively labeled oligopeptides, which correspond to or mimic a natural ligand for some particular receptor (Boerman et al. 2000). Relatively well developed is the use of labeled variants of an octreotide, an analog of the peptide hormone somatostatin. This hormone mainly functions by inhibiting the secretion of other hormones. The ^{111}In-labeled-DTPA-octreotide is successfully used for imaging of neuro-endocrine and gastro-entero-pancreatic tumors, breast carcinoma, and lymphoma. Besides their advantageous diagnostic use, octreotides are in therapy trials, for instance in the form of ^{111}In-labeled DTPA-octreotide and ^{90}Y-labeled DOTA-Tyr3-octreotide. This use demands internalization of the receptor-ligand complex and long-term retention. Other peptides that have potential for imaging receptors on tumor cells include vasoactive intestinal peptides such as cholecystokinin, bombesin, and calcitonin (Boerman et al. 2000). Among hormones or analogues that have been labeled and found useful for tumor imaging are ^{18}F-labeled fluoro-estradiol, for instance for estrogen receptor-carrying breast tumors (Silverman et al. 1998) and ^{131}I- or ^{123}I-labeled meta-iodo-benzylguanidine (MIBG) for neural crest tumors (Shapiro et al. 1995). Still another class of compounds for tumor receptor imaging encompasses pharmaceutical agents, e.g., ^{18}F- or ^{131}I-labeled tamoxifen analogues for estrogen receptor-carrying breast tumors (Silverman et al. 1998). Similarly, angiogenesis (and apoptosis, too) are now amenable to ligand-receptor imaging. Indeed, targeted tumor-receptor imaging in the broadest sense is currently one of the most rapidly evolving nuclear medical techniques in clinical oncology (Kim and Yang 2001).

The special case of neuroreceptor stimulation by neurotransmitters at neuronal synapses results in conversion to an electrical signal. Neurotransmission in the living brain is being extensively investigated with some clinically important results for diagnosis and therapy planning, as discussed later in this book (see Chap. 18). Labeled neurotransmitters are crucial in functional brain mapping in vivo regarding motor, sensory, or pure mental activity. Following initial animal studies (Firnau et al. 1976), neurotransmitter imaging in the human brain began with the dopamine receptors (Wagner et al. 1983). Many such investigations now also pertain to measuring receptor availability and, dependent on post-synaptic receptor activity, to on-site concentration of neurotransmitters such as dopamine, serotonin, acetylcholine, in neurological and psychiatric disease or intoxication states

(Laruelle 2000), as discussed in more detail in Chaps. 18, 19. Useful information has accumulated with radioactive analogues of neurotransmitters, which behave according to the "occupancy model" of straightforward competition between natural ligand and labeled analogues. However, some of the labeled analogues do not follow the degree or kinetics of binding expected from the "occupancy model," which has led attention to effects of the cycling of receptors between internalization and externalization and perhaps of other factors which may determine substrate binding. Such investigations complement the data obtained by FDG imaging of regional cerebral glucose metabolism or by corresponding functional MRI of local cerebral blood flow, as well as by intracerebral magneto-encephalography (MEG). Indeed, these various approaches in unison today open synoptically different kinds of views on the living brain and occupy a major research field of applied neurobiology, with revolutionizing consequences regarding understanding of neural and mental function.

Promising radioactively labeled ligands for receptors on leukocytes, monocytes, and lymphocytes are under investigation for imaging infection/inflammation, and on thrombi for detecting deep vein thrombosis and pulmonary emboli (Boerman et al. 2000; see Chaps. 10, 11). Beta-adrenergic receptor density in the myocardium in heart failure, ischemic heart disease, non-ischemic cardiomyopathy, and after heart transplantation has been evaluated with ^{123}I-labeled MIBG and other tracers (Syrota et al. 1995; Bengel et al. 2002).

Receptor imaging is likely to be extended to all body tissues and many diseases and be directed to questions regarding receptor sites, density, function, and regulatory susceptibility. As suggested by the multiplicity of involved genes and/or gene mutations in determining for instance estrogen receptor activity in breast cancer (Gruvberger et al. 2001) and insulin receptor activity in insulin resistance (Stern 2000), genetically determined nuances of individual responses to labeled ligands can be expected for various other kinds of receptors as well. Other possible modifications of receptor activity come from "epigenetic" reactions, such as acetylation, phosphorylation, or methylation of amino acids in nuclear histones, for example, in the context of the new "histone code" hypothesis (Maher 2001). Moreover, symbiotic effects on a specified target from exposure to differing levels of various natural ligands circulating in the blood need consideration.

Present day advances in this type of imaging are rapid and diversified and will undoubtedly open many new biochemical and cell biological systems to in vivo observation with reference to genetic control.

1.6
Probing for Gene Expression In Vivo

Over the past decade, in vivo gene expression has become a major research area with profound implications to clinical diagnosis and therapy, especially in oncology. The analytical techniques largely rely on the two-tissue compartment model. With the decoding of the human genome, new impetus is given to this field, which like hardly any other nuclear medical development relies on the intimate interdisciplinary interaction between cell and molecular biology and non-invasive in vivo imaging procedures, including radiopharmaceutical work.

Gene expression in cells and tissues may be measured by the activity of gene products, as referred to in various sections above, for instance, in the context of ligand-receptor imaging. In addition, the expression and efficiency of groups of correlated genes may become observable by combining various techniques discussed above for in vivo measurements of circuits of biochemical reactions dependent on these genes. The effect of a signaling pathway on one or more biochemical reactions within a functional circuit may thus eventually be unraveled in vivo and correlated with gene expression.

A direct way to analyze gene expression in vivo employs radioactively labeled probes that may bind to genes or their transcribed RNA. Such labeled markers need complementary binding to the DNA or RNA nucleotide chain, and thus consist of chains of a limited number of nucleotides. Such relatively small single-chain oligonucleotides serve as "antisense" markers defining precise nucleotide sequences. Initial work with such radioactively labeled antisense single oligonucleotides in vitro and animal experiments show the feasibility of the approach for diagnosis and therapy, as extensively presented in Chaps. 28, 30. A serious limitation is the relative instability of oligonucleotides through the action of nucleases; as a consequence, one tries to increase oligonucleotide resistance against nucleases by attaching metabolically blocking end groups. Another limitation lies in the intracellular barriers and the relatively small numbers of DNA or RNA targets in cells. This causes rather weak signals for imaging unless the target number is amplified, which is still an unsolved problem. The situation is different for enzyme-catalyzed reactions where repetitive binding of labeled substrate as indicator allows a relatively large image signal. Other caveats arise through potentially interfering background radioactivity in whole system images after administration of labeled antisense probes. More confounding factors relate to the fate of large molecular indicators on the way through interstitial spaces to reaction sites, general complications of in vivo measurement of biochemical reaction, as referred to previously. One may expect that these problems will eventually be solved.

The other approach focuses on measuring specific gene transcription products, be they enzymes or receptors, as outlined above. This way is especially effective for assaying the success of gene transfer into cells in the course of attempts at gene-based therapy, be it for tumor cell killing or for altering cellular control or protective systems. Here, the primary concern is to localize the transferred gene and to describe its expression. When the transferred gene is not easily accessible to being observed by its transcription product, the goal is to attach a second gene whose transcription product can report the activity of the primary gene when both genes have the same promoter. Chapter 30 extensively discusses the progress in this field.

In vivo observation of gene expression may limit its focus to a single gene that has been stably transferred and is resident in the genome in a cell system. Or such observation may reveal the need for the *ad hoc* transfer of a gene as a desired therapy for eradicating a tumor or changing cellular function. The latter scenario is in early development, particularly challenging and demanding attention as to selection of genes for transfer, to the construction of gene groups including the reporter gene, to the techniques of transfer, and to the choice of the optimal indicator for analyzing the location and activity of the transferred gene or its reporter gene.

One example should suffice here to illustrate the potency of gene transfer for both tumor therapy and monitoring of gene expression by in vivo imaging of a reporter gene. The most widely used gene in the context of developing gene transfer and in vivo imaging of gene expression codes for herpes-simplex-virus-specific thymidine kinase, HSV-1-tk. The choice of the HSV-1-tk gene arose from the well-known application of acyclovir in the treatment of herpes simplex virus infection. Acyclovir is a purine analogue

(9-[(2-hydroxyethoxy)methyl]guanine) and a substrate for the HSV-1-tk but not for the more substrate-specific thymidine kinase in normal mammalian cells. After its phosphorylation acyclovir becomes substrate to host cell kinases to be phosphorylated to triphosphate. As such, it inhibits virus DNA polymerase and also may be incorporated into virus DNA with subsequent nucleotide chain termination. Nucleotides as phosphorylated substrates do not easily escape cells but remain trapped. An analogue of acyclovir, ganciclovir (9-[(1,3,-dihydroxy-2-propoxy)-methyl]guanine) is again a substrate of HSV-1-tk, yet also to a lesser extent for thymidine kinase in normal mammalian cells. The triphosphate of ganciclovir is toxic to any cell where it is formed and trapped. Thus, transfecting tumor cells first with the HSV-1 virus, and then exposing them to ganciclovir will overwhelmingly affect the tumor host cells. Since the HSV-1-tk is less substrate-specific than thymidine kinase in normal mammalian cells, it also accepts thymidine analogues as substrates for phosphorylation.

Therefore, two classes of substrates have been extensively studied to observe in vivo the expression of the HSV-1-tk gene by monitoring its transcription product: one is the group of guanine derivatives, and the other encompasses thymidine and its analogues, both labeled, for example, with ^3H for autoradiography, or with ^{123}I for single-photon-tomography, or ^{11}C or ^{18}F for PET. The advantage of ^{123}I, and ^{18}F is, of course, their relatively long physical half-life, which allows for delayed imaging. The guanine analogues used as indicator for HSV-1-tk activity include, for instance, labeled acyclovir (ACV), ganciclovir (GCV), fluoroganciclovir (FGCV), penciclovir (PCV), fluoropenciclovir (FPCV), 9-[3-fluoro-1-hydoxy-2-propoxymethyl] guanine (FHPG), and 9-[4-fluoro-3-(hydoxymethyl)butyl] guanine (FHBG). The thymidine analogue group includes ^{123}I- and ^{18}F-labeled indicators: 3'-deoxy-3'-fluorothymidine (FLT), 5 iodo-2'-deoxyuridine (5-IUdR), and the metabolically more stable 2'-fluoro-2'-deoxy-1-b-D-arabinofuranosyl-5-iodouracil (FIAU), 2'-fluoro-2'-deoxy-5-iodo-1-b-D-ribofuranosyl-uracil (FIRU), 2'-fluoro-2'-deoxy-5-methyl-1-b-D-arabino-furanosyl-uracil (FMAU), and 2'-fluoro-2'-deoxy-5-iodovinyl-1-b-D-ribofuranosyl-uracil (IVFRU). These and other similarly labeled substrate analogues are still under study and being compared in various cell systems in vitro and in vivo with different advantages and disadvantages regarding their metabolic fate in the whole organism, degree of background signaling, and cell specificity in the context of using the HSV-1-tk as reporter gene in linkage with a primary gene and a common promoter for both in a DNA construct for transfer.

The choice of the DNA construct, or "cassette," varies with the goal of the transfer. Besides the choice of primary gene for transfer, various reporter genes may offer advantages. In addition, the promoter may be chosen to be cell specific and exclude gene expression in certain cells. Or the promoter may function either continuously or upon a defined external signal such as a drug or low-dose irradiation.

Transfection of cells relies on various vectors well known in cell biology. They include a virus-type vector, liposomes, plasmids, or purposefully guiding peptides. Important is the proper binding of the vector to the cell membrane with subsequent internalization. These types of vector, as long as they operate, usually do not affect the functioning of the transferred gene complex.

Transfection may use cells in vitro to be transferred into body tissue, or may directly aim at cells in the body. For experimental and some therapeutic reasons, in vitro transfected cells are injected into the site of envisaged action. Reporter genes have thus been experimentally transferred to cells in different body sites and shown to function.

In summary, a gene or genetic construct with its reporter gene in a complying vector must enter the target cells to become integrated into the host genome or be metabolized rapidly. The transferred gene alone or its reporter gene should not already be present in the cells under observation and should be specific for the transfected cells. The expression of the primary gene or, if present, its reporter gene should transcribe a product that has a biological residence time long enough to be measured. In vivo imaging gains from an appropriately labeled indicator substrate to bind irreversibly to the gene product such as enzyme, receptor, or transporter, without disturbing cellular homeostasis. Of course, any of the above-mentioned methods for in vivo measuring a biochemical reaction may be useful.

No doubt the possibility of in vivo observing the expression of transferred genes, its duration, and location in cells and tissues will bring new technologies and insights and will immensely benefit clinical research and practice both in diagnosis and therapy. Molecular biology research and application in living tissues as complex adaptive systems is becoming a reality.

1.7
Conclusion

This condensed overview and review of the present state of art in measuring biochemical reactions as they occur in living systems presents the various approaches that allow the linking of gene expression to cell function at the molecular level. Justification, goals, complicating factors, and solutions are given in a synopsis. Some of these techniques are currently more in use than others, but the power of all is evident. The following chapters in this book give more details than the sections in this chapter, in the context of different research and clinical demands. Common to nearly all procedures is the multiple parameter analysis, whether it applies to dynamic imaging of whole body or its regions of interest, or to in vitro analyses of body fluids, excreta, and/or exhaled breath. The various parameters of measurement may combine tracers, substrates and tissue compartments. In whatever interrelationship, the simultaneous attention to different parameters appears to be nearly indispensable for in vivo quantification of biochemical reactions.

Increased investments in cost and labor into further developing these approaches appear to be worthwhile in that they promise to lead to the understanding of fundamental biological functions in living systems, in fact bring molecular biology and cell biology into intimate contact with clinical medicine. They allow the biomedical researcher and physician to respond to the challenge of in vivo functional genomics and proteomics and thus open new diagnostic and therapeutic dimensions.

1.8
References

Antar MA, Spohr G, Herzog HH et al (1986) 15-(ortho-123-I-phenyl)-pentadecanoic acid, a new myocardial imaging agent for clinical use. Nucl Med Commun 7:683–696

Armbrecht JJ, Buxton DB, Schelbert HR (1990) Validation of [1-11-C]acetate as a tracer for noninvasive assessment of oxidative metabolism with positron emission tomography in normal, ischemic, postischemic, and hyperemic canine myocardium. Circulation 81:1594–1605

Ballare E, Persana L, Lania AG et al (2001) Mutation of somatostatin receptor type 5 in an acromegalic patient resistant to somatostatin analog treatment. J Clin Endocrinol Metab 86:3809–3814

Beckurts TE, Shreeve WW, Schieren R et al (1985) Kinetics of different ^{123}I- and ^{14}C-labeled fatty acids in normal and diabetic rat myocardium in vivo. Nuclear Med Commun 6:415–424

Bengel FM, Permanetter B, Ungerer M et al (2002) Alterations of the sympathetic nervous system and metabolic performance of the cardiomyopathic heart. Eur J Nucl Med Mol Imaging 29:198–202

Blaufox MD (1996) Becquerel and the discovery of radioactivity: early concepts. Semin Nucl Med XXVI:145–154

Boerman OC, Oyen WJG, Corstens FHM (2000) Radio-labeled receptor-binding peptides: a new class of radiopharmaceuticals. Semin Nucl Med 30:195–208

Brown MA, Marshall DR, Sobel BE et al (1987) Delineation of myocardial oxygen utilization with carbon-11 labeled acetate. Circulation 76:687–696

DeGrazia JA, Ivanovich P, Fellows H et al (1965) A double isotope method for measurement of intestinal absorption of calcium in man. J Lab Clin Med 66:822–829

Delbeke D, Martin WH, Patton JA, Sandler M (eds) (2002) Practical FDG imaging, a teaching file. Springer, Berlin Heidelberg New York

De Meutter RC, Shreeve WW (1963) Conversion of DL-lactate-2-C^{14} or 3-C^{14} or pyruvate-2-C^{14} to blood glucose in humans: effects of diabetes, insulin, tolbutamide and glucose load. J Clin Invest 42:525–533

Ebert A, Feinendegen DL, Czech N et al (1993) Erfassung des Lipidstoffwechsels und der hepatozellulären Viabilität mittels 15-(para-123-J-Phenyl)-Pentadecansäure (pPPA) und 15-(ortho-131-J-Phenyl)-Pentadecansäure (oPPA) (abstract). Nuklearmedizin 32a:105

Feinendegen LE (1993) Single photon metabolic imaging in cardiology. In: Zaret BL, Beller GA (eds) Nuclear cardiology, state of the art and future directions, chap 24. Mosby Year Book, St Louis, Mo

Feinendegen LE (2000) Myocardial imaging of lipid metabolism with labeled fatty acids. In: Dilsizian V (ed) Myocardial viability: a clinical and scientific treatise, chap 16. Futura, Armonk NY

Feinendegen LE, Ritzl F (1971) Insulin metabolism determination in vivo using iodine-125 and chromium-51 double labeling. Nucl Med (Stuttg) 9:748–751

Feinendegen LE, Heiniger HJ, Friedrich G et al (1973) Differences in reutilization of thymidine in hemopoietic and lymphopoietic tissues of the normal mouse. Cell Tissue Kinet 6:573–585

Feinendegen LE, Vyska K, Freundlieb C et al (1981) Non-invasive analysis of metabolic reactions in body tissues, the case of myocardial fatty acids. Eur J Nucl Med 6:191–200

Feinendegen LE, Henrich MM, Kuikka JT et al (1995) Myocardial lipid turnover in dilated cardiomyopathy: a dual in vivo tracer approach. J Nucl Cardiol 2:42–52

Feinendegen DL, Ohlenschlaeger U, Grossmann K et al (1996) Lipid metabolism in the liver studied in vivo with two isomers of labeled fatty acid analogs. J Nucl Med 37:1841–1845

Feinendegen LE, Herzog II, Thompson KH (2001) Cerebral glucose transport implies individualized glial cell function. J Cereb Blood Flow Metab 21:1160–1170

Firnau G, Garnett ES, Chan PK et al (1976) Intracerebral dopamine metabolism studied by a novel radioisotope technique. J Pharm Pharmacol 28:584–585

Fowler JS, Wolf AP, MacGregor RR et al (1988) Mechanistic positron emission tomography studies: demonstration of a deuterium isotope effect in the monoamine oxidase-catalyzed binding of {C-11}-deprenyl in living baboon brain. J Neurochem 51:1524–1534

Freundlieb C, Hoeck A, Vyska K et al (1978) Use of ω-123-I-labeled heptadecanoic acids for non-invasively measuring myocardial metabolism. In: Woldring M, Schmidt HAE (eds) Proceedings of the 15th international meeting of the Society of Nuclear Medicine, Groningen, 1977. Schattauer, Stuttgart, pp 216–219

Freundlieb C, Hoeck A, Vyska K et al (1980) Myocardial imaging and metabolic studies with (17-123-I) iodoheptadecanoic acid. J Nucl Med 21:1943–1950

Friedrich G, Feinendegen LE, Heiniger HJ (1972) Studies on the incorporation of exogenic DNA in mammalian cells. Hoppe Seylers Z Physiol Chem 353:705–706

Gallagher BM, Ansari A, Atkins H et al (1977) Radiopharmaceuticals XXVII. 18F-labeled 2-deoxy-2-fluoro-d-glucose as a radiopharmaceutical for measuring regional myocardial glucose metabolism in vivo: tissue distribution and imaging studies in animals. J Nucl Med 18:990–996

Gjedde A, Diemer NH (1983) Kinetic analysis of the uptake of glucose and some of its analogs in the brain using the single capillary model: comments on some points of controversy. In: Lambrecht RM, Rescigno A (eds) Lecture notes in biomathematics 48: Tracer kinetics and physiological modeling. Springer, Berlin Heidelberg New York, pp 387–410

Gruvberger S, Ringner M, Chen Y et al (2001) Estrogen receptor status in breast cancer is associated with remarkably distinct gene expression patterns. J Cancer Res 61:5979–5984

Herzog H, Lele VR, Kuwert T et al (1990/1991) Changed pattern of regional glucose metabolism during yoga meditative relaxation. Neuropsychobiology 23:182–187

Hevesy G (1913) Radioelements as indicators in chemistry and physics. Chem News 108:166–167

Hevesy G (1948) Radioactive indicators. Interscience, New York

Hoeck A, Spohr G, Schmitz M et al (1986) 17-iodine-123 iodoheptadecanoic acid for metabolic liver studies in humans. J Nucl Med 27:1533–1539

Kaiser KP, Geuting B, Grossmann K et al (1990) Tracer kinetics of 15-(ortho-123-/131-I-phenyl)-pentadecanoic acid (oPPA) and 15-(para-123/131-I-phenyl)-pentadecanoic acid (pPPA) in animals and man. J Nucl Med 31:1608–1616

Kallie RN, Shreeve WW, Joubert SM (1968) Studies in primary hyperuricaemia III. The conversion of ^{14}C to $^{14}CO_2$ from glucose-1-^{14}C and glucose-6-^{14}C in hyperuricaemia and gout. S African Med J 42:473–476

Kelly DP, Mendelsohn NJ, Sobel BE et al (1993) Detection and assessment by positron emission tomography of a genetically determined defect in myocardial fatty acid utilization (long-chain acyl-Co-A dehydrogenase deficiency). Am J Cardiol 71:738–744

Knapp FF Jr, Kropp J (1995) Iodine-123-labeled fatty acids for myocardial single photon emission tomography: Current status and future perspectives. Eur J Nucl Med 22:361–381

Kim EE, Yang DJ (eds) (2001) Targeted molecular imaging in oncology. Springer, Berlin Heidelberg New York

Landau BR, Chandramouli V, Schumann WC et al (1995) Estimates of Krebs cycle activity and contributions of gluconeogenesis to hepatic glucose production in fasting health subjects and IDDM patients. Diabetologia 38:831–838

Laruelle M (2000) Imaging synaptic neurotransmission with in vivo binding competition techniques: a critical review. J Cereb Blood Flow Metab 20:423–451

Logan J, Fowler JD, Volkow ND et al (1990) Graphical analysis of reversible radioligand binding from time-activity measurements applied to [N-11C-methyl]-(-)cocaine PET studies in human subjects. J Cereb Blood Flow Metab 10:740–747

Machulla H-J, Stoecklin G, Kupfernagel CH et al (1978) Comparative evaluation of fatty acids labeled with C-11, Cl-34m, Br-77, and I-123 for metabolic studies of the myocardium: concise communication. J Nucl Med 19:298–302

Machulla H-J, Marsmann M, Dutschka K et al (1980) Biochemical concept and synthesis of a radioiodinated phenylfatty acid for in vivo metabolic studies of the myocardium. Eur J Nucl Med 5:171–173

Maher BA (2001) Researchers focus on histone code. Scientist 15:15–16

Meyers DK, Feinendegen LE (1975a) Incorporation of thymidine and iododeoxyuridine into the DNA of mouse tissues. Can J Physiol Pharmacol 53:1014–1022

Meyers DK, Feinendegen LE (1975b) Double labeling with [3H]thymidine and [125I]iododeoxyuridine as a method for determining the fate of injected DNA and cells in vivo. J Cell Biol 67:484–488

Meyers DK, Feinendegen LE (1976) DNA turnover and thymidine re-utilization in mouse tissues. Cell Tissue Kinet 9:215–221

Nyhan WL (1984) Nonketotic hyperglycinemia. In: Nyhan WL (ed) Abnormalities in amino acid metabolism in clinical medicine, chap 34. Appleton-Century-Croft, New York, pp 333–351

O'Brien KO, Zaveleta N, Caulfield LE et al (1999) Influence of prenatal iron and zinc supplements on supplemental iron absorption, red blood cell incorporation, and iron status in pregnant Peruvian women. Am J Clin Nutr 69:509–515

Patlak CS, Blasberg RG, Fenstermacher JD (1983) Graphical evaluation of blood-to-brain transfer constants from multiple-time uptake data. J Cereb Blood Flow Metab 3:1–7

Phelps ME, Huang SC, Hoffman EJ et al (1979) Tomographic measurement of local cerebral glucose metabolic rate in humans with (F-18) 2-fluoro-2-deoxy-D-glucose: validation of method. Ann Neurol 6:371–388

Rader DJ, Schaefer JR, Lohse P et al (1993) Increased production of apolipoprotein A-1 associated with elevated plasma levels of high-density lipoproteins, apolipoprotein A-1 and lipoprotein A1 in a patient with hyperalphalipoproteinemia. Metab Clin Exp 42:1429–1434

Reivich M, Kuhl D, Wolf A et al (1979) The (^{18}F)fluorodeoxyglucose method for the measurement of local cerebral glucose utilization in man. Circ Res 44:117–127

Rensberger B (1996) Life itself: exploring the realm of the living cell. Oxford University Press, Oxford

Ritzl F, Feinendegen LE (1971) In vivo determination of site and rate of insulin catabolism using the double tracer technique with 51-Cr and 131-I. In: Dynamic studies with radioisotopes in medicine. Proceedings of the Symposium on Dynamic Studies with Radioisotopes in Clinical Medicine and Research. International Atomic Energy Agency, Vienna, Austria, pp 57–68

Ritzl F, Feinendegen LE, Schnippering HG (1974) A double isotope technique for estimating insulin degradation in vivo. Nucl Med (Stuttg) 13:85–97

Ritz P, Coward WA (1995) Doubly labeled water measurement of total energy expenditure. Diabetes Metab 21:241–251

Rodwell VW (1996) Enzymes: kinetics. In: Murray RK, Granner DK, Mayes PA et al (eds) Harper's biochemistry, 24th edn, chap 9. Appleton and Lange, Stamford, Conn, pp 75–90

Sasaki Y (1995) Carbon-14 and Carbon-13 breath tests. In: Wagner HN Jr, Szabo Z, Buchanan JW (eds) Principles of nuclear medicine, 2nd edn, chap 40. Saunders, Philadelphia, pp 958–965

Schelbert HR, Henze E, Sochor H et al (1986) Effects of substrate availability on myocardial C-11 palmitate kinetics by positron emission tomography in normal subjects and patients with ventricular dysfunction. Am Heart J 111:1055–1064

Schoenheimer R (1946) The dynamic state of body constituents. Harvard University Press, Cambridge, Mass, pp 1–78

Schroeder H, Schelbert HR (2000) Positron emission tomography for the assessment of myocardial viability: noninvasive approach to cardiac pathophysiology. In: Dilsizian V (ed) Myocardial viability: a clinical and scientific treatise, chap 17. Futura, Armonk, NY

Schultze B, Gregoire F, Hughes WL (1964) Renal uptake of pancreatic ribonuclease after intravenous injection in mice and rats. Technical Report, Brookhaven National Laboratory, Upton, NY, BNL-8683

Shapiro B, Gross MD, Sisson JS (1995) Neural crest tumors. In: Wagner HN Jr, Szabo Z, Buchanan JW (eds) Principles of nuclear medicine, 2nd edn, chap 33. Saunders, Philadelphia, pp 665–680

Shreeve WW, Cerasi E, Luft R (1970) Metabolism of (2-^{14}C) pyruvate in normal, acromegalic and growth hormone-treated human subjects. Acta Endocrinol 65:155–169

Shreeve WW, Tashjian AJ, Oji N et al (1971) Formation of $^{14}CO_2$ and ^3HOH from glucose-1-^{14}C-1-^{3}H during oral cortisone glucose tolerance tests in obese patients. Metab Clin Exp 20:280–292

Shreeve WW, Schieren R, Machulla HJ et al (1984) Hepatic uptake and fate of ^{123}I- and ^{14}C-fatty acids in normal and ethanolic mice. Nucl Med Commun 5:519–524

Silverman DHS, Hoe CK, Seltzer MA et al (1998) Evaluating tumor biology and oncological disease with positron-emission tomography. Semin Rad Oncol 8:183–196

Smith SM, Wastney ME, Nyquist LE et al (1996) Calcium kinetics with microgram stable isotope doses and saliva sampling. J Mass Spectrom (CMB) 31:1265–1270

Sokoloff L, Reivich M, Kenney C et al (1977) The [14C]deoxyglucose method for the measurement of focal cerebral glucose utilization: theory, procedure and normal values in the conscious and anesthetized albino rat. J Neurochem 28:897–916

Stern MP (2000) Strategies and prospects for finding insulin resistance genes. J Clin Invest 106:323–327

Stryer L (1995) Enzymes: basic concepts and kinetics. In: Stryer L, Biochemistry, 4th edn, chap 8. Freeman, New York, pp 181–206

Syrota A, Merlet P, Delforge J (1995) The heart: clinical neurotransmission. In: Wagner HN Jr, Szabo Z, Buchanan JW (eds) Principles of nuclear medicine, 2nd edn, chap 37, sect 2. Saunders, Philadelphia, pp 759–773

Taylor SI (1995) Diabetes mellitus. In: Scriver CR, Beaudet AL, Sly WS, Valle D (eds) The metabolic and molecular bases of inherited disease I, 7th edn, chap 21. McGraw-Hill, NY, pp 843–896

Van Eenige MJ, Visser FC, Duwel CMB et al (1987) Analysis of myocardial time activity curves of I-123-heptadecanoic acid I. Curve fitting. Nucl Med 26:241–247

Wagner HN Jr (1995) Nuclear medicine: what it is, what it does. In: Wagner HN Jr, Szabo Z, Buchanan JW (eds) Principles of nuclear medicine, 2nd edn, chap 1. Saunders, Philadelphia, pp 1–8

Wagner HN, Burns HD, Dannals RF et al (1983) Imaging dopamine receptors in the human brain by positron tomography. Science 221:1264–1266

Wagner HN Jr, Szabo Z, Buchanan JW (eds) (1995) Principles of nuclear medicine, 2nd edn. Saunders, Philadelphia

CHAPTER 2

The Human Genome and Disease

EBERHARD PASSARGE

2

Contents

also see Waterston 2002), *Drosophila* (Adams et al. 2000), *C. elegans* (The *C. elegans* Sequencing Consortium 1998), yeast (Goffeau et al. 1996), and microbes (Doolittle 2002); websites are included in the publications cited.

The branch of human genetics primarily concerned with human diseases, medical genetics, benefits directly from the rapid advances in genomic sciences. Human genetic disorders can now be defined and diagnosed at a level of precision beyond imagination just a decade ago. This chapter reviews the principal results and open questions in viewing human diseases against the background of the unfolding knowledge of the human genome.

2.1
Introduction

The exploration of the human genome yields new information about the structure and function of the genes of man, their evolutionary origin, the genomic endowment of *Homo sapiens* in relation to other organisms, the causes of diseases, and other subjects. Human genetics today derives important insights from the study of model organisms such as the mouse and rat, the fruit fly (*Drosophila melanogaster*), a nematode (*Caenorhabditis elegans*), yeast (*Saccharomyces cerevisiae*), many bacteria and their plasmids, and also plants. The complete DNA sequence of these organisms provides an understanding of the pathogenicity of microbial organisms, opens new ways to develop therapies and vaccines and to understand the functional consequences of mutations. Data on sequenced organisms are readily available for man (International Human Genome Sequencing Consortium 2001; Venter et al. 2001), mouse (draft of 96% of the genome of the C57BL/6J mouse strain posted on 9 May, 2002 at Genbank (www.ncbi.nlm.nih.gov;

2.2
Components of the Nuclear Human Genome

The human genome (and that of other mammals) contained in the cell nucleus (as opposed to the small 16,569-bp mitochondrial genome) consists of 3 billion base pairs (bp) of DNA (3000 Mb) per haploid set of chromosomes (chromosome 1–22, the X- and the Y-chromosome). However, only about 3% contains coding information and about 7% are related to regulatory functions. Most of the DNA (90%) is not related to genes. Forty-five percent of the human genome is composed of DNA derived from previous transposon and virus invasions. These are represented as various types of repetitive sequences: tandemly repeated DNA (satellite DNA) and interspersed genome-wide repeats (long interspersed nuclear elements, LINEs, accounting for 21%; short interspersed nuclear elements, SINEs, accounting for 13%; long terminal repeats, LTRs, derived from retroviruses, accounting for 8%; and DNA-transposons, accounting for 3%. All of this is reviewed by

Strachan and Read 1999, Lewin 2000, Lodish et al. 2000, Alberts et al. 2002, and Brown 2002. LINEs and SINEs are referred to as retroelements (or retrotransposons) because they are thought to have arisen by transposition involving an RNA intermediate (retrotransposition). This is a noteworthy feature of the human genome because of their similarities to retroviruses. Alu sequences are the most abundant type of SINE sequences in the human genome. They occur about once every 3 kb and account for about ten percent of the mass of the human genome (Batzer and Deininger 2002). They derive their name from the restriction enzyme AluI, which has a recognition site within the sequence. Alu repeats have a total length of about 300 bp. They have a bipartite structure with a 31-bp insertion in the 5′ half. In contrast, the mouse genome does not contain Alu sequences.

A distinctive feature of the human genome is its great interindividual variability. At least once every 600 base pairs, humans differ in that one may have one nucleotide (e.g., C) at a position, while others have a different nucleotide at this position (e.g., G). This is called a single nucleotide polymorphism (SNP). Although theoretically four alternatives for each of the four types of nucleotides are possible, most exist in just two variants. SNPs are target sites in attempts to elucidate possible predisposition for particular complex diseases or variant pharmacologic responses to certain chemical compounds (pharmacogenomics; for a review see Roden and George 2002). The reason for the great interest SNPs have created is the expectation that they can be used as markers to identify genes that predispose individuals to common, complex disorders by detecting linkage disequilibrium (LD). This is a deviation from the random distribution of marker alleles (SNPs) and disease-predisposing alleles. SNPs alleles located close enough to disease-predisposing alleles will be inherited together at least through a few generations (linkage), whereas all other alleles will segregate in a random manner. Thus, SNP alleles could serve as markers to identify loci relevant to the causes of particular disorders. Presumably this could be a basis for developing new therapeutic drugs aimed at individual disease predisposition (for review see Syvänen 2001). A map of 1.42 million SNPs has been published for the human genome (Sachidanandam et al. 2001).

Another feature of the human genome is the widespread presence of tandemly arranged DNA repeats, e.g., repeats of CA, with the number of CAs varying from 2 to 8 per site per chromosome (variable number of tandem repeats, VNTRs). Since these are individual, hereditary variations, they can be used to trace their origin from the parents and grandparents of any individual. If located near or within a disease-causing allele of a gene, they can be utilized in the diagnosis by genotyping (haplotype analysis) even when the relevant mutation cannot be shown directly (indirect DNA diagnosis).

2.3
Impact of the Human Genome Project

The reports of the DNA sequence of the human genome (International Human Genome Sequencing Consortium 2001; Venter et al. 2001) revealed a surprisingly low number of genes, only about 34,000 instead of the expected 80,000 (up to 140,000 were thought to be possible). Thus, in size the human genome is not very much greater than that of Drosophila (about 14,000 genes), C. elegans (about 19,000), yeast (6000), or E. coli (4300).

Thus, the number of genes per se does not seem to determine the overall biology of an organism. Presumably it is the regulation of genes and their expression pattern during development, the formation of functional networks, and other features that are equally important. One mechanism contributing to the functional complexity of the human genome may be alternative splicing of the transcript (RNA). Indeed, about 40–60% of human genes have alternative splice forms (Graveley 2001; Mironov et al. 1999; Modrek et al. 2001; Modrek and Lee 2002). However, open questions remain about their true functionality and about the regulation of splicing.

The complete sequence of human chromosomes 14, 20, 21, and 22 has greatly contributed to the identification of genes involved in disease (Deloukas et al. 2001; Heilig et al. 2003). For example, human chromosome 20 contains 727 genes in 59,187,298 bp, corresponding to 12 genes per 1 million base pairs (chromosome 21: 6.7 and chromosome 22: 16.3). The relatively low number of genes on chromosome 21 presumably is the reason for survival to birth in 1 of 4 fetuses with trisomy 21, whereas trisomies 20 and 22 do not occur in liveborn infants.

The density of genes along a stretch of DNA or on a whole chromosome differs considerably. The average gene density in the human genome is low, i.e., about one gene per 100 kb (100,000 bp) compared to one gene in 9 kb in *Drosophila* or in 7.5 kb in *Ciona intestinalis*, a 550 million year deuterostome (Dehal et al. 2002). Seven genes may be present along a 6-Mb region, whereas another 1.1-Mb region may contain 17 genes. The human genome contains 234 gene-poor sections ranging from 620,000 to 4 million base pairs ("deserts") which account for about 9% of the genome.

Aside from transfer RNAs (tRNA) and ribosomal RNAs (rRNA), other RNAs that do not function as messenger RNA (mRNA) have been found. In prokaryotes they are referred to as small RNAs (sRNA) and in eukaryotes as noncoding RNA (ncRNA). They are involved in a wide variety of fundamental processes, such as transcriptional regulation, replication, RNA processing, and translation. Because they cannot be identified by an open reading frame as coding genes, it is difficult to determine how many ncRNAs are encoded by a genome (Hannon 2002). Thus, many questions will remain after the complete sequence of the human genome is available.

2.4
Comparative Genomics

Genomes of different organisms are related because of their evolutionary origin from a common ancestor at different stages of the development of life on earth. Many human genes have a homologous counterpart in such different organisms as the fruit fly, a small nematode, yeast, bacteria, and even plants. In particular, human disease genes (i.e., genes known to cause a human disease when mutated to a nonfunctional variant) involved in important physiological functions related to cell growth and differentiation, neurological functions, and aging, have been shown to be similar in other organisms. In many cases the human gene has been found owing to its similarity to a gene in another organism.

The human genome and that of the mouse share about 200 homology segments along their chromosomes. These are stretches of a chromosome that carry homologous genes in the same sequence in both organisms. The human chromosome 19 has 9 mouse homology segments, the mouse chromosome 16 has 7 human homology segments.

2.5
Human Genetic Diseases

Diseases result from environmental influences interacting with the individual genetic makeup – the genome and its genes. Usually there is no access to individual components of this interaction. We just see the result as a disease diagnosed. This group of diseases is referred to as complex diseases, also named multifactorial or multigenic disorders. The individual contributions of the genetic components are visible by a difference between the population frequency and the frequency in relatives, which can be used to establish the risk of occurrence in first and second degree relatives. This group predominates in all medical areas, because it involves such important groups as cardiovascular diseases, diabetes mellitus, cancer, psychiatric disorders, age-related disorders, and others (Rees 2002; Strohman 2002).

In contrast, about one thousand human diseases have been recognized at the molecular level to result from a pathological change in a single gene (monogenic diseases). All manifestations of the disease (the phenotype) result from a specific interruption of a metabolic or signal pathway or a cell-specific functional deficiency (the genotype). About another 2,000 monogenic disorders are defined at the clinical level within varying degrees of precision. A complete catalogue is available as *Mendelian Inheritance in Man* (MIM, McKusick 1998, and online OMIM, McKusick 1998). The monogenic diseases are individually rare, with some exceptions in different populations, but together they are frequent enough to contribute in a major way to morbidity and mortality (Table 2.1).

The use of the McKusick catalogue is indispensable when dealing with monogenic human disorders. The main reason is the genetic heterogeneity of most of

Table 1. Categories and Frequency of Genetically Determined Diseases

Category of disease	Frequency per 1000 individuals
Complex (multifactorial) disorders	70–90
Monogenic diseases (total)	4.5–15
Autosomal dominant	2–9.5
Autosomal recessive	2–3.5
X-chromosomal	0.5–2
Chromosomal aberrations	5–7

them: they may be clinically indistinguishable (same or similar phenotype), but result from mutations in different genes of the same or a related pathway or in different alleles of the same gene. As these may differ in the mode of Mendelian inheritance, failure to recognize genetic heterogeneity may result in erroneous risk assessment, because autosomal dominant, autosomal recessive, or X-chromosomal inheritance is not distinguished.

Human chromosomal disorders are another category of disease. They result from a specific aberration of the chromosome number (aneuploidy: trisomy or monosomy) or structure (partial deficiency or duplication). They always involve several organ systems and tissues, because even small structural changes involve many genes. The human genome is markedly sensitive to gene dosage. Any deficiency or duplication of a chromosomal segment results in developmental disturbance. This probably is the reason for the limited spectrum of numerical aberrations in human liveborn infants: just trisomy 13, trisomy 18, and trisomy 21, each with a characteristic phenotype and impaired developmental range, and variable numbers of the X- and the Y-chromosome.

Many human diseases occur in a non-hereditary form due to one or several somatic mutations in a particular cell type or tissue. All human tumors result from somatic mutations. They differ with respect to the genes involved, number, sequence, and type of mutations, time of occurrence in relation to the cell cycle, the age of the individual, and other aspects. In a variable, small proportion of individuals, the first, tumor-predisposing mutation may be present in the germline, either due to a new mutation or resulting from transmission of the mutation present in one of the parents (hereditary form). For example, a heritable predisposing mutation in one of the two main genes involved in breast cancer (*BRCA1* and *BRCA2*) is present in 5–10% of patients with breast cancer. In practice, it is quite difficult to assess the risk conferred by many of the different tumor-predisposing mutations that are found in different patients (for review see Vogelstein and Kinzler 2001). The principles and practice of diagnosis and management of human genetic disorders are presented in depth in several multivolume, multiauthor books (King et al. 1992; Jameson 1998; Gilbert-Barness and Barness 2000; Scriver et al. 2001; Rimoin et al. 2002), some of the underlying principles and concepts, in single author books (Weatherall 1991; Childs 1999; Passarge 2001) and the references cited therein.

2.6
Genomic Disorders

The human genome contains numerous areas that are duplicated either on the same or a different chromosome (segmental duplications). If located near each other or arranged tandemly, this may interfere with the precise alignment of homologous chromosomes during meiosis. As a result, complementary deletions and duplications may arise after unequal crossing-over and separation at anaphase. Region-specific low-copy repeats are involved in a number of diseases that are transmitted in a Mendelian mode of inheritance (Emmanuel and Shaikh 2001; Guttmacher and Collins 2002; Stankewiecz and Lupski 2002). For example, in autosomal dominant neurofibromatosis (NF1) 5–20% of patients have an interstitial microdeletion of 1.5 Mb at region 1, band 1.2 of the long arm (q) of chromosome 17 (17q11.2). The proximal breakpoints of the common deletion cluster at two directly oriented low-copy repeats of 85 kb (85,000 bp), whereas the distal breakpoints vary. Unequal crossing-over at this region results in the deletion. Another important group of so-called genomic disorders are rearrangements of subtelomeric regions.

2.7
Transcriptome and Proteome

The transcriptome is the initial product of the overall expression of a genome, corresponding to a collection of RNA molecules derived from protein-coding genes required for the biological functions of a cell. The proteome corresponds to the final product of genome expression. It includes all proteins present in a cell at a given stage of development and function. The proteome links the genome and its expression to the biochemical functions. Thus, it assumes a central role or our understanding of cellular functions and life (functional genomics). When the sequence of nucleotides of the genome of an organism has been determined, its transcriptome and proteome will assume the center stage of investigation.

2.8
Ethical and Societal Issues (ELSI)

ELSI was an important component of the Human Genome Project from the beginning in the early 1990s. It anticipated that the new knowledge expected would need to be channeled in a way to protect the confidentiality of individuals and avoid discrimination based on data obtained by genetic testing. It was recognized early on that the public should be informed properly about the goals and the consequences of the Human Genome Project.

A particularly important aspect is the application of predictive genetic tests. These are defined as diagnostic procedures, usually based on DNA analysis, aimed at detecting a genetic disorder or its predisposition before it becomes manifest. Thus, information about a person's future health status can be obtained years or in some cases decades before onset of a disease. Although currently this is possible only under relatively narrow circumstances, it raises important questions about the goals of a predictive genetic diagnosis, its basis of decision, individual choice and confidentiality, reliability of the test result, and other considerations. Recommendations and guidelines by major scientific societies – such as the American Society of Human Genetics or the European Society of Human Genetics – and legal requirements demand that informed consent has be obtained by proper medical (genetic) counseling prior to any predictive genetic test. One main aspect of the counseling is to determine whether the result of the test would benefit the individual. If this is not the case or is doubtful, for example because the test result can not be used for meaningful medical intervention, the test should not be done or should be postponed. The overriding principle is protection of the individual from a breach in confidentiality and misuse of genetic data by third parties (Burke 2002; for a review see German Research Council 2000).

2.9
Technical Considerations

The analysis of the human genome would not be possible without major advances in automated DNA sequencing, various strategies in mapping genes, molecular biology, and bioinformatics during the past decade. Large-scale DNA sequencing has been completely automated. It involves four base-specific fluorophores, one for each of the four DNA bases. These are attached to dideoxynucleotide triphosphates (ddNTP). These are analogues of the normal dNTPs, but lack a hydroxyl group at the 3′ and the 2′ carbon position. When incorporated into the growing DNA chain they cannot participate in the normal phosphodiester bonding at its 3′ carbon. Thus, at the site of its incorporation the DNA chain synthesis is interrupted. This results in a DNA fragment whose size is determined by the position of the particular base in one of four parallel base-specific reactions. The DNA fragments, each specified by the terminal position of a ddNTP following random chain termination, are separated according to size in a small electrophoretic capillary. Each ddNTP is labeled with a base-specific fluorescent dye. The electrophoretic migration of the labeled fragments are recorded by a laser beam in a fixed position, inducing a fluorescent signal for each of the four bases. Four example, if ddATP is labeled with a green dye, ddCTP blue, ddGTP yellow, and ddTTP red, the sequence CTAAGTACG would result in visible peaks of the colors blue-red-green-green-yellow-red-green-blue-yellow. The sequence obtained is recorded. Subsequently it can be compared with any of the known sequences stored in large data banks, and conclusions about its possible functional properties can be drawn.

Prior to sequencing, various techniques to amplify small amounts of DNA, to arrange overlapping DNA fragments in the correct order, to map genes, to search for mutations, and to reach other investigative goals are available (for review see Strachan and Read 1999).

Microarrays have proven to be useful for the simultaneous study of expression patterns of hundreds of genes. A microarray or DNA chip is an assembly of DNA probes fixed on a fine grid of small surfaces to analyze the expression of genes represented in cDNA (complementary DNA) prepared from mRNA. Two basic types of microarrays are DNA clones or PCR products of genes (relevant DNA fragments isolated and amplified by the polymerase chain reaction) and microarrays of oligonucleotides synthesized in situ on a suitable surface. In either case, labeled RNA probes can be hybridized to the targets. Tens of thousands of probes can be examined on a high density glass slide (chip) of just 1.28×1.28 cm (Strachan and Read 1999; Brown 2002; Petricoin et al. 2002).

The human genome can be visualized at the chromosomal level (molecular cytogenetics). Fluorescence in situ hybridization (FISH) rests on the principle of denaturing metaphase chromosome preparations (making the DNA single-stranded) and then hybridizing them with a chromosome- or site-specific single-stranded fluorochrome-labeled DNA probe. At the site of hybridization this will elicit a signal visible by dark field microscopy. The advantage is the greatly increased resolution compared to conventional chromosome preparation. As an alternative, one or more entire chromosomes can be specifically stained (so-called whole chromosome painting) and studied for structural aberrations. Multiplex FISH and spectral karyotyping are two other important advances with high resolution and a detection rate of small structural aberration down to about 5-10 Mb (for further information see Schröck et al. 1996; Speicher et al. 1996; Passarge 2001).

- Genomes of other organisms: (www.ncbi.nlm.nih.gov/Entrez/Genome/org.html).
- UK Human Genome Mapping Project Resource Center: (www.hgmp.mrc.ac.uk/).
- German Human Genome Project: (www.dhgp.de).
- European Bioinformatics Institute: (www.ebi.ac.uk).
- Genbank: (www.ncbi.nlm.nih.gov).
- DNA Database of Japan: (www.nig.oc.jp/home.html).
- Medline: (www.ncbi.nlm.nih.gov/PubMed/).

(For additional information about websites see Strachan and Read 1999; Passarge 2001; Brown 2002).

2.10
Conclusions

When the gaps remaining in the sequence of the human genome are closed, by about 2003, we will be left with many genes of unknown function. Functional genomics will reach the mainstream in the quest for interaction of genes and gene systems, structure and function of gene products, individual predisposition to complex disorders, and other challenges of the future.

■ **Acknowledgement.** I thank Professor Bernhard Horsthemke for helpful comments about the manuscript.

2.11
Appendix

Selected Websites Providing Access to Genomic Data

- Human Genome Organization (HUGO): (www.gene.ucl.ac.uk/hugo/).
- The National Center for Biotechnology Information: (www.ncbi.nlm.nih.gov/).
- National Center for Human Genome Research: about the Human Genome Project: (www.nhgri.nih.gov/HGP/), about Genes and Disease: (www.ncbi.nlm.nih.gov/disease/).
- Network of databases: (www.ncbi.nlm.nih.gov/Database/).

2.12
References

Adams MD (2000) et al. The genome sequence of *Drosophila melanogaster*. Science 287:2185–2195

Alberts B, Johnson A, Lewis JJ et al (2002) Molecular biology of the cell, 4th edn. Garland, New York

Batzer MA, Deininger PL (2002) Alu repeats and human diversity. Nature Rev Genet 3:370–379

Brown TA (2002) Genomes, 2nd edn. Bios, Oxford

Childs B (1999) Genetic medicine. A logic of disease. Johns Hopkins University Press, Baltimore

Burke W (2002) Genetic testing, N Engl J Med 347:1867–1875

Dehal P, Satou Y, Campbell RK (2002) The draft genome of *Ciona intestinalis*: Insights into chordate and vertebrate origins. Science 298:2157–2167

Deloukas P, Matthews LH, Ashurst J et al (2001) The DNA sequence and comparative analysis of human chromosome 20. Nature 414:865–871

Doolittle RF (2002) Microbial genomes multiply. Nature 416:697–698

Emmanuel BS, Shaikh TH (2001) Segmental duplications: an expanding role in genomic instability and disease. Nature Rev Genet 2:791–800

German Research Council (2000) Human Genome Research and Predictive Genetic Diagnosis: possibilities – limitations – consequences. Statement by the Senate Commission on Genetic Research of the German Research Council, DFG, report 2, 20 June 1999. Wiley-VCH, Weinheim, pp 107–137

Gilbert-Barness E, Barness L (2000) Metabolic diseases. Foundation of clinical management, genetics, and pathology. Eaton Publishing Co., Natick, Mass.

Goffeau A, Barrell BG, Bussey H et al (1996) Life with 6000 genes. Science 274:562–567

Graveley BR (2001) Alternative splicing: increasing diversity in the proteomic world. Trends Genet 17:100–107

Guttmacher AE, Collins FS (2002) Genomic medicine – a primer. N Engl J Med 347:1512–1520

Hannon GJ (2002) RNA interference. Nature 418:244–251

International Human Genome Sequencing Consortium (2001) Initial sequencing and analysis of the human genome. Nature 409:860–921

Jameson JL (ed) (1998) Principles of molecular medicine. Humana Press, Totowa, NJ

King R, Rotter J, Motulsky AG (eds) (2002) The genetic basis of common disorders, 2nd edn. Oxford University Press, Oxford

Lewin B (2000) Genes VII. Oxford University Press, Oxford

Lodish H, Berk A, Zipurska SL et al (2000) Molecular cell biology (with animated CD-ROM), 4th edn. Freeman, New York

McKusick VA (1998) Mendelian inheritance in man. A catalog of human genes and genetic disorders, 12th edn. Johns Hopkins University Press, Baltimore (www.ncbi.nlm.nih.gov/Omim)

Mironov AA, Ficket JW, Gelfand MS (1999) Frequent alternative splicing of human genes. Genome Res 9:1288–1293

Modrek B, Lee C (2002) A genomic view of alternative splicing. Nature Genet 30:13–19

Modrek B, Resch A, Grasso C, Lee C (2001) Genome-wide analysis of alternative splicing using human expressed sequence data. Nucleic Acids Res 29:2850–2859

Passarge E (2001) Color atlas of genetics, 2nd edn. Thieme, Stuttgart New York

Petricoin EF III, Hackett JL, Lesko LJ (2002) Medical applications of microarray technologies: a regulatory science perspective. Nature Genet Suppl 32:474–479

Rees J (2002) Complex disease and the new clinical sciences. Science 296:698–701

Rimoin DL, Connor JM, Pyeritz RE, Korf RB et al (eds) (2002) Principles and practice of medical genetics, 4th edn. Churchill-Livingstone, Edinburgh

Roden DM, George AL Jr (2002) The genetic basis of variability in drug response. Nature Rev Drug Discov 1:37–44

Sachidanandam R, Weissman D, Schmidt SC et al (2001) A map of human genome sequence variation containing 1.42 million single nucleotide polymorphisms. Nature 409:928–933

Schröck E, du Manoir S, Veldman T et al (1996) Multicolor spectral karyotyping of human chromosomes. Science 273:494–497

Scriver CR, Beaudet AL, Sly WS et al (eds) (2001) The metabolic and molecular bases of inherited disease, 8th edn. McGraw-Hill, New York

Speicher MR, Ballard SG, Ward DC (1996) Karyotyping human chromosomes by combinatorial multi-fluor FISH. Nature Genet 12:368–375

Stankewiecz P, Lupski JR (2002) Genome architecture, rearrangements and genomic disorders. Trends Genet 18:74–82

Strachan T, Read AP (1999) Human molecular genetics, 2nd edn. Bios, Oxford

Strohman R (2002) Maneuvering in the complex path from genotype to phenotype. Science 296:701–703

Syvänen A-C (2001) Accessing genetic variation: genotyping single nucleotide polymorphisms. Nature Rev Genet 2:930–942

The C. elegans Sequencing Consortium (1998) Genome sequence of the nematode C. elegans: a platform for investigating biology. Science 282:2012–2018

Venter JC, Adams MD, Myers EW et al (2001) The sequence of the human genome. Science 291:1304–1351

Vogelstein B, Kinzler KW (2001) The genetic basis of human cancer, 2nd edn. McGraw-Hill, New York

Waterston RH (2002) Mouse genome sequencing consortion: Initial sequencing and comparative analysis of the mouse genome. Nature 420:520–562

Weatherall DJ (1991) The new genetics and clinical practice, 3rd edn. Oxford University Press, Oxford

CHAPTER 3

Functional Genomics and Proteomics: Basics, Opportunities and Challenges

3

Nikolai Kley, Stefan Schmidt, Vivian Berlin, Hannes Loferer, Jeno Gyuris

Contents

3.1 Introduction

Genomics and proteomics are changing our understanding of biology. To date, the greatest impact has come from DNA sequencing projects, which recently culminated in the unveiling of the near-complete 3.2-billion base-pair sequence of the human genome (International Human Genome Sequencing Consortium 2001; Venter et al. 2001). We have thus advanced from having only limited information about the genetic details of biology to possessing an immense amount of structural information about individual genes. The complete genome sequences of more than 60 species are now available in databases, and many more are expected to become available in the near future. These information resources alone will have a significant impact on biomedical research. Even more importantly, blueprints of genomes can provide the basis for the integration of complex data sets derived from a wide range of studies in genomics, functional genomics, and proteomics. The resulting increase in genetic and biological information will have an even greater impact on biomedical research and the way medicine is practiced.

In general terms, genomics refers to the generation of information about genes and genomes by systematic approaches that can be performed at an industrial scale, such as the sequencing and physical mapping of genes. A richness of information that is relevant to basic, applied, and medical sciences can be derived from blueprints of genomes. This includes, but is not necessarily limited to, information about: (1) the identity and modular structure of genes and their encoded proteins; (2) the classification of gene and protein families, and their evolutionary relationships; (3) the molecular basis of evolution (i.e., changes that have led to speciation and existing phylogeny);

(4) the linkage of specific genetic markers to inherited traits and disease susceptibility; and (5) the molecular basis of susceptibility/responsiveness to drug treatment (pharmacogenomics).

However, only limited information about the biological function of genes and proteins can be extracted from sequence information alone, despite the diverse bioinformatic queries to which genome sequences can be subjected at present. Indeed, the surprisingly low number of genes predicted to be encoded by the human genome (currently estimated to be 30,000–40,000 genes) implies that regulatory processes such as differential gene expression, alternative splicing, and post-transcriptional and post-translational regulatory processes must significantly contribute to the complexity of molecular events underlying cellular processes and specification. Indeed, it is estimated that, on average, ten variants of each protein exist (which includes splice variants and post-translationally modified variants). This raises the complexity of the proteome significantly beyond that of the genome. If one includes differential expression as a dynamic parameter, the number of possible states of the proteome are yet several orders of magnitude more complex. This is what underlies the true complexity of cellular functions. A major challenge ahead will be to use genomic information resources to understand the functions of all genes and the proteins they encode. Understanding how genes and proteins collaborate and interact to carry out cellular processes – that is, understanding how complex biological systems operate ("systems biology") – will be among the most difficult challenges ahead.

Functional genomics refers to the systematic generation and analysis of information about what genes do, and involves a broad range of technologies and experimental approaches that strive to integrate genomic information with information gathered from gene-driven experimentation. These include genome-wide analysis of gene expression (analysis of the transcriptome) as well as the analysis of phenotypic effects associated with induced changes in gene expression. Operating on a large scale, functional genomics is a truly multidisciplinary science that integrates genomics, genetics, molecular and cellular biology, computer science (bioinformatics), and engineering and automation technologies. Proteomics integrates a similarly diverse range of technologies but focuses on the large scale analysis of the functions of protein products encoded by genomes – and includes (1) the analysis of cellular protein expression, protein modifications, and protein-protein interactions; (2) the mapping of signaling

pathways; and ultimately (3) the development of a proteome network of cellular signaling pathways.

Chemical genomics and proteomics refers to the systematic analysis of the impact of organic small molecule drugs on the genome and proteome. This chemistry-oriented analysis is focused on improving our understanding of the molecular basis of the mechanism of action of drugs on a genome- and proteome-wide level. Current strategies include: (1) imaging the dynamics of the impact of small molecules, and their metabolites, on the expression of genomes (transcriptome analysis) and proteomes; (2) uncovering cellular targets and pathways that underlie cellular responses to small molecules; and (3) the rational design of small molecules for the selective perturbation of endogenously expressed or selectively engineered proteins. Insights gained from the use of these strategies can afford a better understanding of what drugs do and how they act. Thus, collectively, chemical genomics and proteomics strive to improve our understanding of the target specificity of a compound, reveal early in the discovery process potential safety issues with respect to identified target and off-target interactions and effects, highlight potential new uses of compounds, and guide medicinal chemistry to the synthesis of drugs with more selective and improved properties. Furthermore, in the near future, chemical genomics and proteomics data may be integrated with pharmacogenomic data on allelic variances occurring in populations and, thereby, improve the prediction of drug response profiles. Chemical genomics and proteomics will also promote the emergence of small molecules that can be used as tools to probe the function of genes and proteins. To quote Richard Klausner, director of the NCI: "The discovery, modification and annotation of small molecules in terms of their ability to probe and perturb biological targets will be one of the central tools of the post-genomic era". Chemical genomics emulates all the principles of genetics, but rather than relying on genetic mutations to dissect function, it uses small molecules. These may prove particularly useful in probing the function of proteins in vivo.

The exploitation of genomics, functional genomics, and proteomics to elucidate physiological and pathological processes will play an important role in the future practice of medicine. In this chapter we review the basic principles as well as recent technological advances in functional genomics and proteomics, and examine how progress in these areas affects our understanding of systems biology and the molecular basis of disease, and how it stimulates the development of novel research tools, diagnostics, prognostics, and therapeutics.

3.2
Functional Genomics: Technologies and Applications

3.2.1
Genome Profiling

3.2.1.1
Gene Expression Profiling

The traditional gene-by-gene approach will not suffice to meet the sheer magnitude of the challenge of understanding biological systems with 40,000 or more genes. It will be necessary to take "global views" of biological processes, which requires a simultaneous monitoring of the activity and regulation of as many cellular components as possible. Global analysis of gene expression represents an important piece of such a puzzle. Imaging transcriptional programs can reveal how global gene expression is remodeled during changes in cell growth, physiology, pathology, or environment, and can provide important information about gene function. Already, the recent development of technologies that enable gene expression studies at large scale have begun to have a profound impact on biological research, pharmacology, and medicine – as exemplified, for instance, by:

1. Determination of the tissue and cell type specificity of gene expression. A gene with expression restricted to a certain cell type or tissue is unlikely to be involved in the pathology of a disease affecting another cell type or tissue, unless a secreted protein is involved. Knowledge of the tissue (and cell) specificity of gene expression is also critical for assembling biological pathways (co-expression) and validating suitable targets for therapeutic intervention.
2. Cell-signaling studies. Alteration of gene expression is a critical step in most cellular signaling pathways. Gene-expression studies have been designed to identify genes whose expression depends on a cell state or on the functions of specific components of signaling or transcription apparatuses. For instance, as studies in yeast indicate, functions of a gene can be predicted from the expression profile of a cell with a mutation in that gene.
3. Composite gene expression studies to infer the function(s) of a gene on the basis of its co-regulation with other genes. Large-scale gene expression profiling and gene clustering studies have shown, for both prokaryotes and eukaryotes, that genes that regulate specific cellular processes (e.g., RNA splicing, cell cycle progression, glycolysis, and other metabolic pathways) are often co-regulated. Consequently, when analyzed in the context of the pattern of expression of thousands of genes, the similarity of the behavior of a gene to that of other genes with known function can provide clues to biological function ("guilt by association").
4. Determination of gene expression patterns in disease. Diversion of normal physiology is frequently accompanied by a panoply of histological and biochemical changes, including changes in gene expression patterns. The up or downregulation of gene activity can be either the cause of the pathophysiology or the result of disease. Consequently, genes may be identified which (1) can serve as diagnostic markers, (2) cause disease and can be targeted for therapeutic intervention, and/or (3) are expressed as a consequence of disease and can lead to strategies aimed at alleviation of symptoms.

   Cancer is a good example for which the utility of gene expression has been amply demonstrated in the discovery of novel therapeutic targets as well as in the classification of cancers. Gene expression profiles will become a valuable tool with which pathologists and oncologists can obtain a more global quantitative approach in the classification of cancers and the prediction of outcomes. It will allow physicians to follow the progression of disease even when there is no histological evidence of change. Furthermore, in conjunction with gene mutation and gene polymorphism analysis, the ability to detect differences in human cancers by the difference in expression profiles is likely to aid in the selection of appropriate therapies.
5. Gene expression studies in disease models in inbred animals. Detailed profiling of gene expression in model systems can yield important insights into cellular, animal, and human physiology critical to the discovery and validation of therapeutic targets.
6. Gene expression studies in pathogens. With the entire genome sequence of many pathogens now available, new approaches toward understanding the molecular basis of pathogenesis can be gained.

For instance, gene expression studies can provide insights into the biology of acute infection vs. latency in vivo, virulence factors, and host response to pathogens. It should also be possible to obtain signatures for pathogens that are diagnostic, even when the etiologic agent is not known.

7. Gene expression in response to drug treatment. Gene expression studies can give important insights into the mechanism of action of small molecule drugs and drug resistance mechanisms. They can also be used to delineate and predict adverse events based on the identification of genes with toxicity potential. In conjunction with traditional toxicity studies, the identification of deregulated gene products that can be used as surrogate markers will further improve the drug development process.

It is apparent that gene expression analysis is an important functional genomics tool that has many applications in the basic and applied medical sciences. It is thus not surprising that many gene expression technologies have been developed over the years. These include techniques that generally depend on DNA sequencing (such as sequencing of EST libraries or SAGE elements), or on PCR-based differential display methods, microarrays, or gene traps (which allow direct measurements of gene activity in intact cells). Microarrays have many advantages over other gene expression technologies, which are often laborious and insensitive, and they seem likely to become a standard tool of both molecular biology research and clinical diagnostic research. The most significant advantage of microarrays is the ability to analyze the expression of thousands of genes in a parallel and systematic way, enabling the investigator to take a more "global view" of biological processes as they are reflected in signature gene expression changes. Genome sequencing efforts will provide the necessary information to display a large number of genes on high-density arrays. Industrialization of chip production and free market competition will reduce costs over time and make such tools more widely available to the scientific community. Due to the importance of microarray technology, we will focus in this review on a brief description of currently used array-based profiling technologies. In vivo profiling methods based on gene traps will also be addressed, as they represent an important avenue of gene expression analysis in intact cells and in in vivo model systems.

3.2.1.1.1
Microarrays

DNA microarrays provide a simple and natural yet systematic and comprehensive vehicle for exploring the genome. The power and universality of DNA microarrays as experimental tools derive from the exquisite specificity and affinity of complementary base-pairing. A DNA copy of an individual gene provides an ideal reagent for specific and quantitative detection and measurement of the sequence of the gene, even in an extremely complex mixture. Expression analysis is carried out by hybridizing RNA/cDNA to the immobilized DNA. Several different DNA microarray technologies are currently in use (Fig. 3.1).

In one type of array, DNA is printed and subsequently immobilized on solid supports (Cheung et al. 1999). The first such solid support described was nylon, and nylon remains widely used. This array boasts superior sensitivity with its utilization of radioactively labeled probes, and is relatively inexpensive (Duggan et al. 1999). The small amount of RNA sample required for hybridization makes this technology particularly suitable for the analysis of expression profiles in microdissected disease tissues. However, glass supports have distinct advantages as well: DNA probes can be covalently attached onto glass surfaces, glass is a durable material, it produces low background as compared to nylon membranes when fluorescence detection is used, and it is nonporous so that hybridization volumes can be kept to a minimum. Thus, despite the higher sensitivity of nylon-based filter arrays, the use of nonporous solid supports is becoming a more popular technology with time. It has facilitated miniaturization and utilizes the safer fluorescence-based signal detection methods. The method pioneered by Pat Brown and colleagues uses arrays where DNA is printed on glass microscope slides using a robotic "arrayer" (Schena et al. 1995; Shalon et al. 1996; Eisen and Brown 1999). Information about the relative abundance of genes in two DNA or RNA samples is obtained through the labeling of such samples with different fluorescent dyes. These are then mixed and hybridized to the arrayed DNA spots. After hybridization, the fluorescence of each dye is measured separately. The ratio of signals reflects the relative abundance of the sequence of each gene in the two RNA or DNA samples.

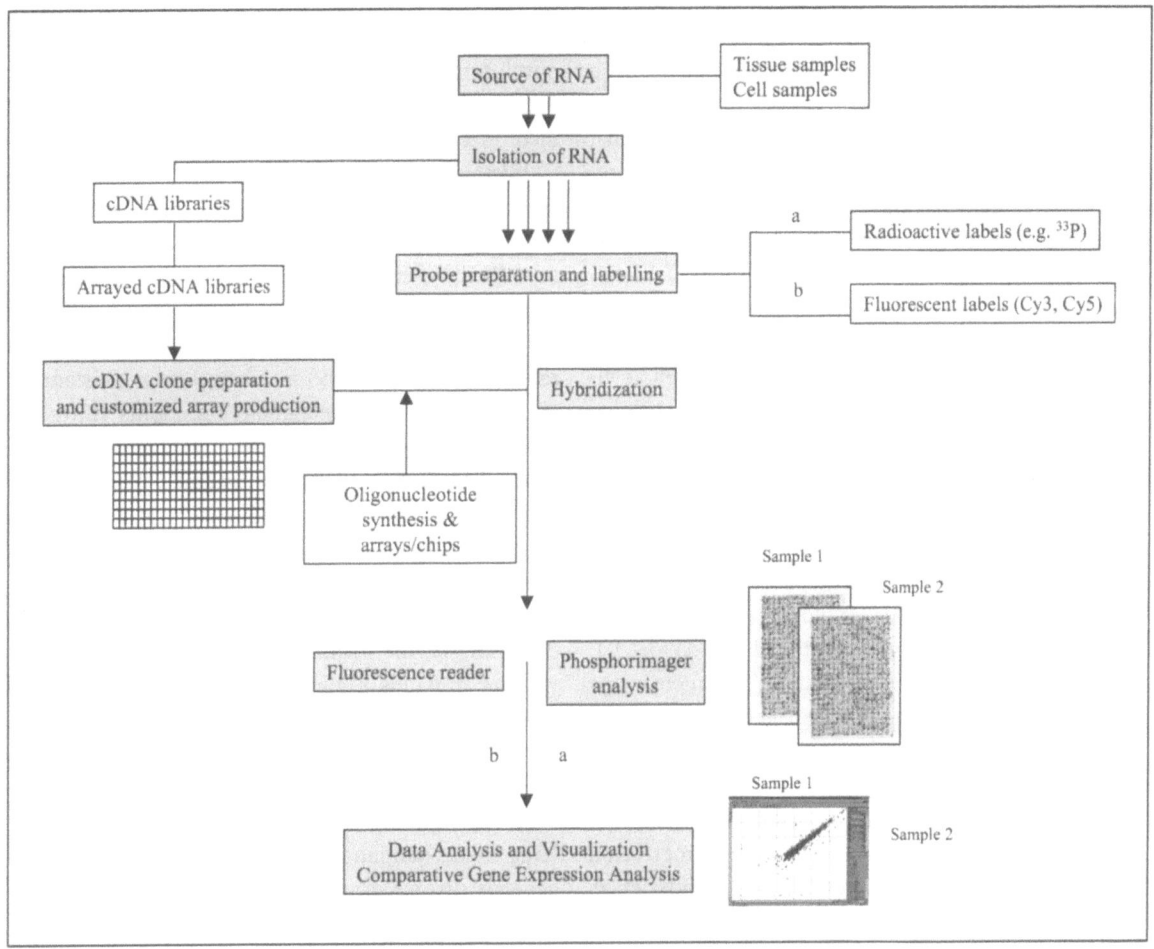

Fig. 3.1. Array-based gene expression monitoring. A *flow path* schematically represents the interacting components required for monitoring RNA abundance levels using DNA microarrays generated by robotic arraying of DNA (e.g., cDNA arrays from PCR products) or by in situ oligonucleotide synthesis (oligonucleotide arrays, Affymetrix chip technology). (*a*) cDNA arrays produced on nylon filter membranes are hybridized with radioactively labeled probes (generally ^{33}P-cDNA). Subsequent to image analysis, the signal intensities (adjusted using various quality control measures such as background subtraction and multiplication with a normalization factor – for instance, normalization to the average of all probed genes, or normalization to the total signal of some housekeeping genes) obtained by hybridization of probes generated from different samples can be arranged in a data matrix containing expression values for each probed gene and for each condition (sample type). Various types of data analyses may be performed, such as pairwise comparisons when two samples are compared, or multiple-pairwise comparisons or clustering analyses (hierarchical and partitioning clustering methods) when more samples are used in the analysis. (*b*) cDNA arrays on glass supports or oligonucleotide arrays (Affymetrix chips) are usually hybridized with probes generated using fluorescent dyes (usually Cy3 and Cy5). From two different conditions, cDNAs are labeled with the two different dyes, and the two samples are co-hybridized to a single array (as opposed to two matching arrays as required for radioactive probes from different samples). Subsequently, the array is scanned at two different wavelengths to detect the relative transcript abundance for each condition

Other arrays utilize oligonucleotides rather than DNA or cDNA arrays (Fodor et al. 1991; Kharpko et al. 1991; Southern et al. 1992). Among these, the method for high density spatial synthesis of oligonucleotides, as developed by Steve Fodor and colleagues (and commercialized by Affymetrix), is the best known (Fodor et al. 1991; Lockhard et al. 1996). This method exploits photolithography technology (an adaptation from computer chip production) in parallel oligonucleotide synthesis, and has been used so far to produce chips with 400,000 or more distinct oligonucleotides.

The emergence of microarray technologies, although still in their infancy, has greatly facilitated large scale gene expression studies. It is important to note that large scale gene expression profiling does not simply entail a scaling up of numbers of events that can be measured simultaneously. As alluded to previously, large data sets harbor information about patterns and systematic features, which can support the building of a picture of complex systems. Bioinformaticians are challenged with evolving tools to facilitate the search and display of "hidden" features. These include sophisticated tools for statistical analysis as well as visualization tools that display data in a easily comprehensible manner. In turn, these should facilitate the formulation of experimentally testable hypotheses. For a description of some such analysis tools, the interested reader is referred to recent reviews covering this topic (Wittes and Friedman 1999; Eisen et al. 1998).

Microarrays have been employed in many of the types of investigations briefly outlined in Sect. 3.2.1.1. For instance, in the basic research arena, microarrays have been applied to the study of genome-wide patterns of gene expression in response to a variety of imposed stimuli. Many early studies have focused on the yeast model system and revealed waves of gene expression that correlate with cell cycle progression and various environmental conditions (De Risi et al. 1997; Lashkari et al. 1997; Wodicka et al. 1997; Spellman et al. 1998). The results from such studies clearly indicate that gene expression data reflect information about gene function and even about the physical association of gene products. Numerous clusters of co-expressed genes, representing diverse expression patterns across even a limited set of conditions, were strikingly coherent in their cellular functions (Eisen et al. 1998). Other investigations have explored whether expression profiles of mutant cells can be used to classify the functions of previously uncharacterized genes (Hughes et al. 2000). Microarrays have also been applied in many other studies involving mammalian cells. These include, for instance; (1) the genome-wide analysis of the mitotic cell cycle (Cho et al. 1998); (2) the study of the transcription response of human fibroblasts to serum (Iyer et al. 1999); (3) the study of the response to activation of a specific gene, such as the MYC oncogene; (4) the large scale identification of secreted and membrane-associated proteins (Diehn et al. 2000); (5) the differential analysis of normal and disease tissues (Perou et al. 1999); and (6) the analysis of tumor subtypes (see below). Clearly, large scale gene expression stud-

ies will help to reveal cellular circuitry at an unprecedented level of detail and lead to the discovery of novel targets for therapeutic intervention. Another application is the study of cell response to drug treatment (see Sect. 3.4.1).

Microarray studies have also been used to improve clinical disease diagnosis. For example, microarrays are already providing insights into cancer that would be difficult, if not impossible, to obtain by a gene-by-gene approach. They have been used to identify specific subtypes of a variety of cancers, including leukemias (Golub et al. 1999) and lymphomas (Alizadeh et al. 2000), cutaneous malignant melanoma (Bittner et al. 2000), breast cancer (Perou et al. 1999) and colon cancer (Alon et al. 1999). For instance, comparison of profiles of acute myeloid leukemia (AML) and acute lymphoblastic leukemia (ALL), two hematologic malignancies that are often difficult to differentiate by standard pathological examination of the diseased cells, allows the distinction between these cancers without previous knowledge of these classes (Golub et al. 1999). Importantly, an automatically derived call predictor (based on identified marker genes) is able to determine the class of new leukemia cases. Another example is highlighted by studies of diffuse large B-cell lymphomas (DLBCL), a clinically heterogeneous disease group. Clustering analyses of microarray data resulted in the identification of marker genes that allowed the definition of two DLBCL groups: one had the signature of B-cells from germinal centers, the other the signature of activated B-cells (Alizadeh et al. 2000). The clinical outcome of patients carrying the activated B-cell-like signature was much worse than that of patients with the germinal center B-cell signature. These studies, although still preliminary, indicate how large scale gene expression studies may improve the resolution of cancer subtypes. Similar kinds of approaches to other diseases are also being undertaken (Friddle et al. 2000) that should accelerate our understanding of disease etiologies, and thus the development of mechanism-based therapeutics and tools for better diagnosis and prognosis (Khan et al. 2001).

3.2.1.1.2

Non-Array Based Gene Expression Methods: In Vivo Analysis of Gene Expression

Microarrays and other sequencing- or PCR-based methods for quantifying gene expression are limited to the use of RNA isolated from cells or tissues. Certain experimental paradigms, however, require measurement of gene expression in intact cells or whole organisms.

Methods that utilize gene traps have been successfully used in this context (Brown et al. 1998). These are primarily based on the use of retroviral vector systems that support the integration of reporter genes into transcriptionally active chromosomal regions. Integration may result in knocking out the targeted gene, and this strategy has been used to generate large embryonic stem (ES) cell libraries for the generation of transgenic knock-out animal model systems (Zambrowicz et al. 1998). However, integration may also be utilized to measure the activity of the gene into which the transgene construct has integrated (Whitney et al. 1998; Ishida and Leder 1999; Akiyama et al. 2000; Medico et al. 2001; Mitchell et al. 2001). Inclusion of appropriate splice acceptor and splice donor sites in the transgene constructs puts expression of the transgene reporter under the control of the endogenous promoter of the affected gene, thereby providing an in situ readout of endogenous gene activity. Numerous reporters, primarily drug-resistance genes or genes encoding fluorescent proteins such as green fluorescent protein (GFP), have been used in this context. Retroviral as well as plasmid-based gene reporter traps have been used to generate complex cell populations with numerous integration sites. Such cell populations may be analyzed to profile gene expression at a genome-wide level. Fluorescent reporter probes enable real time analysis of gene expression using such systems and will be useful in the dissection of signaling pathways and in the study of the effects of environmental agents (Whitney et al. 1998).

The development of noninvasive methods for measuring reporter activity would further expand the use of such systems to in vivo analysis of single or composite gene expression. Such an advance would have applications in a wide range of fields, including the development of animal model systems using gene "knock-in" strategies and gene therapy settings. The recent development of PET reporter genes may provide such an avenue. Positron emission tomography (PET) is a noninvasive imaging modality used to study biochemical and biological process in vivo. PET probes incorporate a positron-emitting isotope attached to a molecule of interest. Such tracers are usually positron-labeled ligands for receptors or positron-labeled enzyme substrates. Retention of PET probes in living tissue generally occurs after ligand binding or after conversion of a substrate to "trapped" product(s), and can be measured and

quantified directly from PET scanner images. This approach has already found application in the development of PET reporters for measuring gene expression. For instance, a PET-reporter-gene approach based on herpes simplex virus type 1 thymidine kinase (HSV1-tk) trapping of positron-labeled substrates in cells expressing HSV1-tk has been reported (Gambhir et al. 1998, 1999). Others have shown that coupling target and reporter expression through internal ribosome entry sites (IRES) could be used for quantitation of the expression of virtually any gene of interest from the respective transgene in intact cells (Yu et al. 2000). Such bi-cistronic reporter systems may be of particular use in gene therapy.

In addition to clinical applications, transgenic and "knock-in" animal models incorporating PET reporters could be used to measure endogenous gene expression in many experimental settings. Thus, integrating large scale experiments that identify novel genes of interest with strategies for in vivo monitoring of gene expression changes, will further facilitate linking genotypes and phenotypes. Further exploitation and development of PET reporter systems should prove useful for both basic and clinical research.

3.2.1.2
Mapping DNA-Protein Interactions

It will be useful to augment global gene expression data with information about the function of the factors that directly regulate genes. Again, DNA arrays have shown promise in mapping the binding sites of proteins to target DNA sequences. Two complementary techniques have already been used, chromatin immunoprecipitation and DNA adenine methyltransferase identification (DamID) (Ren et al. 2000; Iyer et al. 2001; Van Steensel et al. 2001). These techniques can provide genome-wide biochemical information about DNA-protein interactions.

3.2.1.3
DNA Variations

The elucidation of DNA variations has been central to the discovery of genetic mutations underlying many inherited and sporadic diseases. Polymorphic DNA variations are also used as markers of genes and genomes with which researchers perform genetic analysis in outbred species where matings are not controlled. In the context of sequencing of the human genome, a huge collection of single nucleotide polymorphisms (SNPs) is being discovered, and these SNPs are anticipated to provide a basis for more efficient mapping of disease genes and for understanding the role of gene variations in altered gene function, susceptibility to disease, and the response of individual patients to drug treatment (Chakravarti 2001). The general expectation is that SNPs will have a significant impact on basic and clinical research (Risch 2000; Roses 2000), although some fundamental questions, such as the role of common genetic variants in causing human disease, and the nature of variations within and between populations, still need to be answered.

Many different methodologies are being explored in the analysis of DNA variations, a detailed discussion of which is beyond the scope of this review. However, although not strictly within the realm of functional genomics, gene variance analysis will have a significant impact on the determination of functional differences between polymorphic variants; that is, how these result in phenotypic polymorphisms. Microarrays, as described earlier in this review, have had a great impact on large-scale gene expression profiling. Microarrays can also be used to study DNA, primarily for identifying and genotyping mutations and polymorphisms (Brown and Botstein 1999; Hacia 1999; Lockhard and Winzeler 2000).

Array-based characterization and identification of novel DNA variants have largely been performed using oligonucleotide arrays, exploiting the ability to perform custom synthesis at high density. Thus, oligonucleotides with defined substitutions can be used to scan a target sequence for mutations. Such arrays have been used to detect variants in the HIV genome, human mitochondria, and various disease genes such as the p53 tumor-suppressor gene. Array technology is likely to become an important tool in diagnostics, as in genotyping cancer subtypes, thereby aiding the oncologist in the selection of appropriate therapies. An impact will no doubt soon be seen of SNPs upon the discovery of the molecular basis of complex polygenic diseases and drug susceptibility (pharmacogenomics).

3.2.2
Phenotype-Based Functional Genomics

Data emerging from the functional genomics approaches described in the previous sections largely relate to the surveying of molecular differences between biological samples and the placement of identi-fied targets into known signaling pathways. Although large scale studies facilitate the prediction of gene function, functional characterization of any candidate target will ultimately need to be studied in more detail utilizing a spectrum of different technologies and biological model systems.

The goal of phenotype-based functional genomics is to accelerate this process through the identification of targets on the basis of their function in the regulation of specific cellular phenotypes. Thus, target discovery and validation are addressed in parallel. In other words, a combinatorial screening paradigm first identifies elements that induce certain phenotypes, and the genes pertaining to such elements are then identified – a reversed order of discovery from that usually pursued in functional genomics (conceptually similar to principles followed in reverse genetics).

Classical genetic analysis in genetically tractable organisms has been among the approaches traditionally taken to study gene function. This involves the introduction of mutations and the subsequent observation of the mutant phenotype. Complex biological processes have been dissected by genetic screens in which a large collection of random mutations are introduced into the genome of the model organism and mutants with the desired altered phenotype are selected and analyzed. Classical genetics, combined with molecular genetics and genomics, will remain a powerful tool for analyzing the workings of complex biological systems.

Mammalian cells are widely used as model systems to gain insights into normal and disease-related processes. However, with the exception of mutagenesis, none of the classical genetic tools are applicable to mammalian cell systems. Somatic mammalian cells are asexual diploids, and it is impossible to perform genetic crosses for the generation of well-characterized cell strains – the cornerstone of classical genetic analysis. Functional genomics tools have been developed to address these limitations.

3.2.2.1
Induced Phenotypes and Gene Function

In an effort to emulate all the principles of classical genetics, various functional genomics tools have been developed to dissect the function of individual genes or perform genome-wide target discovery screens in mammalian cells. The central element of these tools is the cDNA itself. Individual cDNAs, or a library of cDNAs representing the entire genetic content of a cell or an organism, are inserted into vector systems that allow efficient gene transfer and expression in a variety of mammalian cells. The most prominent of these vector systems are based on retroviruses because of their versatility, efficiency, and ease of manipulation (Morgenstein and Land 1990; Hannon et al. 1999). The expression of a cDNA in sense orientation leads to overexpression of the gene product in the target cell, mimicking the effect of a "gain of function" mutation. The antisense expression of a cDNA leads to the expression of an RNA molecule whose sequence is complementary to the mRNA. The binding of antisense RNA to the target mRNA interferes with its translation, leading to the knock-out or knock-down of gene activity. To improve the efficiency of inhibition, small antisense fragment libraries can be generated. Small antisense RNA fragments may inhibit gene expression more effectively than complete antisense mRNA (Nellen and Sczakiel 1996). The anti-sense expression of a cDNA or cDNA fragments intends to mimic the effect of a "loss of function" mutation (Deiss and Kimchi 1991; Holzmayer et al. 1992; Gudkov and Roninson 1997).

The overexpression or antisense-mediated inhibition of the expression of a gene or its encoded protein results in a reversible alteration of the cellular phenotype, and this can provide valuable insights into gene function. Ample examples in the scientific literature demonstrate that the overexpression of sense and anti-sense cDNAs results in induced phenotypic changes that reflect functional properties of the gene of interest (see, for example, Gallagher et al. 1997; Carnero et al. 2000; Gudkov et al. 1999; Xu et al. 2001). Genome-wide genetics screens are conducted with complex sense or antisense cDNA expression libraries to isolate genes whose "gain of function" (sense libraries) or "loss of function" (antisense libraries) induces the desired phenotypic change. Other approaches may use combinatorial peptide expression libraries, if the discovery of a peptide surrogate ligand with target inhibitory function is desired. Typically, a screen begins with the transfer of a library into the target cells. The transduced cell population is grown under selective conditions to allow for the enrichment of cells that display the phenotype of interest. Cells with the altered phenotype are collected and the library-derived genetic element (cDNA or antisense fragment) whose expression is responsible for the induction of the phenotype is recovered and amplified. This enriched sub-library of functional elements may be used in multiple rounds of phenoty-

pic selections (cycle cloning). Cycle cloning is important in increasing the validity of the screen since mammalian cells are inherently heterogeneous and prone to spontaneous reversions, and many interesting phenotypes are leaky. Ultimately, functional screens will select the relevant genes that are rate-limiting in the control of a biological process, and can lead to the identification of novel targets and unpredicted mechanisms. In the past decade, functional genetic screens have been used with great success to study complex biological processes such as tumor suppression, apoptosis, cellular senescence and drug resistance (see, for example, Gudkov et al. 1993; Deiss et al. 1995; Garkavtsev et al. 1996; Sun et al. 1998; Hudson et al. 1999; Mahon and Whitehead 2001).

3.2.2.2
Applications of Phenotype-Based Functional Genomics: Drug Discovery and Diagnostics

The sequencing of the human genome revealed the existence of thousands of potential drug targets. Functional genomics is being used to validate these potential targets in order to establish a causative relationship between the target and the disease of interest.

Target validation studies have multiple goals. First, they attempt to link the target mechanistically to the disease and demonstrate the role of the target in pathogenesis. Second, they attempt to demonstrate that the inhibition of the target in diseased cells results in a desired therapeutic effect. These experiments may indicate the efficacy of a future drug developed against the target. A third goal is to determine whether the inhibition of the target has a toxic effect on normal cells.

Phenotype-based functional genomics screens achieve these goals through the use of cellular disease models and antisense-mediated inhibition of target expression. Antisense inhibition of gene function mimics the effect of a drug that would bind to and inhibit the target protein. Such an approach can be taken at a genome-wide level or at the level of specific gene families that are of special interest for drug discovery. Using bioinformatic search tools, genes encoding proteins with the same predicted biochemical activity (gene families) can be identified in genomic databases. Gene families encoding protein kinases, ion channels, G protein coupled receptors, and proteases can be identified by the presence of characteristic motifs in the predicted protein sequences. The challenge for functional genomics is to match individual members of gene families to disease indicators and demonstrate that the targeting of those particular members would be therapeutically beneficial. This task requires the simultaneous validation of hundreds of genes in multiple disease models.

Phenotype-based approaches have utility not only in the discovery and validation of drug targets but also in the development of diagnostic tools. For instance, the identification of specific cell surface markers for diseased tissues would facilitate the development of imaging agents for early detection of pathological states. Similarly, the identification of proteins secreted by the diseased tissue would form the basis for the development of novel blood tests for early disease detection. These efforts can be aided by phenotype-based functional studies designed to find membrane-bound or secreted proteins. Such an approach takes advantage of the fact that membrane targeting of both secreted and cell surface proteins requires a short, amino-terminal hydrophobic peptide called a signal peptide. The sequence of the signal peptide uniquely identifies the secreted proteins. The goal is the isolation of cDNA fragments from a library made from the diseased tissue that encode signal peptides using a signal sequence trap (Tashiro et al. 1993; Kojima and Kitamura 1999).

The signal trap is based on the use of truncated membrane-bound cell-surface molecules such as CD4, CD8, or CD95. These proteins are rendered incapable of localizing to the surface of the cell if they lack a functional signal sequence. The fusion of a cDNA fragment that encodes functional signal peptides will restore the localization of the truncated proteins to the surface of the cell. Cells expressing the marker on the surface can be isolated using marker-specific antibodies and cell-sorter or magnetic beads. After the recovery of the cDNAs from the positive cells, a catalogue of membrane-bound and secreted proteins can be compiled from normal and diseased tissues. Candidate diagnostic markers that are abundant in the diseased sample can be further validated for the development of novel diagnostics using additional techniques.

3.3
Proteomics: Technologies and Applications

As a counterpart to functional genomics, proteomics addresses the challenge of the dissection of the function of proteins. Proteins are more difficult to work

with than DNA, which explains the comparably slower development of large scale technology platforms that would facilitate the analysis of proteins at a proteome wide level. Nonetheless, significant developments in this area have been achieved in the past few years. Technologies have emerged that facilitate more sensitive and larger scale profiling of changes in protein expression, and the mapping of protein-protein interactions and signaling pathways. As already noted in the introduction of this chapter, large scale analysis of the regulation and dynamic interactions of components of the proteome will be indispensable in the functional characterization and annotation of the genome, and will have an impact on all areas of basic and medical sciences.

Here we discuss recent developments in proteomics technologies and their impact on profiling the dynamic regulation of the proteome, the mapping of "interactomes" (protein-protein interaction networks), the development of novel tools to probe the function of proteins in vivo, some of which are predicted to aid in the development of imaging agents, and the biology underlying physiological and pathological processes.

3.3.1
Proteome Expression Profiling

3.3.1.1
2D-PAGE-MALDI-MS

The most frequently used method for protein separation and identification is the two-dimensional polyacrylamide gel electrophoresis (2D-PAGE, 2D-GE) followed by mass spectrometry (MS; Gygi and Aebersold 2000; Fig. 3.2). This classical profiling method consists of several steps, beginning with sample preparation. No single method of sample preparation can be applied universally due to the diverse nature of samples that are analyzed, in contrast to the more uniform sample preparation methods involving RNA or DNA. Methods for sample preparation include stepwise precipitation, immuno-affinity isolation and/or subcellular fractionation. Following sample preparation, proteins are separated in two dimensions according to two independent and distinct properties, isoelectric point and molecular weight. Recently, advances in 2D-GE technology have resulted in improved resolution of proteins with polyacrylamide gels (Goerg et al. 2000). To date, 2D-GE is still the most powerful method used to separate com-

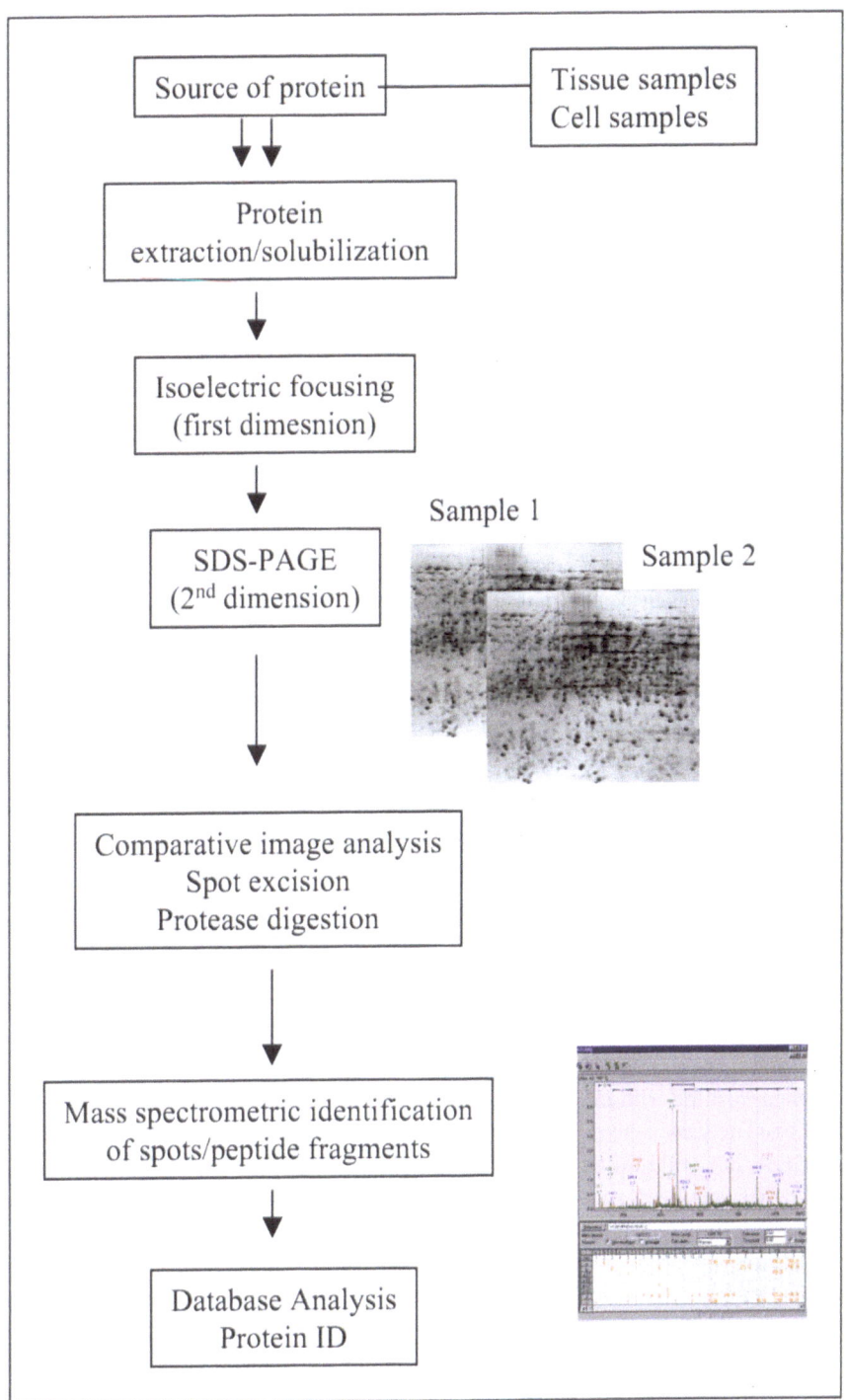

Fig. 3.2. Protein expression monitoring. A *flow path* schematically represents the use of two-dimensional gel electrophoresis for visualizing proteins and their relative expression levels. This approach may be used to compare the expression levels of proteins in two or more different samples/conditions (e.g., normal and diseased). Proteins are solubilized and the protein mixture applied to a "first dimension" gel strip that separates proteins based on their isoelectric points. Subsequently, the strip is subjected to reduction and alkylation and applied to a "second dimension" SDS-PAGE gel. Proteins are then separated on the basis of size. Gels are fixed and proteins visualized by silver staining. Visualized spots are then recorded and quantified. Spots are excised, tryptic digests generated, and peptides analyzed by mass spectrometric analysis

plex protein mixtures. Separated proteins are visualized by chromophoric staining, isotopic labeling, or with fluorescent dyes, and further examined by image analysis. This process is only semi-automated and still requires manual editing of critical positions. In order to achieve statistically significant detection of protein expression differences between two different samples, 2D-GE has to be repeated several times. Individual protein spots of interest are excised, alkylated, reduced, and digested with trypsin. The resulting peptides are mixed with a large excess of UV-absorbing matrix, dried on a spot, ionized by a pulsed laser, then extracted by an electrical field into the mass analyzer. This method, known as matrix-assisted laser desorption-ionization mass spectroscopy (MALDI-MS), creates a peptide mass fingerprint. This signature of a protein is calibrated against the defined internal standard of trypsin auto-cleavage products and compared to a database of tryptic peptides created by virtual cleavage of stored protein sequences. State-of-the-art software is capable of including various potential post-translational modifications into the search algorithm to improve the confidence level of protein identification. Confidence in the database search results is directly correlated with mass accuracy. Recently, methods have been established to automate the processes of spot identification, cutting, digestion, and MS-analysis. The coupling of 2D-GE and MALDI-MS is well established now (Pandey and Mann 2000). However, it still suffers from limitations in reproducibility, sensitivity, and automation potential. MS data also have shortcomings in that peptide masses are often not sufficient to unambiguously identify a protein, and these shortcomings are of particular relevance when working with organisms whose genomes are not completely sequenced. The range of applications of 2D-GE is thus still limited (Pandey and Mann 2000). Still, 2D-GE has been successfully applied to profile expression profiles of fractions of the proteome of diverse species, including human. Complementary technologies, such as those described below, may circumvent some of the limitations inherent in 2D-GE approaches and provide larger scale solutions for the future.

3.3.1.2
Multidimensional LC-MS/MS

Methods to characterize complex protein mixtures without the need of pre-purification increase the efficiency of protein identification and avoid difficulties associated with 2D-GE. The most prominent limitations

of 2D-GE (slow spot excision, low detection limit, low loading capacity, and biased applicability), can be partially overcome by automatically separating proteins by multidimensional liquid chromatography. This method also applies orthogonal separation principles based on different physicochemical parameters. In multidimensional liquid chromatography, the crude lysate (which may even contain solubilized membrane proteins) is first digested as a complex mixture of proteins and then subjected to a series of liquid chromatographic (LC) steps before analysis with an electrospray ionization mass spectrometer (ESI-MS). Coupling the injection-capillary to a low-flow chromatographic system permits direct on-line analysis. The electrospray source operates at atmospheric pressure, making interfacing relatively simple (Rowley et al. 2000). Recently, a biphasic microcapillary column integrating directly overlaid beds of strong anionic exchange beads with reversed phase particles was applied successfully to separate the eukaryotic 80S ribosome and the yeast proteome (Link et al. 1999; Washburn et al. 2001). The combination of fast data acquisition with sophisticated search algorithms capable of subtracting identified peptides before the next round of database searching, such as SEQUEST, leads to a higher sample throughput than can be achieved with any other method (Yates et al. 1995). The success rate of this approach is increasing with the completion of sequenced genomes. Current systems do not yet achieve the resolving power of 2D-GE but have the potential to improve in the future.

3.3.1.3
Isotopic Methods

Isotopic labeling methods facilitate better quantitative measurements of dynamic changes in protein expression. One approach involves stable isotopic dilution, and utilizes heavy isotopes as internal standards that can easily be differentiated by MS from non-labeled samples (de Lenheer et al. 1985). This method usually introduces the isotopic label before protein extraction, which prohibits its application to biopsied tissue samples. Another promising method for quantification consists of three modules and is called isotope-coded affinity tag (ICAT). The elements are: an affinity tag, a linker to incorporate stable isotopes, and a reactive group specific to thiol groups present at cysteines. Proteins from two samples are denatured, reduced, and labeled with either the heavy or the light variant of the isotope. Then the samples are combined, digested, and isolated by affinity chromatography directed against the affinity tag. The column is coupled to a tandem MS and ratios of heavy and light versions of the peptides are determined. Additionally, the identity of the peptide can be revealed by MS/MS sequencing. The major drawbacks of this method are: (1) the attachment of additional mass (complicates database searches), and (2) the biased selectivity for cysteine-containing peptides (Gygi et al. 1999v). The method nevertheless holds great promise for differential protein profiling at a larger scale and for discovering novel disease proteins and markers.

3.3.1.4
Protein Chip

Expression profiles can also be analyzed by replacing the tedious and complex process of 2D-GE with a sample preparation on a chip. This technology reduces the complexity of the protein mixture by separating the proteins into sets of proteins with common properties. Protein Chip arrays contain various affinity surfaces, allowing separation according to various protein characteristics such as charge, hydrophobicity, and metal binding. The different surfaces bind subsets of the protein mixture. After differential washing and removal of unbound material, the bound proteins are analyzed in a time-of-flight mass spectrometer (TOF MS) (Fung et al. 2001). The technology has been used for disease profiling in cancer and to define biomarkers significant for different disease statuses (von Eggeling et al. 2000). The Protein Chip (Ciphergen, Fremont, Calif., USA) system promises to become a powerful proteomics tool for highlighting the differences in protein expression profiles directly from complex lysates (Senior 1999), but major technological improvements are still needed for broader applications.

3.3.1.5
Protein Arrays

Protein arrays are inspired by the success of DNA arrays and are likely to become another important tool for monitoring protein expression profiles and for the larger scale analysis of biochemical properties of proteins. Successful implementation of protein arrays for expression profiling requires access to protein and ligand libraries, suitable solid supports, immobilization techniques, and sensitive detection devices. Protein binding molecules can be derived from small molecules, oligonucleotides, aptamers, antibodies, and macromolecules such as phages (Borrebaeck 2000). Immo-

bilization supports include filter membranes [polyvinylidene difluoride (PVDF) Nitrocellulose], and glass, plastic, metal or silicone surfaces. Detection may be based on electromagnetic waves (absorption, reflection, transmission, fluorescence, luminescence, phospho-imaging), label-free techniques such as surface plasmon resonance, monitoring of molecular parameters such as electrochemical properties (conductivity), atomic force microscopy (Jones et al. 1998), or mass spectrometry (Borrebaeck et al. 2001). The most advanced example of this approach has been the investigation of the presence and relative abundance of 115 different antigens on an antibody array by differential labeling of extracts with Cy3 and Cy5 fluorescent dyes (Haab 2001). When dealing with a limited number of target molecules, a miniaturized sandwich-enzyme-linked immunosorbent assay (ELISA) which requires a pair of antibodies, one for binding and one for detecting the antigen, may be used (Mendoza et al. 1999; Joos et al. 2000; Huang 2001). Currently, the greatest obstacles for protein arrays are the limited stability of the immobilized molecules, the limited availability of monospecific antibodies (which might be overcome by recombinant and synthetic antibody generation techniques; see Sect. 3.3.3.1) and the high variance of affinities and on/off rates. Protein array technologies are still in their infancy but will no doubt grow to become larger scale proteome profiling tools as more technical advances are made in the near future.

3.3.2
Profiling of Macromolecular Interactions: Elucidation of Proteome Networks

Molecular interactions are essential to many biological processes. Constitutive and induced noncovalent associations of proteins and multienzyme complexes play a major role in the regulation of cellular machineries implicated in diverse processes such as DNA replication, transcription, translation, and metabolic and signal transduction pathways. Much of modern biological research is concerned with the identification of proteins involved in such cellular processes, with determining their function, and with elucidating how, when, and where they interact with other proteins involved in specific biochemical pathways. Proteomics technologies that facilitate the large scale mapping of protein interactions will pave the way toward the establishment of maps of cellular circuitry and functions of the proteome.

3.3.2.1
Molecular Interaction Screening Technologies

Biochemical separation technologies have been successfully applied to the study of macromolecular complexes, although they are limited to in vitro analysis of such complexes (Pandey and Mann 2000). Developments in mass spectrometry technology are greatly facilitating the scaling up of such studies (Shevchenko et al. 1996). Proteomics has also led to development of methods for the detection and identification of protein-protein or other macromolecular interactions in intact cells. The most widely known and used technologies that support large scale screen-

ing of interactions are the yeast two hybrid (Y2H) and phage display technologies (for more detailed description of the latter, see Sect. 3.3.3).

The Y2H system is a powerful method for the in vivo analysis of protein-protein interactions in intact cells, and uses yeast as a surrogate host system (Fields and Song 1989; Chien et al. 1991; Gyuris et al. 1993; Vidal and Legrain 1999). It is based on the reconstitution of a transcription factor complex, which is mediated by the interaction of two fusion proteins (Fig. 3.3 A). These fusion proteins consist of two interacting protein entities fused to either a DNA-binding component or transcription activation domain component. An interaction-driven reconstitution of transcription activation

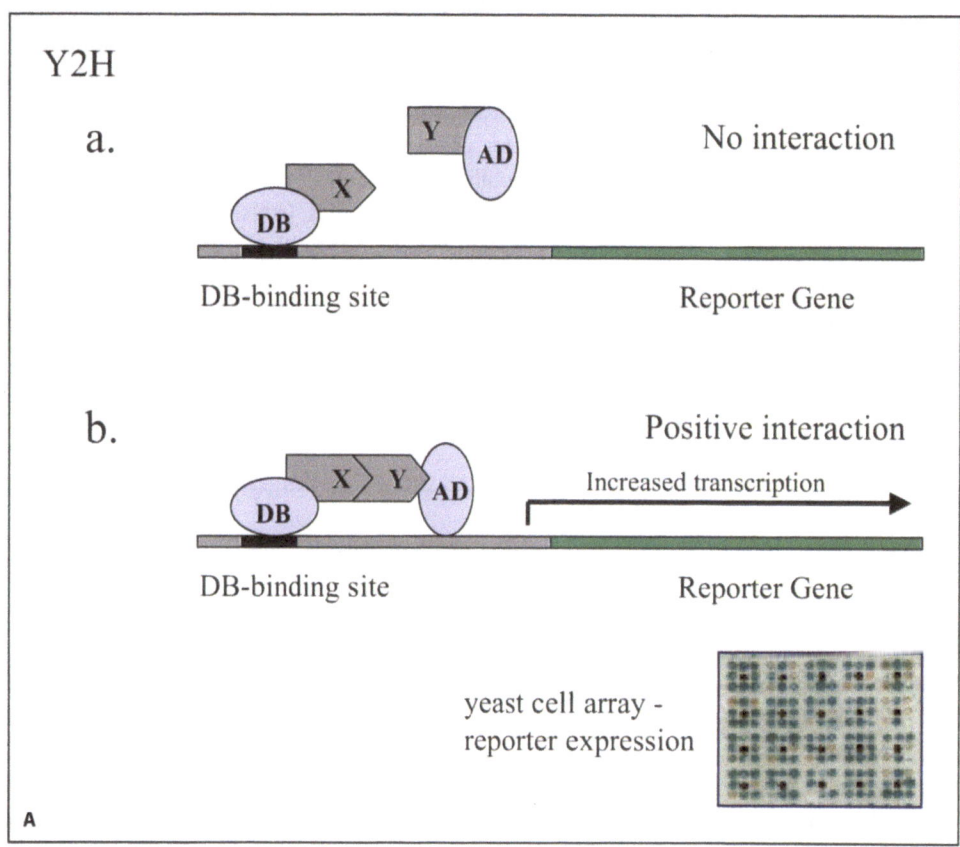

Fig. 3.3 A–C. Cell-based protein-protein interaction monitoring. Variations of two-hybrid themes. (A) The classic yeast-two-hybrid (Y2H) assay. Reconstitution of an active transcription factor complex using two hybrid proteins. One hybrid protein encodes a DNA binding domain (DB) fused to a protein X (bait). The second hybrid protein encodes a transcription activation domain (AD) fused to a protein Y (prey). Protein Y may be encoded by a cDNA library when library screens are performed with the objective to identify a prey protein that binds to a bait protein X of interest. Interaction of the two hybrid proteins results in increased transcription of a reporter gene. Increased reporter expression/activity therefore reflects a positive protein-protein interaction. Inset: Yeast cells expressing hybrid protein may be arrayed and grown on filter membranes, and reporter expression monitored by image analysis in ways similar to image analyses performed with DNA arrays (see Fig. 3.1). When LacZ is used as a reporter, yeast cells turn blue-greenish upon induction of transcription.

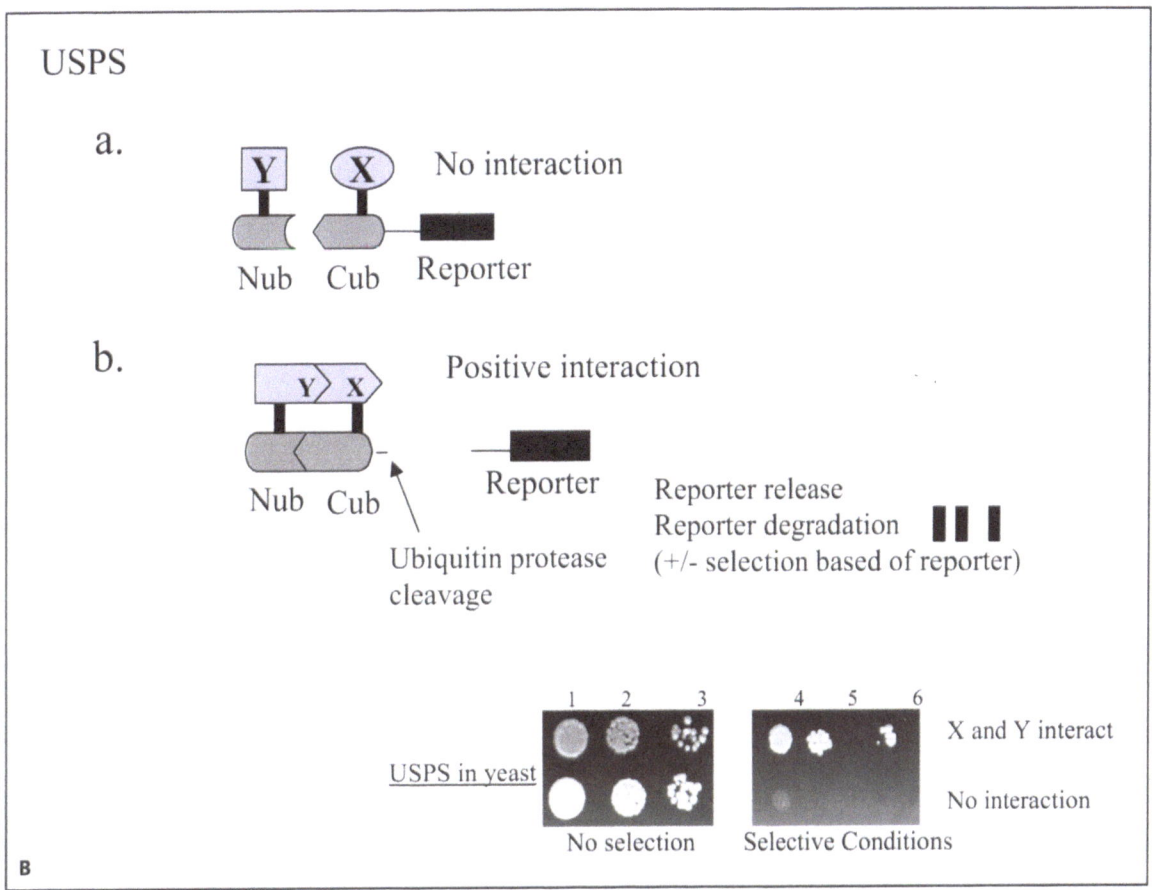

Fig. 3.3. B Ubiquitin split protein system (USPS). C-terminal domain of ubiquitin is fused to protein X (bait) and a reporter. N-terminal domain of ubiquitin (usually some mutant form of Nub with lower affinity for Cub) is fused to protein Y (prey). Interaction of bait and prey proteins brings Nub and Cub fragments in close proximity and drives reconstitution of ubiquitin. This is now recognized by one or more endogenous ubiquitin protease, resulting in cleavage at the C-terminal end and release of the fused reporter. Depending on the nature of the N-terminal residue of the cleaved reporter, the reporter may remain intact or targeted for degradation by processes governed by the N-end rule of protein degradation. Consequently, depending on choice of reporter (and cloning strategy), positive or negative selection schemes may be used to monitor protein-protein interactions. *Inset*: shows use of USPS in yeast employing a reporter (ura3) that undergoes degradation upon release. Yeast cells expressing ura3 reporter are sensitive to the drug 5-FOA (i.e., no growth under selective conditions when no interaction occurs, *right panel – row 2 – columns 5–6*). In contrast, if X and Y interact, resulting in degradation of reporter, cells become resistant to drug treatment (positive selection, *right panel-row 1 – columns 5–6*). *Columns 1–3 and 4–6* show decreasing amounts of spotted yeast cells. C see p. 54

activity leads to induction of the expression of a responsive reporter (for a more detailed review on Y2H, see Mendelsohn and Brent 1999; Vidal and Legrain 1999). Variations on the Y2H theme have been used to study protein-protein, protein-RNA (SenGupta et al. 1996), protein-peptide (Yang et al. 1995), protein-single chain antibody (Visintin et al. 1999), and protein-small molecule interactions (Licitra et al. 1996). Despite the great utility of this technology in interaction mapping, it also has limitations. It is limited to the analysis of soluble proteins or soluble domains of proteins in the nucleus of yeast cells; i.e., it is not applicable to many signaling events, and not suitable for the analysis of full length transcription factors or membrane proteins. Recent developments utilizing repression domains may, however, make the Y2H system more appropriate to the study of transcription factors (Hirst et al. 2001). Additionally, interactions are analyzed in yeast cells, an environment extraneous to proteins derived from other host systems.

Driven by the limitations of the aforementioned systems, other technologies have been developed. These include the ubiquitin split protein sensor technology (USPS) (Johnsson and Varshavsky 1994; Stagl-

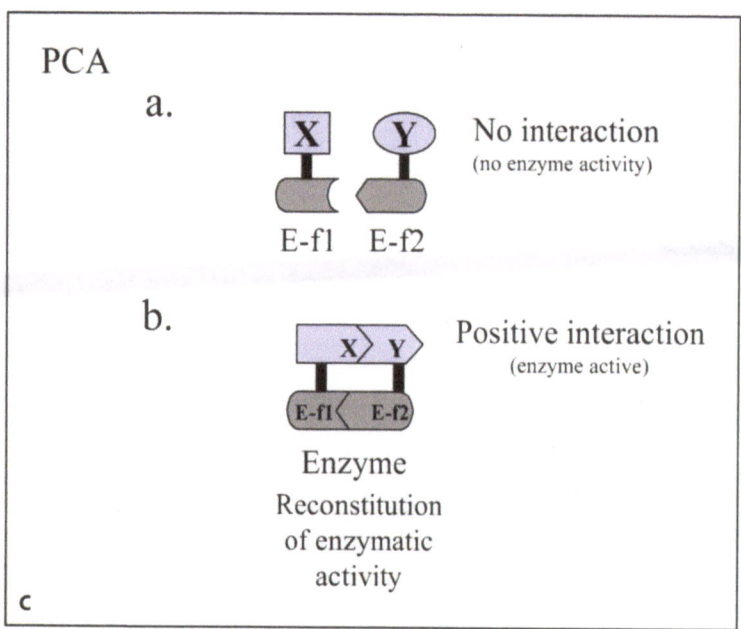

Fig. 3.3. C Protein complementation assay (PCA). Conceptually similar to the USPS technology. In this scenario, an active enzyme is reconstituted through "forced" interaction of N-terminal (Ef1) and C-terminal fragments (Ef2) of the enzyme. Interaction is driven by the interaction of proteins X (bait) and Y (prey) fused to the two enzyme fragments. Consequently, changes in enzyme activity reflect a change in interaction of hybrid proteins

jar et al. 1998; Wittke et al. 1999; Rojo-Niersbach et al. 2000), and other two component protein fragment complementation assays (PCAs). The USPS technology (Fig. 3.3 B) is based upon a protein-protein interaction/dimerization-assisted reassembly of fragmented ubiquitin, which results in susceptibility of the reconstituted ubiquitin to cellular ubiquitin-specific proteases. These recognize the intact ubiquitin structure and can cleave the protein between the C-terminus of ubiquitin and a C-terminally fused reporter protein. Interaction can be measured either as a release of reporter activity or loss in reporter activity. USPS interaction screening can be performed in yeast as well as mammalian and plant cells. It is also suitable for the analysis of membrane proteins and full length transcription factors. The USPS technology promises to become yet another high throughput interaction screening system for the mapping of cellular circuitry.

Conceptually similar to USPS is a PCA technology (Fig. 3.3 C) based on dihydrofolate reductase (DHFR) (Pelletier et al. 1998; Remy et al. 1999). PCA-DHFR is based on the reconstitution of two DHFR fragments through induced proximity mediated by interacting proteins fused to the DHFR fragments. In contrast to

USPS, this assay is based on the reconstitution of an enzymatic activity. Interactions can be detected either by scoring for cellular growth using specific culture conditions for DHFR-deficient cells, or by using a fluorescence-labeled methotrexate (MTX) as a reporter probe. MTX is capable of binding to DHFR with a 1:1 stoichiometry. This system is also applicable to diverse host systems, including bacterial systems (Hochschild and Dove 1998; Joung et al. 2000).

There are also other interaction technologies based on resonance energy transfer between reporter proteins with fluorescent or bioluminescent properties, and these may be used to measure interactions in real time. These technologies are known as FRET (fluorescence resonance energy transfer) and BRET (bioluminescence resonance energy transfer). Green fluorescent protein (GFP) and variants thereof (e.g., YFP and RFP; yellow and red fluorescent proteins) have been used in FRET to monitor protein-protein interactions by measuring energy transfer between these reporter molecules (Devi 2000). Energy transfer occurs when proteins fused to these reporters interact and bring the reporter molecules in close physical proximity to one another. The same principle is used in BRET, using yet other variants of GFP (Devi 2000).

3.3.2.2
Applications of Molecular
Interaction Screening Technologies

The interaction technologies described in the previous section have so far been primarily used in cell culture systems to characterize or identify novel protein-protein interactions. However, USPS and PCA could in principle be extended to incorporate a number of different reporters, some of which could even support detection by non-invasive imaging techniques. This could allow for real time in vivo monitoring of protein-protein interactions, an objective that is currently being pursued by various investigators in the field.

The molecular interaction screening systems discussed here have been applied to the identification of multiple types of interactions. These include protein-protein interactions, peptide-protein interactions, single chain antibody-protein target interactions, and small molecule-target interactions (see Sect. 3.4.2). Consequently, they can also be utilized to find new surrogate ligands, as can also be achieved with phage display approaches (see Sect. 3.3.3), for probing the function of genes and proteins. Such surrogate ligands may be used, for example, in the development of novel peptide or antibody-based imaging agents, and in the mapping of cellular signaling pathways and proteome networks.

To date, application of molecular interaction screening technologies to large scale mapping of cellular signaling pathways has been pioneered by the Y2H technology, primarily because Y2H is well established, and because yeast is a model system that can easily be adapted and manipulated for the screening of large and complex cDNA expression libraries. It has been directed toward a mapping of "interactomes" of yeast (Ito et al. 2000, 2001; Uetz et al. 2000), *C. elegans* (Walhout et al. 2000), and the infectious agents *H. pylori* (Rain et al. 2001) and vaccinia virus (Mc Craith et al. 2000). Characterization of other model system "interactomes," such as the *Drosophila* "interactome," is underway. Numerous studies utilize Y2H and other interaction technologies in the characterization of human protein interactions and the mapping of cellular signaling pathways, with the goal of establishing maps of the human "interactome."

Much is to be gained by proteomics efforts to develop refined methods for measuring macromolecular interactions and for the large scale mapping of "interactomes." They represent cornerstone activities in (1) the elucidation of the function and regulation of genes, proteins, and signaling pathways: (2) the discovery and validation of drug discovery targets; and (3) the development of assays for monitoring specific pathway activities in multiple experimental settings (see also Sect. 3.5). Integration of data sets emerging from genomics and functional genomics research with data sets obtained from large scale proteomics studies will lead to a better understanding of the molecular basis of cellular processes and will facilitate the manipulation of such processes for the development of therapeutic agents.

3.3.3
Probing the Proteome: Discovery of Surrogate Ligands

Having the complete sequence of an entire genome in databases is not sufficient to elucidate the biological function of the encoded genes. There is no strict linear relationship between genes and encoded proteins since the protein product of one gene might be present in many isoforms and post-translationally modified versions in the same cell. Therefore, probing protein function inside the cell requires the ability to differentiate individual proteins, isoforms, and modified derivatives from one another. This can be achieved through the use of surrogate ligands that bind with high specificity and affinity to the protein of interest. Most often, surrogate ligands are either antibodies or peptides that are isolated from large combinatorial libraries. In the simplest case, surrogate ligands are used to detect the protein target inside the cell and in complex biological mixtures. In more complex situations, the surrogate ligands not only bind to the target protein but also modulate its activity. They may inhibit the target's biochemical activity, prevent its interaction with regulatory subunits, or interfere with its interactions with other cellular proteins.

3.3.3.1
Protein-Antibody Interaction Screening Technologies

In the immune system, vast libraries of antibodies and T-cell receptors recognize virtually any foreign entity. When the interaction of a specific antibody with an antigen occurs, proliferation of the antibody-producing cells is stimulated, leading to the selective amplification of that particular antibody-producing clone. Because of this design, the immune system efficiently selects for antibodies capable of recognizing virtually any molecular shape.

This central feature of the immune system has been exploited to generate polyclonal and monoclonal antibodies against specific targets. The theoretical and practical issues of producing polyclonal and monoclonal antibodies are discussed in great detail elsewhere (Harlow and Lane 1998). Early methods of antibody generation and isolation were time-consuming and hampered by low throughput. The emergence of antibody display methods and the availability of antibody display libraries has overcome these shortcomings and made possible the simultaneous isolation of antibodies that specifically recognize dozens or even hundreds of targets.

This breakthrough was made possible by two technical developments. First, it was demonstrated that libraries of randomly recombined heavy and light chains can generate antibodies with defined specificity (Huse et al. 1989). Second, it was shown that foreign DNA fragments can be inserted into the *gene III* of filamentous phages (such as f1 and M13), which encodes phage coat protein pIII, to create a fusion protein with the foreign sequence at the N-terminus (Smith 1985). The fusion protein is incorporated into the virion, which retains infectivity and displays the foreign protein in a form that can bind to other proteins. The phage displaying a protein with the desired binding specificity to an immobilized target can be extracted from the total population of recombinant phages by affinity selection. The recovered, "affinity-enriched" phage population can be amplified and subjected to another round of selection. This process of selective enrichment for proteins with the desired binding specificity is called "panning," and is made possible by the linkage of the genotype (the foreign DNA fragment inside the virion) to the phenotype (the binding affinity of the displayed protein domain).

Antibody libraries displayed on the surface of phage are either single-chain (scFv) or fragment-antigen-binding (Fab) libraries. In the Fab libraries, gene fragments encoding entire light- and heavy-chain antibody binding regions are amplified by PCR from immune cells, combined and displayed on the surface of phage. In single-chain antibody libraries, only the variable domains of the light chain (V_L) and heavy chain (V_H) are amplified, spliced together and displayed. Only these V_H and V_L regions mediate antigen recognition in the Fab fragment.

A typical phage display antibody library contains approximately 10^{10}–10^{11} antibody fragments, which corresponds to the number of different binding specificities and affinities. Antibodies that recognize an immobilized target are enriched from this library by cycles of selective enrichment, ultimately leading to the isolation of a subset of antibodies that may recognize different epitopes on the target and have different binding affinities (for review see Barbas et al. 2001). This process of selective enrichment is highly amenable to automation, allowing the parallel screening of the library with a large number of different antigens. This feature makes antibodies a very valuable tool to study the entire proteome.

3.3.3.2
Combinatorial Peptide Libraries and Surrogate Ligand Discovery

Surrogate peptide ligands, like antibodies, specifically recognize target proteins and can be isolated from combinatorial peptide libraries. In these libraries, random oligonucleotides are inserted into a filamentous phage *gene III* or *gene VIII* allowing the fusion of the encoded random peptides to the N-terminus of the coat proteins pIII or pVIII, respectively. Peptides displayed as pIII fusions can be present in up to 5 copies per virion because there are 5 copies of pIII protein on each virion. The number of pVIII fused peptides can vary from dozens to hundreds per virion because there are around 3000 pVIII proteins on every virion. Typically, higher-binding-affinity peptides can be isolated from the pIII library because of the avidity effect.

Peptide libraries may differ in the length of the random region and may contain fixed residues that impose predetermined structural constraints on the peptides, such as disulfide bridged loops or α-helices. The peptides may also be displayed on the surface of folded proteins or protein domains (for review, see Smith and Petrenko 1997; Barbas et al. 2001).

Random peptide libraries are searched for peptides that bind to the target protein using cycles of affinity selection as described for antibodies. Peptide ligands that recognize the target can be found because only a limited number of critical residues are required for the specific interaction between the peptide and the target. Peptides usually bind at or near the natural ligand-binding site or active site of the target, thereby acting as natural ligand agonists or antagonists.

Random antibody or peptide libraries can also be enriched through binding to intact cells and used to identify ligands for cell surface receptors. These surrogate antibody or peptide ligands can be used to discriminate between different cell types with distinct peptide binding signatures, employed as reagents for receptor identification and cloning, or as tools to induce receptor-mediated phenotypic changes in target cells.

Recently, another type of random peptide library has been developed for the intracellular expression of peptides. The expressed peptides ("perturbagens") may perturb the biological activity of their intracellular targets, leading to detectable phenotypic change. Peptides are recovered from cells that exhibit the desired phenotype, and are used in subsequent experiments to isolate their target(s). The peptide/target pair could form the basis for mechanism-based drug discovery efforts (Capronigro et al. 1998).

3.3.3.3
Implications for Diagnostics, Prognostics and Therapy

Surrogate ligands may form the basis for the development of diagnostic and prognostic agents if alterations in the concentration, localization, or modification of a target are linked to disease development and/or therapeutic outcome. Using these ligands, in vitro assays can be developed to detect and measure the target in tissue biopsies or body fluids. The linkage of the surrogate ligands to an imaging agent may allow the development of a new generation of in vivo diagnostic and prognostic tools. Antibody and peptide ligands may also be used for the generation of protein chips for the parallel detection of tens to hundreds of proteins in one and the same biological sample.

Surrogate ligands may also modulate the biological activity of their target molecules, and may therefore provide a route for the development of therapeutic agents. Antibodies have an excellent biodistribution profile and are stable in vivo. This property, combined with the relatively straightforward generation of fully human antibodies, makes antibodies a suitable drug candidate if the goal is to target cell surface receptors and secreted proteins. The promise of antibodies as therapeutic agents has been substantiated by the recent success of antibody drugs such as herceptin and rituxan. Currently, there are hundreds of antibodies in development for multiple therapeutic indications.

Peptides are less stable in vivo than antibodies, and may require chemical modification and/or the development of sustained release formulations for the delivery of the peptide drug. Nonetheless, there are currently over ten peptide drugs on the market and many more are in development. Surrogate peptide ligands provide useful information for nonpeptidic drug development as well, since the interaction of the peptide with the target provides critical structural information that could be exploited either by screening for nonpeptidic molecules with the same binding specificity, or by a more systematic chemistry effort to incorporate the critical interactions revealed by the peptide binding to the target into synthetic small molecules.

3.4
Chemical Genomics and Proteomics

As briefly discussed in the introduction of this chapter, chemical genomics and proteomics refer to the systematic analysis of the impact of organic small molecules on the genome and proteome (MacBeath 2001). The power of this approach is that global effects of drug treatment can be studied. In the same way that systems biology using genomics is supplanting studies of individual genes, chemical genomics could potentially replace the analysis of drugs and small molecules using conventional assays that measure individual cellular parameters.

Chemical genomics may address a number of questions: What biological processes are affected by drug treatment? What is the mechanism of action and what is the precise molecular target(s) through which a compound achieves a therapeutic effect? What additional targets are affected? Are these additional targets associated with toxicity, or with other potential therapeutic applications? Chemical genomics can also be applied to the design of small molecules that can probe the function of gene and proteins in vitro and in vivo.

Multiple strategies that seek to meet such objectives have been developed over the last years, and integrate many aspects of biology and chemistry. Here we focus primarily on (a) large scale gene expression profiling studies, and (b) the search and characterization of targets of small molecules.

3.4.1
Impact of Small Molecules on Genome Expression

One branch of chemical genomics utilizes differential-gene-expression technologies to analyze the biological effects of drug treatment. The focus here is on DNA microarrays (described in Sect. 3.2.1.1). RNA, extracted from cultured cells or from tissues of animals, is reverse transcribed, labeled, and used to probe whole genome arrays or arrays containing sequences that represent a subset of the genome. Differences in gene expression, which may be caused by drug treatment or genetic mutation, are monitored. Expression profiles of test compounds are compared to those of reference compounds with known mechanisms of action and to those of mutations in genes of known function. By pattern matching, the cellular pathways and even the molecular targets of compounds of unknown mechanism can be revealed. The more comprehensive the reference data set, the greater the probability that compounds can be accurately categorized.

In one example, expression profiling in yeast revealed that the molecular target of dyclonine, a topical anesthetic, is ERG2, a gene required for ergosterol biosynthesis. The human homolog of ERG2 is the sigma receptor, a neurosteroid-interacting protein that regulates potassium conductance (Hughes et al. 2000), suggesting that this receptor may represent the human target of dyclonine. This study highlights how a non-human model organism may be utilized to elucidate the molecular mechanism of a therapeutic agent in human cells.

Good drugs are specific for a biological target and have minimal effects on other targets. A major objective in developing new drugs is to determine if a compound is specific for the intended target. Chemical genomics provides a comprehensive approach to the analysis of genome-wide consequences of a given perturbation are analyzed, as well as a method for identifying unanticipated activities of drugs – i.e., off-target effects. Using DNA microarray technology, the expression levels of thousands of genes are measured simultaneously, giving a "fingerprint" of a compound's activity at the genomic level. Genetic and pharmacologic inhibition of gene function can result in extremely similar changes in gene expression, thus providing a method for confirming a potential target. The pattern of gene expression in cells containing a deletion of the potential target, when studied in drug-treated vs. untreated cells, can also provide information about off-target effects. This approach has been used in studies of the immunosuppressants FK506 and cyclosporin, confirming calcineurin as the molecular target of these drugs and identifying off-target effects of FK506 (Marton et al. 1998).

A major problem in the pharmaceutical industry is the high failure rate of compounds that enter development. Fifty percent of drugs are lost between nomination for development and clinical use. There is an 80% failure after compounds enter the clinic due to lack of efficacy, toxicity issues, or competition, to give an overall failure rate of 90%. Efforts are underway to use genomics to predict compound toxicity. This involves the generation of expression profiles for various tissues (e.g., liver and kidney) from animals treated with compounds of known toxicity. The relationship between the expression profiles can then be compared to traditional measures of toxicity. The goal of this approach is to develop predictive tools for the assessment of toxicity at an early stage in the drug discovery process to decrease the failure rate of drugs that enter development.

In summary, large scale gene expression analysis is a powerful approach for monitoring and decoding the cellular pathways affected by drug treatment. Such analysis relies on a reference database of profiles generated from compounds of known mechanism and toxicity, or from mutations in genes of known function. Pattern recognition is employed to determine what cellular pathways are affected by chemical perturbation. This tool has the potential to rapidly assess the specificity and selectivity of compounds in the drug discovery process, identify compounds with novel mechanism of action, and define new therapeutic applications for compounds that have unanticipated effects.

3.4.2
Small Molecule-Target Interactions

Classic approaches for detection of ligand-protein receptor interactions have relied primarily on in vitro biochemical methods, including radiolabeled ligand binding, affinity chromatography, and photo-cross-linking. These methods are often laborious and time-consuming, and suffer from the requirement of obtaining sufficient material for purification, peptide sequence analysis, and subsequent cloning of the cDNAs encoding these targets. Technological advances in solid support technology (Shimizu et al. 2000) and protein separation and analysis methods (see Sect. 3.3.1) are improving the range of applications of biochemical methods. However, higher throughput and more generally applicable methods for the detection of interactions between organic small molecules and their targets are necessary in or-

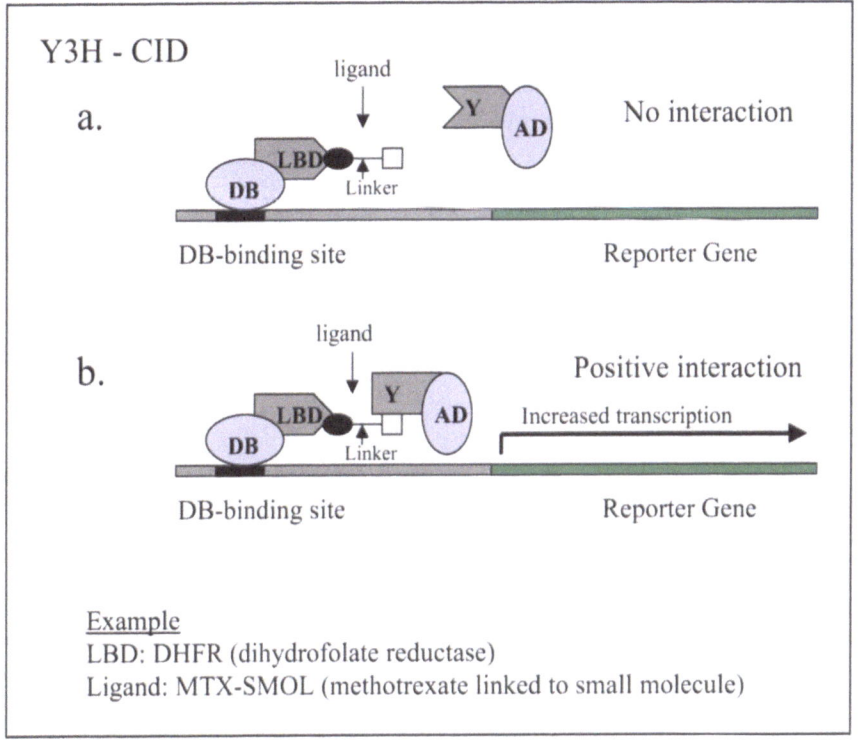

Y3H - CID

a.

ligand

No interaction

LBD
DB
Linker

Y
AD

DB-binding site

Reporter Gene

b.

ligand

Positive interaction

LBD
DB
Linker

Y
AD

Increased transcription

DB-binding site

Reporter Gene

Example
LBD: DHFR (dihydrofolate reductase)
Ligand: MTX-SMOL (methotrexate linked to small molecule)

Fig. 3.4. Monitoring of small molecule-protein target interactions. A schematic representation of a variation of the Y2H system (see Fig. 3.3 A), adapted for the analysis of the interaction of proteins as mediated by a cell permeable small molecule (compound induced dimerization, CID). A bifunctional small molecule is synthesized that incorporates an entity with known target binding properties (e.g., methotrexate, MTX; •-) fused via a linker to a compound of interest (*square with light gray shading*). Interaction of the chimeric small molecule with the ligand binding domain (*LBD*; e.g., DHFR) of the *DB*-fusion protein and the prey target molecule (*Y*) of the *AD*-fusion protein, promotes CID and induction of the transcription of the reporter gene. An AD-cDNA library may be used to screen for prey proteins that interact with a small molecule of interest

der to speed up the drug discovery process. Here we describe a few cell-based technologies that conceptually derive from two-hybrid interaction screening technologies (see Sect. 3.3.2).

One method that highlights the concept of two-hybrid interaction screening technologies as applied to assaying compound-mediated protein dimerization is the yeast three-hybrid (Y3H) technology (Licitra and Liu 1996). It shares many properties with its predecessor, the Y2H technology (Fig. 3.3A), but differs in that interaction between fusion proteins is mediated by a bridging small molecule (Fig. 3.4). The bridging small molecule is a bifunctional synthetic ligand that is used to induce protein dimerization. In brief, one component of the bifunctional ligand is an entity with known target binding properties (for example, methotrexate, which binds with high affinity to the target protein DHFR; Lin et al. 2000), and the other component is the compound of interest. These two components are linked by a specifically designed linker. Interaction

with a target protein (the "prey") results in a compound-mediated dimerization (recruitment) of two distinct target fusion proteins (e.g., DHFR fused to a DNA-binding protein, DB-DHFR, and the prey fused to a transcription activation domain, prey-AD), forming a complex that bears the properties of a functional transcription factor (e.g., a complex such as: DB-DHFR×MTX-cpd×Target-AD). This transcription factor can activate specific reporter genes integrated into the yeast genome. Thus, binding of a small molecule to its target protein may be monitored via the activation of reporters that promote the growth of yeast cells (when auxotrophic markers are used) or can be assayed using colorimetric assays (e.g., LacZ assay). Other systems that are designed for protein-protein interaction screening (see Sect. 3.3.2) may also be adapted for the analysis of small molecule-target interaction. This is a very powerful demonstration of how integration of proteomics and chemistry can lead to development of novel platforms that accelerate drug discovery.

Cellular interaction screening systems for monitoring drug-target interactions will find uses in many different settings. They provide more general and rapid methods for large scale screening of cDNA libraries for target proteins (target discovery). They are not hampered by scarcity of biological samples. They can be employed in high throughput mutagenesis studies to discover receptors with higher affinity to small molecule ligands (target evolution), profile the interaction of small molecules with target polymorphic variants (pharmacogenomics/proteomics), study structure-function relationship, and develop new types of high throughput small molecule screening assays for drug discovery. The synthesis of new kinds of synthetic small molecule libraries that are compatible with three-hybrid systems has already been reported (Koide et al. 2001).

Recently, an interesting approach for engineering small molecules that disrupt protein function has been reported. This approach can be used to investigate the function of genes that, until now, have defied functional analysis because of inherent limitations in the genetic manipulation of mammalian cells (Bishop et al. 2000). It integrates biology, structural biology and chemistry. Shokat and colleagues focused on a class of enzymes known as kinases. They mutated a selected kinase gene to change the site on the enzyme that binds ATP. The result was an enlarged ATP-binding pocket that did not, however, affect the enzyme's normal function. By adding various bulky chemical groups to small molecules already known to inhibit kinases, they identified analogs of such inhibitors that specifically fit into the enlarged active site and selectively inhibited the mutagenized variant but not the wild-type form of the enzyme. A nonspecific kinase inhibitor was thus transformed into a highly specific one. This allowed them to study the function of certain kinases in model systems such as yeast, in which the wild-type form of the kinase of interest was replaced by the mutant inhibitor-responsive variant. The concept of "drug-target fitting" has previously been applied to proteins other than kinases. Clackson et al. (1998) reported the synthesis of analogs of the FK506 immunosuppressive agent that bind mutant forms of FKBP with higher affinity than wild-type FKBP (Clackson et al. 1998). Such analogs are being used in gene therapy settings to induce small molecule dimerizer-induced protein-protein interactions that activate various types of biological processes in a controlled fashion (Clackson 2000).

Other approaches to the analysis of small molecule-target interactions take advantage of recent advances in array technology developed for various functional genomics applications such as gene expression profiling. Schreiber and his colleagues, for instance, have created microarrays printed with libraries of small molecules (MacBeath et al. 1999). These arrays may be interrogated with diverse target populations, including purified proteins or even cell extracts, to detect known as well as novel types of interactions. Although still in their infancy, the further development of such array technology platforms will likely result in opportunities to profile the mechanism of action of small molecule drugs.

3.5
Functional Genomics and Proteomics: Implications for Molecular and Nuclear Medicine

In the context of the revelation of the blueprint of genomes, functional genomics and proteomics have begun to have an impact on many disciplines in basic science. Obtaining more "global views" of the multiplicity of components encoded by genomes, and how they interact and are regulated at the level of the proteome, is accelerating annotation of the genome, the understanding of gene function, the dissection of molecular signaling pathways, and the validation of drug discovery targets. Such enlarged views accelerate our understanding of the molecular basis of cellular processes, and how these regulate physiological and pathological phenotypes. Integration of data sets obtained from large scale "survey" profiling studies with those from more focused gene and protein function studies that biologists are traditionally trained to seek will constantly refine the picture of cellular circuitry with a goal of unraveling the many mysteries underlying organ and whole body system functions.

The impact of functional genomics and proteomics is not limited to basic science research. These disciplines have already begun to have an impact on medical sciences in numerous ways, including the discovery of novel drug targets (target discovery), the introduction of new guiding principles in medicinal chemistry for the development of novel mechanism-based therapeutics (drug discovery), the anticipation of potential side effects of drugs based on an understanding of their molecular mechanism of action (toxicogenomics), and the establishment of new guiding principles in the stratification of patient population on the basis of an understanding that drug response is influenced by genotype (pharmacogenomics). Functional genomics and proteomics research is leading to the discovery of many novel types of markers for disease diagnosis and prognosis. For instance, the emerging large number of new tumor-cell markers, discovered in part using microarray-based gene expression profiling, are likely to allow physicians to eventually tailor cancer therapies to their individual patients. Further, response to a specific therapy may be predicted not only on the basis of the patient's genotype, but also by the expression pattern of a set of specific markers. It is still some way from translating clues provided by gene and protein expression studies into validated diagnostics or prognostic tools, but the path is clearly visible. As certain functional genomics and proteomic tools become more refined and more widely used in the clinical arena, they will become an integral part of up-to-date medicine.

Functional genomics and proteomics are also providing the basic and medical scientist with an increasing number of new research tools for probing genomes and proteomes. These include small molecules, peptides, and antibodies, many of which are relevant to the field of nuclear medicine. Two major foci of nuclear medicine – non-invasive imaging of organs and the destruction of cancer cells, which rely largely on the localization of radioactive compounds within organs of the body – will benefit from the current genomics and proteomics revolution. Several benefits derive from technological developments as described in this review alone. A sampling follows. (1) Profiling-mediated discovery of novel tissue-specific markers, in particular cell surface markers, will spur the development of more specific imaging agents and therapeutics. (2) Non-invasive reporter assays (such as PET reporters; see Sect 3.2.1.1.2), when utilized in conjunction with genome-manipulation technologies (e.g., transgene-expression technologies and directed-gene "knock-in" methodologies), should enable the in vivo monitoring of gene expression. Such efforts would have direct implications for gene therapy as well. (3) The discovery of numerous new types of receptor/target ligands, including peptide- and antibody-based ligands (for both extracellular and intracellular targets) could form the basis of new imaging and therapeutics agents. (4) A better characterization of the target interaction landscape of small molecule drugs should improve the analysis and understanding of drug target interaction imaging data. (5) The discovery of signaling pathways and new enzymes within these pathways, in conjunction with

new strategies in assay development, should lead to the development of new agents that enable the monitoring of the activity of specific enzymes and pathways in vivo. For example, new tools have recently been developed for the imaging of metalloprotease activity in vivo.

Cancer studies have shown a positive correlation between cancer progression and expression of extracellular proteinases, such as the metalloproteinases (MMPs). Malignant cells depend on these proteinases in order to disrupt basement membranes, invade neighboring tissues and metastasize to different organs. A method that would allow in vivo measurement of MMP activity would be highly desirable, in particular for monitoring the therapeutic efficacy of any anti-MMP drug.

Bremer et al. (2001) have developed a method that is based on non-immunogenic MMP substrates that accumulate in tumors and can be used as enzymatic reporter probes. Cleavage by MMPs converts the probe into a fluorescent product, which is detected using near-infrared fluorescence imaging (NIRF), a newly developed technology. Similar methods may be followed for the analysis of other proteases, although this is more difficult for intracellular proteases. However, whole cell assays for the monitoring of the cellular activity of cellular proteases have been and are being developed: for example, in substrate-based fluorescence assays for apoptotic caspases the cleaved product is detected on the basis of an induced fluorescence signal. Such caspase assays are useful in monitoring the activity of cellular circuits converging on this important class of enzymes, and it is possible that future technological developments will translate the use of this kind of assay into development of in vivo imaging agents.

Further progress in the development of technologies that enable the monitoring of the association or dissociation of macromolecules, such as protein-protein interactions, may also lead to novel means by which to measure the activity of signaling pathways in intact cells. For instance, as previously discussed in this review, interaction technologies based on the induced proximity of fluorescent or bioluminescent reporter molecules are already being used in monitoring protein-protein interactions in intact cells. Similarly, protein fragment complementation assays, in which an interaction between macromolecules promotes the reconstitution of reporter activity, may in the future be used to monitor in vivo activity of specifically induced interactions and signaling pathways. If successfully applied to reporters that can be monitored with noninvasive imaging methods, such proteomics technologies could prove useful for applications in nuclear medicine. Technologies based on small-molecule-induced dimerization of macromolecules are already being applied in gene therapy settings (Clackson 2000).

Significant advances in small molecule or macromolecule delivery technologies have been made in recent years. For instance, peptide sequences encoded by diverse proteins (such as HIV-Tat and antennapedia), have been shown to be effective signals in shuttling peptides, and even proteins, into mammalian cells (Ford et al. 2001). Applications of these peptides for in vivo delivery have also been reported. Efforts inspired by these discoveries have led to the design of synthetic molecules that can be linked to other small molecules of interest and promote their uptake into cells (Wender et al. 2000). Such developments promise to have significant implications for basic science research and drug delivery strategies. As discussed previously in this review, functional genomics and proteomics-based combinatorial functional screens can lead to the discovery of target specific peptide surrogate ligands, which in combination with delivery and labeling technologies, may be used in target function studies, and, possibly, as future imaging agents for intracellular targets.

3.6

Functional Genomics and Proteomics: Opportunities and Challenges

The potential for functional genomics and proteomics is almost unlimited. Practiced at smaller scale, research that utilizes genomics and proteomics technologies is already generating important information about gene and protein functions and the molecular basis of disease, whereas large-scale "surveying" and "cataloging" studies are currently giving rise to complex information we cannot yet fully understand. Integrating the newly emerging "global view" approach, or systems biology, with more traditional gene-by-gene function studies will be critical in the functional mapping of the human genome. Potential applications in clinical medicine have been discussed throughout this review.

An interdisciplinary spirit will be necessary in order to fully reap the benefits from the genomics revolution. Geneticists will need to communicate with

chemists, physiologists with physicists, cell biologists with computer scientists, and so on. Especially challenging will be the development of coherent and uniform genomic information storage platforms and of tools that allow the scientist to investigate and understand relationships among diverse data elements. Genomics research is utterly dependent on computer science. Thus, biologists will need to understand the concepts behind relational databases, concepts that include the reduction of a vast amount of information to defined types of entities and their enumeration, and to relations between such entities – concepts that date only from the 1970s (Codd 1998). The evolving multidisciplinary genomic and proteomic sciences will define a new scientific activity that integrates basic and applied sciences, an important objective of which is improved diagnosis, treatment, and prevention of disease.

■ **Acknowledgements.** We would like to thank Dr. Margaret Lee Kley for critical reading of the manuscript and many helpful suggestions, and Zephyr Secher for help in preparing the manuscript.

3.7
References

Akiyama N, Matsuo Y, Sai H et al (2000) Identification of a series of transforming growth factor bet-responsive genes by retrovirus-mediated gene trap screening. Mol Cell Biol 20:3266–3273

Alizadeh AA, Eisen MB, Davis RE et al (2000) Distinct types of diffuse large B-cell lymphoma identified by gene expression profiling. Nature 403:503–511

Alon U, Barkai N, Notterman DA et al (1999) Broad patterns of gene expression revealed by clustering analysis of tumor and normal colon tissues probed by oligonucleotide arrays. Proc Natl Acad Sci USA 96:6745–6750

Barbas CF, Burton DR, Scott JK, Silverman GJ (2001) Phage display: a laboratory manual. Cold Spring Harbor Laboratory Press, Cold Spring Harbor, New York

Bishop AC, Ubersax JA, Petsch DT et al (2000) A chemical switch for inhibitor-sensitive alleles of any protein kinase. Nature 407:395–401

Bittner M, Meltzer P, Chen Y et al (2000) Molecular classification of cutaneous malignant melanoma by gene expression profiling. Nature 406:536–540

Borrebaeck CA (2000) Antibodies in diagnostics – from immunoassays to protein chips. Immunol Today 21:379–382

Borrebaeck CA, Ekstrom S, Hager AC et al (2001) Protein chips based on recombinant antibody fragments: a highly sensitive approach as detected by mass spectrometry. Biotechniques. 2001 30:1126–1132

Bremer C, Tung CH, Weissleder R (2001) In vivo molecular target assessment of matrix metalloproteinase inhibition. Nature Med 7:743–748

Brown PO, Botstein D (1999) Exploring the new world of the genome with DNA arrays. Nat Genet 21:33–37

Brown SD, Nolan PM (1998) Mouse mutagenesis-systematic studies of mammalian gene function. Hum Mol Genet 7:1627–1633

Capronigro G, Abedi MR, Hurlburt AP et al (1998) Transdominant genetic analysis of a growth control pathway. Proc Natl Acad Sci USA 95:7508–7513

Carnero A, Hudson JD, Hannon GJ et al (2000) Loss-of-function genetics in mammalian cells: the p53 tumor suppressor model. Nucleic Acids Res 28:2234–2241

Chakravarti A (2001) To a future of genetic medicine. Nature 409:822–823

Cheung VG, Morley M, Aguilar F et al (1999) Making and reading microarrays. Nat Genet 21:15–19

Chien C, Bartel PL, Sternglanz R et al (1991) The two hybrid system: a method to identify and clone genes for proteins that interact with a protein of interest. Proc Natl Acad Sci USA 88:9578–9582

Cho RJ, Campbell MJ, Winzeler EA et al (1998) A genome-wide transcription analysis of the mitotic cell cycle. Mol Cell 2:65–73

Clackson T (2000) Regulated gene expression systems. Gene Ther 7:120–125

Clackson T, Yang W, Rozamus LW et al (1998) Redesigning an FKBP-ligand interface to generate chemical dimerizers with novel specificity. Proc Natl Acad Sci USA 95:10437–10442

Codd EF (1998) A relational model of data for large shared data banks. 1970. MD Comput 15:162–166

De Leenheer AP, Lefevere MF, Lambert WE, Colinet ES (1985) Isotope-dilution mass spectrometry in clinical chemistry. Adv Clin Chem 24:111–161

De Risi JL, Iyer V, Brown PO (1997) Exploring the metabolic and genetic control of gene expression on a genomic scale. Science 278:680–686

Deiss LP, Kimchi A (1991) A genetic tool used to identify thioredoxin as a mediator of a growth inhibitory signal. Science 252:117–120

Deiss LP, Feinstein E, Berissi H et al (1995) Identification of a novel serine/threonine kinase and a novel 15-kD protein as potential mediators of the gamma interferon-induced cell death. Genes Dev 9:15–30

Devi LA (2000) G-protein-coupled receptor dimmers in the lime light. Trends Pharmcol Sci 21:324–326

Diehn M, Eisen MB, Botstein D et al (2000) Large-scale identification of secreted and membrane-associated gene products using DNA micorarrays. Nat Genet 25:58–62

Duggan DJ, Bittner M, Chen Y et al (1999) Expression profiling using DNA microarrays. Nat Genet 21:10–14

Eisen MB, Brown PO (1999) DNA arrays for analysis of gene expression. In: Weissman SM (ed) cDNA preparation and characterization. Methods in enzymology. Academic Press, San Diego, pp 179–205

Eisen MB, Spellman PT, Brown PO et al (1998) Cluster analysis and display of genome-wide expression patterns. Proc Natl Acad Sci USA 95:14863–14868

Fields S, Song OK (1989) A novel genetic system to detect protein-protein interactions. Nature 340:245–246

Fodor SPA, Read JL, Pirung MC et al (1991) Light-directed, spatial addressable parallel chemical synthesis. Science 251:767–773

Ford KG, Souberbielle BE, Darling D, Farzaneh F (2001) Protein transduction: an alternative to genetic intervention? Gene Ther 8:1–4

Friddle CJ, Koga T, Rubin EM et al (2000) Expression profiling reveals distinct sets of genes altered during induction and regression of cardiac hypertrophy. Proc Natl Acad Sci USA 97:6745–6750

Fung ET, Thulasiraman V, Weinberger SR, Dalmasso EA (2001) Protein biochips for differential profiling. Curr Opin Biotechnol 12:65–69

Gallagher WM, Cairney M, Schott B et al (1997) Identification of p53 genetic suppressor elements which confer resistance to cisplatin. Oncogene 14:185–193

Gambhir SS, Barrio JR, Wu L et al (1998) Imaging of adenoviral-directed herpes simplex virus type 1 thymidine kinase gene expression in mice with ganciclovir. J Nucl Med 39:2003–2011

Gambhir SS, Barrio JR, Phelps ME et al (1999) Imaging adenoviral-directed reporter gene expression in living animals with positron emission tomography. Proc Natl Acad Sci USA 96:2333–2338

Garkavtsev I, Kazarov A, Gudkov AV et al (1996) Suppression of the novel growth inhibitor p33ING1 promotes neoplastic transformation. Nat Genet 14:415–420

Golub TR, Slonim DK, Tamayo P et al (1999) Molecular classification of cancer: class discovery and class prediction by gene expression monitoring. Science 286:531–537

Goerg A, Obermaier C, Boguth G et al (2000) The current state of two-dimensional electrophoresis with immobilized pH gradients. Electrophoresis 21:1037–1053

Gudkov AV, Zelnick CR, Kazarov AR et al (1993) Isolation of genetic suppressor elements, inducing resistance to topoisomerase II-interactive cytotoxic drugs, from human topoisomerase II cDNA. Proc Natl Acad Sci USA 90:3231–3235

Gudkov AV, Roninson IB (1997) Isolation of genetic suppressor elements (GSEs) from random fragment cDNA libraries in retroviral vectors. Methods Mol Biol 69:221–240

Gudkov AV, Roninson IB, Brown R (1999) Functional approaches to gene isolation in mammalian cells. Science 285:299

Gygi SP, Aebersold R (2000) Mass spectrometry and Proteomics. Curr Opin Chem Biol 4:489–494

Gygi SP, Rochon Y, Franza BR, Aebersold R (1999a) Correlation between protein and mRNA abundance in yeast. Mol Cell Biol 19:1720–1730

Gygi SP, Rist B, Gerber SA et al (1999b) Quantitative analysis of complex protein mixtures using isotope-coded affinity tags. Nat Biotechnol 17:994–999

Gyuris J, Golemis E, Chertkov H et al (1993) Cdi1, a human G1 and S phase protein phosphatase that associates with Cdk2. Cell 75:791–803

Haab BB (2001) Advances in protein microarray technology for protein expression and interaction profiling. Curr Opin Drug Discov Devel 4:116–123

Hacia J (1999) Resequencing and mutational analysis using oligonucleotide microarrays. Nat Genet 21:42–47

Hannon GJ, Sun P, Concklin DS et al (1999) MaRX: an approach to genetics in mammalian cells. Science 283:1129–1130

Harlow E, Lane D (1998) Using antibodies: a laboratory manual. Cold Spring Harbor Laboratory Press, Cold Spring Harbor, New York

Hirst M, Ho C, Sabourin L et al (2001) A two-hybrid system for transactivator bait proteins. Proc Natl Acad Sci USA 98:8726–8731

Hochschild A, Dove S (1998) Protein-protein contacts that activate and repress prokaryotic transcription. Cell 92:597–600

Holzmayer TA, Pestov DG, Roninson IB (1992) Isolation of dominant negative mutants and inhibitory antisense RNA sequences by expression selection of random DNA fragments. Nucleic Acids Res 20:711–717

Huang R (2001) Detection of multiple proteins in an antibody-based protein microarray system. J Immunol Methods 255:1–13

Hudson JD, Shoaibi MA, Maestro R et al (1999) A proinflammatory cytokine inhibits p53 tumor suppressor activity. J Exp Med 190:1375–1382

Hughes TR, Marton MJ, Jones AR et al (2000) Functional discovery via a compendium of expression profiles. Cell 102:109–126

Huse WD, Sastry L, Iverson SA et al (1989) Generation of a large combinatorial library of the immunoglobulin repertoire in phage lambda. Science 246:1275–1281

International Human Genome Sequencing Consortium (2001) Initial sequencing and analysis of the human genome. Nature 409:860–921

Ishida Y, Leder P (1999) RET: a polyA-trap retrovirus vector for the reversible disruption and expression monitoring of gene in living cells. Nucleic Acids Res 27:35

Ito T, Tashiro K, Muta S et al (2000) Toward a protein-protein interaction map of the budding yeast: a comprehensive system to examine two-hybrid interactions in all possible combinations between the yeast proteins. Proc Natl Acad Sci USA 97:1143–1147

Ito T, Chiba T, Ozawa R et al (2001) A comprehensive two-hybrid analysis to explore the yeast protein interactome. Proc Natl Acad Sci USA 98:4569–4574

Iyer VR, Eisen MB, Ross DT et al (1999) The transcriptional program in the response of human fibroblasts to serum. Science 283:83–87

Iyer VR, Horak CE, Scafe CS et al (2001) Genomics binding sites of the yeast cell cycle transcription factors SBF and MBF. Nature 409:533–538

Jones VW, Kenseth JR, Porter MD et al (1998) Microminiaturized immunoassays using atomic force microscopy and compositionally patterned antigen arrays. Anal Chem 70:1233–1241

Joos TO, Schrenk M, Hopfl P et al (2000) A microarray enzyme-linked immunosorbent assay for autoimmune diagnostics. Electrophoresis 21:2641–2650

Johnsson N, Varshavsky A (1994) Split ubiquitin as a sensor of protein interactions in vivo. Proc Natl Acad Sci USA 91:10340–10344

Joung JK, Ramm EI, Pabo CO (2000) A bacterial two-hybrid selection system for studying protein-DNA and protein-protein interactions. Proc Natl Acad Sci USA 97:7382–7387

Khan J, Wei JS, Ringner M et al (2001) Classification and diagnostic prediction of cancers using gene expression profiling and artificial neural networks. Nature Med 7:673–679

Kharpko KR, Khorlin AA, Ivanov IB et al (1991) Hybridization of DNA with oligonucleotides immobilized in gel: a convenient method for detecting single base substitutions. Mol Biol 25:581–591

Koide K, Finkelstein JM, Ball Z, Verdine GL (2001) A synthetic library of cell-permeable molecules. J Am Chem Soc 123:398–408

Kojima T, Kitamura T (1999) A signal sequence trap based on a constitutively active cytokine receptor. Nat Biotechnol 17:487–490

Lashkari DA, De Risi JL, McCusker JH et al (1997) Yeast genome micorarrays for parallel genetic and gene expression analysis of the yeast genome. Proc Natl Acad Sci USA 94:13057–13062

Licitra EJ, Liu JO (1996) A three-hybrid system for detecting small ligand-protein receptor interactions. Proc Natl Acad Sci USA 93:12817–12821

Lin H, Abida WM, Sauer RT et al (2000) Dexamethasone-Methotrexate: an efficient chemical inducer of protein dimerization in vivo. J Am Chem Soc 122:4247–4248

Link AJ, Eng J, Schieltz DM et al (1999) Direct analysis of protein complexes using mass spectrometry. Nat Biotechnol 17:676–682

Lockhart DJ, Winzeler EA (2000) Genomics, gene expression and DNA arrays. Nature 405:827–836

Lockhart DJ, Dong H, Byrne MC et al (1996) Expression monitoring by hybridization to high density oligonucleotide arrays. Nat Biotechnol 14:1675–1680

MacBeath G (2001) Chemical genomics: what will it take and who gets to play? Genome Biol 2:2005

MacBeath G, Koehler AN, Schreiber SL (1999) Printing small molecules as micorarrays and detecting protein-ligand interactions en masse. J Am Chem Soc 121:7967–7968

Mahon GM, Whitehead IP (2001) Retrovirus cDNA expression library screening for oncogenes. Methods Enzymol 332:211–221

Marton MJ, De Risi JL, Bennett HA et al (1998a) Drug target validation and identification of secondary drug target effects using DNA microarrays. Nat Med 4:1293–1301

Mc Craith S, Holtzman T, Moss B, Fields S (2000) Genome-wide analysis of vaccinia virus protein-protein interactions. Proc Natl Acad Sci USA 97:4879–4884

Medico E, Gambarotta ZG, Gentile A et al (2001) A gene trap vector system for identifying transcriptionally responsive genes. Nat Biotechnol 19:579–582

Mendelsohn AR, Brent R (1999) Protein interaction methods: towards an endgame. Science 284:1948–1950

Mendoza LG, McQuary P, Mongan A et al (1999) High-throughput microarray-based enzyme-linked immunosorbent assay (ELISA). Biotechniques 27:778–780; 782–788

Mitchell K, Pinson KI, Kelly OG et al (2001) Functional analysis of secreted and transmembrane proteins critical for mouse development. Nat Genet 28:241–249

Morgenstern JP, Land H (1990) Advanced mammalian gene transfer: high titre retroviral vectors with multiple drug selection markers and a complementary helper-free packaging cell line. Nucleic Acids Res 18:3587–3596

Nellen W, Sczakiel G (1996) In vitro and in vivo action of antisense RNA. Mol Biotechnol 6:7–15

Pandey A, Mann M (2000) Proteomics to study genes and genomes. Nature 405:837–846

Pelletier JN, Campbell-Valois FX, Michnick SW(1998) Oligomerization domain-directed reassembly of active dihydrofolate reductase from rationally designed fragments. Proc Natl Acad Sci USA 95:12141–12146

Perou CM, Jeffrey SS, Van De Rijn M et al (1999) Distinctive gene expression patterns in human mammary epithelial cells and breast cancers. Proc Natl Acad Sci USA 96:9212–9217

Rain JC, Selig L, De Reuse H et al (2001) The protein-protein interaction map of Helicobacter pylori. Nature 409:211–215

Remy I, Michnick SW (1999) Clonal selection and in vivo quantitation of protein interactions with protein-fragment complementation assays. Proc Natl Acad Sci USA 96:5394–5399

Ren B, Robert F, Wyrick JJ et al (2000) Genome-wide location and function of DNA binding proteins. Science 290:2306–2309

Risch NJ (2001) Searching for genetic determinants in the new millennium. Nature 405:847–856

Rojo-Niersbach E, Morley D, Heck S, Lehming N (2000) A new method for the selection of protein interactions in mammalian cells. Biochem J 348:585–590

Roses AD (2000) Pharmacogenetics and the practice of medicine. Nature 405:857–856

Rowley A, Choudhary JS, Marzioch M et al (2000) Applications of protein mass spectrometry in cell biology. Methods 20:383–397

Schena M, Shalon D, Davis RW, Brown PO (1995) Quantitative monitoring of gene expression patterns with a cDNA microarray. Science 270:467–470

SenGupta DJ, Zhang B, Kraemer B et al (1996) Three-hybrid system to detect RNA-protein interactions in vivo. Proc Natl Acad Sci USA 93:8496–8501

Senior K (1999) Fingerprinting disease with protein chip arrays. Mol Med Today 5:326–327

Shalon D, Smith SJ, Brown PO (1996) A DNA micro-array system for analyzing complex DNA samples using two-color fluorescent probe hybridization. Genome Res 6:639–645

Shevchenko A, Jensen ON, Podtelejnikov AV et al (1996) Linking genome and proteome by mass spectrometry: large-scale identification of yeast proteins from two dimensional gels. Proc Natl Acad Sci USA 94:14440–14445

Shimizu N, Sugimoto K, Tang J et al (2000) High-performance affinity beads for identifying drug receptors. Nat Biotechnol 18:877–881

Smith GP (1985) Filamentous fusion phage: novel expression vectors that display cloned antigens on the virion surface. Science 228:1315–1317

Smith GP, Petrenko VA (1997) Phage display. Chem Rev 97:391–410

Southern EM, Maskos U, Elder JK (1992) Analyzing and comparing nucleic acid sequences by hybridization to arrays of oligonucleotides: evaluation using experimental models. Genomics 13:1008–1017

Spellman PT, Sherlock G, Zhang MQ et al (1998) Comprehensive identification of cell-cycle-regulated genes of the yeast Saccharomyces cerevisiae by microarray hybridization. Mol Biol Cell 9:3273–3297

Stagljar I, Korostensky C, Johnsson N et al (1998) A genetic system based on split-ubiquitin for the analysis of interactions between protein in vivo. Proc Natl Acad Sci USA 95:5187–5192

Sun P, Dong P, Dai K et al (1998) p53-independent role of MDM2 in TGF-beta1 resistance. Science 282:2270–2272

Tashiro K, Tada H, Heilker R et al (1993) Signal sequence trap: a cloning strategy for secreted proteins and type I membrane proteins. Science 261:600–603

Uetz P, Giot L, Cagney G et al (2000) A comprehensive analysis of protein-protein interactions in Saccharomyces cerevisiae. Nature 403:623–627

Van Steensel B, Delrow J, Henikoff S (2001) Chromatin profiling using targeted DNA adenine methyltransferase. Nat Genet 27:304–308

Venter JC, Adams MD, Myers EW et al (2001) The sequencing of the human genome Science 291:1304–1351

Vidal M, Legrain P(1999) Yeast forward and reverse n'n-hybrid systems. Nucleic Acids Res 27:919–929

Visintin M, Tse E, Axelson H et al (1999) A. Selection of antibodies for intracellular function using a teo-hybrid in vivo system. Proc Natl Acad Sci USA 96:11723–11728

Von Eggeling F, Davies H, Lomas L et al (2000) Tissue-specific microdissection coupled with ProteinChip array technologies: applications in cancer research. Biotechniques 29:1066–1070

Walhout AJ, Sorcella R, Lu X et al (2000) Protein interaction mapping in C. elegans using proteins involved in vulval development. Science 287:116–122

Washburn MP, Wolters D, Yates JR III (2001) Large-scale analysis of the yeast proteome by multidimensional protein identification technology. Nat Biotechnol 19:242–247

Wender PA, Mitchell DJ, Pattabiraman K et al (2000) The design, synthesis, evaluation of molecules that enable or enhance cellular uptake: peptoid molecular transporters. Proc Natl Acad Sci USA 97:13003–13008

Whitney M, Rockenstein E, Cantin G et al (1998) A genome-wide functional assay of signal transduction in living mammalian cells. Nat Biotechnol 16:1329–1333

Wittes J, Friedman H (1999) Searching for evidence of altered gene expression: a comment on statistical analysis of microarray data. J Natl Cancer Inst 91:400–401

Wittke S, Lewke N, Mueller S et al (1999) Probing the molecular environment of membrane proteins in vivo. Mol Cell Biol 10:2519–2530

Wodicka L, Dong H, Mittman M et al (1997) Genome-wide expression monitoring in Saccharomyces cerevisiae. Nat Biotechnol 15:1359–1367

Xu X, Leo C, Jang Y et al (2001) Dominant effector genetics in mammalian cells. Nat Genet 27:23–29

Yang M, Wu Z, Fields S (1995) Protein-peptide interactions analyzed with the yeast-two-hybrid system. Nucleic Acids Res 23:1152–1162

Yates JR III, Eng JK, McCormack AL, Schieltz D (1995) Method to correlate tandem mass spectra of modified peptides to amino acid sequences in the protein database. Anal Chem 67:1426–1436

Yu Y, Annala AJ, Barrio JR et al (2000) Quantification of target gene expression by imaging reporter gene expression in living animals. Nat Med 6:933–937

Zambrowicz BP, Friedrich GA, Buxton EC et al (1998) Disruption and sequence identification of 2000 genes in mouse embryonic stem cells. Nature 392:608–611

CHAPTER 4

How Proteins Speak with One Another In Cell Signaling   4

EDMOND H. FISCHER

Contents

This chapter will focus on cell signaling and signal transduction, that is, the extremely complex series of reactions by which cells receive and react to external signals (for reviews, see Hunter 2000; Pawson and Nash 2000). Particularly, it will deal with those processes that are regulated by protein phosphorylation. The topic is of significance considering that approximately 15 to 20% of the entire human genome (at least 6400 genes) is devoted to the expression of the molecules implicated.

4.1
Cell Signaling

However, for those who might not be conversant with the field, let us try to explain what is meant by "cell signaling" on the basis of a concept everybody is familiar with, namely, a television set. A TV set has a picture tube and requires a source of energy to activate all the components it contains. Biological cells are powered by nutrients: carbohydrates, amino acids, fats, that will be converted into ATP, the universal energy currency that all living organisms use.

In the back of the set, one finds a couple of receptors to capture external electromagnetic signals received from the antenna, the satellite or the VCR. These signals will be transduced by a few hundred elements (transistors, capacitors, resistors, etc.) all acting in concert to give a physical response, that is, an image on the picture tube. These few hundred elements, by speaking with one another, will allow for the generation of tens or hundreds of thousands of interactions by which not only the picture can be formed, but its tint, color intensity, brightness, sharpness, and sound can be modulated at will. This is what is called "signaling" or "signal transduction" – the conversion of various external stimuli into different kinds of responses.

Now, television sets have evolved over the last 75 years or so. By contrast, biological cells have evolved over more than 3.5 billion years. They contain well over 100,000 components and the number of interactions that can be generated among them must be something in the order of 10^{20}. This is what allows cells to become autarkic, i.e., self-sufficient, to synthesize their own elements, to regulate their own development, replicate, repair their damages, and even program their own death at the appropriate time.

Contrary to TV sets that have only a couple of receptors in the back to capture electromagnetic signals, cells have thousands of different receptors responding to tens of thousands of different signals: hormones, growth factors, neurotransmitters, pharmacological agents, etc. Also electromagnetic radiations such as light in the visual system (or ultraviolet light and heat among various stress signals), odorants in the olfactive system, osmotic pressure, oxidative stresses or hypoxia, and so forth. Animals may have close to 1000 different olfactive receptors (about 300 in humans) that can recognize about 20,000 different odorants. This is what allows a dog, for instance, to follow the trail of a single person in a large crowd by computing the response generated by the combination of a few odorant molecules floating around in the atmosphere, or a honey-bee to be attracted by the right flower.

But cells differ from TV sets in an even far more important way in that in higher organisms, millions or billions of cells come together to form tissues and organs. The transition that occurred from single cell to multicellular organisms was without question one of the most determining and consequential events in biological evolution (Eigen). Because before that, and for more than three billion years, unicellular organisms had to compete with one another for food, micronutrients, light, vital space, what ever. But when cells came into contact with one another, for the first time, they had to cooperate for the good of the whole. They had to share their resources, synchronize their growth and behavior in response to internal or external demands. To do so, they had to adhere to one another either directly or through the network of matrix proteins with which they are surrounded, using a great variety of cell adhesion molecules by which a cross-talk among them could be established. Since messages must be constantly sent back and forth in these systems, every cell not only has to sense what happens in all the others but to define what should happen in all the others. This revolution occurred very recently in geological times: something like 530 million years ago, at the pre-cambrian/cambrian border. But this is what triggered the explosion of reactions that led, within an amazingly short span of time – perhaps 5, 10, or 15 million years - to the establishment of all the species one finds inscribed in the fossil record today (Miklos 1993).

So, before that switch and for 3.5 billion years, one witnessed a slow evolution marked by countless small victories followed by countless failures. Until that day when, for the first time, cells began to communicate and cooperate with one another for the good of the whole, only then could they develop and evolve into the more and more complex organisms that led finally to the appearance of man.

Of course, the regulation of such complicated cellular processes requires a myriad of commands, positive and negative, that have to be tightly coordinated to keep all the reactions that take place under control. Particularly, to make sure that no crucial event should occur at an inappropriate time or out of phase, we know today that most of the reactions that are needed to orchestrate these processes, the switches that have to be turned on and off, rely on

protein phosphorylation. However, before entering on the subject, it seems appropriate to recall where we were 50 years ago in this area because the advances that have occurred since then are truly without precedent.

4.2
Yesterday

To begin with, 50 years ago, the title of this chapter mentioning "cell signaling" and a cross-talk among proteins would not have been understood. Although endocrinology was already firmly established as a discipline, it remained purely at the phenomenological, mostly intact animal, level. Some circulating hormones had been identified: they were known to carry messages from one organ to the other and result in some kind of physiological response. Insulin was known as the messenger of the pancreas that affected carbohydrate mechanism, just like epinephrine was known as the messenger sent by the adrenals to prepare an animal for what the physiologist Cannon some 70 years ago called "fight-or-flight," as when a person suddenly confronted by a stressful or emergency situation experiences a "shot of adrenaline" or an "adrenaline rush." But the action of these messengers stopped at the cell membrane and what happened next was totally unknown. It was inconceivable that hormones could act in a cell-free system until Earl Sutherland came along with his stunning discovery of cAMP as a second (intracellular) messenger for the action of epinephrine (Sutherland and Rall 1960).

Then too, there was a fundamental difference in the way science was conducted 50 years ago as compared to today. At that time and, in fact, since the days of Claude Bernard in the mid-1800s (the French physiologist who first described the glycogenic function of the pancreas) one first observed a physiological phenomenon and then tried to identify the enzymes involved. Whereas today, by and large, it is the other way around: new molecules are first identified (and with the amazing advances that have occurred in many genome sequencing projects, literally thousands are being uncovered) and then, by overexpressing them or by knocking them out, one tries to define their function. In other words, and to paraphrase Luigi Pirandello, in the last 20 years we have gone from "Six Functions in Search of an Enzyme" to "Thousands of Enzymes and Proteins in Search of a

Function." This is something analogous to what has been called "reverse endocrinology," where orphan receptors are first identified and then one tries to find the ligands for them in order to define their function.

Finally, 50 years ago, very little was known about cell regulation and absolutely nothing, of course, about a possible involvement of protein phosphorylation in these processes. In fact, this is how phosphoproteins were described around 1950 in a textbook by Hawk, Oser, and Summerson (Hawk et al. 1948), a classical source of biochemical information at that time:

These proteins contain phosphate bound in ester linkage to the hydroxy amino acids serine and threonine. They include casein from milk, ovovitellin from egg yolk, and other proteins associated with the feeding of the young; also the proteolytic enzyme pepsin.

But in the mid-1950s, it was shown that protein phosphorylation could serve as a means to regulate the activity of glycogen phosphorylase, the enzyme that catalyzes the first step in the degradation of glycogen.

4.3
Today: Regulation by Protein Phosphorylation

We know today that reversible protein phosphorylation is the most prevalent mechanism by which physiological processes can be regulated. It is involved in the control of metabolism, gene transcription/translation, transport, the immune response, cell development and differentiation, cell cycle, apoptosis, etc. In fact, it would be difficult to find a cellular event that is not directly or indirectly modulated by protein phosphorylation. Quantitatively, most of these reactions occur on serine or threonine or both. However, one of the most exciting developments in this field was the realization, just over 20 years ago, that the phosphorylation of proteins on tyrosines was intimately implicated in cell transformation and oncogenesis, bringing into play a multitude of kinases of cellular or viral origin, or linked to receptors for mitogenic hormones and growth factors. This work originated with the discovery that the product of the src gene responsible for the powerful oncogenicity of Rous sarcoma virus was a protein kinase designated as pp60^{src} (Brugge and Erickson 1977; Collett and Erickson 1978; Levinson et al. 1978). Second, the unexpected finding showed that this enzyme, unlike all previously known kinases, phosphorylated its protein substrates exclusively on tyrosyl residues (Hunter and Sefton 1980). This is the particular aspect of signal transduction that will be dealt with in the rest of this chapter.

All growth factor receptors are tyrosine kinases; approximately 60 of them are known, distributed in 20 subfamilies (Fig. 4.1). They have a similar general architecture, namely, a single transmembrane segment separating the catalytic domain inside from an external moiety displaying a considerable diversity of classical motifs to recognize the different ligands: cysteine-rich segments; fibronectin type III (FNIII) repeats; immunoglobulin-like loops; EGF-like, factor VIII-like, cadherin-like domains; kringles; acidic-, proline- or leucine-rich segments; and so forth (Schlessinger and Ullrich 1992). The ligands for these receptors are nearly all circulating molecules (although, for instance, FGF needs to bind to proteoglycans such as heparin in order to be active) except for the Eph family of receptors that recognize membrane-bound ligands such as the ephrins.

The Eph receptors constitute the largest family of receptors; more than a dozen have been identified. They all seem to be concentrated – if not exclusively found – in the nervous system. They are not mitogenic; that is, they do not induce cell proliferation or differentiation like most other growth factor receptors. Rather, their primary role is to help developing axons reach their proper targets, perhaps by crawling along microtubule fibers. During embryonic development, nerve growth cones must explore their local environment and make the proper choices in order to go where they are supposed to go. Pathfinding involves very sophisticated mechanisms with soluble and membrane-bound molecules in the target area acting as attractants or repellents (Barinaga 1995). In some instances, like in the optic tectum of the chicken brain, the ligands are distributed along a gradient of concentration, from back to front and from top to bottom, providing a grid of position coordinates that will direct the axons toward their designated targets (Tessier-Lavigne 1995; Tessier-Lavigne and Goodman 1996; Flanagan and Vanderhaeghen 1998; Holland et al. 1998).

We begin to understand today how signal is transduced by these receptors. The most important factor is not so much that the activated receptors can phosphorylate target substrates but that they can undergo autophosphorylation, i.e., phosphorylate themselves

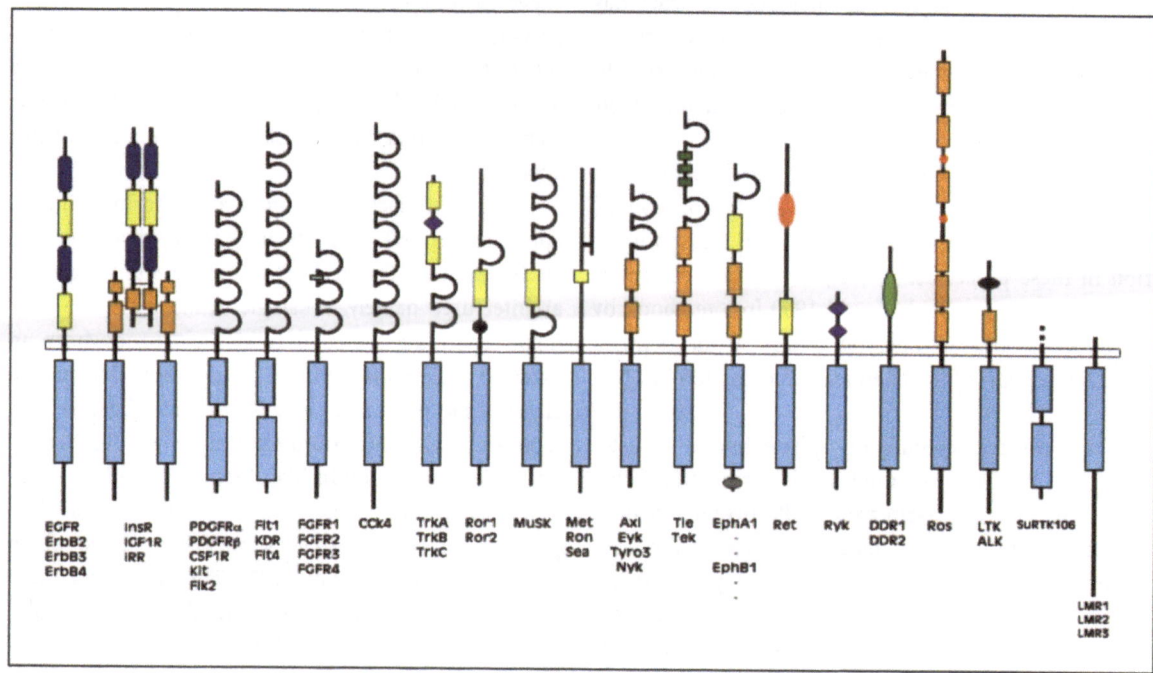

Fig. 4.1. Growth factors receptors with protein tyrosine kinase activity. Courtesy of Jossie Schlessinger, with permission

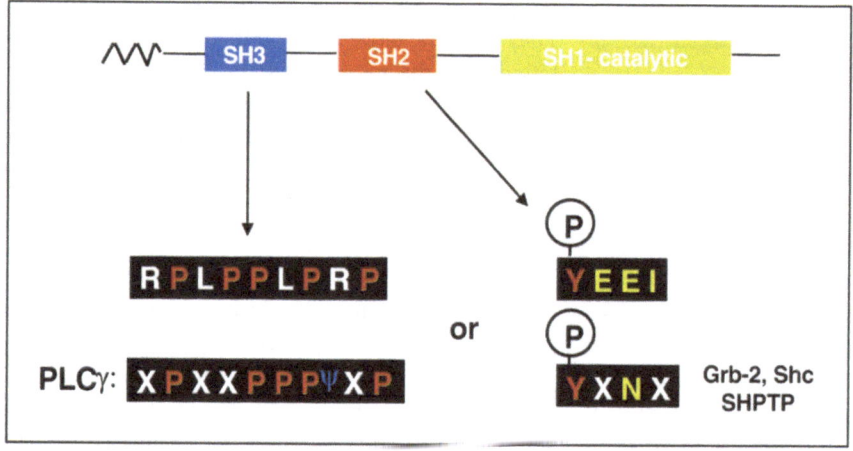

Fig. 4.2. With its three src-homology domains SH1, SH2, and SH3, this represents pp60^{src}

on tyrosyl residues. The phosphotyrosyl groups thus generated can then serve as docking sites for the attachment of adapter proteins by virtue of the fact that they contain structural modules designated src-homology 2 (SH2) domains, which display a high affinity for phosphotyrosyl residues.

This work originated from the observation that p60^{src}, the product of the src gene of Rous sarcoma virus, displayed several highly conserved sequences that were present in many other proteins (Fig. 4.2). First, as to be expected, in the catalytic domain, which is structurally related to the catalytic domain of essentially all other protein kinases, and which was called "src-homology domain 1" or SH1. But then, two other highly conserved regions were also found to be present in a great number of either catalytic or non-catalytic proteins. These were termed "src-homology domain 2" (SH2) and "src homology domain 3" (SH3). As it turned out, SH2 domains mediate protein-protein interaction by binding strongly to particular phosphotyrosyl residues within a specific sequence. By contrast, SH3 domains have a high affinity for short proline-rich segment (Pawson 1995; Kuryian and Cowburn 1997).

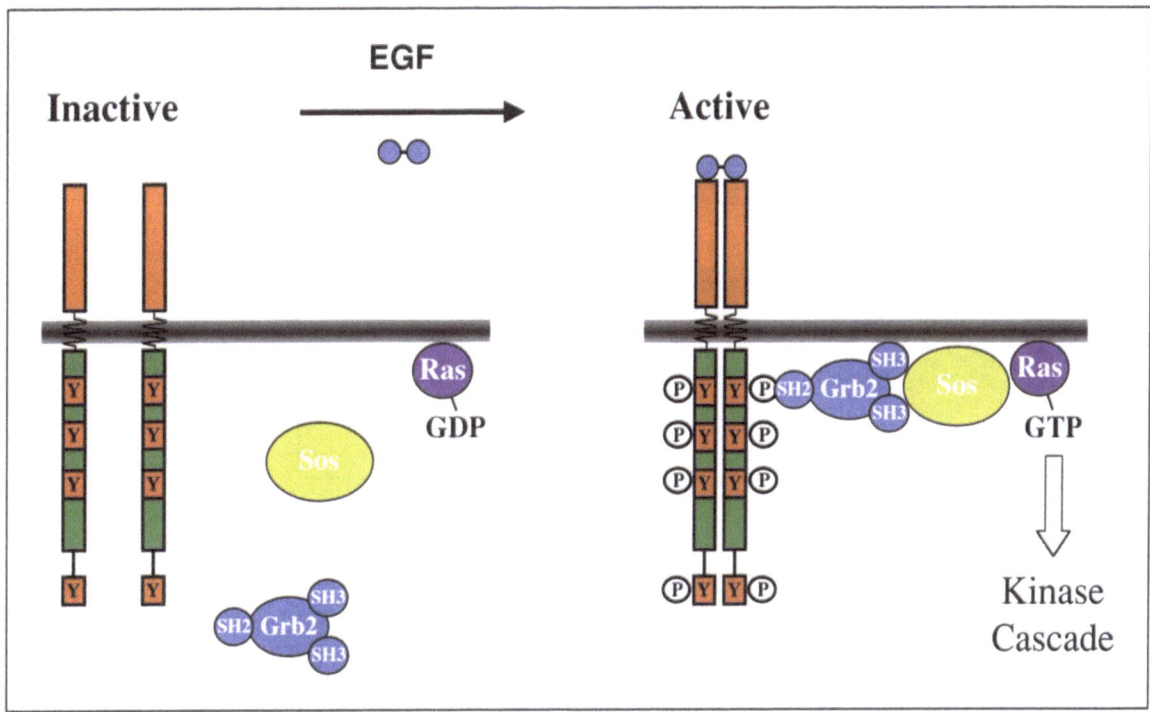

Fig. 4.3. Model of signal transduction down from the EGF receptor, with the recruitment of Grb-2 and SOS by interaction of SH2 and SH3 domains

This is the way it works, for instance, during stimulation of the MAP kinase pathway down from the EGF receptor (Fig. 4.3). In the resting state, the receptor is monomeric and inactive and the adaptor proteins remain unattached, free in the cytoplasm. Upon ligand binding, the receptor dimerizes and undergoes trans-phosphorylation on tyrosyl residues (Schlessinger 1988, 1993). These will then recruit adaptor proteins such as Grb-2 (for Growth factor receptor bound) which has no catalytic activity of its own. The change in conformation in Grb-2 allows its SH3 domain to interact with and activate the next linker protein, namely, SOS. SOS is a guanine nucleotide exchange factor (GEF) that catalyzes the exchange of GDP for GTP in Ras, a small guanine nucleotide-binding protein, thereby initiating a whole cascade of events leading to the activation of the MAP kinase pathway and gene expression. Note that the whole complex made up of the growth factor receptor, Grb-2 and SOS plays a role analogous to that of the 7-transmembrane (G-coupled or "serpentine") receptors whose function is also to activate large G proteins such as Ga, Gs, Go, etc., that transduce their signal.

4.4
Protein Domains that Control Signal Transduction

Today, many dozens of such binding modules have been identified in hundreds of enzymes, adaptor, regulatory, or scaffolding proteins, transcription factors, etc. All display a modular architecture that allows them to interact with one another, or to bind to various compartments of the cell, in a vast, tinker-toy sort of way (Pawson 1995; Pawson and Scott 1997; Buday 1999). A few such domains are described below (Fig. 4.4).

The SH2 domain binds to phosphotyrosyl residues within a specific sequence extending to only a few residues downstream. For instance, the sequence YxNx serves as the recognition motif for the adaptor molecule Grb-2.

The SH3 domain recognizes proline-rich sequences (such as xPxxP) often found in more than one copy per adaptor protein.

WW (an approximately 27-residue long segment sandwiched between two tryptophanyl residues) also binds to proline-rich stretches but of different sequences (e.g., PPxY or PPxP) with recent evidence

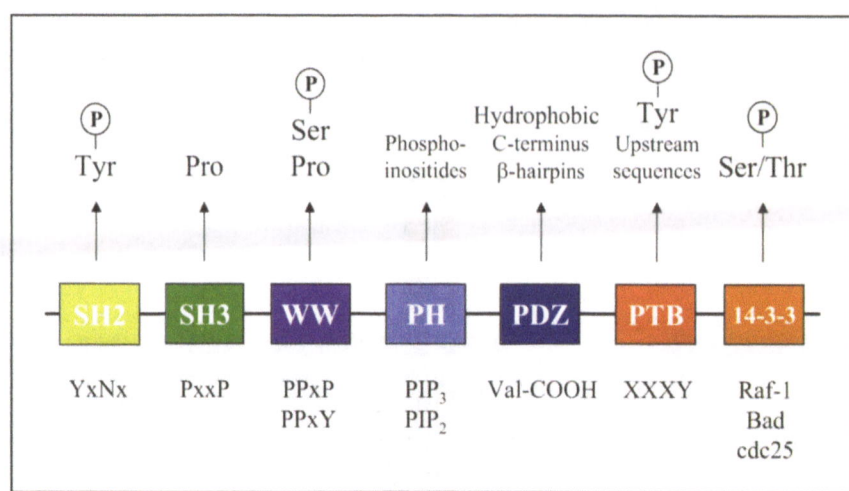

Fig. 4.4. Some of the protein modules that control signal transduction

that it also interacts with phosphoseryl or phospho-threonyl residues (Sudol 1996; Lu et al. 1999).

Contrary to the other binding modules that promote protein-protein interaction, PH (for pleckstrin homology) domain promotes interaction with lipids, i.e., they bind selectively to the anionic head groups of specific phosphoinositides such as PIP2 and (mainly) PIP3. It is mostly responsible for recruiting signaling proteins to the plasma membrane.

PTB (for phosphotyrosine binding domain) also recognizes phosphotyrosyl residues but displays a strong affinity for upstream sequences to such an extent that it will bind to a tyrosine-containing segment even if it is not phosphorylated.

PDZ (from the post-synaptic density protein PSD-95, disc large and zona occludens-1) is a small, four residue motif containing serine or threonine and pops off when these are phosphorylated. The PDZ module binds either to hydrophobic C-termini (such as Val-COO⁻) or to sharp β-hairpin turns present on other proteins or within itself, allowing it to undergo head-to-tail oligomerization. It is often found in a string of multiple copies: for instance, PSD-95 has three; the scaffolding protein InaD that tethers several proteins involved in the Drosophila phototransducing cascade has five. Its main function is to organize or cluster ion channels or receptors on the membrane and bridge them to the cytoskeleton (Songyang et al. 1997; Fanning and Anderson 1999; Hillier et al. 1999). There are, of course, innumerable other interacting domains that are not detailed herein such as 14-3-3, WD40, EH, EVH1, FYVE, FHA, SAM, LIM, etc.

Let's examine a few examples of how these modules are utilized. The SH2 domain, for instance, in addition to promoting protein-protein interaction by binding to specific phosphotyrosyl residues, serves as a conformational switch designed to regulate the activity of innumerable enzymes such as pp60[src] (where it was first identified) and all other src kinases; the syk family of tyrosine kinases found in B- and T-cells; the SHP family of tyrosine phosphatases; phospholipase Cγ, etc.. All these enzymes are maintained in an inactive conformation in the resting state because their SH2 domain interacts either with a phosphotyrosyl residue that serves as a negative determinant (this is the case for Y527 in p60[src] or Y505 in p56[lck]) or with an autoinhibitory motif even when it is not phosphorylated (as seen for the SHP tyrosine phosphatases or the tyrosine kinase ZAP-70). In both instances, these internal interactions shield the catalytic site. Activation results from the dephosphorylation of the inhibitory site or by having the enzyme come into contact with another protein displaying a higher affinity for that particular SH2 module. Such protein then switches its allegiance to the new partner thus unmasking the catalytic site. In all, then, the SH2 domain must be viewed as an internal conformational switch that turns on or off the activity of an enzyme, just like the addition or removal of a phosphate group serves as an external molecular switch to activate or inhibit an enzyme (Hof et al. 1998)

Finally, phosphotyrosyl-binding modules (such as the SH2 or PTB domains) might have yet another function, that is, to protect the phosphotyrosyl residues with which they interact from unspecific or unwanted dephosphorylations. The interaction would maintain for a while that protein in whatever active or inactive state it might be. For instance, the SH2 domain of PLC-γ protects the phosphotyrosyl resi-

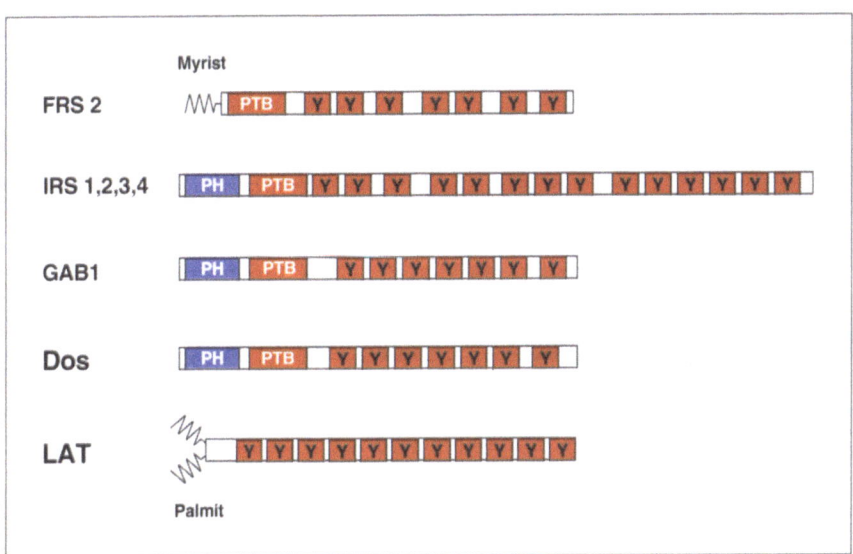

Fig. 4.5. Role of the pleckstrin-homology (*PH*) module in recruiting proteins to the membrane

dues of the EGF and PDGF receptors with which it interacts from the action of tyrosine phosphatases (Rotin et al. 1992).

As indicated earlier, the main function of the PH domain (which has now been found in about 500 different signaling or regulatory proteins), just like that of the myristyl or palmityl groups, is to recruit various molecules to the membrane, often in close proximity to a receptor (Fig. 4.5). This is the case, for instance, for the FGF-receptor substrate 2 (FRS-2), the insulin receptor substrates IRS-1,2,3, etc. These contain a great number of phosphotyrosyl residues and serve as annex or additional docking sites for the attachment of adaptor proteins or enzymes that have SH2 domains. In a coordinated reaction upon receptor stimulation, the docking molecules are recruited to the membrane through their PH domains, and then the PTB module (which has a very high affinity for a specific sequence on the receptor, such as NPXY) slaps them against the receptor, thereby allowing the phosphorylation of their tyrosyl residues.

The importance of the PH module is seen with the Bruton kinase (Btk) that plays such an essential role in B-cell signaling. A single R28C point mutation in its PH domain interferes with the binding of Btk to the membrane and leads to gross functional abnormalities resulting in x-linked agammaglobulinemia.

Clustering of receptors, which is one of main function of the PDZ domain, is essential, indeed, because efficient synaptic transmission demands that the appropriate neurotransmitter be released at high concentration precisely above the postsynaptic site where it is supposed to act. A single neuron may undergo

thousands of such synaptic contacts with other neurons releasing a variety of neurotransmitters. Thus, neurons must possess the ability not only to anchor these receptors, but also to sort them out and fix them at the proper synaptic site. At the neuromuscular junction, for instance, the nicotinic acetylcholine receptors (AcChRs) are clustered to the tune of $10,000/\mu m^2$ by the anchoring protein rapsin which, itself, is hooked onto the membrane by a N-terminal myristyl group. It contains 8 TPR (tetratrico, i.e., 34 residues-long) peptide repeats, a cysteine-rich domain containing a zinc-ring finger motif presumably responsible for AcChR binding and a putative site for Ser phosphorylation (whose function is not yet clear). Disruption of the rapsin gene abolishes AcChR clustering and disrupts the synaptic localization of several other muscle proteins such as syntrophin, utrophin, dystrophin, and various other dystroglycans (Colledge and Foehner 1998; Hillier et al. 1999).

The complexity of the regulation of signaling pathways through adaptor or transducing molecules is well demonstrated by the proto-oncogene p95 Vav. It is important because its multi-modular architecture allows it to interact with or bridge multiple pathways. With its nine structural domains, it looks like one of those thick Swiss army knives that can do just about anything (Fig. 4.6). Its main variant is expressed exclusively in cells of hematopoietic origin. It becomes tyrosine phosphorylated and activated following stimulation of the T- or B-cell receptor, growth factor receptors, and co-precipitates with them or their associated proteins. It is said to be phosphorylated by Bcr-Abl in chronic myelogenous leukemia. Indeed, it

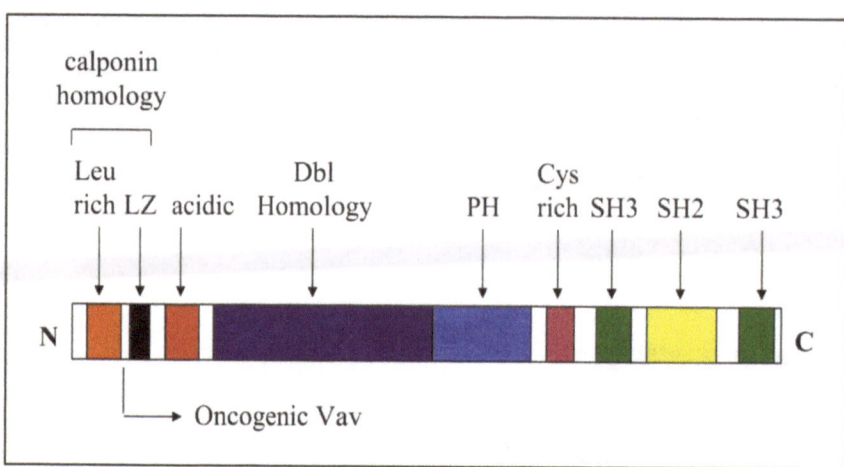

Fig. 4.6. Multi-modular structure of the proto-oncogene molecule p95 Vav

contains: (1) a domain homologous to Calponin (a family of proteins involved in the regulation of smooth muscle contraction) that serves an inhibitory function because when part of it is deleted, Vav becomes oncogenic; (2) an acidic region; (3) a DH domain (homologous to the human dbl oncogene that serves as a GDP-GTP exchange factor for small G proteins of the Rho, Rac, cdc42 family), invariably followed by a PH domain; (4) a cysteine-rich domain with two putative Zn-fingers; and finally (5) one SH2 and two SH3 modules (Bustelo 1996; Romero and Fischer 1996).

Incidentally, it is increasingly clear today that all proteins have this kind of modular architecture. They must be viewed as mosaics of structural motifs (sometimes no larger than three or four residues), as necklaces made up of many different beads whose selection and disposition ultimately define their function. There is no question that, as daily we gain increasing information on the sequence of proteins and the structural homology that exists among them, dozens of such new motifs will be uncovered. Their identification and the determination of their specific function should provide us with a better understanding of how proteins are constructed, how they operate, how they have evolved, and ultimately how they have learned to speak with one another in the course of time.

Since receptors may have seven or eight tyrosyl residues that can be phosphorylated (Heldin 1995), they can initiate several signaling pathways by triggering a vast, tinker toy-like system of protein-protein interactions. In a much over-simplified way, then, receptors can be viewed somewhat like the hand-operated telephone switchboard of old in which the operator linked a call coming from the outside to the proper

recipient by inserting the plug in the appropriate hole. Likewise, receptor stimulation by an external signal may result in the initiation of various transducing pathways leading to gene expression, cytoskeletal rearrangement, cell proliferation or differentiation, apoptosis, etc.

Paul Valery, the French poet, author, mathematician and sometimes philosopher of sciences, once said: "*Ce qui est simple est faux; ce qui est compliqué est incompréhensible.*" Well, the analogy given above is a good example of something simple and wrong because, obviously, no one receptor alone will trigger all these processes. In most cases, it is only when several receptors act together, inducing a combinatorial system of signals, that selectivity and specificity can be introduced and that a particular physiological response can be generated (Fambrough et al. 1999). In the immune system, for instance, when the B- or T-cell receptors are engaged, they integrate their signals with those of many other surface antigens to bring about cell activation (Clark and Ledbetter 1994). In other words, there is a high degree of degeneracy and redundancy in the signaling process, and this is why knock-out experiments often give ambiguous answers with weak or unrecognizable phenotypes.

Elucidation of some of the major signaling pathways (mitogenic, immune, apoptotic, and others) has represented a major challenge. Every one of them brings into play an extremely intricate array of enzymes, adaptor, anchoring, regulatory, and scaffolding proteins, all interacting with one another under the influence of a maze of internal or external conditions according to their cellular distribution and localization, the particular stage of cell development, and so on. Most of the components of those extremely complicated circuitries have been identified through their ability to rescue a variety of mutant cells defective at a particular step of the cascade system.

4.5
Receptor Mutations and Disease

Mutations in any one of these receptors may generate pathological conditions of varying degrees of severity (Table 4.1). For instance, mutations in the FGF receptor may lead to various kinds of skeletal disorders resulting in achrondroplastic dwarfism and cranial and facial anomalies including tower-shaped skull (as seen in Apert's and Crouzon's syndromes), downslanting and protruding eyes due to shallow orbits, and general facial asymmetries as a consequence of the premature fusion of the bones of the skull (Muenke and Schell 1995).

Mutations in the c-kit receptor or in its ligand, the Stem cell (or Steel) factor, interferes with the development of thymocytes and certain skin cell populations including melanoblasts, resulting in localized depigmented areas devoid of melanocytes (Fleischman 1993). While these mutations are in the external domain of the receptor, substitution of one of the intracellular tyrosyl residues responsible for signal trans-

duction impairs the survival of male germ cells and ovarian follicle development in mouse, even though it does not affect the tyrosine kinase activity of the receptor (Blume-Jensen et al. 2000; Kissel et al. 2000).

Among the best studied receptor aberrations are those involving the insulin receptor leading to various forms of insulin-resistant diabetes (NIDDM). More than a hundred of these have been identified. Some will prevent the expression of the receptor, others its transport from the ER/Golgi to the plasma membrane or will increase the rate of its degradation following internalization. Mutations in the external domain will interfere with ligand binding while those in the kinase domain might block its function. The most severe forms may lead to leprechaunism, resulting in early death. As expected, these subjects are severely glucose intolerant in spite of huge amounts of circulating insulin in the blood. In fact, in a desperate but futile attempt to overcome the deficiency, the organism can produce concentrations of insulin 100 times above normal (Taylor 1992).

But the most dramatic of these receptor alterations are those that lead to oncogenicity (Cantley et al. 1991). Indeed, receptor or cytoplasmic tyrosine kinases represent a high proportion of all oncoproteins, which underscores the essential role tyrosine phosphorylation plays in regulating cell function. Oncogenicity may result either from the overexpression of

Table 4.1. Diseases associated with mutations in growth factor receptors

EGF receptor	Mutated, amplified and overexpressed in cancer
ErbB2, ErbB3, ErbB4	Amplified and overexpressed in breast cancer
PDGF receptor β	Activated by chromosomal translocation in leukemia
CSF-1 receptor	Ectopic expression in cancer
Kit (SCF receptor)	Inactivated by mutation in hereditary piebaldism
	Activated by mutation in leukemia
FGF receptor 1	Activated by mutation in craniosynostosis syndromes that affect skull/bone
FGF receptor 2	Development (e.g. Crouzon and Pfeiffer syndromes, thanatophoric dysplasia)
FGF receptor 3	FGFR1/FGFR3 activated by chromosomal translocation in leukemia/myeloma
Met (HGF receptor)	Activated by mutation, amplified, overexpressed in sporadic/hereditary cancer
Axl	Overexpressed in myeloid leukemias
Rse/Sky	Overexpressed in breast cancer
Ret (GDNF receptor)	Inactivated by mutation in hereditary intestinal disease
	Activated by mutation in hereditary cancer syndromes
Alk	Activated by chromosomal translocation in leukemia
Trk, TrkC	Activated by chromosomal translocation in cancer
Tie2 (angiopoietin R)	Activated by mutation in hereditary vascular dysmorphogenesis disease
Insulin receptor	Inactivated by mutation in non-insulin dependent diabetes

Compiled by Jossi Schlessinger and Tony Hunter, with permission

the normal, non-mutated gene, or mutations that would render the encoded enzymes constitutively active. The first to be discovered was v-erb B from an avian erythroblastosis retrovirus (Yamamoto et al. 1983; Downward et al. 1984; Ullrich et al. 1984). It results from a massive truncation of the external domain of the EGF receptor plus a few other point mutations and a deletion at the C-terminus. But mutations can also be very discrete, such as the replacement under the influence of carcinogens of a single valine by glutamic acid in the transmembrane domain of the EGF-related neu/her2 receptor leading to neuroblastomas in rats (Bargmann et al. 1986). Its human homolog, albeit with other mutations in the catalytic domain, has also been implicated in the progression of human mammary carcinomas (Slamon et al. 1987). Other mutations may involve fusion with segments of other genes, particularly those that encode proteins that can easily undergo dimerization. Many have leucine repeats that predict a coiled-coil configuration as found in tropomyosin. This is the case, for instance, with the oncogenic form of the Trk family of nerve growth factor receptors originally isolated from a colon carcinoma. It is a chimeric molecule made by the fusion of the 221 N-termini residues of non-muscle tropomyosin (or of Tpr, a large coiled-coil protein that localizes on the cytoplasmic surface of the nuclear pore complex), also found in the oncogenic forms of the Met receptor and Raf. In fact, all these mutations are oncogenic because they allow the mutated structures to associate with one another, to dimerize and, therefore, to undergo spontaneous, unrestricted transphosphorylation and activation.

4.6
Protein Kinases and Phosphatases: Friends or Foes?

With accumulating evidence implicating tyrosine phosphorylation in cell transformation, it is hardly surprising that many groups would become interested in the protein tyrosine phosphatases that catalyze the reverse reaction, with the hope that their overexpression might block or reverse transformation. This turned out not to be the case.

A protein tyrosine phosphatase (PTP) was first isolated in homogeneous form from human placenta (Tonks et al. 1988 a, b). Surprisingly (and contrary to what had been expected, since essentially all protein kinases are homologous), the amino acid sequence of the enzyme showed no homology with any of the previously known protein phosphatases. But a search of the database revealed a close structural relationship with a surface antigen already well-known by the immunologists, namely, the leukocyte common antigen CD45 (Charbonneau et al. 1988).

CD45 comprises a broad family of heavily O- and N-glycosylated membrane-spanning molecules found in all hematopoietic cells except mature erythrocytes (Fig. 4.7). Six to eight isoforms are known due to the alternative splicing of three exons close to the N-terminus; these are differentially expressed on leukocyte subsets. They have a single transmembrane segment. Their intracellular moiety is highly conserved and contains two, internally homologous domains of about 30 kDa each. It is those two domains that are structurally related to the placenta PTP1B that had been isolated. CD45 had been implicated in the regulation of lymphocyte function, including signaling through the T- or B-cell receptor, cytotoxicity, proliferation, and differentiation (Thomas 1999).

But of course the important question is the role CD45 might play in lymphocyte activation. A T-cell reacts with an antigen-presenting cell through its T-cell receptor (TCR)/CD3 complex (Fig. 4.8). The conventional TCR is made up of the variable alpha-beta immunoglobulin heterodimer whose function is to recognize antigenic peptides presented by the major histocompatibility complexes (MHC) while the function of the invariant CD3 complex is to transduce the signal. Interaction of the receptor with class 1 or class 2 MHCs triggers a number of reactions. It would be too complicated to go into the details of these reactions. Suffice it to say that the first thing that happens upon cell stimulation is an immediate activation of the src tyrosine kinases $p56^{lck}$ and $p59^{fyn}$, following the dephosphorylation of an inhibitory phosphotyrosyl residue close to the C-terminus. This allows for the autophosphorylation of a second tyrosyl residue within the catalytic domain that is required for activity. The activated src kinase catalyzes the phosphorylation of the CD3 complex on so-called ITAM motifs (for immunoreceptor, tyrosine-based activation motifs); three of these are present on the zeta chains, perhaps for amplification of the signal. This then allows the recruitment, tyrosine phosphorylation and activation of ZAP-70 (a CD3 zeta-activated protein tyrosine kinase of the syk family containing 2 SH2 domains) and many other proteins downstream including PLCγ. This then triggers phos-

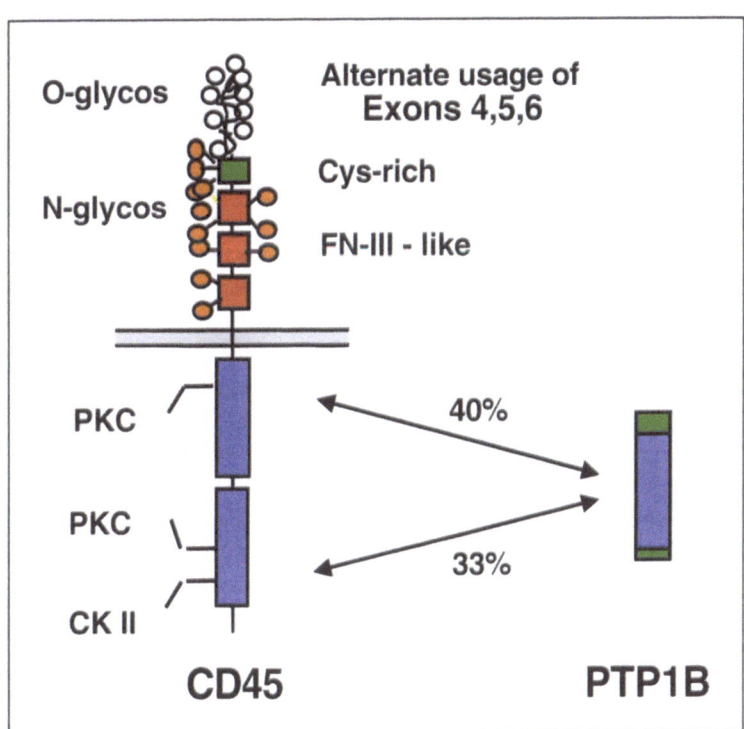

Fig. 4.7. Structure of the leukocyte common antigen CD45 and its relationship to the human placenta tyrosine phosphatase PTP1B

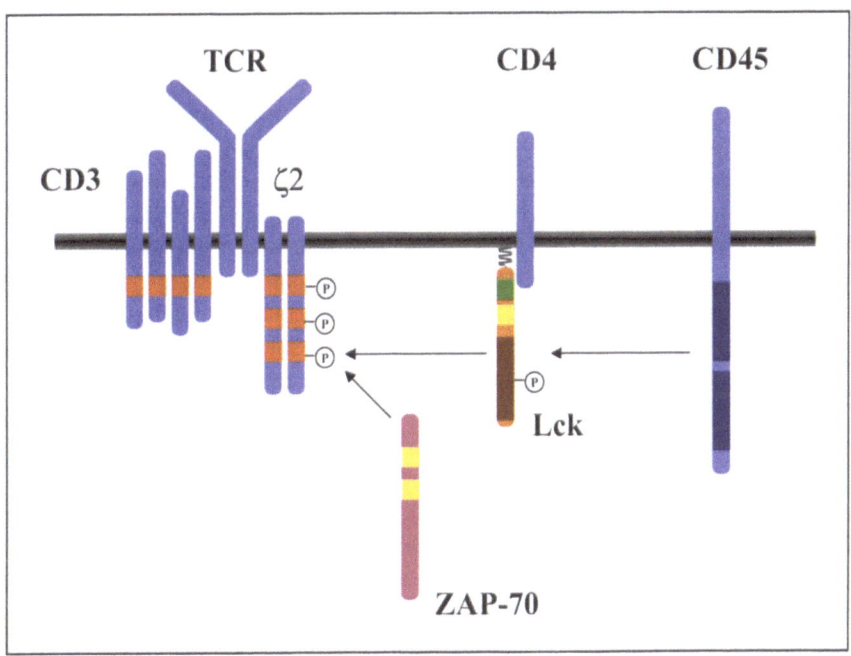

Fig. 4.8. Involvement of CD45 in the immune response

phatidyl inositol (PI) turnover, IL-2 production, transcription of many genes, and finally cell proliferation and differentiation. But for the immune signaling through the T-cell receptor to occur, one needs CD45: mutant forms of a human leukemia T-cell line lacking CD45 fail to signal through the T-cell receptor and overall tyrosine phosphorylation is almost abolished (Pingel and Thomas 1989; Koretzky et al. 1991). Whatever, what these data already indicate is that one needs a tyrosine phosphatase to activate the tyrosine kinases involved (Trowbridge and Thomas 1994). But its mechanism of action is not fully understood – in fact, to such an extent that one isn't sure whether the phosphatase is always inactive in the resting state and activated during the immune reaction, or always active until inactivated by cell-cell interaction (Desai et al. 1993; Majeti et al. 1998; Weiss and Schlessinger 1998).

4.7
Receptor Protein Tyrosine Phosphatases

Since then, a great variety of receptor forms of tyrosine phosphatases have been identified (Fig. 4.9). Just like the growth factor receptors, they have a single transmembrane segment, separating the catalytic domains inside, mostly two in tandem as in CD45, and external segments displaying once again a considerable structural diversity for ligand recognition. But what is striking here, and really unexpected, is that all these structures display the structural characteristics, all the hallmarks of cell adhesion molecules. Some, such as the LARs have Ig-like and FNIII repeats, as found in the cell adhesion molecules belonging to the immunoglobulin super family that include the N-CAMs, Ng-CAMs, neurofascin, contactin, L1, fasciclin 2 and 3, etc. These molecules promote neuron outgrowth; they participate in axonal guidance, fasciculation and synapse formation, and are responsible for the generation of tissue pattern and form during embryonic development (Edelman 1985, 1993; Kallunki et al. 1998). Indeed, contrary to growth factor receptors that respond preponderantly to circulating hormones and ligands, as of now no circulating ligand has been detected for transmembrane tyrosine phosphatases. This implies, of course, that they must participate in – or be regulated by – cell-cell or cell-matrix interaction. It further raises the exciting possibility that they might be directly implicated in contact inhibition, which plays such a crucial role in malignant transformation.

The properties of only a few of these receptors shall be mentioned. RPTPα, for instance is known to play an important role in focal adhesion and in the activation of src RPTPε, with a very short extracellular domain, is expressed in three forms due to alternative splicing: one of these (an intracellular form with no transmembrane domain) appears to participate in the regulation of a voltage-gated K^+ channel that affects Schwann cell myelination (Peretz et al.

2000). R-PTP κ & μ undergo homophilic interaction. They contain on the outside 4 FNIII repeats, an Ig-like domain and, at the N-terminus, a 150-residue globular MAM motif (for Meprin, an enterokinase; A5, a neuronal antigen of Xenopus, and receptor μ). The intracellular portion of RPTPμ has been said to interact with the cadherin/catenin cell-adhesion complex one finds at adherens junction. Phosphorylation of cadherin reduces its adhesive properties, a condition potentially contributing to transformation and metastasis. Complex formation with PTPμ (and probably with LAR) could maintain the system in its normal, adhesive, dephosphorylated state (Brady-Kalnay et al. 1994, 1998; Zontag et al. 1995).

LAR (for Leucocyte common Antigen-Related) also binds to focal adhesions where it participates in cytoskeletal reorganization, probably through interaction with its binding protein Trio (Debant et al. 1996). Trio, like vav, represents one of those multi-modular/multifunctional proteins similar to those one encounter in the regulation of signaling pathways, which underlies the enormous complexity and pleiotropicity of these systems. It has eight Spectrin-like repeats, not one but two DH motifs, each paired with its distinct PH domain, an SH3 domain, an Ig-like loop, and a third catalytic domain with Ser/Thr kinase activity similar to that of MAP kinase. Many other proteins (such as the proto-oncogene product Vav described earlier) have this paired DH/PH motif. What is unique here is that Trio has two of them with specific functions: the first signals through Rac-1 to elicit actin reorganization probably by binding to filamin, while the second signals through Rho-A. But one still doesn't really understand how the LARs transduce their signals.

Some receptors have surprising modules serving as adhesion motifs. Such is the case for RTP β (or ζ or γ) which exists in three forms: two transmembrane structures with either a very long or a short external segment, and a third soluble form lacking the transmembrane and catalytic domains. All three forms contain at their N termini a FNIII motif linked to a globular molecule almost identical to carbonic anhydrase (CAH) except that it contains no Zn and is enzymatically inactive (Krueger and Saito 1992; Levy et al. 1993). The two transmembrane forms are preponderantly found in the embryonic nervous system while the soluble form exists in large amounts in the adult brain: it was already known as a chondroitin sulfate-related proteoglycan termed "phosphacan" and shown to interact with N-CAM, Ng-CAM, and the extracellular matrix (ECM) protein tenascin (Milev et

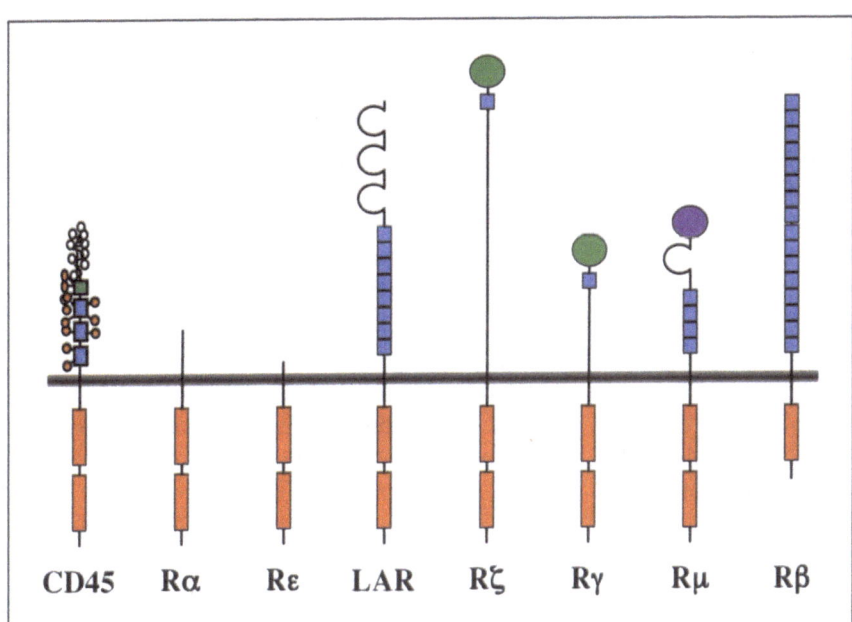

Fig. 4.9. Receptor protein tyrosine phosphatases

CD45   Rα   Rε   LAR   Rζ   Rγ   Rμ   Rβ

al. 1994). The ligand for these receptors was reported to be contactin (Peles et al. 1995).

Contactins (F3/F11) represent a subgroup of compounds belonging to the immunoglobulin superfamily of cell-adhesion molecules. They contain six Ig and four FNIII domains and are expressed during development in a restricted manner on the surface of specific neuronal cells, to which they are attached by a glycosyl phosphatidyl inositol (GPI) anchor. They are able to elicit either positive or negative signals in response to various stimuli: positive when they interact with membrane-bound adhesion molecules such as the N-CAMs or the ECM protein tenascin, leading to axonal growth and differentiation, negative when they interact with ECM molecules such as Janusin (Restrictin) resulting in neuronal repulsion (Vaughan et al. 1994).

Interaction of contactin with R PTP-β requires the CAH domain. However, one doesn't know the manner or even the direction in which the signal is being transduced, a deficiency of such degree that the classical notion of ligands and receptors collapses here. Both Rβ and contactin can exist in membrane-bound and soluble forms so that the soluble external domain of Rβ can act as a ligand for the contactin receptor, just as the soluble form of contactin can serve as a ligand for the phosphatase receptor. Then, of course, the two receptors can come into contact with one another in the course of cell-cell interaction. As a result, the type and direction of the response could conce-

ivably be switched from one to the other at various stages of neuronal development (Peles et al. 1995).

Why this receptor would use a carbonic anhydrase-like motif for this interaction is not known, but it does not seem to be a unique invention. Vaccinia virus has on its surface the same kinds of CAH-like modules and it is possible that it uses these structures to allow the virion to bind to the host cell, perhaps by associating with molecules or receptors related to contactin.

In addition to the diversity in structure and function introduced by alternate exon usage, there is that produced by limited proteolysis of the gene products. Indeed, many tyrosine phosphatase receptors are cleaved in the external juxtamembrane region by different endopeptidases. The functional significance of this process is still unclear and both extremes are conceivable, that is, the external, soluble ligand-binding segment could serve a dominant negative function in blocking receptor signaling while, at the opposite end, the truncated phosphatase might become constitutively active. The basis for this latter assumption comes from the fact that there is some evidence indicating that CD45 might be active in the resting state and, contrary to the kinases, become inactivated by dimerization upon ligand binding (Desai et al. 1994; Majeti et al. 1998). Cleavage within the intracellular juxtamembrane segments would release the catalytic domain from its membrane anchor (Gil-Henn et al. 2001).

4.8
Intracellular Protein Tyrosine Phosphatases

Likewise, intracellular protein tyrosine phosphatases (PTPs) display a great diversity of structures, either preceding or following a highly conserved catalytic core (Fig. 4.10) (Hunter 1995; Neel and Tonks 1997). These are undoubtedly involved in the localization and regulation of the enzymes. The placenta phosphatase (PTP1B) and a human T-cell phosphatase (TCPTP) have a markedly hydrophobic tail that anchors them to the ER. TCPTP also displays a nuclear localization motif (Lorenzen et al. 1995) that takes over when the C-terminal hydrophobic stretch is eliminated in an alternatively spliced form (Champion-Arnaud et al. 1991; Mosinger et al. 1992).

Some PTPs have segments homologous to cytoskeletal proteins such as band 4.1, ezrin, and talin, which probably allows them to interact with the cytoskeleton (Yang and Tonks 1991). Others (the SHP family of enzymes) contain two SH2 domains. On the one hand, these serve to maintain the enzymes in an inactive state by interacting with an internal, non-phosphorylated, auto-inhibitory domain (Hof et al. 1998). On the other hand, they allow interaction with growth factor receptors. Like many other phospha-

tases, they can serve both a positive and negative function; that is, they can either enhance or repress a cellular response (Shen et al. 1991; Zhao et al. 1995). SHP-1 and SHP-2 are the enzymes that are recruited to the membrane by those receptors that negatively regulate signal transduction by the TCR, such as the KIRs (Killer Inhibitory Receptors), CTLA-4, or the FcγRIIB that negatively regulates BCR signaling. Their ablation is embryonically lethal. Two PTPs contain PEST sequences (i.e., sequences enriched in proline, glutamic acid, serine, and threonine). PTP-PEST translocates to the membrane following ECM stimulation of fibronectin and regulates focal adhesion or disassembly, cell migration, and cytokinesis in fibroblasts.

An interesting tyrosine phosphatase was shown to be expressed as the gene product of a virulence plasmid from bacteria of the genus *Yersinia* (which includes *Y. pestis*, responsible for the bubonic plague – the black death – that wiped out a good segment of the human population in times past). The enzyme is involved in the virulence of the organism because when it is replaced it by a dead enzyme, the bacterium is no longer pathogenic. It should be noted that *Yersinia* itself contains no tyrosine kinase and no phosphotyrosine; the phosphatase serves absolutely no internal purpose. Its only mission is to trigger a catastrophic set of tyrosine

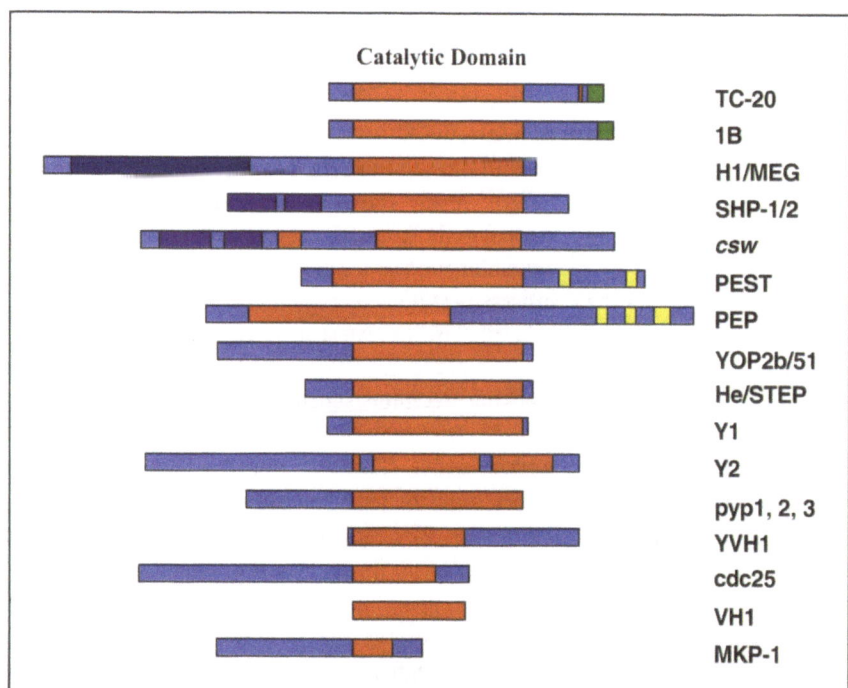

Fig. 4.10. Intracellular protein tyrosine phosphatases

Catalytic Domain

TC-20
1B
H1/MEG
SHP-1/2
csw
PEST
PEP
YOP2b/51
He/STEP
Y1
Y2
pyp1, 2, 3
YVH1
cdc25
VH1
MKP-1

dephosphorylations in the host cells that obliterates its immune defenses and induces massive pathogenicity (Bliska et al. 1991; Stone and Dixon 1994).

Phosphatases that have small catalytic domains tend to have dual specificity, that is, they can act both on tyrosine and serine/threonine phosphates. That is the case for the vaccinia enzyme, for the MAP kinase phosphatase MKP, and cdc25 that regulates the cell cycle.

4.9
Conclusions, and What About Tomorrow?

Here are some of the conclusions that can be drawn from the data on hand:

First, phosphatases cannot be viewed simply as providing an "off" switch in an "on/off" kinase/phosphatase cyclic system. They are not merely scavenger enzymes, acting unidirectionally and there only to remove the phosphate groups introduced by the kinases. In certain cases, they can act synergistically with the kinases to enhance a cellular response (e.g., by activating the src family of kinases whose activities are repressed by phosphorylation at the C-terminus); in others, they can counteract kinase action. The factors determining whether a phosphatase is to enhance or oppose a kinase reaction would seem to depend less on its state of activity than on its regulatory properties and subcellular localization. This suggests that if one wanted to call upon these enzymes to control transformation, one should try to tinker with their regulatory/localization segments – or whatever binding proteins they might be attached to – rather than with their catalytic domains. Displacement of these enzymes from where they are meant to bind would seem a more promising approach than trying to modulate their catalytic activity.

Second, the structural properties of the receptor tyrosine phosphatases that display all the characteristics of cell adhesion molecules are so different from those of the receptor tyrosine kinases that they must have a mission of their own in cell-cell or cell-matrix interaction.

And finally, and perhaps most importantly, where does one go from here? What are some of the main problems that confront us today and that remain to be solved? Admittedly, several of the major signaling pathways have now been elucidated and the structure and properties of many of the molecules involved have been characterized. Some of their physiological functions have been defined. But these molecules are only the words the cells use to perform their daily chores. We know quite a few of these words; we recognize probably some of the phrases they spell out, the scattered fragments of a few sentences they employ to elicit a particular response. But we still don't understand the language they have to use to communicate with one another, to coordinate and keep under control all the reactions that take place, to guaranty the specificity and selectivity these systems require, or to determine whether a given signal should result in cell growth rather than differentiation, or in cell death. We do not know how cells maintain and preserve the fidelity of the signaling process.

The problem is further complicated by the fact that during the several billion years over which cells have evolved, they have had all the opportunities in the world to put in place the vast array of secondary or parallel pathways, shunts, compensatory mechanisms, feed-back loops, and fail-safe systems through which they regulate their growth and development, protect themselves against all sorts of adversities, and program their own death when the time comes. We don't understand the interactivity, the cross-talk that must exist to sort out all the reactions that take place.

Even more importantly, we don't understand the cross-talk that must exist among cells, how they communicate with one another in order to synchronize their behavior in response to internal or external signals. This cross-talk, this sharing of information was crucial for the establishment of such sophisticated networks of communication as seen, for instance, during embryonic development and organogenesis, in our immune system, or in the infinitely more complex central nervous system where a thousand billion cells speak with one another through more than a million billion synapses, leading ultimately to the generation of memory and thought and consciousness. Solving these problems will be one of the major challenges that will confront the biologists in the years to come.

4.10
Acronyms

BCR	B cell receptor
CAH	carbonic anhydrase
CD45	leukocyte common antigen
DH	Dbl-homology domain

EGF	epidermal growth factor
FGF	fibroblast growth factor
FNIII	fibronectin type III
GEF	guanine nucleotide exchange factor
Grb-2	growth factor receptor-bound module
LAR	leukocyte common antigen-related
MAP	microtubule-associated or mitogen-activated proteins
N-CAMs, Ng CAMs	neural or neuron-glial cell adhesion molecules
PH	pleckstrin homology
PI	phosphatidyl inositol
PLCγ	phospholipase Cγ
PTB	phosphotyrosine-binding domain
PTP	protein tyrosine phosphatase
Ras, Rac-2, RhoA	small GTP-activated proteins
R PTP	receptor PTP
SH2	src-homology 2 domain
SH3	src-homology 3 domain
SOS	"son of sam" a GEF
TCR	T-cell receptor
ZAP-70	CD3 zeta-activated protein kinase

4.11
References

Bargmann CI, Hung M-C, Weinberg RA (1986) Multiple independent activations of the neu oncogene by a point mutation altering the transmembrane domain of p185. Cell 45:649–657

Barinaga M (1995) Receptors find work as guides. Science 269:1668–1670

Bliska JB, Guan K-L, Dixon JE et al (1991) Tyrosine phosphate hydrolysis of host proteins by an essential Yersinia virulence determinant. Proc Natl Acad Sci USA 88:1187–1191

Blume-Jensen P, Jiang G, Hyman R et al (2000) Kit/stem cell factor receptor-induced activation of phosphatidylinositol 3'-kinase is essential for male fertility. Nat Genet 24:157–162

Brady-Kalnay SM, Rimm DL, Tonks NK (1994) The receptor protein tyrosine phosphatase PTPμ associates with cadherins and catenins in vivo. Adv Prot Phosphatases 8:227–257

Brady-Kalnay SM, Mourton T et al (1998) Dynamic interaction of PTPμ with multiple cadherins in vivo. J Cell Biol 141:287–296

Brugge JS, Erickson RL (1977) Identification of a transformation-specific antigen induced by an avian sarcoma virus. Nature 269:346–347

Buday L (1999) Membrane-targeting of signaling molecules by SH2/SH3 domain-containing adaptor proteins, Biochem Biophys Acta 1442:187–204

Bustelo XR (1996) The VAV family of signal transduction molecules. Crit Rev Oncogen 7:65–88

Cantley LC, Auger KR, Carpenter C et al (1991) Oncogenes and signal transduction. Cell 64:281–302

Champion-Arnaud P, Gesnet M-C et al (1991) Activation of transcription via AP-1 or CREB regulatory sites is blocked by protein tyrosine phosphatases. Oncogene 6:1203–1209

Charbonneau H, Tonks NK, Walsh KA et al (1988) The leukocyte common antigen (CD45): a putative receptor-linked protein tyrosine phosphatase. Proc Natl Acad Sci USA 85:7182–7186

Clark EA, Ledbetter JA, How B (1994) T cells talk to each other. Nature 367:425–428

Colledge M, Foehner SC (1998) To muster a cluster: anchoring neurotransmitter receptors at synapses. Proc Natl Acad Sci USA 95:3341–3343

Collett MS, Erickson RL (1978) Protein kinase activity associated with the avian sarcoma virus src gene product. Proc Natl Acad Sci USA 75:2021–2024

Debant A, Serra-Pages C, Seipel K et al (1996) The multidomain protein Trio binds the LAR transmembrane tyrosine phosphatase, contains a protein kinase domain, and has separate rac-specific and rho-specific guanine nucleotide exchange factor domains. Proc Natl Acad Sci USA 93:5466–5471

Desai DM, Sap J, Schlessinger J et al (1993) Ligand-mediated negative regulation of a chimeric transmembrane receptor tyrosine phosphatase. Cell 73:541–554

Downward J, Yarden Y, Mayes E et al (1984) Close similarity of epidermal growth factor receptor and v-erb-B oncogene protein sequences. Nature 307:521–527

Edelman GM (1985) Cell adhesion and the molecular processes of morphogenesis. Annu Rev Biochem 54:135–169

Edelman GM (1993) A golden age for adhesion. Cell Adhesion Commun 1:1–7

Fambrough D, McClure K, Kaziauskas A et al (1999) Diverse signaling pathways activated by growth factor receptors induce broadly overlapping, rather than independent, sets of genes. Cell 97:727–741

Fanning AS, Anderson JM (1999) Protein modules as organizers of membrane structure. Curr Opin Cell Biol 11:432–439

Flanagan JG, Vanderhaeghen P (1998) The ephrins and Eph receptors in neural development. Annu Rev Neurosci 21:309–345

Fleischman RA (1993) From white spots to stem cells: the role of the Kit receptor in mammalian development. TIG 9:285–289

Gil-Henn H, Volohonsky G, Elson A (2001) Regulation of protein tyrosine phosphatases α and ε by calpain-mediated proteolytic processing. J Biol Chem 276: 31772–31779

Hawk PB, Oser BL, Summerson WH (1948) Practical physiological chemistry, 12th edn. Blakiston, Philadelphia

Heldin C-H (1995) Dimerization of cell surface receptors in signal transduction. Cell 80:213–223

Hillier BJ, Christopherson KS, Prehoda KE et al (1999) Unexpected modes of PDZ domain scaffolding revealed by structure of nNOS-synthophin complex. Science 284:812–815

Hof P, Pluskey S, Dhe-Paganon S et al (1998) Crystal structure of the tyrosine phosphatase SHP-2. Cell 92:441–450

Holland SJ, Peles E, Pawson T et al (1998) Cell-contact-dependent signalling in axon growth and guidance: Eph receptor tyrosine kinases and receptor protein tyrosine phosphatase beta. Curr Opin Neurobiol 8:117–127

Hunter T (1995) Protein kinases and phosphatases: the yin and yang of protein phosphorylation and signaling. Cell 80:225–236

Hunter T (2000) Signaling – 2000 and beyond. Cell 100: 113–127

Hunter T, Sefton BM (1980) Transforming gene product of Rous sarcoma virus phosphorylates tyrosine. Proc Natl Acad Sci USA 77:1311–1315

Kallunki P, Edelman GM, Jones FS (1998) The neural restrictive silencer element can act as both a repressor and enhancer of L1 cell adhesion molecule gene expression during postnatal development. Proc Natl Acad Sci USA 95:3233–3238

Kissel H, Tomokhina I, Hardy MP et al (2000) Point mutation in Kit receptor tyrosine kinase reveals essential roles for Kit signaling in spermatogenesis and oogenesis without affecting other Kit responses. EMBO J 19:1312–1326

Koretzky GA, Picus J, Schultz J et al (1991) Tyrosine phosphatase CD45 is required for T-cell antigen receptor and CD2-mediated activation of a protein tyrosine kinase and interleukin 2 production. Proc Natl Acad Sci USA 88:2037–2041

Krueger NX, Saito H (1992) A human transmembrane protein-tyrosine-phosphatase, PTPζ, is expressed in brain and has an N-terminal receptor domain homologous to carbonic anhydrases. Proc Natl Acad Sci USA 89:7417–7421

Kuryian J, Cowburn D (1997) Modular peptide recognition domains in eukaryotic signaling. Annu Rev Biophys Biomol Struct 26:259–288

Levinson AD, Opperman H, Levintow L et al (1978) Evidence that the transforming gene of avian sarcoma virus encodes a protein kinase associated with a phosphoprotein. Cell 15:561–572

Levy JB, Canoll PD, Silbennoinen O et al (1993) The cloning of a receptor-type protein tyrosine phosphatase expressed in the central nervous system. J Biol Chem 268:10573–10581

Lorenzen JA, CY Dadabay, Fischer EH (1995) COOH-terminal sequence motifs target the T cell protein tyrosine phosphatase to the ER and nucleus. J Cell Biol 131:631–643

Lu PJ, Shou XZ, Shen M et al (1999) Function of WW domains as phosphoserine- or phosphothreonine-binding modules. Science 283:1325–1328

Majeti R, Bilwes AM, Noel JP et al (1998) Dimerization-induced inhibition of receptor protein tyrosine phosphatase function through an inhibitory wedge. Science 279:88–91

Miklos GLG (1993) Emergence of organizational complexities during metazoan evolution: perspectives from molecular biology, palaeontology and neo-Darwinism. Mem Assoc Australas Paleontols 15:7–41

Milev P, Friedlander DR, Sakurai T et al (1994) Interactions of the chondroitin sulfate proteoglycan phosphacan, the extracellular domain of a receptor-type protein tyrosine phosphatase, with neurons, glia, and neural cell adhesion molecules. J Cell Biol 127:1703–1715

Mosinger B Jr, Tillman U, Westphal H et al (1992) Cloning and characterization of a mouse cDNA encoding a cytoplasmic protein-tyrosine-phosphatase. Proc Natl Acad Sci USA 89:499–503

Muenke M, Schell U (1995) Fibroblast-growth-factor receptor mutations in human skeletal disorders. TIG 11:308–313

Neel BH, Tonks NK (1997) Protein tyrosine phosphatases in signal transduction. Curr Opin Cell Biol 9:293–304

Pawson T (1995) Protein modules and signaling networks. Nature 373:573–580

Pawson T, Nash P (2000) Protein-protein interactions define specificity in signal transduction. Genes Dev 14:1027–1047

Pawson T, Scott JD (1997) Signaling through scaffold, anchoring, and adaptor proteins. Science 278:2075–2080

Peles E, Nativ 1995, Campbell PL et al (1995) The carbonic anhydrase domain of receptor protein tyrosine phosphatase β is a functional ligand for the axonal cell recognition molecule contactin. Cell 82:251–260

Peretz A, Gil-Henn H, Sobko H et al (2000) Hypomyelination and increased activity of voltage-gated K^+ channels in mice lacking protein tyrosine phosphatase ε. EMBO J 19:4036–4045

Pingel JT, Thomas ML (1989) Evidence that the leukocyte-common antigen is required for antigen-induced T lymphocyte proliferation. Cell 58:1055–1065

Romero F, Fischer S (1996) Structure and function of *Vav*. Cell Signal 8:545–553

Rotin D, Margolis B, Mohammadi M et al (1992) SH2 domains prevent tyrosine dephosphorylation of the EGF receptor: identification of Tyr 992 as the high-affinity binding site for SH2 domains of phospholipase Cγ. EMBO J 11:559–567

Schlessinger J (1988) Signal transduction by allosteric receptor oligomerization. Trends Biochem Sci 13:443–447

Schlessinger J (1993) How receptor tyrosine kinases activate Ras. Trends Biochem Sci 18:273–275

Schlessinger J, Ullrich A (1992) Growth factor signaling by receptor tyrosine kinases. Neuron 9:383–391

Shen S-H, Bastien L, Posner BI et al (1991) A protein-tyrosine phosphatase with sequence similarity to the SH2 domain of the protein-tyrosine kinases. Nature 352:736–739

Slamon DJ, Clark GM, Wong SG et al (1987) Human breast cancer: correlation of relapse and survival with amplification of the HER-2/neu oncogene. Science 235:177–182

Sonyang Z, Fanning AS, Fu C et al (1997) Recognition of unique carboxyl-terminal motifs by distinct PDZ domains. Science 275:73–77

Stone RL, Dixon JE (1994) Protein-tyrosine phosphatases. J Biol Chem 50:31323–31326

Sudol M (1996) Structure and function of the WW domain. Prog Biophys Mol Biol 65:113–132

Sutherland EW, Rall TW (1960) The relation of adenosine-3′,5′-phosphate and phosphorylase to the actions of catecholamines and other hormones. Pharmacol Rev 12:265–299

Taylor SI (1992) Molecular mechanisms of insulin resistance: lessons from patients with mutations in the insulin receptor gene. Diabetes 41:1473–1490

Tessier-Lavigne M (1995) Eph receptor tyrosine kinases, axon repulsion, and the development of topographic maps. Cell 82:345–348

Tessier-Lavigne M, Goodman CS (1996) The molecular biology of axon guidance. Science 274:1123–1133

Thomas ML (1999) The regulation of antigen-receptor signaling by protein tyrosine phosphatases: a hole in the story. Curr Opin Immunol 11:270–276

Tonks NK, Diltz CD, Fischer EH (1988a) Purification of the major protein-tyrosine-phosphatases of human placenta. J Biol Chem 263:6722–6730

Tonks NK, Diltz CD, Fischer EH (1988b) Characterization of the major protein tyrosine phosphatases of human placenta. J Biol Chem 263:6731–6737

Trowbridge IS, Thomas ML (1994) CD45: an emerging role as a protein tyrosine phosphatase required for lymphocyte activation and development. Annu Rev Immunol 12:85–116

Ullrich L, Coussens L, Hayflick JS et al (1984) Human epidermal growth factor receptor cDNA sequence and aberrant expression of the amplified gene in A431 epidermoid carcinoma cells. Nature 309:418–425

Vaughan L, Weber P, D'Alessandri L et al (1994) Tenascin-contactin/F11 interactions: a clue for a developmental role? Perspect Dev Neurobiol 1:43–52

Weiss A, Schlessinger J (1998) Switching signals on or off by receptor dimerization. Cell 94:277–280

Yamamoto T, Nishida T, Miyajima N et al (1983) The erbB gene of avian erythroblastosis virus is a member of the src gene family. Cell 35:71–78

Yang Q, Tonks NK (1991) Isolation of a cDNA clone encoding a novel human protein tyrosine phosphatase with homology to the cytoskeletal-associated proteins band 4.1, ezrin and talin. Proc Natl Acad Sci USA 88:5949–5953

Zhao Z, Shen S-H, Fischer EH (1995) Structure, regulation, and function of SH2 domain-containing protein tyrosine phosphatases. Adv Prot Phos 9:297–317

Zontag GCM, Koningstein GM, Jiang Y-P et al (1995) Homophilic interactions mediated by receptor tyrosine phosphatases gamma and kappa. J Biol Chem 270:14247–14250

CHAPTER 5

Labeled Agents as Tracers and Carriers: Theory and Practice

5

WILLIAM C. ECKELMAN

Contents

5.1

Background

In vivo imaging with radiotracers is the most sensitive method for studying chemistry at easily saturable sites. Both single photon emitting radiotracers and positron emitting radiotracers have been developed to measure biochemical changes at receptors and enzymes. With the recent completion of the structural sequence of the genome, the discussion now turns to what role molecular imaging can play in applying this genetic information to alterations in chemistry in the body. These phenotypes, i.e., measurable attributes such as a change in enzyme activity or a change in receptor concentration compared to the norm, are characteristic of the individual and should yield important information in studying the effects of therapy. Given that the phenotypic expression prod-

uct is the key to the progression of the disease and its therapy, this has been and will remain the target for radiotracer studies. Although mRNA has been proposed as a target, there is not always a straightforward correlation between mRNA and the phenotypic expression product. In addition, the limited number of sites will make targeting difficult. It has recently been proposed that, even under optimistic conditions of all mRNA being bound to a radioligand, the amount of radioactivity in the target tissue will be overwhelmed by the background activity (Taylor and Budinger 2000).

5.1.1

Individual Phenotypic Variation

Where imaging is likely to be most useful is in dividing phenotypic variations into discrete, non-overlapping categories. Balaban (1998) has recently set criteria for a special case where the genotype and the phenotype could be linked in a causal relationship (eugenics): (a) If the phenotypic differences are sharply discrete (the area where imaging is mostly like to contribute), (b) the phenotypic differences are correlated with not more than three gene differences, and (c) there are no changes due to developmental conditions. The classic example of a human phenotype that does not vary much as a function of the environment is cystic fibrosis. Cystic fibrosis is an example of cell surface protein mislocalization where the cAMP-regulated chloride-conductance channel protein is expressed not at all or inadequately on the cell surface. One that is very dependent on the environment is phenylketonuria and, in fact, reducing the intake of phenylalanine can alter the pathological phenotype.

The examples that fulfill these three criteria are relatively few, and will likely remain so as long as methods of detection remain static. But the number is likely to increase if better definition of the phenotypic properties can be achieved through molecular imaging. In addition, for the large number of phenotypic expression products that are influenced by environmental conditions, imaging can monitor the expression of the phenotype.

5.1.2
Complex Adaptive Systems

Recently, there has been increasing interest in characterizing complex adaptive systems. There has been much controversy on what constitutes a complex adaptive system (CAS) and there are few examples in biology (Cowan 1994). A CAS is often defined by what it is not. One could imagine that ordered systems, where every point in time and space is the same as every other point in time and space, are not complex. Neither are random systems (Bak 1994). Complexity comes, for example, when moving from a system described by a linear equation to one described by a quadratic equation. The latter system can have more than one equilibrium point and the solution depends on initial conditions and perturbations (Saunders 1993). One example is the second order reaction in receptor binding, where the bound fraction increases slowly with concentration of ligand, then in a linear fashion, and then at a plateau value. In analyzing such data, there can be multiple solutions for the affinity constant and receptor concentration, given the initial values chosen, etc. (Rodbard and Feldman 1975). Major areas of study in complex adaptive systems are mental processes such as cognition and consciousness (Morowitz and Singer 1993). Complex interactions can best be monitored in vivo using radiopharmaceuticals to probe the complexity of phenotypic changes.

5.1.3
Carlsson's Hypothesis on the Balance of Activating and Inhibiting Processes

Many diseases have been originally characterized as related to a single neurotransmitter alteration. For example, biological schizophrenia has been related to changes in dopamine for decades. More recently, the data are best described as a result of complex interactions between facilitating and inhibitory mechanisms with adaptive characteristics (Carlsson et al. 1997). The working hypothesis is that the dopamine system and the glutamatergic system are antagonistic, with the dopamine system being excitatory and the glutamate system inhibitory. Experiments using amphetamine to increase the extracellular concentration of dopamine show that more D2 receptor radioligand is blocked in the brains of schizophrenic patients than in controls (Breier et al. 1997). Perhaps this indicates that the glutamatergic system is not responsive and is not able to activate its glutamate-mediated negative feedback in the same way that normal subjects do. In this respect, Breier et al. found that ketamine, a NMDA glutamate site substrate, was as effective as amphetamine in increasing dopamine concentration such that the binding of a radioligand for the D2 receptor was blocked. In this situation, schizophrenic patients also showed a greater decrease in binding of the D2 receptor radioligand compared to normal controls (Breier et al. 1998). Carlsson et al. (1997) have also found other receptor systems involved in the activating and inhibiting processes.

5.1.4
Endocytosis and Recycling of G Protein Coupled Receptors as an Example of a Complex Adaptive System

Endocytosis and recycling of receptors can be considered a complex adaptive system (Koenig and Edwardson 1997). In the two-compartment model, the rate of change of receptors on the surface is equal to the rate of recycling minus the rate of endocytosis. The endocytosis is accelerated by the presence of agonist. The steady-state number of surface receptors equals the rate of recycling times the sum of the receptors on the surface and in the endosomes divided by the sum of the rates for recycling and endocytosis. However, a more complete model takes into account the degradation of receptors inside the cell and delivery to the surface after new synthesis. The calculation of the fit for this two-tissue compartment model with an independent receptor synthesis rate is best done by iteration, using nonlinear curve-fitting techniques. The initial values of the four rate constants and the initial values of the surface receptor concentration, the endosomal receptor concentration, and the extracellular agonist concentration are key, given that multiple combinations of rate constants are possible.

5.1.5
Receptor Cross-Talk

There is a vast literature on the activation of one receptor causing the release of neurotransmitter from the synapse of another receptor system. Imperato et al. (1993a) have published a series of experiments to measure the increase in neurotransmitter. For example, D1 and D2 receptor agonists cause an increase or a decrease in ACh output, respectively, in the striatum. This effect can differ depending on the target tissue and the concentration of agonist. They have recently reported that in the hippocampus, both D1 and D2 agonists increase ACh extracellular concentration at high doses. At doses of 0.025 and 0.050 mg/Kg of LY 171555, a D2 receptor agonist, basal ACh release was less than 100%. At doses of 0.1, 0.5, and 1 mg/kg the release was greater than 100% in a dose-dependent fashion. They confirmed the decrease in ACh release in the striatum at all doses. With the D1 agonist, SKF 38393, they observed increases of ACh release at doses between 1 to 10 mg/Kg in both the hippocampus and the striatum, in agreement with previous work. Apomorphine, a mixed D1/D2 agonist, had an effect similar to that observed for LY 171555. The antagonist, SCH 23390, prevented the increased ACh release induced with SKF 38393 and LY 171555. Apomorphine showed the same low dose, high dose behavior as LY 171555. The maximal increase in ACh concentration was by a factor of 2.

The same group extended these studies to experiments in the striatum using dopamine antagonists (Imperato et al. 1994b). They found that D2 receptor antagonists increase ACh release in the striatum whereas D1 antagonists decrease ACh release. Raising extracellular dopamine using either amphetamine or cocaine increases the concentration of ACh in the striatum. Blockade of the D1 receptor with SCH 23390 blocks the ACh release. However, the D2 antagonist, raclopride, facilitates the release of ACh. On the other hand, the D1 antagonists, SCH 23390 and SCH 39166 decreased the release to a low of 65%.

The role of cocaine and amphetamine on ACh release in the hippocampus was studied in further detail by Imperato et al. (1993b). The injection of cocaine intraperitoneally caused an increase in ACh to 290% of basal release at 40 min with a half-life of about 60 min. Procaine, a local anesthetic, acts at the dopamine uptake sites with lower affinity than cocaine and produces increases in striatal dopamine as measured by microdialysis that were 50% of those found

for cocaine (Woodward et al. 1995). Therefore, similar but weaker effects on increase in ACh would be expected.

Likewise, d-AMP cause a similar maximal increase. SCH 23390 prevented the increase with both compounds. The γ-aminobutyric acid (GABA) mediated transmission also exerts a tonic modulation on ACh. Flumazenil, an antagonist, increased the basal ACh release to about 200%, whereas the agonist diazepam decreased the release to about 60% of baseline (Imperato et al. 1994a).

The question of how these receptor systems communicate with each other has been described for the GABA$_A$ receptor system and the D5 dopamine receptor system (Liu et al. 2000). Agonist-dependent, protein-protein complex formation between the C-terminal tail domain of D5 receptors and IL2 of GABA$_A$ receptor g2 subunits permits the rapid, reciprocal modulation of GABA and D5 receptor-mediated events. Given the extensive co-localization of GABA$_A$ and the D5 dopamine receptor, dopamine might preferentially activate the D5 receptor, and this could modulate inhibitory synaptic inputs in the GABA$_A$ system. This is interesting given the proposal by Carlsson et al. (see above) that D1 dopamine activity may not be effectively modulated by the GABA system resulting in the pathophysiology of both the positive and negative symptomatology of schizophrenia. The groups at New York University and Brookhaven National Laboratory have characterized functional interactions of neurotransmitter systems by assaying drug-induced displacement of specific receptor radioligands in monkeys using positron emission tomography (PET) (Schloesser et al. 1996).

Many diseases are thought to be the result of neurotransmitter deficiencies, e.g., acetylcholine deficiency in Alzheimer's disease and dopamine deficiency in Parkinson's disease. Monitoring the increase of neurotransmitter brought about by either cross-receptor mechanisms, as discussed above, or by stimulation of release in the same receptor system is only possible using high-specific-activity radiotracers and external imaging.

5.1.6
The Tracer Principle

George Charles de Hevesy is usually considered the first to identify the tracer principle (Myers 1979). In 1923, he used 10.6 hour lead-212 to study the uptake of solutions in bean plants. Although lead is generally considered toxic, he was able to use small, non-toxic amounts because of the sensitivity of the radioactivity techniques. The following year he carried out the first experiment in animals using Bismuth-210 to label and follow the circulation of Bi in rabbits after IM injection of bismuth-containing antisyphilitic drugs. In a later de Hevesy book (with Fritz Paneth), the tracer method was introduced as the use of radioelements as indicators. An indicated element contains a small amount of one of its isotopes, which serves as an indicator for the purpose of detection or measurement (Hevesy and Paneth 1938). Martin D. Kamen in the three editions of his book entitled *Isotopic Tracers in Biology* chronicled the rapid progress in applying the newly discovered tracer principle to uses in clinical research (Kamen 1998). Some of the important "firsts" in humans were the determination of the speed of diffusion and peripheral circulation using ^{24}Na by Blumgart and Weiss in 1927 and thyroid metabolism by Hamilton and Stone in 1937 (as reported in Kamen 1998). Uptake, retention, and excretion of radiolabeled phosphate (^{32}P) and radiolabeled iodide (^{131}I) provided valuable information about the selectivity of proposed therapeutic regimens. Radioisotopes were also valuable in applications in hematology. Data on red cell survival, iron physiology, and blood volume were some of the important contributions. In the early 1940s, ^{32}P, followed by ^{35}S and ^{131}I, were used to label antigens and antibodies. In the process of studying the behavior of ^{131}I, Berson and Yalow developed the sensitive system for measuring blood components known as radioimmunoassay (Yalow and Berson 1960).

5.2
Examples of the Use of Information on the Genotype for Imaging Studies of the Phenotype

5.2.1
Isotopic Fluoro Deoxyglucose (FDG) in von Gierke Disease

Glucose-6-phosphatase (G6Pase) (Chen and Burchell 1995), most abundantly found in liver and kidney, catalyzes the last step of both gluconeogenesis and glycogenolysis. This enzyme is a key protein in the regulation of glucose homeostasis. Its function is to dephosphorylate glucose-6-phosphate (G6P), so that free glucose can be transported out of cells and re-

leased into the blood. Thus, the activity of this enzyme controls liver glucose production. Defects of the G6Pase enzyme system cause glycogen storage disease type 1 (GSD-1), also known as von Gierke disease, and manifests with severe hypoglycemia and hepatomegaly caused by accumulation of glycogen. In GSD-1 the G6Pase enzyme system is associated with the endoplasmic reticulum and has multiple components (Chen and Burchell 1995). Current understanding of this system postulates that G6P is transported into the microsome by a G6P translocase (G6PT) (Arion et al. 1980). Once inside the microsome G6P is dephosphorylated by the G6Pase catalytic unit and finally phosphate and glucose are transported out of the microsome via specific transport proteins (T2 and T3, respectively; Fig. 5.1). The cDNA and gene for G6PT have been recently described (Lin et al. 1998; Pan et al. 1999). Full characterization of this system is still pending, but it has been shown that patients labeled "GSD-1a" have mutations in the G6Pase gene that inactivate or greatly reduce G6Pase activity (Pan et al. 1998).

FDG is [^{18}F]-2-fluoro-2-deoxyglucose, a glucose analog currently utilized for PET imaging studies in humans. FDG uptake has been used for many years to measure in vivo regional glucose utilization (Phelps et al. 1979). This tracer competes with glucose for phosphorylation by hexokinase or glucokinase. After it is phosphorylated, FDG-6-phosphate (FDG6P) does not undergo glycolysis. In tissues with elevated glucose metabolic rates, such as tumors, with low or absent dephosphorylating activity, FDG6P is trapped inside cells. This property allows for imaging of areas with increased FDG retention and has been extensively applied to visualize, stage, and monitor progression of tumors (Conti et al. 1996). On the other hand, in organs such as the liver, FDG is taken up and rapidly released, presumably by the presence and activity of the G6Pase enzyme (Gallagher et al. 1978). Preliminary data indicate that in patients with GSD-1, FDG is more avidly retained in the liver compared to normal subjects (Gjedde 1995). The different steps in FDG metabolism, developed from the seminal work by Sokoloff using [^{14}C]deoxyglucose, can be described by the two tissue compartment model (Sokoloff et al. 1977; Reivich et al. 1979). Kinetic analysis of dynamic PET-FDG studies allows estimation of the rate constants for each step. The current understanding of FDG metabolism is that the k_4 parameter is tightly linked to the presence and activity of the G6Pase enzyme system. Therefore, in vivo measure-

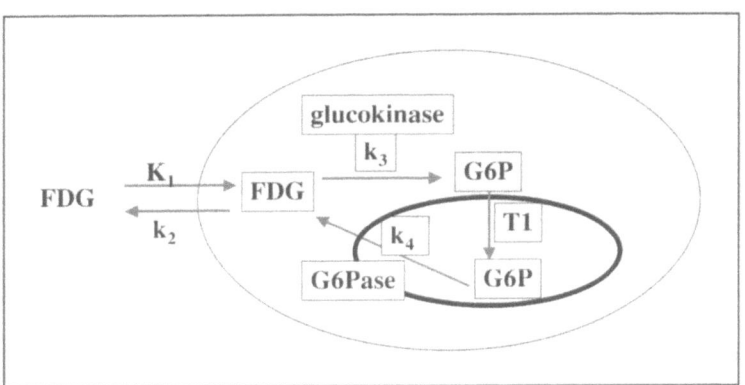

Fig. 5.1. Schematic representation of the two tissue compartment model for FDG

ment of k_4 for FDG may provide a non invasive quantitative method to evaluate liver G6Pase function and also be a tool to monitor the effectiveness of gene therapy in patients with GSD type 1.

To further characterize this issue, the group at NIH has analyzed the FDG turnover properties of cultured cells and verified if a correlation exists between G6Pase activity and whole cell FDG release kinetics (Caraco et al. 2000). Three different cell lines have been studied: A431, derived from a human epidermoid carcinoma; Ho-15 cells, derived from the liver of a G6Pase knockout mouse (Lei et al. 1996); and hepatocytes derived from a normal mouse as control (WT10). In the first two cell lines isolated clones that overexpress the G6Pase catalytic unit by transfection were isolated. Overexpression of the G6Pase catalytic unit in a G6Pase knockout mouse liver cell line (Ho-15) and in a cell line derived from a human epidermoid carcinoma (A431) was also achieved. Both cell lines were permanently transfected with a plasmid containing the full coding sequence for mouse (Ho-15-D3) or human (A431-AC3) liver G6Pase catalytic unit under a viral promoter. G6Pase enzyme activity was measured in all clones and in mouse liver tissue and in normal mouse hepatocytes (WT10) as controls. FDG release rates were determined in cells transfected with the G6Pase catalytic unit gene, in mock transfected cells of both cell lines, and WT10. A431-AC3 showed only a slight increase in release rate compared to the control clone (30%). On the other hand, the Ho-15-D3 clone showed three-fold higher FDG release rates compared to the mock transfected cells. Release rates in WT10 cells were similar to Ho-15-D3. Intracellular radioactivity during the release phase was found to be virtually all in the form of FDG6P in all cell lines tested. Glucose-6-phosphate transporter (G6PT) RNA was found to be expressed at higher levels in Ho-15 cells compared to

A431. This study shows that activity of the G6Pase enzyme system can be quantified in whole cells by measuring FDG release and that it requires adequate levels of the other components of the G6Pase enzyme system such as the G6P transporter. FDG-PET may be utilized in patients with GSD-1 to characterize the disease and monitor the effectiveness of gene therapy.

5.2.2
Alzheimer's Disease and the Muscarinic System

The current state of molecular genetics of Alzheimer's disease (AD) focuses on four genes: the β-amyloid precursor protein (β-APP) gene, presenilin 1 gene, presenilin 2 gene, and the apolipoprotein E (APOE) gene (St. George-Hyslop 2000). The ε_4 allele of the (APOE) gene is associated with an increased risk for late-onset AD. Only 10% of the AD cases appear to be transmitted as an autosomal dominant trait. Approximately 30–50% of the population risk for AD can be attributed to genetic factors with a final risk of 38% by age of 85 years. The characteristic neuropathology is neurofibrillary tangles and amyloid plaques. The end result of the phenotypic changes is a result of an alteration in the processing of β-APP favoring the production of the potentially toxic $A\beta_{42}$ protein.

Few of the imaging studies to date have been based on these gene products, although binding to $A\beta_{42}$ protein has been reported (Skovronsky et al. 2000). There has also been a report on the use of [^{18}F]FDG in a sub-population of AD patients with the APOE gene (Kuhl et al. 1999) although this is an indirect measure of the phenotype. The authors found changes in FDG distribution before clinical manifestations. However, the majority of the effort in radiopharmaceutical development has been to design ligands based on deficits found at autopsy studies of AD patients.

The primary focus of our work has been the development of M2 subtype selective cholinergic ligands, based on the observation that this subtype is lost in the cerebral cortex in Alzheimer's disease (Quirion et al. 1989; Aubert et al. 1992; Rodriguez-Puertas et al. 1997). Postmortem quantitation of muscarinic subtypes indicated a selective loss of M2 subtype in cortical regions while the M1 subtype was preserved. Thus, an M2 selective ligand labeled with a positron-emitting radionuclide would allow determination of M2 subtype concentrations in the living human brain. This may allow study of the progression of Alzheimer's dementia, early diagnosis, and non-invasive monitoring of drug therapies.

The muscarinic acetylcholine receptor (mAChR) was one of the earliest receptors studied with in vivo techniques. As early as 1973, Farrow and O'Brien described the use of [H-3]atropine to define the mAChR. With the development of higher affinity and higher specific activity compounds such as [H-3]QNB, receptor distribution was mapped in isolated tissue (Yamamura et al. 1974). The first ligand to map mAChR in humans in vivo was the radioiodinated form of QNB, 3-R-quinuclidinyl 4-S-iodobenzilate (RS IQNB) (Eckelman et al. 1984). The use of receptor-binding radiotracers for external brain imaging differs from their use in vitro. In order for a radiotracer to be useful in vivo, its distribution must be driven by the local receptor concentration rather than by local blood flow or membrane transport properties, so that the images obtained primarily reflect receptor binding. A comprehensive kinetic analysis of RS IQNB in rats was performed by Sawada et al. (1990). It showed that the uptake in cerebrum is essentially irreversible during the first 360 min after intravenous administration and that the rate of RS IQNB tissue uptake depends on transport across the blood-brain barrier (BBB) as well as on the rate of binding to the receptor. However, the later data (after $\sim$24 h) are sensitive to receptor concentration. Clinical studies with RS IQNB have indicated that it is responsive to changes in receptor concentration at late times postinjection (Weinberger et al. 1991, 1992). However, it does not demonstrate significant muscarinic subtype selectivity.

Several investigators have pursued muscarinic receptor imaging with PET and single-photon emission computed tomography (SPECT), and this topic has been the subject of a review (Maziere 1995). As with

RS IQNB, most tracers for cholinergic receptors have not demonstrated subtype selectivity. Alternatively, ligands that are subtype-selective in vitro typically do not cross the BBB. To develop imaging agents with subtype selectivity, two approaches have been used. In one approach, a ligand that is selective in vitro, AF-DX 116 [11-(((2-(diethylamino)-methyl)-1-piperidinyl) acetyl)-5-11-dihydro-6H-pyrido (2,3β) (1,4) benzodiazepin-6-one], was modified to increase its BBB permeability while maintaining subtype selectivity (Gitler et al. 1992, 1993; Doods et al. 1993). In the second approach, non-subtype selective IQNB [1-azabicyclo[2.2.2]octan-3-yl (a-hydroxy-a-4-iodophenyl)-benzeneacetate], which crosses into the brain, was modified to increase its subtype selectivity (McPherson et al. 1993, 1995).

In addition to muscarinic receptors, ligands for the nicotinic receptors (Gatley et al. 1998; Horti et al. 1998) and acetylcholinesterase (Kuhl et al. 1999) have been investigated. Acetylcholinesterase (Tavitian et al. 1993; Namba et al. 1994; Kilbourn et al. 1996) has been a target for radioligand development, because its levels also change in Alzheimer's disease. In addition, the vesicular acetylcholine transporter has recently been studied in patients with Alzheimer's disease using [Iodine-123]iodobenzovesamicol (Kuhl et al. 1999). The probe for acetylcholinesterase ([^{11}C]PMP) has been used in Alzheimer's patients and found to show a decrease that is more uniform and broadly distributed than the focal defects found with [^{18}F]FDG (Kuhl et al. 1999).

We have also been working extensively with a muscarinic agonist based on a series first proposed by Sauerberg et al. (1992) of Novo-Nordisk. These ligands contain a thiadiazolyl moiety attached to various heterocycles, including tetrahydropyridine. Two of these compounds, xanomeline and butylthio-TZTP, demonstrated M1 selectivity and have been labeled with carbon-11 and studied with PET (Farde et al. 1996). Another compound (3-(propylthio)-1,2,5-thiadiazol-4-yl)-tetrahydro-1-methylpyridine (P-TZTP) is M2 selective. In our NovaScreen assay using brain and heart tissue, this compound showed a K_i of 23 nM for M1 and 1.5 nM for M2 (Fig. 5.2).

We expanded upon this work and described the radiosynthesis and preliminary biodistribution of 3-(3-(3-[^{18}F]fluoropropyl)thio)-1,2,5-thiadiazol-4-yl)-1,2,5,6-tetrahydro-1-methylpyridine ([^{18}F]FP-TZTP) (Kiesewetter et al. 1995). [^{18}F]FP-TZTP showed a K_i of 7.4 nM for M1 and 2.2 nM for M2; it did not bind to M3 receptors or other biogenic amine receptors. In

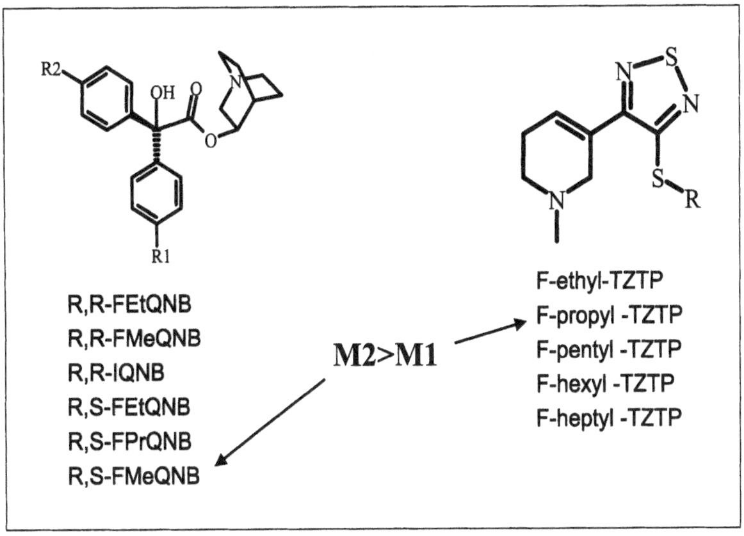

Fig. 5.2. Structures of [^{18}F]-labeled muscarinic ligands

vivo studies with [^{18}F]FP-TZTP in rat brain showed rapid uptake early, but the net efflux was also fast. Autoradiography using no carrier added [^{18}F]FP-TZTP confirmed the uniform distribution of radioactivity characteristic of the M2 pattern of localization. At 1 h after injection, co-injection of P-TZTP at 5, 50, and 500 nmol inhibited [^{18}F]FP-TZTP uptake in a dose-dependent manner. The difference in brain regions between each dose level was significant ($P < 0.01$), except for the 5 nmol value in medulla ($P < 0.06$). The brain distribution of the agonist [^{18}F]FP-TZTP was unaffected by coinjection of 5, 50, or 500 nmol of the antagonist RR-IQNB. Likewise, the distribution of [^{18}F]FP-TZTP was unaffected by coinjection of 500 nmol of the M2 selective antagonist RS-FMeQNB except in cerebral cortex and hippocampus, and there the difference was significant ($P < 0.03$). Binding in the heart was low by 15 min, so earlier time points were studied. At 5 minutes, we observed 55% inhibition of uptake in the heart with co-injection of P-TZTP.

A major concern in PET studies is metabolism of the radioligand to radioactive metabolites. In order to quantify receptors accurately, the time course of the parent compound, [^{18}F]FP-TZTP, in blood must be determined. In addition, it is important to verify that any radioactive metabolites do not cross the BBB. Metabolism of [^{18}F]FP-TZTP in vivo is rapid. In rat, only 5% of plasma radioactivity was in the parent compound by 15 min after injection. One metabolite was almost as lipophilic as the parent compound as measured by thin-layer chromatography (TLC), suggesting that it might cross the BBB. However, the parent compound was found to represent greater than

95% of extracted radioactivity in rat brain through 30 min, and greater than 90% at 45 and 60 min. In summary, our experiments in the rat with [^{18}F]FP-TZTP, a reversible M2 radioligand with predominantly parent compound in the brain, provided strong support for its use as a PET tracer.

PET studies in isoflurane-anesthetized rhesus monkeys were performed to assess the in vivo behavior of [^{18}F]FP-TZTP (Carson et al. 1998a). Control studies ($n = 11$) were performed first to characterize tracer kinetics and to choose an appropriate mathematical model for the in vivo behavior of the tracer. Application of a model requires measurement of the arterial input function for parent compound. Parent compound comprised $48 \pm 9\%$, $28 \pm 6\%$, and $13 \pm 3\%$ of plasma radioactivity at 15, 30, and 90 min, respectively. [^{18}F]FP-TZTP time-activity curves (TACs) in brain were well-described by a one-compartment model with three parameters: uptake rate constant K_1, total volume of distribution V, and a global brain-to-blood time delay Δt. Models with additional parameters could not be used because reliable parameter estimates could not be obtained. Tracer uptake in the brain was rapid, with K_1 values of 0.4 to 0.6 ml/min ml^{-1} in gray matter. The volume of distribution V represents total tracer binding, i.e., free, nonspecifically bound, and specifically bound. V values were very similar (22–26 ml/ml) in cortical regions, basal ganglia, and thalamus, but were significantly lower (16 ml/ml, $P < 0.01$) in the cerebellum. In comparing these V values with the receptor distribution reported for rat and monkey, there is an excellent match with the M2 distribution. In rats, the concentration of M2 receptors in cortical structures, basal ganglia, and thalamus is highly uniform, and is approximately 50% lower in cerebellum (Li et al. 1991). In rhesus monkey, the distribution of M2 receptors is also uniform in cortex, basal ganglia, and thalamus (Flynn and Mash 1993). This pattern is unlike that of M1 receptors (the muscarinic subtype to which this tracer shows threefold lower in vitro affinity) for which basal ganglia >cortex >thalamus (Wall et al. 1991; Flynn and Mash 1993).

Pre-blocking studies were used to measure non-specific binding. Pre-administering 200–400 nmol/kg of non-radioactive FP-TZTP produced a dramatic reduction in total binding of $\sim 50\%$ in cerebellum and 60–70% in other gray matter regions. Similar blockade was seen in analogous rat studies (Kiesewetter et al. 1995). This reduction was highly significant in all regions ($P < 0.001$) and the regional distribution of V values after pre-blocking became nearly uniform. Specific binding values were then determined by the difference between control and pre-blocked V values. In addition to pre-blocking studies, displacement of [^{18}F]FP-TZTP with 80 nmol/kg of FP-TZTP at 45 min post injection caused a distinct increase in net efflux with decreases of 20%, 36%, and 41% in cerebellum, cortex, and thalamus, respectively (Kiesewetter et al. 1999).

The sensitivity of [^{18}F]FP-TZTP binding to changes in brain acetylcholine was assessed by administering physostigmine, an acetylcholinesterase inhibitor, by i.v. infusion beginning 30 min before tracer injection. Physostigmine produced a 35% reduction in cortical specific binding ($P < 0.05$), consistent with increased competition from acetylcholine. The studies in monkeys indicated that [^{18}F]FP-TZTP should be useful for the in vivo measurement of muscarinic receptors.

In collaboration with NIMH investigators, we recently began studies in normal human volunteers, with the eventual goal of studying patients with Alzheimer's disease (Carson et al. 1999). The initial analysis, based on data from six young control subjects, concentrated on the determination of the appropriate kinetic model for [^{18}F]FP-TZTP in humans. In plasma, parent compound represented 68 ± 8, 41 ± 9, and $14 \pm 4\%$ of radioactivity at 20, 40, and 120 min, respectively. The plasma free fraction (f_p) was $5.9 \pm 1.2\%$. A model with one tissue compartment produced an excellent fit for the full 120 min of data, so that the additional parameters of a two-compartment model were unidentifiable. K_1 values in gray matter regions were high, 0.36–0.56 ml/min ml^{-1}, and showed excellent correlation with cerebral blood flow.

V values, representing total tissue binding, were very similar in cortical regions, basal ganglia, and thalamus, but were significantly higher ($P < 0.01$) in amygdala. Unlike the results in the monkey, binding in cerebellum was similar to that in the cerebral cortex. V values correlated with f_p and normalization of V by f_p reduced the coefficient of variation of V from 24% to 16%. The methodology and results from our analysis of [^{18}F]FP-TZTP data in young controls provide the basis for ongoing studies in elderly controls and patients with Alzheimer's disease.

5.2.3
Alcoholism and 5-HT$_{1A}$

Many genes involved in serotonin metabolism have allelic variants including the 5-HT$_{1A}$ receptor. Our work in the serotonin system has concentrated primarily on the 5-HT$_{1A}$ receptor because it has been implicated in alcoholism (Fletcher et al. 1993). A 5-HT$_{1A}$ variant lowers agonist-mediated receptor down-regulation rate and desensitization (Nielsen et al. 1998). In alcoholics, autoradiographic analysis of the 5-HT$_{1A}$ receptor in the human brain showed a 38–86% reduction in various regions of the frontal-parietal cortex (Dillon et al. 1991).

Similar to the muscarinic cholinergic project one of our hypotheses is that subtype selective ligands will prove more useful than compounds without subtype specificity. In addition, our second hypothesis for the 5-HT$_{1A}$ system is that the use of two subtype selective ligands with different receptor affinities will provide more sensitivity to serotonergic processes. The use of a high affinity ligand should provide more accurate measures of serotonin receptor density, suitable for comparisons between patient groups. However, lower affinity ligands are also useful. While contrast between receptor-rich and receptor-poor regions is reduced for ligands with lower affinity, they tend to reach equilibrium more rapidly, and thus are more sensitive to measuring dynamic changes in neurotransmitter concentration. Thus, our objectives for the 5-HT$_{1A}$ system involve the development and application of both high and lower affinity tracers.

Although in vivo studies of the 5-HT$_{1A}$ system have been underway for some time, it has been difficult to isolate its actions fully because of the large number of 5-HT receptor subtypes (Alexander and Peters 1999) and the difficulty in characterizing the agonist and antagonist properties of potential ligands with physiological models (Jain 1988). Studies with the 5-HT$_{1A}$ receptor agonist 8-OH-DPAT [8-hydroxy-2-(di-n-propylamino)tetralin] showed it to be selective and defined many 5-HT$_{1A}$ functional responses. However, many antagonists used in earlier studies were nonselective and some of the ligands initially characterized as antagonists were subsequently shown to have partial agonist properties and often high affinity to other biogenic amine receptor systems (Fletcher et al. 1993). In addition, most ligands showed relatively weak binding to the 5-HT$_{1A}$ receptor and would not be useful as PET radioligands. Thus, the need for new, highly selective antagonists to characterize 5-HT$_{1A}$ receptors motivated our work.

The first high affinity, subtype selective, silent antagonist was WAY 100635 (WAY) (Khawaja et al. 1995). A carboxamide derivative was radiolabeled with Iodine-123 by Kung et al. (1994) and was shown to have appropriate properties for in vivo imaging. Laporte et al. (1994) and Hume et al. (1994) showed that tritiated WAY had appropriate binding characteristics in mouse and rat brain, respectively, and Mathis et al. (1994) obtained similar results with [O-methyl-^{11}C]WAY in rhesus monkeys with PET. However, Osman et al. showed that in monkey, a radioactive metabolite of this tracer, ([O-methyl-^{11}C]WAY 100634, N-{2-[4-(2-methoxyphenyl)-piperazino]ethyl}-2-pyridinamine), crossed the BBB but did not bind to the 5-HT$_{1A}$ receptor (Osman et al. 1996; Pike et al. 1995), thus reducing image contrast. As a result, subsequent labeling attempts concentrated on incorporating the radionuclide at the carboxamide moiety, i.e., the carbonyl carbon of the cyclohexanecarboxamide (Pike et al. 1996). This compound showed reduced nonspecific binding and higher target to nontarget ratios in nonhuman primates and humans, and produced only radiolabeled carboxylic acid metabolites, which did not appear to cross the BBB. Human studies with this tracer are underway elsewhere (Farde et al. 1998; Gunn et al. 1998; Ito et al. 1999).

Our goal is to develop a series of Fluorine-18 labeled radiopharmaceuticals that are 5-HT$_{1A}$ subtype-specific and have a range of binding affinities. We first synthesized three fluorinated derivatives of WAY 100635, N-{2-[4-(2-methoxyphenyl)-piperazino]ethyl}-N-(2-pyridyl)cyclohexanecarboxamide using various acids in place of the cyclohexanecarboxylic acid (CHCA) in the reaction scheme (Lang et al. 1999a). These three acids are: 4-fluorobenzoic acid (FB), trans-4-fluorocyclohexanecarboxylic acid (FC), and 4-

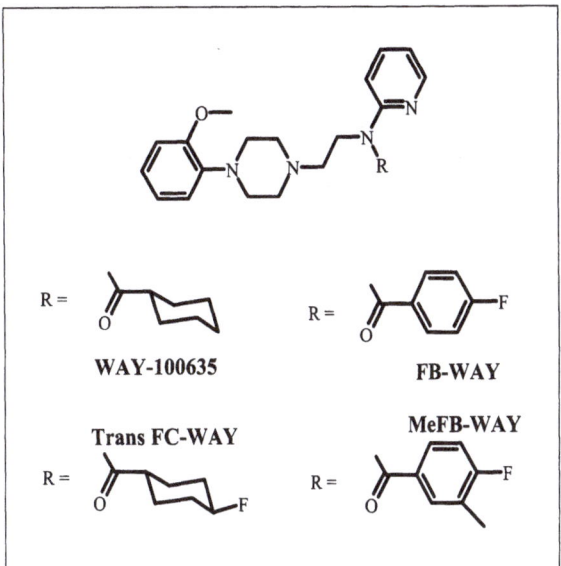

Fig. 5.3. Structures of 5-HT$_{1A}$ antagonists

fluoro-3-methylbenzoic acid (MeFB). They were labeled with ^{18}F and their biological properties were evaluated in rats and compared with those of [carbonyl-^{11}C]WAY100635 ([^{11}C]WAY) (Fig. 5.3).

For all compounds there was a good correlation (r > 0.97) between the differential uptake ratio (DUR, %ID/g×body weight (g)100^{-1}) in individual rat brain regions at 30 min after injection and the concentration of receptors as determined by in vitro quantitative autoradiography. However, specific binding ratios [region of interest/cerebellum] for cerebral cortex and hippocampus were higher for [^{11}C]WAY100635 and [^{18}F]FCWAY compared to the other compounds. This difference in affinity was also apparent in the pharmacokinetic profiles of these compounds. The lower affinity compounds, [^{18}F]FBWAY and [^{18}F]MeFBWAY, showed more rapid equilibrium between brain and blood, as reflected by highly similar brain and blood clearance rates. For the higher affinity compounds, [^{18}F]FCWAY and [^{11}C]WAY, clearance from brain was slower than from blood. Based on these data, we chose [^{18}F]FCWAY to pursue as our high affinity ^{18}F labeled 5-HT$_{1A}$ ligand, to provide better statistics and quantification than [^{11}C]WAY for static measurement of 5-HT$_{1A}$ receptor distribution.

The lower affinity phenylcarboxamide ligands, [^{18}F]FBWAY and [^{18}F]MeFBWAY, had similar specific binding ratios (~5 in hippocampus and ~1.5 in cortex). However, preblocking studies using 50 nmol of WAY showed that the specific binding of [^{18}F]MeFB-WAY could be more effectively blocked. Since our goal with a lower affinity ligand is to be sensitive to dynamic changes in receptor occupancy, we chose [^{18}F]MeFBWAY to pursue as our lower affinity Fluorine-18 labeled 5-HT$_{1A}$ ligand.

In order to completely characterize the chemistry and biology of FCWAY, we prepared both the 4-cis and 4-trans forms and studied their properties in vitro and in vivo (Lang et al. 1999b). Fluorine-18 cis FCWAY was prepared starting from the 4-trans nitro-benzenesulfonate of cyclohexanecarboxylic acid pentamethylbenzyl ester in a procedure similar to that used for preparing ^{18}F trans FCWAY. An in vitro binding assay showed that cis FCWAY had a weaker affinity for 5-HT$_{1A}$ receptor than that of trans FCWAY. Cis FCWAY is also a less lipophilic compound than trans FCWAY based on reversed phase high-performance liquid chromatography (HPLC). In vivo metabolism studies and in vitro studies using isolated rat hepatocytes showed that these two compounds had similar metabolite profiles. However, in vivo rat brain uptake of cis ^{18}F FCWAY at 30 min was much lower than that of trans ^{18}F FCWAY; the specific binding ratio for hippocampus was 19.3 for ^{18}F trans FCWAY and 3.6 for ^{18}F cis FCWAY. Coinjection of 200 nmol WAY 100635 produced a similar percent of blocking for both compounds. These results indicated that the conformation of the ligand could greatly affect its affinity for the 5-HT$_{1A}$ receptor and uptake in the brain.

[Fluorine-18]MeFBWAY was chosen as a lower affinity ligand for studies of dynamic changes in serotonin. Initial studies in monkey with [^{18}F]MeFBWAY showed a lower ratio of specific to nonspecific binding than was anticipated based on the original rat studies. Therefore, in an attempt to find a radioligand with intermediate properties between [^{18}F]FCWAY and [^{18}F]MeFBWAY, we synthesized eight new 4-fluoro-phenylcarboxamide derivatives with various substituted pyridines and pyrimidines replacing the pyridine ring (Lang et al. 1999c). Five of these compounds were labeled with Fluorine-18 and their binding properties evaluated in rats at 30 min with or without coinjection of 200 nmol of WAY. The results indicate that further study of the compound with the pyrimidine replacing the pyridine as a PET ligand for mapping 5-HT$_{1A}$ receptors in vivo is warranted.

In PET studies of [^{11}C]WAY in rhesus monkeys (Carson et al. 1997), we had found that it was slow to reach equilibrium in regions with high specific binding, and that because of its short half-life (20.4 min), the reliability of the later data was poor, particularly

the blood metabolite data. Therefore, we evaluated [^{18}F]FCWAY (Lang et al. 1999a) in rhesus monkeys with PET (Carson et al. 1998b). The goal of these studies was to further develop [^{18}F]FCWAY as a radioligand to measure 5-HT$_{1A}$ receptors in humans. Studies included control experiments to develop an appropriate kinetic model, and pre-blocking studies to demonstrate specific binding and estimate the level of nonspecific binding. Separate experiments with injection of the labeled metabolites of [^{18}F]FCWAY were also performed to determine if significant brain accumulation occurred. Control [^{18}F]FCWAY studies were performed to assess the regional binding patterns. Like [^{11}C]WAY, [^{18}F]FCWAY showed a binding pattern consistent with 5-HT$_{1A}$ binding, i.e., highest binding in frontal and medial temporal cortex, intermediate binding in other cortical regions, low binding in thalamus and caudate nuclei, and minimal uptake in cerebellum. By 60–90 min postinjection, the frontal-to-cerebellum binding ratio was ~10:1. Preblocking studies with WAY 100635 (200 nmol/kg) administered 5 min before [^{18}F]FCWAY administration showed complete removal of specific binding, i.e., highly uniform low binding levels in all regions.

The plasma-free fraction of [^{18}F]FCWAY was high compared to many other tracers: e.g., 49±7%, versus 31±6% for [^{11}C]WAY. Metabolite studies of plasma radioactivity using TLC and HPLC were performed to determine the appropriate input function. The primary metabolites were identified as [^{18}F]fluorocyclohexanecarboxylic acid ([^{18}F]FC) and [^{18}F]F$^-$. Studies were performed to assess the magnitude of brain uptake of these metabolites (Carson et al. 1998b). It was found that [^{18}F]FC would contribute no more than 7% of the signal in the cerebellum and less in other regions with specific binding. Similar magnitudes of errors were found in previous studies of the metabolites of [^{11}C]WAY. Brain uptake of [^{18}F]F$^-$ could not be detected, but partial volume averaging of fluoride in monkey skull produced biases of 4–13% in tissue concentration data at 120 min. This latter effect is expected to be much smaller in human data due to size differences. Tracer kinetic modeling analysis of the time-activity data corrected for metabolite contamination showed that the data were well described by a model with two tissue compartments and four parameters. A model with one tissue compartment produced particularly poor fits for regions with little specific binding, and was generally inadequate in all cases. The total volume of distribution (V) was the most useful measure of receptor binding. In control studies, V values were 33 in frontal cortex, 12 in basal ganglia, 8 in thalamus, and 4 in cerebellum. Across brain regions, V estimates for [^{18}F]FCWAY had a correlation of 0.99 with those measured for [^{11}C]WAY. V values in the preblocked [^{18}F]FCWAY studies were low and very uniform across all regions. Clinical studies with [^{18}F]FCWAY have begun in patients with epilepsy and preliminary indications are that [^{18}F]FCWAY receptor binding is decreased in the epileptogenic temporal lobe.

5.3
Conclusion

The next step in the evolution of PET and SPECT is to base studies on the recent anatomical description of the genome and the linkages with disease. The major contributions of molecular imaging will most likely be to better identify the phenotypic differences within a disease category and to monitor the progression of the protein expression product as a function of the disease.

Given the complex adaptive systems that are involved in most biological properties, the second goal is to develop imaging paradigms that account for multiple interactions rather than looking at a single protein product in isolation. It is imperative that these studies be pursued given the uniqueness of PET and SPECT in monitoring easily saturated binding sites using the tracer principle.

5.4
References

Alexander SPH, Peters JA (1999) 1999 receptor and ion channel nomenclature supplement. Elsevier Trends Journal, Cambridge, UK

Arion WJ, Lange AJ, Walls HE et al (1980) Evidence for the participation of independent translocation for phosphate and glucose 6-phosphate in the microsomal glucose-6-phosphatase system. Interactions of the system with orthophosphate, inorganic pyrophosphate, and carbamyl phosphate. J Biol Chem 255:10396–10406

Aubert I, Araujo DM, Cecyre D et al (1992) Comparative alterations of nicotinic and muscarinic binding sites in Alzheimer's and Parkinson's diseases. J Neurochem 58:529–541

Bak P (1994) Self-organized criticality: a holistic view of nature. Addison-Wesley, Reading, MA

Balaban E (1998) Eugenics and individual phenotypic variation: to what extent is biology a predictive science? Sci Context 11:331–356

Breier A, Su T-P, Saunders R et al (1997) Schizophrenia is associated with elevated amphetamine-induced synaptic dopamine concentrations: evidence from a novel positron emission tomography method. Proc Natl Acad Sci USA 94:2569–2574

Breier A, Adler CM, Weisenfeld N et al (1998) Effects of NMDA antagonism on striatal dopamine release in healthy subjects: application of a novel PET approach. Synapse 29:142–147

Caraco C, Aloj L, Chen LY et al (2000) Cellular release of [18F]2-fluoro-2-deoxyglucose as a function of the glucose-6-phosphatase enzyme system. J Biol Chem 275: 18489–18494

Carlsson A, Hansson LO, Waters N et al (1997) Neurotransmitter aberrations in schizophrenia: new perspectives and therapeutic implications. Life Sci 61:75–94

Carson RE, Schmall B, Endres CJ et al (1997) Kinetic analysis of the 5-HT$_{1A}$ antagonist [carbonyl-^{11}C]WAY-100635. J Nucl Med 38:80P–81P

Carson RE, Kiesewetter DO, Jagoda E et al (1998a) Muscarinic cholinergic receptor measurements with [^{18}F]FP-TZTP: control and competition studies. J Cereb Blood Flow Metab 18:1130–1142

Carson RE, Lang L, Watabe H et al (1998b) Evaluation of a new F-18 labeled analog of the 5-HT$_{1A}$ antagonist WAY 100635 for PET. J Nucl Med 39:135P

Carson RE, Kiesewetter DO, Connelly K et al (1999) Kinetic analysis of the muscarinic cholinergic ligand [F-18]FP-TZTP in humans. J Nucl Med 40:30P

Chen Y-T, Burchell A (1995) Glycogen storage diseases. In: Scriver CR, Beaudet AL, Sly WS, Valle D (eds) The metabolic and molecular bases of inherited disease. McGraw-Hill, New York, pp 935–966

Conti PS, Lilien DL, Hawley K et al (1996) PET and [^{18}F]-FDG in oncology: a clinical update. Nucl Med Biol 23:717–736

Cowan GA (1994) Complexity: metaphors, models, and reality. Addison-Wesley, Reading, MA

Dillon KA, Gross-Isseroff R, Israeli M et al (1991) Autoradiographic analysis of serotonin 5-HT1A receptor binding in the human brain postmortem: effects of age and alcohol. Brain Res 554:56–64

Doods H, Entzeroth M, Ziegler H et al (1993) Characterization of BIBN 99: a lipophilic and selective muscarinic M2 receptor antagonist. Eur J Pharmacol 242:23–30

Eckelman WC, Grissom M, Conklin J et al (1984) In vivo competition studies with analogues of quinuclidinyl benzilate. J Pharm Sci 73:529–533

Farde L, Suhara T, Halldin C et al (1996) PET study of the M1-agonists [^{11}C]xanomeline and [^{11}C]butylthio-TZTP in monkey and man. Dementia 7:187–195

Farde L, Ito H, Swahn CG et al (1998) Quantitative analyses of carbonyl-carbon-11-WAY-100635 binding to central 5-hydroxytryptamine-1A receptors in man. J Nucl Med 39:1965–1971

Farrow JT, O'Brien RD (1973) Binding of atropine and muscarine to rat brain fractions and its relation to the acetylcholine receptor. Mol Pharmacol 9:33–40

Fletcher A, Clife IA, Dourish CT (1993) Silent 5-HT1A receptor antagonists: utility as research tools and therapeutic agents. TIPS 141:441–448

Flynn DD, Mash DC (1993) Distinct kinetic binding properties of N-[^{3}H]-methylscopolamine afford differential labeling and localization of M1, M2, and M3 muscarinic receptor subtypes in primate brain. Synapse 14:283–296

Gallagher BM, Fowler JS, Gutterson NI et al (1978) Metabolic trapping as a principle of radiopharmaceutical design: some factors responsible for the biodistribution of [^{18}F] 2-deoxy-2-fluoro-D-glucose. J Nucl Med 19:1154–1161

Gatley SJ, Ding YS, Brady D et al (1998) In vitro and ex vivo autoradiographic studies of nicotinic acetylcholine receptors using [^{18}F]fluoronochloroepibatidine in rodent and human brain. Nucl Med Biol 25:449–454

Gitler MS, Reba RC, Cohen VI et al (1992) A novel m2-selective muscarinic antagonist: binding characteristics and autoradiographic distribution in rat brain (published erratum appears in Brain Res 594:359). Brain Res 582:253–260

Gitler MS, Cohen VI, De La Cruz R et al (1993) A novel muscarinic receptor ligand which penetrates the blood brain barrier and displays in vivo selectivity for the m2 subtype. Life Sci 53:1743–1751

Gjedde A (1995) Glucose metabolism. In: Wagner H, Szabo Z, Buchanan JW (eds) Principles of nuclear medicine. Saunders, Philadelphia, pp 54–71

Gunn RN, Sargent PA, Bench CJ et al (1998) Tracer kinetic modeling of the 5HT1A receptor ligand [carbonyl-11C] WAY-100635 for PET. Neuroimage 8:426–440

Hevesy G, Paneth F (1938) A manual of radioactivity, 2nd edn. Oxford University Press, London

Horti AG, Scheffel U, Koren AO et al (1998) 2-[^{18}F]Fluoro-A-85380, an in vivo tracer for the nicotinic acetylcholine receptors. Nucl Med Biol 25:599–603

Hume SP, Ashworth S, Opacka-Juffry J et al (1994) Evaluation of [O-methyl-^{3}H]WAY100635 as an in vivo radioligand for 5-HT$_{1A}$ receptor. Eur J Pharmacol 271:515–523

Imperato A, Obinu MC, Gessa GL (1993a) Stimulation of both dopamine D1 and D2 receptors facilitates in vivo acetylcholine release in the hippocampus. Brain Res 618:341–345

Imperato A, Obinu MC, Gessa GL (1993b) Effects of cocaine and amphetamine on acetylcholine release in the hippocampus and caudate nucleus. Eur J Pharmacol 238:377–381

Imperato A, Dazzi L, Obinu MC et al (1994a) The benzodiazepine receptor antagonist flumazenil increases acetylcholine release in rat hippocampus. Brain Res 647:167–171

Imperato A, Obinu MC, Gessa GL (1994b) Does dopamine exert a tonic inhibitory control on the release of striatal acetylcholine in vivo? Eur J Pharmacol 251:271–279

Ito H, Halldin C, Farde L (1999) Localization of 5-HT1A receptors in the living brain using [carbonyl-^{11}C]WAY-100635: PET with anatomic standardization technique. J Nucl Med 40:102–109

Jain RK (1988) Determinants of tumor blood flow. Cancer Res 48:2641–2658

Kamen, MD (1998) Isotopic tracers in biology: an introduction to tracer methodology, 3rd edn. Academic Press, New York

Khawaja X, Evans N, Reilly Y et al (1995) Characterization of the binding of [^{3}H]WAY100635, a novel 5-hydroxytryptamine 1a receptor antagonist, to rat brain. J Neurochem 64:2716–2726

Kiesewetter DO, Lee J, Lang L et al (1995) Preparation of ^{18}F-labeled muscarinic agonist with M2 selectivity. J Med Chem 38:5–8

Kiesewetter DO, Carson RE, Jagoda EM et al (1999) In vivo muscarinic binding of 3-(alkylthio)-3-thiadiazolyl tetrahydropyridines. Synapse 31:29–40

Kilbourn MR, Snyder SE, Sherman PS et al (1996) In vivo studies of acetylcholinesterase activity using a labeled substrate, N-[^{11}C]methylpiperidin-4-yl propionate ([^{11}C]PMP). Synapse 22:123–131

Koenig JA, Edwardson JM (1997) Endocytosis and recycling of G protein-coupled receptors. Trends Pharmacol Sci 18:276–287

Kuhl DE, Koeppe RA, Minoshima S et al (1999) In vivo mapping of cerebral acetylcholinesterase activity in aging and Alzheimer's disease. Neurology 52:691–699

Kung M-P, Zhuang Z-P, Frederick D et al (1994) In vivo binding of [^{123}I] 4-(2'-methoxyphenyl)-1-[2'-(N-2''-pyridinyl)-p-iodobenzamido-]ethylpiperazine, p-MPPI, to 5-HT$_{1A}$ receptors in rat brain. Synapse 18:359–366

Lang L, Jagoda E, Eckelman WC (1999a) New F-18 labeled WAY 100635 derivatives for the 5HT$_{1A}$ receptor. J Nucl Med 40:37P

Lang L, Jagoda E, Sassaman MB et al (1999b) Comparison of F-18 labeled cis and trans 4-fluorocyclohexane derivatives of WAY 100635. J Nucl Med 40:37P

Lang LX, Jagoda E, Schmall B et al (1999c) Development of fluorine-18-labeled 5-HT$_{1A}$ antagonists. J Med Chem 42:1576–1586

Laporte A-M, Lima L, Gozlan H et al (1994) Selective in vivo labeling of brain 5-HT$_{1A}$ receptors by [^{3}H]WAY 100635 in the mouse. Eur J Pharmacol 271:505–514

Lei KJ, Chen H, Pan CJ et al (1996) Glucose-6-phosphatase dependent substrate transport in the glycogen storage disease type-1a mouse. Nat Genet 13:203–209

Li M, Yasuda RP, Wall SJ et al (1991) Distribution of M2 muscarinic receptors in rat-brain using antisera selective for M2 receptors. Mol Pharmacol 40:28–35

Lin B, Morris DW, Chou JY (1998) Hepatocyte nuclear factor 1alpha is an accessory factor required for activation of glucose-6-phosphatase gene transcription by glucocorticoids. DNA Cell Biol 17:967–974

Liu F, Wan Q, Pristupa ZB et al (2000) Direct protein-protein coupling enables cross-talk between dopamine D5 and gamma-aminobutyric acid A receptors. Nature 403:274–280

Mathis CA, Simpson NR, Mahmood K et al (1994) [^{11}C]WAY 100635: a radioligand for imaging 5-HT$_{1A}$ receptors with positron emission tomography. Life Sci 55:403–407

Maziere M (1995) Cholinergic neurotransmission studied in vivo using positron emission tomography of single photon emission computerized tomography. Pharmacol Ther 66:83–101

McPherson DW, Dehaven-Hudkins DL, Callahan AP et al (1993) Synthesis and biodistribution of iodine-125-labeled 1-azabicyclo[2.2.2]oct-3-yl alpha-hydroxy-alpha-(1-

iodo-1-propen-3-yl)-alpha-phenylacetate. A new ligand for the potential imaging of muscarinic receptors by single photon emission computed tomography. J Med Chem 36:848–854

McPherson DW, Lambert CR, Jahn K et al (1995) Resolution and in vitro and initial in vivo evaluation of isomers of iodine-125-labeled 1-azabicyclo[2.2.2]oct-3-yl alpha-hydroxy-alpha-(1-iodo-1-propen-3-yl)-alpha-phenylacetate: a high-affinity ligand for the muscarinic receptor. J Med Chem 38:3908–3917

Morowitz HJ, Singer JL (eds) (1993) The mind, the brain, and complex adaptive systems. Addison-Wesley, Reading

Myers WG (1979) Georg Charles de Hevesy: the father of nuclear medicine. J Nucl Med 20:590–594

Namba H, Irie T, Fukushi K et al (1994) In vivo measurement of acetylcholinesterase activity in the brain with a radioactive acetylcholine analog. Brain Res 667:278–282

Nielsen DA, Virkkunen M, Lappalainen J et al (1998) A tryptophan hydroxylase gene marker for suicidality and alcoholism. Arch Gen Psychiatry 55:593–602

Osman S, Lundkvist C, Pike VW et al (1996) Characterization of the radioactive metabolites of the 5-HT1A receptor radioligand, [O-methyl-^{11}C]WAY100635, in monkey and human plasma by HPLC: comparison behavior of an identified radioactive metabolite with parent radioligand in monkey using PET. Nucl Med Biol 23:627–634

Pan CJ, Lei KJ, Chen H et al (1998) Ontogeny of the murine glucose-6-phosphatase system. Arch Biochem Biophys 358:17–24

Pan CJ, Lin B, Chou JY (1999) Transmembrane topology of human glucose 6-phosphate transporter. J Biol Chem 274:13865–13869

Phelps ME, Huang SC, Hoffman EJ et al (1979) Tomographic measurement of local cerebral glucose metabolic rate in humans with (F-18)2-fluoro-2-deoxy-D-glucose: validation of method. Ann Neurol 6:371–388

Pike VW, McCarron JA, Lammertsma AA et al (1995) First delineation of 5-HT1A receptors in human brain with PET and [^{11}C]WAY-100635. Eur J Pharmacol 283:R1–R3

Pike VW, McCarron JA, Lammertsma AA et al (1996) Exquisite delineation of 5-HT1A receptors in human brain with PET and [carbonyl-^{11}C]WAY-100635. Eur J Pharmacol 301:R5–R7

Quirion R, Aubert I, Labchak PA et al (1989) Muscarinic receptor subtypes in human neurodegenerative disorders; focus on Alzheimer's disease. Trends Pharmacol Sci [Suppl]:80–84

Reivich M, Kuhl D, Wolf A et al (1979) The [^{18}F]fluorodeoxyglucose method for the measurement of local cerebral glucose utilization in man. Circ Res 44:127–137

Rodbard D, Feldman HA (1975) Theory of protein-ligand interaction. Methods Enzymol 36:3–16

Rodriguez-Puertas R, Pascual J, Vilaro T et al (1997) Autoradiographic distribution of M1, M2, M3 and M4 muscarinic receptor subtypes in Alzheimer's disease. Synapse 26:341–350

Sauerberg P, Olesen PH, Nielsen S et al (1992) Novel functional M1 selective muscarinic agonists. Synthesis and structure-activity relationships of 3-(1,2,5-thiadiazoyl)-1,2,5,6-tetrahydro-1-methylpyridines. J Med Chem 35:2274–2283

Saunders PT (1993) The organism as a dynamic systems. Addison-Wesley, Reading, MA

Sawada Y, Hiraga S, Francis B et al (1990) Kinetic analysis of 3-quinuclidinyl 4-[^{125}I]iodobenzilate transport and specific binding to muscarinic acetylcholine receptor in rat brain in vivo: implications for human studies. J Cereb Blood Flow Metab 10:781–807

Schloesser R, Simkowitz P, Bartlett EJ et al (1996) The study of neurotransmitter interactions using positron emission tomography and functional coupling. Clin Neuropharmacol 19:371–389

Skovronsky DM, Zhang B, Kung MP et al (2000) In vivo detection of amyloid plaques in a mouse model of Alzheimer's disease. Proc Natl Acad Sci USA 97:7609–7614

Sokoloff L, Reivich M, Kennedy C et al (1977) The [^{14}C] deoxyglucose method for the measurement of local cerebral glucose utilization: theory, procedure, and normal values in the conscious and anesthetized albino rat. J Neurochem 28:897–916

St. George-Hyslop P (2000) Molecular genetics of Alzheimer's disease. Biol Psychiatry 47:183–199

Tavitian B, Pappata S, Planas AM et al (1993) In vivo visualization of acetylcholinesterase with positron emission tomography. Neuroreport 4:535–538

Taylor S, Budinger T (2000) Prospective on the potential of imaging gene expression. Ernest Orlando Lawrence Berkeley National Laboratory LBNL-44311

Wall SJ, Yasuda RP, Hory F et al (1991) Production of antisera selective for m1 muscarinic receptors using fusion proteins: distribution of m1 receptors in rat brain. Mol Pharmacol 39:643–649

Weinberger DR, Gibson R, Coppola R et al (1991) The distribution of cerebral muscarinic acetylcholine receptors in vivo in patients with dementia. A controlled study with ^{123}IQNB and single photon emission computed tomography. Arch Neurol 48:169–176

Weinberger DR, Jones D, Reba RC et al (1992) A comparison of FDG PET and IQNB SPECT in normal subjects and in patients with dementia. J Neuropsychiatry Clin Neurosci 4:239–248

Woodward JJ, Compton DM, Balster RL et al (1995) In-vitro and in-vivo effects of cocaine and selected local-anesthetics on the dopamine transporter. Eur J Pharmacol 277:7–13

Yalow R, Berson S (1960) Immunoassay of plasma insulin in man. J Clin Invest 39:1157

Yamamura HI, Kuhar MJ, Greenberg D et al (1974) Muscarinic cholinergic receptor binding: regional distribution in monkey brain. Brain Res 66:541–546

CHAPTER 6

Instrumentation for Measuring Metabolism

6

PETER F. SHARP, ANDY WELCH, STEPHEN McCALLUM

Contents

6.1
Introduction

The use of a radioactive label to facilitate the measurement of the in vivo distribution of a pharmaceutical has long been recognised as a powerful analytical technique. The first biological application of a tracer was reported in 1923 by Hevesy (1923), who employed a radioisotope of lead to investigate the metabolism of that element in plants. He showed how it was possible to measure quantitatively the uptake and distribution of the lead in different parts of the plant. Since the detection process is more than a million times more sensitive than chemical and physical methods, only minute quantities of lead were needed and so toxic effects could be avoided. Of course, the application was limited as lead is not a normal constituent of biological systems. One of the earliest studies on man was reported by Blumgart and Yens (1927), who measured the arm-to-arm circulation time using a 100 MBq of radium-C. The beta emissions were detected by a Wilson cloud chamber.

Early progress was severely restricted by the limited choice of radionuclides. In the 1930s, just 2 years after Joliot and Curie had produced the first artificial radionuclides, John Lawrence was performing investigations using ^{131}I produced on his brother Ernest's cyclotron. However, for the technique to become widely available large quantities of radionuclides were required and this was finally rectified as a result of developments which had taken place in connection with the use of atomic energy for military purposes during the Second World War. By 1946, Mitchell was able to list 36 artificially produced radionuclides which appeared to be of interest for biomedical investigations.

Nuclear medicine, the medical imaging technique using radiopharmaceuticals, has proved to be a powerful diagnostic tool, providing images of function rather than anatomy as produced by X-ray based approaches. Nuclear medicine was more suited to digital imaging than X-rays, as the relatively poor spatial resolution did not require fine image digitisation and the low photon densities, typically a factor of 10^5 lower, made fewer demands on the technology. Indeed, by the late 1960s digital nuclear medicine was taking its place as a clinical tool. The earliest studies were dynamic tracer studies which provided relative change in the pharmaceutical distribution as a function of time. By the early 1970s computers had enabled three-dimensional tomographic imaging to be performed, predating Hounsfield's development of the X-ray CT by several years. Known as single photon emission computed tomography (SPECT), it is now a routine technique in nuclear medicine, particularly for studies of the brain.

It is when accurate quantitative information about the radiopharmaceutical distribution is required that the greatest problems arise. The accuracy of quantitative measurements is fundamentally limited by the attenuation of the low energy photons by body tissues making it difficult to relate the density of detected photons to the concentration of the radiopharmaceutical in an organ. This is further exacerbated by the presence of a significant amount of scattered radiation in the images, the poor energy resolution of the scintillation detector predominantly used in nuclear medicine making scatter discrimination difficult, and limited spatial resolution. Imaging using positron emitters, positron emission tomography (PET), can overcome some of the attenuation problems, but compared with single photon imaging it is an expensive technique.

6.2
The Gamma Camera

The effectiveness of radionuclide imaging depended upon the development of sensitive and reliable detectors of gamma radiation. During the 1930s and 40s radiation detectors were based on gas ionisation – the ionisation chamber, the gas proportional counter, and the Geiger-Mueller (GM) counter. Kallman laid the foundations of modern nuclear medicine when he developed the scintillation detector – a scintillation crystal coupled to a photomultiplier – in the early 1940s, as this offered by far superior sensitivity to the GM tube. Early approaches relied on the manual positioning of the detector over the area of interest and the manual plotting of the distribution of the radiopharmaceutical, but automatic imaging became possible with the introduction of the rectilinear scanner by Cassen in 1951 (Cassen et al. 1951). Calcium tungstate crystals fitted with focusing collimators were moved mechanically in a raster pattern over the organ. The image was built up of ink dots on paper using a pen linked to the detector; a high count rate led to an increased dot density. The pen was replaced subsequently with an intensity modulated light source which was used to create an image onto standard X-ray film, or the use of colour coding. Since these systems measured the concentration of radioactivity on a point-wise basis, each image required an acquisition time of several minutes and so the capacity to do dynamic studies was severely limited.

It was the introduction of the gamma camera by Hal Anger in 1958 (Anger 1958) that revolutionised radionuclide imaging (Fig. 6.1). The gamma camera consists of a large area scintillation crystal; almost all cameras still use thallium activated sodium iodide, viewed by an array of between 60 and 90 photomultiplier tubes (PMTs) (Fig. 6.2). Typically the device has a rectangular field of view of 60×50 cm.

The modern camera system now often has two (or more) detector heads, permitting simultaneous views from different angles and, when used in conjunction with the ability of the detectors to rotate around the patient, reduces the time required for tomographic imaging (see Sect. 6.3). By scanning the camera along the length of the patient single view whole body imaging can be performed (Fig. 6.3).

The location of the scintillation is computed by analysing the relative strengths of the signals from the PMTs. In modern cameras this is performed digitally. However, the device suffers from two fundamental problems: the need to use collimation to form the image in the crystal and the small number of light photons available to compute the spatial and energy co-ordinates.

6.2.1
Collimation

The collimator consists of a lead sheet perforated by several thousand holes. In the most widely used design of collimator these holes run parallel and orthogonal to

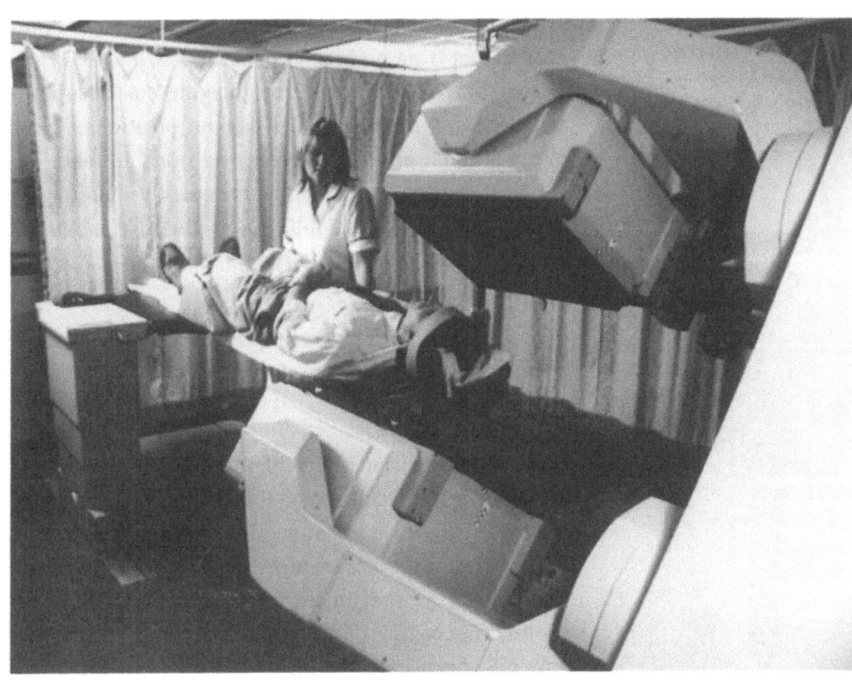

Fig. 6.1. A dual headed Anger gamma camera

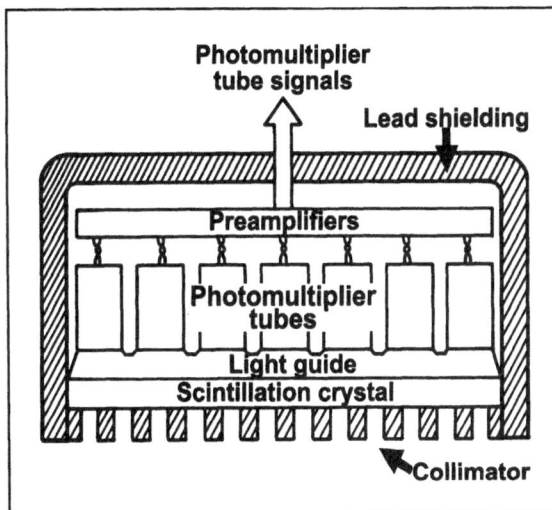

Fig. 6.2. Cross section through the detector head of an Anger gamma camera

the crystal face. An image is produced as only those gamma rays travelling almost parallel to the axis of a hole will be able to hit the scintillation crystal, those at an angle being absorbed by the lead septa between the holes. Thus, image formation is by exclusion of gamma rays rather than by focussing provided by the lens of a photographic camera. The result is that only a very small proportion of the incident photons is actually used to produce the image, typically 0.1%. Thus,

nuclear medicine images are severely photon limited, a situation made worse as the dose of radiopharmaceutical that can be administered to a patient is limited for reasons of radiation safety. A typical nuclear medicine image might contain a total of 10^5 photons.

The geometry of the collimator holes also limits the spatial resolution of images (Anger 1964). Making the acceptance angle of a hole smaller, by decreasing its diameter or increasing its length, improves spatial resolution, but will also decrease sensitivity. Hole shape has little effect on collimator performance (Muehllehner et al. 1976). This trade-off between resolution and sensitivity is unavoidable; the only way in which resolution can be improved is by excluding the more obliquely incident gamma rays.

Hole geometry dictates that resolution will deteriorate with distance from the collimator, while sensitivity is not affected. Thus, the best quality images are obtained with the patient positioned as close as possible to the collimator.

Designs of collimator other than the parallel hole have been used. In the converging collimator the holes converge towards the patient so that gamma rays diverging from the patient are detected and an enlarged image formed in the detector. Such a collimator, with its magnified image, provides the capability of high resolution imaging. It offers the advantage of higher sensitivity than the parallel hole collimator of similar resolution and less rapid degradation of resolution with

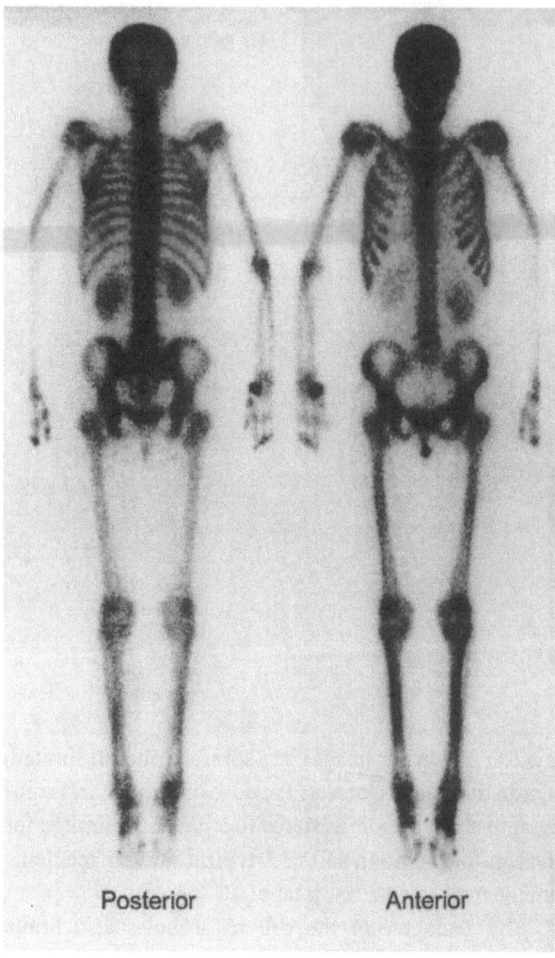

Fig. 6.3. Whole body bone images from a scanning gamma camera

given by the ratio of the distances from detector to pinhole to that from pinhole to patient (Sharp et al. 1985). This is particularly useful for imaging small, superficially located organs, which can be positioned close to the pinhole, and also for imaging small animals. Resolution degrades with depth less quickly than for the parallel hole collimator, but due to the diverging field of view it has less sensitivity at depth.

6.2.2
Detector

Given the poor sensitivity of the collimator, it is essential that the scintillation detector does not introduce a further loss of counts. In addition, the geometrical relationship between the crystal and PMTs must be such as to maximise the transfer of light from scintillation to the photocathodes, so minimising the amount of noise on the output signal from which the location and energy associated with scintillation will be computed. For gamma cameras thallium activated sodium iodide is used universally. Its performance characteristics are given in Table 6.1.

In single photon imaging most studies are performed using gamma rays of energy between 100 and 200 keV. Typically, camera crystals have a thickness of between 3/8 and 5/8 of an inch and will stop over 90% of photons by the photoelectric interaction. However, the conversion efficiency of the crystal is only about 10%, yielding some 5000 light photons for each 150 keV gamma detected. Given that only half will directly hit the PMTs and that these are distributed amongst some tens of tubes, then only about 60 photons hit each tube. With a photocathode conversion efficiency of 25% only about 15 photoelectrons will be produced giving a statistical variability in this signal of about 25%. The consequence of this is poor spatial and energy resolution.

The energy resolution of the scintillation crystal at the energy used in imaging is at best 9%, a value that is now achieved in modern cameras. In order to collect a sufficiently large number of unscattered photons it is necessary to employ an energy window around the photopeak. Typically, this has a value of ±10% of the photopeak energy, i.e. from 126 to 154 keV for ^{99m}Tc. Given that for this 140 keV gamma ray from ^{99m}Tc the Compton edge is 51 keV, then a significant number of photons scattered through quite large angles are going to have energies that will overlap the photopeak window. Approximately 30% of the

depth (Moyer 1974). Its disadvantages are that since magnification varies with depth, it is difficult to estimate size, spatial distortion is produced both by the magnification effect and variation of resolution in a plane due to hole angulation and, finally, sensitivity varies across a plane.

Diverging collimators are used to minify an image, so that a large organ can be imaged within the limited field of view of the crystal (Anger 1958; Muehllehner 1969). These have holes diverging away from the patient. Both efficiency and resolution deteriorate with increasing distance from the collimator face and, as with the converging collimator, vary across a plane.

The final collimator regularly used with the camera is the pinhole collimator. This is similar to the optical pinhole except that the pinhole is made from lead and has a diameter of a few millimetres. An inverted image is formed in the detector with a magnification

Table 6.1. Scintillation crystals

Scintillator	Chemical name	Density (gm/cm^3)	λ_{max} (nm)	Photons (MeV)	Decay time (ns)
Sodium iodide	NaI (Tl)	3.67	415	38,000	230
Bismuth germanate	$Bi_4Ge_3O_{12}$ (BGO)	7.13	480	8,200	300
Barium fluoride	BaF_2	4.89	195, 220	1,800	0.8
Gadolinium orthosilicate	$Gd_2SiO_5(Ce)$ (GSO)	6.70	430	10,000	30–60
Lutetium orthosilicate	$Lu_2SiO_5(Ce)$ (LSO)	7.4	420	30,000	40

photons detected in the photopeak region will have been scattered.

In addition, the noisiness of the PMT signal will lead to random variability in the computed spatial co-ordinates of the scintillation. This intrinsic resolution of the gamma camera is about 3.5 mm for modern cameras.

6.2.3
Calculation of Signal Position

Until the 1990s, the calculation of the spatial co-ordinates of the scintillation position relied on the analogue fixed weighting technique. By weighting the output from each PMT by a value related to the spatial location of the tube, the sum of these signals could be linearly related to signal position (Anger 1958; Barrett and Swindell 1981). However, this approach took no cognisance of the reliability of the signal, spatial resolution. The use of threshold preamplifiers, which eliminated low intensity, and hence noisy, signals from the computation proved to be a solution (Kulberg et al. 1972). A detailed analysis of how resolution and spatial linearity could be jointly optimised (Tanaka et al. 1970; Hiramoto et al. 1971) led to the development of the delay-line arithmetic.

However, analogue arithmetic is intrinsically unreliable and slow. The rate at which successive pulses could be processed and mixed by the position electronics was limited to a few tens of thousands per second. Ultimately, the processing rate was limited by the decay time of the scintillation. With the availability of fast digital processing of signals a variety of correction circuits were introduced to rectify some of the short-comings of the analogue approach. Corrections for spatial

non-linearity (Muehllehner et al. 1980), the variation of energy signal with scintillation position and pulse processing time have all been implemented (Lewellen et al. 1989). Obviously, no correction can be implemented for the random variations in signal, only for those resulting from systematic errors. A variety of proposals have been made for the presence of scatter in the photopeak window (La Fontaine et al. 1986; DeVito et al. 1989; Ogawa et al. 1991; King et al. 1992; Pretorius et al. 1993), but none are satisfactory.

Digital electronic signal location was first proposed in the 1980s, but it was over 10 years later before it appeared in commercial devices. While details of how it is implemented are sparse, the general principle involves the use of a calibration whereby the relative strengths of the signals from each group of PMTs are compared with precalibrated values (Genna et al. 1981). Thus, pulse mixing is avoided. By using only signals from those PMTs closest to the scintillation the effect of the noisiest signals can be reduced. Digital pulse positioning is now the standard on cameras as it achieves an intrinsic resolution comparable with analogue cameras.

6.3
Tomographic Imaging

Tomography refers to the process of reconstructing slices through the object (from the Greek words *temnein*, meaning 'to cut', and *graphia* meaning 'writing'; Webb 1990). The key to tomographic imaging is the acquisition of projection data (e.g. standard planar images) at various positions around the body. In nuclear medicine these data are assumed to be integrals of the activity along lines through the object and corrections are applied to account for errors such as absorption and scatter of the gamma rays. The orientation of the lines is determined by the hardware, i.e. the collimator in SPECT and the location of the detectors in PET.

The main advantages of tomography over planar imaging are: firstly, that it is possible to locate the position of various structures in the image within the body. This localisation can have a direct diagnostic benefit in some circumstances, for example for determining whether a hot lesion is in the bone or the surrounding tissue. Secondly, tomography improves image contrast by "removing" the effect of activity in overlying and underlying tissue, so facilitating the visualisation of lesions that may not be seen on the planar image.

Tomographic images are created by reconstructing the line integral projection data. Virtually all of the methods currently in use for reconstruction of medical images from projections fall into one of two classes; analytic and iterative. The analytic methods are based on those originally derived for use in radioastronomy and electron microscopy (Bracewell 1956; Bracewell and Riddle 1967; DeRosier and Klug 1968), although the problem of reconstructing images from projections was initially addressed much earlier (Radon 1917). These algorithms treat the object and projection data as continuous variables, which are linked via their Fourier transforms and include the filtered backprojection technique, which is the most widely used reconstruction algorithm in commercial SPECT and PET scanners. The big advantage of the filtered backprojection method over the iterative techniques (which will be discussed later) is that it is faster. However, it suffers from two main drawbacks for reconstructing nuclear medicine data. Firstly, the filter used amplifies high frequencies, which results in very noisy images when reconstructing the low-count data typical in nuclear medicine studies. In practice, the noise properties of the algorithm can be improved by modifying the filter with a window function that is "rolled-off" at high frequencies, but this is at the expense of resolution. Secondly, the algorithm can only be applied to a particular geometry, i.e. one in which the projection data are line integrals along sets of parallel lines (although it can be modified to cope with fan-beam data for example). A complete set of parallel projections is required over 180°. These are usually stored in a two-dimensional array (indexed by the radial distance of the projection from the centre of the object and its angle) known as a si-

nogram (see also Sect. 6.5.3). In practice, although a parallel geometry is used, the data are not perfect line integrals. For example, the gamma rays may be absorbed or scattered before being detected and in SPECT the finite size of the collimator holes results in a resolution that falls off with distance.

Iterative algorithms start from the assumption that both the object and the measured data are discrete and that the data are a linear sum of the object elements with weighting factors being used to represent the probability that an event emitted at one location is detected at another. This representation is inherently flexible in that it can be used to represent any geometry and the weights can be modified to incorporate physical effects such as attenuation, scatter and geometric response (see, for example, Bouwens et al. 2001). The aim is to find the object that best fits the measured data assuming some model for the imaging system (i.e. a set of weights). The definition of "best fit" can include some knowledge of the statistical properties of the data leading to a class of algorithms known as maximum likelihood techniques, which include the popular 'expectation maximisation maximum likelihood' (EM-ML) method (Lange and Carson 1984; Shepp and Vardi 1982). Iterative methods such as the EM-ML algorithm have much better noise properties than the filtered backprojection method and are, therefore, becoming more popular for reconstructing nuclear medicine images. However, despite modifications, such as the use of ordered subsets (Hudson and Larkin 1994), which significantly improves the rate of convergence of the algorithm, they are inherently slower. In order to reconstruct an image the data must be projected and backprojected at least once (and usually more than once). With the filtered backprojection algorithm the data are only backprojected once. Since projection and backprojection take a similar amount of time the iterative algorithms take at least twice as long as the filtered backprojection method to produce an image and usually an order of magnitude longer. However, as computational speed has increased, these issues have become less important and the algorithms have gained acceptance for clinical use.

6.3.1
Extension to Three Dimensions

Most of the reconstruction algorithms used in nuclear medicine are two-dimensional algorithms i.e. the object/image is assumed to be two-dimensional and sets of one-dimensional projection data are used to reconstruct the image. These methods are applied to three-dimensional objects by reconstructing a stack of two-dimensional slices with each slice being treated independently of the others. However, in some cases [e.g. cone-beam collimators for SPECT (Jaszczak et al. 1986) or PET scanners without septa] the projection data contain information from multiple transaxial slices so three-dimensional algorithms are required.

As described above, the iterative algorithms are inherently flexible and can easily be extended to cope with three-dimensional data. However, three-dimensional projection and backprojection operations are much slower than two-dimensional ones so the disadvantages of the iterative algorithms in terms of time required to produce an image are more acute than in the two-dimensional case.

The analytic algorithms can be extended to three dimensions. However, in practice very few nuclear medicine scanners acquire sufficient data for a full three-dimensional reconstruction. If a complete set of two-dimensional data is also acquired [as is the case with ring-PET systems (see Sect. 6.5.4.5)], then a sufficient three-dimensional data set can be produced by first reconstructing the two-dimensional data and then projecting through this image to fill in the missing three-dimensional data (Kinahan and Rogers 1989). However, this technique requires at least one three-dimensional projection and backprojection operation so in practice it may not be significantly faster than the iterative methods, which also have the advantage of better noise properties.

An alternative to a full three-dimensional reconstruction is to rebin the three-dimensional data into a two-dimensional data set. This approach is popular in PET where an exact method based on the two-dimensional Fourier transform of the projection data has been developed (Defrise et al. 1997).

6.4
Alternative Imaging Devices

The gamma camera, while now a well developed instrument, has a number of serious shortcomings. A number of attempts have been made to address these, although none have yet been a commercial success. Table 6.2 gives a list of the main limitations of the gamma camera and devices that have been developed to address them.

Table 6.2. Alternative imaging devices

Limitation of the gamma camera	Alternative devices
Need for collimation	Compton effect camera Coded aperture imaging Slat collimator
Poor energy resolution Limited field of view	Semi-conductor camera Multi-wire proportional chambers
Limited count-rate performance	Multi-crystal cameras
Spatial resolution Maneuvrability	Coded aperture imaging Solid state camera Position-sensitive PMT cameras

6.4.1
Coded Aperture Imaging

Collimation produces a one-to-one mapping of the radiopharmaceutical distribution onto the scintillation crystal by only allowing photons travelling in one direction to interact with the crystal; the consequence, however, is poor sensitivity. In a parallel hole collimator the mapping is life size, while other collimator geometries permit magnification or minification of the image. In coded aperture imaging the collimator is replaced with a mask, again consisting of material which is opaque to gamma rays but with a series of apertures in it. The photon source throws a shadow of the mask onto the detector. Provided that the size and position of the shadow is uniquely related to the position of the photon emitter then it will be possible to reconstruct this coded image into a true image.

The fundamental advantage of this approach is that the aperture can have a large proportion of its area transparent to gamma rays – as much as 50% in some systems – and thus offer very high sensitivity. The disadvantage is that the image must be decoded before it is recognisable and this decoding stage can lead to a severe deterioration in the signal-to-noise ratio (SNR) of the image.

The earliest coded aperture to be studied in detail was the Fresnel zone plate (Barrett 1972). It consists of a series of concentric annuli made from alternatively transparent and opaque material, so giving an effective sensitivity of 50%. The gamma rays throw a shadow of the zone plate onto the detector, the position of the ring pattern giving the x and y co-ordinates of the photon source while its magnification depends on the distance of the radiation source from the aperture. Thus, potentially, the system allowed images to be decoded in three dimensions.

Decoding of such images could be done either by digital or analogue means. One property of the zone plate is that it focuses a beam of coherent light; thus the simplest decoding system required the coded image to be reproduced as a transparency and a beam of laser light shone through it onto a photographic film or camera. Simply moving the transparency would bring different imaging planes into focus. Digital decoding is carried out by autocorrelation between the coded image and a digital mask of the zone plate.

While attractive for its simplicity this approach had one major limitation: since noise was spread across the whole of the image, rather than being locally determined by photon density, the SNR was dependent upon the number of resolution areas in the image. For a given SNR, the choice was between having a high spatial resolution, albeit over a limited field of view or a larger image field but at lower spatial resolution (Barrett and De Meester 1974). Also, the conventional zone plate allowed some 50% of incident photons through as a DC component that provided a veiling glare on the coded image. Various forms of the zone plate were tested in order to overcome these problems (Barrett et al. 1972; Crippin 1976), but none were acceptable.

Other types of coded apertures have been used, such as non-redundant pinhole arrays (Chang et al. 1974), pinhole arrays in which the transmission through each pinhole was modulated by shutters (Macovski 1974) and time coded apertures (Koral et al. 1975). Recently, Accorsi and colleagues have looked at the optimal coded aperture pattern (Accorsi et al. 2001a,b). They conclude that these apertures may offer advantages of high resolution gamma camera imaging of small animals, particularly if only a small part of the field of view is used.

6.4.2
Semiconductor Cameras

While the conventional NaI(Tl) scintillation detector is very effective at stopping the low energy gamma rays widely used in nuclear medicine, its poor efficiency for converting gamma rays into photons leads to poor energy resolution and intrinsic resolution.

Semiconductor materials have a small bandgap and therefore only require a few electron volts to form an electron-hole pair compared with 300 eV for a photon with NaI(Tl), thus giving much lower Poisson noise. Unlike the scintillation detector, there is a direct conversion of energy to charge. Thus, detectors based on semiconductors typically have much better energy resolution, of 1%–3%.

Early attempts at constructing a semiconductor camera were hampered by the need to cool the detector during operation and the crystals were expensive to grow and of limited size (Kaufman et al. 1978). However, there are a number of semiconductor materials of a wide bandgap that allow them to act as room temperature gamma ray detectors, of these cadmium telluride (CdTe) and cadmium zinc telluride (CZT) detectors (bandgaps 1.47 and 1.7 eV, respectively) have attracted the most attention. CdTe and CZT have relatively high density (~ 6 g/cm^3) and practically the same mass absorption coefficient as NaI(Tl) giving a good stopping power for gamma rays, but lower photopeak efficiency. Such detectors have achieved an energy resolution of 5–7 keV at 140 keV, when used in a camera configuration (Eisen et al. 1996; Scheiber et al. 1999).

Such detectors also offer the potential advantages of high intrinsic resolution, since detector element sizes of a few millimetres can be achieved, and high countrate capability, as they avoid the pulse processing and scintillation decay time problems encountered with the gamma camera. A number of different detector configurations and read-out schemes have been used. The use of arrays of individual detectors is probably not practical due to the cost of wiring and electronics. The diode matrix (Detko 1969) consists of grooves cut in the n and p layers to produce electrically isolated strips running orthogonally across the detector surface which, unlike a multiple diode array, can be read out by simple electronic circuits. As the pixel size is made smaller, however, there is a reduction in trapping. Separate readouts for each pixel can be achieved using an integrated-circuit chip directly connected to the semiconductor array (Barber et al. 1997).

The main difficulties with semiconductor detectors is achieving high quality, constant resolution at room temperature, and the problem of carrier recombination and trapping which results in incomplete charge collection (Mestais et al. 2001). The presence of zinc in the CZT detector gives it lower leakage currents but makes crystal growing more complex (Fougeres et al. 1999). To achieve a stopping power of at least 75% the crystal has to have a thickness of about

5 mm; at this thickness cadmium telluride has a poor hole mobility compared to its electron mobility. As a consequence pulse height depends on the depth in the crystal at which the gamma ray interacts, so degrading energy resolution (Eisen and Shor 1998; Verger et al. 2001).

A number of CdTe-based gamma cameras have been constructed, although the field of view is still limited to a maximum of about 15×15 cm giving image quality at least as good as that of the modern gamma camera (Eisen et al. 1996; Scheiber et al. 1999; Shor et al. 1999; Scheiber and Giakos 2001). A prototype CZT camera has been developed by Matherson and colleagues with a spatial resolution of about 1.5 mm (Matherson et al. 1998).

6.4.3
The Compton Effect Camera

The camera measures not just the spatial location of the scintillation but also its energy. Since the main scattering process is Compton there is a relationship between the energy lost by a photon and the angle through which it is scattered. In theory this information could be used to calculate the point of emission of a gamma photon (Todd et al. 1974).

Only gamma rays that have not already been scattered in the patient can be used. The process requires the photon to be scattered in the detector system, the energy lost and spatial location to be recorded at the first interaction and the spatial location of the second interaction to be recorded. Knowing the two points of interaction, the trajectory of the photon after its first scattering can be measured. The energy lost in the first interaction can be used to calculate the angle through which the gamma ray had been scattered. This information does not tell us the location of the point of emission of the photon, but the surface of a cone on which that point must lie (Fig. 6.4). By calculating many such conical surfaces their intersection yields the distribution of the photons (for example, Singh and Doria 1983; Herbert et al. 1987).

The main problems are that in order to accurately calculate the scatter angle, it must be possible to accurately measure the energy lost at the first interaction. This means that the first detector must be a semiconductor detector. The location of the interactions must also be known accurately.

Singh (Singh and Brechner 1990) developed a prototype system using a 4×4 Ge detector and an uncol-

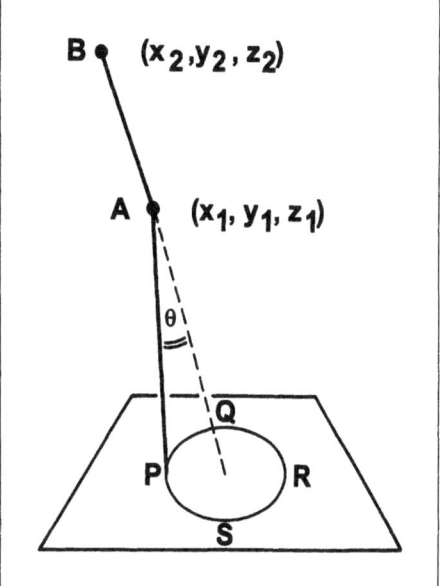

Fig. 6.4. A gamma ray first interacts with a detector at **A** where it is Compton scattered through an angle θ and then stopped in a detector at **B**. The spatial locations of the interactions (x_1, y_1 and x_2, y_2) and the energy lost in the interaction, z, are recorded. From this data the origin of the gamma ray can be computed as on the surface of the cone of semi-angle θ which intersects the plane at PQRS

limated gamma camera as the second detector. Le Blanc and colleagues (1999) have constructed a camera using a single silicon pad, of size $3 \times 3 \times 0.1$ cm^3, pixellated into 1.2×1.2-mm^2 elements, located at the centre of an annular ring of 1/2 inch thick NaI(Tl) detectors viewed by PMTs. They envisage their system being used to image ^{113m}In, which with a 391 keV gamma has the advantage of undergoing less scatter in the patient.

6.4.4
Slat Collimators

Slat collimators consist of a series of parallel slats made from material that is highly attenuating to gamma rays, separated by narrow gaps of transparent material (Keyes 1975). They limit the angle of acceptance of gamma rays in one direction but not in the other and so behave like a one-dimensional version of the parallel hole collimator. The acquired image will be a one-dimensional planar projection of a two-dimensional image. To produce a two-dimensional image the collimator must be rotated around the axis perpendicular to the gamma camera crystal, thus collecting a series a one-dimensional projections at different angles. The resulting data set is similar to that acquired in SPECT, except in one rather than two dimensions, so to form the two-dimensional image a standard tomographic reconstruction algorithm can be used.

A number of designs have been proposed (Keyes 1975; Gindi et al. 1982; Lodge et al. 1995). While the geometric efficiency is about seven times higher than that of a parallel hole collimator of a similar spatial resolution (Webb et al. 1992, 1993), the noise introduced in the reconstruction process leads to a similar SNR. However, using slat collimators as part of a SPECT system with both the collimator and detector rotating is reported to give an improvement of SNR by a factor of nearly two.

One of the problems of the slat collimator is that its response is not spatially invariant across the field of view, and any irregularity in slat spacing leads to image distortion. Efficiency also decreases with distance from the collimator.

6.4.5
Small Field of View Cameras

For a number of applications, in particular scintimammography, cardiac imaging, and small animal imaging, there is no requirement for the large field of view found in conventional cameras. Indeed, it can be a disadvantage as it prevents the camera from being positioned close to the area of interest. While a simple scaling down of the camera is one solution, it has opened the possibility of other designs of cameras. To address this, two camera designs in particular have been explored, those using position sensitive PMTs (PSPMT) and those using photodiodes.

6.4.5.1
Position-sensitive Photomultiplier Tube Camera

As in the conventional camera, a scintillation crystal is optically coupled to a PMT. However, in this case the photons produce electrons from the photocathode that in turn liberate electrons from mesh dynodes, so forming an electron cloud. The location of the centroid of the resulting charge distribution corresponds to the position of the photoelectrons and hence the gamma ray (Pani et al. 1998). The centroid is determined either by resistive chains connected to a cross-wire anode or by measuring the anode output wire by wire (Williams et al. 2000).

The field of view depends upon the size of PMT used and varies between 60×60 and 110×110 mm. Both single NaI(Tl) crystal (Kim et al. 2000) and multicrystal arrays CsI(Na) crystals (Williams et al. 2000) have been used. Using detectors with a multicrystal array gives a better intrinsic resolution than a single crystal design (Pani et al. 1997). Typical values for system resolution are 2–3 mm at the collimator face, but energy resolution is worse (10–30%), in part due to the non-uniform response of the PMT photocathode.

6.4.5.2
Solid State Cameras

The replacement of the PMT by a solid state device leads to a much smaller detector configuration. One commercial device uses a multicrystal CsI(Tl) array of 64×64 crystals each 3×3 mm, with each crystal being optically coupled to a silicon photodiode. CsI(Tl) has a higher density and atomic number than NaI(Tl), thus giving a better stopping power. The system has an intrinsic resolution of 3 mm, similar to that of a conventional camera, and has a high count-rate capability as spatial information is derived from an array of crystal detectors rather than the pulse mixing of the conventional gamma camera. However, CsI(Tl) has a poorer energy resolution than NaI(Tl), only 13% (Narita et al. 2001).

6.4.6
Multiwire Proportional Chamber

Multiwire proportional chamber (MWPC) devices are essentially gas-filled ionisation chambers in which orthogonal arrays of wire planes are used to calculate the spatial location of the event. Their main advantage is that they can be manufactured as large area detectors at relatively low cost; however, this is offset by the extremely poor sensitivity of the gas versus solid detector. Those systems currently being developed are aimed at PET imaging using pairs of large area chambers. While the requirement to detect 511-keV photons exacerbates the problem of the poor stopping power, their large area gives a more efficient geometry for intercepting the pairs of gammas.

Chepel and colleagues (1997) are developing liquid xenon filled detectors. The light from the scintillation is viewed by PMTs to generate a coincidence signal while the MWPC provides high resolution spatial information, of the order of 1 mm. Ott and co-workers (Ott 1997; Visvikis et al. 1997) have used a MWPC filled with a photosensitive vapour, tetrakis(dimethylamino)ethylene (TMAE). An array of barium fluoride scintillation crystals in front of the MWPC stop the gammas, so producing the light that generates the electrons in the chamber. A detection efficiency of 24% for a single detector has been reported for a 40 cm field of view and with a spatial resolution of 6 mm. Whether such systems have a clinical role is not yet clear.

6.5
Positron Emission Tomography (PET)

6.5.1
Introduction

When a positron is emitted from within the body it travels a short distance (of the order of 1 mm) before combining with an electron. When the positron and electron combine they are both destroyed and two gamma rays are produced. In order to satisfy conservation of energy and momentum such an annihilation event results in the emission of two gamma rays, each with an energy of 511 keV, which are emitted at virtually 180° to each other. It is the detection of these two gamma rays that forms the basis of positron emission tomography (PET). If two 511-keV gamma rays are detected at the same time (i.e. in coincidence), then it is a reasonable assumption that an annihilation event has taken place somewhere along the line that connects the two detection locations. This line is known as the line of response (LOR). Additionally, the total number of coincident events acquired by a pair of detector elements in a fixed period of time gives an estimate of the integral of the activity along the LOR that joins the two detectors. By measuring the integral of the activity along sets of LORs we can acquire projection data that can be reconstructed to produce an image of the distribution of the positron emitting radionuclide within the object.

6.5.2
PET Hardware

The technical requirements for the PET imager are more stringent than for the conventional gamma camera as coincidence events must be measured over a wide field of view at an extremely fast rate. Modern dedicated PET scanners consist of a large number of scintillation detectors arranged into multiple rings surrounding the patient. With this arrangement multiple image slices or 3D volume data can be acquired without moving the patient. As with a gamma camera, scintillation light is detected by PMTs. However, unlike a gamma camera which uses one large crystal, dedicated PET systems use an array of many thousands of small crystals. The technical challenge is to detect not only where a scintillation event took place but also when that event took place and how much energy was deposited.

Figure 6.5 shows a schematic of the full PET detection system. Starting at the left hand side, individual scintillation crystals are grouped into square blocks or larger plate arrays. These arrays are viewed by a smaller number of PMTs whose output signals are amplified and fed individually to pulse shaping circuits and summed together to form the input of the constant fraction discriminator (CFD). The four signals from the pulse shapers are processed by the position and energy processor (PEP) which calculates the crystal of interaction and the total energy of the event. The output signal from the CFD is used by the PEP to form an accurate timing marker. The output from the PEPs is fed into the coincidence processors which determine whether a coincidence has taken place within a predefined coincidence time window.

6.5.2.1
Scintillation Crystal

The requirements of the scintillation detector for the PET system are similar to those for the gamma camera. In this case, however, the crystals must be effective at stopping 511-keV gamma rays and the rise and fall time of the light pulse must be fast to ensure good coincidence timing.

Table 6.1 summarises the properties of the principal types of crystals that have been used for PET (Melcher 2000). Sodium iodide with thallium doping NaI(Tl) has the best light output but the worst stopping efficiency. Bismuth germanate (BGO) is the most available dense material but has the lowest light output and the longest decay time. BaF_2 is the fastest but has very low density and efficiency. Lutetium orthosilicate (LSO) is nearly ideal but is expensive and is not yet readily available. Gadolinium orthosilicate (GSO) is good but is only available in small crystals.

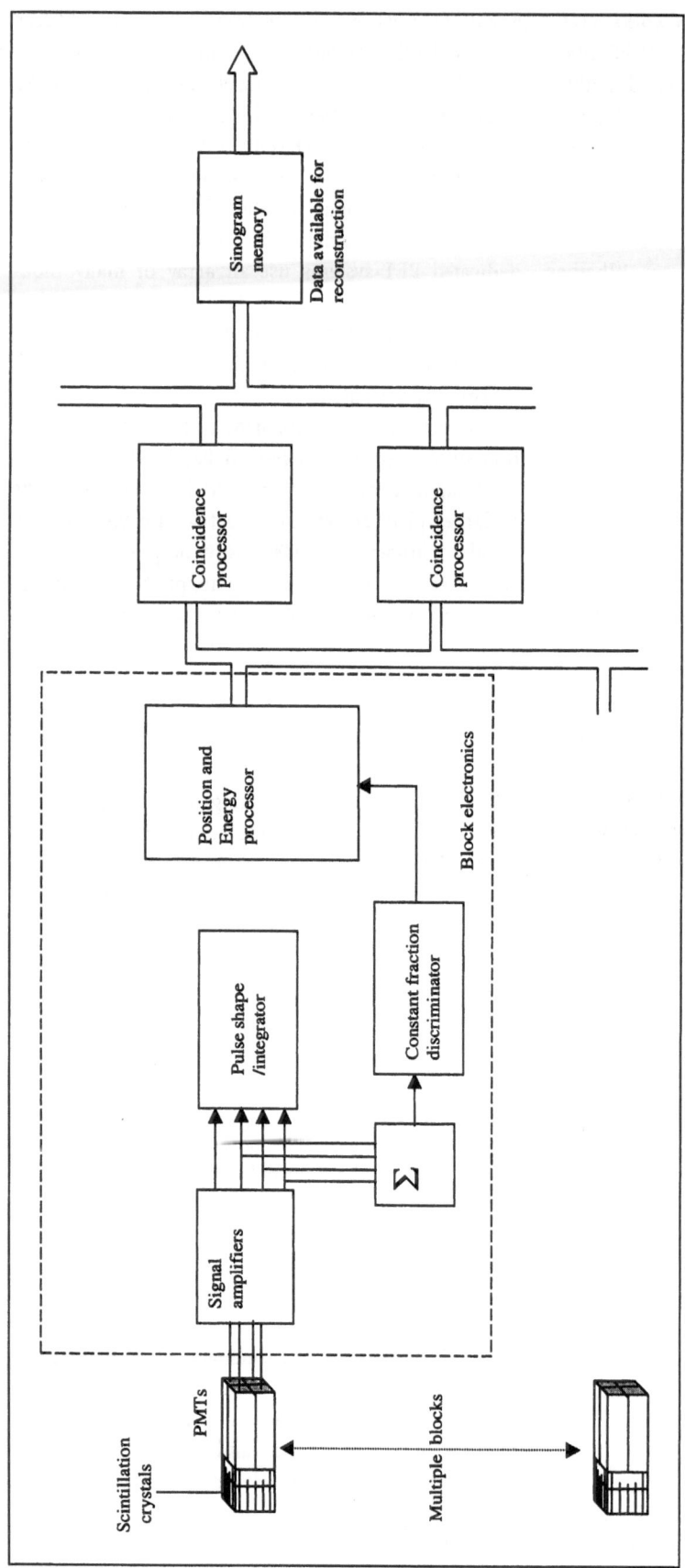

Fig. 6.5. Simplified block diagram of PET imager detection system

BGO has been the crystal of choice in most commercial scanners, chosen for its high efficiency. For the same thickness of crystal BGO is 1.6 times more efficient than NaI(Tl) for single photon events; coincidence events require two detectors and coincidence efficiency is the square of the individual efficiencies. So for PET BGO is 2.6 times more efficient than NaI(Tl).

Energy resolution is important for rejection of scattered events. The processing electronics of a PET imager can impose an energy window. Any events with an energy lying within this window are considered to have originated from a 511-keV interaction and those outside the window are considered as having been scattered.

6.5.2.2
Detector Module

Ideally, a PET imager would have a large array of very small crystals each connected to a PMT arranged in a ring around the patient. Unfortunately due to technical and cost limitations this is not currently possible, modern scanners having well over 9000 detector crystals. Instead a number of techniques have been adopted using a smaller number of PMTs to monitor a number of scintillation crystals.

One manufacturer uses an 8×8 array of BGO coupled to four photomultiplier tubes (Casey and Nutt 1986; see Fig. 6.6). The array is made from a single block of crystal that is cut to separate the individual crystals. As shown in Fig. 6.6a the cuts, or slots, do not run the whole length of the crystal but get progressively deeper towards the edge. This has the effect of differentially distributing the light to the four PMTs to produce a unique signature for an interaction in each of the crystals. The length of the slots increases further away from the centre of the crystal array, with the effect of progressively concentrating the light produced by a single crystal into predominantly one PMT. The crystal of interaction is decoded by comparing the signal intensity from all four PMTs.

With another manufacturer each detector block consists of a 6×6 matrix of discrete crystal elements which are viewed by two dual PMTs. Various coupling compounds and variable degrees of crystal surface roughness are used on the individual crystals of each block detector. Hence each one of the crystal elements has a unique PMT signal signature.

An alternative to the block detector is the plate detector (Wong et al. 1999). Here a large array of small crystals (>1800 in the full ring) are viewed by a much smaller number of PMTs (as few as 280). Between the crystals and the PMTs is a light guide that distributes the light between tubes. Scintillation events in individual crystals can be located by comparing the signal intensities in PMTs within the vicinity of the event, in a similar way to the Anger logic of a gamma camera.

6.5.2.3
Detection Electronics

The electronics connected to the PMTs must decode the position, the energy and the time at which the scintillation occurred. Figures 6.6b and 6.7 show how these functions are performed in a block detector with four PMTs. The signals from the four tubes are amplified before being fed into an integrator. The integrator "cleans up" the signals from the PMTs. After integration the four signals are converted via fast A to D converters. The ratios of the four values can then be calculated digitally to determine the crystal of interaction. The sum of the four integrator outputs is proportional to the energy deposited by the event.

Accurately timing when the scintillation occurred is vital for coincidence detection. The constant fraction discriminator (CFD) activates each time a pulse of light of sufficient amplitude is detected by the PMTs. In the case of the block detector shown in Fig. 6.7 the input to the circuit is the sum of the four PMT outputs and is taken before the integration stage to ensure maximum speed. The CFD technique is used to reduce the effect of amplitude and noise induced inaccuracies on the timing signal. The CFD timing signal is used by processing electronics to tag each event with a time accurate to a few nanoseconds.

6.5.2.4
Coincidence Electronics

Coincidence events must be processed extremely rapidly and with high accuracy. To form coincidence detector pairs, the data from each detector module is compared with the detector module directly opposite and a number of adjacent modules either side (see Sect. 6.5.2.5). The system imposes a coincidence window of typically 12 ns, i.e. two events are only considered in coincidence if they occur within 12 ns of each other. As coincidence events are detected they are sorted and stored into a sinogram.

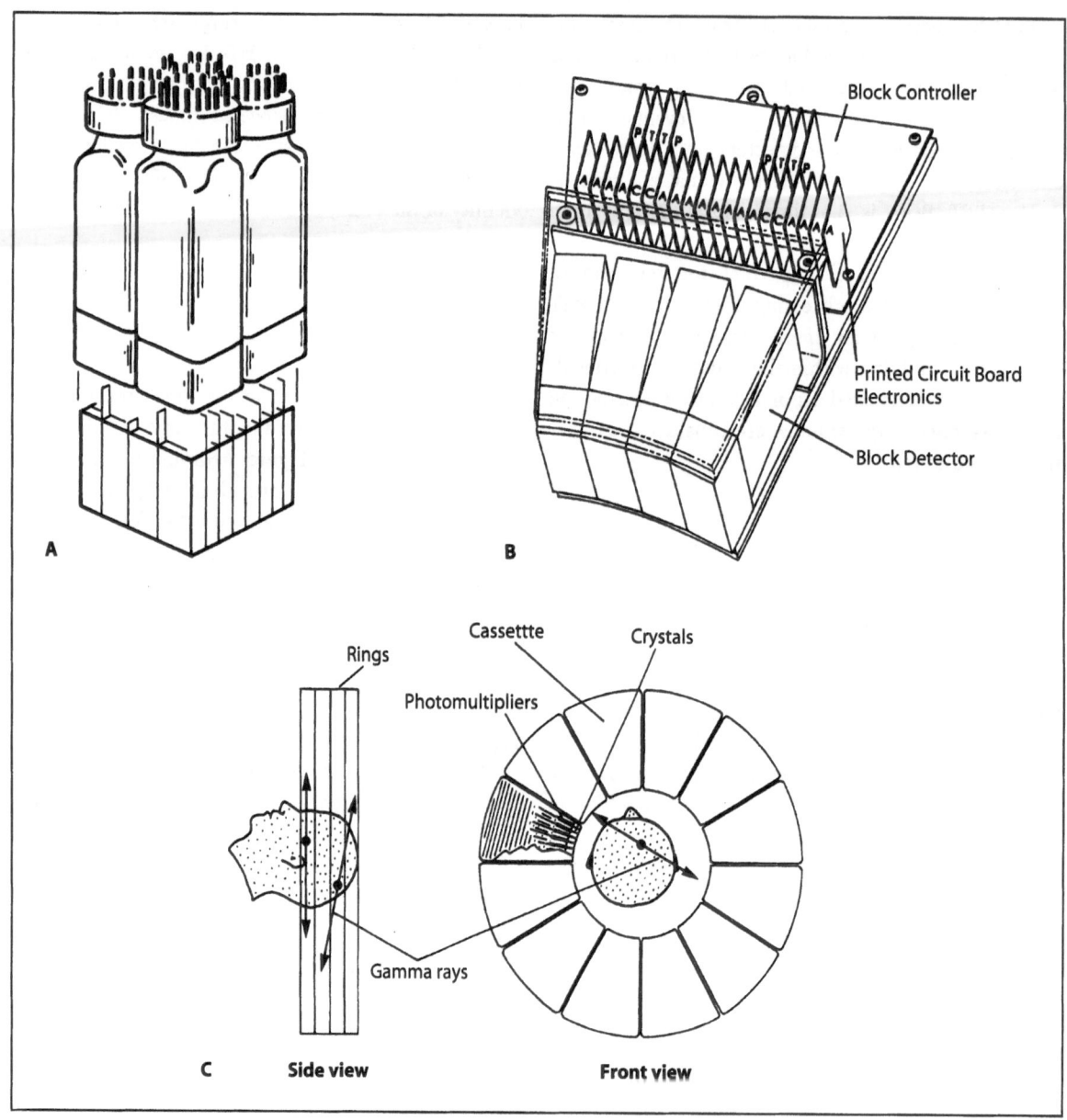

Fig. 6.6. A Block BGO detector (note that in this version the block detector was a 4×8 array of crystal elements). **B** A group of blocks is arranged into a bucket detector sharing electronics. **C** The buckets are arranged into a ring of detectors around the patient

6.5.2.5
Detector Organisation

Figure 6.8 shows a cross section through a ring that is eight detectors wide. Direct image planes are formed between opposing rings. To increase efficiency indirect planes are defined by events occurring in crystal pairs offset from each other by one crystal, known as a ring difference of one. Thus a further seven planes are defined. Efficiency can be increased further by including coincidence between detectors with ring differences of two and three (Fig. 6.8 c). In this case these events are stored in the existing 15 planes. The ring difference will vary depending on the axial position of the plane in the rings. Planes at the end of the rings can only have a ring difference of one and hence the efficiency of the outer planes is low. Towards the centre of the axial field of view the ring difference can increase, improving the quality of the images.

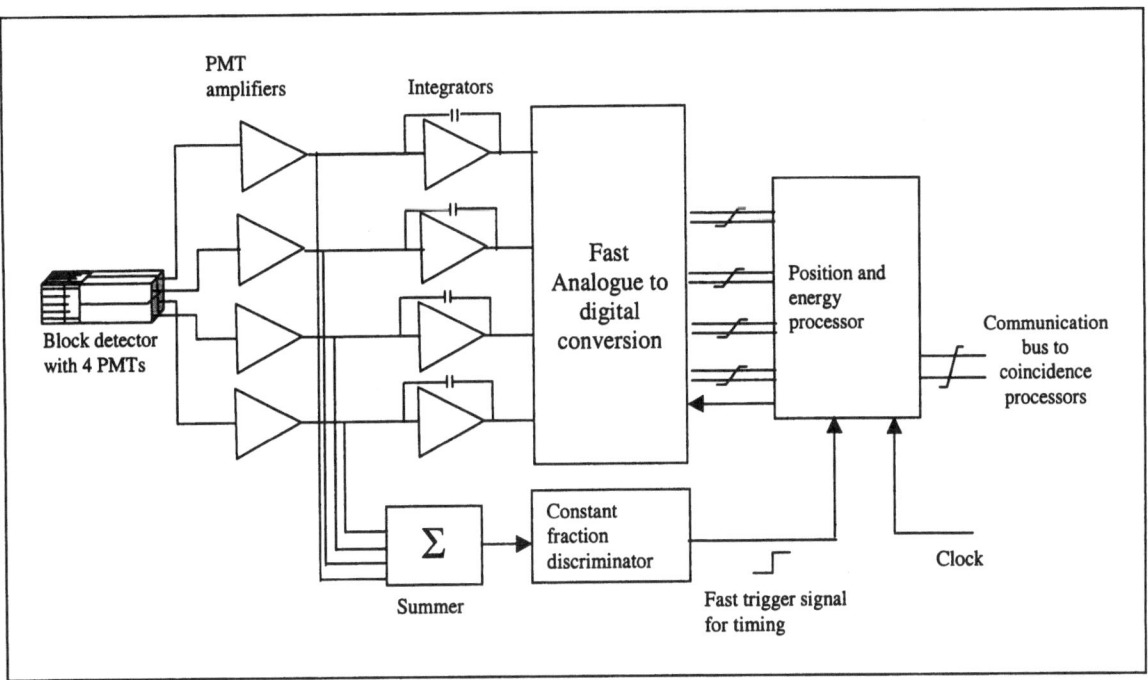

Fig. 6.7. Position and energy electronics

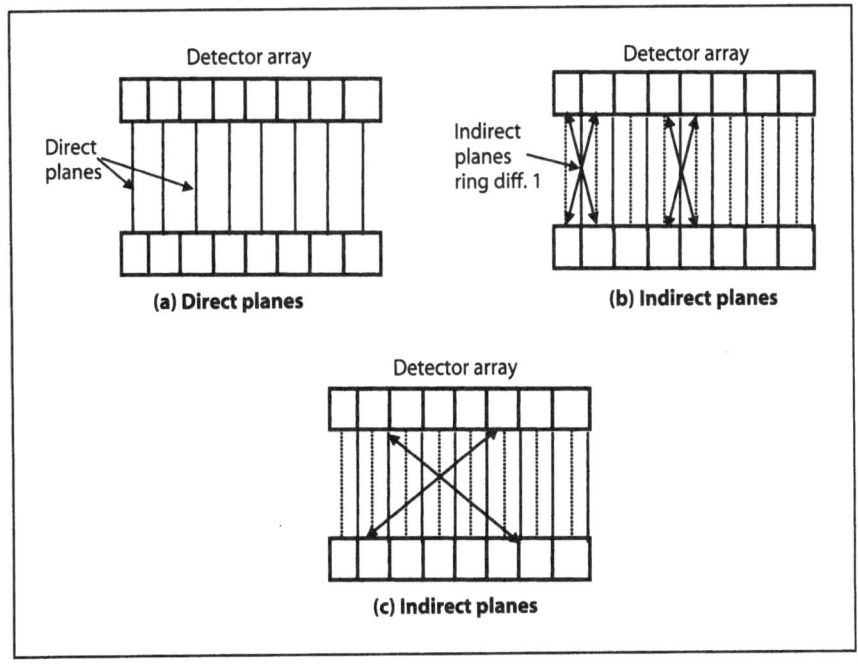

Fig. 6.8. Imager planes

PET systems designed for whole body imaging typically have a field of view of 60 cm in the radial direction. To allow room for transmission sources and septa (see Sects. 6.5.4.2 and 6.5.4.5) the diameter of the imager ring is typically 80 cm. As the field of view is limited to the central region of the detector ring then the number of opposing detectors that any one detector can be in coincidence with is reduced.

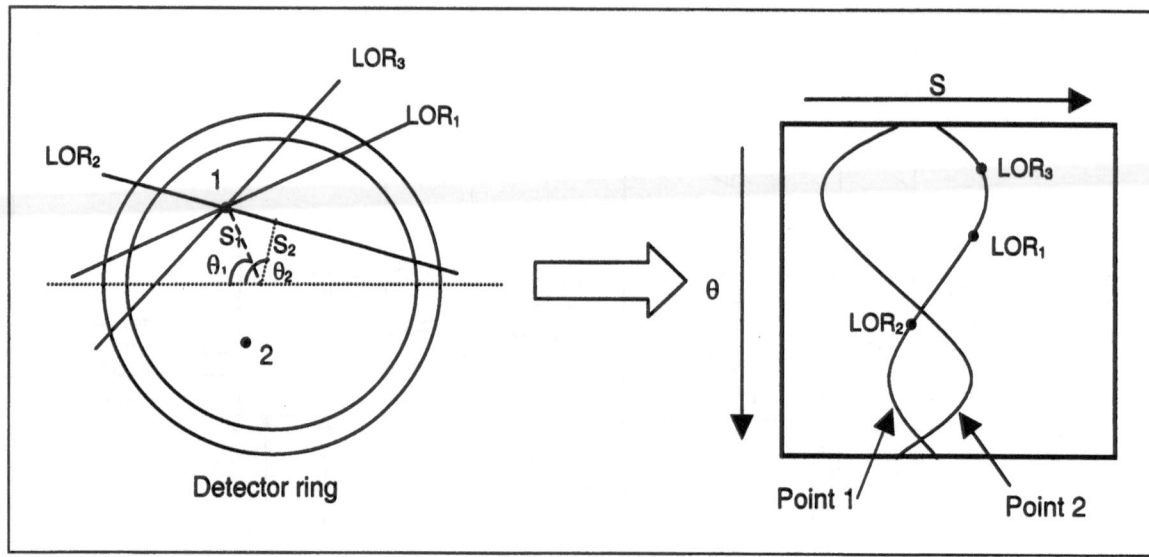

Fig. 6.9. Sinogram formation

6.5.3
Sinogram Formation

Raw data from the scanner are processed into a data format known as a sinogram. Each LOR is uniquely identified by the angle (θ) and radius (s) of a perpendicular line back to the centre of the field of view (Fig. 6.9a) and maps to a single point in the sinogram. The reason for the name "sinogram" becomes clear if we consider all possible LORs passing through a single point and map these to the sinogram. As shown in Fig. 6.9b the resultant trace has a sinusoidal form.

6.5.4
Corrections

6.5.4.1
Normalisation/Uniformity Correction

There are a number of non-uniformities inherent to a PET scanner which have to be corrected. These arise from the geometry of the detector arrays or variations in efficiency between detectors.

Great care is taken in the manufacture of detector crystals but inevitably in a system containing thousands of crystals their efficiency will vary. Factors such as PMT gain, temperature and variations in the electronic circuitry will also have an effect. Crystal efficiency can be measured using a uniform phantom, rod sources or plane source. For each individual de-

tector it is assumed that all LORs will have approximately the same coincidence photon flux and that fan sums do not vary greatly from the average fan sum. The fan sum is the sum of all the LOR between the crystal in question and the opposing crystals.

The need for geometric profile correction or arc correction arises because in order to build up the sinogram, data from a radial array of detectors is projected onto a linear data set. At the centre of the flat array the effective width of the detectors increases and the effective detector density decreases. The effect is to reduce the count rate towards the centre of each projection.

The final normalisation correction is known as crystal interference. This again is due to the geometric problems of mapping radial data acquired from a ring of detectors, onto a square image array. Photons obliquely incident on a crystal not only have a lower probability of being stopped by the crystal, but may also interact with an adjacent crystal. The combined effect of these two factors is measured during the normalisation process after the crystal efficiency and the geometric arc correction have been performed.

6.5.4.2
Attenuation Correction

One of the main advantages of PET over SPECT is the ability to correct for attenuation in the body (Huang et al. 1979). The attenuation distribution can be measured with relative ease by performing a transmission scan using a source which is rotated around the imager field of view with, in one case, the patient present (the transmission scan) and, in the second case, without the patient (the blank scan) (Huesman et al. 1988). These images are formed in a similar fashion to an X-ray CT scan. The attenuation factors are then calculated by taking the ratio of the blank and transmission scans.

In the majority of scanners ^{68}Ge, which has a 271 day half-life and decays into a short-lived positron emitter, is used in the source. These transmission sources are formed into rods, which are housed within lead enclosures on the imager gantry. Recent trends have seen the replacement of the positron source with a rod or point source of a single photon emitter usually ^{137}Cs. ^{137}Cs has a half-life of 30 years and emits a monoenergetic gamma ray at 662 keV. The ^{137}Cs source is placed in the field of view and collimated so only detectors opposite the source are illuminated. Single sources have a number of advantages over positron sources in regard to cost, half-life and efficiency (Dekemp and Nahmias 1994); however, as the gamma emission is at 662 keV rather than 511 keV, small corrections have to be made to the values of the attenuation.

6.5.4.3
Scatter and Scatter Correction

Scattered coincidence events occur when one of the photons released from a positron-electron annihilation is deflected due to Compton scattering. The energy window imposed on the detector will help reduce the effect of scatter; however, problems still exist, particularly when imaging in the 3D imaging mode (see Sect. 6.5.4.5). A number of techniques have been developed to correct for this problem including:

- Using septa between image planes (see below).
- Multiple energy windows (Grootoonk et al. 1996). This method imposes a second lower energy window and assumes that the number of scatters measured in the lower energy window is proportional to those measured in the upper energy window.
- Fitting to data outside object (Cherry et al. 1993). Here it is assumed that the object being imaged can be identified and confined to a region on the sinogram. Signal arising from outside this region is assumed to be scatter and a curve is fitted to

the scatter distribution based on the regions that contain only scattered events.
- Convolution methods (Bailey and Meikle 1994). This method uses a point source to measure a scatter function which is convolved with the image to arrive at a scatter distribution. The distribution is scaled and subtracted from the image.
- Calculation/simulation (Ollinger 1996). A calculation or Monte Carlo simulation is used to model the scatter using the emission and attenuation image as the input data.

6.5.4.4
Randoms and Random Correction

Random coincidence events occur when two photons arising from different positron decays hit the detector ring within the timing window (Hoffman et al. 1981). The conventional way to correct for these events is to employ a second coincidence window with a delayed input. The first coincidence unit records true plus random events. The second coincidence unit can only measure random events because of the short delay introduced to one input. If many events are recorded then statistically the number of randoms recorded by the delayed coincidence window will be the same as those recorded in the normal coincidence window. Random correction is made by subtracting the counts measured in the delayed window from those measured by the trues window.

6.5.4.5
Septa, 2D and 3D Mode

Most modern scanners can operate in either 2D or 3D mode (Michel et al. 1991). In the 2D mode a tungsten septa (collimator) is moved into position between each of the crystal rings. The septa helps to reduce the number of scatter and random events by limiting the acceptance angle of each ring (see Fig. 6.10). However with the septa in place many coincidences are not counted and much information is lost. In the 3D mode the septa are removed; this increases the sensitivity of the system by a factor of about three (Bendriem and Townsend 1998) and data can now be reconstructed as a fully 3D data set. The disadvantages of 3D mode include higher random counts, increased by a factor of two or three, an increased scatter fraction, detector dead time problems at high count rates and increased data processing

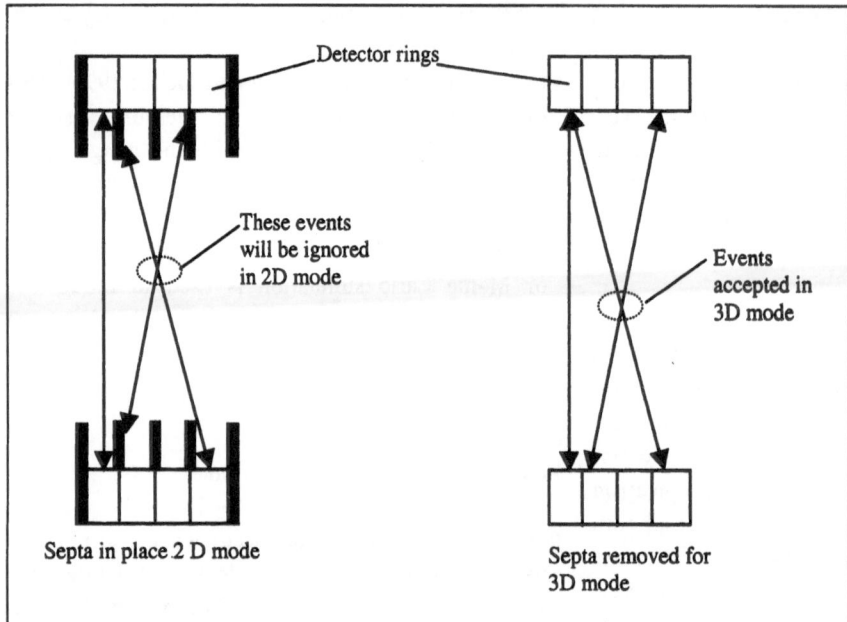

Fig. 6.10. Septa in the 2D and 3D mode, showing cross section of scanner ring (only four rings are shown for simplicity)

time. The 3D mode is often used for brain imaging where there is much less scatter present.

6.5.4.6
Dead Time Correction

The electronics and scintillation detectors take a short time to process each event, the so-called dead time. For block detectors this dead time is approximately 1 μs. As with the gamma camera, there is a maximum count rate a PET system can handle depending on the constraints of the detectors and speed of the processing electronics. If the activity increases then the detector response will begin to fall off due to a pile up of pulses in the detector. A correction is made by the system to compensate for the reduction in count rate until the peak count rate is reached.

6.5.4.7
Decay Correction

Decay correction is applied to account for the radio-nuclide decay during a scanning cycle. Some imaging protocols demand that the patient stays on the imager couch for up to 2 h, during which time a significant amount of radioactive decay can occur. Decay correction simply scales the intensity of each image frame to a reference frame.

6.5.5
Dual Headed Gamma Camera PET

An alternative to a dedicated full PET system is to use a dual headed gamma camera specially modified for PET. The attractions of using gamma cameras for coincidence imaging are mainly economic, but also the fact that they can still be used as conventional cameras makes such systems very versatile. However, such versatility inevitably leads to compromise and the PET performance of these systems has been poor compared to dedicated ring PET systems (Yutani et al. 1999). Increased demand for these systems has meant that manufacturers have been motivated to develop and improve performance (Lewellen et al. 1999; Vandenberghe et al. 2001). State of the art cameras now include transmission sources and have improved detection electronics that can handle multiple events simultaneously.

6.6
Quantitation

Essential to the development of molecular medicine is the ability to measure the behaviour of the radiopharmaceutical in the body. An understanding of the biokinetics of the tracer will depend upon the complexity of the tracer molecule. In this PET usually has an advantage over single photon imaging as the radionuclides are more biologically compatible. A detailed description of modelling and its problems can be found in, for example, Peters and Myers (1998).

Fundamental to studying biokinetics is the ability to accurately quantify the amount of radiopharmaceutical, i.e. the activity per unit mass of tissue in an organ or region of interest. The main problem in relating the detected count rate with the amount of radiotracer present is attenuation of the radiation. To address this problem we must be able to correct both for the absorption of radiation in the patient's tissue and the contamination of image count density by scattered radiation.

As mentioned in Sect. 6.5.1, the fact that PET imaging uses two back-to-back photons means that attenuation is dependent on the total pathway travelled by the two photons, i.e. the thickness of the patient, not on the depth of the organ. It also depends on the effective attenuation coefficient of the material through which the radiation has passed. The acquisition of a transmission scan, as well as the emission scan, allows these parameters to be estimated (see Sect. 6.5.4.2).

A similar approach has been attempted with single photon imaging. Conjugate counting techniques have been used in both planar and SPECT. The geometric mean of opposite (conjugate) views is largely dependent on body thickness, weakly dependent on source thickness, and independent of source depth (Graham and Neil 1974). The arithmetic mean has also been applied. In SPECT projection data can be corrected after reconstruction using the Chang technique (Chang 1978). This approach works well for both uniform and non-uniform attenuation distributions. SPECT reconstruction in situations where there is non-uniform attenuation can also be tackled using the iterative reconstruction techniques (see Sect. 6.3).

While the iterative reconstruction methods can, in principle, include a scatter correction, most methods do not consider scatter. Given the poor energy resolution of the gamma camera, scatter cannot be ignored if accurate quantitation is to be achieved. While there are a number of approaches that have been taken to address this issue (see Sect. 6.2.1.3), in practice none have produced satisfactory results.

In recent years SPECT has emulated PET by using transmission imaging to compensate for attenuation effects. This has been achieved by irradiating the patient using, principally, a scanning line source, although other configurations are used (Hendel et al.

2002). Most sources are filled with ^{153}Gd which has an emission at 100 keV, close to that of ^{99m}Tc, but its long half-life avoids the need to refill the line source frequently. Recently a system using a rotating X-ray tube has also been introduced. The high photon flux gives good quality attenuation maps and also co-registered anatomical and functional images. In the context of cardiac imaging, which has proved to be the main application to date, the value of transmission corrected imaging in improving diagnostic accuracy has yet to be fully defined (Hendel et al. 2002) and while its value in quantitation is encouraging, further work is still required (Tamaki et al. 2001).

One final consideration is that the accuracy of quantitation will also be affected by the spatial resolution of the imaging system. In tomographic imaging this blurring of the image leads to the partial volume effect, the effect being complicated as the reconstructed resolution is both non-stationary and non-isotropic. It makes quantification unreliable for objects smaller than about three to four times the system resolution (Rosenthal et al. 1995).

Unfortunately, given the above issues it is still not possible to answer the question of just how accurately we can measure the absolute concentration of a radiopharmaceutical in tissue. A number of recent papers have tried to address it for specific clinical problems, for example Gonzalez et al. (2001), Blake et al. (2001) and Kalki et al. (1997) and a good discussion of the issues in PET quantitation is given by Bailey et al. (1998). At present quantitative imaging has proved to be a valuable tool in PET, but still remains a desirable but elusive aim for SPECT.

6.7
References

Accorsi R, Gasparini F, Lanza RC (2001a) Optimal coded aperture patterns for improved SNR in nuclear medicine imaging. Nucl Instr Methods Phys Res A 474:273–284

Accorsi R, Gasparini F, Lanza RC (2001b) A coded-aperture for high-resolution nuclear medicine planar imaging with a conventional Anger camera: experimental results. IEEE Trans Nucl Sci 48:2411–2417

Anger HO (1958) Scintillation camera. Rev Sci Instr 29:27–33

Anger HO (1964) Scintillation camera with multichannel collimators. J Nucl Med 5:515–531

Bailey DL, Meikle SR (1994) A convolution-subtraction scatter correction method for 3D PET. Phys Med Biol 39:411–424

Bailey DL, Gilardi M-C, Grootoonk S, Kinahan PE, Nahmias C, Ollinger J, Townsend DW, Trebossen R, Zito M (1998) Quantitative procedures in 3D PET. In: Bendriem B, Townsend D (eds) The theory and practice of 3D PET. Kluwer, Dordrecht, pp 55–109

Barber HH, Apotovsky BA, Augustine FL, Barrett HH, Dereniak EL, Doty FP, Eskin JD, Hamilton WJ, Marks DG, Matherson KJ, Venzon JE, Woolfenden JM, Young ET (1997) Semiconductor pixel detectors for gamma ray imaging in nuclear medicine. Nucl Instr Methods Phys Res A 395:421–428

Barrett HH (1972) J Fresnel zone plate imaging in nuclear medicine. J Nucl Med 13:382–385

Barrett HH, De Meester GD (1974) Quantum noise in Fresnel zone plate imaging. Appl Opt 13:1100–1109

Barrett HH, Swindell W (1981) Radiological imaging. The theory of image formation, detection and processing, vol 1. Academic, New York, pp 262–268

Barrett HH, De Meester GD, Wilson DT, Farmelant MH (1972) Recent advances in Fresnel zone plate imaging. In: Medical radioisotope scintigraphy, vol 1. IAEA, Vienna, pp 269–281

Bendriem B, Townsend D (1998) The theory and practice of 3D PET. Kluwer, Dordrecht

Blake GM, Park-Holohan SJ, Cook GJR, Fogelman I (2001) Quantitative studies of bone with the use of F-18 fluoride and Tc-99m-methylene diphosphonate. Semin Nucl Med 31:28–49

Blumgart HL, Yens OC (1927) Studies on the velocity of blood flow. 1. The method utilised. J Clin Invest 4:1–13

Bouwens L, Van de Walle R, Nuyts J, Koole M, D'Asseler Y, Vandenberghe S, Lemahieu I, Dierckx RA (2001) Image-correction techniques in SPECT. Comp Med Imag Graphics 25:117–126

Bracewell RN (1956) Strip integration in radio-astronomy. Ast J Phys 9:198–217

Bracewell RN, Riddle AC (1967) Inversion of fan-beam scans in radio-astronomy. Astrophys J 150:427–434

Casey ME, Nutt RA (1986) A multicrystal 2-dimensional BGO detector system for positron emission tomography. IEEE Trans Nucl Sci 33:460–463

Cassen B, Curtis L, Reed C, Libby R (1951) Instrumentation for I^{131} use in medical studies. Nucleonics 9:46–50

Chang LT (1978) A method for attenuation correction in radionuclide computed tomography. IEEE Trans Nucl Sci NS-25:638–643

Chang LT, Kaplan SN, Macdonald B, Perez-Mendez V, Shiraishi L (1974) A method of tomographic imaging using a multiple pinhole coded aperture. J Nucl Med 15:1063–1065

Chepel V, Lopes MI, Kuchenkov A, Marques RF, Policarpo AJPL (1997) Performance study of a liquid xenon chamber for PET. Nucl Instr Methods Phys Res A 392:427–432

Cherry SR, Meikle SR, Hoffman EJ (1993) Correction and characterisation of scattered events in three-dimensional PET using scanners with retractable septa. J Nucl Med 34:671–678

Crippin DDM (1976) Single sideband methods in Fresnel zone plate imaging. In: Raynaud C, Todd-Pokropek A (eds) Information processing in scintigraphy. Service de Documentation du CEN Saclay, no 76-003, pp 446–454

De Rosier DJ, Klug A (1968) Reconstruction of three-dimensional structures from electron micrographs. Nature 217:130–134

Defrise M, Kinahan PE, Townsend DW, Michel C, Sibomana M, Newport D (1997) Exact and approximate rebinning algorithms for 3D PET data. IEEE Trans Med Imag 16:145–158

Dekemp RA, Nahmias C (1994) Attenuation correction in PET using single-photon transmission measurement. Med Phys 21:771–778

Detko J (1969) Semiconductor diode matrix for isotope localization. Phys Med Biol 14:245–253

DeVito RP, Hamill JJ, Treffert JD, Stoub EW (1989) Energy weighted acquisition of scintigraphic images using finite spatial filters. J Nucl Med 30:2029–2035

Eisen Y, Shor A (1998) CdTe and CdZnTe materials for room temperature X-ray and gamma-ray detectors. J Crystal Growth 184/185:1302–1312

Eisen Y, Shor A, Gilath C, Tsarbarim M, Chouraqui P, Hellman C, Lubin E (1996) A gamma camera based on CdTe detectors. Nucl Instr Methods Phys Res A 380:474–478

Fougeres P, Siffert P, Hageali M, Koebal J, Regal R (1999) CdTe and $Cd_{1-x}Zn_xTe$ for nuclear detectors: facts and fictions. Nucl Instr Methods Phys Res A 428:38–44

Genna S, Pang S-C, Smith A (1981) Digital scintigraphy: principles, design and performance. J Nucl Med 22:365–371

Gindi GR, Arendt J, Barrett HH, Chiu MY, Ervin A, Giles CL, Kujoory MA, Miller EL, Simpsons RG (1982) Imaging with rotating slit apertures and rotating collimators. Med Phys 9:324–339

Gonzalez DE, Jaszczak RJ, Bowsher JE, Akabani G, Greer KL (2001) High resolution absolute SPECT quantitation for I-131 distributions used in the treatment of lymphoma: a phantom study. IEEE Trans Nucl Sci 48:707–714

Graham LG, Neil R (1974) In vivo quantitation of radioactivity using the Anger camera. Radiol 112:441–442

Grootoonk S, Spinks TJ, Sashin D, Spyrou NM, Jones T (1996) Correction for scatter in 3D brain PET using a dual energy window method. Phys Med Biol 41:2757–2774

Hendel RC, Corbett JR, Cullom S, DePuey G, Garcia EV, Bateman T (2002) The value and practice of attenuation correction for myocardial perfusion SPECT imaging: a joint position statement from the American Society of Nuclear cardiology and the Society of Nuclear Medicine. J Nucl Med 43:273–280

Herbert T, Leahy R, Singh M (1987) Maximum likelihood reconstruction for a prototype electronically collimated single photon emission system. SPIE 767:77–83

Hevesy G (1923) The absorption and translocation of lead by plants. Biochem J 17:439

Hiramoto T, Tanaka E, Nohara N (1971) A scintillation camera based on delay-line time conversion. J Nucl Med 12:160–165

Hoffman EJ, Huang SC, Phelps ME, Kuhl DE (1981) Quantitation in positron emission computed-tomography. 4. Effect of accidental coincidences. J Comput Assist Tomogr 5:391–400

Huang S-C, Hoffman EJ, Phelps ME, Kuhl DE (1979) Quantitation in positron emission computed tomography. 2. Effects of inaccurate attenuation correction. J Comput Assist Tomogr 3:804–812

Huesman RH, Derenzo SE, Cahoon JL, Geyer AB, Moses WW, Uber DC, Vuletich T, Budinger TF (1988) Orbiting transmission source for positron tomography. IEEE Trans Nucl Sci 35:735–739

Hudson HM, Larkin RS (1994) Accelerated image-reconstruction using ordered subsets of projection data. IEEE Trans Med Imag 13:601–609

Jaszczak RJ, Floyd CE, Manglos SH, Greer KL, Coleman RE (1986) Cone beam collimation for single photon emission computed tomography: analysis simulation and image reconstruction using filtered backprojection. Med Phys 13:484–489

Kalki K, Blankenspoor SC, Brown JK, Hasegawa BH, Dae MW, Chin M, Stilson C (1997) Myocardial perfusion imaging with a combined X-ray CT and SPECT system. J Nucl Med 38:1535–1540

Kaufman L, Lorenz V, Hosier K, Hoenninger J, Hattner RS, Okerlund M, Price DC, Shames DM, Swann SJ, Ewins JH, Armantrout GA, Camp DC, Lee K (1978) Two detector, 512 element high purity germanium camera prototype. IEEE Trans Nucl Sci NS-25:189–195

Keyes WI (1975) The fan-beam gamma camera. Phys Med Biol 20:489–493

Kim JH, Choi Y, Joo KS, Sihn BS, Chong JW, Kim SE, Lee KH, Choe YS, Kim BT (2000) Development of a miniature scintillation camera using NaI(Tl) scintillator and PSPMT for scintimammography. Phys Med Biol 45:3481–3488

Kinahan PE, Rogers JG (1989) Analytic 3D image reconstruction using all detected events. IEEE Trans Nucl Sci 36:964–968

King MA, Hademenos GJ, Glick SJ (1992) A dual photopeak window method for scatter correction. J Nucl Med 33:605–612

Koral KF, Rogers WL, Knoll GF (1975) Digital tomographic imaging with time-modulated pseudorandom coded aperture and Anger camera. J Nucl Med 16:402–431

Kulberg GH, Muehllehner G, van Dijk N (1972) Improved resolution of the Anger scintillation camera through the use of threshold preamplifiers. J Nucl Med 13:169–171

La Fontaine R, Stein MA, Graham LS, Winter J (1986) Cold lesions: enhanced contrast using asymmetric photopeak windows. Radiology 160:255–260

Lange K, Carson R (1984) EM reconstruction algorithms for emission and transmission tomography. J Comput Assist Tomogr 8:306–316

LeBlanc JW, Clinthorne NH, Hua C, Rogers WL, Wehe DK, Wilderman SJ (1999) A Compton camera for nuclear medicine applications using [113m]In. Nucl Instr Methods Phys Res A 422:735–739

Lewellen TK, Bice AN, Pollard KR, Zhu J-B, Plunkett ME (1989) Evaluation of a clinical scintillation camera with pulse tail extrapolation electronics. J Nucl Med 30:1554–1558

Lewellen TK, Miyoaka RS, Swan WL (1999) PET imaging using dual-headed gamma cameras: an update. Nucl Med Commun 20:5–12

Lodge MA, Binie DM, Flower MA, Webb S (1995) The experimental evaluation of a prototype slat collimator for single photon emission computed tomography. Phys Med Biol 40:427–448

Macovski A (1974) Gamma ray imaging using modulated apertures. Phys Med Biol 19:523–533

Matherson K, Barber H, Barrett H, Eskin J, Dereniak E, Marks D, Woolfenden J, Young E (1998) Progress in the development of large area modular 64×64 CdZnTe imaging arrays for nuclear medicine. IEEE Trans Nucl Sci 45:354–358

Melcher CL (2000) Scintillation crystals for PET. J Nucl Med 46:1051–1055

Mestais C, Baffert N, Bonnefoy J, Chapuis A, Koenig A, Monnet O, Buffet PO, Rostaing JP, Sauvage F, Verger L (2001) A new design for a high resolution, high efficiency CZT gamma detector. Nucl Instr Methods Phys Res A 458:62–67

Michel C, Bol A, Spinks T, Townsend D, Bailey D, Grootoonk S, Jones T (1991) Assessment of response function in 2 PET scanners with and without interplane septa. IEEE Trans Med Im 10:240–248

Mitchell JS (1946) Applications of recent advances in nuclear physics to medicine. Br J Radiol 19:481–487

Moyer RA (1974) A low energy multi-hole converging collimator compared with a pinhole collimator. J Nucl Med 15:59–64

Muehllehner G (1969) A diverging collimator for gamma ray imaging cameras. J Nucl Med 10:197–201

Muehllehner G, Dudek J, Moyer R (1976) Influence of hole shape on collimator performance. Phys Med Biol 21:242–250

Muehllehner G, Colsher JG, Stoub EW (1980) Correction for field nonuniformity in scintillation cameras through removal of spatial distortion. J Nucl Med 21:771–776

Narita H, Kawaida Y, Ooshita I, Itoh, T, Tsuchida D, Fukumitsu N, Mori Y, Makimo M (2001) Evaluation of efficiency of a multi-crystal scintillation camera Digirad 2020tc Imager using solid state detectors. Jpn J Nucl Med 38:355–362

Ogawa K, Harata Y, Ichihara T, Kubo A, Hashinoto S (1991) A practical method for position dependent Compton-scattered correction in single photon emission CT. IEEE Trans Med Imag 10:408–412

Ollinger JM (1996) Model-based scatter for fully 3D PET. Phys Med Biol 41:153–176

Ott RJ (1997) Position sensitive detectors for medical imaging. Nucl Instr Methods Phys Res A 392:396–401

Pani R, Pellegrini R, Scopinaro F, Soluri A, De Vincentis, G, Pergola A, Iacopi F, Corona A, Filippi S, Ballesio PL (1997) Scintillating array gamma camera for clinical use. Nucl Instr Methods Phys Res A 392:295–298

Pani R, Pellegrini R, Soluri A, De Vincentis G, Scafe R, Pergola A (1998) Single photon emission imaging by position sensitive PMT. Nucl Instr Methods Phys Res A 409:524–528

Peters AM, Myers MJ (1998) Physiological measurements with radionuclides in clinical practice. Oxford University Press, Oxford

Pretorius PH, van Rensburg AJ, van Aswegen A, Lotter MO, Serfontein DE, Herbst CP (1993) The channel ratio method for scatter correction for radionuclide image quantitation. J Nucl Med 34:330–335

Radon J (1917) Über die Bestimmung von Functionen durch ihre Integralwerte längs gewisser Mannigfaltigkeiten. (On the determination of functions from their integrals along certain manifolds.) Ber Vrbhandl Sachs Akad Wiss Leipzig, Math-Phys KL 69:262–277

Rosenthal MS, Cullom J, Hawkins W, Moore SC, Tsui BMW, Yester M (1995) Quantitative SPECT imaging: a review and recommendations by the focus committee of the Society of Nuclear Medicine Computer and Instrumentation Council. J Nucl Med 36:1489–1513

Scheiber C, Giakos GC (2001) Medical applications of CdTe and CdZnTe detectors. Nucl Instr Methods Phys Res A 458:12–25

Scheiber C, Eclancher B, Chambron J, Prat V, Kazandjan A, Jahnke A, Matz R, Thomas A, Warren S, Hage-Hali M, Regal R, Siffert P, Karman M (1999) Heart imaging by cadmium telluride gamma camera. European Program "BIOMED" consortium. Nucl Instr Methods Phys Res A 428:139–149

Sharp PF, Dendy PP, Keyes WI (1985) Radionuclide imaging techniques. Academic, London, pp 55–59

Shepp LA, Vardi Y (1982) Maximum likelihood reconstruction for emission tomography. IEEE Trans Med Imag 1:113–122

Shor A, Eisen Y, Mardor I (1999) Optimum spectroscope performance from CZTγ and X-ray detectors with pad and strip detectors. Nucl Instr Methods Phys Res A 428:182

Singh M, Doria D (1983) An electronically collimated gamma camera for single photon emission computed tomography, part II. Image reconstruction and preliminary experimental measurements. Med Phys 10:428–435

Singh M, Brechner RR (1990) Experimental test-object study of electronically collimated SPECT. J Nucl Med 31:178–186

Tamaki N, Kuge Y, Tsukamoto E (2001) The road to quantitation of regional myocardial uptake of tracer. J Nucl Med 42:780–781

Tanaka E, Hiramoto T, Nohara N (1970) Scintillation cameras based on new pulse arithmetic. J Nucl Med 11:542–547

Todd PW, Nightingale JM, Everett DB (1974) A proposed gamma camera. Nature 251:132–134

Vandenberghe S, D'Asseler Y, Koole M, Van de Walle R, Lemahieu I, Dierckx RA (2001) Physical evaluation of 511 keV coincidence imaging with a gamma camera. IEEE Trans Nucl Sci 48:98–105

Verger L, Biotel M, Gentet MC, Hamelin R, Mestais C, Mongellaz F, Rustique J, Sanchez G (2001) Characterization of CdTe and CdZnTe detectors for gamma ray imaging applications. Nucl Instr Methods Phys Res A 458:297–309

Visvikis D, Ott RJ, Wells K, Flower MA, Stephenson R, Batemen JE, Connolly J (1997) Performance characterisation of large-area BAF2-TMAE detectors for use in a whole body clinical PET camera. Nucl Instr Methods Phys Res A 392:414–420

Webb S (1990) From the watching of shadows: the origins of radiological tomography. Hilger, Bristol

Webb S, Bindi DM, Flower MA, Ott RJ (1992) Monte Carlo modelling of the performance of a rotating slit-collimator for improved gamma-camera imaging. Phys Med Biol 37:1095–1108

Webb S, Flower MA, Ott RJ (1993) Geometric efficiency of a rotating slit collimator for improved gamma-camera imaging. Phys Med Biol 38:627–638

Williams MB, Goode AR, Galbis-Reig V, Majewski S, Weisenberger AG, Wojcik R (2000) Performance of a PSPMT based detector for scintimammography. Phys Med Biol 45:781–800

Wong W-H, Yokoyama S, Uribe Jorge, Baghaei H, Li H, Wang J, Zhang N (1999) An elongated position sensitive block detector designed using PMT quadrant-sharing configuration and asymmetric light partition. IEEE Trans Nucl Sci 46:542–545

Yutani K, Tatsumi M, Shiba E, Kusuoka H, Nishimura T (1999) Comparison of dual-head coincidence gamma camera FDG imaging with FDG PET in detection of breast cancer and axillary lymph node metastasis. J Nucl Med 40:1003–1008

CHAPTER 7

Modelling Metabolite and Tracer Kinetics

ALBERT H. GJEDDE

7

Contents

7.1
Role of Kinetics

7.1.1
Kinetics and Molecular Biology

The purpose of kinetic analysis of living matter is to obtain quantitative measures of the rate of molecular reactions. Quantitative approaches were uncommon in biology and medicine prior to the second half of the 19th century and only slowly gained ground against traditionally qualitative considerations. The competition between quantitative and qualitative perspectives is felt even today.

The struggle reflects the changing views of disease in the medical sciences in which a disorder originally was thought to represent a major imbalance among qualitatively different matters of nature and life including the four elements (water, air, fire, and earth) and the four cardinal fluids (blood, phlegm, yellow bile, and black bile).

This is no longer considered the case. The imbalances underlying disease appear to be due to minute but specific errors that are now known to create the effects of disease by turning open thermodynamic systems implemented in biochemical and physiological compartments into closed systems that must ultimately fail because entropy rises in closed systems as order is replaced by disorder. Thus, it is a fundamental observation that truly closed systems eventually become incompatible with life.

The concept of imbalance is quantitative, as is the injunction of living matter to respond to exigencies with moderation. Thus, measurement is the modern practice, although it is tied to an increasing understanding of the limits of certainty. Competing with this understanding is the rise of information technology, according to which the quantitative properties of the components of living systems could be less important, implying that only their structural relations are informative. Thus, there is a current sense that the tide of scientific philosophy is returning in the direction of the holistic and qualitative. To some extent, molecular biology itself may be the stimulus for this redirection because it is seen as emphasizing structural (i.e., qualitative) rather than quantitative relations. Only the practice of meticulous kinetic analysis can correct this misunderstanding.

7.1.2
Kinetics and Genomics

Living matter is distinguished from nonliving matter primarily by its ability to maintain steady states of incredibly complex molecular compartments far from true equilibrium. The entropy freed by the realization of this enormous potential resides in a remarkably inert and robust molecule called deoxy-ribonucleic acid (DNA). However, the DNA molecule itself does nothing; its entire and completely passive role is to be decoded. Its power to elicit action derives from the ability of other molecules in living tissue to decipher its instructions at the right time and place.

As the decoder must understand the message of at least the opening sections of the manual (the rest can be learned in due course) in advance of the decoding, the fundamental goal of metabolite and tracer kinetic analysis in biology and medicine is to describe and quantify the processes in their entirety from conception to termination of the organism. For example, it is estimated that at the peak of neuronal proliferation during human gestation, as many as 250,000 new brain cells of identical composition are created every minute. Yet metabolite concentrations everywhere remain inside carefully regulated limits. A snapshot of any one cell would produce an unremarkable image; only the proper tracer kinetic analysis could reveal the astounding dynamics of the metabolite fluxes contributing to this development.

7.1.3
Kinetics and Proteomics

The rate of molecular reactions is typically constrained by proteins. An important measure of health is steady state, in which proteins maintain the concentrations of metabolites while the molecular fluxes adjust to local and global requirements. Most importantly, the composition of living matter remains constant in steady state (hence the name), and a momentary glimpse reveals none of the dynamics of the underlying molecular fluxes. The further the steady state is from a state of equilibrium, the greater is the work required to maintain it and the greater are the fluxes controlled by the proteins. Only few processes are near equilibrium, and they typically do not interfere with the regulation of the important molecular fluxes of living matter.

7.1.4
Tracer Kinetics

While concentrations normally do not change outside tightly controlled limits, past attempts to understand the underlying dynamics by perturbing a system often removed the system from its normal state and sometimes failed specifically to reveal the normal dynamic properties of its kinetics. The introduction of suitably flagged ("labeled") and hence identifiable representatives of the native molecules, called "tracers," accomplishes a virtual perturbation without disturbing the steady state of the system, provided the quantity of tracer is kept too low to change the system's properties. Methods of doing just that form the core of the tracer kinetic analysis of biological processes.

7.2
Definition of Relaxation Constants

The key to tracer kinetic analysis is the concept of compartment, a group of atoms or molecules that behave in such an identically predictable manner that the introduction of a few additional but labeled atoms or molecules does not change their behavior significantly. Compartments may be large or small, but they are fundamental abstractions, regardless of their size. As such they can be said to defy the very concept they were created to represent because they require that the members be at equilibrium with each other and hence allow no interactions among each other.

By being relegated to the interfaces between compartments, the kinetic processes studied by tracer kinetic analysis are discontinuous and hence fundamentally at variance with the real nature of kinetic processes, which must be continuous ("distributed"). The mathematically abstract compartments can of course be widely dispersed and physically intermixed with other compartments, but the definition does not allow the members of individual compartments to interact.

Nonlinear modelling of distributed processes is possible in theory, provided the measurements have the necessary power of resolution, but this is so rarely the case that distributed models often arise as assemblies of commensurately diminished compartments and as such can be regarded as extensions of the linear compartments considered here. However, the resulting nonlinear kinetics will not be examined

in this text. The compartment is proof of Niels Bohr's dictum that the measurement must invalidate the measured because it ignores the quantum nature of the distribution. Its saving grace is its usefulness to the practical problems of biology.

7.2.1
Single Compartment

A group of molecules, however confined or identified, forms a single kinetic compartment when the molecules' rate of escape from the compartment (also called "decay" or "relaxation") is constant and proportional to the number of remaining molecules. In this sense, the compartment is an abstraction never found in real life. However, in many cases, although not in all, it is an excellent approximation when the rate of exchange among the members is much faster than the rate of relaxation. This condition is believed to be fulfilled when the compartment is well stirred. Thus, if the stirring is not vigorous enough, the compartmentation is said to fail, and the compartment collapses into smaller compartments or noncompartmental distributions. When the product of a chemical reaction forms a single kinetic compartment, the simplest precursor-product relationship obeys a linear differential equation of the form (Gjedde 1995a)

$$\frac{dm}{dt} = j - k\,m \tag{7.1}$$

where j is the time-variable precursor influx function and k is the relaxation or rate constant. How, or to which location, the product escapes is immaterial to the definition. In this respect, the compartment is a one-dimensional construct with time as the only dimension of change. The relaxation constant is the solution to Eq. 7.1:

$$k = \frac{j}{m} - \frac{1}{m}\left(\frac{dm}{dt}\right) \tag{7.2}$$

where m is the mass of molecules. The equation shows that the relaxation constant k is the ratio between the influx and the compartment content at steady state when $dm/dt = 0$.

It is important to emphasize that the relaxation creates the compartment. Neither membranes nor anatomical subdivisions contribute to the compartmentation, except to the extent that they establish or allow

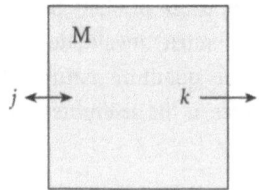

Fig. 7.1. Model of single compartment **M** occupied by the quantity *m*. Note that net fluxes are represented by bidirectional arrows and rate constants and clearances by unidirectional arrows

the relaxation. Nor is the relaxation necessarily the result of a single mechanism. Frequently, the relaxation occurs because multiple so-called first-order mechanisms combine linearly to form a single process.

The mechanisms of passive transport across membranes, enzymatic steps of metabolic pathways, and radioactive decay all generate compartments when they decay at a constant rate (k) such that the loss of members per unit time is proportional to the number of remaining members. Often, but not always, the number of remaining members is proportional to the concentration of the members in an aqueous medium or other solvent. The proportionality depends on several physical and chemical factors, the simplest being the volume of the solvent, but it is not always possible to judge whether they have remained constant to the satisfaction of the required proportionality between number and concentration of members.

The single isolated compartment may exist because the supply of substance is nonexistent, instantaneous, infrequent, or unpredictable, i.e., when the function j is zero or a Dirac delta function or when it is given only numerically.

For this compartment the transient solution to Eq. 7.1 is the convolution integral

$$m = e^{-kT} \left[m(0) + \int_0^T j\, e^{kt}\, dt \right] \tag{7.3}$$

where e^{-kt} is the *impulse response function*, which accounts for the monoexponentially changing loss of the original member molecules as well as for the monoexponentially changing disappearance of new members for whom a correction must be made for the fact that they arrive at different times and hence have different risks of being expelled. The steady-state solution to the contents of the compartment is

$$M = \frac{J}{k} \tag{7.4}$$

in keeping with the defintiion of a compartment, leading to the steady-state solution to the contents of the compartment:

$$J = kM \tag{7.5}$$

where variables at steady state are given uppercase symbols.

7.2.2
Two Compartments

When two or more compartments occupy positions in series, the products have precursors that themselves form single compartments (Fig. 7.2). Applying the definition, Eq. 7.1 becomes

$$\frac{dm_2}{dt} = k_1\, m_1 - k_2\, m_2 \tag{7.6}$$

where m_1 refers to the precursor compartment and m_2 to the product compartment. Note that m_1 and m_2 represent numbers of molecules, usually in units of *mol* when normalized against Avogadro's number ($6 \cdot 10^{23}$). When clearance (which has unit of flow) is the precursor's mechanism of decay, the quantity of molecules can be converted to concentration in the precursor solvent with a known volume (V_1),

$$\frac{dm_2}{dt} = \frac{V_1\, k_1\, m_1}{V_1} - k_2\, m_2\,, \tag{7.7}$$

yielding

$$\frac{dm_2}{dt} = K_1\, c_1 - k_2\, m_2 \tag{7.8}$$

and

$$\frac{dm_1}{dt} = V_1\, \frac{dc_1}{dt} \tag{7.9}$$

where K_1 symbolizes the product $V_1 k_1$ and is known as a clearance with unit of flow and c_1 is the concentration of the precursor. The clearance is the flux from compartment 1 to compartment 2 relative to the concentration. The total flux into the two compartments (and all other compartments fed by this closed system) is

$$j_0 = \sum_{i=1}^{2} \left[\frac{dm_i}{dt} \right] = V_1\, \frac{dc_1}{dt} + K_1\, c_1 - k_2\, m_2 \tag{7.10}$$

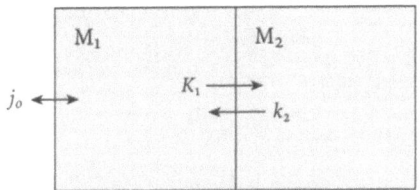

Fig. 7.2. Model of two compartments exchanging contents consisting of the product (m_2) of a reaction and the precursor with the concentration c_1 (m_1)

The distinction between quantity and concentration of substrate is the reason for the distinction between the rate constant k_1 and the clearance K_1. Using concentrations instead of masses requires knowledge of the relative solvent volumes of the two compartments. For example, associating the rate constant k_1 with the concentration c_1 creates an inconsistency, unless m_1 and m_2 occupy the same volume, because the apparent concentration c_1 is the concentration of m_1 in the solvent of M_2, i.e., m_1/V_2:

$$\frac{dc_2}{dt} = \frac{K_1}{V_2} c_1 - k_2 c_2 \tag{7.11}$$

which is different from the mass balance relationship:

$$\frac{dm_2}{dt} = \frac{K_1}{V_1} m_1 - k_2 m_2 \tag{7.12}$$

The transiently nonlinear solution to the differential Eq. 7.8 of this relationship is again the convolution integral

$$m_2 = e^{-k_2 T}\left[m_2(0) + k_1 \int_0^T m_1 e^{k_2 t}\, dt \right] \tag{7.13}$$

where k_1 is the rate constant of the reaction that converts the precursor to the product and k_2 is the rate constant for the reaction that reconverts the product to the precursor. With the substitution of c_1 for m_1, the convolution integral of Eq. 7.8 is

$$m_2 = e^{-k_2 T}\left[m_2(0) + K_1 \int_0^T c_1 e^{k_2 t}\, dt \right] \tag{7.14}$$

which leads to the combined solution for m_1 and m_2 when $m_2(0) = 0$

$$\sum_{i=1}^{2} m_i = V_1 c_1 + K_1 \int_0^T c_1 e^{-k_2(T-t)}\, dt \tag{7.15}$$

and to the combined solution for $n-1$ recipient compartments supplied by the same delivery compartment:

$$\sum_{i=1}^{n} m_i = V_1 c_1 + \sum_{h=1}^{n-1} K_{1_h} \int_0^T c_1 e^{-k_{2_h}(T-t)}\, dt \tag{7.16}$$

A transient multilinear solution to the net flux across the interface is obtained by integration of Eq. 7.8

$$j_1 = K_1 c_1 - k_2 m_2 = K_1 c_1 - K_1 k_2 \int_0^T c_1\, dt$$
$$+ (k_2)^2 \int_0^T m_2\, dt \tag{7.17}$$

where j_1 is the flux between compartments M_1 and M_2, which yields the total accumulation in the two compartments at time T. By simple integration when $m_1(0) = m_2(0) = 0$,

$$m_2 = K_1 \int_0^T c_1\, dt - k_2 \int_0^T m_2\, dt \tag{7.18}$$

and, after the substitution of $\sum_{i=1}^{2} m_i - V_1 c_1$ for m_2, also the total content as a function of the concentration in compartment 1 (M_1),

$$\sum_{i=1}^{2} m_i = V_1 c_1 + (K_1 + k_2 V_1) \int_0^T c_1\, dt$$
$$- k_2 \int_0^T \sum_{i=1}^{2} m_i\, dt \tag{7.19}$$

For $k_2 = 0$, the equation

$$\sum_{i=1}^{2} m_i = V_1 c_1 + K_1 \int_0^T c_1\, dt \tag{7.20}$$

linearizes to the Rutland-Gjedde-Patlak equation (Gjedde 1981):

$$\sum_{i=1}^{2} v_i = V_1 + K_1 \int_0^T \frac{c_1}{c_1(T)}\, dt \tag{7.21}$$

where $v_i(T)$ is an expanding "virtual" volume and $\int_0^T c_1 dt / c_1$ is an expanding "virtual" time, useful to physiologists (Sarna et al. 1977). For $k_2 > 0$, the steady-state solution to Eq. 7.8 is

$$M_2 = \frac{k_1}{k_2} M_1$$

while the steady-state solution to Eq. 7.15 is

$$\sum_{i=1}^{2} M_i = \left[V_1 + \left(\frac{K_1}{k_2} \right) \right] C_1 \qquad (7.22)$$

where V_1 is the distribution volume of the molecules in compartment M_1 and K_1/k_2 defines an additional volume V_e as a "partition volume." The size of the partition volume is dictated by the magnitudes of K_1 and k_2 and is different from the distribution volume of the molecules in compartment M_2 (V_2). The total volume of distribution is therefore $V_1 + (K_1/k_2)$.

The steady-state flux between the compartments is nil:

$$J_1 = K_1 C_1 - k_2 M_2 = 0 \qquad (7.23)$$

and shows that this "system" of two compartments does nothing more than regulate the magnitude of compartment M_2: the steady-state size of M_2 follows passively from the unidirectional fluxes across the interface, as shown by Eq. 7.12.

7.2.3
Two Compartments with Sink

To regulate a steady-state flux without changing the size of compartment M_2, a diversion must be established for the efflux. The receptacles of the efflux from a compartment are known as sinks. A compartment may have multiple sinks and still fulfill the criteria of compartmental behavior, as shown in the model underlying the linked differential equations

$$\frac{dm_1}{dt} = V_1 \frac{dc_1}{dt} \qquad (7.24)$$

and

$$\frac{dm_2}{dt} = K_1 c_1 - (k_2 + k_3) m_2 \qquad (7.25)$$

where K_1 as the product $V_1 k_1$ is the clearance with unit of flow and c_1 is the concentration of the precursor in the volume V_1. The total flux into the system of compartments, and all other compartments fed by it, is given by an equation identical to Eq. 7.10:

$$j_0 = V_1 \frac{dc_1}{dt} + K_1 c_1 - k_2 m_2 \qquad (7.26)$$

The content of M_2 is given by the transient nonlinear solution

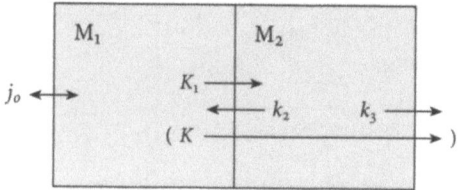

Fig. 7.3. Model of precursor and product compartments with sink. At steady state, the sink defines a single steady-state clearance, K, equal to $K_1 k_3/(k_2 + k_3)$, which reaches K_1 when $k_3 \gg k_2$

$$m_2 = e^{-(k_2+k_3)T} \left[m_2(0) + K_1 \int_0^T c_1 e^{(k_2+k_3)t} \, dt \right] \quad (7.27)$$

which yields the combined solution for $m_2(0) = 0$:

$$\sum_{i=1}^{2} m_i = V_1 c_1 + K_1 \int_0^T c_1 e^{-(k_2+k_3)(T-t)} \, dt \qquad (7.28)$$

The flux into the sink at any time is given by the transient multilinear solution

$$j_2 = k_3 m_2 = K_1 k_3 \int_0^T c_1 \, dt - k_3 (k_2 + k_3)$$
$$\times \int_0^T m_2 \, dt \qquad (7.29)$$

such that for $\sum_{i=1}^{2} m_i = m_2 + V_1 c_1$,

$$\sum_{i=1}^{2} m_i = V_1 c_1 + [K_1 + V_1 (k_2 + k_3)] \int_0^T c_1 \, dt$$
$$- (k_2 + k_3) \int_0^T \sum_{i=1}^{2} m_i \, dt \qquad (7.30)$$

At steady state, the solution to Eq. 7.25 is

$$M_2 = \left(\frac{K_1}{k_2 + k_3} \right) C_1 \qquad (7.31)$$

which yields the steady-state flux through the system, when the content of M_2 is inserted into Eq. 7.29:

$$J_0 = k_3 M_2 = KC_1 \qquad (7.32)$$

where K is the "net" clearance of the precursor from compartment M_1 to compartment M_2:

$$K = K_1 \left(\frac{k_3}{k_2 + k_3} \right) = \frac{K_1}{1 + \frac{k_2}{k_3}}. \qquad (7.33)$$

7.2.4
Three Compartments

The previous systems were open with a throughput. Closed compartments under normal circumstances require a return outlet, as shown in the model below, where

$$\frac{dm_1}{dt} = V_1 \frac{dc_1}{dt} , \qquad (7.34)$$

$$\frac{dm_2}{dt} = K_1 c_1 + k_4 m_3 - (k_2 + k_3) m_2 \qquad (7.35)$$

and

$$\frac{dm_3}{dt} = k_3 m_2 - k_4 m_3 \qquad (7.36)$$

where the total flux into the three compartments is given by the same equation as Eqs. 7.10 and 7.26:

$$j_0 = \sum_{i=1}^{3} \left[\frac{dm_i}{dt} \right] = V_1 \frac{dc_1}{dt} + K_1 c_1 - k_2 m_2 \qquad (7.37)$$

The linked differential Eqs. 7.34–7.36 command particular attention because they underlie major applications of tracer kinetic analysis. They have been solved for both $k_4 = 0$ and $k_4 > 0$. The solutions appear complex but are easily obtained by Laplace transformation into second-order polynomials.

■ **Transiently Nonlinear Solution to Reversible Accumulation.** For $k_4 > 0$, the transient nonlinear solution to the coupled Eqs. 7.34–7.36 is obtained in two steps. The general solution to the differential Eq. 7.36 is the ordinary convolution integral

$$m_3 = e^{-k_4 T} \left[m_3(0) + k_3 \int_0^T m_2 e^{k_4 t} dt \right] \qquad (7.38)$$

where k_3 is the rate constant of the reaction that converts the precursor in M_2 to the product in M_3 and

k_4 is the rate constant for the reaction that reconverts this product to its precursor. When Eq. 7.38 is inserted, the solution to Eq. 7.35 for $m_2(0) = m_3(0) = 0$ is

$$m_2 = K_1 \left[\left(\frac{q_2 - k_4}{q_2 - q_1} \right) \int_0^T c_1 e^{-q_2(T-t)} dt - \left(\frac{q_1 - k_4}{q_2 - q_1} \right) \right.$$
$$\left. \times \int_0^T c_1 e^{-q_1(T-t)} dt \right] \qquad (7.39)$$

where

$$q_1 = \frac{k_2 + k_3 + k_4 - \sqrt{(k_2 + k_3 + k_4)^2 - 4k_2 k_4}}{2} \qquad (7.40)$$

and

$$q_2 = \frac{k_2 + k_3 + k_4 + \sqrt{(k_2 + k_3 + k_4)^2 - 4k_2 k_4}}{2} \qquad (7.41)$$

which in turn changes the solution to Eq. 7.38 to

$$m_3 = \left[\frac{K_1 k_3}{q_2 - q_1} \right] \left(\int_0^T c_1 e^{-q_1(T-t)} dt \right.$$
$$\left. - \int_0^T c_1 e^{-q_2(T-t)} dt \right) \qquad (7.42)$$

such that the sum of the contents of M_1, M_2, and M_3 now is

$$\sum_{i=1}^{3} m_i = V_1 c_1 + K_1 \left(\frac{q_2 - (k_3 + k_4)}{q_2 - q_1} \right)$$
$$\times \int_0^T c_1 e^{-q_2(T-t)} dt + K_1 \left(\frac{(k_3 + k_4) - q_1}{q_2 - q_1} \right)$$
$$\times \int_0^T c_1 e^{-q_1(T-t)} dt \qquad (7.43)$$

which expresses the general property of linear systems that

$$\sum_{i=1}^{3} m_i = V_1 c_1 + \sum_{h=1}^{2} K_{1_h} \int_0^T c_1 e^{-k_{2_h}(T-t)} dt$$

where $K_{1_2} = K_1[q_2 - (k_3 + k_4)]/[q_2 - q_1]$, $K_{1_1} = K_1 \times [(k_3 + k_4) - q_1]/[q_2 - q_1]$, $k_{2_1} = q_2$, and $k_{2_2} = q_1$.

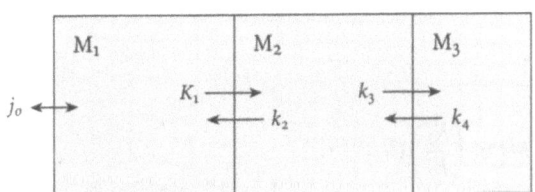

Fig. 7.4. Closed system of three compartments

■ **Transiently Nonlinear Solution to Irreversible Accumulation.** For $k_4 = 0$, the nonlinear solution to Eq. 7.43 reduces to the one given by Sokoloff et al. (1977) for $m_2(0) = m_3(0) = 0$:

$$\sum_{i=1}^{3} m_i = V_1 c_1 + \frac{K_1 k_2}{k_2 + k_3} \int_0^T c_1 e^{-(k_2 + k_3)(T-t)} dt$$
$$+ \frac{K_1 k_3}{k_2 + k_3} \int_0^T c_1 \, dt \qquad (7.44)$$

for which it should be kept in mind that closed systems without backflux have no obvious biological role, although they may exist under experimental circumstances for certain tracers. The solution reduces to

$$\sum_{i=1}^{3} m_i = V_1 c_1 + K \left[\int_0^T c_1 \, dt + \left(\frac{k_2}{k_3} \right) \right.$$
$$\left. \int_0^T c_1 e^{-(k_2 + k_3)(T-t)} dt \right] \qquad (7.45)$$

where the "net" clearance is

$$K = K_1 \left(\frac{k_3}{k_2 + k_3} \right) = \frac{K_1}{1 + \frac{k_2}{k_3}} = K_{1_2} \qquad (7.46)$$

as defined in Eq. 7.33. Gjedde (1982) showed that this equation, when normalized against the concentration c_1 at time T, defines a continuously increasing apparent volume of distribution in the manner of the Rutland-Gjedde-Patlak solution to Eq. 7.19 (Gjedde 1982):

$$\sum_{i=1}^{3} v_i = \left[V_1 + \left(\frac{K_1 k_2}{k_2 + k_3} \right) \int_0^T \left[\frac{c_1(t)}{c_1(T)} \right] e^{-(k_2 + k_3)(T-t)} dt \right]$$
$$+ K \int_0^T \frac{c_1(t)}{c_1(T)} dt \qquad (7.47)$$

if compartments M_1 and M_2 reach a steady balance ("secular" equilibrium), depending on the magnitudes of k_2 and k_3.

■ **Transiently Linear Solution to Reversible Accumulation.** For $k_4 > 0$, the transient multilinear solution to Eqs. 7.34–7.36 was given by Evans et al. (1987) for $m_2(0) = m_3(0) = 0$:

$$\sum_{i=1}^{3} m_i = a_1 c_1 + a_2 \int_0^T c_1 \, dt + a_3 \int_0^T \int_0^u c_1 \, dt \, du$$
$$+ a_4 \int_0^T \sum_{i=1}^{3} m_i \, dt$$
$$+ a_5 \int_0^T \int_0^u \sum_{i=1}^{3} m_i \, dt \, du \qquad (7.48)$$

where

$a_1 = V_1$

$a_2 = K_1 + V_1(k_2 + k_3 + k_4)$

$a_3 = K_1(k_3 + k_4) + k_2 k_4 V_1$

$a_4 = -(k_2 + k_3 + k_4)$

$a_5 = -k_2 k_4$

■ **Transiently Linear Solution to Irreversible Accumulation.** For $k_4 = 0$, the multilinear solution to Eqs. 7.34–7.36 was first given by Blomqvist (1984). It is obtained here by setting $k_4 = 0$ in Eq. 7.48 for $m_2(0) = m_3(0) = 0$:

$$\sum_{i=1}^{3} m_i = V_1 c_1 + [K_1 + V_1(k_2 + k_3)] \int_0^T c_1 \, dt$$
$$+ K_1 k_3 \int_0^T \int_0^u c_1 \, dt \, du$$
$$- (k_2 + k_3) \int_0^T \sum_{i=1}^{3} m_i \, dt. \qquad (7.49)$$

■ **Steady-State Solution to Reversible Accumulation.** When $k_4 > 0$, the steady-state flux (J_0) through the system is zero because of the absent sink. The transient solutions describe the time course of the approach of each of the compartments toward steady state. The net steady-state flux of zero is associated with the following distinct sizes of each compartment, depending on the magnitudes of the rate constants:

$$M_3 = \left(\frac{k_3}{k_4} \right) M_2 = \left(\frac{K_1 k_3}{k_2 k_4} \right) C_1 \qquad (7.50)$$

and

$$M_2 = \left(\frac{K_1}{k_2} \right) C_1 \qquad (7.51)$$

yielding the sum of the three compartments:

$$\sum_{i=1}^{3} M_i = \left[V_1 + \frac{K_1}{k_2} \left(1 + \frac{k_3}{k_4} \right) \right] C_1 \qquad (7.52)$$

where M_i is the mass of molecules. The equation shows that the important effect of this closed system of three compartments is to maintain the content of one or more compartments at a specific level dictated by the magnitude of the relaxation constants. It is important to note that only the relative magnitudes of the constants matter. The steady-state volume of distribution is $\sum_{i=1}^{3} M_i/C_1$, according to Eq. 7.52.

■ **Pseudosteady-State Solution to Irreversible Accumulation.** When $k_4 = 0$, the content of the last of the compartments (M_3) continues to increase in accordance with Eq. 7.47 and thus never reaches true steady state. When $T \gg (k_2 + k_3)^{-1}$, the "virtual" volume of distribution of Eq. 7.47 approaches a line of slope K and ordinate intercept $V_1 + K_1 k_2/(k_2 + k_3)^2$ with time. The pseudosteady-state solution to Eq. 7.47, also known as the Rutland-Gjedde-Patlak plot (Gjedde 1982; Patlak et al. 1983), is

$$\sum_{i=1}^{3} v_i(T) \cong \left[V_1 + \frac{K_1 k_2}{(k_2 + k_3)^2} \right] + K \int_0^T \frac{c_1(t)}{C_1(T)} dt \qquad (7.53)$$

7.2.5
Three Compartments with Sink

An open system with a throughput can be created by adding another rate constant to compartment M_3:

$$\frac{dm_1}{dt} = V_1 \frac{dc_1}{dt} , \qquad (7.54)$$

$$\frac{dm_2}{dt} = K_1 c_1 + k_4 m_3 - (k_2 + k_3)m_2 \qquad (7.55)$$

and

$$\frac{dm_3}{dt} = k_3 m_2 - (k_4 + k_5)m_3 \qquad (7.56)$$

with the total flux into the compartments (and all other compartments fed by this open system) equal to

$$j_0 = V_1 \frac{dc_1}{dt} + K_1 c_1 - k_2 m_2 \qquad (7.57)$$

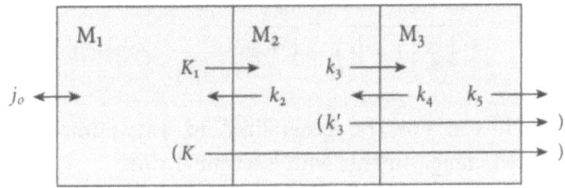

Fig. 7.5. Model of multiple compartment with multiple sinks. Steady-state k_3' equals $k_3 k_5/(k_4 + k_5)$ and reaches k_3 when $k_5 \gg k_4$. Steady-state K equals $K_1 k_3'/(k_2 + k_3')$ and reaches K_1 when $k_3' \gg k_2$ at near equilibrium

The time course of the sum of the magnitudes of all three compartments is a complex solution to Eqs. 7.54–7.56. An easier approach is the numerical solution obtained by simply adding the analytical solutions for the individual compartment, but the number of coefficients makes the solution less useful, and it will not be given here. More important is the net flux through the system at steady state.

7.2.5.1
Generalized Steady-State Solution

The steady-state magnitudes of the compartments are

$$M_3 = \left[\frac{k_3}{k_4 + k_5} \right] M_2 = \left(\frac{K_1}{k_2 + k_3'} \right) \left[\frac{k_3}{k_4 + k_5} \right] C_1 \quad (7.58)$$

and, of course,

$$M_2 = \left(\frac{K_1}{k_2 + k_3'} \right) C_1 \qquad (7.59)$$

such that

$$\sum_{i=1}^{3} M_i = \left(V_1 + \left(\frac{K_1}{k_2 + k_3'} \right) \left[1 + \left(\frac{k_3}{k_4 + k_5} \right) \right] \right) C_1 \qquad (7.60)$$

where the apostrophe ("prime") of the apparent relaxation constant k_3' refers to its definition as a composite of the rate constants of several compartmental interfaces affecting compartment M_3:

$$k_3' = k_3 \left(\frac{k_5}{k_4 + k_5} \right) \simeq k_3 \text{ for } k_5 \gg k_4 \qquad (7.61)$$

The steady-state flux through the system is

$$J_0 = k_5 M_3 = K C_1 \qquad (7.62)$$

where

$$K = \frac{K_1}{1 + \frac{k_2}{k_3}\left(1 + \frac{k_4}{k_5}\right)} = \frac{K_1}{1 + \frac{k_2}{k_3^2}} \quad (7.63)$$

which can now be generalized to any number of nested compartments with a terminal sink:

$$K = \frac{K_1}{1 + \frac{k_2}{k_3}\left(1 + \frac{k_4}{k_5}\left[1 + \ldots \left(1 + \frac{k_{2n-2}}{k_{2n-1}}\right)\right]\right)} \quad (7.64)$$

where n is the number of compartments ending with a sink described by the rate constant k_{2n-1}. The corresponding size of the terminal compartment is given by

$$M_n = \left(\frac{K}{k_{2n-1}}\right) C_1 \quad (7.65)$$

and the combined magnitude of all compartments is

$$\sum_{i=1}^{n} M_i = \left[V_1 + K \left(\frac{1}{k_3'} + \frac{1}{k_5'} + \frac{1}{k_7'} + \ldots + \frac{1}{k_{2i-1}'}\right.\right.$$
$$\left.\left. + \ldots + \frac{1}{k_{2n-1}}\right)\right] C_1 \quad (7.66)$$

where

$$k_{2i-1}' = \frac{k_{2i-1}}{1 + \frac{k_{2i}}{k_{2i+1}}\left[1 + \frac{k_{2i+2}}{k_{2i+3}}\left(1 + \ldots \left[1 + \frac{k_{2n-2}}{k_{2n-1}}\right]\right)\right]} . \quad (7.67)$$

7.2.6
Four or More Compartments

A serial system without a throughput can be created by adding additional compartment (M_4-M_n), such that the system consists of the delivery compartment

$$\frac{dm_1}{dt} = V_1 \frac{dc_1}{dt} , \quad (7.68)$$

the precursor compartment

$$\frac{dm_2}{dt} = K_1 c_1 + k_4 m_3 - (k_2 + k_3) m_2 , \quad (7.69)$$

and multiple product compartments in series, including

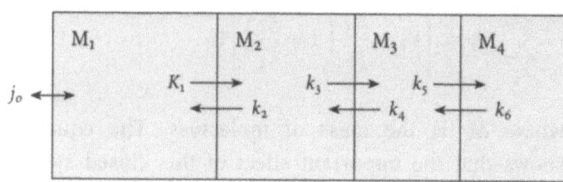

Fig. 7.6. Model of closed system of multiple exchanging compartments

$$\frac{dm_3}{dt} = k_3 m_2 + k_6 m_4 - (k_4 + k_5)m_3 \quad (7.70)$$

and

$$\frac{dm_4}{dt} = k_5 m_3 - k_6 m_4 , \quad (7.71)$$

and ending with the nth compartment

$$\frac{dm_n}{dt} = k_{2n-3} m_{n-1} - k_{2n-2} m_n , \quad (7.72)$$

where the total flux into the four or more compartments generalizes to

$$j_0 = \sum_{i=1}^{n} \left[\frac{dm_i}{dt}\right] = V_1 \frac{dc_1}{dt} + K_1 c_1 - k_2 m_2 . \quad (7.73)$$

7.2.6.1
Convolution of Impulse Response Function as Generalized Transient Solution

The impulse response function is the function that defines the changes of the compartmental contents with time when convolved with the input or "forcing" function, which defines the contents of the delivery compartment, be it the vascular bed or another (reference) tissue. In the case of compartments defined by linear first-order differential equations, the ratio between the integrals of the tissue curve and the forcing function, extrapolated to infinity, yields the ratio between the impulse response function coefficients K_{1_h} and k_{2_h}. Linear systems theory shows that the generalized transiently nonlinear solution to the linked differential Eqs. 7.68–7.72, for $\sum_{i=1}^{n} m_i(0) = 0$, is the convolution of the impulse response function given in Eq. 7.16 with the vascular "forcing" function (Gunn et al. 2001):

$$\sum_{i=1}^{n} m_i = V_1 c_1 + \sum_{h=1}^{n-1} K_{1_h} \int_0^T c_1 e^{-k_{2_h}(T-t)} dt \quad (7.74)$$

where the coefficients K_{1_h} are defined by the equation

$$K_1 = \sum_{h=1}^{n-1} K_{1_h} \qquad (7.75)$$

and the relaxation constants k_{2_h} are defined by the steady-state volume of distribution according to the equation

$$V = \sum_{i=1}^{n} V_i = V_1 + \sum_{h=1}^{n-1} \frac{K_{1_h}}{k_{2_h}} = V_1 + \int_0^\infty r_1(t)\, dt \qquad (7.76)$$

where V is the total steady-state volume of distribution and the impulse response function r_1 is defined as

$$r_1(t) = \sum_{h=1}^{n-1} K_{1_h} e^{-k_{2_h} t} \qquad (7.77)$$

The relaxation constants k_{2_h} individually as well as collectively have no simple biological meaning (see Eq. 7.43). They are, in a sense, *descriptive* rather than *determinant* of the system. The relationship between the parameters of the impulse response function and the relaxation constants of each compartment must be worked out for each constellation of compartments. For the current example of at least four compartments in a series, the relationship is worked out below.

7.2.6.2
Generalized Steady-State and Pseudo-Steady-State Solutions

■ **Reversible Accumulation.** For $k_{2n-2} > 0$, the steady-state flux through the closed system is zero ($J_0 = 0$), but the steady-state contents of the compartments are

$$M_4 = \left(\frac{k_5}{k_6}\right) M_3 = \left(\frac{K_1 k_3 k_5}{k_2 k_4 k_6}\right) C_1$$

and

$$M_3 = \left(\frac{K_1 k_3}{k_2 k_4}\right) C_1$$

and

$$M_2 = \left(\frac{K_1}{k_2}\right) C_1$$

such that the sum of the magnitudes of the four compartments results in a chain of nested coefficient ratios

$$\sum_{i=1}^{4} M_i = \left[V_1 + \frac{K_1}{k_2}\left(1 + \frac{k_3}{k_4}\left[1 + \frac{k_5}{k_6}\right]\right)\right] C_1 \qquad (7.78)$$

that can be generalized to a closed system of n compartments in series:

$$\sum_{i=1}^{n} M_i = \left[V_1 + \frac{K_1}{k_2}\left(1 + \frac{k_3}{k_4}\left[1 + \frac{k_5}{k_6}\left(1 + \cdots \right.\right.\right.\right.$$
$$\left.\left.\left.\left. + \left[1 + \frac{k_{2n-3}}{k_{2n-2}}\right]\right)\right]\right)\right] C_1 \qquad (7.79)$$

where the last, or nth, compartment, relative to the first compartment, has the size

$$M_n = \left(\frac{K_1 k_3 k_5 \ldots k_{2n-3}}{k_2 k_4 k_6 \ldots k_{2n-2}}\right) C_1$$

where n is the number of compartments. The generalization shows that one important function of a closed system is to maintain the steady-state content of one or more compartments at a level regulated by the relaxation constants of all preceeding compartments in the series. For example, the last compartment may represent a neurotransmitter in the synaptic cleft.

The impulse response function yields the total content of the compartments in terms of its own relaxation constants according to Eq. 7.73:

$$\sum_{i=1}^{n} M_i = \left[V_1 + \sum_{h=1}^{n-1} \frac{K_{1_h}}{k_{2_h}} \right] C_1 \qquad (7.80)$$

which yields the relationship between the individual compartmental relaxation constants and the powers of the exponentials of the impulse response function by substitution of Eq. 7.79.

For $k_{2n-2} = 0$, the last compartment never reaches a steady state. The compartment continues to expand while the preceding compartments eventually reach secular equilibria. For $n = 4$, the steady-state solution to this process is the Rutland-Gjedde-Patlak equation (Eq. 7.53):

$$\sum_{i-1}^{4} v_i(T) = V_1 + K\left(\frac{1}{k_3'} + \frac{1}{k_5}\right) + K \int_0^T \frac{c_1(t)}{c_1(T)}\, dt \qquad (7.81)$$

where the apparent relaxation constant k_3' is

$$k_3' = k_3 \left(\frac{k_5}{k_4 + k_5} \right) \tag{7.82}$$

and

$$K = \frac{K_1}{1 + \frac{k_2}{k_3}\left(1 + \frac{k_4}{k_5}\right)} = \frac{K_1}{1 + \frac{k_2}{k_3'}} \tag{7.83}$$

which can be generalized to n compartments

$$\sum_{i=1}^{n} v_i(T) = \left[V_1 + K \left(\frac{1}{k_3'} + \frac{1}{k_5'} + \frac{1}{k_7'} + \ldots + \frac{1}{k_{2i-1}'} \right. \right.$$
$$\left. \left. + \ldots + \frac{1}{k_{2n-3}} \right) \right] + K \int_0^T \frac{c_1(t)}{c_1(T)}\, dt \tag{7.84}$$

where

$$k_{2i-1}' = \frac{k_{2i-1}}{1 + \frac{k_{2i-2}}{k_{2i-1}}\left[1 + \frac{k_{2i}}{k_{2i+1}}\left(1 + \ldots \left[1 + \frac{k_{2n-4}}{k_{2n-3}}\right]\right)\right]} \tag{7.85}$$

and

$$K = \frac{K_1}{1 + \frac{k_2}{k_3}\left(1 + \frac{k_4}{k_5}\left[1 + \ldots \left(1 + \frac{k_{2n-4}}{k_{2n-3}}\right)\right]\right)} \tag{7.86}$$

The impulse response function defines the net clearance K in terms of the "unidirectional" clearance of the irreversibly accumulating compartment $K_{1_{n-1}} = K$.

7.2.7
Multiple Compartments in Series and in Parallel

With a few exceptions (Gunn et al. 2001), the impulse response function defines any constellation of linear compartmental systems, which can be described by first-order linear differential equations (Gunn et al. 2002), regardless of the structure. Thus, compartments in parallel also follow the general principles underlying the steady-state solutions. The impulse response function describes the constellation, but its biological significance must be extracted from an analysis of the relationship between the relaxation constants of each compartment.

7.2.7.1
Generalized Steady-State Solution of Closed System of Branching Compartments

In a closed ("reversible") system, the steady-state flux through the system is of course zero ($J_0 = 0$), but the solution to the steady-state magnitude of the combined compartments, generalized for any closed system of several parallel branches extending n steps from the origin, shows that parallel compartments are additive, while serial compartments are nested:

$$\sum_{i=1}^{n} M_i = \left[V_1 + \frac{K_1}{k_2}\left(1 + \frac{k_{3_1}}{k_{4_1}}\left[1 + \frac{k_{5_{1_1}}}{k_{6_{1_1}}}\ldots\right]\right) + \frac{k_{3_2}}{k_{4_2}}\right.$$
$$\left. \times \left[1 + \frac{k_{5_2}}{k_{6_2}}\left(1 + \ldots \left[1 + \frac{k_{(2n-3)_2}}{k_{(2n-2)_2}}\right]\right)\right]\right] C_1 \tag{7.87}$$

where the additional subscripts refer to the branch in question. The last compartment of the n_{2nd} branch, relative to the first compartment, has the size

$$M_{n_2} = \left(\frac{K_1\, k_{3_2}\, k_{5_2} \ldots k_{(2n-3)_2}}{k_2\, k_{4_2}\, k_{6_2} \ldots k_{(2n-2)_2}} \right) C_1 \tag{7.88}$$

where n is the number of steps the terminus is removed from the source. The generalization shows that one important effect of a closed system is to maintain the steady-state content of one or more compartments at a level regulated by the rate constants of the involved mechanisms. For example, the last compartment may represent a neurotransmitter and its level in the synaptic cleft. This interpretation introduces the general topic of interpretation of the rate constants.

7.2.7.2
Generalized Steady-State Solution of Open System of Branching Compartments

For an open system of n compartments in one branch ("2"), the net flux out of the branch is dictated by the net clearance to the sink of the branch:

$$K_{1_2} = \frac{K_1}{1 + \frac{k_2}{k_{3_2}}\left(1 + \frac{k_{4_2}}{k_{5_2}}\left[1 + \ldots \left(1 + \frac{k_{(2n-2)_2}}{k_{(2n-1)_2}}\right)\right]\right)} \tag{7.89}$$

where $K = K_{1_2}$ is the net clearance by branch 2 and n is the number of compartments leading to the sink governed by the rate constant k_{2n-1_2}. The correspond-

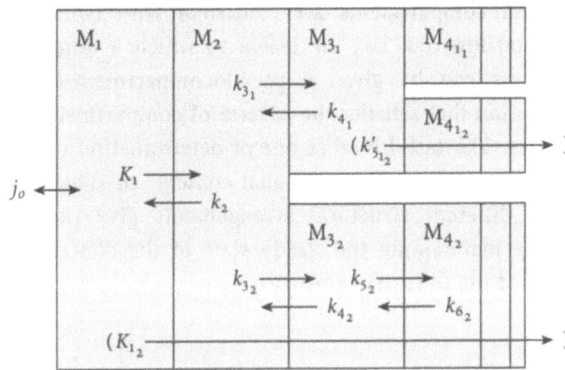

Fig. 7.7. Model of multiple compartments in series and in parallel. Compartments in parallel are additive; compartments in series are nested. Branches 1_2 and 2 are open ("irreversible"); branch 1_1 is closed ("reversible"). Net clearances for open systems are shown in brackets. Note that apostrophes ("primes") refer to composite rate constants or clearances that traverse several membranes

ing magnitude of the nth compartment in this branch (open system) is given by

$$M_{n_2} = \left(\frac{K_{1_2}}{k_{(2n-1)_2}}\right) C_1 \qquad (7.90)$$

and the combined magnitude of all compartments (excluding the sink) in this open branch is

$$\sum_{i=1}^{n} M_{i_2} = \left[V_1 + K'_{1_2}\left(\frac{1}{k'_{3_2}} + \frac{1}{k'_{5_2}} + \frac{1}{k'_{7_2}} + \dots\right.\right.$$
$$\left.\left. + \frac{1}{k'_{(2i-1)_2}} + \dots + \frac{1}{k_{(2n-1)_2}}\right)\right] C_1 \qquad (7.91)$$

where

$$k'_{(2i-1)_2} = \frac{k_{(2i-1)_2}}{1 + \frac{k_{(2i)_2}}{k_{(2i+1)_2}}\left[1 + \frac{k_{(2i+2)_2}}{k_{(2i+3)_2}}\left(1 + \dots\left[1 + \frac{k_{(2n-2)_2}}{k_{(2n-1)_2}}\right]\right)\right]}$$

$$(7.92)$$

where the apostrophe refers to the composite nature of the rate constant, which traverses multiple membranes.

7.3
Interpretation of Relaxation Constants

The biological reality is that concentrations vary in space but not normally in time. The compartmental kinetics must be translated from the space-invariant and time-variant model systems to the space-variant but time-invariant real pathways. Note that time-invariant (steady-state) variables, as above, are given as uppercase symbols.

In many cases, the relaxation constant k represents one of just a few processes including flow, diffusion, membrane permeability, enzyme reaction, facilitated diffusion across membranes, or receptor binding. Depending on the processes known to establish the compartments occupied by a metabolite or by tracer molecules, relaxation constants can be estimated and interpreted in specific ways. In the case of tracer molecules, the molecules trace a specific set of processes, often because the tracer has been designed for the purpose of revealing the kinetics of just these processes.

7.3.1
Flow

When the flow of a solvent links two compartments, the relaxation of the downstream compartment M_2 is due to the washout of the molecules dissolved in the effluent. This is an example of an open vascular system. The products $k_1 V_1$ and $k_2 V_2$ both equal the flow rate F, and k_2 equals the flow through the system per unit volume of the second compartment. Thus, the process establishes a proper compartment when F is constant:

$$k_2 = F/V_2 \qquad (7.93)$$

as shown in the model above.

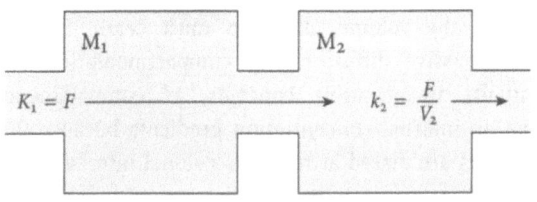

Fig. 7.8. Convection model of precursor and product compartments perfused by solvent

The direct use of this model is limited to the dissolution of inert flow tracers in the bloodstream, which allow the estimation of flow rate, volume of distribution, and mean transit time ("bolus tracking"), as in Ostergaard et al. (1998).

7.3.2
Passive Diffusion

Diffusion is a fundamental process of elimination of concentration differences and release of entropy leading to a flux in the direction of a concentration gradient. In the unidirectional case, according to Fick's first law, the one-dimensional flux is given by (Gjedde 1995a)

$$J = -DA \frac{dC}{dx} \tag{7.94}$$

where D is the diffusion coefficient, A is the cross-sectional area of the volume through which the unidirectional diffusion occurs, and dC/dx is the *gradient* (G), equal to the slope of the concentration curve or plane at the coordinate x. In a closed system, the concentration gradient is eventually dissipated until $J = 0$ when $dC/dx = 0$. The system is at *equilibrium* when net flux is no longer taking place and dC/dt reaches zero. If the diffusing substance undergoes no change other than the diffusion itself, the temporal rate of elimination of the concentration gradient is proportional to the spatially negative rate of change of the gradient:

$$\frac{dc}{dt} = -D \frac{dG}{dx} \tag{7.95}$$

where G is the gradient, such that

$$\frac{\partial c}{\partial t} = +D \frac{\partial^2 c}{\partial x^2} \tag{7.96}$$

where the diffusion coefficient is the proportionality factor. Thus, when the substance is uniformly distributed in the volume, diffusion must cease. In this sense, passive diffusion and compartmentation are mutually incompatible concepts, as compartments have no internal concentration gradients because the gradients are placed at two-dimensional interfaces between the compartments.

The purpose of the treatment below is to establish the extent to which varying concentration profiles within compartments are consistent with compartmental kinetics, i.e., the extent to which a diffusion process can be given a pseudocompartmental description that satisfies the criteria of compartment kinetics. The task becomes one of determination of the concentration profiles and total contents of substance that different structural arrangements give rise to while maintaining the steady state of the concentration of the diffusing substances.

7.3.2.1
Steady-State One-Dimensional Diffusion

In the simplest steady-state condition, the rate of change of the diffusing substance with respect to time is nil ($\partial c_1/\partial t = 0$), but the rate of change of the diffusing substance with respect to distance (one-dimensional space) is significant ($\partial c_1/\partial x \neq 0$). This condition can be evaluated by application of the diffusion equation to the case of one-dimensional diffusion in one direction (x) through a volume ("box") in which no concentration gradients exist in the orthogonal directions y and z and substance is lost or gained only at the ends of the box (Fig. 7.9).

When the concentration of the substance is kept constant, the spatial rate of change of the gradient is nil:

$$\frac{d^2C}{dx^2} = 0 \tag{7.97}$$

and the result is a constant (or uniform) gradient:

$$G(x) = \frac{dC}{dx} = \alpha \tag{7.98}$$

with a linearly stationary concentration profile:

$$C(x) = \alpha x + C(0) \tag{7.99}$$

with a slope of the magnitude α:

$$\alpha = \frac{C(x) - C(0)}{x} \tag{7.100}$$

and a steady-state flux for a box of length $x = L$ of

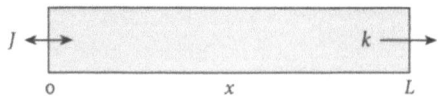

Fig. 7.9. Single compartment with simple passive one-dimensional diffusion due to Brownian motion

$$J_0 = -DA\frac{C(L) - C(0)}{L} = -DA\frac{\Delta C}{L} \qquad (7.101)$$

as well as a steady-state content of the "box" of

$$M = LA\frac{C(L) + C(0)}{2} = LA\overline{C} \qquad (7.102)$$

where $\overline{C}$ is the average steady-state concentration of the compartment. Then, the effective relaxation constant of the efflux is

$$k = \frac{J}{M} = -\left(\frac{D\,\Delta C}{L^2\,\overline{C}}\right) \qquad (7.103)$$

where $L^2/(2D) = \tau_D$ is the characteristic time constant of diffusion along the distance L obtained from the equation derived by Einstein (1908), such that

$$k = -\left(\frac{\Delta C}{2\,\tau_D\,\overline{C}}\right) \qquad (7.104)$$

where the relaxation constant of the substance undergoing passive diffusion is the inverse of the time constant of diffusion if the concentration $C(0)$ is kept constant by quantitative removal to the external medium. From the opposite perspective, the product of k and τ_D can be said to determine the difference between the concentrations at the two extremes of the diffusion path:

$$\Delta C = -2\,k\,\tau_D\,\overline{C} \qquad (7.105)$$

where τ_D is proportional to the square of the length of the diffusion path. Thus, a large concentration difference is favored by long diffusion paths, and, conversely, a small or negligible concentration difference is favored by very short diffusion paths, once the relaxation constant of the compartment is given.

7.3.2.2
Steady-State Passive Diffusion with Concentration-Dependent Loss

Unless the external concentrations are kept constant, equilibrium will eventually ensue. Equilibrium is anathema to the biological *steady state* in which $dC_1/dt = 0$ because the gradient is preserved by energy-requiring processes that maintain the flux by establishing constant concentrations at both ends of the diffusion path. Alternatively, substance may be removed continuously from the diffusion path by a con-

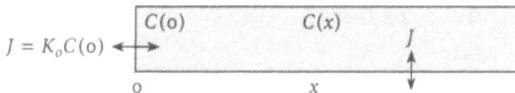

Fig. 7.10. Single compartment with simple passive one-dimensional diffusion through infinitely long tube with concentration-dependent loss

centration-dependent process occurring in the directions orthogonal to the direction of passive diffusion (x). If the volume is sufficiently narrow in the orthogonal directions (infinitely long narrow tube with a cross-sectional area of A) and the product of the rate constant of removal and the time constant of diffusion is very small, no concentration gradients exist in the orthogonal directions.

The spatial rate of change of the gradient is now a function of the rate of removal of the substance along the length of the tube:

$$\frac{d^2C}{dx^2} = \left(\frac{k}{D}\right)C(x) \qquad (7.106)$$

where x is the distance along the length of the tube and k is the relaxation constant of a concentration-dependent loss in the orthogonal y and z directions. At steady state, the gradient is the monoexponential function satisfying Eq. 7.106:

$$G(x) = \frac{dC}{dx} = -C(0)\,e^{-x\sqrt{k/D}}\,\sqrt{k/D} \qquad (7.107)$$

with the concentration profile

$$C(x) = C(0)\,e^{-x\sqrt{k/D}} \qquad (7.108)$$

as well as the total content of the infinitely long tube

$$M = A\int_0^{\infty} C(x)\,dx = A\,C(0)\,\sqrt{D/k} \qquad (7.109)$$

where M is the steady-state content of the infinitely long tube in which the concentration declines to zero. The entire loss of substance therefore occurs in the orthogonal directions, i.e., through the wall of the tube. The efflux through the wall of this virtual compartment is given by

$$J = k\,A\int_0^{\infty} C(x)\,dx = A\,C(0)\,\sqrt{Dk} \qquad (7.110)$$

which confirms that k is indeed the steady-state relaxation constant of the compartment

$$k = \frac{J}{M} = D \left(\frac{A\,C(0)}{M} \right)^2 \qquad (7.111)$$

where the term $M/[A\,C(0)]$ defines an effective diffusion distance in the tube. As the compartment is in steady state and no substance is lost or gained at the end of the tube, the flux into the origin of the tube must be of the same magnitude as the flux through the wall. The influx is determined by the value of the gradient at $x = 0$:

$$J = -D\,A\,G(0) = A\,C(0)\,\sqrt{Dk} \qquad (7.112)$$

such that Eq. 7.108 becomes also

$$C(x) = C(0)\,e^{-x\,kA\,C(0)/J} \qquad (7.113)$$

in which the spatial rate constant of the monoexponential decline is a function of the rate constant of loss or gain through the wall of the tube and the apparent clearance of substance by diffusion into the tube. We now define a clearance as the volume of the precursor compartment that per unit time is cleared of the substance by the diffusion into the infinitely long tube:

$$C(x) = C(0)\,e^{-x\,kA/K_0} \qquad (7.114)$$

where K_0 is defined as the clearance $J/C(0)$, equal to $A\sqrt{Dk}$.

7.3.2.3
Steady-State Passive Diffusion
with Concentration-Dependent Loss
and Concentration-Independent Gain
(Definition of Bidirectionality Factor f_b)

Substance can also be added to the diffusion path by a concentration-independent transfer in the directions orthogonal to the direction of passive diffusion. Such a process could be a second compartment continuously feeding the diffusion path.

The spatial rate of change of the gradient is now a function of the rate of removal of the substance along the length of the tube:

$$\frac{d^2 C_1}{dx^2} = \left(\frac{k}{D} \right) [C_1(x) - C_2] \qquad (7.115)$$

where x is the distance along the length of the tube, k is the relaxation constant of the concentration-de-

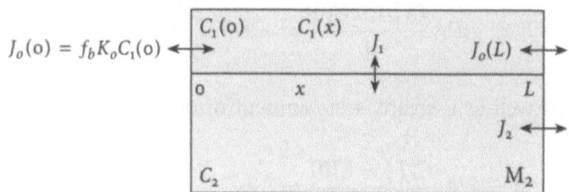

Fig. 7.11. System of two compartments with simple passive one-dimensional diffusion through tube of finite length with concentration-dependent loss to, and concentration-independent gain from, surrounding mantle compartment

pendent loss, and $I = kC_2$ is the constant rate of gain by flux into the diffusion path. The concentration C_2 is a derived variable, defined as the ratio $I/k\,(k \neq 0)$, which is assumed not to undergo change. The treatment does not assume that the rate of gain, I, necessarily is by means of the permeability symbolized by k. At steady state, the gradient is given by the appropriate boundary conditions (see Fig. 7.11):

$$G(x) = \frac{dC_1}{dx} = \sqrt{k/D}\,[C_2 - C_1(0)]\,e^{-x\sqrt{k/D}} \quad (7.116)$$

which when integrated yields the concentration profile

$$C_1(x) = C_2 + (C_1(0) - C_2)e^{-x\sqrt{k/D}} \qquad (7.117)$$

that approaches C_2 rather than the zero of the previous condition. The total influx of substance at the origin of the tube can be evaluated from the gradient at $x = 0$:

$$J_0(0) = -DA\,G(0) = A\sqrt{Dk}\,(C_1(0) - C_2) \qquad (7.118)$$

which relates the influx to the concentration of the diffusing substance at the entrance and to the net gain or loss of substance by transfer in the directions orthogonal to the diffusion path. Let the clearance of substance into the diffusion path from both external sources be K_0' to distinguish it from the clearance previously defined as K_0:

$$K_0' = \frac{J_0(0)}{C_1(0)} = A\left(1 - \frac{C_2}{C_1(0)}\right)\sqrt{Dk} = f_b\,A\sqrt{Dk} = f_b\,K_0$$
$$(7.119)$$

where the fraction f_b, equal to $1 - [C_2/C_1(0)]$, depends on the concentration difference. The fraction f_b accounts for the reduction of the clearance ("backflux") by the bidirectional transfer, such that the influx is

$$J_0(0) = f_b K_0 C_1(0) \tag{7.120}$$

for which the concentration profile along the diffusion path is

$$C_1(x) = C_1(0)\left[1 - f_b\left(1 - e^{-xkA/K_0}\right)\right] \tag{7.121}$$

and the efflux of substance at $x = L$ likewise becomes

$$J_0(L) = -DA\,G(L) = A\,(C_1(0) - C_2)\,e^{-L\sqrt{k/D}}\,\sqrt{Dk}$$
$$= f_b K_0 C_1(0)\,e^{-kLA/K_0} \tag{7.122}$$

where the product LA is the volume of the tubular diffusion path. The net flux in the directions orthogonal to the diffusion path can now be calculated as the difference

$$J_1 = J_0(0) - J_0(L) = f_b K_0 C_1(0)\left(1 - e^{-kLA/K_0}\right) \tag{7.123}$$

which is zero when f_b is zero.[1]

7.3.3
Properties of Delivery Compartment

7.3.3.1
Diffusion-Limited Membrane Permeability

Membrane permeability is a special case of passive diffusion without loss or gain in which the dimensions are fixed but the concentrations on the two sides of the membrane may vary. In this treatment, the membrane is the lining of the narrow tube discussed above.

The diffusion flux is described by Eq. 7.101, modified to include the concentrations of the diffusing

---

[1] The solution is not defined for k equal to zero because the gradient expressed in Eq. 7.116 then has the simpler solution

$$G(x) = \frac{dC}{dx} = \frac{I}{D}x + H_1 \tag{7.124}$$

where $C_2 \neq I/k$, which yields the concentration profile

$$C(x) = \frac{I}{2D}x^2 + H_1 x + H_2 \tag{7.125}$$

where H_1 and H_2 depend on the conditions. This formula can be applied to the concentration-independent consumption of oxygen diffusing unidirectionally through a tissue.

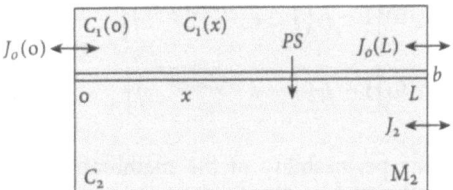

Fig. 7.12. Two compartments separate by membrane with diffusion-limited permeability through wall of thickness b of tube of finite length L

substance in the medium surrounding the membrane, rather than in the membrane itself:

$$dJ = -D'dS\,\frac{C_1(x,b) - C_1(x,0)}{b} = -D'dS\,\frac{\Delta C_1}{b} \tag{7.126}$$

where b is the width of the membrane. The ratio D'/b is defined as the permeability coefficient P, the prime referring to the possibly different solubilities of the diffusing substance in the membrane and the surrounding media, and S is defined as the surface area of the membrane to distinguish that area from A, the cross-sectional area of the tube. Thus,

$$dJ = -PdS\,\Delta C_1 \tag{7.127}$$

where PdS (as a clearance) has unit of flow. Let a tube of length L and inner volume $V_1 = AL$ be lined by a membrane of thickness b and surface area S for which $C_1(x,b) = C_2$ for all x. The relaxation constant of the efflux through the membrane is then given by Eq. 7.93:

$$k = P\,dS/dV_1 = PS/V_1 \tag{7.128}$$

such that together with Eq. 7.121 it yields the concentration profile along the length of the tube

$$C_1(x) = C_1(0)\left[1 - f_b\left(1 - e^{-(x/L)(PS/K_0)}\right)\right] \tag{7.129}$$

where $LA = V_1$, the volume of the entire length of the tube. The relation expresses the monoexponential decline of the concentration along the length of the tube when diffusion in the tube is accompanied by permeation of the membrane lining the tube. For $x = L$, the concentration declines to the expression

$$C_1(L) = C_1(0)\left[1 - f_b\left(1 - e^{-PS/K_0}\right)\right] \tag{7.130}$$

where, as above, $K_0 = A\sqrt{Dk} = \sqrt{DAPS/L}$, such that

$$C_1(L) = C_1(0)\left[1 - f_b\left(1 - e^{-\sqrt{PSL/(DA)}}\right)\right]$$

$$= C_1(0)\left[1 - f_b\left(1 - e^{-L\sqrt{2P/(Dr)}}\right)\right] \quad (7.131)$$

where P is the permeability of the membrane, D is the diffusion coefficient of the substance diffusing through the tube, and r is the radius of the tube. It turns out that this expression can be generalized also to the case of convection through the tube in which the clearance into the proximal end of the tube, and hence the diffusion, is replaced by convection.

When the tube is of finite length, the relaxation constant of the compartment formed by the tube no longer equals just PS/V_1. The content of the compartment is

$$M_1(L) = A \int_0^L C_1(0)\left(1 - f\left[1 - e^{-\frac{xPS}{LK_0}}\right]\right) dx$$

$$= V_1 C_1(0)\left[1 - f_b\left(1 - \frac{1 - e^{-PS/K_0}}{PS/K_0}\right)\right] \quad (7.132)$$

where $M_1(L)$ is the total mass of the substance in the tube of finite length L in which the concentration declines to $C_1(L)$. The decay of the compartment therefore occurs both through the wall and at the end of the tube. The net flux across the wall of the tube is given by

$$J_1 = \frac{PSA}{V_1} \int_0^L (C_1(x) - C_2)\, dx$$

$$= f_b C_1(0) K_0 \left(1 - e^{-PS/K_0}\right) \quad (7.133)$$

while the efflux from the end of the tube is given by

$$J_0(L) = f_b K_0 C_1(0)\, e^{-PS/K_0} \quad (7.134)$$

such that the total flux, equal to the flux at the entry, is

$$J_0(0) = f_b K_0 C_1(0) \quad (7.135)$$

which yields the true relaxation constant of the entire contents of the compartment formed by the tube of finite length:

$$k_0 = \frac{J_0(0)}{M_1} = f_b\left[\frac{K_0}{V_1}\right]$$

$$\times \left(\frac{PS}{PS(1 - f_b) + f_b K_0 \left(1 - e^{-PS/K_0}\right)}\right) \quad (7.136)$$

which approaches $f_b K_0 / V_1$ when the PS product declines relative to the magnitude of the clearance (observe the equivalence with Eq. 7.93), and $k = PS/V_1$ when the PS product is large relative to K_0 and f_b is unity (no gain by influx from compartment M_2). The effective relaxation constant of the loss to compartment M_2 is

$$k_1 = \frac{J_1}{M_1} \quad (7.137)$$

such that

$$k_1 = f_b\left[\frac{K_0}{V_1}\right]\left(\frac{1 - e^{-PS/K_0}}{1 - f_b\left(1 - \frac{1 - e^{-PS/K_0}}{PS/K_0}\right)}\right)$$

$$= f_b\left[\frac{K_0}{V_0}\right]\left(1 - e^{-PS/K_0}\right) \quad (7.138)$$

where V_0 is an apparent volume defined as $M_1/C_1(0)$, which expresses the relaxation of the diffusion path in terms of a virtual compartment with a uniform concentration equal to $C_1(0)$. This convention converts the diffusion path to a virtual compartment, which conforms to the fundamental definition.

7.3.3.2
Flow-Limited Membrane Permeability

Flow-limited membrane permeability is a special case of flow-dependent exchange between two compartments when solvent flows through the precursor compartment and a fixed fraction of the solvent is cleared of the solute by permeation of the solute into the second compartment. The two compartments are separated by a membrane that is impermeable to the solvent but not to the solute (semipermeable membrane). A typical example of this system is the capillary with its surrounding tissue mantle.

The exchange of substance between the capillary and the tissue is assumed to be nonenergy-requiring (passive) and thus to proceed only in the direction of the solute-concentration difference between the two compartments. The system is considered to be closed if the solute in the product compartment decays exclusively by return to the precursor compartment.

As described above, the contents of the capillary bed do not reside in a true compartment because the relaxation constant for the flux through the wall, by means of which a fraction of the solute ($E = J_1/J_0(0)$) enters   the   exchange   compartment,   equals

$k_1 = f_b K_0 (1 - e^{-PS/K_0})/V_0$ rather than PS/V_1. Nonetheless, as derived above (Eq. 7.138), it is possible to reduce the capillary to the status of a compartment by showing when it decays as required of a compartment.

The clearance of a fraction of the contents of the delivery compartment to the exchange compartment is a function both of the permeability-surface area product of the membrane through which the transfer occurs and the flow of the solvent in the tube lined by this membrane, as derived by Crone (1963). Crone derived the clearance on the condition that no more than a negligible amount of capillary solute actually accumulates in the tissue compartment (i.e., when $C_1(x, b) = 0$ for all x), for example, because the tissue volume of distribution is very large or because the solute is quantitatively consumed in the tissue. In this condition, $f_b = 1$, where $1 - f_b$ is the concentration ratio between tissue and capillary at the site of arterial entry.

The clearance equations can be derived for the bidirectional case in which f_b is less than unity. According to one treatment, the flow of a solute through the delivery compartment, combined with the transaxial loss of solute to the surrounding mantle, can be represented by an apparent diffusion coefficient. Let the delivery compartment be a narrow tube of length L, cross-sectional area A, volume V_1, wall permeability-surface area PS, and flow F. The apparent diffusion coefficient of the solute is then

$$D_{app} = \left(\frac{F}{A}\right)^2 \frac{V_1}{PS} \tag{7.139}$$

such that the fundamental flux equation for the loss of solute at any distance along the capillary is given by an expression that is formally analogous to the diffusion equation (Eq. 7.94)

$$J_1 = -D_{app} A \frac{dC_1}{dx} = -\left(\frac{F}{A}\right)^2 \frac{V_1 A}{PS} \frac{dC_1}{dx} \tag{7.140}$$

where the spatial rate of change of the solute gradient in the delivery compartment at steady state is given by

$$\frac{d^2 C_1}{dx^2} = \left(\frac{PS}{FL}\right)^2 (C_1(x) - C_2) = \left(\frac{PS}{FL}\right)^2 (C_1(x) - [1 - f_b] C_1(0)) \tag{7.141}$$

such that the gradient is

$$\frac{dC_1}{dx} = \left(\frac{PS}{FL}\right) \left[(f_b - 1) C_1(0) - f_b C_1(0) e^{-xPS/(FL)} \right] \tag{7.142}$$

and the concentration profile of the solute in the delivery compartment is

$$C_1(x) = f_b C_1(0) e^{-x \, PS/(FL)} + (1 - f_b) C_1(0) \tag{7.143}$$

7.3.3.3
Capillary Model of Bidirectional Flux

In the living organism, the delivery compartment is usually (but not always) a capillary. In those cases, replace $C_1(0)$ by C_a, the symbol of the arterial concentration, x by L, and $C_1(L)$ by C_v, the symbol of the venous concentration. The fundamental equations of the capillary exchange then emerge from Eq. 7.143 as

$$C_v = C_a \left[1 - f_b \left(1 - e^{-PS/F} \right) \right] \tag{7.144}$$

in which the net flux across the capillary wall is given by Eqs. 7.123 and 7.133 for $C_1(0) = C_a$ and $K_0 = F$:

$$J_1 = f_b C_a F \left(1 - e^{-PS/F} \right) \tag{7.145}$$

and the convective flux from the outside into the tube

$$J_0(0) = F C_a \tag{7.146}$$

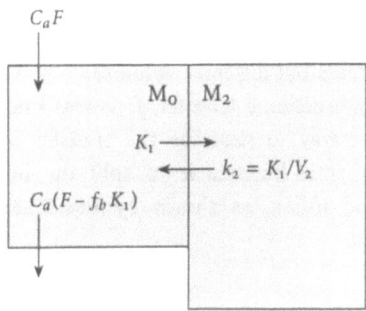

Fig. 7.13. Compartmental model of capillary and tissue compartments in which the tissue compartment clears precursor from solvent flowing through the capillary compartment by means of semipermeable membrane. M_0 is a virtual compartment with the smaller (virtual) volume V_o, uniform concentration $C_a = C_1(0)$, and a substance content of M_1. The fraction $1 - f_b$ is the ratio between concentrations of solute in tissue and capillary compartments

whence the extraction fraction associated with this unidirectional transfer arises as the ratio between the flux into the capillary and the efflux through the wall:

$$E = J_1/J_0(0) = f_b\left(1 - e^{-PS/F}\right) \tag{7.147}$$

which defines the steady-state extraction fraction. For the case of zero tissue concentration, $f_b = 1$ yields

$$E_0 = 1 - e^{-PS/F} \tag{7.148}$$

The total actual content of the capillary compartment is given by Eq. 7.132:

$$M_1 = V_1 C_a\left[1 - f_b\left(1 - \frac{1 - e^{-PS/F}}{PS/F}\right)\right] = V_0 C_a \tag{7.149}$$

in which V_0 defines a virtual volume of the delivery compartment with the uniform concentration C_a. The virtual delivery compartment represents the redistribution of M_1 in a volume, V_0, so small that the concentration is uniformly $C_1(0)$. The redistribution shows that M_1 fulfills the criteria of a compartment with a volume of V_0:

$$V_0 = V_1\left[1 - f_b\left(1 - \frac{1 - e^{-PS/F}}{PS/F}\right)\right] \tag{7.150}$$

where the real average concentration in the physical volume V_1 is

$$\overline{C}_1 = C_a\left[1 - f_b\left(1 - \frac{1 - e^{-PS/F}}{PS/F}\right)\right] \tag{7.151}$$

such that the "virtual" and the real compartments have the same contents but different volumes.

In the case of bidirectional transfer, f_b is less than unity. The simplest way to describe the transfer in terms of compartmental kinetics is to split the net flux into efflux and influx, as shown by Eqs. 7.135 and 7.145, such that

$$J_1 = f_b C_a F E_0 = f_b C_a K_1 \tag{7.152}$$

where K_1 is an apparent clearance, also known as the capillary diffusion capacity, in the direction of transfer from delivery compartment to exchange compartment. It is also the ratio between a fictive unidirectional flux through the wall of the capillary and the arterial concentration, obtained by setting $f_b = 1$:

$$K_1 = F\left(1 - e^{-PS/F}\right) = F E_0 \tag{7.153}$$

such that the same fictive unidirectional flux is given by

$$\overrightarrow{J_1} = K_1 C_a \tag{7.154}$$

where the relaxation constant k_1 for this fictive unidirectional flux is

$$k_1 = \frac{\overrightarrow{J_1}}{M_1} = F\left(1 - e^{-PS/F}\right)/V_0 = K_1/V_0. \tag{7.155}$$

The fictive unidirectional flux in the opposite direction is then

$$\overleftarrow{J_1} = (1 - f_b)\, C_a K_1 \tag{7.156}$$

with the relaxation constant given by the fractional clearance

$$k_2 = \frac{\overleftarrow{J_1}}{M_2} = \frac{F}{V_2}\left(1 - e^{-PS/F}\right) = K_1/V_2 \tag{7.157}$$

where M_2 equals $(1 - f_b)\, C_a V_2$, indicating that the compartment exists when the flow, volume, and permeability-surface area terms are at steady state, as shown in the model of Fig. 7.13.

Very large values of the PS term relative to F turn the clearance compartment into a flow-limited compartment in which the magnitude of the flow determines the influx to the compartment. On the other hand, very low values of the PS product relative to F turn the compartment into a diffusion-limited compartment in which the magnitude of the permeability determines the influx to the product compartment. The ratio between the parameters K_1 and K_1/V_2 is the volume of the solvent V_2 in the product compartment.

Thus, depending on the magnitude of the PS product, the rate constant is an index either of the flow through the system relative to its volume or of the permeability-surface area product of the interface, also relative to the volume of the system.

7.3.4
Protein-Ligand Interaction

In their simplest form, the three processes of enzyme reaction, facilitated diffusion across membranes, and receptor binding are variations on the same underly-

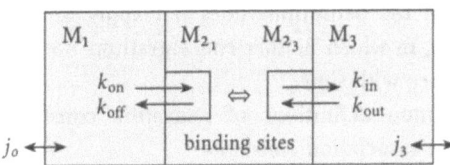

Fig. 7.14. Model of protein-ligand interaction. Unbound ligands (m_1 and m_3) in respective compartments (M_1 and M_3) interact with sites on protein (M_{2_1} and M_{2_3}). Maximum number of available sites is B_{max}

ing theme. The theme received its steady-state formulation by Briggs & Haldane (1925) as an extension of the original equilibrium solution of Michaelis & Menten (1913). The key mechanism is the binding of a ligand to sites on the protein. The interacting compartment M_2 is normally either open, when it acts as a catalyst or transporter, or closed, when it acts predominantly as a receptor.

7.3.4.1
Competitive Interaction

The protein releases the bound ligand (m_2) at a constant rate to the same or a different compartment, either as the intact precursor (m_1) or as a transformed or translocated product (m_3), thus freeing the protein for new occupation, according to the general formula

$$\frac{dm_2}{dt} = (j_1 + j_2) - (k_2 + k_3)\, m_2 \qquad (7.158)$$

where j_1 and j_2 represent the fluxes across the interfaces between M_1 and M_{2_1} and between M_3 and M_{2_3}.

The size of the protein compartment is a function of the magnitude of these coefficients, some of which combine to define the interaction of the protein with two or more competitors for the same sites:

$$\frac{dm_2}{dt} = \left[\frac{k_{on}\,(B_{max} - m_2)}{V_1}\right] m_1 + \left[\frac{k_{out}\,(B_{max} - m_2)}{V_3}\right]$$
$$\times\, m_3 - (k_{off} + k_{in})\, m_2 \qquad (7.159)$$

where m_1 and m_3 are the masses of competitors in compartments M_1 and M_3, and m_2 is the sum of all competitors attached to the binding sites. Equation 7.159 is usually rearranged to allow an evaluation of whether M_2 has the characteristics of a compartment.

Combination of the variables of Eq. 7.159 from this perspective yields an equation with m_2 as the independent variable:

$$\frac{dm_2}{dt} = \left[\frac{k_{on}\, B_{max}}{V_1}\right] m_1 + \left[\frac{k_{out}\, B_{max}}{V_3}\right] m_3$$
$$- (k_{off} + k_{in})\, m_2 - (k_{off} + k_{in}) \left(\left[\frac{k_{on}}{k_{off} + k_{in}}\right]\right.$$
$$\left.\times \left[\frac{m_1\, m_2}{V_1}\right] + \left[\frac{k_{out}}{k_{off} + k_{in}}\right]\left[\frac{m_3\, m_2}{V_3}\right]\right) \qquad (7.160)$$

which shows that the bound ligand constitutes a simple compartment only when m_1 and m_3 remain constant (or effectively nil). The terms $(k_{off} + k_{in})/k_{on}$ and $(k_{off} + k_{in})/k_{out}$ define the Michaelis half-saturation concentrations K_{M_1} and K_{M_3}, and the terms k_{off}/k_{on} and k_{in}/k_{out} the dissociation constants K_{d_1} and K_{d_3}, e.g.,

$$K_{M_1} = \frac{k_{off} + k_{in}}{k_{on}} = K_{d_1} + \frac{k_{in}}{k_{on}} \qquad (7.161)$$

which in turn shows that K_M is always greater than K_d when there is significant translocation of substrate.

■ **Competitive Inhibition.** For the steady state of $dm_2/dt = 0$, Eq. 7.160 yields the famous Michaelis-Menten equation for competition between two ligands (Dixon 1953):

$$M_2 = B_{max}\left[\frac{C_1}{K_{M_1}\left(1 + \frac{C_3}{K_{M_3}}\right) + C_1}\right.$$
$$\left. + \frac{C_3}{K_{M_3}\left(1 + \frac{C_1}{K_{M_1}}\right) + C_3}\right] \qquad (7.162)$$

where $C_1 = M_1/V_1$ and $C_3 = M_3/V_3$. The ratio M_2/B_{max} is the combined occupancies (σ) of the ligands

$$\sum \sigma = \frac{\frac{C_1}{K_{M_1}} + \frac{C_3}{K_{M_3}}}{1 + \frac{C_1}{K_{M_1}} + \frac{C_3}{K_{M_3}}} = \frac{\chi_1 + \chi_3}{1 + \chi_1 + \chi_3} = \frac{\sum \chi}{1 + \sum \chi} \qquad (7.163)$$

where χ_1 is the ratio C_1/K_{M_1}, the normalized or relative concentration of the ligand in the precursor compartment, and χ_3 is the ratio C_3/K_{M_3}. The occupancy of ligand m_1 is then

$$\sigma_1 = \frac{\chi_1}{1 + \sum \chi} \qquad (7.164)$$

which expresses the effect of the competition between the two ligands when the occupation by one ligand causes the occupancy of other ligands to decline.

The bound ligand relative to the unbound ligand at steady state can be expressed as a binding potential (p_B), obtained from Eq. 7.162, as in Gjedde et al. (1986):

$$p_{B_1}(\chi_1) \equiv \frac{M_{2_1}}{M_1} = \frac{\sigma_1\,B_{\max}}{M_1} = \frac{B_{\max}}{V_1\,K_{M_1}\,(1 + \sum \chi)}$$
$$= \frac{p_{B_1}(0)}{1 + \sum \chi} \qquad (7.165)$$

which indicates that a change of the binding potential reflects any change of $B_{\max}$, C_1 or C_2. This equation linearizes to the Eadie-Hofstee equation (Eadie 1952):

$$M_{2_1} = B_{\max} - [V_1\,K_{M_1}\,(1 + \chi_3)]\,p_{B_1}(\chi_1) \qquad (7.166)$$

according to which competitive interaction increases the magnitude of the Michaelis constant in proportion to the magnitude of competitor concentrations, χ_3, provided they are constant.

Depending on its function, the protein exerts its function when it unites with the occupant, either by reacting with other molecules or by translocating the product to another compartment. If the translocated product is identical to the precursor, the process is transport with a maximum velocity of $k_{in}\,B_{\max}$, symbolized by $T_{\max}$. If the product is different from the precursor, the protein is an enzyme catalyst with a maximum reaction rate $k_{in}\,B_{\max}$, symbolized by $V_{\max}$.

Many processes retain transitional features, combining transport and catalysis, or binding and transport. The significance of the transitional processes depends on the relationship between the rate constants k_{off} and k_{in}. Little or no transport or catalysis takes places when k_{in} is equal or close to zero. On the other hand, little or no binding occurs when k_{off} is equal or close to zero. Thus, whether binding takes precedence over translocation or transformation depends on the relative rates of product release.

In principle, protein-ligand interaction does not fulfill the basic requirement of compartmental kinetics because the rate of relaxation depends on the concentration of the decaying substance. However, under special circumstances of the relaxation, depending on the magnitudes and time courses of the ligand concentrations, it is possible to define relaxation constants that reflect these circumstances. The following definitions are valid only when the ligand concentrations are effectively constant or nil relative to their Michaelis constants or when the total ligand occupancy of the receptors is constant. The distinc-

tion between the definitions does not apply to true steady states, in which neither concentrations nor occupancies vary with time.

The common definitions of relaxation constants based on the association constants

$$k_1 = \left(1 - \sum \sigma\right) \frac{k_{on}\,B_{\max}}{V_1} \quad \text{and}$$
$$k_4 = \left(1 - \sum \sigma\right) \frac{k_{out}\,B_{\max}}{V_3} \qquad (7.167)$$

are valid only when $\sum \sigma$ is constant or effectively nil and the concentration of any ligand varies as a function of time; in those cases, the dissociation constants are

$$k_2 = k_{off} \quad \text{and} \quad k_3 = k_{in} \qquad (7.168)$$

while, *alternatively*, the less common definitions based on the dissociation constants

$$k_2 = (1 + \chi_1)k_{off} \quad \text{and} \quad k_3 = (1 + \chi_3)k_{in} \qquad (7.169)$$

are valid only when the concentrations of all ligands are constant and $\sum \sigma$ varies; in that case, the definitions based on the association constants are

$$k_1 = \frac{k_{on}\,B_{\max}}{V_1} \quad \text{and} \quad k_4 = \frac{k_{out}\,B_{\max}}{V_3} \qquad (7.170)$$

which cover the common test-tube cases of buildup of bound substance from a constantly maintained medium but do not easily apply to living matter in which changes of concentrations lead to changes of binding rather than the reverse. It is important to keep in mind that the equations can be extended for any number of ligands in any combinations, some of which may be of constant concentration while others are effectively nil. At normal steady state, all compartments have constant magnitudes. In that situation, only the ratios between the relaxation constants have meaning, and neither set of definitions is valid.

■ **Competitive Activation.** In competitive activation, the affinity of the binding sites for a ligand rises when another ligand binds to the same sites, i.e., the dissociation rates k_{off} and k_{in} decline. In the simplest way, the action is explained by the occupancy of the secondary ligand preventing the dissociation of the primary ligand for as long as the secondary ligand is bound. The occupancy of the ligand is then

$$\sigma'_1 = \frac{C_1}{K_{M_1}(1-\sigma_3)+C_1} = \frac{C_1}{K'_{M_1}+C_1}$$

$$= \frac{\chi_1(1+\chi_3)}{1+\chi_1(1+\chi_3)} = \frac{\chi'_1}{1+\chi'_1}$$

where $K'_{M_1} = K_{M_1}/(1+\chi_3)$ and $\chi'_1 = \chi_1(1+\chi_3)$. The action of the secondary ligand changes the mass of the bound ligand relative to the unbound ligand at steady state expressed as the binding potential

$$p_{B_1}(\chi_1) = \frac{M_{2_1}}{M_1} = \frac{\sigma'_1 B_{max}}{M_1} = \frac{B_{max}}{V_1 K_{M_1}(1+\chi_1-\sigma_3)}$$

$$= \frac{p_{B_1}(0)}{1+\chi_1-\sigma_3} \qquad (7.171)$$

which predicts an increase of the binding potential in the presence of the secondary ligand. This equation also linearizes to the Eadie-Hofstee equation (Eadie 1952):

$$M_{2_1} = B_{max} - [V_1 K_{M_1}(1-\sigma_3)]p_{B_1}(\chi_1) \qquad (7.172)$$

according to which the competitive activation reduces the magnitude of the Michaelis constant in proportion to the magnitude of activator concentration, χ_3, provided they are constant.

7.3.4.2
Noncompetitive Interaction

In noncompetitive interaction, the occupation of the sites by the noncompetitively interacting ligand is not influenced by the changing concentrations of other ligands, either because their concentrations are too low relative to their Michaelis constants ("tracers") or because the noncompetitively interacting ligand permanently renders particular sites unavailable to occupation by other ligands, i.e., lowers the association rate for other ligands so much that their concentrations are now insignificant relative to their noncompetitively elevated Michaelis constants. There are two kinds of noncompetitive interaction – inhibition and activation.

■ **Noncompetitive Inhibition.** In noncompetitive inhibition, a fraction of the protein sites are permanently occupied by the inhibitor, regardless of the concentration of other ligands, or they are rendered permanently unavailable for occupation by any ligand. If the permanently unavailable sites are M_{2_3},

the steady-state binding potential, obtained from Eq. 7.162, is

$$p_{B_1}(\chi_1) = \frac{M_{2_1}}{M_1} = \frac{\sigma_1(1-\varsigma_3)B_{max}}{M_1} = \frac{(1-\varsigma_3)B_{max}}{V_1 K_{M_1}(1+\chi_1)}$$

$$= \frac{1-\varsigma_3}{1+\chi_1} p_{B_1}(0) \qquad (7.173)$$

where ς_3 is now the occupancy of the noncompetitively interacting ligand, which is independent of the concentrations of other ligands. This equation also linearizes to the Eadie-Hofstee equation:

$$M_{2_1} = (1-\varsigma_3)B_{max} - V_1 K_{M_1} p_{B_1}(\chi_1) \qquad (7.174)$$

which confirms that noncompetitive inhibition reduces the number of available binding sites but leaves the affinity of the available sites intact.

As above, the corresponding definitions of relaxation constants for mixed competitive and noncompetitive interaction, i.e.,

$$k_1 = (1-\sigma_1)\frac{k_{on}(1-\varsigma_3)B_{max}}{V_1} \quad \text{and} \quad k_4 \cong 0$$

$$(7.175)$$

are valid only when σ_1 is constant or effectively nil and the concentration of any ligand varies as a function of time; in those cases, the dissociation constants are

$$k_2 = k_{off} \quad \text{and} \quad k_3 = k_{in}. \qquad (7.176)$$

Alternatively, the less common definitions based on the dissociation constants, i.e.,

$$k_2 = (1+\chi_1)k_{off} \quad \text{and} \quad k_3 = (1+\chi_1)k_{in}$$

are valid only when the concentrations of all ligands are constant and σ_1 varies; in that case, the definitions based on the association constants are

$$k_1 = \frac{k_{on}(1-\varsigma_3)B_{max}}{V_1} \quad \text{and} \quad k_4 \cong 0$$

which, as above, can be extended for any number of ligands in any combinations, some of which may be of constant concentration while others are effectively nil.

■ **Noncompetitive Activation.** In noncompetitive activation, the interaction is coupled to the presence of an activator that gives the ligand access to the binding sites. For example, transport of some solutes against a concentration gradient is directly or indirectly coupled to the breakdown of ATP. In these cases, the binding potential is given by

$$p_{B_1}(\chi_1) = \frac{\sigma_1 \, \varsigma_3 \, B_{\max}}{M_1} = \frac{\varsigma_3 \, B_{\max}}{V_1 \, K_{M_1} \, (1 + \chi_1)}$$
$$= \frac{\varsigma_3 \, p_{B_1}(0)}{1 + \chi_1} \qquad (7.177)$$

where ς_3 is now the occupancy of the agonist. The interaction of G-proteins with G-protein-linked receptors is of this kind, being activated by the primary receptor ligand, e.g., dopamine. Likewise, the Eadie-Hofstee plot becomes

$$M_{2_1} = \varsigma_3 B_{\max} - [V_1 \, K_{M_1}] \, p_{B_1}(\chi_1) \qquad (7.178)$$

and the relaxation constant is

$$k_1 = \frac{(1 - \sigma_1) \, k_{\mathrm{on}} \, \varsigma_3 \, B_{\max}}{V_1} \, .$$

7.3.5
Receptor Binding

The protein is said to function primarily as a receptor for a single ligand type when $k_{\mathrm{off}} \gg k_{\mathrm{in}}$, and $K_{M_1} \cong K_{d_1}$ (see Eq. 7.161). Thus, when the translocation of the ligand ("internalization") is minimal, only association and dissociation occur. This is the situation originally conceived by Michaelis & Menten (1913), who assumed that the entity later termed the Michaelis constant would equal the dissociation constant $k_{\mathrm{off}}/k_{\mathrm{on}}$ of the protein-ligand complex because of near-equilibrium between M_1 and M_2. Subsequently it was shown that this kind of near-equilibrium exists mostly in the cases of receptor binding. The interaction between a single ligand and the receptor protein is a simple exchange between two compartments as shown in Fig. 7.15 (adopted from Fig. 7.2).

The main applications of the protein-ligand association equations for receptor binding are the determination of the maximum number of receptor sites, the affinity of the sites for the ligand, or the ligand's occupancy of the receptors. These determinations all involve the binding potential. The binding potential is obtained from the steady-state receptor binding

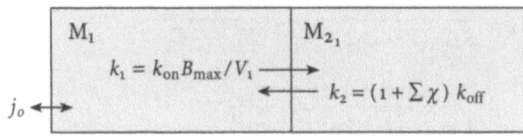

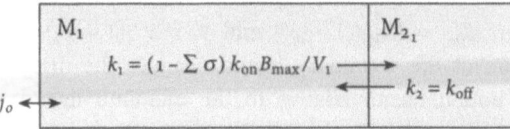

Fig. 7.15. Partial steady states of free and bound ligand compartments established by receptor protein occupied by ligand m_1. Depending on individual steady states, either top or bottom panel defines relaxation constants. Note that receptor protein may also be occupied by other ligands. Top panel requires constant ligand concentrations, bottom panel constant receptor occupancy

given by the product of the steady-state ligand concentration and volume of distribution (Eq. 7.21):

$$M_{2_1} = V_1 \left(\frac{k_1}{k_2}\right) C_1 = \frac{k_{\mathrm{on}} B_{\max} \, C_1}{k_{\mathrm{off}} \, (1 + \sum \chi)}$$
$$= \frac{B_{\max} \, C_1}{K_{d_1} \, (1 + \sum \chi)} = \frac{B_{\max} \, \chi_1}{1 + \sum \chi} = \sigma_1 \, B_{\max} \qquad (7.179)$$

which is the formulation of the Michaelis-Menten equation for the binding of a ligand to its receptor. The ratio k_1/k_2 is the ligand's binding potential, according to Eq. 7.165:

$$p_{B_1}(\chi_1) = \frac{k_1}{k_2} \qquad (7.180)$$

and its steady-state volume of distribution is $V_1 \, (1 + [k_1/k_2])$. The maximum number of available sites, $B_{\max}$, is related to the binding potential by rearrangement of Eq. 7.166:

$$M_{2_1} = B_{\max} - [V_1 \, K_{d_1}(1 + \chi_3)] \, p_{B_1}(\chi_1) \qquad (7.181)$$

which is the Eadie-Hofstee equation of a line with slope $-V_1 \, K_{d_1}(1 + \chi_3)$ and ordinate intercept $B_{\max}$, but only when the concentrations of all ligands other than m_1 are constant or effectively nil. The slope reflects the dissociation constant of the binding, the volume of distribution of the unbound ligand, and the concentrations (constant or nil) of all other ligands interacting with the receptors. The presence of other ligands affects the affinity, and therefore the binding potential, of the ligand in question.

7.3.6
Facilitated Diffusion

A protein can facilitate the diffusion of the ligand across a membrane when k_{in} is not negligible relative to k_{off}. The ligand usually remains intact during the translocation but in principle may also undergo a chemical change. This process is important when the unassisted diffusion is slow, i.e., when the protein spans a membrane in which the diffusion coefficient of the ligand is low. The association and release obey the same equation as receptor binding, but the predominant decay of the compartment occurs by release of the intact ligand to a compartment other than the one whence it originated. The translocation is unidirectional if the binding sites are accessible from only one compartment ("side"), here symbolized by M_1. Spontaneous conformational change of the transporter protein makes the transporter accessible from alternating sides, but the basic model has two compartments with a sink as shown in Fig. 7.16 (adopted from Fig. 7.3).

The flux and hence the relaxation constant depend on the concentration of the ligand. The maximum transport rate T_{max} is defined as $k_{in}B_{max}$. Hence the rate constants are

$$k_1 = \left(1 - \sum \sigma\right)\frac{k_{on} B_{max}}{V_1} = \left(1 - \sum \sigma\right)\frac{k_{on} T_{max}}{k_{in} V_1}$$

$$= \frac{(1 - \sum \sigma)T_{max}}{V_1 (K_{M_1} - K_{d_1})} \tag{7.182}$$

where the magnitude of $\sum \sigma$ depends on the simultaneous accessibility of the transporter from both sides of the membrane. The steady-state relaxation constant is a function of the partition coefficient as expressed in Eq. 7.33:

$$k_1' = k_{in}\, p_{B_1}(\chi_1) = \frac{(1 - \sum \sigma)k_{in} B_{max}}{V_1 K_{M_1}}$$

$$= \frac{(1 - \sum \sigma)T_{max}}{V_1 K_{M_1}} \tag{7.183}$$

provided $\sum \sigma$ reflects uniformly distributed ligands in true compartments. Because the presence of the transporter establishes an effective permeability for the ligand, the steady-state relaxation constant of compartment M_1 defines an apparent permeability in the direction of the net transport, P_1':

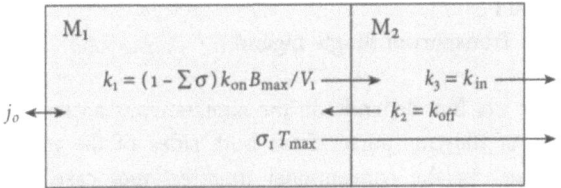

Fig. 7.16. Steady-state model of metabolite interaction with facilitating transport protein (M_2). When p_{B_1} is binding potential, steady-state k_1' equals $k_{in}\, p_{B_1}$ and reaches $k_1 = k_{on}B_{max}/V_1$ when $k_{in} \gg k_{off}$

$$P_1'S = V_1\, k_1' = V_1\, k_{in}\, p_{B_1} = \left(1 - \sum \sigma\right)\frac{T_{max}}{K_{M_1}} \tag{7.184}$$

where the binding potential is now given by

$$p_{B_1} = \frac{k_1}{k_2 + k_3} \tag{7.185}$$

such that the apparent permeability-surface area product now is

$$P_1'S = V_1\left(\frac{k_1 k_3}{k_2 + k_3}\right) \tag{7.186}$$

and the steady-state flux is a simple function of the ligand's occupancy

$$J_1' = P_1'SC_1 = \left(1 - \sum \sigma\right)\left[\frac{T_{max}}{K_{M_1}}\right]C_1 = \sigma_1\, T_{max} \tag{7.187}$$

where the evaluation of the influence of the ligand's concentration, i.e., the magnitude of the occupancy, uses the linear rearrangement of the Eadie-Hofstee equation (Eq. 7.166):

$$J_1' = T_{max} - \left[V_1 K_{M_1}\left(1 + \sum \chi - \chi_1\right)\right]k_1'$$

$$= T_{max} - \left[K_{M_1}\left(1 + \sum \chi - \chi_1\right)\right]P_1'S \tag{7.188}$$

where T_{max} is the ordinate intercept and $-V_1 K_{M_1}(1 + \sum \chi - \chi_1)$ is the slope, provided the interactions are competitive.

7.3.6.1
Net Transport of Single Ligand

The net flux depends on the simultaneous accessibility of the transporter from both sides of the membrane. In the conventional *unidirectional* case, the transporter sites are accessible only from one side of the membrane at a time and hence enjoy a higher occupancy of the ligand on that side, while in the *bidirectional* case, the transporter sites are accessible from both sides at the same time and hence have lower degrees of saturation from either side.

In the case of the unidirectional transporter, the transporter sites can be accessed by several ligands but only from the same side of the membrane at any one time. If only a single ligand is present in the precursor compartment, as in the case of glucose and the GLUT1 transporter (Gjedde 1992), Eq. 7.184 is modified to

$$P_1' S = \frac{(1 - \sigma_1) T_{max}}{K_{M_1}} = \frac{T_{max}}{K_{M_1} + C_1} \qquad (7.189)$$

and the flux of the uniformly distributed substrate, facilitated by the action of the unidirectional transporter, equals

$$J_1' = V_1 k_1' C_1 = P_1' S C_1 = \sigma_1 T_{max} \qquad (7.190)$$

and in the opposite direction, provided $k_{off} = k_{in}$, i.e., T_{max} is the same in the two directions,

$$J_3' = V_3 k_4' C_3 = P_3' S C_3 = \frac{T_{max} C_3}{K_{M_3} + C_3} = \sigma_3 T_{max} \qquad (7.191)$$

which implies a different apparent permeability, dictated by the magnitude of C_3 relative to C_1, and the difference in turn affects the magnitude of the distribution volume. The net transport is the difference between the fluxes in the two directions

$$J_1' = (\sigma_1 - \sigma_3) T_{max} = f_b' \sigma_1 T_{max} \qquad (7.192)$$

where $1 - f_b'$ is now the ratio between the different occupancies of the ligand on the two sides of the interface, established by the saturable nature of the binding to the transporter, rather than simply being the ratio between the concentrations, as in Eq. 7.119.

Facilitated diffusion is in principle passive, i.e., nonenergy-requiring. In case of active, energy-requir-

ing transport against a concentration gradient, the ligand can be translocated by association with sites activated by another ligand, which itself may or may not be translocated. The occupancy of the ligand in question may be low, yet net transport proceeds against a concentration gradient because more sites are made available for the ligand at the low concentration than for the ligand at the high concentration on the opposite side of the interface. The activation is reflected in the definition of f_b', when $\sigma_1 \varsigma_1 > \sigma_3 \varsigma_3$

$$f_b' = 1 - \frac{\sigma_3 \varsigma_3}{\sigma_1 \varsigma_1}$$

in which the low occupancy of M_1 (σ_1) is compensated by the high occupancy (ς_1) of the activating ligand.

7.3.6.2
Flow-Limited Net Transport of Single Ligand

Equation 7.189 raises the issue of the distribution of the ligand in the delivery compartment, when this compartment is the vascular bed. Not only are the apparent permeabilities different for the two directions of transport but they also vary along the length of the capillary because of the decline of the ligand concentration, as the ligand is delivered to the exchange compartment. Gjedde (1980) treated the case of the varying apparent permeability of the blood-brain barrier to glucose and showed that the effect can be ignored when the extraction is low. The analysis showed that the decrease of the ligand concentration, as blood passes along the capillaries, has little influence on the magnitude of the apparent permeability when the ratio $J_1' / [F(K_{M_1} + C_1)]$ is less than 0.1, where F is blood flow, i.e., when $f_b' \sigma_1 P_1' S / F < 0.1$. Thus, it is approximately correct to modify Eq. 7.192 for the apparent permeability generated by the facilitated diffusion, as illustrated above:

$$J_1' \cong f_b' C_a F \left(1 - e^{-P_1' S / F}\right) = f_b' K_1' C_a \qquad (7.193)$$

where C_a is the arterial ligand concentration and K_1' is the apparent clearance established by the presence of the permeability symbolized by $P_1' S$, the prime referring to the net transport through the two membranes of the endothelium.

The error incurred by ignoring the effect of the changing substrate concentration is revealed by the

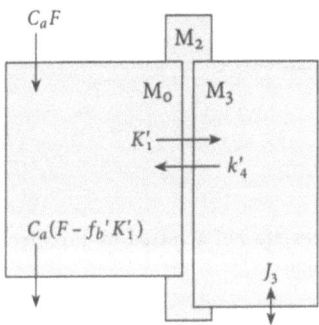

Fig. 7.17. Compartmental model of capillary and tissue compartments in which tissue compartment clears solute such as glucose from plasma flowing through capillary compartment by means of unidirectional GLUT1 transporters. The fraction $1 - f_b{}'$ is the ratio between effective concentrations of glucose in tissue and capillary compartments caused by higher apparent permeability in direction from tissue to capillary because of lower glucose concentration. Primes refer to net transport through membranes of M_2

formula for the average capillary concentration of the ligand (Eq. 7.151):

$$\overline{C}_1 = C_a \left[1 - f_b{}' \left(1 - \frac{1 - e^{-P_1'S/F}}{P_1'S/F} \right) \right] \qquad (7.194)$$

where the relationship between $\overline{C}_1$ and C_a depends on the magnitudes of $f_b{}'$ and the $P_1'S/F$ ratio. The ratio of the occupancies of the ligand on the two sides of the interface can be inferred from the ratio between the "unidirectional" and net fluxes at steady state:

$$f_b{}' = \frac{J_3}{C_a K_1'} = \frac{K'}{K_1'} \qquad (7.195)$$

where K' is the net clearance, which is approximately 0.5 for the transport of D-glucose across the human blood-brain barrier, indicating that the occupancy on the tissue side of the interface is half of that in plasma and the concentration in the tissue water one third of that in plasma (Gjedde 1995).

7.3.6.3
Flow-Limited Net Transport of Multiple Ligands

Equation 7.193 is valid also when multiple ligands are present in the precursor compartment, as in the case of the transporter of large neutral amino acids (Le-Fauconnier 1992). Changes of the apparent permeabilities on the two sides are then buffered by the large number of ligands. The net flux is given by the equation

$$J_1 \cong K_1' C_a \left(1 - \frac{\sigma_3}{\sigma_1} \right) = f_b{}' K_1' C_a \qquad (7.196)$$

where, to recapitulate, $1 - f_b{}'$ is the ratio between the occupancies on the two sides of the interface (the prime referring to facilitated rather than simple diffusion), K_1' is the rate of clearance from the virtual compartment M_0 through compartment M_2 to compartment M_3 in the direction from vessel to tissue, and C_a is the arterial concentration of the ligand, equal to the uniform concentration of the ligand in M_0. The net translocation ceases when the occupancies are the same on the two sides. Multiple ligands, on the other hand, may influence the translocation indirectly by changing the occupancy of the ligand on one or both sides.

7.3.7
Enzymatic Reactions

The protein is said to be an enzyme when the product is different from the precursor and physical translocation of the product is not a main function of the protein. The action of the protein may or may not involve translocation, and the sites usually are accessible by both product and precursor, although not at the same time. The model is adopted from Fig. 7.5. The maximum rate of the enzymatic reaction (V_{max}) is $k_{in} B_{max}$. Thus,

$$k_1 = \left(1 - \sum \sigma \right) \frac{k_{on} B_{max}}{V_1} = \left(1 - \sum \sigma \right) \frac{k_{on} V_{max_1}}{k_{in} V_1}$$

$$= \frac{(1 - \sum \sigma) V_{max_1}}{V_1 (K_{M_1} - K_{d_1})} \qquad (7.197)$$

where the magnitude of $\sum \sigma$ depends on the number of substrates and inhibitors accessed by the enzyme. The steady-state relaxation constant in the direction from precursor to product is a function of the affinity of the enzyme for the precursor, as expressed in Eq. 7.167:

$$k_1' = \frac{(1 - \sum \sigma) k_{in} B_{max}}{V_1 K_{M_1}} = \frac{(1 - \sum \sigma) V_{max_1}}{V_1 K_{M_1}} \qquad (7.198)$$

provided the precursor is uniformly distributed in the compartment. Likewise, the steady-state relaxation constant in the direction from product to precursor is a function of the affinity of the enzyme for the product:

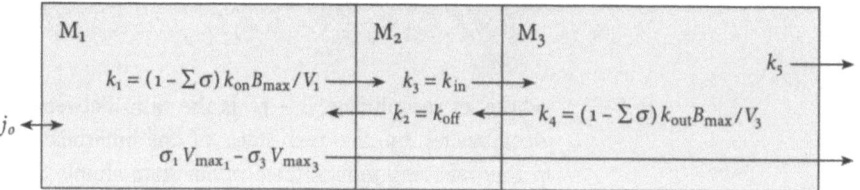

Fig. 7.18. Model of precursor and product compartments separated by enzyme (M_2) occupied by precursor (m_1) and product (m_3). Reaction sites are not accessed by precursor and product at the same time

$$k_4' = \frac{k_{off} B_{max}}{V_3 K_{M_3}} = \frac{(1 - \sum \sigma) V_{max_3}}{V_3 K_{M_3}} \qquad (7.199)$$

provided the product is also uniformly distributed in the compartment. The relaxation constant for an enzymatic reaction is often referred to as the enzyme activity, i.e., the maximum rate relative to the K_M and corrected for volume of distribution (V) and presence of competitors ($\sum \sigma$). The actual rate of the enzymatic reaction is then

$$J_1' = \sigma_1 V_{max_1} - \sigma_3 V_{max_3} = f_b' \, \sigma_1 V_{max_1} \qquad (7.200)$$

where $1 - f_b'$ is now the ratio $\sigma_3 V_{max_3} / [\sigma_1 V_{max_1}]$, that is, not only a function of the affinity of the enzyme and concentration of the ligand but also of the maximum rate in both directions of the transformation. This is the general form of the Michaelis-Menten equation.[2]

If the net flux is close to zero, this relationship may be difficult to establish. The reaction then is said to be near equilibrium. In this case, the ratio of the precursor and the substrate is a simple function of the ratio between the dissociation constants at steady state

$$\frac{C_3}{C_1} = \frac{K_{M_3} V_{max_1}}{K_{M_1} V_{max_3}} = \frac{k_{in} \, k_{on}}{k_{out} \, k_{off}} = \frac{K_{d_3}}{K_{d_1}} \qquad (7.201)$$

which may differ for different precursors and substrates, as both K_M and V_{max} magnitudes vary among

precursors and substrates. At near equilibrium, fluxes in the two directions are similar in magnitude as well as much greater than the net reaction rate. This contrasts with nonequilibrium steady state, when concentrations are constant and fluxes in the two directions are very different.

7.4
Determination of Relaxation Constants

7.4.1
Stimulus-Response Relations

The purpose of the kinetic analysis is to deconvolve the characteristic constants of the impulse response function from the relationship between a perturbation of the steady state of a system of compartments and the time course of its return to steady state. The perturbation of the steady state usually consists in an induced change of the contents of the delivery compartment ("stimulus"). The subsequent return to steady state ("response") is recorded as timed measurements of the contents of the entire system under investigation. Under certain limited circumstances, the use of tracers allows investigators to estimate the magnitude of relaxation constants and derived variables (clearance and volume of distribution) by deconvolving the impulse response function from one or more solutions of the differential equations that predict the response of the compartments to the stimulus.

Generally, tracers are labeled molecules that enter compartments without affecting the magnitude of the relaxation constants that define the compartments. Tracers are designed to have the same physical properties as the native molecules but to exist in such low concentration that the total number of molecules in the compartment can be considered. This characteristic is not always fulfilled, however, because the tracer molecules may differ from the native molecules in

---

[2] Although f_b' varies with the occupancies of precursor and product, magnitudes of V_{max_1} and an apparent K_{M_1} can be related by the Eadie-Hofstee equation (Eadie 1952; Hofstee 1952),

$$J_1' = V_{max_1} - \left[V_1 K_{M_1} \left(1 + \left[\frac{C_3}{K_{d_3}} \right] \right) \right] k_1'$$

where V_{max_1} is the ordinate intercept and $-V_1 K_{M_1}(1 + [C_3/K_{d_3}])$ the slope, provided C_3 is constant. The relationship implies that the bidirectional exchange influences the apparent Michaelis constant.

their chemical properties, particularly with respect to the magnitude of affinities and dissociation and Michaelis constants. For this reason, it is often necessary to carefully distinguish between relaxation constants that follow the exchanges of the native molecules and relaxation constants that describe the exchanges of the tracers.

When tracer molecules enter a compartment, the underlying assumption is that they are chemically indistinguishable from the native members of the compartment from the point of view of the mechanism responsible for the relaxation. Below, the tracer molecules in any compartment are symbolized by m^* and their concentration in the compartment by c^*.

By definition, tracers rarely reach steady state. Eventually, they disappear by decay or washout, unless they are continuously supplied, or neither decay nor escape from the compartments where they end up. The transient solutions are therefore useful to the estimation of relaxation constants, provided the predicted solutions of the differential equations are sufficiently accurate. The significance of this accuracy is the object of the regression analysis.

The methods fall into three categories: the single-bolus nonlinear deconvolutions, the steady-state programmed infusions, and the multiple-time graphical analyses. The deconvolutions are accurate but potentially imprecise and offer no directly verifiable results. The infusions are designed to reach steady state in extended periods of time. The multiple-time graphical analyses combine the advantage of the bolus injection with the visibility of the infusion results.

7.4.2
Regression Analysis

7.4.2.1
Regression Versus Function Analysis

The purpose of regression analysis is to establish the relations between two sets of variables, in the current context between the timed measurements of tracer concentration or content in a delivery compartment, say M_1, and the timed measurements of tracer concentration in a number of tissue compartments, M_2 to M_n, often including the delivery compartment. The relationship between the two sets of variables is established either in the form of a model with a number of constants or parameters whose magnitudes are unknown (model-based regression) or in the form of a class of equations (basis-functions) with known coefficients applicable to a system of compartments whose number and structure must be deduced from the data (data-based function analysis). The regression relation is based on the variations of both the "true" values of the variables and the random errors to which the observations of these variables are subject. The functional relations, on the other hand, depend on the variation of "true" values of these variables only, stripped of the random errors associated with their observation.

■ **Regression Analysis.** Conventional regression analysis assumes a model solution and estimates the parameters most consistent with the known model. The regression analysis indicates with a certain statistical probability the range of values of the parameters that are consistent with the chosen model. As the relationship in principle is either nonlinear or linear (i.e., uni- or multilinear), regression analyses are likewise either nonlinear or linear (i.e., uni- or multilinear). Nonlinear regression analysis is still an imperfect art with many pitfalls, while linear regression analysis is simpler both conceptually and mathematically. While nonlinear regression analysis proceeds by iteration until a given threshold of accuracy is reached, linear regression analysis often can be completed in a single analytical step. The disadvantage of linear regression, on the other hand, is that it can be subject to important bias, which must be taken into account.

The importance of the parameter estimates must be established by statistical tests, which are frequently misinterpreted. The problem is that a parameter estimate may have been established within satisfactorily narrow limits yet may express no more than the noise inherent in the measured variables. Another parameter estimate may have been established within fairly wide limits yet may represent a very accurate estimate of the biological variability of the measured variables. For these and other reasons, it is important to distinguish between precision and accuracy, also in terms of their practical applicability. Accuracy may be useless if the variables indeed are subject to unavoidable biological variation, while precision may be useful as an indicator of the difference among several sets of data, despite the bias of the estimates.

■ **Basis-Function Analysis.** Data-driven basis-function analysis assumes a set of parameters of known magnitude and chooses the model configuration most consistent with the data. This approach is a search for the optimal structural relations among compartments that predict the form of the response function. As an intriguing consequence, it is possible to fit combinations of models to the data. These model solutions are "probabilistic" in the sense that the resulting model structure can be established only with a finite probability of consistency with the data. The analysis is not an actual deconvolution but rather a segmentation of the data in probabilistic clusters of model structure.

7.4.2.2
Model-Based Deconvolution

In seeking the regression relation between two sets of tracer measurements, $c^*(t)$ and $m^*(t)$, the values of the dependent variable m^* are presumed to be randomly distributed around the regression function, so that its *expected* values are known to be a specific function of the *observed* values of the independent variable c^*. The expected values of the dependent variable m^* are the function $\overline{m}(c^*(t), t, p_1, \ldots, p_n)$, where $p_1, \ldots, p_n$ refers to the n parameters of the functional relation. If the distribution of $m^*(T)$ (where T is a time of observation) around $\overline{m}(c^*(T), T, p_1, \ldots, p_n)$ is normal for all T, the method of least squares is the most efficient means of estimating the parameters. The sum of squares to be minimized is

$$S_{sq} = \sum_{i=a}^{b} [m^*(T_i) - \overline{m}(c^*(T_i), T_i, K_1, k_2, \ldots, k_n)]^2$$

(7.202)

where a and b define a list of measurements chosen to ensure sufficient degrees of freedom in relation to the number of parameters. The minimization is carried out by means of differential calculus, but it can be performed by elementary algebra in linear and multilinear regression in a way that simultaneously yields the estimates of the parameters and the minimal sum of the squares. These methods will not be discussed here, as a large number of commercially available programs efficiently carry out the necessary computations.

In regression analyses of model parameters, it is common to refer to the "virtual" first compartment, M_0, as the *delivery compartment* and the second compartment, M_2, as the *precursor* or *exchange compartment*. The subsequent compartments, M_3-M_n, are then the *product compartments*. The concentration of the tracer precursor, $c_1^*(0)$, in the delivery compartment is c_a^*, which denotes the arterial concentration when the delivery compartment is the capillary bed. The tracer content of the precursor compartment is then m_e^*, and the content of all the subsequent compartments is m_p^*. The sum of all compartments, $V_0 c_a^* + m_e^* + m_p^*$, is then m^*. For three compartments, the tracer content obeys Eq. 7.43. The regression analysis then yields the parameter estimates, which minimize the sum of squares, and determines the residual sum of squares.

7.4.2.3
Data-Driven Basis-Function Analysis

In general, linear first-order differential equations define a system with an infinite number of compartments distinguished by the magnitude of their exponentials. An experimentally relevant subset of the compartment can be identified by means of least-squares fitting of the basis functions to the data (Gunn et al. 2001). The convolution of the impulse response function with the vascular or tissue-forcing function leads to the actually observed tissue curve predicted by Eq. 7.16. The underlying impulse response function is assumed to be represented as a cluster of elements picked from a sufficiently large ("overcomplete") pool of pairs of preselected coefficients (K_{1_h}) and relaxation constants (k_{2_h}) assumed to include all of the actual compartments of the system. As standard least-squares analysis does not apply when the number of parameters exceeds the number of measurements, addition of a penalty function reduces the number of permitted elements to one that minimizes the penalty function. The penalty function depends on the variability of the data, expressed in the magnitude of the regularization term μ. The variability coefficient can be determined by any one of several methods of "denoising" or "smoothing" (Gunn et al. 2002). Greater variability of the data allows fewer elements in the cluster of compartments and prevents both overfitting and underfitting of the data (Shao 1993; Hjorth 1994):

$$S_{sq} = \sum_{i,h=a,1}^{b,n-1} \left([m^*(T_i) - \overline{m}(c^*(T_i), T_i, K_{1_h}, k_{2_h})]^2 + \mu_h\right)$$

(7.203)

where μ is the regularization term.

7.4.3
Deconvolution of Response Function by Differentiation

In the cases of more than three or four compartments, the number of parameters is so great that regression analysis fails to identify a meaningful and unique set of estimates. In these cases, it is possible to use the regression analysis to identify a precursor compartment because it regulates the influx to all of the subsequent compartments.

Using the substitution m^* for the total contents of the precursor and product compartment, c_a^* for $c_1(0)$, and V_0 for the volume of the "virtual" capillary compartment defined in Eq. 7.150, the total flux of a labeled substance into any closed system of compartments is given by Eq. 7.73:

$$\frac{dm^*}{dt} = V_0 \frac{dc_a^*}{dt} + K_1 c_a^* - k_2 m_e^* \qquad (7.204)$$

where asterisks denote the labeled substance ("tracer"). When tracers are introduced into living tissue, the tracer precursor and its products distribute in separate compartments, which cannot be probed individually. The purpose of the regression is to determine the individual time courses of the precursor and the products. Among other applications, this regression is used to determine true steady-state values of a binding potential (p_B) and the corresponding steady-state magnitude of bound tracer. The regression is based on the claim that the tracer flux into the tissue is driven by the concentration difference between the tracer concentration in the capillary circulation and the tissue with which the capillary is directly exchanging (precursor compartment). Equation 7.204 can be solved for the tracer content of the precursor (exchange) compartment:

$$m_e^* = V_e \left[c_a^* - \frac{1}{K_1} \left(\frac{dm^*}{dt} - V_0 \frac{dc_a^*}{dt} \right) \right] \qquad (7.205)$$

where $V_e = K_1/k_2$ is the partition volume. To obtain an estimate of the quantity of exchangeable tracer, it is necessary to determine the derivatives of m^* and c_a^* and the magnitudes of V_0, V_e, and K_1.

To differentiate the function underlying the observed values m^*, the function must first be optimized by any suitably realistic formula, which minimizes the sum of squares but also has the property that the observed values maintain normal distribution around the optimized values of this formula. A de-

gree of smoothing must be performed to allow differentiation with minimal variability. Equations 7.43 and 48 are recommended for this smoothing. With Eq. 7.43, the smoothing is accomplished by ordinary six-parameter nonlinear regression to the measured pairs of m^* vs. c_a^*. With Eq. 7.48, the best estimate of the expected ("smoothed") value $\overline{m}$ is calculated by an appropriate method of "denoising." One approach is to apply a differentiable function directly:

$$\overline{m} = a_1 c_a^* + a_2 \int_0^T c_a^* \, dt + a_3 \int_0^T \int_0^u c_a^* \, dt \, du$$
$$+ a_4 \int_0^T m^* \, dt + a_5 \int_0^T \int_0^u m^* \, dt \, du \qquad (7.206)$$

where $\overline{m}$ is the optimized value and the coefficients are those defined for Eq. 7.48.

■ **Analytic Differentiation.** The nonlinear smoothing by means of Eq. 7.43 has the advantage that the derivatives of the function are analytically known in advance. Using the parameter estimates $\overline{k}_2$, $\overline{k}_3$, and $\overline{k}_4$, the expected quantity of tracer precursor is calculated from the relationship given in Eq. 7.39:

$$\overline{m}_e = V_e \overline{k}_2 \overline{\zeta}(t) \qquad (7.207)$$

where V_e is the partition volume measured in a reference region and $\overline{\zeta}(t)$ is the expected function $\overline{m}_e/K_1$ calculated from Eq. 7.39:

$$\overline{\zeta}(t) = \left(\frac{\overline{q}_2 - \overline{k}_4}{\overline{q}_2 - \overline{q}_1} \right) \int_0^T c_a^* e^{-\overline{q}_2 (T-t)} \, dt - \left(\frac{\overline{q}_1 - \overline{k}_4}{\overline{q}_2 - \overline{q}_1} \right)$$
$$\times \int_0^T c_a^* e^{-\overline{q}_1 (T-t)} \, dt \qquad (7.208)$$

where $\overline{q}_1$ and $\overline{q}_2$ are calculated from the estimates $\overline{k}_2$, $\overline{k}_3$, and $\overline{k}_4$ by means of Eqs. 7.40 and 7.41.

■ **Numerical Differentiation.** Simple differentiation of Eq. 7.206 (Wong et al. 1998) yields the flux into the precursor and product compartments required for determination of $\overline{m}_e$:

$$\overline{j}_1 = \frac{dm^*}{dt} - a_1 \frac{dc_a^*}{dt} = a_2 c_a^* + a_3 \int_0^T c_a^* \, dt$$
$$+ a_4 m^* + a_5 \int_0^T m^* \, dt \qquad (7.209)$$

such that Eq. 7.205 becomes

$$\overline{m}_e = V_e \left[c_a^* - \frac{\overline{j}_1}{a_1 a_4 + a_2} \right] \tag{7.210}$$

where V_e is known from separate measurement of the tracer's partition volume (K_1/k_2) in a reference region.

■ **Determination of Precursor-Product Ratio (Binding Potential).** The tracer product is the precursor tracer subtracted from the total:

$$\overline{m}_p = \overline{m} - \overline{m}_e - a_1 c_a^*$$

where $\overline{m}$ is the time-activity curve in the region of interest, smoothed according to Eq. 7.206 (Wong et al. 1998). At the peak of $\overline{m}_p$, $d\overline{m}_p/dt$ is zero and $\overline{M}_p$

is at steady state (transient equilibrium). The precursor-product ratio (steady-state binding potential), p_B, is the ratio between $\overline{M}_p$ and $\overline{M}_e$ at this time only (indicated by uppercase symbols):

$$p_B = \frac{\overline{M}_p}{\overline{M}_e} = \frac{\overline{M} - V_0 C_a^*}{\overline{M}_e} - 1 \tag{7.211}$$

where V_0 is the estimate of a_1 in the case of the multilinear regression by Eq. 7.206, as shown in Wong et al. (1998). The quantity of tracer product at the peak is determined as the ratio between $\overline{M}_p$ and the specific activity A:

$$M_p = \frac{\overline{M}_p}{A} \ .$$

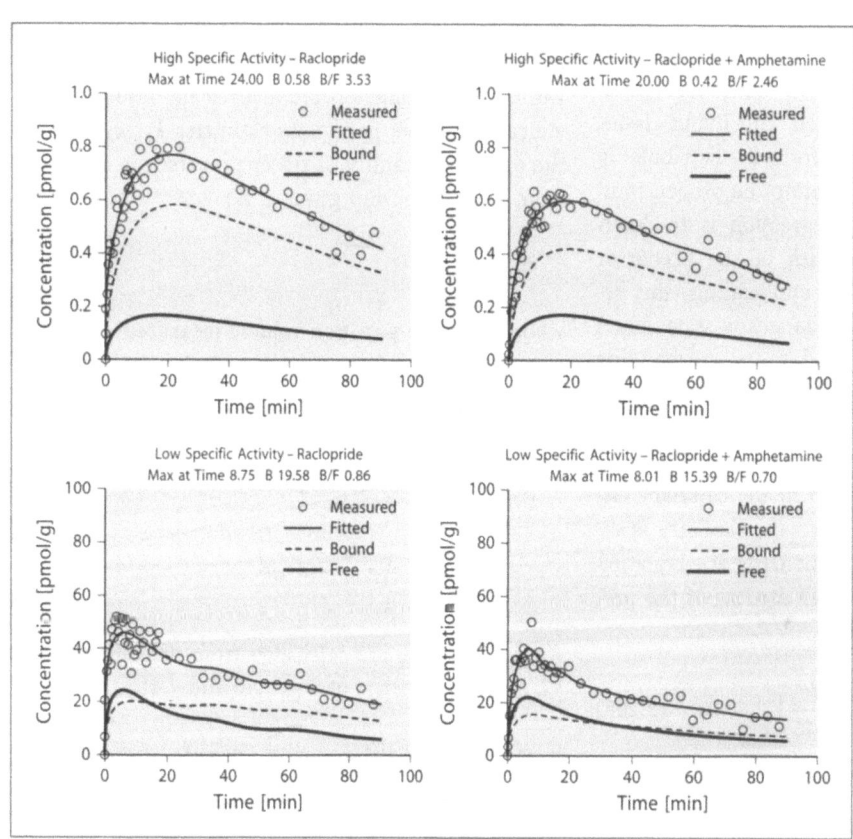

Fig. 7.19. Precursor and product curves at high (top panels) and low (bottom panels) specific activies before (left panels) and after (right panels) competitor (dopamine-released amphetamine) action. Resulting estimates for the upper left-hand panel are listed in Table 7.1. From Wong et al. 1998 (TREMBLE)

7.4.4
Deconvolution of Response Function by Temporal Transformation (Rutland-Gjedde-Patlak and Related Plots)

7.4.4.1
Constant Forcing Function

The deconvolution of the response function by means of temporal transformation is based on the properties of Eq. 7.74 when c_a^* is maintained constant (i.e., symbolized by C_a^*):

$$v^*(T) \equiv \sum_{i=1}^{n} \frac{m_i^*(T)}{C_a^*} = V_0 + \sum_{h=1}^{n-1} \left[\frac{K_{1_h}}{k_{2_h}} \left(1 - e^{-k_{2_h}T} \right) \right]$$

where $v^*(T)$ is the apparent volume of distribution as a function of time (T). The equation simplifies to the equation

$$v^*(T) = V - \sum_{h=1}^{n-1} \left[\frac{e^{-k_{2_h}T}}{k_{2_h}} \right]$$

where V is the total steady-state volume of distribution. The behavior of the equation depends on the relationship between k_{2_h} and T. For values of $T \ll k_{2_h}^{-1}$ for all h,

$$v^*(T) = V_0 + T \sum_{h=1}^{n-1} K_{1_h} = V_0 + K_1 T$$

which has the properties of a straight line and is a special case of Eq. 7.21 for constant c_1. For values of $T \gg k_{2_h}^{-1}$ for all h,

$$v^*(T) = V_0 + \sum_{h=1}^{n-1} \frac{K_{1_h}}{k_{2_h}} = V = V_0 + \sum_{h=1}^{n-1} V_h$$

which is a steady-state plateau equivalent to the level defined by Eq. 7.22. For intermediate values of T, $k_{2_h} \ll T^{-1}$ for $h < g$ and $k_{2_h} \gg T^{-1}$ for $h \geq g$, the intermediate equation results:

$$v^*(T) = V_0 + \sum_{h=1}^{g-1} V_h + T \sum_{h=g}^{n-1} K_{1_h} = V_0$$
$$+ \sum_{h=1}^{g-1} V_h + K T \qquad (7.212)$$

which is a straight line with an ordinate intercept and a slope of K, formally equivalent to Eq. 7.84. The

equivalence indicates that normalization of the integrated forcing function compensates for nonlinearity of the forcing function. The normalization has unit of time and represents a temporal transformation.

The use of the transformed time as independent variable and the volume v^* as the dependent variable changes the time course of the response and allows inference ("graphical analysis") to be made of the steady-state clearance of the tracer.

7.4.4.2
Reparameterization of Independent and Dependent Variables for Variable Forcing Function

In analogy with Eqs. 7.21 and 7.84, the primary temporal transformation creates a new time variable in the case of a variable forcing function:

$$\Theta^*(T) = \frac{\int_0^T c_a^* \, dt}{c_a^*(T)}$$

as does the secondary temporal transformation

$$\Theta'^*(T) = \frac{\int_0^T \int_0^u c_a^* \, dt \, du}{\int_0^T c_a^* \, dt} \qquad (7.213)$$

Also judging from the equivalence with Eqs. 7.21 and 7.84, the primary dependent variable is an apparent partition volume:

$$v^*(T) = V_0 + \sum_{i=2}^{n} v_i^* = \sum_{i=1}^{n} \frac{m_i^*}{c_a^*}$$

and the secondary dependent variable an apparent partition volume that approaches the steady-state volume of distribution, i.e., the integral of the impulse response function, when time approaches infinity:

$$v'^*(T) = V_0 + \sum_{i=2}^{n} v_i'^* = V_0 + \sum_{i=2}^{n} \frac{\int_0^T m_i^* \, dt}{\int_0^T c_a^* \, dt} \qquad (7.214)$$

which has the special significance of being the ratio of the total areas under the curves of the response and the stimulus. The fact that this ratio approaches the steady-state volume of distribution as T approaches infinity is the *stimulus-response theorem* (Lassen & Perl 1979).

7.4.4.3
Graphical Analysis
of Primary Temporal Transformation

For n compartments, the expected results of the multilinear solution to Eq. 7.204 are given by Eq. 7.74:

$$m^*(c_a^*, m_e^*) = V_0\, c_a^* + K_1 \int_0^T c_a^*\, dt - k_2 \int_0^T m_e^*\, dt$$

where m_e^* is the tracer content of the precursor compartment. Equation 7.216 rearranges to

$$\frac{m^*}{c_a^*} = V_0 + K_1 \left(1 - \left[\frac{k_2 \int_0^T m_e^*\, dt}{K_1 \int_0^T c_a^*\, dt}\right]\right) \int_0^T \left[\frac{c_a^*(t)}{c_a^*(T)}\right] dt$$

and thus to the temporally transformed equation

$$v^*(T) = V_0 + K_1\, \Theta^* \left(1 - \left[\frac{v_e'^*(T)}{V_e}\right]\right) \qquad (7.215)$$

where $v_e'^*(T)$ is the apparent partition volume at time T.

■ **Two Compartments.** For two compartments, Eq. 7.215 reduces to the Rutland-Gjedde-Patlak Eq. 7.21 for all times T for which $v_e^*(T) \ll V_e$

$$v^*(T) = V_0 + K_1\, \Theta^*. \qquad (7.216)$$

■ **Multiple Compartments.** With continued incremental accumulation of at least one product (m_p^*), Gjedde (1982) showed that Eq. 7.215 reduces to Eq. 7.84 also for multiple compartments, provided steady state exists between the delivery and precursor compartments. Thus, when it is true that the ratio $v_e^*(T)/V_e$ maintains a constant ratio less than unity but greater than zero, it is also true that

$$m_p^*(T_2) - m_p^*(T_1) = K\left(\Theta^*(T_2) - \Theta^*(T_1)\right) \quad (7.217)$$

where K is the net clearance of the precursor to at least one of several product compartments, relative to the concentration of the precursor in the circulation. The general Rutland-Gjedde-Patlak equation (Eq. 7.81) is then valid:

$$v^*(T) = V_0 + \sum_{i=2}^n V_i + K\, \Theta^* = V_n + K\, \Theta^* \quad (7.218)$$

where V_n is a "virtual" precursor volume of distribution. The precursor volume is "virtual" because it

does not actually contain the precursor from the onset of the uptake but from the onset of the steady state (secular equilibrium) between the delivery and precursor compartments.

The lumped variables (Eqs. 7.213 and 7.214) define linear relationships applicable to results from in vivo or ex vivo tomography, including autoradiography, positron emission tomography, and magnetic resonance spectroscopy.

7.4.4.4
Graphical Analysis
of Secondary Temporal Transformation

The apparent partition volume $v'^*(T)$ is the ratio of the areas under the response function and the forcing (stimulus) function curves (AUC). The analysis extends the graphical analysis, which can be useful also in the absence of measurements of a proper vascular forcing function. It also extends the graphical analysis by integrating delivery and precursor-product curves prior to the normalization. As with the simple ratios, the independent variable has unit of time, and the new dependent variable is the ratio between the integrals of the total content of the precursor and product compartments and the integral of the tracer concentration of the delivery compartment.

■ **Vascular Forcing Function.** The approach to the ratio of integrated compartments employs the temporally transformed independent variable Θ'^*, equal to $\int_0^T \int_0^u c_a^*\, dt\, du / \int_0^T c_a^*\, dt$. The AUC analysis is applicable to models with any number of compartments, but the interpretation of the ratios of the areas under the curves depends on the model.

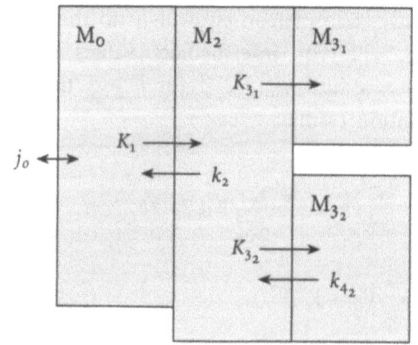

Fig. 7.20. Model of multiple compartments with two parallel compartments and a vascular delivery compartment

The model shown in Fig. 7.20 is an example of a model that can be analyzed by the AUC analysis. M_0 is the virtual delivery compartment in which the tracer concentration is c_a^*. It masks the real delivery compartment, which does not fulfill the fundamental compartment criteria as discussed above. M_2 is the precursor compartment, and M_{3_1} and M_{3_2} are the irreversible and reversible accumulation compartments, respectively. It is further assumed that the relaxation constants k_{3_2} and k_{4_2} are of such great magnitude that compartments M_2 and M_{3_2} maintain an approximately constant ratio (secular equilibrium), equal to a binding potential, p_B. The constant ratio has the effect of reducing the three compartments M_2, M_{3_1}, and M_{3_2} to two with the relaxation constants k_2' and k_{3_1}':

$$k_2' = \frac{k_2}{1 + p_B} \quad \text{and} \quad k_{3_1}' = \frac{k_{3_1}}{1 + p_B} \qquad (7.219)$$

where p_B is the ratio k_{3_2}/k_{4_2}.

In this model, the total uptake of tracer is given by the Blomqvist equation (Eq. 7.49):

$$m^* = V_0 c_a^* + \left[K_1 + V_0(k_2' + k_{3_1}') \right] \int_0^T c_a^* \, dt$$
$$+ K_1 k_{3_1}' \int_0^T \int_0^u c_a^* \, dt \, du - (k_2' + k_{3_1}') \int_0^T m^* \, dt \qquad (7.220)$$

where the symbols have their usual meaning. Equation 7.220 rearranges to the ratio of the areas under the curves as a function of time T:

$$v'^*(T) = \frac{\int_0^T m^* \, dt}{\int_0^T c_a^* \, dt} = \left[\frac{K_1}{k_2' + k_{3_1}'} + V_0 \right] + \frac{K_1 k_{3_1}'}{k_2' + k_{3_1}'}$$
$$\times \left(\frac{\int_0^T \int_0^u c_a^* \, dt \, du}{\int_0^T c_a^* \, dt} \right) - \left[\frac{m^* - V_0 c_a^*}{(k_2' + k_{3_1}') \int_0^T c_a^* \, dt} \right] \qquad (7.221)$$

where the ratio of the areas under the time-activity curves is v'^*, $K_1 k_{3_1}'/(k_2' + k_{3_1}')$ is K, the net clearance of the ligand, $\int_0^T \int_0^u c_a^* \, dt \, du / \int_0^T c_a \, dt$, is the new time variable Θ'^*, and $(K_1 \int_0^T c_a^* \, dt - m)/[(k_2' + k_{3_1}') \int_0^T c_a^* \, dt]$ represents the monoexponential approach to the time-variable ordinate-intercept $V_g(1 + p_B)$, equal to $K_1 k_2'/(k_2' + k_{3_1}')^2$, where V_g is defined as $K_1 k_2/(k_2 + k_{3_1})^2$:

$$v'^* = K \Theta'^* + V_0 + V_g(1 + p_B)(1 - e^{-\alpha \Theta'^*}) \qquad (7.222)$$

which defines an early phase of slope K/α and ordinate intercept V_0, a late phase of slope K and ordinate intercept $V_0 + V_g$, and a monoexponentially changing transition between the two phases. For $k_{3_1} = 0$, Eq. 7.220 shrinks to

$$v'^* = V_0 + V_e \, (1 + p_B) \, (1 - e^{-\alpha \Theta'^*}) \qquad (7.223)$$

where $V_0 + V_e (1 + p_B)$ is the steady-state partition volume V, such that $p_B = [(V - V_0)/V_e] - 1$. For $k_{3_1} = 0$ and $p_B = 0$, Eq. 7.223 further reduces to

$$v'^* = V_0 + V_e (1 - e^{-\alpha \Theta'^*}) \qquad (7.224)$$

which yields the partition volume $V = V_e + V_0$ at steady state.

■ **Tissue (Indirect) Forcing Function.** Although safe, the use of arterial sampling can be an inconvenience to experimental subjects and, with patients, impossible. Several methods have been proposed for the estimation of binding potentials using tracer-uptake curves in regions of little or no specific binding.

The reference-tissue area-under-curves (AUC) ratio analysis extends the ordinary AUC ratio analysis by integrating the reference region and region-of-interest tracer-uptake curves. The reference region takes the place of the combined delivery and precursor compartments by reversely estimating the tracer concentration in the delivery compartment that yielded the

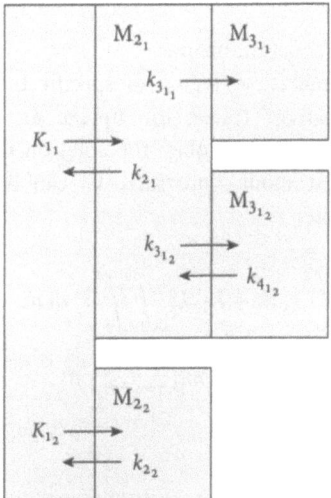

Fig. 7.21. Model of four compartments with two parallel compartments in region of interest, and two compartments in reference region

tracer content of the reference-region precursor compartment. The new independent variable still has unit of time, and the the new dependent variable is the ratio between the integrals of the regions of interest and the integral of the reference-region tracer-uptake curves.

In the region of reference, the term $m_{2_2}^*$ denotes the tracer content of the precursor compartment of the reference region, M_{2_2}, as a function of time. The term $m_{[2_1+3_{1_1}+3_{1_2}]}^*$ denotes the tracer content of the precursor and product compartments of the region of interest as a function of time. In the reference region, let the tracer uptake as a function of time be given by Eq. 7.17, provided V_0 can be claimed to be negligible:

$$m_{2_2}^* = K_{1_2} \int_0^T c_a^* \, dt - k_{2_2} \int_0^T m_{2_2}^* \, dt \qquad (7.225)$$

where the symbols have their usual meaning. Note that no initial volume of distribution was assumed in this region. The equation was rearranged to yield an expression of the integral of c_a^*:

$$\int_0^T c_a^* \, dt = \frac{m_{2_2}^*}{K_{1_2}} + \frac{k_{2_2}}{K_{1_2}} \int_0^T m_{2_2}^* \, dt \qquad (7.226)$$

which is integrated once more to yield

$$\int_0^u \int_0^T c_a^* \, dt \, dT = \frac{\int_0^u m_{2_2}^* \, dT}{K_{1_2}} + \frac{k_{2_2}}{K_{1_2}} \int_0^u \int_0^T m_{2_2}^* \, dt \, dT \qquad (7.227)$$

for insertion into the equation below.

In the region of interest, a region of specific but reversible accumulation of tracer, the uptake as a function of time is expected to obey the solution to the three-compartment model, provided V_0 can be claimed to be negligible:

$$m_{[2_1+3_{1_1}+3_{1_2}]} = K_{1_1} \int_0^T c_a^* \, dt + K_{1_1} \, k_{3_1}' \int_0^T \int_0^u c_a^* \, dt \, du$$
$$- (k_2' + k_{3_1}') \int_0^T m_{[2_1+3_{1_1}+3_{1_2}]}^* \, dt \qquad (7.228)$$

where again the symbols have their usual meaning (for example $k_2' = k_{2_1}/(1 + p_B)$ where p_B is the binding potential of the tracer for rapidly reversible accumulation in a compartment other than the one re-

sponsible for the relaxation constant $k_{3_{1_1}}$). Insertion of Eqs. 7.226 and 7.227 yields

$$m_{[2_1+3_{1_1}+3_{1_2}]} = R_1 m_{2_2}^* + R_1 (k_{2_2} + k_{3_1}') \int_0^T m_{2_2}^* \, dt$$
$$+ R_1 \, k_{2_2} \, k_{3_1}' \int_0^T \int_0^u m_{2_2}^* \, dt \, du$$
$$- (k_2' + k_{3_1}') \int_0^T m_{[2_1+3_{1_1}+3_{1_2}]}^* \, dt \qquad (7.229)$$

where R_1 symbolizes the K_{1_1}/K_{1_2} ratio.

To compute the ratio of the contents of tissue of interest relative to the reference tissue as a function of time, transformed time variables are generated from the reference-tissue curve in analogy with the definitions of Θ^* and Θ'^* (Eq. 7.213):

$$\theta^* = \left[\frac{\int_0^T m_{2_2}^* \, dt}{m_{2_2}^*} \right] \qquad (7.230)$$

and

$$\theta'^* = \left[\frac{\int_0^T \int_0^u m_{2_2}^* \, dt \, du}{\int_0^T m_{2_2}^* \, dt} \right] \qquad (7.231)$$

and dividing by the integral of the tracer accumulation in the reference region, the following version of Eq. 7.221 is obtained:

$$\rho^* = \frac{\int_0^T m_{[2_1+3_{1_1}+3_{1_2}]}^* \, dt}{\int_0^T m_{2_2}^* \, dt} = \left(\frac{k_{2_2} \, k_{3_{1_1}}}{k_{2_2} + k_{3_{1_1}}} \right) \theta'^*$$
$$+ R_1 \left(\frac{k_{2_2} + k_{3_1}'}{k_2' + k_{3_1}'} \right) \left[1 - \frac{\left(\frac{m_{[2_1+3_{1_1}+3_{1_2}]}^*}{R_1 \, m_{2_2}^*} - 1 \right)}{(k_{2_2} + k_{3_1}') \, \theta^*} \right] \qquad (7.232)$$

where ρ^* is defined as the ratio between the contents of the region of interest and the reference region as a function of the transformed time variable. The following substitutions are based on compartment modelling, which will reduce Eq. 7.232 to a more manageable form. For the ratio between the rates of washout from the region of interest and the reference region, the following linked differential equations are solved:

$$\frac{dm^*_{2_1}}{dt} = -(k'_2 + k'_{3_1})\, m^*_{2_1}$$

$$\frac{dm^*_{[3_{1_1}+3_{1_2}]}}{dt} = k'_{3_1}\, m^*_{2_1}$$

and

$$\frac{dm^*_{2_2}}{dt} = -k_{2_2}\, m^*_{2_2} \qquad (7.233)$$

from which follows the solution for $m^*_{[2_1+3_{1_1}+3_{1_2}]}$ $= m^*_{2_1} + m^*_{[3_{1_1}+3_{1_2}]}$,

$$\frac{m^*_{[2_1+3_{1_1}+3_{1_2}]}}{m^*_{2_2}} - R_1 = R_1 \left[\left(\frac{k'_{3_1}}{k'_2 + k'_{3_1}}\right) e^{k_{2_2} T} \right.$$
$$\left. + \left(\frac{k'_2}{k'_2 + k'_{3_1}}\right) e^{(k_{2_2}-k'_2-k'_{3_1})\, T} - 1 \right] \qquad (7.234)$$

The second substitution is the result of the washout of $m^*_{2_2}$, such that

$$\theta^* \cong \frac{e^{k_{2_2} T} - 1}{k_{2_2}} \qquad (7.235)$$

where k_{2_2} is the washout rate or fractional clearance of tracer from the reference region, which is valid when Eq. 7.233 holds. With the substitutions, the following intermediate equation arises:

$$\rho^* = \left(\frac{k_{2_1} k_{3_{1_1}}}{k_{2_1} + k_{3_{1_1}}}\right) \theta'^* + R_1\, \frac{k_{2_2} + k'_{3_1}}{k'_2 + k'_{3_1}}$$
$$\times \left[1 - \frac{k_{2_2}\, k'_2}{(k_{2_2} + k'_3)(k'_2 + k'_{3_1})} \right.$$
$$\times \left(\frac{1 - e^{-(k_{2_2}-k'_2-k'_{3_1})\, T}}{1 - e^{-k_{2_2}\, T}}\right) e^{-(k'_2+k'_{3_1})\, T}$$
$$\left. - \frac{k_{2_2}\, k'_{3_1}}{(k_{2_2} + k'_{3_1})(k'_2 + k'_{3_1})} \right] \qquad (7.236)$$

By making the substitution

$$\left(\frac{k_{2_2} - k'_2 - k'_{3_1}}{k_{2_2}}\right) e^{-\alpha\, \theta'^*}$$
$$\cong \left(\frac{1 - e^{-(k_{2_2}-k'_2-k'_{3_1}\, T}}{1 - e^{-k_{2_2}\, T)}}\right) e^{-(k'_2+k'_{3_1})\, T} \qquad (7.237)$$

the operational equation is derived:

$$\rho^* = \left(\frac{k_{2_1} k_{3_{1_1}}}{k_{2_1} + k_{3_{1_1}}}\right) \theta'^* - \frac{R_1\, k_{2_2}\, k'_{3_1}}{(k'_2 + k'_{3_1})^2} + \frac{R_1}{k'_2 + k'_{3_1}}$$
$$\times \left[k_{2_2} + k'_{3_1} - k'_2\left(\frac{k_{2_2}}{k'_2 + k'_{3_1}} - 1\right) e^{-\alpha\, \theta'^*} \right] \qquad (7.238)$$

■ **Irreversible Accumulation.** For $p_B = 0$, Eq. 7.238 reduces to

$$\rho^* = \left(\frac{k_{2_1} k_{3_{1_1}}}{k_{2_1} + k_{3_{1_1}}}\right) \theta'^* + R_1 \left[\frac{k_{3_{1_1}}}{k_{2_1} + k_{3_{1_1}}} \right.$$
$$+ \left(\frac{k_{2_1}}{k_{2_1} + k_{3_{1_1}}}\right) e^{-\alpha\, \theta'^*} \right]$$
$$+ \left(\frac{k_{2_1}}{k_{2_1} + k_{3_{1_1}}}\right)^2 \left[1 - e^{-\alpha\, \theta'^*} \right] \qquad (7.239)$$

which is symmetrical in k_{2_1} and $k_{3_{1_1}}$ when R_1 is close to unity and θ'^* exceeds a certain minimum. Therefore, the relation does not accurately identify the individual estimates of the two rate constants, which must be inferred from other information.

The steady-state ordinate intercept is an indicator of the relative magnitude of the two rate constants, as

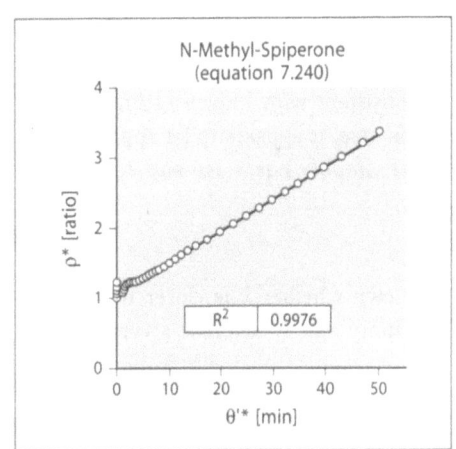

Fig. 7.22. Reference-region area-under-curves ratio analysis according to Eq. 7.240 of tracer N-methylspiperone (NMSP) uptake into human striatum in vivo, assumed to be mediated by high-affinity binding to dopamine receptors. Slope is 0.0460 min^{-1} (s.e. ± 0.0005), R_1 is 1.11 ± 0.01, and α is 0.52 ± 0.10 (DILBERT)

shown on the left panel of the graph: a lower intercept is indicative of rate constants of similar magnitudes. The closer to unity the steady-state ordinate intercept is, the more dissimilar the magnitudes of the rate constants are. Thus, for $k_{3_{1_1}} \ll k_{2_1}$,

$$\rho^* \simeq k_{3_{1_1}} \, \theta'^* + 1 - (1 - R_1) \, e^{-\alpha \theta'^*}. \qquad (7.240)$$

On the other hand, for $k_{2_1} \ll k_{3_{1_1}}$, as shown for the tracer N-[^{11}C]methylspiperone (NMSP) in Fig. 7.22, the slope loses its sensitivity to k_3, but the steady-state ordinate intercept remains close to R_1:

$$\rho^* \simeq k_{2_1} \, \theta'^* + R_1 \left(1 + \frac{k_{2_1}}{k_{3_{1_1}}} \left[\frac{k_{2_1}}{k_{3_{1_1}}} + \left(1 - \frac{k_{2_1}}{k_{3_{1_1}}} \right) \right. \right.$$
$$\left. \left. \times \, e^{-\alpha \theta'^*} \right] \right) \qquad (7.241)$$

Thus, only for significantly reduced binding of NMSP, e.g., by blockade of the binding by a pharmacologically active dose of an antagonist, is it possible to show any relation of the slope to the receptor density.

Judged from other studies (Gjedde & Wong 2001), Fig. 7.22 shows that the slope is not too far from the known value of $k_2 = K_1/V_e = 0.1/3$ min^{-1} for NMSP and thus has lost its sensitivity to the magnitude of k_3. A partial but inadequate solution to the problem of estimating k_3 under these circumstances is to use the estimate of α, which is close to the value of $k_{2_1} + k_3$, as shown in Eq. 7.237. Assuming this to be the case, k_{2_1} is close to 0.051 min^{-1} and k_3 close to 0.47 min^{-1}, almost an order of magnitude greater. In principle, the ratio could be the inverse, although this would not be consistent with known estimates of K_1 for NMSP. In this case, it appears to be approximately correct to further simplify Eqs. 7.240 and 7.241 to

$$\rho^* \simeq k \, \theta'^* + 1 \qquad (7.242)$$

where it is unknown whether k is closer to k_{2_1} or to $k_{3_{1_1}}$ when R_1 is unity and θ'^* exceeds a certain minimum.

■ **Reversible Accumulation.** When it is known a priori that the tracer is subject to completely reversible accumulation, the relaxation constant $k_{3_{1_1}}$ is zero. The comprehensive Eq. 7.238 then reduces to

$$\rho^* = 1 + p_B \left(1 - e^{-\alpha \theta'^*} \right) + (R_1 - 1) \, e^{-\alpha \theta'^*} \qquad (7.243)$$

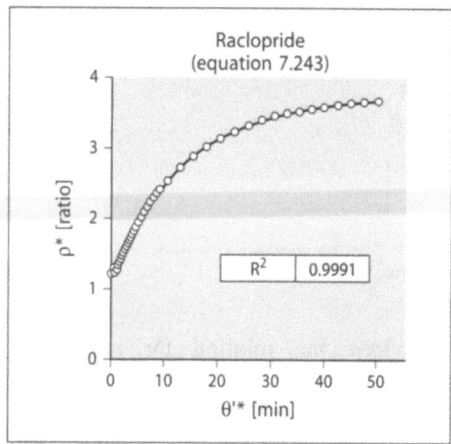

Fig. 7.23. Reference-region area-under-curves ratio analysis of reversible tracer raclopride uptake into human striatum in vivo, assumed to be mediated by medium-affinity binding to dopamine receptors. Binding potential estimate is 2.8 (s.e., ±0.01), R_1 is 1.08±0.01, and α is 0.072±0.001 (ERLiBiRD)

which for $R_1 = 1$ becomes

$$\rho^* = 1 + p_B \left(1 - e^{-\alpha \theta'^*} \right) \qquad (7.244)$$

which is shown in the graph above (Fig. 7.23) for raclopride. The results of applying this analysis to multiple subjects are listed in Table 7.4.

Although the result is visually satisfying, there is reason to be cautious about the many assumptions underlying the analysis and plot, known colloquially as "ERLiBiRD" (Estimation of Reversible Ligand Binding and Receptor Density).

7.4.5
Deconvolution of Response Function by Linearization (Logan and Related Plots)

The descriptive graphical analyses reduce the number of apparent variables by lumping the primary variables. They provide precise and often reasonably accurate illustrations of transient kinetic processes of brain uptake and metabolism. One great advantage of the methods is the ease with which they reveal key features of kinetic processing, which can be invisible in more conservative presentations of uptake data. A disadvantage is the compounding comminution of dependent and independent variables, which yields a potential bias (Logan et al. 2001).

The apparent clearance as a function of time of circulation is defined as

$$\kappa_i^* = \frac{m_i^*(T)}{\int_0^T c_a^* \, dt},$$

and the apparent residence time in the precursor or exchange compartment has previously been defined as

$$\theta_i^* = \frac{\int_0^T m_i^* \, dt}{m_i^*(T)}$$

which are useful when they can be determined directly from the timed measurements of c_a^* and m^* as functions of time, as described above.

7.4.5.1
Vascular Reference

■ **Single-Compartment Linearization.** For the single-compartment tracer model, the linear plots arising from the application of the lumped variables to Eq. 7.17 roughly divide into negative-slope and positive-slope plots. In this model, $m^* = m_{2_1}^*$. The *negative-slope* plots include the "clearance" plot (Eq. 37, in Gjedde 1982; Eq. 3, in Cumming et al. 1993), which relates the ratio between the product and the substrate integral to the ratio between the product and substrate integrals, as in the equation

$$\kappa_e^*(T) = K_{1_1} - k_{2_1} \, v_e'^*(T) \tag{7.245}$$

where $v_e'^*(T)$ (Eq. 7.215) is the apparent partition volume. The reciprocal "ratio" plot (Gjedde et al. 2000)

$$v_e'^*(T) = V_e - \frac{1}{k_{2_1}} \, \kappa_e^*(T) \tag{7.246}$$

where V_e is the precursor's partition volume. The *positive-slope* plots include the "reciprocal clearance" plot (Eq. 36, in Gjedde 1982), which relates the ratio between the precursor integral and the product to the ratio between the product integral and the product, the latter being equal to the transit time of the product through the product pool, as in the equation

$$\frac{1}{\kappa_e^*(T)} = \frac{\theta_e^*(T)}{V_e} + \frac{1}{K_{1_1}} \tag{7.247}$$

where $1/\kappa^*$ is a reciprocal clearance and k_{2_1}/K_{1_1} is the reciprocal partition volume. The equation was reversed by Logan et al. (1990) as the Logan plot:

$$\theta_e^*(T) = \frac{V_e}{\kappa_e^*(T)} - \frac{1}{k_{2_1}} \tag{7.248}$$

where the positive slope is the partition volume. The third positive-slope equation is the "reciprocal time" plot (Eq. 2, in Reith et al. 1990), which relates the ratio between the precursor and the product integral to the ratio between the integrals of the precursor and product:

$$\frac{1}{\theta_e^*} = \frac{K_{1_1}}{v_e'^*(T)} - k_{2_1} \tag{7.249}$$

where the slope is the rate constant of the unidirectional clearance. The most commonly used of these pseudolinear relations are the Logan plot for the estimation of steady-state volumes of distribution and the plots of Gjedde (1982) and Cumming et al. (1993) for the estimation of the rates of enzymatic reaction. The plots were originally derived to allow information from multiple experiments to be jointly analyzed because the ratios eliminate differences of scale among the experiments. The advent of positron emission tomography made the use of the unilinearized plots obsolete, except in cases in which the number of regressions made the non- or multilinear regressions computationally prohibitive.

■ **Two-Compartment Linearization.** For the two-compartment models, Eqs. 7.76 and 7.221 are replaced by Eq. 7.18:

$$\frac{\int_0^T m^* \, dt}{\int_0^T c_a^* \, dt} = \left(\frac{K_{1_1}}{k_{2_1}} + V_0 \right) - \left(\frac{m^* - V_0 \, c_a^*}{k_{2_1} \, \int_0^T c_a^* \, dt} \right) \tag{7.250}$$

which yields the negative-slope "ratio" plot (Gjedde et al. 2000) for times great enough to reduce the term V_0/Θ^* to zero:

$$v^*(T) = (V_e + V_0) - \left(\frac{1}{k_{2_1}} \right) \kappa^* \tag{7.251}$$

where the ordinate intercept is $V_e + V_0$ and the slope is $-1/k_{2_1}$ when the term V_0/v^* approximates zero. The alternative slope or Logan plot follows from the rearrangement of Eq. 7.251:

$$\theta^*(T) = (V_e + V_0) \left(\frac{1}{\kappa^*(T)} \right) - \left(\frac{1}{k_{2_1}} \right) \tag{7.252}$$

which has the slope $V_e + V_0$ and the ordinate intercept $-1/k_{2_1}$.

7.4.5.2
Tissue Reference

The reparameterization allows the deconvolution to proceed by linearization. Thus, for $k_{3_{1_1}} = 0$, provided $K_{1_1}/k_{2_1} = K_{1_2}/k_{2_2}$, Eq. 7.238 reduces to

$$\int_0^T m^*_{[2_1+3_{1_1}+3_{1_2}]} \, dt = (1 + p_B) \int_0^T m^*_{2_2} \, dt \\ - \left[\frac{m^*_{[2_1+3_{1_1}+3_{1_2}]} - R_1 \, m^*_{2_2}}{k'_2} \right] \tag{7.253}$$

which can be linearized in several ways. For $k_{3_{1_1}} = 0$, it also follows from Eq. 7.234 that

$$1 - R_1 \, \frac{m^*_{2_2}}{m^*_{[2_1+3_{1_1}+3_{1_2}]}} = 1 - e^{-(k_{2_2}-k'_2) \, T} \tag{7.254}$$

such that

$$\int_0^T m^*_{[2_1+3_{1_1}+3_{1_2}]} \, dt = (1 + p_B) \int_0^T m^*_{2_2} \, dt \\ - m^*_{[2_1+3_{1_1}+3_{1_2}]} \left[\frac{1 - e^{-(k_{2_2}-k'_2) \, T}}{k'_2} \right] \tag{7.255}$$

With its three parameters, Eq. 7.255 can be fitted with nonlinear regression analysis, but Eqs. 7.253 and 7.255 can be further rearranged to yield approximately linear relationships shown in Fig. 7.24.

■ **Ordinate-Intercept Plot.** The ordinate-intercept plot is obtained from Eq. 7.256 by subtracting, and dividing with, $\int_0^T m^*_{2_2} \, dt$:

$$\frac{\int_0^T \left(m^*_{[2_1+3_{1_1}+3_{1_2}]} - m^*_{2_2} \right) dt}{\int_0^T m^*_{2_2} \, dt} = p_B - \frac{m^*_{[2_1+3_{1_1}+3_{1_2}]}}{\int_0^T m^*_{2_2} \, dt} \\ \times \left[\frac{1 - e^{-\alpha\tau}}{k'_2} \right] \tag{7.256}$$

where τ is $\int_0^T m^*_{2_2} \, dt / m^*_{[2_1+3_{1_1}+3_{1_2}]}$, which is linear with an ordinate intercept of p_B and a slope of $-1/k'_2$ for times great enough to render the exponential term negligible.

■ **Logan-Type Plot.** Alternatively, Eq. 7.255 can be rearranged to yield the slope plot (reminiscent of the tissue reference-region version of the Logan plot) by dividing with $m^*_{[2_1+3_{1_1}+3_{1_2}]}$:

$$\frac{\int_0^T \left(m^*_{[2_1+3_{1_1}+3_{1_2}]} - m^*_{2_2} \right) dt}{m^*_{[2_1+3_{1_1}+3_{1_2}]}} = p_B \left[\frac{\int_0^T m^*_{2_2} \, dt}{m^*_{[2_1+3_{1_1}+3_{1_2}]}} \right] \\ - \frac{1 - e^{-\alpha\tau}}{k'_2} \tag{7.257}$$

which is linear with a slope of p_B and an ordinate intercept of $-k'_2$ for times great enough to render the exponential term negligible.

7.5
Application of Relaxation Constants

7.5.1
Peroxidation

The original Michaelis-Menten formulation of enzymatic action required approximate steady state between the ligands in compartments M_1 and M_2. An

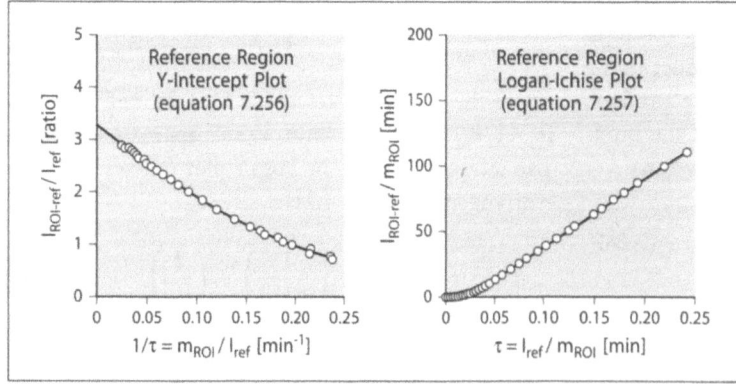

Fig. 7.24. Pseudolinear reference-region plots of binding of tracer (labeled) raclopride to dopamine receptors in striatum are linear only after steady state between precursor and delivery compartments. Averages of 13 healthy volunteers. In both cases, α is 0.30 min^{-1} and k'_2 is 0.68 min^{-1}. Binding potentials are listed in Table 7.4

Table 7.1. Kinetics of enzyme-substrate compound of peroxidase

Symbol	Mean	Unit
k_{on}	0.6	$min^{-1} nM^{-1}$
k_{off}	12	min^{-1}
k_{in}	252	min^{-1}
K_d	20	nM
K_M	440	nM

approximate steady state exists between these two compartments when $k_{in} \ll k_{off}$. Briggs & Haldane (1925) showed that the formulation is also valid when $k_{in} \gg k_{off}$, which appears to be the case for many enzymes. Estimates of the relaxation constants tend to confirm that magnitudes of K_M greatly exceed the magnitudes of K_d for most enzymes. Chance (1943) studied the enzyme peroxidase and observed that the dissociation constant was only 4.5% of the Michaelis constant, as shown in Table 7.1, below, indicating that only this small fraction of the molecules do not change into the product.

This observation means that the maximum rate is different for the precursor and the product.

7.5.2
Dopaminergic Neurotransmission

7.5.2.1
Dopamine Receptor Binding Constants

Analysis of tracer [11C]raclopride binding to dopamine receptors in the putamen region of the neostriatum of healthy volunteer subjects by means of standard nonlinear regression of Eq. 7.43 (assuming V_1 has a value of zero) to measurements of the time courses of tracer accumulation in arterial blood and a region of interest in the brain ("putamen") yielded the estimates listed in Table 7.2

The estimates of $V_e = K_1/k_2$ (Eq. 7.51) and $p_B = V_e [1 + (k_3/k_4)]$ (Eq. 7.52) were calculated from the individual estimates of the clearance and relaxation constants. When calculated from the average estimates, the binding potential (p_B) had the value 3.5, indicating some noise-dependent bias also of the conventional nonlinear regression analysis. It may be appropriate to regard this value as the unbiased "gold" standard.

7.5.2.2
Dopamine Receptor Binding Potential

Other analyses of tracer [11C]raclopride binding to dopamine receptors in the putamen region of the neostriatum of the healthy volunteer subjects yielded slightly different values of the binding potential at the dopamine D_2 and D_3 receptors, to which raclopride binds. The results of the different approaches are shown in Table 7.3.

The table confirms that estimates vary according to the regression method used to obtain the results. Compared to the previously identified "gold" standard of 3.5, the differentiation and linearization methods, the latter applied to population averages, yield the closest results.[3]

7.5.2.3
Comprehensive Model
of Dopaminergic Neurotransmission

To examine the ability of compartmental modelling to reveal the role of individual relaxation constants in the regulation of the turnover of molecules in a specific system of brain biochemistry, we have combined the known elements of dopaminergic neurotransmission in the striatum of the brain. The model consists of the following compartments: the tyrosine compartment being fed from the bloodstream by the blood-brain-barrier-facilitated diffusion transporter, the dopa compartment being fed by tyrosine hydroxylase and decaying both to the bloodstream, to 3-O-methyldopa, and to the main conduit in the form of dopa decarboxylation to dopamine. The intracellular dopamine department is also fed by the dopamine transporters linking the extracellular and intracellular dopamine compartments. One fraction of the intracellular dopamine is transported into and concentrated in the vesicles and subsequently released to the intrasynaptic cleft, whence the dopamine diffuses to the extrasynaptic space. From the extrasynaptic space it returns to the intracellular space by means of the dopamine transporters' facilitated diffusion. Another fraction of the intracellular dopamine is converted to DOPAC by the monoamine oxidase enzyme. DOPAC

---

[3] "Individual subjects" and "population average" refer to results averaged from regression to 13 individual observations ("individual subjects"), or single result from population averaged observation ("population average"). Averaging removes noise and leads to substantially but unrealistically lowered standard errors.

Table 7.2. Tracer raclopride compartmental relaxation constants for binding to dopamine $D_{2,3}$ receptors in putamen of human neostriatum (Gjedde & Wong 1990)

Constant Unit	K_1 (ml g^{-1} min^{-1})	k_2 (min^{-1})	k_3 (min^{-1})	k_4 (min^{-1})	V_e (ml g^{-1})	p_B (ratio)
Mean estimate	0.16	0.49	0.37	0.100	0.34	3.8
Standard error	0.01	0.03	0.02	0.008	0.02	0.2

Table 7.3. Tracer raclopride binding potential at dopamine $D_{2,3}$ receptors in human striatum, estimated by multiple methods

Analysis	Regression	Differentiation		Temporal transformation		Linearization	
Equation	(43)	(208)	(210)	(243)	(243)	(256)	(257)
Material	Individual subjects (n = 13)			Population average			
Mean estimate	3.8	3.3	3.5	3.0	2.9	3.3	3.3
Standard error	0.2	0.1	0.2	0.1	0.01	0.02	0.01

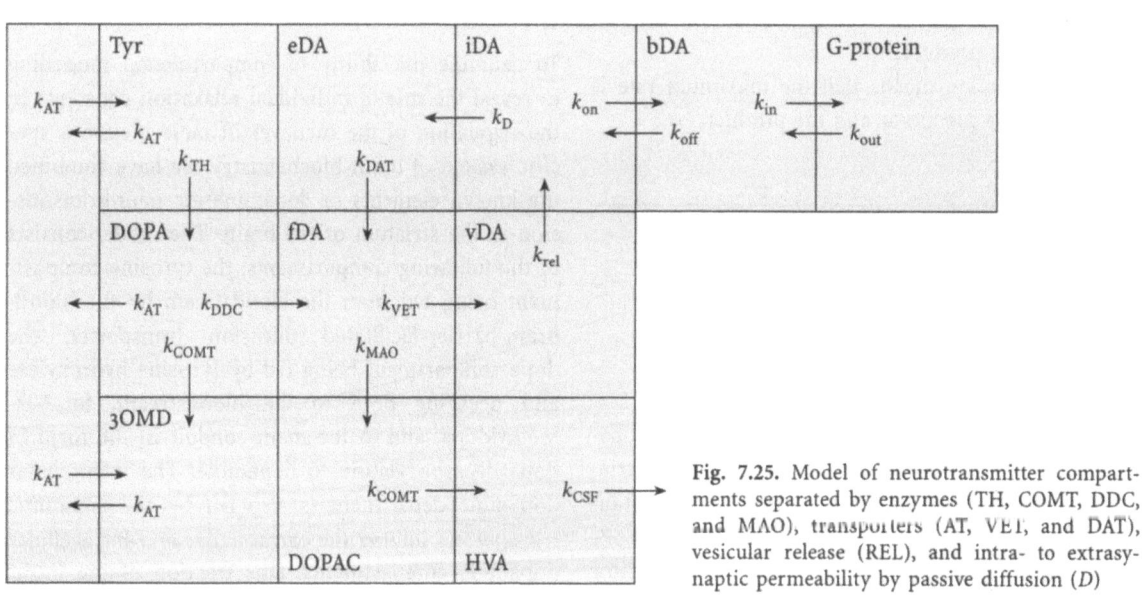

Fig. 7.25. Model of neurotransmitter compartments separated by enzymes (TH, COMT, DDC, and MAO), transporters (AT, VET, and DAT), vesicular release (REL), and intra- to extrasynaptic permeability by passive diffusion (D)

in turn is 3-O-methylated to homovanillic acid (HVA).

At steady state, the system is fully described by two out of three variables, e.g., the relaxation constants and the flux through the system, which together define the compartment contents. To calculate the compartment contents, and thus to describe the system in full, tracer measurements of relaxation constants and net flux through the dopamine metabolic pathway were combined to yield the missing compartment contents. The relaxation constants ap-

plied to the analysis of dopamine turnover are listed in Table 7.4, and the resulting compartment contents and concentrations are listed in Table 7.5.

The values of the relaxation constants were acquired or calculated on the basis of their definitions. Note that the flux-generating steps have relaxation constants of the lowest magnitude of the steps that constitute the pathway. In the definitions below, f_b' symbolizes the bidirectionality of the flux, assuming a value of unity for unidirectional flux and a value of zero for a net flux of zero. The symbol σ refers to the occupancy of the sub-

Table 7.4. Relaxation constants of dopamine turnover

Enzyme or transporter	Relaxation constant	Estimate (min^{-1})
Tyrosine hydroxylase (*Secondary Flux Generator*)	k_{TH}	0.005
DOPA decarboxylase	k_{DDC}	1
Monoamine oxidase	k_{MAO}	20
Dopamine vesicular transporter	k_{VET}	50
Dopamine release (*Primary Flux Generator*)	k_{rel}	0.015
Dopamine diffusion from synapse	k_{D}	125
Dopamine plasma membrane transporter	k_{DAT}	1,000
LNAA transporter (tyrosine)	$k_{\text{AT}_{\text{TH}}}$	0.06
LNAA transporter (DOPA)	$k_{\text{AT}_{\text{DOPA}}}$	0.05
LNAA transporter (3-O-methyl-DOPA)	$k_{\text{AT}_{\text{3OMD}}}$	0.2
Catechol-O-methyl-transferase (DOPA)	$k_{\text{COMT}_{\text{DOPA}}}$	0.03
Catechol-O-methyl-transferase (DOPAC)	$k_{\text{COMT}_{\text{DOPAC}}}$	0.03
Choriod plexus anion transporter	k_{CSF}	0.04

stance and the symbol ς to the occupancy of a noncompetitive activator in the cases of active transport or an agonist in the cases of receptor binding.

■ **Tyrosine.** The large neutral amino acid (LNAA) transporter of the blood-brain barrier transports tyrosine by facilitated diffusion from the circulation. The transporter is 95% saturated with the many LNAA in the circulation. However, this competition does not affect the dopamine turnover, as most of the tyrosine continues into proteins and the entry into the monoamine synthetic pathways is regulated by the activity of tyrosine hydroxylase (TH), which is saturated with tyrosine (Eq. 7.198). Hence the relaxation constant is inversely proportional to the content of tyrosine:

$$k_{\text{TH}} \cong \frac{V_{\text{max}}^{\text{TH}}}{M_{\text{Tyr}}} \tag{7.258}$$

which in brain tissue is close to 60 nmol g^{-1} in brain tissue. The rate of dopamine turnover has been calculated to be 0.3 nmol g^{-1} min^{-1}, equal to the V_{max} of TH. Thus, the magnitude of k_{TH} is close to 0.005 min^{-1} (Cumming et al. 1998).

■ **DOPA.** DOPA (di-hydroxy-phenylalanine) is the product of tyrosine hydroxylation. DOPA is the substrate of aromatic amino acid decarboxylase, also called DOPA decarboxylase (DDC), as well as of the large neutral amino acid transporter of the blood-brain barrier and the enzyme catechol-O-methyltransferase (COMT). DDC is unsaturated, and the DDC reaction is irreversible, with the relaxation constant

$$k_{\text{DDC}} \cong \frac{V_{\text{max}}^{\text{DDC}}}{V_{\text{DOPA}} K_{\text{M}}^{\text{DDC}}} \cdot \tag{7.259}$$

As the enzyme is unsaturated, it is difficult to determine its true activity in vivo. Instead, the relaxation constant of DDC in mammalian brain has been measured directly with labeled substrates of the enzyme in vivo, and its magnitude varies from region to region, but the results are suspected of being low because of loss of tracer. In striatum, the measured value ranges from 0.05 to 0.1 min^{-1}, but the true value may be as high as 1 min^{-1} (Cumming et al. 1995). In addition to the decarboxylation, DOPA is also subject to transport by the LNAA transporter. The relaxation constant of this process, k_{AT}, averaged 0.05 min^{-1} in vivo (Cumming et al. 1993):

$$k_{\text{AT(DOPA)}} = \frac{f_b{'}\ \sigma_{\text{DOPA}}^{\text{LNAA}}\ T_{\text{max}}^{\text{LNAA}}}{M_{\text{DOPA}}} \tag{7.260}$$

where $f_b{'}$ is close to unity because of the unidirectional efflux of DOPA from brain tissue. Finally, DOPA is subject to 3-O-methylation in the liver and other tissues including brain. The relaxation constant of this process averages about 0.03 min^{-1} in vivo (Cumming et al. 1993).

$$k_{\text{COMT(DOPA)}} = \frac{\sigma_{\text{DOPA}}^{\text{COMT}}\ V_{\text{max}}^{\text{COMT}}}{M_{\text{DOPA}}} \cdot \tag{7.261}$$

■ **3-O-Methyl-DOPA (3OMD).** The 3-O-methyl-DOPA (3OMD) generated by COMT in tissues is exchanged with 3OMD in the circulation by means of the neutral amino acid transporter of the blood-brain barrier.

The concentrations of 3OMD in plasma and tissues are likely to be similar because of the comparatively high relaxation constant of the transporter, $k_{AT(3OMD)}$, which has been measured to average about 0.2 min^{-1} (Reith et al. 1990) (Eq. 7.194):

$$k_{AT(3OMD)} = \frac{f_b' \, \sigma_{3OMD} \, T_{max}^{LNAA}}{M_{3OMD}} \, . \tag{7.262}$$

■ **Intracellular dopamine (fDA).** As the product of DDC and dopamine transporter action, intracellular dopamine is the junction of two metabolic paths, one accessing the vesicular transporter (VET) in the membranes of vesicles, driven by proton antiport, the other accessing monoamine oxidase (MAO):

$$k_{MAO} = \frac{\sigma_{fDA}^{MAO} \, V_{max}^{MAO}}{M_{fDA}} \tag{7.263}$$

where the unidirectionality of the reaction renders f_b' equal to unity. The relaxation constant of MAO has been estimated as the ratio between the V_{max} and the K_M in tissue homogenates and was found to be approximately 20 min^{-1} (Azzaro et al. 1985). To match the known dopamine turnover rate of 0.3 nmol g^{-1} min^{-1}, the intracellular dopamine content (outside vesicles) must equal 0.015 nmol g^{-1}, corresponding to an intracellular dopamine concentration of 2 μM. This concentration is consistent with the free intracellular dopamine concentration in giant-squid neurons of the order of 1 μM (Chien et al. 1990). Also acting on the intracellular dopamine, the vesicular dopamine transporter actively transfers dopamine to the vesicles in symport with hydrogen ions, which act as noncompetitive activators (Johnson & Scarpa 1979):

$$k_{VET} = \frac{f_b' \, \sigma_{fDA}^{VET} \, \varsigma_{H^+}^{VET} \, T_{max}^{VET}}{M_{fDA}} \tag{7.264}$$

The relaxation constant of the VET was calculated from the rate of release of dopamine from the vesicles. To be consistent with a lower limit of dopamine release from vesicles (see below), k_{VET} must average at least 50 min^{-1}, acting on the intracellular content of 0.015 nmol g^{-1} for a flux of 0.75 nmol g^{-1} min^{-1}. Then the total influx of dopamine into the free intracellular compartment must be at least 1.05 nmol g^{-1} min^{-1}, of which an amount of 0.75 nmol g^{-1} min^{-1} reenters the vesicles and an amount of 0.3 nmol g^{-1} min^{-1} undergoes monoamine oxidation.

■ **Vesicular dopamine (vDA).** Most of the tissue dopamine, 48 nmol g^{-1}, is believed to reside in vesicles in the dopaminergic terminals. Quanta of the vDA are released to the intrasynaptic space in response to action potential arrival. The relaxation constant of the release of dopamine from vesicles in the baseline state is the ratio between the dopamine release rate of 0.75 nmol g^{-1} min^{-1} and the vDA content of 48 nmol g^{-1}:

$$k_{rel} = \frac{J_{vDA}^{rel}}{M_{vDA}} \tag{7.265}$$

where the baseline state is considered the lower limit of the half-life of total dopamine in the tissues with active dopaminergic neurotransmission. This half-life is close to 45 min (Cumming et al. 1999), corresponding to a rate constant of 0.015 min^{-1}.

■ **Intrasynaptic dopamine (iDA).** Intrasynaptic dopamine is balanced among the processes of vesicular release, receptor binding, and passive diffusion to the extrasynaptic space. The passive diffusion is driven by the concentration gradient established by the dopamine transporters. The compartment decays by diffusion to the extrasynaptic space, with the relaxation constant (see Eq. 7.111)

$$k_D = \frac{A\sqrt{kD}}{V_{iDA}} \tag{7.266}$$

where k is the rate constant of the concentration-dependent removal of the dopamine, catalyzed by the dopamine transporter facing the extrasynaptic space (see below), D is the dopamine diffusion coefficient, A is the area of the interface between the intra- and extrasynaptic compartments, and V_{iDA} is the volume of the intrasynaptic compartment. The dopamine content of the compartment was assumed to be 6 pmol g^{-1} in agreement with the binding potential of unity derived from one study of dopamine binding (see below), corresponding to a relaxation constant of 125 min^{-1} for a baseline dopamine flux of 0.75 nmol g^{-1} min^{-1}.

■ **Extrasynaptic dopamine (eDA).** Extrasynaptic dopamine is a weighted average of diffusion profiles established by the clearance of dopamine from the intrasynaptic space and facilitated diffusion mediated by the dopamine transporters surrounding the extrasynaptic space. The dopamine transport actively transports dopamine in proportion to the sodium-ion gradient. The sodium-ion concentration provides the noncompetitive activation that raises the maximum transport capacity in proportion to the sodium-ion concentration, as expressed operationally by the difference between the magnitudes of the sodium-ion occupancy (ς_{Na^+}) on the two sides of the membrane. The relaxation constant of this compartment is (Eq. 7.138)

$$k_{DAT} = f_b' \frac{A\sqrt{kD}}{V_{eDA}} \left[\frac{1 - e^{-V_{eDA}\sqrt{k/D}/A}}{1 - f_b'\left(1 - \frac{1-e^{-\left[V_{eDA}\sqrt{k/D}/A\right]}}{V_{eDA}\sqrt{k/D}/A}\right)} \right] \tag{7.267}$$

where $1 - f_b'$ is the ratio between the occupancies of the transporter ligand on the two sides of the transporting membrane at the onset of diffusion, i.e., at the interface between the intra- and extrasynaptic spaces (see Eq. 7.138), A is the area of the interface between the intra- and extrasynaptic compartments, V_{eDA} is the volume of the extrasynaptic space, D is the dopamine diffusion coefficient in the extrasynaptic space, and k is the relaxation constant of the facilitated transport, defined as (Eq. 7.128)

$$k = \frac{\overline{P}_{DAT}' S}{V_{eDA}} \tag{7.268}$$

and equal to (Eq. 7.183)

$$k = \frac{\overline{f}_b' \, \overline{\sigma}_{eDA} \, \overline{\varsigma}_{Na^+} T_{max}}{\overline{M}_{eDA}} \tag{7.269}$$

where $\overline{f}_b'$ is the weighted average magnitude of f_b', $\overline{\sigma}_{eDA}$ is the weighted average extracellular dopamine occupancy of the transporter, $\overline{\varsigma}_{Na^+}$ is the weighted average occupancy of the noncompetitive activator (Na$^+$), and $\overline{M}_{eDA}$ is the weighted average eDA content. The magnitude was calculated to be 1000 min^{-1} from the relationships derived below. Because $1 - f_b'$ refers to the ratio between the concentrations of free dopamine in the dopaminergic neurons and dopamine in the intrasynaptic space, f_b' must be close to unity. When $f_b' = 1$, and the term $V_{eDA}\sqrt{k/D}/A$ is replaced by the term λ, Eq. 7.267 reduces to

$$k_{DAT} = \lambda \left(\frac{A\sqrt{kD}}{V_{eDA}} \right) \tag{7.270}$$

where λ is an index of the degree to which the dopamine transport raises the gradient of the diffusion of dopamine in the extracellular space.

■ **Receptor-bound dopamine (bDA).** The dopamine occupancy of dopamine's receptors is a function of the intrasynaptic-concentration and the receptor-dissociation constants, and the amount of bound dopamine is the product of the maximum binding capacity and the occupancy (Eq. 7.165):

$$\rho_{B_{iDA}} = \frac{\sigma_{iDA}' B_{max}^{DAR}}{M_{iDA}} \tag{7.271}$$

where σ_{iDA}' symbolizes the increased binding due to the higher affinity caused by the binding of GTP-free G-protein. The binding of dopamine to G-protein-linked receptors acts as a competitive activation of the G-protein binding, which in turn activates second-messenger systems in the cells Eq. 7.177 as in Gjedde & Wong (2001) and Cumming et al. (2002). The average normal dopamine occupancy was assumed to be 20%. With equal amounts of bound and free dopamine, 6 pmol g^{-1} (Borbely et al. 1999), the binding potential is unity.

■ **DOPAC.** Dihydroxy-phenyl-acetic acid (DOPAC) is the product of the reaction of dopamine with MAO. In turn, DOPAC is the substrate of an irreversible reaction catalyzed by COMT. The relaxation constant of the COMT reaction was argued above to be close to 0.03 min^{-1} in vivo (Cumming et al. 1993).

$$k_{COMT(DOPAC)} = \frac{\sigma_{DOPAC} V_{max}^{COMT}}{M_{DOPA}} \tag{7.272}$$

■ **HVA.** Homovanillic acid (HVA) is the product of the reaction between DOPAC and COMT. HVA is cleared from brain tissue by anionic transport across the choroid epithelium, at a rate estimated to be 0.04 min^{-1} (Cumming et al. 1992).

$$k_{CSF} = \frac{f_b' \, \sigma_{HVA}^{CSF} \varsigma_{OH^-}^{CSF} \, T_{max}^{CSF}}{M_{HVA}}. \tag{7.273}$$

7.5.2.4
Dopamine Turnover

To determine the magnitudes of the relaxation constants and compartments participating in the dopamine turnover, it is necessary to measure the relaxation constants or contents, or both, in vivo. However, not all constants or contents are known yet, in humans or other mammals. The relationships can be used to infer values that subsequently undergo experimental verifi-

Table 7.5. Contents and concentrations of dopaminergic metabolites

Metabolite	Content (nmol g⁻¹)	Concentration (μM)	Volume (ml g⁻¹)
Tyrosine	56	190	0.300
DOPA	0.3	0.4	0.800
3OMD	0.06	0.075	0.800
Free intracellular DA	0.015	2	0.0075
Vesicular DA	48	60,000	0.0008
Intrasynaptic DA	0.006	7.5	0.0008
Extrasynaptic DA	0.00075	0.005	0.150
Total DA	48		
DA Dissociation constant ($K_{d_{DA}}$)		30	
DA Maximum binding capacity (B_{max})	0.03		
Bound DA	0.006		
DOPAC	10		
HVA	7.5		

cation. The values listed in Tables 7.4 and 7.5 are presented as internally consistent examples, to be adjusted in the future, with the aid of the following ratios. The steady-state sum of unbound dopamine in dopaminergec neurons, relative to tyrosine, is

$$\frac{M_{DA}}{M_{Tyr}} = \left(\frac{k_{DDC}}{k_{MAO}}\right)\left[\frac{k_{TH}}{k_{DDC} + k_{AT} + k_{COMT}}\right]$$
$$\left(1 + \frac{k_{VET}}{k_{rel}} + \frac{k_{VET}}{k_D} + \frac{k_{VET}}{k_{DAT}}\right) \quad (7.274)$$

where the free iDA is given by

$$M_{fDA} = \left(\frac{k_{DDC}}{k_{MAO}}\right)\left[\frac{k_{TH}}{k_{DDC} + k_{AT} + k_{COMT}}\right] M_{Tyr}$$
$$= f_f M_{Tyr} , \quad (7.275)$$

the vDA by

$$M_{vDA} = \left(\frac{k_{DDC}}{k_{MAO}}\right)\left[\frac{k_{TH}}{k_{DDC} + k_{AT} + k_{COMT}}\right]$$
$$\times \left(\frac{k_{VET}}{k_{rel}}\right) M_{Tyr} = f_v M_{Tyr} \quad (7.276)$$

and the iDA is given by

$$M_{iDA} = \left(\frac{k_{DDC}}{k_{MAO}}\right)\left[\frac{k_{TH}}{k_{DDC} + k_{AT} + k_{COMT}}\right]$$
$$\times \left(\frac{k_{VET}}{k_D}\right) M_{Tyr} = f_i M_{Tyr} \quad (7.277)$$

such that the bound dopamine is

$$M_{bDA} = \frac{B_{max} M_{Tyr}}{M_{Tyr} + \left(\frac{k_{MAO}}{k_{DDC}}\right)\left[\frac{k_{TH} + k_{AT} + k_{COMT}}{k_{TH}}\right]\left(\frac{k_D}{k_{VET}}\right) K_{d_{DA}}}$$
$$= \frac{B_{max} M_{Tyr}}{M_{Tyr} + (K_{d_{DA}}/f_i)} \quad (7.278)$$

and the eDA is

$$M_{eDA} = \left(\frac{k_{DDC}}{k_{MAO}}\right)\left[\frac{k_{TH}}{k_{DDC} + k_{AT} + k_{COMT}}\right]$$
$$\times \left(\frac{k_{VET}}{k_{DAT}}\right) M_{Tyr} = f_e M_{Tyr} \quad (7.279)$$

which, as confirmation, yields the correct ratio between iDA and eDA:

$$\frac{M_{iDA}}{M_{eDA}} = \frac{k_{DAT}}{k_D} = \lambda \frac{V_{iDA}}{V_{eDA}} \quad (7.280)$$

and hence the ratio between the concentrations of iDA and eDA:

$$\frac{C_{iDA}}{C_{eDA}} = \lambda = 1,500 \quad (7.281)$$

where λ is

$$\lambda = \sqrt{\frac{\overline{\sigma}_{eDA} \varsigma_{Na^+} T_{max}}{\overline{C}_{eDA} DA/L}} \quad (7.282)$$

according to which the ratio rises and falls with the occupancy of the dopamine transporter and the length of the diffusion path. The value of λ suggests that the dopamine transporter accelerates the dopamine diffusion from the intrasynaptic space by a factor of 2.25×10^6.

7.6
Glossary

$*$	Asterisk, denotes labeled tracer
a	Parameter of multilinear regression
A	Area (unit of distance^2)
$\bar{b}$	Bar, denotes calculated rather than measured variable
$B_{\max}$	Maximum binding capacity of protein in ligand-protein interaction (unit is mass)
c	Concentration, time-variable concentration of delivery compartment (unit of mass volume^{-1})
C	Concentration, time-invariant (steady-state) concentration of delivery compartment (unit of mass volume^{-1})
D	Diffusion coefficient (unit of distance^2 time^{-1})
ζ	Dummy differentiation variable equal to $\overline{m}_e/K_1$
f_b	Bidirectionality factor (ratio)
f_b'	Bidirectionality factor for saturable ligand-protein interactions (ratio)
G	Spatial concentration gradient, dC/dx (unit is mass volume^{-1} distance^{-1})
j, j_n	Time-variable flux (unit is mass time^{-1}, n is any number)
J, J_n	Flux, time-invariant (steady-state) flux (unit is mass time^{-1}, n is any number)
k, k_n	Relaxation constant (unit is time^{-1}, n is any number)
k', k_n'	Steady-state relaxation constant representative of multiple compartments (unit is time^{-1}, n is any number)
k_{in}	Dissociation-rate constant (unit of time^{-1})
k_{off}	Dissociation-rate constant (unit of time^{-1})
k_{on}	Association coefficient (unit of mass volume^{-1} time^{-1})
k_{out}	Association coefficient (unit of mass volume^{-1} time^{-1})
K_1	"Unidirectional" clearance (unit is volume time^{-1})
K, K_1'	Net clearance, spanning several compartments (unit is volume time^{-1})
K_d	Dissociation constant (unit of mass volume^{-1})
κ_i^*	Apparent clearance of tracer to compartment i as function of time of circulation defined as $m_i^*(T)/\int_0^T c_a^* \, dt$
χ	Concentration ratio (example $C \, K_M^{-1}$)
K_M	Michaelis half-saturation constant (unit of mass volume^{-1})
L	Diffusion distance, physical length of pathway
λ	Ratio
m, m_n	Mass, time-variable content of any compartment or compartment numbered "n"
m_e^*	Time-variable mass of tracer in exchange compartment
m_p^*	Time-variable mass of tracer in compartments (product compartments) other than delivery and exchange compartments
M, M_n	Mass, time-invariant (steady-state) content of compartment or compartment n
$\mathbf{M}, \mathbf{M}_n$	Compartment, compartment n
μ_h	Penalty function of compartment system of h compartments
P	Permeability (unit of volume distance^2 time^{-1})
p_B	Binding potential (ratio)
q	Parameter of three-compartment model (unit of time^{-1})
r_1	Impulse response function
R_1	Ratio between clearances into region of interest and reference region
ρ^*	Time-varying ratio between integrated masses of tracer of region of interest and reference region
S	Surface area (unit of distance^2)
S_{sq}	Sum of squares to be minimized by regression analysis
σ	Saturation, occupancy, fraction of binding sites occupied by ligand (example $\chi \, (1+\chi)^{-1}$)
ζ_n	Occupancy of noncompetitively interacting ligand n
t	Time
T	Time, integration limit
τ_D	Characteristic time constant of diffusion
$T_{\max}$	Maximum flux mediated by transporter protein
Θ^*	Transformed time, equal to ratio between integral of arterial concentrations and arterial concentrations of tracer
Θ'^*	Transformed time, equal to ratio between twice-integrated arterial concentrations and integrated arterial concentrations of tracer
θ^*	Transformed time, equal to ratio between integral of reference-region mass and reference-region mass of tracer

θ'^* Transformed time, equal to ratio between twice-integrated reference-region mass and integrated reference-region mass of tracer item [u] dummy time variable of integration

v^* Apparent volume of distribution of tracer, defined as time-variable mass-concentration ratio

v'^* Apparent volume of distribution of tracer, defined as time-variable mass-concentration ratio

V Volume of distribution, time-invariant (steady-state) mass-concentration ratio

V_d Physical (aqueous) volume of distribution, time-invariant (steady-state) mass-concentration ratio of mass of substance in compartments and concentration of substance in reference fluid (water) in same compartments

V_e Partition volume, time-invariant (steady-state) mass-concentration ratio of mass of substance in compartments supplied by, and concentration of substance in, delivery compartment, equal to ratio K_1/k_2

V_f Special partition volume, time-invariant (steady-state) mass-concentration ratio of mass of substance in exchange compartment M_2 shown in Fig. 7.3 and concentration of substance in delivery compartment M_1 (M_0 in case of capillary delivery compartment), equal to ratio $K_1/(k_2 + k_3)$

V_g Special partition volume, time-invariant (steady-state) mass-concentration ratio of mass of substance in exchange compartment M_2 shown in Fig. 7.3 and concentration of substance in delivery compartment M_1 (M_0 in case of capillary delivery compartment), equal to ratio $K_1 k_2/(k_2 + k_3)^2$

V_{max} Maximum reaction rate of enzyme catalysis (unit is flux)

x Distance

y Ordinate, axis of orthogonal coordinate system

z Third axis of three-dimensional orthogonal coordinate system.

7.7
References

Azzaro AJ, King J, Kotzuk J, Schoepp DD, Frost J, Schochet S (1985) Guinea pig striatum as a model of human dopamine deamination: the role of monoamine oxidase isozyme ratio, localization, and affinity for substrate in synaptic dopamine metabolism. J Neurochem 45:949–956

Blomqvist G (1984) On the construction of functional maps in positron emission tomography. J Cereb Blood Flow Metab 4:629–632

Borbely K, Brooks RA, Wong DF, Burns RS, Cumming P, Gjedde A, Di Chiro G (1999) NMSP binding to dopamine and serotonin receptors in MPTP-induced parkinsonism: relation to dopa therapy. Acta Neurol Scand 100:42–52

Briggs GE, Haldane JBS (1925) A note on the kinetics of enzyme action. Biochem J 19:338–339

Chance B (1943) The kinetics of the enzyme-substrate compound of peroxidase. J Biol Chem 151:553–577

Chien JB, Wallingford RA, Ewing AG (1990) Estimation of free dopamine in the cytoplasm of the giant dopamine cell of Planorbis corbeus by voltammetry and capillary electrophoresis. J Neurochem 54:633–638

Crone C (1963) The permeability of capillaries in various organs as determined by use of the "indicator diffusion" method. Acta Physiol Scand 58:292–305

Cumming P, Brown E, Damsma G, Fibiger HC (1992) Formation and clearance of interstitial metabolites of dopamine and serotonin in rat striatum: an in vivo microdialysis study. J Neurochem 59:1905–1914

Cumming P, Léger GC, Kuwabara H, Gjedde A (1993) Pharmacokinetics of plasma 6-[^{18}F]fluoro-L-3,4-dihydroxyphenylalanine ([^{18}F]FDOPA) in humans. J Cereb Blood Flow Metab 13:668–675

Cumming P, Kuwabara H, Ase A, Gjedde A (1995) Regulation of DOPA decarboxylase activity in brain of living rat. J Neurochem 65:1381–1390

Cumming P, Ase A, Kuwabara H, Gjedde A (1998) [^{3}H]DOPA formed from [^{3}H]tyrosine in living rat brain is not committed to dopamine synthesis. J Cereb Blood Flow Metab 18:491–499

Cumming P, Hermansen F, Gjedde A (1999) Cerebral dopamine concentrations during levodopa treatment. Neurology 53:1374–1375

Cumming P, Wong DF, Gillings N, Hilton J, Scheffel U, Gjedde A (2002) Specific binding of [(11)C]raclopride and N-[(3)H]propyl-norapomorphine to dopamine receptors in living mouse striatum: occupancy by endogenous dopamine and guanosine triphosphate-free G protein. J Cereb Blood Flow Metab 22:596–604

Dixon M (1953) The determination of enzyme inhibitor constants. Biochem J 55:170–171

Eadie GS (1952) On the evaluation of the constants V_m and K_M in enzyme reactions. Science 116:688

Einstein A (1908) The elementary theory of the Brownian motion. Z f Elektrochemie 14:235–239

Evans AC (1987) A double integral form of the three-compartmental, four-rate-constant model for faster generation of parameter maps. J Cereb Blood Flow Metab 7 (Suppl 1):S453

Gjedde A (1980) Rapid steady-state analysis of blood-brain glucose transfer in rat. Acta Physiol Scand 108:331–339

Gjedde A (1981) High- and low-affinity transport of D-glucose from blood to brain. J Neurochem 36:1463–1471

Gjedde A (1982) Calculation of glucose phosphorylation from brain uptake of glucose analogs in vivo: a re-examination. Brain Res Rev 4:237–274

Gjedde A, Wong DF, Wagner HN Jr (1986) Transient analysis of irreversible and reversible tracer binding in human brain in vivo. In: Battistin L (ed) PET and NMR: new perspectives in neuroimaging and clinical neurochemistry. Alan R Liss, New York, pp 223–235

Gjedde A, Wong DF (1990) Modeling neuroreceptor binding of radioligands in vivo. In: Frost J, Wagner HN Jr (eds) Quantitative imaging of neuroreceptors. Raven Press, New York, pp 51–79

Gjedde A (1992) Blood-brain glucose transfer. In: Bradbury MWB (ed) Handbook of Experimental Pharmacology, Chap. 6a. Springer, Berlin Heidelberg New York, pp 65–115

Gjedde A (1995a) Compartmental analysis. In: Wagner HN Jr, Szabo Z, Buchanan JW (eds) Principles of nuclear medicine, 2nd edn. Saunders, Philadelphia, pp 451–461. J Cereb Blood Flow Metab 20:834-838

Gjedde A (1995b) Glucose metabolism. In: Wagner HN Jr, Szabo Z, Buchanan JW (eds) Principles of nuclear medicine, 2nd edn. Saunders, Philadelphia, pp 54–71

Gjedde A, Gee AD, Smith DF (2000) Basic CNS drug transport and binding kinetics in vivo. In: Begley DJ, Bradbury MW, Kreuter J (eds) The blood-brain barrier and drug delivery to the CNS. Marcel Dekker, New York, pp 225–242

Gjedde A, Wong DF (2001) Quantification of neuroreceptors in living human brain. V. Endogenous neurotransmitter inhibition of haloperidol binding in psychosis. J Cereb Blood Flow Metab 21:982–994

Gunn RN, Gunn SR, Cunningham VJ (2001) Positron emission tomography compartmental models. J Cereb Blood Flow Metab 21:635–652

Gunn RN, Gunn SR, Turkheimer FE, Aston JAD, Cunningham VJ (2002) Positron emission tomography compartmental models: a basis pursuit strategy for kinetic modelling. J Cereb Blood Flow Metab 22:1425–1439

Hjorth JSU (1994) Computer intensive statistical methods validation, model selection and bootstrap. Chapman & Hall, London

Hofstee BHJ (1952) On the evaluation of the constants V_m and K_M in enzyme reactions. Science 116:329–331

Johnson RG, Scarpa A (1979) Protonmotive force and catecholamine transport in isolated chromaffin granules. J Biol Chem 254:3750–3760

Lassen NA, Perl W (1979) Tracer kinetic methods in medical physiology. Raven Press, New York

LeFauconnier JM (1992) Transport of amino acids. In: Bradbury MWB (ed) Physiology and pharmacology of the blood-brain barrier. Springer, Berlin Heidelberg New York, pp 117–150

Logan J, Fowler JS, Volkow ND, Wolf AP, Dewey SL, Schlyer DJ, MacGregor RR, Hitzemann R, Bendriem B, Gatley SJ, Christman DR (1990) Graphical analysis of reversible radioligand binding from time-activity measurements applied to [N-11C-methyl]-(-)-cocaine PET studies in human subjects. J Cereb Blood Flow Metab 10:740–747

Logan J, Fowler JS, Volkow ND, Ding YS, Wang GJ, Alexoff DL (2001) A strategy for removing the bias in the graphical analysis method. J Cereb Blood Flow Metab 21:307–320

Michaelis L, Menten ML (1913) Zur Kinetik der Invertinwirkung. Biochem Z 49:333–369

Ostergaard L, Johannsen P, Host-Poulsen P Vestergaard-Poulsen P, Asboe H, Gee AD, Hansen SB, Cold GE, Gjedde A, Gyldensted C (1998) Cerebral blood flow measurements by magnetic resonance imaging bolus tracking: comparison with [(15)O]H$_2$O positron emission tomography in humans. J Cereb Blood Flow Metab 18:935–940

Patlak CS, Blasberg RG, Fenstermacher JD (1983) Graphical evaluation of blood-to-brain transfer constants from multiple-time uptake data. J Cereb Blood Flow Metab 3:1–7

Reith J, Dyve S, Kuwabara H, Guttman M, Diksic M, Gjedde A (1990) Blood-brain transfer and metabolism of 6-[^{18}F]fluoro-L-Dopa. J Cereb Blood Flow Metab 10:707–719

Sarna GS, Bradbury MW, Cavanagh J (1977) Permeability of the blood-brain barrier after portocaval anastomosis in the rat. Brain Res 138:550–554

Shao J (1993) Linear model selection by cross-validation. J Am Statist Assoc 88:486–494

von Smoluchowski M (1914) The kinetic theory of matter and electricity. Leipzig and Berlin

Sokoloff L, Reivich M, Kennedy C, Des Rosiers MH, Patlak CS, Pettigrew KD, Sakurada O, Shinohara M (1977) The [14C]deoxyglucose method for the measurement of local cerebral glucose utilization: theory, procedure, and normal values in the conscious and anesthetized albino rat. J Neurochem 28:897–916

Wong DF, Sølling T, Yokoi F, Gjedde A (1998) Quantification of extracellular dopamine release in schizophrenics and cocaine use by means of TREMBLE. In: Carson RE, Daube-Witherspoon ME, Herscovitch P (eds) Quantitative functional brain imaging and positron emission tomography. Academic, San Diego, pp 463–468

CHAPTER 8

Compound Delivery and Local Blood Flows

JAMES B. BASSINGTHWAIGHTE

Contents

8.1
Introduction

Genomics and molecular and cell biology have evolved so rapidly in the past few years that there is more information available than can be readily or accurately understood about the behavior of whole cells, intact tissues and organs, and the functioning organism. Even where the biochemical pathways are known, it is not yet possible to fully portray the fluxes along them or their regulation over a physiological range of conditions. Yet these and a host of related functions, at all levels of integration, are precisely what is required to understand and predict the effects of genomic or pharmacological intervention. Because even small systems are complex, nonlinear, and time varying, an untutored intuitive understanding is impossible. In accordance with the goals for the practice of nuclear medicine, one should attempt to gain information of diagnostic and therapeutic importance from the patient while using the most efficient noninvasive techniques.

"Noninvasive" and "in vivo" are no longer words in total conflict with "quantitative" and "cellular," but the key to "conflict avoidance" is the judicious handling of complex situations in which multiple processes are involved. The complexity of the processes of solute delivery and intracellular metabolism can be managed only through detailed understanding of those processes and how they impact the kinetics of tracer or indicator transport through the vascular system, across capillary walls, through cell membranes, and in fluxes through reactions. Only through adequately detailed quantitative portrayal of these events can data acquired by noninvasive imaging techniques be interpreted with confidence in making life-determining decisions.

8.2
The Processes of Delivery and Uptake

Whether tracers are injected for diagnostic purposes or drugs for therapeutic purposes, all the same basic mechanisms are involved: convection in flowing fluids, permeation by diffusion or by transmembrane transporters, diffusion, binding, and, usually, reaction. For an orally administered drug the order is different, but all these mechanisms are involved. The binding to a protein of an enzyme, receptor, or combination of sites is almost inevitable.

8.2.1
Flow, the Convective Process

Generally speaking, despite the fact that blood flow is pulsatile in the arterial system, the flow is stable, not turbulent. Blood viscosity is about five centipoise, which is about five times the viscosity of water. Poiseuille's equation for flow through a straight tube is

$$F = \frac{\pi r_0^4 \Delta P}{8 \eta L} , \qquad (8.1)$$

where F is flow (ml/s), r_0 is the tube radius (cm), ΔP is the pressure difference between the entrance and the exit of the tube (dynes/cm^2 or g cm^{-1} s^{-1}), η is viscosity (poise or dyne s cm^{-2}), and L is the tube length (cm). Vascular resistance equals pressure divided by flow:

$$\text{Resistance} = \frac{\text{pressure}}{\text{flow}} = \frac{\Delta P}{F} = \frac{8 \eta L}{\pi r_0^4} . \qquad (8.2)$$

While the equation assumes that the viscosity is a constant value independent of flow, this is not quite true for blood, the apparent viscosity of which increases somewhat as the flows approach zero. Blood is a thixotropic fluid. Viscosity exerts its retarding influence on flow because the layer at the wall has zero velocity – it sticks to the wall; thus a more highly viscous fluid has more resistance to flow. Equations 8.1 and 8.2 are simply a restatement of Ohm's law: resistance equals voltage over current. In the fluid near the wall the neighboring laminae have increasing velocities, v, as the radial distance, r_0–r, from the wall increases: the shear rate is defined as dv/dr. The shear rate at the wall is given by

$$\left. \frac{dv}{dr} \right|_{r=r_0} = \frac{\Delta p \cdot r_0}{2 \eta L} , \qquad (8.3)$$

where the symbols are defined as in Eq. 8.1. The average shear rate across the cylinder is two thirds of this. At the center of the tube, the shear rate at the axis is zero and increases gradually as it approaches the wall. The shear stress for a cylinder is $\tau_{\text{mean}} = 4 \eta \bar{v}/r_0$, where $\bar{v}$ is the mean velocity.

For very small vessels, less than 300 μm diameter, the apparent viscosity diminishes as diameter narrows and shear rate increases, a phenomenon known as the Fahraeus-Lindqvist effect (Fahraeus and Lindqvist 1931), in accordance with Eqs. 8.1 and 8.3. This was shown by Chien et al. (1966), as diagrammed in Fig. 8.1. The diminution in apparent viscosity with

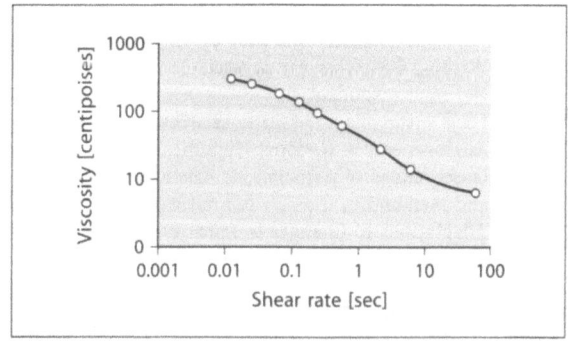

Fig. 8.1. Decrease in apparent viscosity of blood (hematocrit ratio, 51.7%) at increasing rates of shear, both plotted on logarithmic scales. The shear rate refers to the relative velocity of one layer of fluid with respect to that of the adjacent layers and is directionally related to the rate of flow (From Chien et al. 1966, with permission)

increasing shear has a number of explanations. One is that the red blood cells line up with the laminae; another is that in small vessels the red cells are readily deformed and take on a parachute shape just as they do when going through capillaries; a third explanation is that the red cell is not only flexible but also has a fluid content that can rotate inside the cell. All these mechanisms reduce the effective thickness of the red cell so that the laminae are thinner.

8.2.1.1

Driving Forces: Hydrostatic Pressure, Osmotic Pressure, Water Flux

Bulk flow is driven by a hydrostatic pressure gradient as defined in, for example, Eq. 8.1. Laminar flow, usually known as Newtonian flow or Poiseuille flow, is less resistant than turbulent flow. A measure of this is given by the Reynolds number, the dimensionless number that is the ratio of inertial forces to viscous forces:

$$\mathrm{Re} = \frac{2\rho r_0 \bar{v}}{\eta} = \frac{2\rho r_0 (F/\pi r_0^2)}{\eta} = \frac{2\rho F}{\eta \pi r_0} , \qquad (8.4)$$

where Re is the Reynolds number, ρ is fluid density (g/ml), r_0 is the tube radius, and $\bar{v}$ is the average velocity (cm/s), which is equivalent to the flow, F, divided by the cross-sectional area, πr_0^2. Reynolds numbers in microcirculation are usually less than 1 but, in the ascending aorta and even in other arteries during exercise, may approach 2000. At about a critical Reynolds number of 2100 to 2400 the flow becomes turbulent, destroying its laminar nature and increasing its resistance. Acceleration of the flow tends to stabilize it, even if the Reynolds numbers are high; deceleration of the fluid column tends to destabilize it. Backflow in the ascending aorta at the end of each systole renders it effectively turbulent, a kind of mixing chamber.

Thus, the first and most important driving force for not only blood flow along the vessels but for water transport across membranes is a hydrostatic pressure gradient. Of equal or greater importance, however, are osmotic gradients. These are due to differences in total solute concentrations on the two sides of a membrane. Water, by virtue of its higher chemical activity in the less concentrated of the two solutions, will, without any hydrostatic pressure difference, move down its activity gradient into the more concentrated solution, e.g., across a cell mem-

brane from the interstitium into the cytosol of a cell. Producing small metabolites in excess to raise cytosolic osmolarity as a muscle cell does during exercise causes the cell to swell and the muscle to become turgid. This is a diffusional process equivalent to the convection of the solvent. It is the interaction of convective and diffusive driving forces and fluxes that governs exchanges at the capillary wall and across cell membranes.

8.2.2

Permeation and Diffusion

8.2.2.1

Permeation in Transcapillary Exchange

Consider the situation, diagrammed in Fig. 8.2, where there are two well-stirred tissue compartments having volumes V_1 and V_2 (ml/g) with concentrations C_1 and C_2 (molar). The conductance of the membrane separating the two compartments is given by PS, the permeability-surface area product, which is the product of the permeability P (cm/s) times the surface area S (cm^2/g) of the membrane per gram of tissue. The exchange is attributable to the diffusion of the solute through the membrane, so permeability is really equivalent to a diffusion coefficient, D (cm^2/s), in the membrane divided by the membrane thickness, Δx (cm). Fluxes (moles/s) across the membrane go in both directions, and PS (ml s^{-1} g^{-1}) is the conductance for the unidirectional flux. The concentrations (moles/ml) change as a result of a net flux as expressed in Eqs. 8.5a and 8.5b:

$$V_1 \frac{dC_1}{dt} = PS(C_2 - C_1) , \qquad (8.5\,\mathrm{a})$$

$$V_2 \frac{dC_2}{dt} = PS(C_1 - C_2) . \qquad (8.5\,\mathrm{b})$$

$$P = D/(\Delta x) . \qquad (8.6)$$

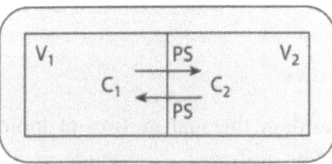

Fig. 8.2. Fluxes of solute across a membrane between two stirred tanks

A passive system like this will equilibrate to give equal concentrations on the two sides of the membrane if the PS is not zero. When there is flow through compartment 1 (of Fig. 8.2, not diagrammed) Eqs. 5a and 5b become

$$V_1 \frac{dC_1}{dt} = F(C_{in} - C_1) - PS(C_2 - C_1) , \qquad (8.7\,a)$$

$$V_2 \frac{dC_2}{dt} = PS(C_2 - C_1) , \qquad (8.7\,b)$$

where C_{in} is the concentration at the inflow and C_1 is the outflow concentration.

Because capillary-tissue exchange occurs over the full length of the capillary, and the length, in most tissues, is about 200 times the diameter, about 5 μm, there is normally a gradient in intracapillary concentrations from inflow to outflow. The stirred tank approximation in Eqs. 8.7a and 8.7b is therefore not valid because the inflow and outflow concentrations are, inappropriately, considered to be instantaneously equalized, so Eqs. 8.7a and 8.7b are replaced by

$$V_1 \frac{\partial C_1}{\partial t} = -FL \frac{\partial C_1}{\partial x} + PS(C_2 - C_1) , \qquad (8.8\,a)$$

$$V_2 \frac{\partial C_2}{\partial t} = PS(C_1 - C_2) , \qquad (8.8\,b)$$

where these partial differential equations (PDE) differ from the ordinary differential equations (ODE) in Eqs. 8.7a and 8.7b by accounting for the intracapillary axial gradient $\partial C_1/\partial x$ along the length as a function of position x. The boundary conditions for the PDEs are that $C_1 = C_{in}$ at the inflow at $x=0$, and $C_{out}=C_1$ (at $x=L$, the end of the capillary). It is apparent that the PDE representation is as simple as the ODE representation and has all the same parameters except for an effective length L. (In practice the distance x is considered as x/L, a normalized, dimensionless length, thereby reducing the parameters of the PDE to be only those of the ODE but preserving the all-important accounting for the gradient in concentrations.

8.2.2.2
Diffusion

Diffusion is due to random thermal motion of molecules and particles. The diffusion coefficient is increased by raising the temperature, which reduces the viscosity, and is lower for large molecules than for small ones because the hindrance is greater; for spherical molecules D is well described by

$$D = \frac{RT}{6\pi a\eta N_A} , \qquad (8.9)$$

where R is the gas constant (98,500 liters×atmospheres moles^{-1} deg^{-1}), T is temperature (degrees Kelvin), a is molecular radius (cm), η is viscosity (poise), and N_A is Avogadro's number, 6.03×10^{23} molecules per mole. Typical diffusion coefficients are around 10^{-5} or lower. For example, the diffusion coefficient for water in aqueous solution is 2.3×10^{-5} cm^2/s, but in myocardial tissue it is 2.17×10^{-6} cm^2/s, an order of magnitude lower (Safford et al. 1978). The striking lowering of diffusion coefficients in tissue, even for highly diffusive solutes that penetrate membranes quickly, is due to the hindrance to diffusion. First, the viscosity is higher because the cytosol is a semifluid gel. Second, inside the cell and in the interstitium are hindering structures, high molecular-weight proteins (collagen, actin, myosin), and many subcellular organelles.

When a solute binds to proteins, diffusion is retarded yet further, as shown in Eq. 8.10. If the protein is mobile, its diffusion as the solute protein complex contributes to the diffusion of the free solute, as in Eq. 8.10:

$$D'_S = aD_S + (1 - a)D_{SB} , \qquad (8.10)$$

where the effective diffusion coefficient is D'_S, D_S is the diffusion coefficient for the free solute and D_{SB} is that for the solute-protein site complex, a is the fraction of free solute, and $(1-a)$ is the fraction bound. If the protein is immobile, the second term is zero. This is pretty much the case for calcium inside cells where the effective diffusion coefficient for calcium at peak concentrations of about 10^{-5} M is about 0.2×10^{-5} cm^2/s and falls to about 1/400th of the free diffusion coefficient at concentrations around 1×10^{-7} M (Safford and Bassingthwaighte 1977). If the solute is mainly in the bound form, as is oxygen in the presence of hemoglobin, then $(1-a)$ is large and the diffusion rate is dominated by that of the hemoglobin-oxygen complex, HbO$_2$, facilitated diffusion.

The water space available for solute diffusion in structured media is reduced because of molecular exclusion as well as the tortuosity of the available paths. Molecular exclusion is the name given to the situation in which no two molecules can be centered at the same location; likewise a protein in the cytosolic water in the neighborhood of a collagen molecule cannot be centered on the collagen and cannot diffuse between closely juxtaposed collagens. (The diffusing molecule cannot be centered on the structural protein; thus its center must be at least one molecular radius away from the structure.) The result is that only a fraction of the cytosolic water is available for mobile solutes; for large proteins the fraction is substantially reduced. In the interstitial space of the heart, only about 50% of the water is available for albumin simply because the spaces between structuring molecules are not much bigger than the albumin itself. Molecular exclusion is the prime reason for the reduction in intracellular and interstitial diffusion coefficients. The combination of lowered water-space availability and tortuosity of paths taken by diffusing molecules, combined further with retention of some molecules by binding to specific proteins, can make intracellular diffusion coefficients very low. For example, Luxon and Weisiger (1992) found diffusion coefficients of tri-iodothyronine to be 3×10^{-8} cm^2/s and for palmitate to be about 4×10^{-9} cm^2/s (Luxon and Weisiger 1993); both tri-iodothyronine and palmitate bind to intracellular proteins in liver cells.

8.2.2.3
Hindrance to Diffusion in Transcapillary Exchange

Some insight into the variety of mechanisms for solute movement across the capillary wall between blood and the interstitial space is given by Fig. 8.3, which illustrates the various means of transcapillary exchange, and Fig. 8.4, which illustrates the periendothelial pathway.

Figure 8.3 shows that the route for hydrophilic solutes is through the interendothelial clefts but for lipid-soluble substances is a combination of through the clefts and across the endothelial cells themselves. In closed capillaries the junctional regions are narrow, with interendothelial cleft widths of about 17 nm from cell membrane to cell membrane; these are functionally only about 9 nm wide due to the presence of glycocalyx and interendothelial cell gap junctional connections. In open or fenestrated capillaries the spaces are larger but still usually contain glycocalyx, which inhibits the passage of charged proteins; for example, negatively charged albumin doesn't cross the glomerular membrane. Large proteins cannot penetrate the clefts of closed capillaries but cross the barrier slowly via incorporation into vesicles and vesicular transport to the opposite surface or possibly, and certainly rarely, through chains of linked vesicles.

A possible mode of transport for lipid-soluble substances is a periendothelial path via dissolution in the endothelial luminal surface plasmalemma followed by diffusion within the phospholipid bilayer around the edges of the cell to the abluminal surface of the endothelial cell, followed by release into the interstitial fluid (ISF) space. The difficulties associated with attributing much flux to this route are shown in Fig. 8.4 and described in the figure legend. In brain capillaries the chains of gap junctional connections between neighboring cells probably form an impenetrable barrier to intramembranous pericellular diffusion.

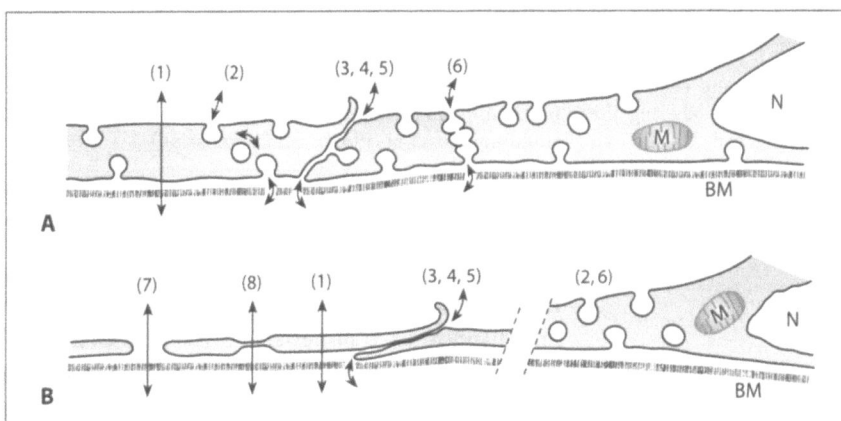

Fig. 8.3. Transcapillary exchange routes in the capillary endothelial barrier (adapted from Renkin 1977, with permission from the American Heart Association). **A** Closed endothelium. **B** Fenestrated endothelium. The numbered routes are: (1) Transendothelial (for water and small non-polar solutes). (2) Vesicular, including shuttling across the cell (for macro- molecules and particles). (3) Lateral diffusion within the cell membrane (lipids only). (4, 5) Diffusion and convection through the aqueous cleft, a few of which are large (5). (6) Aqueous passage via chains of vesicles. (7) Open fenestrae (intestine, kidney cortex, liver)

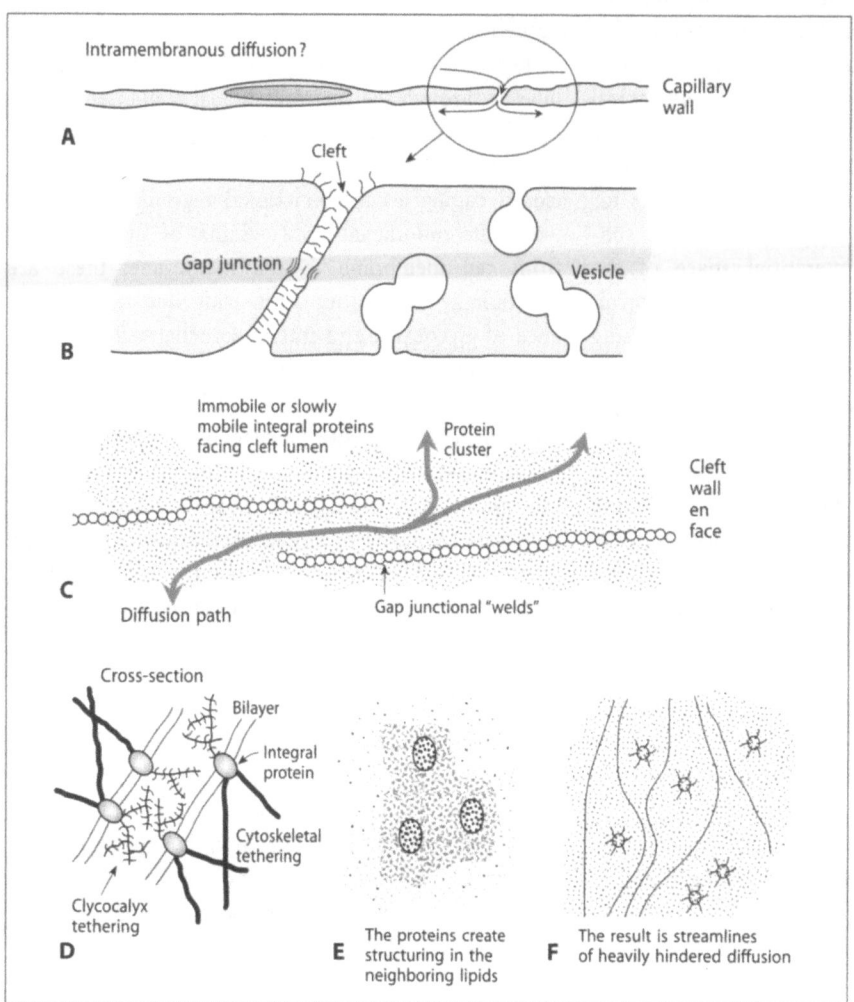

Fig. 8.4 A–F. Factors in the passive diffusion of solutes from plasma to the interstitium via intramembranous diffusion in the plasmalemma of the endothelial cells. A Overall view of cross section of capillary endothelium. B Magnification of cleft showing gap junction linking neighboring endothelial cells and the glycocalyx within the cleft and tethering the membrane proteins to which they are attached. C Cleft wall *en face* with immobile strings of gap junctions around which intramembranous diffusion must occur along the "diffusion path." D Tethering of glycocalyx within cleft reduces mobility of membrane proteins and the membrane lipids near them. E Structuring in neighborhood of proteins reduces phospholipid mobility. F Lines of fastest intramembrane diffusion are farthest from the proteins (from Baosingthwaighte et al. 1989b, with permission)

8.2.2.4
Permeation Across Cell Membranes

Lipid-soluble molecules traverse cell membranes relatively easily since they can dissolve in the fatty-acid tails of the phospholipid bilayer. However, hydrophilic molecules do not permeate the bilayer, and even water has a very low rate of permeation across a pure phospholipid bilayer (known as a black lipid membrane). Consequently, charged molecules, which are normally surrounded by water molecules, *waters of hydration*, and hydrophilic molecules don't cross membranes unless the membrane is equipped with ion channels or transporters to facilitate transmembrane exchange. Ion channels are proteins allowing bidirectional transport of ions when the channel is open; the driving force for this is the electrochemical gradient: the channel conductances are dependent on transmembrane voltage and on the time course of voltage changes. They tend to be selective for specific ions, so we talk of the Na-channel, the K-channel, and the Ca-channel. However, the Na-channel allows passage to Li and Tl, and the Ca-channel allows Sr. None is perfectly selective (Hille 2001). There are several K-channels with different time and voltage dependencies.

For neutral hydrophilic solutes the transmembrane electrical gradient has no influence, but there are large sets of specialized transporters for such things as purine nucleosides, amino acids, sugars, and so on. All of these specialized transport mechanisms are saturable because the transporter is designed to have a high affinity for a particular solute and can usually carry only one at a time. (Ionic channels do not normally exhibit the saturability of their binding sites.) The effective permeability via carrier-mediated transport is reduced by raising the concentration of the solute species, as described in Eq. 8.11:

$$P = \frac{P_{max}}{1 + C_S/K_m} \,, \tag{8.11}$$

where P is the permeability (cm/s), P_{max} is the maximum permeability at low concentrations where the transporter is unsaturated (cm/s), C_S is the solute concentration (molar), and K_m is the effective Michaelis constant for the transporter binding and translocation (molar). When the concentration is equal to the apparent K_m, the permeability is $0.5 P_{max}$. Transporters typically have fairly high K_ms. For the purine nucleosides they are about 100 μM, almost 10,000 times as high as the ambient free concentrations of the substrate, so they are apparently designed to handle the fluxes over very wide ranges of concentrations. Even water needs a transporter, the aquaporin channel (Agre et al. 1993), to allow rapid penetration of cell membranes.

8.2.3
Volumes of Distribution

Here we start with the simplest principle of indicator dilution. The injection of a mass, Q_0 (moles), into an unknown volume, V_{dist} (liters), gives, after a thorough mixing, an equilibrium concentration, C_{eq}. If one samples the fluid and measures the concentration, C_{eq}, the volume can be estimated:

$$V_{dist} = \frac{Q_0}{C_{eq}} \,. \tag{8.12}$$

8.2.3.1
The Partition Coefficient

When a solute can distribute itself by diffusional exchange between two regions, V_1 and V_2, the partition coefficient is defined as the ratio of the solubility in one region to that in the other region. When equilibration is by purely passive processes, the solubility ratio equals the concentration ratio:

$$\lambda_{21} = \frac{C_2}{C_1} \quad \text{at equilibrium} \,, \tag{8.13}$$

where λ_{21} is the ratio of solubility in region 2 compared to that in region 1. Thus tissue/plasma partition coefficients are the relative concentrations of solute in the tissue at equilibrium with the plasma, $\lambda_{tiss} = C_{tissue}/C_{plasma}$.

For lipid-soluble substances the tissue/plasma partition coefficients are very high for fatty tissues and relatively low for aqueous tissues. Substances that are concentrated in cells have high tissue/plasma partition coefficients, e.g., about 30 for potassium. Contrarily, when a substance is excluded from some of the tissue, its effective partition coefficient is low. For example, the partition coefficient for albumin in the ISF (interstitial fluid) compared to plasma is about 0.5 simply because of molecular exclusion. Thus partitioning occurs not merely by solubility but also by physical exclusion and by concentrative transport. If there is a binding substance in the extravascular space for a solute, its tissue/plasma partition will be raised by the fraction of the solute bound to the binding site, removing it from free solution. While perhaps we should not really lump all of these varied mechanisms governing intratissue concentrations into the catchall term *partition coefficient*, it is important to appreciate that partitioning is not due to solubility alone. The fact of the matter is that plasma and interstitial and cellular fluid spaces are not very different, and therefore, strictly speaking, the solubility in water is a great influence. So solute partitioning different from passive equilibration is due to the other mechanisms: molecular exclusion, concentrative transport, or binding to medium- or high-affinity sites in the tissue.

Plasma is taken to be the reference fluid in defining volumes of distribution; thus the V_{dist} for a tissue is

$$V_{dist} = \lambda_{tiss} \cdot V_{actual} \,, \tag{8.14}$$

and since tissue includes all components, this implies that the λ_{tiss} is based on the average concentration of all components of the tissue, including blood, relative to the plasma concentration.

8.2.4
Retention Mechanisms

The three main mechanisms for retaining a tracer solute within the tissue are (1) concentrating it within a cell, (2) attaching it to a high-affinity' binding or receptor site, and (3) transforming it by a reaction and retaining the reaction product.

8.2.4.1
Concentrating Processes

A concentrating process, raising the concentration of a solute inside a cell or vesicle to a high level compared with that outside the membrane, is an energy-requiring process. Potassium accumulation inside cells is brought about by the sodium-potassium ATPase, NaK-ATPase, an ionic pump that uses the energy of ATP breakdown to ADP to extrude three Na's for every two Ks brought in. This ATP-powered exchanger or pump creates the negative transmembrane intracellular potential; it concentrates potassium by 30-fold and reduces Na concentration by 15-fold. The actual ionic current flows, especially in excitable cells, are more complex than this and are well summarized in the action potential models of Winslow et al. (1999), Greenstein et al. (2000), Michailova and McCulloch (2001), Luo and Rudy (1994), and Hund et al. (2000).

8.2.4.2
Exchangers and Obligatory Countertransporters

The Na gradient developed by the NaK-ATPase serves also to drive out calcium from the cell. Ca^{2+} efflux is effected through the Na/Ca exchanger by which the inflow of three Na's drives out one Ca and in this mode operates as a net inward current of positive charge. Thus if the NaK-ATPase and the Na/Ca exchanger had equal flux rates, they could balance the Na exchange across the wall. The Na/Ca transporter is an obligatory transporter that can run backwards: it actually creates a net outward current briefly during the plateau phase of the cardiac action potential. The Na/H exchanger is electrically neutral and works to prevent cellular hyperacidity. An obligatory cotransport system is exemplified by the glucose/Na cotransporter (Stein 1986).

8.2.4.3
Binding and Adsorption

The binding of a tracer-labeled solute to a receptor is normally fast compared with its rate of release. In order to estimate the kinetics of the binding and unbinding there must be release, simply in order to get the information. When release is very slow and the uptake fast, the retention of a receptor marker gives a measure of the rate of delivery to the receptor rather than indicating anything about the kinetics of the receptor itself. For example, Little and Bassingthwaighte (1983), in searching for the ideal "molecular microsphere," found that a blocker of serotonin HT_2 receptors, iododesmethylimipramine (IDMI), worked well. In the heart, it was nearly 100% extracted during a single pass through the capillary in going from artery to vein and was well retained by binding to the receptor.

Other mechanisms of intracellular sequestration that sestamibi undergoes (Piwnica-Worms et al. 1992), such as the mitochondrial uptake due to the transmitochondrial potential, are also an important factor in retention. Dependence on membrane potential alone is good only for charged tracers. The nice thing about potassium and rubidium is that there is not only a transmembrane driving force for uptake because of the charge but also a large potassium pool, maintained by the NaK-ATPase, that retains them in the cell. There is still a problem with using ions as flow markers in that they are not lipid-soluble and therefore must traverse specialized channels through which their permeability is limited. Therefore they are not completely extracted during transcapillary passage. It has generally been considered that thallium uses the same mechanism as potassium and rubidium; this is only partly true, as thallium can also enter through the Na channel (Hille 1972) and thallium may actually attach to the external surface of the sarcolemmal membrane, not entering the cells completely (Winkler and Schaper 1979).

The key factor in common for the success of these deposition markers is that their effective volumes of distribution are large, simply because of the binding or because of the fact that they are concentrated in cells by asymmetrical transporters. For potassium the NaKATPase, the "NaK pump," so raises the intracellular concentration of K^+ that it is 150 mM, as opposed to 5 mM outside, so the volume of distribution inside myocytes is 30 times the real volume, i.e., the effective cell-to-plasma partition coefficient is 30 for tracer potassium.

8.2.4.4
Intraregional Binding
Enlarges the Partition Coefficient

The partition coefficient is influenced by binding either in plasma space or in the tissues. Strong binding in the plasma reduces tissue/plasma partition coefficients. For some substances there are specific plasma protein carriers: retinone, testosterone, and other steroids. Fatty acids are approximately 99.94% bound to albumin at 0.4 mM. For a substance that is free in the plasma, remains extracellular, but binds in the interstitium, the interstitium/plasma partition coefficient is increased by the binding. At equilibrium:

$$\lambda_{ISF/pl} = \frac{C_{ISF}\,(\text{free}) + C_{ISF}\,(\text{bound})}{C_{pl}}\,. \tag{8.15}$$

When the concentration of the binding sites and the dissociation constant are known, one can translate this expression into one that gives a partition coefficient as a function of the free concentration in the plasma. Under this circumstance the effective interstitial volume of distribution, V_{ISF}, for this solute is then concentration-dependent, being higher at low concentrations when few of the binding sites are filled and low at high concentrations when the binding sites are saturated. (Such binding will also reduce its diffusion; Safford and Bassingthwaighte 1977.) The virtual volume of distribution, V', for a solute in a region of volume V is increased by the presence of a substance to which it binds and unbinds rapidly. Consider the binding to a protein of concentration B in the region for which the dissociation constant for the solute is K_d. The equilibrium ratio of the total amount in the region to the amount of free solute is dependent on the solute concentration, C, in the region:

$$\frac{V'}{V} = 1 + \frac{B}{K_d + C} = 1 + \frac{B/K_d}{1 + C/K_d}\,. \tag{8.16}$$

In this expression, the 1 represents the ratio in the absence of binding, and the $B/(K_d + C)$ represents the bound solute in the form of a complex with the protein. C is the concentration of free solute. With $C \ll K_d$, $V'/V \simeq 1 + B/K_d$. With $C \gg K_d$, $V' = V$. The impact of binding is shown in Fig. 8.5 for a protein B at two concentrations, 0.5 and 1.0 mM, which has a single binding site with a dissociation constant of 0.1 mM.

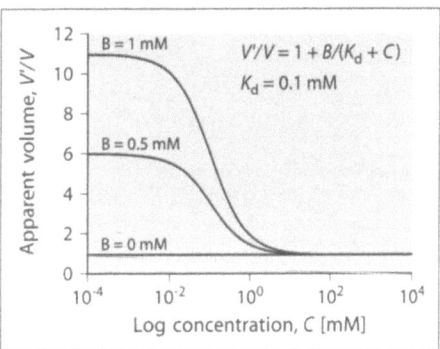

Fig. 8.5. Volume of distribution for a solute that binds to an intraregional site. In the presence of a binding site of concentration B whose dissociation constant, K_d, is 0.1 mM for the solute, the effective equilibrium volume of distribution is increased, particularly at low concentrations. Curves are for $B = 1$, 0.5, and 0 mM

At low solute concentrations the volume of distribution is enlarged by B/K_d, a factor of 11 for $B = 1$ mM.

8.2.4.5
Trapping by Transformation

While serotonin is trapped in endothelial cells by binding to HT$_2$ (hydroxytryptamine) receptor sites, adenosine is also trapped in endothelial cells by another mechanism, namely, enzymatic reaction facilitated by adenosine kinase to form AMP and then ATP. Since the ATP pool is exceedingly large (5 mM) compared to the free adenosine concentration inside cells (5–20 nM), this leads to very long-term retention. In dog hearts the extraction of adenosine is nearly complete (Kroll and Stepp 1996). This means that adenosine is an excellent marker of regional blood flow in the canine heart and probably in other organs including brain. However, this works in dogs because *dog* red blood cells take up adenosine only very slowly (Kroll and Stepp 1996), whereas human and guinea pig red blood cells take up adenosine rapidly through a purine nucleoside transporter. When there is rapid uptake of adenosine into the ATP pool of the red blood cells, it becomes useless as a regional flow marker since the red cells will simply travel through the capillaries without losing their adenosine-labeled ATP to the tissue. This is known as a "red cell carriage effect," whereby the rate of escape from red blood cells limits the tissue uptake (Goresky et al. 1975).

A similar mechanism applies to the uptake and metabolism of fatty acids. They are not much more

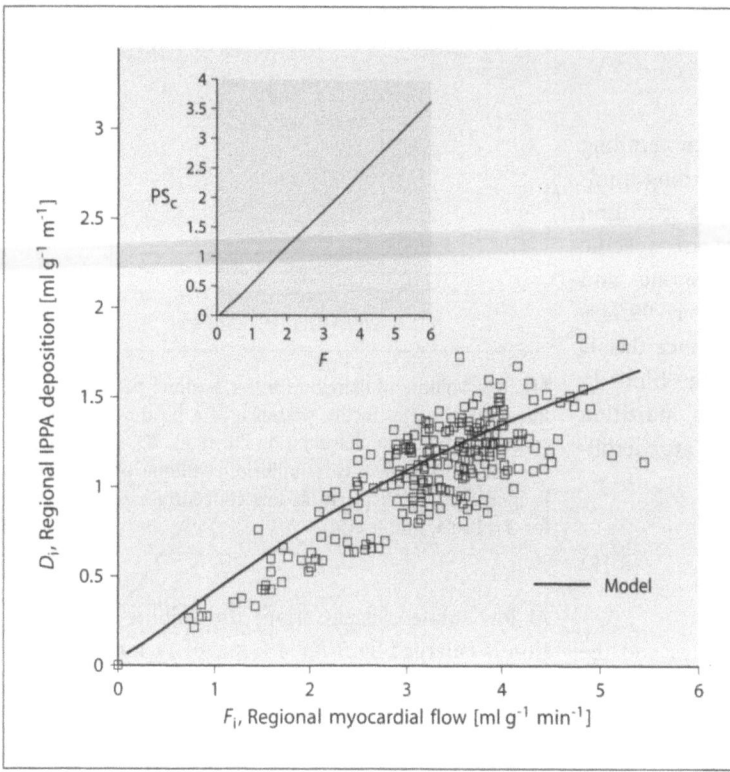

Fig. 8.6. Regional deposition, D_i, of fatty acid (iodopentaphenyldecanoic acid) vs. regional flow, F_i, in a dog heart. The model fit is for the relationship $PS_{c_i} = \overline{PS_c} \cdot [1 + a(F_i/\bar{F} - 1)]$, where $a = 0.9$. This is very close to a pure proportionality that occurs with $a = 1$, $PS_{c_i} = \overline{PS_c} \cdot (F_i/\bar{F})$, that is, that regional permeability to fatty acid is proportional to regional flow (Fig. 8 of Caldwell et al. 1994, with permission)

than half extracted during single passage in normal hearts (Caldwell et al. 1994). In ischemic cardiac regions, because flow is reduced, the extraction locally is higher. ^{131}I-labeled fatty acids were used in identifying ischemic regions in the early years of cardiac imaging because of their retention there as di- and tri-glycerides (Evans 1964). There is in normal hearts evidence (Caldwell et al. 1994) that the uptake of fatty acids is *proportional* to the regional flow (Fig. 8.6). Such proportionality indicates a special means of uptake. There is apparently increased transport capacity in high-flow regions. The basis presumably lies in the regional metabolic needs and accordingly the regulation of transcription of transporter protein for fatty acid, high in regions of high flow and metabolism, whereas low-flow regions have low requirements for fatty acid.

8.3
Flow Heterogeneity and Its Relationship to Local Metabolism

8.3.1
A Heterogeneity of Regional Flows Is Normal

8.3.1.1
Probability Density Functions of Flows in Intact Organs

Regional blood flows in the heart are spatially heterogeneous (Richmond et al. 1973; Buckberg et al. 1971; Yipintsoi et al. 1973; Falsetti et al. 1975; Sestier et al. 1978). Detailed analysis in awake baboons showed that this variation was normal and not due to methodologic error (King et al. 1985). King et al. (1985) showed that probability density functions of flows (Fig. 8.7, left panel) in awake baboons are almost Gaussian. The figure shows that the subsidiary distributions in the right and left ventricles are distinctly different from each other and that the average flow in the right ventricle is only about 70% of that in the left ventricular tissue. The normal heterogeneity has been observed in much detail, as reviewed by Bassingthwaighte et al. (2001).

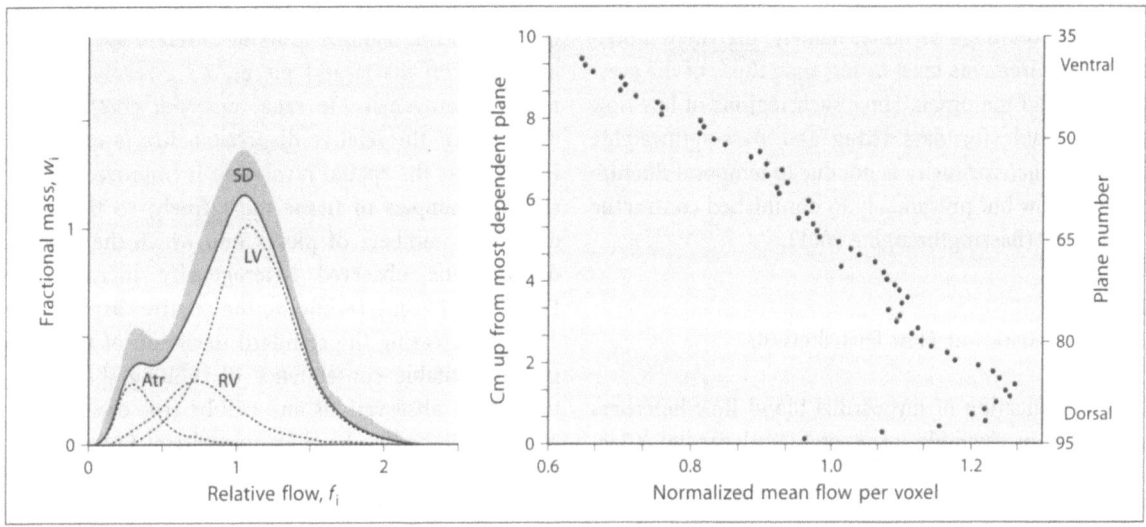

Fig. 8.7. Flow distributions in normal hearts and lungs. *Left panel:* the distribution of relative flows in the hearts of 13 awake baboons, from 13,114 estimates at four to six different times in each animal at a spatial resolution of about 250 tissue pieces per heart. The solid line represents the distribution for the whole heart, while the dotted lines are its left ventricular (LV), right ventricular (RV), and atrial (Atr) components. The standard deviation (SD) for the whole heart curve is shown by the shaded region. The mean cardiac flow per gram is normalized to 1.0. For the LV (70% of cardiac mass) the average relative flow is 1.14, for the RV (20% of mass) 0.81, and atria (10%) 0.41. (From King et al. 1985, with permission.) *Right panel:* mean flow per voxel by isogravitational (coronal) plane. Isogravitational planes are represented by both plane number and distance in centimeters up from most dependent plane. When flow data are analyzed in this way, the classically described pattern of increasing flows down the lung is reproduced (from Glenny and Robertson 1990, with permission)

Similarly wide variation in regional flows is found in the lung, as shown in Fig. 8.7 (right panel, from Glenny and Robertson 1990); the probability density functions for the lung are similar to those in the heart, but the fractal dimension for the regional flows shows that the near-neighbor correlation in the lung is higher than in the heart.

Regional flows in other organs are similarly heterogeneous even in high-flow organs like the liver (Thompson et al. 1959), the kidney (Lumsden and Silverman 1990), and the brain (Sokoloff et al. 1977). Heterogeneity in the kidney and in the brain, unlike the heterogeneity found in heart, lung, and liver, is due in part to the differences in regional metabolism and to the structure of the organ. Regional flows in the brain are about twice as high in gray matter, with its high density of neural cells, as they are in white matter, which is composed mainly of axons and dendrites. Flows in the cortex of the kidney are many times higher than those in the renal medulla; the medullary vasculature is composed of ascending and descending vasa recta, small vessels that are in osmotic balance with the osmotic gradients along the renal medulla, peaking at the tip of the renal papilla. It has been difficult to measure flows in the renal medulla because this is in fact a "portal" situation, with the blood flow coming to the medulla after passing through the capillaries of the juxtamedullary glomeruli. This anatomic arrangement precludes the use of the microsphere technique to measure medullary flow. Washout techniques, for example using 133-xenon (Aukland 1980), give gross underestimates of the medullary flow: the shunting of any diffusible indicator (they used xenon) in the countercurrent exchange region from the venules back into the arterioles prolongs its retention and reduces artifactually the estimate of the local flow.

While the variations between regions in kidney and brain are usually understood, those in the heart are not, for it has been commonly thought that all cardiac regions should have equivalent flow requirements because the whole heart contracts as an electrical syncytium and every cell is excited with every heart beat. Cardiac metabolism is high, and transcapillary oxygen extraction is greater than 50%. Thus, if a low-flow region, for example with flow of 30% of the mean flow, had the same flow as the average for the whole heart, not enough oxygen would be delivered to this region to satisfy its metabolic needs. The stability of the flow in these low-flow regions demands another explanation for their continued viability, namely, that their nutritional requirements must be less than those of the average tissue of the organ. Since such regions of low flow remain viable for days (King and Bassingthwaighte 1989), the heterogeneity is not due to temporal fluctuations in flow but presumably to diminished contractile workloads (Bassingthwaighte 2001).

8.3.1.2
Fractal Statistics of Flow Distributions

The identification of myocardial blood flow heterogeneity as an example of a statistical fractal (Bassingthwaighte 1988) has provided some insight into the basis of flow heterogeneity. First, the demonstration of a fractal dimension of about 1.2 or a Hurst coefficient of 0.8 indicates that the heterogeneity is not spatially random but that near neighbors are similar in their local flows. High-flow regions are surrounded by relatively high-flow regions and low-flow regions have low-flow neighbors, regardless of the spatial resolution used to make the observation. This, coupled with the fact that the low-flow and high-flow regions are stable over time, suggests that the regions are stable with respect to their metabolic requirements, something that will be elucidated below.

Figure 8.8 shows a set of probability density functions from the regional flows at different spatial resolutions. With the largest pieces, 7.2-g resolution, not much heterogeneity is seen, and the coefficient of variation or the relative dispersion, RD, is only 12%. However, as the spatial resolution is improved by cutting the samples of tissue more finely, so that there are larger numbers of pieces into which the heart is divided, the observed heterogeneity increases; in Fig. 8.8 at 170-mg resolution the relative dispersion is 26.5%. Increasing the standard deviation of the flows is an inevitable consequence of refining the resolution of the observations and will be true of any property of a tissue that has heterogeneity at fine scales.

The fractal dimension of the flow heterogeneity is determined by plotting the observed relative dispersion of regional flows at each spatial resolution on a logarithmic scale against the sizes of the pieces of tissue into which the heart has been divided, again on a log scale, as in Fig. 8.9. The observed variation is shown using open circles. In this particular set of experiments the methodologic variation was known for each sample size because duplicate observations were made with differently labeled microspheres, as shown by the diamonds, giving RD_M, the methodologic standard deviation divided by the mean. The spatial variation corrected for the methodologic error is given

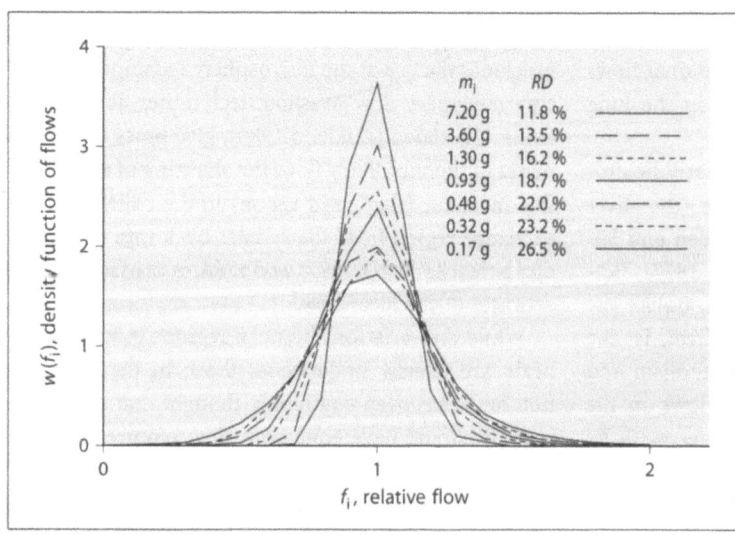

m_i	RD	
7.20 g	11.8 %	
3.60 g	13.5 %	
1.30 g	16.2 %	
0.93 g	18.7 %	
0.48 g	22.0 %	
0.32 g	23.2 %	
0.17 g	26.5 %	

Fig. 8.8. Composite probability density functions for regional flows at seven different element sizes in the left ventricles of 11 awake baboons. The order of the average element sizes, m_i, is the same, from top to bottom, as the heights of the peaks of the density functions. Bin width = 0.1 times mean flow. The spread of the probability density function of regional flows is dependent on the size of the elements used to make the observations, more heterogeneity being revealed as spatial resolution is increased (figure reproduced from Bassingthwaighte et al. 1989a, their Fig. 4, left panel)

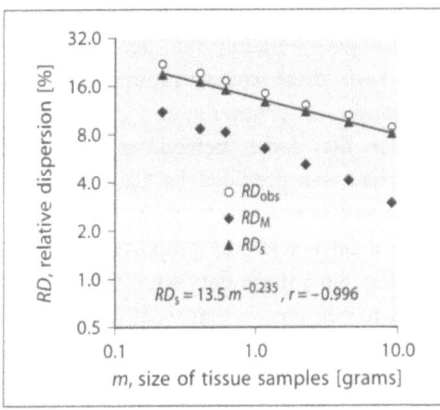

Fig. 8.9. Fractal regression for spatial flow variation in left ventricular myocardium of a baboon. Plotted are the relative dispersion of the observed density function (RD_{obs}), the methodological dispersion (RD_M), and the spatial dispersion (RD_s) at each piece mass calculated using Eq. 8.17 (figure reproduced from Bassingthwaighte et al. 1989a, Fig. 5, left panel)

by the triangles labeled RD_s; the fractal scaling relationship is given by

$$RD(m) = RD(m_0) \cdot (m/m_0)^{1-D} , \qquad (8.17)$$

where m is the mass of the observed tissue elements, m_0 is the chosen reference mass (usually 1 g), and RD is the relative dispersion, which is the standard deviation divided by the mean for the whole heart at each level of resolution. The equation given in the figure, which is the regression line through the solid triangle symbols, shows the exponent, which is the $1-D$ in Eq. 8.17, to be –0.235, thereby showing that the value for the fractal dimension is 1.235. The Hurst coefficient $H = 2-D$, so H is equal to 0.77. These values indicate that the nearest-neighbor correlation coefficient is $r_1 = 0.44$, since

$$r_1 = 2^{2H-1} . \qquad (8.18)$$

This degree of correlation is far different from zero. The average near-neighbor correlation coefficient is from data for the whole heart, so the correlation coefficient 0.44 applies to both high-flow and low-flow regions in the heart. Bauer et al. (2001) found that the fractal dimension of regional myocardial blood flows was lowered by nitroglycerin, decreasing the heterogeneity. In contrast, Austin et al. (1990) found that hypoxia increased heterogeneity and reduced the correlation with the control state.

8.3.1.3
Vascular Geometry and Flow Heterogeneity

Bassingthwaighte and van Beek (1988) postulated that near-neighbor correlation might be a natural result of the fact that the flows are delivered to neighboring regions by near-neighboring branches of the coronary arterial system. This model of a dichotomously branching network gave results similar to the data, something that was later affirmed with quite independent ways of creating coronary vascular network reconstructions (Beard and Bassingthwaighte 2000). Such views cannot be used to assign cause-and-effect relationships. What is clear is that the associations between asymmetry in flow at branch points and spatial flow heterogeneity are strong. Glenny and Robertson (1990) made similar observations in the lung and Grant and Lumsden (1994) in the kidney.

8.3.1.4
Washout Processes Are Fractal, not Multiexponential

Yipintsoi et al. (1970) observed that the washout curves for antipyrine and for tracer water from the heart were never monoexponential but remained concave upward on a semilog plot. Bassingthwaighte and Beard (1995) analyzed the washout of ^{15}O-water injected into the coronary inflow and found that these tended toward power-law relationships. The persisting upward concavity on the semilog plot in the left panel of Fig. 8.10 indicates that these curves could be considered as multiexponential, but the log plot showing how the data tend toward linear relationships when $\log R(t)$, the residual tracer, is plotted against $\log t$, indicates that they are actually power-law functions. In their appendix, Bassingthwaighte and Beard (1995) show how, given enough exponentials, a multiexponential function can be fitted arbitrarily closely to a power-law function.

The power-law washout curve is one showing self-similarity, that is, the fraction of tracer remaining in the organ diminishes at a constant fraction for a given doubling of the time. One should therefore use the power-law approximation rather than the multiexponential approximation when fitting washout curves. The power-law function is more parsimonious, having only two parameters – an intercept and a power-law exponent – whereas the multiexponential fit has two parameters for each exponential that is used. The fractal description is not only more parsimonious

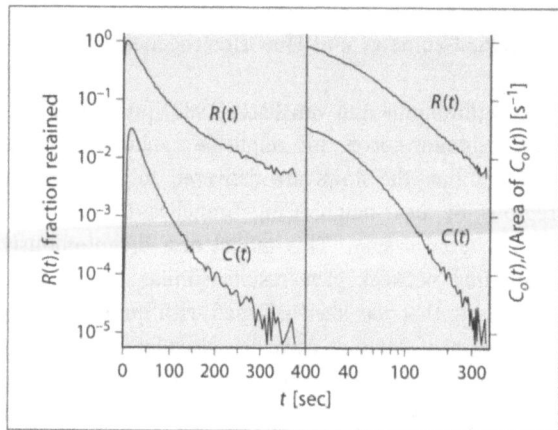

Fig. 8.10. Residue curves, $R(t)$, and outflow concentration-time curves, $C(t)$, from rabbit after injection of ^{15}O-water into the coronary inflow. Both $R(t)$ and $C(t)$ have tails that are powerlaw functions of time, that is, they are straight on log-log plots (right panel) and clearly not monoexponential, being concave up, on semilog plots (left panel) (from Fig. 10.18 of Bassingthwaighte et al. 1994, with permission)

than any multiexponential description, but it also avoids misleading one into interpretations wherein multiexponentials are assigned meanings with respect to the physiology.

8.3.2
Local Metabolism Drives Local Flows and Local Angiogenesis: A Rational Hypothesis

8.3.2.1
Observations of Associations Between Flow and Metabolism

Data on this topic have been obtained in the heart, but it is very likely that all organs show the same behavior. The major substrates for energy metabolism, glucose and fatty acids, are normally delivered to tissues at rates far in excess of their utilization. Net glucose extraction is normally only a few percent, and fatty acid extraction, for example in the heart, is only about one half. It is oxygen that may be in limited supply, particularly in the heart where the effluent PO_2 is only about 25 mmHg compared to the inflow PO_2 of 100 mmHg. Normally the regional concentrations of such substances as creatine kinase, lactate dehydrogenase, ATP, and glycogen, all major players in energy metabolism, are found in constant abundance in normal tissues and have concentrations un-

related to the local flow (Franzen et al. 1988). This is what one would expect – mainly that normal regulatory responses keep these concentrations at steady levels despite changes in turnover rates.

Good evidence that local metabolism and local flows were matched was provided by Caldwell et al. (1994) who observed in the hearts of exercising dogs that the uptake of fatty acid was proportional to the local flow (see Fig. 8.6); these data were from hearts divided into 500 to 600 pieces, that is, at a spatial resolution of better than 0.2% of the organ. Our interpretation of these data is that the transporters and enzymes involved in the first steps of the fatty-acid translocation and transformation have abundances that are proportional to the local flow.

Weiss et al. (1978) showed that venous oxygen concentrations in small venules were low in regions of high flow (measured by microspheres) and computed that more oxygen was used in high-flow regions than in low-flow regions. Van Beek et al. (1999) observed that the turnover of acetate was high in high-flow regions compared to that in low-flow regions. In the setting of their observations in Ringer-perfused rabbit hearts supplied only with acetate, where tricarboxylic acid cycle turnover was attributable only to acetate, they concluded that local flow must be nearly proportional to local oxygen consumption. Decking et al. (2001) made similar observations when pyruvate was the only substrate. The observations on acetate and pyruvate implied that oxygen consumption should be proportional to the local flow, but only the direct observations of Li et al. (1997) demonstrated this explicitly. While we can conclude from these observations that local flows and local metabolism in an organ with high energy consumption (the heart) are proportional to flow, we can only surmise that the same principle holds in other organs. In the brain the high-flow gray-matter regions stand out in a background of low-flow white-matter regions; the high abundance of neuronal bodies in gray matter can be expected to have high energy demand.

8.3.2.2
Adaptations to Changes in Local Workload

Metabolism changes when the local load changes, in the heart. Taegtmeyer and colleagues (Depre et al. 1998) noted that the unloaded rat heart has not only the expression of fetal genes such as *cfos* and TGFβ but also a diminution in glucose uptake as a proportion of the energy source. Diminished glucose uptake has long been recognized to be a feature of left bundle branch block where septal glucose utilization in the interventricular septum is diminished out of proportion to the regional flow (Altehoefer 1998). McGowan et al. (1976) had observed in such patients that there was a septal perfusion deficiency even though the coronary arteries were patent. The work of Prinzen et al. (1990, 1992, 1995) showed that the flows redistribute in the paced heart, diminishing at sites of early activation and contraction against no load and increasing in late-activated sites that have higher contractile loads imposed on them. The work by van Oosterhout et al. (2002) shows that the heart remodels, atrophying in the low-flow, early-activated regions and hypertrophying in late-activated regions where blood flows have increased. These appear to be compensatory changes. Presumably the capillarity as well as the size of the larger vessels increases in the hypertrophying regions, as was earlier observed by Batra et al. (1989).

8.3.2.3
Hypothesized Algorithms for Vascular Growth and Adaptation

Vascular endothelial growth factor, VEGF, is released from endothelial cells when there is a sustained increase in shear rate at the endothelial lining of the vessel (Cowan and Langille 1996). In the acute response to increased shear, nitric oxide, NO, is released, causing smooth muscle to relax and allowing arterioles to dilate. The NO may play a role in stimulating increased expression of VEGF and other factors.

The stimulus for vascular growth, however, comes first from the metabolizing cells of the organ in response to their needs. While it is known that increased metabolic demand is associated with raised interstitial K^+, CO_2, often lactate, and acidic pH, all of which are vasodilatory, it is not known whether it is the parenchymal cells or the endothelial cells that initiate the drive to secrete growth factors. Certainly heightened metabolism increases the flux of purine nucleotides from parenchymal cells; hydrolysis of the nucleotides in the ISF (interstitial fluid) releases adenosine, a vasodilator suspected to stimulate vascular growth factors.

8.3.3
Approaches to Estimating Regional Flows

8.3.3.1
Deposition Techniques

The standard approach is to inject particles that cannot pass through capillaries into the left atrial or left ventricular cavity so that they mix with the blood, to give them time to deposit in the tissues of the body, then to remove the organs or tissues, and finally to estimate regional blood flow from the deposition of the particles, on the assumption that the particle deposition is in proportion to the regional flow. For an individual organ this gives flow relative to the mean flow in the whole organ. The use of macroaggregated albumin particles, with particle sizes mainly above 10 µm, was introduced by Ueda et al. (1964) and Taplin et al. (1964): quantitation using macroaggregated albumin was possible (Richmond et al. 1970, 1973) but was not regarded as a particularly robust method because of the diversity of particle sizes, and albumin macroaggregates were soon replaced by 15-µm-diameter radioactive microspheres (Buckberg et al. 1971; Yipintsoi et al. 1973). This method also demonstrated substantial heterogeneity of regional myocardial blood flows and suggested further that flows to the subendocardial region were higher than flows to the subepicardial region. This approach, in a generalized form, was used to determine the flows to all the organs of the body. (Liver flows, however, were calculated as the sum of the hepatic artery flow and the flows to all the abdominal organs whose drainage was into the portal vein.)

8.3.3.2
The Classic 15-Micron-Diameter Microsphere Method

From the start, the 15-µm-microsphere technique was good, and it got better (Buckberg et al. 1971; Yipintsoi et al. 1973; Baer et al. 1984). The developments are well summarized in a review paper by Heymann et al. (1977).

Buckberg et al. (1971) and Nose et al. (1985) predicted that the radioactive microsphere method should have a methodologic error of the order of 5 to 8%. The experimental observations of King et al. (1985) and of Bassingthwaighte et al. (1990), using duplicate observations, showed errors of 5–7%, so the statistical predictions of Buckberg et al. and Nose et al. were correct.

Deposition techniques of this sort work well because the two fundamental conditions are relatively easily fulfilled: (1) particles should be deposited in tissue in proportion relative to the flow and (2) the deposition of the markers should not obstruct the flow. Even microspheres as large as 15 μm are deposited primarily in capillaries, even though their average diameter is only slightly more than 5 μm. The 15-μm spheres distort the flexible walls of the capillaries as they become lodged. They do not stop in 15-μm-diameter terminal arterioles since these vessels are also flexible. If there is a bias to microsphere distribution, it is that they will behave like stiffened spherical red blood cells rather than like plasma. With larger spheres there is a tendency for preferential deposition in the subendocardium of the heart (Yipintsoi et al. 1973), simply because the penetrating vessels extend straight through the wall toward the endocardium, making a preferential pathway for such particles. The bias is small for 15-μm spheres but large for larger sphere sizes. Hales and Cliff (1977) made movies showing that red blood cells still flowed through capillaries in which 15-μm microspheres were trapped and indeed flow around the spheres. Thus it has been found that color microspheres injected in separate injections some time apart can be found together in the same capillary, indicating the flow continued when one large sphere was trapped but did not really block the flow. With molecular microspheres there is no obstruction.

8.3.3.3
Fluorescent Versus Tracer Microspheres

The use of varicolored microspheres avoided using radioactivity (Kowallik et al. 1991), and it has been found even more convenient to use varicolored fluorescent microspheres (Glenny et al. 1993). The use of fluorescent microspheres had an unexpected advantage, compared with the use of radiotracer microspheres, in that fluorescent microspheres did not leach out over a period of several months, and so fluorescent microspheres should now be recognized as the marker of choice for determining flow distributions when long periods of time precede sacrifice (van Oosterhout et al. 1998; Prinzen and Bassingthwaighte 2000).

8.3.3.4
"Molecular Microspheres"

While the finding of a "molecular microsphere," a molecule that could be nearly 100% extracted from the blood in single passage through the organ and then retained within the organ, was thought impossible, it is in fact quite achievable. The requirement is that the molecule be of a sort that can be readily trapped. One mechanism is attachment to a high-affinity receptor, exemplified first by the IDMI (iodo-desmethylimipramine) (Little and Bassingthwaighte 1983). It showed 96–99% trapping, even at high flows. Any indicator with high lipid solubility, so that it traverses membranes and high-affinity receptor or binding sites within a few microns of the capillary, could be similarly trapped, so one can anticipate that many similar molecular-flow markers will be discovered. The deposition of the molecular microsphere is complete enough that it can be used to evaluate the existing 15-μm-diameter microsphere technique. This was done for the first time in the paper of Bassingthwaighte et al. (1990). Its relatively long retention is due to the high-affinity binding, owing to its slow release from the site, giving it a high volume of distribution as diagrammed in Fig. 8.5. A comparison of the molecular microsphere IDMI with standard tracer microspheres showed that the molecular microsphere had an error of about 2%, whereas the 15-μm spheres had errors of 5 to 7%, just as had been predicted by Buckberg et al. (1971) and Nose et al. (1985). The molecular microsphere, however, could provide much higher resolution since many molecules are deposited in small regions, and so the agent was useful in estimating flows at very high spatial resolution using autoradiography (Stapleton et al. 1995).

This idea can be extended to other molecular microspheres because different ones might well serve different purposes. For example, rotenone and dihydrorotenone (Van Brocklin et al. 1995) are highly lipid-soluble compounds that bind complex I, an enzyme in the respiratory chain inside mitochondria; they are good flow markers because of their quick access to the binding site and long retention. They are better than sestamibi, which is more polar and whose entry into cells is retarded such that sestamibi extraction is not complete during a single transcapillary passage.

8.3.3.5
Incompletely Deposited Tracers
(Tl, Rb, K, Tc-sestamibi, rotenone, etc.)

If a tracer is highly extracted and retained for a long time in the tissue, it can serve as a deposition marker for flow estimation using positron emission tomography (PET) and single photon emission computed tomography (SPECT). Molecules should be better than microspheres since there can be no bias due to the hydrodynamics at branch points (Yen and Fung 1978). Tracer water, D_2O or ^2H_2O, was used by Thompson et al. (1959). The markers with large volumes of distribution, and therefore long retention, were potassium and thallium. More recently sestamibi (Udelson et al. 1994) and rotenone (Van Brocklin et al. 1995) have been considered. Potassium was first used by Love and Burch (1959). None of these markers except water is highly extracted during a single pass through the heart, but the fraction that was extracted was retained for a long enough time that the heart could be removed and diced into small pieces for the measurement of local concentrations and estimation of flow. The transcapillary extraction of potassium measured in skeletal muscle by Renkin (1959a,b) was about 70% extraction. Extraction in the heart was lower, 40 to 70% depending on flow (Tancredi et al. 1975). Retention is good because of the large intracellular volume of distribution, and potassium washout is about 1% per minute. While it was commonly thought that thallium shares this large potassium pool, there is some evidence that its good retention may involve absorption to something on the cell surface (Winkler and Schaper 1992).

8.3.4
Flow Estimation from Indicator
Input-Output Measurements

8.3.4.1 The Fick Method

The practical measurement of flows is based on conservation: material that enters a system must also exit the system. The system must not leak solvent or solute; it must not gain material from another source; it must not store material so its volume, mass, and density remain the same; and it must not destroy, convert, or metabolize the substance whose concentration is being measured to estimate flow. Volume is conserved first. Volume of inflow must equal volume of outflow. Mass of solute is conserved second. The mass that entered must be equaled by the sum of the mass that exits plus that which remains in the system. This is more accurate than simply saying what goes in goes out. In more explicit, mathematical terms,

$$F \int_0^t C_{in} dt = F \int_0^t C_{out} dt + q(t) , \qquad (8.19)$$

where F is flow in ml/min, $C_{in}(t)$ is the inflow concentration-time curve, and $C_{out}(t)$ is the same at the outflow, both in units of mmol/ml, and $q(t)$ is the amount of the solute within the system at time t, minute. This is a wonderfully all-encompassing statement that we shall explore from several angles.

Nontracer, mass-balance techniques for measuring flow make use of steady-state differences due to the addition or subtraction of a solute from the flowing stream. The best-known technique is that described by Fick (1870) in a one-paragraph article. (The article has been translated from German and published by Hoff and Scott 1948.) Consider material being added to the fluid flowing through the system, such as oxygen, and added to the blood flowing through the lungs: in a steady state the total amount of material per unit time entering the organ at all entrances (air and blood stream) is equal to the amount per unit time leaving the organ at all exits:

$$(dq/dt)_{in} = (dq/dt)_{out} . \qquad (8.20)$$

As shown in Fig. 8.11, the amount exiting (or leaving) must be considered the sum of the quantity brought to the organ by the inflowing blood, $F_{in}C_{in}$, plus an amount per unit time, $(dq/dt)_{in}$, added to the blood while it passes through the organ.

In calculating cardiac output by the usual "Fick technique," the lung is chosen as the organ. Inflowing blood is that passing through the pulmonary artery; outflowing blood is that passing through the pulmonary veins or the aorta. The indicator input rate, $(dq/dt)_{in}$ mmol/min, is the rate of addition of material, O_2 or CO_2, from the lung air space to the blood. It is positive for oxygen intake and negative for carbon dioxide output. With steady flow and constant pulmonary blood volume, F_{in} is the same as F_{out}, and the equation may be rewritten as

$$F = \frac{(dq/dt)_{in}}{C_{out} - C_{in}} . \qquad (8.21\,a)$$

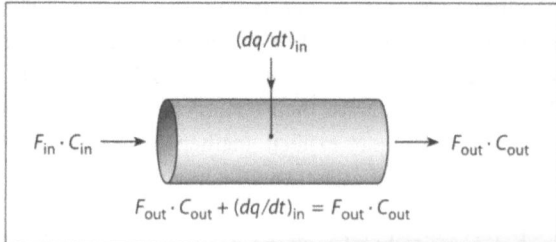

Fig. 8.11. Fick equation. In a steady-state system where mass is conserved the sum of the rates of entry of mass is equal to the rate of exit

In experimental terms, this is

$$F = \frac{\text{oxygen uptake (ml/min)}}{\text{A-V concentration difference (ml O}_2/\text{ml blood)}}, \tag{8.21 b}$$

and a numerical example is

$$F = \frac{200 \text{ ml/min oxygen consumption}}{[20 \text{ ml}/100 \text{ ml} - 15 \text{ ml}/100 \text{ ml}] \times 10(100 \text{ ml/liter})}$$

$$= 4.0 \text{ liters/min} . \tag{8.21 c}$$

Because the steady-state oxygen uptake by the lungs is equal to the use of O_2 by the body, the Fick equation can also be thought of as the arterio-venous, A-V, concentration difference produced by the loss of O_2 into the tissues:

$$F = \frac{(dq/dt)_{\text{out}}}{C_{\text{in}} - C_{\text{out}}} . \tag{8.22}$$

8.3.4.2
Use of the Fick Expression at the Whole-Organ Level

Instead of using it to estimate flow, the usual application is to measure A-V differences in the steady state and from that estimate and the known flow to estimate the intraorgan utilization, dq/dt, of an extracted material, e.g., oxygen, by the organ. By combining external detection methods with outflow, under specified circumstances one can measure the flow by measuring the consumption: the circumstance is that the material must be entirely retained by the organ and detectable externally in quantitative terms. Examples are the uptake of Rb and ^{18}F-deoxyglucose; the former is retained because of the large volume of distribution in the potassium space, and the latter is retained as deoxyglucose-6-phosphate since this com-

pound is not further metabolized and the rate of dephosphorylation is low. With a constant infusion into the organ and C_{in} constant, C_{out} follows to reach a constant but lower level; since $C_{\text{in}} - C_{\text{out}} = F \, dq/dt$, the key to estimating F is to measure dq/dt. Using PET of the heart and ventricular cavity a signal β times dq/dt can be obtained; its calibration factor, β, is the same as that for the inflowing blood, C_{in}, so $\beta = C_{in}/S_{bl}$ where S_{bl} is the signal detected from a whole blood region such as the aortic root or the ventricular cavity. Then $dq/dt = \beta \cdot dS_{\text{tiss}}/dt$, where dS_{tiss}/dt is the steady-state rate of rise of the observed tissue signal. The technique requires sampling the outflow, C_{out}, and making sure this C_{out} represents the average outflow from the whole organ. When this is assured,

$$F_{\text{organ}} = \frac{\beta \cdot dS_{\text{tiss}}/dt}{C_{\text{in}} - C_{\text{out}}} . \tag{8.23}$$

The awkward part of this method is getting the estimate of C_{out} in the steady state; this is straightforward only in an organ with a single, and accessible, venous outflow.

8.3.4.3
Use of the Fick Expression at the Regional Intraorgan Level

To estimate regional intratissue flow, the same approach, Eq. 8.23, can in theory by used, but now the potential for obtaining the local venous blood sample is difficult even with invasive techniques and impossible noninvasively.

8.3.5
Washout and Washin Techniques for External Detection

These techniques derive from the theoretical development described by Zierler (1963, 1965) and are essentially measures of the transit time through the organ: the mean transit time is the volume divided by the flow, where the volume is the volume of distribution for the tracer. The volume of distribution is the plasma-equivalent space into which the tracer distributes during single passage through the organ. For a tracer such as tritiated water, that space is the total water space of the organ. When the volume of distribution for the tracer is known, as it is for water in the heart (0.78 ± 0.01 ml/g) (Yipintsoi et al. 1972), then measuring the mean transit time gives the answer:

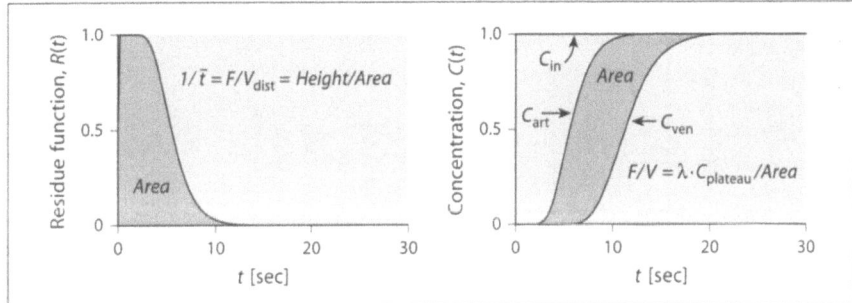

Fig. 8.12. Organ washout and washin curves. *Left panel:* residue analysis. Washout after an impulse injection at $t=0$. The flow per unit volume is the height divided by the shaded area under the curve. *Right panel:* input-output analysis. In the Kety-Schmidt approach, there is delay and dispersion in the system between the site of infusion and the entrance to the organ, deforming $C_{in}(t)$ into $C_{art}(t)$, and again between $C_{art}(t)$ and the concentration at the outflow $C_{ven}(t)$. The flow per unit volume is the height $C_{plateau}$ times the tissue-blood partition coefficient, λ, divided by the area (shaded)

$$F = \frac{V_{dist}}{\bar{t}} , \qquad (8.24)$$

where F, the flow per gram of tissue (ml g^{-1} min^{-1}), is equal to the volume of distribution (ml/g) divided by the mean transit time (min). An example of the washout following a brief pulse injection at $t=0$ into the inflow is shown in Fig. 8.12 for the conditions in which there is no previous background of tracer in the recirculated blood. The shaded area of the curve, divided by the initial height, is equal to the mean transit time. The flow per unit volume $F/V_{dist} = 1/\bar{t}$ is the height divided by the area. The right panel shows that when the input function is a dispersed response to a step increase in concentration due to a steady infusion somewhere upstream to the organ, there is delay and dispersion in the rise of concentration toward the plateau. The area to be measured is that between the concentration-time curves at the inflow and outflow (shaded areas in the figure). This is the technique used by Kety and Schmidt (1948) for measuring brain blood flow using nitrous oxide, which equilibrates between the plasma and the brain tissue with a partition coefficient of about 1. These techniques tend to be compromised to some extent by the presence of recirculation and the intact human, but this can be corrected for fairly accurately by recording the inflow, $C_{art}(t)$ (Bassingthwaighte and Holloway 1976).

Washout techniques work quite well when there is flow-limited exchange between the blood and the tissue for the particular tracers. To have enough detection time it is important that the tracers exchange in flow-limited fashion between blood and tissue, a requirement that is fulfilled by highly lipid-soluble indicators such as antipyrine, xenon, hydrogen, and even tracer water.

8.3.6
Pulse Injection Indicator Dilution Techniques

The various techniques for estimating flow from sudden injection techniques are all based on the conservation of mass, following the developments of Stewart (1897) and Hamilton et al. (1932) for the estimation of cardiac output, assuming a linear stationary system. Linearity and stationarity imply that the output responses are proportional to the magnitude of the inputs and that the responses to similar inputs are the same from one time to the next.

When a mass of indicator traverses a system, its many particles become spread with respect to distance along the vascular system and therefore also spread with respect to time of arrival at the outflow. An impression of the delay and dispersion is gained immediately when one inspects the concentration-time curve at the outflow after making a short injection at the inflow (Fig. 8.13, left panel). The dispersion is produced by several factors acting either together or consecutively: (1) via the wide range of the velocities within in each vessel (with parabolic flow the velocities in the middle of the stream are higher than velocities near the wall); (2) via mixing effects (e.g., in the ventricle where partial mixing and dispersion occur during the diastolic filling phase, in the great vessels during early diastole, or in the eddies that occur at branches in arteries and veins); (3) via molecular diffusion, both radially between central and peripheral laminae in a stream, and axially; and (4) via the variance in pathway lengths, velocities, or mean transit times that generally exist in vascular networks of individual organs or of a system consisting of organs in parallel.

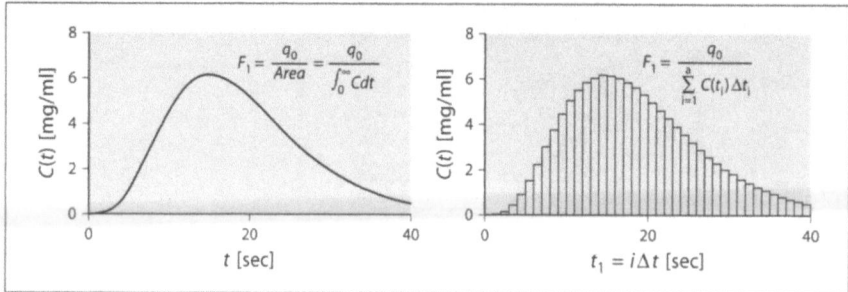

Fig. 8.13. Estimation of flow from the concentration-time curve obtained following a brief injection into the flowing stream. *Left panel:* indicator dilution curve at flow F_1. *Right panel:* histogram representation of the dilution curve of the left panel. The area is approximated by a summation replacing the integral

Furthermore, dispersion occurs when an injection is made into a fluid column; the injectate tends to be fairly widely dispersed (in space) and so contributes to the temporal dispersion observed at the outflow. (This can be seen when one injects ink into water flowing within clear plastic tubing.) In cross-sectional labeling the injectate labels one slice that extends over the whole cross-sectional area of the vascular lumen at the injection site; however, this is theoretically inappropriate. Ideally the labeling is flow-proportional, i.e., the fraction of indicator injected into each element of the cross section at the injection site is proportional to the flow through that element. Similarly, ideally the sampling of the indicator-blood mixture is flow-proportional. (For other cases, see the detailed analysis by Gonzalez-Fernandez 1962 concerning various laminar flow situations.) It is on these assumptions that the pioneering descriptions by Stephenson (1948), Meier and Zierler (1954), and Zierler (1958, 1962b) were based.

When a very short injection, ideally an impulse in put, is made at the entrance to a system, the fraction of the indicator passing the sampling site per unit time is the impulse response of the system, or its transport function, $h(t)$. Conservation of mass requires that the flow times the concentrations at each time add up to the injectate mass, i.e., $q_0 = F \int_0^\infty C(t)dt$, as implied in Fig. 8.13. (For an impulse injection, the output concentration-time curve, $C(t)$, is simply the transport function, $h(t)$, times q_0/F.)

From the area under the concentration-time curve in Fig. 8.13 and the known injected dose, q_0, moles, one can estimate F:

$$F(\text{ml/s}) = \frac{q_0}{\int_0^\infty C(t)dt} = \frac{\text{injected dose (moles)}}{\text{area ((moles/ml)} \times \text{s})} .$$

$$(8.25)$$

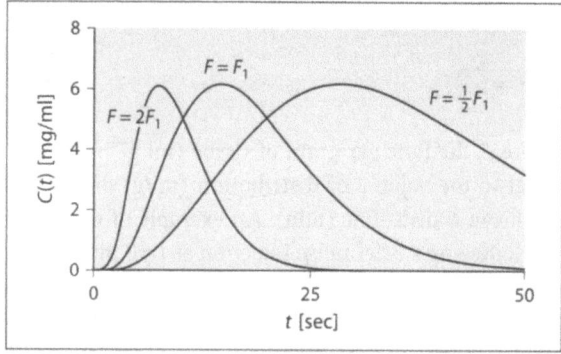

Fig. 8.14. Outflow concentration-time curves at double and half the flow. Same conditions as in Fig. 8.13 except for the changes in flow

The curve $C(t)$ may be obtained as a sequence of samples at discrete time intervals so that $C(t)$ is represented as a histogram (Fig. 8.13, right panel) rather than a smooth curve. For convenience, a smooth curve may be read at a succession of intervals. In either case, the area under the curve is then obtained by adding up the elements of the histogram; the area of the element i is $C(t_i)\Delta t_i$, with t_i being the median time of the interval and $C(t_i)$ the average concentration over the interval. The total area, A, under the histogram composed of n elements is

$$A = \sum_{i=1}^{i=n} C(t_i)\Delta t_i , \qquad (8.26)$$

and $F = q_0/A$ just as for the smooth curve.

The shape of the outflow dilution curve changes when flow is changed. Just as for the outflow responses from a nondispersive pipe, slowing the flow retards the outflow, and increasing the flow leads to the earlier appearance of indicator in the outflow. Fig-

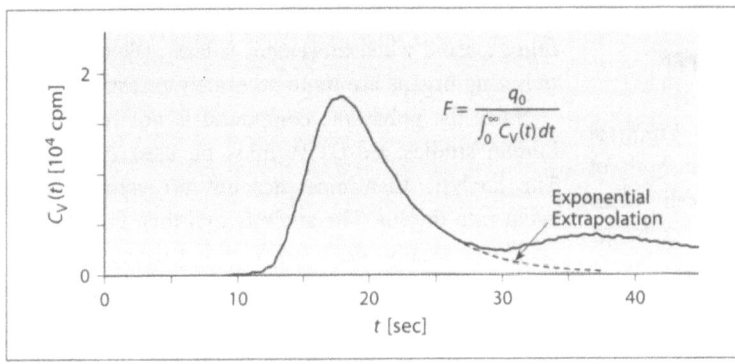

Fig. 8.15. Recorded indicator dilution curve with recirculation. A monoexponential extrapolation is applied to approximate the tail of the primary curve. The equation applies to the "corrected" primary curve (not shown). Semilog plot of the same curve, with a monoexponential curve fitted and used to give an extrapolated estimate of the primary, first-pass indicator

$$F = \frac{q_0}{\int_0^\infty C_V(t)\,dt}$$

Exponential Extrapolation

ure 8.14 illustrates two other important features: (1) the higher the flow, the smaller the area under the outflow curve; and (2) the heights of the peaks are not changed by changing flow. In fact, the three curves in Fig. 8.14 can all be superimposed on one another by scaling the time. Doubling the time scale for $F=2F$, and halving it for $F=\frac{1}{2}F$, superimposes the high and low curves onto that for $F=F_1$. This shows that the curves have "similar" shapes, an important point with respect to the question of whether or not the transport function for the system changed its characteristics by changing the flow. In this case, since the curves can be superimposed, the nature of the events causing dispersion between inflow and outflow did not change.

8.3.6.1
Correction for Recirculation

In the late 1920s and early 1930s, W.F. Hamilton and his colleagues took Stewart's basic approaches to flow estimation and brought them to a new level of practice. Hamilton et al. (1932) recognized that recirculation interfered with obtaining estimates of the area under the concentration-time curve for the "first-pass" indicator. The recirculating tracer mixing with late-emerging, first-pass tracer marred the tail of the dilution curves as in Fig. 8.15. Hamilton et al. noted that the downslopes of the curves were more or less exponential. When they tested an artificial system, a volume of fluid with glass beads in a column, they affirmed by a monoexponential fit that the washout curves were approximately correct. To apply this idea to in vivo systems was simple: replot the dilution curve on semilog paper (plot log $C(t)$ versus t), use a ruler to obtain a monoexponential fit (a straight line on semilog paper) to the segment of the downslope prior to obvious recirculation, and extend the line to

later times, thereby obtaining an approximation to the curve that would have been observed in the absence of recirculation. Then to estimate the area by Eq. 8.26, one sums $C(t_i)$ to the point where the exponential departs from the recorded curve and then adds the sums of the extrapolated points from later times.

An analytical shortcut can be used. Starting at the point, t_0, where the exponential departs from the observed curve we can calculate the area, A_2, beyond this point by using the analytical expression for the area under the exponential tail:

$$A_2 = C(t_0) \cdot \tau = C(t_0)/k . \tag{8.27}$$

The total area is the sum

$$A = A_1 + A_2 , \tag{8.28}$$

or one can approximate the area under the first-pass dilution curve by summing the discretized representation of the primary peak and an analytic expression for the tail:

$$\int_0^\infty C(t)\,dt \simeq \sum_{i=1}^{i=n} C_i \Delta t_i + \int_{t_{n+1/2}}^\infty C(t)\,dt$$

$$= \sum_{i=1}^{i=n} C_i \Delta t_i + \tau \cdot C_{n+1/2} , \tag{8.29}$$

where n is the number of time intervals of duration Δt_i, and the value $C((n+1/2)\Delta t)$ is used, as suggested by Hoffman (1960), to account for the fact that half an interval is accounted for in the sum A_1.

8.3.7
^{15}O-Water for Estimating Regional Flow Using PET

Studies on ^{3}HHO and ^{131}I-antipyrine by Yipintsoi and Bassingthwaighte (1970) indicated that both of these tracers traversed the capillary and cell membranes so rapidly that their blood-tissue exchange was flow-limited in the heart. This makes these tracers ideal for estimating *regional* flows, given that external detection techniques can be applied with adequate spatial and temporal resolution. Flow limitation is not a requirement since all that is required is a priori knowledge of V_{dist} and measurement of the mean transit time. "Ideal" therefore means only that having a flow-limited marker means not having to estimate the fraction of the area under the very long low tails of washout curves that occur when there is partial barrier limitation, or, put another way, that the tails are the shortest possible and are the least masked by recirculating indicator.

Raichle and colleagues (1983), using ^{15}O-water for estimating brain blood flow, used a compartmental model to estimate the mean transit time. Even though the washout curves for tracer water have been shown to be a power-law function, $t^{-\beta}$, rather than an exponential (Bassingthwaighte and Beard 1995), the error is probably only a few percent underestimation. Since the power-law extrapolation of the tail of the residue curves is as easy as an exponential extrapolation, the power-law method appears preferable.

8.3.8
Contrast Agents for Flow Estimation by MRI

Magnetic resonance imaging (MRI) is another external detection technique that can be used for estimating regional blood flows. A bolus of contrast agent injected upstream from the organ becomes dispersed as it traverses the circulation. The methods using MRI to detect regional concentration-time curves for an intravascular agent, polylysine Gd-DTPA (diethylenetriaminepentaacetic acid), are described in detail by Wilke et al. (1995), who gives estimates with about 20% accuracy when the signal is captured by gating the image at once per heart beat (Kroll et al. 1996). However, to obtain this accuracy in the heart, one needs to use spatially distributed rather than compartmental models, account for heterogeneity of flows within ROIs (regions of interest), and sample once

per beat using about 300 ms during the late diastolic filling period when movement is least. (Flows in non-pulsating organs are more accurately measured.)

Since the polylysine compound is not released for human studies, Gd-DTPA must be used in humans. The analysis then must account for escape of the agent into the ISF. The analysis tool then is the multicapillary axially distributed blood-tissue exchange model accounting for the flow heterogeneity and the permeation. The flow estimation from this more complicated MRI signal appears to be about as accurate as for the intravascular agent. There is an additional free parameter, the capillary PS_c, which for Gd-DTPA is similar to that for sucrose and does play a role in optimizing fits of the model to the observed ROI contrast-time curves. For both the polylysine Gd-DTPA and the monomer Gd-DTPA, the models should account for the exchanges of spin-labeled water molecules between the capillary region (and ISF for free Gd-DTPA) containing the contrast agent and those excluding it, the cell water space (and the ISF in the case of the polylysine compound). The effect of the exchange is to dilute the influence of the contrast agent inside its volume of distribution and to enhance its effect outside of that, as modeled by Bauer et al. (1996). This understanding of the effect of proton exchange on MRI signals needs consideration also for noncontrast perfusion imaging (Bauer et al. 2001). The axially distributed models (e.g., Bassingthwaighte et al. 1992, available as BTEX30, a three-region model, or MMID4, a four-region multiple parallel path model accounting for flow heterogeneity, for download from http://nsr.bioeng.washington.edu) can be adapted for this purpose, and GENTEX, a general-purpose multicapillary version, has been used without adaptation.

8.4
Flow-Limited Versus Barrier-Limited Tracers

Gd-DPTA, sestamibi, Tl, K, and Rb are all polar solutes. All but Gd-DTPA can enter cells, albeit slowly. At high flows their entry is limited by the slow rate of transmembrane flux: they are "barrier-limited" in their exchange. Crone (1963) and Renkin (1959) developed an expression to estimate the permeability-surface area product of the capillary wall, PS_c, to such solutes:

$$PS_c = -F \ln (1 - E), \tag{8.30}$$

where F is the flow and E is the fractional extraction occurring during transcapillary passage. The multiple-indicator dilution method of Chinard et al. (1955) was developed further by Crone (1963) to determine E from the difference between the outflow curves for an intravascular tracer and one that partially escaped from the blood into the tissue:

$$E(t) = (h_R - h_D)/h_R, \tag{8.31}$$

where h_R and h_D are the normalized impulse responses at the outflow, R is the intravascular reference tracer, and D is the diffusing or permeating solute. The maximum value of $E(t)$ is usually taken to represent the extraction due to unidirectional flux out of the blood, obtained at an early time before there is any tracer backflux from tissue into the blood. This pioneering but simple expression is compromised by intraorgan-flow heterogeneity and by intravascular

dispersion, tending to underestimate PS_c. With infusions at a constant rate, the technique Renkin used in 1959, the problem is similar, giving even lower estimates of PS_c unless full modeling analysis is undertaken. Figure 8.16 shows the response to a step infusion, a constant-rate infusion for 50 s.

The peak values of the $E(t)$ provide underestimates of the true PS_c using Eq. 8.30, being 0.80 and 0.81 ml g^{-1} min^{-1} for solutes E and D, while the correct values are 1 ml g^{-1} min^{-1}. The underestimation illustrates a shortcoming of the Crone-Renkin expression, Eq. 8.30, due to the dispersion in the vascular space and not merely to the return flux from ISF to plasma, which, of course, also reduces $E(t)$.

The later plateau value of $E_D(t)$ clearly does not give a measure of unidirectional flux out of the capillary and cannot be used to estimate PS_c. Rather it gives a measure of the total conductance from blood to the cellular sink for D, so the total resistance is the

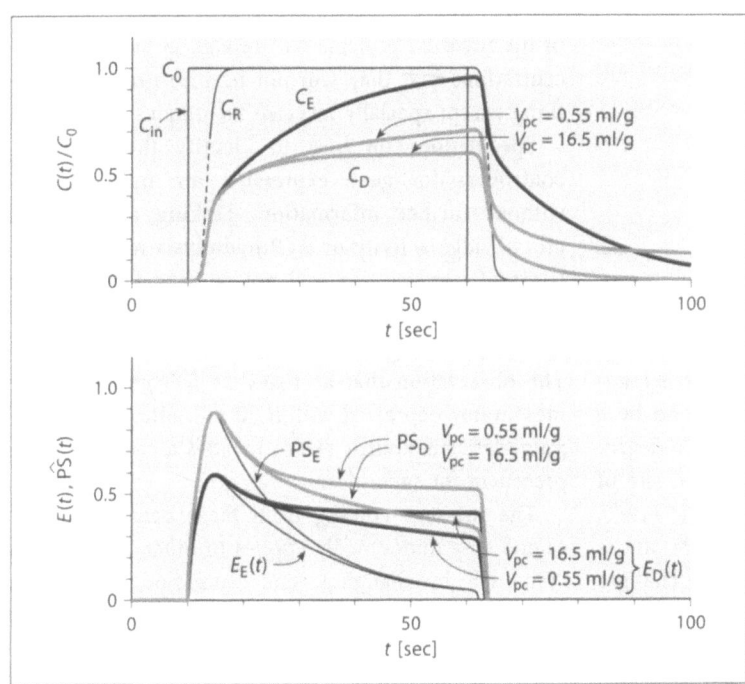

Fig. 8.16. Multiple indicator dilution experiment (modeled) for an input step function, $C_{in}(t)$, at level C_0 from 10 to 60 s. *Upper panel:* $C_R(t)$ is concentration-time curve for the impermeant reference intravascular solute, $C_E(t)$ for the extracellular solute with the same PS_c as for $C_D(t)$, and $C_D(t)$, for the permeating or diffusing solute entering the parenchymal cells of the organ. Parameters were: PS_c (for solutes E and D) = 1 ml g^{-1} min^{-1}, PS_{pc} (for D only) = 1 ml g^{-1} min^{-1}, and V_{cap}, V_{isf} and V_{pc} were 0.06, 0.18 and 0.55 or 16.5 ml g^{-1}.

The axial intravascular dispersion coefficient for all solutes was the same, 10^{-4} cm^2/s. *Lower panel:* the instantaneous extractions for solutes E and D, using Eq. 8.31, are $E_E(t)$ and $E_D(t)$; their peaks were almost the same, at 0.550 and 0.556 at 16.3 s; for $E_D(t)$ the later plateau values at $t = 60$ s were 0.345 with V_{pc} at 0.55 ml g^{-1} and 0.512 with V_{pc} at 16.5 ml g^{-1}. The estimations of $PS_c(t)$ were calculated at each time point for all positive values of $E(t)$ using Eq. 8.30

sum of two resistances in series, $1/PS_c$ and $1/PS_{pc}$, where PS_{pc} is the conductance across the cell membrane:

$$\frac{1}{PS_{tot}} = \frac{1}{PS_c} + \frac{1}{PS_{pc}} \ . \qquad (8.32)$$

If there is no, or at least little, return flux from the cell, then in principle

$$PS_{tot} = -F \ln \left(1 - E_{plateau}\right) , \qquad (8.33)$$

and if PS_c can be estimated from the early part of $E(t)$, then PS_{pc} can also be estimated by Eq. 8.32. In this case, the plateau $E_D(t)$ gave an estimate of $PS_{tot} = 0.34$ and 0.5 when V_{pc} was 0.55 and 16.5 ml/g. With the larger volume of distribution of 16.5 ml/g (which is equivalent to the potassium space within parenchymal cells), the back diffusion from cell to ISF is low and PS_{tot} is 0.5, as Eq. 8.32 predicts it to be. Since the behavior of a tracer being tightly bound by a receptor or incorporated into a cellular constituent is going through a process kinetically equivalent to cellular uptake with retention, these equations apply also to estimating the uptake by receptors or enzymatic reactions.

8.5
Flows, Local Functions, and Gene Expression

One can be easily persuaded that transcription is regulated by cell activity, by external influences and by a variety of intracellular signals. The remarkable degree of remodeling of the heart with an abnormal site of electrical activation seen by Prinzen et al. (1990, 1992, 1995) and van Oosterhout et al. (2002) attests to the heart's capabilities to undergo massive changes in its protein composition. With artificial pacing at the right ventricular outflow tract there was atrophy at the pacing site and hypertrophy at late activated sites amounting to about a 40% increase in wall thickness; there were no changes in blood pressure. Since there are no new cells formed, or cells lost, so far as is known, we attribute the changes in myofilament protein expression to be the result of down-regulation of transcription in those cells which are activated early, and up-regulation in those which are activated late (and are usually prestretched by the contraction of

those activated early). In the hypertrophic regions the cells increase in diameter and in length. The cellular remodeling is on a grand scale. The calculations of the changes in regional stress/strain patterns appear to offer the explanation: the early-activated regions work less hard and the late-activated regions harder, requiring more ATP and more oxygen.

Vascular remodeling was not assessed in this set of projects but is observed normally with regional hypertrophy. Presumably changes in transcription and in endothelial cell division are occurring to allow this. Intercellular communication among myocytes, endothelial cells, fibroblasts, and sympathetic neuronal cells must be occurring. If fetal genes are turned on by unloading the heart (Depre et al. 1998), one can speculate that they act in some permissive fashion. Young et al. (2001) have shown that there is a set of "clock genes" whose mRNA levels peak at different times of day and speculate that these may be permissive of specific responses by transcription or translation events. Tracking down the details of how this occurs is going to require protein array techniques as well as mRNA arrays; it is reasonably likely that some of the involved proteins will remain at such low concentrations that they will not be measurable by anything except specially targeted techniques.

Speculations on how to identify the regulatory controllers for gene expression are pretty useless without further information. Linking actin-myosin cross-bridge activity or ATP hydrolysis rates to regulators of transcription will not be easy. On the other hand, we may certainly expect that hundreds of gene transcription regulators will be changed in concert. The observation that as many as 500 genes become measurably depressed within 30 min after human immunodeficiency virus (HIV) invades a cell provides a precedent for such ideas.

The message coming from these data on cardiac remodeling undoubtedly applies to other tissues: cells "remodel" by changing gene expression in response to demand. What "remodel" means depends on the cell's functions, e.g., more or less acid secretion in gastric epithelium or fewer molecules of glycolytic enzymes if glycolysis is reduced in favor of fatty acid metabolism.

One conceptual approach to understanding these coordinated events is to perform large-scale integrated analysis. One idea is to combine the information from mRNA arrays, protein arrays, and arrays of small metabolite concentration changes. While neither protein-array nor metabolite-array technologies are highly accurate at this point, they are improving rapidly. This combination of measurement technologies can be brought together through large models of cellular metabolic and signaling fluxes using the information on protein and mRNA concentrations together with that on metabolite concentrations. This will require a massive modeling effort as well as a highly coordinated data-collection effort to obtain large amounts of information at the regional level in controlled circumstances.

■ **Acknowledgements.** The author thanks the many co-workers who contributed to the experimental and analytical work and the development of these ideas. Particular thanks go to James Eric Lawson for his help in the preparation of the manuscript. Support has been provided by NIH grants HL19139, HL-T07403, and RR1243. The latter supports a Research Resource Facility for Circulatory Transport, Exchange and Metabolism from which software may be downloaded at http://nsr.bioeng.washington.edu.

8.6
References

Agre P, Preston GM, Smith BL, Jung JS, Raina S, Moon C, Guggino WB, Nielsen S (1993) Aquaporin CHIP: the archetypal molecular water channel. Am J Physiol Renal Physiol 265:F463–F476

Altehoefer C (1998) Editorial: LBBB: challenging our concept of metabolic heart imaging with fluorine-18-FDG and PET. J Nucl Med 39:263–265

Aukland K (1980) Methods for measuring renal blood flow: total flow and regional distribution. Annu Rev Physiol 42:543–555

Austin RE Jr, Aldea GS, Coggins DL, Flynn AE, Hoffman JIE (1990) Profound spatial heterogeneity of coronary reserve: discordance between patterns of resting and maximal myocardial blood flow. Circ Res 67:319–331

Baer RW, Payne BD, Verrier ED, Vlahakes GJ, Molodowitch D, Uhlig PN, Hoffman JIE (1984) Increased number of myocardial blood flow measurements with radionuclide-labeled microspheres. Am J Physiol 246 (Heart Circ Physiol 15):H418–H434

Bassingthwaighte JB (1988) Physiological heterogeneity: fractals link determinism and randomness in structures and functions. News Physiol Sci 3:5–10

Bassingthwaighte JB (2001) The modeling of a primitive "sustainable" conservative cell. Philos Trans R Soc Lond A 359:1055–1072

Bassingthwaighte JB, Holloway GA Jr (1976) Estimation of blood flow with radioactive tracers. Semin Nucl Med 6:141–161

Bassingthwaighte JB, van Beek JHGM (1988) Lightning and the heart: fractal behavior in cardiac function. Proc IEEE 76:693–699

Bassingthwaighte JB, Beard DA (1995) Fractal ^{15}O-water washout from the heart. Circ Res 77:1212–1221

Bassingthwaighte JB, King RB, Roger SA (1989a) Fractal nature of regional myocardial blood flow heterogeneity. Circ Res 65:578–590

Bassingthwaighte JB, Noodleman L, van der Vusse GJ, Glatz JFC (1989b) Modeling of palmitate transport in the heart. Mol Cell Biochem 88:51–58

Bassingthwaighte JB, Malone MA, Moffett TC, King RB, Chan IS, Link JM, Krohn KA (1990) Molecular and particulate depositions for regional myocardial flows in sheep. Circ Res 66:1328–1344

Bassingthwaighte JB, Chan IS, Wang CY (1992) Computationally efficient algorithms for capillary convection-permeation-diffusion models for blood-tissue exchange. Ann Biomed Eng 20:687–725

Bassingthwaighte JB, Liebovitch LS, West BJ (1994) Fractal physiology. Oxford University Press, London

Bassingthwaighte JB, Beard DA, Li Z (2001) The mechanical and metabolic basis of myocardial blood flow heterogeneity. Basic Res Cardiol 96:582–594

Batra S, Kuo C, Rakusan K (1989) Spatial distribution of coronary capillaries: A-V segment staggering. Adv Exp Med Biol 248:241–247

Bauer WR, Hiller K-H, Roder F, Rommel E, Ertl G, Haase A (1996) Magnetization exchange in capillaries by microcirculation affects diffusion-controlled spin-relaxation: a model which describes the effect of perfusion on relaxation enhancement by intravascular contrast agents. Magn Reson Med 35:43–55

Bauer WR, Hiller K-H, Galuppo P, Neubauer S, Köpke J, Haase A, Waller C, Ertl G (2001) Fast high-resolution magnetic resonance imaging demonstrates fractality of myocardial perfusion in microscopic dimensions. Circ Res 88:340–346

Beard DA, Bassingthwaighte JB (2000) The fractal nature of myocardial blood flow emerges from a whole-organ model of arterial network. J Vasc Res 37:282–296

Buckberg GD, Luck JC, Payne BD, Hoffman JIE, Archie JP, Fixler DE (1971) Some sources of error in measuring regional blood flow with radioactive microspheres. J Appl Physiol 31:598–604

Caldwell JH, Martin GV, Raymond GM, Bassingthwaighte JB (1994) Regional myocardial flow and capillary permeability-surface area products are nearly proportional. Am J Physiol Heart Circ Physiol 267:H654–H666

Chien S, Usami S, Taylor HM, Lundberg JL, Gregersen MI (1966) Effects of hematocrit and plasma proteins of human blood rheology at low shear rates. J Appl Physiol 21:81–87

Chinard FP, Vosburgh GJ, Enns T (1955) Transcapillary exchange of water and of other substances in certain organs of the dog. Am J Physiol 183:221–234

Cowan DB, Langille BL (1996) Cellular and molecular biology of vascular remodeling. Curr Opin Lipidol 7:94–100

Crone C (1963) The permeability of capillaries in various organs as determined by the use of the 'indicator diffusion' method. Acta Physiol Scand 58:292–305

Decking UKM, Skwirba S, Zimmerman MF, Preckel B, Thämer V, Deussen A, Schrader J (2001) Spatial heterogeneity of energy turnover in the heart. Pflugers Arch (Eur J Physiol) 441:663–673

Depre C, Shipley GL, Chen W, Han Q, Doenst T, Moore ML, Stepkowski S, Davies PJA, Taegtmeyer H (1998) Unloaded heart in vivo replicates fetal gene expression of cardiac hypertrophy. Nature Med 4:1269–1275

Evans JR (1964) Importance of fatty acid in myocardial metabolism. Circ Res 15:II96–II106

Fahraeus R, Lindqvist T (1931) The viscosity of the blood in narrow capillary tubes. Am J Physiol 96:562–568

Falsetti HL, Carroll RJ, Marcus ML (1975) Temporal heterogeneity of myocardial blood flow in anesthetized dogs. Circulation 52:848–853

Fick A (1870) Über die Messung des Blutquantums in den Herzventrikeln. Verhandl Phys-Med Ges Würzburg 2:36

Franzen D, Conway RS, Zhang H, Sonnenblick EH, Eng C (1988) Spatial heterogeneity of local blood flow and metabolite content in dog hearts. Am J Physiol Heart Circ Physiol 254:H344–H353

Glenny RW, Robertson HT (1990) Fractal properties of pulmonary blood flow: characterization of spatial heterogeneity. J Appl Physiol 69:532–545

Glenny RW, Bernard S, Brinkley M (1993) Validation of fluorescent-labeled microspheres for measurement of regional organ perfusion. J Appl Physiol 74:2585–2597

Gonzalez-Fernandez JM (1962) Theory of the measurement of the dispersion of an indicator in indicator-dilution studies. Circ Res 10:409–428

Goresky CA, Bach GG, Nadeau BE (1975) Red cell carriage of label: its limiting effect on the exchange of materials in the liver. Circ Res 36:328–351

Grant PE, Lumsden CJ (1994) Fractal analysis of renal cortical perfusion. Invest Radiol 29:16–23

Greenstein JL, Wu R, Po S, Tomaselli GF, Winslow RL (2000) Role of the calcium-independent transient outward current I_{to1} in shaping action potential morphology and duration. Circ Res 87:1026–1033

Hales JRS, Cliff WJ (1977) Direct observations of the behaviour of microspheres in microvasculature. Bibl Anat 15:87–91

Hamilton WF, Moore JW, Kinsman JM, Spurling RG (1932) Studies on the circulation. IV. Further analysis of the injection method, and of changes in hemodynamics under physiological and pathological conditions. Am J Physiol 99:534–551

Heymann MA, Payne BD, Hoffman JIE, Rudolph AM (1977) Blood flow measurements with radionuclide-labeled particles. Prog Cardiovasc Dis 20:55–79

Hille B (1972) The permeability of the sodium channel to metal cations in myelinated nerve. J Gen Physiol 59:637–658

Hille B (2001) Ion channels of excitable membranes, 3rd edn. Sinauer Associates, Sunderland, MA

Hoff HE, Scott HJ (1948) Medical progress: physiology (continued). N Engl J Med 239:120–126

Hoffman JIE (1960) Calculation of output central volume and variance from indicator-dilution curves. J Appl Physiol 15:535–538

Hund TJ, Otani NF, Rudy Y (2000) Dynamics of action potential head-tail interaction during reentry in cardiac tissue: ionic mechanisms. Am J Physiol Heart Circ Physiol 279:H1869–H1879

Kety SS, Schmidt CF (1948) The nitrous oxide method for the quantitative determination of cerebral blood flow in man: theory, procedure and normal values. J Clin Invest 27:476–483

King RB, Bassingthwaighte JB (1989) Temporal fluctuations in regional myocardial flows. Pflugers Arch (Eur J Physiol) 413:336–342

King RB, Bassingthwaighte JB, Hales JRS, Rowell LB (1985) Stability of heterogeneity of myocardial blood flow in normal awake baboons. Circ Res 57:285–295

Kowallik P, Schulz R, Guth BD, Schade A, Paffhausen W, Gross R, Heusch G (1991) Measurement of regional myocardial blood flow with multiple colored microspheres. Circulation 83:974–982

Kroll K, Stepp DW (1996) Adenosine kinetics in the canine coronary circulation. Am J Physiol Heart Circ Physiol 270:H1469–H1483

Kroll K, Wilke N, Jerosch-Herold M, Wang Y, Zhang Y, Bache RJ, Bassingthwaighte JB (1996) Modeling regional myocardial flows from residue functions of an intravascular indicator. Am J Physiol Heart Circ Physiol 271:H1643–H1655

Little SE, Bassingthwaighte JB (1983) Plasma-soluble marker for intraorgan regional flows. Am J Physiol Heart Circ Physiol 245:H707–H712

Li Z, Yipintsoi T, Bassingthwaighte JB (1997) Nonlinear model for capillary-tissue oxygen transport and metabolism. Ann Biomed Eng 25:604–619

Love WD, Burch GE (1959) Influence of the rate of coronary plasma flow on the extraction of Rb^{86} from coronary blood. Circ Res 7:24–30

Lumsden CJ, Silverman M (1990) Multiple indicator dilution and the kidney: kinetics, permeation, and transport in vivo. Methods Enzymol 191:34–72

Luo C-H, Rudy Y (1994) A dynamic model of the cardiac ventricular action potential I. Simulations of ionic currents and concentration changes. Circ Res 74:1071–1096

Luxon BA, Weisiger RA (1992) A new method for quantitating intracellular transport: application to the thyroid hormone 3,5,3'-triiodothyronine. Am J Physiol 263:G733–G741

Luxon BA, Weisiger RA (1993) Sex differences in intracellular fatty acid transport: role of cytoplasmic binding proteins. Am J Physiol 265:G831–G841

McGowan RL, Welch TG, Zaret BL, Bryson AL, Martin ND, Flamm MD (1976) Noninvasive myocardial imaging with potassium-43 and rubidium-81 in patients with left bundle branch block. Am J Cardiol 38:422–428

Meier P, Zierler KL (1954) On the theory of the indicator-dilution method for measurement of blood flow and volume. J Appl Physiol 6:731–744

Michailova A, McCulloch A (2001) Model study of ATP and ADP buffering, transport of Ca^{2+} and Mg^{2+}, and regulation of ion pumps in ventricular myocyte. Biophys J 81:614–629

Nose Y, Nakamura T, Nakamura M (1985) The microsphere method facilitates statistical assessment of regional blood flow. Basic Res Cardiol 80:417–429

Piwnica-Worms D, Chiu ML, Kronauge JF (1992) Divergent kinetics of ^{201}Tl and ^{99m}Tc-SESTAMIBI in cultured chick ventricular myocytes during ATP depletion. Circulation 85:1531–1541

Prinzen FW, Bassingthwaighte JB (2000) Blood flow distributions by microsphere deposition methods. Cardiovasc Res 45:13–21

Prinzen FW, Augustijn CH, Arts T, Allessie MA, Reneman RS (1990) Redistribution of myocardial fiber strain and blood flow by asynchronous activation. Am J Physiol (Heart Circ Physiol 28) 259:H300–H308

Prinzen FW, Augustijn CH, Allessie MA, Arts T, Delhaas T, Reneman RS (1992) The time sequence of electrical and mechanical activation during spontaneous beating and ectopic stimulation. Eur Heart J 13:535–543

Prinzen FW, Cherlex EC, Delhaas T, Oosterhout MFM, Arts T, Wellens HJJ, Reneman RS (1995) Asymmetric thickness of the left ventricular wall resulting from asynchronous electric activation: a study in dogs with ventricular pacing and in patients with left bundle branch block. Am Heart J 130:1045–1053

Raichle ME, Martin WRW, Herscovitch P, Mintun MA, Markham J (1983) Brain blood flow measured with intravenous $H_2^{15}O$: II. Implementation and validation. J Nucl Med 24:790–798

Renkin EM (1959a) Transport of potassium-42 from blood to tissue in isolated mammalian skeletal muscles. Am J Physiol 197:1205–1210

Renkin EM (1959b) Exchangeability of tissue potassium in skeletal muscle. Am J Physiol 197:1211–1215

Renkin EM (1977) Multiple pathways of capillary permeability. Circ Res 41:735–743

Richmond DR, Tauxe WN, Bassingthwaighte JB (1970) Albumin macroaggregates and measurements of regional blood flow: validity and application of particle sizing by Coulter counter. J Lab Clin Med 75:336–346

Richmond DR, Yipintsoi T, Coulam CM, Titus JL, Bassingthwaighte JB (1973) Macroaggregated albumin studies of the coronary circulation in the dog. J Nucl Med 14:129–134

Safford RE, Bassingthwaighte JB (1977) Calcium diffusion in transient and steady states in muscle. Biophys J 20:113–136

Safford RE, Bassingthwaighte EA, Bassingthwaighte JB (1978) Diffusion of water in cat ventricular myocardium. J Gen Physiol 72:513–538

Sestier FJ, Mildenberger RR, Klassen GA (1978) Role of autoregulation in spatial and temporal perfusion heterogeneity of canine myocardium. Am J Physiol Heart Circ Physiol 235:H64–H71

Sokoloff L, Reivich M, Kennedy C, Des Rosiers MH, Patlak CS, Pettigrew KD, Sakurada O, Shinohara M (1977) The [^{14}C]deoxyglucose method for the measurement of local cerebral glucose utilization: theory, procedure, and normal values in the conscious and anesthetized albino rat. J Neurochem 28:897–916

Stapleton DD, Moffett TC, Baskin DG, Bassingthwaighte JB (1995) Autoradiographic assessment of blood flow heterogeneity in the hamster heart. Microcirculation 2:277–282

Stein WD (1986) Transport and diffusion across cell membranes. Academic, Orlando, FL

Stephenson JL (1948) Theory of the measurement of blood flow by the dilution of an indicator. Bull Math Biophys 10:117–121

Stewart GN (1897) Researches on the circulation time and on the influences which affect it. IV. The output of the heart. J Physiol 22:159–183

Tancredi RG, Yipintsoi T, Bassingthwaighte JB (1975) Capillary and cell wall permeability to potassium in isolated dog hearts. Am J Physiol 229:537–544

Taplin GV, Johnson DE, Dore EK, Kaplan HS (1964) Suspensions of radioalbumin aggregates for photoscanning the liver, spleen, lung and other organs. J Nucl Med 5:259

Thompson AM, Cavert HM, Lifson N, Evans RL (1959) Regional tissue uptake of D_2O in perfused organs: Rat liver, dog heart and gastrocnemius. Am J Physiol 197:897

Udelson JE, Coleman PS, Metherall J, Pandian NG, Gomez AR, Griffith JL, Shea NL, Oates E, Konstam MA (1994) Predicting recovery of severe regional ventricular dysfunction. Comparison of resting scintigraphy with 201Tl and 99mTc-sestamibi. Circulation 89:2552–2561

Ueda H, Iio M, Kaihara S (1964) Determination of regional pulmonary blood flow in various cardiopulmonary disorders: study and application of macroaggregated albumin (MAA) labelled with I^{131} (I). Jpn Heart J 5:431

Van Beek JHGM, Van Mil HGJ, King RB, de Kanter FJJ, Alders DJC, Bussemaker J (1999) A ^{13}C NMR double-labeling method to quantitate local myocardial O_2 consumption using frozen tissue samples. Am J Physiol Heart Circ Physiol 277:H1630–H1640

Van Brocklin HF, Enas J, Hanrahan S, O'Neil J (1995) Fluorine-18 labeled dihydrorotenone analogs: preparation and evaluation of PET mitochondrial probes. J Lab Comp Radiopharm 37:217–219

Van Oosterhout MFM, Prinzen FW, Sakurada S, Glenny RW, Hales JRS (1998) Fluorescent microspheres are superior to radioactive microspheres in chronic blood flow measurements. Am J Physiol Heart Circ Physiol 275:H110–H115

Van Oosterhout MFM, Arts T, Bassingthwaighte JB, Reneman RS, Prinzen FW (2002) Relation between local myocardial growth and blood flow during chronic ventricular pacing. Cardiovasc Res 53:831–840

Weiss HR, Neubauer JA, Lipp JA, Sinha AK (1978) Quantitative determination of regional oxygen consumption in the dog heart. Circ Res 42:394–401

Wilke N, Kroll K, Merkle H, Wang Y, Ishibashi Y, Xu Y, Zhang J, Jerosch-Herold M, Mühler A, Stillman AE, Bassingthwaighte JB, Bache R, Ugurbil K (1995) Regional myocardial blood volume and flow: First-pass MR imaging with polylysine-Gd-DTPA. J Magn Reson Imaging 5:227–237

Winkler B, Schaper W (1979) Tracer kinetics of thallium, a radionuclide used for cardiac imaging. In: Schaper W (ed) The pathophysiology of myocardial perfusion. Elsevier/North Holland, Amsterdam, pp 102–112

Winkler B, Schaper W (1992) The role of radionuclides for cardiac research. In: Pabst HW, Adam WE, Ell P, Hör G, Kriegel H (eds) Handbook of nuclear medicine, vol 2. Heart. Fischer, Stuttgart, pp 390–407

Winslow RL, Rice J, Jafri S, Marbán E, O'Rourke B (1999) Mechanisms of altered excitation-contraction coupling in canine tachycardia-induced heart failure. II: Model studies. Circ Res 84:571–586

Yen RT, Fung YC (1978) Effect of velocity distribution on red cell distribution in capillary blood vessels. Am J Physiol 235 (Heart Circ Physiol 4):H251–H257

Yipintsoi T, Bassingthwaighte JB (1970) Circulatory transport of iodoantipyrine and water in the isolated dog heart. Circ Res 27:461–477

Yipintsoi T, Tancredi R, Richmond D, Bassingthwaighte JB (1970) Myocardial extractions of sucrose, glucose, and potassium. In: Crone C, Lassen NA (eds) Capillary permeability (Alfred Benzon Symp II). Munksgaard, Copenhagen, pp 153–156

Yipintsoi T, Scanlon PD, Bassingthwaighte JB (1972) Density and water content of dog ventricular myocardium. Proc Soc Exp Biol Med 141:1032–1035

Yipintsoi T, Dobbs WA Jr, Scanlon PD, Knopp TJ, Bassingthwaighte JB (1973) Regional distribution of diffusible tracers and carbonized microspheres in the left ventricle of isolated dog hearts. Circ Res 33:573–587

Young ME, Razeghi P, Taegtmeyer H (2001) Clock genes in the heart: characterization and attenuation with hypertrophy. Circ Res 88:1142–1150

Zierler KL (1958) A simplified explanation of the theory of indicator-dilution for measurement of fluid flow and volume and other distributive phenomena. Johns Hopkins Med J 103: 199–217

Zierler KL (1962) Theoretical basis of indicator-dilution methods for measuring flow and volume. Circ Res 10:393–407

Zierler KL (1963) Theory of use of indicators to measure blood flow and extracellular volume and calculation of transcapillary movement of tracers. Circ Res 12:464–471

Zierler KL (1965) Equations for measuring blood flow by external monitoring of radioisotopes. Circ Res 16:309–321

Part II Whole Body Processes

CHAPTER 9

Endothelium and Compound Transfer 9

CHRISTOPHER A. DAWSON, SAID H. AUDI, GARY S. KRENZ, DAVID L. ROERIG

Contents

9.1
Introduction

One approach to the study of cell function in situ and to the quantitative phenotyping required for functional genomics and genetic circuit analysis is the bolus injection – outflow detection – multiple indicator dilution (MID) method. The MID method is a tracer dilution method used to measure tissue volumes (Goresky 1963; Goresky et al. 1970; Chinard 1975; Bassingthwaighte and Goresky 1984; Dawson et al. 1989, 1992), tissue composition (Dawson et al. 1989, 1992; Roerig et al. 1999), transcapillary transport parameters (Goresky et al. 1970; Bassingthwaighte 1974; Harris et al. 1978; Bassingthwaighte and Goresky 1984; Harris et al. 1987; Dawson et al. 1989; Audi et al. 1996a, 2000; Linehan et al. 1998), enzyme or receptor binding (Maolli et al. 1985; Dawson et al. 1989; Linehan et al. 1998; Roerig et al. 2000), and enzyme kinetics (Goresky et al. 1983, 1993; Riggs et al. 1988; Dawson et al. 1989; Linehan et al. 1998; Audi et al. 2000) within an organ. The fo-

cus of this chapter will be on the application of the MID method to measurement of endothelial transport and metabolism, particularly in the lungs. In comparison to other organs, lung tissue is disproportionately comprised of endothelial cells, and the lungs contain a large fraction (nearly half) of the vascular endothelium of the entire body. Although endothelial functions can be organ-specific, the lung serves as a good model for demonstrating general principles involved in the application of the method.

The MID method involves injection of a bolus containing two or more indicators into the arterial inlet to the organ and the measurement of the indicator concentrations in the organ venous effluent as a function of time following the injection. Typically, at least one indicator is a substance confined to the intravascular volume as it passes through the organ and is, therefore, referred to as the vascular reference indicator. The other indicators, referred to subsequently as test indicators, interact with the tissue in some way that reflects the organ functions of interest. Data analysis focuses on explaining the separation in time and concentration among the indicators resulting from passage through the organ, where, in this context, the concentrations of the injected indicators in the venous effluent are normalized to the injected amounts of the respective indicators. Interpretation of the data includes use of mathematical models based on hypotheses regarding the processes responsible for the separation. The model parameters are typically measures of vascular and tissue-distribution volumes, partition coefficients, and kinetic rate constants for transport between blood and tissue, associations between the indicators and molecules within the tissue, and/or metabolic conversions of the indicators. When metabolic conversions take place, effluent blood analysis includes measurement of both the injected test indicator and its metabolic products. The vascular reference indicator traces blood flow through the organ and thus the fate of those molecules of test indicator that remain in the blood during their entire organ transit. In other words, the effluent vascular reference indicator curve is what the test indicator curve would have been if none of the test indicator had left the intravascular volume. For test indicators that do not enter or interact with blood cells the appropriate vascular reference indicator is a substance carried in the plasma. The situation is more complicated when a test indicator interacts with blood cells (usually red blood cells) as well as tissue (Goresky et al. 1975; Bassingthwaighte and Goresky 1984; Pang et al. 1995). Then, both plasma and red blood cell reference indicators are needed to account for the fact that, in general, red blood cell transit time through an organ is shorter than the plasma transit time. The MID method is somewhat more invasive than tracer imaging methods, and it does not provide spatial information. However, for processes that take place in a time frame on the order of the organ capillary transit time, it can provide considerable information about their function within the whole organ. Although the use of radioactive tracers as indicators is particularly pertinent in the context of this text, in comparison to imaging methods there is less constraint on the range of physical and chemical properties that can be exploited for online or offline detection and measurement (Demarino et al. 1998; Audi et al. 2000).

MID methods have been important research tools for studying metabolic functions of various organs including, in particular, the liver (Goresky 1963; Goresky and Rose 1977; Goresky et al. 1983, 1993, 1998; Schwab and Goresky 1996; Pang et al. 1998; Schwab 1998), heart (Bassingthwaighte and Levin 1981; Goresky and Rose 1977; Bassingthwaighte and Goresky 1984; Kuikka et al. 1986; Cousineau et al. 1995; Bassingthwaighte et al. 1998), brain (Bissonnette et al. 1991; Kassissia et al. 1995; Paulson et al. 1982), and lungs (Goresky et al. 1969; Chinard 1975; Dawson et al. 1989, 1992; Audi et al. 1994, 1995, 1996 a, b, 1998 a, b, 1999, 2000, 2001; Dupuis et al. 1994; Roerig et al. 1995, 1999; Linehan et al. 1998; Maolli et al. 1985). The lungs are uniquely situated for in vivo MID studies because of the relatively easy access available for indicator injection and sampling via systemic venous and arterial sites, respectively. However, the lungs are normally provided with a high blood flow relative to the metabolic requirements of the lung cells themselves. One does not expect to find substantial arterial-venous differences or, in the MID context, separation in pulmonary venous effluent con-

centrations between a reference indicator and an indicator for most endogenous substrates and products for intermediary metabolism. Instead, the processes that can be detected and quantified tend to be those in which the lungs are involved in processing the systemic venous blood before it passes to the systemic arterial system and must therefore occur at rates consistent with the high pulmonary blood flow. In addition, for some processes specific properties of xenobiotic chemical probes can be exploited as indicators (Goresky et al. 1983; Dawson et al. 1992; Roerig et al. 1995; Maolli et al. 1985; Audi et al. 2000).

That the pulmonary endothelium interacts with a number of blood constituents as they pass through the lungs has been understood since the work of Vane (1969) and others (Ben-Harari and Bakhle 1980; Morel et al. 1985; Gillis 1988; Dawson et al. 1989; Wiedemann et al. 1990) showed that a number of vasoactive substances carried in the venous blood are either removed from the blood and/or chemically modified by the pulmonary endothelial cells during passage through the pulmonary circulation. This stimulated interest in the use of MID methods for evaluating these "nonrespiratory functions" of the pulmonary endothelium as well as in the development of the means for deciphering the information contained in the pulmonary effluent concentration data (Linehan et al. 1987, 1998; Audi et al. 1994, 1995, 1998 b). One motivation for attempting to understand the information content of these data has been the concept that if these metabolic functions of the pulmonary endothelium could be measured in vivo they might be useful as metabolic phenotypes reflecting the physiological or pathophysiological status of the endothelial cells. This has proven to be a useful concept for experimental studies on the mechanisms of endothelial injury (Gillis and Pitt 1982; Morel et al. 1985; Gillis et al. 1986; Gardaz et al. 1988; Gillis 1988; Merker and Gillis 1988; Dawson et al. 1989). It has also been suggested that, with the appropriate indicators and detection systems, the approach might have clinical diagnostic and prognostic utility in situations where injury to the pulmonary endothelium is a likely outcome (Gillis and Pitt 1982; Johnston et al. 1985; Morel et al. 1985; Gillis et al. 1986; Gardaz et al. 1988; Gillis 1988; Dawson et al. 1989; Wiedemann et al. 1990; Cowen et al. 1992). Thus, the application of the MID approach to the pulmonary endothelium has been an important research tool with potential for clinical application as well.

A key aspect distinguishing MID methods from steady-state indicator dilution methods is that the transient nature of the response provides information for separating individual processes contributing to, but not individually identifiable in, the steady state. The information content of the data can be complex because, in addition to the processes of interest occurring within the tissue, several other factors can influence the amount of a test indicator that is removed and/or modified on passage through an organ. These factors include organ perfusion (e.g., blood flow, perfused capillary surface area, and the distribution of capillary transit times) and reactions taking place in the blood (e.g., plasma protein binding, metabolism by blood-borne enzymes, uptake, and/or binding to formed elements). With the appropriate experimental design and indicator probes, the MID data can contain information about all of these processes. Approaches to deciphering this information, particularly as applied to the pulmonary endothelium, will be the focus of this chapter.

Two representations of the pulmonary venous (or systemic arterial) effluent data obtained after a pulmonary arterial (or systemic venous) bolus injection will be used in what follows. One representation is concentration vs. time following arterial injection (with indicator concentration normalized to the amount injected). The other representation is the cumulative effluent fraction of the amount of an indicator that has emerged from the lung vs. time following its pulmonary arterial injection. While both representations have almost the same information content, they tend to focus attention on different aspects of the MID signal. The differences in effluent time courses between indicators is visually emphasized by the concentration curves, whereas cumulative effluent fractions emphasize differences in indicator recovery at a given time. Figure 9.1 is an example of experimental data from a perfused rabbit lung presented using the two representations. In this case, the injected bolus included the reference indicator [fluorescein isothiocyanate labeled dextran (FITC-Dex)], and either ^{3}HOH or ^{14}C-phenylethylamine (^{14}C-PEA) as the test indicator. The ^{3}HOH rapidly equilibrates throughout the lung water volume accessible via the perfused capillaries as if there were no barriers to its access to the tissue water volume. In contrast, the tissue uptake of ^{14}C-PEA involves transport through the luminal endothelial cell membrane barrier followed by metabolism within the endothelial cells to phenylacetic acid (^{14}C-PAA) via a monoamine oxidase (MAO) catalyzed reaction (Roth and Gillis 1975; Ben-Harari and Bakhle 1980). In addition to examples

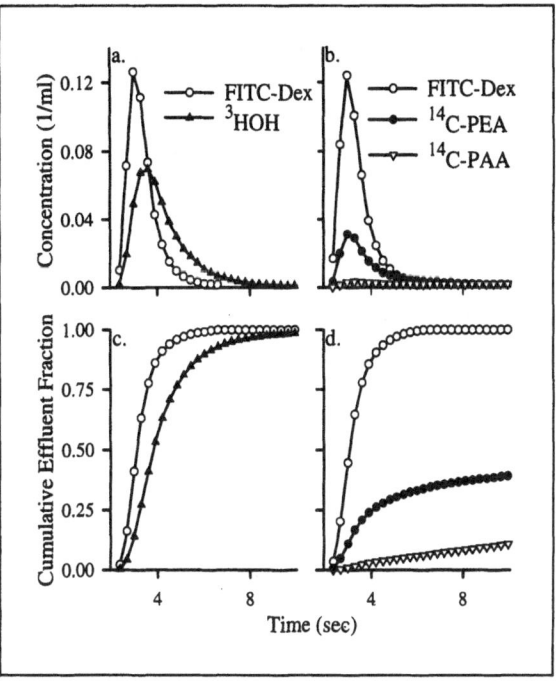

Fig. 9.1. a, b Venous effluent concentration vs. time curves for a vascular reference indicator fluorescein isothiocyanate-labeled dextran (FITC-Dex), ^{3}HOH, ^{14}C-phenylethylamine (^{14}C-PEA) and ^{14}C-phenylacetic acid (^{14}C-PAA) following a bolus injection of FITC-Dex and either ^{3}HOH (**a**) or ^{14}C-PEA (**b**) into the pulmonary artery of an isolated rabbit lung perfused with an artificial plasma solution as described in Audi et al. (2001). The concentrations on this and subsequent graphs are normalized to the amounts of the injected indicators and are thus the fraction of injected dose per milliliter of venous effluent. **c, d** The fraction of the injected indicators (cumulative effluent fraction) that had emerged in the lung effluent by a given time for the same injections as the respective upper panels (**a, b**). Modified version of Fig. 1 in Audi et al. (2001) with permission from the American Physiological Society

based on experimental data such as in Fig. 9.1, model simulations will be used for points that are not readily made with actual data.

9.2
The MID (Multiple Indicator Dilution) Method Applied to Endothelial Uptake and Metabolism

9.2.1
The General MID Model

The following model development results in a means of expressing the information content of the data as well as a method for data parameterization. This development begins with the single capillary element, which includes the capillary lumen and surrounding tissue to which a given test indicator has access (for example, in the case of ^{14}C-PEA, the endothelial cell volume, or for ^{3}HOH the entire tissue water volume). In the model, all interactions between tissue and test indicator take place within the capillary element (or, put another way, the capillary is defined as the place where the test indicator-tissue interactions take place). Pulmonary capillary dimensions (on the order of 0.005 mm in diameter and 1 mm long; Haworth et al. 1991) are such that radial diffusion is sufficiently rapid to prevent radial indicator concentration gradients within the capillary lumen even when the indicator extraction fraction is high. On the other hand, the capillary length and magnitude of the Peclet number (ratio of convection to diffusion velocities; Lassen and Perl 1979) are such that mass transport by axial diffusion flux is not included within the capillary element (Bass and Robinson 1982; Bassingthwaighte and Goresky 1984; Audi et al. 1998b). The number of species included in the capillary element model depends on the number and kind of indicators used, but the species balance equations for the indicators within the capillary element have the following general form.

For the capillary lumen (c):

$$\partial\frac{[U_c]}{\partial t} + W\frac{\partial[U_c]}{\partial x} = \Phi_c + \Phi_m - \frac{\dot{M}}{Q_c} \tag{9.1}$$

For the endothelial membrane (m):

$$\frac{\partial[U_m]}{\partial t} = \Phi_m \tag{9.2}$$

For the extravascular (intracellular volume) (i):

$$\frac{\partial[U_i]}{\partial t} = \frac{\dot{M}}{Q_i} + \Phi_i \tag{9.3}$$

The [U]s denote generic indicator concentration, and x and t are capillary axial distance and time, respectively. W is perfusate velocity, Q_c and Q_i are the vascular and extravascular components of the volume of the capillary element, and Φ_c, Φ_i, and Φ_m denote the relevant chemical kinetics terms in the capillary and extravascular volumes and at the intervening luminal endothelial membrane, respectively. $\dot{M}$ represents the net rate of substrate mass transfer across the capillary surface between Q_c and Q_i. In Eq. 9.1, local changes

in $[U_c]$ due to chemical reactions are balanced by axial convective transport in the capillary. The vascular reference indicator is confined to the capillary lumen, i.e., both $\dot{M}$ and the $\Phi's = 0$. A test indicator might interact only with endothelial surface molecules, in which case Φ_c, Φ_i, $\dot{M}$ and $Q_i = 0$, but $\Phi_m > 0$, or the test indicator might diffuse into the tissue in the radial direction such that Q_i and $\dot{M} > 0$, and $\Phi's \geq 0$. In either case, the test indicator is delayed in time relative to the vascular reference indicator. Selection of the appropriate kinetics terms (the $\Phi's$) is an initial step in formulating the model for a specific test indicator. As an example, for a typical reversible bimolecular association reaction, e.g., $U + B \underset{k_{-1}}{\overset{k_1}{\rightleftharpoons}} UB$ based on the law of mass action, Φ would be $(k_1[U][B]-k_{-1}[UB])$, where k_1 and k_{-1} are the association and dissociation rate constants, respectively, for the interactions of the indicator with some tissue constituent B. For transport across the capillary surface between Q_c and Q_i, $\dot{M}$ would be $(PS_c[U_c]-PS_i[U_i])$, where PS_c and PS_i are, respectively, the permeability surface area products representing the test indicator conductances into and out of the extravascular volume. This representation would include membrane permeation via either passive diffusion or facilitated transport in the concentration range consistent with the linear kinetic model. $PS_c \neq PS_i$ implies active transport, also in the concentration range in which linear kinetics apply.

9.2.2
The MID Model for Pulmonary Endothelial Uptake and Metabolism of Monoamine Oxidase Substrates

To illustrate the approach utilized for quantitative interpretation of MID data, a kinetic model that was developed to account for the pulmonary disposition of the MAO substrate PEA will be considered (Audi et al. 2001). PEA was chosen to illustrate the approach because its endothelial disposition on passage through the lungs involves several steps of the type common to the disposition of other substances of interest. PEA processing is perhaps the most general or prototypical example of a pulmonary endothelial cell metabolic function studied to date using the MID method. In addition, because MAO is primarily located in the outer mitochondrial membrane, which plays a key role in cell signaling pathways, MAO substrates may be useful as mitochondrial outer membrane probes (Goncharova 1983; Ishiwata et al. 1985;

Zhang and Piantadosi 1991; Hauptmann et al. 1996; Boulton et al. 1998; Malorni et al. 1998; Paterson and Tatton 1998). The MID method has been used to study equally or more complex processes in other organs (Kuikka et al. 1986; Goresky et al. 1993, 1998; Cousineau et al. 1995; Schwab and Goresky 1996; Bassingthwaighte et al. 1998; Pang ct al. 1998; Schwab 1998), but the PEA example illustrates some key points.

9.2.2.1
Single Capillary Element

For the PEA model, the single capillary element is composed of a capillary volume, Q_c, and an endothelial volume, Q_i. The spatial and temporal variations in the vascular reference indicator (R), PEA, and its final MAO product PAA within Q_i and Q_c are described by the species balance equations:

$$\frac{\partial[R_c]}{\partial t} + W\frac{\partial[R_c]}{\partial x} = 0 \tag{9.4}$$

$$\frac{\partial[PEA_c]}{\partial t} + W\frac{a_1 Q_c}{a_1 Q_c + Q_1}\frac{\partial[PEA_c]}{\partial x} =$$
$$= \frac{k_{PEA}PEA_i - PS_1[PEA_c]}{a_1 Q_c + Q_1} \tag{9.5}$$

$$\frac{\partial PEA_i}{\partial t} = PS_1[PEA_c] - k_{PEA}PEA_i - (k_{met} + k_{seq})PEA_i \tag{9.6}$$

$$\frac{\partial[PAA_c]}{\partial t} + W\frac{a_2 Q_c}{a_2 Q_c + Q_2}\frac{\partial[PAA_c]}{\partial x} =$$
$$= \frac{k_{PAA}PAA_i - PS_2[PAA_c]}{a_2 Q_c + Q_2} \tag{9.7}$$

$$\frac{\partial PAA_i}{\partial t} = PS_2[PAA_c] - k_{PAA}PAA_i + k_{met}PEA_i \tag{9.8}$$

where $[R_c](x, t)$, $[PEA_c](x, t)$, and $[PAA_c](x, t)$ are the vascular concentrations of the vascular reference indicator, PEA and PAA, respectively, at distance x from the capillary inlet ($x=0$) and time t. $PEA_i(x, t) = (a_3 Q_i)[PEA_i](x, t)$ and $PAA_i(x, t) = (a_4 Q_i)[PAA_i](x, t)$, where $[PEA_i](x, t)$, $[PAA_i](x, t)$ are the endothelial cell concentrations of PEA and PAA, respectively, at dis-

tance x from the capillary inlet and time t; $a_1 = 1 + \frac{[Pr_c]}{K_{eq1}}$ and $a_2 = 1 + \frac{[Pr_c]}{K_{eq2}}$, where K_{eq1} and K_{eq2} are the equilibrium dissociation constants for PEA and PAA associations, respectively, with perfusate albumin, Pr_c; $a_3 = 1 + \frac{[Pr_i]}{K_{eq3}}$ and $a_4 = 1 + \frac{[Pr_i]}{K_{eq4}}$, where Pr_i represents nonspecific intracellular sites of association for PEA and PAA having equilibrium dissociation constants K_{eq3} and K_{eq4}, respectively; W is the average linear flow velocity within Q_c; PS_1 and $k_{PEA} = \frac{PS_1}{a_3 Q_i}$ are the rates of mass transport of PEA in and out of the endothelial cells, respectively; PS_2 and $k_{PAA} = \frac{PS_2}{a_4 Q_i}$ are the rates of mass transport of PAA in and out of the endothelial cells, respectively; $Q_1 = Q_c K_1 [Z_1]$ and $Q_2 = Q_c K_2 [Z_2]$, where Z_1 and Z_2 represent nonspecific sites of association for PEA and PAA on the endothelial surface, respectively, with equilibrium dissociation constants K_1 and K_2, respectively.

The identifiable model parameters are k_{met} (s^{-1}), which is the measure of the rate of PEA deamination by MAO (actually PAA is the product formed by MAO followed by the aldehyde dehydrogenase reaction, which in the model development is considered to be much faster than the MAO reaction and, therefore, not separately identifiable; Audi et al. 2001); k_{seq} (s^{-1}), which is the measure of the PEA sequestration rate within the lung tissue (Audi et al. 2001); PS_1 (ml/ s) and PS_2 (ml/s), which are the endothelial permeability surface area products for PEA and PAA, respectively; k_{PEA} (s^{-1}) and k_{PAA} (s^{-1}), which are measures of the respective rates of PEA and PAA egress from the cells; and the virtual volumes Q_1 (ml) and Q_2 (ml), which are measures of the magnitudes of the rapidly equilibrating cell surface interactions of PEA and PAA, respectively. For PAA injections, Eqs. 9.5–9.8 reduce to Eqs. 9.7–9.8 with PEA_i set to zero, and the number of identifiable parameters reduces to three, namely, PS_2 (ml/s), k_{PAA} (s^{-1}), and Q_2 (ml).

For ^{14}C-PEA model bolus injections the numerical solution (Finlayson 1992) of Eqs. 9.4-9.8 is determined by the initial conditions ($t=0$): $[R](x, t) = [PEA_c](x, t) = [PAA_c](x, t) = 0$, $PEA_i(x, t) = PAA_i(x, t) = 0$, and boundary conditions ($x=0$): $[PAA_c](x, t) = 0$, $PEA_i(x, t) = PAA_i(x, t) = 0$, $[R](x, t) = C_{in}(t)$, and $[PEA_c](x, t) = (1/a_1)C_{in}(t)$, where $C_{in}(t)$ is the capillary input concentration function. For ^{14}C-PAA model bolus injections the numerical solution of Eqs. 9.4 and 9.7-9.8 is determined by the initial conditions ($t=0$): $[R](x, t) = 0$, $PAA_i(x, t) = 0$, and boundary conditions ($x=0$): $PAA_i(x, t) = 0$, $[R](x, t) = C_{in}(t)$, and $[PAA_c](x, t) = (1/a_2)C_{in}(t)$.

9.2.2.2
The Whole Organ Model

9.2.2.2.1
Significance of Capillary Transit-Time Distribution

The distribution of transit times for blood flowing through the various capillary elements comprising the organ contributes to the dispersion of the bolus as it passes from injection to sampling site. This capillary transit-time distribution, $h_c(t)$, can have a substantial impact on the disposition of a test indicator during its passage through an organ independently of the kinetics of vascular and tissue interactions (Goresky 1963; Gonzalez-Fernandez and Atta 1973; Goresky and Rose 1977; Bass and Robinson 1982; Bronikowski et al. 1987; King et al. 1996; Audi et al. 1998 b). In fact, it can be shown that for a given set of kinetic parameters and capillary mean transit time if the individual capillary transit times are distributed, the fractional extraction of a test indicator will be less than if all individual transit times were equal (Gonzalez-Fernandez and Atta 1973). Thus, $h_c(t)$ must be taken into account in the mathematical modeling interpretation of MID data. To construct an organ

model from the single capillary element, the random coupling assumption is invoked (Audi et al. 1998 b; Bronikowski et al. 1987), i.e., each capillary element receives the same input, $C_{in}(t)$, which is dispersed as a result of mixing at the injection site and dispersion in the conducting vessels connecting the capillary elements to the injection system.

The capillary outlet concentrations for the various species, $C_c(t)$'s, are obtained by solving Eqs. 9.4–9.8 for a capillary element having a transit time as long as the longest individual capillary transit time, t_N, in the whole organ. In this representation, transit time and distance traveled along the capillary are in constant proportion for ease of illustration. However, the transit-time distribution resulting from varying flows and geometries are also represented in the same formulation (Audi et al. 1998 b). As shown in Fig. 9.2, the time-concentration curves at all x locations along the capillary with the longest transit time are obtained. These curves are equivalent to the outflow concentrations for all capillaries with transit times between the shortest and longest capillary transit times, i.e., concentrations computed at the jth node ($j\Delta x$ from the capillary inlet, where $\Delta x = L\Delta t/t_N$, $j = 1,\ldots,$ N, Δt is the transit time increment, and L is the length of the

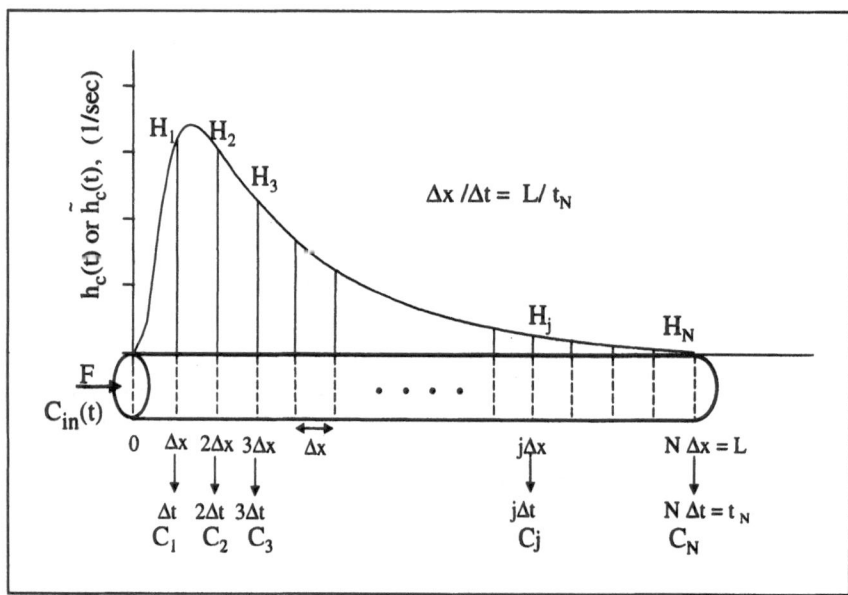

Fig. 9.2. A diagrammatic representation of the capillary transport function ($h_c(t)$ or $\tilde{h}_c(t)$) weighted outputs from capillaries with different transit times ($t_j = j\Delta t$). $C_{in}(t)$ and F are the capillary input function and flow, respectively. Δx and Δt are the space (distance along capillary) and time increments, respectively. N is the total number of different capillary transit times represented in the effluent concentra-

tion curve. L is the length of the capillary having the longest transit time. Model computed concentrations, C_j, are also interpretable as the outflow concentration from capillaries with transit time t_j. The weighting factor H_j represents the flow-weighted fraction of capillaries with transit time t_j. Adapted from Audi et al. (1998 b) with permission from the Biomedical Engineering Society

capillary having the longest transit time) correspond to the outflow concentration-time curve for any capillary having a transit time, $t_j = j\Delta t$ (Audi et al. 1998 b). For a given indicator the capillary outflow concentration, $C_c(t)$, is then obtained by summing the concentrations, $C_j(t)$, at all N nodes, each weighted by its corresponding H_j:

$$C_c(t) = \sum_{j=1}^{N} H_j C_j(t) \qquad (9.9)$$

where $H_j = (\Delta t/2)(h_c(t_j - \Delta t/2) + h_c(t_j + \Delta t/2))$ is the flow-weighted fraction of capillaries with transit time t_j such that $\sum_{j=1}^{N} H_j = 1$. A key point regarding the computations that follow from this development is that the outflow curves from the thousands of parallel capillary elements are obtained by solving only one capil-

lary element equation. The organ outflow concentrations for the various species, $C_o(t)$'s, are then obtained by convolving the capillary outflow concentrations with $h_v(t)$, the venous transit-time distribution. Because of the commutativity and associativity of the convolution integral and the applicability of first-order kinetics for the tracer concentrations used, it is not actually necessary to specify $C_{in}(t)$. Instead, only $h_c(t)$ and the reference indicator curve, $C_R(t)$, are needed since $C_R(t) = C_n(t) \cdot h_c(t)$, where $C_n(t) = C_{in}(t) \cdot h_v(t) = (q/F) h_n(t)$ represents the transport of indicators through the noncapillary (arteries, veins, tubing, and injection-sampling system) part of the system, $h_n(t)$ is the nocapillary transit-time distribution, and F and q are the total flow through the organ and the mass of the injected reference indicator, respectively. Thus, with $C_{in}(t)$ replaced with $C_n(t)$, the capillary outflow concentrations, $C_c(t)$'s, become the organ outflow concentrations, $C_o(t)$'s (Audi et al. 1996 a, 1998 a).

The above approach can be used to generate simulated MID data. For example, Fig. 9.3 emphasizes the impact of distributed capillary transit times represented by $h_c(t)$ on the test indicator disposition. This example is of simulations from two organs whose capillary mean transit time, model kinetic parameters, and the total reference indicator dispersion were the same, but each had a different relative contribution of $h_c(t)$ and $C_n(t)$ to the total variation in vascular transit times. For the simulation represented in Fig. 9.3 a, c, $C_n(t)$ accounted for the larger fraction of the variation in the whole organ vascular transit-time distribution, while for the simulation represented in Fig. 9.3 b, d, $h_c(t)$ accounted for the larger fraction of the vascular transit-time variation. The resulting simulated test indicator outflow concentration vs. time curves in Fig. 9.3 a, b are quite different, demonstrating the need to account for $h_c(t)$ in the computation of the kinetic parameters.

9.2.2.2.2
Estimation of Capillary Transit-Time Distribution

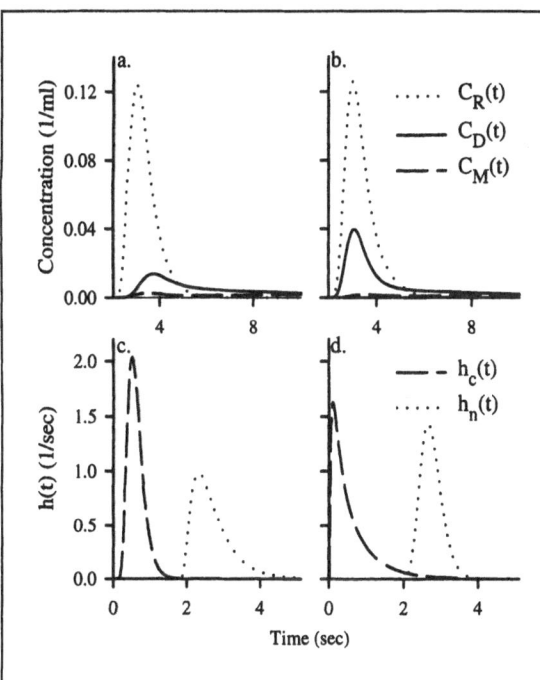

Fig. 9.3. a, b Simulated concentration vs. time organ outflow curves, $C_o(t)$'s, for a reference indicator, $C_R(t)$, a test indicator, $C_D(t)$, and the metabolite of the test indicator, $C_M(t)$, obtained by numerically solving Eqs. 9.1–9.8 with the same set of kinetic parameters ($F = 6.6$ ml/s, $Q_1 = 7.4$ ml, $Q_2 = 5.4$ ml, $k_{met} = 0.09$ s^{-1}, $k_{seq} = 0.033$ s^{-1}, $PS_1 = 18.4$ ml/s, $k_{PEA} = 0.19$ s^{-1}, $PS_2 = 2.6$ ml/s and $k_{PAA} = 0.05$ s^{-1}), but for the two different sets of capillary, $h_c(t)$, and noncapillary, $h_n(t)$, transport functions shown in the respective lower panels (**c, d**). $C_n(t) = (q/F)h_n(t)$, where q is the mass of the injected reference indicator and F is the total flow through the organ

Given this impact of $h_c(t)$, it is important to have a means of obtaining the necessary information regarding $h_c(t)$. One approach has been to measure the perfusion distribution using radiolabeled microspheres (Yipintsoi et al. 1973; King et al. 1985; Bassingthwaighte et al. 1987) and then, under certain assumptions about how the distribution of flow and transit times relate to each other, $h_c(t)$ is estimated (Bassingthwaighte and Levin 1981; Audi et al. 1998 b;

King et al. 1996). However, that approach is not consistent with the goal of nondestructive evaluation of organ function. As an alternative, Goresky (1963) noted that under the assumption that the average cross-sectional area orthogonal to the capillary axis is similar among capillary elements, $h_c(t)$ can be obtained from the indicator dilution data themselves if one of the test indicators rapidly equilibrates between blood and tissue during passage through the parallel noninteracting capillary elements. The rapidly equilibrating indicators are referred to as "flow-limited indicators" (Goresky 1963; Goresky et al. 1970; Bassingthwaighte and Goresky 1984; Dawson et al. 1992). For this class of test indicators, Eqs. 9.1–9.3 reduce to

$$\frac{\partial [U_c]}{\partial t} + W \frac{Q_c}{Q_c + Q_L} \frac{\partial [U_c]}{\partial x} = 0 \qquad (9.10)$$

where Q_L includes the physical extravascular volume accessible to the flow-limited test indicator. For flow-limited hydrophilic compounds, Q_L includes the water volume to which they have access, for example, the total perfused water volume for labeled water (Goresky et al. 1969; Chinard 1975). For lipophilic compounds, Q_L includes any rapidly equilibrating intravascular, luminal surface and/or extravascular interactions with other molecular species that result in a tissue-to-perfusate partition coefficient $\neq 1$ [the latter Q_L is sometimes referred to as a "virtual volume of distribution" (Dawson et al. 1992; Audi et al. 1994, 1995)].

A key assumption used to take advantage of flow-limited indicators for estimating $h_c(t)$ is that the reference indicator curve dispersion can be decomposed into two distinct components. One is noncapillary dispersion due to the injection and transit through arteries, veins, and any connecting tubing (e. g., catheters and/or perfusion tubing). The other is dispersion resulting from the diversity of pathways the blood can take through the tortuous capillary bed. In the noncapillary portion of the system, the dispersion of the flow-limited and vascular reference indicators is the same. However, in the capillaries, the flow-limited indicator transit is retarded in comparison to the reference indicator because it rapidly equilibrates with a larger volume (Q_L). The delay depends on Q_L and results in an increase in the moments of the flow-limited indicator curves relative to those of the reference indicator curve. The delay is distributed in proportion to the capillary transit-time distribution for the reference indicator, resulting in a fixed relationship between the dispersions of the reference and flow-limited indicators

within the capillary bed. The theory (Audi et al. 1994, 1995) leads to Eq. 9.11 relating the mean transit time, $\bar{t}$ the variance, σ^2, and the third central moment, m^3, (referred to subsequently as "the moments"), of the capillary transport function, $h_c(t)$, to the measured moments of the effluent concentration vs. time curves for a vascular reference indicator, $C_R(t)$, and a flow-limited test indicator, $C_F(t)$.

$$\sigma_F^2 - \sigma_R^2 = \left| \left(1 + \frac{\bar{t}_e}{\bar{t}_c}\right)^2 - 1 \right| \sigma_c^2 \; ; \qquad (9.11\,a)$$

$$m_F^3 - m_R^3 = \left| \left(1 + \frac{\bar{t}_e}{\bar{t}_c}\right)^3 - 1 \right| m_c^3 \; ; \qquad (9.11\,b)$$

where $\bar{t}_e = \bar{t}_F - \bar{t}_R$.

The moments of $h_c(t)$ can be estimated if the injected bolus includes, along with the vascular reference indicator, at least two flow-limited indicators having sufficiently different mean transit times. When the moments of $h_c(t)$ obtained using Eq. 9.11 are used as the parameters of a specific probability density function, such as the random walk or lagged normal density function (Bassingthwaighte et al. 1966; Harris and Newman 1970; Bronikowski et al. 1987; Audi et al. 1996, 1998b, c), an estimate of $h_c(t)$ itself results (Audi et al. 1994, 1995, 1998b). For example, for a time-shifted random walk function,

$$\mathrm{RW}(t) = \frac{\sqrt{\frac{\phi\theta}{4\pi(t-t_s)}}}{\theta} \exp\left(\frac{-\phi\left(1 - \frac{(t-t_s)}{\theta}\right)}{4\frac{(t-t_s)}{\theta}} \right) \text{ for } t > t_s$$

$$(9.12)$$

where $\int_0^\infty$; $\mathrm{RW}(t)dt = 1$; $\mathrm{RW}(t)$ is zero for $t \leq t_s$, and ϕ and θ are related to the mean transit time, $\bar{t}_c$, variance, σ_c^2, and third central moment, m_c^3, of $h_c(t)$ by:

$$\bar{t}_c = \theta\left(1 + \frac{2}{\phi}\right) + t_s \; ; \qquad (9.13\,a)$$

$$\sigma_c^2 = \frac{2\phi + 8}{(\phi + 2)^2} (\bar{t}_c - t_s)^2 \; ; \qquad (9.13\,b)$$

$$m_c^3 = \frac{12\phi + 64}{\phi^3} \theta^3 \qquad (9.13\,c)$$

The parameters of $\mathrm{RW}(t)$, namely θ, ϕ, and t_s, are related to the mean transit time and relative dispersion, $\mathrm{RD}_c = \sigma_c/\bar{t}_c$, of $h_c(t)$ via

$$\frac{t_s}{\bar{t}_c} = 1 - RD_c \frac{\phi + 2}{\sqrt{2\phi + 8}} \qquad (9.14)$$

Knowing $h_c(t)$, $C_n(t)$ can also be represented by Eq. 9.12 scaled by q/F, the parameters of which can be specified by iteratively convolving $C_n(t)$ with $h_c(t)$ until the optimal least square fit to $C_R(t)$ is obtained (Audi et al. 1998a, 1999, 2000; Roerig et al. 1999).

Estimation of $h_c(t)$ is useful because the descriptors of $h_c(t)$ (e.g., the moments) are measures of an important aspect of lung vascular function and also because knowledge of $h_c(t)$ allows for identification of the separate contributions of reaction kinetics and perfusion kinematics to test indicator disposition as indicated above. However, it turns out that for estimation of the kinetic parameters, knowledge of the relationship between the moments of $h_c(t)$ is sufficient information (Audi et al. 1998b). Thus, for kinetic parameter estimation, one does not need to know $h_c(t)$ per se but rather the common characteristics (relationship between moments) of the members of a family of effectively similar $h_c(t)$'s, subsequently referred to as $\tilde{h}_c(t)$'s. Knowledge of the common characteristics is useful because of the simplification it provides. That is the moment relationships can be obtained from the outflow concentration curve from a single flow-limited indicator, such as labeled water, ^3HOH, by replacing $\bar{t}_c$ with $\tilde{\bar{t}}_c = \bar{t}_e/(\sqrt{\sigma_F^2/\sigma_R^2} - 1)$ in Eq. 9.11, where $\tilde{\bar{t}}_c$ is an upper bound on the actual capillary mean transit time specified by the data. The actual $\bar{t}_c$ is not obtained, but the practical gain is that by including only a single flow-limited indicator, such as ^3HOH, in the set of injected indicators, one can distinguish the effects of capillary perfusion kinematics from the effects of the kinetics of the tissue interactions for the test indicators included in the bolus (Audi et al. 1998b).

In what follows, $\tilde{h}_c(t)$ obtained as described above from the ^3HOH data such as in Fig. 9.1 will be used to account for the effects of the capillary transit-time distribution in the estimation of parameters for pulmonary endothelial uptake and metabolism of ^{14}C-phenylethylamine.

9.2.2.3
Kinetic Parameter Estimation

Estimation of the model kinetic parameters requires appropriate experimental protocols that yield MID data containing discriminating information for separating the processes of interest. The protocol used for the pul-

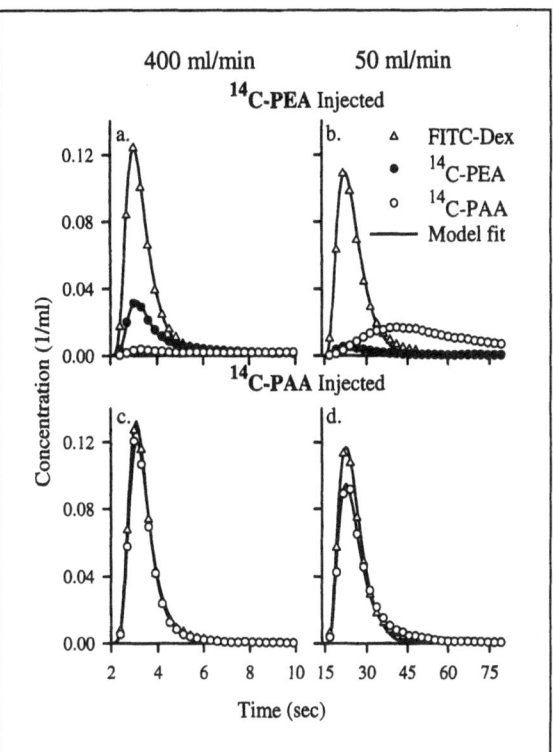

Fig. 9.4a–d. Venous effluent concentration vs. time curves for FITC-Dex, ^{14}C-phenylethylamine (^{14}C-PEA) and ^{14}C-phenylacetic acid (^{14}C-PAA) after bolus injections of FITC-Dex and either ^{14}C-PEA (**a, b**) or ^{14}C-PAA (**c, d**) into the pulmonary artery of rabbit lungs perfused at the two indicated flows. Solid lines are model fits to the data. The model lines do not show when fit is within the extent of the data symbols. Modified version of Fig. 1 in Audi et al. (2001) with permission from the American Physiological Society

monary endothelial PEA uptake and metabolism studies included boluses consisting of the vascular reference indicator FITC-Dex, ^3HOH, and either ^{14}C-PEA or ^{14}C-PAA injected into the pulmonary artery of isolated rabbit lungs perfused at two flows, 400 and 50 ml/min. The typical FITC-Dex, ^{14}C-PEA, and ^{14}C-PAA outflow data are exemplified in Fig. 9.4. Since ^3HOH is a flow-limited indicator, the overall shape of its outflow curve is essentially the same for all these study conditions and is exemplified in Fig. 9.1a.

While each of the PEA model parameters is mathematically identifiable from a single ^{14}C-PEA bolus injection, the high correlation between some parameters can make independent estimation from the data from a single bolus injection unreliable. The means for reducing the correlations between parameters is to vary experimental conditions. For example, measurements made at different flows produce data sets in which the information about the various processes is weighted differently (Audi et al. 2001). This may be anticipated from the data in Fig. 9.4, in which following ^{14}C-PEA injection the relative magnitudes of the effluent ^{14}C-PEA and ^{14}C-PAA concentrations are quite different between the two flows. The PEA transport into the cells tends to be rate-limiting at higher flows, while the rate of intracellular metabolism dominates at lower flows. The effluent ^{14}C-PAA concentration curve following ^{14}C-PEA injection contains information about the ^{14}C-PEA metabolism, but the metabolism rate parameter is highly correlated with k_{PAA}, the parameter governing the rate of ^{14}C-PAA return to the perfusate. This correlation can be broken by utilizing the effluent ^{14}C-PAA concentration curves following the injection of ^{14}C-PAA (Audi et al. 2001). These data contain information about the ^{14}C-PAA transport into and out of the endothelial cells independent of metabolism. Thus, to estimate model parameters from data in Fig. 9.4, the model was fit to all the concentration data obtained following the bolus injection of either ^{14}C-PEA or ^{14}C-PAA and at both flows simultaneously. Additional information, namely, the plasma protein binding equilibrium dissociation constants per unit plasma protein concentration, [Pr$_c$], for PEA and PAA ($\frac{K_{eq1}}{[Pr_c]} = 7.1$, and $\frac{K_{eq2}}{[Pr_c]} = 0.12$, respectively) obtained from a separate ultrafiltration experiment (Audi et al. 1994, 1995, 2001; Dawson et al. 1992), was also included as part of the total data set. The model fit was obtained by carrying out the same procedure as described above for the simulations at each iteration of a Levenberg-Marquardt optimization scheme (Audi et al. 1996a, 1998a, 1999, 2000).

The numerical values of the model kinetic parameters are generally most meaningful in the context of changes in response to manipulations directed at the hypothesized mechanisms of disposition of the test indicator, or in response to changes in physiological or pathophysiological conditions or differences in genotype. With regard to manipulations of mechanisms, this might be through the use of pharmacological inhibitors of selected points in the putative pathways or through the use of genetic variants. The pharmacological approach is exemplified by the data in Figs. 9.5 and 9.6 obtained after adding the MAO inhibitors pargyline and semicarbazide (Roth and Gillis 1975; Ben-Harari and Bakhle 1980) to the lung perfusate and then carrying out the ^{14}C-PEA and ^{14}C-PAA bolus injections. Following MAO inhibition, ^{14}C-PAA was no longer detectable in the venous effluent. The inhibition had relatively little effect on the effluent ^{14}C-PEA curves and virtually no effect on the ^{14}C-PAA curves following ^{14}C-PAA injections. The cumulative effluent fractions

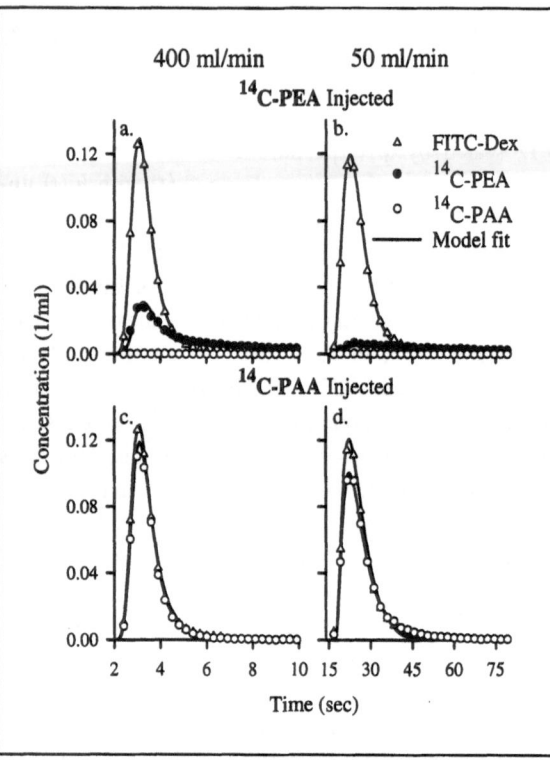

Fig. 9.5 a–d. Venous effluent concentration vs. time curves for FITC-Dex, ^{14}C-phenylethylamine (^{14}C-PEA), and ^{14}C-phenylacetic acid (^{14}C-PAA) after bolus injections of FITC-Dex and either ^{14}C-PEA (**a,b**) or ^{14}C-PAA (**c,d**) into the pulmonary artery of rabbit lungs perfused at two flows following the inhibition of monoamine oxidase with pargyline and semicarbazide. Solid lines are model fits to the data. Modified version of Figs. 1 and 2 in Audi et al. (2001) with permission from the American Physiological Society

in Fig. 9.6 emphasize the effect of MAO inhibition on the recoveries of ^{14}C-PEA and ^{14}C-PAA in the venous effluent. After MAO inhibition the 80-s ^{14}C-PAA recovery following ^{14}C-PEA decreased from nearly 60% to essentially zero. That this decrease in ^{14}C-PAA recovery was due to a loss of MAO activity rather than to a decrease in permeability to ^{14}C-PAA is suggested by the fact that the MAO inhibitors had virtually no effect on ^{14}C-PAA data obtained following the ^{14}C-PAA injection. The increase in ^{14}C-PEA recovery with MAO inhibition, observable by comparing the two upper panels of Fig. 9.6 (Fig. 9.6 a, b), is consistent with an increase in the amount of unmetabolized ^{14}C-PEA available for diffusion back into the perfusate in the absence of its intracellular metabolism. As might be expected from the data in Figs. 9.5 and 9.6, fitting the model to the Fig. 9.5 data resulted in a much lower value for the ^{14}C-PEA metabolism parameter k_{met} (Table 9.1), with little effect on the other parameters. This kind

Table 9.1. Values of model kinetic parameters estimated from data in Figs. 9.4 and 9.5

	PS_1 (ml/s)	k_{PEA} $(10^{-2}/s)$	PS_2 (ml/s)	k_{PAA} $(10^{-2}/s)$	k_{met} $(10^{-2}/s)$	k_{seq} $(10^{-2}/s)$	Q_1 (ml)	Q_2 (ml)	Q_V (ml)	Q_W (ml)
Control	18.4	19.0	2.57	4.7	9.1	3.3	7.4	5.4	4.2	5.7
Inhibitors	18.1	17.0	2.04	4.3	0.0	2.7	10.3	5.3	4.1	5.6

Q_V and Q_W are the pulmonary vascular volume and extravascular lung water volume estimated from $Q_V = Ft_R$ and $Q_W = F(t_F - t_R)$, respectively, where F is the flow, and t_R and t_F are the mean transit times of the concentration vs. time outflow curves of FITC-Dex and ^3HOH, respectively, obtained by fitting Eq. 9.12 to the data. t_R was corrected for the mean transit time of the nonvascular part of the perfusion system (e.g., tubing, catheters) (Audi et al. 1995)

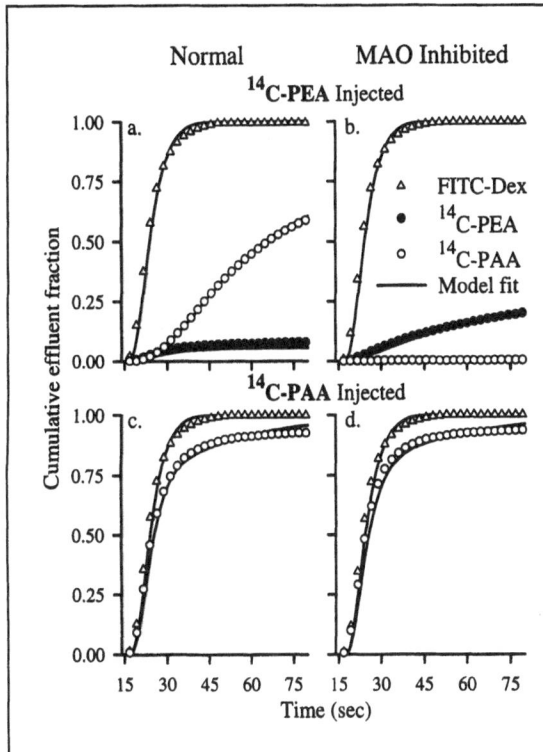

Fig. 9.6. a,b Cumulative effluent fraction representation of the data and model fits in Figs. 9.4b and 9.5b, respectively. **c,d** Cumulative effluent fraction representation of the data and model fits in Figs. 9.4d and 9.5d, respectively. Modified version of Figs. 1 and 2 in Audi et al. (2001) with permission from the American Physiological Society

of result tends to confirm that the model does in fact include representations of the dominant processes involved in the test indicator disposition on passage through the organ to the extent that the outcome was predictable from the known effects of the inhibitors. It also tends to confirm that the experimental protocol is sufficient to allow for calculation of model parameters that quantify the separate processes. The latter is a key question in any parameter-estimation scheme and is further addressed in the next section.

9.2.2.4
Robustness of Estimated Parameters

The model parameters do not contribute equally to the model fit to the data, and their contributions depend on experimental factors such as flow, bolus composition (e.g., ^{14}C-PEA or ^{14}C-PAA), and other manipulations such as the introduction of inhibitors in the above example. This can be appreciated using sensitivity analysis. For the ith model parameter, θ_i, the sensitivity function $S_i(t) = \partial C_o(t)/\partial \theta_i$, where $C_o(t)$ is the calculated test indicator effluent concentration. The change in $C_o(t)$ resulting from a 1% change in θ_i divided by the change in θ_i provides an approximation to $S_i(t)$ (Bassingthwaighte and Chaloupka 1984; Audi et al. 2000). When $S_i(t)$ is multiplied by the value of the parameter estimate, θ_i, the relative amplitude of this function for each respective parameter provides an indication of the relative contribution of the parameter to the model fit to the data at a given time during the evolution of the $C_o(t)$ data. Comparison of the respective shapes of $\theta_i S_i(t)$ reveals the degree to which any pair of these parameters is correlated. Similar shapes indicate high positive correlations, while mirror images indicate high negative correlations. Examples of the sensitivity functions obtained for two of the model kinetic parameters, PS_1 and k_{met}, are given in Fig. 9.7. The sensitivity functions for only these key parameters are displayed for the purpose of illustration because a graph including the sensitivity functions for all combinations of all the parameters would be quite complex. One point evident from Fig. 9.7 is the utility of the flow variation as one means for decreasing a correlation, in this case, between ^{14}C-PEA uptake and metabolism parameters. The dominant role of PS_1 in the fit to the ^{14}C-PEA data following ^{14}C-PEA injection at high flow can be seen in Fig. 9.7a. k_{met}, on the other hand, has most of its influence on the model fit to the ^{14}C-PAA data following the ^{14}C-PAA injected at the low

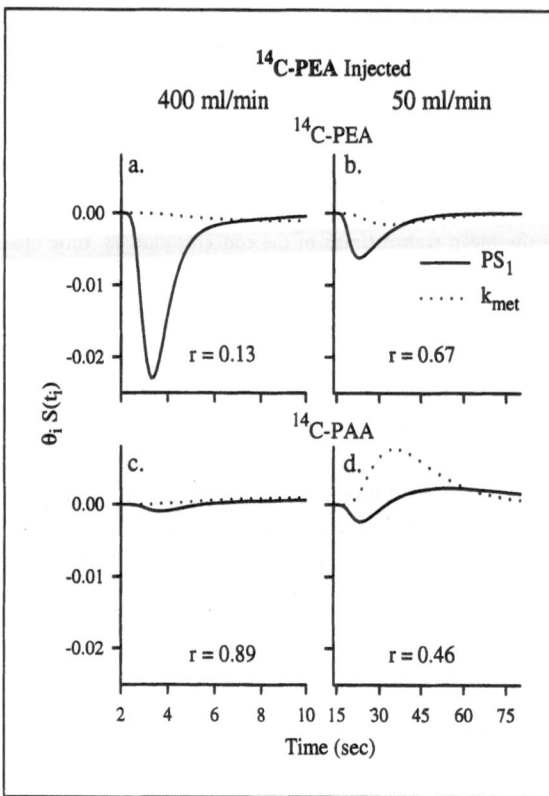

Fig. 9.7 a–d. Sensitivity functions, $\theta S(t)$, for the two parameters PS_1 and k_{met} from the model fit to the ^{14}C-PEA and ^{14}C-PAA concentration data in Fig. 9.4 obtained following the injection of ^{14}C-PEA at two different flows as indicated. r is the correlation coefficient between the PS_1 and k_{met} sensitivity functions

flow, as seen in Fig. 9.7 d. Not only does the relative influence of the parameters on the fits change with flow, but also the relative time at which the influence is strongest changes with flow. In other words, the relative shapes of the sensitivity functions change. This is reflected in the effect of flow on the correlations between parameters. The latter is quantified by the correlation coefficient, r, between the sensitivity functions for PS_1 and k_{met}, which are lowest in the opposite quadrants for the ^{14}C-PEA and ^{14}C-PAA data in Fig. 9.7 a and d, respectively.

9.2.2.5
Interpretation of Estimated Parameters

MID model parameters provide a quantitative description of relevant processes within the portion of the organ that is accessible to test indicators via the circulation. Thus, the parameters are potentially affected not only by changes in the relevant tissue

properties but also by changes in organ perfusion, which may be affected by recruitment, derecruitment, or vascular remodeling, etc., depending on physiological or pathophysiological conditions (Harris et al. 1978; Merker and Gillis 1988; Dawson et al. 1992; Audi et al. 1999; Roerig et al. 1999). Parameters that characterize extensive properties of the perfused tissue, such as a permeability surface area product (PS), reflect any change in the fraction of the tissue volume and/or vascular surface area that is perfused as well as any change in local function. On the other hand, parameters that characterize intensive properties, such as a_j, are independent of changes in the perfused tissue volume. The distinction between parameters representing extensive and intensive properties may be appreciated by noting that mole, calorie, and Becquerel are units of extensive properties for which molar, °C, and half-life seconds are, respectively, units for related intensive properties. Extensive MID parameters can be normalized to appropriate indices of perfused tissue volume and/or capillary surface area to aid in interpretation, as indicated below. In addition, the identifiable parameters are often products of other parameters that are inseparable without some additional information either from some additional test indicator or the application of some other methodology. For the PEA example, K_{eq1} and K_{eq2}, the plasma protein binding equilibrium dissociation constants for PEA and PAA, respectively, were estimated using a separate methodology (ultrafiltration). The PS is an example of a group parameter included in MID models for indicators such as the PEA in which there is a diffusion barrier at the blood-tissue interface. The PS is the product of permeability, which is an intensive property of the barrier conductance, and capillary surface area, which is an extensive property of the conductance. Normalization of PS to other extensive parameters results in intensive parameters that can provide additional insights. Insofar as the perfused surface area is related to the perfused vascular volume, Q_V, normalization of PS to Q_V accounts for changes in PS resulting from gross changes in the fraction of perfused tissue volume. Normalization of PS to total water volume of perfused tissue, Q_W, would provide a similar index unless some change in tissue hydration (edema) accompanied the change in PS (Dawson et al. 1992). Thus, comparison of a change in PS/Q_V with PS/Q_W can provide insight into the nature of the change in PS even without separation of P and S per se. This problem has been of particular interest in the lungs because both edema and perfusion abnormalities are common sequelae of lung injury. Combinations of lipophilic, hydrophilic, and/or amphipathic indicators with barrier or flow-limited distributions into the lungs have been used (Harris et al. 1987; Merker and Gillis 1988; Dawson et al. 1992) to calculate such indices from MID model parameters. Alternatively, nonlinear models that are used to account for the saturation of certain transport or metabolic processes when nontracer concentrations of indicators are injected (Linehan et al. 1987, 1998) or infused (Goresky et al. 1983; Schwartz et al. 2000) have the potential for further subdividing group parameters and revealing changes in intensive properties of the tissue.

9.3
Applications

The approach to modeling and parameterization of data presented here is representative of the kind of "black box" analysis typically required for nondestructive evaluation of organ function from outflow (MID) or residue (imaging) detection indicator methods. This mathematical modeling has two primary functions. One is to provide a concise statement of the hypotheses regarding the processes involved in the indicator disposition. These include hypotheses regarding the subject processes carried out by cells within the intact organ in which the cells are exposed to in situ environments not readily reproduced artificially such as in cell culture. The ability of the mathematical model to fit the data is one test of those hypotheses and as such is a means for understanding the cellular mechanisms involved in the indicator disposition. As an approach to hypothesis testing mathematical modeling contains key elements of any method of hypothesis testing including the caveat that it can reveal only whether or not the hypotheses are consistent with (fit) the data but not whether the hypotheses are in fact the explanation for the data. As a means of hypothesis testing, mathematical modeling does contribute in at least two unique ways. First, it can be applied to hypothesized explanations for systems that are complex enough that, without a formal (mathematical) means of keeping track of the interactions, it can be difficult to predict even the direction of a response. Thus, the mathematical simulations can reveal the consequences of hypotheses that might otherwise be missed. Second, it allows for quantita-

tive as opposed to only directional predictions. This helps to avoid the trap set by plausible explanations that are not actually consistent with the magnitude of a given response. The role of quantification in hypothesis testing is emphasized by Alfred Crosby in *The Measure of Reality* (Crosby 1997): "It was quantification, not aesthetics, not logic per se, that parried Kepler's beloved Platonic solids and goaded him on until he begrudgingly devised his planetary laws."

The other primary function of the mathematical modeling exemplified here is that the calculated parameters can be quantitative measures of the status of the subject processes within the functioning intact organ. For example, the MID approach applied to lung monoamine disposition has been used to determine the oxygen dependency of lung MAO k_{met} (Audi et al. 2001) and the effects of lung injury on monoamine PS_1 (Dawson et al. 1989). Given that the model fits the data, evaluation of the use of a parameter as a quantifier of a specific process focuses on the robustness of the parameter, e.g., whether or to what extent a change is proportional to a change in the function the parameter is purported to represent. As indicated above, the correlation matrix is a key source of information in this regard (Audi et al. 1998a). When this issue is satisfactorily resolved, the model parameters can be and have been used to distinguish functional states, i.e., for example detecting endothelial injury (Dawson et al. 1989) or, in the example shown, for determining the effect of experimental perturbations (e.g., inhibitors of particular steps in the process) on the systems under study. When a genetically determined variation in a model parameter is identified, the parameter can be viewed as a trait or phenotype for genetic analysis. Commonly, the parameters will be complex traits, as are blood pressure, body size, etc., conducive to genetic analysis and identification of genes affecting the processes reflected quantitatively by the model parameter. Even the activity of a single enzyme will generally be governed by several aspects of the cellular environment that are controlled by multiple genes. Given that gene segregation during meiosis is a function of distance between gene loci, the loci of contributing genes can be determined as quantitative trait loci by correlating the segregation of the magnitude of the parameter with that of genetic markers distributed along the chromosomes (Cowley et al. 2000). Thus, with a high-resolution genetic map and a genetically determined parameter (phenotype) cline, the chromosomal loci affecting the magnitude of the parameter can be mapped using various breeding strategies or lineages in the same manner as for any other quantitative phenotype (Cowley et al. 2000).

■ **Acknowledgements.** The authors' research contribution to this chapter was supported by NHLBI grant HL24349 and the Department of Veterans' Affairs.

9.4
References

Audi SH, Krenz GS, Linehan JH et al (1994) Pulmonary capillary transport function from flow-limited indicators. J Appl Physiol 77:332–351

Audi SH, Linehan JH, Krenz GS et al (1995) Estimation of the pulmonary transport function in isolated rabbit lungs. J Appl Physiol 78:1004–1014

Audi SH, Dawson CA, Linehan JH et al (1996a) An interpretation of ^{14}C-urea and ^{14}C-primidone extraction in isolated rabbit lungs. Ann Biomed Eng 24:337–351

Audi SH, Schuster DP, Merker MP et al (1996b) Pulmonary angiotensin converting enzyme ligand binding kinetics. FASEB J 10:A99

Audi SH, Dawson CA, Linehan JH et al (1998a) Pulmonary disposition of lipophilic amine compounds in the isolated perfused rabbit lung. J Appl Physiol 84:516–530

Audi SH, Linehan JH, Krenz GS et al (1998b) Accounting for the heterogeneity of capillary transit times in modeling multiple indicator dilution data. Ann Biomed Eng 26:914–930

Audi SH, Linehan JH, Krenz GS et al (1998c) Lipophilic amines as probes for measurement of lung capillary transport function and tissue composition using the multiple-indicator dilution method. In: Bassingthwaighte JB, Goresky CA, Linehan JH (eds) Whole organ approaches to cellular metabolism. Springer, Berlin Heidelberg New York, pp 517–543

Audi SH, Roerig DL, Ahlf SB et al (1999) Pulmonary inflammation alters the lung disposition of lipophilic amine indicators. J Appl Physiol 87:1831–1842

Audi SH, Olson LE, Bongard RD et al (2000) Toluidine blue O and methylene blue as endothelial redox probes in the intact lung. Am J Physiol Heart Circ Physiol 278:H137–H150

Audi SH, Dawson CA, Ahlf SB et al (2001) Oxygen dependency of monoamine oxidase activity in the intact lung. Am J Physiol (Lung Cell Mol Physiol) 281:L969–L981

Bass L, Robinson PJ (1982) Capillary permeability of heterogeneous organs: a parsimonious interpretation of indicator diffusion data. Clin Exp Pharmacol Physiol 9:363–388

Bassingthwaighte JB (1974) A concurrent flow model for extraction during transcapillary exchange. Circ Res 35:483–503

Bassingthwaighte JB, Levin M (1981) Analysis of coronary outflow dilution curves for the estimation of cellular uptake rates in the presence of heterogeneous regional flows. Basic Res Cardiol 76:404–410

Bassingthwaighte JB, Chaloupka M (1984) Sensitivity functions in the estimation of parameters of cellular exchange. Fed Proc 43:180–184

Bassingthwaighte JB, Goresky CA (1984) Modeling in the analysis of solute and water exchange in the microvasculature. In: Renkin EM, Michel CC (eds) Handbook of physiology, vol IV, sect 2. The cardiovascular system. Microcirculation, part 1. American Physiological Society, Bethesda, MD, pp 549–626

Bassingthwaighte JB, Ackerman FH, Wood EH (1966) Application of the lagged normal density curve as a model for arterial dilution curve. Circ Res 38:398–415

Bassingthwaighte JB, Malone MA, Moffett TC et al (1987) Validity of microsphere depositions for regional myocardial flows. Am J Physiol (Heart Circ Physiol 22) 253:H184–H193

Bassingthwaighte JB, Kroll K, Schwartz LM et al (1998) Strategies for uncovering the kinetics of nucleoside transport and metabolism in capillary endothelial cells. In: Bassingthwaighte JB, Goresky CA, Linehan JH (eds) Whole organ approaches to cellular metabolism. Springer, Berlin Heidelberg New York, pp 163–188

Ben-Harari R, Bakhle YS (1980) Uptake of β-phenylethylamine in rat isolated lung. Biochem Pharmacol 29:489–494

Bissonnette JM, Hohimer AR, Chao CR (1991) Unidirectional transport of glucose and lactate into brain of fetal sheep and guinea-pig. Exp Physiol 76:515–523

Boulton AA, Yu PH, Davis BA et al (1998) Aliphatic N-methylpropargylamines: monoamine oxidase-B inhibitors and antiapoptotic drugs. Adv Pharmacol 42:308–311

Bronikowski TA, Dawson CA, Linehan JH (1987) On indicator dilution and perfusion heterogeneity: a stochastic model. Math Biosci 83:199–225

Chinard FP (1975) Estimation of extravascular lung water by indicator-dilution techniques. Circ Res 37:137–145

Cousineau DF, Goresky CA, Rose CP et al (1995) Effects of flow, perfusion pressure, and oxygen consumption on cardiac capillary exchange. J Appl Physiol 78:1350–1359

Cowen ME, Mulvin D, Howard RB et al (1992) Lung tolerance to hyperthermia by in vivo perfusion. Eur J Cardio Thoracic Surg 6:167–173

Cowley AW Jr, Stoll M, Green AS et al (2000) Genetically defined risk of salt sensitivity in an intercross of brown Norway and Dahls rats. Physiol Genom 2:107–115

Crosby AW (1997) The measure of reality. Cambridge University Press, London, p 229

Dawson CA, Roerig DL, Linehan JH (1989) Evaluation of endothelial injury in the human lung. In: Jenkinson SG (ed) Clinics in chest medicine, vol 10. Saunders, Philadelphia, pp 13–23

Dawson CA, Roerig DL, Rickaby DA et al (1992) Use of diazepam for interpreting changes in extravascular lung water. J Appl Physiol 72:686–693

Demarino S, Olson LE, Pou NA et al (1998) Optical and radioisotope indicator dilution measurements in pulmonary edema. Ann Biomed Eng 26:417–430

Dupuis J, Goresky C, Stewart DJ (1994) Pulmonary removal and production of endothelin in the anesthetized dog. J Appl Physiol 76:694–700

Finlayson BA (1992) Numerical methods for problems with moving fronts. Ravena Park Publishing, Seattle

Gardaz JP, Py P, Suter PM et al (1988) Effects of oleic acid-, alpha-naphthylthiourea-, and phorbol myristate acetate-induced microvascular damage on indexes of pulmonary endothelial function in anesthetized dogs. Am Rev Respir Dis 137:1350–1355

Gillis CN (1988) Pulmonary extraction of PGE1 in the adult respiratory distress syndrome. Am Rev Respir Dis 137:1–2

Gillis CN, Pitt BR (1982) The fate of circulating amines within the pulmonary circulation. Annu Rev Physiol 44:269–281

Gillis CN, Pitt BR, Wiedemann HP et al (1986) Depressed prostaglandin E1 and 5-hydroxytryptamine removal in patients with adult respiratory distress syndrome. Am Rev Respir Dis 134:739–744

Goncharova VA (1983) Content of serotonin, catecholamines and monoamine oxidase activity of the lung in disease. Vopr Med Khim 29:34–39

Gonzalez-Fernandez JM, Atta SE (1973) Maximal substrate transport in capillary networks. Microvasc Res 5:180–198

Goresky CA (1963) A linear method for determining liver sinusoidal and extravascular volumes. Am J Physiol 204:626–640

Goresky CA, Rose CP (1977) Blood-tissue exchange in liver and heart: the influence of capillary transit times. Fed Proc 36:2629–2634

Goresky CA, Cronin RFP, Wangel BE (1969) Indicator dilution measurements of extravascular water in the lungs. J Clin Invest 48:487–501

Goresky CA, Ziegler WH, Bach GG (1970) Capillary exchange modeling. Circ Res 27:739–764

Goresky CA, Bach GG, Nadeau BE (1975) Red cell carriage of label. Its limiting effect on the exchange of materials in the liver. Circ Res 36:328–351

Goresky CA, Bach GG, Rose CP (1983) Effects of saturating metabolic uptake on space profiles and tracer kinetics. Am J Physiol 244 (Gastrointest Liver Physiol 7):G215–G232

Goresky CA, Bach GG, Schwab AJ (1993) Distributed-in-space product formation in vivo: enzymic kinetics. Am J Physiol 264 (Heart Circ Physiol 33):H2029–H2050

Goresky CA, Bach GG, Schwab AJ et al (1998) Liver cell entry in vivo and enzymic conversion. In: Bassingthwaighte JB, Goresky CA, Linehan JH (eds) Whole organ approaches to cellular metabolism. Springer, Berlin Heidelberg New York, pp 297–324

Harris TR, Newman EV (1970) An analysis of mathematical models of circulatory indicator-dilution curves. J Appl Physiol 28:840–850

Harris TR, Brigham KC, Rowlett RD (1978) Pressure, serotonin, and histamine effects on lung multiple-indicator curves in sheep. J Appl Physiol 44:245–253

Harris TR, Roselli RJ, Mauner CR et al (1987) Comparison of labeled propranediol and urea as markers of lung vascular injury. J Appl Physiol 62:1852–1859

Hauptmann N, Grimsby J, Shih JC et al (1996) The metabolism of tyramine by monoamine oxidase A/B causes oxidative damage to mitochondrial DNA. Arch Biochem Biophys 335:295–304

Haworth ST, Linehan JH, Bronikowski TA et al (1991) A hemodynamic model representation of the dog lung. J Appl Physiol 70:15–26

Ishiwata K, Ido T, Yanai K et al (1985) Biodistribution of a positron-emitting suicide inactivator of monoamine oxidase, carbon-11 pargyline, in mice and a rabbit. J Nucl Med 26:630–636

Johnston MR, Christensen CW, Minchin RF et al (1985) Isolated total lung perfusion as a means to deliver organ-specific chemotherapy: long-term studies in animals. Surgery 98:35–44

Kassissia IG, Goresky CA, Rose CP et al (1995) Tracer oxygen distribution is barrier-limited in the cerebral microcirculation. Circ Res 7:1201–1211

King RB, Bassingthwaighte JB, Hales JRS et al (1985) Stability of heterogeneity of myocardial blood flow in normal awake baboons. Circ Res 57:285–295

King RB, Raymond GM, Bassingthwaighte JB (1996) Modeling blood flow heterogeneity. Ann Biomed Eng 24:352–372

Kuikka J, Levin M, Bassingthwaighte JB (1986) Multiple tracer dilution estimates of D- and 2-deoxy-D-glucose uptake by the heart. Am J Physiol 250 (Heart Circ Physiol 19):H29–H42

Lassen NA, Perl W (1979) Tracer kinetic methods in medical physiology. Raven, New York, pp 156–175

Linehan JH, Bronikowski TA, Dawson CA (1987) Kinetics of uptake and metabolism by endothelial cell from indicator dilution. Ann Biomed Eng 15:201–215

Linehan JH, Audi SH, Dawson CA (1998) The uptake and metabolism of substrates by endothelium in the lung. In: Bassingthwaighte JB, Goresky CA, Linehan JH (eds) Whole organ approaches to cellular metabolism. Springer, Berlin Heidelberg New York, pp 427–438

Maolli R, Howell RE, Gillis CN (1985) Kinetics of captopril- and enalapril-induced inhibition of pulmonary angiotensin converting enzyme in vivo. J Pharmacol Exp Ther 234:372–377

Malorni W, Giammarioli AM, Matarrese P et al (1998) Protection against apoptosis by monoamine oxidase A inhibitors. FEBS Lett 426:155–159

Merker MP, Gillis CN (1988) Propranolol and serotonin removal in lung injury. J Appl Physiol 65:2579–2584

Morel DR, Dargent F, Bachmann M et al (1985) Pulmonary extraction of serotonin and propranolol in patients with adult respiratory distress syndrome. Am Rev Respir Dis 132:479–484

Pang KS, Barker F, Simard A et al (1995) Sulfation of acetaminophen by the perfused rat liver: the effect of red blood cell carriage. Hepatology 22:267–282

Pang KS, Goresky CA, Schwab AJ et al (1998) Probing the structure and function of the liver with the multiple-indicator dilution technique. In: Bassingthwaighte JB, Goresky CA, Linehan JH (eds) Whole organ approaches to cellular metabolism. Springer, Berlin Heidelberg New York, pp 325–368

Paterson IA, Tatton WG (1998) Antiapoptotic actions of monoamine oxidase B inhibitors. Adv Pharmacol 42:312–315

Paulson OB, Gyory A, Hertz MM (1982) Blood-brain barrier transfer and cerebral uptake of antiepileptic drugs. Clin Pharmacol Ther 32:466–477

Riggs D, Havill AM, Pitt BR et al (1988) Pulmonary angiotensin-converting enzyme kinetics after acute lung injury in the rabbit. J Appl Physiol 64:2508–2516

Roerig DL, Ahlf SB, Dawson CA et al (1995) First pass uptake in the lung of drugs used during anesthesia. In: Bosnjak AJ, Kampine JP (eds) Advances in pharmacology. Anesthesia in cardiovascular disease, vol 31. Plenum, New York, pp 531–549

Roerig DL, Audi SH, Linehan JH et al (1999) Detection of changes in lung tissue properties with multiple-indicator dilution. J Appl Physiol 86:1866–1880

Roerig DL, Audi SH, Ahlf SB et al (2000) Increases in mitochondrial benzodiazepine receptors and caspase-3 activity in inflamed lungs. FASEB J 14:A195

Roth JA, Gillis CN (1975) Multiple forms of amine oxidase in perfused rabbit lung. J Pharmacol Exp Ther 194:537–544

Schwab AJ, Goresky CA (1996) Hepatic uptake of protein-bound ligands: effect of an unstirred Disse space. Am J Physiol 270 (Gastintest Liver Physiol 33):G869–G880

Schwab AJ (1998) A generalized mathematical theory of the multiple-indicator dilution method. In: Bassingthwaighte JB, Goresky CA, Linehan JH (eds) Whole organ approaches to cellular metabolism. Springer, Berlin Heidelberg New York, pp 369–388

Schwartz LM, Bukowski TR, Ploger JD et al (2000) Endothelial adenosine transporter characterization in perfused guinea pig hearts. Am J Physiol Heart Circ Physiol 279:1502–1511

Vane JR (1969) The release and fate of vasoactive hormones in the circulation. Pharmacologist 35:209–242

Wiedemann HP, Matthay MA, Gillis CN (1990) Pulmonary endothelial cell injury and altered lung metabolic function. Clin Chest Med 11:723–736

Yipintsoi T, Dobbs Jr WA, Sanlon PD et al (1973) Regional distribution of diffusible tracers and carbonized microspheres in the left ventricle of isolated dog hearts. Circ Res 33:573–587

Zhang J, Piantadosi CA (1991) Prevention of H_2O_2 generation by monoamine oxidase protects against CNS O_2 toxicity. J Appl Physiol 71:1057–1061

Chapter 10

Coagulation and Peripheral Vascular Disease 10

Leonard Rosenthall

Contents

10.1
Introduction

Deep vein thrombosis (DVT) of the lower extremities is a serious disease which can occur spontaneously or as a complication of inherited and acquired disorders. These include coagulation abnormalities, presence of cancer and as a complication of trauma and surgical procedures, particularly of the hip (Kierkegaad 1980; Anderson et al. 1991; Salzman and Hirsh 1993). The clinical course of DVT could lead to pulmonary embolism and recurrent episodes of DVT, and serious postphlebitic sequelae in 25–65% of the patients such as intractable edema, debilitating pain, pigmentation and ulceration (Strandness et al. 1983; Salzman and Hirsh 1993; Prandoni et al. 1996). Upper extremity

DVT is more common than previously recognized as a result of more frequent objective testing with non-invasive ultrasonography. The precise incidence of DVT and pulmonary embolism is unknown (Gillium 1987; Cronan 1993). It is estimated that in the USA there are 300,000–600,000 hospitalizations per year for venous thrombosis and pulmonary emboli, and about 50,000 patients die per year from pulmonary embolism. Prevalence in the general population is approximately 2 million cases per year, which is exceeded only by acute coronary syndromes and cardiovascular disease. The incidence of DVT increases exponentially with age (Anderson et al. 1991).

Both DVT and pulmonary embolism can be asymptomatic and undiagnosed.

Thrombi frequently originate in the deep veins of the calf and may then extend superiorly to the popliteal vein and above. Untreated, about 20–30% of the calf thromboses will extend to the proximal veins where they are apt to embolize to the lungs (Kakkar et al. 1969; Anonymous 1975; Douss 1976). Proximal DVT can also occur de novo after major pelvic or hip surgery, in patients with hip fracture, or during pregnancy or the postpartum period (Clarke-Pearson et al. 1984; Kalebo et al. 1990). In a large multicenter study it was shown that silent pulmonary emboli occurred with an estimated frequency of 40–59% in patients with established proximal DVT (Meignan et al. 2000). Silent and symptomatic pulmonary emboli associated with an isolated calf DVT is an uncommon event (Moser and LeMoine 1981; Clarke-Pearson et al. 1984; Dorfman et al. 1987).

The frequency of recurrence in the untreated isolated calf DVT is low (Dorfman et al. 1987). Indirect evidence is also derived from prospective serial impedance plethysmography studies in about 2000 patients, which indicated that less than 2% of the patients developed DVT of those who continued to have negative results during the long-term follow-up

(Wheeler et al. 1982; Huisman et al. 1986). Risk of recurrence after proximal DVT is increased with absent or inadequate treatment, diagnosed cancer, and inherited thrombophilic abnormalities, and it is reported to occur with a frequency of 20–50% (Hull et al. 1979, 1986; Hansson et al. 2000). After an episode of DVT, 30–50% of the patients will develop symptomatology suggestive of recurrent DVT, but only about a third of them will be so afflicted. To identify this subset of true recurrence, objective testing is mandatory in these patients in order to avoid unnecessary anticoagulant therapy (Koopman et al. 1995).

10.2
Diagnosis

Physical examination, impedance plethysmography, contrast venography, and ultrasonography are available methods for thrombi detection. In patients with early evidence of thrombosis the clinical criteria using physical examination and symptoms are seldom definitive. In a study of 102 patients with suspected DVT who had contrast venography, clinical assessment proved to have a sensitivity and specificity of 66% and 55%, respectively (O'Donnell et al. 1980; Well et al. 1995).

10.2.1
Plethysmography

Impedance plethysmography is a noninvasive method based on the principle that the blood volume changes in the calf produced by inflation and deflation of a pneumatic thigh cuff result in changes in the electrical conductivity that can be detected by electrodes placed around the calf. Conductivity increases during inflation and decreases to baseline when the cuff is released. The amplitude of the electrical change is reduced in proximal DVT. Sensitivity is reported to vary from 81% to 100% and specificity from 87% to 100% in proximal symptomatic DVT (Hull et al. 1978; Lanigan et al. 1979; Peters et al. 1982). It is less sensitive in the disclosure of silent proximal DVT, a frequent occurrence after major hip and abdominal surgery. The reported sensitivity and specificity is about 15% and 99%, respectively (Anand et al. 1998). False positive results may occur in any condition that

causes decreased venous emptying or filling. This includes extrinsic venous compression, severe peripheral arterial insufficiency, congestive heart failure and advanced postphlebitic changes, among others.

10.2.2
Ultrasonography

Ultrasonography is a widely used modality for the detection and monitoring of DVT. Duplex ultrasonography refers to a system that allows dual real-time B-mode imaging and Doppler capabilities simultaneously or alternatively. The real time component visualizes the venous channels, whereas Doppler provides acoustic and graphic representation of blood flow owing to a reflected sound beam at an altered frequency which is proportional to the velocity of blood. Color Doppler depicts flow toward the transducer in red and flow away from the transducer in blue. B-mode criteria for the presence of thrombosis are visualization of the thrombi, incompressibility of veins with transducer cutaneous probe pressure, and absent or diminished percentage change in venous diameter with Valsalva maneuvers. The Doppler criteria include visualization of the thrombus as a defect in the blood flow column and abnormal flow responses to external compression at various sites in the lower extremity (Meadway et al. 1975; Barnes et al. 1976; Ouriel et al. 1984). In an exhaustive review of the world literature it was found that for first-time DVT in symptomatic patients the mean sensitivity of proximal DVT and calf DVT was 97% and 62%, respectively. Positive and negative predictive values for symptomatic proximal DVT were 97% and 98%, respectively. In the asymptomatic group, the mean sensitivity of proximal DVT and calf DVT was 62% and 53%, respectively, and the positive and negative predicted values for proximal DVT were 74% and 95%, respectively. The authors also concluded that incomplete venous compressibility with application of probe pressure at the common femoral vein and mid-popliteal vein is the most accurate diagnostic criterion for DVT and that Doppler assessment of blood flow did not improve the diagnostic accuracy for DVT (Kearon et al. 1998). Diagnostic accuracy is challenged in patients with a history of recurrent DVT, peripheral edema and congestive failure.

10.2.3
Contrast Venography

Contrast venography allows the visualization of the entire deep venous system of the lower extremity and offers direct visual evidence of thrombi. It is equally sensitive to both proximal and calf DVT, and it can establish or exclude the diagnosis in one session. Impedance plethysmography and duplex ultrasound, in contrast, are less sensitive to calf DVT and require repeated examinations in patients with an initially normal result. Venography has, therefore, been considered as a gold standard, but it is not free of side effects. Contrast material can cause a painful irritation of the venous endothelium and may induce thrombus formation with reported frequencies as high as 5% (Hull et al. 1981). There is also a small risk of systemic reactions such as transient nausea and flushing, but more serious are the uncommon occurrences of bronchospasm and hemodynamic instability. Contrast administration should be avoided in patients with acute renal failure, chronic renal failure with serum creatinine exceeding 3 mg/dl and a history of idiosyncratic reactions. Fatalities of 1 in 10,000 to 1 in 40,000 have been registered (Cohan and Dunnick 1987). Technical factors may lead to flow artifacts and non-filling of the deep veins which are falsely interpreted as positive for DVT. Previous DVTs can alter the venous anatomy such that the changes may be misinterpreted as acute DVT. False negative results occur when an involved non-visualized vein is missed in the presence of otherwise normal filling of the deep veins. Interobserver variability in assessing the number of abnormalities within a single contrast venogram study, rather than agreement on the final diagnosis, was evaluated by the kappa statistic, which is an index agreement excluding chance alone. It yielded a kappa value of 0.56, where 1 is perfect agreement and 0 is purely chance (Illescas et al. 1990). For interobserver agreement on the absence or presence of DVT in a given individual, a kappa value of 0.93 is reported (de Valois et al. 1990).

10.2.4
Computed Tomography
and Magnetic Resonance Angiography

There is a growing interest in the application of computed tomographic angiography (CTA) and magnetic resonance angiography (MRA) to the diagnosis of DVT of the pelvis and lower limbs. CTA has the attribute of visualizing the pulmonary arteries for emboli and the venous system for thrombi with the same injection of contrast material and MRA does not produce ionizing radiation. Both modalities have reported instances of identifying thromboses of the pelvic veins when contrast venography and duplex ultrasound failed to do so. CTA is generally comparable to contrast venography and ultrasound in the femoral-popliteal region. The precise role of CTA and MRA in the diagnostic algorithm is presently undefined (Reimer and Landwehr 1998; Cham et al. 2000; Kemp et al. 2000; Loud et al. 2000).

10.3
Radionuclide Applications

10.3.1
Fibrinogen Uptake Counting

Fibrinogen labelled with ^{125}I was one of the earliest test agents applied to the diagnosis of DVT. It has been instrumental in providing a large amount of information concerning the epidemiology and natural history of DVT, and function as a means to monitor the efficacy of prophylactic treatment (Flanc et al. 1968; Negus et al. 1968; Kakkar et al. 1969). Labelled fibrinogen is incorporated into actively forming thrombi and the soft X-ray emissions of ^{125}I are detected and counted on the surface of the lower extremity with a scintillation probe 1 day after its intravenous administration. Readings are obtained at several constant points over the calves and thighs, and the percent uptake is related to the counts over the heart. Differences exceeding 20% between adjacent points, or the same level of the contralateral limb, which persist 24 h later are considered abnormal. ^{125}I-fibrinogen leg scanning is 95% sensitive to symptomatic calf DVT within 1 week of onset. In the thigh, sensitivity progressively decreases proximally because of the increase in background count from adjacent large arteries, the radioactivity in the urinary bladder and absorption of the low energy X-rays by overlying soft tissue (Kakkar et al. 1969; Harris et al. 1975). Point counting with ^{125}I-fibrinogen is less sensitive in asymptomatic calf DVT after major hip surgery. The reason for this may be due to the small size of the thrombi resulting from the prophylactic measures taken. It has been shown that sensitivities of 83%,

72% and 40% are obtained for large, medium and small thrombi, respectively (Paiement et al. 1988). False positive results can occur from the operative wound, hematoma, extensive joint or skin inflammation, varicose veins and radioactive urine on the lower extremities of incontinent patients. The 24- to 48-h delay in obtaining the results is a significant disadvantage of ^{125}I-fibrinogen method. A major hazard in the use of human fibrinogen is the theoretical risk of transmitting serum hepatitis and human immunodeficiency virus infections.

10.3.2
Radionuclide Venography

Radionuclide venography is a method that utilizes the gamma ray scintillation camera to visualize venous blood flow dynamically with serial 1- to 3-s images following an intravenous bolus injection of the test agent. This is supplemented by a delayed equilibrium static image at about 5 min post-dose and later as needed depending on the test agent used. The high photon flux necessary to obtain these images is achieved with the ^{99m}Tc radionuclide label. Historically, the most widely used radiopharmaceutical was ^{99m}Tc labelled albumin macroaggregates (MAA), although for the dynamic phase alone ^{99m}Tc as a pertechnetate is sufficient. These agents are injected into a foot vein with a tourniquet in place around the ankle to promote a more complete filling of the deep venous system of the calf. Simultaneous bilateral injections of the feet is an option.

The DVT criteria for MAA are entrapment of the particles by the thrombi, which appear as "hot spots", the presence of venous collaterals, or both (Fig. 10.1). The mean unweighted sensitivity and specificity for the technique are 92% and 78%, respectively, for the entire lower limb (Yao et al. 1973; Henkin et al. 1974; Johnson et al. 1974; Webber et al. 1974; Pollak et al. 1975; Vlahos et al. 1976; Ryo et al. 1977; Bentley and Kakkar 1979) (Fig. 10.1). These data were not analyzed with regard to calf and proximal DVT separately. The number of channels within the calf are poorly resolved by this technique. No figures are available on the efficacy of diagnosing calf DVT in the absence of proximal extension. An advantage of using ^{99m}Tc-MAA is the ability to obtain a perfusion lung scan with same dose after completion of the venography (Rosenthall 1971). Two main disadvantages are the requirement of a pedal injection, which is not always possible in the presence of edema, and a failure to visualize all the deep veins of the calf because the superficial veins are preferentially filled in the absence of a tourniquet above the ankles or when it is ineffectively applied. Under these circumstances the MAA will enter the proximal deep veins at the level of the knee through the saphenous veins and function to interrogate the iliofemoral veins. Previous DVT may cause venous collaterals and particle entrapment by inflammatory disease render false positive results. "Hot spots" may occur as a result of holdup behind valves and areas of blockage, and stasis in varicose veins or venous insufficiency. Exercising the limb will remove many of these accumulations. It has been suggested that a "hot spot" cannot be considered as a true thrombus entrapment unless it remains in the same location on two successive postexercise images (Hayt et al. 1977). The most reliable sign is the presence of venous collaterals.

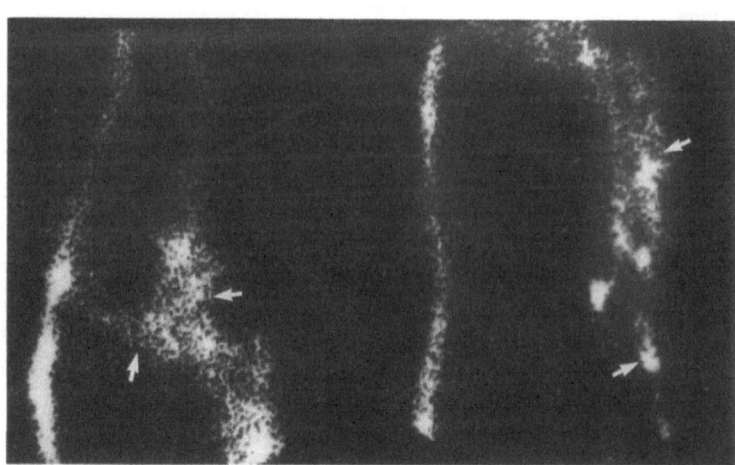

Fig. 10.1. ^{99m}Tc-macroaggregates of albumin (MAA) images. *Left*, bilateral pedal injections portray obstruction in the left iliofemoral region with collateral runoff (*arrows*). *Right*, postperfusion image of the thigh showing focal retentions suggestive of thrombi entrapment of MAA (*between arrows*)

10.3.3
Venous Blood Pool Imaging

This method utilizes ^{99m}Tc labelled red blood cells (RBC) as the test agent. It was developed during the 1970s to obviate the pedal injection that is necessary for radionuclide venography and to enhance visualization of the deep venous system of the calf (Beswick et al. 1979). The patient's own RBC are labelled with ^{99m}Tc in vitro and then reinjected into a peripheral arm vein. Gamma camera images of the lower extremities are procured soon afterward. Venous blood pool imaging is based on the premise that most of the blood volume of the limbs is contained within the venous system, therefore, images of the deep veins can be obtained without significant "cross talk" from neighboring arteries (Fig. 10.2). With DVT, a completely obstructed segment exhibits absent radioactivity. Incomplete obstruction may manifest a relative decrease in radioactivity and diameter, but it is less reliable. The presence of collateral vessels is strong evidence of venous compromise (Fig. 10.3). The reported mean unweighted sensitivity and specificity of ^{99m}Tc-RBC blood pool imaging for the entire limb are 93% and 87%, respectively (Beswick et al. 1979; Singer et al. 1984; Zorba et al. 1986; Robertson et al. 1994). In one prospective study of 110 pa-

tients, IPG and ^{99m}Tc-RBC blood pool scans were compared in patients with clinically suspected first DVT episodes (Leclerc et al. 1988). Patients with abnormal IPG underwent confirmatory contrast venography. Those with normal IPG findings had the test repeated every other day for up to about 14 days. Anticoagulant treatment was withheld in patients with normal IPG results regardless of the ^{99m}Tc-RBC diagnosis. All patients with initial abnormal IPG had the ^{99m}Tc-RBC study on the same day, whereas those with an initial normal IPG received the ^{99mTc}-RBC test sometime during the 14-day follow-up. A scan was considered abnormal if the concentration was decreased by 50% or more compared to the same level of the contralateral limb. "Possibly abnormal" was assigned to concentrations that were decreased by less than 50% in order to accommodate the presence of nonocclusive disease and some cross talk from adjacent arteries. ^{99m}Tc-RBC sensitivity and specificity for the entire lower extremity was 90% and 56%, respectively, when abnormal and possibly abnormal categories were included. For proximal DVT, the sensitivity and specificity were 68% and 88%, respectively. The conclusions from this prospective study were: (1) a normal ^{99m}Tc-RBC scan excludes DVT because of the high sensitivity and negative predictive value; (2) an abnormal result, particularly in the calf region, should

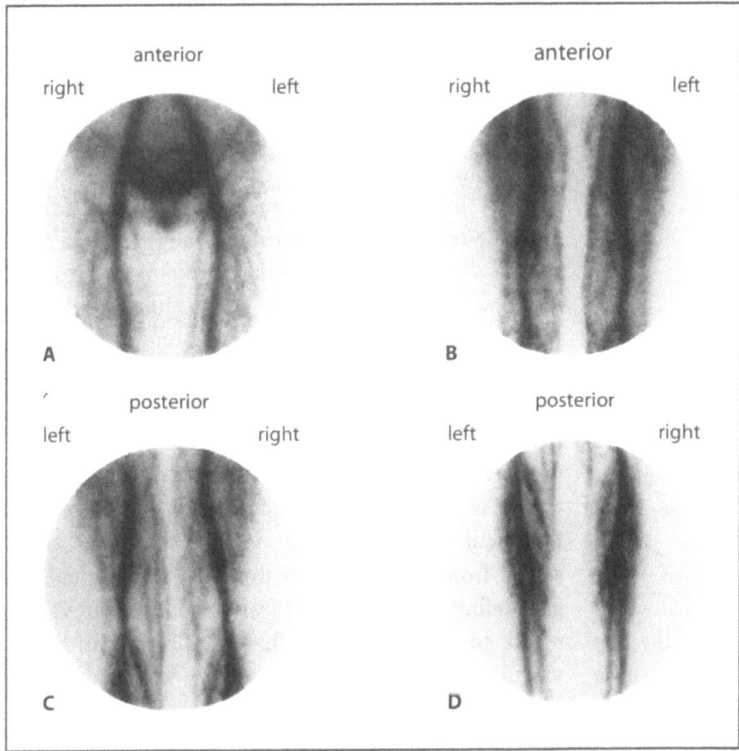

Fig. 10.2 A–D. Normal ^{99m}Tc labelled red blood cell pool (RBC) images of the lower extremities. **A** Anterior thigh. **B** Anterior lower thigh and knee. **C** Posterior lower thigh and knee. **D** Posterior calves

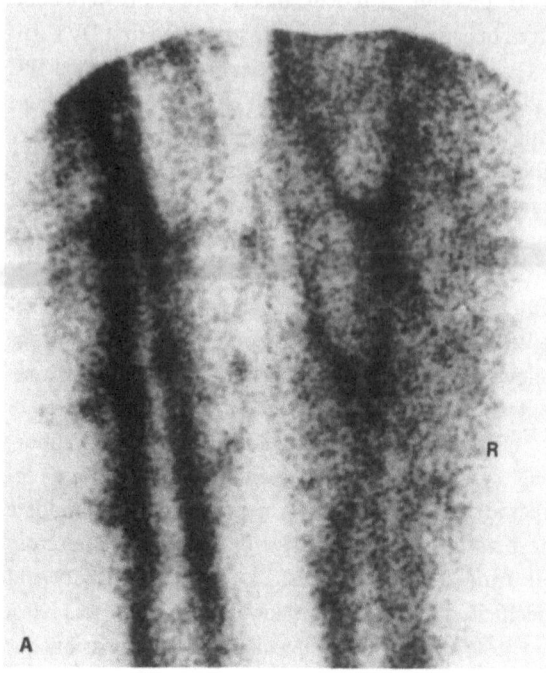

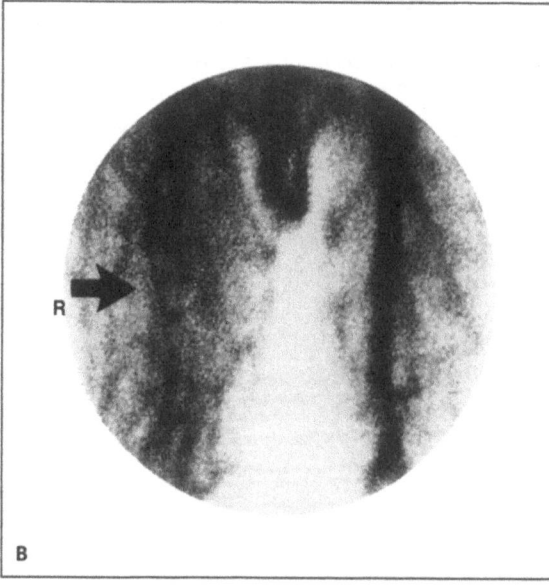

Fig. 10.3 A, B. Extensive right lower extremity deep vein thrombosis. A ^{99m}Tc-RBC images of the posterior calves. Absent posterior tibial and peroneal veins and increased collateral circulation in the right calf. B Anterior thighs. There is extension into the femoral vein of the right thigh (*arrow*)

be confirmed by another method in view of the low specificity and low positive predictive value; (3) false negative results may be obtained in isolated proximal DVT without calf involvement.

A major limitation of radionuclide dynamic and static venous blood pool imaging, contrast venography, compression ultrasound, CT, MRI and plethysmography is that the information provided is related to venous morphology only. They cannot reliably distinguish an acute thrombophlebitis, i.e. within 7 days of onset, which is liable to extend and embolize, from aged, organizing thrombi or chronic venous changes following a remote DVT. Much research has been devoted to the development of radiopharmaceuticals that bind specifically to the various components of the thrombus, particularly acute active thrombi which may require immediate anticoagulant therapy. Many of the test agents produced were found to be efficacious in animal models and some of them have been subjected to clinical trials.

10.4
Radionuclide Thrombus Imaging

10.4.1
Fibrinogen

The labelling of fibrinogen with gamma emitters such as ^{131}I, ^{123}I, ^{67}Ga, ^{111}In and ^{99m}Tc was introduced to permit thrombus imaging and overcome the limitations inherent in ^{125}I counting (Charkes et al. 1974; DeNardo et al. 1977; Jeghers et al. 1978; Layne et al. 1982; Ohmomo et al. 1982; Yamamoto et al. 1988). Unlike ^{125}I-fibrinogen counting which has a relatively poorer accuracy in the detection of iliofemoral venous thrombi, imaging enables visualization of thrombi in both the proximal and distal calf veins. Most of the injected labelled fibrinogen binds nonspecifically to freshly growing thrombi by virtue of its conversion to fibrin, but anticoagulation therapy may interfere with its accretion. In one report, however, accretion in the presence of anticoagulation was noted with the high flux ^{99m}Tc label rather than ^{131}I. Persistent uptake was associated with a significant recurrence of clinical disease and/or pulmonary emboli, and it was used as an indication for more vigorous treatment (DeNardo et al. 1977). Fibrinogen clears slowly from the blood and this high background delays definitive results by at least 24 h, including some of the scans performed with ^{99m}Tc-fibrinogen. Extravascular clotting and inflammatory conditions promote fibrinogen deposition. There are no large quantitative comparisons with contrast venography or duplex ultrasound reported in the literature, but some discordance is to be expected as the two are primarily anatomical methods of thrombus detection, whereas fibrinogen is functional. Fibrinogen has been superseded by labelled antibody and peptide imagining agents.

10.4.2
Platelets

Platelets labelled with ^{111}In-oxine demonstrate a high concentration in fresh thrombi which exceeds that of fibrinogen (Knight et al. 1978). Like fibrinogen, blood clearance of the ^{111}In-platelets is slow and definitive results may be delayed by 24 h or more. Anticoagulation therapy also interferes with its deposition at the site of thrombosis (Moser et al. 1980; Ezekowitz et al. 1986).

[111]In-platelet scintigraphy was studied in two groups of patients following major orthopedic pelvic surgery: asymptomatic patients who were injected 1 day after the surgery and screened for 5 days with the same administered dose; symptomatic patients who were injected a day before or 3 days after contrast venography. Sensitivity and specificity for DVT of the entire lower extremity were 93% and 97%, respectively, in the asymptomatic group, but only 42% and 67% in the symptomatic group. The lower sensitivity in the symptomatic group was attributed to therapy with anticoagulants (Ezekowitz et al. 1986). A follow-up report, comparing contrast venography and ^{111}In-platelet scintigraphy in 65 patients suspect for DVT but not on anticoagulation therapy, stated that the sensitivity and specificity were 42% and 96%, respectively. It was speculated that the low sensitivity was related to inactivity of the thrombus, and the test would only be useful in patients whose symptoms are of recent onset. From the foregoing results, it would appear that anticoagulation and thrombus age weigh against ^{111}In-platelet deposition. Interobserver agreement for interpretation of the images in this study was 92% for contrast venography and 79% for scintigraphy (Farlow et al. 1989). The efficacy of screening asymptomatic patients following abdominal and pelvic surgery with ^{111}In-platelet imaging has been confirmed by others (Clarke-Pearson et al. 1988). Platelet uptake was shown to be associated with embolic events in chronic left ventricular thrombi diagnosed by echo cardiography. Embolism occurred in 7 of 34 (21%) patients with a positive uptake in the ventricular thrombus compared to 2 of 69 (3%) patients without visual platelet uptake

(Stratton and Ritchie 1990). A comparison of contrast venography and ^{99m}Tc-labelled autologous platelets was made in 33 patients clinically suspected of having DVT. Thirteen of the patients were on heparin treatment at the time of scintigraphy. Fifteen of 23 patients with DVT were positive by venography and scintigraphy; 5 of the 8 patients with false negative ^{99m}Tc-platelet studies were on heparin. Nine out of 12 patients without anticoagulation had positive platelet images compared to 2 of 11 patients on heparin therapy, which is a statistically significant difference. All patients who were negative by venography were also negative by ^{99m}Tc-platelet imaging. Sensitivity and specificity of ^{99m}Tc-platelet scintigraphy were 65% and 100%, respectively, in all patients, and 83% and 100% in patients without anticoagulation (Honkanen et al. 1992).

A deterrent to a more widespread application of radiolabelled platelets is its preparation which requires 2–3 h and expertise in handling the extensive in vitro separations necessary for efficient labelling.

10.4.3
Antifibrin Monoclonal Antibodies

Monoclonal antibodies can be fabricated to recognize specific components of thrombin and are highly specific for these epitopes when they are exposed during various stages of thrombus formation and dissolution. Early animal studies demonstrated that whole ^{131}I-antibodies (T2G1s-IgG) directed to fibrin were capable of detecting and imaging venous thrombi at various ages (Rosebrough et al. 1985, 1987, 1988). Results were not definitive until 24 h after the injection of the whole antibody owing to the relatively slow clearance and poorer thrombus penetration of the large molecule. To be clinically helpful in patient management, a diagnostic test of DVT should be clearly evident within several hours. This has been accomplished by using smaller fragments of the antifibrin antibody, the Fab fragments, and ^{99m}Tc as the label (Knight et al. 1989). These smaller molecules render accurate diagnostic images within 1–6 h after intravenous administration. The usual technique is to obtain equilibrium blood pool images soon after injection and again at 2–6 h to observe retention of the test agent within the thrombus. Several monoclonal antifibrin antibodies have been developed and tested in animal models, but few have come to clinical trial. These include T2G1s and 59D8 which recognize the same epitope on the beta chain of human fibrin and

Table 10.1. Detection of venous thrombosis in humans using antifibrin Fab

Reference study	Number	Epitope	Label	% Sensitivity (N)	% Specificity (N)
Jung et al (1989)	52	59D8	In-111	84 (26/31)	81 (17/21)
Lusiani et al (1989)	30	59D8	In-111	78 (14/18)	92 (11/12)
De Faucal et al (1991)	44	59D8	In-111	85 (29/34)	100 (10/10)
Alavi et al (1990)	33	59D8	In-111	97 (28/29)	75 (3/4)
Schaible et al (1992)	256	T2g1s	Tc-99m	79 (77/97)	91(145/159)
Overall	415			83 (174/209)	90 (186/206)

anti-D antibodies which bind in the region of the covalent crosslink between paired D domains. The latter has the potential of disclosing older thrombi as there are more binding sites available for this antibody on each fibrin strand.

Table 10.1 details the performance of the antifibrin Fab studies from five centers. Most of the results with the ^{111}In label were obtained from images taken between 2 and 4 h; later images up to 24 h did not improve the diagnostic accuracy. Overall sensitivity and specificity were 83% and 90%, respectively. An analysis of these studies indicated that sensitivity was higher in the more acute thrombi. When patient symptoms were less than 10 days sensitivity increased from 84% to 92%. Location of the DVT also influences efficacy. Sensitivity and specificity were, respectively, 92% and 96% in the calf; 82% and 83% in the popliteal region; 63% and 96% in the thigh; 18% and 100% in the pelvis (Jung et al. 1989). Similar results were reported by others (Lusiani et al. 1989; De Faucal et al. 1991). It is speculated that the larger thrombi in the proximal veins slow penetration of the antibody into the central portion of the thrombi, and coupled with a rapidly clearing blood background, the uptake and sensitivity are reduced.

The reports regarding the effect of heparin on sensitivity are conflicting. On theoretical consideration, heparin would be expected to slow the rate of new fibrin formation and reduce the number of exposed binding sites for 59D8 and T2G1s antibodies. In one study, using contrast venography for confirmation, all five thrombi in the thigh were identified in 19 nonheparinized patients compared to only seven of 13 proximal thrombi in heparinized patients (Alavi et al. 1990). Another investigation, in which 79% of the patients received heparin, no change in the overall sensitivity was observed, but there was a tendency to a lower target-to-background ratio in the treated patients (Jung et al. 1989).

There are two reports on the clinical use of ^{99m}Tc-labelled antifibrin-DD Fab' monoclonal antibody fragments which specifically target human cross-linked fibrin with high affinity. All the patients had established thrombosis confirmed by contrast venography, duplex ultrasound, or both. In one group there were 20 patients with DVT of the lower extremities, of whom 19 were anticoagulated with heparin. Positive uptakes were registered in all 20 patients for 100% sensitivity, without inference from heparin (Bautovich et al. 1994). The other group consisted of 16 patients with lower limb DVT in three and superficial venous thrombosis in 13 patients who were slated for saphenous vein stripping the day after the radionuclide study. All 25 sites of thrombosis, which were of variable age, that were detected with duplex ultrasound were also imaged with ^{99m}Tc-antifibrin-DD, and confirmed in vitro postoperatively (Ciavolella et al. 1999).

10.4.4
Antiplatelet Monoclonal Antibodies

Antibodies against two major platelets sites have been developed, the fibrinogen receptor glycoprotein IIa/IIIb and GMP-140. GMP-140 is a protein derived from the platelet granule and is expressed on the surface of thrombus activated platelets and not on the resting or circulating platelet, whereas glycoprotein IIa/IIIb is found on both resting and activated platelets. There are multiple epitopes related to the IIa/IIIb complex for which a number of antibodies have been raised (Coller et al. 1983; Stuttle et al. 1988; Taub et al. 1989; Thakur et al. 1987). However, since resting circulating platelets concentrate these antibodies in vivo the blood background activity remains high for several days with the ^{111}In label, much the same as occurs with in vitro ^{111}In-oxine labelled platelets. Animal studies with whole or Fab antibodies have shown that these antibodies are capable of excellent imaging of both venous and arterial thrombi, and binding is not influenced by heparin or aspirin (Oster et al. 1985; Peters et al. 1986; Som et al. 1986). The number of reports on human trials are few and small in size, but they render some insight into the potential of these antibodies. Ten patients with suspected thrombosis were investigated with ^{111}In-F(ab')$_2$ fragments of P252, a monoclonal antibody which recognizes an epitope on the platelet membrane glycoprotein IIa/IIIb complex. Six patients exhibited localization: three lower limb DVT, one aortofemoral graft, one aortic aneurysm and one carotid stenosis (Stuttle et al. 1988). Sequential imaging with ^{111}In-labelled Fab' fragment of the monoclonal P256 was employed in 11 patients undergoing total hip arthroplasty to monitor the development of postsurgical thrombosis. There were six patients with focal uptakes in the lower extremities, and these were confirmed as thrombi in three patients who had contrast venography. Two of the six patients also demonstrated localization of the P256 Fab' antibody in the lungs and in one of these patients it was confirmed as an embolus with a high probability ventilation/perfusion lung scan (Stuttle et al. 1989). A similar occurrence was published as a case history (King et al. 1992). A preliminary trial of ^{111}In-labelled P256 Fab' imaging in 17 patients with acute lower extremity arterial occlusions indicated that it may be helpful in selecting patients for thrombolysis (Berridge et al. 1991).

The advent of small molecular weight peptide radiopharmaceuticals has largely superseded the developmental efforts in antibody components.

10.4.5
Fibrin Fragment E$_1$

Fragment E$_1$ is a plasmin degradation product from cross-linked fibrin. It is a 60-kDa fragment that binds specifically to fibrin polymers in mid-strand (Knight et al. 1983). An important attribute of this molecule is that it yields diagnostic images in 20–60 min and is not appreciably influenced by heparin. Using ^{123}I as a label for fragment E$_1$, a limited clinical trial demonstrated that all five patients with established lower extremity DVT had focally positive scans, three patients without DVT were normal, and two patients with superficial phlebitis exhibited a diffuse uptake pattern along the veins (Knight et al. 1985). A technique for labelling fragment E$_1$ with ^{99m}Tc has been developed (Knight et al. 1992).

10.4.6
Tissue Plasminogen Activator

Tissue plasminogen activator (t-PA) can be produced in large quantities by means of recombinant DNA technology. It binds to fibrin and is rapidly cleared from the blood. Since t-PA can also lyse the fibrin to which it is bound, it may hasten the washout of a positive scan. There is also a catalytic site on t-PA which binds plasminogen and circulating inhibitors that interfere with fibrin binding. Methods to block the catalytic site without inactivating the fibrin binding process have been developed. A clinical trial involving 79 patients with suspected DVT was undertaken using such a modified recombinant tissue plasminogen activator (rt-PA) labelled with ^{99m}Tc. Compared to contrast venography, the reference standard, sensitivity and specificity were 93% and 92%, respectively, for the proximal veins, and 80% and 93%, respectively, for the calf veins (Butler et al. 1996). An investigation of 53 consecutive asymptomatic postarthroplasty patients with the same test agent, ^{99m}Tc-rt-PA, was reported. Compared to contrast venography, there were 15 thrombosed segments which were positive in 14 by ^{99m}Tc-rt-PA. Of the 69 segments not thrombosed by venography, 63 had negative scans for a sensitivity of 93% and a specificity of 91% (Butler et al. 1997).

10.4.7
Peptides

Recent interest in designing radiopharmaceuticals for specific thrombus imaging has been focussed on synthetic peptides. The low molecular weight and small size of the peptides enables rapid background clearance through renal excretion rather than liver concentration and improved penetration of the thrombus. Most of the peptides produced thus far have a high affinity to the glycoprotein IIa/IIIb receptor expressed on activated platelets and can be labelled with ^{99m}Tc. Of the several agents developed and tested in animal models only one has been extensively applied to human trials; the synthetic peptide P280 labelled with ^{99m}Tc. Early in its investigation the peptide was found to bind to acute thrombi and not chronic thrombi which lack activated platelets. In a study consisting of 15 patients with proved DVT by contrast venography, all demonstrated abnormal uptake of ^{99m}Tc-P280 except for four cases of DVT wherein the onset of symptomatology preceded the test by more than 40 days (Lastoria et al. 1995). This also suggests that ^{99m}Tc-P280 may be helpful in monitoring active venous thrombosis as a guide to management.

The results of ^{99m}Tc-P280 (presently called ^{99m}Tc-apticide) imaging of 243 patients derived from a multicenter study has recently been published (Taillefer et al. 2000b). The criteria for selection and performance were onset of signs and symptoms of acute DVT, or a surgical procedure associated with a high risk of DVT, within 10 days of diagnostic imaging, and contrast venography and ^{99m}Tc-apticide scintigraphy in all patients within 3 days of each other. Planar scintigrams were taken at 10 min, 1 h and 2-3 h postdose. Of the 243 patients, 150 (61.7%) were receiving heparin. Overall, masked readings of the ^{99m}Tc-apticide images and contrast venograms yielded sensitivity, specificity, and agreement rates of 73.4%, 67.5% and 69.1%, respectively. In order to preclude venograms with residual anatomic abnormalities from previous insults which could confound interpretation, a subset of 63 patients was chosen with first time DVT, onset of symptomatology 3 days or less, and no history of pulmonary embolism. Analysis of this group yielded sensitivity, specificity and agreement rates of 90.6%, 83.9% and 87.3%, respectively. This implies that contrast venography, an anatomic study, cannot consistently distinguish acute from chronic disease. It follows that ^{99m}Tc-apticide imaging, a functional study, may be falsely positive because the thrombi were too small to produce a visible filling defect on the venogram, and a false negative peptide result may in fact be a false positive reading of the contrast venogram due to previous disease. Another factor was the considerable interobserver variability in the interpretation of the venograms. Heparin had no appreciable influence in the diagnosis of acute DVT by ^{99m}Tc-apticide imaging. The precise role for peptide scintigraphy in the context of other modalities, availability and cost has not yet been fully defined.

Fibronectin is a multifocal glycoprotein that is present in blood and extracellular matrix. It contains several domains with different functional roles including binding to cells, heparin, DNA, collagen and gelatin, fibrin and fibrinogen (Ruoslahti et al. 1981;

Mosher 1990; Ezov et al. 1997). The fibrin binding domain (FBD) possesses both a fibrin binding site and a site for covalent crosslinking to fibrin. Results of animal studies with the human fibronectin molecule labelled with either ^{131}I or ^{111}In were somewhat skeptical about its potential to disclose thrombi and pulmonary emboli in patients (Uehara et al. 1988; Zoghbi et al. 1988). This was much improved by using recombinant FBD as the test agent instead of fibronectin. ^{111}In-FBD was used in a pilot study to investigate its clinical potential in 62 patients, 30 of whom were suspected of lower extremity DVT and 32 were controls. All 30 patients had either impedance plethysmography, duplex ultrasound, contrast venography, or various combinations of the three. The 32 controls were drawn from patients admitted to hospital with infection, rheumatoid arthritis, malignant disease and lung transplant complication. Scintigraphy was performed 18-24 h postdose, and 26 of the 30 patients were on heparin at the time. There were 22 agreements (73%) in the 30 patients between ^{111}In-FBD and one or more of the other modalities. Three of the patients with negative scan results had thrombi that were more than 1 week old, and this may be a limitation of the technique. In three other patients who had negative scan results, but positive

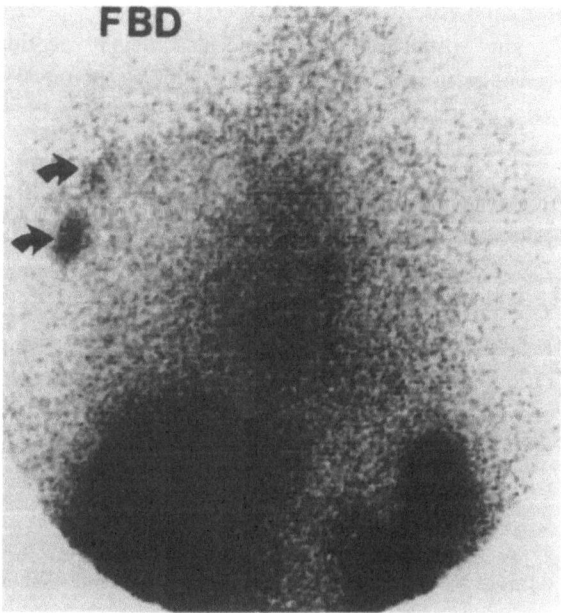

Fig. 10.4. ^{111}In-labelled fibrin binding domain image showing focal uptake in the right axillary vein caused by a recently removed in-dwelling catheter (*arrows*)

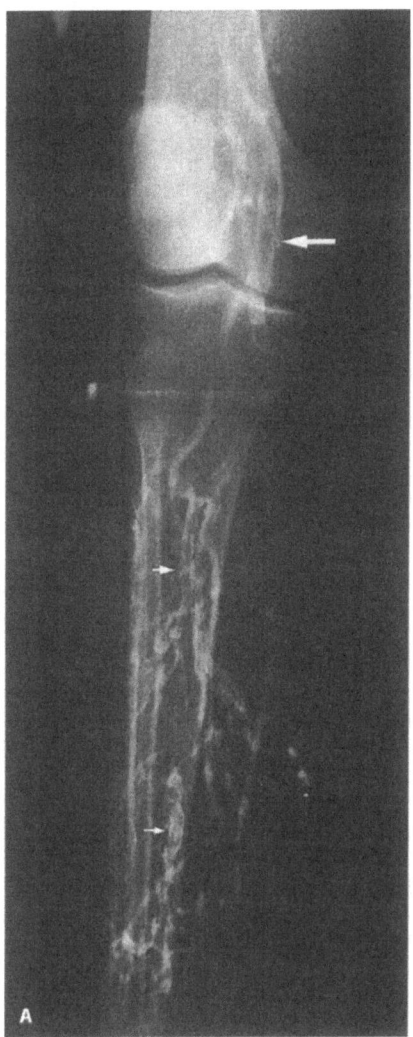

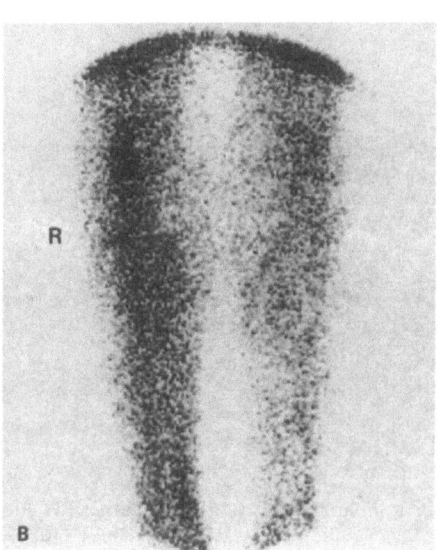

Fig. 10.5 A, B. Deep vein thrombosis in the right lower extremity. A Contrast venogram showing clots in the calf veins with extension into the popliteal and superficial veins (*arrows*). B Abnormal accretion of ^{111}In-FBD in the calf extending proximally, consistent with the presence of fresh thrombi

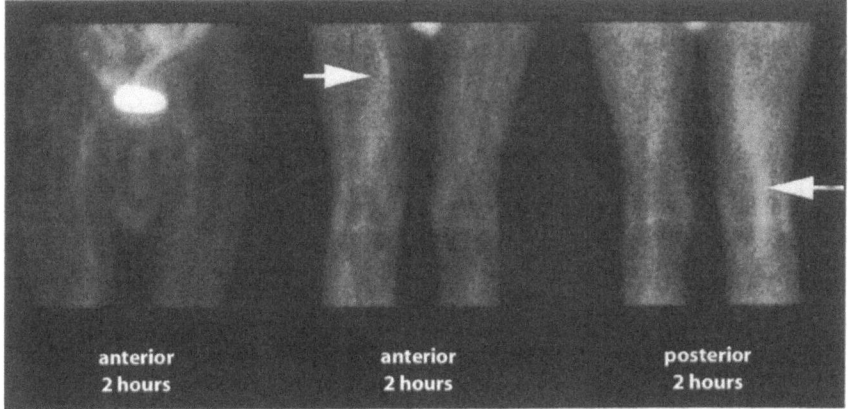

Fig. 10.6. ^{99m}Tc-FBD images of the lower extremities in a patient with acute deep vein thrombosis. There is increased concentration in the right popliteal region extending into the femoral vein (*arrows*)

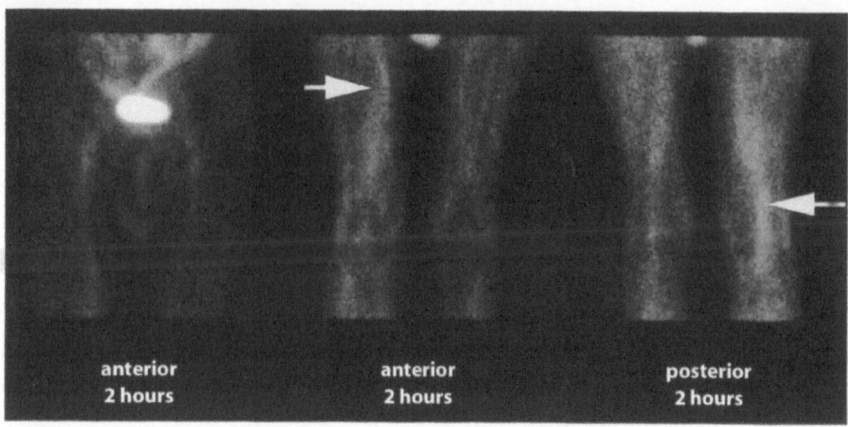

Fig. 10.7. ^{99m}Tc-FBD images of the lower extremities demonstrating accretion in calf with proximal extension into the thigh in a patient with acute deep vein thrombosis

alternative studies, there was a history of previous episodes of DVT and perhaps the latter were, indeed, falsely positive for acute DVT. Normal ^{111}In-FBD images were obtained in the lower extremities of the control patients, except in two patients who exhibited accretion at the sites of insulin injections and focal vessel wall injury (Rosenthall and Leclerc 1995) (Fig. 10.4). A phase II multicenter study of ^{99m}Tc-FBD has been reported in 41 patients with DVT in the lower extremities established by compression ultrasonography and a positive d-dimer test (Taillefer et al. 2000a; Taillefer 2001). Patients with a history of previous DVT or pulmonary embolism were excluded, but heparinization was acceptable. Overall diagnostic sensitivity was about 80% when equivocal scan interpretations were considered normal, but 93% when counted as abnormal. Sensitivity was higher for proximal DVT than distal DVT, 92% versus 68%, respectively (Figs. 10.5, 10.6, 10.7).

References

Alavi A, Palevsky HI, Gupta N et al (1990) Radiolabeled antifibrin antibody in the detection of venous thrombosis: preliminary results. Radiology 175:79–85

Anand S, Wells PS, Hunt D et al (1998) Does this patient have deep vein thrombosis? JAMA 279:1094–1099

Anderson FA, Wheeler HB, Goldberg RJ et al (1991) A population-based perspective on the hospital incidence and case-fatality rates of deep vein thrombosis and pulmonary embolism. The Worcester DVT study. Arch Intern Med 151:933–938

Anonymous (1975) Prevention of fatal postoperative pulmonary embolism by low doses of heparin. An international multicenter trial. Lancet 2:45–51

Barnes RW, Russell HE, Wu KK et al (1976) Accuracy of Doppler ultrasound in clinically suspected venous thrombosis of the calf. Surg Gynec Obstet 143:425–428

Bautovich G, Angelides S, Lee FT et al (1994) Detection of deep venous thrombi and pulmonary embolus with technetium-99m-DD-3B6/22 anti-fibrin monoclonal antibody Fab' fragment. J Nucl Med 35:195–202

Bentley PG, Kakkar VV (1979) Radionuclide venography for the demonstration of the proximal deep venous system. Br J Surg 66:687–690

Berridge DC, Perkins AC, Frier M et al (1991) Detection and characterization of arterial thromboses using a platelet-specific monoclonal antibody (P256 Fab'). Br J Surg 78:1130–1133

Beswick W, Chmiel R, Booth R et al (1979) Detection of deep venous thrombosis by scanning of 99m-technetium-labelled red-cell venous blood pool. Br J Med 1:82–84

Butler SP, Boyd SJ, Parkes SL et al (1996) Technetium-99m-modified recombinant tissue plasminogen activator to detect deep venous thrombosis. J Nucl Med 37:744–748

Butler SP, Rahman T, Boyd SJ et al (1997) Detection of postoperative deep-venous thrombosis using technetium-99m-labeled tissue plasminogen activator. J Nucl Med 38:219–223

Cham MD, Yankelevitz DF, Shaham D et al (2000) Deep venous thrombosis: detection by using indirect CT venography. The Pulmonary Angiography-Indirect CT Venography Cooperative Group. Radiology 216:744–751

Charkes ND, Dugan MA, Maier WP et al (1974) Scintigraphic detection of deep-vein thrombosis with ^{131}I-fibrinogen. J Nucl Med 15:1163–1166

Ciavolella M, Tavolaro R, Di Loreto M et al (1999) Immunoscintigraphy of venous thrombi: clinical effectiveness of a new antifibrin D-dimer monoclonal antibody. Angiology 50:103–109

Clarke-Pearson DL, Synan IS, Coleman RE et al (1984) The natural history of postoperative venous thromboemboli in gynecologic oncology: a prospective study in 382 patients. Am J Obstet Gynecol 148:1015–1054

Clarke-Pearson DL, Coleman RE, Siegal R et al (1988) Indium-111 platelet imaging for the detection of deep vein thrombosis and pulmonary embolism in patients without symptoms after surgery. Surgery 98:98–104

Cohan RH, Dunnick NR (1987) Intravascular contrast media: adverse reactions. AJR 149:665–670

Coller BS, Peerschke EI, Scudder LE et al (1983) A murine monoclonal antibody that completely blocks the binding of fibrinogen to platelets produces a thrombasthenic-like state in normal platelets and binds to glycoprotein IIb and/or IIIa. J Clin Invest 72:325–338

Cronan JJ (1993) Venous thromboembolic disease: the role of ultrasound. Radiology 186:619–650

De Faucal P, Peltier P, Planchon B et al (1991) Evaluation of indium-111-labelled antifibrin monoclonal antibody for the diagnosis of venous thrombotic disease. J Nucl Med 32:785–791

DeNardo SJ, DeNardo GL (1977) Iodine-123 fibrinogen scintigraphy. Semin Nucl Med 7:245–251

Dorfman GS, Cronan JJ, Tuypper TB et al (1987) Occult pulmonary emboli: a common occurrence in deep vein thrombosis. AJR 148:263–266

Douss TW (1976) The clinical significance of venous thrombosis of the calf. Br J Surg 63:377–378

Ezekowitz MD, Pope CF, Sostman HD et al (1986) Indium-111 platelet scintigraphy for the diagnosis of acute venous thrombosis. Circulation 73:668–674

Ezov N, Nimrod A, Parizada B et al (1997) Recombinant polypeptides derived from fibrin binding domain of fibronectin are potential agents for the imaging of blood clots. Thromb Haemost 77:796–803

Farlow DC, Ezekowitz MD, Rao SR, Martinez C, Denny DF et al (1989) Early image acquisition after administration of indium-111 platelets in clinically suspected deep vein thrombosis. Am J Cardiol 64:363–368

Flanc C, Kakkar VV, Clarke MB (1968) The detection of venous thrombosis of the legs using ^{125}I-labelled fibrinogen. Br J Surg 55:742–747

Gillium RF (1987) Pulmonary embolism and thrombophlebitis in the United States 1970–1985. Am Heart J 114:1262–1264

Hansson PO, Sorbo J, Eriksson H (2000) Recurrent venous thromboembolism after deep vein thrombosis: incidence and risk factors. Arch Intern Med 160:769–774

Hayt DB, Blatt CJ, Freeman LM (1977) Radionuclide venography: its place as a modality for the investigation of thromboembolic phenomena. Semin Nucl Med 7:263–281

Harris WH, Salzman E, Athanasoulis C et al (1975) Comparison of the ^{125}I-fibrinogen count scanning with phlebography for detection of venous thrombi after elective hip surgery. N Engl J Med 292:665–667

Henkin RE, Yao JS, Quinn JL et al (1974) Radionuclide venography (RVN) in lower extremity venous thrombosis. J Nucl Med 15:171–175

Honkanen T, Jauhola S, Karppinen K et al (1992) Venous thrombosis: a controlled study on the performance of scintigraphy with 99mTc-HMPAO-labelled platelets versus venography. Nucl Med Commun 13:88–94

Huisman MV, Buller HR, Cate JW ten et al (1986) Serial impedance plethysmography for suspected deep venous thrombosis in outpatients. N Engl J Med 314:823–828

Hull R, Taylor DW, Hirsh J et al (1978) Impedance plethysmography: the relationship between venous filling and sensitivity and specificity for proximal vein thrombosis. Circulation 58:898–902

Hull RD, Delmore T, Genton E et al (1979) Warfarin sodium versus low dose heparin in the long-term treatment of venous thrombosis. N Engl J Med 301:855–858

Hull RD, Hirsh J, Sackett DL et al (1981) Clinical validity of a negative venogram in patients with clinically suspected venous thrombosis. Circulation 64:622–625

Hull RD, Raskob GE, Hirsh J et al (1986) Continuous intravenous heparin compared with intermittent subcutaneous heparin in the initial treatment of proximal-vein thrombosis. N Engl J Med 315:1109–1114

Illescas FF, Leclerc J, Rosenthall L et al (1990) Interobserver variability in the interpretation of contrast venography, technetium-99m red blood cell venography and impedance plethysmography for deep vein thrombosis. J Can Assoc Radiol 41:264–269

Jeghers O, Abramivici J, Jonckheer M et al (1978) A chemical method for labeling of fibrinogen with ^{99m}Tc. Eur J Nucl Med 3:95–100

Johnson WC, Patten DH, Widrich WC et al (1974) Technetium-99m isotope venography. Am J Surg 127:424–428

Jung M, Kletter K, Dudczak R et al (1989) Deep vein thrombosis: scintigraphic diagnosis with In-111 labelled monoclonal antifibrin antibodies. Radiology 173:469–476

Kakkar VV (1977) Fibrinogen uptake test for detection of deep vein thrombosis: Review of current practice. Semin Nucl Med 7:229–244

Kakkar VV, Howe CT, Flanc C et al (1969) Natural history of postoperative deep vein thrombosis. Lancet 2:230–233

Kalebo P, Anthymyr BA, Erichsson BI et al (1990) Phlebographic findings in venous thrombosis following total hip replacement. Acta Radiol 31:259–263

Kearon C, Julian JA, Newman TE et al (1998) Noninvasive diagnosis of deep venous thrombosis. Ann Intern Med 128:663–677

Kemp GK, Wojcik D, Hoehn S et al (2000) Thromboembolic disease: comparison of combined CT pulmonary angiography and venography with bilateral leg sonography. AJR 175:997–1001

Kierkegaad A (1980) Incidence of acute deep vein thrombosis in two districts. A phlebographic study. Acta Chir Scand 146:267–269

King AD, Bell SD, Stuttle AW et al (1992) Platelet imaging of thromboembolism. Natural history of postoperative deep venous thrombosis and pulmonary embolism illustrated using the 111In-labelled platelet-specific monoclonal antibody, P256. Chest 101:1597–1600

Knight LC, Primeau JL, Siegel BA et al (1978) Comparison of In-111 labelled platelets and iodinated fibrinogen for the detection of deep vein thrombosis. J Nucl Med 19:391–394

Knight LC, Mauer AH, Robbins PS et al (1985) Fragment E1 labeled I-123 in the detection of venous thrombosis. Radiology 156:509–514

Knight LC, Mauer AH, Ammar IA et al (1989) Tc-99m antifibrin Fab' fragments for imaging venous thrombi: evaluation in a canine model. Radiology 173:163–169

Knight LC, Abrams MJ, Schwartz DA et al (1992) Preparation and preliminary evaluation of technetium-99m-labeled fragment E1 for thrombus imaging. J Nucl Med 33:710–715

Koopman MM, Buller HR, ten Cate JW (1995) Diagnosis of recurrent deep vein thrombosis. Haemostasis 25:49–57

Lanigan DP, Goitre JJ, Burnham ST et al (1979) Vascular-laboratory diagnosis of clinically suspected acute deep vein thrombosis. Lancet 2:331–334

Lastoria S, Vergara E, Varella P et al (1995) Imaging of thromboembolism by scintigraphy with 99m-technetium-labelled synthetic peptide P280. Radiol Med (Torino) 90:812–819

Layne WW, Hnatowich DJ, Doherty PW et al (1982) Evaluation of the variability of In-111 labelled DTPA coupled to fibrinogen. J Nucl Med 23:627–630

Leclerc JR, Rosenthall L, Wolfson T et al (1988) Technetium-99m red blood cell venography in patients with suspected deep vein thrombosis: a prospective study. J Nucl Med 29:1498–1506

Loud PA, Katz DS, Klippenstein DL et al (2000) Combined CT venography and pulmonary angiography in suspected thromboembolic disease: diagnostic accuracy for deep venous evaluation. AJR 174:61–65

Lusiani L, Zanco P, Visona A et al (1989) Immunoscintigraphic detection of venous thrombosis of the lower extremities by means of human antifibrin monoclonal antibodies labelled with [111]In. Angiology 40:671–677

Meadway J, Nicolaides AN, Walker CJ et al (1975) Ultrasound in diagnosis of clinically suspected deep vein thrombosis. Br Med J 4:552–554

Meignan M, Rosso J, Gauthier H et al (2000) Systematic lung scans reveal a high frequency of silent pulmonary embolism in patients with proximal dep venous thrombosis. Arch Intern Med 160:195–164

Moser KM, Spragg RG, Bender F et al (1980) Study of factors that may condition scintigraphic detection of venous thrombi and primary pulmonary emboli with indium-111 labelled platelets. J Nucl Med 21:1051–1058

Moser KM, LeMoine JR (1981) Is embolic risk conditioned by local deep vein thrombosis? Ann Intern Med 94:439–444

Mosher DF (1990) Fibronectin. Prog Hemost Thromb 5:111–116

Negus D, Pinto DJ, Le Quesne LP et al (1968) ^{125}I-labelled fibrinogen in the diagnosis of deep vein thrombosis and its correlation with phlebography. Br J Surg 55:835–839

O'Donnell T, Abbott W, Athanasoulis C et al (1980) Diagnosis of deep vein thrombosis in the outpatient by venography. Surg Gynecol Obstet 150:69–74

Ohmomo Y, Yokoyama A, Suzuki J et al (1982) ^{67}Ga-labelled human fibrinogen: a new promising thrombus imaging agent. Eur J Nucl Med 7:458–461

Oster Z, Srivastava S, Som P et al (1985) Thrombus radioimmunoscintigraphy: an approach using monoclonal antiplatelet antibody. Proc Natl Acad Sci USA 82:3465–3468

Ouriel K, Whitehouse WM, Zarins CK (1984) Combined use of Doppler ultrasound and phlebography in suspected deep venous thrombosis. Surg Gynecol Obstet 159:242–246

Paiement G, Wessinger SJ, Waltman AC et al (1988) Surveillance of deep vein thrombosis in asymptomatic total hip replacement patients. Impedance phlebography and fibri-

nogen scanning versus roentgenographic phlebography Am J Surg 155:400–404

Peters SH, Schulman ES, Schleimer RP et al (1982) Home diagnosis of deep venous thrombosis with impedance plethysmography. Thromb Haemocyte 48:297–300

Peters A, Lavender J, Needham S et al (1986) Imaging thrombus with radioactive monoclonal antibody to platelets. Br Med J 293:1525–1527

Pollak EW, Webber MM, Victery W et al (1975) Radioisotope detection of venous thrombosis. Venous scan vs fibrinogen uptake test. Arch Surg 110:613–616

Prandoni D, Lensing AWA, Cogo A et al (1996) The long-term clinical course of acute deep venous thrombosis. Ann Intern Med 125:1–7

Reimer P, Landwehr P (1998) Noninvasive imaging of peripheral vessels. Eur Radiol 8:858–872

Robertson PL, Berlangier SU, Goergen SK et al (1994) Comparison of ultrasound and blood pool scintigraphy in the diagnosis of lower limb deep vein thrombosis. Clin Radiol 49:382–390

Rosebrough SF, Kudryk B, Grossman ZD et al (1985) Radioimmunoimaging of venous thrombi using iodine-131 monoclonal antibody. Radiology 156:515–521

Rosebrough SF, Grossman ZD, McAfee JG et al (1987) Aged venous thrombi: Radioimmunoimaging with fibrin specific monoclonal antibody. Radiology 162:575–577

Rosebrough SF, Grossman ZD, McAfee JG et al (1988) Thrombus imaging with indium-111 and iodine-131-labelled fibrin-specific monoclonal antibody and its F(ab')$_2$ and Fab fragments. J Nucl Med 29:1212–1222

Rosenthall L (1971) Combined inferior vena cava, iliac venography and lung scanning with 99mTc-albumin macroaggregates. Radiology 98:623–627

Rosenthall L, Leclerc J (1995) A new thrombus imaging agent: human recombinant fibrin binding domain labeled with In-111. Clin Nucl Med 20:398–402

Ruoslahti E, Engvall E, Hayman EG (1981) Fibronectin: current concepts of its structure and and function. Coll Relat Res 1:95–128

Ryo VY, Siddaligappa-Srikantaswamy MQ, Pinsky S (1977) Radionuclide venography: correlation with contrast venography. J Nucl Med 18:11–17

Salzman EW, Hirsh J (1993) The epidemiology, pathogenesis and natural history of venous thrombosis. In: Coleman RW, Hirsh J, Marder VJ, Salzman EW (eds) Hemostasis and thrombosis: basic principles and clinical practice. Lippincott, Philadelphia, pp 1275–1296

Schaible T, Dewoody K, Weisman H et al (1992) Accurate diagnosis of acute deep vein thrombosis with technetium-99m antifibrin scintigraphy. J Nucl Med 33:848 (abstract)

Singer I, Royal HD, Uren RF et al (1984) Radionuclide plethysmography and Tc-99m red blood cell venography in venous thrombosis: comparison with contrast venography. Radiology 150:213–217

Som P, Oster Z, Zamora PO et al. (1986) Radioimmunoimaging of experimental thrombi in dogs using technetium-99m-labeled monoclonal antibody fragments reactive with human platelets. J Nucl Med 27:1315–1320

Strandness DE, Langlois Y, Cramer M et al (1983) Long-term sequelae of acute thrombosis. JAMA 250:1289–1292

Stratton J, Ritchie J (1990) In-111 platelet imaging of left ventricular thrombi: Predictive value for systemic emboli. Circulation 81:1182–1189

Stuttle AW, Peters AM, Loufti I et al (1988) Use of an antiplatelet monoclonal antibody F(ab')$_2$ fragment for imaging thrombus. Nucl Med Commun 9:647–655

Stuttle AW, Klosok J, Peters AM et al (1989) Sequential imaging of postoperative thrombus using In-111-labelled platelet specific monoclonal antibody P256. Br J Radiol 62:963–969

Taillefer R (2001) Radiolabeled peptides in the detection of deep venous thrombosis. Semin Nucl Med 31:102–123

Taillefer R, Barnes D, Hill J et al (2000a) ^{99m}Tc-FBD (fibrin binding domain of fibronectin) scintigraphy in patients with acute deep vein thrombosis: results of the phase II multicenter trial. J Nucl Med 41:12P (abstr)

Taillefer R, Edell S, Innes G et al (2000b) Acute thromboscintigraphy with ^{99m}Tc-apcitide: results of the phase 3 multicenter clinical trial comparing ^{99m}Tc-apcitide scintigraphy with contrast venography for imaging acute DVT. J Nucl Med 41:1214–1223

Taub R, Gould RJ, Garsky VM et al (1989) A monoclonal antibody against the fibrinogen receptor contains a sequence that mimics a receptor recognition domain in fibrinogen. J Biol Chem 264:259–265

Thakur ML, Thiagarajan P, White F et al (1987) Monoclonal antibodies for specific cell labelling: considerations, preparations and preliminary evaluation. Int J Rad Appl Instrum 14:51–58

Uehara A, Isaka Y, Hashikawa K et al (1988) Iodine-131 labeled fibronectin: potential agent for imaging atherosclerotic lesion and thrombus. J Nucl Med 29:1264–1267

Uphold RE, Knopp R, Santos PAL dos (1980) Radionuclide venography as an outpatient screening test for deep venous thrombosis. Ann Emerg Med 9:613–616

Valois JC de, Schaik CC van, Verzijlbergen F et al (1990) Contrast venography: from gold standard to 'golden backdrop' in clinically suspected deep vein thrombosis. Eur J Radiol 11:131–137

Vlahos L, MacDonald AF, Causer DA (1976) Combination of isotope venography and lung scanning. Br J Radiol 49:840–851

Webber MM, Pollak EW, Victery W et al (1974) Thrombosis detection by radionuclide particle entrapment (MAA): correlation with fibrinogen uptake and venography. Radiology 111:645–650

Well RD, Hirsh J, Anderson DR et al (1995) Accuracy of clinical assessment of deep vein thrombosis. Lancet 34:1326–1330

Wheeler HB, Anderson FA, Cardullo PA (1982) Suspected deep vein thrombosis. Management by impedance plethysmography. Arch Surg 117:206–1209

Yamamoto K, Senda M, Fujita T et al (1988) Positive imaging of venous thrombi and thromemboli with Ga-67 DFO-DAS-fibrinogen. Eur J Nucl Med 14:60–64

Yao JS, Henkin RE, Conn J et al (1973) Combined isotope venography and lung scanning. A new diagnostic approach to thromboembolism. Arch Surg 107:146–151

Zoghbi SS, Sostman HD, Duberg AC et al (1988) Radiolabeled fibronectin for the scintigraphic detection of pulmonary emboli in dogs. Invest Radiol 23:574–578

Zorba J, Schier D, Posmituck G (1986) Clinical value of blood pool radionuclide venography. AJR 146:1051–1055

CHAPTER 11

Molecular Imaging of Infection and Inflammation 11

HUUB J. J. M. RENNEN, OTTO C. BOERMAN, WIM J. G. OYEN, FRANS H. M. CORSTENS

Contents

In response to tissue damage, powerful defense mechanisms are activated, consisting of cells (leukocytes) and plasma proteins (antibodies, complement). Besides, a complex variety of mediators, both vasoactive and chemotactic, is involved in the process. These mediators are generated in the focus of inflammation and amplify the local response by the recruitment of cells and plasma components from the blood. Vasodilatation and increased endothelial permeability are induced to facilitate the extravasation of proteins and cells. Tracers in nuclear medicine such as ^{67}Ga-citrate and radiolabeled polyclonal immunoglobulins, liposomes, and avidin-biotin accumulate in the inflammatory lesion based on a nonspecific process of leakage out of dilated capillaries.

In addition, the expression of adhesion molecules on endothelial cells and leukocytes is stimulated. In this way, leukocytes are arrested on the endothelial lining and migrate actively from the circulation into inflamed tissue. There is a highly complex interplay between cells and mediators, regulating the inflammatory response in time from the point of initiation to the final resolution of inflammation or to a more persisting state of chronic inflammation. Different leukocyte subtypes are involved in this process, presenting various receptors on their outer membranes. In scintigraphic imaging of inflammation, the focus of interest are neutrophilic granulocytes, predominantly active in acute stages, and, to a lesser degree, lymphocytes, predominantly active in chronic stages of the inflammatory process. Direct labeling of isolated neutrophils and reinjecting them is considered the "gold standard" nuclear medicine technique for imaging inflammation, although this technique is very laborious and the handling of potentially contaminated blood poses risks to both laboratory personnel and patient.

Alternatively, labeling granulocytes in vivo (the indirect approach) can be achieved by the use of radiolabeled antibodies or compounds binding to receptors on granulocytes with high affinity. At present, antibodies, antibody fragments, and smaller ligands such as chemotactic mediators (e.g., the interleukins) do play or are starting to play important roles in molecular imaging. Several monoclonal antibodies reactive with antigens expressed on granulocytes (NCA, CD15, CD66 and CD67) have been developed. At least three ^{99m}Tc-labeled anti-granulocyte antibodies have been tested for imaging inflammation in patients: anti-NCA-95 IgG (BW250/183), anti-CD66 Fab' (Immu-MN3, Leukoscan), and anti-CD15 IgM (LeuTech).

In the molecular imaging approach toward inflammation, the most interesting group of inflammation-targeting agents are the chemotactic mediators: relatively small sized proteins and peptides that bind with high affinity to receptors which are abundantly present in the area of inflammation. Chemotactic peptide formyl-Met-Leu-Phe (and its derivatives), cytokines such as interleukin-1, interleukin-2, interleukin-8, and platelet factor 4 (derivatives) have been tested for their potential to image inflammation. Although the development is still in a relatively early phase, further advances could yield safe and easy-to-prepare imaging agents that can substitute radiolabeled white blood cells in the clinic.

In addition, positron emission tomography (PET) with ^{18}F-fluorodeoxyglucose (FDG) takes advantage of the enhanced metabolic requirements of leukocytes in inflammatory foci.

One typical example of the genomic approach has recently entered the field of nuclear medicine: radiolabeled ciprofloxacin binds to DNA gyrase in all dividing bacteria. It is claimed that with this agent it is possible to discriminate between infection and sterile inflammation.

Here, the characteristics and the diagnostic potential of established and experimental radiopharmaceuticals for imaging infection and inflammation are reviewed with a central focus on molecular imaging using radiolabeled mediators.

11.1
Introduction

Major advances in molecular biology have revealed the sequence, structure, and function of genes and proteins, the functions of receptors and their ligands. Molecular imaging exploits this new knowledge and aims to visualize cellular events at the molecular level. Molecular imaging of infection and inflammation aims to visualize molecular events of the inflammatory response. In nuclear medicine the central focus of interest in imaging infection and inflammation is the radiolabeling of neutrophils and, to a lesser degree, of lymphocytes. Radiolabeled autologous leukocytes are considered the "gold standard" nuclear medicine technique for imaging infection and inflammation. Alternatively, new radiolabeled agents under development interact with receptors and other membrane components of cells involved in the inflammatory response. So far, nuclear medicine has offered only one example of a radiopharmaceutical that approximates the genomic approach: radiolabeled ciprofloxacin (Infecton) is a microbiological agent binding to the DNA gyrase enzyme present in all living bacteria, and thereby obstructing the translation of bacterial DNA.

11.2
Pathophysiology of Infection and Inflammation

Inflammation can be described as the reaction of the body to any kind of injury (Roitt 1997). Such injury can vary from trauma to ischemia, to neoplasm, to injury from thermal, immunological, or chemical origin. In case the tissue injury is caused by invading micro-organisms, the term infection is used. Infection simply means "contamination with micro-organisms." When tissue injury occurs, regardless of cause or anatomic site, several defense mechanisms are activated to destroy and remove the noxious stimulus. Figure 11.1 presents a schematic overview of the process. The defensive materials are leukocytes and plasma containing defensive proteins of many kinds, including opsonins, complement factors, and antibodies. Furthermore, a complex variety of soluble mediators is involved in the inflammatory response. These mediators are molecules that are generated in the focus of inflammation and modulate the inflammatory response. Some act directly on the smooth muscle wall surrounding the arterioles to alter blood flow. Others cause contraction of the endothelial cells resulting in transient opening of the inter-endothelial junctions and consequent transudation of plasma. The migration of leukocytes from the bloodstream is facilitated by mediators which upregulate the expression of adhesion molecules on both endothelial and white blood cells. Other mediators direct the leuko-

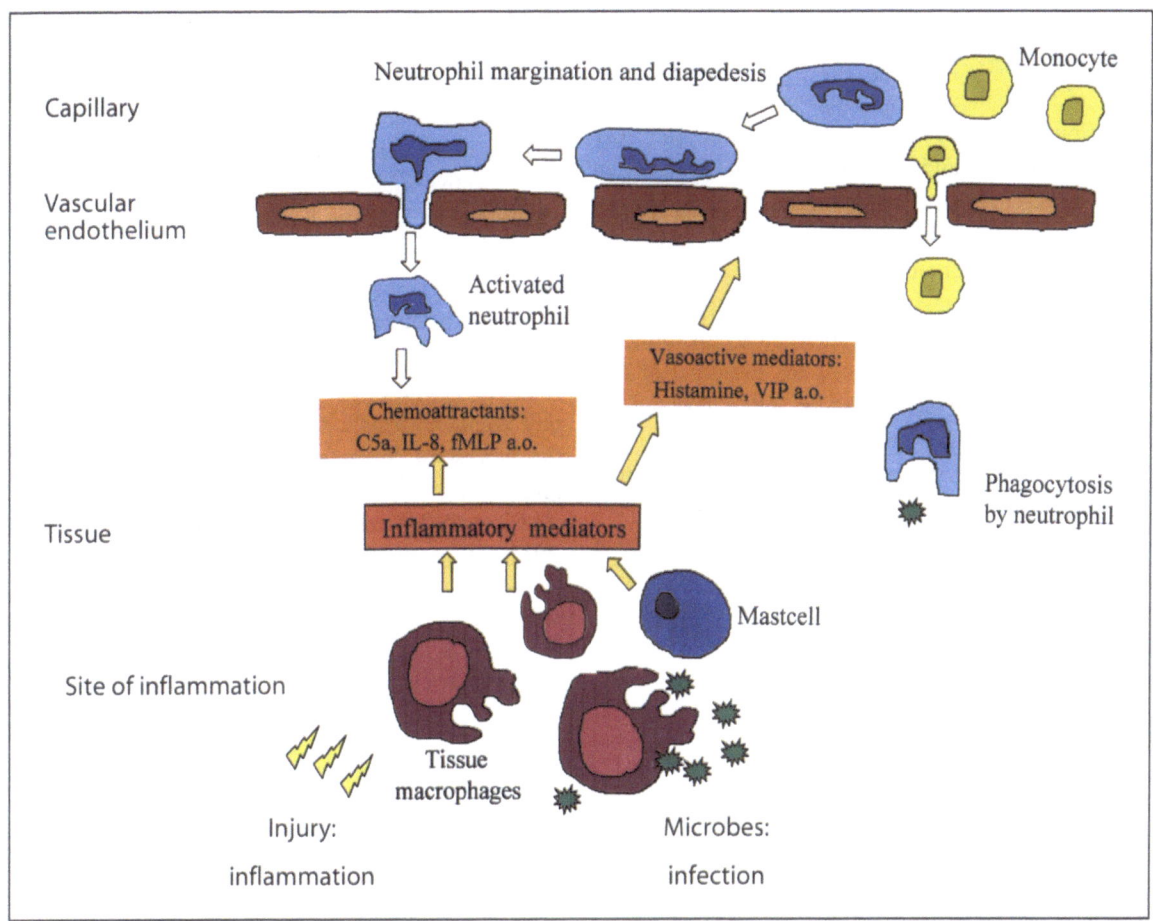

Fig. 11.1. Schematic representation of the inflammatory response. The noxious stimulus of the anti-inflammatory response can be either biotic (infection) or non-biotic (inflammation). In response, inflammatory mediators are generated by the locally present macrophages and mast cells in order to increase tissue perfusion and vascular permeability (vasoactive mediators), and to attract leukocytes to the site of inflammation (chemoattractants). Leukocytes marginate by upregulation of adhesion molecules on both endothelial cells and the leukocytes themselves. They pass through the gap between the endothelial cells and migrate up the chemotactic gradient to the site of inflammation, where they remove the noxious stimulus by phagocytosis

cytes to the inflamed site through chemotaxis up a concentration gradient of these molecules. In this way leukocytes migrate actively from the circulation into inflamed tissue. First, they adhere to the vascular endothelium due to locally enhanced expression of adhesion molecules (rolling, arrest, and adhesion). Subsequently, they pass through the endothelium and the basal membrane (diapedesis) and migrate into the inflammatory focus (chemotaxis). In acute inflammation, the infiltrating cells are predominantly neutrophils. In summary, the pathophysiology of the acute inflammatory response can be characterized by increased blood supply to the region, increased vascular permeability and emigration of leukocytes out of blood vessels into the affected tissues (chemotaxis).

In chronic inflammation, the cellular response is different from that of acute inflammation. Infiltrating cells are predominantly mononuclear cells: lymphocytes, monocytes, and macrophages. Vasodilatation, vascular permeability, and endothelial activation tend to normalize.

11.3
Mediators of the Inflammatory Response

Understanding the role of mediators of the inflammatory response is of particular importance for a number of reasons. Basic, fundamental knowledge of the molecular biology of the inflammatory process is crucial for every medical discipline involved in immunological disorders. Secondly, the design and development of anti-inflammatory drugs is directed not only to antibiotics but also to intervention in the interplay between mediator and inflammatory cell, especially in disorders such as autoimmune diseases where the inflammatory response is activated inappropriately. For instance, a mediator might be overexpressed resulting in a potential tissue-damaging situation. The action of such a mediator could be blocked by introducing appropriate receptor antagonists for receptors of this mediator. Thirdly, the mediators offer proteins of small size (molecular weight <20,000) that bind to receptors on inflammatory cells with high affinity (nanomolar range). For the purpose of scintigraphic imaging of inflammations in nuclear medicine such proteins are attractive candidates, supposing that their biological activity could be circumvented. Here, a concise overview will be given.

11.3.1
The Onset of the Inflammatory Process

One of the very early events in the inflammatory response is the release of histamine by mast cells and basophils and of thrombin by platelets. These products increase both the blood supply in the injured area and the permeability of the venules. So, histamine and thrombin support the nonspecific mechanisms of localization of radiotracers. The same agents also induce upregulation of the expression of the adhesion molecule P-selectin and platelet-activating factor (PAF) on endothelial cells. This facilitates the process of arresting and binding of neutrophils, after which they may infiltrate into the affected area. Neutrophils are made more responsive to chemotactic agents by the influence of complement factor C5a (product of the activated complement system) and leukotriene B$_4$ (LTB4, secreted by mast cells). Besides the variety of endogenous mediators, the inflammatory process can also be triggered by exogenous mediators including the bacterial products formyl-Met-Leu-Phe and lipopolysaccharide (LPS).

11.3.2
The Ongoing Inflammatory Process

Vasodilatation and increased permeability of the venules is induced also by numerous other agents, such as vasoactive intestinal peptide (VIP) and substance P. The macrophages secrete the cytokines interleukin-1 (IL-1) and tumor necrosis factor a (TNF-a), which act at a later stage than histamine or thrombin to stimulate the endothelial cells. Other late products are E-selectin, an adhesion molecule which binds and activates neutrophils, and interleukin-8 (IL-8), a neutrophil chemoattractant. Other chemokines involved in the process of chemotaxis of neutrophils are neutrophil-activating protein 2 (NAP-2), platelet factor 4 (PF4), epithelial-cell-derived neutrophil attractant 78 (ENA-78), granulocyte chemotactic protein 2 (GCP-2), interferon γ-inducible protein (IP-10), and the growth-related gene products GRO-a, -β, -γ. In a later stage, monocytes are attracted to the area of inflammation by mediators such as monocyte chemotactic protein (MCP-1, -2, -3, -4), secreted by various cells activated by IL-1 and TNF-a.

11.3.3
Regulatory Mechanisms

Regulatory mechanisms, both at a humoral and a cellular level, prevent inflammation from getting out of hand. At the humoral level there is a series of complement regulatory proteins such as C1 inhibitor, C4 binding protein, and many others. At the cellular level, prostaglandin (PGE$_2$), transforming growth factor (TGF-β), and glucocorticoids are powerful regulators. The action of pro-inflammatory cytokines is kept in balance by a number of anti-inflammatory cytokines, e.g., the effects of IL-1 are counteracted by the IL-1 receptor antagonist (IL-1RA), blocking IL-1 receptors without induction of signal transduction. Inhibition of cytokine production may be brought about by cytokines such as transforming growth factor (TGF-β), interleukin-4 (IL-4), and interleukin-10 (IL-10). In addition, cytokine activity may be neutralized via the production of high-affinity antibodies to the cytokines that form complexes with them.

11.3.4
Mediators in Chronic Inflammation

Monocytes and lymphocytes are attracted to the site of inflammation by several chemo-attractants, including the lymphocyte chemotactic factor (LCF), the chemokine RANTES (regulated on activation normal T-cell expressed and secreted), macrophage inflammatory proteins 1 (MIP-1α, -1β), and monocyte chemotactic proteins (MCP-1, -2, -3, -4). The immune response is regulated by various cytokines, such as interleukin-2 (IL-2), and (neuro)peptides, such as somatostatin and substance P.

11.4
Developing New Radiotracers

For the development of new infection imaging radiopharmaceuticals the following route can be designed. Once a potential new radiopharmaceutical has passed the laboratory level of synthesis, purification, and quality control, in vitro and in vivo testing starts. In case the potential new radiotracer aims at binding to certain cell types in vivo (e.g. leukocytes, lymphocytes, bacteria), in vitro binding assays with the target cells can give an indication of the potential of the product. Target cells for binding studies can be isolated from the blood, or cell lines can be generated bearing receptors from original cells by cDNA transfection techniques. Reproducibility of receptor binding assays is generally better using these cell cultures bearing the desired transfected receptors. In binding assays the affinity and specificity of the agent for the target cells can be determined.

Subsequently, the new radiotracer can be tested in one of the simple and easy-to-use animal models of infection (for a recent review see Oyen et al. 2001). Mice or rats with thigh muscle infections induced by *S. aureus* or *E. coli* can serve as a good model for these first explorations. Biodistribution data generated in such a model can give a first impression of the imaging potential of the new tracer and allow the first comparison with known agents (Van der Laken et al. 1995, 2000). Data on accumulation in the target, target/non-target ratios, blood clearance, and the primary route of clearance (hepatobiliary/renal) can be obtained.

The next and in most cases last step in preclinical testing will involve more advanced models of infection

in animals of more closely related species like rabbits and dogs. High costs and strict legal regulations will in most cases impede the use of primates for these studies. Moreover, the new information that experiments in primates could reveal does not outweigh the investments in money and energy in general. Besides the thigh muscle infection mentioned above, more complicated models of infection of more clinical relevance can be used: colitis (Rogler and Andus 1998; Neurath et al. 2000), osteomyelitis (Smeltzer et al. 1997; Dams et al. 2000a), endocarditis (Hershberger et al. 2000; Veltrop et al. 2000), meningitis (Koedel and Pfister 1999; Sorensen et al. 2000), and respiratory tract infections (Dei-Cas et al. 1998; Cere and Polack 1999). In many cases these models require more specialized skills and experience. The model of choice depends on the intended application of the new radiotracer. In the rabbit and dog model(s) a direct comparison between the new radiotracer and the "gold standard" ^{111}In- or ^{99m}Tc-labeled purified granulocytes can be carried out.

Once a new radiotracer has successfully passed these stages of laboratory and preclinical research, the agent should be tested in clinical trials in patients suspected of infectious or inflammatory diseases.

11.5
Nonspecific and Specific Radiopharmaceuticals to Image Infection

11.5.1
Nonspecific Radiotracers of Infection

Increased blood supply, increased vascular permeability, and enhanced transudation are processes that can be utilized for nonspecific accumulation of radiotracers. It must be emphasized that all radiopharmaceuticals accumulate to some extent in this nonspecific way at the site of infection (Rennen et al. 2001a). This is of particular importance when evaluating new radiotracers that seem to accumulate in inflammatory foci in a specific way. Nonspecific accumulation can be erroneously interpreted as specific accumulation. Examples of nonspecific radiotracers for detection of infection based on increased vascular permeability are:

- ^{67}Ga-citrate
- Radiolabeled nonspecific immunoglobulins
- Radiolabeled liposomes
- Radiolabeled avidin-biotin.

11.5.2
Specific Radiotracers of Infection

Specific processes of accumulation comprise a number of possible interactions between radiopharmaceutical and target, e.g., receptor binding and antibody-antigen binding. Several pathways can be distinguished in infection imaging based on specific processes of accumulation of the radiopharmaceutical, as shown in Table 11.1. Leukocytes preferentially target infection by chemotaxis and can therefore be used to transport radionuclides to the infected area (Pathway A). They move massively to the site of infection and localize there in great numbers. Detection of infection can be accomplished by direct labeling of leukocytes (ex vivo labeling) or by labeling leukocytes indirectly, i.e., in vivo. Ex vivo labeling requires withdrawal of blood from the patient, purification of leukocytes, labeling, and reinjection of the radiolabeled cells. Autologous leukocytes labeled with ^{111}In or ^{99m}Tc can lead to positive imaging since leukocytes, even after ex vivo labeling, have the capacity to migrate to the inflamed area. In vivo labeling of circulating and marginated leukocytes can be based on antibody-antigen interactions (e.g., radiolabeled antigranulocyte monoclonal antibodies) or on leukocyte receptor binding (e.g., radiolabeled chemotactic peptides, cytokines, and complement factors). Another approach is to target mediators of the inflammatory process in vivo (Pathway B). Pathway C can be regarded as true infection imaging, since radiopharmaceuticals are used that specifically target micro-organisms. PET scans using [18]FDG have a completely different mechanism of accumulation at the inflammatory site: enhanced uptake of radiolabeled glucose by infiltrated granulocytes and tissue macrophages with increased metabolic requirements (Pathway D). The accumulation of [18]FDG in cells with increased glucose metabolism is specific. However, [18]FDG is taken up also by, e.g., tumor cells. So, accumulation of [18]FDG is not specific for infection and inflammation processes as such.

11.6
Detection of Infection by Nonspecific Radiotracers

11.6.1
[67]Ga-Citrate

In clinical practice ^{67}Ga-citrate is used in several pathological conditions, including infection and many skeletal disorders (Lavender et al. 1971; Staab and McCartney 1978). Once injected into the circulation, ^{67}Ga-citrate binds to circulating transferrin. This complex extravasates at the site of infection due to the locally enhanced vascular permeability (Tsan 1985), and is partly bound there to lactoferrin excreted by leukocytes or to siderophores produced by microorganisms (Weiner 1990). The agent is excreted partly via the kidneys (especially during the first 24 h after injection), and via the gastrointestinal tract. Physiological uptake of the radiolabel occurs in liver, bone, bone marrow, and bowel. Although ^{67}Ga-citrate scintigraphy has high sensitivity for both acute and chronic infection and noninfectious inflammation (Palestro 1994), there are several shortcomings that limit its clinical application. The specificity of the technique is low because of physiological bowel excretion and accumulation in malignant tissues and areas of bone modeling (Perkins 1981; Bekerman et al. 1984; Seabold et al. 1989). In addition, the radiopharmaceutical has unfavorable imaging characteristics (long physical half-life and high energy gamma photons), causing high doses of absorbed radiation. Furthermore, optimal imaging often requires delayed recordings up to 72 h. These unfavorable characteristics and the development of newer radiopharma-

Table 11.1. Strategies in infection imaging with specific radiotracers

A. Target leukocytes moving to and present in the focus of infection:
   Direct labeling of leukocytes (ex vivo labeling)
   Indirect labeling of leukocytes (in vivo labeling):
     Antibody-antigen interactions: radiolabeled antigranulocyte monoclonal antibodies/antibody-fragments
     Binding to receptors on leukocytes:
       Radiolabeled chemotactic peptides: formyl-Met-Leu-Phe
       Radiolabeled cytokines: the interleukins, platelet factor 4, granulocyte chemotactic protein 2
       Radiolabeled complement factors: C5a, C5a-des-Arg

B. Target mediators that are already present in or migrate to the site of infection:
   Anti-E-selectin antibodies or anti-E-selectin F(ab')$_2$ fragments

C. Target locally present micro-organisms:
   Radiolabeled ciprofloxacin
   Radiolabeled antimicrobial peptides

D. Target the increased glucose uptake of infiltrated granulocytes and tissue macrophages:
   [18]FDG in positron emission tomography (PET)

ceuticals have narrowed the clinical indication for gallium scintigraphy to certain conditions such as lung infections and chronic osteomyelitis (Seabold et al. 1989; Palestro 1994).

11.6.2
Nonspecific Immunoglobulins

Initially it was hypothesized that human polyclonal immunoglobulin (HIG) was retained in infectious foci due to the interaction with Fc-γ receptors as expressed on infiltrating leukocytes (Fischman et al. 1990). Later studies have shown that radiolabeled HIG accumulates in infectious foci by nonspecific extravasation due to the locally enhanced vascular permeability (Fischman et al. 1992).

For clinical use HIG has been labeled with ^{111}In as well as [99m]Tc. Both agents have slow blood clearance and physiological uptake in the liver, the spleen and the kidneys. The ^{99m}Tc-labeled preparation has the well-known ideal radiation characteristics, while the ^{111}In-labeled preparation allows imaging at time points beyond 24 h postinjection. HIG labeled with [111]In- or ^{99m}Tc has been extensively tested in a large number of clinical studies. It has shown to be excellent in the localization of musculoskeletal infection and inflammation (Fig. 11.2; Nijhof et al. 1997). In addition, good results have been reported in pulmonary infection, particularly in immunocompromised patients (Oyen et al. 1992; Buscombe et al. 1993), and abdominal inflammation (Mairal et al. 1995). In a comparative study, Dams et al. (1998) showed that ^{99m}Tc-HIG labeled with the Tc-99m chelator hydrazinonicotinamid (HYNIC) has in vivo characteristics highly similar to those of ^{111}In-HIG, and in most cases can replace the ^{111}In-labeled compound. Poor sensitivity of radiolabeled HIG is found in the diagnosis of endocarditis and vascular lesions in general, due to long lasting high levels of circulating activity. A drawback is the long time span between injection and final diagnosis (24–48 h).

11.6.3
Liposomes

Liposomes are spheres consisting of one or more lipid bilayers surrounding an aqueous space. They were proposed as vehicles to image infection some 20 years

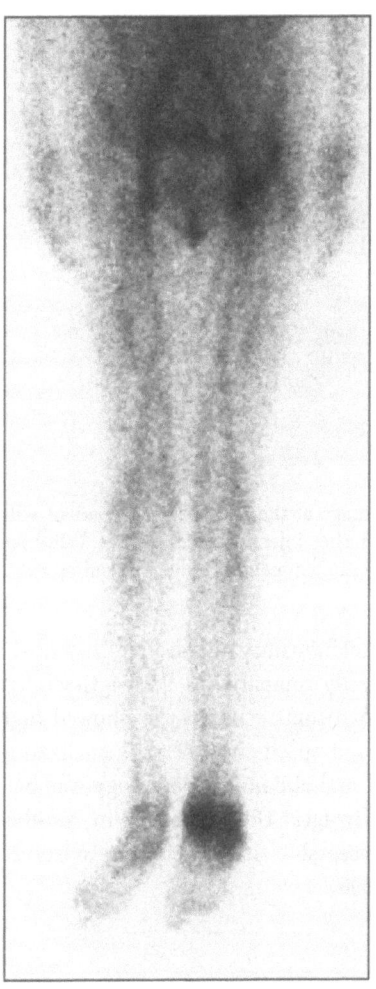

Fig. 11.2. Anterior image below the waist of a 41-year old female patient with focal infection in the left ankle and around the left hip prosthesis, 24 h after injection of 740 MBq ^{99m}Tc-HIG

ago, but the preparations used in those early days were cleared from the circulation very rapidly by the mononuclear/phagocyte system (MPS). However, if the surface of the liposomes is coated with a hydrophilic polymer such as polyethyleneglycol (PEG), they circumvent recognition by the MPS. This leads to a prolonged residence time in the circulation and enhanced uptake in pathological sites (Boerman et al. 1995). Such stabilized PEG-liposomes can be labeled with ^{111}In-oxinate and with ^{99m}Tc using either HMPAO as an internal label or via HYNIC as an external chelator. Labeling is easy and takes only minutes (Laverman et al. 1999) and the first clinical evaluation showed good imaging of focal infection (Fig. 11.3; Dams et al. 2000b). In patients suspected

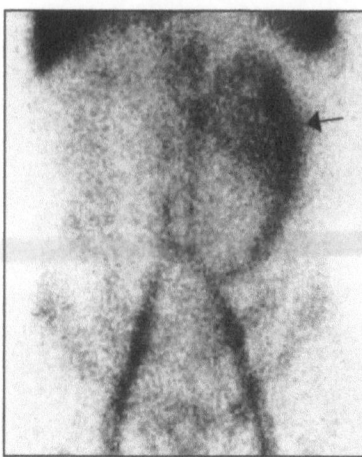

Fig. 11.3. Anterior image of the abdomen in a patient with Crohn's disease, 24 h after injection of 700 MBq ^{99m}Tc-liposomes, demonstrating inflammation in the descending colon

of infectious or inflammatory disease, ^{99m}Tc-PEG-liposomes were directly compared to ^{111}In-IgG-scintigraphy. ^{99m}Tc-PEG-liposome scintigraphy showed high sensitivity (94%) and specificity (89%). Visualization of musculoskeletal and abdominal pathology was better than with ^{111}In-IgG. Unfortunately, in another clinical study unacceptable side-effects were observed (Brouwers et al. 2000).

11.6.4
The Avidin-Biotin System

Avidins are a family of proteins present in the eggs of amphibians, reptiles, and birds. Streptavidin is a member of the same family. Avidin and streptavidin (MW 64 60 kDa) bind to biotin, a compound of low molecular weight that can be radiolabeled, with extremely high affinity ($K_d = 10^{-15}$ M). The avidin-biotin approach is based on the fact that avidin (or streptavidin) will nonspecifically localize at sites of infection due to increased vascular permeability. One or the other of these is injected as a pretargeting agent, followed hours later by a second injection with radiolabeled biotin (Hnatowich et al. 1987). Good diagnostic accuracy was demonstrated in studies of vascular infection (Samuel et al. 1996) and chronic osteomyelitis (Rusckowski et al. 1996; Lazzeri et al. 1999).

11.6.5
Limitations in the Use of Nonspecific Radiotracers to Image Infections

Infectious foci can be visualized with radiotracers without a specific interaction between the agent and a tissue component in the infectious focus in a nonspecific process of localization due to the locally enhanced vascular permeability. Extravasation of (macro)molecules via diffusion is a slow process. Prolonged high blood levels are needed to allow (sufficient) diffusion into the target tissue. However, high blood levels involve relatively high background levels, especially in well-perfused tissues. Secondly, in chronic inflammation the vascular permeability tends to normalize. Furthermore, because nonspecific agents accumulate through a common mechanism of infection and inflammation, these agents cannot distinguish the two conditions. As a result, nonspecific radiotracers are principally limited in their ability to detect (and discriminate between) infections and inflammations. Thus, ^{67}Ga-citrate, radiolabeled HIG, radiolabeled liposomes, and the avidin-biotin approach face the same inherent limitations.

11.7
Detection of Infection by Specific Radiotracers

As outlined in Table 11.1, different strategies in imaging infection using specific radiotracers can be employed. These agents will be presented in more detail.

11.7.1
Direct Labeling of Leukocytes (Ex Vivo Labeling)

Radiolabeled autologous leukocytes were developed in the 1970s and 1980s (McAfee and Thakur 1976; Peters et al. 1986; Peters 1994) and are still considered the "gold standard" nuclear medicine technique for infection and inflammation imaging. After intravenous administration, initial sequestration of the labeled leukocytes occurs in the lungs with subsequent rapid clearance of the activity from there. The radiolabel rapidly clears from the blood as well, and in most cases there is high uptake in granulocytic infiltrates, and a substantial portion of the leukocytes (presumably the old and damaged cells) accumulate in the spleen. Thus, as a radiopharmaceutical, radiolabeled leukocytes are a specific indicator for leukocytic infiltration, but not for infection. McAfee et al. (1976) developed a technique to label autologous leukocytes with ^{111}In using oxinate as a chelator to transfer the radiolabel into the cell. Peters et al. (1986) developed a labeling technique using HMPAO, a lipophilic chelator, that allows efficient labeling of white blood cells with [99m]Tc. In contrast to ^{111}In-oxinate, some of the ^{99m}Tc-HMPAO is released from the leukocytes after injection and subsequently excreted via the renal (within minutes) and hepatobiliary (after hours) routes (Peters 1994). Due to their more optimal radiation characteristics, ^{99m}Tc-labeled leukocytes have replaced those labeled with ^{111}In for most indications. Still, for evaluation of kidney, bladder, and gall bladder infections ^{111}In labeling is preferred. The excellent performance of radiolabeled leukocytes for imaging infection and inflammation was demonstrated in a series of studies; indeed for such foci sensitivity was found to exceed 95% (Datz 1994; Peters 1994; Kipper 1999; Wolf et al. 2001). There has been some concern that more chronic infections could be missed with labeled leukocyte scans, because such infections generate a smaller granulocyte response than acute infections. However, a study in 155 patients showed that the sensitivity of labeled leukocytes for chronic infections (86%) was not significantly different from the sensitivity for detection of acute infections (90%) (Datz and Thorne 1986). Labeled leukocytes proved to be useful also in the assessment of the extent and severity of acute exacerbations of ulcerative colitis (Bennink et al. 2001). Because the presence and severity of disease correlated well with endoscopic and histologic findings in this study, it was concluded that leukocyte scintigraphy could assess disease extent without the need for colonoscopy. With regard to diagnostic accuracy there is no pressing need for a better imaging agent than labeled autologous leukocytes. However, the preparation of this radiopharmaceutical is laborious, requires specialized equipment, and is potentially hazardous. Isolating and labeling a patient's white blood cells takes a trained technician approximately 3 h. In addition, the required handling of potentially contaminated blood could lead to transmission of blood-borne pathogens such as HIV and HBV (Lange et al. 1990).

11.7.2
Indirect Labeling of Leukocytes (In Vivo Labeling)

A gross division can be made in radiotracers that bind leukocytes by receptor binding (relatively small molecules; MW < 20 kDa) and radiotracers that bind leukocytes by antibody-antigen interaction (relatively large molecules). The antibodies range in molecular weight from 50 kDa (antibody fragments) through 150 kDa (IgG) to 900 kDa (IgM).

11.7.2.1
Antigranulocyte Antibodies and Antibody Fragments

Ever since it became clear that radiolabeled autologous leukocytes could visualize infectious foci, investigators have tried to develop a methodology to label leukocytes in circulation or in infectious foci. Instead of isolating the white blood cells from a patient and labeling the cells ex vivo, these methods aimed at labeling white blood cells in vivo. Such labeling procedures can be easier and do not require handling of potentially contaminated blood. The use of radiolabeled monoclonal antibodies against surface antigens present on granulocytes was one of the first attempts at in vivo labeling of leukocytes. Several monoclonal antibodies reactive with antigens expressed on granulocytes (NCA, CD15, CD66, and CD67) have been developed. At least three ^{99m}Tc-labeled anti-granulocyte antibodies have been tested for infection imaging: anti-NCA-95 IgG (BW250/183) (Becker et al. 1989, 1992; Segarra et al. 1991; Schubiger et al. 1989; Papos et al. 1996; Krause et al. 1999; Gyorke et al. 2000), a Fab'-fragment of IgG directed against NCA-90 (Immu-MN3, leukoscan: anti-CD66; Becker et al. 1994a, 1996; Barron et al. 1999; Gratz et al. 2000), and anti-SSEA-1 IgM (LeuTech: anti-CD15) (Thakur et al. 1996, 2001; Gratz et al. 1998; Kipper et al. 2000; Rypins and Kipper 2000). Each of these anti-granulocyte antibodies allowed accurate delineation of infection.

It was soon realized that the in vivo behavior of these labeled anti-granulocyte antibody preparations did not mimic the behavior of radiolabeled leukocytes. In general, blood clearance of the IgG preparations was much slower, giving a high background radioactivity that decreases slowly with time. For that reason the time interval between injection of the labeled antibodies and the acquisition of images is relatively long in order to get good target-background ratios. Furthermore, no initial lung entrapment was seen and splenic uptake was much lower, while the preparations based on antibody fragments (Fab, Fab') had a much higher renal excretion. Similarly, the IgM antibody had a much higher liver uptake as compared to the ex vivo labeled white blood cells. Becker et al. (1989) showed that less than 10% of the radiolabeled BW250/183 antibody in the blood was actually associated with granulocytes. These observations indicated that the anti-granulocyte antibody approach for infection imaging, although feasible, did not represent a method to label white blood cells in vivo. It is now generally accepted that radiolabeled anti-granulocyte antibodies localize in infectious foci mainly by nonspecific extravasation as a result of locally enhanced vascular permeability, and that binding of the antibody to infiltrated leukocytes in the inflamed tissue may contribute to the retention of the radiolabel in the focus (Becker et al. 1989; Gyorke et al. 2000). Perhaps an exception should be made for anti-SSEA-1 IgM (LeuTech: anti-CD15; Thakur et al. 1996, 2001). This antibody recognizes CD-15 antigens on granulocytes with high affinity ($K_d = 10^{-11}$ M) and the in vivo granulocyte binding exceeds 50%, pointing toward more specific processes of accumulation in infected tissue. Recently, a ^{99m}Tc-labeled anti-CD15 IgM monoclonal antibody has shown promising results in patients with equivocal appendicitis (Kipper et al. 2000).

The anti-granulocyte antibody-based radiopharmaceuticals visualized infectious foci in patients with a sensitivity of 80–90% (Becker et al. 1994b). Scintigraphy employing ^{99m}Tc-BW250/183 was useful in the evaluation of vascular graft infection and prosthetic heart valve infection. Good results were also obtained in the evaluation of patients with inflammatory bowel disease (Becker et al. 1992, 1994b), although the agent appeared to be less accurate compared to labeled leukocytes (Segarra et al. 1991; Papos et al. 1996). Pulmonary infections – with the exception of lung abscesses – were not visualized. Peripheral bone infections were adequately visualized, but the sensitivity decreased as the focus approached the spine (Becker et al. 1994a). Due to the relatively slow blood clearance of the agent, a 24-hour postinjection scan is generally necessary for correct localization of the inflammatory focus. A major disadvantage of the murine monoclonal antibodies is that they may induce human antimouse antibodies (HAMA), which can result in altered biodistribution after subsequent injections (Sakahara et al. 1989; Becker et al. 1994b). In this respect, the use of antibody fragments instead of the whole antibody seems to be more advantageous, since such fragments have ap-

Table 11.2. Criteria for an ideal radiotracer

- Efficient accumulation and good retention in inflammatory foci
- Rapid clearance from the background
- No accumulation in non-target organs
- No toxicity, no immune response
- Early diagnostic imaging
- Ready availability and low cost
- Low radiation burden
- Easy, low-hazard preparation
- Differentiation between infection and non-microbial inflammation

peared to be less immunogenic (Becker et al. 1994b). In addition, antibody fragments show faster blood clearance and may thus provide earlier diagnosis. The ^{99m}Tc-labeled antigranulocyte Fab'-fragment (Leukoscan) has been registered in Europe as an infection imaging agent (Becker et al. 1996). Further clinical studies will help to define the utility of these new agents in clinical practice.

11.7.2.2
Radiolabeled Mediators

It appears that small receptor-binding peptides could fulfill most of the requirements for an ideal radiotracer for imaging inflammation and infection as summarized in Table 11.2. The binding to receptors, which are abundant in infectious and inflammatory tissue, theoretically may allow specific uptake in these tissues, while their smaller size facilitates rapid clearance from the background resulting in high target-to-background ratios in a relatively short time. Recently, much attention has been directed to labeling of mediators of the inflammatory response (chemotactic peptides, cytokines, among others) for the purpose of imaging infection and inflammation.

11.7.2.2.1
Chemotactic Peptides

Like anti-granulocyte antibodies, peptides with high affinity for receptors expressed preferentially on granulocytes could be suitable for targeting granulocytes in vivo. A wide variety of peptides that bind to receptors as expressed on white blood cells has been tested for the detection of infection. One of the first receptor-binding peptides that was studied for its ability to image infectious foci was the chemotactic peptide formyl-Met-Leu-Phe. This tripeptide, N-terminally formylated, is a chemotactic factor produced by bacteria. It binds to receptors on granulocytes and monocytes with high affinity ($K_d = 10$–30 nM). The first work on this agent was reported more than twenty years ago. Zoghbi et al. (1981) labeled f-Met-Leu-Phe and investigated its in vivo characteristics. They found that even low doses of peptide induced a transient granulocytopenia. In 1991, Fischman et al. described the synthesis of four DTPA-derivatized chemotactic peptide analogues and their labeling with [111]In. All peptides maintained biologic activity and receptor binding affinity. The peptides were tested in rats with *Escherichia coli* soft tissue infections. All analogues showed preferential localization in the focal infection within 1 h after injection. In a comparative study in rabbits with *E. coli* infections, it was demonstrated that localization of infection using ^{99m}Tc-labeled f-Met-Leu-Phe was superior to that of ^{111}In-labeled leukocytes (Babich et al. 1993). However, although a high specific activity ^{99m}Tc-labeling method was applied, a peptide dose as low as 10 ng/kg still resulted in a drop of the peripheral leukocyte counts (Fischman et al. 1993). Several antagonists were developed to circumvent this undesirable biologic activity. However, these antagonists had lower uptake in the infectious focus, most likely due to reduced affinity for the receptor (Pollak et al. 1996; Babich et al. 1997). In summary, rapid imaging of infection and inflammation is feasible with radiolabeled chemotactic peptides, but their undesired biologic side effects would seem to impede further clinical development.

11.7.2.2.2
Cytokines

Labeled cytokines are a potential class of protein radiopharmaceuticals of small molecular weight (< 20 kDa). Cytokines act through an interaction with specific cell-surface receptors expressed on known cell populations. Binding affinities are usually high (nanomolar range). Cytokine receptors are expressed at low levels on non-excited cells, but their expression can be upregulated during activation in situations such as inflammation and infection.

11.7.2.2.2.1
Interleukin-1 (IL-1)

IL-1 binds receptors as expressed mainly on granulocytes, monocytes, and lymphocytes with high affinity. Studies in mice with focal *Staphylococcus aureus* infections showed specific uptake of radioiodinated IL-1 at the site of infection (Van der Laken et al. 1995).

Using IL-1 receptor blocking antibodies, it could be demonstrated that accumulation of the agent in the infectious foci was due to binding to the IL-1 type II receptor (Van der Laken et al. 1997). Unfortunately, the biologic effects (e.g., hypotension, headache) of IL-1 even at very low doses (10 ng/kg) precluded clinical application of radiolabeled IL-1. Therefore, the naturally occurring IL-1 receptor antagonist (IL-1ra) was tested as an imaging agent. This equally-sized (17 kDa) protein binds IL-1 receptors with similar high affinity but lacks any biologic activity. In a comparative study in rabbits with focal *E. coli* infections, the abscess uptake of radioiodinated IL-1ra was half that of radioiodinated IL-1 (Van der Laken et al. 1998). Barrera et al. (2000) tested ^{123}I-IL-1ra in patients with rheumatoid arthritis. Inflamed joints were nicely visualized, but major retention of the radiolabel in the intestinal tract indicated that this agent can not be used to visualize infectious and inflammatory lesions in the abdomen.

11.7.2.2.2.2
Interleukin-2 (IL-2)

Chronic inflammation is characterized by infiltration of the target tissue by lymphocytes. These infiltrates have been successfully targeted with radiolabeled IL-2. The IL-2 is considered to bind specifically to IL-2 receptors as expressed on activated T-lymphocytes. In a study in an animal model of human autoimmune diabetes mellitus, Signore et al. (1992) showed that lymphocytic infiltration in the pancreas could be visualized with ^{123}I-labeled IL-2 5–15 min after injection. A method was developed for preparation of a ^{99m}Tc-IL-2 radiopharmaceutical with a high specific activity (Chianelli et al. 1997). Studies in patients with chronic inflammatory conditions, including insulin-dependent diabetes, Hashimoto thyroiditis, Graves' disease, Crohn's disease, celiac disease, and other autoimmune diseases, demonstrated localization of [123]I- or ^{99m}Tc-labeled IL-2 at the site of lymphocytic infiltration (Signore 1999; Signore et al. 2000a, b). In cases of Crohn's disease, the intestinal ^{123}I-IL-2 uptake, as assessed by the number of positive ROIs, positively correlated with the Crohn's disease activity index (Signore et al. 2000a). A study in patients with celiac disease demonstrated a positive correlation between the number of lymphocytes infiltrating the jejunal mucosa, histologically determined, and the jejunal accumulation of ^{123}I-IL-2 as measured by gamma camera (Signore et al. 2000b). The results suggested that radiolabeled IL-2 might be a suitable agent for in vivo targeting of mononuclear cell infiltration in autoimmune diseases.

11.7.2.2.2.3
Interleukin-8 (IL-8)

IL-8 is a small protein (8.5 kDa) belonging to the CXC subfamily of the chemokines, or chemotactic cytokines, in which the first two cysteine residues are separated by one amino acid residue. IL-8 binds to receptors on neutrophils with high affinity (0.3–4 nM). Hay and colleagues (1997) studied the in vivo behavior of radioiodinated IL-8 in a rat model with carrageenan-induced sterile inflammations. The uptake peaked at 1–3 h after injection and declined thereafter, while target-to-background ratios remained relatively low. In a pilot study in 11 patients Gross et al. (2001) showed that ^{123}I-IL-8 could visualize inflammatory foci, although the labeling method appeared to have major effects on the in vivo biodistribution of radioiodinated IL-8. The scintigraphic imaging characteristics of IL-8 labeled by the Bolton-Hunter method were clearly superior to IL-8 labeled by the iodogen (or Iodo-Gen) method, despite similar in-vitro cell binding characteristics (Van der Laken et al. 2000). In rabbits with focal E. coli infection, accumulation of ^{123}I-labeled IL-8 in the abscess was rapid and high. The specific activity of this IL-8 preparation was relatively low resulting in a transient drop of peripheral leukocyte counts to 45% after a dose of 25 µg/kg ^{123}I-IL-8, followed by a leukocytosis (170% of preinjection level) over several hours. Recently, a ^{99m}Tc-labeled IL-8 preparation was developed using HYNIC as a chelator. In rabbits with thigh muscle infection induced by E. coli, high abscess uptake of ^{99m}Tc-HYNIC-IL-8 and high abscess to background ratios were obtained compared to those using the radioiodinated preparation (Rennen et al. 2001b). In rabbits with chemically induced acute colitis, visualization of the extent of colonic inflammation was possible within 1 h after injection of ^{99m}Tc-HYNIC-IL-8 (Fig. 11.4; Gratz et al. 2001a). Within 4 h after injection, this agent allowed a meticulous evaluation of the severity of inflammatory bowel disease. The performance of ^{99m}Tc-HYNIC-IL-8 was also evaluated in a rabbit model of acute osteomyelitis (Gratz et al. 2001b). In this model, ^{99m}Tc-labeled IL-8 clearly delineated the osteomyelitic lesion within a few hours after injection. The labeling of IL-8 with ^{99m}Tc using HYNIC as a chelator could be further optimized by

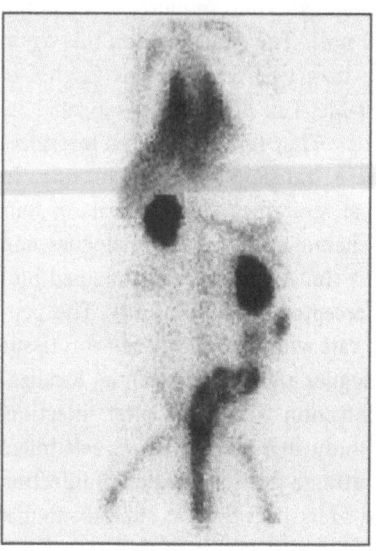

Fig. 11.4. Scintigraphic image of a rabbit with experimental colitis in the descending colon, 4 h after injection of 18.5 MBq ^{99m}Tc-labeled Interleukin-8

the use of alternative coligands to stabilize the ^{99m}Tc-protein complex (Rennen et al. 2001c). The most optimal infection imaging characteristics in a rabbit model of E. coli induced thigh muscle infection were found for preparations using nicotinic acid and tricine as coligands. This formulation combined a high specific activity and high in-vitro stability with high abscess/muscle ratios and high abscess/background ratios. In this way, protein doses to be administered could be lowered substantially for this highly bioactive protein.

11.7.2.2.2.4
Platelet Factor 4 (PF-4)

PF4 is, like IL-8, a member of the CXC chemokines and has been called the "body's heparin neutralizing agent." The imaging characteristics of ^{99m}Tc-labeled PF4 were compared to those of IL-8 in a rabbit model of soft tissue infection (Rennen et al. 2001d). Abscess uptake of PF4 was considerably lower and liver uptake considerably higher. Thus, ^{99m}Tc-labeled native PF4 is an unfavorable alternative for infection imaging. The peptide P483H, synthesized at Diatide Inc., contains the heparin-binding region of PF4, complexed with heparin, and a lysine-rich sequence to facilitate rapid renal clearance and reduce liver uptake. In a rabbit model of infection, ^{99m}Tc-P483H clearly delineated the infectious foci as early as 4 h after injection. The transient neutropenia observed with IL-8 and f-Met-Leu-Phe was not encountered after IV injection of P483H (Moyer et al. 1996). Also studied in patients has been ^{99m}Tc-P483H as an imaging agent for scintigraphic detection of infection and inflammation, with fair results (sensitivity 82–86%, specificity 77–81%; Palestro et al. 1999, 2001). However, in some patients excessive thyroid uptake was observed, suggesting the release of ^{99m}Tc from the agent in vivo.

11.7.2.2.2.5
Other Cytokines

Theoretically, other neutrophil binding mediators such as NAP-2, ENA-78, GCP-2, IP-10, the GRO-proteins (GRO-α, -β, -γ), and the MCP-proteins (MCP-1, -2, -3, -4) could also be used as infection imaging agents. In a preclinical study the imaging characteristics of a ^{99m}Tc-labeled GCP-2 derivative were compared to those of IL-8 (Rennen et al. 2001d). Here again, IL-8 proved to be the superior agent: abscess uptake and abscess to muscle ratios were four times as high as for the GCP-2 derivative. Blankenberg et al. (2001) tested ^{99m}Tc-labeled MCP-1 for imaging inflammation in a rat model of sterile inflammation induced by turpentine. Abscess uptake and abscess-to-muscle ratios were relatively low, showing no conclusive proof of specific uptake of this agent.

11.7.2.2.3
Complement Factors

The complement anaphylatoxin C5a and its natural metabolite C5a-des-Arg are involved in several stages of the inflammatory process. C5a-des-Arg differs from C5a only in lacking the C-terminal Arg-residue of C5a. Both act on a common receptor on different cell types, including neutrophils and monocytes. The receptor-binding affinity of C5a exceeds that of C5a-des-Arg by 1 to 2 orders of magnitude. and the latter has considerably reduced biologic potency compared to that of C5a. A study with ^{99m}Tc-labeled C5a and C5a-des-Arg in rabbits with *E. coli* induced thigh muscle infections showed that ^{99m}Tc-labeled C5a could rapidly visualize the inflammatory focus with high uptake of the radiolabel in the affected muscle (Rennen et al. 2001e). The antagonist C5a-des-Arg showed low uptake in infection. The ideal combination of reduced biological potency and preferable imaging characteristics was not found in C5a-des-Arg.

11.7.3
Detection of Infection by Targeting Adhesion Molecules: Anti-E-Selectin Antibodies or Anti-E-Selectin F(ab')$_2$ Fragments

E-selectin is an endothelial adhesion molecule exclusively expressed on the luminal surface of activated endothelial cells and capable of binding to leukocytes. Radiolabeled anti-E-selectin monoclonal antibodies (Keelan et al. 1994) or anti-E-selectin F(ab')$_2$ fragments (Jamar et al. 1995) have been successfully used to image arthritis and chronic inflammatory bowel disease (Bhatti et al. 1996, 1998; Chapman et al. 1996; Jamar et al. 1997).

11.7.4
Detection of Infection by Radiolabeled Antibiotics

11.7.4.1
Ciprofloxacin (Infecton)

None of the agents discussed above can make a differential diagnosis between infection and inflammation as they accumulate in the focus due to a common feature of infection and inflammation. Ciprofloxacin, a fluoroquinolone antimicrobial agent, binds to the DNA gyrase enzyme present in all dividing bacteria, even to those resistant to ciprofloxacin. In this way, ciprofloxacin introduces the genomic approach in nuclear medicine. Fluoroquinolones are thought not to bind to dead bacteria nor to accumulate in non-microbial inflammatory processes such as Crohn's disease. Originally, it was claimed that with these ^{99m}Tc-labeled agents it is possible to discriminate between infection and sterile inflammation (Vinjamuri et al. 1996). First clinical studies showed high accuracy in the detection of bacterial infection. Data on the efficacy of imaging using a ^{99m}Tc-ciprofloxacin derivative called "Infecton" in patients with suspected infective disorders revealed sensitivities between 70 and 94% and specificities between 83 and 93% (Hall et al. 1998; Sundram et al. 2000; Sonmezoglu et al. 2001). Repeat studies on patients following antibiotic treatment over a long period of time seemed to be very useful in deciding on termination of this treatment (Sundram et al. 2000). However, results with Infecton obtained by another group pointed towards good sensitivity but low specificity in a study in rabbits with infected and non-infected knee prostheses

(Sarda et al. 2000a) and in patients after joint arthroplasty or surgical osteosynthesis (specificity only 31%) (Sarda et al. 2000b).

11.7.4.2
Antimicrobial Peptides

Neutrophil defensins (human neutrophil peptides, HNP) are stored in the granules of neutrophils. In addition to their direct antimicrobial activity, the peptides have chemo-attractive activity for various monocytes and lymphocytes. It has been hypothesized that their cationic charge facilitates binding of these peptides to various micro-organisms. HNP-1 was labeled with ^{99m}Tc using a direct method by reducing the disulfide bridges of the molecule. Using this agent in experimental thigh infections in mice, abscess-to-background ratios were low and decreased with time (Welling et al. 1998, 1999). In the peritoneal cavity of infected mice, ^{99m}Tc-HNP-1 bound to bacteria rather than to leukocytes. However, this agent needs extensive optimization and tailoring to become able to distinguish between bacterial infection and sterile inflammation.

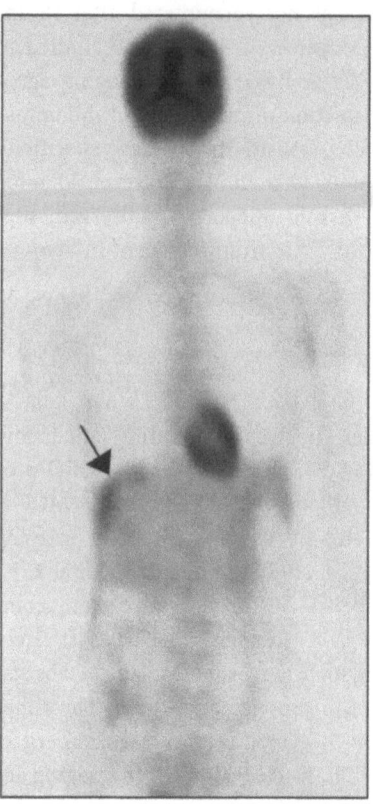

Fig. 11.5. Coronal view of a ^{18}F-FDG PET scan of a 46-year old female patient with a postsurgical perihepatic abscess

11.7.5
Detection of Infection by Positron Emission Tomography (PET) Using ^{18}FDG

The use of ^{18}F fluorodeoxyglucose positron emission tomography (^{18}FDG-PET) has become increasingly important for differentiating malignant from benign tumors, for tumor staging, and for evaluating treatment efficacy in cancer patients (Nunan and Hain 2000; O'Doherty 2000). As early as 1931, Warburg demonstrated an increased glucose metabolism in malignant tumors in vitro. Accumulation of ^{18}FDG in malignant tumors in vivo is due mainly to their increased glucose metabolism. However, during the staging and follow-up of malignant tumors, false-positive findings occasionally were found to occur, chiefly because of infectious or granulomatous processes (Strauss 1996). Ever since Tahara et al. (1989) first demonstrated high ^{18}FDG uptake in human abdominal abscesses in 1989, ^{18}FDG has been shown to accumulate in various inflammatory processes. Infection imaging with ^{18}FDG-PET is based on the fact that granulocytes and macrophages utilize glucose as an energy source. When activated through infection,

metabolism and thus ^{18}FDG-uptake increases. The usefulness of ^{18}FDG-PET to image infections has amply been demonstrated by a number of patient studies (Ichiya et al. 1996; O'Doherty et al. 1997; Guhlmann et al. 1998a,b; Sugawara et al. 1998; Kalicke et al. 2000; Stumpe et al. 2000; Zhuang et al. 2000). Studies of ^{18}FDG-PET have been undertaken in a wide variety of bone and soft-tissue infections of bacterial, tuberculous, and fungal origins (Fig. 11.5). Sensitivities and specificities generally exceeded 90%. Especially successful use of ^{18}FDG-PET has been recorded in cases of osteomyelitis (Guhlmann et al. 1998a, b; Kalicke et al. 2000; Stumpe et al. 2000; Zhuang et al. 2000). The high spatial resolution allows differentiation between osteomyelitis or inflammatory spondylitis and infection of the soft tissue surrounding the bone (Kalicke et al. 2000). High spatial resolution and rapid accumulation into infectious foci are significant advantages over conventional imaging techniques such as with labeled leukocytes. However, low specificity due to indiscriminative uptake in any cell type with high glycolytic activity is a serious limitation. For example, [18]FDG-PET cannot discriminate between tumor and inflammatory lesions. Moreover, [18]FDG uptake in infectious foci is affected by serum glucose levels and by conditions such as diabetes mellitus. Cost is another issue; [18]FDG-PET is rather expensive, and hence prospective cost-effectiveness studies in various patient populations have to be carried out.

11.8
Toward an Ideal Infection Imaging Agent

A lot of radiopharmaceuticals for the diagnosis of infection have been developed in the last decades, each with its own advantages and disadvantages. Criteria for the ideal radiotracer are summarized in Table 11.2. As opposed to nonselective nonspecific localization, selective and irreversible localization in inflammatory foci is an absolute prerequisite. Binding with high affinity to inflammatory cells or to the causative noxious stimuli appears to be a requirement for high selective uptake. Selectivity could be addressed in preclinical in vitro and in vivo binding experiments and in blocking experiments.

High blood pool activity is undesirable because it increases background and complicates the imaging of lesions in well vascularized tissues. In general, agents of larger size, such as antibodies and liposomes, stay longer in and are cleared more slowly from the circulation than are agents of smaller size, such as the chemotactic peptides, the cytokines and complement factors, which are rapidly cleared from the circulation.

Low uptake in non-target organs is particularly important. For instance, high uptake in the liver (or in the bowel in general) would make detection of infection in the vicinity of these organs difficult. Agents of larger size tend to clear predominantly via the hepatobiliary route, whereas smaller agents do so renally (Rennen et al. 2001a). On a molecular level this is reflected in the degree of lipophilicity, or inversely hydrophilicity, of a compound. For instance, PF-4 was found to be cleared by the liver to a substantial degree. P483, a PF4 derivative, was given a hydrophilic lysine-rich sequence to facilitate rapid renal clearance and to lower hepatobiliary clearance.

Absence of toxicity is a crucial requirement when one deals with radiopharmaceuticals derived from biologically active compounds, such as radiolabeled chemotactic peptides, cytokines, and complement

factors. For a diagnostic tool, side-effects, even mild ones, are undesirable. So far, the only solution to reduce unwanted biological activity is to increase the specific activity (MBq/mg). Attempts to introduce radiolabeled receptor antagonists for imaging of inflammation to replace their biologically potent counterparts have failed so far. Radiolabeled IL-1ra proved to be no alternative for IL-1, just as radiolabeled C5a-des-Arg was no alternative for C5a. In general, every mutation in, addition to, or deletion in a high-affinity binding and highly bioactive compound, will alter both characteristics at the same time: binding affinity and bioactivity. A radiolabeled agent with reduced binding affinity will show reduced biological potency together with suboptimal imaging characteristics. The solution to reduce side-effects is to be found in radiochemistry. In the proper choice of the linking agent, which couples radiometal to protein, a substantial gain in specific activity can be achieved. In case of the bifunctional coupling agent HYNIC, the proper choice of coligands for complexation and stabilization of the radiometal-complex was found to be essential for both specific activity and biodistribution. It is a challenge to radiochemistry to reduce the doses of radiolabeled bioactive proteins below the threshold of biological activity.

Another safety issue is mentioned in the section on radiolabeled antigranulocyte antibodies. On repeated administration of radiolabeled murine monoclonal antibodies the human immune system may respond with the production of human antimouse antibodies, which results in an altered biodistribution of subsequently injected antibodies and, possibly, in clinical side-effects. In most cases, diagnosis of inflammatory disorders is possible with a single procedure. In case of repeated diagnoses and expected immune response, the physician may have to switch to an alternative radiopharmaceutical for imaging inflammation.

Early diagnostic imaging is preferable but not always necessary, as in the case of osteomyelitis or in infected bone/joint prostheses. A one-day protocol is favorable in cases of severe and acute illness such as lung infections. For a one-day protocol the small sized agents are preferred because of their fast pharmacokinetics.

Universal availability, low cost, and low radiation burden unquestionably favor ^{99m}Tc as radiometal. Production costs of peptides and proteins depend on size and complexity of the compound and scale of the production process. At present, solid-phase peptide synthesis enables chemical synthesis of proteins of over 100 amino acid residues. Recombinant DNA technology enables synthesis of even larger proteins of 300 amino acid residues or more.

Radiopharmaceuticals that can be labeled in one step using ^{99m}Tc without the need for further purification are particularly interesting. The preparation of radiolabeled isolated leukocytes is laborious and presents serious hazards of viral contamination for both personnel and patient. These are the principal reasons for replacing labeled leukocytes by suitable alternatives.

Differentiation between infection and non-microbial inflammation can be achieved only with radiolabeled agents that directly target the micro-organisms. Radiolabeled antibiotics are the first choice to accomplish this task. However, recent studies with radiolabeled antibiotics showed a lack of specificity in targeting, which impedes a clear discrimination between infection and non-microbial inflammation. Further studies are needed to elucidate the issue of specificity.

11.9
Conclusions

Although radiolabeled leukocytes can accurately delineate inflammatory foci, there is a clear need for an imaging agent that can be prepared instantaneously. Ideally, this agent should be an agent that can be labeled off-the-shelf with ^{99m}Tc and should have at least the diagnostic performance of radiolabeled leukocytes. In the molecular imaging approach, the most interesting group of inflammation targeting agents are the chemotactic mediators. This group includes several small hydrophilic proteins and peptides that bind with high affinity to receptors abundantly present in the area of inflammation. Some of these mediators combine high uptake in infectious and inflammatory foci with rapid background clearance. Several chemotactic peptides and cytokines (IL-1, IL-2, IL-8, PF-4, among others) are currently under investigation. Potential biologic effects is still the major drawback in the use of these molecules for scintigraphic imaging. This problem has to be addressed on a radiochemical rather than a biochemical level. Attempts to introduce radiolabeled receptor antagonists for imaging of inflammation to replace their biologically potent counterparts have failed so far. Development of high-specific-activity labeling of agonists appears to be a more fruitful approach. There is an ongoing progress in the selection and optimiza-

tion of these agents, although to date their development is still in a relatively early phase. In future studies an extensive clinical evaluation is necessary in order to elucidate whether these receptor-specific small proteins and peptides are suitable as new radiopharmaceuticals for scintigraphic detection of infection and inflammation. In addition, [18]FDG-PET may prove to be as useful in the rapid detection and management of human infections as it is for management of malignant diseases.

11.10
References

Babich JW, Graham W, Barrow SA et al (1993) Technetium-99m-labeled chemotactic peptides: comparison with indium-111-labeled white blood cells for localizing acute bacterial infection in the rabbit. J Nucl Med 34:2176–2181

Babich JW, Dong Q, Graham W et al (1997) A novel high affinity chemotactic peptide antagonist for infection imaging. J Nucl Med 38:268P

Barrera P, Van der Laken CJ, Boerman OC et al (2000) Radiolabelled interleukin-1 receptor antagonist for detection of synovitis in patients with rheumatoid arthritis. Rheumatology 39:870–874

Barron B, Hanna C, Passalaqua AM et al (1999) Rapid diagnostic imaging of acute, nonclassic appendicitis by leukoscintigraphy with sulesomab, a technetium 99m-labeled antigranulocyte antibody Fab' fragment. LeukoScan Appendicitis Clinical Trial Group. Surgery 125:288–296

Becker W, Borst U, Fischbach W et al (1989) Kinetic data of in vivo labelled granulocytes in humans with a murine Tc-99m-labelled monoclonal antibody. Eur J Nucl Med 15:361–366

Becker W, Saptogino A, Wolf F (1992) The single late Tc 99m granulocyte antibody scan in inflammatory diseases. Nucl Med Commun 13:186–192

Becker W, Bair J, Behr T et al (1994a) Detection of soft-tissue infections and osteomyelitis using a technetium-99m labelled anti granulocyte monoclonal antibody fragment. J Nucl Med 35:1436–1443

Becker W, Goldenberg DM, Wolf F (1994b) The use of monoclonal antibodies and antibody fragments in the imaging of infectious lesions. Semin Nucl Med 24:142–153

Becker W, Palestro CJ, Winship J et al (1996) Rapid imaging of infections with a monoclonal antibody fragment (Leukoscan). Clin Orthop 329:263–272

Bekerman C, Hoffer PB, Bitran JD (1984) The role of gallium-67 in the clinical evaluation of cancer. Semin Nucl Med 14:296–323

Bennink R, Peeters M, D'Haens G et al (2001) Tc-99m HMPAO white blood cell scintigraphy in the assessment of the extent and severity of an acute exacerbation of ulcerative colitis. Clin Nucl Med 26:99–104

Bhatti M, Chapman P, Jamar F et al (1996) Immunolocalization of active inflammatory bowel disease (IBD) using a monoclonal antibody against E-selectin. J Nucl Med 37 [Suppl 5]:114

Bhatti M, Chapman P, Peters M et al (1998) Visualising E-selectin in the detection and evaluation of inflammatory bowel disease. Gut 43:40–47

Blankenberg FG, Tait JF, Blankenberg TA et al (2001) Imaging macrophages and the apoptosis of granulocytes in a rodent model of subacute and chronic abscesses with radiolabeled monocyte chemotactic peptide-1 and annexin V. Eur J Nucl Med 28:1384–1393

Boerman OC, Storm G, Oyen WJG et al (1995) Sterically stabilized liposomes labeled with ^{111}In to image focal infection in rats. J Nucl Med 36:1639–1644

Brouwers AH, Jong DJ de, Dams ETM et al (2000) Tc-99m-PEG-liposomes for the evaluation of colitis in Crohn's disease. J Drug Target 8:225–233

Buscombe JR, Oyen WJG, Grant A et al (1993) Indium-111-labeled human polyclonal immunoglobulin: identifying focal infection in patients positive for human immunodeficiency virus (HIV). J Nucl Med 34:1621–1625

Cere N, Polack B (1999) Animal pneumocystosis: a model for man. Vet Res 30:1–26

Chapman PT, Jamar F, Keelan ET et al (1996) Use of a radiolabeled monoclonal antibody against E-selectin for imaging of endothelial activation in rheumatoid arthritis. Arthritis Rheum 39:1371–1375

Chianelli M, Signore A, Fritzberg AR et al (1997) The development of technetium-99m-labelled interleukin-2: a new radiopharmaceutical for the in vivo detection of mononuclear cell infiltrates in immune-mediated diseases. Nucl Med Biol 24:579–586

Dams ET, Oyen WJ, Boerman OC et al (1998) Technetium-99m labeled to human immunoglobulin G through the nicotinyl hydrazine derivative: a clinical study. J Nucl Med 39:119–1124

Dams ET, Nijhof MW, Boerman OC et al (2000a) Scintigraphic evaluation of experimental chronic osteomyelitis. J Nucl Med 41:896–902

Dams ETM, Oyen WJG, Boerman OC et al (2000b) Tc-99m-PEG-liposomes for the scintigraphic detection of infection and inflammation: Clinical evaluation. J Nucl Med 41:622–630

Datz FL (1994) Indium-111-labeled leucocytes for the detection of infection: current status. Semin Nucl Med 24:92–109

Datz FL, Thorne DA (1986) Effect of chronicity of infection on the sensitivity of the In-111-labeled leucocyte scan. Am J Roentgenol 147:809–812

Dei-Cas E, Brun-Pascaud M, Bille-Hansen V et al (1998) Animal models of pneumocystosis. FEMS Immunol Med Microbiol 22:163–168

Fischman AJ, Rubin RH, White JA et al (1990) Localization of Fc and Fab fragments of nonspecific polyclonal IgG at focal sites of inflammation. J Nucl Med 31:1199–1205

Fischman AJ, Pike MC, Kroon D et al (1991) Imaging focal sites of bacterial infection in rats with indium-111-labeled chemotactic peptide analogs. J Nucl Med 32:483–491

Fischman AJ, Fucello AJ, Pellegrino-Gensey JL et al (1992) Effect of carbohydrate modification on the localization of human polyclonal IgG at focal sites of bacterial infection. J Nucl Med 33:1378–1382

Fischman AJ, Rauh D, Solomon H et al (1993) In vivo bioactivity and biodistribution of chemotactic peptide analogs in nonhuman primates. J Nucl Med 34:2130–2134

Gratz S, Behr T, Herrmann A et al (1998) Intraindividual comparison of 99mTc-labelled anti-SSEA-1 antigranulocyte antibody and 99mTc-HMPAO labelled white blood cells for the imaging of infection. Eur J Nucl Med 25:386–393

Gratz S, Rennen HJJM, Boerman OC et al (2001a) Rapid imaging of experimental colitis with ^{99m}Tc-Interleukin-8 in rabbits. J Nucl Med 42:917–923

Gratz S, Raddatz D, Hagenah G, et al (2000) Tc99m-labelled antigranulocyte monoclonal antibody Fab' fragments versus echocardiography in the diagnosis of subacute infective endocarditis. Int J Cardiol 75:75–84

Gratz S, Rennen HJJM, Boerman OC et al (2001b) ^{99m}Tc-Interleukin-8 for imaging acute osteomyelitis. J Nucl Med 42:1257–1264

Gross MD, Shapiro B, Fig LM et al (2001) Imaging of human infection with ^{131}I-labeled recombinant human Interleukin-8. J Nucl Med 42:1656–1659

Guhlmann A, Brecht-Krauss D, Suger G et al (1998a) Chronic osteomyelitis: detection with FDG PET and correlation with histopathologic findings. Radiology 206:749–754

Guhlmann A, Brecht-Krauss D, Suger G et al (1998b) Fluorine-18-FDG PET and technetium-99m antigranulocyte antibody scintigraphy in chronic osteomyelitis. J Nucl Med 39:2145–2152

Gyorke T, Duffek L, Bartfai K et al (2000) The role of nuclear medicine in inflammatory bowel disease. A review with experiences of aspecific bowel activity using immunoscintigraphy with 99mTc anti-granulocyte antibodies. Eur J Radiol 35:183–192

Hall AV, Solanki KK, Vinjamuri S et al (1998) Evaluation of the efficacy of 99mTc-Infecton, a novel agent detecting sites of infection. J Clin Pathol 51:215–219

Hay RV, Skinner RS, Newman OC et al (1997) Scintigraphy of acute inflammatory lesions in rats with radiolabelled recombinant human interleukin-8. Nucl Med Commun 18:367–378

Hershberger E, Coyle EA, Kaatz GW et al (2000) Comparison of a rabbit model of bacterial endocarditis and an in vitro infection model with simulated endocardial vegetations. Antimicrob Agents Chemother 44:1921–1924

Hnatowich D, Virzi F, Rusckowski M (1987) Investigation of avidin and biotin for imaging investigation. J Nucl Med 28:1294–1302

Ichiya Y, Kuwabara Y, Sasaki M et al (1996) FDG-PET in infectious lesions: the detection and assessment of lesion activity. Ann Nucl Med 10:185–191

Jamar F, Chapman PY, Harrison AA et al (1995) Inflammatory arthritis: imaging of endothelial cell activation with an In-111 labeled F(ab')$_2$fragment of anti-E-selectin monoclonal antibody. Radiology 194:843–850

Jamar F, Chapman PT, Manicourt DH et al (1997) A comparison between 111In-anti-E-selectin mAb and 99Tcm-labelled human non-specific immunoglobulin in radionuclide imaging of rheumatoid arthritis. Br J Radiol 70:473–481

Kalicke T, Schmitz A, Risse JH et al (2000) Fluorine-18 fluorodeoxyglucose PET in infectious bone diseases: results of histologically confirmed cases. Eur J Nucl Med 27:524–528

Keelan E, Chapman P, Binns R et al (1994) Imaging vascular endothelial activation: an approach using radiolabeled monoclonal antibodies against the endothelial cell adhesion molecule E-selectin. J Nucl Med 35:276–281

Kipper SL (1999) The role of radiolabeled leukocyte imaging in the management of patients with acute appendicitis. Q J Nucl Med 43:83–92

Kipper SL, Rypins EB, Evans DG et al (2000) Neutrophil-specific 99mTc-labeled anti-CD15 monoclonal antibody imaging for diagnosis of equivocal appendicitis. J Nucl Med 41:449–455

Koedel U, Pfister HW (1999) Models of experimental bacterial meningitis. Role and limitations. Infect Dis Clin North Am 13:549–577

Krause T, Reinhardt M, Nitzsche E et al (1999) Photopenic lesions in bone marrow scintigraphy using technetium-99m labeled antigranulocyte antibody without known tumour. Nuklearmedizin 38:85–89

Lange JMA, Boucher CAB, Hollak CEM et al (1990) Failure of zidovudine prophylaxis after accidental exposure to HIV-1. N Eng J Med 323:915–916

Lavender JP, Lowe J, Barker JR et al (1971) Gallium 67 citrate scanning in neoplastic and inflammatory lesions. Br J Radiol 44:361–366

Laverman P, Dams ETM, Oyen WJG et al (1999) A novel method to label liposomes with Tc-99m via the hydrazino nicotinyl derivative: a comparison with Tc-99m-HMPAO-labeled PEG-liposomes. J Nucl Med 40:192–197

Lazzeri E, Manca M, Molea N et al (1999) Clinical validation of the avidin/indium-111 biotin approach for imaging infection/inflammation in orthopaedic patients. Eur J Nucl Med 26:606–614

Mairal L, Lima PD, Martin Comin J et al (1995) Simultaneous administration of ^{111}In-human immunoglobulin and ^{99m}Tc-HMPAO labelled leucocytes in inflammatory bowel disease. Eur J Nucl Med 22:664–670

McAfee JG, Thakur ML (1976) Survey of radioactive agents for the in vitro labeling of phagocytic leucocytes. I. Soluble agents. II Particles. J Nucl Med 17:480–492

Moyer BR, Vallabhajosula S, Lister-James J (1996) Technetium-99m-white blood cell-specific imaging agent developed from platelet factor 4 to detect infection. J Nucl Med 37:673–679

Neurath M, Fuss I, Strober W (2000) TNBS-colitis. Int Rev Immunol 19:51–62

Nijhof MW, Oyen WJ, Kampen A van et al (1997) Evaluation of infections of the locomotor system with indium-111-labelled human IgG scintigraphy. J Nucl Med 38:1300–1305

Nunan TO, Hain SF (2000) PET in oncology II – other tumours. Nucl Med Commun 21:229–233

O'Doherty MJ (2000) PET in oncology I – lung, breast, soft tissue sarcoma. Nucl Med Commun 21:224–229

O'Doherty MJ, Barrington SF, Campbell M (1997) PET scanning and the human immunodeficiency virus-positive patient. J Nucl Med 38:1575–1583

Oyen WJG, Claessens RAMJ, Raemaekers JMM et al (1992) Diagnosing infection in febrile granulocytopenic patients with indium-111 labeled human IgG. J Clin Oncol 10:61–68

Oyen WJG, Boerman OC, Corstens FHM (2001) Animal models of infection and inflammation and their role in experimental nuclear medicine. J Microbiol Meth 47:151–157

Palestro CJ (1994) The current role of gallium imaging in infection. Semin Nucl Med 24:128–141

Palestro CJ, Tomas MB, Bhargava KK et al (1999) Tc-99m P483H for imaging infection: phase 2 multicenter trial results. J Nucl Med 40:15

Palestro CJ, Weiland FL, Seabold JE et al (2001) Localizing infection with a technetium-99m-labeled peptide: initial results. Nucl Med Commun 22:695–701

Papos M, Nagy F, Narai G et al (1996) Anti-granulocyte immunoscintigraphy and [99mTc]hexamethylpropyleneamine-oxime-labeled leucocyte scintigraphy in inflammatory bowel disease. Dig Dis Sci 41:412–420

Perkins PJ (1981) Early gallium-67 abdominal imaging: Pitfalls due to bowel activity. Am J Roentgenol 136:1016–1017

Peters AM (1994) The utility of [99mTc]HMPAO-leucocytes for imaging infection. Semin Nucl Med 24:110–127

Peters AM, Danpure HJ, Osman S et al (1986) Preliminary clinical experience with 99mTc-hexamethylpropyleneaminoxime for labelling leucocytes and imaging infection. Lancet ii:945–949

Pollak A, Goodbody AE, Ballinger JR (1996) Imaging inflammation with Tc-99m labelled chemotactic peptides: analogues with reduced neutropenia. Nucl Med Commun 17:132–139

Rennen HJ, Makarewicz J, Oyen WJ et al (2001a) The effect of molecular weight on nonspecific accumulation of (99m)Tc-labeled proteins in inflammatory foci. Nucl Med Biol 28:401–408

Rennen HJJM, Boerman OC, Oyen WJG et al (2001b) Specific and rapid scintigraphic detection of infection with Tc-99m-labeled interleukin-8. J Nucl Med 42:117–123

Rennen HJ, Van Eerd JE, Oyen WJ et al (2001c) Scintigraphic detection of infection with ^{99m}Tc-labeled interleukin-8: effects of coligand on its in vivo characteristics. J Nucl Med 42:73 (abstract)

Rennen HJ, Boerman OC, Frielink C et al (2001d) Three ^{99m}Tc-labeled CXC-cytokines for infection imaging: IL-8, GCP-2 and PF-4. J Nucl Med 42:72 (abstract)

Rennen HJ, Boerman OC, Oyen WJ et al (2001e) Infection imaging with ^{99m}Tc-labeled C5a and C5a-antagonists. Eur J Nucl Med 28:1043 (abstract)

Rogler G, Andus T (1998) Cytokines in inflammatory bowel disease. World J Surg 22:382–389

Roitt IM (1997) Essential immunology, 9th edn. Blackwell Scientific, Oxford

Rusckowski M, Paganelli G, Hnatowich D et al (1996) Imaging osteomyelitis with streptavidin and indium-111-labeled biotin. J Nucl Med 37:1655–1662

Rypins EB, Kipper SL (2000) Scintigraphic determination of equivocal appendicitis. Am Surg 66:891–895

Sakahara H, Reynolds JC, Carrasquillo JA et al (1989) In vitro complex formation and biodistribution of mouse antitumor monoclonal antibody in cancer patients. J Nucl Med 30:1311–1317

Samuel A, Paganelli G, Chiesa R et al (1996) Detection of prosthetic vascular graft infection using avidin/indium-111-biotin scintigraphy. J Nucl Med 37:55–61

Sarda L, Saleh-Mghir A, Peker C et al (2000a) Infecton imaging in a rabbit model of prosthetic joint infection due to Staphylococcus aureus. Eur J Nucl Med 27:971 (abstract)

Sarda L, Lebtahi R, Genin R et al (2000b) Value of Infecton scintigraphy for the detection of postoperative osteo-articular infection: preliminary results of a clinical study. Eur J Nucl Med 27:970 (abstract)

Schubiger PA, Hasler PH, Novak-Hofer I et al (1989) Assessment of the binding properties of Granuloszint. Eur J Nucl Med 15:605–608

Seabold JE, Nepola JV, Conrad GR et al (1989) Detection of osteomyelitis at fracture nonunion sites: comparison of two scintigraphic methods. Am J Roentgenol 152:1021–1027

Segarra I, Roca M, Baliellas L et al (1991) Granulocyte specific monoclonal antibody technetium-99m-BW 250/183 and indium-111 oxine labelled leucocyte scintigraphy in inflammatory bowel disease. Eur J Nucl Med 18:715–719

Signore A (1999) Interleukin-2 scintigraphy: an overview. Nucl Med Commun 20:938

Signore A, Chianelli M, Toscano A et al (1992) A radiopharmaceutical for imaging areas of lymphocytic infiltration, ^{123}I-interleukin-2 labeling procedure and animal studies. Nucl Med Commun 13:713–722

Signore A, Chianelli M, Annovazzi A et al (2000a) ^{123}I-interleukin-2 scintigraphy for in vivo assessment of intestinal mononuclear cell infiltration in Crohn's disease. J Nucl Med 41:242–249

Signore A, Chianelli M, Annovazzi A et al (2000b) Imaging active lymphocytic infiltration in coeliac disease with iodine-123-interleukin-2 and the response to diet. Eur J Nucl Med 27:18–24

Smeltzer MS, Thomas JR, Hickmon SG et al (1997) Characterization of a rabbit model of staphylococcal osteomyelitis. J Orthop Res 15:414–421

Sonmezoglu K, Sonmezoglu M, Halac M et al (2001) Usefulness of 99mTc-ciprofloxacin (infecton) scan in diagnosis of chronic orthopedic infections: comparative study with 99mTc-HMPAO leukocyte scintigraphy. J Nucl Med 42:567–574

Sorensen KN, Sobel RA, Clemons KV et al (2000) Comparison of fluconazole and itraconazole in a rabbit model of coccidioidal meningitis. Antimicrob Agents Chemother 44:1512–1517

Staab EV, McCartney WH (1978) Role of gallium-67 in inflammatory disease. Semin Nucl Med 8:219–234

Strauss LG (1996) Fluorine-18 deoxyglucose and false-positive results: a major problem in the diagnostics of oncological patients. Eur J Nucl Med 23:1409–1415

Stumpe KD, Dazzi H, Schaffner A et al (2000) Infection imaging using whole-body FDG-PET. Eur J Nucl Med 27:822–832

Sugawara Y, Braun DK, Kison PV et al (1998) Rapid detection of human infections with fluorine-18 fluorodeoxyglucose and positron emission tomography: preliminary results. Eur J Nucl Med 25:1238–1243

Sundram FX, Wong WY, Ang ES et al (2000) Evaluation of technetium-99m ciprofloxacin (Infecton) in the imaging of infection. Ann Acad Med Singapore 29:699–703

Tahara T, Ichiya Y, Kuwabara Y et al (1989) High [18F]-fluorodeoxyglucose uptake in abdominal abscesses: a PET study. J Comput Assist Tomogr 13:829–831

Thakur ML, Marcus CS, Henneman P et al (1996) Imaging inflammatory diseases with neutrophil-specific technetium-99m-labeled monoclonal antibody anti-SSEA-1. J Nucl Med 37:1789–1795

Thakur ML, Marcus CS, Kipper SL et al (2001) Imaging infection with LeuTech. Nucl Med Commun 22:513–519

Tsan MF (1985) Mechanism of gallium-67 accumulation in inflammatory lesions. J Nucl Med 26:88–92

Van der Laken CJ, Boerman OC, Oyen WJG et al (1995) Specific targeting of infectious foci with radioiodinated human recombinant interleukin-1 in an experimental model. Eur J Nucl Med 22:1249–1255

Van der Laken CJ, Boerman OC, Oyen WJG et al (1997) Preferential localization of systemically administered radiolabeled interleukin-1a in experimental inflammation in mice by binding to the type II receptor. J Clin Invest 100:2970–2976

Van der Laken CJ, Boerman OC, Oyen WJG (1998) Imaging of infection in rabbits with radioiodinated interleukin-1 (a and β), its receptor antagonist and a chemotactic peptide: a comparative study. Eur J Nucl Med 25:347–352

Van der Laken CJ, Boerman OC, Oyen WJG et al (2000) Radiolabeled interleukin-8: scintigraphic detection of infection within a few hours. J Nucl Med 41:463–469

Veltrop MH, Bancsi MJ, Bertina RM et al (2000) Role of monocytes in experimental Staphylococcus aureus endocarditis. Infect Immun 68:4818–4821

Vinjamuri S, Hall AV, Solanki KK et al (1996) Comparison of ^{99m}Tc Infecton imaging with radiolabelled white-cell imaging in the evaluation of bacterial infection. Lancet 347:233–235

Warburg O (1931) On the origin of cancer cells. The metabolism of tumors. Smith, New York, pp 129–169

Weiner R (1990) The role of transferrin and other receptors in the mechanism of 67 Ga localization. Int J Rad Appl Instrum B 17:141–149

Welling MM, Hiemstra PS, van den Barselaar MT et al (1998) Antibacterial activity of human neutrophil defensins in experimental infections in mice is accompanied by increased leucocyte accumulation. J Clin Invest 102:1583–1590

Welling MM, Nibbering PH, Paulusma-Annema A et al (1999) Imaging of bacterial infections with 99mTc-labeled human neutrophil peptide-1. J Nucl Med 40:2073–2080

Wolf G, Aigner RM, Schwarz T (2001) Diagnosis of bone infection using 99m Tc-HMPAO labelled leukocytes. Nucl Med Commun 22:1201–1206

Zhuang H, Duarte PS, Pourdehand M et al (2000) Exclusion of chronic osteomyelitis with F-18 fluorodeoxyglucose positron emission tomographic imaging. Clin Nucl Med 25:281–284

Zoghbi S, Thakur M, Gottschalk A (1981) Selective cell labelling: a potential radioactive agent for labeling of human neutrophils. J Nucl Med 22:32

CHAPTER 12

Element Metabolism and Body Composition

12

WALTON W. SHREEVE, RICHARD N. PIERSON JR.

Contents

12.1
Common Mineral Elements

12.1.1
Introduction

Life in its very beginnings surely included the conjunction of mineral ions with the simplest organic compounds and with subsequent nucleic acids and proteins to determine the form and function of most cellular components and processes. A separation of various cations and anions accompanied the creation of membranes and the development of the cell. Electrochemical and electromotive forces were thus established with selective, often linked, passage of inorganic and organic molecules through channels or pores in membranes (Beyenbach 1990; Alberts 1994). Thereby, a system of bioenergetics developed and the fundamental, unique nature of intracellular life was determined. Disturbances or break-downs in the dynamic systems dependent on location, movement and functions of mineral ions underlie many diseases.

Early biomedical researchers using radioactive or stable isotope tracers had a number of isotopes of inorganic elements available by which to evaluate total body amounts of ions, flux across membranes, combination with organic compounds and localization in organs and tissues (Hevesy 1948; Edelman et al. 1952). Physiological differences, as well as pathological changes in the above parameters, have long been objectives of studies with tracers. Several overt genetic diseases are the results of dysfunction of mineral ionic processes, and the elucidation and diagnosis of some of these with radiotracers is an important part of the history of nuclear medicine. Other disorders arise mainly due to environmental circumstances. This chapter describes the ways in which tracer and related techniques have helped – and could help more in the future – to understand the function and dysfunction of mineral elements in the body at the cellular level, as well as that of mineral and some organic elements at the level of major body compartments (extra- and intracellular water, lean mass, fat mass, bone mass), where the integral effect of ionic function/dysfunction is expressed.

12.1.2
Cations

12.1.2.1
Sodium

Because of the separation of concentration of the two free ions, i.e., Na^+ essentially in the extracellular water (ECW) and K^+ essentially in the intracellular water (ICW), there is a chemical gradient for each across the cellular outer membrane. In the case of sodium there is also a voltage gradient (70–90 mV) from positive extracellular to negative intracellular sites. Special ion channels in the cell membrane are used to exchange ions with the aid of energetic mechanisms to maintain the electrochemical gradients. The exchange processes perform useful cell functions such as exchange of Na^+ for K^+ and for movement of other ions such as Ca^{2+}, Mg^{2+} and Cl^-. Furthermore, nature has linked the influx of sodium to the co-transport of glucose and some amino acids (Beyenbach 1990) (Fig. 12.1). To maintain osmotic balance, the ATPase "pump" in the outer cell membrane transports three Na^+ atoms out while "pulling" two K^+ ions from an extra- to intracellular location, against gradients in each case. The importance of the maintenance of electrochemical gradients in the body is indicated by the fact that 35–40% of the total energy expenditure of cells is required for this function (Murray et al. 1996). A special function of Na^+ (and K^+) ion flow is that which occurs through the voltage-gated Na^+ channel in a localized area of nerve ending after some stimulus which depolarizes the plasmalemma (cell membrane) very transiently before local ion restoration, while the depolarization proceeds in rapid stages along the axon to carry its "message" as an electric current (Bolsover et al. 1997).

Of total body sodium (TBNa) about 25–30% is "bound" in bone not significantly exchangeable (with a radiotracer, such as ^{24}Na or ^{22}Na) within 24 h (Ellis et al. 1976). Another 10–15% of TBNa is in bone, dense connective tissue and cartilage, but is part of the exchangeable Na (Na_e) within 24 h (the usual equilibration time for measurement of Na_e) (Edelman and Liebman 1959). The rest of Na_e is in plasma and interstitial lymph fluid except for about 2–5% of Na_e which is normally intracellular (Edelman and Liebman 1959; Pierson et al. 1982) – as would be required by processes described above. This small, but critical, fraction itself increases with the age of adults at a rate of almost 2% per year in males and about 0.6% per year in females, possibly due to a gradual decline in (Na^+-K^+) ATPase (Pierson et al. 1982), with attendant decline in resting membrane potential.

Various disease states are characterized by abnormal distributions or amounts of sodium, which can be detected and quantified by tests available to nuclear medicine. Edema formation results in gross increases (20–100%) of Na_e (mostly extracellular) in patients with overt liver, kidney or heart disease, while Addison's disease is associated with major depletion of Na_e (Edelman and Liebman 1959). Substantial increases of Na_e by ^{24}Na measurement (20% above control), which remained in patients weeks or months after treatment of congestive heart failure, when no clinical edema was present (Carroll et al. 1965), have been attributed more likely to binding to bone or connective tissue than to intracellular binding. Others (Segawa et al. 1989) have detected increased Na_e in patients following even the mildest degree of myocardial infarction (MI) while noting progressively more abnormal Na_e with more serious MI, though no heart failure was evident. Suspected instances of excessive intracellular penetration of sodium in disease gave rise decades ago to the term "sick cell syndrome" (Flear and Singh 1973).

Other examples of abnormal Na_e include observations of increase in patients with clinically occult recurrent depression, with significant restoration upon treatment with lithium, as seen with ^{24}Na (Cox et al. 1977), erratic changes in patients with ileostomy, seen with ^{22}Na (Brevinge and Jacobson 1994), and diverse abnormalities in patients with hypertension, as tested with ^{24}Na (Davies et al. 1973). In the latter study there was abnormally high Na_e relative to plasma renin/angiotensin in patients with chronic renal failure, but much less abnormal ratios in patients with malignant hypertension, except for subjects with "intractable hypertension". Such diversity suggests possible value in correlating abnormalities of Na_e with different genetic types of hypertension.

More clinical use of isotope dilution of ^{24}Na or ^{22}Na for recognizing or evaluating acute abnormalities (patients on dialysis, post-surgery or trauma, acute renal failure) or chronic disease (liver or heart disease, obesity, affective disorders, malnutrition, etc.) might well be justified. The short $T_{1/2}$ of ^{24}Na (15 h) allows for frequent repeats of the test, although the longer-lived ^{22}Na ($T_{1/2\,eff} = 11$ days) (Table 12.1) is currently more widely available. In either case the radiation dose from beta emissions (1.0–2.0 mSv whole body) is not prohibitive.

A new and powerful method for the study of regional changes in sodium concentration is NMR

Table 12.1. Radionuclides used, or appropriate for, clinical studies of element metabolism and/or body composition

Physical data				Applications	Tracer amount (MBq)/ study	Absorbed dose (mSv)	
Element	Isotope	Energy/type	$T_{1/2}$ (p, eff)			Whole body	Local (organ)
Americium	^{241}Am	60 KEVγ	–	Bone density by SPA	–	–	0.05 (limb)
Barium	^{135m}Ba	268 KEVγ	29 h (p)	Surrogate for Ca in bone imaging	37–74	3.0–6.0	40–80 (large intestine)
Bromine	^{82}Br	0.6–1.4 MEVγ; 150 KEVβ^-	36 h (p)	Surrogate for Cl (ECW)	0.9–1.8	0.56–1.1	
Calcium	^{45}Ca	257 KEV β^-	162 d (eff)	Intestinal uptake and body retention	0.15–0.48	0.61–2.0	5.3–17.0 (bone)
	^{47}Ca	1.30 MEVγ; 0.69 MEVβ^-	4.9 d (eff)	Intestinal uptake and body kinetics	0.18–0.37	0.36–0.72	3.11 (bone)
	^{49}Ca	3.08 MEVγ; 2.18 MEVβ^-	8.7 m (p)	Bone mass by DGNAA	–	3.0–6.0	–
Chlorine	^{38}Cl	1.6–2.2 MEVγ; 1.4 MEVβ^-	38 m (p)	ECW by DGNAA	–	3.0–6.0	–
Chromium	^{51}Cr	320 KEVγ	27.7 d (p)	Blood volume, distribution and kinetics	1.0	0.1	1–5 (b.m., spleen)
Copper	^{64}Cu	1.34 MEVγ; 578 KEVβ^-; 511 KEVβ^+	12.8 h (p) 9.9 h (eff)	Intestinal uptake; body kinetics and locale	18.5	0.7	12.5 (spleen)
	^{67}Cu	93 & 185 KEVγ; 390–580 KEVβ^-	60 h (p)	Intestinal uptake; body kinetics and locale	3.7	0.5	9.0 (spleen)
Gadolinium	^{153}Gd	44 and 100 KEVγ	242 d (p)	Bone density by DPA	–	0.05–0.15	–
Hydrogen	^{3}H	18.6 KEVβ^-	12.3 d (eff)	Body water amount	2–6	0.04–0.12	–
Indium	^{113m}In	393 KEVγ	99 m (p)	Plasma volume, distribution, and kinetics	0.05–2.0	0.004–0.01	
Iodine	^{123}I	159 KEVγ	13.3 h (p)	Thyroid uptake	3.7–7.4	0.06–0.12	13–26 (thyroid)
	^{125}I	35 KEVγ, 27–33 KEV x-rays	60 d (p)	1) Plasma volume and kinetics	0.2–0.4	0.3	–
				2) Bone density by SPA	–		0.02 (limb)
	^{131}I	364 KEVγ; 606 KEVβ^-	8.0 d (p)	Thyroid uptake	1.5–3.7	0.14–0.35	55–130 (thyroid)
Iron	^{55}Fe	7 KEV x-ray	2.7 y (p)	Intestinal uptake	0.11–0.44	0.15–0.60	0.8–3.0 (spleen)
	^{59}Fe	1.0 and 1.3 KEVγ; 273 and 464 KEVβ^-	45 d (p)	1) Intestinal uptake	0.15	1.5	6 (spleen)
				2) Body amounts and kinetics	0.2–0.75	1.8–7.2	7–28 (spleen)
Krypton	^{79}Kr	44–606 KEVγ; 511 KEVβ^+	35 h (p)	Amount and distribution of fat	37	0.014	0.02 (lungs)
Magnesium	^{28}Mg	32–1340 KEVγ; 460 KEVβ^-	22 h (p)	Intestinal uptake; body amount, distribution and kinetics	1.1–5.5	0.6–3.0	–

Table 12.1 (continued)

Physical data					Tracer amount (MBq)/ study	Absorbed dose (mSv)	
Element	Isotope	Energy/type	$T_{1/2}$ (p, eff)	Applications		Whole body	Local (organ)
Manganese	^{52m}Mn	1.43 MEVγ; 378 KEVβ^-; 511 KEVβ^+	22 m (p)	Cardiac uptake			
	^{56}Mn	847 KEVγ; 1.04 and 2.85 MEVβ^-	2.6 h (p)	Body distribution and kinetics	0.56–0.74	0.03–0.04	0.8 (liver)
Nitrogen	^{15}N	10.8 MEVγ	10^{-15} s (p)	Protein mass by PGNA	–	0.8	–
Potassium	^{40}K	1.46 MEVγ; 1.35 MEVβ^-	1.25×10^9 (p)	Body potassium amount	–	–	–
	^{42}K	1.5 MEVγ; 2.0 and 3.5 MEVβ^-	12 h (p)	Body amount and kinetics	3.5	0.77	–
	^{43}K	370–610 KEVγ	22 h (p)	Body amount, distribution and kinetics	18.5	4.5	–
Sodium	^{22}Na	1.27 MEVγ; 511 KEVβ^+	2.6 y (p) 11 d (eff)	Body amount and kinetics	0.05–0.35	0.15–2.0	–
	^{24}Na	1.4 and 2.8 MEVγ; 1.4 MEVβ^-	15 h (p)	Body amount and kinetics	2–5	1.0–3.0	–
Strontium	^{87m}Sr	388 KEVγ	2.8 h (p)	Surrogate for Ca in bone imaging	3.7–37.	0.02–0.20	1–10 (bone)
Sulfur	^{35}S	167 KEVβ^-	87 d (p)	1) Body distribution for ECW	1.8–3.7	1.3–2.6	–
				2) Uptake by isolated cells	–	–	–
Technetium	^{99m}Tc	141 KEVγ	6 h (p)	1) Plasma volume, distribution, kinetics	0.4–40	0.2–20	–
				2) Thyroid uptake	74–148	0.25–0.50	2.5–5.0 (large intestine)
Thallium	^{201}Tl	135 + 167 KEVγ	73 h (p)	Surrogate for K in muscle uptake	56–74	3,6–4,8	6–9 (gonad)
Xenon	^{127}Xe	172–375 KEVγ	36 d (p)	Amount and distribution of fat	185–370	0.12–1.2	–
	^{133}Xe	81 KEVγ; 346 KEVβ^-	5.3 d (p)	Amount and distribution of fat	185–370	0.14–1.4	–
Zinc	^{62}Zn	42 and 590 KEVγ; 511 KEVβ^+	9.3 h (p)	Localization in prostate	37	4–8	120 (prostate)
	^{65}Zn	1.12 MEVγ; 511 KEVβ^+	244 d (p)	Intestinal uptake and body kinetics	0.02–0.22	0.06–1.2	3.0 (liver)
	^{69m}Zn	574 KEVγ; 439 KEVβ^-	13.8 h (p)	Intestinal uptake and body kinetics	0.37		

Data obtained from original articles (see textual references), textbooks (e.g., Harbert and DaRocha 1984a, b; Wagner et al. 1995) and handbooks (e.g., Syed and Hosain 1982; Loevinger et al. 1991; SNM Procedure Guidelines Manual 1999).
SPA, single photon absorptiometry; p, physical; eff, effective; ECW, extracellular water; h, hour; d, day; y, year; b.m., bone marrow; DGNAA, delayed gamma neutron activation analysis; DPA, dual photon absorptiometry; PGNA, prompt-gamma neutron activation.

imaging of natural ^{23}Na. Although lower in sensitivity than NMR of water protons and at much lower concentration, special optimizing of imaging parameters can provide images of ^{23}Na with a resolution of 7×7×7 mm for human hearts at 1.5 T (Parrish et al. 1997) with much higher resolutions available at higher tesla. Besides heart imaging, there may be also some practical applications of imaging of ^{23}Na in the brain (Winkler 1990). Nuclear medicine is appropriately involved in this methodology in which findings have a basis in metabolism.

12.1.2.2
Potassium

The principal intracellular cation, potassium, is maintained by energy-requiring processes at a steep concentration gradient between intra- and extracellular locations. The (Na$^+$-K$^+$) ATPase pump and other ionic transport mechanisms exchange minerals across outer cell membranes so fast that nuclear medicine conveniently uses the extent of uptake of the potassium analogue, ^{201}Tl, from plasma by areas of heart and skeletal muscle to signify blood flow, which is presumed to be the limiting factor in the ^{201}Tl uptake.

Though rapid uptake of labeled K or analogues may simply show initial blood flow, various influences operate to determine rates and sites of local and whole body distribution. Nuclear medicine has an array of different methods by which to evaluate these factors. There is a considerable time factor in the process of equilibration of labeled potassium (^{42}K) from plasma to those tissues (or cellular parts) containing potassium. The final equilibration in normal tissues is asymptotic and requires 48–72 h. Early studies in rabbits and cats using ^{42}K (Fenn et al. 1941/1942) provided time curves of uptake in various tissues which showed early, but transient, concentration by heart muscle, liver, kidney and lung with slower sustained accumulation by skeletal muscle, marrow and other tissues with brain, red cells, nerve and bone slowest. In normal human subjects measurements of exchangeable potassium (K_e) with ^{42}K in successive plasma and urine samples has been validated by total body potassium (TBK) by whole body counting of natural ^{40}K (Jasani and Edmonds 1971) (Table 12.1), indicating that K_e was 85% of TBK at 24 h, 90% at 48 h, but in some cases was incomplete at 70 h. Two components of mixing with half-times of 7 and 77 h were defined – the slow component (15–18%) was attributed to red cells and brain.

The rate of equilibration into various organs or the whole body varied from normal in certain disease states. This is shown, for instance, by the finding of normal TBK in chronically diabetic subjects by the method of whole body counting of ^{40}K, which has no time factor of equilibration (Ellis et al. 1974), but decreased K_e in diabetics by isotopic dilution of ^{42}K measured 48 h post-injection (Telfer 1966). Supportive of the latter finding is the reduction of (Na$^+$-K$^+$) ATPase activity in erythrocyte membranes of patients with non-insulin-dependent diabetes mellitus (NIDDM) vs. control subjects and more abnormality in hyperkalemic than in normokalemic patients with NIDDM (Mimura et al. 1992). By contrast, a study of isotope dilution of ^{42}K from plasma to whole body in normal subjects vs. those with at least one other abnormal K parameter (serum level, renal excretion, exchangeable amount) suggested more rapid membrane transport in the "abnormal" group in a three-compartment model (Neumann et al. 1980).

Influences acting on potassium movement include that of insulin to promote transfer of K from extra- to intracellular sites, and aldosterone to cause secretion by the kidney and ion transport in sweat glands, intestinal mucosa and salivary glands (Murray et al. 1996). In turn, a slight increase in serum concentration of K$^+$ stimulates production of aldosterone. Known roles for K$^+$ occur in synaptic transmission and gastric HCl secretion (Alberts et al. 1994; Smith et al. 1983), which are primarily membrane functions, and extend to a wide range of intracellular functions which probably depend on the physiochemical relationship between the different states and transitions between K$^+$ and protein-bound K. The general constancy of the TBK/TBN (total body nitrogen) ratio in healthy subjects suggests a close alliance of K$^+$ with proteins. Yet the concentration of K$^+$ in plasma is maintained carefully within a tight range at considerable expense of TBK if need be (Edelman and Liebman 1959; Jeejeebhoy 1990). Characteristics of TBK in health and disease are further discussed in Sect. 12.3.

The radioisotope, ^{43}K, has a longer physical $T_{1/2}$ and lower (though still high) set of gamma energies compared to ^{42}K (Table 12.1). Therefore, ^{43}K would be preferable not only for a more extended time study of kinetics in body fluids, but for external regional (organ) counting and low-resolution imaging of potassium distribution. Equilibrium imaging of ^{43}K in skeletal muscle has been carried out (Miyamoto et al. 1975). Unfortunately, neither ^{42}K nor ^{43}K is presently readily available. While K_e by measure with the surrogate, ^{201}Tl, can differ unpredictably from the K_e using ^{43}K (MacKay et al. 1978), ^{201}Tl has been used not only for kinetic changes to be seen in heart muscle, but also in skeletal muscle, e.g., in lower limbs with peripheral vascular disease (Shreeve et al. 1978; Segall et al. 1990). Much further study of this type could be done to evaluate physical fitness, at one end of the spectrum, as well as conditions of muscular impairment (genetic dystrophies, atrophies due to surgery, immobilization, etc.) at the other end. The recent availability of the NMR technique for measuring protein-bound K as a component of intracellular K could open further vistas (Elia 1997; Parrish et al. 1997).

12.1.2.3
Calcium

Calcium is an essential constituent of bone and 98% of total body calcium (TBCa) is in bone. The other 2% in soft tissue has highly important functions, however, as will be discussed. The evaluation of bone mass by measurement of TBCa by neutron activation analysis and by other static techniques is a subject of a later section of this chapter. Here, the more dynamic aspects of calcium metabolism will be considered, and reasons given for continued or renewed interest in the application of available radiotracer techniques.

Early studies (1960s and 1970s) used the beta-emitting ^{45}Ca and/or the beta- and gamma-emitting ^{47}Ca (Table 12.1) to reveal the absorption and retention characteristics of calcium given by mouth or by vein and with different carrier loads (Tran and Chaudhuri 1982; Lutwak 1969). Measurements were made by various choices or combinations of plasma, urine and fecal specimens and whole body counting. One study (DeGrazia et al. 1965) used both tracers (one orally, one intravenously) with measurement of their ratio in the urine as a relatively simple evaluation of absorption. Another study (Cohn et al. 1965) analyzed the kinetics of calcium metabolism with data from blood, urine and feces plus whole body counting after intravenous injection. Modeling of data suggested a pool of calcium (plasma, extra- and intracellular) equilibrated in 1 h, a second pool (exchangeable bone) at equilibrium by 3 days (similar in size to pool 1) and a third pool of very slowly exchanging, "deep" bone. Another reason, besides the investigation of intestinal, hormonal or bone diseases, for studies of this type is a need to ascertain the de-

position and turnover in bone of potentially hazardous bone-seeking elements such as ^{90}Sr and ^{226}Ra.

Absorption of ingested calcium ranges from 25–60% in humans (Lutwak 1969; Sjoberg et al. 1970; Tran and Chaudhuri 1982). Differences are due partly to differing methods but they are in major part dependent on carrier load. There is a negative correlation between the mass of carrier ingested and per cent absorption. This is so for normal subjects at all ages; however, in elderly females (but not males) with osteoporosis there can be fixed absorption with changing intake, different etiologies, and perhaps genetic types, of osteoporosis. There is a gradual decrease in absorption with age (Avioli et al. 1965; Lutwak 1969). Various malabsorptive diseases (Lutwak 1969) and ileal shunt operations for obesity (Dano and Christiansen 1974) cause substantial decrease in absorption, whereas absorption is increased in Paget's disease (Lutwak 1969). Individual variation among normal (and osteoporotic) subjects in the degree of retention of ingested calcium at equivalent loads (Lutwak 1969; Sjoberg et al. 1970; Tran and Chaudhuri 1982), together with a long-known association of family history with osteoporosis, suggests that a renewed search for genotypic/phenotypic links by the above techniques with labeled calcium tracers (see also Chap. 14, Sect. 14.3.1.1.1) might be rewarding.

In soft tissue the concentration (about 2 mM) of total (protein-bound plus free-ionized) calcium within cells is only slightly less than the total concentration (2.5 mM) in plasma and extracellular fluid. However, in the extracellular spaces the concentration of ionized Ca^{2+} is about 1 mM, whereas the intracellular concentration is only 10^{-6} to 10^{-8} mM (Cohn and Roth 1983; Murray et al. 1996). Thus, a steep electrochemical gradient exists for Ca^{2+} across the cell membrane. Channels exist for movement down the gradient, and the opening of these channels is linked in a variety of cell types to an electrical impulse (nerve) or to the action of specific agonists on specific receptors in outer cell membranes or those of intracellular organelles (Bolsover et al. 1997). In conjunction with the electrical impulse or agonist-receptor interaction, ionized calcium enters the cytosol (raising its concentration ten-fold) and binding, in many cases, to a protein, calmodulin. This Ca-protein complex then typically stimulates functions specific for the cell type. In this way, Ca^{2+} is a non-specific, but vital, "second messenger". As has been phrased (Bolsover et al. 1997), its job is just to say "do it" – whether this be the release of neurotransmitters, the contraction of skeletal, cardiac or

smooth muscle, blood platelet coagulation, glycogenolysis, activity of ATP-generating enzymes, or gene
transcription and likely a variety of other cell functions. Under normal, healthy conditions the momentary increased amount of intracellular Ca^{2+} is promptly
extruded by a Ca^{2+}-ATPase pump or a Ca^{2+}/Na^+ ion exchanger. There is some evidence that Ca^{2+} can also be a
first messenger, since an extracellular calcium-sensing
receptor has been recognized to exist on cells from
parathyroid, thyroid, kidney, and brain (Hebert and
Brown 1996; Hory et al. 1998). This may give new
meaning to the understanding of hyper- and hypocal-
cemia, including mutant genotypes for the calcium receptor protein in one or another organ (Chattopadhyay
et al. 1996).

The generalized nature and transience of the role
of Ca^{2+} in intracellular activity, the very low concentration at which the action takes place, and the unfavorable physical characteristics of the beta-gamma
emitter, ^{47}Ca, discourage a consideration of external
imaging or counting of calcium changes in soft tissue. In some pathological states, however, the removal
of excess intracellular Ca^{2+} may be impaired with resultant accumulation and possibly increased Ca binding to proteins. This occurs in a variety of disease
states and in ischemia and injured tissues. There is a
striking demonstration of the latter by autoradiography of ^{45}Ca after blunt injury brain contusion in rats
(Nadler et al. 1995). After the availability of ^{99m}Tc-
phosphates in the 1970s, these agents have repeatedly
shown hyper-concentration in diseased or damaged
tissue at various sites, including liver, spleen, lung,
kidney, skin, skeletal and heart muscle, brain, and
blood vessels (Brill 1981). Many neoplasms, particularly more malignant types and metastatic sites, attract ^{99m}Tc-phosphates, seemingly on the basis of mi-
cro- (or macro-) calcification, as in the case of soft
tissues. Hypercalcemia, due to multiple myeloma
(Eagel et al. 1988) or to hyperparathyroidism (Padhy
et al. 1990), is associated with increased deposition of
^{99m}Tc-phosphates in various non-osseous organs.
Since calcium has no radioisotopes appropriate for
imaging, it is worth considering whether other alkaline earth metals have radioisotopes which could act
as surrogates of Ca^{2+} and reveal further or different
information than that gained from ^{99m}Tc-phosphates.
Earlier trials of bone-scanning with ^{135}Ba or ^{87m}Sr
(Table 12.1) (Hosain and Dunson 1982) suggests that
one or another of these nuclides might represent
Ca^{2+} in regard to its accumulation and turnover in
pathological soft tissues.

12.1.2.4
Magnesium

Magnesium is the fourth most abundant cation in the
body, the second most common divalent cation (after
calcium), and the second most common intracellular
cation (after potassium). Because of the frequent occurrence of Mg deficiency and its seeming involvement in a number of common diseases (see below), it
is unfortunate that nuclear medicine has not much
explored the potential of labeled Mg. However, ^{28}Mg
(Table 12.1) has been and could further be useful, as
can the newer method of evaluating concentrations of
Mg by the character of the MR spectrum of ATP.

Like calcium, magnesium is present in bone (and
teeth), but only about half of total body Mg is located
there. Furthermore, about 1/3 of skeletal Mg is said
to be readily exchangeable and to serve as a reservoir
for maintaining Mg in extracellular space and intracellular soft tissue (Nadler and Rude 1995). Of the
half of Mg in soft tissue (about 10–12 gm), a very minor amount is in the extracellular space; most is in
cytosol or stored in intracellular organelles. After
intravenous injection of ^{28}Mg in man (Avioli and
Berman 1966) compartmental modeling of plasma
data appeared to indicate boundaries representing ex-
tra- and intracellular spaces for exchangeable Mg
with 80% in the "intracellular" site. Others (Silver et
al. 1960) found with similar techniques three compartments with half-times of 1, 3 and 35 h. Skeletal
muscle, which holds most of soft tissue Mg, has a relatively slow turnover and in small animal studies
appeared to have heterogeneous components (Avioli
and Berman 1966).

The normal serum Mg concentration is 0.7–
0.9 mM/L of which most (60–75%) is ionized; intracellular total Mg concentration is 1–3 mM/L, but 10%
or less is ionized (Al-Ghamdi et al. 1994; Nadler and
Rude 1995). Because of the difference between extra-
and intracellular Mg as per cent ionized, there is a
moderately lower electrochemical potential of Mg^{2+}
inside the cell, thus a natural gradient towards influx
from outside the cell. Channels and carriers exist for
movement in both directions across outer cell membranes, but energetic mechanisms are required for efflux. Most evidence indicates the operation of a $Na^+/$
Mg^{2+} exchanger for extruding Mg^{2+} with the likelihood of involvement also of primary active transport
by an ATP pump (Beyenbach 1990) (Fig. 12.1). The
intracellular Mg^{2+} concentration is said to be maintained at the expense of ECF and bone reservoirs (Al-

Ghamdi et al. 1994). However, there are various clinical and experimental instances of decreased total body or intracellular (e.g., skeletal muscle) Mg despite normal serum levels (Flink 1986; Al-Ghamdi et al. 1994; Nadler and Rude 1995; Gustafson et al. 1996).

Magnesium is involved in a great variety of intracellular and membrane functions. More than 260 enzymes require Mg^{2+} for activation (Cohn and Roth 1983; Beyenbach 1990). Mg^{2+} stabilizes DNA, initiates its synthesis and promotes gene transcription, regulates muscle contraction, hormone synthesis and target action, and has multiple effects on ion traffic across the cell membrane (Beyenbach 1990). Mg^{2+} competes with Ca^{2+} for binding sites and helps sequester Ca^{2+} in sarcoplasmic reticulum (Al-Ghamdi et al. 1994). By preventing excessive cytosolic Ca^{2+} undue vasoconstriction as well as muscular hypercontractility is prevented. Mg^{2+} is needed to prevent hypokalemia as well as intracellular deficiency of K^+ (Al-Ghamdi et al. 1994), accomplished by its influence on membrane transport, and on ion binding at intracellular sites.

Although Mg is widely present in food sources and its intestinal absorption (normally 30–50%) varies inversely with intake (Graham et al. 1960; Nadler and Rude 1995), deficiency can be consequent to malnutrition, or to a variety of gastrointestinal disorders. While magnesium deficiency is supposedly unusual in healthy individuals with a normal caloric intake (Al-Ghamdi et al. 1994; Nadler and Rude 1995), the average consumption in food and water may be marginal for some groups in modern society (Rylander 1996). Insufficient intake is one of the reasons for the frequent occurrence (as found with [28]Mg tracer studies) of Mg deficiency in chronic alcoholism (Flink 1986; Al-Ghamdi et al. 1994; Nadler and Rude 1995). A variety of drugs (e.g., loop diuretics, chemotherapeutic and immunosuppressive agents) lead to depletion by renal wasting of Mg (Nadler and Rude 1995). Evidence continues to accumulate that the common

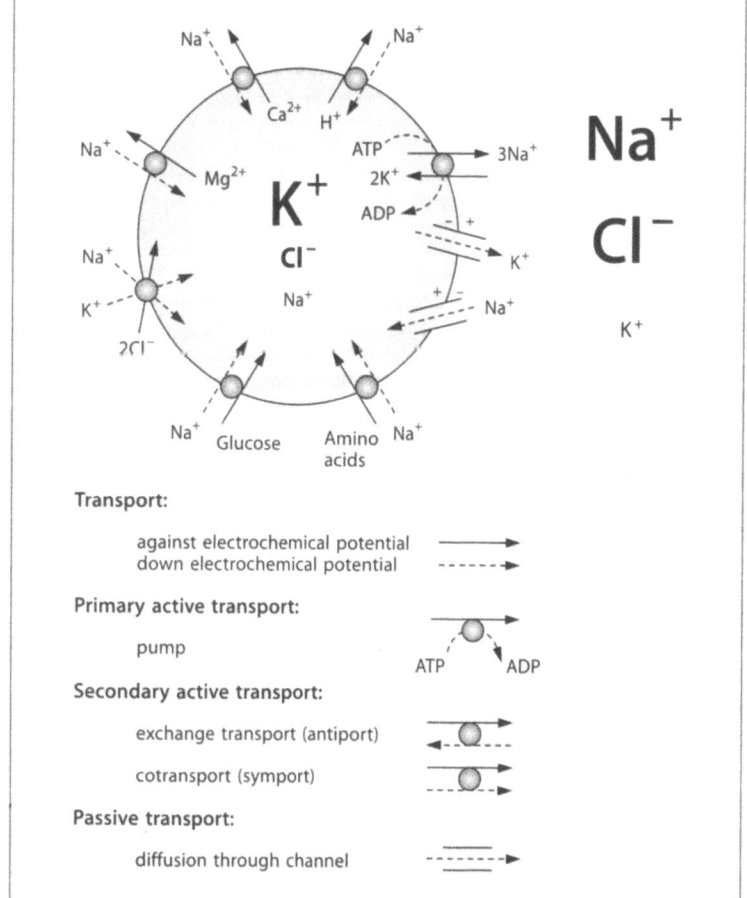

Fig. 12.1. Relationships for transport of Na^+, K^+, other inorganic ions and some small organic molecules across the plasma membrane. Coupling, types of transport, movement down or against electrochemical gradients and some linkage to energy pumps are indicated. (From Beyenbach 1990, with permission)

and important diseases of NIDDM, hypertension, obesity, coronary arterial disease and atherosclerosis are all characterized by high incidences of Mg deficiency (Al-Ghamdi et al. 1994; Nadler and Rude 1995; Satake et al. 1996). A factor common among these conditions is insulin resistance, which has been found to be closely associated with intracellular Mg deficiency (Resnick et al. 1990; Resnick 1992). The pathogenesis of Mg deficiency among these different diseases can be multi-factorial, but one of the factors is likely to be increased urinary loss. There are several known special, relatively rare genetic types of renal wasting of Mg (Al-Ghamdi et al. 1994) and it seems quite possible that the above common diseases may have more subtle, genetically-influenced renal Mg wasting. Further correlation of genes disposing to Mg deficiency with the phenotypic cluster of NIDDM, hypertension, insulin resistance, etc., seems indicated.

The above considerations make a strong argument for renewal of tests with tracers for Mg which can reveal Mg deficiency. The lack of convenient, available isotopes has been deplored (Beyenbach 1990; Al-Ghamdi et al. 1994). ^{28}Mg has been available only from a limited source (essentially Brookhaven National Laboratory) and at high cost with no expectation of change unless the volume of demand increases. The stable isotope, ^{26}Mg, has been measured by in vitro neutron activation analysis in double isotope (^{28}Mg-^{26}Mg) studies of intestinal absorption (Schwartz et al. 1978). The possibility of using ^{54}Mn as a surrogate for Mg has been suggested (Beyenbach 1990) because of the longer $T_{1/2p}$ (310 days) than for ^{28}Mg (22 h) (Table 12.1). In some sites (potentially an area for nuclear medicine), a fundamentally different technique is available, that of ^{31}P-NMR spectroscopy of the different phosphoryl groups of ATP. The chemical shifts of the α, β, and γ-phosphoryls of ATP depend on the state of ATP complex formation with Mg^{2+} and the separation between α and β depends on the fraction of ATP complexed with magnesium (Resnick et al. 1990; Resnick 1992). By this means the intracellular concentration of free Mg^{2+} has been measured in human erythrocytes in vitro (Resnick et al. 1990) and in different human muscles in vivo (Ryschon et al. 1996; Widmaier et al. 1996). A significant negative correlation was found between the intracellular concentration of Mg^{2+} in muscle and serum levels of Mg. Lymphocyte Mg is said to be more consistently reduced than the plasma level in conditions known to cause Mg depletion (Al-Ghamdi et al. 1994) and Mg in mononuclear cells more reflective of Mg in skeletal and cardiac muscle than Mg in erythrocytes

(Elin 1987; Satake et al. 1996). A further advantage of ^{31}P-NMR spectroscopy is the ability to measure intracellular pH; a decrease in the latter is a further characteristic of Mg^{2+} deficiency (Resnick et al. 1990).

12.1.2.5
Iron

Iron in the body is generally involved in oxidative functions, including transport of O_2 by hemoglobin in red cells, transfer of electrons in oxidative pathways by cytochromes and oxidation of particular compounds (e.g., tryptophan, xanthine). The most common disorder of iron (Fe) is that of Fe-deficiency anemias caused by malnutrition, malabsorption and/or chronic blood loss. On the other hand, increased iron storage which can occur in older males and post-menopausal women may increase vulnerability to ischemic heart disease and myocardial infarction (Sullivan 1989; Uchida 1995). The genetic disorder of hemochromatosis – with a fairly common prevalence among Caucasians of 1:300/400 for homozygotes and 1:7/12 for heterozygotes – is a more specific disease of excessive iron storage with crippling effects on liver, heart and pancreas (Bothwell et al. 1995; Uchida 1995). The disorder involves hyperactivity of the reduction of Fe^{3+} to Fe^{2+} with resultant excessive intestinal absorption (Raja et al. 1996). Absorption has been studied by the combined use of two radioisotopes, ^{59}Fe and ^{55}Fe (Viteri and Kohaut 1997) (Table 12.1) or of two stable isotopes, ^{54}Fe and ^{57}Fe (Barrett et al. 1997). In vivo kinetics of disappearance of iron from plasma, incorporation into red cells and storage and turnover in different tissues (liver, spleen, bone marrow) has been evaluated with ^{59}Fe in various types of anemia with distinctive characteristics in each, as well as in hemochromatosis (Pollycove and Tono 1975; Price and McIntyre 1984).

12.1.3
Anions

12.1.3.1
Chloride

In nature chloride is the ubiquitous, major anion balancing the cation, sodium, as exemplified in human and other mammalian extracellular fluids (ECF). Early studies utilized information from isotope dilution of radiotracers for chloride to define the ECF, but it was realized that the application was imperfect, because labeled chloride (or bromide) occupied the intracellular space of many organs, serving also as an important intracellular anion. The latter finding has become more significant since the recognition of specific Cl ion channels and their function and dysfunction. Thereby, nuclear medicine techniques may share in the evaluation of the "model" genetic disease, cystic fibrosis, and possibly other disorders (e.g., in the CNS).

Early studies used ^{38}Cl, ^{36}Cl and, as a surrogate for Cl, ^{82}Br (Table 12.1) (Manery and Haege 1941; Gamble and Robertson 1952; Nicholson and Zilva 1960; Pierson et al. 1978). For most soft tissues of man and other mammals (dog, rat, rabbit) equilibration of labeled chloride or bromide between plasma and a space definable as extracellular water (ECW) (synonymous with ECF) occurred within 1 h. However, certain tissues, i.e., testes, brain and pyloric mucosa (Manery and Haege 1941), intestine, skin (Pierson et al. 1978) or red blood cells (Nicholson and Zilva 1960) showed amounts of chloride which were either slowly or never penetrated by a corresponding tracer. Slow penetration of brain fluids, like that of ^{24}Na, was presumed due to the blood-brain barrier. An unusual concentration in RBCs is explainable as due to the well-known "chloride shift" in and out of the interior of red cells which facilitates transfer of CO_2 from oxidizing tissues to lung. Apparent intracellular chloride in RBCs (of man) are, however, quite small (2–3% of total body Cl), as are the amounts in thyroid, salivary glands or gastric juice (Nicholson and Zilva 1960; Pierson et al. 1978). Nevertheless, comparisons with other tracers of ECW indicate that about 23% of the space penetrated in the first few hours by radio-bromide or radio-chloride is intracellular, and that tissues such as stomach, intestine, brain and skin are the main sites of the intracellular halide (Pierson et al. 1978). Further discussion of various radioactive indicators for the ECW and their clinical application is contained in Sect. 12.3.2.1.2.

Until the 1980s there was little clinical interest in chloride ion channels that provide intracellular access – which is down a chemical gradient but dominated by the adverse voltage gradient. The lack (or rarity) of pathology of previously recognized functions, e.g., the chloride shift in RBCs, gastric secretion of HCl, or active re-absorption in the distal ascending loop of Henle in the kidney, suggested little need for special clinical (or nuclear medical) attention to chloride channels in membranes. One important disease which has focused attention on chloride channels is cystic

fibrosis. European folklore associated salty-tasting skin of a baby to a likelihood of dying young. By the mid-twentieth century a disease of thick mucous secretions from the lungs (inviting infection) associated with pancreatic insufficiency was termed "cystic fibrosis of the pancreas" or "mucoviscidosis". Shortly thereafter the connection with salty sweat was made. Family studies clearly indicated an autosomal recessive disease affecting usually Caucasian populations, in which the incidence of 1/2500 births make it the most common fatal disease of such genetic type (Welsh et al. 1995; Ross 1996; Bolsover et al. 1997). Cystic fibrosis (CF) is a multi-organ disease of epithelial secretory glands, in which faulty electrolyte transport somehow promotes hypersecretion of glycoproteins, which may also be abnormal in chemical content (e.g., oversulfated) (Welsh et al. 1995). Altered electrical properties of CF respiratory epithelium were found to be associated with abnormalities of both sodium and chloride transport. Sweat glands were found to secrete initially normal sodium and chloride, but in a more peripheral part of the gland the normal re-absorption of Cl fails to occur. Because of electrolytic attraction, Na is also retained abnormally in the sweat. The primacy of abnormality of Cl absorption is indicated by the greater electronegativity of CF sweat compared to normal sweat. In the lung and pancreas the lack of properly functioning Cl ion channels appears to cause an initial anion deficit in the secretions. Meconium ileus is seen both early and late; bronchiectasis, respiratory failure, male infertility, biliary cirrhosis and diabetes are late manifestations. Presently with supportive therapy the life expectancy is in early adulthood (Bolsover et al. 1997).

In the 1980s the above findings on pathogenesis paralleled an intensive genetic hunt which is a model in the history of genetic identification of disease (Welsh et al. 1995; Bolsover et al. 1997). The genetic search was at first difficult because there was inadequate knowledge of what the gene does in the phenotype. No abnormality in heterozygotes (1/25–40 white persons) could be found. Extensive trials with restriction endonuclease enzymes with analysis of many fragment length polymorphisms finally revealed a particular DNA variant inherited with the gene. Further screening with genomic clones of cDNA from sweat glands located the gene and identified one primary triplet codon (delta F508) which is missing in 70% of patients. However, to date 300–350 mutations of the same gene (on chromosome 7) have been discovered, although only seven or eight have significantly high incidence. This multiplicity helps to explain the considerably variable pattern of clinical expression of the disease and attendant difficulty in diagnosis (Welsh et al. 1995; Bolsover et al. 1997). The protein coded by the gene is now recognized to be a chloride ion membrane channel, which may be deficient in production or defective in protein processing, regulation or conductance. These different defects correspond to classes of mutation and to severer or milder forms of disease. There is a looser correlation of specific mutations (including delta F508) with lung disease than with pancreatic insufficiency (PI). Approximately 80–90% of homozygotes have PI and are classified as severe CF. The possibility of gene insertion into pulmonary endothelium is in active development (Tanne 1999).

A number of studies with ^{99m}Tc- or otherwise labeled particles inhaled by CF patients have demonstrated abnormal distribution patterns of varying severity with impairment or unusual variability of clearance (Laube et al. 1989). Such imaging techniques are also used to evaluate the efficacy and effect of delivery of therapeutic agents, such as gentamycin (Ilowite et al. 1987) or rhDNAse (Laube et al. 1996). Imaging for PI was started in the 1960s with ^{75}Se-selenomethionine and later has been performed with ^{11}C-L-methionine (Syrota et al. 1981), ^{11}C-DL-valine and ^{11}C-DL-tryptophan (Hubner et al. 1979). These studies show that PI can be well demonstrated by evidence of low initial pancreatic uptake, rapid fall-off of activity, poor response to secretin stimulation by external imaging, and/or low appearance of tracer in duodenal aspirate. Biliary cirrhosis is demonstrable as a greatly increased T1/2 of hepatic clearance without lower retention as seen with ^{99m}Tc-IDA (Keefe et al. 1988). It may be that a combination of various nuclear medicine tests would be useful not only to diagnose disease initially but to characterize different patterns of multi-organ involvement which are the phenotypic expressions of particular mutant genotypes.

Genetic typing by linkage analysis can now be done for pre-natal diagnosis of homozygous CF and also to detect heterozygotes. Screening for the latter has been recommended where there is a positive family history, but the desirability of general population screening is controversial (Welsh et al. 1995). The most obvious somatic test for CF is electrolytic analysis of sweat, since this is the most common abnormality of classical CF (Welsh et al. 1995). The identification of heterozygotes has seemingly not been tested. Chemical (or pH) analysis is available but may have drawbacks, e.g., inaccu-

racy due to evaporation, or variable hydration, or lactic acid content of sweat. A substantial volume of sweat, e.g., up to 14 l/day, may be lost with copious sweating (Smith et al. 1983). Such sweat may normally contain as much as 75 mmol NaCl/l, but there is normally lesser concentration of salt in more moderate sweating and virtually none in insensible perspiration. However, homozygotes for CF (and possibly heterozygotes) would show excessive content of Na or Cl even with minor sweating and, indeed, homozygotes have shown Addisonian-like crisis with heavy sweating. There is the possibility that an abnormally high concentration of labeled chloride or bromide in the skin/sweat could be observed by in vitro analysis of sweat. Also, abnormal expansion of the apparent chloride space by plasma analysis in the first hours after i.v. injection of labeled halide could signify abnormal loss into sweat, if other (e.g., renal) factors are excluded.

There is increasing recognition of the concept that chloride ion channels are the conduit for cellular influx of a variety of small molecular compounds, e.g., neutral or acidic amino acids and lactic acid (suggesting anion selectivity) and polyols (Kirk et al. 1994; Pasantes-Morales et al. 1994). In the central nervous system the action of two amino acids, gamma-amino butyric acid (GABA) and glycine, as neurotransmitters has been linked to a chloride ion channel in a specific way. These compounds are thought to be the major transmitters for mediating fast inhibition in the vertebrate CNS. By opening Cl ion channels in the post-synaptic receptor, hyperpolarization is maintained, thus opposing the action of excitatory stimuli which operate through cationic influx to depolarize the membrane potential. Tranquilizing drugs (benzodiazepines and barbiturates) potentiate the action of GABA (Alberts et al. 1994). Certain radioligands for PET or SPECT are directed toward the benzodiazepine receptors (Pike et al. 1993). Another class of drugs, the picrotoxins, are believed to act through binding to sites within the Cl ion channel itself and the potential for imaging has been investigated by labeling with ^{11}C, ^{18}F or ^{123}I (Snyder et al. 1995).

12.1.3.2
Phosphate

The major part of body phosphate is in mineral form in structural materials such as bone and teeth, but a minor part (15–20%) in soft tissues has long been known to be mainly present as important intracellular phosphorylated organic compounds. The early availability of the beta-emitter, [32]P (Table 12.1) aided biomedical researchers in the 1930s and 1940s to gain evidence (mostly from animal tissues) on the turnover and other characteristics of the large variety of phosphorylated compounds (nucleic acids, phospholipids, intermediary substrates, energy-transferring nucleotides) then known. The particularly rapid accumulation of inorganic phosphate (Pi) into bone marrow, liver and spleen, the close link of phosphorylation to oxidative metabolism, the phenomena of intra- or inter-molecular transfer of phosphate groups without the intermediation of inorganic phosphate were among the characteristics elucidated with the aid of [32]P (Hevesy 1948). Knowledge of the phosphorylation of proteins or amino acids (tyrosine, threonine, serine, aspartate, glutamate) at free hydroxyl or carboxyl groups and the consequent activation or deactivation of enzymes, transporters, etc., was gradually revealed in research of later decades and continues to unfold (Bolsover et al. 1997).

The beta emission of ^{32}P was first applied in the 1930s for the treatment of leukemia, but was unsuccessful. Because of increased bone mineral turnover adjacent to bone marrow metastases (and possibly by direct incorporation into tumor cells) ^{32}P-phosphate was much more effective in the relief of bone pain from prostatic or mammary cancer and was used widely from the 1940s through the 1980s (Silberstein 1993). Since then other bone-seeking radionuclides have superseded [32]P. The latter was used from an early time for suppression of excessive red cell production in polycythemia vera and continues to remain an option for the treatment of this disease, particularly in elderly patients (Sostre 1995), in whom the induction of leukemia is not a major concern.

The advent of in vivo nuclear magnetic resonance spectroscopy (MRS) of the natural, non-radioactive isotope, ^{31}P, which began in the 1970s and developed further in the 1980s, has provided a new and exciting possible direction for nuclear medicine. While MR imaging of protons in water has moved rapidly into clinical usage, ^{31}P-MRS (and ^{1}H-MRS) have been much slower to develop clinically. The magnetic sensitivity of ^{31}P is only 1/10 that of ^{1}H and the far lower biological concentrations (1/500–1/1000) (Willcott et al. 1983) makes detection far more difficult and spatial localization much more crude. Yet the information on the energy state and other metabolic characteristics of high, medium and low energy-containing compounds of phosphorus in organic combination or as free phosphate makes this technique highly valuable and worthy of strong pursuit of clinical applica-

tions. In the 1990s much more was done, mostly by adapting clinical MRI units. The nuclear medicine practitioner can appropriately have an active, or even leading, role in ^{31}P spectroscopy, in which there is emphasis on biochemical interpretations and intracellular kinetic changes. (For the same reasons, nuclear medicine personnel can well be involved in ^{1}H-MRS and the still-emerging natural abundance and tracer ^{13}C spectroscopy.)

Spectra of ^{31}P effectively display relative, and can be calibrated to measure absolute, concentrations of phosphorus compounds of varying energy content and organic association. This includes the three separately seen phosphates of the common "energy coin" of cells, ATP (or, more generally, NTP, nucleotide triphosphates). The energy contents (energy of hydrolysis) of ATP and ADP are intermediate between those of higher energy compounds (e.g., creatine phosphate, phosphoenolpyruvate) and lower energy compounds (e.g., hexose phosphates) (Murray et al. 1996). Concentrations of ATP tend to remain fairly constant in cells under widely different conditions of energy availability. In states of good oxygen and substrate supply and mitochondrial integrity high-energy compounds ("phosphagens"), mainly phosphocreatine (PCr), are high in concentration, while under conditions of ischemia or other deprivation, phosphagens are low in concentration and inorganic phosphate (P$_i$) is typically elevated. In current practice, seven distinct peaks in the spectrum are generally recognized, i.e., three phosphorus atoms of ATP (NTP), PCr, P$_i$, phosphomonoesters (PME) and phosphodiesters (PDE) (Fig. 12.2). (Sometimes diphosphodiesters (DPDE), mostly UDP-hexosamines, is another peak.) The PME and PDE may be sugar phosphates or phospholipids or both, depending on the tissue being analyzed. The pH of the tissue is revealed by the degree of separation or "chemical shift" of the Pi peak relative to that of PCr. The degree of binding to Mg can also cause chemical shifts (see Sect. 12.1.2.4). At present the spatial resolution of useful spectra of ^{31}P is on the order of 1–10 cc volumes (voxels), using appropriately configured and localized surface coils (Ross et al. 1997; Richardson et al. 1997) (Fig. 12.2).

The earliest studies of in vivo ^{31}P spectroscopy focused on relatively large muscle masses under conditions of rest, exercise and recovery (Chance et al. 1983). These studies typically used small-bore magnets of high magnetic field strength, e.g., 4.5 T; advances in technology now permit use of wide-bore magnets with 1.5 T, although higher strengths are

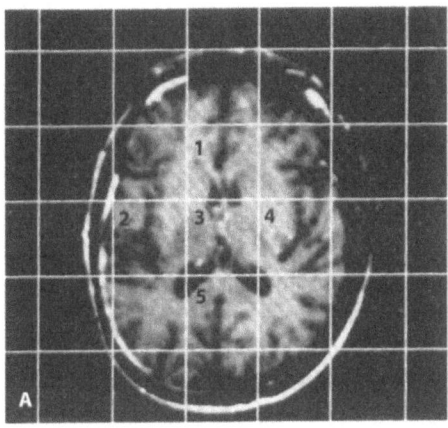

Fig. 12.2 A, B. Representative magnetic resonance (MR) data from a 33-year-old control male. **A** A T$_1$-weighted transverse image (SE 600/22) at the level of the basal ganglia, showing the voxels selected at this level. The same configuration of voxels was selected from the transverse image 4 cm higher, at the level of the centrum semiovale. **B** A representative ^{31}P MR spectrum, from voxel no. 3 of (**A**). (Reprinted from Richardson et al. 1997, with permission)

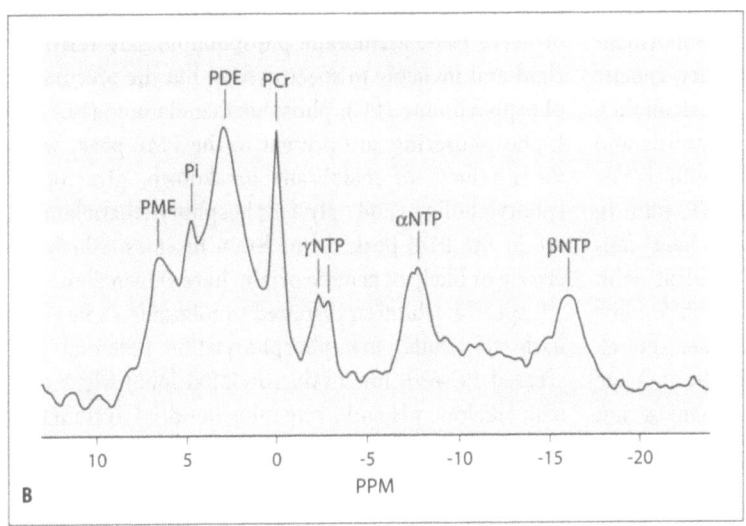

preferable. For muscle studies limbs can be inserted into small-bore magnets and exercise stress applied with in situ ergometers. The disorder earliest studied was McArdle's disease (genetic deficiency of muscle glycogen phosphorylase), showing expected, abnormal findings upon exercise, including the unique lack of fall of intracellular pH (Chance et al. 1983; Kent-Braun et al. 1994). Other genetic disorders, such as Duchennes's muscular dystrophy, phosphofructokinase deficiency, and mitochondrial myopathies have shown characteristic changes (Kent-Braun et al. 1994). Post-viral and other cases of chronic muscular fatigue and myalgia have shown abnormalities. Effects of attempted therapy of these various muscular disorders can be followed by ^{31}P-MRS. As opposed to the study and diagnosis of disease, the effect of muscle training in normal human subjects has been repeatedly studied with ^{31}P-MRS and shown clearly increased potential for oxidative metabolism, particularly at high exercise levels (Kent-Braun et al. 1990,

1994). In connection with recent suggestions that creatine administration will improve muscle performance even in normal persons, a finding by ^{31}P-MRS (Frassineti et al. 1996) that oral phosphocreatinine results in a smaller depletion of PCr at high work rates and smaller acidification should stimulate further such study not only in normal but in muscle-diseased states.

Techniques of limb (particularly lower limb) rest/exercise/recovery studies with appropriate magnets and ^{31}P-MRS have also been valuable for demonstrating abnormalities in peripheral vascular disease (Chance et al. 1983; Schunk et al. 1997). Findings have included noticeable changes even at rest (elevated PME/βATP, PDE/βATP, Pi/PCr+P$_i$), increased acidosis on exercise and (most prominently) long-delayed post-exercise recovery of the ratio of PCr/P$_l$ (Chance et al. 1983; Kent-Braun et al. 1994; Schunk et al. 1997). Application of surface coils at different levels on an affected leg might aid in locating the appropriate point for surgical intervention (Chance et al. 1983). The technique may reveal not only macro- but microangiopathy (Schunk et al. 1997).

Heart muscle is quite accessible to ^{31}P-MRS by means of surface coils placed on the chest. This was first done on an infant who had cardiac hypertrophy and familial evidence of congenital cardiomyopathy; such a study at that time was possible only by placing the body entirely within a small-bore, high field strength magnet (Whitman et al. 1985). The subject showed marked departure from normal in the baseline PCr/Pi ratio of heart and also skeletal muscle. There were general metabolic abnormalities (e.g., fasting hypoglycemia) and nutritional treatment showed improvement in the spectroscopic abnormalities. The area of cardiomyopathies, whether genetic (e.g., diabetic) or environmental (e.g., alcoholic), should be a fruitful one for diagnosis, prognosis and evaluation of treatment. Adult patients with aortic valve disease have shown decreases of PCr/P_i ratio in heart muscle commensurate with degree of heart failure. The abnormality is much more evident with pressure overload (aortic stenosis) than with volume overload (aortic incompetence) (Neubauer et al. 1997). Heart muscle which is ischemic from acute myocardial infarction promptly shows changes not only in PCr/P_i ratio but gradual decrease of ATP, while after restoration of blood flow the extent of correction of spectroscopic abnormality can signify the amount of remaining damaged tissue (de Roos and van der Wall 1994). Likewise, severe chronic coronary artery disease (>70% LAD stenosis) is detectable by ^{31}P spectra from the anterior myocardium with evidence of efficacy from re-vascularization (de Roos and van der Wall 1994).

Both liver and kidneys can be viewed by ^{31}P-MRS, either with surface coils or with whole body coils. Patients with diffuse liver disease, such as hepatitis and cirrhosis, tend to have normal ratios of various phosphorus metabolites, but general decreases in absolute amounts, including ATP (Meyerhof et al. 1989). Intracellular pH was more acidic than normal in cirrhosis, more alkaline in hepatitis. Both rats and humans showed low maximum PME peaks after intravenous fructose load during biliary obstruction, continuing after biliary drainage (and after normalization of serum bilirubin) (Kamuma et al. 1997). These findings indicate that fructose phosphorylation is inhibited. A test in healthy human volunteers of provocation by concomitant administration of fructose and ethanol showed a significant reduction of PME degradation,

suggesting that fructose metabolism is slowed by ethanol after the initial phosphorylation step (Boesch et al. 1997). A study (Heindel et al. 1997) of patients with pathologic changes after kidney transplant demonstrated differentiation from uncomplicated kidney transplants by increased ratio of P_i/a-ATP and reduced pH. Kidneys with tubular necrosis had low PME/PDE ratio and could be further differentiated from rejection reaction by pH (lower in rejection than in tubular necrosis).

^{31}P-MRS of the brain is promising not only because of information on bioenergetics, but because of phosphorylated intermediates from glycolysis and of phospholipids involved in formation of the myelin sheaths of nerve cells. Membrane phospholipids are relatively rigid and invisible to spectroscopy, but the precursors, phosphocholine (PC), phosphoethanolamine (PE) and L-phosphoserine, are present in the PME peak, while the products of membrane breakdown, glycerophosphorylcholine and glycerophosphorylethanolamine, are in the PDE peak. Some brain diseases, which are clearly or likely of genetic origin, have shown abnormal ^{31}P spectra. Children disposed to migraine show (similarly to adults) low phosphorylation potential (decreased PCr/P_i), low brain (occipital lobe) Mg^{2+} content, alkalotic pH and likely mitochondrial dysfunction even in interictal periods (Lodi et al. 1997). Studies in a few subjects with epilepsy have variously indicated low PCr and/or high P_i, low PME and alkaline pH shift (Novotny 1995). Low PME are also seen in demyelinating disorders, bipolar disorders and schizophrenia with some sub-types suggested (Kato et al. 1994; Hinsberger et al. 1997). The low PME are evidently not due simply to loss of brain volume. The common (5% of population) and presumably genetic disorder of dyslexia has some biological characteristics similar to schizophrenia, but, in contrast to the latter, has increased PME on ^{31}P spectra. This suggests a different kind of metabolic abnormality of phospholipids and different genetic factors (Richardson et al. 1997). The clinical heterogeneity of dyslexia has yet to be correlated adequately with genetic sub-types, and ^{31}P-MRS could possibly help with the sorting out of this disorder. A new (and, so far, rare) inborn error of metabolism is that of creatine deficiency caused by deficiency of hepatic guanidinoacetic acid methyltransferase. There occurs in early childhood a dystonic-dyskinetic syndrome with motor and mental developmental delay. The ^{31}P-spectra of the brain is strikingly abnormal (very low PCr; high GAP); beneficial effects of therapy with oral creatine can be documented by the ^{31}P pattern (Schulze et al. 1997). Another brain disorder originating in the liver is that of hepatic encephalopathy, usually associated with alcoholic liver disease. Study by ^{31}P-MRS found (in basal ganglia) reductions of PME/ATP and PDE/ATP, which, together with evidence from ^{1}H-MRS, indicated altered cerebral phospholipid metabolism (Taylor-Robinson et al. 1996). Because both PME and PDE peaks may contain glycolytic intermediates, interpretations of findings in this (and other such) studies must remain guarded until future studies can better define contents of the ester peaks.

Various tumors show abnormal ^{31}P spectra, which differ depending on tumor type (Daly and Cohen 1989; Rutter et al. 1995) and which are useful for staging degree of malignancy, possibly (in some cases) for prognosis of treatment, and certainly for detecting and following response to treatment (Maris et al. 1985; Carswell 1987; Daly and Cohen 1989; Maldonado et al. 1998). A characteristic most common in malignant tumors is high PME peaks (probably PC and PE) relative to others (ATP, Pi) and some also show high PDE or DPDE; most likely the high phosphoesters signify active cell proliferation with changes in key directional enzymes, which, in turn, reflect differences in gene expression (Daly and Cohen 1989). Some tumors exhibit basically high energy states (high PCr, low Pi), while others before treatment have an opposite pattern of bioenergetics. Intracellular pH values may be high or low. There is a very early response to effective treatment, often preceding clinical or other imaging evidence by days, weeks or even months (Carswell 1987; Daly and Cohen 1989). The initial response may be a rapid increase of P_i with substantial drop in high-energy phosphates. This is followed by a return toward pretreatment pattern, but the long-term effectiveness of treatment is generally indicated by the curtailment of the previously high PME and PDE values (Daly and Cohen 1989; Maldonado et al. 1998).

12.2
Uncommon or Trace Mineral Elements

12.2.1
Introduction

Some 50–60 trace elements may be found in the human body but only a few are "essential", having known biochemical functions, while still fewer are as-

sociated with distinct pathology or clinical disease (Underwood 1971). The multi-valent property of some trace cationic metals appears to add stability to the polyanionic protein molecule and in so doing confers essential co-enzyme activity for many enzymes. When such an element is firmly bound and uniquely required the complex is termed a metalloenzyme. In another, larger group the metal can be readily removed or replaced by another, and these are designated metal-enzyme complexes. It is notable that some trace metals (nickel, chromium, cobalt and manganese) are firmly bound to RNA, which may also stabilize the three-dimensional structure and thus play a role in the synthesis of proteins in general (Underwood 1971; Cohn and Roth 1983).

Trace cations which have important biochemical functions include copper, zinc, manganese and (less clearly) nickel. Copper and zinc have been linked to particular, and essentially genetic, diseases, while manganese has been most noted clinically for its environmental toxicity. Biomedical studies of all three metals have depended extensively on studies with radioisotopes.

Among anions, iodide is classical, in endocrinology and nuclear medicine, for providing early and effective application of radiotracers. Because so much is elsewhere recorded, only recent findings, limited essentially to the inorganic ion, will be discussed. Without a gamma-emitting isotope, metabolism of sulfate has received little attention in nuclear medicine, but its biological importance calls for mention of some tracer studies.

12.2.2
Cations

12.2.2.1
Copper

Nuclear medicine has had available two radioisotopes of copper (^{64}Cu and ^{67}Cu), with which useful studies and diagnosis of disabling genetic disorders of either excess or deficiency of body copper have been done.

Copper is a vital trace element which has been found associated with at least 12 enzymes, most of which have oxidative functions (Cohn and Roth 1983). Deficiency of copper produces two devastating clinical entities, known as Mencke's disease and occipital horn syndrome, whereas an excess of copper, in Wilson's disease, has widespread adverse effects but appears at a later age, has more subtle and (usually) chronic manifestations and is more susceptible to treatment. Much is known about the genetics of each of these three diseases (Danks 1995; Murray et al. 1996). The deficiency diseases are due to mutations on chromosome X and may be allelic mutants or on closely linked foci; the mutant gene for Wilson's disease, which is inherited as an autosomal recessive, is found on chromosome 13 in the region 13q 14.

In Mencke's disease there is a mutation for a copper-binding P-type ATPase, which is thought to direct the efflux of copper from cells. This is mainly expressed as defective intestinal absorption of Cu, leading to a deficiency in which the newly- or recently-born show depigmented steely hair, bloated, saggy-skinned facies, arterial degeneration, neuronal degeneration mainly in the cerebellum, and other organ defects. Death usually occurs before 2 years of age. Occipital horn syndrome may be a milder form of the same defect, since it shows similar clinical changes but survival into adult life is usual. In Mencke's disease serum concentrations of Cu and ceruloplasmin (the a_2-globulin which carries 90% of serum Cu) are both low. ^{64}Cu given orally is poorly absorbed, but is cleared normally from serum after intravenous injection. Disturbances of handling of ^{64}Cu in cultured cells, including fibroblastic, amniotic and chorionic cells, can be definitive for diagnosis (Danks 1995).

Wilson's disease (WD), also called hepatolenticular degeneration, presents a still better opportunity for useful application of copper radiotracers, which can play a decisive role in diagnosis of a disease that can be difficult to distinguish from other, mostly hepatic, diseases. Although this genetic disease usually appears in childhood or young adulthood, it may not do so until middle age (Danks et al. 1990). The two main disturbances of copper metabolism in WD are a reduction in the rate of incorporation into ceruloplasmin and reduction in biliary excretion of copper. Copper accumulates substantially (ten-fold normal) in the liver and subsequently in other organs, such as brain, particularly the basal ganglia. Total serum Cu may not be abnormal due to the combination of low circulating ceruloplasmin and high free copper.

The genetic mutation now appears to involve a gene encoding a copper-binding P-type ATPase, which is expressed in liver and thereby different from that of Mencke's disease. Because an important function of ceruloplasmin is to oxidize Fe^{2+} to Fe^{3+}, hemolytic anemia is often seen. Neurologic manifestations are common. Bone and joint disorders may oc-

cur. These and other clinical features reflect deranged activity of many known copper-containing enzymes, including tyrosinase, lysyl oxidase, cytochrome oxidase, dopamine β-hydroxylase and superoxide dismutase (Cohn and Roth 1983; Danks 1995).

The homozygous condition of WD is rare, i.e., one case per 50,000–100,000, in most Western countries, but more frequent in some island or other isolated, in-bred populations. Heterogeneity is likely in view of variability of disease effects and age of onset. Although heterozygotes (about 1/100 persons) do not have clinical manifestations, reduction of serum ceruloplasmin and copper are found in 20%; a larger percentage (80–90%) show changes in handling of radioactive copper.

Considering the physical characteristics of ^{64}Cu and ^{67}Cu (Table 12.1), the latter, with longer half-life, is more desirable, but is less available. Analyses have included rates of plasma disappearance and accumulation in liver and brain, rates of oral absorption (to evaluate effectiveness of therapy with zinc, which inhibits absorption), rates of urinary excretion and whole body disappearance rates. A further test without radiocopper is measurement of copper content in liver biopsy by neutron activation. The ultimate test for WD is demonstration of negligible uptake of radiocopper into ceruloplasmin. At first rapid and then slowing, but continuing decline of total serum radioactivity over 24–48 h (no normal resurgence of radiocupric-ceruloplasmin) is diagnostic (Sternlieb and Scheinberg 1979; Wahner and Goldstein 1979; Danks et al. 1990; Danks 1995). Other abnormal findings in WD with ^{64}Cu or ^{67}Cu are high 24/4 h ratios of hepatic content by external imaging and counting (Czerniak and Zwas 1976) and prolonged whole body turnover (Willvonseder et al. 1974; Chaudhuri and Tran 1982) (Fig. 12.3) with heterozygotes intermediate between normal and homozygous WD in these parameters. Efficacy of treatment of WD with D-penicillamine can be well evaluated with radiocopper (Willvonseder et al. 1974) (Fig. 12.3). An interesting observation is a high uptake of ^{64}Cu into the frontal brain of schizophrenics, with or without WD (Czerniak and Zwas 1976).

The damaging effect of increased deposition of copper in the basal ganglia of the brain in WD, which may be manifested as parkinsonism, dystonia or cerebellar ataxia, has been evaluated with ^{18}F-FDG and ^{18}F-DOPA, both of which show decreased striatal uptake with the former also showing general cerebral hypometabolism (Snow and Calne 1995).

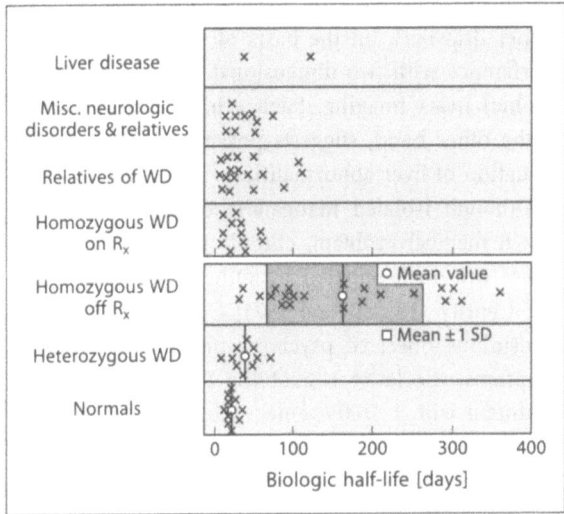

Fig. 12.3. Biological half-life of ^{67}Cu in whole body of patients with treated and untreated Wilson's disease (WD) and in related or other subjects. (Reprinted from Willvonseder et al. 1974, with permission)

12.2.2.2
Zinc

Gamma emitters (^{65}Zn and ^{69m}Zn) have been useful in one known genetic disorder and these isotopes of zinc (or the positron/gamma emitting ^{62}Zn) might be applied in other ways as suggested below.

Zinc is normally present in the human body in such abundance (ca. 2 g/70 kg body weight) as to make it one of the trace elements present in highest amount – about half that of iron, 10–15 times that of copper and more than 100 times that of manganese (Underwood 1971). As in the case of both copper and manganese, a large number and diversity of enzymes contain zinc. Enzymes appearing to require zinc include carbonic anhydrase, several dehydrogenases (alcohol, lactate, malate, glutamate), alkaline phosphatase, aldolase, carboxypeptidase, RNA and DNA polymerase, reverse transcriptase and others (Underwood 1971; Cohn and Roth 1983). One would therefore expect that zinc deficiency would produce widespread pathologic effects. This is evidenced in the deficiency disease of acrodermatitis enteropathica (see below).

In animals and humans addition of zinc to the diet increases the rate of wound healing, and zinc therapy is claimed to be beneficial in some cases of atherosclerosis (Underwood 1971) – the two phenomena could be linked in terms of repair of arterial wall trauma (Underwood 1971). Perhaps relatedly, the aortic wall has a relatively high concentration and turnover of ^{65}Zn (Strain et al. 1964). Zinc is involved in the production and function of insulin (Underwood 1971), which could relate to problems of wound healing in diabetics.

Radiozinc, i.e., ^{65}Zn and ^{69m}Zn (Table 12.1), have been useful in showing absorption and the different accumulation rates, relative concentrations, and turnover rates of zinc in various organs and tissues. Rapid accumulation and turnover occurs in liver, kidney, pancreas, heart muscle and spleen (Strain et al. 1964; Underwood 1971). Skeletal muscle, red blood cells, bone and CNS are sites of slower uptake and longer retention. Skin, nails and hair and the reproductive organs are particularly vulnerable to zinc deficiency, as seen in acrodermatitis enteropathica. The latter is an inherited recessive autosomal disorder, in which the predominant signs, usually starting in infancy, are diarrhea, growth retardation, dermatitis, alopecia, recurrent infections, mood disorders and anemia (Van Wouwe 1989). The primary problem, as demonstrated with ^{65}Zn and ^{69m}Zn, is a defect in active transport across the duodenal mucosa. The clinical course can be variable and the disease sometimes not recognized until adulthood. The disease can be confused with cystic fibrosis, coeliac disease or other dermatologic or mental disorders. Absorption of radiozinc is the only test, done in vitro by biopsy or in vivo by counting of whole body, serum or urine, which gives, according to most studies, no overlap with controls or false positives with other conditions (Van Wouwe 1989).

Primary zinc deficiency occurs among young men in Iran and Egypt, who have subsisted on diets of whole wheat, corn and beans containing high phytates which bind zinc and prevent absorption. Hypogonadism and dwarfism have been ascribed to zinc deficiency in the affected subjects (Underwood 1971).

A relatively prolonged retention of radiozinc in the prostate gland has led some workers to attempt imaging with hopes of recognizing either benign prostatic hypertrophy or carcinoma of the prostate. However, such attempts with ^{65}Zn (Chisholm et al. 1974), ^{69m}Zn (Fruhling and Coune 1975) or the positron- and gamma-emitter, ^{62}Zn (Yano and Budinger 1977) have not been encouraging. The study with ^{62}Zn (in animals) was also evaluated for pancreas imaging (zinc in islet cells is at higher concentration than in exocrine cells) and used zinc chelates with various amino acids besides ZnCl$_2$. Some potential for emission computed tomography was suggested.

12.2.2.3
Manganese

Preliminary studies with labeled manganese suggest promise in the study of heart, and possibly liver disease. Other potential may exist for evaluation of Parkinson's disease.

Manganese is an intracellular mineral present only in trace amounts but is important as a necessary constituent of the pivotal carbohydrate-utilizing enzyme, pyruvate carboxylase. Other enzymes requiring or usually containing Mn are arginase, prolidase (imidodipeptidase), mitochondrial superoxide dismutase, enzymes performing synthesis of glycoproteins and proteoglycans, and various enzymes involved in metabolism of carbohydrates, fats, and amino acids (Underwood 1971; Cohn and Roth 1983; Phang and Scriver 1995; Murray et al. 1996). Intestinal absorption and excretion (via the bile) are responsive to stores of body manganese in ways to maintain homeostasis (Cotzias 1962). As with other minerals, manganese is present mostly in nuts, cereals, vegetables and fresh fruits (Underwood 1971; Cohn and Roth 1983). Manganese deficiency has not been documented in humans, but experimental Mn deficiency in animals produces skeletal abnormalities, impaired growth and reproductive function and ataxia of the newborn (Underwood 1971). The skeletal problems are believed secondary to disturbances in cartilage formation, based upon malfunction of certain polymerase and transferase enzymes (Underwood 1971).

A series of early studies in man and animals used [54]Mn and [56]Mn to establish various parameters of metabolism of manganese (Cotzias 1962). Among the observations were those of relatively rapid turnover as well as high concentration of Mn in mitochondria. This led to an attempt to measure total body mitochondrial mass by external scanning (mainly of the liver) and a correlated blood disappearance curve (Cotzias 1962). A more specific study of heart muscle uptake of [54]Mn was also based on the high abundance of mitochondria in that organ (Chauncey et al. 1977). Myocardium-to-blood ratios of tracer in rats was higher than for ^{201}Tl and in dogs higher than reported for ^{201}Tl by others. In the dogs experimental myocardial ischemia reduced uptake by 17–75%. While this appeared encouraging for myocardial imaging, the authors rightly indicated that neither [54]Mn nor [52]Mn have suitable physical characteristics, but suggested that [52m]Mn (Table 12.1) might be used with an acceptable radiation dose. An observed uptake of 30–50% of dose in liver was considered a further drawback on the basis of radiation dose and interference with two-dimensional (though not tomographic) heart imaging. Such a high hepatic uptake, on the other hand, suggests possible applications for evaluation of liver abnormalities.

Although isolated manganese deficiency is not a known medical problem, chronic poisoning by manganese ore dusts inhaled by miners is a long-established entity (Underwood 1971; Cotzias 1962). There is insidious onset of psychotic and extra-pyramidal symptoms, the latter resembling Parkinson's disease. Treatment with L-DOPA ameliorates the symptoms of chronic manganese poisoning much as it does those of Parkinsonian patients (Mena et al. 1970). Study with [54]Mn of miners with manganese poisoning indicated much slower loss of radiotracer from the whole body and from the area of the liver than for a group of "healthy" manganese miners, despite lesser tissue concentrations of manganese in the poisoned miners (Cotzias 1962). It was therefore concluded that chelation therapy would not likely be effective. The difference in turnover of [54]Mn between healthy and poisoned miners plus known striking variations in individual susceptibility to toxicity of Mn (Cotzias 1962) suggest modulation by genetic factors. The individual variations found among miners exposed equally to manganese further suggests that more study with radioactive manganese might help delineate predilection to or stages or types of spontaneous Parkinson's disease.

12.2.3
Anions

12.2.3.1
Iodide

Much evidence has been gathered and literature written to tell the story of the early availability of a radioisotope of iodine, ^{131}I, with physical characteristics (Table 12.1) appropriate for both diagnosis and therapy, for a specific target, the thyroid gland, vulnerable to various abnormalities of hypo- or hyper-function (benign or malignant). Later, the use of ^{123}I and [99m]TcO$_4$ have become even more appropriate for diagnostic imaging and counting. These applications have long served as a paradigm for the successful practice of nuclear medicine, and also have been fundamental to understanding the basic biochemistry and physiology of the thyroid gland. This chapter will consider only the trapping of inorganic iodide by the thyroid gland and some other tissues; organification of iodine and other aspects of thyroid hormone (Medeiros-Neto 1996) are addressed in Chap. 15.

Although only the first in a series of several steps, the trapping of circulating iodide by a specialized ion channel protein, known as the Na^+/I^- symporter (NIS), appears to be normally the limiting factor in the rate of production of hormone (Vassart et al. 1995). This places high importance on the testing of thyroidal concentration of radioisotopes of iodine (and $^{99m}TcO_4$) and focuses attention on new information about the NIS, the gene which codes for it, and other influences upon it. Among these are some which may change the balance of uptake of ^{131}I by normal and neoplastic tissue in ways favorable to more effective therapy, as will be described. Thus, the NIS deserves to be a significant topic in nuclear medicine.

Still the major world-wide cause of hypothyroid goiter is insufficient intake of iodine, for which the RDA is 200 µg/day, but which in inland areas lacking iodized salt or oil or proper food sources may be much less. If present from birth the result can be cretinism with mental retardation, arrested physical development and possible deaf-mutism. Congenital hypothyroidism occurs sporadically in about 1:4,000 births (Medeiros-Neto 1996; Pohlenz et al. 1998) and is usually due to thyroid agenesis or dysgenesis. However, about 15% of cases are due to defects of thyroid hormonogenesis which are inherited in an autosomal recessive manner. Defects have been recognized as occurring at any of various metabolic or molecular sites, e.g., TSH availability, TSH receptor, iodide trapping, intracellular peroxidase and further organification and binding by thyroglobulin. A trapping defect has so far been identified in only about 40 cases (in 28 families) world-wide (Vassart et al. 1995; Pohlenz et al. 1998). Subjects may have either partial or profound defects, suggesting genetic variability. Where there is co-existent iodide deficiency there is the potential for interaction between genetic potential and the environment in determining the phenotypic expression (Medeiros-Neto 1996). Subjects usually (except in childhood) exhibit goiter, which is often multinodular and hyperplastic with a propensity to neoplasia. With the recent cloning of the human cDNA which encodes for the NIS protein (Smanik et al. 1996), there is the potential for identifying specific genetic mutations for corresponding phenotypes. This is currently being done for obvious clinical cases, including an interesting case of hypothyroidism due to compound heterozygous condition with inheritance of different mutant forms from the mother and the father (Pohlenz et al. 1998). Others are more standard examples of homozygotes with inheritance of one or another specific recessive genetic defect (Matsuda and Kosugi 1997; Pohlenz et al. 1997). One of these (Matsuda and Kosugi 1997) is notable in that the patient, who had a huge goiter but was euthyroid, was found to have 100-fold higher than normal mRNA in thyroid cells, suggesting compensation by over-expression.

It might now be asked whether there could be subclinical hypothyroid subjects who have inherited from both parents genes with relatively minor damage to the NIS gene and whose radioiodine (or pertechnetate) uptake is marginal or in the "low normal range". Furthermore, it could be suggested that a closer study of the early rate of thyroidal uptake of ^{123}I (after bolus i.v. injection) in addition to the usual 24-h value (or analogous study of $^{99m}TcO_4$) might be revealing of a particular defect in the NIS in the clinical setting.

Details of information about the NIS continue to accumulate, particularly with the recent availability of its cloned cDNA. It is well-established that the NIS is linked to the action of adenyl cyclase (and TSH) and utilizes the energy of Na^+/K^+-ATPase to cause co-transport of Na^+ through the basolateral membrane down an electrical gradient with I^- against the gradient, such as to concentrate the anion to levels of 20–40:1 vs. plasma concentration (Vassart et al. 1995; Medeiros-Neto 1996). The NIS is one of a "family" of Na^+/K^+ ATPase-dependent co-transporters and shows homology, e.g., to the glucose transporter (Dai et al. 1996). The NIS protein has 12 putative domains (transmembrane segments) with both the amino and C termini located on the intracellular side of the membrane (Dai et al. 1996). Whereas the thyroid membrane is the most vital site of the NIS, it is also found to some extent in salivary glands, stomach, some portions of intestine, skin, hair, mammary gland, placenta and ovary (Grass 1962; Vassart et al. 1995). There is less capacity for concentration of iodine by these other tissues and none, except for the lactating mammary gland, is able to organify the iodine (Gross 1962). The latter fact permits the use of saliva/serum ratios for radioiodine or $^{99m}TcO_4$ to be used for evaluating the function of the NIS better than the usual test for thyroidal uptake of radioiodine, since the latter (and possibly even that of

^{99m}TcO$_4$) is affected by organic binding (Vassart et al. 1995; Medeiros-Neto 1996).

Various influences are now known to impinge on the NIS. The strong component of autoimmune processes in Grave's disease, Hashimoto's thyroiditis and other thyroid disorders has, until recently, been supposed to be due to auto-antibodies to thyroid peroxidase, TSH receptor and/or thyroglobulin. Newer evidence (Raspe et al. 1995; Endo et al. 1996a; Ajjan et al. 1998) has been presented to show the occurrence of auto-antibodies to NIS in a large percentage of sera from patients with Grave's disease or with Hashimoto's thyroiditis. Different studies have identified different epitopes (sections) of the NIS protein as responsible for generating the antibodies. In Hashimoto's thyroiditis the typical (and expected) effect of the isolated antibody is to inhibit uptake of ^{125}I by cultured cells expressing NIS (Endo et al. 1996b). In thyroid tissue from Grave's disease the NIS mRNA was increased almost four-fold, along with the mRNA for thyroid peroxidase and thyroglobulin (but not TSH receptor) (Saito et al. 1997). Thus, increased expression of NIS may contribute to the development of Grave's disease. Further study of the antibody-specific sites on the NIS (such as that which has implicated the sixth extracellular loop of the protein; Endo et al. 1996b) will help elucidate the molecular mechanism of iodine binding and possibly lead to new modes of treatment of thyroid disease.

Other recently identified influences may have more relevance or promise in nuclear medicine. One series of studies (Schmutzler et al. 1997) has shown that the compound, retinoic acid (RA), a metabolite of Vitamin A with differentiation-inducing properties, will cause in normal thyroid cells a down-regulation of NIS mRNA while in a line of follicular carcinoma cells in vitro there is up-regulation, though without increased iodide uptake. Nevertheless, in pilot clinical studies after 5–6 weeks of treatment with RA about half of the patients with different forms of thyroid carcinoma, that had lost the ability to accumulate iodide, again showed iodide uptake in tumor tissue. Others (Grünwald et al. 1998) have confirmed the redifferentiation effect of RA to cause increased uptake of ^{131}I in metastases from follicular Hurthle cell carcinoma. These studies hold promise for improving the sensitivity for recognizing as well as treating some thyroid carcinoma with ^{131}I. Another, perhaps more futuristic, hope is contained in a study of transfection of the NIS gene into malignantly transformed rat thyroid cells which did not previously concentrate iodide, but after transfection accumulated ^{125}I 60-fold. Tumors grown in vivo from these cells injected into rats accumulated up to 27% of ^{125}I administered (Shimura et al. 1997). The short biological half-life (6 h) of iodine in the tumor cells did not allow adequate radiation dose from ^{131}I to change significantly the tumor volume in rats. Nevertheless, in human subjects with likely slower iodide kinetics this approach with adequate doses of ^{131}I directed into radio-sensitive tumors could provide a novel form of treatment as well as helping to localize metastases by imaging and predicting therapeutic advantage.

12.2.3.2
Sulfate

Nature was not generous to nuclear medicine in provision of useful isotopes of sulfur. Only the weak beta emitter, ^{35}S (Table 12.1) is available. It has been used almost entirely for biomedical research. In lieu of sulfur, the gamma emitting ^{75}Se has been used clinically in the form of ^{75}Se-selenomethionine, which localizes in pancreas and liver. Sulfate in the body, derived from sulfur-containing amino acids and attached to carbohydrate moieties of proteoglycans and glycolipids, is widely distributed into many organs and secretions and has important functions. Whereas in vivo diagnostic studies are generally impractical there are some in vitro applications of ^{35}S, including a recently-developed "beta camera", which hold promise for the future evaluation of clinical disease.

In the component, glycosaminoglycans (GAG), proteoglycans contain sulfate attached co-valently to various hexoses and hexosamines, which are contained typically in long polysaccharide chains composed of repeating disaccharide units (Shreeve 1974; Murray et al. 1996). In proteoglycans (in contrast to glycoproteins) the amount of polysaccharide is usually much greater than the protein component. The sulfates are attached through hydroxyl or amine groups at C-4 or C-6 positions, though there can be (e.g., in dermatan sulfate) hypersulfation with additional sulfyl groups (at C-2 and C-3). The strong acidity of the sulfyl and carboxyl groups (the latter in contained uronic acids) confers strong electronegativity which attracts mono- and di-valent cations and by osmotic force much water, producing the viscosity characteristic of mucous secretions, joint surfaces, etc. The loose areolar stroma of subcutaneous tissue, fibrous ligaments, vascular walls, collagenous tissue in bone, cartilage and skin, synovial sac and fluid of joint surfaces, sclera and cornea of the eye – all contain various GAG. The major ones are chondroitin-4-sulfate, chondroitin-6-sulfate, dermatan sulfate, keratan sulfate and heparan sulfate. A kind of GAG with different (i.e., anti-coagulant) properties is heparin, which is paradoxically more sulfated than those GAG which bear the term, sulfate. Of likely genetic and possibly clinical significance is that the chemical diversity of the GAG allows for far greater potential for polymorphism and microheterogeneity than is the case for protein (Shreeve 1974).

The glycolipids comprise a family of compounds which provide hydrophilic and very hydrophobic components in the same molecule. One type of glycolipid, i.e., ceramide hexosides (cerebrosides) are composed of an amino alcohol (sphingosine), a long-chain fatty acid and one or more hexosides (usually galactose). When a sulfate is attached at the C-3 position of the hexose, a "sulfatide" is formed (Shreeve 1974). The myelin sheaths of white matter of the brain and spinal cord have high sulfatide content, but sulfatides are also found in gray matter of brain, kidney, liver, lung, heart, skeletal muscle, RBCs, spleen, respiratory, intestinal and gastric mucosas and various other body fluids. Glycolipids occur particularly in the outer leaflet of the plasma membrane where the carbohydrate moiety (and the sulfate moiety) are likely factors in cell surface receptors. Other compounds in the body which may be found sulfated are the steroids, catecholamines, 5-hydroxytryptamines (serotonin) and some proteins (at tyrosine and glycosyl substituents) (Kolodny and Fluharty 1995).

Inorganic sulfate is hardly a dietary staple, so sulfur must be introduced normally into the body in the form of the sulfur-containing amino acids, methionine and cysteine. Indeed, the sulfate ion is excluded from intracellular access to major tissues, since ^{35}S-sulfate given orally or parenterally appears initially restricted to a space definable as the ECW (see Sect. 12.3.2.1.2). There is also evidence (in the adult rat) for a blood-joint barrier to ^{35}S-sulfate in vivo (Kleine and Baumann 1991). A key reaction to generate "active sulfate" is the addition of the adenosyl moiety of ATP to the sulfur atom of methionine, from which is derived 3-phospho-adenosine-5-phosphosulfate (PAPS). For cysteine there are other catabolic pathways which can produce sulfate at the intra-mitochondrial site (Schepartz 1973). PAPS is the immediate donor of "active sulfate" in the sulfotransferase reactions to form sulfated GAG and sulfatides and steroid sulfates (Shreeve 1974; Kolodny and Fluharty 1995; Murray et al. 1996). Despite the restrictions on utilization of sulfate administered in vivo, ^{35}S-sulfate

given i.v. is incorporated into the GAG of normal and neoplastic human chondrocytes in vivo (Mayer et al. 1978) and into chondrocytes in tissue culture (Quatela et al. 1993; Lammi et al. 1994; Verschure et al. 1996). The rate of synthesis of sulfated GAG in articular cartilage is affected by the availability of inorganic sulfate (Kelly 1998). Sulfur-35 may be found in the sulfatides in the myelin sheath of rats after in vivo injection of ^{35}S-sulfate (Shreeve 1974).

Studies so far do not suggest in vivo localization of ^{35}S-sulfate which would encourage tissue sampling for clinical diagnosis of diseases involving either proteoglycans or glycolipids. However, technical developments in recent years have resulted in detailed analyses of urinary products and the contents of joint fluids and intra-articular cartilage as well as metabolism of cultured fibroblasts or chondrocytes ex vivo. A disease which involves an inherited defect in desulfation of cerebroside sulfate (by the enzyme, arylsulfatase A) is that of metachromatic leukodystrophy. The disease, while fortunately uncommon, is neurologically very disabling and may appear in late infantile, early or late juvenile, or adult stages of life, depending on the pattern of inheritance of two different mutations in the gene for arylsulfatase A (Kolodny and Fluharty 1995). The most sensitive test has been the cerebroside sulfate loading test, as carried out with cultured skin fibroblasts and metachromatic staining. Possibly an even better test, which can correlate with the degree of severity (age of onset) of the disease, is that of performance of the cerebroside loading test with ^{35}S-cerebroside sulfate and measurement of the relative amounts of intracellular and extracellular ^{35}S over a period of several days. An inverse relationship between the intracellular/extracellular ratio of ^{35}S and age of onset of disease was found (Porter et al. 1971; Kolodny and Fluharty 1995). Other sulfatase deficiencies, which cause abnormal accumulation of the amounts of GAG, are responsible for the various forms of mucopolysaccharidosis. One such disease, known as multiple sulfatase deficiency, appears to occur in two different types (mild and severe), which have been distinguished by the amount of ^{35}S-sulfate incorporated into intracellular mucopolysaccharides in cultured skin fibroblasts (Eto et al. 1982; Kolodny and Fluharty 1995). In view of the observation that the result of various sulfatase deficiencies is a particularly noticeable increase in urinary excretion of typical sulfatides or sulfated GAG (Kolodny and Fluharty 1995), consideration could be given to the possibility that tolerable in vivo doses of ^{35}S-sulfate would provide a tracer for both the amount and kind of excreted, labeled glycolipids or proteoglycans in the various associated clinical entities.

In contrast to the characteristics of obvious heritability, extensive early pathology but rarity of occurrence in the diseases mentioned above, there is the group of arthritides, which are far more common and less disastrous, but which also involve disturbances of metabolism of sulfated GAG, i.e., in the epiphyseal bone ends and cartilaginous surfaces of joints. In this group there have been various studies of uptake, turnover and localization of ^{35}S-sulfate in isolated cartilage and/or cells of synovial fluid from normal and arthritic human subjects (Ostensen et al. 1991; Quatela et al. 1993; Flechtenmacher et al. 1996; Verschure et al. 1996) or other mammalian species (Weiss et al. 1986; Lammi et al. 1994; Guingamp et al. 1997). These studies have provided varied information such as inhibition of synthesis of GAG by pressure effects (Lammi et al. 1994) and preferred selection of types of cyanoacrylate tissue adhesives for cartilage graft fixation (Quatela et al. 1993). Biochemical changes in the GAG of isolated tissue may be found in early, low-grade osteoarthritis (Ratcliffe et al. 1996) or osteonecrosis of the femoral head (Iwase et al. 1998) before clinical or imaging diagnosis is clear. Findings with ^{35}S-sulfate and from non-isotopic studies indicate that in mild or moderate osteoarthritis (Verschure et al. 1996) (Fig. 12.4) there is increased anabolism in cells with reversion to a fetal or post-natal type of cartilage, whereas in advanced osteoarthritis and in rheumatoid arthritis (Ostensen et al. 1991) there is a decreased level of chondrocyte activity.

It might be supposed that the relatively benign nature of the arthritides would not merit the degree of invasiveness (even at planned arthroscopy) of sampling of cartilage or even synovial fluid and the labor of analysis with ^{35}S and other laboratory methods, except in limited research studies. Perhaps this is so, but recent developments in quite new and interesting directions of treatment plus superior methods both for chondrocyte culture and for imaging of ^{35}S are incentives not only for more research but possibly for direct clinical application. Moreover, arthritis (particularly osteoarthritis) becomes ever more common

and disabling in the demographics of an expanding population and length of life of the elderly. While the commonness of primary osteoarthritis suggests mainly the universally shared genetics of aging, there is, in fact, evidence (Felson et al. 1998; Hirsch et al. 1998) for familial inheritance, which adds further reason for distinction of biochemical differences.

New kinds of treatment include the use of recombinant human osteogenic protein-1, which is capable of inducing both new cartilage and new bone formation and even (in animal models) of restoring large bone defects (Flechtenmacher et al. 1996). In another approach adenovirus-mediated gene transduction to chondrocytes of transforming growth factor-β1 and of heat shock protein 70 enhanced expression of mRNA of type II collagen and proteoglycan core protein (major components of articular cartilage) and suppressed the expression of an mRNA which can degrade proteoglycan (Arai et al. 1997). Ways to administer these agents locally for tissue repair capability, and means to control unwanted effects, are suggested.

A third kind of therapy, presently popular with easy availability, is oral administration of glucosamine sulfate and/or chondroitin sulfate (CS). As found with ^{3}H-labeled CS given orally to rats and dogs (Conte et al. 1995) this compound, with a relatively low molecular mass of 14 Kd, showed about 70% absorption of radioactivity, though only about 8.5% was in original form with the rest as depolymerized products. Amounts of activity found in synovial fluid and joint cartilage were not much different than in other tissues (trachea, eye, muscle, lung) but in joint cartilage, trachea and eye it stayed longer. When (unlabeled) CS was given for several days to normal and osteoarthritic human subjects, biochemical parameters of GAG in synovial fluid changed in ways consistent with clinical advantage (Conte et al. 1995). As reported by others from controlled studies (Morreale et al. 1996), there is a slow but sustained clinical response to oral CS in osteoarthritic subjects which can last for up to 3 months after the end of treatment. Various evidence for positive effects of either glucosamine sulfate or CS has been summarized (Kelly 1998).

Fig. 12.4A–D. Zonal distribution of in situ autoradiography of ^{35}S-sulfate incorporated into isolated human cartilage after incubation for 4 hours with 14.8×10^4 Bq Na$_2$^{35}SO$_4$ (magnification 20×). **A** Normal cartilage (from knee) with low levels of ^{35}S-sulfate incorporated (PG synthesis) into chondrocytes of surface zone (*S*) with distinctly higher amounts in middle (*M*) and deeper (*D*) zones (surface at *arrow point*). **B** Mild osteoarthritis (OA) cartilage shows no zonal variation. **C** Moderate OA cartilage shows high PG synthesis in both upper and deeper layers. **D** Severe OA cartilage shows very low levels of chondrocyte incorporation of ^{35}S-sulfate in any strata. (From Verschure et al. 1996, with permission)

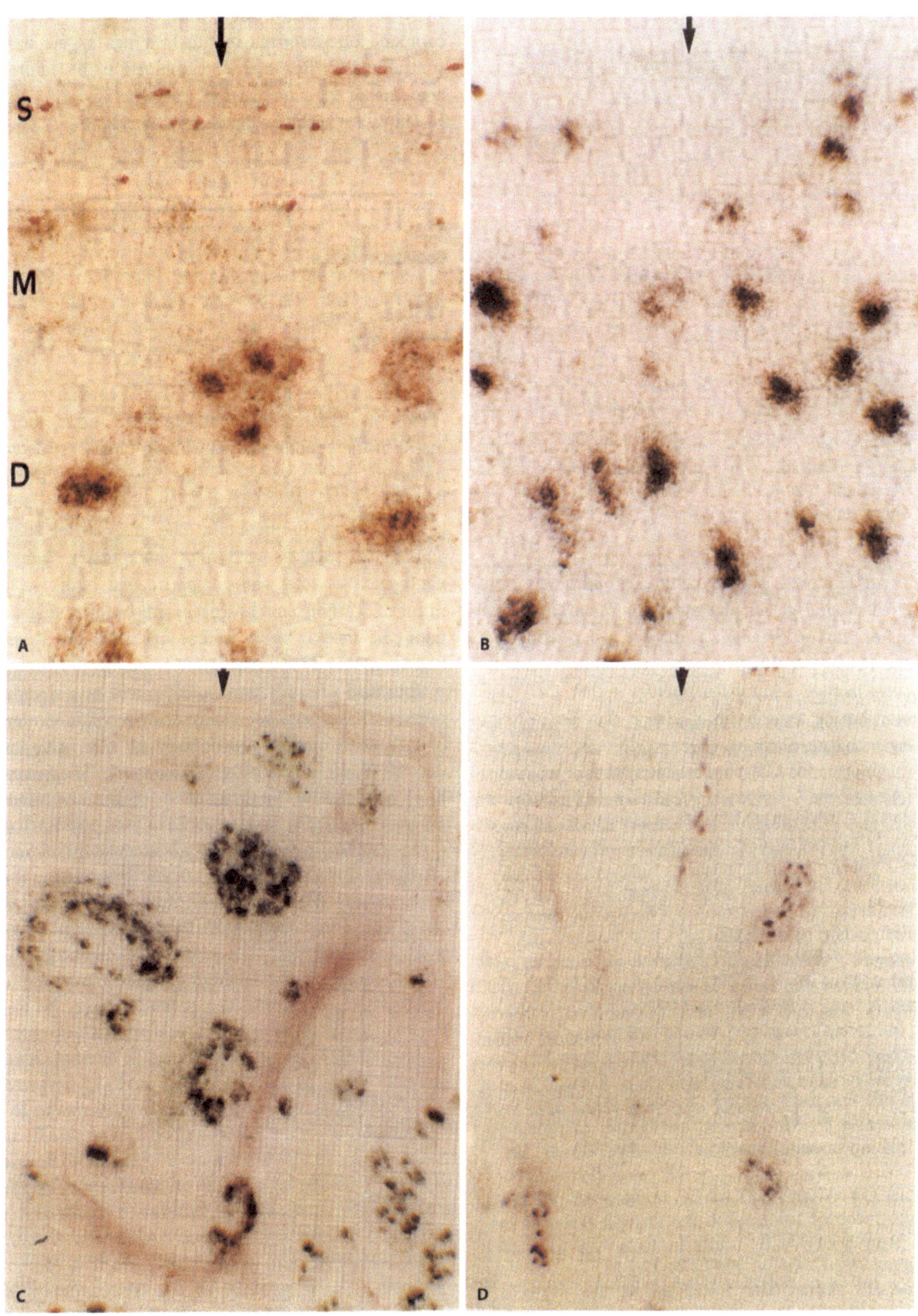

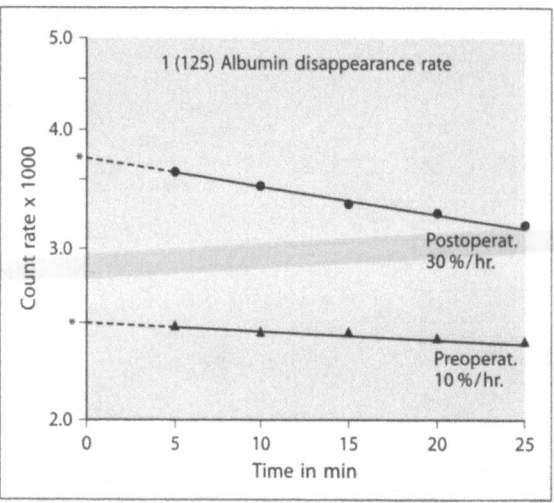

Fig. 12.5. Rate of disappearance of ^{125}I-albumin from plasma in pre-operative and post-operative periods – note increase from 10% to 30% per hour loss. (From Alberts 1974, with permission)

An improved way to culture chondrocytes, clustered around gelled alginate (glycuranan) "semi-solid" or "hollow" beads, was devised a decade or more ago (Guo et al. 1989) in order to maintain chondrocytes in their natural morphology rather than metamorphosing towards fibroblasts as they do in monolayer culture. Even in agarose culture without beads the morphology of chondrocytes is well-maintained (Ostensen et al. 1991) and cells tend to grow in aggregates or clusters. Various studies have used autoradiography of cartilage slices (Fig. 12.5) or of GAG separated by electrophoresis on agarose gel (Lammi et al. 1994). A newer and more facile analysis (in hours rather than many days) could be the use of the "beta camera", which has been developed in recent years (Ljunggren and Strand 1990; Ljunggren et al. 1993) for imaging of ^{35}S, ^{14}C, ^{131}I, ^{90}Y and even such emitters as ^{68}Ga, ^{18}F1 and ^{201}T1. This miniature camera (with a resolution of 500 microns) could presumably delineate ^{35}S in clusters or beads of cultured chondrocytes vs. that in the extracellular medium, thus making possible kinetic studies on cultured cells in situ, as well as quicker, possibly easier distinction of separated biochemical compounds than by autoradiography.

There have been efforts to use the affinity for ^{35}S-sulfate of polysaccharide-containing tumors, e.g., chondrosarcoma or chordoma, to deliver safely a significant radiation dose to such tumors (Mayer et al. 1978). Although repression of tumor growth (especially for chondrosarcoma) was obtained, the concomitant long-term marrow toxicity has appeared to contraindicate its use. A combination of limited amounts of ^{35}S and external radiotherapy was suggested as a possibly workable alternative.

12.3
Body Composition

12.3.1
Introduction

As stated earlier, certain major body compartments (fluids in different components, lean mass, bone mass) in the equilibrium state are in large part determined by the energetic and other mechanisms which define locations, concentrations and amounts of inorganic elements in the body. The analyses of static amounts of such elements, plus other major components such as total and compartmental body water, nitrogen, carbon and hydrogen, are brought to bear upon the field (also, of course, including fat mass) known as "body composition". From the beginning of this field of study both isotopic and non-isotopic methods have been used. There is no single clinical or laboratory specialty identified with this field. Researchers and practitioners interested in measuring body composition have included nutritionists, physiologists, internists and surgeons, as well as nuclear medicine physicians and radiologists.

For decades much interest in body composition has centered, by necessity, upon physiological differences among major categories of healthy persons, i.e., dependent upon gender, ethnicity, age, body habitus, physical activity, etc., in order to build definitions of normal values and their ranges. Much data of this type has been accumulated and can be related in extenso to fundamental genetic constitutions, which in turn may be correlated to hormonal or other somatic determinants from the genetic make-up. The disease states to which body composition studies apply include these related to over-nutrition (obesity) or under-nutrition (e.g., AIDS, various starvation states) and the therapeutic measures designed to ameliorate these problems. Chronic disorders of bone (osteoporosis) and other organs (heart, kidney, endocrine glands) are also candidates for study. More acute traumatic disruptions of normal body composition have also been objects of study in this field.

12.3.2
Compartments

12.3.2.1
Water

Maintenance of the amounts and compartmental distributions of water in the body is vital to the transport of nutrients and oxygen to tissues, removal of waste, support of cell volume and function, thermal regulation and gene expression (Haussinger 1996; Wang et al. 1999).

12.3.2.1.1
Plasma

Maintenance of plasma and blood volume is so fundamental that no major genetic change (e.g., in hormonal balance or renal regulation) which would allow significant chronic hypo- or hypervolemia from an early age is likely to persist in the gene pool (the rare disorder, diabetes insipidus, is an exception). However, there are acute, temporary disruptions of blood volume (e.g., hemorrhage, traumatic shock, post-surgical state, acute environmental change in temperature or altitude) and some chronic changes in later age (e.g., polycythemia vera, nephrotic syndrome, hemodialysis, protein-losing enteropathy) in which measurement of plasma volume may guide treatment (Price and McIntyre 1984). The total blood volume (red cells plus plasma) has long been judged to be about 7% of body weight; a recent study indicates that it is more accurately related to the fat-free (mostly skeletal muscle) mass (Hunt et al. 1998).

Nuclear medicine has long had reliable, feasible radiotracers: albumin labeled with ^{131}I or ^{125}I, ^{99m}Tc albumin or ^{113m}In-transferrin (Table 12.1), for measuring plasma volume by isotope dilution (Alberts 1974; Price and McIntyre 1984). This is also true for the other major blood component, the red cell mass, measurable by ^{51}Cr or ^{99m}Tc-labeled red cells. In the past three decades there has been decreasing use of plasma (or RBC) radiotracers. This is due to reliance upon non-tracer methods, e.g., central venous pressure for the functional assessment of the relationship between venous volume (capacitance) and cardiac function. The non-isotopic tracer method, Evans blue dye (T-1824), is now often employed for plasma volume, perhaps to avoid radiation effects or due to lack of available radio-laboratory facilities. The very low

radiation dose (Kerr et al. 1967) (Table 12.1) should be readily acceptable.

Advantages characterize the use of labeled albumin to measure plasma volume (or labeled RBC for blood volume) compared to other methods. First, data indicate that reliance upon central venous pressure (Alberts 1974; Johansen et al. 1998a) or calculation from hematocrit/hemoglobin values (Johansen et al. 1998b) to assess plasma volume indirectly can be misleading compared to direct measurement. The repetitive use of Evans blue dye may pose technical problems (Alberts 1974). More significantly, the use of the gamma-emitting ^{99m}Tc-albumin (or ^{113m}In-transferrin or ^{99m}Tc-RBCs) can provide evaluation of changes in regional distribution of plasma (blood) volume by external counting. An integrated system using ^{99m}Tc-RBC and four probes located on the lower limbs or over the lung fields and liver has been devised for such evaluation (Baccelli et al. 1995). Besides analyzing such medical problems as postural hypotension, heart failure or venous insufficiency in the lower limbs, this technique could add another dimension to the recent interests in plasma/blood volume change in prolonged microgravity in extended space missions (Johansen et al. 1997; Hsieh et al. 1998), bed rest and water immersion deconditioning (Greenleaf 1997), intense exercise (Boulay et al. 1995), prolonged endurance training (Pickering et al. 1997), marked altitude changes (Grover et al. 1998) or other environmental stress situations. Further information could come from measuring regional changes of both plasma volume and RBC mass (either consecutively with ^{99m}Tc label or simultaneously with two nuclides of different energy, e.g., ^{99m}Tc-RBC and ^{113m}In-transferrin), since effects on "plasma" radioactivity could signify change in capillary integrity (leakage of albumin) in addition to change in vascular volume. It is well established that there is a normal transcapillary escape rate (TER) of albumin from the circulation of about 10%/h. This is usually successfully ignored by taking the plasma sample at 10 min post-injection of tracer, though correction by obtaining samples at 10, 20 and 30 min with extrapolation to zero time is advised when very accurate results are required, particularly with expected increased escape rate, as in post-operative states (Fig. 12.5). Increased TER of albumin can be seen also in more chronic disease such as juvenile diabetes (Parving et al. 1976) or essential hypertension (Parving and Gintelberg 1973). More recently, a decreased TER of albumin, which was correlated with a slight but definite increase in plasma volume, was noted in normal subjects 24 h after intense exercise (Haskell et al. 1997). Thus, the parameter of TER of albumin is variable under physiological as well as pathological circumstances and probably subject to both environmental and genetic determinants. Studies of this phenomenon are an example of applications in which tracer methods are either the best, or the only, means applicable to measuring important physiological functions.

12.3.2.1.2
Extracellular Water

Changes in the total extracellular water (ECW) are generally not as critical hemodynamically as those of plasma (blood) volume and characteristically persist on a more chronic basis. In addition to obvious causes of grossly decreased ECW (hemorrhage, shock, profound diarrhea), or causes of grossly increased ECW (cardiac, hepatic or renal failure, starvation), there is an assortment of hormonal influences which may become unbalanced and lead to moderate, persistent change (usually an increase) of ECW. Increased ECW can impose a burden on the heart, kidney and other internal organs as well as the interstitial tissue compartment. Several radio-tracers or chemical tracers have been used in the past to measure ECW by isotope dilution, as has the special technique of neutron activation analysis (NAA). Currently there is a major trend toward replacement of the dilution techniques, and extension to wider field conditions, by the method of bio-electric impedance analysis (BIA).

Measurements by tracer dilution after intravenous injection of tracers, which may be isotopic (^{82}Br, ^{77}Br, ^{24}Na, ^{22}Na, ^{14}C-sucrose or -inulin, ^{35}S-sulfate) or non-isotopic (chemical sucrose, inulin or bromide), have given estimates of the ECW in healthy, non-obese men or women which range from 16% to 34%, depending on the tracer used (Yasumura et al. 1983). For reasons discussed in Sects. 12.1.2.1 and 12.1.3.1, the sodium and bromide tracers provide the higher values in this range, and require some kind of informed correction. The lowest value, and apparently the most accurate method (when corrected for early urinary loss and definitely requiring extrapolation to zero time from multiple plasma samples) is that using ^{35}S-sulfate, based on the estimated concentrations of intracellular potassium and sodium which match the terms of the Nernst equations and known osmolalities (Ryan et al. 1956). Among radioisotopes for dilution tracers only ^{35}S-sulfate (Ernest et al. 1992), and

occasionally ^{82}Br have been in use in recent times. The very different technique of "delayed gamma" neutron activation analysis (DGNAA) (Yasumura et al. 1983; Cohn 1995a) measures, by whole body counting of the gamma emissions from activated Na and Cl (plus Ca) in a single procedure, the total body sodium (TBNa) and total body chloride (TBCl). The technique obviates the vagaries of tracer dilution (incomplete mixing, reversible or irreversible loss to compartments or chemical entities not being measured) but requires correction for non-ECW Na or Cl (helped by the simultaneous Ca measurement). DGNAA is costly, involves radiation burdens in the range of 500 mrem (5 mSv), and is presently performed in only two institutions in the USA and in a few other sites around the world. With proper corrections TBCl is a robust and superior method for ECW; TBNa (non-bone) is closely proportional to ECW. The single frequency BIA method, which measures total body water (see below), or the multi-frequency BIA, which in theory can divide between ECW and intracellular water (ICW) in healthy persons is presently an appealing alternative to more complex methods, but can give misleading results in pathological conditions (Thomas et al. 1998; Johnson et al. 1996). Present techniques of BIA, while being relatively simple in application and inexpensive, have too large a coefficient of variation ($\pm$10%) for research level use. Even in healthy persons there can be significant differences (individual or total group) between results from tracer dilution and those from BIA (Armstrong et al. 1997). Results of study of a large number of hospitalized patients indicated that sensitivity of BIA for diagnosing dehydration or over-hydration, using deuterium-labeled water and bromide dilution as reference techniques, was just 14% and 17%, respectively, though responsiveness of serial measurements to intra-individual changes was good (Olde Rikkert et al. 1997).

There are no pronounced physiological differences in ECW (relative to body weight) in regard to gender, race, or age, as gauged by DGNAA for TBCl or TBNa (Ellis 1990). However, a difference perhaps worth noting is a slight increase in TBCl and TBNa with age in black males, whereas no such difference appeared in the group of white males. This may relate to the higher incidence of hypertension in black men, particularly with advancing age. The genetic components contributing to hypertension are complex, and likely operate through interplay among various hormonal systems affecting blood pressure. Conducive to increased arterial pressure are the renin-angiotensin-aldosterone system (RAAS), arginine vasopressin, endothelin (Jespersen 1997), catecholamines, and glucocorticoids (Whitworth et al. 1995), while opposing these, both for blood pressure and potential body accumulations of sodium, chloride and water, are atrial natriuretic peptide (ANP) and brain natriuretic peptide (BNP) (both from the heart) and prostaglandin E_2 from the kidney (Hunt et al. 1996; Jespersen 1997). In one of these influences, the RAAS, an interesting change has been noted with age. There is reduced adrenal responsiveness to angiotensin II and a more pronounced reduction in blood pressure in response to angiotensin-converting enzyme (ACE) inhibitors in old than in young hypertensive patients (of white race) (Belmin et al. 1994). In this connection an insertion/deletion polymorphism in the ACE gene has been described and a deletion/deletion genotype of the gene reported to be associated with longevity (despite an association of the same polymorphism as a risk factor for myocardial infarction). Much interest, and indeed controversy, for decades has surrounded the question of salt sensitivity in at least some subtypes of hypertensive disease of humans. Future prospective studies of the genotype/phenotype connections (possibly with salt loads or dietary manipulation) might well be undertaken using appropriate indicator dilution tracers for Na or Cl or the DGNAA technique. Only with use of sophisticated measurements of the ECW and ICW by nuclear techniques, plus awareness of hormonal and drug changes (Trovato et al. 1998), can important questions about genesis and types of hypertension be addressed.

In the case of chronic renal diseases, such as the nephrotic syndrome (Jespersen 1997) and renal failure requiring hemodialysis (Trovato et al. 1998; Leypoldt and Cheung 1998) the effects of various influences (diet, fluid intake, hormones, medication) on the ECW can be judged by one or another of the isotope tracer or NAA techniques. While study of the Na_e with ^{24}Na in patients recovered from cardiac failure has suggested lingering excess of ECW (Carroll et al. 1965) (see Sect. 12.1.2.1), more recent study with chemical bromide found no difference from normal controls (Steele et al. 1998) or (in contrast to the ^{24}Na study) a condition of "intracellular edema" with decreased ECW, increased ICW, normal TBK and thereby decreased Ki and resting membrane potential (Sackner-Bernstein et al. 2000). Further investigation and direct comparison of different methods may be indicated. Evaluation of hepatic cirrhosis with hypoalbuminemia and varying degrees of ascites can benefit from measurement of ECW, particularly in the settings of various therapies (Jespersen 1997).

Both acute and chronic environmental stresses lead to situations in which the amount of water and electrolytes in the ECW may be markedly abnormal, either absolutely or relative to other water compartments. Among acute changes, excess of ECW has been documented for extensive burns (Murton et al. 1998; Zdolsek et al. 1998), sepsis, such as bacteremia (Schwenk et al. 1998) and peritonitis (Plank et al. 1998), and a deficiency of ECW ("hypoxia diuresis") from acute high altitude exposure (Jain et al. 1980). Patients who present to major surgery with weight loss and hypoalbuminemia, or develop this post-surgically, are said to have "protein energy malnutrition" (PEM) (Hill 1992). If such patients (adults) have had more than 20% weight loss they may have a kind of "adult kwashiorkor" with at least minimal clinical edema but a marked expansion of the ECW. While PEM may be encountered in some surgical patients, it can be found on a large scale among adults and children in areas of the world afflicted with semi-starvation. Here also, measurement of the ECW could be useful in nutrition rescue efforts. The simpler field methods (e.g., BIA) would likely be used, but may yet require calibration and verification with tracer methods.

12.3.2.1.3
Total and Intracellular Water

The intracellular water (ICW) is more remote to measurement than the ECW. Potassium is the one element which is almost entirely (98%) confined to the aqueous phase of the cell interior. After equilibration with the 2% in ECW (including plasma), either ^{42}K or ^{43}K can be sampled from plasma. This technique has both advantages and disadvantages (see Sect. 12.1.2.2). More commonly used now is the whole body counting of natural ^{40}K (Table 12.1) for total body potassium (TBK). If one assumes a uniform, normal concentration of intracellular K (145–150 meq/l in muscle, 125 meq/l in the "average" cell) in men or women of a large range of ages and ethnicities, as to be expected in healthy subjects or those with moderate disease without significant abnormality of electrolyte/water balance, TBK may be used to estimate ICW. However, data (Sutcliffe 1996) indicate that, while in a large assortment of patients with various diseases there is fairly good correlation of TBK with total body water (TBW), there is a closer correlation of TBK with total body nitrogen (TBN), representing body cell mass. Changes in cellular potassium concentration in disease are characteristically in the direction of a decreased concentration (Monk et al. 1996).

A more direct way of deriving ICW is measurement of TBW by ^3H_2O, ^2H_2O or $H_2{}^{18}O$ and subtraction of ECW obtained by one of the techniques already discussed. The method with tritiated water (Vaughan and Boling 1961; Pierson et al. 1982) remains the easiest for most laboratories to perform and involves a very small radiation dose (Table 12.1), whereas the stable isotope analyses (usually by means of mass spectrometry or infrared spectroscopy) are much less commonly available, but allow the use of ^2H or ^{18}O, or the two combined for additional information (see below). Measurement of electrical impedance at the high, as well as low, end of the frequency range in multi-frequency BIA will provide (again by difference) the ICW, but variations in ratio of ECW/ICW require special formulae and the technique can be applied only with caution in subjects with disturbed water distribution (Duerenburg et al. 1989; Bedogni et al. 1996).

The ICW is usually preserved in disease more assiduously than the ECW, although in such conditions as marasmic kwashiorkor or acute pancreatitis (Hill 1992) the ICW may become depleted while there is an increase in the ECW. This is usually associated with extensive loss of fat-free mass (Bedogni et al. 1996; Baarends et al. 1997). However, there is evidence that after major trauma or sepsis the loss of ICW may precede or, at least, exceed the loss of body protein (Finn et al. 1996). Such a finding places greater emphasis on the need for measuring ICW in severe, acute or wasting disease and for evaluating corrective therapy, accordingly. Conversely to the usual increase of ECW/ICW ratio in disease, there can be a situation of acute diuresis, e.g., as caused by the drug, acetazolamide (Brechue et al. 1990), in which the ECW is depleted with a simultaneous actual increase in the ICW. A situation of pathological diuresis occurs in the rare disease, diabetes insipidus (DI), which may be caused either by deficiency of the neurohypophyseal hormone, arginine vasopressin (AVP), or by defects in the AVP-dependent receptors in the renal collecting ducts. The nephrogenic type of DI is congenital and is discussed in Chap. 25.

As mentioned above, there are special advantages of measuring TBW by the method of simultaneous double labeling with ^2H and ^{18}O. By this technique the energy expenditure in the resting state, or with free-living activity, can be done by collections and analysis of water from blood or urine rather than by more cumbersome and interfering methods of analysis of respiratory gas exchange. The main premise of the method is that the O atoms in expired CO_2 have isotopically equilibrated with the O atoms in body water. Thus, ^2H is eliminated from the body only as water, while the ^{18}O is eliminated both as water and as CO_2. The difference between the elimination rates of the two isotopes in water indicates the CO_2 production and (with some knowledge or estimate of the respiratory quotient) the energy expenditure (Schoeller et al. 1986; Roberts 1989; Ritz and Coward 1995). The energy intake can also be estimated. There can be definite clinical value in measuring resting energy expenditure (REE), or varying expenditure with moderate activity, in disease states, particularly for prediction of wasting tendency and the prescription of nutritional needs (Azcue et al. 1997). Genetic subtypes of disease may be recognized early by differences in REE, as in the case of cystic fibrosis (Thomson et al. 1996). Measurement of energy expenditure may help to evaluate functional thyroid states and to classify types of obesity (genetic or otherwise) and the metabolic effects of weight loss programs. Few nuclear medicine establishments now have capability of or access to stable isotope mass spectrometry or other equipment required for high precision results, but there may be sharing with a large chemistry department, an entirely practical arrangement, since the specimens can be readily transported to a distant laboratory site. These are promising clinical developments, not only for ^2H and ^{18}O but for other stable isotopes.

As yet unmentioned is one of the advantages earliest recognized for the measurement of TBW, by whatever method. This is the derivative calculation of lean body mass (LBM) and from that the body fat mass as the difference between body weight and LBM (see Sects. 12.3.2.2 and 12.3.2.3).

12.3.2.2
Lean Mass

The lean body mass (LBM) [differing from the fat-free mass (FFM) only by inclusion of the adipocyte in the LBM] includes the metabolically-active body cell mass (BCM) plus such non-cellular, non-fat parts as bone, tendons, teeth and cartilage and the inert layers of skin. To measure the masses of these entities, and to study the general kinetics of the main constituent, protein, in the LBM or BCM, nuclear medicine provides various techniques which range from simple to complex in performance. Relatively simple is the derivation of LBM from measurement of TBW, or the BCM by derivation from total body potassium (TBK). Total body nitrogen (TBN), from which derives estimates for total protein (as being 16% N), is measurable directly with the special equipment used for "prompt-gamma" neutron activation analysis (PGNAA). Protein mass and turnover are measured together in more elaborate techniques employing (usually) stable isotope-labeled amino acids.

The longest-used and most enduring method for calculating FFM is based on the assumption that the average hydration of the FFM is 73% water. Measurement of the TBW in the total body has been discussed in the preceding section. The value of 0.73 ± 0.03 as the fraction of water in the FFM of healthy individuals is a remarkably constant value across many mammalian species, verified by repeated chemical analyses in animals and human cadavers (Pace and Rathbun 1945; Wang et al. 1999). Although the number of 73% is reliable in healthy adult persons (or post-infant children) and some with various diseases, it cannot be safely assumed in patients with possible or likely over-hydration, e.g., malnutrition or post-surgical complications (Hill 1992; Plank et al. 1998) or obesity (Pierson et al. 1976), or those with under-hydration due to prolonged diarrhea, ileostomy, etc. The value of TBK (measured presently by counting natural ^{40}K with reproducibility within 3% in human subjects over a wide range of size and shape) can estimate the BCM by assuming constant proportions of water and potassium. The LBM (FFM) is also estimated using formulae based on assumed concentrations of potassium (Sutcliffe 1996). Assumptions may not be valid in subjects with deranged water balance, as in some diseases, or among the very young or very elderly. When measurements are required essentially for fluid compartments, BIA is commonly used for deriving LBM (as above) from high-frequency BIA for TBW. Comparative values and limitations of BIA have been discussed in Sect. 12.3.2.1. Another non-isotopic method now available for evaluating lean tissue (if whole body studies are available) is dual X-ray absorptiometry (DXA) of lean vs. fat tissues (Krall and Dawson-Hughes 1995). A much more direct and therefore more reliable indicator of the total metabolic cell mass is based on body protein measurements, done either from potassium or by PGNAA for TBN. Certain special equipment and neutron generators (cyclotrons or, more commonly now, isotope sources, e.g., ^{238}Pu-

Be or ^{252}Cf) are required. These are mainly located in a few large medical centers or special institutions, but such locations have steadily increased in number and applications in recent years (Hill 1992; Cohn 1995a; Sutcliffe 1996). The precision of measurement of TBN by the usual technique is presently about 2–3% for a radiation dose (at skin surface) of 0.40–0.80 mSv from combined gamma emissions and absorbed neutrons which are assigned a quality factor of 10 or (more conservatively) 20. Whereas the usual type of PGNAA for TBN gives the above modest dose from a scan time of 30–40 min, a further new application of the "associated particle technique" (Mitra et al. 1995; Sutcliffe 1996), holds promise of a ten-fold lower dose (e.g., 0.03 mSv) with shorter scan time (15 min) and a current precision of about 4%. Moreover, this newer technique simultaneously can measure (with a D-T generator) TBN, total body carbon (TBC) for estimating fat mass (see Sect. 12.3.2.3) and total body oxygen (TBO) from which TBW can be derived. Lastly can be mentioned the much different way of evaluating the metabolically active protein (particularly muscle) mass, as well as rates of protein synthesis and proteolysis, by analyzing from multiple blood and urine samples the kinetics of certain labeled amino acids (e.g., leucine or methylhistidine) after intravenous administration (Rathmacher et al. 1995).

Many of the studies of BCM and FFM to date have had as their purpose the cataloguing of physiological differences in normal subjects, recognizing the large influences of gender, ethnicity and age. While such differences are often self-evident by casual observation, body composition techniques have provided the means to quantify the differences and even to add qualitative distinctions. The relevance of these studies for understanding the degree of genetic control has been recently emphasized by evidence within families of a higher degree of heritability of LBM than even of body fat (Forbes et al. 1995; Krall and Dawson-Hughes 1995). Much attention has been paid to multiple genes disposing to obesity; comparatively little comparable attention or knowledge seemingly exists for genes controlling LBM.

Gender differences among healthy adults for parameters of BCM and FFM have been reported for white North Americans (USA) using TBK by ^{40}K (Pierson et al. 1974), for white and black North Americans (USA) using TBK by ^{40}K and PGNAA for TBN (Ellis 1990), and for Chinese subjects measuring TBW by the "falling drop technique" using ^{2}H$_2$O (precision of 2%) (Jiang et al. 1991). Using a normalization of K/height, the earlier (1974) study (of about 1,800 males and 1,300 females) showed a ratio of about 60% more K in young males/females. With age, both males and females decline from peak at age 25–35, such that the larger male TBK has decreased to a ratio of about 1.3 by age 70. The later (1990) study of another large group of North American whites (about 110 males and 175 females) showed similar ratios (1.66 and 1.42, respectively). Analysis of data for TBN in that study indicates (corrected for height) a male/female ratio of 1.29, a difference which increases (to 1.4) by age 70. The differences based on TBK reflect more directly than TBN the skeletal muscle mass, which is known to diminish more rapidly, starting at an earlier age in males than in females (Gallagher et al. 1997; Kehayias et al. 1997). After age 60 females may lose FFM more rapidly than males (Flynn et al. 1989). Gender differences for black North Americans (Ellis 1990) were presented only for mean age (about 50 years) of a much smaller total number of subjects and indicate 50% higher K/height and 11% higher TBN/height in males than in females. In the study of normal Chinese subjects (also small in number and grouped together in age, i.e., average 36 years for males, 26 years for females) the BCM/height was 48% greater for males than females. The SD about the means of the various measurements among whites and blacks in the USA was generally about 10–15%, though only 5% for Chinese. The extent of the SD has been supposed to be due to differences in fatness and therefore "fat-obligated skeletal muscle" as well as differences in physical activity (Pierson et al. 1974), but some basic individual genetic differences are not ruled out. The useful studies of LBM in heterozygous and dizygous twins (Forbes et al. 1995; Arden and Spector 1997) suggest that about half of the variance among individuals is environmental in origin, the other half attributable to genetic determinants.

As indicated above, with aging there is a natural loss of LBM and particularly muscle mass ("sarcopenia") to the extent of about 15–25% from the third to the eighth decade in generally healthy men and a similar (or greater) rate of decline from about the time of menopause in women (Flynn et al. 1989; Poehlman et al. 1995; Harper 1998). The ways in which degrees of fatness affect the BCM and its changes with age (based on TBK measurement) have been analyzed and discussed (Pierson et al. 1974). The main proximate causes of decline of skeletal muscle mass with age appear to be changes in physical activity and reduction in the activity of growth hormone (GH) (Harper 1998). By acting directly, or indirectly by enhancing insulin-like growth factor-1 (IGF-1), GH increases LBM through stimulation of protein synthesis and reduction in protein oxidation (Ho et al. 1996). Maintenance of physical activity in later periods of life increases IGF-1, and offsets the sarcopenia of aging, giving objective evidence of its efficacy (Fielding 1995; Horber et al. 1996; Fiatarone Singh 1998). Administration of GH is effective in healthy elderly men and also, but less so, in women (Marcus et al. 1993). Because of several side effects and expense, GH might best be considered as an option for selected elderly patients suffering from catabolic illness (Corpas et al. 1993). The effect of non-hormonal interventions, such as the use of oral creatine to increase fat/bone-free mass and muscle performance in young athletes, was demonstrated (Kreider et al. 1998). The casual use of creatine by healthy older persons to re-create their youth has been applied without risk/benefit analysis. The use of androgenic steroids in men also stimulates a protein anabolic effect, but is ill-advised because of the increased risk of prostate cancer. Likewise, the use of other anabolic hormones (rHGH, erythropoietin) by young superathletes and "wannabes" (Begley and Brant 1999) is associated with risks.

In regard to ethnic or racial differences, various studies (Meneely et al. 1963; Ellis 1990; Aloia et al. 1997) indicate by measurement of TBK and/or TBN that the BCM (FFM) of American (USA) black male subjects (men and women) is about 5–10% higher than for comparable whites. This is most readily interpreted as indicating more skeletal muscle mass in blacks. An interesting observation from one study (Ellis 1990) is that, unlike the difference in TBK, the TBN was no greater in black men than white men, while the difference of TBN was like that of TBK in black vs. white women. This would suggest that in black men there is unusually high TBK/TBN in skeletal muscle or some shift in balance of protein between skeletal muscle (with high K concentration) and visceral organs (with lower K concentration). When female Chinese subjects were compared with a matched group of North American whites (Jiang et al. 1991), there was possibly lower BCM and LBM in the Chinese, but this could be well attributed to more fatness in the whites and thereby more fat-obligated skeletal muscle. Another ethnic difference (whether genetic or environmental) comes from comparison of Caucasian, African-American, Mexican-American and Native American women of

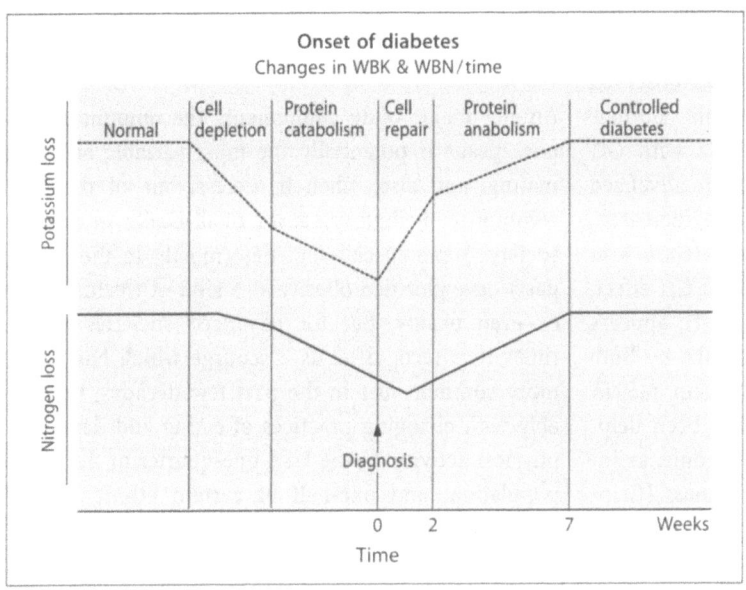

Fig. 12.6. Illustration of the general pattern of change in whole body potassium (WBK) and whole body nitrogen (WBN) in patients at the onset of diabetes and following treatment. (From Walsh et al. 1976, with permission)

younger (20–30 years) and older (40–50 years) age. There was a much greater decrease in TBK with age in all three other groups compared with Caucasian women (Thomas et al. 1997).

The hazards of major losses of BCM, and particularly muscle mass, in conditions of acute stress (trauma, sepsis, post-major surgery) or in prolonged wasting illness or semi-starvation has received increasing clinical attention in the past 20 years (Cuthbertson 1979; Hill 1992; Chang et al. 1998). There is well-documented early loss of BCM, especially skeletal muscle, following extensive burns or fracture (even moderate) (Cuthbertson 1979), acute onset of diabetes (Walsh et al. 1976), inflammation such as peritonitis (Plank et al. 1998) and in complicated post-operative patients or those who come to surgery under adverse conditions (Hill 1992; Monk et al. 1996). Combined measurements of TBK and TBN indicate that there is in acute diabetes (Walsh et al. 1976) (Fig. 12.6) and in cancer (Cohn et al. 1981) a greater and faster loss of K than N, probably indicating predominant loss of muscle with its high K concentration relative to other protein (visceral organs, connective tissue). Deficits of LBM and/or BCM in more chronic diseases have been recognized by body composition studies in patients with renal failure including those on dialysis (Johansson et al. 1998; Mitch 1998), growth-hormone deficiency (Christ et al. 1997), Parkinson's disease (Poehlman et al. 1995), Crohn's disease with short-bowel syndrome (Azcue et al. 1997; Ellegard et al. 1997), celiac disease (Smecuol et al. 1997), rheuma-

toid arthritis (Westhovens et al. 1997), patients on long-term glucocorticoids (Reid et al. 1996), and (prominently early in the disease) AIDS patients (Kotler et al. 1989). A significant finding is that men with HIV infection and weight loss have high serum and intramuscular concentrations of a myostatin, a glycoprotein which limits muscle growth and for which the controlling gene is well-characterized (Gonzalez-Cadavid et al. 1998). Whether increased myostatin contributes to wasting in other chronic illness is an open question.

The availability of body composition facilities can contribute importantly to the evaluation of results of efforts to treat deficiency or loss of BCM or FFM or muscle mass, whether from acute or chronic cause. The treatments evaluated are generally of two types, i.e., the composition of nutritional protein or amino acid (a.a.) supplements (especially glutamine, arginine, branched-chain a.a. or metabolites) and anabolic drugs or agents, e.g., growth hormone and anabolic steroids (Hill 1992; Nissen and Abumrad 1997; Chang et al. 1998). The availability in recent years of recombinant human GH (rHGH) has allowed wider application. Its use in the treatment of GH deficiency in adults (Cuneo et al. 1991; Christ et al. 1997; Johansson et al. 1997) has had a general success by evidence of measurements of LBM and BCM. GH has also been used with positive results in patients with short bowel syndrome due to Crohn's disease (Ellegard et al. 1997), patients with AIDS and weight loss (Schambelan et al. 1996) or various hospitalized patients with post-surgical complications or trauma

(Garlick et al. 1998). GH acts partially through stimulation of hepatic production of IGF-1, and the combined administration of GH and IGF-1 was observed to be more anabolic than either one alone (Bernais and Keller 1996). The treatment of AIDS with GH does have a "down side" – patients with advanced disease appear to react paradoxically with increased, rather than decreased, lean body wasting (Garlick et al. 1998). The usual anabolic mechanism of GH effect, as studied by 6-h infusion of methionyl GH, appears to involve an increase in amino acid uptake by limb muscle and increase in myosin heavy-chain mRNA levels (Fong et al. 1989). Testosterone has been helpful in the treatment of AIDS wasting syndrome, as indicated by increases in LBM and muscle mass (Grinspoon et al. 1998). Testosterone has reversed the deleterious effects of glucocorticoids on lean tissues in men (Reid et al. 1996). Other androgenic steroids have reduced nitrogen loss after surgical procedures and following trauma (van Wayjen 1993). Insulin reverses, of course, the loss of TBK and TBN in acute onset of juvenile diabetes (Walsh et al. 1976) (Fig. 12.6), but has not proved useful in other general wasting conditions. There is an interesting suggestion of developing agents to block genetic expression of myostatin in AIDS or possibly other conditions of muscle wasting (Gonzalez-Cadavid et al. 1998).

Some of the studies cited above used only the isotope or nuclear techniques, while others included, or used instead, those methods (anthropometry, BIA, DXA, X-ray CT, hydrogen MRI) which are not usually or properly in the province of nuclear medicine. The non-nuclear techniques can analyze regional differences in the body (including ratios of lean/fat tissue), with some extrapolation for whole body, whereas the current nuclear techniques presently measure only, but more directly, global body composition. Once facilities are in place, the nuclear techniques achieve reliable results in longitudinal studies with a variety of kinds of hospitalized patients, or in larger populations with nutritional disorders. Such studies are relatively comfortable for patients and involve moderate radiation doses. In the future, with the more efficient and sensitive "associated particle" or gamma resonance absorptiometry techniques for nitrogen, certain regional body analyses may become feasible with very low radiation dose and possible 3-D imaging. These techniques could be very useful for relatively acute disorders (e.g., fractured limb) or more chronic disabilities (e.g., paralysis from CVA or spinal cord injury) (Spungen et al. 2000).

12.3.2.3
Fat Mass

Among major body components the amount of adipose tissue is potentially the most variable and fluctuating, and also, when in excess, one of the most common contributory causes of ill-health in modern society. Some special societies (mostly in the distant past) have glorified obesity as a sign of wealth, health or even beauty, but for advanced societies in our times it is recognized as a scourge which has grown more common just in the past few decades, presumably with changing practices of eating and degrees of physical activity. In the USA one-quarter of the global population, and one-half of certain ethnic populations, at virtually all ages, is now considered obese, and other societies around the globe are becoming similarly affected (Taubes 1998). Besides being highly correlated with dangerous degenerative disorders such as diabetes, hypertension and atherosclerosis, the psychological effects are distressingly negative. For the obese person feelings range from embarrassment to self-contempt, while the public is inclined to view the condition with distaste if not revulsion and generally as a sign of laziness and/or gluttony. Such a view may ignore or overlook the considerable evidence that obesity is importantly determined by genetic make-up as well as by life-style practices. The nuclear medicine physician needs to have some knowledge about the roles and interactions of environmental, genetic and metabolic factors in order to advise and help in the evaluation and management of obesity.

For measurement of fat mass (FM) nuclear medicine has techniques from simple to complex in terms of availability of technical equipment and personnel, cost, radiation exposure and state of development for clinical use. Most of the methods are indirect in that they calculate body FM by difference between body weight and other measured body components. Most methods depend on a primary measurement of total body water, which may serve (in the simplest two-compartment model) to calculate lean body mass (LBM) by using the Pace constant, 0.73, as the mean for hydration fraction of the LBM. Radioactive tritiated water (^3H_2O) is often the method of choice for ease of performance and cost, but the stable isotopes, ^2H and/or ^{18}O, may be preferred in some sites and for some reasons (Jebb and Elia 1993; Bosaeus et al. 1996) (see Sect. 12.3.2.1.3). In the obese subject difficulty arises in the use of the hydration constant, because there is typically excess ECW relative to ICW, e.g., in very obese males a mean ECW/ICW of 0.74 vs. 0.42 in normals (Pierson et al. 1976). The difficulty may be addressed by measurement of the ECW by a dilution tracer, e.g. ^{35}S-sulfate or chemical Br (see Sect. 12.3.2.1.2) with concomitant whole body counting of ^{40}K (Pierson et al. 1976). Use of ^{40}K alone as a measurement of body fat involves further questionable constancy assumptions for estimating LBM or body cell mass (BCM). Greater accuracy (and other information) is obtained by increasing the number of compartments measured: TBN (and hence protein) by PGNAA (becoming more available; see Sect. 12.3.2.2) and TBCa (hence bone) by DGNAA (less available, more costly, and with more radiation exposure; see Sects. 12.3.2.1.2 and 12.3.2.4). In contrast to any of the above methods, the FM is evaluated more directly by the nuclear technique of inelastic neutron scattering (INS) which can measure total body carbon (TBC) (abundantly present in fat) with correction for non-fat carbon by measurement of TBN (Sutcliffe 1996; Wang et al. 1998). The technique for INS requires special facilities, but delivers only a small radiation dose (0.5 mSv). As described in Sect. 12.3.2.2, there is the promise of the "associated particle technique" for measuring TBC and TBN plus also total body oxygen (TBO), from which body water is calculable; this technique involves a trivial radiation dose (0.03 mSv) for total body assay and would allow regional assay with moderately higher dose (Mitra et al. 1995).

All of the above techniques measure total amounts of body components and are not presently practicable (though hopeful by the associated particle or gamma resonance techniques) for regional analysis, which is quite important in evaluating types of obesity, e.g., central (particularly visceral) or peripheral (essentially subcutaneous). In theory, the commonly-used techniques of BIA (Sect. 12.3.2.1.2) and DXA (Sect. 12.3.2.2) have such capability. However, the special prediction formulae needed for BIA of FM in obese persons are presently uncertain in application (Bosaeus et al. 1996), and, even for subjects of normal size, BIA gives low values for FM compared with other techniques (Bosaeus et al. 1996; Wang et al. 1998). DXA provides a more direct measure of FM and compares well with other techniques (except BIA), but accuracy falls off with increasing tissue thickness and abnormal hydration. Besides the "associated particle technique", regional analysis may also be done by the method of gamma resonance absorp-

tiometry (GRA), which also has the theoretical capacity to measure whole body N (demonstrated), C, Ca and O (Vartsky and Goldberg 1997). GRA involves a very low dose of radiation, which would allow long regional counting with increasing statistical validity. None of the foregoing techniques can provide organ analysis of fat content, which is usefully provided now by the non-nuclear techniques of CT and MRI to delineate local FM from FFM. A radioisotopic technique which has this capability, in addition to measuring total body fat, is based on the high solubility of noble gases (xenon, krypton) in fat. As shown with ^{127}Xe or ^{79}Kr (Susskind et al. 1978) (Table 12.1), the slowest component of wash-out of either of these gases from the whole body correlates highly with total body fat, as judged indirectly by the TBK method (^{40}K) for LBM. Regional profiling of activity from the positron emitter, ^{79}Kr, indicated distinct shifts in body location with time after initial inhalation, with eventual display of central vs. peripheral adiposity (Fig. 12.7). Other studies with ^{133}Xe have shown high correlation of gas retention in human liver by external imaging/counting compared with measurement of fat in liver biopsies (Ahmad et al. 1979; Rogus and Blumenthal 1981). A study combining ^{133}Xe gas and ^{131}I-iodo-antipyrine in washout measurements analyzed the increasing fat content of skeletal muscle with age (Lindbjerg 1967). Renewal of such studies with labeled inert gases combined with present technology for anatomical definition (SPECT, PET, CT, MRI) could provide more precise "denominators" for local fat content in the body, while also providing information on total body FM by whole-body counting of wash-out rates.

The amount and location of fat in the body is the result of continuing interplay of environmental and genetic factors. There is a complex mixture of signals emanating from endocrine organs, digestive tract, central nervous system, adipose tissue and muscle. Receptors in those and other organs receive signals by neuronal or circulatory routes, and respond by upward or downward regulation, generally in directions tending to maintain a balance or "set-point" for each individual's degree of fatness (Gladwell 1998). Traditionally there is the effect of some hormones, e.g., estrogens, glucocorticoids and insulin, to induce to fatness, whereas others, e.g., growth hormone, thyroid hormone and catecholamines, induce the opposite. A recently-discovered hormone produced by adipose tissue is leptin, a protein which is the product of the "OB gene" and which acts on CNS receptors to stimulate sa-

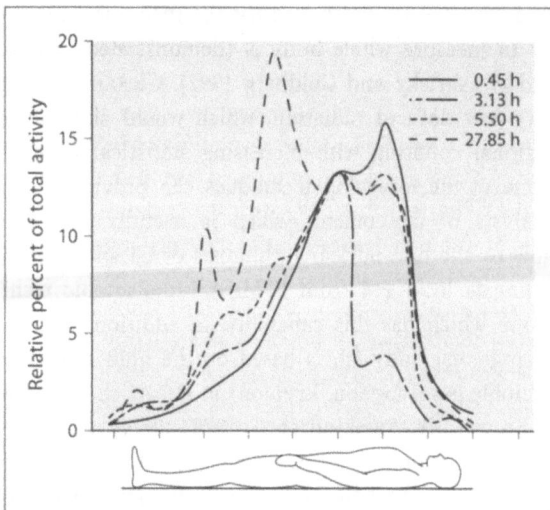

Fig. 12.7. Typical profile distribution of ^{79}Kr in man, based on coincidence counting mode of the whole body counter at Brookhaven National Laboratory. (Reprinted from Susskind et al. 1978, with permission)

tiety centers, is inhibitory to insulin secretion, and increases energy expenditure, all actions which tend to limit fat accumulation (Fehman et al. 1997). Tumor necrosis factor-alpha (Fernandez-Real et al. 1997) and neuropeptide-Y (Erickson et al. 1996) are two compounds which modulate production and CNS action of leptin, respectively. Serum leptin levels vary in proportion to the amount of body fat. A peptide hormone, ghrelin, from the stomach and duodenum induces hunger and also diposes metabolically to obesity; efficacy of gastric by-pass for morbid obesity may relate partly to decreased ghrelin (Cummings et al. 2002). Cholecystokinin and serotonergic and dopaminergic neurotransmitters are among other agents affecting the hunger and satiety centers (Legato 1997; Wang et al. 2002). There is also the effect of "uncoupling proteins" (UCP), which divert energy of substrates from ATP formation (and fat formation) to generation of heat (Gura 1998). Two alleles of a human UCP gene correlate with two phenotypes, as indicated by more or less increase in percentage body fat over a 12-year period (Oppert et al. 1994). Many such chances exist for derangements in these multiple control points, which account for obesity that has been correlated with the growing list of obesity-related genes (Weigle and Kuijper 1996; Perusse and Chagnon 1997; Comuzzie and Allison 1998).

Obesity is a condition in which there is a wide range of phenotypes within a "normal" population. The phenotypes occur in a seeming continuum, and the line between normal and abnormal is arbitrary. This continuum and lack of definite abnormal margin is surely due to the multiplicity of genetic polymorphisms/mutations bearing upon the condition. The definition of "normal" is commonly based on body mass index (BMI), wherein a range of the index from 21–25 (kg/ m^2) has been viewed as being within normal, 25–27 as "overweight", 27–30 as "obese", and >30 as "morbid obese" with good evidence for disease association of BMI > 25 (Legato 1997). BMI as an indicator is not ideal due to variations in lean body (usually muscle) mass. Percentage of weight as FM is more directly pertinent and often used, as measured by various means (see above). In a typical Western population the "normal" FM of women is about 20–40% and that of men is about 10–30% with the range dependent on age, race, method of analysis (Pierson et al. 1997; Wang et al. 1998) and other factors. Another significant difference between men and women, is regional fat distribution. Men are inclined to have central (particularly visceral) accumulation of fat, whereas the distribution in women is generally more peripheral, e.g., subcutaneous in buttocks and limbs. According to much evidence, central adiposity has a much stronger linkage to the pathologies of hypertension, diabetes, atherosclerosis and coronary heart disease ("metabolic syndrome x"), and also is more genetically determined, than peripheral adiposity. After menopause women are more disposed to central obesity and its accompanying risk factors. A gradual increase in weight (likely fat) with age up to 60 or 70 years is common. This could relate to a gradual disruption of the relationship between fat content and leptin with aging (Moller et al. 1998). In regard to racial differences, black women (BW) have less fat, as a percentage of body weight, but more upper body fat, than white women (WW) (Gasperino 1996). Although this would imply greater disease risk for BW, other findings indicate that BW with upper body fat are less disposed than WW to insulin resistance and dyslipidemia (Dowling et al. 1995). Various observations indicate that non-Caucasian populations, e.g., Pima Indians of Southwest United States (Gladwell 1998) and First Nation Canadians (Katzmarzyk and Malina 1998), particularly after adoption of Western life styles of little physical activity and "fast-food" eating, are highly prone to obesity, especially that of the central type (Fujimoto et al. 1995). Research to identify loci for specific genotypes for obesity in such non-Caucasian populations is being actively pursued (Norman et al. 1998).

Pathologically extreme or "morbid" obesity is in most cases related to an unfortunate massive concurrence or expression of a variety of obesity-disposing genes which, when less expressed, account for moderate obesity in the general population. So far in just a few families has severe obesity from childhood been identified with defects of the OB gene causing true leptin deficiency (low serum leptin) despite the markedly elevated fat mass, or else mutation of the gene for the leptin receptor (Campfield et al. 1998). Of course, severe obesity may also have strong environmental components, i.e., physical inactivity and indulgent overeating. There is the hazard of the disabling obesity limiting the physical activity with the institution of a vicious cycle. In any case, to the "metabolic syndrome x" may be added in the excessively obese person such pathologies as arthritis, heart failure, sleep apnea, elephantiasis with skin changes and (quite commonly) hepatic steatosis progressing occasionally to fibrosis and cirrhosis (Marceau et al. 1999). The challenges to the practical utilization of different nuclear or non-nuclear methods for measuring body composition in extreme obesity are evident and documented (Allison et al. 1998). A small fraction of the obese population have diverse clinical syndromes, often with hormonal defects such as hypogonadism or isolated growth hormone deficiency. These are rare in occurrence, and in some cases autosomal recessive (Bouchard and Perusse 1988; Perusse and Chagnon 1997). In most of these syndromes the genetic defect is not known. An exception is the Prader-Willi syndrome with an abnormality on chromosome 15, a prevalence of about 1:25,000 and the characteristic of uncontrollable hyperphagia. Some chromosomal aneuploidies, such as Down's syndrome and Turner's syndrome, have excess body fat, which is resistant to change (Bouchard and Perusse 1988).

Monitoring of intentional weight loss, particularly in the very obese and when major losses are involved, should include measurement of various body components – not only fat, but FFM, by measuring water, protein and minerals (bone), when possible. To that end, use of some nuclear techniques (for TBW, EWC, TBK, TBN, TBCa) are very helpful. The important difference between central visceral vs. central or peripheral subcutaneous fat requires either CT or MRI or, as suggested earlier, possibly a renewed application of the labeled inert gas technique for measuring localization (and change) in regional internal fat as well as change in total FM. Strategies for weight loss include generally four modalities which are, in usual order of importance and proper use: diet, exercise, biochemical agents (mainly drugs) and gastrointestinal surgery. Kinds of diet (low-carbohydrate vs. low-fat) are controversial and beyond the scope of detailed discussion here. Relatively high protein will obviously help preserve LBM. Calorie restriction is paramount over kinds

of food content in achieving weight loss. It is axiomatic, with much supporting evidence, that the more rigorous the diet the more likely will be loss of various components of the FFM. A very strict diet can lead to excessive loss of FFM (both muscle and bone). It is said that FFM loss should compose no more than 30% of weight loss; also, that in the obese some of the loss of muscle (of less good quality) may be desirable (Marks and Rippe 1996). That the proportional associated loss of FFM, as well as variability in fat loss, is partly under genetic control is indicated by considerable differences among individual subjects who were in a fairly homogeneous group of mostly black, middle-aged women of similar obesity (Albu et al. 1992). Exercise is ordinarily a very useful supplement to diet. While aerobic exercise during dieting may not increase the amount of weight loss, such exercise can limit the loss of FFM (Andersen et al. 1999), improve cardiorespiratory fitness during dieting (Utter et al. 1998) and limit the reduction of resting energy expenditure (Wadden et al. 1997). However, people differ considerably in their cardiovascular response to exercise, as well as in their spontaneous physical activity and in energy expenditure during activity (Vogel 1999). Genetic factors in each of these differences are suspected. Free-activity energy expenditure, measured by the dual stable isotope (^{2}H/^{18}O)-labeled water method (see Sect. 12.3.2.1.3), can be used to evaluate individual responses to exercise. Whereas various medications have been used for decades to accompany diet and exercise to promote fat loss, in recent years there has been particularly active development of such agents based on the growing knowledge of the hormonal, enzymatic and other various metabolic factors which may dispose to obesity, as revealed in some cases by genetic studies. Thus, drugs which affect satiety (through serotonergic, vagal or noradrenergic neurons, leptin or opioid receptors), fat absorption or biosynthesis, or gastric emptying rate are currently being used or are in development (Legato 1997; Campfield et al. 1998). Effects of drugs acting on insulin production, i.e., metformin (Paolisso et al. 1998) and diazoxide (Alamzadeh et al. 1998), to cause weight and fat loss in non-diabetic obese subjects emphasize the close relationship of obesity and diabetes. Among hormones possibly to be exploited, attention has focused recently on growth hormone and IGF-1, which aid fat loss with minimal loss of FFM (Johannsson et al. 1997; Thompson et al. 1998). Gastrointestinal surgery (reserved generally for morbid obesity) is well-known to have side-effects, but nevertheless can bring about significant fat loss with relative preservation of LBM (Gahtan et al. 1997; Legato 1997).

Patterns of loss of weight and FM differ according to gender and race and thereby reinforce the evidence for genetic factors controlling the amount and distribution of body fat. On weight loss programs men lose more intra-abdominal (visceral) fat whereas women lose more subcutaneous fat (Wirth and Steinmetz 1998). Unlike men, women may require exercise as well as dieting to decrease waist-hip ratio, a marker for visceral fat (Legato 1997). However, in inter-racial studies, obese white women (WW), who basically have higher visceral/subcutaneous abdominal fat ratios then obese black women (BW), mobilize more visceral and less subcutaneous fat than BW (Van der Merwe et al. 1996). Obese BW have lower resting energy expenditure (REE) than WW (Jakicic and Wing 1998) and yet have greater percentage drops in REE upon weight loss than do WW (Foster et al. 1999). These differences may help explain lesser degrees of fat loss in BW than in WW on weight loss programs.

The difficulty of maintaining weight loss is notorious, with return to previous high levels occurring in 90% or more in many studies. This is commonly attributed to the "set-point" theory (see above) which gains evidence from metabolic changes that occur either upon underfeeding or overfeeding of normal or overweight subjects and which operate, in either case, to resist weight change and are conducive to returning to previous weight (Gibbs 1996). Much other evidence, as discussed above, which reveals the particular vulnerability of certain racial groups, by reason of having more "thrifty genes", to become readily obese upon adoption of Western life styles (decreased physical activity, increased eating of high-density foods) has led to the idea of a "settling-point". The latter takes account of the environment interacting with the genetic predisposition. The settling-point theory allows the possibility that deliberate, drastic and prolonged change in life-style after weight loss might reset the set-point at a new weight (Gibbs 1996; Gladwell 1998).

At the opposite extreme from obesity is the paucity of fat mass in the much less common disorders of anorexia nervosa (AN) and bulimia nervosa (BN). Body composition changes in AN upon treatment and weight gain have been studied (Polito et al. 1998; Orphanidou et al. 1997). This subject is discussed further in Sect. 12.3.2.4 below, since the long-term clinical significance of AN and BN (in body composition) is greatest in this area (Walsh and Devlin 1998).

12.3.2.4
Bone Mass

In the matter of measurement and evaluation of bone mass, nuclear medicine has an important past, a rather specialized and minimal present (unless one includes DXA) and a possibly new and interesting future. Whatever techniques may be employed will have a worthy goal in recognizing bone loss, and aiding in its prevention and treatment, in various diseases, especially age-related osteoporosis, a major health (and financial) threat for families and society.

Radiographic photodensitometry on X-ray films has limitations of standardization and repeatability and is confined to the appendicular skeleton. A quantum improvement was made in the 1970s–1980s with the more sensitive and quantitative techniques of single photon absorptiometry (SPA) using ^{125}I or ^{241}Am as external sources, and then (better) dual photon absorptiometry (DPA) using (most often) ^{153}Gd. These techniques are described in detail elsewhere (Wahner et al. 1983, 1984). SPA can measure the bone mineral density (BMD, gm/cm^2) of distal sites of the appendicular skeleton only and therefore mainly cortical bone, except for the mostly trabecular site in the calcaneus, while DPA can evaluate the central skeleton, particularly the important proximal femur and the vertebrae, which contain mostly the metabolically active trabecular bone, where fractures due to osteoporosis are more common. In the late 1980s SPA and DPA were largely displaced by dual X-ray absorptiometry (DXA) because of much higher photon flux, absence of source decay, lower cost of maintenance and more convenience. DXA, like DPA, is applicable to both central and appendicular skeletal sites. DXA also has a precision (1–2%) which is better than that of DPA (2–3%), but the highest precisions are accomplished by utilization of the same machine (and, still better, the same operator), since comparisons among even the same type machines from the same manufacturer can give results as disparate as ±5% (Formica 1998). Differences between instruments from different manufacturers may be up to 15% (Tothill et al. 1994). The modest radiation doses from DXA are similar to those from SPA and DPA (Table 12.1). While not a nuclear technique, DXA is performed in some nuclear medicine locations as a "hold-over" from use of SPA/DPA. Also in the 1980s the technique of quantitative CT of bone was found useful to delineate trabecular from cortical bone in the same image; the technique is used presently in research settings in which the high cost and radiation exposure are tolerable. Another more recent non-nuclear technique is that of quantitative ultrasound, which is promising (especially for the heel), but has some questions of accuracy and precision (Galt 1999).

The considerably different technique of neutron activation of natural ^{48}Ca to produce gammas from ^{49}Ca in local sites (Zaichick and Morukov 1998) or whole body (Cohn 1995a, b) was begun in the 1960s and continues, mostly in clinical research applications, to the present time. The precision of measurements of TBCa by delayed gamma (DGNAA) is high: 1–2% (Wahner et al. 1984; Cohn 1995) and accuracy is within ±5%, according to recent comparisons of data between two different facilities in the US (Ma et al. 1999). This technique has served as a standard for calibration of bone densitometry devices, and has provided data on differences in TBCa among normal subjects of differing age, sex, ethnicity and life-style (physical activity) plus other subjects with osteoporosis. However, the cost, required instruments and technical expertise and moderately high radiation dose (see Sect. 12.2.3.4) are barriers to wide-spread application. Since the total skeleton is only 20% trabecular bone, the measurement of TBCa theoretically should be relatively insensitive for recognizing osteoporosis and fracture risk, but in practice high sensitivity and specificity have been demonstrated (Cohn et al. 1986). A promising newer technique with very low radiation dose, and a potential for high precision and accuracy, as well as imaging with about 2-cm resolution at regional sites or for the whole body, is that of gamma resonance absorptiometry (GRA), which uses a compact proton accelerator to produce incident gamma rays resonant with calcium (as well as others related to nitrogen) (Cohn 1995a). GRA is currently in research development. Still another technique bearing upon evaluation of bone metabolism and mass is that of employing radioactive or stable isotopes of calcium (Goans et al. 1995; Kurz et al. 1995; O'Brien et al. 1998) (see Sect. 12.1.2.3 and Table 12.1) to measure intestinal absorption and turnover in bone, which can evaluate major factors in determining vulnerability to osteoporosis.

Osteoporosis is defined as a degree of bone loss more than 2.5 SD below the peak bone mass (the young adult norm) ("T-score"). Osteopenia is a lesser degree of bone loss (−1.0 to −2.5 SD). Osteoporosis involves loss of organic matrix or osteoid portion (mostly collagen) as well as mineral. There is disturbance in micro-architecture with thinning of trabeculae in cancellous bone and resultant weakening (osteomalacia). Non- or minimal-trauma fractures often occur in vulnerable sites (vertebrae, upper femur and wrist).

There are several diseases or environmental conditions which cause osteopenia and osteoporosis (Smith 1986) (see below). Many quantitative studies have generated a list of risk factors: estrogen deficiency or resistance, steroid use, cigarette smoking, alcohol abuse, thyroid disease, hyperparathyroidism, epilepsy medication, leanness, low calcium and/or low vitamin D intake, or low physical activity. Other predisposing diseases and conditions are Cushing's disease, acromegaly, prolonged heparin therapy, thalassemia major, lipid storage disease and diabetes (Wasserman and Barzel 1987; Wonke et al. 1998). The mechanisms, including the genotype-phenotype correlations, in these disorders are increasingly under study.

Osteoporosis in the aged is so common (40% of women after age 50) and progressive in both women and (at a slower rate) in men, that it has until recent years been ascribed to general aging. While some familial incidence has long been recognized, this could be attributed to cultural factors, including physical activity and low-calcium diet as well as to genetic factors. Elderly subjects, with or without osteoporosis, show decline of intestinal absorption of calcium and increased blood levels of parathyroid hormone (Wasserman and Barzel 1987; O'Brien et al. 1998). In the last two decades more intensive family studies of first-degree relatives and twin-pairs have more specifically implicated genetics in osteoporosis by correlating BMD and other signs with an array of polymorphisms in several candidate genes, including those for vitamin D receptors (VDR), estrogen receptors (ER), calcitonin receptors (CTR), Type 1 collagen (COLIA1 and COLIA2), apolipoprotein E, transforming growth factor-beta and interleukin-6 (Ralston 1998; Taboulet et al. 1998; Wood and Fleet 1998). (Interestingly, the heterozygous condition for the CTR polymorphism conveys higher BMD than does either of the homozygous types.) From these studies it is presently estimated that 50–90% of variability in bone density is determined by a genetic component, depending on the site measured. Differences in axial skeletal sites appear to be more genetically influenced than those of distal sites. While many studies present evidence of significant correlations of the major candidate genes with phenotypic expression, others find little or no evidence. This suggests that osteoporosis has a complex polygenic origin and/or that age or environmental factors modulate or counteract genetic expression. Occurrences of gene-gene (or allele-allele) interaction are exemplified in particular observations: the co-presence of mutated VDR and ER genes has an effect on BMD not seen with either one alone (Deng et al. 1998) and the combined effect of a VDR-5' start codon polymorphism and a kind of VDR-3' allele has a greater effect than the occurrence of the VDR-3' allele only (Ferrari et al. 1998). Environmental factors such as dietary intake of Ca^{2+} or Vitamin D (or available sunlight to produce the latter), amount and kind of exercise, or postmenopausal estrogen administration can well interact with genetic expression. Thus, postmenopausal women who are positive, negative or mixed (heterozygous) for a certain VDR-3' allele will all respond differently to Ca^{2+} supplementation with respect to change in BMD. Also, homozygotes for the unfavorable allele may show a lower fractional absorption of ^{45}Ca than homozygotes without such an allele on low Ca^{2+} intake but not on high Ca^{2+} intake (Ferrari et al. 1998; Eisman 1998). In a study of the kinetics of stable ^{42}Ca (O'Brien et al. 1998) middle-aged and elderly women from osteoporotic families (unlike control counterparts) were unable to maintain positive calcium balance (due to higher bone resorption) even on high Ca^{2+} intake (1500 mg). These and other data suggest the need for supplementation with both Ca^{2+} and Vitamin D in elderly women.

In osteoporosis, as in many diseases, an ounce of prevention is worth a pound of cure. In the early menopausal period in any woman pre-disposed by body habitus, family history, or other risk factors, the use of estrogen replacement plus high Ca^{2+} and vitamin D intake can be highly effective as shown by either densitometry or DGNAA measurements (Aloia et al. 1994; Ziegler et al. 1995). Deficiency of androgens may also play a role in the pathophysiology in men or women and should be considered in the equation of management. Exercise programs, especially resistance training, have proven value in either prevention or rehabilitation (Wasserman and Barzel 1987; Nelson et al. 1991; Harrison et al. 1993). Other useful treatments may include calcitonin, fluoride (Murray et al. 1990) or certain drugs, e.g., alendronate or raloxifene, one of the new class of "SERMS" (selected estrogen receptor modulators) (Muchmore 1999). Attention is being directed to diagnostic evaluation and preventive measures for susceptible persons early in life. The growing evidence for specific genetic factors has spurred this attention, along with the recognition that peak bone mass in early adulthood strongly determines the reservoir of bone mass in later decades (Eisman 1998). For example, it has been shown that the lesser incidence of osteoporo-

sis and related fractures in older black than in age-matched white women is mainly due to the higher peak bone mass of blacks (Wasserman and Barzel 1987). Daughters of women with osteoporosis have less good response to increasing Ca^{2+} intake than do controls (O'Brien et al. 1998) and this may be related to the presence of unfavorable VDR alleles (Eisman 1998; Wood and Fleet 1998). Daughters of mothers with low BMD have low BMD (Barthe et al. 1998; Ferrari et al. 1998). The young (particularly females) may be put at risk by environmental as well as by genetic factors. Besides the obvious cases of malnutrition due to prolonged illness or famine, anorexia nervosa (AN) and the related bulimia nervosa are strong risk factors, especially when associated with amenorrhea. AN usually occurs at an age crucial for development of peak bone mass, which may be diminished by low Ca^{2+} intake, low estrogen secretion, and increased levels of cortisol (Walsh and Devlin 1998). Estrogen lack may be related to hypogonadism or to dearth of adipose tissue, which is involved in converting precursors to estradiol (Wasserman and Barzel 1987) (and which may help explain the low incidence of osteoporosis in obese persons). Another factor in AN may be a decrease in the weight-bearing loads of ordinary physical activity, which maintains bone mass at all ages. A critical decrease in peak bone mass puts even recovered AN patients at jeopardy for osteoporosis in later life. An entity somewhat related to AN is that of the "female athlete triad" of disordered eating, amenorrhea and bone loss, which may be seen, for instance, (despite exercise) among young gymnasts, cross-country runners and ballet dancers (Villarosa 1999). A well-known possible cause of early osteoporosis in an elite few is that of astronauts, who must carry out special exercise programs to avoid bone loss (Zaichick and Morukov 1998). Temporary but pronounced bone loss may occur at regional sites of immobilization due to injury or surgery, and may be evaluated by regional densitometry or calcium measurement.

One of the more vexing problems with osteoporosis is the difficulty in predicting fracture risk from lifestyles, biophysical measurements or genetic information. Bone strength is determined not only by bone density but by bone quality, bone geometry and microdamage, while neuromuscular function, affecting susceptibility to falls, is another important factor (Greenfield 1992). Although certain VDR genotypes correlate with bone mass, no clear association is found between VDR polymorphisms and osteoporotic fracture (Eisman 1998; Ralston 1998). Although Ca^{2+} absorption and bone deposition are influenced by variation in VDR genes, wide differences in Ca^{2+} intake do not correlate with fracture risk, possibly due to ethnic or environmental differences (Eisman 1998). Calcitonin receptor polymorphism has been correlated to fracture risk (Taboulet et al. 1998). Linkage has been observed between the COLIA1 Sp1 polymorphism and vertebral bone mass and fracture (Keen et al. 1999; Sainz et al. 1999), whereas the association between the collagen genotype and femoral BMD is inconsistent or absent (Ferrari et al. 1998; Ralston 1998; Wood and Fleet 1998), suggesting a genetic fracture risk independent of bone mass. (Such findings pose a challenge to nuclear medicine to devise ways to recognize kinetic or other abnormalities of bone collagen.) Findings about the vertebrae do not necessarily predict fracture of the hip (Wahner et al. 1983) and there is a difference in time pattern between hip and vertebral fracture – for hip fractures there is no apparent increased rate at the time of menopause, unlike fractures of the spine or wrist (Wasserman and Barzel 1987). Such differences as above have led to speculation that spinal osteoporosis and age-related hip fractures may be two different osteoporotic syndromes (Wahner et al. 1983). Very likely more genetic or other information will be discovered which will help explain present puzzling discrepancies. A genome-wide scan on a larger number of individuals in nuclear families in China showed evidence of linkages to forearm BMD of polymorphisms on chromosomes 2 and 13, and identified further candidate genes such as calmodulin 2, pro-opiomelanocortin, COL4A1 and COL4A2 (Niu et al. 1999).

12.4
References

Ahmad M, Perillo RP, Sunwoo YC et al (1979) Xenon-133 retention in hepatic steatosis-correlation with liver biopsy in 45 patients: concise communication. J Nucl Med 20:397–401

Ajjan RA, Findlay C, Metcalfe RA et al (1998) The modulation of the human sodium iodide symporter activity by Grave's disease sera. J Clin Endocrinol Metab 82:1217–1221

Al-Ghamdi SMG, Cameron EC, Sutton RAL (1994) Magnesium deficiency: pathophysiologic and clinical overview. Am J Kidney Dis 24:737–752

Alberts SN (1974) Blood volume in clinical practice. In: Rothfeld B (ed) Nuclear medicine in vitro, chap 5. Lippincott, Philadelphia, pp 52–68

Alberts B, Bray D, Lewis J et al (1994) Molecular biology of the cell, 3rd edn. Garland, New York

Albu J, Smolowitz J, Lichtman S et al (1992) Composition of weight less in severely obese women: a new look at old methods. Metab Clin Exp 41:1068–1074

Alemzadeh R, Langley G, Upchurch L et al (1998) Beneficial effect of diazoxide in obese hyperinsulinemic adults. J Clin Endocrinol Metab 83:1911–1915

Allison DB, Nathan JS, Albu JB et al (1998) Measurement challenges and other practical concerns when studying massive obese individuals. Int J Eat Disord 24:275–284

Aloia JF, Vaswani A, Ma R et al (1997) Comparisons of body composition in black and white premenopausal women. J Lab Clin Med 129:294–299

Aloia JF, Vaswani A, Yeh JK et al (1994) Calcium supplementation with and without hormone replacement therapy to prevent postmenopausal bone loss. Ann Intern Med 120:97–103

Andersen RE, Wadden TA, Bartlett SJ (1999) Effects of lifestyle activity vs. structured aerobic exercise in obese women: a randomized trial. J Am Med Assoc 281:335–340

Arai Y, Kubo T, Kobayashi K et al (1997) Adenovirus vector-mediated gene transduction to chondrocytes: in vitro evaluation of therapeutic efficiency of transforming growth factor-1 and heat shock protein 70 gene transduction. J Rheumatol 24:1787–1795

Arden NK, Spector TD (1997) Genetic influences on muscle strength, lean body mass and bone mineral density: a twin study. J Bone Miner Res 12:2076–2081

Armstrong LE, Kenefick RW, Castellani JW et al (1997) Bioimpedance spectroscopy technique: intra-, extracellular and total body water. Med Sci Sports Exerc (MG8) 29:1651–1663

Avioli LV, Berman M (1966) Mg^{28} kinetics in man. J Appl Physiol 21:1688–1694

Avioli LV, McDonald JE, Lee SW (1965) The influence of age on the intestinal absorption of ^{47}Ca in women and its relation to ^{47}Ca absorption in post-menopausal osteoporosis. J Clin Invest 44:1960–1967

Azcue M, Rashid M, Griffiths A et al (1997) Energy expenditure and body composition in children with Crohn's disease: effect of enteral nutrition and treatment with prednisolone. Gut 41:203–208

Baarends EM, Schols AM, van Marken Lichtenbelt WD (1997) Analysis of body water compartments in relation to tissue depletion in clinically stable patients with chronic obstructive pulmonary disease. Am J Clin Nutr 65:88–94

Baccelli G, Pacenti P, Terrani S et al (1995) Scintigraphic recording of blood volume shifts. J Nucl Med 36:2022–2031

Barrett JFR, Whittaker PG, Williams JG et al (1997) Absorption of non-haem iron from food during normal pregnancy. Br Med J 309:79–82

Barthe N, Basse-Cathalinat B, Meunier PJ (1998) Measurement of bone mineral density in mother-daughter pairs for evaluating the family influence on bone mass acquisition. Osteoporos Int 8:379–384

Bedogni G, Merlini L, Ballestrazzi A et al (1996) Multifrequency bioelectric impedance measurements for predicting body water compartments in duchenne muscular dystrophy. Neuromusc Disord 6:55–60

Begley S, Brant M (1999) The real scandal. Newsweek Magazine, 15 Feb 1999, pp 48–51

Belmin J, Levy BI, Michel JB (1994) Changes in the renin-angiotensin-aldosterone axis in later life. Drugs Aging 5:391–400

Bernais K, Keller U (1996) Metabolic actions of growth hormone: direct and indirect. Baillieres Clin Endocrinol Metab 10:337–352

Beyenbach KW (1990) Transport of magnesium across biological membranes. Magnesium Trace Elem 9:233–254

Boesch C, Elsing C, Wegmuller H et al (1997) Effect of ethanol and fructose on liver metabolism: a dynamic 31 phosphorus magnetic resonance spectroscopy study in normal volunteers. Magn Reson Imaging 15:1067–1077

Bolsover SR, Hyams JS, Jones S et al (1997) From genes to cells. Wiley, New York

Bosaeus I, Johannsson G, Rosen T et al (1996) Comparison of methods to estimate body fat in growth hormone deficient adults. Clin Endocrinol 44:395–402

Bothwell TH, Charleton RW, Motulsky AG (1995) Hemachromatosis. In: Scriver CR, Beaudet AL, Sly WS, Valle D (eds) The metabolic and molecular bases of inherited disease, vol II, 7th edn. McGraw-Hill, New York, pp 2237–2269

Bouchard C, Perusse L (1988) Heredity and body fat. Annu Rev Nutr 8:259–277

Boulay MR, Song TM, Serresse O et al (1995) Changes in plasma electrolytes and muscle substrates during short-term maximal exercise in humans. Can J Appl Physiol 20:89–101

Brechue WF, Stager JM, Lukaski HC (1990) Body water and electrolyte responses to acetazolamide in humans. J Appl Physiol 69:1397–1401

Brevinge H, Jacobson L (1994) Total exchangeable sodium related to body composition in patients with conventional or reservoir ileostomy. Scand J Gastroenterol 29:160–165

Brill DR (1981) Radionuclide imaging of nonneoplastic soft tissue disorders. Semin Nucl Med XI:277–288

Campfield LA, Smith FJ, Burn P (1998) Strategies and potential molecular targets for obesity treatment. Science 280:1383–1387

Carroll HJ, Gotterer R, Altschuler B (1965) Exchangeable sodium, body potassium and body water in previously edematous cardiac patients. Circulation XXXII:185–192

Carswell H (1987) Glimpses into spectroscopy's future reveals promising clinical utility. Diagnostic Imaging, May, pp 157–165

Chance B, Eleff S, Leigh JP et al (1983) Phosphorus NMR. In: Partain CL, James AE, Rollo FD, Price RR (eds) Nuclear magnetic resonance (NMR) imaging. Saunders, Philadelphia, pp 399–415

Chang DW, DeSanti L, Demling RH (1998) Anticatabolic and anabolic strategies in critical illness: a review of current treatment modalities. Shock 10:155–160

Chattopadhyay N, Mithal A, Brown EM (1996) The calcium-sensing receptor: a window into the physiology and pathophysiology of mineral ion metabolism. Endocr Rev 17:289–307

Chaudhuri TK, Tran N (1982) Copper metabolism and Wilson's disease. In: Spencer RA (ed) CRC handbook series in clinical laboratory science, sect A. Nuclear medicine. CRC Press, Boca Raton, pp 294–296

Chauncey DM Jr, Schelbert HR, Halpern SE et al (1977) Tissue distribution studies with radioactive manganese: a potential agent for myocardial imaging. J Nucl Med 18:933–936

Chisholm GD, Short MD, Ghanadian R et al (1974) Radiozinc uptake and scintiscanning in prostatic disease. J Nucl Med 15:739–742

Christ ER, Carroll PV, Russell-Jones DL et al (1997) The consequences of growth hormone deficiency in adulthood, and the effects of growth hormone replacement. Schweiz Med Wochenschr 127:1440–1449

Cohn RM, Roth KS (1983) Metabolic disease: a guide to early recognition. Saunders, Philadelphia

Cohn SH (1995 a) Body composition. In: Wagner HN Jr, Szabo Z, Buchanan JW (eds) Principles of nuclear medicine, 2nd edn, chap 28. Saunders, Philadelphia, pp 475–482

Cohn SH (1995 b) Whole body counters. In: Wagner HN Jr, Szabo Z, Buchanan JW (eds) Principles of nuclear medicine, chap 16. Saunders, Philadelphia, pp 298–305

Cohn SH, Bozzo SR, Jesseph JE et al (1965) Formulation and testing of a compartmental model for calcium metabolism in man. Radiat Res 26:319–333

Cohn SH, Gartenhaus W, Sawitsky A et al (1981) Compartmental body composition of cancer patients by measurement of total body nitrogen, potassium and water. Metab Clin Exp 30:222–229

Cohn SH, Aloia JF, Vaswani AN et al (1986) Women at risk for developing osteoporosis: determination by total body neutron activation analysis and photon absorptiometry. Calcif Tissue Int 38:9–15

Comuzzie AG, Allison DB (1998) The search for human obesity genes. Science 280:1374–1377

Conte A, Volpi N, Palmieri L et al (1995) Biochemical and pharmacokinetic aspects of oral treatment with chondroitin sulfate. Arzneimittelforschung 45:918–925

Corpas E, Harman SM, Blackman MR (1993) Human growth hormone and aging. Endocr Rev 14:20–29

Cotzias G (1962) Manganese. In: Comar CL, Bronner F (eds) Mineral metabolism: an advanced treatise, vol II, chap 33. The elements, Part B. Academic, New York, pp 404–442

Cox RJ, Pearson RE, Brand HL (1977) Lithium in depression: a biochemical study. Gerontology (FPB) 23:219–235

Cummings DE, Weigle DS, Frayo RS et al (2002) Plasma ghrelin levels after diet-induced weight loss or gastric by-pass surgery. N Engl J Med 346:1623–1630

Cuneo RC, Solomon F, Wiles CM et al (1991) Growth hormone treatment in growth hormone-deficient adults I: effects on muscle mass and strength. J Appl Physiol 70:688–694

Cuthbertson DP (1979) The metabolic response to injury and its nutritional implications: retrospect and prospect. J Parent and Ent Nutr 3:108–130

Czerniak P, Zwas ST (1976) Usage de radiocuivre et du radiozinc pour le diagnostic psychiatrique. Ann Med Psychol (Paris) 134:566–576

Dai G, Levy O, Carrasco N (1996) Cloning and characterization of the thyroid iodide transporter. Nature 379:458–460

Daly PF, Cohen JS (1989) Magnetic resonance spectroscopy of tumors and potential in vivo clinical applications: a review. Cancer Res 49:770–779

Danks DM (1995) Disorders of copper transport. In: Scriver CR, Beaudet AL, Sly WS, Valle D (eds) The metabolic and molecular bases of inherited disease II, 7th edn. McGraw-Hill, New York, pp 2211–2236

Danks DM, Metz G, Sewell R et al (1990) Wilson's disease in adults with cirrhosis but no neurological abnormalities. Br Med J 301:331–332

Dano P, Christiansen C (1974) Calcium absorption and bone mineral contents following intestinal shunt absorption in obesity. Scand J Gastroenterol 9:775–779

Davies DL, Beevers DG, Briggs JD et al (1973) Abnormal relation between exchangeable sodium and the renin-angiotensin system in malignant hypertension and in patients with chronic renal failure. Lancet 1:683–686

De Roos A, van der Wall EE (1994) Evaluation of ischemic heart disease by magnetic resonance imaging and spectroscopy. Radiol Clin North Am 32:581–593

DeGrazia JA, Ivanovich P, Fellows H et al (1965) A double isotope method for measurement of intestinal absorption of calcium in man. J Lab Clin Med 66:822–829

Deng HW, Li J, Li JL et al (1998) Change of bone mass in postmenopausal Caucasian women with and without hormone replacement therapy is associated with Vitamin D receptor and estrogen receptor genotypes. Human Genet 103:576–585

Dowling HJ, Fried SK, Pi-Sunyer FX (1995) Insulin resistance in adipocytes of obese women; effects of body distribution and race. Metabolism 44:987–995

Duerenberg P, van der Kooy K, Leenen R et al (1989) Body impedance is largely dependent on the intra- and extracellular water distribution. Eur J Clin Invest 43:845–853

Eagel BE, Stier SA, Wakem C (1988) Non-osseous bone scan abnormalities in multiple myeloma associated with hypercalcemia. Clin Nucl Med 13:869–873

Edelman IS, Leibman J (1959) Anatomy of body water and electrolytes. Am J Med 27:256–277

Edelman IS, Olney JM, James AH et al (1952) Body composition: studies in the human being by the dilution principle. Science 115:447–454

Eisman JA (1998) Genetics, calcium intake and osteoporosis. Proc Nutr Soc 57:187–193

Elia M, Ward LC (1997) What is needed in metabolic research? In proceedings, Serona symposium on quality of the body cell mass: body composition in the third millenium, Ft. Lauderdale, FL, 27 Feb–2 Mar, 1997, Springer, New York, pp 219–232

Elin RJ (1987) Assessment of magnesium status. Clin Chem 33:1965–1970

Ellegard L, Bosaeus I, Nordgren S et al (1997) Low-dose recombinant human growth hormone increases body weight and lean body mass in patients with short bowel syndrome. Ann Surg 225:88–96

Ellis KJ (1990) Reference man and woman more fully characterized. Variations on the basis of body size, age, sex and race. In: Schrauzer GN (ed) Biological trace element research, vols 26 and 27. Humana Press, Totawa, New Jersey, pp 385–400

Ellis KJ, Shukla KK, Cohn SH et al (1974) A predictor for total body potassium in man based on height weight sex and age: applications in metabolic disorders. J Lab Clin Med 83:716–721

Ellis KJ, Vaswani A, Zanzi I et al (1976) Total body sodium and chlorine in normal adults. Metab Clin Exp 25:645–654

Endo T, Kogai T, Nakazato M et al (1996a) Autoantibody against Na^+/I^- symporter in the sera of patients with autoimmune thyroid disease. Biochem Biophys Res Commun 224:92–95

Endo T, Kaneshige M, Nakazato M et al (1996b) Autoantibody against thyroid iodide transporter in the sera from patients with Hashimoto's thyroiditis possesses iodide transport inhibitory activity. Biochem Biophys Res Commun 228:199–202

Erickson JC, Hollopeter G, Palmiter RD (1996) Attenuation of the obesity syndrome of ob/ob mice by the loss of neuropeptide Y. Science 274:1704–1707

Ernest D, Hartman NG, Deane CP et al (1992) Reproducibility of plasma and extracellular fluid volume measurements in critically ill patients. J Nucl Med 33:1468–1471

Eto Y, Tokoro T, Liebaers I et al (1982) Biochemical characterization of neonatal multiple sulfatase deficient (MSD)

disorder cultured skin fibroblasts. Biochem Biophys Res Commun 106:429–434

Fehman HC, Peiser C, Bode P et al (1997) Leptin: a potent inhibitor of insulin secretion. Peptides 18:1267–1273

Felson DT, Couropmitree NN, Chaisson CE et al (1998) Evidence for a Mendelian gene in a segregation analysis of generalized radiographic osteoarthritis: the Framingham Study. Arthritis Rheum 41:1064–1071

Fenn WD, Noonan TR, Mullins LJ et al (1941/42) The exchange of radioactive potassium with body potassium. Am J Physiol 135:149–163

Fernandez-Real JM, Gutierrez C, Ricart W et al (1997) The TNF-alpha gene Nco-I polymorphism influences the relationship among insulin resistance, per cent body fat, and increased serum leptin levels. Diabetes 46:1468–1472

Ferrari S, Rizzuli R, Bonjour J-P (1998) Heritable and nutritional influences on bone mineral mass. Aging Clin Exp Res 10:205–213

Fiatarone Singh MA (1998) Combined exercise and dietary intervention to optimize body composition in aging. Ann NY Acad Sci 854:378–393

Fielding RA (1995) The role of progressive resistance training and nutrition in the preservation of lean body mass in the elderly. J Am Coll Nutr 14:587–594

Finn PJ, Plank LD, Clark MA et al (1996) Progressive cellular dehydration and proteolysis in critically ill patients. Lancet 347:654–656

Flear CTG, Singh CM (1973) Hyponatremia and sick cells. Br J Anaesth 45:976–994

Flechtenmacher J, Huch K, Thonar EJ-M A et al (1996) Recombiant human osteogenic protein-1 is a potent stimulator of the synthesis of cartilage proteoglycans and collagens by human articular chondrocytes. Arthritis Rheum 39:1896–1904

Flink EB (1986) Magnesium deficiency in alcoholism. Alcohol Clin Exp Res 10:590–594

Flynn MA, Nolph GB, Baker AS et al (1989) Total body potassium in aging humans: a longitudinal study. Am J Clin Nutr 50:713–717

Fong Y, Rosenbaum M, Tracey KJ et al (1989) Recombinant growth hormone enhances muscle myosin heavy-chain mRNA accumulation and amino acid accrual in humans. Proc Natl Acad Sci USA 86:3371–3374

Forbes GB, Sauer EP, Weitkamp LR (1995) Lean body mass in twins. Metabolism 44:1442–1446

Formica CA (1998) Standardization of BMD measurements. Osteopor Int 8:1–3

Foster GD, Wadden TA, Swain RM et al (1999) Changes in resting energy expenditure after weight loss in obese African American and white women. Am J Clin Nutr 69:13–17

Frassineti C, Iotti S, Lodi R et al (1996) Effect of oral phosphocreatinine on human skeletal muscle shown by in vivo ^{31}P-NMR. In Vivo 10:429–433

Fruhling J, Coune A (1975) Radiozinc as a scintigraphic agent for the human prostate. J Nucl Med 16:495–497

Fujimoto WY, Bergstrom RW, Boyko EJ et al (1995) Susceptibility to development of central adiposity among populations. Obes Res 3 [Suppl 2]:179S–186S

Gahtan V, Goode SE, Kurto HZ et al (1997) Body composition and source of weight loss after bariatric surgery. Obes Surg 7:184–188

Gallagher D, Visser M, DeMeersman RE et al (1997) Appendicular skeletal muscle mass: effects of age, gender and ethnicity. J Appl Physiol 83:229–239

Galt JR (1999) Instrumentation and quality control in bone measurements. In: Continuing education course: bone densitometry and osteoporosis. 46th Ann Meeting Soc of Nucl Med, Los Angeles, Calif, June 1999

Gamble JL Jr, Robertson JS (1952) Volume of distribution of radioactive chloride in dogs; comparisons with sodium, bromide and inulin spaces. Am J Physiol 171:659–667

Garlick PJ, McNurlan MA, Bark T et al (1998) Hormonal regulation of protein metabolism in relation to nutrition and disease. J Nutr 128:356S–359S

Gasperino J (1996) Ethnic differences in body composition and their relation to health and disease in women. Ethn Health 1:337–347

Gibbs WW (1996) Gaining on fat. Sci Am August, pp 88–94

Gladwell M (1998) The Pima paradox. The New Yorker Magazine, 2 Feb 1998

Goans RE, Abrams SA, Vieira NE et al (1995) A three-hour measurement to evaluate bone calcium turnover. Bone 16:33–38

Gonzalez-Cadavid NF, Taylor WE, Yarasheski K et al (1998) Organization of the human myostatin gene and expression in healthy men and HIV-infected men with muscle wasting. Proc Natl Acad Sci USA 95:14938–14943

Graham LA, Caesar JJ, Burgen ASV (1960) Gastrointestinal absorption and excretion of Mg^{28} in man. Metabolism Clin Exp 9:646–659

Greenfield MA (1992) Current status of physical measurements of the skeleton. Med Phys 19:1349–1357

Greenleaf JE (1997) Exercise thermoregulation with bed rest, confinement and immersion deconditioning. Ann NY Acad Sci 813:741–750

Grinspoon S, Corcaran C, Askari H et al (1998) Effects of androgen administration in men with the AIDS wasting syndrome. A randomized, double-blind, placebo-controlled trial. Ann Intern Med 129:18–26

Gross J (1962) Iodine and bromine. In: Comar CL, Bronner F (eds) Mineral metabolism: an advanced treatise, vol II, chap 29. The elements, part B. Academic, New York, pp 221–265

Grover RF, Selland MA, McCullough RG et al (1998) Beta-adrenergic blockade does not prevent polycythemia or decrease in plasma volume in men at 4300 m altitude. Eur J Appl Physiol 77:264–270

Grünwald F, Pakos E, Bender H et al (1998) Redifferentiation therapy with retinoic acid in follicular thyroid cancer. J Nucl Med 39:1555–1558

Guingamp C, Gegout-Pottie P, Philippe L et al (1997) Mono-iodoacetate-induced experimental osteoarthritis: a dose-response study of loss of mobility, morphology, and biochemistry. Arthritis Rheum 40:1670–1679

Guo J, Jourdian GW, MacCallum DK (1989) Culture and growth characteristics of chondrocytes encapsulated in alginate beads. Connect Tissue Res 19:277–297

Gura T (1998) Uncoupling proteins provide new clue to obesity's causes. Science 280:1369–1370

Gustafson T, Boman K, Rosenhall L et al (1996) Skeletal muscle magnesium and potassium in asthmatics treated with oral beta 2-agonists. Eur Respir J 9:237–240

Harbert J, DaRocha AFG (eds) (1984a) Textbook of nuclear medicine, vol I. Basic science, 2nd edn. Lea and Febiger, Philadelphia

Harbert J, DaRocha AFG (eds) (1984b) Textbook of nuclear medicine, vol II. Clinical applications, 2nd edn. Lea and Febiger, Philadelphia

Harper EJ (1998) Changing perspectives on aging and energy requirements: aging, body weight, and body composition in humans, dogs and cats. J Nutr 128:2627S–3631S

Harrison JE, Chow R, Dornan J et al (1993) Evaluation of a program for rehabilitation of osteoporotic patients: a 4-year follow-up. Osteoporosis Int 3:13–17

Haskell A, Nadel ER, Stachenfeld NS et al (1997) Transcapillary escape rate of albumin in humans during exercise-induced hypervolemia. J Appl Physiol 83:407–413

Haussinger D (1996) Regulation of cell function by level of hydration (in German). Naturwissenschaften 83:264–271

Hebert SC, Brown EM (1996) The scent of an ion: calcium-sensing and its roles in health and disease. Curr Opin Nephrol Hypertens 5:45–53

Heindel W, Kugel H, Wenzel F et al (1997) Localized ^{31}P MR spectroscopy of the transplanted human kidney in situ shows altered metabolism in rejection and acute tubular necrosis. J Magn Reson Imaging 7:858–864

Hevesy G (1948) Radioactive indicators: their application in biochemistry, animal physiology and pathology. Interscience, New York

Hill GL (1992) Body composition research: Implications for the practice of clinical nutrition. J Parent Ent Nutr 16:197–218

Hinsberger AD, Williamson PC, Carr TJ et al (1997) Magnetic resonance imaging volumetric and phosphorus 31 magnetic resonance spectroscopy measurements in schizophrenia. J Psychiatr Neurosci 22:111–117

Hirsch R, Lethbridge-Cejku M, Hanson R et al (1998) Familial aggregation of osteoarthritis: data from the Baltimore Longitudinal Study on Aging. Arthritis Rheum 41:1227–1232

Ho KK, O'Sullivan AJ, Hoffman DM (1996) Metabolic actions of growth hormone in man. Endocrine J 43:S57–S63

Horber FF, Kohler SA, Lippuner K et al (1996) Effect of regular physical training on age-associated alterations in body composition in men. Eur J Clin Invest 26:279–285

Hory B, Rousanne MC, Drueke TB et al (1998) The calcium receptor in health and disease. Exp Nephrol 6:171–179

Hosain P, Dunson GL (1982) Radioactive complexes as radiopharmaceuticals. In: Spencer RP (ed) CRC handbook series in clinical laboratory science, sect A. Nuclear medicine, vol. 2. CRC Press, Boca Raton, pp 143–157

Hsieh ST, Ballard RE, Murthy G et al (1998) Plasma colloid osmotic pressure increases in humans during simulated microgravity. Aviat Space Environ Med 69:23–26

Hubner KF, Andrews GA, Buonocore E et al (1979) Carbon-11-labeled amino acids for the rectilinear and positron tomographic imaging of the human pancreas. J Nucl Med 20:507–513

Hunt BE, Davy KP, Jones PP et al (1998) Role of central circulatory factors in the fat-free mass-maximal aerobic capacity relation across age. Am J Physiol 275:H1178–H1182

Hunt PJ, Espiner EA, Nicholls MG et al (1996) Differing biological effects of equimolar atrial and brain natriuretic peptide infusions in normal man. J Clin Endocrinol Metab 81:3871–3876

Ilowite JS, Gorvoy JD, Smaldone GC (1987) Quantitative deposition of aerosolized gentamycin in cystic fibrosis. Am Rev Respir Dis 136:1445–1449

Iwase T, Hasegawa Y, Ishiguro N et al (1998) Synovial fluid cartilage metabolism marker concentrations in osteonecrosis of the femoral head compared with osteoarthrosis of the hip. J Rheumatol 25:527–531

Jain SC, Bardhan J, Swamy YV et al (1980) Body fluid compartments in humans during acute high-altitude exposure. Aviat Space Environ Med 51:234–236

Jakicic JM, Wing RR (1998) Differences in resting energy expenditure in African-American vs. Caucasian overweight females. Int J Obes Relat Metab Disord 22:236–242

Jasani BM, Edmonds CJ (1971) Kinetics of potassium distribution in man using isotope dilution and whole-body counting. Metab Clin Exp 20:1099–1106

Jebb SA, Elia M (1993) Techniques for the measurement of body composition: a practical guide. Int J Obes Relat Metab Disord 17:611–621

Jeejeebhoy KN (1990) Mechanism of reduction of total body potassium in malnutrition. In: Yasumura S et al (eds) Advances in in vivo body composition studies. Plenum Press, New York

Jespersen B (1997) Regulation of renal sodium and water excretion in the nephrotic syndrome and cirrhosis of the liver. Dan Med Bull 44:191–207

Jiang Z, Yang N, Chou C et al (1991) Body composition in Chinese subjects: comparison with data from North America. World J Surg 15:95–102

Johannsson G, Måron P, Lönn L et al (1997) Growth hormone treatment of abdominally obese men reduces abdominal fat mass, improves glucose and liprotein metabolism, and reduces diastolic blood pressure. J Clin Endocrinol Metab 82:727–734

Johansen LB, Gharib C, Allevard AM et al (1997) Haematocrit, plasma volume and noradrenaline in humans during simulated weightlessness for 42 days. Clin Physiol 17:203–210

Johansen LB, Pump B, Warberg J et al (1998a) Preventing hemodilution abolishes natriuresis of water immersion in humans. Am J Physiol 275:879–888

Johansen LB, Videbaek R, Hammerum M et al (1998b) Underestimation of plasma volume changes in humans by hematocrit/hemoglobin method. Am J Physiol 274:R126–R130

Johanssen AC, Samuelsson O, Haraldsson B et al (1998) Body composition in patients treated with peritoneal dialysis. Nephr Dial Transplant 13:1511–1517

Johanssen G, Rosen T, Bengtsson BA (1997) Individualized dose titration of growth hormone (GH) during GH replacement in hypopituitary adults. Clin Endocr 47:571–581

Johnson DW, Thomas BJ, Fleming SJ et al (1996) Monitoring of extracellular and total body water during haemodialysis using multifrequency bio-electrical impedance analysis. Kidney Blood Press Res 19:94–99

Kamuma O, Asano T, Ikehara H et al (1997) Assessment of energy status and fructose metabolism of liver in obstructive jaundice by ^{31}P magnetic resonance. J Gastroenterol Hepatol 12:740–744

Kato T, Shioiri T, Murashita J et al (1994) Phosphorus-31 magnetic resonance spectroscopy and ventricular enlargement in bipolar disorder. Psychiatr Res Neuroimaging 55:41–50

Katzmarzyk PT, Malina RM (1998) Obesity and relative subcutaneous fat distribution among Canadians of First Nation and European ancestry. Int J Obes Relat Metab Disord 22:1127–1131

Keefe EB, Lieberman DA, Krishnamurthy S et al (1988) Primary biliary cirrhosis: Tc-99m-IDA planar and SPECT scanning. Radiology 166:143–148

Keen RW, Woodford-Richens KL, Grant SF et al (1999) Association of polymorphism at the type 1 collagen (COLIA1) locus with reduced bone mineral density, increased fracture risk, and increased collagen turnover. Arthritis Rheum 42:285–290

Kehayias JJ, Fiatarone MA, Zhuang H et al (1997) Total body potassium and body fat: relevance to aging. Am J Clin Nutr 66:904–910

Kelly GS (1998) The role of glucosamine sulfate and chondroitin sulfates in the treatment of degenerative joint disease. Altern Med Rev 3:27–39

Kent-Braun JA, McCully KK, Chance B (1990) Metabolic effects of training in human: a ^{31}P-MRS study. J Appl Physiol 69:1165–1170

Kent-Braun JA, Miller RG, Weiner MW (1994) Magnetic resonance spectroscopy studies of human muscle. Radiol Clin North Am 32:313–335

Kerr RM, Dubois JJ, Holt RR (1967) Use of ^{125}I- and ^{51}Cr-labeled albumin for the measurement of gastrointestinal and total albumin catabolism. J Clin Invest 46:2064–2082

Kirk K, Horner HA, Elford BC et al (1994) Transport of diverse substrates into malaria-infected erythrocytes via a pathway showing functional characteristics of a chloride channel. J Biol Chem 269:3339–3347

Kleine TO, Baumann HJ (1991) On the detection of a blood-joint barrier for radiosulfate and (3H) glucosamine in single joints of aging rats. Z Gerontol 24:303–310

Kolodny EH, Fluharty AL (1995) Metachromatic leukodystrophy and multiple sulfatase deficiency: sulfatide lipidosis. In: Scriver CR, Beaudet AL, Sly WS, Valle D (eds) The metabolic and molecular bases of inherited disease, vol II, 7th edn. McGraw-Hill, New York, pp 2693–2739

Kotler DP, Tierney AR, Wang J et al (1989) Magnitude of body cell mass-depletion and the timing of death from wasting in AIDS. Am J Clin Nutr 50:444–447

Krall EA, Dawson-Hughes B (1995) Soft tissue body composition: family resemblance and independent influences on bone mineral density. J Bone Miner Res 10:1944–1950

Kreider RB, Ferreira M, Wilson M et al (1998) Effects of creatine supplementation on body composition, strength and sprint performance. Med Sci Sports Exerc (MG8) 30:73–82

Kurz P, Tsobanelis T, Roth P et al (1995) Differences in calcium kinetic pattern between CAPD and HD patients. Clin Nephrol 44:255–261

Lammi MJ, Inkinen R, Parkkinen JJ et al (1994) Expression of reduced amounts of structurally altered aggrecan in articular chondrocytes exposed to high hydrostatic pressure. Biochem J 304:723–730

Laube BL, Links JM, LaFrance ND et al (1989) Homogeneity of bronchopulmonary distribution of ^{99m}Tc-aerosol in normal subjects and in cystic fibrosis patients. Chest 95:822–830

Laube BL, Auci RM, Shields DE et al (1996) Effect of rhDNAse on airflow obstruction and mucociliary clearance in cystic fibrosis. Am J Respir Crit Care Med 153:752–760

Legato MJ (1997) Gender-specific aspects of obesity. Int J Fertil 42:184–197

Leypoldt JK, Cheung AK (1998) Evaluating volume status in hemodialysis patients. Adv Ren Replace Ther 5:64–74

Lindbjerg IF (1967) The variations in fat content of various skeletal muscles. Scand J Clin Lab Invest [Suppl] 93:14–17

Ljunggren K, Strand S-E (1990) Beta camera for static and dynamic imaging of charged-particle emitting radionuclides in biologic samples. J Nucl Med 31:2058–2063

Ljunggren K, Strand S-E, Ceberg CP et al (1993) Beta camera low activity tumor imaging. Acta Oncol 32:869–872

Lodi R, Montagna P, Soriani S et al (1997) Deficit of brain and skeletal muscle bioenergetics and low brain magnesium in juvenile migraine: an in vivo ^{31}P magnetic resonance spectroscopy interictal study. Pediatr Res 42:866–871

Loevinger R, Budinger TF, Watson EE (1991) Mird primer for absorbed dose calculations, revised edn. Society of Nuclear Medicine, New York

Lutwak L (1969) Tracer studies of intestinal calcium absorption in man. Am J Clin Nutr 22:771–785

Ma R, Ellis KJ, Yasumura S et al (1999) Total body-calcium measurements: comparison of two delayed-gamma neutron activation facilities. Phys Med Biol 44:N113–N118

Mackay A, Davies DL, Horton PW (1978) Is thallium-201 of use in the measurement of total exchangeable potassium in man? Eur J Clin Invest 8:261–262

Maldonado X, Alonso J, Giralt J et al (1998) [31]Phosphorus magnetic resonance spectroscopy in the assessment of head and neck tumors. Int J Radiat Oncol Biol Phys 40:309–312

Manery JF, Haege LF (1941) The extent to which radioactive chloride penetrates tissues and its significance. Am J Physiol 134:83–93

Marceau P, Biron S, Hould FS et al (1999) Liver pathology and the metabolic syndrome x in severe obesity. J Clin Endocrinol Metab 84:1513–1517

Marcus R, Holloway L, Butterfield G (1993) Clinical uses of growth hormone in older people. J Reprod Fertil Suppl 46:115–118

Maris JM, Evans AE, MacLaughlin AC et al (1985) [31]P Nuclear magnetic resonance spectroscopic investigation of human neuroblastoma in situ. N Engl J Med 312:1500–1505

Marks BL, Rippe JM (1996) The importance of fat-free mass maintenance in weight loss programmes. Sports Med 23:273–281

Matsuda A, Kosugi S (1997) A homozygous missense mutation of the sodium/iodide symporter gene causing iodide transport defects. J Clin Endocrinol Metab 82:3966–3971

Mayer K, Pentlow KS, Marcove RC et al (1978) Sulfur-35 therapy for chondrosarcoma and chordoma. In: Spencer RP (ed) Therapy in nuclear medicine, chap 16. Grune and Stratton, New York, pp 185–192

Medeiros-Neto G (1996) Clinical and molecular advances in inherited disorders of the thyroid system. Thyroid Today XIX:1–13

Mena I, Court J, Fuenzalida S et al (1970) Modification of chronic manganese poisoning: treatment with L-DOPA or 5-OH tryptophane. N Engl J Med 282:5–10

Meneely GR, Heyssel RM, Ball COT (1963) Analysis of factors affecting body composition determined from potassium content in 915 normal subjects. Ann NY Acad Sci 110:271–281

Meyerhoff DJ, Baska MD, Thomas AM et al (1989) Alcoholic liver disease: quantitative image-guided P-31 MR spectroscopy. Radiology 173:393–400

Mimura M, Makino H, Kanatsuka A et al (1992) Reduction of erythrocyte (Na[+]+K[+]) ATPase activities in non-insulin-dependent diabetic patients with hyperkalemia. Metab Clin Exp 41:426–430

Mitch WE (1998) Mechanisms causing loss of lean body mass in kidney disease. Am J Clin Nutr 67:359–366

Mitra S, Wolff JE, Garrett R et al (1995) Application of the associated particle technique for the whole-body measurement of protein, fat and water by 14 MEV neutron activation analysis – a feasibility study. Phys Med Biol 40:1045–1055

Miyamoto AT, Mishkin FS, Maxwell TM (1975) Noninvasive study of extremity perfusion by potassium 43 scanning. Arch Surg 110:58–63

Moller N, O'Brien P, Nair KS (1998) Disruption of the relationship between fat content and leptin levels with aging in humans. J Clin Endocrinol Metab 83:931–934

Monk DN, Plank LD, Franch-Areas G et al (1996) Sequential changes in the metabolic response in critically injured patients during the first 25 days after blunt trauma. Ann Surg 223:395–405

Morreale P, Manopulo R, Galati M et al (1996) Comparison of the antiinflammatory efficacy of chondroitin sulfate and diclofenac sodium in patients with knee osteoarthritis. J Rheumatol 23:1385–1391

Muchmore DB (1999) SERMS: a new option for the prevention of osteoporosis. Endotrends 6:8–10

Murray RK, Granner DK, Mayes PA et al (1996) Harper's biochemistry, 24th edn. Appleton and Lange, Stamford, CT

Murray TM, Harrison JE, Bayley TA et al (1990) Fluoride treatment of postmenopausal osteoporosis: age, renal function, and other clinical factors in the osteogenic response. J Bone Miner Res 5 [Suppl 1]:S27–S35

Murton SA, Tan ST, Prickett TC et al (1998) Hormone responses to stress in patients with major burns. Br J Plast Surg 51:388–392

Nadler V, Biegon A, Beit-Yannai E et al (1995) [45]Ca accumulation in rat brain after closed head injury: attenuation by the novel neuroprotective agent HU-211. Brain Res 685:1–11

Nadler JL, Rude RK (1995) Disorders of magnesium metabolism. Endocr Metab Clin North Am 24:623–641

Nelson ME, Fisher EC, Dilmanian FA et al (1991) A 1-yr walking program and increased dietary calcium in postmenopausal women: effects on bone. Am J Clin Nutr 53:1304–1311

Neubauer S, Horn M, Pabst T et al (1997) Cardiac high-energy phosphate metabolism in patients with aortic valve disease assessed by ^{31}P-magnetic resonance spectroscopy. J Invest Med 45:453–462

Neumann J, Wüstenberg P-W, Esther G (1980) Kompartmentanalyse des Kaliumhaushaltes – mathematische Modellierung des [42]K-Tracerkinetic. Ber Ges Inn Med 12:273–274

Nicholson JP, Zilva JF (1960) Estimation of extracellular fluid volume using radiobromine. Clin Sci 19:391–398

Nissen SL, Abumrad NN (1997) Nutritional role of the leucine metabolite – hydroxy-methylbutyrate (HMB). J Nutr Biochem 8:300–311

Niu T, Chen C, Cordell H et al (1999) A genome-wide scan for loci linked to forearm bone mineral density. Hum Genet 104:226–233

Norman RA, Tataranni PA, Pratley R et al (1998) Autosomal genomic scan for loci linked to obesity and energy metabolism in Pima Indians. Am J Human Genet 62:659–668

Novotny EJ Jr (1995) Overview – the role of NMR spectroscopy in epilepsy. Magn Reson Imaging 13:1171–1173

O'Brien KO, Abrams SA, Liang LK et al (1998) Bone turnover response to changes in calcium intake is altered in girls and adult women in families with histories of osteoporosis. J Bone Miner Res 13:491–499

Olde Rikkert MG, Deurenberg P, Jansen RW et al (1997) Validation of multi-frequency bioelectrical impedance analysis in detecting changes in fluid balance of geriatric patients. J Am Geriatr Soc (H6V) 45:1345–1351

Oppert JM, Vohl MC, Chagnon M et al (1994) DNA polymorphism in the uncoupling protein (UCP) gene and human body fat. Int J Obes Relat Metab Disord 18:526–531

Orphanidou CI, McCargar LJ, Birmingham CL et al (1997) Changes in body composition and fat distribution after short-term weight gain in patients with anorexia nervosa. Am J Clin Nutr 65:1034–1041

Ostensen M, Veiby OP, Raiss R et al (1991) Responses of normal and rheumatic human articular chondrocytes cultured under various experimental conditions in agarose. Scand J Rheumatol 20:172–182

Pace N, Rathbun EN (1945) Studies in body composition III. The body water and chemically combined nitrogen content in relation to fat content. J Biol Chem 158:685–691

Padhy AK, Gopinath PG, Amini AC (1990) Myocardiol, pulmonary, diaphragmatic, gastric, splenic, and renal uptake of Tc-99m-MDP in a patient with persistent, severe hypercalcemia. Clin Nucl Med 15:648–649

Paolisso G, Amato L, Eccelente R et al (1998) Effects of metformin on food intake in obese subjects. Eur J Clin Invest 28:441–446

Parrish TB, Fieno DS, Fitzgerald SW et al (1997) Theoretical basis for sodium and potassium MRI of the human heart. Magn Reson Med 38:653–661

Parving H-H, Noer I, Deckert P-E et al (1976) The effect of metabolic regulation on microvascular permeability to small and large molecules in short-term juvenile diabetics. Diabetologia 12:161–166

Parving H-H, Gintelberg F (1973) Transcapillary escape rate of albumin and plasma volume in essential hypertension. Circ Res 32:643–651

Pasantes-Morales H, Murray RA, Sanchez-Olea R et al (1994) Regulatory volume decrease in cultured astrocytes II. Permeability pathway to amino acids and polyols. Am J Physiol 266:C172–C178

Perusse L, Chagnon YC (1997) Summary of human linkage and association studies. Behav Genet 27:359–372

Phang JM, Yeh GC, Scriver CR (1995) Disorders of proline and hydroxyproline metabolism. In: Scriver CR, Beaudet AL, Sly WS, Valle D (eds) The metabolic and molecular bases of inherited disease I, 7th edn. McGraw-Hill, New York, pp 1125–1146

Pickering GP, Fellmann N, Morio B et al (1997) Effects of endurance training on the cardiovascular system and water compartments in elderly subjects. J Appl Physiol 83:1300–1306

Pierson RN Jr, Lin DHY, Phillips RA (1974) Total body potassium in health: effects of age, sex, height and fat. Am J Physiol 226:206–212

Pierson RN Jr, Wang J, Yang MU et al (1976) The assessment of human body composition during weight reduction: evaluation of a new model for clinical studies. J Nutr 106:1694–1701

Pierson RN Jr, Price DC, Wang J et al (1978) Extracellular water measurements: organ tracer kinetics of bromide and sucrose in rats and man. Am J Physiol 235:F254–F264

Pierson RN Jr, Wang J, Colt EW et al (1982) Body composition measurements in normal man: the potassium, sodium, sulfate and tritium spaces in 58 adults. J Chron Dis 35:419–428

Pierson RN Jr, Wang J, Thornton JC (1997) Measurement of body composition: applications in hormone research. Horm Res 48 [Suppl 1]:56–62

Pike VW, Halldin C, Crouzel C et al (1993) Radioligands for PET studies of central benzodiazepine receptors and PK (peripheral benzodiazepine) binding sites – current status. Nucl Med Biol 20:503–525

Plank LD, Connolly AB, Hill GL (1998) Sequential changes in the metabolic response in severely septic patients during the first 23 days after the onset of peritonitis. Ann Surg 228:146–158

Poehlman ET, Toth MJ, Fishman PS et al (1995) Sarcopenia in aging human: the impact of menopause and disease. J Geront A Biol Sci Med Sci 50:73–77

Pohlenz J, Medeiros-Neto G, Gross JL et al (1997) Hypothyroidism in a Brazilian kindred due to iodide trapping defect caused by a homozygous mutation in the sodium/iodide symporter gene. Biochem Biophys Res Comm 240:489–491

Pohlenz J, Rosenthal IM, Weiss RE et al (1998) Congenital hypothyroidism due to mutations in the sodium/iodide symporter: identification of a nonsense mutation producing a downstream cryptic 3′ splice site. J Clin Invest 101:1028–1035

Polito A, Cuzzolaro M, Raguzzini A et al (1998) Body composition changes in anorexia nervosa. Eur J Clin Nutr 52:655–662

Pollycove M, Tono M (1975) Studies of the erythron. Semin Nucl Med 5:11–61

Porter MT, Fluharty AL, Trammell J et al (1971) A correlation of intracellular cerebroside sulfatase activity in fibroblasts with latency in metachromatic leukodystrophy. Biochem Biophys Res Commun 44:660–666

Price DC, McIntyre PA (1984) The hematopoietic system. In: Harbert J, DaRocha AFG (eds) Textbook of nuclear medicine, vol II. Clinical applications, 2nd edn. Lea and Febiger, Philadelphia, pp 535–605

Quatela VC, Futran ND, Frisina RD (1993) Effects of cyanoacrylate tissue adhesives on cartilage graft viability. Laryngoscope 103:798–803

Raja KB, Pountney S, Bomford A et al (1996) A duodenal mucosal abnormality in the reduction of the Fe(III) in patients with genetic haemochromatosis. Gut 38:765–769

Ralston SH (1998) Do genetic markers aid in risk assessment? Osteopor Int [Suppl] 1:S37–S42

Raspe E, Costagliola S, Ruf J et al (1995) Identification of the thyroid Na$^+$/I$^-$ cotransporter as a potential autoantigen in thyroid autoimmune disease. Eur J Endocrinol 132:399–405

Ratcliffe A, Flatow EL, Roth N et al (1996) Biochemical markers in synovial fluid identify early osteoarthritis of the glenohumeral joint. Clin Orthop Relat Res 330:45–53

Rathmacher JA, Flakoll PJ, Nissen SL (1995) A compartmental model of 3-methylhistidine metabolism in humans. Am J Physiol 269:E193–E198

Reid IR, Wattie RN, Evans MC et al (1996) Testosterone therapy in glucocorticoid-treated men. Arch Intern Med 156:1173–1177

Resnick LM (1992) Cellular ions in hypertension, insulin resistance, obesity and diabetes: a unifying theme. J Am Soc Nephrol 3:S78–S85

Resnick LM, Gupta RK, Gruenspan H et al (1990) Hypertension and peripheral insulin resistance: possible mediating role of intracellular free magnesium. Am J Hypertens 3:373–379

Richardson AJ, Cox IJ, Sargentoni J et al (1997) Abnormal cerebral phospholipid metabolism in dyslexia indicated by phosphorus-31 magnetic resonance spectroscopy. NMR Biomed 10:309–314

Ritz P, Coward WA (1995) Doubly labeled water measurement of total energy expenditure. Diabet Metab 21:241–251

Roberts SB (1989) Use of the doubly labeled water method for measurement of energy expenditure, total body water, water intake, and metabolizable energy intake in humans and small animals. Can J Physiol Pharmacol 67:1190–1198

Rogus J, Blumenthal SA (1981) Variations in dietary intake after by-pass surgery for obesity. J Am Diet Assoc 79:437–441

Ross BD, Bluml S, Cowan R et al (1997) In vivo magnetic resonance spectroscopy of human brain: the biophysical basis of dementia. Biophys Chem 68:161–172

Ross DW (1996) Introduction to molecular medicine, 2nd edn, chap 6. Genetic diseases. Springer, Berlin Heidelberg New York

Rutter A, Hugenholtz H, Saunders JK et al (1995) One-dimensional phosphorus-31 chemical shift imaging of human brain tumors. Invest Radiol 30:359–366

Ryan RJ, Pascal LR, Inoye T et al (1956) Experiences with radiosulfate in the estimation of physiologic extracellular water in healthy and abnormal man. J Clin Invest 35:1119–1130

Rylander R (1996) Environmental magnesium deficiency as a cardiovascular risk factor. J Cardiovasc Risk 3:4–10

Ryschon TW, Rosenstein DL, Rubinow DR et al (1996) Relationship between skeletal muscle intracellular ionized magnesium and measurements of blood magnesium. J Lab Clin Med 127:207–213

Sackner-Bernstein J, Toma C, Haynes P et al (2000) Evidence for intracellular edema in euvolemic heart failure patients. Workshop on body composition in basic and clinical research and the emerging technologies, Brookhaven National Laboratory, Upton, NY, 14 Dec 2000

Sainz J, Van Tournout JM, Sayre J et al (1999) Association of collagen type 1 alpha 1 gene polymorphism with bone density in early childhood. J Clin Endocrinol Metab 84:853–855

Saito T, Endo T, Kawaguchi A et al (1997) Increased expression of the Na^+/I^- symporter in cultured human thyroid cells exposed to thyrotropin and in Grave's thyroid tissue. J Clin Endocrinol Metab 82:3331–3336

Satake K, Lee J-D, Shimuzu H et al (1996) Relation between severity of magnesium deficiency and frequency of anginal attacks in men with variant angina. J Am Coll Cardiol 28:897–902

Schambelan M, Mulligan K, Granfeld C et al (1996) Recombinant human growth hormone in patients with HIV-associated wasting. A randomized, placebo-controlled trial. Ann Intern Med 125:873–882

Schepartz B (1973) Regulation of amino acid metabolism in mammals. Saunders, Philadelphia

Schmutzler C, Winzer R, Meissner-Weigl J et al (1997) Retinoic acid increases sodium/iodide symporter mRNA levels in human cancer cell lines and suppresses expression of functional symporter in nontransformed FRTL-5 rat thyroid cells. Biochem Biophys Res Commun 240:832–838

Schoeller DA, Ravussin E, Schutz Y et al (1986) Energy expenditure by doubly labeled water: validation in humans and proposed calculation. Am J Physiol 250 (Reg Integr Comp Physiol 19):R823–R830

Schulze A, Hess T, Wevers R et al (1997) Creatine deficiency syndrome caused by guanidinoacetate methyl transferase deficiency: diagnostic tools for a new inborn error of metabolism. J Pediatr 131:626–631

Schunk K, Romaneehsen B, Mildenberger P et al (1997) Dynamic phosphorus-31 magnetic resonance spectroscopy in arterial occlusive disease. Invest Radiol 32:651–659

Schwartz R, Spencer H, Wentworth RA (1978) Measurement of magnesium absorption in man using stable ^{26}Mg as a tracer. Clin Chim Acta 87:265–273

Schwenk A, Ward LC, Elia M et al (1998) Bioelectrical impedance analysis predicts outcome in patients with suspected bacteremia. Infection 26:277–282

Segall GM, Lennon SE, Stevick CD (1990) Exercise whole-body thallium scintigraphy in the diagnosis and evaluation of occlusive arterial diagnosis in the legs. J Nucl Med 31:1443–1449

Segawa I, Otokida K, Kato M (1989) Total body water, total exchangeable sodium and potassium in patients with convalescent acute myocardial infarction one to two months after onset. Jpn Circ J 53:1215–1220

Shimura H, Haraguchi K, Miyazaki A et al (1997) Iodide uptake and experimental ^{131}I therapy in transplanted undifferentiated thyroid cancer cells expressing the Na^+/I^- symporter gene. Endocrinology 138:4493–4496

Shreeve WW (1974) Physiological chemistry of carbohydrates in mammals. Saunders, Philadelphia

Shreeve WW, Giwa L, Ficek MA, et al (1978) Effects of exercise or surgery on uptake of thallium-201 into legs in peripheral vascular disease. J Nucl Med 19:709

Silberstein EB (1993) The treatment of painful osseous metastases with phosphorus-32-labeled phosphates. Semin Oncol 20 [Suppl 2]:10–21

Silver L, Robertson JS, Dahl LK (1960) Magnesium turnover in the human studied with Mg^{28}. J Clin Invest 39:420–423

Sjoberg HE, Reizenstein P, Arman E (1970) Retention of orally administered ^{47}Ca in man measured in a whole-body counter. Scand J Clin Lab Invest 26:67–71

Smanik PA, Liu Q, Furminger TL et al (1996) Cloning of the human sodium iodide symporter. Biochem Biophys Res Commun 226:339–345

Smecuol E, Gonzalez D, Mautelen C et al (1997) Longitudinal study on the effect of treatment on body composition and anthropometry of celiac disease patients. Am J Gastroenterol 92:639–643

Smith EL, Hill RL, Lehman IR et al (1983) Principles of biochemistry: mammalian biochemistry, 7th edn. McGraw-Hill, New York

Smith LH (1986) Diseases of bone and bone mineral metabolism. In: Andreoli TE, Carpenter CCJ, Plum F, Smith LH Jr (eds) Cecil essentials of medicine, sect IX. Saunders, Philadelphia

SNM procedure guidelines manual (1999)

Snow BJ, Calne DB (1995) Movement disorders. In: Wagner HN Jr, Szabo Z, Buchanan JW (eds) Principles of nuclear medicine, 2nd edn. Saunders, Philadelphia, pp 557–564

Snyder SE, Kume A, Jung Y-W et al (1995) Synthesis of carbon-11, fluorine-18, and iodine-125-labeled $GABA_A$-gated chloride ion channel blockers: substituted 5-tert-butyl-2-phenyl-1,3-dithianes and -dithiane oxides. J Med Chem 38:2663–2671

Sostre S (1995) Treatment of polycythemia vera. In: Wagner HN, Szabo Z, Buchanan JW (eds) Principles of nuclear medicine, 2nd edn. Saunders, Philadelphia, pp 722–727

Spungen AM, Wang J, Pierson R (2000) Soft tissue body composition differences in monozygotic twins discordant for spinal cord injury. J Appl Physiol 88:1310–1315

Steele IC, Young IS, Stevenson HP et al (1998) Body composition and energy expenditure of patients with chronic cardiac failure. Eur J Clin Invest 28:33–40

Sternlieb I, Scheinberg IH (1979) The role of radiocopper in the diagnosis of Wilson's disease. Gastroenterology 77:138–142

Strain WH, Huegin F, Lankau CA Jr et al (1964) Zinc-65 retention by aortic tissue of rats. Int J Applied Radiat Isotopes 15:231–237

Sullivan JL (1989) The iron paradigm of ischemic heart disease. Am Heart J 117:1177–1188

Susskind H, Ellis KJ, Atkins HL et al (1978) Studies of whole-body retention and clearance of inhaled noble gases. Prog Nucl Med 5:13–34

Sutcliffe JF (1996) A review of in vivo experimental methods to determine the composition of the human body. Phys Med Biol 41:791–833

Syed IB, Hosain F (1982) Radiation dose considerations from radiopharmaceuticals. In: Spencer RP (ed) CRC handbook series in clinical laboratory science, sect A. Nuclear medicine, vol II. CRC Press, Boca Raton, pp 169–183

Syrota A, Dop-Ngassa M, Cerf M et al (1981) ^{11}C-methionine for evaluation of pancreatic exocrine function. Gut 22:907–915

Taboulet J, Frenkian M, Frendo JL et al (1998) Calcitonin receptor polymorphism is associated with a decreased fracture risk in post-menopausal women. Hum Mol Genet 7:2129–2133

Tanne JH (1999) US trial of gene therapy for cystic fibrosis. Brit Med J 318:1096

Taubes G (1998) As obesity rates rise, experts struggle to explain why. Science 280:1367–1368

Taylor-Robinson SD, Sargentoni J, Oatridge A et al (1996) MR imaging and spectroscopy of the basal ganglia in chronic liver disease: correlation of T_1-weighted contrast measurements with abnormalities in proton and phosphorus-31 MR spectra. Metab Brain Dis 11:249–267

Telfer N (1966) Exchangeable potassium in diabetes. Metab Clin Exp 15:502–508

Thomas BJ, Ward LC, Cornish BH (1998) Bioimpedance spectrometry in the determination of body water compartments and clinical significance. Appl Radiat Isot (BQG) 49:447–455

Thomas KT, Keller CS, Holbert KE (1997) Ethnic and age trends for body composition in women residing in the U.S. Southwest II: total fat. Med Sci Sports Exerc (MG8) 29:90–98

Thompson JL, Butterfield GE, Gylfadottir UK et al (1998) Effects of human growth hormone, insulin-like growth factor I, and diet and exercise on body composition of obese postmenopausal women. J Clin Endocrinol Metab 83:1477–1484

Thomson MA, Wilmott RW, Wainwright C et al (1996) Resting energy expenditure, pulmonary inflammation and genotype in the early course of cystic fibrosis. J Pediatr 129:367–373

Tothill P, Avenell A, Reid DM (1994) Precision and accuracy of measurements of whole-body mineral: comparisons between Hologic, Lunar and Norland duel energy x-ray absorptiometers. Br J Radiol 67:1210–1217

Tran N, Chaudhuri TK (1982) Calcium absorption. In: Spencer RP (ed) CRC handbook series in clinical laboratory science, sect A. Nuclear medicine. CRC Press, Boca Raton, pp 297–304

Trovato GM, Iannetti E, Carpentieri G (1998) Nifedipine and extracellular water in dialysis arterial hypertension. Rec Prog Med 89:438–443

Uchida T (1995) Overview of iron metabolism. Int J Hematol 62:193–202

Underwood EJ (1971) Trace elements in human and animal nutrition, 3rd edn. Academic, New York

Utter AC, Nieman DC, Shannonhouse EM et al (1998) Influence of diet and/or exercise on body composition and cardiorespiratory fitness in obese women. Int J Sports Nutr 8:213–222

Van der Merwe MS, Wing JR, Celgow LH et al (1996) Metabolic indices in relation to body composition changes during weight loss on Dexfenfluramine in obese women from two South African ethnic groups. Int J Obes Relat Metab Disord 20:768–776

Van Wayjen RG (1993) Metabolic effects of anabolic steroids. Wien Med Wochenschr 143:368–375

Van Wouwe JP (1989) Clinical and laboratory diagnosis of acrodermatitis enteropathica. Eur J Pediatr 149:2–8

Vartsky D, Goldberg MB (1997) Body composition studies – neutron activation analysis vs. nuclear resonance absorption. Proc SPIE Int Soc Opt Engin 2867:383–386

Vassart G, Dumont JE, Refetoff S (1995) Thyroid disorders. In: Scriver CR, Beaudet AL, Sly WS, Valle D (eds) The metabolic and molecular bases of inherited disease, vol II. McGraw-Hill, New York, pp 2883–2928

Vaughan BE, Boling EA (1961) Rapid assay procedures for tritium-labeled water in body fluids. J Lab Clin Med 57:159–164

Verschure PJ, van Marle J, Joosten LAB et al (1996) Localization of insulin-like growth factor-1 receptor in human normal and osteoarthritic cartilage in relation to proteoglycan synthesis and content. Br J Rheumatol 35:1044–1055

Villarosa L (1999) Tri-fold ailment stalks female athletes. The New York Times, 22 June 1999

Viteri FE, Kohaut BA (1997) Improvement of the Eakins and Brown method for measuring ^{59}Fe and ^{55}Fe in blood and other iron-containing materials by liquid scintillation counting and sample preparation using microwave digestion and ion-exchange column purification of iron. Anal Biochem 244:116–123

Vogel S (1999) Why we get fat. Discover Magazine, April 1999

Wadden TA, Vogt RA, Andersen RE et al (1997) Exercise in the treatment of obesity: effects of four interventions on body composition, resting energy, expenditure, appetite and mood. J Consult Clin Psychol 65:269–277

Wagner HN Jr, Szabo Z, Buchanan JW (eds) (1995) Principles of nuclear medicine, 2nd edn. Saunders, Philadelphia

Wahner HW, Goldstein NR (1979) Radiocopper kinetic studies in the diagnosis of Wilson's disease. J Nucl Med 20:679

Wahner HW, Dunn WL, Riggs BL (1983) Noninvasive bone mineral measurements. Semin Nucl Med XIII:282–289

Wahner HW, Dunn WL, Riggs BL (1984) Assessment of bone mineral, part 2. J Nucl Med 25:1241–1253

Walsh BT, Devlin MJ (1998) Eating disorders: progress and problems. Science 280:1387–1390

Walsh CH, Soler NG, James H et al (1976) Studies in whole body potassium and whole body nitrogen in newly diagnosed diabetics. Quart J Med 178:295–301

Wang G-J, Volkow N, Fowler JS (2002) The role of dopamine in motivation for food in humans: implications for obesity. Expert Opin Ther Targets 6:601–609

Wang ZM, Duerenberg P, Wang W et al (1999) Fat-free body mass hydration: review and critique of a classic body composition constant. Am J Clin Nutr 69:833–841

Wang ZM, Duerenberg P, Guo SS et al (1998) Six-compartment body composition model: Inter-method comparisons of total body fat measurement. Int J Obes Relat Metab Disord (BTX) 22:329–337

Wasserman SHS, Barzel US (1987) Osteoporosis: the state of the art in 1987: a review. Semin Nucl Med XVII:283–292

Weigle DS, Kuijper JL (1996) Obesity genes and the regulation of body fat content. Bioassays 18:867–874

Weiss A, Raz E, Silbermann M (1986) Effects of systemic glycocorticoids on the degradation of glycosaminoglycans in the mandibular condylar cartilage of newborn mice. Bone Miner 1:335–346

Welsh MJ, Tsui L-C, Boat TF et al (1995) Cystic Fibrosis. In: Scriver CL, Beaudet AL, Sly WS, Valle D (eds) Metabolic and molecular bases of inherited disease, vol III, 7th edn. McGraw-Hill, New York, pp 3799–3878

Westhovens R, Nijs J, Taelman V et al (1997) Body composition in rheumatoid arthritis. Br J Rheumatol 36:444–448

Whitman GJR, Chance B, Bode H et al (1985) Diagnosis and therapeutic evaluation of a pediatric case of cardiomyopathy using phosphorus-31 nuclear magnetic resonance spectroscopy. J Am Coll Cardiol 5:745–749

Whitworth JA, Brown MA, Kelly JJ et al (1995) Experimental studies on cortisol-induced hypertension in humans. J Hum Hypertens 9:395–399

Widmaier S, Hoess T, Jung WI et al (1996) 31P NMR studies of human soleus and gastrocnemius show differences in the J gamma beta coupling constant of ATP and in intracellular free magnesium. MAGMA 4:47–53

Willcott MR, Cook JP, Ford JJ et al (1983) NMR in Chemistry. In: Partain CL, James AE, Rollo FD, Price RR (eds) Nuclear magnetic resonance (NMR) imaging. Saunders, Philadelphia, pp 45–59

Willvonseder R, Goldstein NP, Tauxe WN (1974) Long-term body retention of radiocopper (^{67}Cu) and the diagnosis of Wilson's disease. Mayo Clinic Proc 49:387–393

Winkler SS (1990) Sodium-23 magnetic resonance brain imaging. Neuroradiology 32:416–420

Wirth A, Steinmetz B (1998) Gender differences in changes in subcutaneous and intra-abdominal fat during weight reduction: an ultrasound study. Obes Res 6:393–399

Wonke B, Jensen C, Hanslip JJ et al (1998) Genetic and acquired predisposing factors and treatment of osteoporosis in thallasemia major. J Pediatr Endocrinol Metab 11 [Suppl 3]:795–801

Woods RJ, Fleet JC (1998) The genetics of osteoporosis: Vitamin D receptor polymorphisms. Annu Rev Nutr 18:233–258

Yano Y, Budinger TF (1977) Cyclotron-produced ZN-62: its possible use in prostate and pancreas scanning as a Zn-62 amino acid chelate. J Nucl Med 18:815–821

Yasumura S, Cohn SH, Ellis KJ (1983) Measurement of extracellular space by total body neutron activation. Am J Physiol 244 (Regulatory Integrative Comp Physiol 13): R36–R40

Zaichick VY, Morukov BV (1998) In vivo bone mineral studies on volunteers during a 370-day antiorthostatic hypokinesia test. Appl Radiat Isot 49:691–694

Zdolsek HJ, Lindahl OA, Angquist KA et al (1998) Non-invasive assessment of intercompartmental fluid shifts in burn victims. Burns (AFC) 24:233–240

Ziegler R, Scheidt-Nave C, Scharla S (1995) Pathophysiology of osteoporosis: unresolved problems and new insights. J Nutr 125 [Suppl 7]:2033S–2037S

CHAPTER 13

The Biology of Aging and Related Nuclear Medicine Studies 13

Richard N. Pierson Jr., Walton W. Shreeve

"That time of life though may in me behold
when yellow leaves, or none, or few do hang,
on those bare choirs where late
the sweet birds sang." Sonnet [53]
William Shakespeare

"The days of our years are three score years and ten;
and if by reason of strength they be fourscore years,
yet is their strength labor and sorrow;
for it is soon cut off and we fly away."
Psalms 90:10

Contents

13.1
Introduction

In this chapter we shall explore the convergences between the sciences of aging and the methods of molecular nuclear medicine. First we shall provide a "modern" glimpse of the state of the art in aging research in 2002, with an acute sense that the recent rate-of-change trajectory in this field, projected forwards, promises that the scenery will change radically, and frequently. Our domain-mapping will emphasize the molecular levels of understanding, of analysis, and of research via tracer studies, where this little specialty has its methods-oriented history of past and future contribution. Indeed the tissue-inherent, gene-driven vital processes to be considered are fundamentally molecular in their expression, and genetic in their transmission. At the molecular level, until recently a more visible/discoverable domain than the genetic, the physiology of aging may be described as an integration between two overlapping processes. First is the *degradative*: (a) free radical destructive reactions (targeting both chemistry and genes), (b) molecular cross-linking, (c) deteriorating immune function, (d) telomere shortening, leading to cell and cell-line death, by apoptosis and perhaps by other mechanisms, and (e) the influences of active "senescence" genes. Each of these processes operates by damaging a gene, which is needed for cellular regeneration. These degrading forces interact with *reparative* influences: (a) free radical scavenging, (b) genetic deletion, and (c) genetic or somatic repair. A genetically programmed number of cell-renewals (80 has been suggested by Hayflick and Moorehead 1961 for mammals) is contained in many genetic instruction sets. These agencies are indeed not necessarily independent. Of even greater importance, "apoptosis" or the death of one cell, cell line, or organ, may enhance the survival of the host organism, most obviously when a malignant cell or cell-line is eradicated.

The study of aging has come to require, in the past decade, a reduction to basics: elucidation of structural biology, at levels from cm to angstroms, is a creature of the current era of scientific discovery, growing from and abetting perhaps three centuries of natural philosophy, in which methods were serially and progressively invented to explore and enrich biological hypotheses in finer detail. In the present decade, many maps have been created to describe our terrestrial universes, both the biological and the physical. The critical fourth dimension for both is time, a dimension which is, at least for biology, non-relativistic and manageable: ranging from seconds to years (for the physical universe, relativity is needed to deal with the 10^6 to 10^9 year scale). If we can adduce to the analysis of the experimental studies of structures, and of the interactions, which invest structures with life (the physiologies), as freeze-frame snapshots, to which by "streaming" we may add the linear dimension of time, we have a practical description, if not a definition, of aging. Observations about aging are as old as the recorded word, reflecting the prejudice of the philosopher-poets that "the proper study of mankind is man" (Pope 1733). Scientists have both preceded and followed philosophers, using the tools of quantitative biology to address this ancient concern, with proliferating and cascading series of research studies, extending the time-dimension of analysis (from the age of the subject: one-cell organism, fruit fly, rodent, or human) to virtually every discovery of an enzyme system, a cytokine level, a membrane-kinetic control mechanism, etc., to a study of the effects of life-stage on its levels, on its function, and on its relevance to cell death, organ death, or, in the case of cancer, cellular immortality. If by chance the enzyme, cytokine, or hormone level is associated with longevity in one or another species, the "association" is tried out for a relationship to the biology (biogerontology), or the "cause", of aging, and, especially in recent years, to its reversal. In this setting, cause and effect may or may not be analytically separated. According to the principle that there is "no sweeter fat than sticks to my own bones" (Whitman 1855), the fascination with aging (and its reversal) may increase as a function of the age of the investigator – and investigators, like all human subspecies, are living and working longer.

Short-lived organisms (compared to humans), from bacteria to fruit flies to mice have been studied with ease, by observation and intervention, with great benefit. Nuclear medicine methods have been applied to these species, occasionally leading the way to clinical studies. Our focus in this chapter will lean towards the "higher", or at least the more evolved orders, where strong homologies in genetic codes, from bacteria through man, render anthropometric conclusions more credible, building on the recognition that basic mechanisms are more fruitfully explored at the bench, with controlled studies of short-lived creatures readily studied through consecutive generations, with

isolation from external/environmental forces. With the recent discovery that large gene sequences in humans are identical to, and presumably inherited from, bacterial species of vastly longer survival and development (10^9 vs. 10^5 years), the study of older, possibly simpler, genetic codes may be rewarding in mapping the patterns, and eventually the causalities, of cell programming, and cell and organ survival.

Two critical dimensions, time and location, must be considered. The dimension of time, while uni-directional, operate in cycles, creating windows of opportunity, vulnerability, and inertness, best exampled by the G-S phases of replication which describe the mitotic process, where the time spent in one or another phase is highly determining of the resultant vector of a given intervention according to the time in a cycle when an ionization or a chemical event occurs: the difference between 10^{-2} and 10^{-9} second half-lives respectively of H_2O_2 and hydroxyl radicals entirely determines the physical range, the locations, in which their toxic oxidative effects will be expressed. Location also identifies the specificity of chemical reactions, often oxidations and anti-oxidant repairs, but also other chemical and physical events, to membrane-guarded, fat or water-soluble, intracellular or intra-mitochondrial sites, a fact which, for example, makes nuclear DNA, much more actively guarded by abundant antioxidants, far less susceptible than mitochondrial DNA to unrepaired oxidative injury.

Living organisms, and their subunits down to the membrane, protein, and genetic levels, are in metabolic flux, with active metabolic "work" required to maintain each unit, whereby the loss of a given structure, in a given cell, in a given organ, so long as it can be renewed/repaired/replaced, results in survival of the cell line and the organism. Genetic studies define a *cellular proliferation potential*, with specific genes identified for specific cell lines, which may operate through the telomere system to render some cell lines (usually human cancers) immortal, while others receive instructions for a fixed number of cell divisions, followed by apoptosis, or cell death. Since "aging" and "passage of time" are functionally synonymous, the *random somatic mutation* theory and its first cousin, *chromosome damage rate*, relate prolonged life spans to interventions, which block, reverse, or eradicate mutations. These theories regarding mechanisms are highly supported by radiation studies in which the correlations of radiation dose (and dose rate) to chromosome breaks are quantitative, and these effects may be blocked or reversed by a variety of interventions. The *free radical* concept is a mechanism within these theories, since interactions with free radicals may result in chromosome breaks, and gene-transmissible defects. The very large number of chromosome breaks/day, estimated at 10^6 for humans, "most" of which are repaired or discarded, introduces a numerically stochastic filter which allows the laws of probability to be invoked when impugning "risk", a concept of the greatest importance to the research physicians who will persuade themselves, their staffs, their patients and their "normal volunteers", that they may reasonably agree to being exposed to experimental regimens involving mitogenic chemical or radiative interventions.

The numbers of chromosome-altering events caused by chemical species greatly outnumbers those caused by radiation, prior to any additions to the natural backgrounds of either chemical or ionizing events. With regards to ionizing radiation, the exposure to radiation required in a research protocol is usually a small multiple of that from cosmic and other natural sources (3 mSv/year at sea level in the United States). The wide variations in background "natural" chemical and ionizing radiation exposure by geographic site (by a factor of at least 30 with radiation), with no apparent variations in somatic effects (the most visible of which is the rate of malignancy), are among the skeins of evidence, which support the "minimal effect" hypothesis for radiation.

These then are the arguments for the structure of this chapter: aging provides the fourth dimension, time, to the three-dimensional maps of structural biology, with their superimposed physiologies. The clock may run fast or slow for a given event-window (the swollen chromosomes in metaphase, 3% of a typical cell cycle, are a briefly "large" target), but it runs in one direction. And the metabolic turnover rates for a given membrane or synthesis may be measured in seconds, days, or years, creating another hierarchy of probabilities for damage or intervention for a given external or internal influence. The aging clock is surely not simple, but it is probably stochastic.

Tracer imaging and non-imaging methods make up the disciplines of nuclear medicine. This chapter does not explore except by example, in Sect. 13.5, the explicit procedures to which nuclear/radioisotopic techniques apply; these will change and adapt as the research goes forward. The domains in which our methods may be applied are both too large and too

fully described elsewhere to be included in this chapter. The "molecular" focus has been particularly well served by radioactive and other nuclear tracer methods, which predominate in the service of research agendas. The fundamental advantages of nuclear methods for tracer studies originate in the chemical identity between isotopes, and the large Avogadro denominator of $\sim 10^{23}$ atoms in a gram molecular weight, providing favorable tracer/tracee ratios in the order of 10^{-15} or better, which permit distinctly odd or unlikely molecules to be injected into a complex and often multi-staged metabolic process as tracers, with a vanishingly low probability of chemical or biological disruption. In the case of an isotopic tracer, the "isotope effect", whereby the tracer is either metabolized or measured atypically based on variant nuclear mass, is rarely significant except for some applications of isotopes of hydrogen and possibly carbon (Chase and Rabinowitz 1959). Tracers with short T1/2 are particularly apposite to the study of metabolic rates and kinetics. The characteristics of many frequently used tracers are listed in Chap. 11.

13.2
Models for Aging

13.2.1
Rate of Living

This physiologic model (Table 13.1) dates from the 19th and early 20th centuries (Weissman 1889; Pearl 1928), and is fundamentally physiologic, envisaging a built-in, species-specific, energy "reservoir", which may be drawn down slowly or rapidly, but which is limited. While this model does address the wide variation of species-specific life spans, the capacity to explain this observation by several other models based on experiment had made it a "transient" by the second half of the 20th century.

13.2.2
Evolutionary

Teleology has had expression among theories of aging. *Programmed senescence* as a means to population control, favoring survival only until reproductive maturity, followed by death-reduction of segments of

Table 13.1. An epistemology for the study of aging

Principles	Mechanisms	Physical targets	Organ systems	Interventions
Genetic controls (immortality, senescence, Hayflick limit)	Nuclear DNA, mitochondrial DNA	Chromosomes, telomeres		Gene deletions and insertions
Physiologies	Oxidative stress, free radicals, anti-oxidative apoptosis	Genes, cells	All (brain, heart)	Scavengers
	Nutrition	Membranes (turnover, maintenance, repair)	Liver, muscle, endocrine	Essentials, caloric restriction
	Radiation, chemical	Genes		
Immune systems Endocrine systems	Cellular, circulating Hormones (tonic)			Stem cell systems

a population, which primarily consume resources, has been expressed using the Pacific salmon as an extreme example, in which death coincides with the end of the single reproductive cycle. Thus wild animals rarely live to a "natural", or genetically imposed death, dying after their prime, after their reproductive powers have waned, but before any genetically imposed decay, as a result of extrinsic mortality (Kirkwood and Austad 2000). Williams (1957) proposed a life-history trade-off, termed a *disposable soma*, based on optimized allocation of resources between somatic maintenance and reproduction. A short-lived species will have little benefit from somatic maintenance, while a long-lived species will benefit from a greater investment in building and maintaining a durable soma.

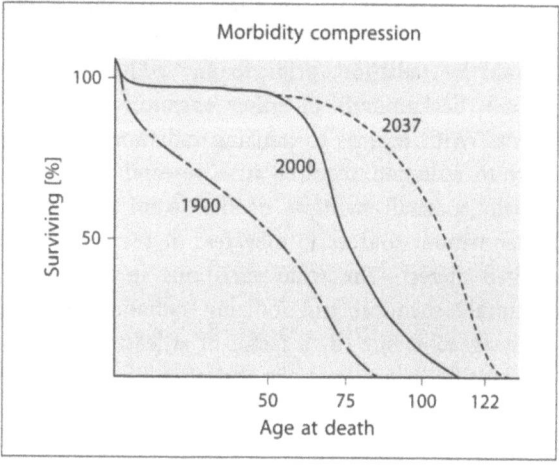

Fig. 13.1. Morbidity compression describes the effects of life span on removing the effects of illness, predation and natural forces, developed by Fries (1980)

13.2.3
Stochastic

"Aging is the accumulation of adverse changes that increase the risk of death" (Harman 1998). This model is intensely driven by data: from observation, from experiment, and from interventions designed to tweak each way-station in the biologic, chemical, and ultimately the molecular-biologic chains of events. The current decade has well prepared this path. Analytically, by this model aging is defined in the absence of disease, labeling as exogenous the pathologies of bacterial or toxic invasion, and the endogenous deleterious influences – arthritis, arteriosclerosis, and cancer are examples – often considered as programmed-in to "normal" cell timetables, but which have widely been analytically separated from aging. After filtering ex-

perience/observation to study aging separate from disease, the net time-ordered "changes" of aging are a weighted summation of genetic and environmental elements which operate to increase the risk that a cell, an organ, or an individual will die at a certain age, providing a "residual" estimate for a built-in longevity. Dissecting away the random causes of death (infection, accident, predation, etc.) to define biologic life expectancy, a concept quite different from Darwinian "survival" in a competitive tooth-and-claw environment, exposes an "age rectangle" (Fig. 13.1) of ideal survival (the "piston function" of normal red cell survival is an excellent medical model), when the distractions of competition and disease are (analytically) removed. An historic look at human death rates drives this model: average human life expectancy (here the stochastic element) has increased greatly and continuously over two centuries, resulting from reductions in the expressions of environmental and disease factors. As risk- and disease-reduction strategies (more "public health" than "medical" interventions are noted) approach a limit, the term "morbidity compression" is demonstrated; the exponential chance of death with age asymptotically approaches a limit, leaving increasingly exposed an intrinsic upper boundary, perhaps in the vicinity of the 122 years recorded in 1997 in France. For both philosophers and biologists, it is this intrinsic limit, the *"maximum potential life span – MLSP"*, which studies of aging have addressed. The mechanisms most compatible with this model converge on an "accumulation of toxic products" pathway, usually, in the current view, the ROS common pathway (13.3.1).

13.2.4
Heredity

Longevity by heredity has been proposed as a model. In men over age 85 certain components of the immune system may have influence: expression of HLA-DR5 was higher, and HR-B40 lower, than in controls matched for all other characteristics (Lagaay et al. 1991). The differences were greater in women than in men. The statistical significances of the differences are not given. In mice a similar finding is reported for the major histocompatibility complex (MHC) H2 antigen.

Recent studies of several hundred centenarians who also had at least two long-lived (>90) siblings have implicated a short stretch on chromosome 4 which may contain an anti-oxidant support gene which is of high frequency in this rare "experiment of nature" subgroup (Perls 2002). Weaving a connection between other observations on aging: the telomere/telomerase mechanisms; the powerful effectiveness of energy-reduction interventions; a hyperinsulinemia/ROS interaction; and selected ROS-immune-system modifications, may provide convergence towards a unified theory, which would serve the many scientists who seek to understand with more precision the intermediary biophysical and predictive correlates of aging. It also stimulates the metaphysical, who seek clues to immortality. Both genetic (Perls 2002) and stochastic (random occurrence of "redundant capacity and physiological reserve"; Hayflick 1998) models have been developed. The existence of species (some tortoises, fish, and the American lobster), which never reach a fixed size in adulthood,

and may never age, presents a challenge to biogerontologists.

13.3
Key Concepts for Aging

13.3.1
Reactive Oxygen Species (ROS)

The free radical theory of aging was proposed by Harman (1956), stating the premise that a single common process, modifiable by genetic and environmental factors, could account for aging and death in all living species, and that the free electron associated with oxidation reactions in tissue could best explain the chain of molecular events leading to deterioration (aging) of cells, systems, and tissues, to slowed or inefficient reactions, and ultimately to the death of the organism (see Tables 13.1 and 13.2). In 1972 mitochondrial DNAs, *mtDNA*, were proposed as the principal and most effective intermediary target. An inborn aging process, defined as the inexorable accumulation of irreversibly damaged proteins and lipids, caused by very transient superoxide radicals and H_2O_2 in mitochondria, in the course of normal metabolism, damages the molecular machinery of intermediary metabolism, until the various defense mechanisms have been overmatched (Harman 1998). The specific organ, cell-type, and intracellular structural targets of free-radical damage are many and varied, have been reviewed in excellent detail (Beckman and Ames 1998), commented on by many others, and are summarized in this chapter. The integrity and survival of both the individual and of the species depends on capacity to maintain homeostasis against the oxidative stresses of their environments, many of which are mediated via the generation of free radicals. The mobilization and functional integration of specific defense components include membrane-associated and cytosolic scavenging and repair enzymes, which dampen the effects of these destructive final-pathway agents by reducing their toxic life-times, limiting their access to the most sensitive micro-sites, and repairing the lesions induced, extending from instant successful prevention of protein-damage through rapid scavenging, to failure, via genetic damage and loss of membrane integrity. Thus the miracles of energy efficiency accomplished in an oxygen-based metabolism resemble the Promethean gift of fire, bearing the

Table 13.2. Natural and natural antagonistic systems

Natural systems	Natural antagonistic systems
Superoxide dismutase, catalase, glutathione peroxidases, cyclic nitroxides, uric acid, PBN N-tert-butyl-phenylnitrone	*NADPH oxidases (PMN), nitric oxide synthase (NOS), NADH dehydrogenase (Complex 1), ubiquitone-cytochrome c-reductase complex 3, advanced glycation product (AGE)*
Spin traps Vitamin E, alpha tocopherol Vitamin C Beta carotene Alpha lipoic acid Melatonin Metallothionein Polyphenols Endorphin Homocysteine Cytoxins Enkephalin Ubiquitin Phenolic anti-oxidants	May favor apoptosis Nuclear factor (NF) κB p53 Activation Heat shock response
Chelating agents EDTA Phytic acid	
Signaling pathways Extracellular signal-regulated kinase (ERK) c-Jun amino-terminal kinase (JNK) p38 Mitogen-activated protein kinase (MAPK)	
Environmental	*Ozone? Atmospheric NO, CO, etc.*
RADIATION – natural: hormesis via stress up-regulation, via stress-induced up-regulation	*RADIATION – unnatural: therapy, diagnostic, industrial*
Exercise	*Cigarette smoke*
Diet – caloric reduction: Cruciferous vegetables, genistein, oltipraz, anethol dithiolthione, sulforaphane, quinone reductase, glutathione-S synthetase	*Amino acids easily oxidized* *Histidine, lysine*

seeds of potential destruction entirely intertwined within the magical chemistries of its central agency, oxygen metabolism. The self-sustained protective components which have evolved are defined as "anti-oxidant defense systems" (Davies 1986; Heffner and Repine 1989). The "enemy" is broadly defined as the ROS.

An articulate minority (Maynard-Smith 1963; Sohal 1993) have challenged the ROS thesis, interpreting the evidence both to diminish the likely scope of ROS damage, and to upgrade assessment of the capacity of antioxidant and repair chemistries. The argument questions whether the structural damage in tissues of senescent animals is of sufficient severity and specificity to explain the measured loss of function, since some nearby enzyme systems are unaffected, and the overall concentrations of oxidatively damaged molecules are present in only trace quantities. Sohal directs an alternate hypothesis to an effect of ROS on quantitative expression of genes starting from the early embryonic stages of development. Reviews by Greenstock (1993) and Knight (1998) provide balanced commentaries across the pro-con ROS rift, with the review by Beckman and Ames (1998) providing a most complete and dispassionate summary. Furthermore, a great and increasing (2002) majority expression from the literature on caloric restriction, apoptosis, mtDNA, atherogenesis and cancer induction, link their data and their arguments to elements of the ROS hypothesis.

13.3.1.1
Mechanisms of ROS

Atomic electron structure is thermodynamically stable when orbital electrons are paired with antiparallel spin, resulting in no net spin (Yu 1994). A free radical is a molecule or an atom with an unpaired spin in its outer orbit. Molecular oxygen is a double radical, having two unpaired electrons, each situated at a different Pi orbital, but having the same spin quantum number, located in a parallel spin configuration. When O oxidizes another molecule by accepting a pair of electrons, these pairs must be aligned with parallel spin. The major route for metabolism of molecular O involves complete reduction to H_2O by accepting four electrons. However with one-electron reduction, several free radicals and H_2O_2 are formed, in the following sequences:

		Lifetime
$O_2 + e^- \rightarrow O_2^{-*}$	*Superoxide* radical	10^{-6} s
$O_2 + H_2O \rightarrow HO_2^* + OH^-$	Hydroperoxyl radical	10^{-2} s
$HO_2 + e^- + H \rightarrow H_2O_2$	Hydrogen peroxide	10^{-6} s
$H_2O_{2++}e^- \rightarrow OH^* + OH^-$	*Hydroxyl* radical	10^{-9} s

The *superoxide* is the best-known free radical, with many potential cascades generating other species of

ROS. The *hydroxyl* radical is potentially the most potent agent, but has such a short half-life that a time-space radius-of-interaction of only about 2 molecular diameters limits its range. *Hydrogen peroxide* has been widely studied, and while not itself an ROS, it readily passes biological membranes, being un-ionized, and spawns ROS by interacting with redox-active transitional metals such as Cu and Fe.

The nature of cellular damage by free radicals is categorized into three stages: fragmentation, aggregation, and susceptibility to proteolytic digestion. (Typical of such damage is the *cross-linking* and protein degradation that occurs to ocular lens proteins resulting in cataracts, a well recognized accompaniment of aging.) A parallel three-part process in lipid peroxidation is described as initiation, propagation, and termination (Girotti 1985), a sequence associated with formation of diene bonds, in the presence of redox-active metals such as Fe and Cu. Aldehydes, which are both cytotoxic and mutagenic, are frequently encountered in the degradative cascades. They can be measured in the expired air, a tracking method applied in mice and humans. While many subcellular targets have been identified, the bilipid membranes are prominently susceptible targets. Accumulation of malondialdehyde, associated with lipofuscin materials by histology, is a marker for the age-related loss of mitochondrial oxidation capacity. Even carbohydrates undergo oxidative degradation, an example being the formation of dicarbonyl compounds and H_2O_2.

The *ubiquitin*-proteasome pathway, particularly the 20S *proteasome*, controls the degradation of most eukaryocytic cells (Davies 2001). Both short- and long-lived proteins: tumor suppressors, transcription factors, and cell cycle proteins, are cleaved, modified, folded, and transported by this ubiquitous and multitasked protein. Ubiquitin's highly conserved 76 amino acids are identical in insects, fish, and humans. It is well known currently in relation to protein degradation, chromatin structure, and heat-shock, and has anabolic as well as catabolic functions. Activity declines with age, and is increased by caloric reduction (Gaczynska et al. 2001). It has "turned up" in new intra- and extracellular settings at frequent intervals in the past 3 years; a growth industry is suggested.

ROS modifications of cellular and genomic structures in the damage pathways have been abundantly studied. The mitochondria play a particularly sensitive role, being major producers of free radicals, poorly endowed with DNA repair processes, and preferential targets for many xenobiotic chemical carcinogens. Here converge the radiative and chemical pathways to ROS production, and the well recognized mutual enhancement of carcinogenesis when both insults combine. While the DNA double helix confers a major degree of protection from oxidative injury via its template-repair mechanism, the ubiquitous and multifarious availability of ROS provides a sufficiently diverse set of challenges, which assure a final failure of defenses.

A second major source of abnormally aggregated "aging" proteins is via advanced glycation end-products (*AGE's*) derived from the Maillard reaction, whereby the carbonyl (aldehyde or ketone) of the reducing sugar binds to an amino group (Cerami 2001), resulting in a slow transformation via still-reversible Amadori products, to a heterogeneous array of irreversible products, the AGE's, some of which cause intra- and intermolecular *cross-links*. HgbA1c is an example of an Amadori product, with a T1/2 of 28 days, serving as a very reliable index of average intracellular glucose over several weeks. Collagen and elastin are prime examples of structural proteins with long turnover times, slow to develop and slow to clear, with uncoupling of the elastin laminin receptor. This results in structural rigidity in membranes, and in the endothelia of arterioles, strong functions of age, which have been traced directly to cross-linked AGE proteins, explaining the premature arteriosclerosis in diabetics, an impressively reversible process under experimental treatment with several compounds, for example, *p*henacyl*t*hiazolium *b*romide (PTB), and *d*imethyl*p*renacy*t*hiazolium *c*hloride (DPTC), in animal models.

13.3.1.2
Defenses

The litany of defense mechanisms, which are designed to prevent, counteract, or repair oxidative toxicities has become the most active domain in the Aging field. Primary native defenses include *superoxide dismutase* (*SOD*) in Mn, Cu, and Zn complexes, because it prevents further generation of free radicals, acting as a catalyst. The rate of SOD dismutation is 10^4 greater than that of chemical dismutation, its rate of formation regulated through biosynthesis, being increased in the presence of high oxygen tensions. *Catalase* is similar to SOD in function and distribution, playing its major role in converting H_2O_2 to H_2O. Glutathione peroxidase combines with selenium in some circumstances. Vitamins E, C, and A serve different roles according to their tissue distributions. Vitamin E, its dominant form being alpha tocopherol, is fat soluble, and is in high concentration in lipid-rich membranes, in mitochondria, and in endoplasmic reticulum. Protection against lipid peroxyl and alkoxyl radicals occurs because of its access to key intracellular and intramitochondrial sites. Vitamin E supplementation must be provided at a higher level when a diet high in fish oil is taken, especially when exercise will increase the oxidation of these unsaturated fats (Meydani 1992). Vitamin C is hydrophilic, is active in plasma lipid peroxidation, and indeed serves to re-activate oxidized vitamin E. Vitamin A and the related beta carotenes are potent scavengers of ROS at low partial pressures of oxygen. Glutathione in the reduced form is ubiquitous and abundant in all mammalian cell systems, being on average present in 0.5 mM intracellular concentrations, rising 20-fold to 10 mM when stimulated. Uric acid has been recognized as a potent antioxidant in both intracellular and extracellular sites, and its levels correlate with *maximum life span potential.*

Secondary defenses include lipolytic and proteolytic enzymes, which serve by preferentially degrading oxidatively-altered lipids and proteins. These functions are particularly relevant in the context of aging, since the accumulation of damaged proteins correlates highly with age-ordered diminution of function, being the center of the hypothesis of Orgel (1963), which proposed these buildups as a central correlate and marker for aging, a hypothesis not borne out by subsequent research, but stimulating for its linking of the ROS system specifically to aging. The cellular strategy to combat injury to protein molecules by proteolytic enzyme systems seems analogous to that in lipids, in the dual processes of protection and repair, the primary defense being free-radical scavenging, (leaving the ROS less available to damage proteins), and the secondary being either the elimination of damaged molecules, or their restoration by repair. Further investigations identify the *proteasome*, a multicatalytic proteinase complex of 600–700 kDa, and another proteolytic complex called MOP (Salo et al. 1988).

Exercise at high levels produces a two-fold increase in O^- production, while whole-body increase in O_2 metabolism is 20-fold. The normal homeostasis between production and clearing is perturbed, and anti-oxidant functions will fail to keep up, or if they compensate at young ages, they may fail at older ages. Aged rat heart and liver show 40% higher O_2 production at rest. There is some evidence that sus-

Table 13.3. Enhancing nature's systems: the interventions which have been discovered by the "experiments of nature", in which antioxidant treatments derive from utilizing known chemistries, often vitamins. These interventions are associated minimally with the risks of allergic responses and toxicities. In addition, being natural substances, minimal requirements for approval by regulatory agencies

Spin traps:
Butylated hydroxytoluene
Butylated hydroxyanisole
MEA 2-mercaptoethylamine
Ethoxyquin
Lazaroids (2-methylaminochromans)
21-Aminosteroids

tained exercise regimens result in hypertrophy of antioxidant capacity.

Greenstock (1993) has discussed the radiation-aging question within the framework of ROS kinetics. A linkage between ROS pathways and the energy-reduction (favorable) and the cross-linking (unfavorable) motifs is recognized (Finkel and Holbrook 2000).

Table 13.3 lists the interventions, which have been turned up by the "experiments of nature", in which antioxidant treatments derive from utilizing known chemistries, often vitamins. These interventions are associated minimally with the risks of allergic responses and toxicities, and, being natural substances, minimal requirements for approval by regulatory agencies.

Several "by design" interventions with antioxidant potential have been studied, which propose to improve on natural antioxidant systems, but which must survive clinical trial for regulatory review. Pharmaceutical interventions with DHEA, Q10, thioproline, flavonoid, and the metals Fe, Sr, Zn have been approved for human use. The use of growth hormone, and many "look-alike" hormones which have some activity in the GH-IGF pathway, has received US FDA approval for treating specific diseases, but not for "treating" aging, leaving a market for Mexican and other more international sites which advertise hormone-intervention approaches to longevity.

Repair mechanisms are scavengers, antioxidants, able to combat the free radical-superoxide load. These vary in fat-water phase solubility, size, reaction rate, avidity, recharge chemistry, capacity to hand-off electrons to one another, and most importantly, in location within the cell and its membranes. Within the nucleus, histones dominate. In the mitochondria, Mn-SOD (Mori et al. 1998), catalase, glutathione peroxidase, cytochrome c, succinate, and coenzyme Q (ubiquinone) have major natural presence (Wei 1998). The spontaneous decline with age, and also in several of the "natural aging experiments" in the premature genetic aging diseases, of natural antioxidants has presented strong research evidence favoring the ROS theory, and experimental interventions to explore ROS theories are legion. The spin-trapping agent PBN (phenylbutylnitrone) (Davies 1986; Mori et al. 1998) resulted in life-extension by 25% in a short-lived strain of mice. By extension, the study of oxidation and protection kinetics in many organisms and tissues has produced the venerable "rate of living" theory in which the mitochondrial rate of ROS production is inversely related to *MLSP* (Rubner 1908; Pearl 1928; Harman 1972).

The free radical-superoxide load is reduced by *caloric restriction* (see Sect. 13.4.1), and by various chemical interventions, which may be endogenous, (these are numerous), or by therapeutic interventions. These have been studied, as part of the scene of natural processes; they have been discovered, first by observation and then by experiment; and they have been designed, as in the modern tradition of the pharmaceutical industries, which have the capability to plan to intervene. Whether they come by study, discovery, or design, they become substrates for the popular industry of seeking "Fountains of Youth", "Elixirs of Life", or other expressions of the missions to prolong life, delay death, and seek "Immortality". The perhaps ultimate approach in therapeutic design for the 21st century will be the intervention of genetic transfection, which might inject "super genes" which might improve the resistance of substrates to oxidation, or might upgrade the internal anti-oxidant environment.

The protective or toxic influences of Fe and Cu were recognized early (Koechlin 1952) and repeatedly (Stocks et al. 1974); a single ferritin molecule can carry 4,500 Fe atoms, a potent scavenger for unbound iron, when binding sites are only partly occupied, but when saturated, ferritin (and transferrin) become sources of free iron, and further oxidative damage. Ceruloplasmin has a parallel role for copper, and Cu and Fe transfers may interact (Gutteridge 1988). Selenium has unique chemical properties related to its multivalent potential (−2, 0, +4, +6), and an impressive role in cancer epidemiology (Schamberger and Frost 1969). It is closely associated with the glutathione system, and is found in highest concentrations in nuclei, and in descending prominence in cytosol, mitochondria, and microsomes, which are major action sites both in carcinogenesis and in cancer-protective reactions. Evidences supporting the free radical hypothesis in aging present both molecular and mechanistic theses. Antioxidant feeding experiments to suppress free-radical half-lives have resulted in maximum life span extensions of 12–20% in mice, fruit flies, nematodes, and rotifers. Further, Cutler (1974) demonstrated a positive correlation between "life span energy potential" and interventions with dietary scavengers. Liu et al. (2002) have recently patented the paired supplements acetyl-L-carnitine and alpha-lipoic acid to enhance the enzyme carnitine acetyltransferase, with reversals in memory loss and physical performance in rat colonies.

Therapeutic strategies are often considered in the aging literatures. Wide current acceptance of the ROS mechanisms outlined above, largely by a channeling of the other active theories into pathways incorporating ROS kinetics, dictates that enhancement of antioxidant defenses should be studied for therapeutic potential. Questions: first, can effective therapeutics address ROS damage? Second, would enhancing these mechanisms retard aging, or prevent aging-associated diseases? The development of new SOD- and catalase-mimetic drugs have succeeded in mice and Caenorhabditis elegans. "Hormesis" – induced by modest stresses in the radiation, exercise, and heat-shock protein scenarios have been effective in C. elegans and Drosophila. Identification of longevity-influencing genes, both in laboratory animals and in mammals, strengthens the motivation to extend the inquiry into a mechanistic connection between oxidants, stress, and aging.

The chain of causality for the mtDNA as intermediary mechanism for aging has now extended to finding somatic mtDNA rearrangement mutations in post-mitotic tissues in brain and skeletal muscle in humans, as well as in a wide range of tissues in animals. The mitochondrial genome contains 37 genes, all of them involved with synthesizing units of the respiratory chain complex, 13 of them as polypeptide components, 22 as transfer RNAs, and two as ribosomal RNAs (Cottrell et al. 2001). Establishing a connection between gross histological changes such as the "ragged red fibers" in muscle, with chemical species representing evidence of toxic oxidation (8-hydroxy deoxyguanosine as example), with degraded mutant mtDNA, has extended the range of research methods adduced to the ROS-mtDNA argument. These multi-specimen, multi-mechanism studies have been widely applied to caloric-reduction studies, with general confirmation of the cause-effect chain (Wallace 2001).

Immune systems of both the cell-mediated (T cell) and humoral (B cell) types have been widely studied in mice and humans (Effros 2001). The simple observation that centenarians show well-preserved immune function compared to younger cohorts (Franceschi et al. 2000) has stimulated studies showing age-ordered increases in natural killer (NK) cells, and qualitative shifts in antibody responses, which have been termed "regulatory changes". Poor survival is associated with decreased T cell proliferative response, high CD8+, low CD4+, and low B cell counts. The MHC is proposed as the genetic locus governing immune function.

B cell antigen-driven kinetics in young-normals show a very high rate of point mutations in the variable regions of Ig genes (Miller and Kelsoe 1995). The T1/2 of mature B cells increases with age, while a ten-fold decrease of marrow and splenic émigrés is measured. T cell clonal expansions characterize successful aging, favoring CD8+ (intracellular pathogen-related) over CD4+ (related to extracellular pathogens) species. Recent recognition of the CD28 species notes that it is the CD28– elements which dominate in the elderly. Also, a shift from the Th1 to the Th2 cytokines, with greater expression of autoantibodies and less cellular immunity, accompanies aging, but is counteracted by caloric-reduction (Pahlavani et al. 1997). The Fas system, a potent *apoptosis* inducer, is diminished with aging. Mucosal immune systems, with the GI tract representing about half of the total, show reduced IgA secretion with age.

Thymic involution, a well recognized physical concomitant of early growth, is now thought not to apply to thymic function, with the anatomically diminished gland being highly functional during aging (Rodewald 1998). *Proteasome* functions decline; these are involved in the intermediary steps around *apoptosis* and scavenging of residuals, producing peptides which may become antigen-presenting vehicles, more likely with age to elicit aberrant and autoimmune responses, which increase with age.

13.3.2
Reactive Nitrogen Species (RNS)

Nitrous oxide (NO) is generated by NO synthase (NOS) acting on L-arginine. Analogous to the ROS cascade, NO, itself a relatively unreactive radical, is the parent molecule in a redox cascade, producing NO_3^-, NO_2^-, and $ONOO^-$, peroxynitrite; nitrates are considered harmless, but the reduced forms have a high potential for destructive interactions on proteins in the mitochondria. Aging may influence this pathway in the opposite direction than the ROS analogue; NOS expression in skeletal muscle declines with age (Reid and Durham 2002). NO diffuses readily across cell membranes acting readily with guanylate cyclase to convert a linear triphosphate to the cyclic 3', 5' form. These species have the capacity to affect signaling pathways by modifying the tyrosine residues in receptor molecules, affecting neurotransmission, vasodilation, and immune function. Dietary vegetables contribute primarily harmless nitrates; 60–90% of nitrogen absorbed from meats is in the potentially toxic nitrite form.

The very short biological half lives of the reactive N species contribute to the more recent discovery, by more modern techniques, of this parallel chemical pathway to protein degradation, and the buildup of strategically located toxic byproducts, which characterize, perhaps cause, the alterations in protein function which are the chemical definitions of aging (Drew and Leeuwenberg 2002).

13.3.3
Mitochondrial DNA, *mtDNA*

More than 90% of whole-body oxygen consumption takes place by oxidative phosphorylation in the mitochondria, which, compared with nuclear DNA, lack protective histones, have a much lower rate of repair, and present a much larger, less antioxidant-protected target, with a large opportunity for transmission of altered genetic material (Richter et al. 1988). Studies of function, intermediary metabolism, biochemistry, and recently the genetics of the mitochondria have progressed rapidly, and their roles in disease (particularly atherogenesis), cancer, and aging, have received intense study in the past 15 years. Indeed a *mitochondrial theory of aging* has appeared. Research 50 years ago in species of flies, rodents, and birds, has progressed in the past 20 years to bats, primates, and humans, becoming more experimental with the recent adoption of knockout mice techniques in which the responses to interventions could test antioxidant hypotheses. Rates of mitochondrial free radical production vary across six species by a factor of two, $r = -0.91$, with an inverse associated *maximum life span potential* (MLSP) (Calder and Newsholme 1993).

Mitochondrial studies of the reactions within the electron transport chain, which serially energize complexes I–V, ultimately phosphorylating ADP to ATP, show the stages at which proton leaks, and various levels of oxidative damage, may occur (see Sect. 13.3.1). The effects on animal life span of a given mitochondrial metabolic rate were attributed to different rates of "free radical leaks", when 1–3% of the oxygen used in this transport chain is diverted to free radical production (Linnane et al. 1998). Recognition of the propinquity of mtDNA to the principal generator of free radicals, the "inner mitochondrial membrane", and then attention to the different substrates, the carbohydrate, protein, and lipid sequences, in the mitochondrial cytosols and membranes, set the stage for placing many determinants of cell health, longevity, and repair in the domain of the mitochondria, which are now seen as the industrial centers of the life and commerce of all cells. Increased mutation rates in DNA are caused by free radicals (Mori et al. 1998). Indeed this chain of causality was widely hypothesized in the 1960s and 1970s, considered unclear/unproved in the 1980s, and now seems widely adopted by the turn of the century. The toxic effect of O^- or OH^- radicals in their capacity to cause chromosome breaks has long been well recognized, occurring non-specifically as responses to radiation, chemical effects, and other causes of ROS. The mtDNA is 10–20 times more vulnerable to mutation than nuclear DNA (Fraga et al. 1990); 24 large-scale deletions and two-point mutations of mtDNA occurring and accumulating (Ikebe et al. 1990) in the mitochondria have been identified. The chain of causality between events taking place at different loci in the mitochondria, mutations in mtDNA, and the life span of the organism is not simple; experiments in different tissues of the same animal show very different metabolic rates, and different metabolic efficiencies, as measured by free radical leaks. While it is possible that one or another organ (heart, brain, kidney and lung have been measured) may be critical in determining total-organism median life span – *MLSP* (a "*critical organ*" theory), only the simple association is claimed; the metabolic "efficiency", not the metabolic rate, is ascribed a causal (inverse) relationship. The liver has been described as immortal; the heart and brain are the weak links (Linnane et al. 1998).

By contrast with mtDNA, effects on chromosomes occurring within the nucleus are potentially carried forward to subsequent cell generations, but the susceptible chromosome represents a target almost infinitely

small in volume, arguably less than 0.1% of tissue volume. When the chromosomes swell during metaphase, their fractional volume increases by a large factor, increasing the "target" size for a transient ROS. However, the mitochondria represent an enormously larger target volume, and given the same "background" load of radiative, chemical, or other influences, the availability of free radicals is likely to be a function of mitochondrial mass. To the extent that the aging organism accumulates free radicals, or has exposure, however brief, to free radicals, any toxic effects would also be cumulative, absent scavenging or anti-oxidant counter effects which might suspend, relieve, or cancel the toxic effects of the free radicals. The accumulation of oxidative DNA damage is higher when free radical production is higher, or DNA repair is lower.

Genetic targets: the post-mitotic mitochondria of the "irreversibly differentiated" cell maintains a large gene population different from the nuclear DNA by virtual lack of intron spacer material, and by a (relatively) poorly developed resident intrinsic scavenger/ antioxidant system, specialized for excision and recombination repair. While cyclic mitosis provides a purging and rejuvenation opportunity for nuclear DNA committed to germ line preservation, the mitochondrial DNA have a more transient mission, perhaps related to the thesis that post-reproductive organisms are best allowed to age and die out, preserving limited resources for future, evolutionarily potent generations, a thesis stated by Weissman (1889), and echoed by many since. Miquel (1998) summarizes the connections between the ROS pathways and its protagonists, both the oxygen-stress and the scavenger/ antioxidant systems, and the genetics which control their dispositions and their efficiencies. He draws in many of the arguments presented under the ROS (Sect. 13.3.1) and caloric reduction (CR) (Sect. 13.4.1) sections, repeats the basic arguments, and brings forward in summary the mechanics of the genetic engines which preserve and transmit structure. His summary defines the characteristics of different organs by their degree of potential renewal (low in CNS, skeletal and cardiac muscle, high in liver and immune systems), in terms of their mitotic potentials, their susceptibilities to permanent ROS-initiated cumulative dysfunction, and the degree to which an intervention (with external antioxidant medication), or strategy (such as CR or exercise), might tilt a balance.

In Drosophila there is a 35% increase in life-span coincident with mutation of the "Methuselah" gene, the protein product of which is not known. Mutations in another gene, called the INDY (I'm not dead yet) gene (which codes for a protein that transports and recycles metabolic products), results in coming closer than any previous report to the Fountain of Youth, by causing a doubling of life expectancy, with no apparent price paid in vigor, fertility, or exercise potential (Rogina et al. 2000). Of special interest, the protein is highly homologous with a membrane protein common to the animal kingdom, including humans. The apparent mechanism of action is to mimic the effects of caloric reduction – without the caloric reduction. A second gene, *sir2*, invades the nucleus to reprogram a ribosome, which then also mimics caloric reduction; this gene fascinates its discoverers because it is highly homologous to the gene which causes Werner's syndrome, paradoxically a premature aging syndrome which functions at just the opposite phenotype (Guarente 2000).

13.3.4
Telomeres

Most cultured mammalian cells do not proliferate indefinitely, but achieve a status of "terminal proliferation arrest" (Reddel 1998), also called "replicative senescence", after a preset number of doublings, 50–90 for the human fibroblast cell lines studied by Hayflick and Moorehead (1961). Fundamentally a protection against the development of cancers – a special case of cellular "immortality" – is a requisite of successful aging strategies. Thus aged cells, whose roles do not require their replacement for the survival of the organism, may live for a long period, having lost the ability to respond to stimuli to proliferate. In 1990 telomeres, linear arrays of DNA which "cap" the ends of all chromosomes and protect against degradation, were described (Harley 1991), leading to the proposal that this might be an internal timing mechanism for cellular aging. Telomeres are long repeat sequences of 5′ $(TTAGGG)_n$–3′; with each cell division 30–200 base pairs (between 5 and 30 repeats) are lost, resulting in telomere shortening. Once the telomeres become sufficiently short, they lose the ability to mask the end of the chromosome, preventing it from being recognized as a broken DNA molecule. This degradative loss of sequences may be repaired by "activated telomerase", a ribonucleoprotein reverse transcriptase, HERT, that uses an internal RNA as template to elongate the G-rich 3′ ends of the telomeres. This intervention, accomplished by infection with a recombinant retrovirus in fibroblast experiments, decreased functional p53 gene activity, and restored proliferation potential for a finite number of cycles, indicating plasticity in the inherent longevity instruction set, which allows for variation. P53 is a transcription factor and a mediator of the senescence response, whose absence allows cells to live significantly longer. It was proposed by Lane (1992) that p53 is a tumor-suppressor, the loss of which is essential to the immortalization of a cancer cell line. By this model, p53 acts as a "molecular policeman", monitoring the integrity of the genome at the G1 checkpoint, able to switch off replication of defective DNA allowing time for its repair, but able to trigger apoptosis in the event of failed repair. Thus p53 knockout mice have high spontaneous tumor rates.

Discovered in the past year (Shay and Wright 2001), limitations of the two-dimensional cell-culture environment to represent in vivo replication biology provided incorrect interpretation scenarios on which much of the 1990s conclusions, drawn from rodent studies, were based, requiring modifications of the "telomere story" recognizing species differences; a science in flux. The results of two new studies support the hypothesis that a cell's conventional defense arsenal, which includes DNA repair pathways and antioxidant enzymes, is adequate to protect against the accumulation of mutations and development of cancer during the short life span of small organisms – rodents. In contrast, larger and longer-lived species had to evolve replicative senescence to ensure an increased protection that their longevity necessitated: a biological clock, the telomere, and its defense-repair alter ego, the telomerase system. *Telomerase* is a ribonucleoprotein present to a small extent in normal cells, and in high concentration in cancer cell lines, which uses an internal RNA as template, providing a mechanism for telomere lengthening, or "stabilization" (de Lange and Jacks 1999). This occurs at the G-rich 3′ end of telomeres. Telomere length maintenance is a necessary, but not sufficient, accompaniment to cell-strain immortality. Telomerase loss is a sufficient, but not a necessary condition for cell senescence.

This combination of mechanisms results in markers of gene expression typical of younger cells, without the hallmarks of malignant transformation (Bodnar et al. 1998). The conditions for achieving "immortality" for a cell line have been extensively studied, being the conditions which produce a cancer, a growth-circumstance which allows a particular cell-line to escape the "normal" limit on growth.

The interactions of telomerase with cancer have received recent attention (de Lange 1999). The power of telomerase to render a cell line immortal raises the specter of inducing a malignancy; the distinction between the human telomerase reverse transcriptase (*HERT*), which it turns out does not induce cancers, and oncogene activation (as with Myc), which is highly associated with malignancy, has been demonstrated (Harley 1991). Thus the dream survives that HERT might become a fountain of youth, bypassing the *replicative senescence* (Campisi 1996, 1997), which applies to all normal cells via the telomere doubling limitation.

Capacity to respond to stress, defined in many and diverse ways, has been associated with telomere length in mice (Jazwinski 1996) and humans (Jurivich et al. 1997). Shay has recently suggested that short-lived rodents (whose telomeres are much longer than those in humans) use a conventional arsenal of DNA repair mechanisms and antioxidants over their 2-year life span, and do not require the replication-counting system, which has evolved in long-lived humans (Shay and Wright 2001). Interaction of the p53 protein, p53 gene activity, telomere function, and an additional list of related genes: ING1, p33^{ing1}, and p21, are introduced by Reddel (1998).

13.3.5
Membrane Function

The cell membrane of the living cell requires active metabolic support to sustain the membrane potential, which enables the separation of an intracellular environment markedly different in K^+ and Na^+ concentrations and pH, and other aspects of the internal milieu. Integrity of membrane function, in the construct of Linnane et al. (1998), is dependent on the mitochondrial genome as the principal source of the instruction set and the mechanics for this basically respiratory function, which produces Na-K ATP synthetase, the fundamental respiratory engine. This genome is a closed circular molecule of just 16,569 base pairs, which are of two types. One type codes for the 13 proteins which make up this engine, the other for the two ribosomal and 22 transfer RNA genes which mediate the synthetase process. There is little "spacer" DNA in the genome, most of which is therefore an active target for destructive oxidative events. Each cell contains over 1000 copies of the mtDNA molecule, providing a robust source of "institutional memory" support, but the mutations, which accumulate over long periods eventually deplete the efficiency of this basic respiratory function.

Linnane has championed the role of the coenzyme Q_{10}, which can substitute for pyruvate to provide redox support for cell respiration at the critical cell membrane sites, through a plasma membrane oxidoreductase. Both mitochondrial respiration and glycolysis decline with age, and are improved by intervention with Q_{10}. A series of experiments in animals, and biopsy-supported measurements in young and aged humans, support the observation that with advanced age, less than 5% of the mtDNA in skeletal muscle retained the full-length mtDNA instruction set, the residual being defective from deletion and mutation products. In animals, and in humans with certain premature aging syndromes, treatment with exogenous coenzyme Q_{10} reversed this respiratory defect, and was associated with clinical improvement.

Clinical observations in the era of the treatment of AIDS with AZT, an agent which causes degeneration of mitochondrial function, have noted myopathies as an occasional side effect in humans, and in experimental rodents, as a predictable cause of muscle strength decline, which was then reversed by provision of dietary Q_{10}. Observations on mitochondrial function in rat hepatocytes showed a similar age-ordered decline in membrane potential, which could be reversed by dietary supplements of L-carnitine and a hydrophilic antioxidant chemical, phenylbutylnitrone.

These observations of Linnane et al. (1998) and Ames et al. (1993; Ames and Shigenaga 1992) provide tissue-specific support for the broad hypothesis of Harman, developed over 30 years, that free radical production is a concomitant of aging, and that interventions which scavenge free radicals at critical time-phases of cell cycles may achieve morbidity-compression and life-extension, in a variety of cellular and animal models. These observations support the "rate of living" theory of aging enunciated in the early 20th century by Rubner (1908) and Pearl (1928). This theory is derived from observations that longevity correlates with the total calories burned according to a "life energy potential", in which a faster rate of metabolism would associate with a shorter life span, assuming a higher production of oxidative free radicals, and a correlation between free-radical production and mortality. Correlations are excellent within species, although by this system, birds are four times as efficient as large mammals.

The lower free radical production of bird vs. mammalian mitochondria strongly supports the free radical theory of aging. It also supports the concept that the inverse correlation observed between mitochondrial free radical production and maximum life span potential (MLSP) in mammals is mechanistically related to their MLSPs, instead of being a simple correlate of their metabolic rates. While there are many inter-species exceptions, the inverse relationship between mitochondrial free radical production and MLSP occurs in all the animal species studied to date, for mammals and birds. The rate of free radical production, not the rate of oxygen consumption, correlates with the MLSP and with the rate of aging. More than 90% of oxygen consumption is in mitochondria, in which a very high information density, lacking protective histones and polyamines, assures a much lower repair rate than in nuclear DNA. Approximately 1–3% of this oxygen metabolism is diverted to superoxide radical formation. Further, mtDNA lies in close vicinity to the main generator of free radical production, the inner mitochondrial membrane. The comparative survival rates of different species indeed correlate highly with the physical distance of mtDNA from this potent source of free radicals, suggesting that physical distance, and perhaps a sheath of absorbing fat molecules, may provide an insulating layer. The observation that high levels of DNA repair and low rates of free radical production near DNA sites result in low states of steady-state oxidative DNA damage has been repeated in many mammalian species.

Tissue specificities are critical in studying the systems listed above. Muscle systems, skeletal and cardiac, have been easily accessed, and have been well and profitably studied. Neural systems are also considered strongly at risk, and likely to be relevant to aging, although more difficult to access in humans. Hepatocytes by contrast are considered relatively immortal.

The investigative pathways in aging have had a rich contribution from the application of interventions, which have the double benefit of dissecting and exploring causalities, and of suggesting potential strategies for treatment/prevention.

13.3.6
Apoptosis

Apoptosis, or timed cell death, has the Greek root "falling off", as in autumnal leaves. This mechanism triggers cells to self-destruct as a natural, timely event, offsetting cell proliferation, to be distinguished from *necrosis*, which infers injury. It is a genetically programmed event initiated by a variety of internal or external stimuli, which activate a suicide pathway

arousing the proteins that destroy the cell's structural proteins and genetic material. Menstruation represents a benign example, and cancer a failure of apoptosis. Morphological states begin with a shriveling and pulling away from other cells, bubble-like formations on the cell surface, and condensation of nuclear chromatin, followed by dissolution and scavenging. The tumor-suppressor gene p53 is an essential intermediary. When groups of cells are involved, presumably responding to a "class action" signal, there results histological "coagulation necrosis". Examples of developmental involution are very many, as ontogeny repeats phylogeny in the embryonic period. The immediate mechanism is loss of control over cell volume, as the membrane loses the capacity to pump in K and extrude Na.

An earlier view of apoptosis considered it as the destruction of essential and irreplaceable cells. The current view sees rather the elimination of damaged and dysfunctional cells, which may then be replaced by cell proliferation, maintaining homeostasis, by turnover of fresh cells for old, regulating cell number. Control of cancer cells, an excellent example being the prostate, where hyperplasia and cancer are a virtual certainty with increasing age, occurs by apoptosis, the controlled removal of malignant cells targeted for destruction by immunologic means. Wang (1997) identifies four groups of apoptosis genes: signal generators, signal processors, activators, and substrates, for the apoptotic pathway.

Two genes, ced3 and ced4, identified in C. elegans, have mammalian homologs bax and bcl-2, in the much larger group of mammalian *caspases*, which engage in a ritual coordinated cascade (Hampton and Orrenius 1998), via cytochrome c, to prevent DNA replication and repair (Warner 1999). The checkpoint protein p53 is activated by DNA damage, and "decides" whether to engage in repair or apoptosis. This process is particularly key in regulating T lymphocytes, which have the potential to engage in auto-destructive activity in auto-immune states, but which are frequently involved in apoptosis regimes, in which the Fas gene is frequently a partner. The teleologies of apoptosis were summarized by Jacobson et al. (1997) as "sculpting" body parts, deleting unwanted structures, controlling cell numbers, and eliminating nonfunctional cells. In Alzheimer's, ALS, Parkinson's, and Huntington's diseases apoptosis is implicated as the active agency describing the pathological loss of needed cells, there being animal models for each in which genetic origins are well known.

The ischemia-reperfusion sequence in atherosclerotic myocardial disease is associated with apoptosis. Immunological aging has been related to changes in T lymphocytes, 95% of which are constructively destroyed by apoptosis when they arise in response to self-antigens. With aging the apoptotic pathways may be suppressed, allowing a deterioration in this self-preserving homeostatic cycle, with the (teleologically inappropriate) survival of damaged cells. Lymphopenia and decline in T cell function are seen in conjunction with instances in which a neural-immune lesion is identified; decline in circulating lymphocytes is frequently associated with aging. Warner (1987) posed four questions: Are there age-related changes in apoptosis? What molecular mechanisms are altered to produce these changes? Can detection and elimination of damaged cells be enhanced? Can apoptosis be delayed in non-renewable tissues? Higher incidences of both autoimmune diseases and cancer in the elderly are attributed to failures in the balance between forces favoring cell proliferation, and apoptotic removal (Ginaldi et al. 2000).

An understanding of the place of apoptosis, the death and removal of cells, in the regulation of the mechanisms of survival has a primary purpose unrelated to aging and the death of the organism. Indeed the most powerful defense against cancer cell lines is their recognition by the policing functions, some of them immunologic, and many depending on the p53 gene, with its great power to recognize and protect normal cells, marking for execution by apoptosis the cancer cells. Thus elevated from the "devil" to the "angel" list largely during the past decade, apoptosis and its regulation are developing a strong literature in the many individual-organ subspecialties of "Aging", which have not been individually addressed in this chapter.

13.3.7
Cellular Senescence

Cellular senescence is a mechanism different from apoptosis which provides a fail-safe mechanism to prevent the proliferation of cells at risk for malignant transformation. Replicative senescence is mediated via telomeres as previously described, providing a biological clock for a life-span signal (Itahana et al. 2001). P53 does not increase during *replicative* senescence, but is a critical regulator in *cellular* senescence. It is a protein whose activity as a tumor suppressor, via p53 acting as a check-point control at G1 for DNA damage, was discovered in 1991. "Successful" cancers produce a p53-inactivating binding protein, with the functional effect of inactivating its guardian role, which is normally expressed by its capacity to bind to specific sequences of DNA in the genome. This has the same effect as would a p53 point-mutation. Thus p53 acts as a "molecular policeman", monitoring the integrity of the genome. If DNA is damaged, p53 accumulates and switches off replication to allow time for repair. If repair fails, p53 may trigger cell suicide by apoptosis. With inactivation of p53, abnormal genetic material replicates, and cancer spreads.

13.4
Interventions

13.4.1
Caloric Restriction

Caloric restriction (CR), also titled "Dietary" and "Energy" restriction, has developed an impressive bibliography. CR is the only known way to delay the aging process, and to extend *the mean* and *maximal lifespans*. McCay et al. (1935) and Weindruch et al. (1979) consistently demonstrated prolongation of both by 30–50% (Luft 1994; Sohal et al. 1994) in rats and mice. Suggestive evidence has been developed in other short-lived mammalian species, and prospective studies in primates are in process since 1987 at the National Institute of Aging (NIA) in Baltimore, and at the University of Wisconsin. Regimens optimized at 60% of the calories in control groups (Masoro et al. 1991), initiated either after weaning, or at mid-life, produced body weights plateauing at levels 60–70% of control weights, with body compositions showing slightly lower fat. Health-neutral effects include decreases in plasma cholesterol, triglycerides, and adipocyte responsiveness to hormones, and in Fischer rats, rates of nephropathy, cardiomyopathy, and neoplasms were diminished. Average blood glucose in successful animal colonies is 15% lower than in controls. Body temperatures are lower. Body weight is inversely, and brain weight directly, correlated with length of life. Less salutary responses include lower fertility, decreased CNS neurotransmitter functions, lens crystallins, and some humoral immune functions, with enhancement of cell-mediated immunity

(Youngman et al. 1992). Although lower body fats are constant findings, fat levels did not correlate with length of life. Early studies (Rubner 1908; Pearl 1928; Sacher 1968) relate an associated decreased metabolic rate as causal, although later studies clearly dissociate MR. Sapolsky et al. (1986) hypothesized an intermediary "glucocorticoid cascade hypothesis" in which a loss of hippocampal neurons is associated with hyperadrenocorticism. Masoro et al. (1991) did not confirm a hyperadrenocortical state, but did note higher cortisol levels. Interactions between insulin, the insulin/growth hormone interaction, and an entity called by Parr "reserve capacity" connects with CR, in that lower caloric exposure reduces the level of insulin exposure (Parr 1997).

The *Proteasome* is a highly regulated intracellular enzymatic proteolytic system responsible for the repair of degraded proteins, usually when they are damaged by free radical activities. These are located in the cytosol and nuclei of mammalian cells, being rich in the endoplasmic reticulum and plasma membranes (Coux et al. 1996; Rivett 1998). They possess three separable enzymatic activities, one trypsin-like for the carboxyl side of basic amino acids, one chymotrypsin-like for hydrophobic AA, and a peptidylglutamyl-peptide-hydrolase for acidic AA (Merker et al. 2001). Merker discusses the regulation of the protease system. A complementary glycation mechanism (Cerami 1985) states that excessive glycation of proteins and nucleic acids is detrimental to longevity because of enhanced cross-linking in critical organs, and decreased enzymatic activity. Free radical production is decreased, and lipid peroxidation in hepatocytes is much lower with CR. These data incorporate the *free radical theory* mechanics proposed by many presented above. Pieri (1991) studied cell membranes in CR colonies: in splenic lymphocytes, and in the liver plasma membrane, higher fluidity and lower viscosity were associated with lower levels of ROS. CR data provide auxiliary evidence from the much older lipofuscin observations and data (Hammer and Braum 1988; Yin 1996), which correlate the rate of development of these "aging pigments" with CR interventions. Thus CR acts at many levels: reduced tissue levels of ROS, increased scavenger activity, protection of membranes, reduced levels of lipid peroxidation and higher removal of byproducts (Kristal et al. 1994, 1998). Mechanisms may include not only reduction of DNA damage, but also production of favorable growth factors and stress proteins (Mattson et al. 2001). While earlier students of CR suggested that the "rate of life" theory might interact with their observations, the (modest) positive effects of exercise on life-spans opposes this explanation (Ramsey et al. 2001).

Extension of experimental data from rodents, with ~2-year life spans, to longer-lived species has been an essential, if not holy, grail for students of aging of the late 20th century. From 1987 several groups at the NIA, and later studies from the primate colony in Wisconsin, have studied squirrel (Ingram et al. 1990), rhesus (Kemnitz et al. 1993), and cynomolgus (Cefalu et al. 1997) monkeys (average lifespan 15–30 years); CR to 70% of control was better tolerated than the slightly more effective reduction to 60%, achieving a 20% reduction in body weight. Studies in larger animals include pigs, sheep, and cattle. These studies conclude, inter alia that the responses to *starvation* are very different from *programmed undernutrition* at the –30% to –40% caloric reductions projected from rodent studies.

Human studies for as long as 1 year – 16 were summarized in 2001 – have been undertaken, with hypocaloric diets given for multiple week experiments, primarily to obese subjects. Five studies in humans, for relatively short periods, mostly begun late in life, serve at most as guides for planning for more definitive studies, for which early experience has generated great enthusiasm. The biosphere experiment had four subjects on hypocaloric diets for over 2 years (Ramsey et al. 2001). Experiments of nature, such as an Okinawan population having natural low-caloric intakes, have been analyzed. The variables putatively associated with life span, termed "aging biomarkers", include blood glucose and insulin/glucose ratios, transaminase, alkaline phosphatase, albumen, globulin, BUN/creatinine ratio, DHEA, and lipids. Some theoreticians are inclined to accept slope/rate data acquired in vivo, while others have insisted that full life-span observations will be required (Roth et al. 1999). Other mechanisms studied include an enhanced capacity to repair UV-induced DNA damage (Weraarchakul et al. 1989), and enhanced capacity to express a gene responsible for inducing the heat-shock protein Hsp70, itself a promoter of survival from ROS. Development of guidelines/maps, both for outcomes/measurement strategies, and for socio-legal envelopes, which might protect the investigators and the subjects, will be needed.

Mitochondrial proton leak describes a pro-oxidant process at the mitochondrial membrane by which a dissipation of proton-motive force and uncoupling of substrate oxidation from ADP phosphorylation occurs, dependent on membrane potential, being exponentially higher with increased membrane potential. Proton leak is a major contributor to resting O_2 consumption (~25%), and it continues to contribute at higher respiratory rates. This is a major factor in determining efficiency of ATP synthesis, and in overall rates of energy expenditure. CR could decrease proton leak by influencing membrane lipid composition towards the linoleic, and away from the long chain (Brookes et al. 1998). A strong influence of CR on levels of uncoupling proteins, also highly correlated with proton leak, supports this chain of causality as a potential mechanistic link via polyunsaturates between CR and aging (Fleury 1997; Brookes et al. 1998). The proton-leak explorations have gone much further: analyses across species based on body size, ectothermy and thyroid hormone levels, result in an integrated mechanistic theory with seven steps (Ramsey et al. 2001). CR induces changes in the mitochondrial membranes favoring linoleic acid, while reducing long chain polyunsaturates and uncoupling proteins. Proton leak diminishes, decreasing O_2 consumption. Further membrane changes decrease membrane surface areas and permeability, decrease electron transport chain activity, reducing ROS formation and resulting damage to proteins lipids and DNA, thus "improving" the cellular redox state, and increasing life span. It is corollary to this proposed mechanism that initiation of CR at any age could be effective, a hypothesis supported by other observations of late-induction of CR (Aspnes et al. 1997), and by finding supportive CR-induced genetic modifications in skeletal muscle (Lee et al. 1999), although another study in rats suggested that late-induction of CR did not match the benefits of early diet change (Lass et al. 1998), and yet another in rhesus monkeys showed beneficial results on gene expression with middle-aged induction of CR (Kayo et al. 2001).

13.5
Molecular Nuclear Medicine in Studies of Aging

13.5.1
Introduction

Both experimental and clinical studies have used radionuclide methods to evaluate the processes of aging: the effects of oxidative radical (ROS) damage to DNA, and telomere shortening. Tracer imaging and non-imaging studies have monitored the rates of progression, both by age-group comparisons, and by longitudinal studies of individuals. The gradual erosion of membrane integrity, with its vital separation of intracellular and extracellular environments, may be recognized by change in whole body, or in discrete organ contents of inorganic elements, usually K+ and Na+, representing intracellular and extracellular domains. Measurements have been made by isotope dilution, neutron activation or counting of natural ^{40}K (see Chap. 11). Many standard and research-level nuclear methods record the multiple pathological processes which typically increase with age, which may to various extents be attributable to aging per se, and yet may also be consequent to adverse environmental circumstances. While comprehensive coverage of such tests, and listing of the variety of the physiologies they address, is not appropriate here, a few outstanding examples are illustrative: arterial atherosclerosis, osteoporosis, osteoarthritis, and Alzheimer's disease have had major contributions from molecular-oriented tracer studies, and these contributions will grow. These common "diseases" and "conditions" – let us recognize the challenge of separating these categories when we can – all have profiles of genetic predisposition. Frequently the predispositions – diabetes is an example – show clinical and pathological manifestations early in life, raising perhaps the ultimate question in molecular/genomic medicine, "might we insert a genetic repair?" The genomic equivalents for the phenotypes due to aging are turning up, with new additions monthly, perhaps weekly.

13.5.2
Arterial Atherosclerosis

Pathological and clinical evidence (including the progeroid Werner syndrome) have implicated innate aging processes as etiologic in atherosclerosis. The connection has recently been given biochemical support by the finding of telomere shortening in leukocyte DNA of patients with severe coronary artery disease compared to age-matched controls (Samani et al. 2001). In regard to genotypes, there are polymorphisms in both nuclear (Castro et al. 2000) and mitochondrial (Matsunaga et al. 2001) DNA, which have been associated with coronary stenosis or medial-intimal thickening of carotid arteries, respectively. Chronic systemic inflammation (related to autoimmunity, a genetic condition) is accompanied by the presence of activated T lymphocytes and macrophages in atherosclerotic plaques (Brod 2000) and this situation appears to be related to aging (Franceschi et al. 2000; Krause and Clark 2001). Monocytes and macrophages are attracted to arterial lesions by monocyte chemo-attractant peptide-1 (MCP-1). Experimental arterial lesions (in rabbit iliac artery or aorta) show pronounced accumulation of ^{125}I-MCP-1 by autoradiography (Ohtsuki et al. 2001). MCP-1 labeled otherwise (e.g., ^{123}I or ^{131}I) is suggested as a useful tracer for imaging. Abnormal lipoprotein (LP) levels in plasma have long been considered as risk factors for atherosclerosis and are found increasingly with advance in age. Studies with radio-iodinated lipoproteins (Counsell et al. 1991; Virgolini et al. 1991; Lupattelli et al. 1999), or peptide fragments of LP (Hardoff et al. 1993) have shown some promise for imaging of peripheral arterial lesions in experimental animals, and in more limited studies in humans. Strauss has aptly commented (Ohtsuki et al. 2001) that, while peripheral sites of plaque formation may well be imaged externally, coronary sites are so small that detection (with any of the candidate agents) may require an intravascular radionuclide detection device. Cardiac metabolism of positron-labeled substrates has been compared between young and old human subjects: with age, glucose replaces fatty acids as fuel under fasting conditions, but the aged may not *increase* the use of glucose (the "emergency" fuel) in response to catecholamine (dobutamine) stimulus as do young persons (Herrero et al. 2001). This may indicate a significant deficiency under stressful conditions.

13.5.3
Osteoporosis

Osteoporosis has succeeded hypertension as "the silent killer" in the 21st century. Advancing age of survival well beyond the menopausal loss of estrogen stimulus to bone formation has unearthed and defined a "disease" process (an earlier teleology-dominated ethos had labeled post-menopausal women as dispensable to the survival of the species). To separate this condition from "normal aging", the distinction is greatly influenced (as it was with hypertension 75 years ago) by the fact that effective treatments are available. While various environmental influences (diet, drugs, exercise) can modulate progression, genetic and ethnic differences appear to predominate (Kobyliansky et al. 2000). Polymorphisms in several candidate genes have

been correlated to familial predisposition, to deficiencies of bone mineral density (BMD), and to impaired calcium absorption (see Chap. 11). There is reason to differentiate *types* of osteoporosis (structural integrities of hip, spine, and wrist have been differentially defined) by their effects early or later in life, degree of dependence on collagen metabolism and structure, or by degree of involvement of calcium absorption, all of which can depend on genetic make-up (Braga et al. 2000; see Chap. 11). At the research and discovery frontiers, nuclear techniques are available to make these phenotypic distinctions and genetic correlations. At the clinical frontier, the single and dual photon absorptiometry (DPA) methods of the 1980s have been succeeded by dual X-ray absorptiometry (DXA) as a two-dimensional approach to estimating bone mineral density.

Whose turf? Some nuclear medicine facilities which began with DPA have carried on the clinical imperative to operate DXA instruments. Recently DXA has been applied to making images of the lumbar and lower thoracic spine from different directions (called "instant vertebral assessment" or IVA, a somewhat primitive effort to make the measurement three-dimensional), with results showing greater sensitivity for osteoporosis than BMD (Boughton 2002). A rarely available and expensive technique, delayed-gamma neutron activation (DGNA), measures total body calcium and was early and successfully applied to recognizing osteoporosis (Aloia et al. 1994; see Chap. 11), and as a criterion method to calibrate DXA instruments. Another useful nuclear tracer technique has been evaluation of intestinal absorption of Ca^{++} with one or more calcium isotopes (radioactive or stable), one being given orally and the other intravenously (see Chaps. 11 and 27). A technique still in experimental stage is that of gamma resonance absorptiometry (GRA), which requires a small proton accelerator and can measure concentration of Ca^{++} (and other elements) in small regions and further provide a 3-D image (Wielopolski et al. 2000). In the future, besides detecting osteoporosis and monitoring effects of drugs and other established interventions, the techniques mentioned above may be used to judge the efficacy of mesenchymal stem cells (MSC), which have already been found to augment healing of bone fractures (Justesen et al. 2001). Animal studies have shown that MSC given intravenously target bone. Human studies show that the number and differentiation potential of MSC are unchanged with age and osteoporosis.

13.5.4
Osteoarthritis

There is a strong association between age and the incidence of osteoarthritis (OA). The etiology is often a complex mixture of environmental influences (e.g., trauma, diet, obesity, muscle weakness) and of "innate" aging processes. Lately there has been an increasing emphasis on the role of inflammation. Evidence for this includes the association of OA with increases of particular cytokines [e.g., interleukins (IL)] in chondrocytes or matrix of cartilage (Johnson et al. 2001a; Mazzetti et al. 2001; Pincus 2001; Gouze et al. 2002). The increase of cytokines is donor age-dependent. Transforming growth factor-beta and IL-1 beta, acting through transglutaminases (and nitric oxide), appear to promote calcification in matrix (Johnson et al. 2001a), which has subsequent damaging effects, including stimulation of more cytokines and of genes causing expression of proteolytic enzymes (Misra 2000). In support of "innate aging", as opposed to the "disease" model in pathogenesis of OA, there are correlations between age and replicative senescence in chondrocytes from patients with OA; the evidence includes increase of senescence-associated beta-galactosidase activity, decrease of mitotic activity (shown with ^{3}H-thymidine), and shortening of telomere length (Martin and Buckwalter 2001). Study of the effects of oxidizing agents producing ROS in cultured chondrocytes (shown by leakage from matrix of ^{35}S-labeled components), and the inhibiting effect of anti-oxidants, points to an important mechanism of matrix degradation and the possible therapeutic value of anti-oxidants for OA (Tiku et al. 1999). However, reactive nitrogen radicals [NO(-)(2)] appear to be more detrimental than ROS in OA, whereas the opposite is found with rheumatoid arthritis (RA) (Mazzetti et al. 2001). Although OA is often viewed as a disease due to general aging abetted by environmental insults, there is growing knowledge of associated polymorphisms of particular genes (Ikeda et al. 2001; Ingvarsson et al. 2001) and of distinctive familial incidences (Ingvarsson et al. 2001; see Chap. 11). In current nuclear medicine practice the use of planar or SPECT imaging with conventional bone radiopharmaceuticals can detect OA early and lead to timely therapy (De Maeseneer et al. 1999; Sarikaya et al. 2001; see Chap. 25). New directions are suggested by several pathways. ^{35}S-sulfate has been applied to show, by autoradiography of cultured human chondrocytes, variations in metabolic activity in different layers of joint surface cartilage de-

pending on stage of OA (Verschure et al. 1996; see Chap. 11). Demonstration of uptake in abnormal cartilage in rabbits with ^{75}Se-labeled methylamino-ethyl selenide diiodide indicates a potential new imaging agent with strong avidity for cartilage due to its binding to chondroitin sulfate (Yu et al. 1999). Future possibilities for gene therapy (of either OA or RA) are suggested by the effect of an adenovirus-transfected gene for a cytokine repressor to reduce drastically the severity of arthritis when injected periarticularly into mice with induced arthritis (Shouda et al. 2001). Associated incorporation of a reporter gene could use nuclear imaging to verify and monitor the therapeutic effect.

13.5.5
Alzheimer's Disease

In view of the rapidly increasing incidence of Alzheimer disease (AD) in the very, very old (some have proposed up to 50% by age 90) it seems highly likely that fundamental aging processes are an important component of the condition. Oxidative damage from production of reactive oxygen species has been linked to the neurotoxic effects of beta amyloid precursor protein (betaAPP), well-recognized as an intermediate in formation of amyloid plaques in the brain in AD, when APP is chelated to metals, particularly Cu (Multhaup et al. 1997). Beta-amyloid (Abeta), aided by Cu, itself appears able to generate free radicals (Multhaup et al. 1997; Kontush 2001). Other findings bolster the evidence incriminating ROS: there is a reduction of an endogenous NO synthase inhibitor, asymmetrical dimethyl arginine (ADMA), in the CSF of patients with AD and (less so) in aging normal controls (Abe et al. 2001). Paradoxically, Abeta has a putative role as an anti-oxidant for brain lipoprotein, but with increased production (perhaps an overexpression) of betaAPP, becomes a pro-oxidant (Kontush 2001). Another theory (different theories and research directions abound in this field) involves activation of microglia (special macrophages of the CNS) to secrete neurotoxic compounds, including free radicals, cytokines and nitric oxide (Multhaup et al. 1997; Dukic-Stefanovic et al. 2001). In aging brain advanced glycation endproducts (AGEs) accumulate; AGEs cause cross-linking of cytoskeletal proteins, exert chronic oxidative stress on neurons, and activate glial cells. AGEs are attenuated by certain inhibitors, which have been suggested as treatment for AD (Dukic-Stefanovic et al. 2001). Cytokines promote inflammation, which is invoked to account, at least in a secondary role, for AD, as it is for other diseases of aging (Krause and Clark 2001; see above). There are claims that polymorphic forms of cytokines are particularly responsible (Mrak and Griffin 2000; Kolsch et al. 2001). Primary brain aging is typically found to be characterized by activation of proteolytic systems (Nixon 2000). Still another theory would put soluble Abeta in the position of attempting rescue of a fuel-deprived cell by inducing changes in the glucose-fatty acid cycle (Heininger 2000). Another research study indicates that brain cells in areas involved in AD (hippocampus, cerebral cortex), which are normally quiescent, appear in AD to "emerge from senescence" and undergo cell cycling, at which time they are presumably vulnerable to degeneration and, perhaps, apoptosis (Raina et al. 2001). The phenomenon of molecular misreading of DNA into mRNA (to cause "+1 proteins") is a process found generally in various tissues in the elderly, both in AD and normal subjects, but there has been specific identification of involvement of certain genes, i.e., for betaAPP and for ubiquitin-B, which are linked to AD (van Leeuwen et al. 2000). These +1 proteins accumulate in brain tissues considered to show the neuropathological hallmarks of AD.

Compared to the diverse evidence, beliefs and opinions as to the importance of various phenomena of intrinsic neurological aging in the causation of AD, there is relatively good agreement about at least some of the genes, mutations of which account for different ages of onset, and probably other distinctions of types, of AD. Particular polymorphisms of the presenilin-1 gene (associated with secretase enzymes converting betaAPP to Abeta), the similar presenilin-2 gene, and the betaAPP gene (located on chromosomes 14, 1 and 21, respectively) characterize familial, early-onset AD. Familial or sporadic late-onset AD is commonly found to correlate with homozygosity (less so for heterozygosity) of the varepsilon 4 allele of the gene for apolipoprotein E (ApoE4) (Farlow 1998; Rogaev 1999; Boss 2000; Kardaun et al. 2000; Morris 2000; Molero et al. 2001). There is a lesser inverse association of AD with ApoE2 (Kardaun et al. 2000). ApoE4 is a risk factor rather than a deterministic gene – i.e., some elderly patients with AD do not have the ApoE4 allele, while the latter may be present in very old subjects who do not get AD. There is incomplete identical twin concurrence. There may be different sub-groups of AD which begin >65 years: late-onset and late-late onset (Wu et al. 1998). Further genetic factors for AD are suspected to be due to polymorphisms in IL-1 among cytokines (see

above), and are likely in the ROS-prone, mutation-prone "second genome" of the mitochondria (Graeber et al. 1998). Many researchers suspect a greater heterogeneity of phenotypes of AD than has yet been fully recognized by clinical means (Rogaev 1999; Khachaturian 2000; Morris 2000). Mild cognitive impairment (MCI) is regarded as a possible prelude to AD. Neuropsychological and genetic assessment (prevalence of ApoE4) in older persons with subjective memory complaints reveals a significant separation of those with only age-associated memory impairment (AAMI) from cases identified as MCI, with a significantly higher incidence of ApoE4 in MCI than in AAMI (Bartres-Faz et al. 2001). The various considerations of heterogeneity and of pre-clinical involvement should give encouragement to nuclear medicine to extend its participation in diagnosing and monitoring AD and in correlating phenotypes with developing knowledge of genotypes.

For the past 20 years nuclear medicine has with considerable success used first planar, then SPECT, and eventually PET to establish a role for nuclear imaging in search of a biological marker for AD. Some of these studies have also compared nuclear imaging findings with the occurrence of known predisposing genes. An FDG/PET study of normal, non-demented persons of various ages, with no memory impairment, showed that even a single copy of the ApoE4 allele had an association with hypometabolism in the classical AD brain regions (temporoparietal, posterior cingulate), and a combination of the two markers (ApoE4 allele and hypometabolism by FDG) predicted cognitive decline (Small et al. 2000). Others report on an association of ApoE4 and regional hypometabolism using FDG/PET (Mielke et al. 1998; Fox and Rossor 1999) or hypoperfusion using ^{133}Xe/SPECT (Tanaka et al. 1998). Using ^{99m}Tc-HMPAO/SPECT to study members of a pedigree for the presenilin-1 (PS-1) gene, it was found that, not only did the early-onset AD subjects show decreased perfusion, but also those who had the PS-1 mutation but were asymptomatic (Johnson et al. 2001a). A preliminary study of asymptomatic carriers of the mutated betaAPP gene suggests that reduction of regional cerebral blood flow (rCBF) is a superior predictor of AD (Julin et al. 1998). Different clinical sub-types within the diagnosis of AD have been shown to be distinctive also in brain regions of hypoperfusion (Galton et al. 2000; Cappa et al. 2001) or hypometabolism (Schroder et al. 2001). Both ^{99m}Tc-HMPAO/SPECT (Charpentier et al. 2000) and FDG/PET (Hoffman et al. 2000; Nagata et al. 2000) have been successfully applied to distinguish AD from other dementias. One study of rCBF (with ^{133}Xe) and clinical follow-up (Nobili et al. 2001) indicated definite prognostic value for AD patients, although in another group of subjects (with aging-associated cognitive decline but not yet AD at the time of study with ^{99m}Tc-HMPAO/SPECT), outcome could not be predicted (McKelvey et al. 1999).

Cerebral cholinergic deficits have been described both in AD, and as consequent to normal aging. Memory and attention are cognitive functions particularly dependent on the cholinergic system. Various radio-labeled (^{123}I, ^{11}C) ligands for muscarinic acetylcholine (ACH) receptors have been imaged in vivo with animal models or humans with or without AD (e.g., Yoshida et al. 1998; Herholz et al. 2000; Nobuhara et al. 2000; Tanaka et al. 2001; Zubieta et al. 2001). Areas of deficient concentration of ACH agents in AD appear to be more global than the more specific areas defined by hypoperfusion or hypometabolism. Studies making direct comparisons do not so far show more sensitivity of ACH agents for recognizing or characterizing AD than does FDG/PET. There has been a study of histamine receptors in brains of AD patients and controls by in vivo imaging of a receptor ligand (^{11}C-doxepin) with PET (Higuchi et al. 2000). Findings include decreased binding particularly in frontal and temporal areas, good correlation to severity of AD by neuropsychological score, and a greater rate of binding decrease with progression of disease than of decrease in rCBF. Search for biochemical markers of AD in serum has not generally been rewarding, but it is reported (Gottfries et al. 1998) that high levels of serum homocysteine occur in 39% of patients with mild cognitive impairment. This could suggest that some dynamic changes of regional intracerebral anomaly of amino acid metabolism, measurable by in vitro isotope tracer studies (possibly of the involved one-carbon transfer), could be linked to AD or other dementias.

The demonstrated potential of non-invasive nuclear tracer-imaging methods to study regional cerebral metabolic pathways, build-up of abnormal metabolites, and deposition of toxic materials, has utilized blood flow indicators (functional MRI, ^{133}Xe washout), location of the binding of chemical indicators (examples – Tc tracers), non-specific location of regional glucose metabolism (FDG-PET), localization of key receptor ligands, and, most exquisitely with regards future potential, the tracing of specific high-order metabolic pathways by which an Alzheimer's defect may be identified. In each of these methodologic pathways, we shall find greatest benefit when the same methods may be applied to show the reversal of an abnormal process associated with a targeted intervention. In the case of studies of the primate or human brain, more than in any other example in the available biological models for the study of aging, the definitive studies will be in large-brained, highly-evolved humans, in whom non-invasive, non-destructive methods may be applied to the study of interventions, followed by cognitive outcomes-testing of subtle functions unlikely to be achieved in animal models.

13.6
References

Abe T, Tohgi H, Murata T et al (2001) Reduction in asymmetrical dimethylarginine, an endogenous nitric oxide synthase inhibitor, in the cerebrospinal fluid during aging and in patients with Alzheimer's disease. Neurosci Lett 312:177–179

Aloia JF, Vaswani A, Yeh JK et al (1994) Calcium supplementation with and without hormone replacement therapy to prevent post-menopausal bone loss. Ann Intern Med 120:97–103

Ames BN, Shigenaga MK (1992) Oxidants are a major contributor to aging. Ann NY Acad Sci 663:85–96

Ames BN, Shigenaga MK, Hagan TM (1993) Oxidants, antioxidants, and the degenerative diseases of aging. Proc Natl Acad Sci USA 90:7915–7922

Aspnes LE, Lee CM, Weindruch R et al (1997) Caloric restriction reduces fiber loss and mitochondrial abnormalities in aged rat muscle. FASEB II:573–581

Bartres-Faz D, Junque C, Lopez-Alomar A et al (2001) Neuropsychological and genetic differences between age-associated impairment and mild cognitive impairment entities. J Am Geriatr Soc 49:985–990

Beckman KB, Ames BN (1998) The free radical theory of aging matures. Physiol Rev 78:547–581

Bodnar AG, Ouellette M, Frolkis M et al (1998) Extension of life-span by introduction of telomerase into normal human cells. Science 279:349–352

Boss MA (2000) Diagnostic approaches to Alzheimer's disease. Biochim Biophys Acta 1502:188–200

Boughton B (2002) Instant vertebral assessment. Radiol Today 3:9–11

Braga V, Mottes M, Mirandola S et al (2000) Association of CTR and COLIA1 alleles with BMD values in peri- and postmenopausal women. Calcif Tissue Int 67:361–366

Brod SA (2000) Unregulated inflammation shortens human functional longevity. Inflamm Res 49:561–570

Brookes PS, Buckingham JA, Tenreiro AM et al (1998) The proton permeability of the inner membrane of liver mitochondria from ectothermic and endothermic vertebrates and from obese rats: correlations with standard metabolic rate and phospholipid fatty acid composition. Comp Biochem Physiol 119B:325–334

Calder PC, Newsholme EA (1993) Influence of antioxidant vitamins on fatty acid inhibition of lymphocyte proliferation. Biochem Mol Biol Int 29:175–183

Campisi J (1996) Aging cancer: the double edged sword of replicative senescence. Cell 84:497–500

Campisi J (1997) Replicative senescence: an old lives' tale? J Am Geriatr Soc 45:482–488

Cappa A, Calcagni ML, Villa G et al (2001) Brain perfusion abnormalities in Alzheimer's disease: comparison between patients with focal temporal lobe dysfunction and patients with diffuse cognitive impairment. J Neurol Neurosurg Psychiatry 70:22–27

Castro E, Edland SD, Lee L et al (2000) Polymorphisms at the Werner locus: II. 1074Leu/Phe, 1367Cys/Arg, longevity, and atherosclerosis. Am J Med Genet 95:374–380

Cefalu WT, Wagner JO, Wang ZQ et al (1997) A study of caloric restriction and aging in cynomolgus monkeys. J Gerontol Biol Sci 52:B10–B19

Cerami A (1985) Hypothesis. Glucose as a mediator of aging. J Am Geriatr Soc 33:626–634

Cerami A (2001) Pharmaceutical intervention of advanced glycation end products discussion 212–216, 217–220 Review 2001. Novartis Found Symp 235:202–212

Charpentier P, Lavenu I Defebvre L et al (2000) Alzheimer's disease and frontotemporal dementia are differentiated by discriminant analysis applied to (99m)Tc HMPAO SPECT data. J Neurol Neurosurg Psychiatry 69:661–663

Chase GD, Rabinowitz JL (1959) Radioisotopes in the Biological Sciences. In: Principles of radioisotope methodology, chap 11. Burgess, Minneapolis, pp 305–328

Cottrell DA, Blakely EL, Johnson MA et al (2001) Mitochondrial DNA mutations in disease and aging. Novartis Foundation Symposium. Wiley, Chichester, pp 234–246

Counsell RE, Schwendner SW, DeForge LE et al (1991) Lipoproteins as carriers for organ-imaging radiopharmaceuticals. Target Diagn Ther 5:251–314

Coux O, Tanaka K, Goldberg AL (1996) Structure and functions of the 20S and 26S proteasomes. Annu Rev Biochem 65:801–847

Cutler RG (1974) Redundancy of information content in the genome of mammalian species as a protective mechanism determining aging rate. Mech Ageing Dev 2:381

Davies KJ (1986) Intracellular proteolytic systems may function as secondary antioxidant defenses: an hypothesis. J Free Rad Biol Med 2:155–173

Davies KJ (2001) Degradation of oxidized proteins by the 20S proteasome. Biochimie 83:301–310

de Lange T (1999) Unlimited mileage from telomerase? Science 283:947–949

de Lange T, Jacks T (1999) For better or worse? Telomerase inhibition and cancer. Cell 98:273–275

De Maeseneer M, Lenchik L, Everaert H et al (1999) Evaluation of lower back pain with bone scintigraphy and SPECT. Radiographics 19:901–912

Drew B, Leeuwenberg C (2002) Aging and the role of the reactive nitrogen species. Ann NY Acad Sci 959:66–81

Dukic-Stefanovic S, Schinzel R, Riederer P et al (2001) AGES in brain ageing: AGE-inhibitors as neuroprotective and anti-dementia drugs? Biogerontology 2:19–34

Effros RB (2001) Ageing and the immune system. Novartis Found Symp. Wiley, New York, pp 130–139

Farlow MR (1998) Etiology and pathogenesis of Alzheimer's disease. Am J Health Syst Pharm 55 [Suppl 2]:S5–S10

Finkel T, Holbrook NJ (2000) Oxidants, oxidative stress and the biology of ageing. Nature 408:239–247

Fleury C (1997) Uncoupling protein-2: a novel gene linked to obesity and hyperinsulinemia. Nat Genet 15:269–272

Fox NC, Rossor MN (1999) Diagnosis of early Alzheimer's disease. Rev Neurol (Paris) 155 [Suppl 4]:S33–S37

Fraga CG, Shigenaga MK, Park JW et al (1990) Oxidative damage to DNA during aging: 8-hydroxy-2deoxyguanosine in rat organ DNA and urine. Proc Natl Acad Sci USA 87:4533–4537

Franceschi C, Bonafe M, Valensin S et al (2000) Inflamm-aging. An evolutionary perspective on immunosenescence. Ann NY Acad Sci 908:244–254

Gaczynska M, Osmulski PA, Ward WF (2001) Caretaker or undertaker? The role of the proteasome in aging. Mech Ageing Dev 122:235–254

Fries JF (1980) Aging, natural death, and the compression of morbidity. N Engl J Med 303:130–135

Galton CJ, Patterson K, Xuereb JH et al (2000) Atypical and typical presentations of Alzheimer's disease: a clinical, neuropsychological, neuroimaging and pathological study of 13 cases. Brain 123:484–498

Ginaldi L, DeMartinis M, D'Ostilio A et al (2000) Cell proliferation and apoptosis in the immune system in the elderly. Immunol Res 21:31–38

Girotti AW (1985) Mechanisms of lipid peroxidation. J Free Radic Biol Med 1:87–95

Gottfries CG, Lehmann W, Regland B (1998) Early diagnosis of cognitive impairment in the elderly with the focus on Alzheimer's disease. J Neural Transm 105:773–786

Gouze JN, Bianchi A, Becuwe P et al (2002) Glucosamine modulates IL-1-induced activation of rat chondrocytes at a receptor level, and by inhibiting the NF-kappa B pathway. FEBS Lett 510:166–170

Graeber MB, Grasbon-Frodl E, Eitzen UV et al (1998) Neurodegeneration and aging: role of the second genome. J Neurosci Res 52:1–6

Greenstock CL (1993) Radiation and aging: free radical damage, biological response, and possible antioxidant intervention. Med Hypotheses 41:473–482

Guarente L (2000) Sir2 links chromatin silencing, metabolism, and aging. Genes Dev 14:1021–1026

Gutteridge JMC, Halliwell B (1988) The antioxidant proteins of extracellular fluids. In: Chow CK (ed) Cellular antioxidant defense mechanisms. CRC Press, Boca Raton, pp 1–23

Hammer C, Braum E (1988) Quantification of age pigments (lipofuscin). Comp Biochem Physiol B 90:7–17

Hampton MB, Orrenius S (1998) Redox regulation of the caspases during apoptosis. Ann NY Acad Sci 854:328–335

Hardoff R, Braegelmann F, Zanzonico P et al (1993) External imaging of atherosclerosis in rabbits using an 123I-labeled synthetic peptide fragment. J Clin Pharmacol 33:1039–1047

Harley CB (1991) Telomere loss: mitotic clock or genetic time-bomb? Mutat Res 256:271–282

Harman D (1956) Aging: a theory based on free radical and radiation chemistry. J Gerontol 11:298–300

Harman D (1972) The biologic clock: the mitochondria? J Am Geriatr Soc 20:145–147

Harman D (1998) Extending functional life span. Exp Gerontol 33:95–112

Hayflick L (1998) How and why we age. Exp Gerontol 33:639–653

Hayflick L, Moorehead PS (1961) The serial culture of human dip cell strains. Exp Cell Res 25:585–621

Heffner JE, Repine JE (1989) Pulmonary strategies of antioxidant defense. Am Rev Respir Dis 140:531–544

Heininger K (2000) A unifying hypothesis of Alzheimer's disease. IV. Causation and sequence of events. Rev Neurosci 11:213–328

Herholz K, Bauer B, Wienhard K et al (2000) In-vivo measurements of regional acetylcholine esterase activity in degenerative dementia: comparison with blood flow and glucose metabolism. J Neural Transm 107:1457–1468

Herrero P, Vedala G, Kowalski R et al (2001) Impact of aging on dobutamide-induced changes in myocardial substrate utilization. J Nucl Med 5 [Suppl]:52P

Higuchi M, Yanai K, Okamura N et al (2000) Histamine H(1) receptors in patients with Alzheimer's disease assessed by positron emission tomography. Neuroscience 99:721–729

Hoffman JM, Welsh-Bohmer KA, Hanson M et al (2000) FDG PET imaging in patients with pathologically verified dementia. J Nucl Med 41:1920–1928

Ikebe S, Tanaka M, Ohno K et al (1990) Increase of deleted mitochondrial DNA in the striatum in Parkinson's disease and senescence. Biochem Biophys Res Commun 170:1044–1048

Ikeda T, Mabuchi A, Fukuda A et al (2001) Identification of sequence polymorphisms in two sulfation-related genes, PAPSS2 and SLC26A2, and an association analysis with knee osteoarthritis. J Hum Genet 46:538–543

Ingram DK, Cutler RG, Weindruch R et al (1990) Dietary restriction and aging: the initiation of a primate study. J Gerontol Biol Sci 45:B148–B163

Ingvarsson T, Stefansson SE, Gulcher JR et al (2001) A large Icelandic family with early osteoarthritis of the hip associated with a susceptibility locus on chromosome 16p. Arthritis Rheum 44:2548–2555

Itahana K, Dimri G, Campisi J et al (2001) Regulation of cellular senescence by p53. Eur J Biochem 268:2784–2791

Jacobson MD, Weil M, Raff MC (1997) Programmed cell death in animal development. Cell 88:347–354

Jazwinski SM (1996) Longevity, genes, and aging. Science 273:54–59

Johnson K, Hashimoto S, Lotz M et al (2001a) Interleukin-1 induces pro-mineralizing activity of cartilage tissue transglutaminase and factor XIIIa. Am J Pathol 159:149–163

Johnson KA, Lopera F, Jones K et al (2001b) Presenilin-1-associated abnormalities in regional cerebral perfusion. Neurology 56:1545–1551

Julin P, Almkvist O, Basun H et al (1998) Brain volumes and regional cerebral blood flow in carriers of the Swedish Alzheimer amyloid protein mutation. Alzheimer Dis Assoc Disord 12:49–53

Jurivich DA, Qiu L, Welk JF (1997) Attenuated stress responses in young and old human lymphocytes. Mech Ageing Dev 94:233–249

Justesen J, Stenderup K, Kassem MS (2001) Mesenchymal stem cells. Potential use in cell and gene therapy bone loss caused by aging and osteoporosis. Ugeskr Laeger 163:5491–5495

Kardaun JW, White L, Resnick HE et al (2000) Genotypes and phenotypes for apolipoprotein E and Alzheimer disease in the Honolulu-Asia aging study. Clin Chem 46:1548–1554

Kayo T, Allison DB, Weindruch R et al (2001) Influence of aging and caloric restriction on the transcriptional profile of skeletal muscle from rhesus monkeys. Proc Natl Acad Sci USA 98:5093–5098

Kemnitz JW, Weindruch R, Roecker EB et al (1993) Dietary restriction and aging: design, methodology, and preliminary findings from the first year of the study. J Gerontol 48:B17–B26

Khachaturian ZS (2000) Toward a comprehensive theory of Alzheimer's disease – challenges, caveats and parameters. Ann NY Acad Sci 924:184–193

Kirkwood TBL, Austad SN (2000) Why do we age? Nature 408:233–238

Knight JA (1998) Free radicals: their history and current status in aging and disease. Ann Clin Lab Sci 28:331–346

Kobyliansky E, Karasik D, Belkin V et al (2000) Bone aging: genetics versus environment. Ann Hum Biol 27:433–451

Koechlin BA (1952) Preparation and properties of serum and plasma proteins XVIII. The β1-metal-combining protein of human plasma. J Am Chem Soc 74:2649–2653

Kolsch H, Ptok U, Bagli M et al (2001) Gene polymorphisms of interleukin-1alpha influence the course of Alzheimer's disease. Ann Neurol 49:818–819

Kontush A (2001) Amyloid-beta: an anti-oxidant that becomes a pro-oxidant and critically contributes to Alzheimer's disease. Free Radic Biol Med 31:1120–1131

Krause KH, Clark RA (2001) Geneva Biology of Ageing Workshop 2000: phagocytes, inflammation, and ageing. Exp Gerontol 36:373–381

Kristal BS, Chen J, Yu BP (1994) Sensitivity of mitochondrial transcription to different free radical species. Free Radic Biol Med 16:323–329

Kristal BS, Yu BP (1998) Dietary restriction augments protection against induction of the mitochondrial permeability transition. Free Rad Biol Med 24:1269–1277

Lagaay AM, D'Amaro J, Ligthart GJ et al (1991) Longevity and heredity in humans. Ann NY Acad Sci 62:78–89

Lane DP (1992) p53, guardian of the genome. Nature 358:15–16

Lass A, Sohal BH, Weindruch R et al (1998) Caloric restriction prevents age-associated accrual of oxidative damage to mouse skeletal muscle mitochondria. Free Radic Biol Med 25:1089–1097

Lee CK, Klopp RG, Weindruch R et al (1999) Gene expression profile of aging and its retardation by caloric restriction. Science 285:1390–1393

Linnane AW, Kovalenko S, Gingold EB (1998) Age associated cellular bioenergetic degradation and amelioration therapy. The universality of bioenergetic disease. Ann NY Acad Sci 854:202–213

Liu J, Killilea DW, Ames BN (2002) Age-associated mitochondrial oxidative decay: improvement of carnitine acetyltransferase substrate-binding affinity and activity in brain by feeding old rats acetyl-L-carnitine and/or R-alpha-lipoic acid. Proc Natl Acad Sci USA 19:99:1876–1881

Luft R (1994) The development of mitrochondral medicine. Proc Natl Acad Sci USA 9:8731–8738

Lupattelli G, Siepi D, Palumbo B et al (1999) Vascular 131I-low-density lipoprotein imaging in patients with hypertension and early atherosclerotic lesions in the carotid artery. Vasa 28:185–189

Martin JA, Buckwalter JA (2001) Telomere erosion and senescence in human articular cartilage chondrocytes. J Gerontol A Biol Med Sci 56:B172–179

Masoro EJ, Shimokawa I, Yu BPl (1991) Retardation of the aging process in rats by food restriction. Ann NY Acad Sci 621:337–352

Mattson MP, Duan W, Lee J et al (2001) Suppression of brain aging and neurodegenerative disorders by dietary restriction and environmental enrichment: molecular mechanisms. Mech Ageing Dev 122:757–778

Matsunaga H, Tanaka Y, Tanaka M et al (2001) Antiatherogenic mitochondrial genotype in patients with type 2 diabetes. Diabetes Care 24:500–550

Maynard-Smith J (1963) Temperature and the rate of aging in poikilotherms. Nature 199:400–402

Mazzetti I, Grigolo B, Pulsatelli L et al (2001) Differential roles of nitric oxide and oxygen radicals in chondrocytes affected by osteoarthritis and rheumatoid arthritis. Clin Sci 101:593–599

McCay CM, Cromwell MF, Maynard LA (1935) The effect of retarded growth upon the length of life span and upon the ultimate body size. J Nutr 10:63–79

McKelvey R, Bergman H, Stern J et al (1999) Lack of prognostic significance of SPECT abnormalities in non-demented elderly subjects with memory loss. Can J Neurol Sci 26:23–28

Merker K, Stolzing A, Grune T (2001) Proteolysis, caloric restriction, and aging. Mech Ageing Dev 122:595–615

Meydani M (1992) Vitamin E requirement in relation to dietary fish oil and oxidative stress in elderly. EXS 62:411–418

Mielke R, Zerres K, Uhlhaas S et al (1998) Apolipoprotein E polymorphism influences the cerebral metabolic pattern in Alzheimer's disease. Neurosci Lett 254:49–52

Miller C, Kelsoe G (1995) Ig VH hypermutation is absent in the germinal centers of aged mice. J Immunol 155:3377–3384

Miquel J (1998) An update on the oxygen stress-mitochondrial mutation theory of aging: genetic and evolutionary implications. Exp Gerontol 33:13–126

Misra RP (2000) Calcium and disease: molecular determinants of calcium crystal deposition diseases. Cell Mol Life Sci 57:421–428

Molero AE, Pino-Ramirez G, Maestre GE (2001) Modulation by age and gender of risk for Alzheimer's disease and vascular dementia associated with the apolipoprotein E-varepsilon4 allele in Latin-Americans: findings from the Maracaibo Aging Study. Neurosci Lett 307:5–8

Mori A, Utsumi K, Liu J et al (1998) Oxidative damage in the senescence-accelerated mouse. Ann NY Acad Sci 854:239–250

Morris JC (2000) The nosology of dementia. Neurol Clin 18:773–788

Mrak RE, Griffin WS (2000) Interleukin-1 and the immunogenetics of Alzheimer disease. J Neuropathol Exp Neurol 59:471–476

Multhaup G, Ruppert T, Schlicksupp A et al (1997) Reactive oxygen species and Alzheimer's disease. Biochem Pharmacol 54:533–539

Nagata K, Maruya H, Yuya H et al (2000) Can PET data differentiate Alzheimer's disease from vascular dementia? Ann NY Acad Sci 903:252–261

Nixon RA (2000) A protease activation cascade in the pathogenesis of Alzheimer's disease. Ann NY Acad Sci 924:117–131

Nobili F, Copello F, Buffoni F et al (2001) Regional cerebral blood flow and prognostic evaluation in Alzheimer's disease. Dement Geriatr Cogn Disord 12:89–97

Nobuhara K, Halldin C, Hall H et al (2000) Z-IQNP: a potential radioligand for SPECT imaging of muscarinic acetylcholine receptors in Alzheimer's disease. Psychopharmacology (Berl) 149:45–55

Ohtsuki K, Hayase M, Akashi K et al (2001) Detection of monocyte chemoattractant protein-1 receptor expression in experimental atherosclerotic lesions: an autoradiographic study. Circulation 104:203–208

Orgel LE (1963) The maintenance of the accuracy of protein synthesis and its relevance to aging. Proc Natl Acad Sci USA 49:517–521

Pahlavani MA, Harris MD, Richardson A (1997) The increase in the induction of IL-2 expression with caloric restriction is correlated to changes in the transcription factor NFAT. Cell Immunol 180:10–19

Parr T (1997) Insulin exposure and aging theory. Gerontol 43:182–200

Pearl R (1928) The rate of living. Knopf, New York

Perls TT (2002) Genetic and environmental influences on exceptional longevity and the AGE Nomogram. Ann NY Acad Sci 959:1–13

Pieri C (1991) Food restriction slows down age-related changes in cell membrane parameters. Ann NY Acad Sci 621:353–362

Pincus T (2001) Clinical evidence for osteoarthritis as an inflammatory disease. Curr Rheumatol Rep 3:524–534

Pope A (1733–1734) An essay on man II

Raina AK, Pardo P, Rottkamp CA et al (2001) Neurons in Alzheimer disease emerge from senescence. Mech Ageing Dev 123:3–9

Ramsey JJ, Harper MF, Weindruch R (2001) Restriction of energy intake: energy expenditure and aging. Free Radic Biol Med 29:946–968

Reddel RR (1998) Genes involved in the control of cellular proliferative potential. Ann NY Acad Sci 854:8–19

Reid MB, Durham WI (2002) Generation of reactive oxygen and nitrogen species in contracting skeletal muscle. Ann NY Acad Sci 959:108–116

Richter C, Park JW, Ames BN (1988) Normal oxidative damage to mitochondrial and nuclear DNA is extensive. Proc Natl Acad Sci USA 85:6465–6467

Rivett AJ (1998) Intracellular distribution of proteasomes. Curr Opin Immunol 10:110–114

Rodewald HR (1998) The thymus in the age of retirement. Nature 396:630–631

Rogaev EI (1999) Genetic basis for Alzheimer's disease and other dementia and prospects of molecular diagnosis (in Russian). Vestn Ross Akad Med Nauk 1:33–39

Rogina B, Reenan RA, Nilsen SP et al (2000) Extended lifespan conferred by cotransporter gene mutations in Drosophila. Science 290:2137–2140

Roth GS, Ingram DK, Lane MA (1999) Calorie restriction in primates: will it work and how will we know? J Am Geriatr Soc 47:896–903

Rubner M (1908) Das Problem der Lebensdauer und seine Beziehungen zum Wachstum und Ernährung. Munich, Oldenburg, pp 150–204

Sacher GA (1968) Molecular versus systemic theories on the genesis of ageing. Exp Gerontol 3:265–271

Salo DC, Lin SW, Pacifici RE et al (1988) Superoxide dismutase is preferentially degraded by a proteolytic system from red blood cells following oxidative modification by hydrogen peroxide. Free Radic Biol Med:335–339

Samani NJ, Boultby R, Butler R et al (2001) Telomere shortening in atherosclerosis. Lancet 358:472–473

Sapolsky RM, Krey LC, McEwen BS (1986) The neuroendocrinology of stress and aging: the glucocorticoid cascade hypothesis. Endocr Rev 7:284–301

Sarikaya I, Sarikaya A, Holder LE (2001) The role of single photon emission computed tomography in bone imaging. Semin Nucl Med 31:3–16

Schamberger RJ, Frost DV (1969) Possible protective effects of selenium against human cancer. Can Med Assoc J 100:682

Schroder J, Buchsbaum MS, Shihabuddin L et al (2001) Patterns of cortical activity and memory performance in Alzheimer's disease. Biol Psychiatry 49:426–436

Shay JW, Wright WE (2001) Aging. When do telomeres matter? Science 291:839–840

Shouda T, Yoshida T, Hanada T et al (2001) Induction of the cytokine signal regulator SOCS3/CIS3 as a therapeutic strategy for treating inflammatory arthritis. J Clin Invest 108:1781–1788

Small GW, Ercoli LM, Silverman DH et al (2000) Cerebral metabolic and cognitive decline in persons at genetic risk for Alzheimer's disease. Proc Natl Acad Sci USA 97:5696–5698

Sohal RS (1993) The free radical hypothesis of aging: an appraisal of the current status. Aging Clin Exp Res 5:3–17

Sohal RS, Agarwal S, Candas M et al (1994) Effect of age and caloric restriction on DNA oxidative damage in different tissues of C57BL/6 mice. Mech Ageing Dev 76:215–224

Stocks J, Gutteridge GM, Sharp RJ et al (1974) The inhibition of lipid peroxidation by human serum and its relationship to serum proteins and alpha-tocopherol. Clin Sci 47:223–233

Tanaka N, Fukushi K, Shinotoh H et al (2001) Positron emission tomographic measurement of brain acetylcholinesterase activity using N-[(11)C]methylpiperidin-4-yl acetate without arterial blood sampling: methodology of shape analysis and its diagnostic power for Alzheimer's disease. J Cereb Blood Flow Metab 21:295–306

Tanaka S, Kawamata J, Shimohama S et al (1998) Inferior temporal lobe atrophy and APOE genotypes in Alzheimer's disease. X-ray computed tomography, magnetic resonance imaging and Xe-133 SPECT studies. Dement Geriatr Cogn Disord 9:90–98

Tiku ML, Gupta S, Deshmukh DR (1999) Aggrecan degradation in chondrocytes is mediated by reactive oxygen species and protected by antioxidants. Free Radic Res 30:395–405

Van Leeuwen FW, Fischer DF, Kamel D et al (2000) Molecular misreading: a new type of transcript mutation expressed during aging. Neurobiol Aging 21:879–891

Verschure PJ, Van Marle J, Joosten LAB et al (1996) Localization of insulin-like growth factor-1 receptor in human normal and osteoarthritic cartilage in relation to proteoglycan synthesis and content. Br J Rheum 35:1044–1055

Virgolini I, Rauscha F, Lupattelli G et al (1991) Autologous low-density lipoprotein labeling allow characterization of human atherosclerotic lesions in vivo as to presence of foam cells and endothelial coverage. Eur J Nucl Med 18:948–951

Wallace DC (2001) A mitochondrial paradigm for degenerative diseases and aging Novartis Foundation Symposium. Wiley, Chichester, pp 247–263

Wang E (1997) Regulation of apoptosis resistance and ontogeny of age-dependent diseases. Exp Gerontol 32:471–484

Warner HR (1999) Apoptosis: a two-edged sword in aging. Ann NY Acad Sci 887:1–11

Warner HR, Butler RN, Sprott RL, et al (eds) (1987) Modern biological theories of aging. Raven Press, New York

Wei YH (1998) Oxidative stress and mitochondrial DNA mutations in human aging. Proc Soc Exp Biol Med 217:53–63

Weindruch R, Kristie JA, Cheney KE et al (1979) Influence of controlled dietary restriction on immunologic function and aging. Fed Proc 38:2007–2016

Weissman A (1942) The duration of life. In: Poulton EB (ed) New paths in genetics. Harper, New York

Weraarchakul NR, Strong R, Wood WG et al (1989) The effect of aging and dietary restriction on DNA repair. Exp Cell Res 181:197–204

Whitman W (1855) from 'Song of myself'

Wielopolski L, Vartsky D, Pierson R et al (2000) Gamma resonance absorption. New approach in human body composition studies. Ann NY Acad Sci 904:229–235

Williams GC (1957) Pleiotropy, natural selection, and the evolution of senescence. Evolution 11:398–411

Wu Z, Kinslow C, Pettigrew KD et al (1998) Role of familial factors in late-onset Alzheimer disease as a function of age. Alzheimer Dis Assoc Disord 12:190–197

Yin D (1996) Biochemical basis of lipofuscin, ceroid, and age pigment-like fluorophores. Free Radic Biol Med 21:871–888

Yoshida T, Kuwabara Y, Ichiya Y et al (1998) Cerebral muscarinic acetylcholinergic receptor measurement in Alzheimer's disease patients on 11C-N-methyl-4-piperidyl benzilate – comparison with cerebral blood flow and cerebral glucose metabolism. Ann Nucl Med 12:35–42

Youngman LD, Park JY, Ames BN (1992) Protein oxidation associated with aging is reduced by dietary restriction of protein or calories. Proc Natl Acad Sci USA 89:9112–9116

Yu BP (1994) Cellular defenses against damage from reactive oxygen species. Physiol Rev 74:139–162

Yu SW, Shaw SM, Van Sickle DC (1999) Radionuclide studies of articular cartilage in the early diagnosis of arthritis in the rabbit. Ann Acad Med Singapore 28:44–48

Zubieta JK, Koeppe RA, Frey KA et al (2001) Assessment of muscarinic receptor concentrations in aging and Alzheimer disease with [11C]NMPB and PET. Synapse 39:275–287

CHAPTER 14

In Vitro Indicators of Metabolism of Natural Compounds

WALTON W. SHREEVE

Contents

14.1
Introduction

Measurements of the concentrations of radioactive or stable isotopes in body fluids (e.g., blood, plasma), excreta (e.g., urine, breath, feces, saliva) or biopsied tissue after in vivo administration constituted a major part of nuclear medicine in the early years of the specialty (roughly, the third quarter of the 20th Century). This era was mostly research-oriented and pre-clinical. Yet, clinical applications of in vitro tests were developing for the purposes of recognizing problems of gastrointestinal absorption of minerals, vitamins and energy constituents of the diet and the fate of elements or compounds after enteral or parenteral administration. Amounts and kinetics of body constituents were evaluated in ways which were, and in various cases still are, achievable only, or best, with the isotopic tracers. Another important facet of radiotracer use was the development of radioimmunoassay, by which minute concentrations of proteins (especially hormones) in body fluids would be measured with much benefit to clinical medicine. Autoradiography of uptake of weak beta emitters into isolated cells after in vivo or in vitro exposure also developed in this period, as did initial work with in vitro blood cell tagging for volume assays and for imaging.

In the past two or three decades the exceptional progress of in vivo imaging techniques and applications has diverted attention from many in vitro tests, although in large nuclear medicine services "wet-bench" laboratories are in use for support activities such as labeling of white cells or antibodies, evaluating in-put functions from blood concentrations of radiotracers, or measuring excreted radioactivity to evaluate internal dosimetry. Some in vitro measurements of radionuclides, such as for radioimmunoassay, have been transferred to non-nuclear medicine sites, such as clinical pathology or pharmaceutical company laboratories. Some former in vitro tests with radiotracers have been superseded by other methods, but some still valuable tests have languished for lack of attention and/or resources. It bears reminding that many in vitro assays can be done with only moderately expensive single or multi-channel crystal counters (for gammas) and liquid scintillation counters (for betas). Moreover, new applications are continually developing to utilize the same resources.

Both radioactive and stable isotopes have been used to provide information on kinetics of labeled compounds, as measured in bodily fluids, excreta or biopsied tissue. In several cases the most precise and direct diagnosis of enzymatic defects of genetic type are provided by this means with relatively simple and easy techniques. Other such data can help focus on the most likely vulnerable phenotypic sites in multifactorial disease of partly genetic origin and possibly reveal genetic sub-types. In future the stable isotopes ought to receive greater use for various reasons. While absorbed radiation doses from administration of metabolically-useful radioisotopes (^{14}C, ^{3}H, ^{18}F, ^{11}C, ^{13}N, ^{15}O, etc.) are tolerable in diseased adult patients and even in limited numbers of, and doses for, normal adult volunteers, use of stable isotopes would much enlarge the base of data on normal subjects without any concern about ionizing radiation. This is a particularly important point for the study of children, both normal and abnormal, in whom the benefits of linking biochemical phenotype to genotype could have great advantages. Another asset of stable isotopes is the possibility of using them together with corresponding radioisotopes in dual-isotope studies, which have special power to define sites of biochemical diathesis in disease.

14.2
Techniques

14.2.1
Radioisotope Tracers

14.2.1.1
Instrumentation

For many years the well-type scintillation counter, usually with NaI (Tl) solid crystal detector(s) (either single or multi-sample) and the liquid scintillation counter (usually multi-sample) with radioactive source dissolved (or intimately suspended) with the scintillating agent, have been the major means of counting, respectively, higher-energy gamma- or beta-, alpha- and low-energy gamma-emitters in vitro. The instrumentation, reagents and performance of these counters have been lately reviewed (McIntyre and Saha 1995). Most models have multiple pulse-height analyzers for evaluating simultaneously two or more radiotracers, e.g., ^{14}C and ^{3}H (Okita et al. 1957; Shigeta and Shreeve 1964; Wolfe 1992), ^{45}Ca and ^{47}Ca (Heaney et al. 1989) or ^{58}Co and ^{57}Co (Price and McIntyre 1984; Lindgren et al. 1997). A relatively new

development is that of a hybrid combination of a liquid and a NaI (Tl) scintillation monitor which facilitates the analysis of a mixture of beta- and beta-gamma-emitters, such as ^{3}H, ^{51}Cr and ^{125}I (Fujii et al. 2000). While developed to analyze radioactive waste solutions, such a counter could be used for biological fluids in various applications. Another new technique employs a miniaturized microplate scintillation counter, which is suited to miniaturized separation techniques and is especially sensitive to low concentrations of radioactive compounds within complex biological fluids (Boernsen et al. 2000). Still another new variation is that of a flow-through scintillation counter designed to be on line with high performance liquid chromatography for assay of radioactive amino acids in plasma (Ahmed et al. 1998). Liquid and solid scintillation counters will be needed not only for the types of metabolic studies discussed in this chapter but for in vitro measurements for dosimetry of alpha- or beta-emitting radiotherapeutic agents (Salako and De Nardo 1997; Cremonesi et al. 1999) and for the presence of labeled drugs and drug metabolites in plasma and excreta (Boernsen et al. 2000).

Much of the description in this chapter includes the use and measurement of ^{14}C-labeled compounds. The simplest and most clinically feasible of these applications is the assay of $^{14}CO_2$ in breath after passage through a drying agent and then directly into scintillation vials containing a precisely measured amount of hyamine which determines stoichiometrically the amount of $^{14}CO_2$ trapped, thereby enabling a calculation of the specific activity after liquid scintillation counting (LSC) (Kaihara and Wagner 1968; Shreeve 1984; Sasaki 1995). A related device is that of a plastic scintillation filament detector system, which measures the rate of total CO_2 output as well as $^{14}CO_2$ and is portable (Lorenz et al. 1978). Several applications of these simple techniques are possible, but use has been very limited, due partly to concern about the radiation dose from the very long-lived ^{14}C. Most such tests require about 1–10 µCi (37–370 kBq), when the administered compound is mostly metabolized to CO_2 (the compound of usual interest) within one to a few days and analysis is by LSC. The corresponding radiation burden, assuming long-term (years) retention of 2–10% of administered dose is variously estimated (depending on kind of compound and organs most affected) as 1–20 mrem (0.01–0.20 mSv)/µCi (Yap et al. 1975; Pedersen and Marqverson 1981; Landau and Shreeve 1991). The risk of genetic damage due to transmutation of ^{14}C to ^{14}N is said to be about equal to that of the radiation effect (Yap et al. 1975). Most of the radiation and transmutation effects come from the small fraction of long-retained radiotracer or its metabolites. The dose rate is therefore quite low and the concern regarding radiation has probably been over-emphasized for the situations where there is important diagnostic information to be obtained in seriously-diseased, non-pregnant, adult patients. For some other kinds of studies (e.g., evaluation of protein synthetic rates, gluconeogenesis or lipogenesis) larger amounts of ^{14}C may be required and the risk/benefit ratio is clearly higher if LSC is the method employed. Potentially on the horizon for clinical application is a quite different kind of assay of ^{14}C, i.e., accelerator mass spectrometry (AMS), which is several orders of magnitude more sensitive than LSC. This technique was developed about 20 years ago, essentially for carbon dating, but is beginning to be applied to compounds in biological fluids (Vogel and Turteltaub 1998; Barker and Garner 1999). AMS is a type of tandem isotope ratio mass spectrometry in which initially negatively charged atomic and small molecular ions are attracted to a collision cell at very high positive potential, thereby stripping off multiple electrons and destroying all molecules, leaving only nuclear ions to be identified as nuclear isobars (Vogel and Turteltaub 1998). The AMS process is far more efficient for measurement of ^{14}C than LSC, detecting 10-fold lower concentrations and 100-fold more mass amounts (Gilman et al. 1998). Tracer studies can be done by administering amounts of ^{14}C in biological organic compounds which deliver no more than 1 µSv (0.1 mrem) radiation dose in the first 1000 h (Vogel and Turteltaub 1998). A life-time integrated dose of only 11 µSv was calculated from a study of metabolism of labeled folic acid, in which 100 nCi were administered and >75% retained beyond 40 days (Buchholz et al. 1999). Tritium and ^{41}Ca are other long-lived radioisotopes which are measurable by AMS. Relatively small AMS spectrometers designed for biological tracing are still not "bench-top" (requiring 100 sq. ft.) and cost about one million dollars. However, a centralized commercial company is already offering services using a biomedical AMS (Barker and Garner 1999). AMS could open up the permitted use of ^{14}C (and some other radiotracers) in a variety of disease states, physiological and genetic variants, and normal control subjects.

The likelihood of performance of autoradiography (Lear 1995; Clark and Klein 1996), radioimmunoassay (Yalow 1995) or ligand assays (Ekins 1995) in nuclear medicine departments depends on special research interests within the department or collegial interests within an institution or hospital, but well-diversified departments should be prepared to respond to such interests. Besides conventional autoradiography for content and location of ^{14}C or ^{3}H by development of single X-ray film, there are other techniques by which to quantitate amounts of separated compounds (or their location in isolated cellular studies) by electron microscopic autoradiography (Carpentier et al. 1986; Englert et al. 1995) or to provide separate counts of ^{14}C and ^{3}H in double isotope studies by a two-way imaging plate (Yamane et al. 1995). Another new technique shortens development time (hours instead of weeks) as well as quantitates data with high sensitivity by digital autoradiography (DAR) (Ludanyi et al. 1999). DAR might well be applied to the analysis of uptake and/or retention of ^{14}C- or ^{3}H-labeled compounds incubated with isolated blood (or tissue) cells in vitro, such as described for monocytes in Sect. 14.3.3.4 and for platelets in Sect. 14.3.4.1.3. A further possibility is the use of the miniaturized "beta camera" (Ljunggren and Strand 1990; Ljunggren et al. 1993) (see Chap. 12, Sect. 12.2.3.2), which is designed for rapid, kinetic assay of in vitro studies with ^{14}C and other beta-, or low-energy gamma-, or positron-emitters (though not for the very low energy of ^{3}H). However, the relatively low resolution (500 µm) of the instrument would require separation of cells from extracellular medium, e.g., by centrifugation, for analysis.

14.2.1.2
Preparation of Compounds for Analysis

Large nuclear medicine services should have the "wet-bench" laboratory capacity, physical facilities and materials, and expertise of personnel for doing initial "clean-up" of plasma, urine, feces, cerebrospinal fluid, etc., to isolate labeled compounds and their metabolites. Such procedures would include deproteination of plasma or serum (Berrish et al. 1995), passage of the product through anion and/or cation exchange columns (De Meutter and Shreeve 1963; Wolfe 1992; Thornburn et al. 1995), oxidation of specific compounds (urea, glutamate, glucose, lactate) to CO_2 (Landau et al. 1995), extraction of albumin from plasma (Ballmer et al. 1993), hydrolysis of isolated proteins to constituent amino acids (Demant et al. 1996) and some kinds of derivatization of compounds for analysis (De Meutter and Shreeve 1963). A few specialized services might also perform ultracentrifugation of lipids or complex derivatization of compounds to prepare them for special chromatography (depending on particular research interests), but in most cases collaboration with a basic science laboratory would be needed to extend to sizeable clinical applications.

Of central importance to the separation, isolation and assay of radioactive or stable isotope-labeled compounds in body fluids or excreta is the principle of chromatography (Greek for "color writing"), which was first named as such in 1906 by the Russian botanist, Michael Tswett, when he separated constituents of a chloroplast mixture (Wolfe 1992). In essence chromatography separates substances by their differential migration through a two-phase system, consisting of a mobile phase and a stationary phase. The stationary phase tends to retard movement of a compound by a physical characteristic (e.g., pore size or electrical charge) or chemical attribute (molecular polarity or non-polarity), while the mobile phase opposes by its differential polarity and/or a physical driving force, which can be merely gravity, or applied pressure, or by electrochemical attraction. The mobile phase is usually liquid for radioisotopic tracers, while for stable isotope-labeled compounds it is more likely gaseous (as are the derivatized compounds being separated), so as to be conveniently on line with a mass spectrometer. The substances ("analytes") separate based upon the particular balance of opposing attractions between the mobile and stationary components for each individual substance. "Reverse phase" liquid chromatography is more common now and somewhat of a misnomer, since it merely indicates that the mobile phase is more polar than the stationary phase, which is the reverse of the original phases used by Tswett (Wolfe 1992). "High performance" liquid chromatography (HPLC) consists of modern extensions of Tswett's original column technique by adding "bells and whistles" such as pressure pumps for aiding movement of the mobile, multiple capillary columns to increase the area of the stationary phase, etc. (Wolfe 1992). For radioisotope analysis the timed eluates from the stationary phase (column/plate) are counted in well counters or liquid scintillation counters and identified by reference to behavior of labeled standards.

Practical for most nuclear medicine "wet" labs are techniques such as ion exchange columns (to remove inorganic or organic cations and/or anions) and ordinary, room-pressure thin-layer chromatography (TLC), most commonly with silica gel plates, which can separate lipids, carbohydrates or amino acids

(Wu et al. 1974; Wolfe 1992; Clark and Klein 1996). The performance and range of application of TLC has been much increased by introduced forced-flow (i.e., overpressure) layer chromatography which enables the use of viscous solvent systems and superfine sorbent particles (stationary phase), greatly shortens separation time, and increases efficiency of separation (Ludanyi et al. 1999). Compounds which are highly polar or have very large mass, so can be separated but not readily eluted, can be detected by X-ray film development or other methods of imaging of the plate (see Sect. 14.2.1.1).

14.2.2
Stable Isotope Tracers

14.2.2.1
Instrumentation

The basic instrument for most metabolic studies with stable isotopes is the mass spectrometer. The latter is complex, labor-intensive, moderately costly (though no more than "an NMR machine"; Halliday and Rennie 1982), so is unlikely to be acquired by more than a few nuclear medicine departments in the foreseeable future. Yet, sharing arrangements can be made for the assay of samples collected (and possibly initially "cleaned-up") at a clinical (nuclear medicine) site. This may become important since (as indicated throughout Sect. 14.3; see also Chap. 12) the use of stable isotopes is a large and growing component of human metabolic studies, some of which are already at the clinical level and others likely to develop. Nuclear medicine practitioners should, therefore, at least know about the fundamental nature and characteristics of mass spectrometer(s) (MS). For MS the compound(s) to be measured must be in a gaseous state and be caused to be ionized positively, in which state they are then attracted to a collecting plate, where amounts of ions impinging on certain locations of the plate, depending on mass and degree of ionization of a compound, are measured. A magnetic field interposed between the ionization site and the collecting site differentially bends the streams of compounds in a mixture depending on their mass/ion ratios. Any molecular compound (or kind of atom) enriched with one or more atoms of a rare (natural or artificial) isotope (the tracer) is thus distinguished from those without any (or less or more of) the tracer. The chamber from ion source to collecting plate is under high vacuum, assuring a gaseous state even for large molecules.

There are two general types of mass spectrometry (Halliday and Rennie 1982; Wolfe 1992). One is the isotope-ratio mass spectrometer (IRMS), which is usually employed to measure elements of biomedical interest, e.g., $^2H/^1H$, $^{13}C/^{12}C$, $^{18}O/^{16}O$ or $^{15}N/^{14}N$, in their simplest state. Usually these are introduced into the IRMS in the gaseous state of primary elements (H_2, N_2, O_2) or the gaseous oxides of C, N or S. About 10–50 cc of these pure gases are required. The major and minor (tracer) isotopes can be collected for measurement simultaneously. Detection of ten parts of tracer isotope per million is possible, even against the rather high background of 1.1% natural ^{13}C. The other general kind of MS links a gas chromatograph (GC) directly to the MS. The mobile phase of the GC is an inert carrier gas such as helium, H_2, N_2, or methane. The stationary phase is a non-volatile liquid or inert, size-graded solid. Compounds being separated and analyzed are generally of high molecular weight, having been derivatized to increase volatility (see Sect. 14.2.2.2). As compounds are successively eluted from the GC they pass into the MS, where they are ionized by one or other of a range of various techniques (Wolfe 1992), then passed through a quadrupole filter, which has a resonating effect on the ionized compound, depending on mass and ionization, and thereby allows certain masses through to be measured. A kind of combination GC/IRMS includes a pyrolysis chamber between the GC and the IRMS, whereby the volatile derivatives of compounds are combusted to small molecules (CO_2, H_2O, N, S) which are analyzed by the isotope-ratio method (Halliday and Rennie 1982; Wolfe 1992). Significant differences between the IRMS and the "selective ion monitor" (SIM)/GC/MS are the greater sensitivity and precision of IRMS (as much as 1000 times greater sensitivity; Maugeais et al. 1998) but the capacity of SIM/GC/MS to measure much smaller (picogram or nanogram) quantities of sample.

Clinical applications of mass spectrometry (mostly by IRMS) have so far been mainly those involving analysis of $^{13}CO_2$ collected from breath after intravenous or oral administration of a strategically-labeled compound (Halliday and Rennie 1982; Shreeve 1984; Sasaki 1995). Many other kinds of human studies (see Sect. 14.3) could be modified to be more widely applicable and should be preferred over ^{14}C for children, pregnant women and young, healthy adults [with the caveat that availability of AMS (Sect. 14.2.1.1) could make ^{14}C more acceptable]. It is

notable that new models for either GC/IRMS (Maugeais et al. 1998) or LC-MS (Van Eijk et al. 1999) are being described as being of "bench-top" size. An alternative to mass spectrometry for analysis of $^{13}CO_2/^{12}CO_2$ ratios (as from breath tests) is non-dispersive infra-red spectroscopy (NDIRS), a technique which has been available for either ^{14}C or ^3H for more than two decades (Irving 1975; Milstein and Kaufman 1975; Hirano et al. 1979). Recently the performance of this kind of instrument in a bench-top configuration (Fig. 14.1) has been tested by directly comparing results from a ^{13}C-urea breath test for *Helicobacter pylori* infection (see Sect. 14.3.1.2) in which aliquots of timed breath samples were analyzed both by NDIRS and by IRMS (Braden et al. 1999). There was 98–99% agreement in raw "delta" values and linear correlation of the two methods and, using the IRMS as the "gold standard", 98% sensitivity and 99% specificity for NDIRS. The analysis time for one sample was less than 1 min. The breath samples can be collected in aluminized plastic bags for analysis on site (Fig. 14.1) or in glass tubes for shipping to some laboratory center. Although the sensitivity of NDIRS is not as good as with IRMS, for most biomedical applications it is adequate. Furthermore, NDIRS is less expensive than IRMS. An apparatus similar to that of NDIRS is based on laser spectroscopy (LS) (Sasaki 1995). The separate detection of $^{13}CO_2$ and $^{12}CO_2$ may be more precise than with NDIRS. The use of NDIRS or LS seems a good, practical choice for many nuclear medicine services which would like to begin some capacity for analysis of stable isotopes in vitro but cannot or do not want to invest in the more expensive and more complex techniques of MS.

A much more esoteric technique of in vitro analysis of in vivo metabolism of stable isotopes is that of NMR spectroscopy of enrichment of some compounds circulating in plasma or other body fluids after tracer administration. Such examination of plasma glucose for different patterns of molecular distribution of tracer isotope, ^{13}C (isotopomers), have been performed after nasogastric infusion of U-^{13}C-glucose to distinguish types of glycogen storage disease (Kalderon et al. 1989) and after similar infusion of U-^{13}C-fructose to show a typical diagnostic pattern for hereditary fructose intolerance (Gopher et al. 1990). While this technique appears too sophisticated for practice beyond a few specialized locations, it is based on the same main principles as MRS of ^1H, ^{31}P or ^{13}C in vivo and particularly could be utilized as a bridge to in vivo MRS tracing of ^{13}C in the future.

Fig. 14.1. The isotope-selective nondispersive infrared spectrometer (NDIRS) (FANci2, Fischer Analysentechnik, Leipzig, Germany). A series of aluminized breath bags may be sequentially connected to eight inlet ports for measurement and analysis by computer within a few minutes. (From Braden et al. 1999, with permission)

14.2.2.2
Preparation of Compounds for Analysis

Surely the simplest kind of preparation occurs in the tests involving collection and analysis of breath CO_2, since the only "clean-up" procedure is that of eliminating water in the breath (as for ^{14}C) by interposition of a drying agent (calcium chloride, silica gel) in the line leading directly from the patient's mouth to the IRMS. Alternatively, breath can be trapped in a form of vacucontainer and sometimes subsequent freezing CO_2 in liquid N_2 or trapping in NaOH can be performed prior to storage and/or transfer to an MS site. Concerns about contamination of masses 44, 45 or 46 after ionization of trace amounts of organic compounds other than CO_2 in the breath or of fractionation of masses upon release from NaOH have been tested (Wolfe 1992) with the finding of very small differences due to either of the suspected possibilities, but so small as to be insignificant relative to physiological noise levels.

Initial isolation and purification measures for obtaining relatively clean samples of labeled carbohydrates, lipids, amino acids (free or released from protein), various metabolites or end-products (e.g., urea, uric acid, 5-hydroxy-indole acetic acid) in plasma, urine, CSF, etc. can be carried out in some nuclear medicine laboratories by techniques as mentioned for ^{14}C in Sect. 14.2.1.2. Some laboratories may have the capability (or collaborate with specialized laborato-

ries) to separate lipoproteins by ultracentrifugation. Some nuclear medicine sites could develop the means to derivatize the primary, isolated labeled compounds or this step could instead be a function of the facility having the GC/MS capability. The required reagents and materials and procedures for initial extraction from body fluids, further isolation by special techniques (e.g., thin-layer chromatography) and derivatization as needed for GC/MS assay of such compounds as amino acids, glucose, glycerol, urea, free fatty acids, pyruvic and lactic acids, have been well set forth (Wolfe 1992). Since MS must analyze molecules (or fragments thereof) after ionization in the gaseous form, it is important in most cases to attach non-polar molecules to the more polar type compounds which are the objects of tracer study in the body fluids. The aim is to form derivatives which will be much more volatile and hence suitable for separation in a gas chromatograph and convenient for on-line transfer to the evacuated MS chamber. Methods of derivatization generally consist of alkylation, acetylation or silylation. The latter seems to be most often used. Possibly this is due to the affinity of silyl derivatives for the stationary phase in the GC (or GLC), which is likely to be silica gel or liquid methyl silicone (Wolfe 1992). Other choices besides (butylated or methylated) silyl derivatives (Bougneres et al. 1995; McMillan et al. 1996) include isopropylidene (Bougneres et al. 1995) or pentaacetate (Wolfe 1992) derivatives for glucose measurement, orthophthaldialdehyde (Van Eijk et al. 1999) or n-propyl-heptafluoro-butyryl esters (Elias et al. 1999) for amino acid derivatives (in an HPCL/MS system), heptafluorobutyric anhydride-acetone for testosterone (Vierhapper et al. 1997) and triacetate for glycerol (Siler et al. 1998) among others. A trick used in the derivatization process with testosterone was the addition of ^{3}H-testosterone to monitor the extent of recovery of ^{2}H-testosterone in the derivative. One feature of the mass spectrometric analysis of derivatives is fragmentation of some of the primary molecules upon ionization but the pattern of fragmentation tends to be typical for the particular molecule. Many more details of kinds of chromatography, kinds of mass spectrometry, characteristics of derivatives and analysis of results can be found in the comprehensive source by Wolfe (1992).

14.3
Applications

14.3.1
Gastrointestinal Tract and Exocrine Pancreas

14.3.1.1
Primary Absorptive Disorders

14.3.1.1.1
Vitamins and Minerals

A long-used in vitro test in nuclear medicine is the Schilling test for deficiency of gastric "intrinsic factor" (IF), a glycoprotein attaching to vitamin B_{12} (cyanocobalamin) and facilitating its absorption in the distal ileum. As initially practiced, B_{12} in the free, crystalline state or attached to IF is labeled with ^{57}Co or ^{58}Co, administered orally (sequentially or simultaneously), and 24-h urine assayed (Price and McIntyre 1984). Low excretion in phase 1 (free B_{12}) but normal in phase 2 (IF-bound B_{12}) indicates IF deficiency, which can occur as a rare genetic trait, but is more often due to chronic inflammatory or atrophic gastritis or surgical loss of stomach function. Low levels of B_{12} in the diet predispose toward a deficiency of IF and absorption of IF-B_{12}. Some technical changes have improved the sensitivity and also directed attention to ileal absorption. When ingested normally in foodstuffs B_{12} is protein-bound. If protein-bound B_{12} is substituted for free B_{12}, there is more likely positiv-

ity in mild or moderate gastritis, because in the latter states hypochlorhydria and decreased pepsin, occurring more readily than IF deficiency, prevent liberation of B_{12} for IF attachment and are the cause of malabsorption. Modifications include using labeled B_{12} bound to protein such as egg yolk (Lindgren et al. 1997). Even with such modification the test may be insensitive for diagnosing early gastric atrophy compared to evidence of abnormal serum indicators (Lindgren et al. 1997; Snow 1999). Another modification is based on the knowledge that ingested protein-bound B_{12} is actually first transferred to gastric "R-protein", then later transferred by pancreatic proteases in the duodenum to IF. By simultaneously using ^{57}Co-B_{12} bound to hog R-protein and ^{58}Co-B_{12} bound to IF, the occurrence of pancreatic insufficiency is indicated by a low ratio of ^{57}Co/^{58}Co in the urine (DiMagno 1995).

The mechanism of IF-B_{12} absorption in the distal ileum involves an epithelial receptor, cubilin. The gene for cubilin has been mapped to a region of chromosome 10 known to contain a recessive-gene locus for juvenile megaloblastic anemia (Imerslund-Grasbeck disease) (Aminoff et al. 1995). There may be impaired synthesis or processing of cubilin or ligand-binding capacity in this rare disease. Another factor in ileal B_{12} absorption is transfer of the B_{12} to transcobalamin II within the intestinal villus prior to transport in the blood. Clinical and laboratory variations among patients suggests a heterogeneity in I-G disease at the molecular level (Altay et al. 1995; Grasbeck 1997). Indeed, I-G disease may be just the tip of an iceberg, expressing the most damaging kind(s) of alleles of involved genes, while less-damaging alleles lead to occult, sub-clinical deficiency in a larger population, perhaps especially the elderly (Loew et al. 1999). Besides genetic disease there can be acquired causes of ileal malabsorption of IF-B_{12}, in such conditions as ileal resection, exocrine pancreatic insufficiency, chronic inflammatory disorders (Behrend et al. 1995; Loew et al. 1999) and intestinal diversion (Terai et al. 1997).

Absorption of mineral elements has been studied for decades by the use of radioactive or stable isotope tracers with analyses of plasma, urine or other fluids or tissues. Various aspects of metabolism, including intestinal absorption, of particular elements are discussed in Chap. 12. Here are provided brief reviews and further comments on the absorption of Ca, Fe, Cu and Zn, for each of which there are significant pathologies dependent on genetic constitution, and for which in vitro isotope tests can supply valuable diagnostic information.

Like many other elements and compounds, the percentage of ingested calcium which is absorbed is inversely proportional to the amount ingested, but there are differences among individuals which are familial in character and likely predispose some persons to osteoporosis (Favus 1989); absorption decreases with age and there are gender differences (Chap. 12, Sects. 12.1.2.3 and 12.3.2.4; Lutwak 1969). There are specific known genes, transport proteins and ion channels for absorption of Ca^{2+} in the upper small intestine (Barley et al. 1999; Peng et al. 1999; Ames et al. 1999). The gene for the vitamin D receptor appears to be particularly important in its polymorphisms (Nakamura 1997; Ames et al. 1999). Some of these polymorphisms are related to different responsiveness of Ca absorption to serum levels of activated vitamin D [1,25 (OH) D or calcitriol] (Nakamura 1997; Wishart et al. 1997). Estrogens in females and androgens in males have a positive effect on supporting Ca absorption (Maurus et al. 1999). A low-protein diet can impair absorption (Kerstetter et al. 1998). While some of these findings involve dual-stable (or radioactive) isotope administration (oral/i.v.) and measurement in late samples of plasma, urine and/or feces (Lutwak 1969; Kerstetter et al. 1998; Ames et al. 1999; Maurus et al. 1999), others use a single radioactive tracer (^{45}Ca or ^{47}Ca) with measurement in serum at 1 h (Wishart et al. 1997) or at 5 h (Lutwak 1969) post-ingestion. An attractive alternative (for cost reduction, non-invasiveness and minimizing of tracer perturbation of the system) is administration of dual (oral/i.v.) stable isotopes (^{43}Ca and ^{46}Ca) in microgram doses with analyses of saliva (Smith et al. 1996).

Intestinal uptake of iron and transfer to serum transferrin is, like Ca^{2+}, inversely proportional to dietary load and, moreover, is normally inversely proportional to body iron stores. Disruption of this regulation can occur as genetic traits to cause either excessive or deficient iron absorption. In the relatively common genetic disorder, hemochromatosis, a known mutation in a gene (HFE) (found in 90% of homozygotes of European origin) interferes with normal binding of transferrin to a transferrin receptor in the duodenal mucosa, so that iron content of the latter remains low despite excessive body iron stores. This triggers genetic over-expression of a membrane protein transporter of Fe (NRAMP-2), thus causing inappropriately high Fe uptake (Worwood 1998; Anderson

and Powell 1999; Zoller et al. 1999). There is accompanying increased reduction of dietary Fe^{3+} to the Fe^{2+} necessary for uptake and transport (Chap. 12, Sect. 12.1.2.5). Such information has come largely from analyses (some with ^{59}Fe) of duodenal biopsy specimens. Possibly useful for clinical studies are the dual-isotope (oral/i.v.) methods with radioisotopes (Saylor and Finch 1953) or stable isotopes (Barret et al. 1997) of iron. Nevertheless, caution is advisable when applying such methods in developed hemochromatosis, because of distortion of isotopic iron concentrations by the large hepatic Fe stores (Price and McIntyre 1984). Paired stable isotopes (^{57}Fe orally and ^{58}Fe i.v.) have been used with analyses of isotope ratios in RBC in pregnant women to show that Fe in unfortified diets is inadequately absorbed, and, therefore, Fe supplementation during pregnancy is clearly important (O'Brien et al. 1999). ^{57}Fe has been used to define a rare, human familial occurrence of isolated iron malabsorption, the genetics of which are unknown, but not the same as for rodent models of microcytic anemia with known specific mutation in the NRAMP-2 gene (Pearson and Lukens 1999). The large, world-wide incidence of anemia related to poor nutrition has prompted the development of genetically-engineered rice, which has new characteristics that support iron absorption; such a development might well utilize dual-Fe isotope tracer studies to guide these nutritional efforts in public health.

In regard to the essential trace minerals, copper and zinc, there are in each case rare, recessive genetic traits causing deficiency of intestinal absorption and resultant devastating multi-organ pathology. Mutations of gene(s) on chromosome X (controlling a Cu-binding ATPase) cause either Mencke's disease, in which there is usually early mortality, or a lesser related disorder, occipital horn syndrome, with usual survival into adult life. Oral ^{64}Cu is poorly absorbed, while parenteral ^{64}Cu shows relatively normal kinetic parameters (though parenteral Cu is inadequate for therapy). Definitive diagnosis is available from the characteristics of retention of ^{64}Cu by cultured fibroblasts or amniotic or chorionic cells (Danks 1995). In the zinc-deficiency disease, acrodermatitis enteropathica, patients may live to adulthood and clinical diagnosis can be difficult. There is low zinc-binding capacity in the duodenum with likely defect of a transport-facilitating ligand. Low in vivo retention of ^{65}Zn or ^{69m}Zn is demonstrable by whole body counting, while in vitro tests of absorption by levels and specific activities of radio-zinc in plasma or 24-h urine samples can show the greatest difference from controls with no overlap (Van Wouwe 1989).

14.3.1.1.2
Macronutrients

Problems of absorption of energy-containing foodstuffs may be divided into those of maldigestion and essential malabsorption. Maldigestion is usually due to pancreatic insufficiency. The latter has various possible causes (e.g., liver or biliary tract disease) but is most often due to chronic pancreatitis, which in developed countries is usually associated with cystic fibrosis, alcoholism or diabetes, though in under-developed countries may be the result of protein-calorie malnutrition (Bass et al. 1986). Pancreatic insufficiency and also some other conditions (short-bowel syndrome, regional ileitis) display steatorrhea, but this may be present only in advanced disease and is cumbersome to measure quantitatively. Nuclear medicine has developed tests of digestion of fat by measuring the output of labeled CO_2 in the breath after ingestion of carbon (^{14}C or ^{13}C)-labeled fat (Sects. 14.2.1 and 14.2.2). Validity of such tests (and others described below) depend on the correctness of the supposition that the digestive process (and/or intestinal absorption) is the limiting factor in the rate of formation of CO_2 and that post-absorptive metabolism is normal. This may not be so in conditions such as diabetes, hyperlipemia, obesity, hyperthyroidism or liver disease. Early use of ^{14}C-labeled-triolein, tripalmitin or trioctanoin (Reba and Salkeld 1982) has been widely superseded by the adoption of the synthetic mixed-triglyceride (MTG), 1-3-distearyl, 2-^{14}C- or (now more often) ^{13}C-octanoyl glycerol, since the rate-limiting step occurs more definitely in the hydrolysis of the long-chain ester bonds, which then allows rapid absorption and metabolism of the labeled octanoic acid (Van Trappen et al. 1989). Various means have been used to validate or improve the specificity of the ^{14}C- or ^{13}C-TG breath test, or provide alternative in vitro isotopic tests for pancreatic insufficiency. Validation comes from observing restoration of normality of output of labeled CO_2 after enzyme therapy (Goff 1982; Weaver et al. 1998). There has been a double-tracer study with administration of ^{14}C-tripalmitin/^{3}H-palmitic acid followed by analysis of the $^{3}H/^{14}C$ ratio in serum lipids, which was more accurate for diagnosis than the breath $^{14}CO_2$ (Adlung et al. 1975). Another multi-tracer test has compared the absorption of bentiromide [con-taining tyrosylated para-aminobenzoic acid (PABA)], $^{13}C_6$-PABA and xylose by measuring with stable isotope methodology each of these substances in the serum at 1 h after ingestion (Deutsch et al. 1995). Conversion of bentiromide to PABA specifically requires chymotrypsin. The addition of xylose further tests intestinal absorption (see below). In this study patients with cystic fibrosis (hence pancreatic insufficiency) could be differentiated from those with celiac sprue (hence diminished intestinal absorption) as well as both groups from normal. It seems that the CO_2 breath tests (with either ^{14}C- or ^{13}C MTG) are the most practical for cost effectiveness, ease of performance, patient comfort, and fairly good sensitivity (Weaver et al. 1998). A comparison with non-isotopic methods found that one of the latter tests, i.e., fecal elastase 1 concentration, shows moderately higher sensitivity and specificity than the ^{13}C-MTG test for mild pancreatic insufficiency (Löser et al. 1998).

Digestion is directly associated with absorption in the case of disaccharidases, which may show specific, genetic variation or be non-specifically deficient due to diffuse mucosal injury. Best known of the specific types is isolated lactase "deficiency", which, since the condition is present in the majority of the world's population (mostly Asians, American Indians, some Africans and some Southern Europeans) is better designated as low lactose digestion capacity (LDC) (after weaning) vs. persistence of high capacity (HDC) (Bass et al. 1986; Flatz 1995). Genotype studies suggest the occurrence of two or more alleles of the involved gene(s) on chromosome 2 with the HDC gene(s) dominant, the LDC gene(s) recessive and control at the transcriptional level (Flatz 1995). However, there is also some evidence for phenotypic adaptation to milk-drinking. A ^{14}C-lactose/CO_2 breath test to determine the cause of the non-specific symptoms of abdominal pain, bloating and diarrhea was early described and later refined (Sasaki 1995). The hydrogen breath test, based on delivery of excessive lactose to the colon and thereby excessive production of H_2 by colonic bacteria, is currently preferred over CO_2 breath tests (Rings et al. 1994; Carroccio et al. 1998). For some subjects with unusual milk intolerance there may be a need to distinguish LDC from milk allergy (or from other kinds of intestinal disease). The carbon-labeled lactose breath test could provide more specificity than the H_2 breath test, particularly since the latter is also positive in other circumstances, e.g., bacterial intestinal overgrowth (see below). The specificity of the $^{14}C/^{13}C$-lactose intolerance test could be increased by comparing extent and time of oxidation to that of labeled lactulose, a disaccharide not absorbed but oxidized by colonic bacteria, or labeled galactose, a product of lactase. There could be a role for a ^{13}C-sucrose breath test, rather than the H_2 breath test (Reba and Salkeld 1982), in the genetic trait of sucrase-isomaltase deficiency, which is symptomatic from infancy and generally rare, although (interesting genetically) prevalent in Eskimos. Glucose-galactose malabsorption is a rare genetic disease present from infancy; this could be diagnosed most non-invasively by CO_2 output from appropriate ^{13}C-labeled sugars.

A classical test for mucosal absorptive deficiency from various causes (Bass et al. 1986) is chemical measurement of D-xylose in urine or plasma after load of D-xylose (a little-metabolized sugar). This test might also be done with carbon-labeled D-xylose.

14.3.1.2
Bacterial Abnormalities

Normally the colonic bacteria are excluded from the small intestine but may invade when there are motility disorders, as in scleroderma, amyloidosis or diabetes mellitus, when the bowel is shortened as in ileal resection, or in blind loops from surgery or disease (diverticulae, fistulae). One consequence of this bacterial overgrowth syndrome is interference with the normal enterohepatic circulation of bile salts, which are reabsorbed mainly in the distal ileum. Colonic bacteria, when present in the small intestine, abnormally deconjugate the bile salts, liberating constituent taurine or glycine. From the 1970s an often-used test for bacterial overgrowth has been the observation of excessive, or abnormally early, production of $^{14}CO_2$ in the breath from ^{14}C-glycine-cholate and a corresponding test with ^{13}C label is described (Sasaki 1995). However, some conditions, such as celiac (non-tropical) sprue (a heritable trait of mucosal allergy to ingested glutens in cereals) may cause false-negative results due to failure of damaged mucosa to absorb the liberated glycine (Shreeve 1984) and false-positives can occur when upper intestinal contents (including unabsorbed bile salts) are delivered unusually early to the colon. The important factor of time of output of labeled breath CO_2 can be evaluated by external monitoring of the in vivo transit of ^{99m}Tc-colloid or ^{99m}Tc-DTPA ingested with the labeled bile salts (Shreeve 1984). The isotopic breath test preferred by some over that with glycine-cholate is the ^{14}C- or ^{13}C-D-xylose test, since D-xylose is absorbed mostly in the upper small intestine and can more spe-

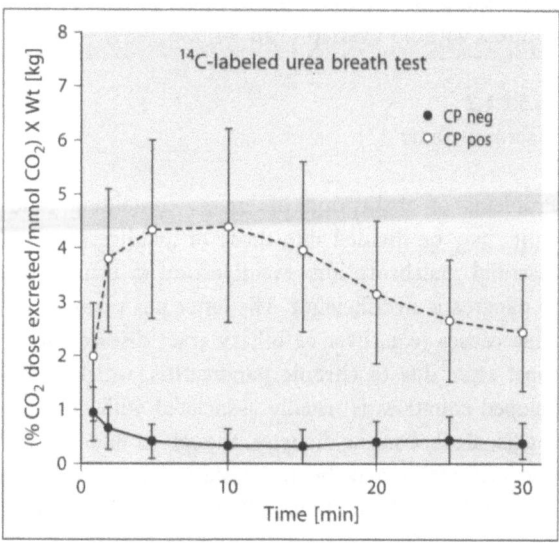

Fig. 14.2. Graphed results of $^{14}CO_2$ output by 16 patients positive for infection of gastric mucosa by *Campylobacter pylori* (*CP*) as defined by characteristic bacteria detected on gram-stained smear, culture or histology of biopsy specimens and 16 patients negative by such criteria. (From Marshall and Surveyor 1988, with permission)

cifically identify bacterial overgrowth in that region, rather than early dumping into the colon (Toskes and Donaldson 1993). There is also evidence that the ^{14}C-xylose breath test is superior to the H_2 breath test (after lactulose or glucose load) for diagnosis of bacterial overgrowth (Toskes and Donaldson 1993).

Another intestinal bacterial problem is that of infestation of the stomach and duodenum by *Helicobacter pylori*. The majority of the world's population, in various areas of the globe, is infected with this pathogen, which is a main contributing cause of peptic ulcer, atrophic gastritis and, sometimes, gastric cancer. Beside environmental and bacterial variation, genetic factors of the host are part determinants of infection and disease. Allele differences of the tumor necrosis factor (TNF)-alpha gene have been identified as risk factors (Kunstmann et al. 1999) and various studies indicate the importance of genetic variations of immune response (B and T lymphocytes). In the 1980s (Marshall and Surveyor 1988) (Fig. 14.2) an effective and now widely-used CO_2 breath test was developed on the basis that, unlike the gastric mucosa (and other resident bacteria), *H. pylori* contains urease. Thus, in *H. pylori*-positive subjects ingestion of ^{14}C- or ^{13}C-urea is followed by typically rapid appearance of tracer in the breath CO_2. Various conditions and protocol for conducting this test have been described (Raju et al. 1994; Sasaki

1995; Suto et al. 1999). Results of the breath test have helped to identify the degree of efficacy of one of the drugs (omeprazole) used to eradicate *H. pylori*, depending on polymorphisms in a host gene for cytochrome activity (Tanigawara et al. 1999).

While rates of gastric emptying of solids or liquids are often evaluated by scintigraphy in vivo, breath tests of CO_2 after ingestion of carbon-labeled acetate or octanoate are accurate indicators of gastric-emptying (Lembcke 1997). Features of gastric emptying (including genetic factors) and relevant in vivo tracer tests are discussed further in Chap. 24.

14.3.2
Liver

14.3.2.1
Proteins and Amino Acids

Synthesis and export of several plasma proteins is a major activity of the liver. Albumin is synthesized in greatest amount, derives only from the liver and has various functions (maintenance of osmotic pressure, carriage of fatty acids, steroids, metals, drugs, etc.). Inadequate synthesis is the known, or suspected, cause of hypoalbuminemia in liver disease, malnutrition, inflammatory states, hormone deficiencies (e.g., Type I diabetes) and the rare recessive genetic disorder of analbuminemia. Low serum albumin may also be due to excessive losses as in albuminuria (e.g., nephrotic syndrome) or protein-losing gastroenteropathy. Kinetics of turnover have been evaluated with various radionuclides attached to serum albumin; the most common tracer is ^{125}I-albumin. Recent use of the latter helped to identify the acute-phase response of the liver to "inflammation", with consequent decrease of hepatic synthesis of albumin, as the main cause of hypoalbuminemia in hemodialysis patients (Kaysen et al. 1995). Relatively slow turnover and varying shifts of distribution of albumin (intra- vs. extra-vascular) may not allow conclusion about momentary rates of hepatic synthesis from studies with labeled albumin per se (Friedman et al. 1996). One method for assessing current rates of synthesis employs injection of $Na_2{}^{14}CO_3$, which provides hepatic ^{14}C-arginine as a common precursor for both albumin and urea. A separate study (with [14]C- or ^{13}C-urea) must be done to determine kinetics and body mass of urea, which allows calculation of albumin synthesis from comparison of specific activities of albumin and urea. By this method low albumin synthesis

rates have been demonstrated in cirrhosis (Sherlock and Dooley 1993) and in various inflammatory states (Moshage et al. 1987). The method is complex, lengthy and more suitable for research than clinical use. A different approach has been that of administering stable isotope-labeled amino acids, e.g., ^{15}N-glycine, ^{13}C-leucine, or [2]H-phenylalanine (preferably in flooding doses), by which the rate of albumin synthesis is calculated according to the rate of change in enrichment of tracer in albumin (Ballmer et al. 1993; McMillen et al. 1996). By this method with ^{13}C-leucine there was good correlation of albumin synthetic rate with clinical scores of stage of cirrhosis (Ballmer et al. 1993). Use of the method (with [2]H-phenylalanine) has also included simultaneous measurement of synthesis of fibrinogen, a positive (increased) acute-phase protein (APP), and albumin, a negative (decreased) APP, to further sharpen the evaluation of the hepatic APP response to inflammation or injury (McMillen et al. 1996). The coordinated increased transcription of mRNA for positive APP and decreased mRNA for albumin is triggered by interleukin (IL)-1 or IL-6 from monocytes and the APP produced contain other immuno-active interleukins (Moshage 1987); this identifies both the cause and role of increased APP production as part of the overall immune response to bodily insult. The technique with stable isotopic amino acids involves minimal subject time (2 h), comparatively little radiation (^{125}I-albumin for plasma volume) and holds clinical promise, if the analyzing facilities are available.

14.3.2.2
Lipoproteins and Lipids

Because of much evidence linking hyperlipemia and high serum lipoproteins (LP) with atherosclerosis, coronary artery disease, diabetes, etc., many studies have been done with radioactive and stable isotopes (Packard 1995) in this metabolic area, which involves hepatic production and secretion of LP and other lipids, e.g., "free fatty acids" (FFA, attached to albumin). Some studies in humans have explored the effects of dietary differences. Particular increase of conversion of ingested ^{14}C-sucrose to serum triglyceride (TG)-fatty acids following exposure to high-sucrose (vs. high starch) diet pointed to increased hepatic lipogenesis (Wu et al. 1974). Whether or not this involved up-regulation of gene(s) is a question. Sustained hyperglycemia (by infusion) caused decreased splanchnic oxidation of ^{13}C-labeled fatty acids and increase in hepatic secretion of labeled VLDL-TG

(Sidossis et al. 1998). Other studies have used various tracers to investigate plasma lipid or LP kinetics in subjects disposed to different kinds of dyslipidemia. The latter are typically familial, usually multi-genic (in rare types, monogenic) traits, which are often much affected by environmental or disease states (alcoholism, diabetes, etc.) (Havel and Kane 1995). Abnormalities of serum levels and metabolism of LP can involve "alpha" or "beta"-LP, which are associated, respectively, with HDL or LDL/VLDL categories of LP, according to content of lipids and, hence, density. Both hyper- and hypobetalipoproteinemias are characterized by disease manifestations (Havel and Kane 1995). A study of familial hypobetalipoproteinemia with two tracers together, i.e., bolus injection of ^{2}H-palmitate and primed constant infusion of ^{2}H-leucine, showed much decrease of hepatic secretion of both TG and apo B-100 (protein moiety) in VLDL, but particularly the latter, suggesting VLDL particles unusually rich in TG (Elias et al. 1999). In a proband of a kindred with the trait of hyperalphalipoproteinemia (and high serum level of the "good" HDL cholesterol) the combined use of ^{125}I-labeled apo A plus endogenous labeling of the protein by primed constant infusion of ^{13}C-phenylalanine demonstrated markedly increased production of apo-A-1 (Rader et al. 1993). The mechanisms of action of two different kinds of serum cholesterol-lowering drugs (Schonfeld et al. 1998; Winkler et al. 1999) have been investigated in patients with mixed hyperlipidemias, using endogenous labeling with stable isotopes. A relatively short-term (only 10-min) infusion of ^{14}C-palmitate indicated more rapid turnover of FFA in plasma of obese subjects than in non-obese (Riemens et al. 1998). With infused ^{14}C-palmitate the plasma FFA were found to provide normally 80% or more of the TGFA in VLDL in the post-absorptive state, but less than 50% in alcoholic liver disease or in subjects on clofibrate therapy (Barter et al. 1972). Because of the extensive time involvement for most subjects, as well as the complexity of laboratory analyses, there has been little encouragement to adapt the above kinds of studies for larger clinical use. Some findings with breath tests could be more adaptable clinically. One is the observation of markedly diminished oxidation of orally-administered ^{14}C-octanoate to breath ^{14}CO$_2$ in human hepatic cirrhosis (Rabinowitz et al. 1978). Another is the recent demonstration of low ^{13}CO$_2$ production from ^{13}C-oleate contained in a chylomicron remnant-like emulsion when administered i.v. to subjects with Type III dyslipidemia but not found with

Type I (Redgrave et al. 2001). The test signifies the particularly pro-atherogenic defect of hepatic clearance of remnants in Type III (homozygosity for apolipoprotein E$_2$ alleles) (Watts et al. 1998). The importance of the dyslipidemias merits further effort to provide clinical applications of tracers.

14.3.2.3
Carbohydrates

The liver is the main site of formation of glucose and an important site of its storage (as glycogen) or utilization under load conditions. Other special monosaccharides, e.g., galactose and fructose, are metabolized primarily by the liver. Particular genetic disorders of these special sugars have been well characterized. Galactosemia stems from a group of relatively rare autosomal recessive traits which cause defects in one or another enzyme (uridyl transferase, kinase or epimerase) and for which corresponding mutations and their loci are identified (Shreeve 1974; Segal and Berry 1995). Besides acute symptoms from ingestion of milk by infants there is long-term damage to the nervous system and other organs with growth impairment, even after early diagnosis and corrective measures. Knowledge of parental genetic linkage and intrauterine diagnosis (amniotic cells) can be critically useful. Carbon-labeled galactose has long been helpful in diagnosis – in some ways with isolated red cells, but also by rate of appearance of labeled CO$_2$ in the breath after in vivo administration. Recent use of this technique with ^{13}C-galactose (now tending to replace ^{14}C-galactose) has further quantitated and extended previous information on different results among different genetic variants, including recognition of lesser severity of the mutant type in black galactosemics than in whites with the disease (Segal and Berry 1995; Berry et al. 1997) (Table 14.1). Oxidation within normal range in vivo does not guarantee absence of clinical disease but deficiency of oxidation can tell about type of mutation and severity of disease (Berry et al. 2000).

While chemical assay (blood or urine) in an in vivo galactose tolerance test has long been used to evaluate hepatic cirrhosis (or hepatitis), a breath test with oral load of ^{14}C- or ^{13}C-galactose was devised as a simpler test with similar diagnostic value (Shreeve et al. 1976; Caspary and Schaffer 1978). This test shows more distinction between cirrhotics and normal subjects than most common "liver function" tests. Performing the test in two stages, i.e., labeled

Table 14.1. Results of test of $^{13}CO_2$ in breath (percentage of dose expired at 1 h, 5 h) after intravenous bolus of 1-^{13}C-galactose given to subjects with classic galactosemic genotype, other genetic variants with galactosemia, heterozygotes or normal controls. All subjects received 7 mg labeled galactose/kg, except for patient 6 (classic galactosemic), one female and one male adult control subject, who received 1 mg 1-^{13}C-galactose/kg. (Modified from Berry et al. 1997, with permission)

Subjects	Age (years)	Sex	$^{13}CO_2$ 1 h	$^{13}CO_2$ 5 h	Genotype
Classic galactosemic					
1	31	F	0.17	3.1	Q188R/Q188R
2	30	M	0.03	2.2	Q188R/Q188R
3	22	F	0.05	2.6	Q188R/Q188R
4	16	F	0.14	3.2	Q188R/Q188R
5	7	F	0.31	3.6	Q188R/Q188R
6	28	F	0.05	1.2	Q188R/unknown
7	6	M	0.03	2.2	Q188R/unknown
African-American variant					
1	12	F	2	19	S135L/S135L
Duarte variant					
1	6	F	3	27	N314D/Q188R
2	29	F	4	32	N314D/N314D
Q188R Heterozygote					
1	35	F	3	28	Q188R/+
S135L Heterozygote					
1	37	F	4	31	S135L/+
Control					
6 Adults	18–37		3–6	27–47	+/+
4 Children	6–13		4–5	21–40	+/+

galactose with and without an acute dose of ethanol, may add sensitivity (Shreeve 1987). The galactose oxidation test could be useful in judging the results of liver transplant or the newer "bridge to transplant" by injection of frozen liver cells. An alternative to the galactose oxidation test is the breath test with ^{14}C- or ^{13}C-dimethyl-amino-antipyrine ("aminopyrine") (Schoeller et al. 1982). Demethylation and conversion to labeled CO_2 occurs in the liver by the microsomal mixed-function oxidase system. Because of this mechanism the aminopyrine breath test is unreliable in patients on barbiturates or other drugs or steroids. On the other hand, there is interference with the galactose breath test if the patient has diabetes (Shreeve 1987) or, of course, galactosemia. About the same distinctions between liver-diseased and normal subjects appear to be found with either the galactose or the aminopyrine breath tests. No studies have been performed with both tests in the same subjects. Genetic

characteristics (other than above) having a bearing upon outcome of either of these tests have not been explored. The breath test with 1-^{14}C-galactose has also been used to assess the degree of liver damage in "Indian childhood cirrhosis" (DaCosta et al. 1976). This disease has been traced to excessive copper in foods (made up in copper vessels) given to infants with neonatal cholestasis (which itself may have genetic propensities) (Danks 1995).

Hereditary fructose intolerance (HFI) is another rare recessive genetic trait, which is due to one or other of three or four known point mutations in the gene (on chromosome 9) for aldolase B in liver, kidney cortex and intestine (Gitzelmann et al. 1995). There are acute gastrointestinal symptoms and marked hypoglycemia after ingestion of sucrose or fructose or i.v. load of fructose. Infants are at risk for death. There can be long-term liver deterioration and poor growth. Signs and symptoms are non-specific, so that many cases may go undiagnosed. Hypoglycemia is due to inhibition of both gluconeogenesis and glycogenolytic enzymes by accumulated fructose-1-phosphate (Shreeve 1974). Diagnosis is established by liver biopsy and biochemical or (more sensitively) radioisotopic assay (Shin et al. 1983) for fructaldolase, or possibly by analysis of DNA from blood leukocytes. Cessation or diminution of glucose production (extent possibly dependent on genotype) can be evaluated by plateauing of slope of specific activity of previously-injected ^{14}C-glucose after fructose administration (Shreeve 1974). The method requires isolation of glucose from multiple blood samples. Rate and metabolic pathways of conversion of U^{13}-C-fructose to glucose after nasogastric infusion have been evaluated by the isotopomer analysis of ^{13}C distribution in glucose using in vitro NMR spectroscopy (Gopher et al. 1990) (see Sect. 14.2.2). Potential diagnostic value is claimed. A breath test for labeled CO_2 from ^{14}C- or ^{13}C-fructose is not feasible due apparently to compensatory fructose utilization by adipocytes in HFI (Gitzelmann et al. 1995). The technique of NMR spectroscopy of isolated plasma samples after infusion of U-^{13}C-glucose (Kalderon et al. 1989) has been used to distinguish Type I from Type III glycogen storage disease (GSD), which is another heterogeneous group of inborn errors of carbohydrate metabolism (Shreeve 1974). The isotopomer analysis with ^{13}C shows different glucose carbon recycling in GSD I vs. GSD III and is suggested as a non-invasive diagnostic test (Gopher et al. 1990). Other tests designed to measure hepatic gluconeogenic activity relative to other metabolic pathways, applicable to diabetes and other endocrine disorders, are discussed in Sect. 14.3.3.4.

14.3.3
Endocrine Pancreas

One of the outstanding achievements of early nuclear medicine was the development of radioimmunoassay (RIA) by Berson and Yalow. Among the several hormones and other proteins/polypeptides present in trace concentrations in plasma, insulin was the first target of measurement by RIA. There were somewhat unexpected and important consequences for the understanding of diabetes mellitus. Measurement of insulin and its precursors/components remains a vital aspect of early recognition and management of diabetes. RIA of other pancreatic islet cell products (glucagon, amylin) and of islet-cell antibodies are also of current interest. Other studies over the last four or five decades have sought to understand derangements of metabolism in diabetes by following the fate of key substrates (carbohydrates, fats, amino acids) labeled with carbon, hydrogen or other isotopes. Isolation and analysis of products in blood, breath, urine, etc. have not usually been "streamlined" for wide clinical application, but could be so. Both the hormonal and the metabolic assays can now be applied anew with a view to better correlation with the genetic make-up of individuals or groups as can be learned from the newly-revealed genome.

Diabetes, as defined by fasting hyperglycemia and glycosuria or "pre-diabetes" by impaired glucose tolerance is not any one disease (or even two) but rather a symptom complex, much as is anemia, which likewise has a multitude of possible causes. "Secondary diabetes" is a term given to the occurrence of diabetic signs or symptoms in conjunction with a more definitive, and usually more prominent, disease. In a panoply of more than 60 such diseases (typically recognized as genetic) (Rotter et al. 1990) the diabetic trait ranges from evident Type I, insulin-dependent diabetes (IDDM) to some kind of insulin resistance, which mimics the far more common, maturity-onset, Type II, non-insulin dependent diabetes (NIDDM). In some of these cases of secondary diabetes, e.g., cystic fibrosis, there are aspects suggestive of both IDDM and NIDDM (Hardin et al. 1999).

14.3.3.1
Antibodies in IDDM

There is abundant evidence that both Type I and Type II "idiopathic" diabetes are polygenic in origin and that, also, both are affected by environmental conditions. The most well-known marker and likely cause of IDDM is development of anti-islet-cell antibodies to antigens of the beta cell, usually insulin and/or glutamic acid decarboxylase (Taylor 1995). Certain heritable haplotypes of the HLA antigens in the major histocompatability (MHC) cluster of genes appear to predispose to the various islet-cell antibodies. Other kinds of autoimmune disease may accompany IDDM. That the genetic component of IDDM is tempered by environmental influences is indicated by studies of close family members. Thus, concordance of IDDM within pairs of monozygotic twins is only 30–50% (Rotter et al. 1990; Taylor 1995). Suspected environmental factors include previous viral infection and feeding of cow's milk, containing bovine insulin and albumin, either of which can trigger immunity (Taylor 1995; Vaarala et al. 1999). Furthermore, there may be rearrangements of genes in cells of the immune system, which confounds the expected orderly inheritance (Taylor 1995). The incidence of IDDM varies widely throughout the world (from lowest in Japan to highest in Scandinavia), suggesting genetic cause, though environmental factors could be operating. Certain combinations of different HLA antigens (e.g., DR3/DR4 compound heterozygotes) appear to dispose individuals to earlier onset and more family linkage of IDDM than one or other antigen alone (Rotter et al. 1990; Kobayashi et al. 1993). On the other hand, the expression of a particular allele of one HLA antigen appears to protect against the development of IDDM (Gianani et al. 1996). Serum levels of predisposing antibodies are relatively high at the time of onset of IDDM and subsequently decline. Indeed, high antibodies may be present years before clinical or other laboratory evidence and presage the occurrence of IDDM. Thus, the RIA for various antibodies to HLA antigens is highly valuable in the prediction of possible future IDDM among family members of a proband with the disease.

14.3.3.2
Insulin/Proinsulin

The early measurement of total serum immunoreactive insulin (IRI) provided another basic distinction between Type I and Type II diabetes, i.e., low fasting and very low response of insulin to glucose stimulus in Type I (presumably due to damage by islet-cell antibodies), but, in sharp contrast, normal or even elevated serum insulin in fasting or fed state in Type II, though with time the latter shows inadequate insulin output. While total IRI remains useful in judging stage of disease or response to drugs, attention has shifted to separate RIA of the precursor, proinsulin (PRO) and to the C-peptide split off in the conversion of proinsulin to insulin (and secreted stoichiometrically with insulin). In patients with IDDM on exogenous insulin, analysis of C-peptide is a superior indicator of insulin production (Taylor 1995). Due to the low molecular weight, C-peptide reactivity (CRP) is conveniently measured in the urine of children with excretion best expressed relative to creatinine concentration (Kuzuya et al. 1978). Proinsulin has a lower biological activity than insulin, and, therefore, measurement of total IRI may over-estimate the level of effective insulin. Proinsulin-like components (proinsulin and conversion intermediates) may constitute an exorbitant fraction of total IRI in both IDDM and NIDDM (De Fronzo et al. 1992; Kahn and Halban 1997). The fraction tends to increase in parallel with the degree of hyperglycemia. The PRO/IRI ratio is a marker for high risk of developing disease in post-puberty first degree relatives of patients with IDDM (Rodriguez-Villar 1997). Nondiabetic Mexican-Americans with a parental history of Type II diabetes have excessive proinsulin relative to insulin compared to subjects without such a history (Haffner et al. 1995). First degree relatives of NIDDM subjects lack the normal oscillations in insulin secretion following intravenous glucose (Rotter et al. 1990; De Fronzo et al. 1992). Some progress has been made in relating these abnormal patterns of proinsulin processing and insulin production to specific mutations in the genome (Tager 1984; Steiner et al. 1995). Presently known mutations in the insulin gene appear to decrease the attachment of insulin to peripheral insulin receptors (Rotter et al. 1990; Taylor 1995).

14.3.3.3
Etiology of NIDDM

The predominant abnormality present in NIDDM and in subjects with impaired glucose tolerance (IGT), who may or may not ever develop NIDDM, is insulin resistance with accompanying hyperinsulinemia. Insulin resistance without clinical diabetes or even IGT is actually quite common in the general, more elderly

population. Polymorphisms of the insulin receptor gene (on chromosome 19) are known in considerable detail (Rotter et al. 1990; Taylor 1995) and are logical causes of insulin resistance. Autoantibodies to the insulin receptor gene can arise and contribute to one kind of NIDDM (Taylor 1995; Gorden 1997). A mutation of the glucokinase gene appears responsible for "maturity-onset diabetes of the young" (MODY) in some patients. Yet even within the category of MODY there is clinical diversity and likely polygenicity (Rotter et al. 1990). (While MODY may be a specific genetic group, NIDDM generally is now appearing much more commonly in younger, even juvenile subjects.) Some mutations of mitochondrial DNA account for a small sub-set of NIDDM which is transmitted through the maternal line (Suzuki et al. 1994; Taylor 1995). There is general consensus that some genetic disorder of insulin production and release must be a factor additional to peripheral insulin resistance to precipitate clinical NIDDM. "Exhaustion" of the beta cell as a consequence of over-stimulation by hyperglycemia has been a popular theory, but there is current supposition of a further genetic trait predisposing to the gradual decline of insulin output.

A variety of polygenic patterns in NIDDM is indicated by the considerable variety of kind and incidence of clinical disease seen according to ethnic differences. Populations in Mexico, East Asia, Asian India, Polynesia, and (more tellingly) these groups (and Southwest Native Americans) when living in or by standards of the Western world exhibit NIDDM more prominently and at earlier ages than the Caucasian population (Taylor 1995). Most of these ethnic groups with high incidence of NIDDM also have common occurrence of obesity, which, as is well known, is even more common within the diabetic groups (including Caucasian). While cause and effect are not easy to separate, rodent models, as well as clinical and laboratory evidence in humans, suggest a commonality of genes disposing to both IDDM and obesity. Further strong evidence for polygenicity and critical grouping(s) of genes to produce clinical NIDDM is a concurrence of almost 100% for the disease within monozygotic twins and high incidences of one or another phenotypic marker for the disease among first-degree relatives. The familial linkages are much stronger than for IDDM. Nevertheless, despite detailed knowledge of mutations of genes for both production and sites of peripheral action of insulin, virtually all such known mutations are either low in incidence or low in penetrance to phenotype, so it is believed that the majority of genetic abnormalities required to produce clinical NIDDM remain undiscovered (Rotter et al. 1990; Gorden 1997). Further progress on this front depends on finer phenotypic classification of sub-sets of NIDDM (types of Type II).

14.3.3.4
Metabolism Tracers

Besides contributing radioimmunoassay of insulin and proinsulin to the understanding of phenotypes of diabetes, nuclear medicine may utilize radioactive or stable isotope tracers to delineate biochemical abnormalities by in vitro assays with isolated blood cells or by measurement of tracer turnover/appearance in products of blood, breath or urine after in vivo administration of tracer. Circulating monocytes share the diabetic diathesis and a test for binding of ^{125}I-insulin to isolated monocytes showed an inverse relationship between capacity for binding and the fasting insulin concentration or area under the insulin curve after glucose stimulation (Olefsky and Reaven 1977). More detailed studies with the same tracer agent and blood cell type (Trischitta et al. 1989) have indicated that cells from obese NIDDM patients have a slower redistribution of insulin receptors from the surface to the cell interior after exposure to unlabeled insulin. In such down-regulated cells from obese NIDDM patients or obese nondiabetics there is slower recovery of binding capacity, slower release of internalized insulin and seemingly less intracellular degradation of insulin than for non-obese, nondiabetic control subjects. As in NIDDM, monocytes from patients with untreated IDDM have a decreased rate of internalization of ^{125}I-insulin (Carpentier et al. 1986). It was suggested that the "purpose" of this phenomenon is to preserve the insulin signal despite down-regulation of cell surface receptor.

Early studies of turnover of ^{14}C-glucose (Shreeve et al. 1956; Searle et al. 1959; Shreeve 1966) attempted to better typify diabetes than the crude phenotypic indicator of IGT, which could be due either to increased (or abnormally maintained) hepatic glucose output (HGO) or to decreased glucose removal (or both). Such studies generally suggested that control of HGO is a more dominant factor, or earlier response to insulin, than decreased removal. More recent studies have tended to employ glucose labeled with deuterium and/or tritium (Consoli and Nurjhan 1990; Berrish et al. 1995; Landau et al. 1995; Thorburn et al. 1995; Diraison et al. 1998) or deuterated

water (Landau 1997). In addition, some of these studies have used carbon-labeled acetate, lactate and/or bicarbonate with isolation and analysis of beta-hydroxy-butyrate in plasma or of glutamine (trapped by phenylacetate) and urea in the urine. This allows more specific evaluation of the contribution of gluconeogenesis to HGO relative to rates of oxidation in hepatic Krebs cycle. Some of these studies (Consoli and Nurjhan 1990; Thorburn et al. 1995) have supported the earlier evidence for increased HGO and gluconeogenesis in NIDDM, while others (Berrish et al. 1995; Diraison et al. 1998) are in contradiction. Different results could be due either to different experimental methods or to different kinds (genotypes?) or disease stages of patients in experimental groups. In any case, these sorts of tracer studies do not seem promising for large clinical application, due to complexity of laboratory analyses as well as involved protocol for subjects.

Another approach has been the analysis of alternative fates of ^{14}C contained in administered three-carbon intermediates, lactate or pyruvate, or in glycerol (Shreeve 1966). Such studies showed markedly increased conversion of tracer to blood glucose and decreased conversion to expired CO_2 in diabetic subjects with a prompt effect of insulin or tolbutamide to restore the balance toward normal. The amount of ^{14}C in blood glucose can be affected by the reactions of glycogenesis or glycolysis. The latter could be increased by hyperglucagonemia, but this has not appeared to be a major factor (De Meutter and Shreeve 1963). Increased hepatic glucose influx or increased size of glucose space are other factors possibly affecting specific activity of ^{14}C in glucose. However, the reciprocal relationship between ^{14}C in blood glucose and in expired CO_2 derived from three-carbon intermediates (Fig. 14.3) drew attention to possible direct effects of insulin on the liver and enzymes (pyruvate kinase, phospho-enol-pyruvate carboxykinase, pyruvate dehydrogenase) likely responsible for the alternative fates of gluconeogenesis from or oxidation of the three-carbon compounds. Subsequent evidence on effect of insulin (either positive or negative) on gene expressions which closely control the activation or deactivation of these enzymes for pyruvate metabolism (Taylor 1995) help explain the earlier tracer findings. Extension of such tracer studies to larger patient groups might preferably use the stable isotope, ^{13}C, at least for glucose analysis, since required amounts of ^{14}C, by current analytic techniques, are several-fold the customarily approved doses for clinical application (Landau and Shreeve 1991). Acceptability of ^{14}C

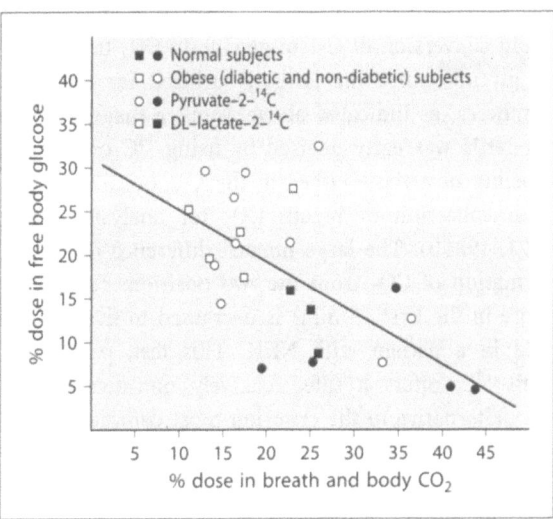

Fig. 14.3. Comparative disposition of ^{14}C between CO_2 and glucose at 1 h after single intravenous injection of a trace amount of labeled pyruvate or lactate into fasting normal-weight and obese human subjects (American Caucasian, mixed sex). (From Shreeve et al. 1968, with permission of the American Society of Clinical Nutrition)

could change with development of more sensitive analytic techniques (Sect. 14.2.1.1) or according to new evidence on tolerance to low doses of ionizing radiation (Feinendegen and Pollycove 2001) (see Chap. 34). Conversions to expired CO_2 are measurable by presently accepted quantities of ^{14}C and involve simple, practical analyses (Sect. 14.2.1.1). A possible approach to defining specific metabolic sites or regions of action of insulin would employ the two-carbon tracers, ^{14}C and ^{13}C, to measure simultaneously CO_2 production from two closely related compounds, e.g. pyruvate and acetate or fructose and glucose. An abnormal ratio of CO_2 formation from the two hexoses could suggest defects in a glucokinase gene or the glucose transporter system. It seems possible that different mutations of the insulin (pre-proinsulin) gene (Steiner et al. 1995) could cause differential effects on the manifold metabolic actions of insulin in the post-insulin receptor response. This possibility could be explored by comparing the degree of abnormality in different metabolic systems with the aid of selected labeled substrates. Another use for the above-described kinds of studies is evaluation of the sites and effectiveness of anti-diabetic drugs (Wiernsperger and Bailey 1999). Early diagnosis of diabetes by the use of well-chosen tracers given in vivo and practicable versions of in vitro analysis of

products (e.g., breath CO_2, blood glucose) is another potential goal.

Some studies with labeled compounds other than those involved in glucose metabolism may offer an opportunity for tests of diabetes by in vitro analyses after in vivo administration. In obese and obese NIDDM human subjects there is a mild galactosemia and logically there could be intolerance to a galactose load (Shreeve 1974). After intravenous galactose load containing ^{14}C-galactose there was, indeed, reduced production of $^{14}CO_2$ in patients with IDDM relative to normal subjects (Shreeve 1987). The degree of abnormality was no more than seen in the fasting blood glucose level, but could be more specific than the latter by signifying a biochemical disorder (causing high NADH) in the liver. The effect of anti-diabetic drugs possibly to cause hepatic damage by inhibition of the microsomal detoxifying enzymes has been investigated by the ^{14}C-aminopyrine breath test with findings that indicate potential damage by tolbutamide (or the latter when combined with any other such inhibitory drug; Kaminska-Gelwas and Scoczynski 1991) but no effects of glyburide (Juan et al. 1990). Disturbances of fat metabolism in diabetes might be followed by in vitro analyses of change of isotope tracer concentrations in blood fatty acids, triglycerides or lipoproteins, but no simple, practical test has been suggested. Using ^3H-leucine and glycemic and insulinemic clamps with kinetic modeling, it has been noted that patients with NIDDM had a very low (3–9%) suppression of hepatic production of VLDL1 apo B compared with 50% decrement in normal subjects (Malmström et al. 1997). The findings may help explain hypertriglyceridemia in NIDDM. The experimental protocol is laborious and little suited, as such, for general clinical application.

14.3.3.5
Insulin in Breast Cancer

Recent findings (Altman 2000) suggest a use for radioimmunoassay of insulin in a disease other than diabetes, i.e., breast cancer. Women with such cancer and hyperinsulinemia have about an eight-fold incidence of recurrence of disease (grade and stage) after standard therapy compared to women with the disease but normal insulin levels. The hyperinsulinemia was found among normal-weight and thin, as well as obese women.

14.3.4
Nervous System

14.3.4.1
Amino Acids

14.3.4.1.1
Glycine

Glycine is one of the three major amino acid neurotransmitters in the brain [the others are glutamic acid and gamma-amino-butyric acid (GABA)], and also plays an important role in neurotransmission in the spinal cord. In the brain the effect of glycine is excitatory, serving as a co-agonist for the N-methyl-D-aspartate (NMDA) glutamate receptor, while in the brain stem and spinal cord glycine is predominantly inhibitory. These opposite effects underlie the signs and symptoms of non-ketotic hyperglycinemia (NKH), a rare autosomal recessive trait causing, in the homozygous state, devastating neurological disease in early infancy and usually early death (Nyhan 1984b; Hamosh et al. 1995). In the classical phenotype early (usually neonatal) signs include lethargy, hypotonia, refusal to feed, seizures, ophthalmoplegia, apneic episodes and hiccups. Those who survive the first few months have severe mental retardation. Different degrees of severity in the same or different families and other aberrant cases suggest genetic heterogeneity. The general world incidence is not known, though in Finland, where it appears to be most common, the incidence is 1/12,000 births.

In NKH there is a defect in the glycine cleavage system (GCS), which accounts normally for the major metabolic pathway of glycine. The GCS is a four-peptide complex found in the inner mitochondrial membrane of the liver, kidney, brain and placenta. The gene for one of these enzymes (P-protein) maps to chromosome 9 and a missense mutation has been defined; the genomic location of others is not known, but DNA and RNA for the next involved enzyme (H-protein) have been cloned (Hamosh et al. 1995). The GCS carries out the decarboxylation of glycine with C-2 of glycine being transferred as a one-carbon unit to tetra-hydrofolic acid. From there the C-2 of glycine is used in the synthesis of purines, serine and other compounds. Although glycine has several different metabolic fates (synthesis of proteins directly or via serine, gluconeogenesis, formation of creatine, glutathione, conjugates and porphyrins) the clinical effects of NKH appear to be essentially neurological. When

the GCS functions normally there is a much more rapid conversion of C-1 of glycine to CO_2 than of C-2 of glycine, since the latter is utilized for metabolic syntheses, as indicated above. Thus, a diagnostic test for NKH was early devised by using ^{14}C or ^{13}C in a labeling of glycine either in the C-1 or C-2 position with collection of breath CO_2 for analysis (Nyhan 1971, 1984b). The large normal difference in rates of formation of CO_2 from the two positions (30-fold or more in the first 15 min) is decreased to five- or tenfold in a patient with NKH. This test, particularly with ^{13}C, offers a safe, relatively non-invasive and easy alternative to the criterion most commonly used, i.e., the ratio of glycine concentration in cerebrospinal fluid (CSF) to that in plasma. The breath test is not valuable in recognizing heterozygotes because of overlap of results with normal subjects. Isotopic assay of metabolism of labeled glycine in isolated transformed lymphocytes might recognize the carrier state (Hamosh et al. 1995). The breath test could be used to judge the efficacy of therapy with benzoate, which disposes of glycine as hippuric acid in the urine, affording improvement in some subjects, especially when combined with other therapy (Hamosh et al. 1995).

Another neurological disease associated with glycine metabolism is Lesch-Nyhan (L-N) syndrome. This autosomal recessive disease is caused by one or more (possibly several) mutations of a gene on the X chromosome which codes for the enzyme, hypoxanthine-guanine phosphoribosyl transferase (HGPRT). HGPRT is present in virtually all tissues and is a key enzyme in the "salvage pathway", which reconverts the purine bases to mononucleotides. In humans up to 90% of free purines are recycled in this manner (Rossiter and Caskey 1995). In the absence or even partial deficiency of HGPRT there is a substrate-engendered increase in de novo synthesis of purines, which incorporates glycine as a whole plus one-carbon units from C-2 of glycine into the basic molecular structure. Consequent to the excess synthesis of purines there is an over-production of uric acid with hyperuricemia, gouty arthritis and nephropathy due to urate stones. The uric acid abnormalities appear late (often not until adulthood), whereas within the first year of life delay in motor development, hypotonia, choreoathetosis and/or spasticity are seen. Later, mental retardation, growth retardation, dysarthria, aggressive behavior and (most distinctively) self-mutilation are characteristic. The self-mutilation usually appears around age 2 or 3 years and may finally make the clinical diagnosis. The connection between the neurological disorders and purine metabolism is still not clear, but animal studies and some clinical evidence suggest subtle changes in dopamine function (Rossiter and Caskey 1995). The brain may be most severely affected because of its particular limitation on compensatory de novo purine synthesis, yet other tissues are affected, as indicated by the occurrence of megaloblastic anemia and frequent infections (Galjaard 1980a).

The full phenotype of L-N syndrome has an incidence of only one in approximately 300,000 births, occurs with equal frequency in a wide range of racial groups and there is no greater incidence within inbred populations (Rossiter and Caskey 1995). The genetic factors are quite heterogeneous. Thus, some subjects are found to have no detectable HGPRT activity in erythrocytes, lymphocytes or skin fibroblasts but no neurological manifestations. Conversely, some patients with L-N syndrome have residual HGPRT activity (10–15% of control) (Galjaard 1980a). Some patients who have familial primary gout have low HGPRT and this may be found in subjects with or without family history of L-N syndrome. It is said that mutations at the HGPRT locus are so heterogeneous that almost every affected family (with either gout or L-N syndrome or both) carries a different alteration (Rossiter and Caskey 1995). (However, partial HGPRT deficiency occurs in less than 5% of all cases of gout.)

Radiotracing with either labeled glycine or labeled purines can aid in the post-natal or pre-natal diagnosis of L-N syndrome or detection of female carriers. Some, but not all, L-N patients have elevated serum uric acid and urinary excretion from birth. Over-production in L-N (or subjects with only gout from partial HGPRT deficiency) can be verified by i.v. administration of ^{14}C- or ^{13}C-glycine and urine collection for 7 days – normally about 0.1% of isotopic carbon appears in uric acid, but in L-N syndrome up to 2.5% (Nyhan 1971). This method can also indicate increased rate of de novo purine synthesis in female carriers (Galjaard 1980a). Biochemical assay for HGPRT is based on incubation of cells (amniotic, skin fibroblasts, hair follicles) with (8-^{14}C)-hypoxanthine and separation of labeled end-products by electrophoresis or chromatography. Micromethods with small numbers (2,000–8,000) of cultured cells have used a combination of ^{14}C-hypoxanthine and ^{3}H-adenine to allow evaluation by direct measurement (without product separation) by liquid scintilla-

tion counting of the ^{14}C/^{3}H ratio. Use of skin fibroblast and ^{3}H-hypoxanthine have enabled detection of HGPRT deficiency by autoradiography of single cells in culture for detection of female carriers (Galjaard 1980a,b).

14.3.4.1.2
Phenylalanine

A group of recessive disorders now known as the hyperphenylalaninemias (HPA) was first exemplified, in the 1930s, by the recognition of phenylketonuria (PKU) due to excessive phenylpyruvic acid in the urine of infants or young children with mental retardation. This was soon associated with elevated serum levels of phenylalanine and subsequently an enzymic defect of phenylalanine hydroxylase (PAH) in the liver (Nyhan 1984a; Scriver et al. 1995). In the absence (or very low value) of PAH phenylalanine is transaminated to phenylpyruvic acid. There are geographical and ethnic differences of incidences in this uncommon but neurologically devastating disorder: 1 in 10,000–20,000 births in Caucasian and Asian populations generally with variants in Yemenite Jews, Scots and Gypsies (1:40 in the latter). The clinical manifestations in classical PKU include 1/3 incidence of cerebral palsy and 1/4 with seizures besides the typical increasing mental retardation and demyelination in the absence of early and continuing treatment by dietary phenylalanine restriction.

PKU has been studied more than any other inborn error of metabolism. There are various reasons for this. Firstly, the relatively easy and effective treatment by restriction of phenylalanine intake justifies widespread testing of infants (usually by competitive inhibition of bacterial growth by phenylalanine or a fluorometric method) (Nyhan 1984a). Secondly, there has been gradually unfolding knowledge of a large degree of heterogeneity of allelic types of classical, symptomatic PKU as well as lesser, more benign cases of non-PKU hyperphenylalaninemia. There are over 100 mutations (missense, nonsense, deletions, insertions) of the PAH gene on chromosome 12. Most probands are compound heterozygotes (Bartholome et al. 1984; Scriver et al. 1995). Furthermore, the finding that tetrahydrobiopterin (BH$_4$) is a co-factor for PAH and that at least four enzymes of production or regeneration of BH$_4$ determine its availability and therefore the activity of PAH adds more complexity to the phenotypic array of PKU or non-PKU hyperphenylalaninemia (Scriver et al. 1995; Blau et al.

1996). Thirdly, there is the intriguing inter-organ connection between PAH deficiency in the liver and the clinical and pathological manifestations virtually all in the CNS. One explanation is the BH_4 requirement of other amino acid hydroxylases, i.e., tyrosine and tryptophan hydroxylases, which, of course, determine the availability of the important neurotransmitters, dopamine, norepinephrine, epinephrine and serotonin (also thyroxine and melanin). But this explains only a minor fraction (1–2%) of HPA (Scriver et al. 1995). There is competition among amino acids for transport into the brain and the very high levels of phenylalanine may inhibit uptake of tryptophan, tyrosine and other amino acids or cause their sequestration in parenchymal tissues (Scriver et al. 1995). Another explanation is suggested by the inhibition of 5-hydroxytryptophan decarboxylase and glutamic acid decarboxylase (generator of the neurotransmitter, GABA) by phenylalanine and its metabolites (Nyhan 1984a). There are extensive genomic and proteomic homologies among the aromatic amino acid hydroxylases (Scriver et al. 1995) and one could posit a feed-back inhibition of genetic expression of any of the hydroxylases by a high cellular concentration of phenylalanine. The same might be said about the various decarboxylases, particularly since the demonstration of a general aromatic L-amino acid decarboxylase deficiency as another inborn error affecting neurotransmitter synthesis (Hyland et al. 1992).

A few nuclear tracer tests have been employed in the HPA and others could be suggested. Assay of the conversion of infused heptadeuterated phenylalanine to labeled plasma tyrosine showed a direct correlation to the more invasive analysis of hepatic PAH activity by needle biopsy with similar tracer test in vitro (Scriver et al. 1995). Furthermore, the assay of in vivo metabolites is more physiological and can identify variants, e.g., cases due to BH_4 deficiency (Bartholome et al. 1975; Scriver et al. 1995). Study by prolonged infusion of (ring-^{2}H)-phenylalanine and (1-^{13}C)-tyrosine with analysis by gas chromatography/mass spectrometry of curves of plasma concentration of labeled amino acids (Clarke and Bier 1982) provided superior accuracy of rate of in vivo PAH activity, but the method seems too cumbersome for clinical use. Of more practical use would be the measurement of rate of delivery to body water of tritium or deuterium from injected ring-labeled phenylalanine (Milstein and Kaufman 1975) or the rate of appearance in respiratory CO_2 of ^{14}C or ^{13}C from uniformly-labeled phenylalanine given orally (Lehmann

et al. 1986). Besides readily diagnosing classic homozygous PKU, these tracer techniques may detect PKU heterozygotes (Lehmann et al. 1986) and differentiate between heterozygotes for PKU and homozygotes for HPA (Milstein and Kaufman 1975). Studies assessing in vivo conversions of relevant amino acids or intermediates [tyrosine, L-DOPA, 5-hydroxytryptophan (5-HTP), glutamic acid] to end-products [homovanillic acid (HVA) from dopamine, vanillylmandelic acid from epinephrine or norepinephrine, 5-hydroxyindole acetic acid (5-HIAA) from 5-HTP, GABA from glutamate] in urine or (better) in CSF would be of interest. Breath CO_2 analyses after administration of targeted carbon-labeled precursors has particular appeal for handy clinical application and could be of value in identifying heterozygotes for PKU or HPA, as well as the possible effect of such heterozygosity on availability of catecholamine or serotonin neurotransmitters.

14.3.4.1.3
Tryptophan

Tryptophan (TP) is crucial for CNS function because of its conversion through 5-hydroxytryptophan (5-HTP) and then decarboxylation to 5-hydroxytryptamine (5-HT), also called serotonin. The latter appears to act mainly as a central inhibitory synaptic neurotransmitter in many regions of the brain (Van Woert et al. 1977). Evidence from ^{11}C-TP or ^{11}C-HTP in monkeys (Hartvig et al. 1992), ^{14}C-HTP in cats (Miyakoshi et al. 1980) and ^{3}H-tryptamine in post-mortem human brain (Mousseau and Butterworth 1994) indicates a broad spread of radioactivity with highest concentration in the striatal region, especially hypothalamus, thalamus, hippocampus, raphe nucleus and adjacent areas. Neuropsychiatric diseases exhibiting abnormal concentrations or kinetics of serotonin in brain and/or abnormal amounts of end-product, 5-HIAA, in CSF (Willmer 1985) include schizophrenia, obsessive-compulsive disorder, social anxiety disorder, bipolar disorder or monopolar depression, alcoholism and drug addiction. Neurologic disorders involved can be post-anoxic intention myoclonus from various causes (Van Woert et al. 1977), Parkinson's disease (Mousseau and Butterworth 1994), autism (Tanguay 2000) and others.

Discussion of the many sorts of mental or neurological disorders with suspected involvement of the serotonergic neurotransmitter system, as listed above, is beyond the scope of this chapter. One of the disorders, i.e., depression with its attendant mood disorders, social anxiety, decrease of sexual desire and other pleasure-seeking, insomnia, tendency to alcoholism and to suicidal ideation or behavior, has been particularly linked to dysfunction of the serotonergic system. Endogenous bipolar or unipolar depression with seemingly slight or no apparent environmental cause is recognized as pathological and (while variable) much more common than the other "inborn errors" discussed in this section above. Genotypic and organic phenotypic correlates have been actively sought and found. Incentive for this search is partly fueled by the success of pharmacological treatment so far and the promise of more success as finer biochemical knowledge unfolds.

Two areas of knowledge about serotonin availability and action in the brain have been pursued in parallel. One has to do with the pre-synaptic transporter (re-uptake) activity and the post-synaptic receptor activity. Genotyping for the involved proteins has been extensively done. There has been a focus on a promoter region for the 5-HT transporter gene (5-HTTLPR), which shows polymorphisms with either a short (s) allele or long (l) allele. The degree of occurrence of the s allele (in which there is less transcription of the transporter gene) correlates positively with the severity and repetitiveness of suicide attempts (Gorwood et al. 2000). Allelic differences do not correlate with depression per se, but there is a co-morbid effect of dependence and alcohol-dependence on suicidal behavior. Others (Mann et al. 2000) do find a correlation of 5-HTTLPR genotype with major depression. Still others (Hallikainen et al. 1999) have also found a higher frequency of the s allele among more impulsive and violent-type alcoholics than among non-violent alcoholics or matched controls. The s allele has been incriminated as being a risk factor for seasonal affective disorder (Rosenthal et al. 1998). Yet, there is conflicting evidence (from autoradiography of brain samples with ^{3}H-analogues) that a higher frequency of the l allele in the regulatory region of the transporter gene is associated with suicide in depressed subjects (Du et al. 1999). Interestingly, there are allelic differences of 5-HTTLPR between alcohol-tolerant and alcohol-intolerant subjects (Heinz and Goldman 2000). At very least, heterogeneity of both genotypes and phenotypes among depressives and alcoholics is suggested. In regard to the 5-HT receptor gene/protein, both quantitative autoradiography of post-mortem brain (pre-frontal cortex) of suicides vs. non-suicides (Mann et al. 2000) and PET imaging studies of patients with major depression (Yatham et al. 2000) indicate decreased 5-HT receptor binding, although there are also negative findings in attempts to link 5-HT receptor polymorphisms with bipolar disorder (Arranz et al. 1997; Massat et al. 2000).

The other line of investigation and knowledge about serotonin involvement in depression and related disorders has to do with metabolism of TP, the amino acid precursor, and the series of metabolic reactions leading to and from 5-HT. It is in this area that some in vitro tracer studies may be able to characterize in vivo metabolic disorder and possibly to categorize genetic types of depressions. There is a report (Bennett et al. 2000) of significant polymorphism of intron 7 of the gene for tryptophan hydroxylase (the rate-limiting enzyme for serotonin synthesis) among suicide victims. Another genetic type of neuroaffective disorder appears to be related to variations in the promoter region of the MAO-A gene (Manuck et al. 2000). When given systemically TP is mainly subject to use in protein synthesis and is mainly degraded by hepatic tryptophan pyrrolase in the "kynurenine pathway". Only a small fraction is used for production of 5-HT in the brain and TP so used competes with other neutral amino acids in passage across the blood-brain barrier. There is much evidence, however, that the amount of TP that gets to the brain is critical for inducing neuroaffective disorder. Thus, a well-known device for testing for, and sometimes revealing, tendency to depression is imposing a TP-depleted diet. Chronic alcoholics may have an increase in the kynurenine pathway with the resultant deprivation of TP for the production of 5-HT in the brain leading to depression (Badawy 1999). Very different to the fate of tryptophan is that of 5-hydroxy-tryptophan (5-HTP) when given systemically. This product by hydroxylation of TP readily crosses the blood-brain barrier without competition of other amino acids and can be more efficiently directed to serotonin production in the brain. This is aided by concomitant administration of a peripheral aromatic amino acid decarboxylase inhibitor (Van Woert et al. 1977; Willmer 1985). It has been shown that 5-HTP is much more effective in treating depression than is TP.

In the brain 5-HT is readily converted by oxidation to 5-HIAA and the latter is present in much greater concentration in the brain and CSF than is 5-HT. Many studies have shown that low 5-HIAA in the lumbar CSF (with or without the use of probenecid to inhibit absorption of 5-HIAA) correlates with depression, though almost as many studies do not show

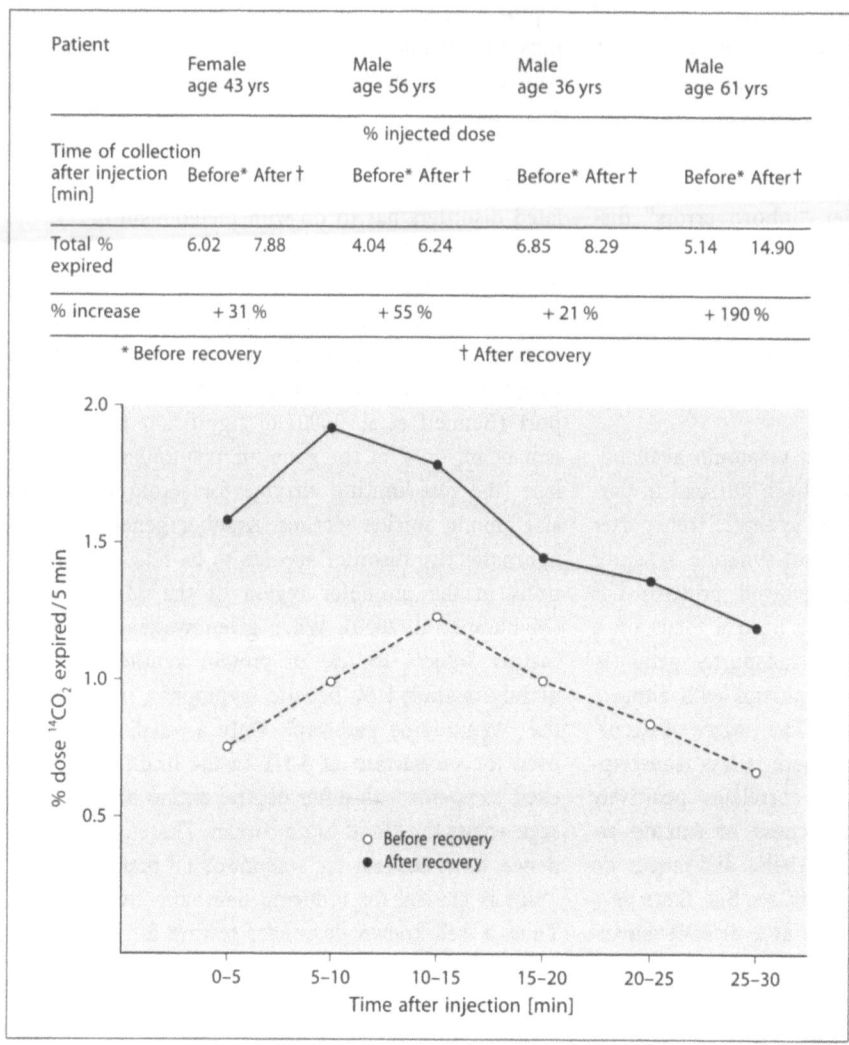

Patient	Female age 43 yrs		Male age 56 yrs		Male age 36 yrs		Male age 61 yrs	
Time of collection after injection [min]			% injected dose					
	Before*	After†	Before*	After†	Before*	After†	Before*	After†
Total % expired	6.02	7.88	4.04	6.24	6.85	8.29	5.14	14.90
% increase	+31%		+55%		+21%		+190%	

* Before recovery   † After recovery

Fig. 14.4. Expiration of $^{14}CO_2$ following injection of 5-HTP-1-^{14}C in patients before and after recovery from a depressive illness – average results in four patients. (Modified from Coppen et al. 1965, with permission)

a correlation (Willmer 1985). The association with depression appears to be a trait-marker (present with or without a depressive period) rather than a state-marker (present only with symptoms). It is said that lack of correlation may be due to a bimodal distribution of findings, which could be caused either by generally higher 5-HIAA in CSF of women than in men (now a well-documented difference) or by a heterogeneity of sub-groups of depressive types, some of which may be more related to a paucity of dopamine or norepinephrine than of 5-HT (Willmer 1985). Clearly, there would be differences in the pharmacologic approach to genotypes differing in the type of monoamine dysfunction disposing to depression (Charney 1998). A low 5-HIAA in CSF is presumed to indicate a "slow serotonin turnover" and, by implication, an inadequate serotonin neuroactivity. One might conceive that systemic administration of ^{14}C- or ^{13}C-labeled 5-HTP followed by properly-timed sampling of 5-HIAA in CSF for tracer activity could be a more dynamic, and perhaps superior, way to detect depression (or traits of impulsivity, hyperaggressiveness, suicidal behavior, irritability, etc.) than measurement of 5-HIAA levels only. Multiple sampling of lumbar CSF would seem to be beyond ethical allowance. More practical, and possibly useful, would be sampling of urine for labeled 5-HIAA after administration of labeled 5-HTP together with a peripheral decarboxylase inhibitor. There is a very early and little-noted report of decreased oxidation of 1-^{14}C-HTP to $^{14}CO_2$ among just four patients during a period of depression compared to much higher oxidation rate upon recovery (Coppen et al. 1965) (Fig. 14.4). This study seems never to have been repeated and surely needs verification, since it involves the very simple breath technique easily used for other purposes in nuclear medicine tests (see other sections of this chapter and Chap. 12). A problem with this early finding is that it appears to indicate a state-dependent marker rather than one of trait-dependence, as generally found for 5-HIAA levels in CSF.

Another approach to an in vitro tracer test for depression would exploit the association between abnormalities of serotonin activity/handling by blood platelets and conditions of depression. There is a large body of evidence that indicates abnormalities of uptake or binding of serotonin or related compounds and also of genetic differences akin to those found in neuroaffective disorders (Willmer 1985; Greenberg et al. 1999; Du et al. 2000; Franke et al. 2000). The use of platelets as a surrogate for neuronal activity (particularly pre-synaptic terminals) has a valid basis dependent on a particular specific enzyme in common ("neuron-specific" enolase), which suggests a common embryological origin for the two cell types (Willmer 1985). Measurements of 5-HT uptake by platelets (Greenberg et al. 1999) could be done using labeled 5-HT. The repeated findings of reduction of ^{3}H-imipramine binding sites observed in platelets taken from depressed patients (Willmer 1985) could be adapted to streamlined techniques appropriate for wider clinical application in nuclear medicine laboratories.

14.4
References

Adlung J, Grazikowske H, Uthgenannt H (1975) Über ein neues Verfahren in der Diagnostik der Maldigestion. Schweiz Med Wochenschr 105:134–140

Ahmed LS, Moorehead H, Leitch CA et al (1998) Determination of the specific activity of sheep plasma amino acids using high-performance liquid chromatography: comparison study between liquid scintillation counter and on-line flow-through detector. J Chromatogr B Biomed Sci Appl 710:27–35

Altay C, Cetin M, Gumruk F et al (1995) Familial selective vitamin B$_{12}$ malabsorption (Immerslund-Grasbeck syndrome) in a pool of Turkish patients. Pediatr Hematol Oncol 12:19–28

Altman LK (2000) High level of insulin linked to breast cancer's advance. NY Times, 24 May 2000

Ames SK, Ellis KJ, Gunn SK et al (1999) Vitamin D receptor gene Fok1 polymorphism predicts calcium absorption and bone mineral density in children. J Bone Miner Res 14:740–746

Aminoff M, Tahranainen E, Grasbeck R et al (1995) Selective intestinal malabsorption of vitamin B$_{12}$ displays recessive mendelian inheritance: assignment of a locus to chromosome 10 by linkage. Am J Hum Genet 57:824–831

Anderson GJ, Powell LW (1999) Haemochromatosis and control of intestinal iron absorption. Lancet 353:2089

Arranz MJ, Erdmann J, Kirov G et al (1997) 5HT2A receptor and bipolar affective disorder: association studies in affected patients. Neurosci Lett 224:95–98

Badawy AA (1999) Tryptophan metabolism in alcoholism. Adv Exp Med Biol 467:265–274

Ballmer PE, Walshe D, McNurlan MA et al (1993) Albumin synthesis rates in cirrhosis: correlation with Child-Turcotte classification. Hepatology 18:292–297

Barker J, Garner RC (1999) Biomedical applications of accelerator mass spectrometry – isotope measurements at the level of the atom. Rapid Commun Mass Spectrom 13:285–293

Barley NF, Prathalingam SR, Zhi P et al (1999) Factors involved in the duodenal expression of the human calbindin-D9K gene. Biochem J 341:491–500

Barrett JFR, Whittaker PG, Williams JG et al (1997) Absorption of non-haem iron from food during normal pregnancy. Br Med J 309:79–82

Barter PJ, Nestel PJ, Carroll KF (1972) Precursors of plasma triglyceride fatty acids in humans. Effects of glucose consumption, clofibrate administration and alcoholic fatty liver. Metab Clin Exp 21:117–124

Bartholome K, Lutz P, Biekel H (1975) Determination of phenylalanine hydroxylase activity in patients with phenylketonuria and hyperphenylalaninemia. Pediatr Res 9:899–903

Bartholome K, Olex K, Trefz F (1984) Compound heterozygotes in hyperphenylalaninemia. Hum Genet 65:405–406

Bass NM, Smith LH Jr, Van Dyke RW (1986) Gastrointestinal diseases. In: Andreoli TE, Carpenter CCJ, Plum F, Smith LH Jr (eds) Cecil essentials of medicine. Saunders, Philadelphia, pp 260–338

Behrend C, Jeppesen PB, Mortensen PB (1995) Vitamin B$_{12}$ absorption after ileorectal anastomosis for Crohn's disease: effect of ileal resection and time span after surgery. Eur J Gastroenterol Hepatol 7:397–400

Bennett PJ, McMahon WM, Watabe J et al (2000) Tryptophan hydroxylase polymorphisms in suicide victims. Psychiatr Genet 10:13–17

Berrish TS, Hetherington CS, Alberti KGMM et al (1995) Peripheral and hepatic insulin sensitivity in subjects with impaired glucose tolerance. Diabetologia 38:699–704

Berry GT, Nissim I, Gibson JB et al (1997) Quantitative assessment of whole body galactose metabolism in galactosemic patients. Eur J Pediatr 156 [Suppl 1]:S43–S49

Berry GT, Singh RH, Mazur AT (2000) Galactose breath testing distinguishes variant and severe galactose-1-phosphate uridyltransferase genotypes. Pediatr Res 48:323–328

Blau N, Thony B, Spada M et al (1996) Tetrahydrobiopterin and inherited hyperphenylalaninemias. Turk J Pediatr 38:19–35

Boernsen KO, Floeckher JM, Bruin GJ (2000) Use of a microplate scintillation counter as a radioactivity detector for miniaturized separation techniques in drug metabolism. Anal Chem 72:3956–3959

Bougneres P-F, Rocchiccioli F, Nurjhan N et al (1995) Stable isotope determination of plasma lactate conversion into glucose in fasting infants. Am J Physiol 268:E652–E659

Braden B, Caspary WF, Lembcke B (1999) Nondispersive infrared spectrometry for $^{13}CO_2/^{12}CO_2$ measurements: a clinically feasible analyzer for stable isotope breath tests in gastroenterology. Z Gastroenterol 37:477–481

Buchholz BA, Arjomand A, Dueker SR et al (1999) Intrinsic erythrocyte labeling and attomole pharmacokinetics tracing of ^{14}C-labeled folic acid with accelerator mass spectrometry. Anal Biochem 269:348–352

Carpentier J-L, Robert A, Grunberger G et al (1986) Receptor-mediated endocytosis of polypeptide hormones is a regulated process: inhibition of ^{125}I-iodoinsulin internalization in hypoinsulinemic-diabetes of rat and man. J Clin Endocrin Metab 63:151–155

Carroccio A, Montalto G, Cavera G et al (1998) Lactose intolerance and self-reported milk intolerance: relationship with lactose maldigestion and nutrient intake. J Am Coll Nutr 17:631–636

Caspary WF, Schaffer J (1978) ^{14}C-D-galactose breath test for evaluation of liver function in patients with chronic liver disease. Digestion 17:410–418

Charney DS (1998) Monoamine dysfunction and the pathophysiology and treatment of depression. J Clin Psychiatry 59 [Suppl 14]:11–14

Clarke JTR, Bier DM (1982) The conversion of phenylalanine to tyrosine in man. Direct measurement by continuous intravenous tracer infusions of L-(ring-$^2H_5$) phenylalanine and L-(1-^{13}C) tyrosine in the postabsorptive state. Metab Clin Exp 38:999–1005

Clark T, Klein O (1996) Thin-layer radiochromatography. In: Sherma J, Fried B (eds) Handbook of thin-layer chromatography, 2nd edn, chap 12. Dekker, New York, pp 341–359

Consoli A, Nurjhan N (1990) Contribution of gluconeogenesis to overall glucose output in diabetic and nondiabetic men. Ann Med 22:191–195

Coppen A, Shaw DM, Malleson H (1965) Changes in 5-hydroxytryptophan metabolism in depression. Br J Psychiatry 111:105–107

Cremonesi M, Ferrari M, Zoboli S et al (1999) Biokinetics and dosimetry in patients administered with ^{111}In-DOTA-Tyr^3-octreotide: implications for internal radiotherapy with ^{90}Y-DOTATOC. Eur J Nucl Med 26:877–886

DaCosta H, Shreeve WW, Merchant S (1976) Radiorespirometric study of carbohydrate metabolism in childhood liver disease. J Nucl Med 17:218–219

Danks DM (1995) Disorders of copper transport. In: Scriver CR, Beaudet AL, Sly WS, Valle D (eds) The metabolic and molecular bases of inherited disease II, 7th edn, chap 68. McGraw-Hill, New York, pp 2211–2236

De Fronzo RA, Bonadonna RC, Ferrannini E (1992) Pathogenesis of NIDDM. A balanced overview. Diabetes Care 15:318–368

Demant T, Packard CJ, Demmelmair H et al (1996) Sensitive methods to study human apolipoprotein B metabolism using stable isotope-labeled amino acids. Am J Physiol 270:E1022–E1036

De Meutter RC, Shreeve WW (1963) Conversion of DL-lactate-2-C^{14} or -3-C^{14} or pyruvate-2-C^{14} to blood glucose in humans: effects of diabetes, insulin, tolbutamide and glucose load. J Clin Invest 42:525–533

Deutsch JC, Santhosh-Kumar CR, Kolli VR (1995) A noninvasive stable-isotope method to simultaneously assess pancreatic exocrine function and small bowel absorption. Am J Gastroenterol 90:2182–2185

DiMagno EP (1995) Laboratory assessment of pancreatic impairment. In: Haubrich WS, Schaffner F, Berk JE (eds) Bockus gastroenterology, 5th edn, vol 4, chap 149. Saunders, Philadelphia, pp 2835–2850

Diraison F, Large V, Brunengraber H et al (1998) Non-invasive tracing of liver intermediary metabolism in normal subjects and in moderately hyperglycaemic NIDDM subjects. Evidence against increased gluconeogenesis and hepatic fatty oxidation in NIDDM. Diabetologia 41:212–220

Du L, Faludi G, Palkovits M et al (1999) Frequency of long allele in serotonin transporter gene is increased in depressed suicide victims. Biol Psychiatry 46:196–201

Du L, Bakish D, Lapierre YD et al (2000) Association of polymorphism of serotonin 2A receptor gene with suicidal ideation in major depressive disorder. Am J Med Genet 96:56–60

Ekins RP (1995) Ultrasensitive ligand assays. In: Wagner HN Jr, Szabo Z, Buchanan JW (eds) Principles of nuclear medicine, 2nd edn, chap 13. Saunders, Philadelphia, pp 247–266

Elias N, Patterson BW, Schonfeld G (1999) Decreased production rates of VLDL triglycerides and apo B-100 in subjects heterozygous for familial hypobetalipoproteinemia. Arterioscler Thromb Vasc Biol 19:2714–2721

Englert D, Roessler N, Jeavons A et al (1995) Microchannel array detector for quantitative electronic radioautography. Cell Mol Biol 41:57–64

Favus MJ (1989) Intestinal calcium absorption: Have we absorbed enough from research to have a test for the patient? J Bone Miner Res 4:461–462

Feinendegen LE, Pollycove M (2001) Biologic responses to low doses of ionizing radiation: detriment versus hormesis, part 1. Dose responses of cells and tissues. J Nucl Med 42:17N–27N

Flatz G (1995) The genetic polymorphism of intestinal lactase activity in adult humans. In: Scriver CR, Beaudet AL, Sly WS, Valle D (eds) The metabolic and molecular bases of inherited disease III, chap 150. McGraw-Hill, New York, pp 4441–4450

Franke L, Schewe J, Muller B et al (2000) Serotonergic platelet variables in unmedicated patients suffering from major depression and healthy subjects: relationship between 5HT content and 5HT uptake. Life Sci 67:301–305

Friedman LS, Martin P, Muñoz SJ (1996) Liver function tests and the objective evaluation of the patient with liver disease. In: Zakim D, Boyer TD (eds) Hepatology, a textbook of liver disease, 3rd edn, vol 1, chap 28. Saunders, Philadelphia, pp 791–833

Fujii H, Matsuno K, Takiue M (2000) Hybrid radioassay of multiple radionuclide mixtures in waste solutions by using liquid and NaI (Tl) scintillation monitors. Health Physics 79:294–298

Galjaard H (1980a) Diagnosis of genetic metabolic diseases. In: Galjaard H (ed) Genetic metabolic diseases, early diagnosis and prenatal analysis, chap. 2. Elsevier/North Holland Biomedical Press, Amsterdam, pp 422–436

Galjaard H (1980b) Practical experience with prenatal diagnosis of genetic metabolic diseases. In: Galjaard H (ed) Genetic metabolic diseases, early diagnosis and prenatal analysis, chap. 4. Elsevier/North Holland Biomedical Press, Amsterdam, pp 670–673

Gianani R, Verge CF, Moromisato-Gianani RI et al (1996) Limited loss of tolerance to islet autoantigens in ICA + first degree relatives of patients with type 1 diabetes expressing the HLA dqbl*0602 allele. J Autoimmun 9:423–425

Gilman SD, Gee SJ, Hammock BD et al (1998) Analytical performance of accelerator mass spectrometry and liquid scintillation counting for detection of ^{14}C-labeled atrazine metabolites in human urine. Anal Chem 70:3463–3469

Gitzelmann R, Steinman B, Van den Berghe G (1995) Disorders of fructose metabolism. In: Scriver CR, Beaudet AL, Sly WS, Valle D (eds) The metabolic and molecular bases of inherited disease I, 7th edn. McGraw-Hill, New York, pp 905–934

Goff JS (1982) Two-stage triolein breath test differentiates pancreatic insufficiency from other causes of malabsorption. Gastroenterology 83:44–46

Gopher A, Vaisman N, Mandel H et al (1990) Determination of fructose metabolic pathways in normal and fructose-intolerant children: a ^{13}C NMR study using (U-^{13}C) fructose. Proc Natl Acad Sci USA 87:5449–5453

Gorden P (1997) Non-insulin dependent diabetes – the past, present and future. Ann Acad Med Singapore 26:326–330

Gorwood P, Batel P, Ades J et al (2000) Serotonin transporter gene polymorphisms, alcoholism, and suicide behavior. Biol Psychiatry 48:259–264

Grasbeck R (1997) Selective cobalamin malabsorption and the cobalamin-intrinsic factor receptor. Acta Biochim Pol 44:725–733

Greenberg BD, Tolliver TJ, Huang SJ et al (1999) Genetic variation in the serotonin transporter promoter region affects serotonin uptake in human blood platelets. Am J Med Genet 88:83–87

Haffner SM, Stern MP, Miettinen H et al (1995) Higher proinsulin and specific insulin are both associated with a parental history of diabetes in nondiabetic Mexican-American subjects. Diabetes 44:1156–1160

Halliday D, Rennie MJ (1982) The use of stable isotopes for diagnosis and clinical research. Clin Sci 63:485–496

Hallikainen T, Saito T, Lachman HM et al (1999) Association between low activity serotonin transporter promoter genotype and early onset alcoholism with habitual impulsive violent behavior. Mol Psychiatry 4:385–388

Hamosh A, Johnston MV, Valle D (1995) Nonketotic hyperglycinemia. In: Scriver CR, Beaudet AL, Sly WS, Valle D (eds) The metabolic and molecular bases of inherited disease, I, 7th edn, chap 37. McGraw-Hill, New York, pp 1337–1348

Hardin DS, LeBlanc A, Para L et al (1999) Hepatic insulin resistance and defects in substrate utilization in cystic fibrosis. Diabetes 48:1082–1087

Hartvig P, Lindner KJ, Tedroff J et al (1992) Brain kinetics of ^{11}C-labelled L-tryptophan and 5-hydroxy-L-tryptophan in the rhesus monkey. A study using positron emission tomography. J Neural Transm Gen 88:1–10

Havel RJ, Kane JP (1995) Introduction: structure and metabolism of plasma lipoproteins. In: Scriver CR, Beaudet AL, Sly WS, Valle D (eds) The metabolic and molecular bases of inherited disease II, 7th edn, chap 56. McGraw-Hill, New York, pp 1841–1851

Heaney RP, Recker RR, Stegman MR et al (1989) Calcium absorption in women. Relationships to calcium intake, estrogen status and age. J Bone Miner Res 4:469–475

Heinz A, Goldman D (2000) Genotype effects on neurodegeneration and neuroadaptation in monoaminergic neurotransmitter system. Neurochem Int 37:425–432

Hirano S, Kanamatsu T, Takagi Y et al (1979) A simple infrared spectroscopic method for the measurement of expired ^{13}CO$_2$. Anal Biochem 96:64–69

Hyland K, Surtees RAH, Rodeck C et al (1992) Aromatic L-amino acid decarboxylase deficiency: clinical features, diagnosis and treatment of a new inborn error of neurotransmitter amine synthesis. Neurology 42:1980–1988

Irving CS (1975) Measurement of ^{13}CO$_2$/^{12}CO$_2$ abundance by nondispersive infrared heterodyne ratiometry as an alternative to gas isotope ratio mass spectrometry. Anal Chem 58:2172–2178

Juan D, Molitch ME, Johnson MK et al (1990) Unaltered drug metabolizing enzyme systems in type II diabetes mellitus before and during glyburide therapy. J Clin Pharmacol 30:943–947

Kahn SE, Halban PA (1997) Release of incompletely processed proinsulin is the cause of disproportionate proinsulinemia of NIDDM. Diabetes 46:1725–1732

Kaihara S, Wagner HN Jr (1968) Measurement of intestinal fat absorption with carbon-14 labeled tracers. J Lab Clin Med 71:400–411

Kalderon B, Korman SH, Gutman A et al (1989) Estimation of glucose carbon recycling in children with glycogen storage disease: a ^{13}C NMR study using (U-^{13}C) glucose. Proc Natl Acad Sci USA 86:4690–4694

Kaminska-Galwas B, Sroczynski J (1991) Evaluation of hepatic microsomal enzyme activity using C-14-labeled aminopyrine breath test in patients with diabetes mellitus type 2 treated with tolbutamide. Pol Arch Med Wewn 86:142–148

Kaysen GA, Rathore V, Shearer GC et al (1995) Mechanisms of hypoalbuminemia in hemodialysis patients. Kidney Int 48:510–516

Kerstetter JE, O'Brien KO, Insogna KL (1998) Dietary protein affects intestinal calcium absorption. Am J Clin Nutr 68:859–865

Kobayashi T, Tamemoto K, Nakanishi K et al (1993) Immunogenetic and clinical characterization of slowly progressive IDDM. Diabetes Care 16:780–788

Kunstmann E, Epplen C, Elitok E et al (1999) Helicobacter pylori infection and polymorphisms in the tumor necrosis factor region. Electrophoresis 20:1756–1761

Kuzuya T, Matsuda A, Sakamoto Y et al (1978) C-peptide immunoreactivity (CPR) in urine. Diabetes 27 [Suppl 1]:210–215

Landau BR (1997) Stable isotope techniques for the study of gluconeogenesis in man. Horm Metab Res 29:334–336

Landau BR, Shreeve WW (1991) Radiation exposure from long-lived beta emitters in clinical investigation. Am J Physiol 261:E415–E417

Landau BR, Chandramouli V, Schumann WC et al (1995) Estimates of Krebs cycle activity and contributions of gluconeogenesis to hepatic glucose production in fasting healthy subjects and IDDM patients. Diabetologia 38:831–838

Lear JL (1995) Autoradiography. In: Wagner HN Jr, Szabo Z, Buchanan JW (eds) Principles of nuclear medicine, 2nd edn. Saunders, Philadelphia, pp 267–274

Lehmann WD, Fischer R, Heinrich HC et al (1986) Metabolic conversion of L-(U-^{14}C) phenylalanine to respiratory ^{14}CO$_2$ in healthy subjects, phenylketonuric heterozygotes and classic phenylketonurics. Clin Chim Acta 157:253–266

Lembcke B (1997) Atemtests bei Darmerkrankungen und in der gastroenterologischen Funktionsdiagnostik. Praxis 86:1060–1067

Lindgren A, Bagge E, Cederblad A et al (1997) Schilling and protein-bound cobalamin absorption tests are poor instruments for diagnosing cobalamin malabsorption. J Intern Med 241:477–484

Ljunggren K, Strand SE (1990) Beta camera for static and dynamic imaging of charged-particle emitting radionuclides in biologic samples. J Nucl Med 31:2058–2063

Ljunggren K, Strand SE, Ceberg CP et al (1993) Beta camera low activity tumor imaging. Acta Oncol 32:869–872

Loew D, Wanitschke R, Schroedter A (1999) Studies on vitamin B$_{12}$ status in the elderly-prophylactic and therapeutic consequences. Int J Vitam Nutr Res 69:228–233

Lorenz E, Brookeman VA, Mauderli W (1978) Plastic scintillation filament detector system for ^{14}CO$_2$ breath-analysis tests. Med Phys 5:195–198

Loser C, Brauer C, Aygen S et al (1998) Comparative clinical evaluation of the ^{13}C-mixed triglyceride breath test as an indirect pancreatic function test. Scand J Gastroenterol 33:327–334

Ludanyi K, Vekey K, Szunyog J et al (1999) Application of overpressured layer chromatography combined with digital autoradiography and mass spectrometry in the study of derameiclane metabolism. J AOAC Int 82:231–238

Lutwak L (1969) Tracer studies of intestinal calcium absorption in man. Am J Clin Nutr 22:771–785

Malmström R, Packard CJ, Caslake M et al (1997) Defective regulation of triglyceride metabolism by insulin in the liver in NIDDM. Diabetologia 40:454–462

Mann JJ, Huang YY, Underwood MD et al (2000) A serotonin transporter gene promoter polymorphism (5-HTTLPR) and prefrontal cortical binding in major depression and suicide. Arch Gen Psychiatry 57:729–738

Manuck SB, Flory JD, Ferrel RE et al (2000) A regulatory polymorphism of the monoamine oxidase-A gene may be associated with variability in aggression, impulsivity, and central nervous system serotonergic responsivity. Psychiatry Res 95:9–23

Marshall BJ, Surveyor I (1988) Carbon-14 urea breath test for the diagnosis of Campylobacter pylori associated gastritis. J Nucl Med 29:11–16

Massat I, Souery D, Lipp O et al (2000) A European multicenter association study of HTR2A receptor polymorphism in bipolar affective disorder. Am J Med Genet 96:136–140

Maugeais C, Ouguerram K, Krempf M et al (1998) Kinetic study of ApoB100 containing lipoprotein metabolism using amino acid labeled with stable isotopes: methodological aspects. Clin Chem Lab Med 36:739–745

Maurus N, Hayes VY, Vieira NE et al (1999) Profound hypogonadism has significant negative effects on calcium balance in males: a calcium kinetic study. J Bone Miner Res 14:577–582

McIntyre WJ, Saha GB (1995) Radiation detection. In: Wagner HN Jr, Szabo Z, Buchanan JW (eds) Principles of nuclear medicine, 2nd edn. Saunders, Philadelphia, pp 235–242

McMillan DC, Slater C, Preston T et al (1996) Simultaneous measurement of albumin and fibrinogen synthetic rates in normal fasted subjects. Nutrition 12:602–607

Milstein S, Kaufman S (1975) Studies on the phenylalanine hydroxylase system in vivo. An in vivo assay based on the liberation of deuterium or tritium into the body water from ring-labeled L-phenylalanine. J Biol Chem 250:4782–4785

Miyakoshi N, Tanaka M, Shinda H (1980) Autoradiographic studies on distribution of L-3,4-dihydroxyphenylalanine (L-DOPA)-14C and L-5-hydroxytryptophan (L-5-HTP)-14C in the cat brain. Jpn J Pharmacol 30:795–805

Moshage HJ, Janssen JAM, Franssen JH et al (1987) Study of the molecular mechanism of decreased liver synthesis of albumin in inflammation. J Clin Invest 79:1635–1641

Mousseau DD, Butterworth RF (1994) The (^{3}H) tryptamine receptor in human brain: kinetics, distribution and pharmacologic profile. J Neurochem 63:1052–1059

Nakamura T (1997) The importance of genetic and nutritional factors in responses to vitamin D and its analogs in osteoporotic patients. Calcif Tissue Int 60:119–123

Nyhan WL (1971) The use of isotopic tracers in the study of genetic disorders of metabolism in children. In: Klein PD, Roth LJ (eds) Proceedings of a seminar on the use of stable isotopes in clinical pharmacology. Argonne National Laboratory, pp 53–70

Nyhan WL (1984a) Phenylketonuria. In: Abnormalities in amino acid metabolism in clinical medicine, chap. 14. Appleton-Century-Crofts, Norwalk, CT, pp 127–148

Nyhan WL (1984b) Nonketotic hyperglycinemia. In: Abnormalities in amino acid metabolism in clinical medicine, chap. 34. Appleton-Century-Crofts, Norwalk, CT, pp 333–351

O'Brien KO, Zaveleta N, Caulfield LE et al (1999) Influence of prenatal iron and zinc supplements on supplemental iron absorption, red blood cell iron incorporation, and iron status in pregnant Peruvian women. Am J Clin Nutr 69:509–515

Okita GT, Kabara JJ, Richardson F et al (1957) Assaying compounds containing H^3 and C^{14}. Nucleonics 15:111–114

Olefsky JM, Reaven GM (1977) Insulin binding in diabetes. Relationships with plasma insulin levels and insulin sensitivity. Diabetes 26:680–688

Packard CJ (1995) The role of stable isotopes in the investigation of plasma lipoprotein metabolism. Baillieres Clin Endocrinol Metab 9:755–772

Pearson HA, Lukens JN (1999) Ferrokinetics in the syndrome of familial hypoferremic microcytic anemia with iron malabsorption. J Pediatr Hematol Oncol 21:412–417

Pedersen NT, Marqverson J (1981) Metabolism of ingested ^{14}C-triolein: estimation of radiation dose in tests of lipid assimilation using ^{14}C and ^{3}H-labeled fatty acids. Eur J Nucl Med 6:327–329

Peng JB, Chen XZ, Berger UV et al (1999) Molecular cloning and characterization of a channel-like transporter mediating intestinal calcium absorption. J Biol Chem 274:22739–22746

Price DC, McIntyre PA (1984) The hematopoietic system. In: Harbert J, DaRocha AFG (eds) Textbook of nuclear medicine, vol II, chap 23. Clinical applications, 2nd edn. Lea and Febiger, Philadelphia, pp 535–605

Rabinowitz JL, Hansell J, Staeffen J (1978) Rate of (C-14)-octanoate oxidation to (C-14) carbon dioxide as a test for cirrhosis. J Nucl Med 19:689

Rader DJ, Schaefer JR, Lohse P et al (1993) Increased production of apolipoprotein A-1 associated with elevated plasma levels of high-density lipoproteins, apolipoprotein A-1 and lipoprotein A-1 in a patient with familial hyperalphalipoproteinemia. Metab Clin Exp 42:1429–1434

Raju GS, Smith MJ, Morton D et al (1994) Mini-dose (1-μCi) ^{14}C-urea breath test for the detection of Helicobacter pylori. Am J Gastroenterol 89:1027–1031

Reba RC, Salkeld J (1982) In vitro studies of malabsorption and other GI disorders. Semin Nucl Med XII:147–155

Redgrave TG, Watts GF, Martins IJ et al (2001) Chylomicron remnant metabolism in familial dyslipidemias studies with a remnant-like emulsion breath test. J Lipid Res 42:710–715

Riemens SC, Dullart RPF, Franssen EJF et al (1998) Measurement of free fatty acid kinetics during non-equilibrium tracer conditions in man: implications for the estimation of the rate of appearance of free fatty acids. Eur J Clin Invest 28:108–114

Rings EH, Grand RJ, Buller HA (1994) Lactose intolerance and lactose deficiency in children. Curr Opin Pediatr 6:562–567

Rodriguez-Villar C, Conget I, Casamitjana R et al (1997) High proinsulin levels in late pre-IDDM stage. Diabetes Res Clin Pract 37:145–148

Rosenthal NE, Mazzanti CM, Barnett RL et al (1998) Role of serotonin transporter promoter repeat length polymorphism (5-HTTLPR) in seasonality and seasonal affective disorder. Mol Psychiatry 3:175–177

Rossiter BJF, Caskey CT (1995) Hypoxanthine-guanine phosphoribosyltransferase deficiency: Lesch-Nyhan syndrome and gout. In: Scriver CR, Beaudet AL, Sly WS, Valle D (eds) The metabolic and molecular bases of inherited disease, II, 7th edn, chap 50. McGraw-Hill, New York, pp 1679–1706

Rotter JL, Vadheim CM, Rimoin DL (1990) Genetics of diabetes mellitus. In: Rifkin H, Porte D Jr (eds) Diabetes mellitus: theory and practice, 4th edn, chap 24. Elsevier, Amsterdam, pp 378–413

Salako QA, De Nardo SJ (1997) Radioassay of yttrium-90 radiation using the radionuclide dose calibrator. J Nucl Med 38:723–726

Sasaki Y (1995) Carbon-14 and carbon-13 breath tests. In: Wagner HN Jr, Szabo Z, Buchanan JW (eds) Principles of nuclear medicine, 2nd edn, chap 40. Saunders, Philadelphia, pp 958–965

Saylor L, Finch CA (1953) Determination of iron absorption using two isotopes of iron. Am J Physiol 172:372–376

Schoeller DA, Baker AL, Monroe SP et al (1982) Comparison of different methods of expressing results of the aminopyrine breath test. Hepatology 2:455–462

Schonfeld G, Agailar-Salina C, Elias N (1998) Role of 3-hydroxy-3-methylglutaryl coenzyme A reductase inhibitors ("statins") in familial combined hyperlipidemia. Am J Cardiol 81:43B–46B

Scriver CR, Kaufman S, Eisensmith RC et al (1995) The hyperphenylalaninemias. In: Scriver CR, Beaudet AL, Sly WS, Valle D (eds) The metabolic and molecular bases of inherited disease, I, 7th edn, chap 27. McGraw-Hill, New York, pp 1015–1048

Searle GL, Mortimore GE, Buckley RE et al (1959) Plasma glucose turnover in humans as studied with ^{14}C glucose. Influence of insulin and tolbutamide. Diabetes 8:167–173

Segal S, Berry GT (1995) Disorders of galactose metabolism. In: Scriver CR, Beaudet AL, Sly WS, Valle D (eds) The metabolic and molecular bases of inherited disease, I, 7th edn, chap 25. McGraw-Hill, New York, pp 967–1000

Sherlock S, Dooley J (1993) Assessment of liver function. In: Sherlock S, Dooley J (eds) Diseases of the liver and biliary system, 9th edn, chap 2. Blackwell, London, pp 17–32

Shigeta Y, Shreeve WW (1964) Fatty acid synthesis from glucose-1-H^3 and glucose-1-C^{14} in obese-hyperglycemic mice. Am J Physiol 206:1085–1090

Shin YS, Moro V, Doliwa H et al (1983) A radioisotopic method for fructose-1-phosphate aldolase assay that facilitates diagnosis of hereditary fructose intolerance. Clin Chem 29:1955–1958

Shreeve WW (1966) Effects of insulin on the turnover of plasma carbohydrates and lipids. Am J Med 40:724–734

Shreeve WW (1974) Physiological chemistry of carbohydrates in mammals. Saunders, Philadelphia, pp 1–318

Shreeve WW (1984) Labeled carbon breath analysis. In: Harbert J, DaRocha AFG (eds) Textbook of nuclear medicine, vol I, chap 16. Basic science. Lea and Febiger, Philadelphia, pp 351–362

Shreeve WW (1987) Impaired oxidation of carbon-labeled galactose by alcoholic or diabetic liver in vivo. Nuklearmedizin 26:159–166

Shreeve WW, Baker N, Miller M et al (1956) ^{14}C studies in carbohydrate metabolism II. The oxidation of glucose in diabetic human subjects. Metabolism 5:22–34

Shreeve WW, Hoshi M, Oji N et al (1968) Insulin and the utilization of carbohydrates in obesity. Am J Clin Nutr 21:1404–1418

Shreeve WW, Shoop JD, Ott DG et al (1976) Test for alcoholic cirrhosis by conversion of [^{14}C]- or [^{13}C] galactose to expired CO_2. Gastroenterology 71:98–101

Siddossis LS, Mittendorfer B, Walser E et al (1998) Hyperglycemia-induced inhibition of splanchnic fatty acid oxidation increases hepatic triacylglycerol secretion. Am J Physiol 275:E798–E805

Siler SQ, Neese RA, Parks EJ et al (1998) VLDL triglyceride production after alcohol ingestion, studied using (2-$^{13}C_1$) glycerol. J Lipid Res 39:2319–2328

Smith SM, Wastney ME, Nyquist LE et al (1996) Calcium kinetics with microgram stable isotope doses and saliva sampling. J Mass Spectrom (CMB) 31:1265–1270

Snow CF (1999) Laboratory diagnosis of vitamin B_{12} and folate deficiency. Arch Int Med 159:1289–1298

Steiner DF, Tager HS, Nanjo K et al (1995) Familial syndromes of hyperproinsulinemia and hyperinsulinemia with mild diabetes. In: Scriver CR, Beaudet AL, Sly WS, Valle D (eds) The metabolic and molecular bases of inherited disease, I, 7th edn. McGraw-Hill, New York, pp 897–904

Suzuki S, Hinokio Y, Hirai S et al (1994) Diabetes with mitochondrial gene tRNALYS mutation. Diabetes Care 17:1428–1432

Suto H, Azuma T, Ito S et al (1999) Evaluation of endoscopic 13C-urea breath test for assessment of Helicobacter pylori eradication. J Gastroenterol 34 [Suppl II]):67–71

Tager HS (1984) Abnormal products of the human insulin gene. Diabetes 33:693–699

Tanguay PE (2000) Pervasive developmental disorders: a 10-year review. J Am Acad Child Adolesc Psychiatry 39:1079–1095

Tanigawara Y, Auyama N, Kita T, Shirakawa K et al (1999) CYP2C19 genotype-related efficacy of omeprazole for the treatment of infection caused by Helicobacter pylori. Clin Pharmacol Ther 66:528–534

Taylor SI (1995) Diabetes mellitus. In: Scriver CR, Beaudet AL, Sly WS, Valle D (eds) The metabolic and molecular bases of inherited disease I, 7th edn, chap 21. McGraw-Hill, New York, pp 843–896

Terai A, Okada Y, Shichiri Y et al (1997) Vitamin B_{12} deficiency in patients with urinary intestinal diversion. Int J Urol 4:21–25

Thorburn A, Litchfield A, Fabris S et al (1995) Abnormal transient rise in hepatic glucose production after oral glucose load in non-insulin-dependent diabetic subjects. Diab Res Clin Pract 28:127–135

Toskes PP, Donaldson RM Jr (1993) Enteric bacterial flora and bacterial overgrowth syndrome. In: Sleisinger MH, Fordtran JS (eds) Gastrointestinal disease, pathophysiology/diagnosis/management, 5th edn, chap 53. Saunders, Philadelphia, pp 1106–1118

Trischitta V, Brunetti A, Chiavetta A et al (1989) Defects in insulin-receptor internalization and processing in monocytes of obese subjects and obese NIDDM patients. Diabetes 38:1579–1584

Vaarala O, Knip M, Paronen J et al (1999) Cow's milk formula feeding induces primary immunization to insulin in infants at genetic risk for type 1 diabetes. Diabetes 48:1389–1394

Van Eijk HMH, Rooyakkers DR, Soeters PB et al (1999) Determination of amino acid enrichment using liquid chromatography-mass spectrometry. Anal Biochem 271:8–17

Van Trappen GR, Rutgeerts PJ, Ghoos YF et al (1989) Mixed triglyceride breath test: A noninvasive test of pancreatic lipase activity in the duodenum. Gastroenterology 96:1126–1134

Van Woert MH, Rosenbaum D, Howieson J et al (1977) Long-term therapy of myoclonus and other neurologic disorders with L-5-hydroxytryptophan and carbidopa. N Engl J Med 296:70–75

Van Wouwe JP (1989) Clinical and laboratory diagnosis of acrodermatitis enteropathica. Eur J Pediatr 149:2–8

Vierhapper H, Nowotny P, Waldhausl W (1997) Determination of testosterone production rates in men and women using stable isotope/dilution and mass spectrometry. J Clin Endocr Metab 82:1492–1496

Vogel JS, Turteltaub KW (1998) Accelerator mass spectrometry as a bioanalytical tool for nutritional research. From a conference on mathematical modeling in experimental nutrition, held 17–20 Aug 1997. Plenum, New York, pp 397–410

Watts GF, Mamo JC, Redgrave TG (1998) Postprandial dyslipidemia in a nutshell: food for thought. Aust NZ J Med 28:816–823

Weaver LT, Amarri S, Swart G (1998) ^{13}C mixed triglyceride breath test. Gut 43 [Suppl 3]:S13–S19

Wiernsperger NF, Bailey CJ (1999) The antihyperglycemic effect of metformin: therapeutic and cellular mechanisms. Drugs 58 [Suppl]:31–39; 75–82

Willmer P (1985) Depression: a psychobiological synthesis. Wiley, New York, pp 3–877

Winkler K, Schafer JR, Klima B et al (1999) Lifibrol enhances the low density apolipoprotein B-100 turnover in patients with hypercholesterolemia and mixed hyperlipidemia. Atherosclerosis 144:167–175

Wishart JM, Horowitz M, Need HG et al (1997) Relations between calcium intake, calcitriol, polymorphisms of the vitamin D receptor gene, and calcium absorption in premenopausal women. Am J Clin Nutr 65:798–802

Wolfe RR (1992) Radioactive and stable isotope tracers in biomedicine. Principles and practice of kinetic analysis. Wiley-Liss, New York

Worwood M (1998) Haemochromatosis. Clin Lab Haematol 20:65–75

Wu C-H, Hoshi M, Shreeve WW (1974) Human plasma triglyceride labeling after high-sucrose feeding. I. Incorporation of sucrose-U-^{14}C. Metab Clin Exp 23:1125–1140

Yalow RS (1995) Radioimmunoassay. In: Wagner HN Jr, Szabo Z, Buchanan JW (eds) Principles of nuclear medicine, 2nd edn, chap 13. Saunders, Philadelphia, pp 243–246

Yamane Y, Ishede N, Kagaya Y (1995) Quantitative double-tracer autoradiography with tritium and carbon-14 using imaging plates: application to myocardial metabolic studies in rats. J Nucl Med 36:518–524

Yap SH, Hafkenscheid CM, Goossens IC et al (1975) Estimation of radiation dosage and transmutation effects of ^{14}C involved in measuring rate of albumin synthesis with ^{14}C-carbonate. J Nucl Med 16:642–648

Yatham LN, Liddle PF, Shiah IS (2000) Brain serotonin 2 receptors in major depression: a positron emission tomography study. Arch Gen Psychiatry 57:850–858

Zoller H, Pietrangelo A, Vogel W et al (1999) Duodenal metal-transporter (DMT-1, NRAMP-2) expression in patients with hereditary haemochromatosis. Lancet 353:2120–2123

Part III Organ Systems

CHAPTER 15

Endocrinology 15

JUNJI KONISHI, BEN A. DWAMENA, MILTON D. GROSS, BRAHM SHAPIRO,
TAKASHI MISAKI, MASAO FUKUNAGA, J.C. SISSON, HONG-YOE OEI, MARION DE JONG, ERIC P. KRENNING

Contents

15.1
General Introduction

JUNJI KONISHI

Endocrine organs are generally small in size but have the unique function of synthesizing and secreting specific hormones. Since the early introduction of radioiodine for both the diagnosis and therapy of thyroid disorders, nuclear medicine has played an important role in clinical practice as well as in research in endocrinology. In particular, radionuclide imaging using tumor-seeking radionuclides, radioreceptor ligands, or radiolabeled precursors can detect functioning tumors and dysfunctional states in various endocrine organs, including the pituitary, thyroid, parathyroid, and adrenal glands. The radionuclide-imaging approach to clinical problems related to the pituitary gland has recently been drastically redefined through the development of the high-performance multi-headed gamma camera, single photon emission computed tomography (SPECT) systems, and dedicated positron emission tomography (PET) machines. Advances in molecular biology have occurred that permit the characterization and successful radiolabeling of a number of pituitary hormonal/hypothalamic neurotransmitter ligands and their cell surface receptors. In addition, PET radiotracers have become available for in vivo evaluation of various aspects of tumor metabolism and blood flow.

Cognizant of remarkable roles of radiopharmaceutical imaging of pituitary lesions in current investigative algorithms, we discuss in this chapter how the integration of the aforementioned developments can now provide opportunities for the use of somatostatin and dopamine receptor-based radiopharmaceuticals and for metabolic imaging with various PET tracers: first to localize the source(s) of humoral hypersecretion when anatomical pituitary imaging results are indeterminate; second, to delineate occult pituitary lesions in ectopic extrapituitary tissues and to screen the whole body for clinically occult neuroendocrine lesions in neurocristopathic syndromes like MEN I (pituitary, pancreatic and parathyroid lesions); third, to noninvasively characterize lesions for the selection and monitoring of therapy with long-acting somatostatin analogues and dopamine agonists; and fourth, to distinguish postoperative changes from residual or recurrent tumor. The biochemical basis, clinical indications, validated or proposed scanning protocols, and published results of clinical studies of receptor-based SPECT and PET as applied to pituitary gland abnormalities are reviewed.

The ability of the thyroid gland to accumulate iodide via Na/I symporter (NIS) has long provided the basis for diagnostic scintigraphy of the thyroid with radioiodine, and also for ^{131}I treatment of Graves' disease and metastatic thyroid cancer. Since the cloning of cDNA of rat NIS, the genomic organization of human NIS has been elucidated. Spontaneous NIS mutation leads to a congenital defect in iodide transport. Gene therapy experiments have been tried by transducing the NIS gene into various types of cells. Clinical indications for radioiodine uptake studies, scintigraphy, and use of tumor-seeking agents are reviewed.

The molecular characterization of thyroid peroxidase (TPO) and TSH receptor has revealed their role as thyroid antigens in autoimmune thyroid diseases. The cloning of these antigens as well as thyroglobulin has made the analysis of antigenic determinants possible. Measurement of thyroid autoantibodies using radioassay systems is contributing to the diagnosis and treatment of autoimmune thyroid disorders. Somatic mutations of the TSH receptor are observed in hyperfunctioning thyroid adenomas. Oncogenes such as ret/PTC have been shown to be involved in the development of thyroid tumors. Recombinant hTSH has been applied for the imaging of metastatic thyroid cancers.

Nuclear imaging methods such as ^{201}Tl-^{99m}TcO$_4$ subtraction and double-phase ^{99m}Tc-methoxyisobutyl-isonitrile (MIBI) scintigraphy are utilized for the localization of the hyperfunctioning parathyroid gland, especially in persistent or recurrent hyperparathyroidism (although detection rates greatly vary). In addition, PET studies with ^{18}F-FDG and L-^{11}C-methionine have been used for the evaluation of glucose or methionine metabolism as well as for localizing parathyroid adenomas and hyperplasia. Recent advances in molecular biology and genetics have made it possible to clarify some of abnormalities in parathyroid disorders. In the future, molecular nuclear imaging, which employs the principles of molecular biology and genetics, will also contribute to in vivo visualization of biochemical processes of hyperfunctioning parathyroid glands.

Nuclear imaging methods that use tumor-avid radionuclides, radio-labeled iodocholesterol (NP-59), and meta-iodobenzylguanidine (MIBG) have demonstrated functioning tumors and dysfunction in the adrenal cortex, adrenal medulla, and adrenergic neurons. For example, the correlation of image patterns with values of specific hormones can determine the nature of hyperplasia and neoplasia with high degree of accuracy. Similar methods using refined radiopharmaceuticals can be applied to further define molecular abnormalities of the adrenal cortex in familial Cushing's syndrome, different types of aldosteronism, and congenital adrenal hyperplasia and hypoplasia. Currently, some radiopharmaceuticals are available to dissect the steps in synthetic pathways of hormone, neurotransmitter, and norepinephrine, helping delineate the individual abnormalities that cause superphysiologic actions or failure of function in the adrenal medulla and sympathetic nervous system. Molecular and genetic changes can be directly correlated to scintigraphic alterations, or at least they can be inferred from correlated data. Reviewed finally is the application of radiolabeled somatostatin analogues for the receptor scintigraphy and radionuclide therapy of gastroenteropancreatic neuroendocrine tumors, which has opened a new era in molecular nuclear medicine.

15.2
Radiopharmaceutical Imaging of the Pituitary Gland

BEN A. DWAMENA, MILTON D. GROSS, BRAHM SHAPIRO

15.2.1
Introduction

15.2.1.1
Historical Background

Depiction of the pituitary gland in health and disease states has been dominated for the last 25 years by the sectional anatomical imaging of CT and MRI (Chakeres et al. 1989; Chen and Kucharczyk 1989; Donovan and Nesbit 1996; Naidich and Russell 1999). These modalities can demonstrate the pituitary gland and lesions arising from it with exquisite spatial resolution (Korogi and Takahashi 1995). However, they frequently fail to reveal the source of functional disorders such as pituitary microadenomas (defined as < 10 mm in size) and hyperplasia with hormone hypersecretion (Freda and Post 1999; Freda and Wardlaw 1999). Pituitary imaging utilizing radiopharmaceutical-based procedures also has a long history and, in fact, predates CT and MRI. However, earlier tracer studies of pituitary and other

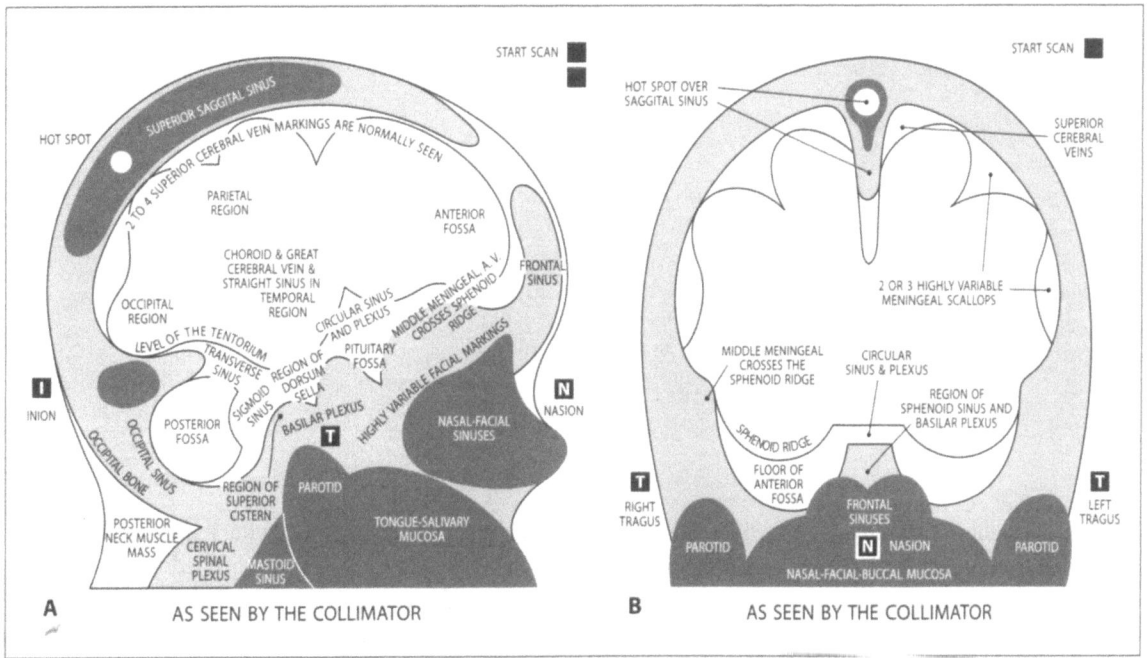

Fig. 15.1 A, B. Diagrammatic depiction of conventional planar rectilinear brain scans. Anterior projection (**A**) and lateral projection (**B**). Reproduced with permission from Brucer (1971)

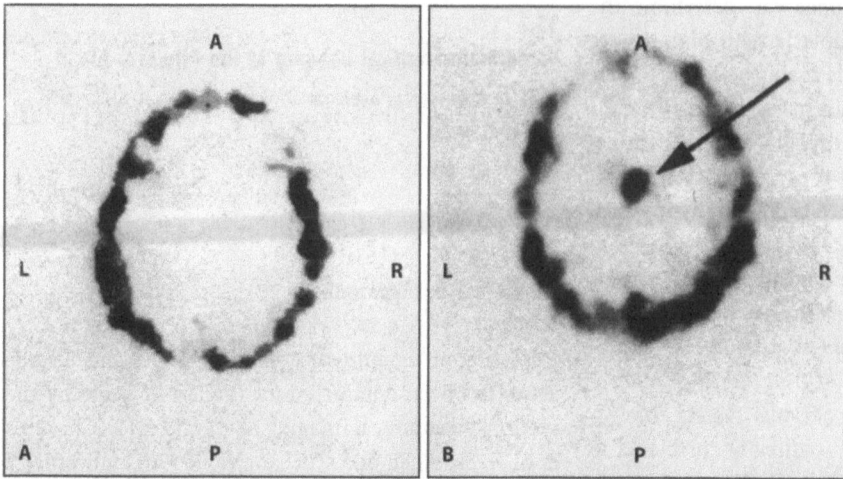

Fig. 15.2 A, B. Early ^{99m}Tc-pertechnetate (15 mCi) SPECT image of pituitary region. Normal subject (**A**) and patient with large pituitary adenoma (**B**, *arrow*). Both scans are parallel to the outer canthal-meatal line (OCM) and 2 cm above it. (Reproduced with permission from Britton and Shapiro 1981)

Table 15.1. Pituitary hypersecretory hormonal disorders

Hormones	Syndrome
Thyrotropin Stimulating Hormone (TSH)	Rare and unusual form of hyperthyroidism characterized by elevated TSH, diffuse goiter, absence of Graves' eye disease and elevated radioiodine uptake.
Growth Hormone (GH)	Acromegaly in adults, gigantism in children. GH levels are not suppressible by glucose load. Effects of GH are primarily mediated by insulin-like growth factor I (IGF-1).
Prolactin (Prl)	Effects include amenorrhea and galactorrhea in women and impotence and infertility in men. Autonomous hypersecretion by adenomas must be distinguished from hypersecretion secondary to interruption of inhibitory dopaminergic signals via the hypothalamic/hypophyseal axis, and also from the effects of drugs.
Adrenocorticotropin secreting Hormone (ACTH)	Cushing's disease results from autonomous stimulation of both adrenal cortices by ACTH secreted from the pituitary. This is characterized by measurable ACTH, elevated urinary cortisol suppressibility by high doses of dexamethasone, and stimulation by CRF. Pituitary Cushing's disease must be distinguished from primary adrenal disease and ectopic ACTH syndrome.
Follicle Stimulating Hormone/Luteinizing Hormone (FSH/LH)	Such lesions are rare (1%) and usually clinically silent but may occasionally cause gonadotropin hypersecretion.
Non-functioning tumors	Modern radioimmunoassay and immunohistochemistry of hormones reveal a small subset of tumors that are truly non-hypersecretory except for the a subunit.

intracranial tumors were largely limited due to the lack of specific functional tracers other than those which depicted breakdown of the blood-brain-barrier (Fig. 15.1; Britton and Shapiro 1981). Normal pituitary has an embryological origin from Rathke's pouch (an upward invagination of the foregut) that grows to meet the base of primordial brain, and which subsequently forms the hypothalamus, pituitary stalk, and posterior hypophysis and thus, is not protected by the BBB.

Although the resolution of the then available rectilinear scanner and gamma camera was limited, the early nuclear imaging was already able to depict macroadenomas, particularly those with suprasellar extension. Radiopharmaceuticals employed at that time included Hg^{137}-chloromeridin, I^{131}-HSA and subsequently Tc^{99m}-O$_4$, Tc^{99m}-DTPA which were endocrinologically not functional. The rectilinear scanner was advantageous for the imaging of pituitary tumors because it had a focusing collimator with the greatest resolution at depth. One of the earliest functional radiopharmaceuticals used for pituitary imaging was Se^{75}-selenomethionine which was an index of protein and peptide synthesis (Fig. 15.2). Imaging was achieved using a first generation rotate-translate tomographic scanner as early as 1978 (Britton and Shapiro 1981). But scan speeds were too slow and count rates too low for practical characterization.

15.2.1.2
Recent Developments

Despite the rather inauspicious beginning, recent brisk developments in instrumentation, molecular biology, and radiopharmaceuticals have greatly stimulated the application of radionuclide imaging for solving clinical problems related to the pituitary gland. Indeed, the construction of modern rotating gamma cameras, multi-head SPECT, and dedicated PET, advances in molecular biology – particularly the characterization of many pituitary hormonal/hypothalamic neurotransmitter ligands and their cell surface receptors (Moyse et al. 1985; Muhr et al. 1986a, b) – and the availability of radiopharmaceuticals and the techniques for labeling recently described hormonal and neurotransmitter ligands (e.g., the entire group of somatostatin-receptor-binding ligands) with them are some examples of the good progress made. Furthermore, PET radiotracers have permitted in vivo evaluation of various aspects of tumor metabolism such as glucose uptake using F^{18}-fluorodeoxyglucose, peptide synthesis using C^{11}-methionine (Muhr et al. 1986a, b; Muhr and Bergstrom 1991), and tumor blood flow using O^{18}-water.

15.2.2
The Current Role of Radiopharmaceutical Imaging of Pituitary Lesions

15.2.2.1
Clinical Context

In placing the radiopharmaceutical imaging of pituitary lesions into an investigative algorithm, it is important to realize that studies of this type are essen-

Table 15.2. Syndromes resulting from impairment of pituitary hormone secretion

Hormone	Syndrome
TSH	Central or secondary hypothyroidism, which is characterized by hypothyroidism with low TSH and thyroid hormone levels and an impaired response to TRH.
GH	Growth hormone deficiency causes stunted growth in children and, in adults, is now increasingly recognized to be a cause of impaired overall physical function, obesity, muscle atrophy, and cardiac dysfunction. GH and IGF-I are low and GH is unresponsive to insulin-induced hypoglycemia, exercise, dopamine, and GnRH.
Prl	Decreased Prl levels seldom occur except in total pituitary infarction (Sheehan's syndrome) causing panhypopituitarism with lactational failure following peripartum hemorrhage. Impaired Prl secretory reserve is disclosed by failure of the hormone levels to increase after administration of a dopamine antagonist such as metaclopropamide.
ACTH	Central or secondary hypoadrenalism characterized by low ACTH and cortisol levels and all the features of Addison's disease except the hyperpigmentation. Cortisol responses to insulin hypoglycemia and synthetic ACTH are impaired.
FSH/LH	Central or secondary hypogonadism, which is characterized by amenorrhea and/or infertility in women and impotence and/or infertility in men. Hormonally, low sex steroid hormones (estradiol in women and testosterone in men) are associated with low gonadotropin levels, which are poorly responsive to GnRH stimulation.

tially adjunctive to comprehensive history and physical examination, to biochemical studies sufficient to characterize the hormonal secretory status, and to MRI imaging with high-level tissue characterization (Korogi and Takahashi 1995). Pituitary and parapituitary lesions typically present with one or both of two broad spectrum clinical features (Tables 15.1–15.4): (1) Humoral effects of autonomous hypersecretion of pituitary hormones caused by adenomas and rare carcinomas or hyperplasia (Table 15.1), or as the consequence of impaired secretion of these hormones due to destruction of secretory capacity or interruption of secretory signals from the hypothalamus, as in damage to the hypothalamic-hypophyseal portal system (Table 15.2); and (2) mechanical or physical effects of tumor mass to many neighboring neurolo-

Table 15.3. Mass effects on adjacent structures by pituitary and parapituitary mass lesions

Directional spread	Structures affected	Effects
Lateral extension	Cavernous sinus	Cavernous sinus obstruction or cranial nerve palsies
	Cranial nerves III, IV, VI	Tumors seldom compress or infiltrate the carotid but aneurysm may be confused with pituitary lesions
	Intracavernous carotid artery	
	Medial surface of temporal lobe	Very large tumors may cause temporal lobe epilepsy
Anterior extension	Olfactory bulbs	Anosmia
	Cribriform plate	CSF rhinorrhea
Inferior extension	Sphenoid sinus	Sphenoid mucocele, CSF rhinorrhea. and sphenoid sinusitis
Superior extension	Diaphragma sellae	Expansion into these suprasellar cisterns is in itself asymptomatic but predicts future optic chiasm involvement.
	Basal CSF cisterns	
	Pituitary stalk	Hypothalamic/pituitary dysfunction
	Optic chiasm	Bitemporal hemianopsia
	Hypothalamus	Hypothalamic pituitary dysfunction, syndrome of increased or decreased appetite, altered thirst somnolence, central fever.
	Third ventricle	Communicating hydrocephalus if compressed
Posterior extension	Occipital bone	This bone is dense and resists infiltration but chordomas of the clivus may involve the pituitary. Non-communicating hydrocephalus if Sylvian aqueduct compressed.
	Brain stem	Long tract signs and cranial nerve palsies
	Posterior hypophysis	Diabetes insipidus

Table 15.4. Parapituitary mass lesions

Craniopharyngiomas
Meningiomas
Optic nerve and optic chiasm gliomas
Hypothalamic gliomas
Cholesteatomas
Germinomas
Teratomas
Chordomas
Arachnoid cysts
Rathke's pouch cysts
Carotid aneurysms
Metastases (lung, breast, melanoma, etc.)
Granulomas   – Sarcoid
              – Tuberculosis
              – Histoplasmosis
              – Histiocytosis, eosinophilic granuloma

gic and other anatomical structures, with or without concurrent hormonal effects (Table 15.3).

15.2.2.2
Indications for Pituitary Scintigraphy

There are a number of indications for radiopharmaceutical imaging in patients with known or suspected pituitary lesions. One of these addresses the localization of the source of humoral hypersecretion, when anatomical pituitary imaging results are indeterminate, in order to delineate occult pituitary lesions from ectopic extrapituitary lesions (Phlipponneau 1994; de Herder 1996; Segu 1997). Another indication for such imaging incorporates the experience that scintigraphy is well suited to whole-body screening for clinically occult neuroendocrine lesions in neuro-cristopathic syndromes like MEN I (which includes pituitary, pancreatic, and parathyroid lesions; Ponssen et al. 1996). Another indication is the in vivo, non-invasive characterization of lesions for the selection and monitoring of therapy with long acting somatostatin analogues and dopamine agonists (Broson-Chazot et al. 1997; Legovini et al. 1997; Ferone et al. 1998; Colao et al. 1999). Finally, radiopharmaceutical imaging can differentiate postoperative changes from residual or recurrent tumors (Lauriero et al. 1998), particularly using radiolabeled somatostatin analogues (which will be described later).

Table 15.5. Radiopharmaceutical agents for pituitary scintigraphy

Radiopharmaceutical	Modality	Mechanism of action	(Potential) Clinical application
^{18}F-FDG	PET	Glucose Metabolism	Primarily, differentiation of postoperative changes from residual or recurrent tumor. Secondarily, differential diagnosis of pituitary adenomas and other parasellar lesions
^{11}C-Methionine	PET	Amino acid metabolism and transport	Differential diagnosis of pituitary adenomas and other parasellar lesions Distinguishing between postoperative changes and residual or recurrent tumors
^{18}F-Fluoroethyl-spiperone	PET	Dopamine receptor agonism	Localization of tumor (pituitary or ectopic) Selection and monitoring of therapy with long acting dopamine agonists
^{11}C-Raclopride	PET	Dopamine receptor agonism	Localization of tumor (pituitary or ectopic) Selection and monitoring of therapy with long acting dopamine agonists
^{111}In-DTPA-Pentetreotide	SPECT	Somatostatin receptor agonism	Localization of tumor (pituitary or ectopic) Selection and monitoring of therapy with long acting somatostatin analogues
^{123}I-IBZM	SPECT	Dopamine receptor agonism	Localization of tumor (pituitary or ectopic) Selection and monitoring of therapy with long acting dopamine agonists
^{123}I-Epidepride	SPECT	Dopamine receptor agonism	Localization of tumor (pituitary or ectopic) Selection and monitoring of therapy with long acting dopamine agonists.

15.2.3
Imaging Techniques (Table 15.5)

15.2.3.1
Single Photon Imaging Techniques

15.2.3.1.1
Somatostatin Receptor Imaging (Maini et al. 1995)

15.2.3.1.1.1
Background

Native somatostatin is a cyclic peptide with 14 and 28 amino acids and widely distributed in the hypothalamus, central nervous system, spinal cord, pancreas, gut, adrenal medulla, and many other tissues. It subserves paracrine, true endocrine, hormonal, and neurotransmitter roles. Somatostatin decreases the secretion of growth hormone, gastrin, insulin, and other hormones; it reduces splanchnic blood flow and decreases pancreatic, gastric, and other exocrine secretions; in addition, it modulates the function of activated leukocytes. Somatostatin receptors, of which there are at least five subtypes whose genes and pro-

teins have now been characterized, are expressed by many normal and neoplastic tissues, most prominently by those of neuroendocrine organs such as the somatotroph cells of the anterior pituitary.

The multiple somatostatin receptor subtypes have varying affinities for different biologic forms of somatostatin and its synthetic analogues. For example, octreotide has a high affinity for somatostatin tissue receptor type 2 (SSTR-2), but not SSTR-1. Some apparent failures in the diagnosis and therapy with octreotide and its radiolabeled preparations may be attributable to SSTR-2 as well as to the imaging technique and true receptor-negative lesions. The heterogeneous distribution of receptor subclasses within tumors has also been documented. The development of octreotide, a relatively long-acting synthetic octapeptide analogue of somatostatin, provided the first practical means for exploiting the actions of somatostatin for treating at least the hypersecretory manifestations of many neuroendocrine tumors.

The advent of ^{123}I-Tyr^{3}-octreotide offered the opportunity to scintigraphically localize tumors with dense somatostatin receptor populations, and possibly to predict their response to treatment. Imaging was reasonably successful but the labeling was technically difficult with drawbacks of limited availability, suboptimal half-life of I-123, pronounced biliary excretion, and persistently high background activity due to circulating degradation products. The advent of ^{111}In-DTPA-D-Phe^1-octreotide (Octreoscan) circumvented many of these problems with its longer half-life, rapid renal clearance (85% of dose recovered by 24 h), and few circulating labeled metabolites (Bakker et al. 1991; Krenning et al. 1992; Boni et al. 1995). The strengths of radiolabeled octreotide are in the localization and staging of neuroendocrine tumors, both pre- and post-operative, and the prediction of tumor response to octreotide (Krenning et al. 1992, 1993 b; Maini et al. 1995; Kwekkeboom et al. 1999).

15.2.3.1.1.2
Imaging Procedure (Figs. 15.3–15.5)

For imaging 3 to 6 mCi is administered intravenously and the pituitary is imaged at 24 h, or occasionally 48 h afterward. Because of normal pituitary uptake, separation between normal and abnormal uptake may be difficult (Colao et al. 1999). Uptake indices using tumor/non-tumor ratios assessed by a standard region of interest can be used but no validated data are available.

Physiologic uptake is seen in the pituitary, thyroid, spleen, gut, liver, kidneys, and urinary tract. SPECT is useful to distinguish pathological from physiological tracer distribution. Concurrent octreotide therapy decreases the sensitivity of ^{111}In-DTPA-D-Phe^1-octreotide imaging in animal models. Thus, when studying hormonally active tumors inquiry should be made about recent or ongoing octreotide pharmacotherapy. If there has been no biochemical or clinical response to octreotide, then the likelihood of visualizing the tumor with Octreoscan is low. Octreotide therapy should be withheld before imaging for 1 to 7 days to reduce competitive inhibition of Octreoscan receptor binding by unlabeled octreotide. However, the cessation of medication has been debated and it is possible to perform octreotide receptor imaging while the patient is on octreotide that cannot be withheld (a circumstance to be documented).

15.2.3.1.1.3
Published Clinical Studies

A wide variety of neuroendocrine and other tumors have been studied (Krenning et al. 1993 a, b; Table 15.6). In most series, scintigraphy revealed lesions

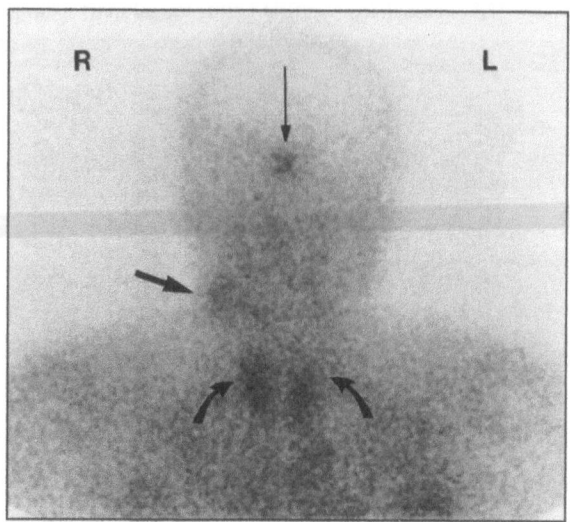

Fig. 15.3. Frontal planar ^{111}In-DTPA-octreotide scan in a patient with MEN-1 shows increased tracer uptake in a growth hormone secreting pituitary tumor (*thin straight arrow*), bilateral early medullary thyroid cancer (*curved arrows*), and faint uptake in right cervical lymph node metastases from the thyroid cancer (*short straight arrow*)

that were either undetected by other modalities or clinically not suspected. The non-neoplastic lesions that accumulate octreotide include recent scars, reactive lymph node hyperplasia, irradiated and bleomycin-exposed lung parenchyma, granulomatous diseases, and rheumatoid tissues. Such accumulation is considered to relate with the presence of somatostatin receptors in activated lymphocytes and other inflammatory cells.

Oppizzi et al. (1998) have suggested that in acromegaly scintigraphy with ^{111}In-pentetreotide (^{111}In-P) visualizes functioning pituitary SS-R coupled to intracellular events that control hormonal hypersecretion and tumor growth. In contrast, despite positive ^{111}In-P imaging in most patients with "clinically nonfunctioning" pituitary adenoma (NFPA), their receptors might have a defective coupling-transduction process since they are not inhibited by octreotide treatment and tumor does not shrink. Oppizzi's group performed pituitary scintigraphy using ^{111}In-P in patients with GH-secreting adenoma and NFPA to evaluate the presence and the functionality of SS receptors (SS-R). Accumulation of ^{111}In-P in the pituitary was expressed as activity ratio (AR), the ratio of radioactivity in the adenoma to that in normal brain tissue. The AR ranged from 1.6 to 2.2 in those without pituitary disease and hence a value lower than 2.2 was arbitrarily considered as "normal." On the other

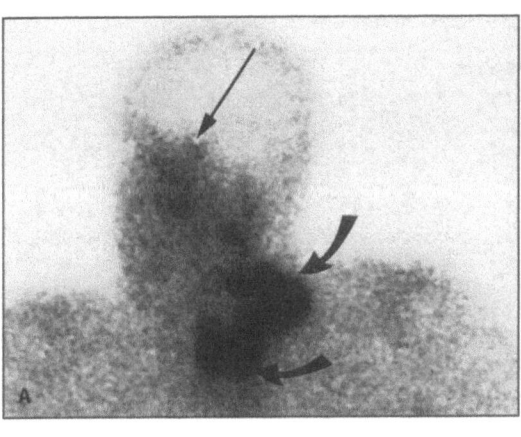
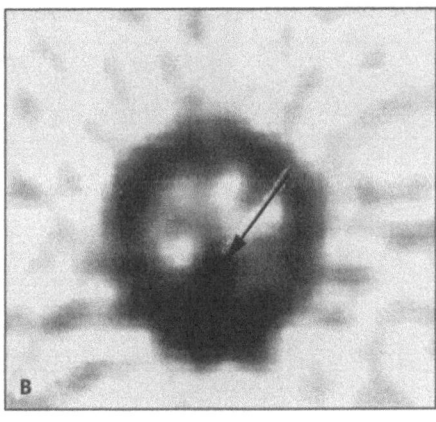

Fig. 15.4 A, B. An ^{111}In-DTPA-octreotide scan in a patient with an unusual crossover syndrome with a growth-hormone-secreting pituitary tumor (*thin straight arrow*) and papillary follicular thyroid cancer with lymph node metastases (*thick curved arrows*) shows positive uptake on an anterior planar image (**A**) and a transaxial SPECT (**B**)

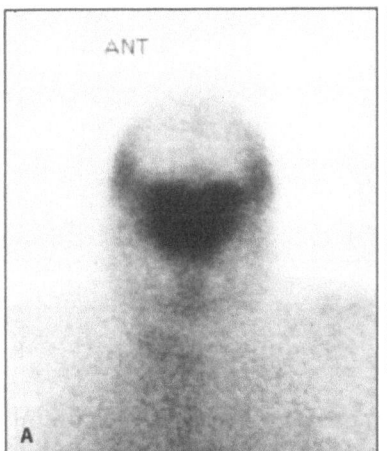
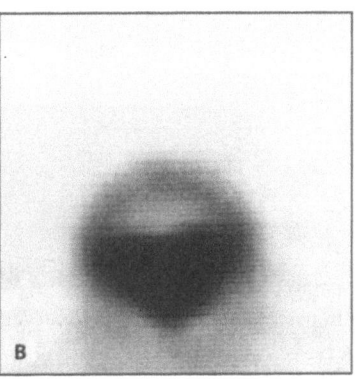
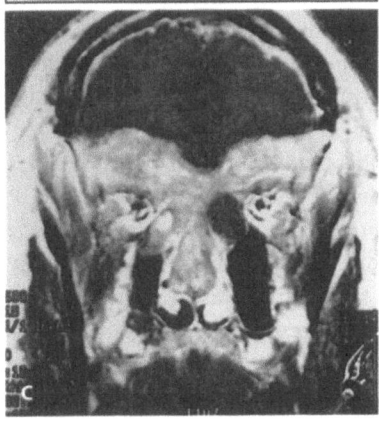

Fig. 15.5 A–C. Giant highly aggressive prolactinoma with extensive spread into the subfrontal, temporal, and periorbital regions, and into the nasal cavity, demonstrating intense receptor binding of ^{111}In-DTP A-octreotide on anterior planar imaging (**A**), anterior coronal SPECT section (**B**) and anterior coronal MRI section (**C**) which is analogous to **B** and demonstrates the subfrontal and nasal spread

hand, a positive accumulation of the radioligand was shown in 15 out of 17 patients with GH-secreting adenoma. The median AR was 3.8 (range 1–6.9; the AR was greater than 2.2 in 14) and ARs were directly correlated ($r = 0.54$; $P = 0.05$) with the suppressibility of plasma GH levels by acute octreotide administration. In two patients who repeated scintigraphy during chronic octreotide treatment, ARs were reduced. In all 22 patients with NFPA an accumulation of ^{111}In-P at the pituitary level was observed and the

Table 15.6. Depiction of radiolabeled somatostatin analogue binding to various tumors[a]

Pituitary tumors				
GH secreting	7/8	(88)	45/46	(98)
TSH secreting	2/2	(100)	2/2	(100)
Non-secretory	6/9	(67)	6/15	(40)
Endocrine pancreatic				
Gastrinoma	10/11	(91)	6/6	(100)
Glucagonoma	1/1	(100)	2/2	(100)
Insulinoma	9/14	(65)	8/11	(72)
Non-functioning	3/3	(100)	4/4	(100)
Exocrine pancreatic	0/18	(0)	0/12	(0)
Carcinoid	37/39	(95)	54/62	(87)
Merkel cell tumors	4/5	(80)	–	
Medullary thyroid carcinoma	20/28	(71)	10/26	(38)
Paraganglioma	50/53	(94)	11/12	(92)
Neuroblastoma	8/9	(89)	15/23	(65)
Small cell lung cancer	56/57	(98)	4/7	(57)
Breast carcinoma	39/52	(75)	33/72	(46)
Lymphocytic lymphoma (all sites)	81/107	(76)	26/30	(87)
Intra-abdominal	1/9	(11)	–	
Extra-abdominal	21/24	(75)	28/32	(88)
Hodgkin's disease	23/24	(96)	2/2	(100)
Glioma				
Grade II	0/5	(0)	–	
Grades III and IV	7/7	(100)	--	
Meningioma	20/20	(100)	54/55	(98)

[a] Compiled from reference sources. Numbers in parentheses represent percentages.

median AR was 3.0 (range 1.5–20; greater than 2.2 in 14). In vitro autoradiography of surgical specimens in six NFPA patients revealed SS-R in four cases with high scintigraphic AR and negative results in two cases with low AR. The scintiscan was repeated during chronic octreotide treatment in five patients with high scores: AR decreased in one patient, increased in three, and did not change in one. Tumor size did not change in any of these patients. A total of eight patients (three GH secreting and five NFPA) had "normal" AR values.

It has been shown that ^{111}In-P scan can be used to image TSH-secreting pituitary adenomas, confirming the presence of somatostatin receptors in these rare tumors. Losa and his group (1997) studied five patients (three women and two men of ages 27–46 years) with TSH-secreting adenomas using ^{111}In-P SPECT. The intensity of ^{111}In-P uptake by tumor was correlated with the degree of TSH suppression after a single subcutaneous administration of octreotide. Except for one patient with acromegaly, all had pure TSH-secreting tumors. One patient was not treated, two treated with octreotide, one with antithyroid drugs, and one with radioiodine. In all, SPECT demonstrated increased uptake of ^{111}In-P by the pituitary adenoma. Administration of octreotide caused a significant reduction in TSH levels.

Legovini and others (1997) have shown that apart from the depiction of pituitary adenomas, ^{111}In-P scintigraphy can predict GH responses to octreotide in unoperated acromegalic patients. In unsuccessfully operated patients, scintigraphy was infrequently positive and did not predict response to octreotide. Based on these results, they suggested that the extensive clinical use of ^{111}In-P scintigraphy is not warranted in acromegaly. They studied ^{111}In-P scintigraphy and hormonal responses to octreotide in 12 acromegalic patients. Nine had active acromegaly before pituitary adenomectomy. Six of these and three others were studied after operation. GH was measured after a single subcutaneous dose of 100 micrograms of octreotide (acute test).

Patients were preoperatively treated with 100 micrograms of octreotide subcutaneously three times daily for 3 months in the same manner as are unsuccessfully operated patients, and GH and IGF-I were then measured. A decrease in hormones greater than 50% of baseline values was considered as a positive response in both the acute test and to chronic treatment. Eight of nine unoperated patients showed an adenoma visualized by scintigraphy and had a positive response to the acute test and to chronic treatment. One showed no evidence of tumor on scintigraphy and did not respond to octreotide. Scintigraphy was negative in all three patients cured by surgery. Six patients still had active disease after adenomectomy but scintigraphy was positive only in one case, and GH responded to octreotide treatment in all patients.

The predictability of somatostatin receptor scintigraphy (SRS) to somatostatin analogue therapy for pituitary adenomas is inconclusive at this time because they used different radiopharmaceutical agents (^{123}I-Tyr3-octreotide and ^{111}In-P) and the numbers of patients were too small in most studies. Plockinger et al. (1997) used ^{111}In-P scintigraphy in 49 patients to correlate SRS results with basal tumor volume as well as volume and hormone response to 3 months of octreotide therapy, to identify tumor remnants after incomplete surgery and to evaluate correlation with immunohistology. Twenty-five patients had a GH-secreting adenoma (GH-A; 15 prior to intended surgery, 10 with persistent or recurrent disease after previous therapy). Twenty-four patients had a NFPA. They reported that ^{111}In-P SRS reflects tumor volume poorly in GH-A and not at all in NF-A. It does not predict the effect of octreotide treatment on the volume of either GH-A and NFPA, nor on the GH concentration in GH-A. Tumor remnants not visible on postoperative MRI cannot be identified with ^{111}In-P SRS.

15.2.3.1.2
Dopamine Receptor Imaging

15.2.3.1.2.1
Background

Dopamine D2 receptors have been demonstrated in vitro in prolactinomas, GH-secreting tumors, and NFPAs. Dopamine receptors can also be demonstrated in vivo using SPECT with ^{123}I-S-[(1-ethyl-2-pyrolidinyl)-methyl]-2-hydroxy-3-iodo-6-methoxybenzamide (IBZM) or with the newer substituted benzamide,

[123]I-epidepride. The affinity of ^{123}I-IBZM for dopamine D_2 receptors is in the nM range compared to an affinity in the pM range for ^{123}I-epidepride. Although not for microprolactinomas or GH-secreting adenomas, ^{123}I-IBZM is a ligand suitable for in vivo imaging of dopamine agonist-sensitive macroprolactinomas. The technique is a potential means of predicting the dopamine agonist-responses of non-functioning pituitary adenomas in vivo. Pituitary scintigraphy with ^{123}I-IBZM may also predict the response to long-term treatment with dopamine agonists such as quinagolide in patients with NFPA, where the lack of pituitary hormone hypersecretion makes monitoring of treatment efficacy difficult. Intense ^{123}I-IBZM uptake in patients with nonfunctioning adenomas predicts a good response to a chronic treatment with quinagolide and cabergoline. Such molecular imaging may be useful in depicting recurrent or residual tumor following surgery when anatomic imaging is made useless by artifacts.

15.2.3.1.2.2
Imaging Procedure

Clinical indications and imaging guidelines of the ^{123}I-IBZM pituitary scintigraphy are currently evolving and no validated protocols are available at present.

15.2.3.1.2.3
Published Clinical Studies

De Herder et al. (1996) evaluated the value of the pituitary dopamine D2 receptor scintigraphy using ^{123}I-IBZM in the diagnosis of patients with pituitary tumors. Imaging was performed in five patients with prolactin (PRL)-secreting macroadenomas, two with PRL-secreting microadenomas, 17 patients with NFPAs, 12 with GH-secreting adenomas and one patient with a TSH-secreting macroadenoma. SPECT showed significant uptake of ^{123}I-IBZM in the pituitary region in three out of five macroprolactinoma patients. These results closely correlated with the positive response of plasma PRL levels to quinagolide (dopamine D_2 receptor agonist). In two scan-positive prolactinoma patients, repeated SPECT during quinagolide therapy showed reduction in the pituitary uptake of ^{123}I-IBZM. The pituitary SPECT was negative in microprolactinoma patients who responded to quinagolide administration. In 4 of 17 patients with NFPA, significant uptake of radioligand in the pituitary region was observed. In two of three scan-positive NFPA patients treated with quinagolide shrinkage of the pituitary tumors occurred. Treatment with quinagolide stabilized tumor growth in other scan-positive patients. Four out of 17 patients with NFPA with negative SPECT were treated with quinagolide. None of the 12 acromegaly patients showed significant uptake of ^{123}I-IBZM. Tumor growth was observed in one patient; no change was seen in three. A sensitivity of GH-secreting adenomas to quinagolide was demonstrated in vivo by an acute test in eight of nine patients and by an in vitro test in six. The pituitary SPECT was negative in patient with TSH-secreting macroadenoma and the tumor did not show response to quinagolide in vivo or in vitro.

Pirker and colleagues (1994) performed SPECT using ^{123}I-IBZM in 15 patients with pituitary tumors. There were five prolactinoma patients with macroadenoma and two acromegalic patients with macroadenoma. Of these, specific binding in the area of adenoma was observed only in one patient with macroprolactinoma who was responsive to dopaminergic treatment. IBZM binding in the striatum was found to be significantly lower in the pituitary tumor patient group as compared to controls.

Ferone and colleagues (1998) performed ^{123}I-IBZM imaging in 14 patients with macroadenoma before starting a long-term treatment with quinagolide. Six patients had NFPA with high circulating alpha-subunit levels, four with PRL-secreting, and four GH-secreting adenomas. In all 14, a significant positive correlation was noted between the degree of ^{123}I-IBZM uptake and clinical response to quinagolide treatment ($r = 0.90$; $P = 0.001$). In particular, the normalization of serum alpha-subunit and PRL levels, respectively, was achieved in three patients with NFPA and in two patients with prolactinoma, who showed intense ^{123}I-IBZM uptake in the pituitary region. In four of these five patients with a positive scan, a significant tumor shrinkage occurred 6–12 months after the beginning of quinagolide treatment. In all patients with GH-secreting adenoma, significant uptake of ^{123}I-IBZM, significant decrease in circulating GH and/or insulin-like growth factor-I levels, and tumor shrinkage were not noted during long-term treatment with quinagolide.

In vivo visualization of dopamine D_2 receptor expression detected by ^{123}I-IBZM scintigraphy was correlated with the response to chronic treatment with quinagolide or cabergoline in ten patients with NFPA (five men and five women ages 25–50) and ten patients with PRL-secreting macroadenomas (three men and seven women ages 22–59) as control. At study entry, an MR scan of the pituitary region and ^{123}I-IBZM pituitary scintigraphy were performed. MR was repeated after 12 months of treatment to evaluate tumor shrinkage. Scintigraphy was negative in six of ten patients with NFPA. Moderate uptake was observed in three patients with prolactinoma and two patients with NFPA, whereas an intense uptake was observed in the remaining seven patients with prolactinoma and two patients with NFPA. Of eight patients with NFPA and high circulating alpha-subunit levels, the acute test was negative in five, positive in three. The acute test was positive in all ten patients with prolactinoma.

After 12 months of treatment with quinagolide and cabergoline, the circulating PRL level was lowered in all ten patients with prolactinoma and normalized in seven. PRL suppression was noted in all ten patients with NFPA. Alpha-subunit level was significantly decreased in nine of ten patients with NFPA, normalized in four of eight. Significant adenoma shrinkage was recorded in four of seven patients with prolactinoma with intense pituitary uptake of ^{123}I-IBZM. Significant adenoma shrinkage was recorded only in two out of ten patients with NFPA with intense pituitary uptake of ^{123}I-IBZM. A significant positive correlation was noted between the intensity of uptake (considered as score) and response to quinagolide or cabergoline treatment (considered as percent hormone suppression) in patients with PRL-secreting adenoma ($r = 0.856$, $P = 0.005$) and in those with NFPA ($r = 0.787$, $P = 0.05$).

The ^{123}I-epidepride SPECT appears superior to ^{123}I-IBZM SPECT for the visualization of dopamine receptor-positive pituitary adenomas. De Herder and colleagues (1999) compared these two for the in vivo imaging of dopamine D_2 receptor in 15 patients with NFPA. Four patients with dopamine agonist-sensitive macroprolactinomas were studied as positive controls. The pituitary uptake was established using a visual scoring system and an uptake index was calculated by dividing the average count rates in the pituitary area by the average count rates in the cerebellum. All four macroprolactinomas showed specific binding of ^{123}I-epidepride, but only one showed specific binding of ^{123}I-IBZM. Specific binding of ^{123}I-epidepride was demonstrated in 9 of 15 clinically nonfunctioning pituitary adenomas (60%), but specific binding of ^{123}I-IBZM was seen in only six patients (40%). The uptake of ^{123}I-epidepride in the pituitary region was consistently higher than that of ^{123}I-IBZM. None of the patients with absence of ^{123}I-epidepride uptake in the pituitary area showed ^{123}I-IBZM uptake.

15.2.3.2
Positron Radiopharmaceuticals

15.2.3.2.1
Fluorodeoxyglucose

15.2.3.2.1.1
Background

Fluorine-18 fluorodeoxyglucose (FDG) is by far the most commonly used PET tracer. The rationale of the [18]FDG PET is based different rates of glucose utilization in normal tissues compared to diseased tissues, in particular to tumors, which often exhibit high metabolic rates. The normal pituitary gland does not accumulate FDG, but most common intrasellar masses, pituitary adenomas, and craniopharyngiomas accumulate FDG to a high degree, even if benign. No established indication for [18]FDG-PET imaging in pituitary adenomas exists at present. In differentiating pituitary adenomas from other sellar and parapituitary lesions, [18]FDG-PET does not provide anatomic information comparable to or have any significant functional imaging advantages over CT and MRI. However, [18]FDG-PET can be used to differentiate postoperative changes from tumor recurrence, nonsecreting adenomas, and rare pituitary carcinoma (De Souza et al. 1990; Muhr and Bergstrom 1991). While recurrent tumor exhibits a high [18]FDG uptake, MRI and CT provide no such reliable information necessary for important management decisions.

15.2.3.2.1.2
Imaging Procedure

Patients must be in the fasting state for at least 4 h to ensure satisfactory tumor uptake of [18]FDG. No other patient preparation is required. Data acquisition is done in the supine position with the head resting in a supportive device to minimize motion artifacts. Data are acquired for 20 min in three-dimensional mode, 30 min after intravenous administration of 180–370 MBq of [18]FDG with transmission scanning for attenuation correction. Image reconstruction is performed with a noniterative algorithm and [18]FDG-PET images may be interpreted visually and/or quantitatively. Visual image analysis provides information of an area with increased, decreased, or normal uptake corresponding to the lesion seen on a neuroradiologic examination by CT or MRI. Quantitative analysis measures absolute metabolic rates and re-

quires multiple arterial or arterialized venous blood samples. Semiquantitative measurements based on tumor-to-white matter or tumor-to-contralateral cortex ratio may be used as well as the standardized uptake values (SUV).

15.2.3.2.1.3
Clinical Studies

A small number of patients with pituitary adenomas have been studied with [18]FDG-PET. Francavilla and colleagues (1991) studied 24 patients with pituitary macroadenomas (32 studies) to assess the glucose utilization of these tumors in vivo. The adenoma metabolic index, which is the ratio of [18]FDG uptake in tumor to its uptake in a whole brain slice, was calculated. Comparisons were made between tumor uptake of [18]FDG and hormone secretion and response to therapies. Macroadenomas were identified visually as foci of increased [18]FDG uptake near the region of the sella. Uptake of [18]FDG was highest in nonfunctional adenomas. Tumors, however, did not exhibit metabolic rates that could characterize the type of hormone produced. Recurrent macroadenomas displayed metabolism similar to that of tumors not operated on, whereas irradiated adenomas showed lower glucose uptake than nonirradiated tumors. Drug therapy with bromocriptine or octreotide also decreased the glucose utilization of tumor. There was no correlation between the amount of hormone produced and the adenoma metabolic index for any group of tumors. According to this study, [18]FDG-PET predicts and defines the growth of pituitary adenomas, particularly when plasma hormone assays and conventional imaging techniques prove inadequate for monitoring patient response to therapy.

Francavilla's group also studied 20 cases of surgically verified pituitary microadenomas (17 with Cushing's disease and three with acromegaly). The [18]FDG-PET results were compared with those of CT and MR imaging and with simultaneous bilateral inferior petrosal sinus sampling (SIPS) in Cushing's disease. PET results showed 12 positive readings and one questionable reading compared to seven positive readings and one questionable reading on CT (18 cases studied) and 13 positive readings and two questionable readings on MRI. This result indicates that five of the positive PET readings were negative or questionable using MRI. PET studies of 20 healthy control subjects showed no false-positive cases, whereas other studies of healthy subjects using contrast-enhanced CT and MRI respectively yielded 20% and 15% positive readings suggesting silent or occult adenomas.

15.2.3.2.2
Other PET Agents

15.2.3.2.2.1
Background

Other possible applications of PET imaging for pituitary adenomas might be the use of ^{11}C-methionine (Bergstrom et al. 1986, 1987a–c) or dopamine D_2-receptor ligands such as ^{11}C-raclopride and ^{18}F-fluoroethyl-spiperone (Muhr et al. 1986a, b). These agents might be used to differentiate pituitary adenomas from other sellar/parasellar masses and to monitor medical therapy when a dopamine agonist is administered before surgical intervention (Bergstrom et al. 1987b). The monoamine oxidase B inhibitor, ^{11}C-1-deprenyl, may be useful in the differential diagnosis between meningiomas and pituitary adenomas (Bergstrom et al. 1992). Limited data are available for review and no validated clinical protocols are presently in use.

15.2.3.2.2.2
Clinical Studies

Bergstrom and colleagues (1991) performed over 400 PET examinations in patients with pituitary adenoma demonstrating that PET with carbon-11-methionine could provide valuable complementary information in the diagnosis of this tumor. PET was able to distinguish viable tumor tissue from fibrosis, cysts, and necrosis. Furthermore, PET with dopamine D_2 receptor ligands was found useful in characterizing the degree of receptor binding, providing information as to the prerequisites for dopamine agonist treatment. The most important finding was very high sensitivity of ^{11}C-methionine PET in the assessment of therapeutic effects. Bergstrom and colleagues (1992) also examined seven patients with clinically nonsecreting pituitary adenoma and five patients with meningioma using ^{11}C-deprenyl and ^{11}C-methionine PET. They demonstrated rapid and high uptake of ^{11}C-L-deprenyl in the pituitary adenomas immediately after the injection. In later phase, the uptake value became at a constant level that was equal to or higher than in normal brain tissue. In the meningiomas, however, the initially high uptake was followed by a marked

decrease with time, reaching a level that was approximately one half of brain tissue level. The study demonstrated that the binding of ^{11}C-deprenyl to monoamine oxidase B in pituitary adenomas was high, whereas it was very low in meningiomas. Operative samples from ten patients with meningioma and five patients with pituitary adenomas were biochemically analyzed for monoamine oxidase B activity using ^{14}C-phenyl-ethylamine as substrate. The nonsecreting pituitary adenoma showed high enzyme activity while secreting adenomas and meningiomas respectively showed about one-tenth and one-thirtieth of enzyme activities of nonsecreting adenomas.

Lucignani and colleagues (1997) investigated whether the uptake of ^{18}F-fluoro-ethyl-spiperone (FESP) with PET was helpful for the differential diagnosis of pituitary adenomas and other parasellar lesions, and for establishing the appropriate therapeutic approach. The population examined comprised 16 preoperative patients with the diagnosis of primary tumor of the sellar/parasellar region. The results showed that PET with ^{18}F-FESP was very useful for differentiating adenomas from craniopharyngiomas and meningiomas. Visual interpretation of images allowed such differentiation at approximately 70 min after tracer injection. Semiquantitative analysis of dynamic PET data confirmed results of the visual interpretation demonstrating that ^{18}F-FESP uptake was consistently at least two- to three-fold higher in nonfunctioning adenomas than in other parasellar tumors at 70 min after the tracer injection, and uptake increased further thereafter. Thus, ^{18}F-FESP PET appears to be quite helpful in those tumors in which differential diagnosis by conventional methods is uncertain.

15.2.4
Future Directions

15.2.4.1
Therapeutic Applications

It would be possible to deliver a therapeutic dose of radiation by a large amount (many hundreds of millicuries) of ^{111}In-DTPA-octreotide, or to use pure beta-emitting radiolabels such as yttrium-90 in place of ^{111}In for lesions shown by diagnostic studies to have plentiful somatostatin receptors. To date, this method has been applied to carcinoids and islet cell tumors. In the pituitary, benign adenomas and circumscribed carcinomas are better treated by surgery and/or local radiotherapy but radiopharmaceutical therapy might have an occasional place for the treatment of very rare but vexing problem of metastatic carcinoma.

15.2.4.2
Potential for Future Imaging

It is conceivable that in the future radiolabeled analogues of hypothalamic-hypophyseal portal regulatory peptides such as thyrotropin-releasing hormone (TRH), gonadotropin-releasing hormone (GnRH), and corticotropin-releasing hormone (CRH) may be developed for scintigraphic use. Such radiopharmaceutical agents, however, must have satisfactory receptor binding capacity, resist in vivo proteolytic enzymes, be suitable for radiolabeling either with ^{123}I via a thyrosine residue or via various bifunctional chelating agents such as DTPA, DOTA, etc., and contain a peptide sequence that binds a radiometal such as the series of radiopharmaceuticals developed by Diatide (Berlex Imaging, www.berlex.com), for example, Acutec for deep venous thrombosis.

The binding of dopamine to its receptors results in an intracellular response that is transmitted by second messenger molecules, such as cAMP, which in turn regulate enzymatic processes within the cell such as protein kinase. Development of radiotracers for second-messenger systems would be an enormous advantage for there are theoretical reasons to expect that pituitary disease and therapeutic interventions will express themselves more clearly at the level of second- and third-messenger systems.

15.3
The Thyroid Gland

JUNJI KONISHI, TAKASHI MISAKI

15.3.1
Thyroidal Uptake of Radioiodine and ^{99m}Tc-Pertechnetate

The thyroid gland concentrates iodide from the serum and organifies and oxidizes it at the apical membrane. This peculiar property of the gland supports molecular nuclear medicine that utilizes radioiodines to play a key role in defining the gland in both physiological and pathological states (Becker et al. 1996a). Iodide is also transported to, but not further metabolized in, the salivary glands, choroid plexus, gastric

Table 15.7. Indications for thyroid uptake measurements and imaging

1. Distinguish causes of thyrotoxicosis
   Graves' disease, Plummer's disease, thyrotoxic multi-nodular goiter
   Painless thyroiditis, subacute thyroiditis, postpartum thyroiditis
   Functioning metastasis of thyroid cancer
   Struma ovarii
   TSH-producing pituitary adenoma
   Pituitary resistance to thyroid hormones
   Hydatidiform mole, choriocarcinoma
   Exogenous thyroid hormone

2. T_3 suppression test

3. Provide data needed for calculation of a therapeutic dose of ^{131}I

4. Reversible hypothyroidism

5. Ectopic thyroid

6. Determine whether a thyroid nodule is functioning

7. Dyshormonogenesis
   Perchlorate discharge test

8. Metastasis of differentiated thyroid cancer

parietal cells, and lactating mammary glands. The measurement of radioiodine uptake by the thyroid, although not a primary test of thyroid function, provides valuable diagnostic information in some clinical situations (Table 15.7).

Another useful agent for thyroid scintigraphy and uptake that reflects transport or trapping activity is ^{99m}Tc-pertechnetate. Because pertechnetate is not organified or retained in the thyroid, peak activity in the gland is reached in 20 to 30 min after its intravenous administration (Atkins and Klopper 1973; Hays and Wesselossky 1973). A number of environmental factors, diseases, and drugs may influence the thyroid uptake of both radioiodine and pertechnetate (Table 15.8). In regions where dietary iodine intake is high, restriction of iodine-containing foods such as seaweeds (kelp in particular) for 2 weeks before the uptake study is required to obtain true value (Becker et al. 1996b).

Most thyrotoxicosis is caused by excessive activity of the thyroid gland, as in Graves' disease, but other causes include inflammatory diseases (subacute thyroiditis, silent thyroiditis, postpartum thyroiditis) and exogenous thyrotoxicosis. In each instance, the radioiodine uptake test may provide useful clues leading to the diagnosis (McDougall 1990; Cavalieri and Gerard 1991). During the thyrotoxic phase of thyroiditis, the uptake value becomes low, usually less than 2% at

Table 15.8. Factors that influence thyroid uptake

	Factor	Diseases
Increased uptake	TSH-receptor stimulating antibody TSH	Graves' disease Hashimoto's disease Dyshormonogenesis TSH-producing pituitary adenoma Thyroid hormone resistance Reversible hypothyroidism Iodine deficiency disorders
	hCG Autonomy (TSH-receptor mutation) Constitutive activation of TSH-receptor	Hydatidiform mole, choriocarcinoma Plummer's disease
Decreased uptake	TSH-receptor blocking antibody	Hypothyroidism due to TSH-receptor blocking antibody
	Suppression of TSH secretion	Subacute thyroiditis Silent thyroiditis Exogenous thyroid hormone Pituitary hypothyroidism
	Destruction due to inflammation	Hashimoto's thyroiditis Subacute thyroiditis Silent thyroiditis Postpartum thyroiditis Primary myxedema
	Radiation	Post ^{131}I-treatment External neck radiation
	Tumor No or small thyroid tissue	Nonfunctioning tumor Post-thyroidectomy Hypoplasia
	Na/I symporter mutation	Iodine trapping defect
Iodine excess	Perchlorate, thiocyanate Anti-thyroid drugs	

24 h. Similarly, exogenous thyrotoxicosis shows the 24-h uptake to be as low as 2–5%.

15.3.2
Molecular Genetics of Thyroid Hormone Synthesis

15.3.2.1
Sodium/Iodide Symporter

Often abbreviated as NIS in recent endocrine literature, the sodium/iodine symporter is deemed the last of major thyroid autoantigens cloned. Despite the critical role it plays in the initial part of thyroidal iodine metabolism, the molecular characterization of this transporter protein was hampered by the technical difficulty of its isolation from plasma membrane. But, Carrasco and her colleagues finally succeeded in fishing out cDNA of rat NIS (Dai et al. 1996) and the human counterpart was cloned by Smanik et al. in the same year. Soon afterward, an iodide transport

defect due to mutation of this gene at diversified points was identified in more than 50 cases in 29 families (Fujiwara et al. 1997; Matsuda and Kosugi 1997; Pohlenz et al. 1998; also see Sect. 11.2.3.1). Because of compensatory overexpression of defective proteins, possibly in conjunction with some yet unknown factors, phenotypic manifestations such as goiter size and extent of hypothyroidism of these mutations vary from patient to patient (Kosugi et al. 1998). Cloning of the NIS gene also sheds a light on such classical phenomena of thyroid physiology as accelerated iodine uptake in Graves' thyroid (Saito et al. 1997) and escape from the Wolf-Chaikoff effect (Eng et al. 1999).

15.3.2.2
Thyroperoxidase (TPO)

TPO is a heme-containing glycoprotein of about 105 kD molecular weight that catalyses the oxidation of iodine, the iodination of tyrosine residues, and the union of iodotyrosines (McLachlan and Rapoport 1992; Taurog 2000). The identity of TPO as a major part of "thyroid microsomal antigen" was established in the mid 1980s (Czarnocka et al. 1985; Hamada et al. 1985; Seto et al. 1987). Autoantibodies to this enzyme seems to inhibit its activities (Kohno et al. 1986; Okamoto et al. 1989), especially when titers are very high. More important is the role played by these antibodies as a marker of thyroid autoimmunity, not only in Hashimoto's thyroiditis but in Graves' hyperthyroidism (McLachlan and Rapoport 1992). Radioimmunometric assays for anti-TPO (Beever et al. 1989) are currently in wide clinical use, contributing to the diagnosis and treatment of autoimmune thyroid diseases (Kasagi et al. 1996). Mutated TPO usually fails to organify iodide resulting in a goitrous hypothyroidism known as "familial iodination defect" (De Vijlder and Vulsma 2000). Recent gene analysis of Pendred's syndrome, another type of iodide organification defect associated with hearing loss, has led to the discovery of a new thyroid-related protein called pendrin (Everett et al. 1997). Defects in iodide oxidation and organification can be detected in vivo by the perchlorate discharge test (Wolff 1964). In patients with defects in iodide oxidation and organification, for example, those with inactivating mutations in thyroid peroxidase, the thyroid radioiodine content decreases by more than 10% soon after perchlorate administration because the iodide not oxidized and organified diffuses back into the circulation.

15.3.2.3
Thyroglobulin (Tg)

The large glycoprotein Tg, molecular weight about 660 kd, serves as the provider of tyrosine residues for thyroid hormone synthesis (Taurog 2000). Mutation in various portion of its gene has been reported to cause diffuse goiter with or without hypothyroidism (Targovnik et al. 1993; Hishinuma et al. 1998). The clinical usefulness of Tg as a tumor marker for differentiated thyroid cancer has been well established (Iida et al. 1991; Ozata et al. 1994). So far, however, there is no reliable and simple laboratory test to distinguish Tg secreted from benign or malignant thyroid tissue, although differences have been reported in iodine content and other chemical characteristics (Tarutani and Ui 1985; Tarutani et al. 1991). In addition, like the anti-TPO, the autoantibody against this large protein molecule is an excellent marker of thyroid autoimmunity. Radioassays of anti-Tg showed its

close correlation to histopathological changes of chronic thyroiditis, even in those with a negative test for conventional particulate agglutination assays (Kasagi et al. 1996).

15.3.3
Molecular Genetics Regarding Regulatory Mechanism of Thyroid Function

After the cloning of the thyrotropin (TSH) receptor gene (Kohn et al. 1995; Parmentier et al. 1989; Nagayama et al. 1989), it has become gradually known that toxic multinodular goiter and some hyperfunctioning thyroid adenomas (Plummer's disease) are attributable to its gain-of-function mutations (Parma et al. 1993; Holzapfel et al. 1997). Predictably, as a mirror image, loss-of-function type aberrations in the gene have been shown to cause hypothyroidism without goiter (Sunthornthepvarakul et al. 1995; Biebermann et al. 1997).

In laboratory medicine, the availability of recombinant human TSH receptor enabled assays for TSH receptor antibodies, the hallmark of Graves' disease (Smith and Hall 1974) and a subtype of autoimmunity-based hypothyroidism (Konishi et al. 1985), with improved sensitivity and specificity compared to conventional assays which use receptors or cells from other species (Morgenthaler et al. 1998; Costagliola et al. 1999). However, the assays that detect TSH receptor antibodies by their inhibition of ^{125}I-TSH binding to receptors cannot discriminate the bioactivity of antibodies. Therefore, in vitro bioassays are still necessary for differentiation of the stimulating and the blocking types of the antibodies (Kasagi et al. 1986).

The triiodothyronine (T_3) suppression test has been utilized to test whether the thyroid function is under the regulation of TSH. After baseline uptake studies, 75–100 µg of T_3 are administered orally for 7 days, and the uptake test or scanning procedure is repeated (Werner 1956). Since the function of normal thyroid tissue decreases if TSH secretion is inhibited, normal subjects demonstrate at least a 50% reduction of the radioiodine uptake. The failure to suppress thyroid uptake in patients with euthyroid bilateral or unilateral exophthalmos is consistent with euthyroid exophthalmic Graves' disease. The failure to suppress thyroid uptake in a patient with a functioning thyroid nodule is consistent with an autonomous thyroid nodule.

15.3.4
Molecular Abnormalities of Thyroid Hormone Action

The nuclear T_3 receptor, a heterodimer homologous to receptors of vitamin D and retinoic acid, regulates synthesis of proteins in cells by binding the specific regulatory site upstream of the encoding sequence for their genes (Anderson et al. 2000). Mutations in various parts of the T_3 receptor gene have been found in thyroid hormone resistance (THR). As molecular analyses make progress, the "selective pituitary THR" described in earlier literatures is not now considered to be separable from the "generalized" one; the reputed distinction of these two subtypes was probably due to the subjectiveness of symptoms (Beck-Peccoz and Chatterjee 1994; Refetoff 2000).

15.3.5
Radioiodine Treatment

Iodine-131 therapy for the management of benign and malignant thyroid disease was introduced in early 1940s. The treatment is indicated in the hyperthyroidism of Graves' disease, multinodular toxic goiter, and toxic adenoma, as well as patients with metastatic thyroid cancer. Whole-body scintigraphy with ^{131}I is done in patients with thyroid carcinoma after surgical treatment to determine whether any remaining normal thyroid tissue or local or distant metastases are present. Because of its longer half-life, ^{131}I allows imaging at 48 to 72 h, when uptake in poorly functioning thyroid tissue is more likely to be detected because background activity is lower than in the earlier phase. The reported incidence of ^{131}I uptake by metastatic differentiated thyroid carcinoma is 60–70% (Charbord et al. 1977; Samaan et al. 1985; Kasagi et al. 1993; Schlumberger et al. 1996). Both papillary and follicular carcinomas have TSH receptors on cell surfaces, providing TSH-dependent ^{131}I uptake. Therefore, treatment should be given when serum TSH concentrations are high. In our hospital, 4–6 GBq (108–162 mCi) [131]I is administered 1–2 months after total thyroidectomy without thyroid hormone replacement. In patients receiving thyroxine (T_4), T_4 is replaced by T_3 5 weeks before ^{131}I treatment. T_3 is withheld 3 weeks prior to treatment and iodine-rich foods are suspended for 20–30 days before and 7 days after treatment.

It has recently been reported that ^{131}I in several millicurie doses used for diagnostic purposes prevents subsequent uptake of therapeutic doses of [131]I (so-called stunning of the functioning thyroid tissues; Park et al. 1994). Iodine-123, with its shorter half life of 13 h and much smaller radiation burden to the thyroid, may be an alternative. Park et al. (1997) reported that the diagnostic accuracy of ^{123}I scan (300 Ci/11 MBq) is essentially equal to that of ^{131}I scan (3–10 mCi/110–370 MBq) for the first postoperative evaluation. Diffuse liver uptake of ^{131}I indicates the presence of a functioning thyroid remnant or metastases (Chung et al. 1997). Therapeutic doses of ^{131}I may reveal distant metastases that are not detected by a tracer dose (< 5 mCi /< 185 MBq) of ^{131}I in some patients who are apparently free of disease except for detectable serum Tg (Pacini et al. 1987). Therefore, in those with detectable Tg a negative ^{131}I whole-body scan with a tracer dose does not necessarily exclude the presence of residual or metastatic tissue that is responsive to treatment.

15.3.6
Radionuclide Imaging of Thyroid Cancer

As noted above, an ^{131}I whole-body scan is indispensable for predicting the efficacy of ^{131}I treatment and also for detecting occult metastases (Schlumberger et al. 1988). Nevertheless, ^{131}I scintigraphy has disadvantages such as dietary restriction of iodide intake and discontinuation of thyroid hormone medications (Becker 1996c). Accumulation of [201]Tl (Iida et al. 1991; Alam et al. 1997), ^{99m}Tc-MIBI (Miyamoto et al. 1997), ^{99m}Tc-tetrofosmin (Lind et al. 1997), and ^{18}F-FDG (Feine et al. 1996) has been reported in thyroid cancer after their use as tumor-seeking agents. Although they may be limited in their ability to differentiate malignant from benign thyroid tumors, their usefulness in localizing thyroid cancer metastases has been evaluated positively. Thyroid hormone medication needs not be discontinued, and imaging can be done immediately after the injection of these agents. Whole-body scans using them may be recommended for metastasis screening in patients who do not undergo total thyroidectomy, since the role of Tg determination in separating patients with metastases from those without recurrence is limited in such patients. FDG-PET has been reported to be superior to other imaging methods (Grunwald et al. 1997).

PET is indicated in ^{131}I-scan-negative patients, especially when regional lymph node metastasis is suspected (Chung et al. 1999).

Radionuclide imaging is used for medullary thyroid cancer (MTC) that originates from parafollicular cells and secretes calcitonin. Eighty percent of MTC occurs sporadically. Familial cases have been reported to crop up as multiple endocrine neoplasia (MEN) type 2 in association with pheochromocytoma (Sipple's syndrome).

Detection of MTC has been reported using ^{99m}Tc-pentavalent dimercaptosuccinic acid (^{99m}Tc(V)-DMSA; Ohta et al. 1984; Miyauchi et al. 1986; Clarke et al. 1988). Rhenium-186 and -188 pentavalent DMSA are available for therapeutic application (Singh et al. 1993). Iodine-131-metaiodobenzylguanidine (^{131}I-MIBG), used mainly to detect pheochromocytoma, does accumulate in some MTCs, but it is not suitable for screening purposes because of low sensitivity (Endo et al. 1984; Keeling and Basso 1988). When uptake of ^{131}I-MIBG is intense in unresectable MTC, radiotherapy with a higher dose is indicated. Indium-111-pentreotide scan can be used in patients with MTC with a higher sensitivity (Krenning et al. 1993; Adams et al. 1998). In general, nuclear imaging plays a limited role in the primary work-up but it may be useful in identifying residual or recurrent disease.

15.3.7
Molecular Biology of Thyroid Cancer

15.3.7.1
Oncogenes

Mutation of oncogenes and tumor-suppressor genes has been found in thyroid malignancy as well as in cancers of other organs. The most notable is the high frequency of rearrangement in ret genes in childhood thyroid cancer after the Chernobyl accident (Smida et al. 1999; Santoro et al. 2000). Other findings related to gene mutation and thyroid cancer include the c-ret protooncogene and familial medullary cancer (MEN type 2; Berndt et al. 1998; Hofstra et al. 1997) and p53 abnormality in some cases of anaplastic thyroid cancer (Blagosklonny et al. 1998; Fagin et al. 1996).

15.3.7.2
Recombinant Human Thyrotropin (rh-TSH) in the Preparation of Diagnostic Radioiodine Scans

The availability of rh-TSH makes radioiodine scintigraphy for metastatic thyroid cancer possible without discontinuing thyroid hormone therapy. Rh-TSH administration has been reported to be a safe and effective means of stimulating radioiodine uptake and raising serum Tg levels. Recently, a multicenter trial of rhTSH was performed in patients undergoing evaluation for thyroid cancer recurrence. In the rh-TSH regimen, 0.9 mg rh-TSH was administered subcutaneously on days 1 and 2 and ^{131}I orally on day 3, with scintiscanning on day 5. Unlike effects experienced after thyroid hormone withdrawal, hypothyroid symptoms were absent after rh-TSH administration. The use of rh-TSH thus becomes a worthwhile alternative to stopping thyroid hormone in patients who undergo studies for thyroid cancer persistence and recurrence (Haugen et al. 1999). Applicability of rh-TSH for ^{131}I -treatment is to be evaluated.

15.3.8
Future Directions

With so many genes cloned and deciphered as described above, the therapy of thyroid carcinoma will definitely be refined using genetic technology. Re-induction of NIS protein in de-differentiated thyroid cancer would restore the effect of radioiodine therapy (Shimura et al. 1997), and may in the future even be applied to tumor cells of non-thyroidal lineage (Nakamoto et al. 2000; Spitzweg et al. 1999). In another approach, re-introduction of p53 gene by means of a viral vector to anaplastic thyroid cancer may render the tumor curable (Narimatsu et al. 1998). As in the past, ever ongoing advances in molecular biology will continuously invigorate nuclear medicine of the thyroid both in academic research and clinical practice.

15.4
Parathyroid Scintigraphy

MASAO FUKUNAGA

15.4.1
Introduction

15.4.1.1
The Parathyroid Gland

The parathyroids are found in all vertebrates higher than fishes. The superior parathyroids, which arise from the fourth pharyngeal pouch, are considered more constant in location, while the inferior parathyroids share their origin with the thymus from the third pharyngeal pouch and may migrate into the anterior mediastinum. Variability in the location and number of parathyroid glands is not rare.

Skilled surgeons can resect 95% or more of hyperfunctioning parathyroid glands during surgery even without a preoperative localization study. However, it is often difficult to identify abnormal parathyroids patients with previous operations (Granberg et al. 1982) due to scarring and loss of normal tissue planes. It may also be difficult to detect ectopic parathyroid. Thus, noninvasive localization such as scintigraphy is used for primary and secondary hyperparathyroidism especially when the condition is persistent or recurrent.

15.4.1.2
Secretion of Parathyroid Hormone (PTH)

The human PTH gene is located in the short arm of chromosome 11, and is expressed only in parathyroid cells. The messenger RNA of PTH is initially transcribed from the PTH gene in the nucleus. It then moves to cytoplasm and is translated into a pre pro PTH of 115 amino acids. This prepro PTH crosses the rough endoplasmic reticulum, and 25 amino acids of the amino-terminal portion are removed yielding a prePTH of 90 amino acids. PrePTH then transfers to the Golgi apparatus, where a mature PTH of 84 amino acids is finally synthesized by cleavage of six amino acids of the amino-terminal portion. The mature PTH is stored in the secretory granules which are mainly released from the cell by exocytosis (Rosenblatt et al. 1989).

PTH secretion is regulated by extracellular concentrations of calcium, $1,25\text{-}(OH)_2D_3$ or phosphate, by increase or decrease in PTH gene expression, and by parathyroid cellular proliferation (Naveh-Many et al. 1995; Slatopolsky et al. 1996). Most these processes are controlled by an extracellular calcium-sensing receptor which is expressed on the cell surface (Brown et al. 1993; Chattopadhyay et al. 1996). Intracellular calcium concentration is the most important regulator of PTH secretion. Other factors, such as isoproterenol, osmolality, prolactin, ATP, fluoride, lithium, potassium, dopamine, glucocorticoid, and estrogen (Silver et al. 1985) may also be responsible for the modulation of PTH secretion (Pocotte et al. 1991). As described previously, a calcium-sensing receptor on parathyroid cells regulates the synthesis and secretion of PTH through a G-protein transducing pathway (Nemeth and Scarpa 1987; Brown et al. 1993, 1995). Like extracellular calcium, phenylalkylamine (R)-N-(3-methoxy-alpha-phenyl ethyl)-3-2(chlorophenyl)-1-propylamine (R-568) can activate the calcium-sensing receptor and inhibit parathyroid cell function (Brown et al. 1991). In fact, this calcimimetic drug R-568 reduces serum PTH and ionized calcium concentrations in postmenopausal women with primary hyperparathyroidism (Silverberg et al. 1997). The finding has encouraged the use of R-568 as an alternative to surgery for primary hyperparathyroidism.

There are several factors involved in the molecular genetics of parathyroid tumors: the activation or overexpression of the protooncogene parathyroid adenoma 1 (PRAD1; Arnold et al. 1989; Arnold 1993)/cyclin D1 of the cyclin D family of cell-cycle regulatory proteins (Motokura et al. 1991) and the RET gene (Mulligan and Ponder 1995), and the inactivation of tumor-suppressor genes such as the RB (retinoblastoma) gene (Cryns et al. 1994), and the MEN1 locus (Friedman et al. 1989; Thakker 1996). Clonal allelic loss of loci on the chromosome arm Ip is a frequent feature of parathyroid adenomas implying that the inactivation of one or more tumor-suppressor genes on Ip contributes to its pathogenesis (Cryns et al. 1995). Acquired transformation is a common occurrence in the refractory parathyroid hyperplasia of chronic renal failure and nonfamilial hyperplasia (Arnold et al. 1995). Expression of both calcium-sensing receptor mRNA and protein is reduced in most primary parathyroid adenomas and secondary hyperplasias (Gogusev et al. 1997).

15.4.2
Radiopharmaceuticals

15.4.2.1
Historical Background

As radiopharmaceutical agents for parathyroid scintigraphy, $^{57}CoB_{12}$ (Sisson and Beierwaltes 1962) and 75-Se-selenomethionine (Potchen and Dealy 1963) were introduced in the early 1960s. Selenomethionine is a biologically active analogue of methionine whose sulfur atom is replaced with selenium atom. Hyperfunctioning parathyroids incorporate methionine or selenomethionine. However, 75-Se-selenomethionine was abandoned because of poor image quality and low detectability of abnormal parathyroid glands. In 1979, Fukunaga et al. reported that ^{201}Tl accumulated in parathyroid adenoma. Accumulation of ^{201}Tl is related to several factors including increased regional blood flow (Strauss et al. 1977), functional activity distribution analogous to that of potassium (Ito et al. 1978), and cellularity. As the uptake mechanism of ^{201}Tl, Na/K ATPase, the $Tl^+-Na^+-2Cl^-$ cotransport and a Ca^{++}-dependent channel have been proposed (Sessler et al. 1986; Nygren et al. 1988). Sandrock et al. (1993) reported that the detectability of abnormal parathyroid glands by $^{201}Tl-^{99m}Tc$ subtraction scintigraphy depended in part on the presence of mitochondria-rich oxyphil cells, which are required for the energy-dependent mechanism of ^{201}Tl uptake. Thallium-201 accumulates in both thyroid and parathyroid tissue but ^{99m}Tc and ^{123}I accumulate in the thyroid gland only. Thus, $^{201}Tl-^{99m}Tc$ subtraction scintigraphy can localize abnormal parathyroid glands (Ferlin et al. 1983). The sensitivity for abnormal parathyroid glands by ^{201}Tl subtraction was 71% in pooled data (McBiles et al. 1995). Gimlette et al. (1986) reported that 25 out of 26 parathyroids of mass >1.0 g were correctly located by $^{201}Tl-^{99m}Tc$ subtraction scintigraphy, but none of 10 parathyroids of mass < 0.3 g were located. The mean parathyroid uptake of ^{201}Tl was 0.018%/g, thyroid 0.01%/g, and muscle 0.0026%/g. False negative results occur primarily in small adenomas or hyperplasia (Fine 1987). On the other hand, false positive results are seen most commonly in concurrent thyroid nodules.

In 1989, Coakley et al. reported that ^{99m}Tc-metoxy-isobutylisonitrile (MIBI) could be used as a new agent for parathyroid imaging. The uptake and washout kinetics of MIBI and ^{201}Tl were compared, and the peak time, clearance half time, difference between thyroid and parathyroid activity, thyroid-to-parathyroid ratio, and uptake per gram of tissue were summarized (McBiles et al. 1995). In addition, ^{99m}Tc-tetrofosmin and ^{99m}Tc (III)-furifosmin (Q12), a myocardial perfusion agent, were applied to parathyroid imaging (Ishibashi et al. 1995; Aigner et al. 1997). A radiolabeled parathyroid-specific monoclonal antibody was also developed for parathyroid imaging (Otsuka et al. 1988), and it was shown to be effective in the nude mouse model. More recently, PET studies with ^{18}F-FDG (FDG) (Neumann et al. 1994; Sisson et al. 1994) and L-^{11}C-methionine (Sundin et al. 1996) have been performed for hyperfunctioning parathyroid glands. Table 15.9 summarizes the radiopharmaceuticals used for parathyroid imaging.

Table 15.9. Radiopharmaceutical for parathyroid imaging

Mechanism of accumulation	Radiopharmaceutical	Authors
Histologic dye	^{131}I-toluidine blue	Zwas et al. 1987
Specific antibody	Radiolabeled parathyroid-specific monoclonal antibody	Otsuka et al. 1988
Perfusion and cell proliferation	^{201}Tl-Cl ^{99m}Tc-MIBI ^{99m}Tc-tetrofosmin ^{99m}Tc (III)-furifosmin (Q12)	Fukunaga et al. 1979 Coakley et al. 1989 Ishibashi et al. 1995 Aigner et al. 1997
Amino acid metabolism	^{75}Se-selenomethionine L-^{11}C-methionine	Potchen et al. 1963 Sundin et al. 1996
Glucose metabolism	^{18}F-FDG	Sisson et al. 1994 Neumann et al. 1994

15.4.2.2
Technetium-99m-Methoxybutylisonitrile (MIBI)

15.4.2.2.1
General Aspects

In 1989, Coakley et al. introduced MIBI as a new agent for parathyroid imaging. Abnormal parathyroids were correctly localized using ^{99m}Tc-MIBI, with a higher parathyroid-to-thyroid uptake ratio compared to other agents. Its physical characteristics for gamma camera imaging is superior to those of ^{201}Tl. Although ^{201}Tl uptake was higher than that of MIBI in both the parathyroid and thyroid tissue, the uptake of MIBI per gram of tissue in the parathyroid was higher than that in the thyroid (O'Doherty et al. 1992). Double-phase scanning, 15 min and 1–2 h after intravenous injection of ^{99m}Tc-MIBI, was reported to be promising for the preoperative detection and localization of parathyroid adenomas (Taillefer et al. 1992). The findings reflected different washout patterns between the early and late phases, namely, persistent uptake in the parathyroid and progressive decrease in uptake in normal thyroid.

15.4.2.2.2
Mechanisms of Accumulation

The multidrug-resistant P-glycoprotein (P-gp) is a plasma membrane protein which appears to function as an efflux transporter of a variety of potent chemotherapeutic agents (Bradley et al. 1988; Juranka et al. 1989). MIBI serves as a substrate for P-gp transport, and it may prove useful for functionally characterizing P-gp expression in human tumors in vivo (Rao et al. 1994). It has also been reported that rapid washout of MIBI from malignant tumors is associated with high levels of P-gp (Rao et al. 1994). The effect of P-gp on the detectability of parathyroid adenomas with ^{99m}Tc-MIBI scintigraphy was evaluated, but no correlation was found between the dual-phase MIBI study and levels of P-gp (Bhatnagar et al. 1998).

Net cellular uptake and retention of MIBI in fibroblasts have been determined from both mitochondrial and plasma membrane potentials (Chiu et al. 1990). As previously stated, detection of abnormal parathyroid glands by ^{201}Tl-^{99m}Tc subtraction scintigraphy partly depends on mitochondria-rich oxyphil cell content (Sandrock et al. 1993). On ^{99m}Tc-MIBI delayed scan, rapid washout of MIBI from parathyroid adenomas with histological absence of oxyphil cells

has been documented (Bénard et al. 1995). A study of 18 parathyroid lesions in 14 patients has shown late retention of ^{99m}Tc-MIBI in double-phase scintigraphy to be related to parathyroid oxyphil cell content (Carpentier et al. 1998). On the other hand, other studies have shown the ^{99m}Tc-MIBI uptake in the parathyroids to be correlated with serum calcium and osteocalcin levels, but not with cell density or the oxyphil cell count (Ishibashi et al. 1997; Staudenherz et al. 1997). In addition, the uptake of ^{99m}Tc-MIBI in hyperfunctioning parathyroid is dependent on gland size, the amount of cellular component, and numbers of chief cells and oxyphil cells (Takebayashi et al. 1999).

15.4.2.2.3
Imaging Procedures

A SNM-approved procedure guideline for parathyroid scintigraphy was published by Greenspan et al. in 1998. The guideline recommends that ^{201}Tl of 75–130 MBq (2–3.5 mCi) and ^{99m}Tc-MIBI of 185–925 MBq (5–25 mCi) are to be used intravenously as radiopharmaceuticals. As thyroid imaging agents in subtraction studies, ^{99m}Tc of 75–150 MBq (2–4 mCi) and 185 MBq (5 mCi) are to be used for ^{201}Tl and MIBI, respectively. Iodine-123 of 7.5–20 MBq (200–500 Ci) can also be used. Planar imaging should include both neck and mediastinum in all cases. There is no uniform agreement regarding which of ^{201}Tl and other thyroid delineating radiopharmaceutical is to be administered first. The procedure of subtraction using ^{99m}Tc-MIBI is similar to that using ^{201}Tl. When a dual-phase study is performed, both early (10 min postinjection) and delayed (1.5 to 2.5 h postinjection) high-count images are to be obtained. Computer processing is used in subtraction studies. After two images have been normalized, the ^{99m}Tc or ^{123}I image is subtracted from the ^{201}Tl or MIBI image. Sources of error include patient motion, image misregistration, a hyperfunctioning parathyroid of less than 500 mg in size, ectopic adenoma, and thyroid lesions (Greenspan et al. 1998).

In general, SPECT more precisely delineates anatomy than does planar imaging. In the detection of residual aberrant hyperfunctional parathyroid glands, ^{99m}Tc-MIBI SPECT has been reported to be superior to planar scintigraphy (Schurrer et al. 1996). Technetium-99m-MIBI SPECT and volume-rendered reproduction, with an intraoperative quick assay of PTH, have improved the results of parathyroidectomy (Sfakianakis et al. 1996). In contrast, it has been shown that delayed imaging and computer subtraction using SPECT is not helpful (Chen et al. 1997). To enhance the localization accuracy of nuclear imaging, gamma camera equipped with an integrated X-ray tube system has been used for patients with endocrine neoplasms including parathyroid adenomas (Martin et al. 2000). The fusion of CT and SPECT images can provide functional and anatomical information of parathyroid adenomas in the mediastinum.

Factor analysis, i.e., the isolation of fundamental kinetics from dynamic image sequence, has been applied to ^{99m}Tc-MIBI parathyroid scintigraphy (Billotey et al. 1994). Three fundamental kinetics including vascular, thyroidal, and parathyroidal kinetics were identified by a 25 min dynamic acquisition following an intravenous bolus injection of ^{99m}Tc-MIBI. A dynamic series of 1 min images was acquired. Factor analysis of dynamic structure (FADS) was specific and more sensitive than the dual-phase method. The two methods respectively took 1 h and 2 h to detect abnormal parathyroids in the cervical area (Billotey et al. 1994). Further, the combination of FADS and SPECT was useful in detecting small parathyroid glands, especially those in a posterior location, inside or outside the thyroid area (Billotey et al. 1996).

15.4.2.2.4
Clinical Studies (Fig. 15.6)

Double-phase ^{99m}Tc-MIBI scintigraphy was positive in 74% of secondary hyperparathyroidism (Piga et al. 1996). Mean intact PTH concentration in serum was significantly higher in ^{99m}Tc-MIBI-positive patients than in negative ones. Comparison of scintigraphic data to functional data strongly suggested that ^{99m}Tc-MIBI scintigraphy revealed no parathyroidal enlargement, but rather demonstrated the presence of hyperfunctioning (autonomous) parathyroid tissue, suggesting tertiary hyperparathyroidism (Piga et al. 1996). The preoperative ^{99m}Tc-MIBI-^{123}I subtraction scintigraphy depicted 95% of solitary adenomas responsible for primary hyperparathyroidism (Hindié et al. 1997). However, the sensitivity was lower in hyperplasia than adenoma (O'Doherty et al. 1992).

Sensitivity was found to be dependent on lesion size, previous neck surgery, and the kind of abnormal parathyroid tissue. Sandrock et al. (1990) reported that the detectability also depended on the anatomic site of the lesion; understandably, detection was easier in typical than in ectopic sites. Anatomic distribution

and number of parathyroid glands greatly vary. In an autopsy study of 503 cadavers (Åkerström et al. 1984), only three glands were found in 3%, and supernumerary glands in 13%. In 71 patients with hyperplasia (Liechty and Weil 1992), mediastinal glands occurred in 13%, supernumerary glands in 11%, and ectopic glands in 10%. In patients with persistent or recurrent hyperparathyroidism, ectopic parathyroids are more prevalent, and in such cases exploration is necessary the for precisely locating the hyperfunctioning tissue.

Weber et al. (1993) reported that ^{99m}Tc-MIBI-^{123}I subtraction imaging was more accurate than ^{201}Tl-^{99m}Tc subtraction and CT in persistent hyperparathyroidism. McIntyre et al. (1994) also observed that among multi-imaging modalities the most sensitive method for preoperative localization was ^{99m}Tc-MIBI (sensitivity 86%) followed by ^{201}Tl-^{99m}Tc (67%), arteriography (63%), venous sampling (52%), CT (42%), MRI (33%), and US (27%). Ultrasound provides the best images for identifying parathyroid glands adjacent to or within the thyroid, while MRI is useful for mediastinal glands. Scanning with ^{99m}Tc-MIBI is suitable for parathyroid tumors situated in both normal and ectopic positions (Rodriquez et al. 1994). Technetium-99m-MIBI SPECT is useful for localizing mediastinal parathyroid adenomas (Teigen et al. 1996).

To assess the viability and function of transplanted parathyroid tissue after total parathyroidectomy in patients with parathyroid hyperplasia, ^{201}Tl or ^{99m}Tc-MIBI scintigraphy was performed (Fronistas et al. 1992; Chen et al. 1995; Kilgore et al. 1996; Thelen et al. 1996). In dynamic studies, not only visual findings of ROI curves but transplant curves to background curves were used to differentiate among the normal, hyperfunctioning, and nonfunctioning states, and rejection (Fronistas et al. 1992). Scintigraphic studies are especially useful for assessment of the vascularity and uptake in transplanted parathyroid tissue in recurrent hyperparathyroidism after parathyroidectomy. Noninvasive imaging modalities such as US, CT, MRI, and nuclear medicine procedures are used for the detection of abnormal parathyroid glands. The sensitivity, specificity, and false positive rate of each modality have been reported in many studies (Krubsack et al. 1989; Miller 1991; Kang et al. 1993; Lee et al. 1996; Mazzeo et al. 1996; McDermott et al. 1996). In comparing advantages and disadvantages of these modalities, care should be taken regarding tissue type (adenoma or hyperplasia), location, size, and prior neck surgery. Each modality has specific diagnostic imag-

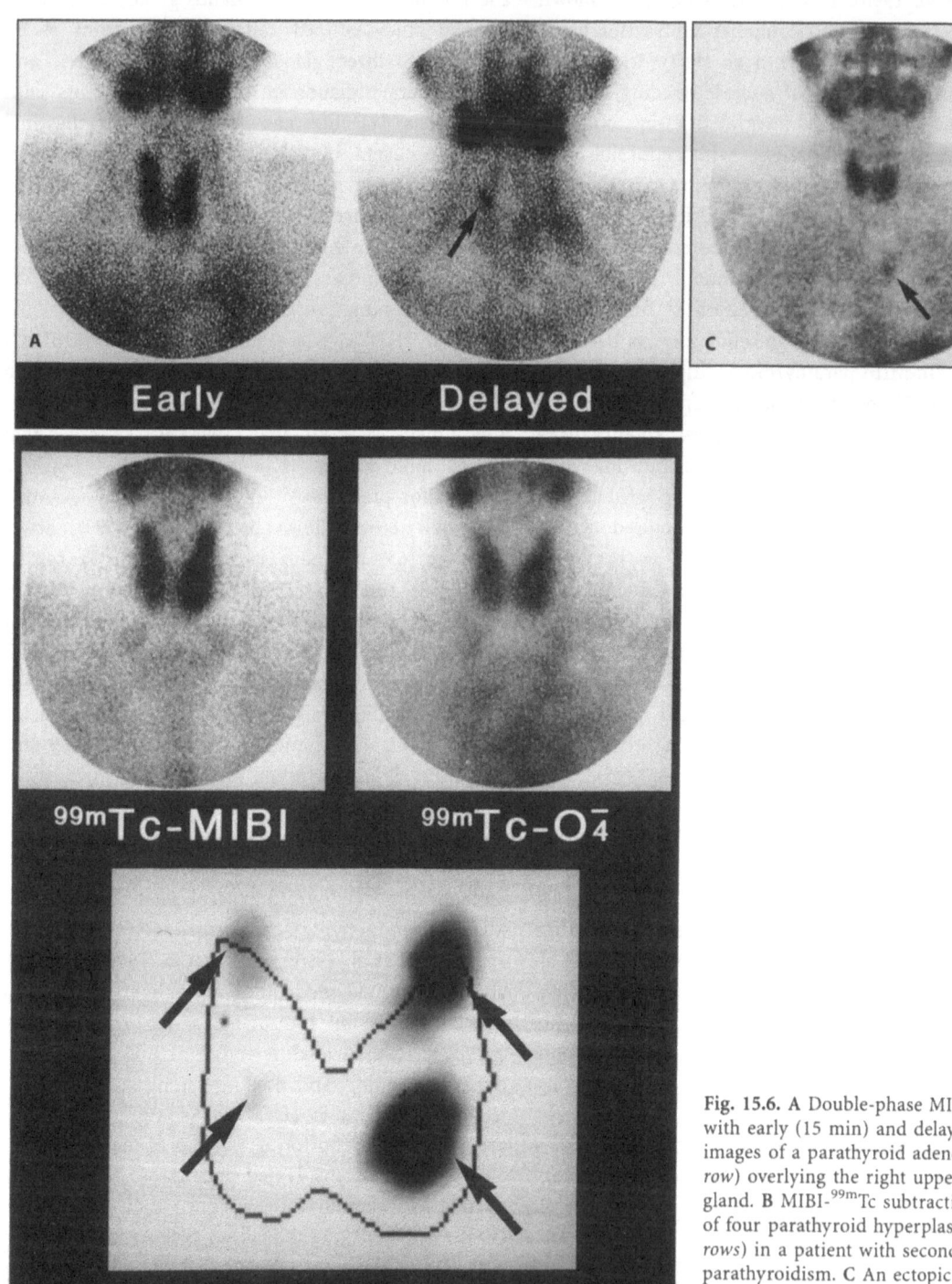

Early   Delayed

^{99m}Tc-MIBI ^{99m}Tc-O$_4^-$

subtraction

Fig. 15.6. A Double-phase MIBI study with early (15 min) and delayed (2 h) images of a parathyroid adenoma (*arrow*) overlying the right upper thyroid gland. B MIBI-^{99m}Tc subtraction image of four parathyroid hyperplasias (*arrows*) in a patient with secondary hyperparathyroidism. C An ectopic parathyroid adenoma (*arrow*) in the mediastinum on the delayed image

ing characteristics. Thus, nuclear imaging depends on the metabolic activity of the gland, whereas CT, US, and MRI identify abnormal parathyroid anatomy. Doppman and Miller (1991) reviewed noninvasive localizing studies of US, ^{201}Tl-^{99m}Tc subtraction scintig-raphy, CT, and MRI in patients with unoperated asymptomatic primary hyperparathyroidism. The sensitive and false positive rates were 65% and 12.5% for US, 55% and 13.5% for ^{201}Tl-^{99m}Tc subtraction scintigraphy, and 74% and 18% for MRI, respectively.

The sensitivity of CT was 63% but the false positive rate was not adequately documented. Goris et al. (1991) also reported the results of meta-analysis for sensitivity. They were 62% ($n = 562$) for ^{201}Tl-^{99m}Tc subtraction scintigraphy, 55% ($n = 276$) for CT, 77% ($n = 56$) for high-resolution CT, 49% ($n = 89$) for US, 57% ($n = 339$) for high-resolution US, and 72% ($n = 77$) for MRI. In addition, imaging modalities were reviewed with regard to sensitivity according to lesion size, the area best or most poorly visualized, resolution, radiation to thyroid availability, operator dependency, risk, and cost (Doppman and Miller 1991). Scintigraphy using ^{99m}Tc-MIBI provided the best specificity and anatomic coverage and, coupled with MRI, the most anatomic detail.

15.4.2.3
Technetium-99m-Tetrofosmin

In 1995, Ishibashi et al. described parathyroid imaging using ^{99m}Tc-tetrofosmin. Used for myocardial perfusion imaging, ^{99m}Tc-tetrofosmin does not require complex preparation (whereas 10 min-incubation at $100\,^{\circ}$C is required for ^{99m}Tc-MIBI). Semiquantitative analyses of parathyroid and thyroid uptake of ^{99m}Tc-MIBI, and ^{99m}Tc-tetrofosmin with respect to background were made by Mansi et al. (1996). The parathyroid/thyroid ratio of ^{99m}Tc-MIBI and ^{99m}Tc-tetrofosmin was demonstrated in a parathyroid adenoma. These ratios were 1.57 at 20 min and 3.00 at 120 min for MIBI and 1.71 at 20 min and 1.33 at 120 min for ^{99m}Tc-tetrofosmin. Image contrast was generally higher with MIBI than ^{99m}Tc-tetrofosmin (Aigner et al. 1996), while ^{99m}Tc-tetrofosmin showed slower elimination from parathyroid adenomas than MIBI in six of ten cases. For the detection of parathyroid adenomas on early images, ^{99m}Tc-tetrofosmin showed the same success rate as ^{99m}Tc-MIBI (Fjeld et al. 1997). In contrast, delayed imaging of ^{99m}Tc-tetrofosmin has no diagnostic impact. Technetium-99m-tetrofosmin was found to accumulate not only in parathyroid adenoma but in hyperplasia (Ishibashi et al. 1997, 1998b). The sensitivity of this agent's parathyroid imaging is higher than that of US or MRI (Ishibashi et al. 1997, 1998a). However, prolonged retention of ^{99m}Tc-tetrofosmin is not dependent on the number of mitochondria-rich oxyphil cells (Ishibashi et al. 1997). Overall sensitivity, specificity, and accuracy of ^{99m}Tc-tetrofosmin-^{99m}Tc subtraction are higher than those of ^{201}Tl-^{99m}Tc subtraction: further, the sensitivity of ^{99m}Tc-tetrofosmin-^{99m}Tc subtraction scintigraphy was found to be 87% for adenoma and 72% for hyperplasia (Apostolopoulos et al. 1998).

15.4.2.4
Fluorine-18-FDG and L-[methyl-^{11}C] Methionine

Aerobic glycolysis is prevalent in many neoplastic tissues, and the rate of glucose metabolism tends to be much higher than in normal tissues (Som et al. 1980). FDG has been found to accumulate in 94% of parathyroid adenoma and 50% of hyperplasia (Neumann et al. 1994). In addition, the diagnostic sensitivity for their detection is higher with FDG-PET (86%) than with dual-phase MIBI SPECT (43%) (Neumann et al. 1996a). The location and extent of parathyroid carcinoma were accurately determined by FDG-PET (Neumann et al. 1996b). However, the sensitivity of FDG-PET has been reported to be low (Sisson et al. 1994; Melon et al. 1995) and its diagnostic efficacy is controversial.

L-[methyl-^{11}C] methionine can be used for the localization of abnormal parathyroid tissues (Sundin et al. 1996). The accumulation of ^{11}C-methionine reflects protein synthesis, transmethylation processes, and transmembranous amino acid transport. The true positive rates are 81% of excised lesions for ^{11}C-methionine metabolites with PET, 57% with CT, and 52% with US. The highest standard uptake value (SUV) was displayed by parathyroid carcinoma and the lowest by hyperplasia.

15.4.3
Future Directions

Nuclear imaging of the parathyroid glands can portray not only in vivo anatomy but also molecular and biochemical alterations (Wagner 1995). For example, the ^{18}F-FDG PET is used for the evaluation of glucose metabolism (Neumann et al. 1994) and L-^{11}C-methionine for methionine metabolism (Sundin et al. 1996) in parathyroid adenoma and hyperplasia. Calcimimetic radiopharmaceuticals which inhibit PTH secretion might be added as a potent agent for the biochemical imaging of the parathyroids (Silverberg et al. 1997). On the other hand, currently available parathyroid scintigraphy has a few important drawbacks to be overcome. First, the detection rate of small adenoma and hyperplasia in asymptomatic subjects and ectopic or supernumerary parathyroid glands, especially in persistent or recurrent cases after parathyroid surgery, is still undesirably low. Second, the differential diagnosis between adenoma, hyperplasia, and carcinoma is still below our expectations and, third, the exclusion of thyroid lesions coexisting with hyperparathyroidism remains a vexing problem.

15.5
The Adrenal Glands

J.C. Sisson

15.5.1
The Adrenal Cortex

15.5.1.1
Available Nuclear Medicine Methods

The adrenal cortex can scintigraphically be imaged using 6 β-iodomethyl-19-norcholesterol (^{131}I-iodocholesterol, called NP-59). NP-59 scintigraphy has been shown to be useful in distinguishing patterns of function within the adrenal glands (Kazerooni et al. 1990). The disorders defined by NP-59 imaging fall into three categories.

ACTH-independent Cushing's syndrome may be caused by benign adenoma, nodular hyperplasia, or carcinoma. The latter generally does not concentrate iodocholesterol, while the other two can be identified by unilateral and bilateral portrayals of adrenal function (Fig et al. 1988).

Primary aldosteronism can be induced either by adenoma or hyperplasia. Iodocholesterol image, especially when normal cortical function is suppressed using an exogenous steroid such as dexamethasone, may show the gland(s) that excrete excessive aldosterone (Conn et al. 1976).

Tumors of the adrenal gland that do not cause hyperfunction are commonly seen on computed tomography. The majority concentrate iodocholesterol, a highly reliable sign of a benign adenoma, whereas those not sequestering the tracer require additional and often invasive investigation (Gross et al. 1988).

15.5.1.2
Molecular Changes in Recognized Forms of Adrenal Dysfunction

15.5.1.2.1
ACTH-Independent Cushing's Syndrome

15.5.1.2.1.1
Unilateral Tumors, Genetic Abnormalities Undefined

■ Adenoma

■ Carcinoma

15.5.1.2.1.2
Bilateral Disorders Manifesting as Cushing's Syndrome

■ **Micronodular Hyperplasia.** Micronodular hyperplasia may be familial. One familial form is the Carney Complex (also manifesting myxomas and skin pigmentation) which is inherited by autosomal dominant mode (Sarlis et al. 1997; Young et al. 1989). In this disorder, evidence indicates that the ACTH receptor is stimulated by an immunoglobulin, but the basic immunologic abnormality has not yet been defined (Young et al. 1989).

■ **Macronodular Hyperplasia.** Little is known of the molecular changes that lead to this disorder (Aiba et al. 1991; Malchoff et al. 1989).

■ **Stimulation by Gastric Inhibitory Polypeptide.** Comprehension of molecular events in this syndrome is still evolving (LaCroix et al. 1992; Reznick et al. 1992; Chabre et al. 1998).

■ **The McCune-Albright Syndrome.** In this syndrome hyperplasia occurs in the adrenal glands, and often in the pituitary, thyroid, and gonadal tissues as well. The basic abnormality may be a genetic change in the G protein stimulating adenyl cyclase, a constitutive stimulation of function (Weinstein et al. 1991).

15.5.1.2.2
Primary Aldosteronism

15.5.1.2.2.1
Unilateral Causes

■ **Adenomas Renin/Angiotensin-Responsive and Glucocorticoid-Responsive.** The basic origin of the adenoma, the classical cause of primary aldosteronism, is still unknown.

■ **Carcinomas**

15.5.1.2.2.2
Bilateral Causes

■ **Hyperplasia, Micronodular and Macronodular.** The molecular aberrations in these disorders remain to be determined.

■ **Glucocorticoid Responsive Functional Aberration.** This is a familial condition inherited by autosomal dominant mode (Rich et al. 1992; Dluhy and Lifton 1999). There is a crossing over of 11-β hydroxylase and aldosterone synthetase genes with the consequence that aldosterone synthetase is stimulated by ACTH in the fasciculata zone of adrenal cells.

■ **Bilateral or Unilateral Production of 21-Deoxyaldosterone and Kelly's M-1 Steroid.** This disorder of aldosterone synthesis produces a clinical picture of aldosteronism but the basic abnormality is synthesis of an excess of a related hormone, 21-deoxyaldosterone. The function in most of these adenomas was sensitive to ACTH (Abdelhamid et al. 1995).

15.5.1.2.3
Congenital Adrenal Abnormalities

15.5.1.2.3.1
Congenital Adrenal Hyperplasia

Most cases arise from mutations that cause deficient function of synthetic enzymes. Deficiency of 21-hydroxylase is inherited by autosomal recessive mode (Nordenstrom et al. 1999; Krone et al. 2000). The enzyme is encoded by two genes in the short arm of chromosome 6 and abnormalities include gene deletions, conversions, and point mutations. Two classic phenotypes are the salt-losing form of infancy and the virilizing form of childhood or later life. However,

the disorder spans a spectrum of phenotypes, and genotyping may help to predict the severity of clinical manifestations.

Deficiency of 11-hydroxylase is the cause of adrenal congenital hyperplasia in about 5% of affected patients and may arise from one of several mutations (Merke et al. 1998).

15.5.1.2.3.2
Congenital Adrenal Hypoplasia

This condition is recognized in two forms; one is inherited by recessive mode and the other is transmitted as x-linked and caused by mutations in the DAX-1 gene (Reutens et al. 1999).

15.5.1.3
Approach to Molecular Diagnosis
Through Nuclear Medicine Techniques

15.5.1.3.1
Evaluation of General Metabolic Processes

Extensions of the use of radio-iodo-cholesterol provide insights into the nature of abnormal function in some adrenal disorders. The high density lipoprotein (HDL) receptor is an important pathway for entry of cholesterol into the adrenal glands and steroidogenesis (Cao et al. 1997; Liu et al. 1997; Krieger 1999). Cholesterol can be labeled within HDL and its entry into the adrenal glands quantified. Kinetics could be more readily measured if a positron emitting nuclide such as ^{18}F was the radioactive tag. The syndromes of congenital adrenal hypoplasia may show distinctive patterns of cholesterol sequestration.

15.5.1.3.2
Evaluation of Specific Abnormalities
and Sites of Abnormality

Radioligands, including antibodies, may be found for individual components or synthetic systems, and quantifications would again reveal the individual disorders as their molecular defects become known.

15.5.2
The Adrenal Medulla

15.5.2.1
Available Nuclear Medicine Methods

Meta-iodobenzylguanidine (MIBG) is known to be concentrated in the adrenal medulla and in adrenergic neurons and tumors. Labeled with ^{123}I, and occasionally with ^{131}I, MIBG enables the scintigraphic portrayal of the small adrenal medullas in human subjects (Nakajo et al. 1983; Lynn et al. 1985). Tissues rich in adrenergic neurons, such as the heart, are also visualized by this technique (Sisson et al. 1987a). Following the pathway of the hormone/neurotransmitter, norepinephrine, the radiopharmaceutical enters adrenergic cells via a specialized transporter of the so-called uptake-1 system. It is then stored in adrenergic vesicles and leaves the cell both by moving against the uptake-1 gradient and by physiologic exocytosis (Jaques et al. 1984; Sisson et al. 1987b, 1991). Some kinetics of MIBG mimic those of norepinephrine, but differences are great, particularly when MIBG is not a substrate for monoamine oxidase and does not interact with adrenergic receptors of effector cells so that there is little or no physiologic response to the injected agent (Wieland et al. 1990). MIBG has been used to locate adrenergic tumors, pheochromocytomas (Shapiro et al. 1985; Sisson and Shulkin 1999), and neuroblastomas (Kimmig et al. 1984; Shulkin and Shapiro 1998) and, in large doses, to treat these neoplasms by selectively delivered radiation (Sisson et al. 1984; Hutchinson et al. 1992; Matthay et al. 1998). It has also provided insights into the adrenergic nervous system of the heart following heart failure, infarctions, and neuropathy (Sisson et al. 1987a; Stanton et al. 1989; Merlet et al. 1999).

15.5.2.2
Molecular Changes in Recognized Diseases of the Adrenergic Nervous System

15.5.2.2.1
Tumors of the Adrenal Medulla (Pheochromocytoma) and Sympathetic Ganglia (Paraganglioma)

Hereditary forms of pheochromocytoma are expressed in about 50% of subjects with MEN types 2a and 2b (Lips et al. 1994) wherein there are mutations in the RET proto-oncogene (Eng 1999). Although the mutations can be detected, the related abnormal protein product(s) are not defined and hence the specific delineation of abnormalities by in vivo methods is not yet possible. MIBG scintigraphy can detect the earliest anatomical change in the medulla, namely, a tumor-preceding hyperplasia. Von Hippel-Lindau disease and neurofibromatosis manifest pheochromocytomas in 10–15% (Eisenhofer et al. 1999) and 1–2% (Gutman et al. 1997) of cases, respectively, but in vivo delineation of specific molecular change is yet not possible. Other germ-line mutations have resulted in pheochromocytomas and paragangliomas (Neumann et al. 2002).

It should be noted that some oncogenes have been identified in neuroblastomas, but the basic nature remains unknown.

15.5.2.2.2
Functional Changes in the Adrenergic Nervous System

Many diseases impair the function of the adrenergic nervous system (Goldstein et al. 1997). Diabetes mellitus is one of the most common causes of degeneration of neuronal function. Some disorders arise in the central nervous system. However, neither a specific molecular defect nor an abnormality in the transport or enzyme system has been defined.

Hyperfunction has not been convincingly described.

15.5.2.3
Approaches to Molecular Diagnosis Through Nuclear Medicine Techniques

Adrenergic tumors carry differing prognoses. Pheochromocytomas, for example, may be benign or malignant. Some of the malignant ones grow rapidly and are fatal within months, while others grow slowly over years and even decades. Similarly, neuroblastomas appear in recognized stages that portend the rate of progression, but individual patients not infrequently defy the predictions for outcome. Through binding to specific molecular structures, radioligands may define prognosis. As yet no abnormalities have been identified in adrenergic neoplasms that correlate with tumor growth.

A number of positron-emitting radiotracers have been developed to dissect components of the adrenergic neurons. For example, fluorodopamine and fluoronorepinephrine (Goldstein et al. 1997) and fluorometaraminol (Wieland et al. 1990) can be labeled with [18]F. Each of these agents mimics some actions of norepinephrine and to some extent reflect the integrity and neuronal function.

Radiopharmaceuticals can be synthesized with ^{11}C as tracer, and the physiology or pharmacology of each agent can be scintigraphically portrayed to elucidate in vivo processes. Hydroxyephedrine labeled with [11]C (HED) is transported into the neuron terminal by the uptake-1 transporter. It is not a substrate for monoamine oxidase; it is sequestered by storage vesicles and then leaks out against the gradients of uptake-1. It is also secreted by physiologic exocytosis (Rosenspire et al. 1990). The hormone ^{11}C-epinephrine differs from HED in that it is a substrate for monoamine oxidase. It more tightly bound to vesicles, a physiologic process that can be depicted in human hearts (Chakraborty et al. 1993). The kinetics of sympathomimetic phenylephrine labeled with ^{11}C is similar to that of HED except for susceptibility to oxidation. However, when deuterium molecules are incorporated into ^{11}C-phenylephrine, the resulting agent resists oxidation and exits the neuron against the uptake-1 gradient at a lower rate than HED (del Rosario et al. 1995). By employing two or more such agents sequentially, relative activities of the uptake-1 system, monoamine oxidase, vesicular storage, and release of norepinephrine from neurons can be deduced. For example, the action of monoamine oxidase in the heart can be measured by quantifying the rate of loss of ^{11}C-phenylephrine and comparing the result to that obtained from ^{11}C-phenylephrine-deuterium. Thus, the probing of the molecular nature and abnormalities within the adrenal medulla and the adrenergic nervous system seems promising.

15.5.3
Overview

Nuclear imaging using radio-iodocholesterol (NP-59) and meta-iodobenzylguanidine (MIBG), can demonstrate functioning tumors and dysfunction of the adrenal cortex, medulla, and adrenergic neurons. By correlating image pattern with appropriate hormone level, the nature of hyperplasia and neoplasia can be determined with high accuracy. Similar methods, using refined radiopharmaceuticals, can be applied to define adrenal cortical molecular alterations in the adrenal cortex such as those which cause familial Cushing's syndrome, different types of aldosteronism, and congenital adrenal hyperplasia and hypoplasia. New radiopharmaceuticals enable the dissection of steps in synthetic pathways of hormones and the neurotransmitter norepinephrine, helping delineate the individual abnormalities that cause super-physiologic actions and failure of function in the adrenal medulla and the sympathetic nervous system. In the future, molecular and genetic changes may be closely correlated with nuclear image findings or at least inferred as data are correlated.

15.6
Gastroenteropancreatic Neuroendocrine Tumors

Hong-Yoe Oei, Marion de Jong, Eric P. Krenning

15.6.1 Introduction

The term "neuroendocrine tumors" comprises a wide variety of rare tumor entities which may originate either from pure endocrine organs (e.g., pituitary adenomas), from pure nerve structures (e.g., neuroblastomas), or from elements of the diffuse (neuro) endocrine system like all endocrine tumors of the gastroenteropancreatic (GEP) tract. These neuroendocrine tumor cells are similar to each other in cytochemical and ultrastructural characteristics; they are capable of uptake and conversion of amine precursors to amines or peptides, or both, which they store in secretory granules in the cytoplasm. However, it had been discovered that other cells throughout the body share this ability of amine precursor uptake and decarboxylation (APUD). The term APUD has more lately been found to be inadequate since several cell types included in the system do not metabolize amines. Furthermore, there is evidence that some APUD cell types are not of neural crest origin, but are derived from endoderm (Delcore and Friesen 1993; Langley 1994; Kirkwood and Debas 1995; Rindi et al. 1999; Tamburrano et al. 1999).

15.6.2
Gastroenteropancreatic (GEP) Tumors

GEP tumors constitute about 2% of all gastrointestinal malignancies. They can be divided into the gastrointestinal submucosal carcinoids and the endocrine tumors of the pancreas that comprise about 10% of all pancreatic tumors (Mignon 1999). The carcinoid tumor, a small slowly growing tumor in the intestine, was named by Oberndorfer in 1907. It was initially considered to be benign and divided into foregut, midgut, and hindgut tumors according to the site of origin (Tiensuu Janson and Öberg 1996). The foregut tumors include primaries located in the lung, stomach, and proximal duodenum; the midgut tumors arise from the rest of the small intestine to the mid transverse colon; and the hindgut tumors are those originating from the distal part of the colon and rectum. Midgut carcinoid tumors are usually argentaffin-positive, while the two other subgroups usually are not. The histological patterns also differ, the midgut carcinoid tumors usually growing in nests separated by connective tissue, while the foregut and hindgut carcinoid tumors tend to grow in a more trabecular pattern. To avoid confusion, the term carcinoid tumor should be used only for the midgut neuroendocrine tumors that cause carcinoid syndrome. The others are better called "neuroendocrine tumors" with the site of origin and dominant hormone production identified, as for instance "gastrin-producing neuroendocrine duodenal tumor." Such a distinction makes

for clearer description of these tumors (Tiensuu Janson and Öberg 1996). GEP tumors consist of cells whose morphofunctional profile partly resembles that of normal GEP endocrine cells – each perhaps to be considered as the potential source of a GEP tumor (Table 15.10; Rindi et al. 1998, 2000). Table 15.11 shows relevant data concerning the most familiar GEP tumors.

15.6.2.1
EC-Cell Carcinoid

Enterochromaffin (EC)-cell carcinoids are most prevalent occupying 75% of all GEP tumors. The clinical syndrome resulting from EC-cell tumors (carcinoid syndrome) may be present in patients with liver metastases as well as in some patients with foregut tumors. The syndrome consists of flushes, diarrhea, carcinoid heart disease with right-sided heart failure, and bronchial constriction. In some patients, the syndrome combined with hypotension and/or severe diarrhea may be life threatening. Clinically, the situation is known as carcinoid crisis. The measurement of 5-hydroxy-indol-acetic acid (5-HIAA) excretion in the urine is the most useful diagnostic test. Approxi-

Table 15.10. Human gastroenteropancreatic endocrine cells

Cell	Main product	Pancreas	Stomach		Small intestine				Large bowel	
			Corpus/ fundus	Antral	Duode- num	Jejunum	Ileum	Appendix	Colon	Rectum
P	Unknown	F	+	+	+	F	F		F	F
EC	5-hydroxytryptamine + peptides	F	+	+	+	+	+	+	+	+
D	Somatostatin	+	+	+	+	F	F	+	F	+
L	GLI+pYY				F	+	+	+	+	+
A	Glucagon	+	A							
PP	Pancreatic peptide	+			A					
B	Insulin	+								
X	Unknown		+							
ECL	Histamine		+							
G	Gastrin			+	+					
CCK	Cholecystokinin				+	+		F		
S	Secretin				+	+				
GIP	Gastric inhibitory polypeptide				+	+		F		
M	Motilin				+	+		F		
N	Neurotensin				F	+		+		

Adapted from Rindi et al. (2000)

A presence of cells in fetus and newborn, *CCK* cholecystokinin, *EC* Enterochromaffin, *ECL* Enterochromaffin-like cell, *F* presence of a few cells, *GIP* gastric inhibitory polypeptide, *GLI* glucagon-like immunoreactants (glicentin, glucagon-37, glucagon-29), *PP* pancreatic polypeptide, *pYY* PP-like peptide with N-terminal tyrosine amide

Table 15.11. Epidemiology and pathology of GEP neuroendocrine tumors. (Adapted from Mignon 1999)

	Incidence (per million population)	Tumors identified at surgery (%)	Patients with metastases (%)	Patients with MEN-1 (%)
EC-cell carcinoids	15	100	100	Rare
Gastrinoma	0.5	52–87	50	18–24
Insulinoma	0.8	80–100	5–10	5–10
VIPoma	Rare	100	40	Occasional
Glucagonoma	Rare	98–100	>70	Occasional
Somatostatinoma	Rare	100	~70	Occasional
ECL-cell carcinoids	Rare	100	30–75	14–25
Non-functioning	0.2	100	60	High
Rare forms	?	Usually	25–100	0–25

mately 75% of patients excrete more than 80 μmol of 5-HIAA per day. The specificity of this finding approaches 100% if the ingestion of substances known to elevate 5-HIAA levels is excluded. The presentation of tumors without carcinoid syndrome is related to the site of origin as well as malignant spread of the tumor. Most small bowel EC-cell tumors are located in the ileum, especially near the ileocecal valve. Primary tumors tend to remain small, but they may spread to regional lymph nodes, and further spread may reach the liver and possibly bone. The incidence of metastasis is related to the size of the primary lesion. Small tumors tend to metastasize more frequently when in the small bowel than in the appendix. Appendiceal carcinoids are usually small, 54% being detected because of signs of acute appendicitis, 46% discovered incidentally (Tiensuu Janson and Öberg 1996; Angeletti et al. 1999; Mignon 1999).

15.6.2.2
Insulinoma

Insulinomas are common pancreatic endocrine neoplasms. They arise from the beta islet cells of the pancreas and are frequently solitary and small (<2 cm), and 10–15% are malignant. Involving the head, body, and tail of the pancreas with equal frequency, distant metastases occur primarily to the liver. These tumors are rarely ectopic but can be multiple, especially when they occur with MEN-1. The most prominent and perhaps most reliable sign of the concomitant hypoglycemia is CNS dysfunction (neuroglycopenia), signs and symptoms of which include diplopia, blurred vision, confusion, abnormal behavior, and amnesia. These may progress to a severe extent to cause seizures, coma, or even permanent brain damage. The catacholamine-response

symptoms triggered by hypoglycemia are sweating, weakness, hunger, tremor, nausea, feeling of warmth, anxiety, and palpitation. Sign and symptoms usually occur in the morning before breakfast or several hours after eating and may be precipitated by exercise. A fasting period of 72 h is considered to be the gold standard. The combination of hypoglycemia symptoms along with blood glucose levels of near or below 2.2 mmol/l, insulin levels of equal or greater 6 μU/ml (43 pmol/l), and elevated C-peptide levels are pathognomonic for insulinoma (Doherty et al. 1991; Grant 1996).

15.6.2.3
Gastrinoma

Gastrinoma is another common pancreatic endocrine tumor. It is located in the pancreas and duodenal wall and rarely in the gastric antrum. Sixty percent are malignant and 20% are associated with MEN type 1 (MEN-1). Gastrinomas are clinically characterized by the Zollinger-Ellison syndrome (ZES). This syndrome is produced by the hypersecretion of gastric acid and pepsin with resultant severe peptic ulcer, indigestion, esophagitis, duodenojejunitis, and/or diarrhea. The diagnosis is made by the measurement of circulating gastrin levels. Gastrinomas are slow-growing and patients can live for years even with multiple metastases if gastric hypersecretion is controlled. In 60–70% of patients, the primary tumor cannot be localized radiographically. The distinctive features of ZES-MEN-1 as compared to sporadic ZES are the impossibility of cure by surgery and relatively low death rate in the presence of liver metastasis (Mignon et al. 1995; Jensen 1996; Hirschowitz 1997; Mignon and Cadiot 1999; Tamburrano et al. 1999).

15.6.2.4
VIPoma

VIPomas, vasoactive intestinal polypeptide (VIP) secreting tumors, are rare tumors arising from the VIP secreting cells. They occur in pancreatic (90%) and extrapancreatic (10%) sites, most of the latter involving the autonomic nervous system and adrenal medulla. Fifty to sixty percent of the pancreatic tumors are malignant, whereas most neural crest tumors are benign. The syndrome is characterized by watery diarrhea, hypokalemia, and achlorhydria. Hypercalcemia and/or hyperglycemia are indicative of the syndrome. Diagnosis is confirmed by high plasma VIP, always in excess of 60 pmol/l. Most VIPomas synthesize at least one neuropeptide such as pancreatic polypeptide (39%), somatostatin (14%), glucagon (7%), or neurotensin (Matuchansky and Rambaud 1995; Park et al. 1996).

15.6.2.5
Glucagonoma

Glucagon-producing tumors of the pancreas are associated with peculiar skin rash called necrolytic migratory erythema, as well as glossitis and angular stomatitis, diabetes mellitus, anemia, and severe weight loss. The mechanism of these cutaneous and mucosal lesions remains incompletely understood, but are at least indirectly related to chronic glucagon hypersecretion. The administration of somatostatin and somatostatin analogues that reduce plasma glucagon level may dramatically cure skin lesions (Sohier et al. 1980; Boden et al. 1986). Other clinical features are venous thrombosis, pulmonary embolism, diarrhea, and complex neurologic symptoms. Diagnosis is established by the demonstration of hyperglucagonemia and hypoaminoacidemia. Plasma glucagon is at least five times higher than the normal fasting concentration, and higher than the level detected in severely ill and metabolically stressed patients. The hypoaminoacidemia is believed to reflect increased hepatic gluconeogenesis induced by glucagon. Diagnosis is almost always achieved at a later stage in the life of the neoplasm, since small tumors do not cause clinical symptoms (Guillausseau and Guillausseau-Scholer 1995).

15.6.2.6
Somatostatinoma

Somatostatinoma is a tumor predominantly composed of somatostatin-producing cells; these are frequently detected as a subpopulation in a variety of GEP tumors. The vast majority of somatostatinomas are located in the duodenum and pancreas. Pancreatic somatostatinomas are usually large at the time of diagnosis. The combination of gallstones, diabetes, diarrhea, and steatorrhea should suggest the possibility of a somatostatinoma. Diagnosis is confirmed by the demonstration of elevated fasting somatostatin concentration, which in normal individuals is less than 110 pg/ml (Sassolas and Chayvialle 1995; Tamburrano et al. 1999). A strong association has been shown between duodenal somatostatinomas and neurofibromatosis type 1 (Von Recklinghausen's disease), a relationship sometimes complemented by adrenal medullary hyperplasia and hyperparathyroidism (Swinburn et al. 1988; Hagen et al. 1992).

15.6.2.7
Gastric Carcinoid

Gastric carcinoids constitute about 0.54% of gastric malignancies and represent about 5.6% of all gastrointestinal neuroendocrine tumors. These tumors are mostly composed of enterochromaffin-like (ECL) cells. Patients with gastric carcinoids present with a variety of nonspecific symptoms and signs including vomiting, dyspepsia, upper gastrointestinal bleeding, and anemia. Three tumor types are distinguished on the basis of pathobiological behavior. Type I is associated with atrophic corporal gastritis (ACG) which accounts for 70–80% of all gastric carcinoids and develops in patients with hypergastrinemia and ACG. Tumors are often multiple and present as small polyps. Rarely, they metastasize to regional and distant sites. Type II is associated with MEN-1, is associated with the Zollinger-Ellison syndrome, and accounts for 6% of ECL-cell carcinoids. These tumors are usually multiple, and are larger in size than the type I (27% > 1.5 cm). Although lymph node metastases occur in 30%, prognosis is usually favorable. Type III is sporadic, arising in the absence of hypergastrinemia and accounting for 14–25% of all ECL-cell carcinoids. These type III tumors are comprised of ECL, enterochromaffin and X cells. This neoplasm is usually solitary in occurrence and 33% are larger than 2 cm. Metastases occur in 75% of patients.

Mechanisms governing the induction of oxyntic glands to type III ECL tumor are unknown, whereas it has been shown in type I and type II that ECL-cell carcinoids develop through a sequence hyperplasia-dysplasia-neoplasia under the influence of hypergastrinemia. The persistence of the promoting action of hypergastrinemia in established ECL-cell tumors has been demonstrated by their regression after antrectomy (Hirschowitz et al. 1992; Rindi et al. 1996; Modlin et al. 1997; Bordi 1999; Läuffer et al. 1999).

15.6.2.8
Nonfunctioning Endocrine Tumors of the Pancreas

These include all pancreatic endocrine neoplasms that are not associated with distinct hormonal syndromes. The reasons why nonfunctioning tumors are clinically silent are manifold. Elevated levels of peptides such as PP, HCG-*a*, HCG-*β* subunits, neurotensin, and chromogranins do not cause specific clinical symptoms, due to low serum concentrations of peptides, biologically inactive precursor forms, the downregulation of peripheral receptors, simultaneous production of an inhibitory peptide such as somatostatin, and the tumor's failure to release synthesized peptide. Nonfunctioning endocrine tumors are relatively common, representing 15–20% of endocrine pancreatic tumors. These tumors are often detected by local expansion of primary the tumor or by liver metastases (Eriksson and Öberg 1995; Mignon 1999).

15.6.2.9
Very Rare GEP Tumors

The relevant features of these are shown in Table 15.12. The most prevalent are ACTHomas, followed by PPomas, GRFomas (tumors of pancreatic polypeptides and growth hormone releasing factor, respectively), neurotensinomas and parathyrinomas.

ACTHomas and GRFomas are often extrapancreatic in location. Diagnostic difficulties in these rare tumors are related to the production of multiple hormones leading to the mimicry of well-recognized syndromes (Delcore and Friesen 1998; Mignon 1999).

15.6.3
Expression of Somatostatin Receptors

In mammals, a common gene encodes somatostatin-14 and somatostatin-28. These bioactive peptides act on various targets and regulate a large number of physiological functions based on its inhibitory effect on secretion, motility, and cell proliferation. Biological effects of somatostatin are mediated by five specific plasma membrane receptors that have been identified in normal and neoplastic tissues and designated as sst1 to sst5. Human sst1–5 are encoded by five genes located at chromosomes 14q13, 17q24, 22q13.1, 20p11.2, and 16p13.3, respectively. Four genes are intronless with the exception of sst2, which is spliced to generate two isoforms named sst2A and sst2B. All five receptors belong to the family of G-protein-coupled receptors and bind both native somatostatin-14 and -28 with a high affinity. However, they show major differences in their affinities for octreotide, a stable somatostatin analogue. Octreotide exhibits a low affinity for sst1 and sst4 (≥ 1 μM), whereas it binds sst2 and sst5 with a high affinity (low nM range) and binds sst3 with moderate affinity (Benali et al. 2000; Epelbaum et al. 1994; Meyerhof 1998; Patel et al. 1996; Reubi et al. 1997).

Table 15.12. The rare forms of GEP neuroendocrine tumors. (Data adapted from Mignon 1999)

Tumor type	Predominant hormone overproduction	Dominant symptomatology
ACTHoma/corticotropinoma	ACTH → hypercortisolemia	Severe Cushing's syndrome
PPoma/pancreatic polypeptidoma	PP → pure PPoma (50% cases)	Pain, weight loss, watery diarrhea
GRFoma (GH releasing factor producing tumor)	GRF (somatoliberin, somatocrin) → GH	Acromegaly + associated syndromes (ZES, Cushing's)
Pancreatic neurotensinoma (NToma)	Neurotensin in pure NTomas	Mimic WDHA syndrome
Pancreatic parathyrinoma (PTHoma)	PTHrP/PTH activity	Ectopic hypercalcemia S
Pancreatic serotoninoma/Carcinoid	Serotonin	Atypical carcinoid syndrome
Pancreatic calcitoninoma	Calcitonin + VIP/PP	WDHA
Pancreatic enteroglucagonoma	Enteroglucagon	Glucagonoma syndrome
Pancreatic cholecystokininoma (CCKoma)	CCK	ZES, negative secretion test
Pancreatic GIPoma	GIP	WDHA
Pancreatic bombesinoma (GRPoma)	GRP-like immuno-activity	Diabetes, flush, hemorrhagic gastritis

ACTH adrenocorticotropic hormone, *GIP* gastric inhibitory polypeptide, *GRF* growth hormone releasing factor, *GRP* gastrin releasing peptide, *PP* pancreatic polypeptide, *PTH* parathyroid hormone, *PTHrP* PTH-related peptide, *VIP* vasoactive intestinal peptide, *WDHA* watery diarrhea, hypokalemia, achlorhydria

Somatostatin receptors are widely distributed throughout many tissues and (over)expressed in a large variety of human cancers including GEP neuroendocrine tumors. The latter characteristic is used in somatostatin-receptor scintigraphy (SRS) and somatostatin-receptor radionuclide therapy (SRRT) using radiolabeled somatostatin analogues. Analyses demonstrate that these tumors express various sst mRNAs, sst2 most frequently. Approximately 85% of the carcinoid tumors examined are positive for sst2. The majority of these also express sst5, with sst1, sst3, and sst4 being less abundant. In contrast to neuroendocrine tumors, sst2 expression is absent in pancreatic adenocarcinoma as well as colorectal carcinomas (Reubi et al. 1994; Buscail et al. 1996; Jaïs et al. 1997; Schaer et al. 1997; Janson et al. 1998; Höcker and Wiedenmann 1999; Benali et al. 2000).

Much progress has been made toward the elucidation of molecular signal transduction, expression, and regulation of somatostatin receptors. However, the biological role and cellular distribution of each receptor subtype is far from being completely understood. The development of sst-specific antibodies and sst-specific peptide ligands as well as "pan somatostatin" analogues, the latter having high affinity for all sst, may improve the diagnostic and therapeutic application of somatostatin. The gene knockout models will help define the specific role of individual receptors in physiological and pathological conditions and the significance of multiple receptor subtypes in the same cell. Further understanding of mechanisms underlying the regulation of sst expression and receptor internalization would allow selective upregulation of somatostatin receptors on the target cell and result in enhanced accumulation of the somatostatin analogue (Schönbrunn 1999; Benali et al. 2000).

15.6.4
Molecular Pathogenesis

Four different genetic syndromes known to be associated with the development of GEP tumors are multiple endocrine neoplasia (MEN-1), von Hippel-Lindau (VHL) disease, neurofibromatosis type 1 (NF-1), and tuberous sclerosis. The gene mutated in each of these diseases has been identified, and the diseases are considered to be caused by the loss of function of a tumor suppressor gene. Menin, the 610 amino acid protein altered in MEN-1, widely occurs in normal tissues but its function is unclear. VHL disease encodes a 213 amino acid protein that plays an important role in regulating cell growth and differentiation. In NF-1 (von Recklinghausen's disease), the altered protein neurofibromin functions in normal cells as a suppressor of the ras signaling cascade. Tuberous sclerosis is caused by mutations in either the 1164 amino acid protein hamartin (TSC1), or the 1807 amino acid protein, tuberin (TSC2) (Feldkamp et al. 1998; Guru et al. 1998; Kaelin and Maher 1998; Young and Povey 1998; Jensen 1999).

Of these four syndromes, MEN-1 is most frequently associated with the development of GEP tumors. Between eighty and one hundred percent of patients with MEN-1 have been found to develop nonfunctioning tumors, most of which are small (< 0.5 cm). Family studies indicated that 54% of affected members developed gastrinomas, 21% insulinomas, and less than 5% the other functional GEP tumors. These tumors are the most common cause of death in MEN-1 patients. Therefore, the possibility of MEN-1 must be explored in all patients with GEP, especially in patients with gastrinomas, GRFomas, and insulinomas. This is essential because the clinical management, the need for additional studies to detect other endocrinopathies, the natural history, the risk of development of carcinoids which may be malignant, and the need for screening of other family members, all differ among patients with as opposed to without MEN-1 (Jensen 1999). In 12–17% of the patients with VHL, a pancreatic neuroendocrine tumor (PNT) has been found to occur, nearly all nonfunctional, although occasional insulinomas have been reported (Libutti et al. 1998). Patients with NF-1 develop duodenal somatostatinomas, but they are seldom associated with the somatostatin syndrome (Mao et al. 1995).

Recent studies have provided evidence of the importance of a number of genes in the molecular pathogenesis of GEP tumors, including the MEN-1 gene, the p16/MTS1 tumor-suppressor gene, the DPC 4/Smad4 gene, a candidate tumor-suppressor gene at 18q21, amplification of the HER-2/neu protooncogene, deletions in chromosome 1, and a possible tumor-suppressor gene on chromosome 3p (Evers et al. 1994; Chung et al. 1997; Muscarella et al. 1998; Bartsch et al. 1999; D'Adda et al. 1999; Ebrahimi et al. 1999; Goebel et al. 1999; Gortz et al. 1999; Rindi et al. 1999; Serrano et al. 2000). Alterations in the MEN-1 gene and the p16/MTS1 tumor-suppressor gene are of particular clinical importance. The loss of heterozygosity at the MEN-1 locus was found in up to 93% of sporadic PNTs and in 26–78% of sporadic carcinoid tumors. Furthermore, mutations in the MEN-1 gene locus have been reported in 18% of sporadic carcinoids and 27–39% of sporadic PNTs. Other researchers found MEN-1 gene mutations in 16 of 51 sporadic gastrinomas and 12 of 31 sporadic PNTs (Zhuang et al. 1997; Hessman et al. 1998; Wang et al. 1998; D'Adda et al. 1999; Goebel et al. 1999; Gortz et al. 1999; Jensen 1999). In gastrinomas and nonfunctioning PNTs, Muscarella et al. (1998) demonstrated genetic alterations in the p16/MTS1 tumor-suppressor gene in 91.7%. This gene encodes p16, an inhibitor of cyclin-dependent kinase 4, which phosphorylates the retinoblastoma gene product Rb to its inactive form. Thus, its failure to function results in the loss of cell cycle inhibition. This study showed the p16/MTS1 gene to be homozygously deleted in 41.7% and methylated in 58.3%, but no identifiable mutations. When confirmed, the inactivating mutations of the p16/MTS1 noted in this study would seem to necessitate an important change in the concepts surrounding the molecular pathogenesis of GEP neuroendocrine tumors (Jensen 2000).

15.6.5
Localization Procedure

Figure 15.7 shows a diagnostic work-up scheme of a suspected neuroendocrine tumor (Öberg 1999). The first step is biochemical diagnosis, and the most important screening marker today is chromogranin A (CgA), which is encoded by a gene on chromosome 14 and elevated in 80–100% of patients. The CgA level reflects tumor load and can be a prognosticating marker. Plasma CgA values might then be supplemented with the analysis of other "general" markers such as pancreatic polypeptide, neuron-specific enolase, and HCG-α and -β subunits. In patients with suspected clinical syndromes, markers such as urinary 5-HIAA, serum gastrin, glucagon, plasma somatostatin, ACTH, and others are chosen according to symptoms (Tiensuu Janson and Öberg 1996; Modlin and Tang 1997; Öberg 1999; Eriksson et al. 2000). Surgery remains the mainstay of treatment if the tumor is resectable. When complete resection is not possible because of metastases or tumor location, as much tumor as possible is still removed. Precise tumor localization and assessment of tumor spread are clearly crucial.

During the years 1998 and 1999, the European Network of Neuroendocrine Tumors (ENET) developed recommendations for imaging neuroendocrine

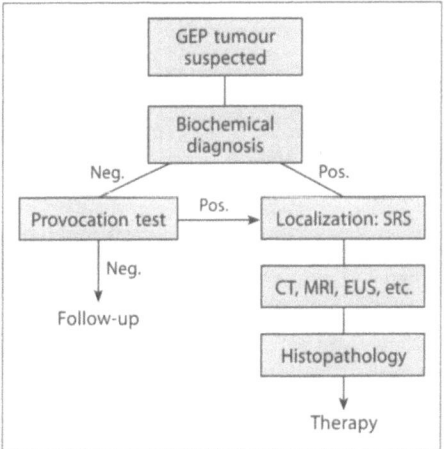

Fig. 15.7. Diagnostic work-up of GEP tumors

tumors, mirroring participants' experiences from the fields of radiology, nuclear medicine, gastroenterology, surgery, and pathology (Ricke and Klose 2000). The central diagnostic modality for the assessment of functional and nonfunctional neuroendocrine tumors of the pancreas is SRS, with a sensitivity and specificity of up to 90% and 80%, respectively. This technique can visualize the primary tumor location and its extent, and also indicate its sensitivity to treatment with a somatostatin analogue (Fig. 15.8). If SRS demonstrates metastasized disease, further diagnostic imaging can be restricted to monitoring therapeutic effect or to taking part in controlled trials. In patients with gastrinoma, endoscopy should be performed to detect lesions in the duodenum, especially when gastrinomas are associated with a MEN-1 (shown to have duodenal manifestations in 80% of cases). Furthermore, endoscopy enables the identification of peptic ulcers in functional gastrinomas. Extended examinations are required to detect lesions in other endocrine organs in MEN-1 patients (Grama et al. 1992; Krenning et al. 1993; Ruszniewski et al. 1995; Schirmer et al. 1995; Cadiot et al. 1996; Zimmer et al. 1996; Lebtahi et al. 1997; Modlin and Tang 1997; Termanini et al. 1997; Bansal et al. 1999; Hiorns and Reznek 2000; Ricke and Klose 2000).

In nonmalignant insulinomas and other tumors of the pancreas, endosonography is regarded as the most sensitive modality. In experienced hands, high-resolution images of the entire pancreas can be obtained with the threshold for detection being 5 mm in size (Proye et al. 1998) at a sensitivity of 79%. Insulinomas rarely express somatostatin receptors, thus limiting the sensitivity of SRS to 50%. However, combined examination

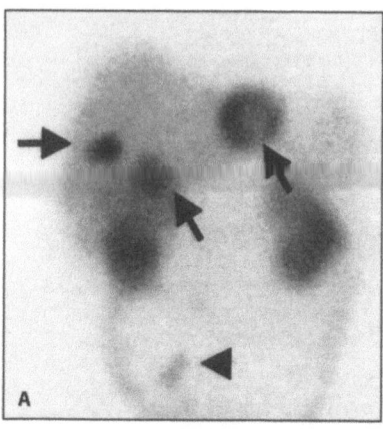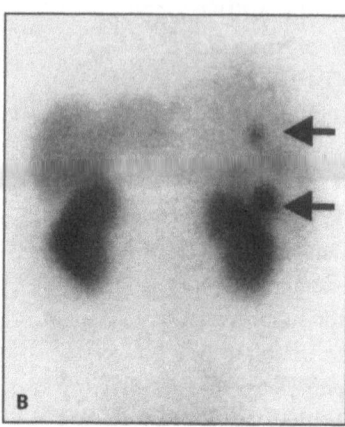

Fig. 15.8 A, B. SRS images obtained from a patient with carcinoid syndrome. The anterior view (**A**) and posterior view (**B**) of the abdomen demonstrates the primary tumor (*arrowhead*) and multiple metastases in the liver (*arrows*). There is physiological accumulation of radioactivity in the liver, spleen, and kidneys, as well as visualization of the colon. This examination also demonstrated distant metastases in the thoracic vertebrae (data not shown)

raises the tumor detectability up to 89%. Since malignant insulinomas often show positive somatostatin receptors, SRS is recommended for their staging. For the local staging of ECL cell carcinoid, endoscopy and endosonography of the stomach are sufficient. In case of suspected malignancy, SRS is the most valuable staging means, with sensitivity of 75% and specificity of 95%. The SRS procedure must be followed by abdominal CT. In patients with ZES, accompanying gastrinomas should be identified (Modlin et al. 1991; Zimmer et al. 1994; Gibril et al. 2000).

For the diagnosis of neuroendocrine tumors of the gut, SRS is the initial diagnostic method of choice. The sensitivity for the detection of foregut, midgut, and hindgut tumors has been reported to be up to 90% with high specificity. Metastasized SRS-positive tumors do not require further imaging study. In patients with negative SRS, the diagnostic approach to foregut tumors depends on the location. CT is the most effective means for diagnosing lung tumors and the combined use of endoscopy and endosonography reinforced with abdominal CT is the best means for tumors in the stomach, duodenum, and pancreas. Midgut tumors with the exception of those in the ascending colon are best visualized by abdominal CT or MRI. A combination of CT and enteroclysis may be recommended. Due to technical limitations, ileoscopy remains optional. For hindgut tumors and tumors of the ascending colon, lower endoscopy is the method of choice. CT scan or MR imaging of the abdomen may complete the study and, if further staging is desired, endosonography is of help. Finally, since hepatic

lesions are occasionally not visible in SRS including single photon emission tomography (SPECT), the screening for liver metastases of neuroendocrine tumors should be completed by abdominal sonography, CT scan, or MR imaging (Thiele et al. 1993; Sugimoto et al. 1995; Semelka et al. 1996; Ricke and Klose 2000).

When noninvasive tests fail to localize functional endocrine tumors identified by biochemistry, intraoperative ultrasound and endoscopic transillumination are valuable alternatives. These imaging diagnoses rely on spatial considerations and, hence, the sensitivity decreases as tumor size diminishes. In contrast, portal venous hormonal assay is completely independent of tumor size. Provocative stimulation of gastrinomas by secretin or of insulinomas by calcium combined with hepatic venous sampling have proven in small series to be better than portal venous sampling alone. The results of intraoperative localization using a gamma probe after administration of presently available radioactive labeled somatostatin analogue are controversial (Imamura et al. 1987; Zeiger et al. 1993; Doppman et al. 1995; Fraker and Alexander 1995; O'Shea et al. 1996; Wängberg et al. 1996; Pelley and Bukowski 1997; Norton 1999).

15.6.5.1
SRS Methods

SRS methods are described in detail elsewhere (Kwekkeboom et al. 2000 a) and are beyond the scope of this chapter. Main factors that determine the ability of SRS to detect a lesion are the size, location, and

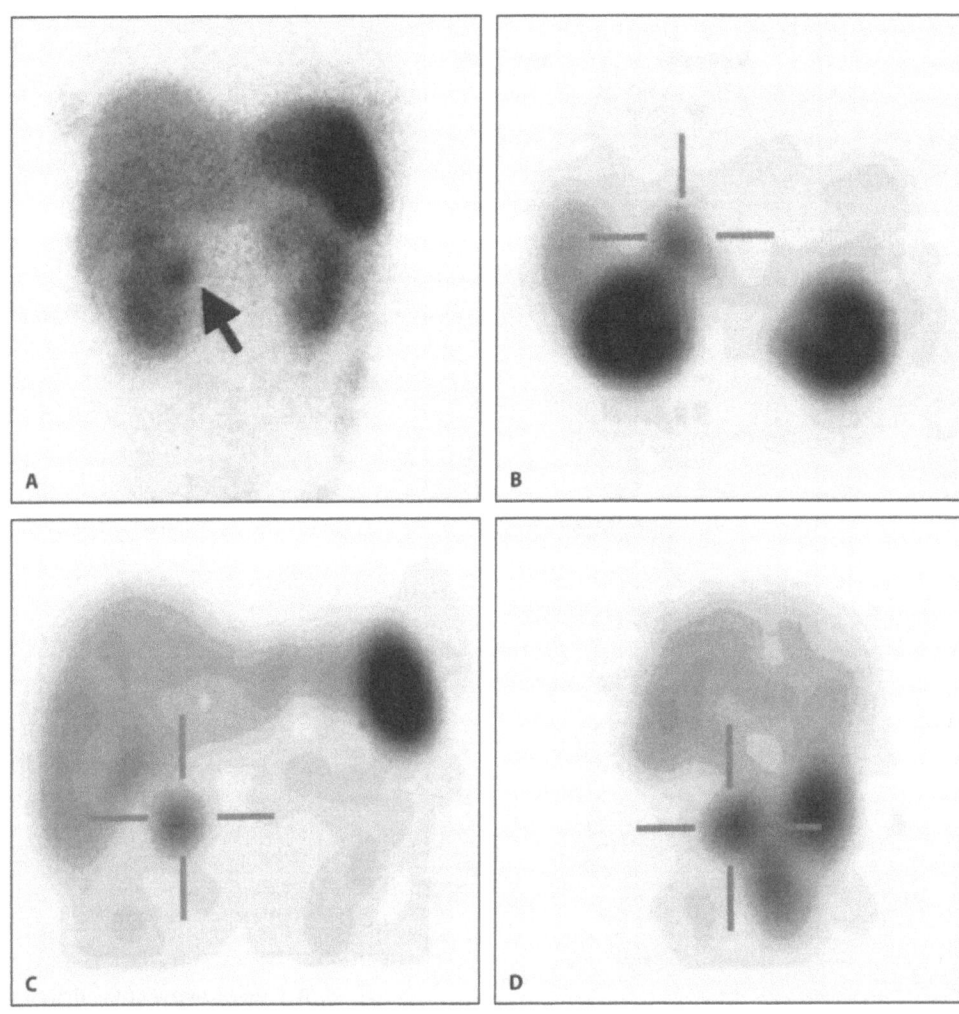

Fig. 15.9 A-D. SRS images obtained from a patient with gastrinoma in the head of the pancreas. On the planar image (A) the lesion is projected on the right kidney mimicking radioactivity in the renal pelvis. The SPECT images (B transverse, C coronal, D sagittal) clearly demonstrate the lesion, which is located ventrally from the right kidney

the receptor density of the lesion. The smallest lesion detectable is of 3 mm diameter. Lesions larger than 2 cm are detected with a sensitivity of 96% (Alexander et al. 1998; Lugtenburg et al. 2001). Sufficient photon counts, SPECT acquisition using a double- or triple-head camera (Fig. 15.9), bowel preparation to reduce background, and proper imaging intervals improve sensitivity (Krenning et al. 1993; McCown et al. 1995; Modlin et al. 1995; Gibril et al. 1996; Kwekkeboom et al. 2000b; Schillaci et al. 2000). When SRS follows other imaging, clinical management has been found to change in 21% to 47% of cases. SRS can identify more bone metastases than bone scan (Fig. 15.10) (Cadiot et al. 1997; Lebtahi et al. 1997; Modlin and Tang 1997; Termanini et al. 1997; Chiti et al. 1998; Frilling et al. 1998; Gibril et al. 1998; Jensen and Gibril 1999; Van Eijck et al. 1999; Jensen 2000), and it is useful for differentiating non functioning GEP tumors from pancreatic adenocarcinomas, which are SRS negative (Pelley and Bukowsky 1997; Van Eijck et al. 1999). However, numerous normal tissues and pathologic processes other than GEP tumors may have high densities of somatostatin receptors and appear as focal lesions on SRS. Gibril et al. (1999) prospectively studied the specificity in 146 gastrinoma patients who underwent 480 examinations. In this study, 12% of patients had false-positive localizations due to thyroid disease, breast disease, granulomatous lung disease, accessory spleen, previous operative sites, renal parapelvic cysts, and procedural problems.

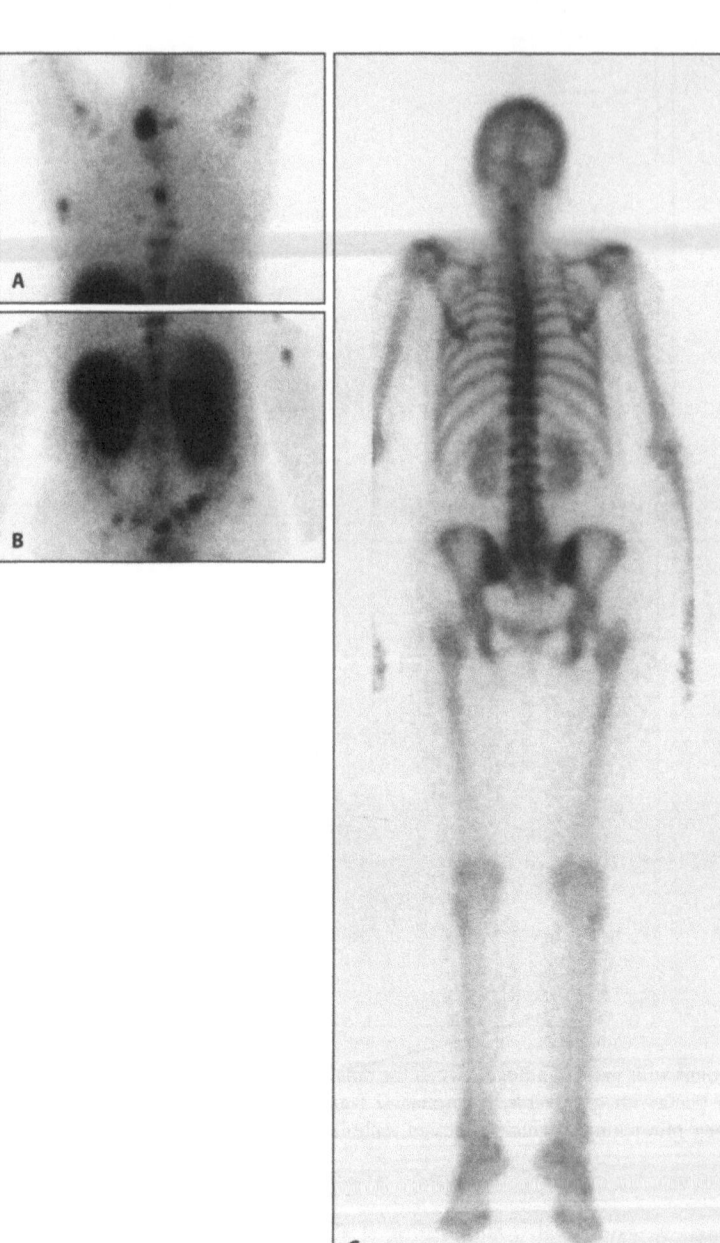

Fig. 15.10. Posterior view of the thorax (A) and abdomen (B) of SRS-images obtained from a patient with metastasized somatostatinoma, and the posterior view of the total body bone scan (C) obtained a week earlier. The SRS demonstrates obvious multiple bone metastases, especially in the vertebral column and pelvis, whereas the bone scan shows no abnormality

However, when SRS results were interpreted in a clinical context, only 2.7% caused management change. SRS should thus always be carefully interpreted and supplemented by other imaging studies (Gibril et al. 1999).

15.6.5.2
CT and MRI Images

CT scan and MR imaging far better delineate the anatomy of tumors and surrounding structures than does SRS, making them indispensable for surgery. Reported sensitivities of CT scan in the detection of GEP tumors vary considerably, reflecting differences between levels of expertise, the age of the scanner, and protocols used among different centers. The advent of helical scanners allows the entire pancreas to be imaged in 3–5 mm slices during a single breath-hold, considerably reducing motion artifacts and enhancing the possibility of visualizing tumor blush on images acquired when the concentration of contrast medium is at its maximum in the arterial phase. CT can also visualize local lymph nodes and hepatic metastases if present. The size detection limit of CT is around 5 mm (Orbuch et al. 1995; Sugimoto et al. 1995; Van Hoe et al. 1995; Gibril et al. 1996; Winter et al. 1996; Pelley and Bukowski 1997; King et al. 1998; Öberg 1999; Hiorns and Reznek 2000). In recent years, the development of MRI techniques that overcome the problems of motion artifact caused by bowel peristalsis, respiratory movement, and vascular motion have significantly improved the quality of MR imaging of the pancreas. The vascular nature of most islet cell tumors means that they enhance markedly after intravenous gadolinium. In clinical centers with experience in imaging pancreatic neuroendocrine tumors, MRI provides better sensitivity than conventional CT. Nevertheless, the majority of authorities agree that dynamic spiral CT is as good as MRI for this purpose (Semelka et al. 1993; Moore et al. 1995; Orbuch et al. 1995; Berger et al. 1996; Gibril et al. 1996; Pelley and Bukowski 1997; Öberg 1999; Hiorns and Reznek 2000; Owen et al. 2001).

15.6.5.3
Other Radionuclide Techniques

Other radionuclide techniques for future use include PET with ^{11}C-5-hydroxytryptophan or ^{18}F-FDG, VIP receptor scintigraphy, and chromogranin A immunoscintigraphy. Using ^{11}C-5-hydroxytryptophan, Orlefors et al. (1998) obtained a higher sensitivity than CT scan and they found a close correlation between its transport rate and 5-hydroxyindole acetic acid excretion rate, showing its potential value in treatment monitoring. Studies with ^{18}F-FDG PET proved valuable for predicting malignancy in relatively poorly differentiated tumors (Adams et al. 1998; Pasquali et al. 1998). Using ^{123}I-VIP, Virgolini et al. (1994, 1996) could visualize 9 of 10 carcinoids and 4 of 4 insulinomas. Moreover, using this radiopharmaceutical they detected colonic and pancreatic adenocarcinomas. The cumbersome peptide labeling with ^{123}I, the relatively high background radioactivity, especially in the thorax, and the limited availability of ^{123}I and VIP in highly purified form are as yet limitations to the widespread use of this type of scintigraphy. The three-step immunoscintigraphy with anti-CgA monoclonal antibody has been introduced by Colombo et al. (1993). With this technique they succeeded in visualizing all of 5 insulinomas, 6 of 8 carcinoids, 1 of 1 gastrinoma and 1 of 1 glucagonoma (Siccardi et al. 1996).

15.6.6
Histopathology and Prognosis

The final diagnosis should always be based on histopathology. The histological features of GEP endocrine tumors are usually distinctive enough to permit their identification. However, the differential diagnosis between some endocrine tumors and exocrine carcinomas and adenocarcinomas may be difficult. Therefore once a GEP tumor is suspected, the diagnosis should rely on known endocrine markers in the tumor cells (Rindi et al. 1998). For general assessment, the techniques of choice include silver impregnation methods such as Grimelius' stain, immunohistochemistry for cytosolic markers of neuroendocrine differentiation such as neuron-specific enolase and the protein gene product 9.5 (PGP 9.5), granular markers associated with large dense core granules, such as chromogranin A, and markers of small synaptic vesicles such as synaptophysin and synaptobrevin (Tiensuu Janson and Öberg 1996; Rindi et al. 1998; Öberg 1999). In addition, GABA and receptors for GABA have been demonstrated in endocrine cell lines deriving from GEP tumors. Along the same lines, two isoforms of ATP-dependent vesicular monoamine transporter proteins (VMAT1 and VMAT2) have recently been identified (Ahnert-Hilger et al. 1993, 1996; Dimaline and Struthers 1996; Erickson et al. 1996). Because the hormonal expressions of GEP endocrine tumors are heterogeneous, the panel of antisera for GEP endocrine tumor analysis should include all the antisera to hormones such as serotonin, gastrin, glucagon, and others that are normally expressed in DES (diffuse endocrine system) cells of the anatomical region from which the tumor arises (Table 15.10). Additional hormones should be tested depending on specific clinical questions or on whether "inappropriate" hormones are expressed by such tumors. The proliferation marker Ki-67 should be included to determine the growth potential (Rindi et al. 1998; Öberg 1999).

Information about biological characteristics of an endocrine tumor is helpful for prognostication. Proposed clinicopathologic classification of neuroendocrine tumors attempts to integrate important clinical and functional findings into a morphologic categorization. Based on this proposal, the WHO Classification Committee gives consideration to three major endocrine tumor categories: well-differentiated endocrine tumor, poorly differentiated endocrine carcinoma, and mixed exocrine-endocrine tumor. The categorization may be applied to neuroendocrine tumors of the pancreas, stomach, duodenum, and others, and supplemented by clinicopathologic classifications. Clinicopathologic classifications define criteria such as size, angioinvasion, and functionality and, based on these criteria, a tumor can be classified as benign, uncertain, or carcinoma with low or high grade malignant potential (Rindi et al. 1998; Klöppel et al. 1999).

15.6.7
Treatment

Surgery is still the only treatment that can cure patients with GEP tumors. Surgical procedures aim at accurate localization and removal of all tumor tissue. These lesions tend to be small and difficult to find, especially true in islet cell tumors. When curative surgery is not achievable, debulking procedures leading to tumor mass reduction must be considered. In an advanced disease state, often with multiple liver metastases, other treatment modalities have to be considered (Ahlman et al. 2000). It is important to realize that patients with GEP tumors not only succumb to tumefaction but also to hormonal overproduction. The aims of medical treatment are to ameliorate symptoms by suppressing hormone production, to reduce tumor growth, and to improve the quality of life. With the exception of insulinoma and gastrinoma, octreotide and other long-acting somatostatin analogues are currently the choice of drugs to control hormone-mediated symptoms. Gastric acid hypersecretion in the Zollinger-Ellison syndrome is best controlled by proton pump inhibitors. In insulinoma, the prevention of hypoglycemia by octreotide is successful in up to 50% of the patients with sstr2 expression (Arnold et al. 2000; Degen and Beglinger 1999). Currently available somatostatin analogues include octreotide, lanreotide, and vapreotide. They predominantly bind to receptor subtypes 2 and 5 and can improve subjective symptoms in over 70% of patients. The success of octreotide therapy is reflected in a patient's quality of life as defined by the ability to perform normal daily activities. A significant recent advance is the development of long acting release forms of octreotide for once-a-month injection and sustained-release forms of lanreotide.

Somatostatin analogues exert antiproliferative effects both in vitro and in vivo predominantly through receptor subtype 2. Available data on growth control indicates that stabilization of tumor growth seems to be the most beneficial effect occurring in up to 50% of patients (Degen and Beglinger 1999; Arnold et al. 2000).

Alpha interferon, an effective antiviral and antitumor agent, has also been applied in the management of classical midgut carcinoids. Up to now, more than 500 patients have been treated with alpha interferon worldwide. The subjective response rate is about 60%. Biochemical responses occurred in 44% and tumor responses in 11% of patients (Öberg 2000). The median survival time from the start of treatment was over three years. The mechanisms of action of alpha interferon is a direct effect on the tumor cells by inhibiting cell proliferation via a cell cycle block in the G1-S phase. The induction of interferon-inducible genes such as p-21, p-27, 2-5-A-synthetase, PKR, IRF-1, IRF-2 contribute to reducing tumor growth potential. Furthermore, alpha interferon has an immunomodulatory effect stimulating natural killer cells and macrophages and possesses anti-angiogenetic effect. Limited data suggest that the combined treatment of octreotide and alpha interferon is superior to the treatment with either agent alone (Arnold et al. 2000; Öberg 2000).

Generally, the efficacy of chemotherapy depends on the primary site and histological differentiation. The standard chemotherapy for pancreatic neuroendocrine tumors is a combination of adriamycin and streptozocin, and less frequently a combination of 5-fluorouracil and streptozocin. Using these combinations, response rates of from 40% to 60% can be achieved. There is no settled standard chemotherapy for carcinoid tumors, but most oncologists use 5-fluorouracil and streptozocin in combination. Results are poor with the response rate being only 20%, and the benefit seldom counterbalances the toxicity of the agents. In undifferentiated or poorly differentiated neuroendocrine tumors, a higher response rates of 41–69% can be achieved by the combined use of etoposide and cisplatin. Unfortunately the chemosensitivity for this combination is of short duration (Rougier and Mitry 2000).

Hepatic arterial chemoembolization seems attractive for unresectable and progressive metastases of GEP tumors confined to the liver, mainly following unsuccessful systemic chemotherapy. Chemoembolization seems more effective than embolization alone or surgical ligation of the hepatic artery, since ischemia may increase the sensitivity of tumor cells to drug activity and anoxia and the slowing of blood flow can increase intra-tumor drug concentrations and dwelling time in tumor cells. This intervention inhibits tumor growth with a mean duration of 6–42.5 months in 33–80% of patients. A decrease of urinary 5-HIAA by more than 50% was observed in 57–91% (Ruszniewski and Malka 2000).

15.6.7.1
Somatostatin Receptor Radionuclide Therapy (SRRT)

A new application of radiolabeled peptides is somatostatin receptor radionuclide therapy, the administration high doses of radiolabeled octreotide analogues such as [^{111}In-DPTA]octreotide, SMT487 (OctreoTher = [^{90}Y-DOTA^0,Tyr^3]octreotide) and [^{177}Lu-DOTA^0,Tyr^3]octreotate. SRRT with somatostatin analogues should be carried out with kidney protection by co-infusion of amino acids: for example, a combination of 25 g lysine and 25 g arginine for 4 h, starting 30 min prior to administration of the radioligand (De Jong et al. 1996).

In a phase 1 trial with cumulative doses of up to 100 GBq [^{111}In-DPTA]octreotide, no major clinical side effects were seen after up to 2 years treatment, except a possible transient decline in platelet counts and lymphocyte subsets. Promising beneficial effects on clinical symptoms, hormone production and tumor proliferation, ascribed to the Auger electrons emitted by ^{111}In, were found. In 21 patients (cumulative dose of more than 20 GBq [^{111}In-DPTA]octreotide) with progressive disease at baseline, eight showed stable disease and six others a reduction in tumor size. There was a tendency towards better results in patients whose tumors had a higher accumulation of the radioligand (Krenning et al. 1999).

Three out of six patients who received higher doses than 100 GBq [^{111}In-DTPA]octreotide developed a myelodysplastic syndrome and/or leukemia. A causal relationship, however, is not certain.

Radionuclide therapy with [^{90}Y-DOTA, Tyr^3]-octreotide started in phase I trials. Yttrium-90 emits β-particles with much longer path lengths compared to those of the Auger electrons of ^{111}In, which is a clear advantage in tumors with heterogeneous cellular receptor distribution. Overall, antimitotic effects have been observed: about 20% partial response and 60% stable disease (n=92) along with complete symptomatic cure of several malignant insulinoma and gastrinoma patients. Maximum cumulative [^{90}Y-DOTA, Tyr^3]octreotide dose was about 26 GBq, without reaching the maximum tolerable dose (Valkema et al. 2000).

New is the use of [^{177}Lu-DOTA, Tyr^3]octreotate, which shows the highest tumor uptake of all tested octreotide analogues so far, with excellent tumor to kidney ratios. SRRT with this analogue in a phase 1 trial started only very recently in our center.

15.7
References

Abdelhamid S, Lewicka S, Vecsei P et al (1995) A new subset of mineralocorticoid hypertension with excess of 21-deoxyaldosterone and Kelly's-M1 steroid: clinical and morphological findings. J Clin Endocrinol Metab 80:737–744

Adams S, Baum RP, Hertel A et al (1998a) Comparison of metabolic and receptor imaging in recurrent medullary thyroid carcinoma with histopathological findings. Eur J Nucl Med 25:1277–1283

Adams S, Baum RP, Hertel A et al (1998b) Metabolic (PET) and receptor (SPET) imaging of well- and less well-differentiated tumours: comparison with the expression of the Ki-67 antigen. Nucl Med Commun 19:641–647

Ahlman H, Wangberg B, Jansson S et al (2000) Interventional treatment of gastrointestinal neuroendocrine tumours. Digestion 62 [Suppl 1]:59–68

Ahnert-Hilger G, Grube K, Kvols L et al (1993) Gastroenteropancreatic neuroendocrine tumors contain a common set of synaptic vesicle proteins and aminoacid neurotransmitters. Eur J Cancer 14:1982–1984

Ahnert-Hilger G, Stadtbäumer A, Strübing C et al (1996) γ-Aminobutyric acid secretion from pancreatic neuroendocrine cells. Gastroenterology 110:1595–1604

Aiba M, Hirayama A, Iri H et al (1991) Adrenocorticotropic hormone-independent bilateral adrenocortical macronodular hyperplasia as a distinct subtype of Cushing's syndrome. Am J Clin Pathol 96:334–340

Aigner RM, Fueger GF, Nicoletti R (1996) Parathyroid scintigraphy: comparison of technetium-99m methoxyisobutylisonitrile and technetium-99m tetrofosmin studies. Eur J Nucl Med 23:693–696

Aigner RM, Fueger GF, Wolf G (1997) Parathyroid scintigraphy: first experiences with technetium (III)-99m-Q12. Eur J Nucl Med 24:326–329

Åkerström G, Malmaeus J, Bergström R (1984) Surgical anatomy of human parathyroid glands. Surgery 95:14–21

Alam MS, Takeuchi R, Kasagi K et al (1997) Values of combined technetium-99m hydroxy methylene diphosphonate and thallium-201 imaging in detecting bone metastases from thyroid carcinoma. Thyroid 7:705–712

Alexander HR, Fraker DL, Norton JA et al (1998) Prospective study of somatostatin receptor scintigraphy and its effect on operative outcome in patients with Zollinger-Ellison syndrome. Ann Surg 228:228–238

Anderson GW, Mariash CN, Oppenheimer JII (2000) Molecular actions of thyroid hormone. In: Braverman LE, Utiger RD (eds) Werner and Ingbar's The thyroid, 8th edn. Lippincott, Williams and Wilkins, Philadelphia, pp 174–195

Angeletti S, Annibale B, Marignani M et al (1999) Natural history of intestinal carcinoids. Ital J Gastroenterol Hepatol 31 [Suppl 2]:S108–S110

Apostolopoulos DJ, Houstoulaki E, Giannakenas C et al (1998) Technetium-99m-tetrofosmin for parathyroid scintigraphy: comparison to thallium-technetium scanning. J Nucl Med 39:1433–1441

Arnold A (1993) Genetic basis of endocrine disease 5. Molecular genetics of parathyroid gland neoplasia. J Clin Endocrinol Metab 77:1108–1112

Arnold A, Kim HG, Gaz RD et al (1989) Molecular cloning and chromosomal mapping of DNA rearranged with the parathyroid hormone gene in a parathyroid adenoma. J Clin Invest 83:2034–2040

Arnold A, Brown MF, Ureña P et al (1995) Monoclonality of parathyroid tumors in chronic renal failure and in primary parathyroid hyperplasia. J Clin Invest 95:2047–2053

Arnold R, Simon B, Wied M (2000) Treatment of neuroendocrine GEP tumours with somatostatin analogues. A review. Digestion 62 [Suppl 1]:84–91

Atkins HL, Klopper JF (1973) Measurement of thyroidal technetium uptake with the gamma camera and computer system. Am J Roentgenol 118:831–835

Bakker WH, Albert R, Bruns C et al (1991) [111In-DTPA-D-Phe1]-octreotide, a potential radiopharmaceutical for imaging of somatostatin receptor-positive tumors: synthesis, radiolabeling and in vitro validation. Life Sci 49:1583–1591

Bansal R, Tierney W, Carpenter S et al (1999) Cost effectiveness of EUS for preoperative localization of pancreatic endocrine tumors. Gastrointest Endosc 49:19–25

Bartsch D, Hahn SA, Danichevski KD et al (1999) Mutations of the DPC4/Smad4 gene in neuroendocrine pancreatic tumors. Oncogene 18:2367–2371

Becker D, Charkes N, Dwarkin H et al (1996a) Procedure guideline for thyroid scintigraphy: 1.0. J Nucl Med 37:1264–1266

Becker D, Charkes N, Dwarkin H et al (1996b) Procedure guideline for thyroid uptake measurement: 1.0. J Nucl Med 37:1266–1268

Becker D, Charkes N, Dwarkin H et al (1996c) Procedure guideline for extended scintigraphy for differentiated thyroid cancer: 1.0. J Nucl Med 37:1269–1271

Beck-Peccoz P, Chatterjee VKK (1994) The variable clinical phenotype in thyroid hormone resistance syndrome. Thyroid 4:225–232

Beever K, Bradbury J, Phillips D et al (1989) Highly sensitive assays of autoantibodies to thyroglobulin and to thyroid peroxidase. Clin Chem 35:1949–1954

Benali N, Ferjoux G, Puente E et al (2000) Somatostatin receptors. Digestion 62 [Suppl 1]:27–32

Bénard F, Lefevre B, Beuvon F et al (1995) Rapid washout of technetium-99m-MIBI from a large parathyroid adenoma. J Nucl Med 36:241–243

Berger JF, Laissy JP, Limot O et al (1996) Differentiation between multiple liver hemangiomas and liver metastases of gastrinomas: value of enhanced MRI. J Comput Assist Tomogr 20:349–355

Bergstrom M, Muhr C, Lundberg PO et al (1986) Amino acid metabolism in pituitary adenomas. Acta Radiol [Suppl] 369:412–414

Bergstrom M, Muhr C, Ericson K et al (1987a) The normal pituitary examined with positron emission tomography and (methyl-11C)-L-methionine and (methyl-11C)-D-methionine. Neuroradiology 29:221–225

Bergstrom M, Muhr C, Lundberg PO et al (1987b) Rapid decrease in amino acid metabolism in prolactin-secreting pituitary adenomas after bromocriptine treatment: a PET study. J Comput Assist Tomogr 11:815–819

Bergstrom M, Muhr C, Lundberg PO et al (1987c) Amino acid distribution and metabolism in pituitary adenomas using positron emission tomography with D-[11C]methionine and L-[11C]methionine. J Comput Assist Tomogr 11:384–389

Bergstrom M, Muhr C, Lundberg PO et al (1991) PET as a tool in the clinical evaluation of pituitary adenomas. J Nucl Med 32:610–615

Bergstrom M, Muhr C, Jossan S et al (1992) Differentiation of pituitary adenoma and meningioma: visualization with positron emission tomography and [11C]-L-deprenyl. Neurosurgery 30:855–861

Berndt I, Reuter M, Saller B et al (1998) A new hot spot for mutations in the ret protooncogene causing familial medullary thyroid carcinoma and multiple endocrine neoplasia type 2A. J Clin Endocrinol Metab 83:770–774

Bhatnagar A, Vezza PR, Bryan JA et al (1998) Technetium-99m-sestamibi parathyroid scintigraphy: effect of P-glycoprotein, histology and tumor size on detectability. J Nucl Med 39:1617–1620

Biebermann H, Schoneberg T, Krude H et al (1997) Mutations of the human thyrotropin receptor gene causing thyroid hypoplasia and persistent congenital hypothyroidism. J Clin Endocrinol Metab 82:3471–3480

Billotey C, Aurengo A, Najean Y et al (1994) Identifying abnormal parathyroid glands in the thyroid uptake area using technetium-99m-sestamibi and factor analysis of dynamic structures. J Nucl Med 35:1631–1636

Billotey C, Sarfati E, Aurengo A et al (1996) Advantages of SPECT in technetium-99m-sestamibi parathyroid scintigraphy. J Nucl Med 37:1773–1778

Blagosklonny MV, Giannakakou P, Wojtowicz M et al (1998) Effect of p53-expressing adenovirus on the chemosensitivity and differentiation of anaplastic thyroid cancer cells. J Clin Endocrinol Metab 83:2516–2622

Boden G, Ryan IG, Eisenschmid BL et al (1986) Treatment of inoperable glucagonoma with the long-acting somatostatin analogue SMS 201-995. N Engl J Med 314:1686–1689

Boni G, Ferdeghini M, Bellina CR et al (1995) [111In-DTPA-D-Phe]-octreotide scintigraphy in functioning and nonfunctioning pituitary adenomas. Q J Nucl Med 39:90–93

Bordi C (1999) Gastric carcinoids. Ital J Gastroenterol Hepatol 31 [Suppl 2]:S94–S97

Bradley G, Juranka PF, Ling V (1988) Mechanism of multidrug resistance. Biochem Biophys Acta 948:87–128

Britton K, Shapiro B (1981) Single photon emission tomography of the pituitary: preliminary communication. J R Soc Med 74:667–669

Broson-Chazot F, Houzard C, Ajzenberg C et al (1997) Somatostatin receptor imaging in somatotroph and nonfunctioning pituitary adenomas: correlation with hormonal and visual responses to octreotide. Clin Endocrinol (Oxf) 47:589–598

Brown EM, Katz C, Butters R et al (1991) Polyarginine, polylysine, and protamine mimic the effects of high extracellular calcium concentrations on dispersed bovine parathyroid cells. J Bone Miner Res 6:1217–1225

Brown EM, Gamba G, Riccardi D et al (1993) Cloning and characterization of an extracellular Ca^{2+}-sensing receptor from bovine parathyroid. Nature 366:575–580

Brown EM, Pollak M, Seidman CE et al (1995) Calcium-ion-sensing cell-surface receptors. N Engl J Med 333:234–240

Brucer M (1971) Vignettes Nucl Med 4:1-14

Buscail L, Saint-Laurent N, Chastre E et al (1996) Loss of sst2 somatostatin receptor gene expression in human pancreatic and colorectal cancer. Cancer Res 56:1823–1827

Cadiot G, Lebtahi R, Sarda L et al (1996) Preoperative detection of duodenal gastrinomas and peripancreatic lymph nodes by somatostatin receptor scintigraphy. Groupe D'etude Du Syndrome De Zollinger-Ellison. Gastroenterology 111:845–854

Cadiot G, Bonnaud G, Lebtahi R et al (1997) Usefulness of somatostatin receptor scintigraphy in the management of patients with Zollinger-Ellison syndrome. Gut 41:107–114

Cao G, Garcia CK, Wyne KL et al (1997) Structure and localization of the human gene encoding SR-BI/CLA-1. J Biol Chem 272:33068–33076

Carpentier A, Jeannotte S, Verreault J et al (1998) Preoperative localization of parathyroid lesions in hyperparathyroidism: relationship between technetium-99m-MIBI uptake and oxyphil cell content. J Nucl Med 39:1441–1444

Cavalieri R, Gerard S (1991) Unusual types of thyrotoxicosis. Adv Intern Med 36:271–286

Chabre O, Liakos P, Vivier J et al (1998) Cushing's syndrome due to a gastric inhibitory polypeptide-dependent adrenal adenoma: insights into hormonal control of adrenocortical tumorigenesis. J Clin Endocrinol Metab 83:3134–3143

Chakeres DW, Curtin A, Ford G (1989) Magnetic resonance imaging of pituitary and parasellar abnormalities. Radiol Clin North Am 27:265–281

Chakraborty PK, Gildersleeve DL, Jewett DM et al (1993) High yield synthesis of high specific activity R-(–)-[^{11}C]epinephrine for routine PET studies in humans. Nucl Med Biol 20:939–944

Charbord P, L'Heritier C, Cukersztein W et al (1977) Radioiodine treatment in differentiated thyroid carcinomas. Treatment of first local recurrences and of bone and lung metastases. Ann Radiol 20:12–13

Chattopadhyay N, Mithal A, Brown EM (1996) The calcium-sensing receptor: a window into the physiology and pathophysiology of mineral ion metabolism. Endocr Rev 17:289–307

Chen CC, Premkumar A, Hill SC et al (1995) Tc-99m sestamibi imaging of a hyperfunctioning parathyroid autograft with Doppler ultrasound and MRI correlation. Clin Nucl Med 20:222–225

Chen CC, Holder LE, Scovill WA et al (1997) Comparison of parathyroid imaging with technetium-99m-pertechnetate/sestamibi subtraction, double-phase technetium-99m-sestamibi and technetium-99m-sestamibi SPECT. J Nucl Med 38:834–839

Chen JC, Kucharczyk W (1989) Hypothalamic-pituitary region: magnetic resonance imaging. Baillieres Clin Endocrinol Metab 3:73–87

Chiti A, Fanti S, Savelli G et al (1998) Comparison of somatostatin receptor imaging, computed tomography and ultrasound in the clinical management of neuroendocrine gastro-entero-pancreatic tumours. Eur J Nucl Med 25:1396–1403

Chiu ML, Kronauge JF, Piwnica-Worms D (1990) Effect of mitochondrial and plasma membrane potentials on accumulation of hexakis (2-methoxyisobutyl-isonitrile) technetium (I) in cultured mouse fibroblasts. J Nucl Med 31:1646–1653

Chung DC, Smith AP, Louis DN et al (1997) A novel pancreatic endocrine tumor suppressor gene locus on chromosome 3p with clinical prognostic implications. J Clin Invest 100:404–410

Chung J-K, Lee YJ, Jeong JM et al (1997) Clinical significance of hepatic visualization on iodine-131 whole-body scan in patients with thyroid carcinoma. J Nucl Med 38:1191–1195

Chung J-K, So Y, Lee JS et al (1999) Value of FDG-PET in papillary thyroid carcinoma with negative ^{131}I whole-body scan. J Nucl Med 40:986–992

Clarke SEM, Lazarus CR, Wraight P et al (1988) Pentavalent [^{99m}Tc]DMSA, [^{131}I]MIBG and [^{99m}Tc]MDP: an evaluation of three imaging techniques in patients with medullary carcinoma of the thyroid. J Nucl Med 29:33–38

Coakley AJ, Kettle AG, Wells CP et al (1989) ^{99m}Tc-sestamibi – a new agent for parathyroid imaging. Nucl Med Commun 10:791–794

Colao A, Lastoria S, Ferone D et al (1999) The pituitary uptake of (111)In-DTPA-D-Phe1-octreotide in the normal pituitary and in pituitary adenomas. J Endocrinol Invest 22:176–183

Colombo P, Paganelli G, Magnani P et al (1993) Immunoscintigraphy with anti-chromogranin A antibodies in patients with endocrine/neuroendocrine tumors. J Endocrinol Invest 16:841–843

Conn JW, Cohen EL, Herwig KR (1976) The dexamethasone-modified adrenal scintiscan in hyporeninemic aldosteronism (tumor versus hyperplasia). A comparison with adrenal venography and adrenal venous aldosterone. J Lab Clin Med 88:841–855

Costagliola S, Morgenthaler NG, Hoermann R et al (1999) Second generation assay for thyrotropin receptor antibodies has superior diagnostic sensitivity for Graves' disease. J Clin Endocrinol Metab 84:90–97

Cryns VL, Thor A, Xu H-J et al (1994) Loss of the retinoblastoma tumor-suppressor gene in parathyroid carcinoma. N Engl J Med 330:757–761

Cryns VL, Yi SM, Tahara H et al (1995) Frequent loss of chromosome arm Ip DNA in parathyroid adenomas. Genes Chromosomes Cancer 13:9–17

Czarnocka B, Ruf J, Ferrand M et al (1985) Purification of the human thyroid peroxidase and its identification as the microsomal antigen involved in autoimmune thyroid diseases. FEBS Lett 190:147–152

D'Adda T, Keller G, Bordi C et al (1999) Loss of heterozygosity in 11q13-14 regions in gastric neuroendocrine tumors not associated with multiple endocrine neoplasia type 1 syndrome. Lab Invest 79:671–677

Dai G, Levy O, Carrasco N (1996) Cloning and characterization of the thyroid iodine transporter. Nature 379:458–460

Degen L, Beglinger C (1999) The role of octreotide in the treatment of gastroenteropancreatic endocrine tumors. Digestion 60 [Suppl 2]:9–14

de Herder WW, Reijs AE, Kwekkeboom DJ, Hofland LJ, Nobels FR, Oei HY, Krenning EP, Lamberts SW (1996) In vivo imaging of pituitary tumours using a radiolabelled dopamine D2 receptor radioligand. Clin Endocrinol (Oxf) 45:755–767

De Herder WW, Reijs AEM, de Swart J et al (1999) Comparison of iodine-123 epipride and iodine-123 IBZM for dopamine D2 receptor imaging in clinically non-functioning pituitary macroadenomas and macroprolactinomas. Eur J Nucl Med 26:46–50

Del Rosario RB, Wieland DM (1995) Synthesis of [^{11}C]-(-)-a,a-dideutero-phenylephrine for in vivo kinetic isotope studies. J Labeled Compds Radiopharm (in press)

Delcore R, Friesen SR (1993) Embryologic concepts in the APUD system. Semin Surg Oncol 9:349–361

Delcore R, Friesen SR (1998) Other rare tumors of the endocrine pancreas. In: Howard J, Idezuki Y, Ihse I, Prinz R (eds) Surgical diseases of the pancreas, 3rd edn. Williams and Wilkins, Baltimore, pp 789–815

De Vijlder JJM, Vulsma T (2000) Hereditary metabolic disorders causing hypothyroidism. In: Braverman LE, Utiger RD (eds) Werner and Ingbar's The thyroid, 8th edn. Lippincott, Williams and Wilkins, Philadelphia, pp 733–742

Dimaline R, Struthers J (1996) Expression and regulation of a vesicular monoamine transporter in rat stomach: a putative histamine transporter. J Physiol (Lond) 490:249–256

Ding YS, Fowler JS, Dewey SL et al (1993) Comparison of high specific activity (-) and (+)-6-[^{18}F]fluoronorepinephrine and 6-[^{18}F] fluorodopamine in baboons: heart uptake, metabolism and the effect of desipramine. J Nucl Med 34:619–629

Dluhy RG, Lifton RP (1999) Glucocorticoid-remediable aldosteronism. J Clin Endocrinol Metab 84:4341–4344

Doherty GM, Doppman JL, Shawker TH et al (1991) Results of a prospective strategy to diagnose, localize, and resect insulinomas. Surgery 110:989–996

Donovan JL, Nesbit GM (1996) Distinction of masses involving the sella and suprasellar space: specificity of imaging features. Am J Roentgenol 167:597–603

Doppman JL, Miller DL (1991) Localization of parathyroid tumors in patients with asymptomatic hyperparathyroidism and no previous surgery. J Bone Miner Res 6 [Suppl 2]:S153–S158

Doppman JL, Miller DL, Chang R et al (1991) Insulinomas: localization with selective intraarterial injection of calcium. Radiology 178:237–241

Doppman JL, Chang R, Fraker DL et al (1995) Localization of insulinomas to regions of the pancreas by intra-arterial stimulation with calcium. Ann Intern Med 123:269–273

Ebrahimi SA, Wang EH, Wu A et al (1999) Deletion of chromosome 1 predicts prognosis in pancreatic endocrine tumors. Cancer Res 59:311–315

Eisenhofer G, Lenders JWM, Linehan WM et al (1999) Plasma normetanephrine and metanephrine for detecting pheochromocytoma in von Hippel-Lindau disease and multiple endocrine neoplasia type 2. N Engl J Med 340:1872–1879

Endo K, Shiomi K, Kasagi K et al (1984) Imaging of medullary thyroid cancer with ^{131}I-MIBG. Lancet 2:233

Eng C (1999) RET Proto-oncogene in the development of human cancer. J Clin Oncol 17:380–393

Eng PHK, Cardona GR, Fang S-L et al (1999) Escape from the acute Wolff-Chaikoff effect is associated with a decrease in thyroid sodium/iodide symporter messenger ribonucleic acid and protein. Endocrinology 140:3404–3410

Epelbaum J, Dournaud P, Fodor M et al (1994) The neurobiology of somatostatin. Crit Rev Neurobiol 8:25–44

Erickson JD, Schäfer MK, Bonner TI et al (1996) Distinct pharmacological properties and distribution in neurons and endocrine cells of two isoforms of the human vesicular monoamine transporter. Proc Natl Acad Sci USA 93:5166–5171

Eriksson B, Öberg K (1995) PPomas and nonfunctioning endocrine pancreatic tumors: clinical presentation, diagnosis, and advances in management. In: Mignon M, Jensen RT (eds) Endocrine tumors of the pancreas. Karger, Basel, pp 208–222 (Frontiers in gastrointestinal research, vol 23)

Eriksson B, Öberg K, Stridsberg M (2000) Tumor markers in neuroendocrine tumors. Digestion 62 [Suppl 1]:33–38

Everett LA, Glaser B, Beck JC et al (1997) Pendred syndrome is caused by mutations in a putative sulphate transporter gene (PDS). Nat Genet 17:411–422

Evers BM, Rady PL, Sandoval K et al (1994) Gastrinomas demonstrate amplification of the HER-2/neu proto-oncogene. Ann Surg 219:596–601

Fagin JA, Tang S-H, Zeki K et al (1996) Reexpression of thyroid peroxidase in a derivative of an undifferentiated thyroid carcinoma cell line by introduction of wild-type p53. Cancer Res 56:765–771

Feine U, Lietzenmayer R, Hanke J-P et al (1996) Fluorine-18-FDG and iodine-131-iodide uptake in thyroid cancer. J Nucl Med 37:1468–1472

Feldkamp MM, Gutmann DH, Guha A (1998) Neurofibromatosis type 1: piecing the puzzle together. Can J Neurol Sci 25:181–191

Ferlin G, Borsato N, Camerani M et al (1983) New perspectives in localizing enlarged parathyroids by technetium-thallium subtraction scan. J Nucl Med 24:438–441

Ferone D, Lastoria S, Colao A et al (1998) Correlation of scintigraphic results using 123I-methoxybenzamide with hormone levels and tumor size response to quinagolide in patients with pituitary adenomas. J Clin Endocrinol Metab 83:248–252

Fig LM, Gross MD, Shapiro B et al (1988) Adrenal localization in the adrenocorticotropic hormone-independent Cushing's syndrome. Ann Intern Med 109:547–553

Fine EJ (1987) Parathyroid imaging: its current status and future role. Semin Nucl Med 17:350–359

Fjeld JG, Erichsen K, Pfeffer PF et al (1997) Technetium-99m-tetrofosmin for parathyroid scintigraphy: a comparison with sestamibi. J Nucl Med 38:831–834

Fraker DL, Alexander HR (1995) The surgical approach to endocrine tumors of the pancreas. Semin Gastrointest Dis 6:102–113

Francavilla TL, Miletich RS, DeMichele D et al (1991) Positron emission tomography of pituitary macroadenomas:

hormone production and effects of therapies. Neurosurgery 28:826–833

Freda PU, Post KD (1999a) Differential diagnosis of sellar masses. Endocrinol Metab Clin North Am 28:81–117

Freda PU, Wardlaw SL (1999b) Clinical review 110: diagnosis and treatment of pituitary tumors. J Clin Endocrinol Metab 84:3859–3866

Friedman E, Sakaguchi K, Bale AE et al (1989) Clonality of parathyroid tumors in familial multiple endocrine neoplasia type I. N Engl J Med 321:213–218

Frilling A, Malago M, Martin H et al (1998) Use of somatostatin receptor scintigraphy to image extrahepatic metastases of neuroendocrine tumors. Surgery 124:1000–1004

Fronistas O, Stavraka-Kakavaki A, Giougi A et al (1992) Evaluation of parathyroid tissue transplants by Tl-201 scintigraphy. Clin Nucl Med 17:954–957

Fujiwara H, Tatsumi K, Miki K et al (1997) Congenital hypothyroidism caused by a mutation in the Na+/I– symporter. Nat Genet 16:124–125

Fukunaga M, Morita R, Yonekura Y et al (1979) Accumulation of ^{201}Tl-chloride in a parathyroid adenoma. Clin Nucl Med 4:229–230

Gibril F, Reynolds JC, Doppman JL et al (1996) Somatostatin receptor scintigraphy: its sensitivity compared with that of other imaging methods in detecting primary and metastatic gastrinomas, a prospective study. Ann Intern Med 125:26–34

Gibril F, Doppman JL, Reynolds JC et al (1998) Bone metastases in patients with gastrinomas: a prospective study of bone scanning, somatostatin receptor scanning, and magnetic resonance image in their detection, frequency, location and effect of their detection on management. J Clin Oncol 16:1040–1053

Gibril F, Reynolds JC, Chen CC et al (1999) Specificity of somatostatin receptor scintigraphy: a prospective study and effects of false positive localizations on management in patients with gastrinomas. J Nucl Med 40:539–553

Gibril F, Reynolds JC, Lubensky IA et al (2000) Ability of somatostatin receptor scintigraphy to identify patients with gastric carcinoids: a prospective study. J Nucl Med 41:1646–1656

Gimlette T, Brownless SM, Taylor WH et al (1986) Limits to parathyroid imaging with thallium-201 confirmed by tissue uptake and phantom studies. J Nucl Med 27:1262–1265

Goebel SU, Heppner C, Burns AL et al (1999) Genotype/phenotype correlation of multiple endocrine neoplasia type 1 gene mutations in sporadic gastrinomas. J Clin Endocrinol Metab 85:116–123

Gogusev J, Duchambon P, Hory B et al (1997) Depressed expression of calcium receptor in parathyroid gland tissue of patients with hyperparathyroidism. Kidney Int 51:328–336

Goldstein DS, Holmes C, Cannon RO III et al (1997) Sympathetic cardioneuropathy in dysautonomias. N Engl J Med 3336:696–702

Goris ML, Basso LV, Keeling C (1991) Parathyroid imaging. J Nucl Med 32:887–889

Gortz B, Roth J, Krahenmann A et al (1999) Mutations and allelic deletions of the MEN1 gene are associated with a subset of sporadic endocrine pancreatic and neuroendocrine tumors and not restricted to foregut neoplasms. Am J Pathol 154:429–436

Grama D, Skogseid B, Wilander E et al (1992) Pancreatic tumors in multiple endocrine neoplasia type 1: clinical presentation and surgical treatment. World J Surg 16:611–618

Granberg P-O, Johansson G, Lindvall N et al (1982) Reoperation for primary hyperparathyroidism. Am J Surg 143:296–300

Grant CS (1996) Insulinoma. Baillieres Clin Gastroenterol 10:645–671

Greenspan BS, Brown ML, Dillehay GL et al (1998) Procedure guideline for parathyroid scintigraphy. J Nucl Med 39:1111–1114

Gross MD, Shapiro B, Bouffard JA et al (1988) Distinguishing benign from malignant euadrenal masses. Ann Intern Med 109:613–618

Grunwald F, Menzel C, Bender H et al (1997) Comparison of ^{18}FDG-PET with ^{131}iodine and ^{99m}Tc-sestamibi scintigraphy in differentiated thyroid cancer. Thyroid 7:327–335

Guillausseau PJ, Guillausseau-Scholer C (1995) Glucagonomas: clinical presentation, diagnosis, and advances in management. In: Mignon M, Jensen RT (eds) Endocrine tumors of the pancreas. Karger, Basel, pp 183–193 (Frontiers in gastrointestinal research, vol 23)

Guru SC, Goldsmith PK, Burns AL et al (1998) Menin, the product of the MEN1 gene, is a nuclear protein. Proc Natl Acad Sci USA 95:1630–1634

Gutmann DH, Aylsworth A, Carey JC et al (1997) The diagnostic evaluation and multidisciplinary management of neurofibromatosis 1 and neurofibromatosis 2. JAMA 278:51–57

Hagen EC, Houben GM, Nikkels RE et al (1992) Exocrine pancreatic insufficiency and pancreatic fibrosis due to duodenal somatostatinoma in a patient with neurofibromatosis. Pancreas 7:98–104

Hamada N, Grimm C, Mori H et al (1985) Identification of a thyroid microsomal antigen by Western blot and immunoprecipitation. J Clin Endocrinol Metab 61:120–128

Haugen BR, Pacini F, Reiners C et al (1999) A comparison of recombinant human thyrotropin and thyroid hormone withdrawal for the detection of thyroid remnant or cancer. J Clin Endocrinol Metab 84:3877–3885

Hays M, Wesselossky B (1973) Simultaneous measurement of thyroid trapping (^{99m}TcO$_4$–) and binding (^{131}I): clinical and experimental studies in man. J Nucl Med 14:785–792

Hessman O, Lindberg D, Skogseid B et al (1998) Mutation of the multiple endocrine neoplasia type 1 gene in non-familial malignant tumors of the endocrine pancreas. Cancer Res 58:377–379

Hindié E, Melliére D, Perlemuter L et al (1997) Primary hyperparathyroidism: higher success rate of first surgery after preoperative Tc-99m sestamibi I-123 subtraction scanning. Radiology 204:221–228

Hiorns MP, Reznek RH (2000) Ultrasound, CT and MRI appearances of pancreatic neuroendocrine tumors and carcinoids. In: de Herder WW (ed) Functional and morphological imaging of the endocrine system. Kluwer, Boston, pp 215–234

Hirschowitz BI (1997) Zollinger-Ellison syndrome: pathogenesis, diagnosis, and management. Am J Gastroenterol 92 [Suppl 4]:44S–48S

Hirschowitz BI, Griffith J, Pellegrin D et al (1992) Rapid regression of enterochromaffinlike cell gastric carcinoids in pernicious anemia after antrectomy. Gastroenterology 102:1409–1418

Hishinuma A, Kasai K, Masawa N et al (1998) Missense mutation (C1263R) in the thyroglobulin gene causes congenital goiter with mild hypothyroidism by impaired intracellular transport. Endocr J 45:315–327

Hocker M, Wiedenmann B (1999) Therapeutic and diagnostic implications of the somatostatin system in gastroenteropancreatic neuroendocrine tumour disease. Ital J Gastroenterol Hepatol 31 (Suppl 2):S139–142

Hofstra RMW, Fattosuro O, Quadro L et al (1997) A novel point mutation in the intracellular domain of the ret protooncogene in a family with medullary thyroid carcinoma. J Clin Endocrinol Metab 82:4176–4178

Holzapfel HP, Fuhrer D, Wonerow P et al (1997) Identification of constitutively activating somatic thyrotropin receptor mutations in a subset of toxic multinodular goiters. J Clin Endocrinol Metab 82: 4229–4233

Hutchinson RJ, Sisson JC, Shapiro B et al (1992) 131-I metaiodobenzylguanidine treatment in patients with refractory advanced neuroblastoma. Am J Clin Oncol 15:226–232

Iida Y, Hidaka A, Hatabu H et al (1991) Follow-up study of postoperative patients with thyroid cancer by thallium-201 scintigraphy and serum thyroglobulin measurement. J Nucl Med 32:2098–2100

Imamura M, Takahashi K, Adachi H et al (1987) Usefulness of selective arterial secretin injection test for localization of gastrinoma in the Zollinger-Ellison syndrome. Ann Surg 205:230–239

Ishibashi M, Nishida H, Kumabe T et al (1995) Tc-99m tetrofosmin. A new diagnostic tracer for parathyroid imaging. Clin Nucl Med 20:902–905

Ishibashi M, Nishida H, Strauss HW et al (1997) Localization of parathyroid glands using technetium-99m-tetrofosmin imaging. J Nucl Med 38:706–711

Ishibashi M, Nishida H, Hiromatsu Y et al (1998a) Comparison of technetium-99m-MIBI, technetium-99m-tetrofosmin, ultrasound and MRI for localization of abnormal parathyroid glands. J Nucl Med 39:320–324

Ishibashi M, Nishida H, Okuda S et al (1998b) Localization of parathyroid glands in hemodialysis patients using Tc-99m sestamibi imaging. Nephron 78:48–53

Ito T, Seyama T, Mizuno T et al (1992) Unique association of p53 mutations with undifferentiated but not with differentiated carcinomas of the thyroid gland. Cancer Res 52:1369–1371

Ito Y, Muranaka A, Harada T et al (1978) Experimental study on tumor affinity of ^{201}Tl-chloride. Eur J Nucl Med 3:81–86

Jacques S Jr, Tobes MC, Sisson JC et al (1984) Comparison of the sodium dependency of uptake of meta-iodobenzyluanidine and norepinephrine into culture bovine adrenomedullary cells. Mol Pharmacol 26:539–546

Jaïs P, Terris B, Ruszniewsk P et al (1997) Somatostatin receptor subtype gene expression in human endocrine gastroentero-pancreatic tumours. Eur J Clin Invest 27:639–644

Janson ET, Stridsberg M, Gobl A et al (1998) Determination of somatostatin receptor subtype 2 in carcinoid tumors by immunohistochemical investigation with somatostatin receptor subtype 2 antibodies. Cancer Res 58:2375–2378

Jensen RT (1996) Gastrinoma. Baillieres Clin Gastroenterol 10:603–643

Jensen RT (1999) Pancreatic endocrine tumors: recent advances. Ann Oncol 10 [Suppl 4]:S170–S176

Jensen RT (2000) Carcinoid and pancreatic endocrine tumors: recent advances in molecular pathogenesis, localization, and treatment. Curr Opin Oncol 12:368–377

Jensen RT, Gibril F (1999) Somatostatin receptor scintigraphy in gastrinomas. Ital J Gastroenterol Hepatol 31 [Suppl 2]:S179–S185

Jong M de, Rolleman EJ, Bernard BF et al (1996) Inhibition of renal uptake of indium-111-DTPA-octreotide in vivo. J Nucl Med 37:1388–1392

Juranka PF, Zastawny PL, Ling V (1989) P-glycoprotein: multidrug-resistance and a superfamily of membrane-associated transport proteins. FASEB J 3:2583–2592

Kaelin WG Jr, Maher ER (1998) The VHL tumour-suppressor gene paradigm. Trends Genet 14:423–426

Kang YS, Rosen K, Clark OH et al (1993) Localization of abnormal parathyroid glands of the mediastinum with MR imaging. Radiology 189:137–141

Kasagi K, Konishi J, Arai K et al (1986) A sensitive and practical assay for thyroid stimulating antibodies using crude immunoglobulin fractions precipitated with polyethylene glycol. J Clin Endocrinol Metab 62:855–862

Kasagi K, Miyamoto S, Endo K et al (1993) Increased uptake of iodine-131 in metastases of differentiated thyroid carcinoma associated with less severe hypothyroidism following total thyroidectomy. Cancer 72:1983–1990

Kasagi K, Kousaka T, Higuchi K et al (1996) Clinical significance of measurements of antithyroid antibodies in the diagnosis of Hashimoto's thyroiditis: comparison with histological findings. Thyroid 5:445–450

Kazerooni EA, Sisson JC, Shapiro B et al (1990) Diagnostic accuracy and pitfalls of [iodine-131]6-Beta-iodomethyl-19-norcholesterol (NP-59) imaging. J Nucl Med 31:526–534

Keeling CA, Basso LV (1988) Iodine-131 MIBG uptake in metastatic medullary carcinoma of the thyroid. Clin Nucl Med 13:260–263

Kerström G, Malmaeus J, Bergström R (1984) Surgical anatomy of human parathyroid glands. Surgery 95:14–21

Kilgore EJ, Teigen EL, Cowan RJ (1996) Imaging of transplanted parathyroid tissue in a patient with recurrent hyperparathyroidism. Clin Nucl Med 21:383–386

Kimmig B, Brandeis WE, Eisenhut M et al (1984) Scintigraphy of a Neuroblastoma with I-131 Meta-iodobenzylguanidine. J Nucl Med 25:773–775

King AD, Ko GT, Yeung VT et al (1998) Dual phase spiral CT in the detection of small insulinomas of the pancreas. Br J Radiol 71:20–23

Kirkwood KS, Debas HT (1995) Neuroendocrine tumors: common presentations of uncommon disease. Compr Ther 21:719–725

Klöppel G, Solcia E, Capella C et al (1999) Classification of neuroendocrine tumours. Ital J Gastroenterol Hepatol 31 [Suppl 2]:S111–S116

Kohn LD, Ban T, Okajima F et al (1995) Cloning and regulation of glycoprotein hormone receptor genes. In: Weintraub BD (ed) Molecular endocrinology: basic concepts and clinical correlations. Raven, New York, pp 133–153

Kohno Y, Hiyama Y, Shimojo N et al (1986) Autoantibodies to thyroid peroxidase in patients with chronic thyroiditis: effect of antibody binding on enzyme activities. Clin Exp Immunol 65:534–541

Konishi J, Iida Y, Kasagi K et al (1985) Primary myxedema with thyrotropin-binding inhibitor immunoglobulins: Clinical and laboratory findings in 15 patients. Ann Intern Med 103:26–31

Korogi Y, Takahashi M (1995) Current concepts of imaging in patients with pituitary/hypothalamic dysfunction. Semin Ultrasound CT MR 16:270–278

Kosugi S, Sato Y, Matsuda A et al (1998) High prevalence of T354P sodium/iodide symporter gene mutation in Japanese patients with iodide transport defect who have heterogeneous clinical pictures. J Clin Endocrinol Metab 83:4123–4129

Krenning EP, Bakker WH, Kooij PP et al (1992) Somatostatin receptor scintigraphy with indium-111-DTPA-D-Phe-1-octreotide in man: metabolism, dosimetry and comparison with iodine-123-Tyr-3-octreotide. J Nucl Med 33:652–658

Krenning EP, Kwekkeboom DJ, Bakker WH et al (1993a) Somatostatin receptor scintigraphy with [111In-DTPA-D-Phe1]- and [123I-Tyr3]-octreotide: the Rotterdam experience with more than 1000 patients. Eur J Nucl Med 20:716–731

Krenning EP, Kwekkeboom DJ, Reubi JC et al (1993b) 111In-octreotide scintigraphy in oncology. Digestion 54:84–87

Krenning EP, Valkema R, Pauwels S et al (1999) Radiolabeled somatostatin analogue(s): peptide receptor scintigraphy and radionuclide therapy. In: Mignon M, Colombel JF (eds) Recent advances in the pathophysiology and management of inflammatory bowel diseases and digestive endocrine tumors. Libbey Eurotext, Paris, pp 220–228

Krieger M (1999) Charting the fate of the 'good cholesterol': Identification and characterization of the high-density lipoprotein receptor SR-BI. Annu Rev Biochem 68:523–558

Krone N, Braun A, Roscher AA et al (2000) Predicting phenotype in steroid 21-hydroxylase deficiency? Comprehensive genotyping in 15 unrelated, well defined patients from southern Germany. J Clin Endocrinol Metab 85:1059–1065

Krubsack AJ, Wilson SD, Lawson TL et al (1989) Prospective comparison of radionuclide, computed tomographic, sonographic, and magnetic resonance localization of parathyroid tumors. Surgery 106:639–646

Kwekkeboom DJ, de Herder WW, Krenning EP (1999) Receptor imaging in the diagnosis and treatment of pituitary tumors. J Endocrinol Invest 22:80–88

Kwekkeboom D, Krenning EP, De Jong M (2000a) Peptide receptor imaging and therapy. J Nucl Med 41:1704–1713

Kwekkeboom DJ, Herder WW de, Krenning EP (2000b) Scintigraphy of pancreatic neuroendocrine tumors and carcinoids. In: de Herder WW (ed) Functional and morphological imaging of the endocrine system. Kluwer, Boston, pp 235–249

Lacroix A, Bolte E, Tremblay J et al (1992) Gastric inhibitory polypeptide-dependent cortisol hypersecretion – a new cause of Cushing's syndrome. N Engl J Med 327:974–980

Langley K (1994) The neuroendocrine concept today. Ann NY Acad Sci 733:1–17

Lauffer JM, Zhang T, Modlin IM (1999) Review article: current status of gastrointestinal carcinoids. Aliment Pharmacol Ther 13:271–287

Lauriero F, Pierangeli E, Rubini G et al (1998) Pituitary adenomas: the role of 111In-DTPA-octreotide SPET in the detection of minimal post-surgical residues. Nucl Med Commun 19:1127–1134

Lebtahi R, Cadiot G, Sarda L et al (1997) Clinical impact of somatostatin receptor scintigraphy in the management of patients with neuroendocrine gastroenteropancreatic tumors. J Nucl Med 38:853–858

Lee VS, Spritzer CE, Coleman RE et al (1996) The complementary roles of fast spin-echo MR imaging and double-phase ^{99m}Tc-sestamibi scintigraphy for localization of hyperfunctioning parathyroid glands. AJR 167:1555–1562

Legovini P, De Menis E, Billeci D et al (1997) 111Indium-pentreotide pituitary scintigraphy and hormonal responses to octreotide in acromegalic patients. J Endocrinol Invest 20:424–428

Lemoine NR, Mayall ES, Wyllie FS et al (1988) Activated ras oncogenes in human thyroid cancers. Cancer Res 48:4459–4463

Libutti SK, Choyke PL, Bartlett DL et al (1998) Pancreatic neuroendocrine tumors associated with von Hippel Lindau disease: diagnostic and management recommendations. Surgery 124:1153–1159

Liechty RD, Weil R (1992) Parathyroid anatomy in hyperplasia. Arch Surg 127:813–816

Lind P, Gallowitsch HJ, Langsteger W et al (1997) Technetium-99m-tetrofosmin whole-body scintigraphy in the follow-up of differentiated thyroid carcinoma. J Nucl Med 38:348–352

Lips CJM, Landsvater RM, Hopener JWM et al (1994) Clinical screening as compared with DNA analysis in families with multiple endocrine neoplasia type 2A. N Engl J Med 331:828–835

Liu J, Voutilainen R, Heikkila P et al (1997) Ribonucleic acid expression of the CLA-1 gene, a human homologue to mouse high density lipoprotein receptor SR-BI, in human adrenal tumors and cultured adrenal cells. J Clin Endocrinol Metab 82:2522–2527

Losa M, Magnani P, Mortini P et al (1997) Indium-111 pentetreotide single-photon emission tomography in patients with TSH-secreting pituitary adenomas: correlation with the effect of a single administration of octreotide on serum TSH levels. Eur J Nucl Med 24:728–731

Lucignani G, Losa M, Moresco RM et al (1997) Differentiation of clinically non-functioning pituitary adenomas from meningiomas and craniopharyngiomas by positron emission tomography with [18F]fluoro-ethyl-spiperone. Eur J Nucl Med 24:1149–1155

Lugtenburg PJ, Krenning EP, Valkema R et al (2001) Somatostatin receptor scintigraphy useful in stage I-II Hodgkin's disease: more extended disease identified. Br J Haematol 112:936–944

Lynn MD, Shapiro B, Sisson JC et al (1985) Pheochromocytoma and the normal adrenal medulla: improved visualization with I-123 MIBG scintigraphy. Radiology 156:789–792

Maini CL, Sciuto R, Tofani A et al (1995) Somatostatin receptor imaging in CNS tumours using 111In-octreotide. Nucl Med Commun 16:756–766

Malchoff CD, Rosa J, DeBold CR et al (1989) Adrenocorticotropin-independent bilateral macronodular adrenal hyperplasia: an unusual cause of Cushing's syndrome. J Clin Endocrinol Metab 68:855

Mansi L, Rambaldi PF, Marino G et al (1996) Kinetics of Tc-99m sestamibi and Tc-99m tetrofosmin in a case of parathyroid adenoma. Clin Nucl Med 21:700–703

Mao C, Shah A, Hanson DJ et al (1995) Von Recklinghausen's disease associated with duodenal somatostatinoma: contrast of duodenal versus pancreatic somatostatinomas. J Surg Oncol 59:67–73

Martin WH, Patton JA, Delbeke D et al (2000) Improved localization of endocrine sites using an integrated CT-SPECT fused imaging. J Nucl Med 41 [Suppl] (abstract):9P

Matsuda A Kosugi S (1997) A homozygous missense mutation of the sodium/iodide symporter gene causing iodide transport defect. J Clin Endocrinol Metab 82:3966–3971

Matthay KK, deSantes K, Hasegawa B et al (1998) Phase I dose escalation of ^{131}I-metaiodobenzylguanidine with autologous bone marrow support in refractory neuroblastoma. J Clin Oncol 16:229–236

Matuchansky C, Rambaud JC (1995) VIPomas and endocrine cholera: clinical presentation, diagnosis, and advances in management. In: Mignon M, Jensen RT (eds) Endocrine tumors of the pancreas. Karger, Basel, pp 166–182 (Frontiers in gastrointestinal research, vol 23)

Mazzeo S, Caramella D, Lencioni R et al (1996) Comparison among sonography, double-tracer subtraction scintigraphy, and double-phase scintigraphy in the detection of parathyroid lesions. AJR 166:1465–1470

McBiles M, Lambert AT, Cote MG et al (1995) Sestamibi parathyroid imaging. Semin Nucl Med 25:221–234

McCown JS, Gordon L, Uflacker RP (1995) In-111 pentetreotide. Superior imaging agent for gastrinomas. Clin Nucl Med 10:896–898

McDermott VG, Fernandez RJM, Meakem TJ III et al (1996) Preoperative MR imaging in hyperparathyroidism: results and factors affecting parathyroid detection. AJR 166:705–710

McDougall IR (1990) The importance of obtaining thyroid uptake measurement in patients with hyperthyroidism. Nucl Med Commun 11:73–76

McIntyre RC, Kumpe DA, Liechty RD (1994) Reexploration and angiographic ablation for hyperparathyroidism. Arch Surg 129:499–505

McLachlan SM, Rapoport B (1992) The molecular biology of thyroid peroxidase: cloning, expression and role as autoantigen in autoimmune thyroid disease. Endocrine Rev 13:192–206

Melon P, Luxen A, Hamoir E et al (1995) Fluorine-18-fluorodeoxyglucose positron emission tomography for preoperative parathyroid imaging in primary hyperparathyroidism. Eur J Nucl Med 22:556–558

Merke DP, Tajima T, Chhabra A et al (1998) Novel CYP11B1 mutations in congenital adrenal hyperplasia due to steroid 11β-hydroxylase deficiency. J Clin Endocrinol Metab 83:270–273

Merlet P, Pouillart F, Subois-Rande JL et al (1999) Sympathetic nerve alterations assessed with ^{123}I-MIBG in the failing human heart. J Nucl Med 40:224–231

Meyerhof W (1998) The elucidation of somatostatin receptor functions: a current view. Rev Physiol 133:55–108

Mignon M (1999) Digestive endocrine tumors: diagnosis and staging. In: Lamberts SWJ (ed) Octreotide: the next decade. BioScientifica, Bristol, pp 133–147

Mignon M, Cadiot G (1999) Natural history of gastrinoma: lessons from the past. Ital J Gastroenterol Hepatol 31 [Suppl 2]:S98–S103

Mignon M, Jaïs P, Cadiot G et al (1995) Clinical features and advances in biological diagnostic criteria for Zollinger-Ellison syndrome. In: Mignon M, Jensen RT (eds) Endocrine tumors of the pancreas. Karger, Basel, pp 223–239 (Frontiers in gastrointestinal research, vol 23)

Miller DL (1991) Pre-operative localization and interventional treatment of parathyroid tumors: when and how? World J Surg 15:706–715

Miyamoto S, Kasagi K, Alam MS et al (1997) Evaluation of technetium-99m MIBI scintigraphy in metastatic differentiated thyroid carcinoma. J Nucl Med 38:352–356

Miyauchi A, Endo K, Ohta H et al (1986) 99mTc(V) dimercaptosuccinic acid scintigraphy for medullary thyroid carcinoma. World J Surg 10:640–645

Modlin IM, Tang LH (1997) Approaches to the diagnosis of gut neuroendocrine tumors: the last word (today). Gastroenterology 112:583–590

Modlin IM, Esterline W, Kim H et al (1991) Enterochromaffin-like cells and gastric argyrophil carcinoidosis. Acta Oncol 30:493–498

Modlin IM, Cornelius E Lawton GP (1995) Use of an isotopic somatostatic receptor probe to image gut endocrine tumors. Arch Surg 130:367–374

Modlin IM, Sandor A, Tang LH et al (1997) A 40-year analysis of 265 gastric carcinoids. Am J Gastroenterol 92:633–638

Moore NR, Rogers CE, Britton BJ (1995) Magnetic resonance imaging of endocrine tumours of the pancreas. Br J Radiol 68:341–347

Morgenthaler NG, Pampel I, Aust G et al (1998) Application of a bioassay with CHO cells for the routine detection of stimulating and blocking autoantibodies to the TSH-receptor. Horm Metab Res 30:162–168

Motokura T, Bloom T, Kim HG et al (1991) A novel cyclin encoded by a bcl1-linked candidate oncogene. Nature 350:512–515

Moyse E, Le Dafniet M, Epelbaum J et al (1985) Somatostatin receptors in human growth hormone and prolactin-secreting pituitary adenomas. J Clin Endocrinol Metab 61:98–103

Muhr C, Bergstrom M (1991) Positron emission tomography applied in the study of pituitary adenomas. J Endocrinol Invest 14:509–528

Muhr C, Bergstrom M, Lundberg PO et al (1986a) Dopamine receptors in pituitary adenomas: PET visualization with 11C-N- methylspiperone. J Comput Assist Tomogr 10:175–180

Muhr C, Bergstrom M, Lundberg PO et al (1986b) In vivo measurement of dopamine receptors in pituitary adenomas using positron emission tomography. Acta Radiol [Suppl] 369:406–408

Mulligan LM, Ponder BAJ (1995) Genetic basis of endocrine disease. Genetic basis of endocrine disease: multiple endocrine neoplasia type 2. J Clin Endocrinol Metab 80:1989–1995

Muscarella P, Melvin WS, Fisher WE et al (1998) Genetic alterations in gastrinomas and nonfunctioning pancreatic neuroendocrine tumors: an analysis of p16/MTS1 tumor suppressor gene inactivation. Cancer Res 58:237–240

Nagayama Y, Kaufman KD, Seto P et al (1989) Molecular cloning, sequence and functional expression of the cDNA for the human thyrotropin receptor. Biochem Biophys Res Commun 165:1184–1190

Naidich MJ, Russell EJ (1999) Current approaches to imaging of the sellar region and pituitary. Endocrinol Metab Clin North Am 28:45–79

Nakajo M, Shapiro B, Copp J et al (1983) The normal and abnormal distribution of the adrenomedullary imaging agent m-[I-131]iodobenzylguanidine (I-131 MIBG) in man. J Nucl Med 24:678–682

Nakamoto Y, Saga T, Misaki T et al (2000) Establishment and characterization of a breast cancer cell line expressing Na+/I– symporters for radioiodide concentrator gene therapy. J Nucl Med 41:1898–1904

Narimatsu M, Nagayama Y, Akino K et al (1998) Therapeutic usefulness of wild-type p53 gene introduction in a p53-null anaplastic thyroid carcinoma cell line. J Clin Endocrinol Metab 83:3668–3672

Naveh-Many T, Rahamimov R, Livni N et al (1995) Parathyroid cell proliferation in normal and chronic renal failure rats. The effects of calcium, phosphate, and vitamin D. J Clin Invest 96:1786–1793

Nemeth EF, Scarpa A (1987) Rapid mobilization of cellular Ca^{2+} in bovine parathyroid cells evoked by extracellular divalent cations. Evidence for a cell surface calcium receptor. J Biol Chem 262:5188–5196

Neumann DR, Esselstyn CB Jr, MacIntyre WJ et al (1994) Primary hyperparathyroidism: preoperative parathyroid imaging with regional body FDG PET. Radiology 192:509–512

Neumann DR, Esselstyn CB, MacIntyre WJ et al (1996a) Comparison of FDG-PET and sestamibi-SPECT in primary hyperparathyroidism. J Nucl Med 37:1809–1815

Neumann DR, Esselstyn CB, Kim EY (1996b) Recurrent postoperative parathyroid carcinoma: FDG-PET and sestamibi-SPECT findings. J Nucl Med 37:2000–2001

Neumann HPH, Bausch B, McWhinney SR et al (2002) Germ-line mutations in non-syndromic pheochromocytoma. N Engl J Med 346:1549–1566

Nordenstrom A, Thilen A, Hagenfeldt L et al (1999) Genotyping is a valuable diagnostic complement to neonatal screening for congenital adrenal hyperplasia due to steroid 21-hydroxylase deficiency. J Clin Endocrinol Metab 84: 1505–1509

Norton JA (1999) Intraoperative methods to stage and localize pancreatic and duodenal tumors. Ann Oncol 10 [Suppl 4]:S182–S184

Nygren P, Gylfe E, Larsson R et al (1988) Modulation of the Ca^{2+}-sensing function of parathyroid cells in vitro and in hyperparathyroidism. Biochim Biophys Acta 968:253–260

Öberg K (1999) Neuroendocrine gastrointestinal tumors – a condensed overview of diagnosis and treatment. Ann Oncol 10 [Suppl 2]:S3–S8

Öberg K (2000) Interferon in the management of neuroendocrine GEP-tumors. Digestion 62 [Suppl 1]:92–97

O'Doherty MJ, Kettle AG, Wells P et al (1992) Parathyroid imaging with technetium-99m-sestamibi: preoperative localization and tissue uptake studies. J Nucl Med 33:313–318

Ohta H, Yamamoto K, Endo K et al (1984) A new imaging agent for medullary carcinoma of the thyroid. J Nucl Med 25:323–325

Okamoto Y, Hamada N, Saito H et al (1989) Thyroid peroxidase activity-inhibiting immunoglobulins in patients with autoimmune thyroid disease. J Clin Endocrinol Metab 68:730–734

Oppizzi G, Cozzi R, Dallabonzana D et al (1998) Scintigraphic imaging of pituitary adenomas: an in vivo evaluation of somatostatin receptors. J Endocrinol Invest 21:512–519

Orbuch M, Doppman JL, Jensen RT (1995) Localization of pancreatic endocrine tumors. Semin Gastrointest Dis 6:90–101

Orlefors H, Sundin A, Ahlstrom H et al (1998) Positron emission tomography with 5-hydroxytryptophan in neuroendocrine tumors. J Clin Oncol 16:2534–2541

O'Shea D, Rohrer-Theus AW, Lynn JA et al (1996) Localization of insulinomas by selective intraarterial calcium injection. J Clin Endocrinol Metab 81:1623–1627

Otsuka FL, Cance WG, Dilley WG et al (1988) A potential new radiopharmaceutical for parathyroid imaging: radiolabeled parathyroid-specific monoclonal antibody II. Comparison of ^{125}I- and ^{111}In-labeled antibodies. Int J Radiat Appl Instrum 15:305–311

Owen NJ, Sohaib SA, Peppercorn PD, Monson JP, Grossman AB, Besser GM, Reznek RH (2001) MRI of pancreatic neuroendocrine tumours. Br J Radiol 74:968–973

Ozata M, Suzuki S, Miyamoto T et al (1994) Serum thyroglobulin in the follow-up of patients with treated differentiated thyroid cancer. J Clin Endocrinol Metab 79:98–105

Pacini F, Lippi F, Formica N et al (1987) Therapeutic doses of iodine-131 reveal undiagnosed metastases in thyroid cancer patients with undetectable serum thyroglobulin levels. J Nucl Med 28:1888–1891

Park HM, Park Y-H, Zhou X-A (1997) Detection of thyroid remnant/metastasis without stunning: an ongoing dilemma. Thyroid 7:277–280

Park HM, Perkins OW, Edmondson JM (1994) Influence of diagnostic radioiodine on the uptake of ablative dose of iodine-131. Thyroid 4:49–54

Park SK, O'Dorisio MS, O'Dorisio TM (1996) Vasoactive intestinal polypeptide-secreting tumours: biology and therapy. Baillieres Clin Gastroenterol 10:673–695

Parma J, Duprez L, Van Sande J et al (1993) Somatic mutations in the thyrotropin receptor gene cause hyperfunctioning thyroid adenomas [see comments]. Nature 365:649–651

Parmentier M, Libert F, Maenhaut C et al (1989) Molecular cloning of the thyrotropin receptor. Science 246:1620–1622

Pasquali C, Rubello D, Sperti C et al (1998) Neuroendocrine tumor imaging: can 18F-fluorodeoxyglucose positron emission tomography detect tumors with poor prognosis and aggressive behavior? World J Surg 22:588–592

Patel YC, Greenwood M, Panetta R et al (1996) Molecular biology of somatostatin receptor subtypes. Metabolism 45 [Suppl 1]:31–38

Pelley RJ, Bukowski RM (1997) Recent advances in diagnosis and therapy of neuroendocrine tumors of the gastrointestinal tract. Curr Opin Oncol 9:68–74

Phlipponneau M, Nocaudie M, Epelbaum J, De Keyzer Y, Lalau JD, Marchandise X, Bertagna X (1994) Somatostatin analogs for the localization and preoperative treatment of an adrenocorticotropin-secreting bronchial carcinoid tumor. J Clin Endocrinol Metab 78:20–24

Piga M, Bolasco P, Satta L et al (1996) Double-phase parathyroid technetium-99m-MIBI scintigraphy to identify functional autonomy in secondary hyperparathyroidism. J Nucl Med 37:565–569

Pirker W, Brucke T, Riedl M et al (1994) Iodine-123-IBZM-SPECT: studies in 15 patients with pituitary tumors. J Neural Transm Gen Sect 97:235–244

Plockinger U, Bader M, Hopfenmuller W et al (1997) Results of somatostatin receptor scintigraphy do not predict pituitary tumor volume- and hormone-response to octreotide therapy and do not correlate with tumor histology. Eur J Endocrinol 136:369–376

Pocotte SL, Ehrenstein G, Fitzpatrick LA (1991) Regulation of parathyroid hormone secretion. Endocr Rev 12:291–301

Pohlenz J, Rosenthal IM, Weiss RE et al (1998) Congenital hypothyroidism due to mutation in the sodium/iodine symporter: identification of a nonsense mutation producing a downstream cryptic 3' splice site. J Clin Invest 101:1028–1035

Ponssen HH, de Herder WW, Bonjer HJ et al (1996) An unusual case of multiple endocrine neoplasia type 1 and the role of 111In-pentetreotide scintigraphy. Nether J Med 49:112–115

Potchen EJ, Dealy JB Jr (1963) Selective isotope labeling of the parathyroid gland. J Nucl Med 4:203

Proye C, Malvaux P, Pattou F et al (1998) Noninvasive imaging of insulinomas and gastrinomas with endoscopic ultrasonography and somatostatin receptor scintigraphy. Surgery 124:1134–1144

Rao VV, Chiu ML, Kronauge JF et al (1994) Expression of recombinant human multidrug resistance P-glycoprotein in insect cells confers decreased accumulation of technetium-99m-sestamibi. J Nucl Med 35:510–515

Refetoff S (2000) Resistance to thyroid hormone. In: Braverman LE, Utiger RD (eds) Werner and Ingbar's The thyroid, 8th edn. Lippincott, Williams and Wilkins, Philadelphia, pp 1028–1043

Reubi JC, Schaer JC, Markwalder R et al (1997) Distribution of somatostatin receptors in normal and neoplastic human tissues: recent advances and potential relevance. Yale J Biol Med 70:471–479

Reubi JC, Schaer JC, Waser B et al (1994) Expression and localization of somatostatin receptor SSTR1, SSTR2, and SSTR3 messenger RNAs in primary human tumors using in situ hybridization. Cancer Res 54:3455–3459

Reutens AT, Achermann JC, Ito M et al (1999) Clinical and functional effects of mutations in the DAX-1 gene in patients with adrenal hypoplasia congenita. J Clin Endocrin Metab 84:504–511

Reznik Y, Allali-Zerah V, Chayvialle JA et al (1992) Food-dependent Cushing's syndrome mediated by aberrant adrenal sensitivity to gastric inhibitory polypeptide. N Engl J Med 327:981–986

Rich GM, Ulick S, Cook S et al (1992) Glucocorticoid-remediable aldosteronism in a large kindred: clinical spectrum and diagnosis using a characteristic biochemical phenotype. Ann Intern Med 11:813–820

Ricke J, Klose KJ (2000) Imaging procedures in neuroendocrine tumours. Digestion 62 [Suppl 1]:39–44

Rindi G, Bordi C, Rappel S et al (1996) Gastric carcinoids and neuroendocrine carcinomas: pathogenesis, pathology, and behavior. World J Surg 20:168–172

Rindi G, Capella C, Solcia E (1998) Cell biology, clinicopathological profile, and classification of gastro-enteropancreatic endocrine tumors. J Mol Med 76:413–420

Rindi G, Candusso ME, Solcia E (1999) Molecular aspects of the endocrine tumours of the pancreas and the gastrointestinal tract. Ital J Gastroenterol Hepatol 31 [Suppl 2]:S135–S138

Rindi G, Villanacci V, Ubiali A (2000) Biological and molecular aspects of gastroenteropancreatic neuroendocrine tumors. Digestion 62 [Suppl 1]:19–26

Rodriquez JM, Tezelman S, Siperstein AE et al (1994) Localization procedures in patients with persistent or recurrent hyperparathyroidism. Arch Surg 129:870–875

Rosenblatt M, Kronenberg HM, Potts JT Jr (1989) Parathyroid hormone. Physiology, chemistry, biosynthesis, secretion, metabolism, and mode of action. In: Degroot LT (ed) Endocrinology, 2nd edn. Saunders, Philadelphia, pp 848–891

Rosenspire KC, Haka MS, Van Dort ME et al (1990) Synthesis and preliminary evaluation of carbon-11-meta-hydroxyephedrine: a false transmitter agent for heart neuronal imaging. J Nucl Med 31:1328–1334

Rougier P, Mitry E (2000) Chemotherapy in the treatment of neuroendocrine malignant tumors. Digestion 62 [Suppl 1]:73–78

Ruszniewski P, Malka D (2000) Hepatic arterial chemoembolization in the management of advanced digestive endocrine tumors. Digestion 62 [Suppl 1]:79–83

Ruszniewski P, Amouyal P, Amouyal G et al (1995) Localization of gastrinomas by endoscopic ultrasonography in patients with Zollinger-Ellison syndrome. Surgery 117:629–635

Saito T, Endo T, Kawaguchi A et al (1997) Increased expression of the Na+/I- symporter in cultured human thyroid cells exposed to thyrotropin and in Graves' thyroid tissue. J Clin Endocrinol Metab 82:3331–3336

Samaan NA, Schultz PN, Haynie TP et al (1985) Pulmonary metastasis of differentiated thyroid carcinoma: treatment results in 101 patients. J Clin Endocrinol Metab 65:376–380

Sandrock D, Merino MJ, Norton JA et al (1990) Parathyroid imaging by Tc/Tl scintigraphy. Eur J Nucl Med 16:607–613

Sandrock D, Merino MJ, Norton JA et al (1993) Ultrastructural histology correlates with results of thallium-201/technetium-99m parathyroid subtraction scintigraphy. J Nucl Med 34:24–29

Santoro M, Thomas GA, Vecchio G et al (2000) Gene rearrangement and Chernobyl related thyroid cancers. Br J Cancer 82:315–322

Sarlis NJ, Chrousos GP, Doppman JL et al (1997) Primary pigmented nodular adrenocortical disease: reevaluation of a patient with Carney complex 27 years after unilateral adrenalectomy. J Clin Endocrinol Metab 82:1274–1278

Sassolas G, Chayvialle JA (1995) GRFomas, somatostatinomas: clinical presentation, diagnosis, and advances in management. In: Mignon M, Jensen RT (eds) Endocrine tumors of the pancreas. Karger, Basel, pp 194–207 (Frontiers in gastrointestinal research, vol 23)

Schaer JC, Waser B, Mengod G et al (1997) Somatostatin receptor subtypes sst1, sst2, sst3 and sst5 expression in human pituitary, gastroentero-pancreatic and mammary tumors: comparison of mRNA analysis with receptor autoradiography. Int J Cancer 70:530–537

Schillaci O, Massa R, Scopinaro F (2000) In-111-pentetreotide scintigraphy in the detection of insulinomas: importance of SPECT imaging. J Nucl Med 41:459–462

Schirmer WJ, Melvin WS, Rush RM et al (1995) Indium-111-pentetreotide scanning versus conventional imaging techniques for the localization of gastrinoma. Surgery 118:1105–1114

Schlumberger MS, Arcangioli O, Piekarski JD et al (1988) Detection and treatment of lung metastases of differentiated thyroid carcinoma in patients with normal chest X-rays. J Nucl Med 29:1790–1794

Schlumberger MS, Challeton C, Vathaire FD et al (1996) Radioactive iodine treatment and external radiotherapy for lung and bone metastases from thyroid carcinoma. J Nucl Med 37:598–605

Schönbrunn A (1999) Somatostatin receptors present knowledge and future directions. Ann Oncol 10 [Suppl 2]:S17–S21

Segu VB, Mahvi DM, Wilson MA, Hale SJ, Warner TF, Meredith M, Shenker Y (1997) Use of In-111 pentetreotide scintigraphy in the diagnosis of a midgut carcinoid causing Cushing's syndrome. Eur J Endocrinol 137:79–83

Semelka RC, Cumming MJ, Shoenut JP et al (1993) Islet cell tumors: comparison of dynamic contrast-enhanced CT and MR imaging with dynamic gadolinium enhancement and fat suppression. Radiology 186:799–802

Schurrer ME, Seabold JE, Gurll NJ et al (1996) Sestamibi SPECT scintigraphy for detection of postoperative hyperfunctional parathyroid glands. AJR 166:1471–1474

Semelka RC, John G, Kelekis NL et al (1996) Small bowel neoplastic disease: demonstration by MRI. J Magn Reson Imaging 6:855–860

Serrano J, Goebel SU, Peghini PL, Lubensky IA, Gibril F, Jensen RT (2000) Alterations in the p16INK4a/CDKN2A tumor suppressor gene in gastrinomas. J Clin Endocrinol Metab 85:4146–4156

Sessler MJ, Greck P, Maul F-D et al (1986) New aspects of cellular thallium uptake: Tl^+-Na^+-$2Cl^-$-cotransport is the central mechanism of ion uptake. J Nucl Med 25:24–27

Seto P, Hirayu H, Magnusson RP et al (1987) Isolation of a complementary DNA clone for the thyroid microsomal antigen: homology with the gene for thyroid peroxidase. J Clin Invest 80:1205–1208

Sfakianakis GN, Irvin GL III, Foss J et al (1996) Efficient parathyroidectomy guided by SPECT-MIBI and hormonal measurements. J Nucl Med 37:798–804

Shapiro B, Copp JE, Sisson JC et al (1985) Iodine-131 metaiodobenzylguanidine for the locating of suspected pheochromocytomas: experience in 400 cases. J Nucl Med 26:576–585

Shimura H, Haraguchi K, Miyazaki A et al (1997) Iodide uptake and experimental ^{131}I therapy in transplanted undifferentiated thyroid cancer cells expressing the Na+/I-symporter gene. Endocrinology 138:4493–4496

Shulkin BL, Shapiro B (1998) Current concepts on the diagnostic use of MIBG in children. J Nucl Med 39:679–688

Siccardi AG, Paganelli G, Pontiroli AE, Pelagi M, Magnani P, Viale G, Faglia G, Fazio F (1996) In vivo imaging of chromogranin A-positive endocrine tumours by three-step monoclonal antibody targeting. Eur J Nucl Med 23:1455–1459

Silver J, Russell J, Sherwood LM (1985) Regulation by vitamin D metabolites of messenger ribonucleic acid for preproparathyroid hormone in isolated bovine parathyroid cells. Proc Natl Acad Sci USA 82:4270–4273

Silverberg SJ, Bone HG III, Marriott TB et al (1997) Short-term inhibition of parathyroid hormone secretion by a calcium-receptor agonist in patients with primary hyperparathyroidism. N Engl J Med 337:1506–1510

Singh J, Reghebi K, Lazarus CR et al (1993) Studies on the preparation and isometric composition of ^{186}Re and ^{188}Re-pentavalent rhenium dimercaptosuccinic acid complex. Nucl Med Commun 14:197–203

Sisson JC, Beierwaltes WH (1962) Radiocyanocobalamine ($Co^{57}B_{12}$) concentration in the parathyroid glands. J Nucl Med 3:160–166

Sisson JC, Shulkin BL (1999) Nuclear medicine imaging of pheochromocytoma and neuroblastoma. Q J Nucl Med 43:217–223

Sisson JC, Shapiro B, Beierwaltes WH et al (1984) Radiopharmaceutical treatment of malignant pheochromocytoma. J Nucl Med 25:197–206

Sisson JC, Shapiro B, Meyers L et al (1987a) Metaiodobenzylguanidine to map scintigraphically the adrenergic nervous system in man. J Nucl Med 28:1625–1636

Sisson JC, Wieland DM, Sherman P et al (1987b) Metaiodobenzylguanidine as an index of the adrenergic nervous system integrity and function. J Nucl Med 28:1620–1624

Sisson JC, Bolgos G, Johnson J (1991) Measuring acute changes in adrenergic nerve activity of the heart in the living animal. Am Heart J 121:1119–1123

Sisson JC, Thompson NW, Ackerman RJ et al (1994) Use of 2-[F-18]-fluoro-2-deoxy-D-glucose PET to locate parathyroid adenomas in primary hyperparathyroidism. Radiology 192:280

Slatopolsky E, Finch J, Denda M et al (1996) Phosphorus restriction prevents parathyroid gland growth. High phosphorus directly stimulates PTH secretion in vitro. J Clin Invest 97:2534–2540

Smanik P, Liu Q, Furminger TL et al (1996) Cloning of the human sodium iodide symporter. Biochem Biophys Res Commun 226:339–345

Smida J, Salassidis K, Hieber L et al (1999) Distinct frequency of ret rearrangements in papillary thyroid carcinomas of children and adults from Belarus. Int J Cancer 80:32–38

Smith BR, Hall R (1974) Thyroid-stimulating immunoglobulins in Graves' disease. Lancet 2:427–431

Sohier J, Jeanmougin M, Lombrail P et al (1980) Rapid improvement of skin lesions in glucagonomas with intravenous somatostatin infusion. Lancet 1:40

Som P, Atkins HL, Bandoypadhyay D et al (1980) A fluorinated glucose analog, 2-fluoro-2-deoxy-D-glucose (F-18): nontoxic tracer for rapid tumor detection. J Nucl Med 21:670–675

Souza B de, Brunetti A, Fulham MJ et al (1990) Pituitary microadenomas: a PET study. Radiology 177:39–44

Spitzweg C, Zhang S, Bergert ER et al (1999) Prostate-specific antigen (PSA) promoter-driven androgen-inducible expression of sodium iodide symporter in prostate cancer cell lines. Cancer Res 59:2136–2141

Stanton M, Tuli M, Radtke N et al (1989) Regional sympathetic denervation after myocardial infarction in humans detected noninvasively using I-123-metaiodobenzylguanidine. J Am Coll Cardiol 14:1519–1526

Staudenherz A, Abela C, Niederle B et al (1997) Comparison and histopathological correlation of three parathyroid imaging methods in a population with a high prevalence of concomitant thyroid diseases. Eur J Nucl Med 24:143–149

Strauss HW, Harrison K, Pitt B (1977) Thallium-201: noninvasive determination of the regional distribution of cardiac output. J Nucl Med 18:1167–1170

Sugimoto E, Lorelius L, Eriksson B et al (1995) Midgut carcinoid tumours: CT appearance. Acta Radiol 36:367–371

Sundin A, Johansson C, Hellman P et al (1996) PET and parathyroid L-[carbon-11] methionine accumulation in hyperparathyroidism. J Nucl Med 37:1766–1770

Sunthornthepvarakul T, Gottschalk ME, Hayashi Y et al (1995) Resistance to thyrotropin caused by mutations in the thyrotropin-receptor gene. N Engl J Med 332:155–160

Swinburn BA, Yeong ML, Lane MR et al (1988) Neurofibromatosis associated with somatostatinoma: a report of two patients. Clin Endocrinol (Oxf) 28:353–359

Taillefer R, Boucher Y, Potvin C et al (1992) Detection and localization of parathyroid adenomas in patients with hyperparathyroidism using a single radionuclide imaging procedure with technetium-99m-sestamibi (double-phase study). J Nucl Med 33:1801–1807

Takebayashi S, Hidai H, Chiba T et al (1999) Hyperfunctional parathyroid glands with ^{99m}Tc-MIBI scan: semiquantitative analysis correlated with histologic findings. J Nucl Med 40:1792–1797

Tamburrano G, Paolini A, Pietrobono D et al (1999) Pancreatic endocrine tumours. Ital J Gastroenterol Hepatol 31 [Suppl 2]:S104–S107

Targovnik HM, Medeiros-Neto G, Varela V et al (1993) A nonsense mutation causes human hereditary congenital goiter with preferential production of a 171-nucleotide-deleted thyroglobulin ribonucleic acid messenger. J Clin Endocrinol Metab 77:210–215

Tarutani O, Ui N (1985) Properties of thyroglobulins from normal thyroid and thyroid tumor on a concanavalin A-sepharose column. J Biochem Tokyo 98:851–857

Tarutani O, Yoshimura H, Ohmori T et al (1991) Enzymatic iodination of thyroglobulins obtained from patients with thyroid disease (in Japanese). Nippon-Naibunpi-Gakkai-Zasshi 67:1186–1196

Taurog AM (2000) Hormone synthesis: thyroid iodine metabolism. In: Braverman LE, Utiger RD (eds) Werner and Ingbar's The thyroid, 8th edn. Lippincott, Williams and Wilkins, Philadelphia, pp 61–85

Teigen EL, Kilgore EJ, Cowan RJ et al (1996) Technetium-99m-sestamibi SPECT localization of mediastinal parathyroid adenoma. J Nucl Med 37:1535–1537

Termanini B, Gibril F, Reynolds JC et al (1997) Value of somatostatin receptor scintigraphy: a prospective study in gastrinoma of its effect on clinical management. Gastroenterology 112:335–347

Thakker RV (1996) Molecular genetics of parathyroid disease. Curr Opin Endocrinol Diabetes 3:521–528

Thelen MH, Kuwert T, Lerch H et al (1996) Double-phase Tc-99m MIBI scintigraphy in secondary hyperparathyroidism relapsing after parathyroidectomy and removal of a parathyroid autograft. Clin Nucl Med 21:609–611

Thiele J, Kloppel R Schulz HG (1993) CT-Sellink – a new method of evaluating the intestinal wall. Rofo Fortschr Geb Rontgenstr Neuen Bildgeb Verfahr 159:213–217

Tiensuu Janson EM, Öberg KE (1996) Carcinoid tumours. Baillieres Clin Gastroenterol 10:589–601

Valkema R, Jamar F, Jonard P et al (2000) Targeted radiotherapy with 90Y-SMT487 (OctreoTher): a phase I study. J Nucl Med 41:111P

Van Eijck CHJ, De Jong M, Breeman WAP et al (1999) Somatostatin receptor imaging and therapy of pancreatic endocrine tumors. Ann Oncol 10 [Suppl 4]:S177–S181

Van Hoe L, Gryspeerdt S, Marchal G et al (1995) Helical CT for the preoperative localization of islet cell tumors of the pancreas: value of arterial and parenchymal phase images. Am J Roentgenol 165:1437–1439

Virgolini I, Raderer M, Kurtaran A et al (1996) 123I-vasoactive intestinal peptide (VIP) receptor scanning: update of imaging results in patients with adenocarcinomas and endocrine tumors of the gastrointestinal tract. Nucl Med Biol 23:685–692

Virgolini I, Raderer M, Kurtaran A et al (1994) Vasoactive intestinal peptide-receptor imaging for the localization of intestinal adenocarcinomas and endocrine tumors. N Engl J Med 331:1116–1121

Wagner HN Jr (1995) Molecular nuclear medicine: from genotype to phenotype via chemotype. J Nucl Med 36 (Suppl):2S–4S

Wang EH, Ebrahimi SA, Wu AY et al (1998) Mutation of the MENIN gene in sporadic pancreatic endocrine tumors. Cancer Res 58:4417–4420

Wängberg B, Forssell-Aronsson E, Tisell L-E et al (1996) Intraoperative detection of somatostatin-receptor-positive neuroendocrine tumours using indium-111-labelled DTPA-D-Phe[1]-octreotide. Br J Cancer 73:770–775

Weber CJ, Vansant J, Alazraki N et al (1993) Value of technetium 99m sestamibi iodine 123 imaging in reoperative parathyroid surgery. Surgery 114:1011–1018

Weinstein LS, Shenker A, Gejman PV et al (1991) Activating mutations of the stimulatory G protein in the McCune-Albright syndrome. N Engl J Med 325:1688–1695

Werner SC (1956) Response to triiodothyronine as an index of persistence of disease in the thyroid remnant of patients in remission from hyperthyroidism. J Clin Invest 35:57–61

Wieland DM, Brown LE, Tobes MC et al (1981) Imaging the primate adrenal medulla with [^{123}I] and [^{131}I] meta-iodobenzylguanidine: concise communication. J Nucl Med 22:358–364

Wieland DM, Rosenspire KC, Hutchins GD et al (1990) Neuronal mapping of the heart with 6-[^{18}F]fluorometaraminol. J Med Chem 30:956–964

Winter TC III, Freeny PC, Nghiem HV (1996) Extrapancreatic gastrinoma localization: value of arterial-phase helical CT with water as an oral contrast agent. Am J Roentgenol 166:51–52

Wolff J (1964) Transport of iodide and other anions in the thyroid gland. Physiol Rev 44:45–90

Young J, Povey S (1998) The genetic basis of tuberous sclerosis. Mol Med Today 4:313–319

Young W Jr, Carney JA, Musa BU et al (1989) Familial Cushing's syndrome due to primary pigmented nodular adrenocortical disease. N Engl J Med 321:1659–1664

Zeiger MA, Shawker TH, Norton JA (1993) Use of intraoperative ultrasonography to localize islet cell tumors. World J Surg 17:448–454

Zhuang Z, Vortmeyer AO, Pack S et al (1997) Somatic mutations of the MEN1 tumor suppressor gene in sporadic gastrinomas and insulinomas. Cancer Res 57:4682–4686

Zimmer T, Ziegler K, Liehr RM et al (1994) Endosonography of neuroendocrine tumors of the stomach, duodenum, and pancreas. Ann NY Acad Sci 733:425–443

Zimmer T, Stolzel U, Bader M et al (1996) Endoscopic ultrasonography and somatostatin receptor scintigraphy in the preoperative localisation of insulinomas and gastrinomas. Gut 39:562–568

Zwas ST, Czerniak A, Boruchowski S et al (1987) Preoperative parathyroid localization by superimposed iodine-131 toluidine blue and technetium-99m pertechnetate imaging. J Nucl Med 28:298–307

Noninvasive Assessment of Inherited Cardiac Disease Using Positron Emission Tomography*

16

SHILPI EPSTEIN, STEVEN R. BERGMANN

Contents

16.1
Introduction

Despite the steady decrease in morbidity and mortality due to cardiovascular disease over recent decades, heart disease remains the leading cause of death in Western societies. The reduction of morbidity and mortality has been attributed primarily to improved detection and treatment of coronary artery disease and hypertension with limited progress in the prevention and treatment of nonischemic cardiomyopathies, a heterogeneous group of diseases that ultimately leads to death secondary to congestive heart failure or cardiac arrhythmias (Podrid et al. 1992; Ho et al. 1993). Despite advances in our knowledge of genetic mechanisms, the biochemical processes that underlie the nonischemic cardiomyopathies, a subset of which has been identified as inherited cardiomyopathies (Kelly and Strauss 1994), remain poorly understood.

Of the many approaches for the measurement of enzyme function and intermediary metabolism, positron emission tomography (PET) is unique in that it

permits noninvasive assessment of regional myocardial perfusion and metabolism with a sensitivity unparalleled by any other imaging approach (Bergmann et al. 1989; Tamaki et al. 1989; Bergmann 1990; Schwaiger et al. 1990). Accordingly, it is uniquely suited for the delineation of the metabolic changes that occur in the heart (as well as other organs in the body) in inherited disease. Although not yet tested for inherited diseases of the heart, strategies using "PET reporter genes" have been shown to enable delineation of expression of administered genes (Gambhir et al. 1998; Blasberg and Tjuvajev 1999). This combined capability of delineating not only genetic expression but also phenotypic consequences should enable noninvasive imaging approaches such as PET to become important in the diagnosis of inherited diseases of the heart and other organs, as well as to provide an objective approach to test and evaluate therapeutic strategies. This chapter will focus on the use of PET for the delineation of biochemical abnormalities that occur with inherited forms of cardiac fatty acid defects as a paradigm for the investigation of phenotypic expression of inherited disease.

16.2
Normal Myocardial Metabolism

Cardiac muscle is dependent on aerobic metabolism for the production of energy, which it requires for contractile function and for maintenance of electrochemical gradients. Under aerobic conditions, abundant long-chain fatty acids serve as a major source of energy. Each fatty acid molecule serves as a rich source of energy (ATP) via mitochondrial β-oxidation. Under normal conditions, fatty acids are the substrate of choice for energy production in the

* Supported in part by grants from the Department of Energy (DE-FG02-97ER62433) and from the Jacob and Hilda Blaustein Foundation (JB 980049).

heart, their oxidation providing 60 to 70% of the energy required by the heart (Bing 1965; Neely and Morgan 1974; Camici et al. 1989). Fatty acids traverse the sarcolemal membrane with the assistance of multiple membrane-bound transporter proteins. Once fatty acids are intracellular, they are thio-esterified, and transported through the mitochondrial membrane via a carnitine mediated transport system (Fig. 16.1; Bing 1965; Neely and Morgan 1974; Camici et al. 1989). Once inside the mitochondrial matrix, β-oxidation consists of four enzyme driven reactions: dehydrogenation, hydration, dehydrogenation, and thiolytic cleavage. The enzymes required are specific to the reaction they catalyze, as well as to the length of the carboxylic acid chain of the fatty acid being oxidized: short, medium, long, or very long. Not unexpectedly, with recent strides in genetic testing, alterations in nearly all steps of myocardial fatty acid metabolism have been identified with increasing frequency (Pollitt 1989; Hale and Bennett 1992; Kelly and Strauss 1994; Largillerie et al. 1995; Boles et al. 1998; Saudubray et al. 1999; Wanders et al. 1999).

Depending on flow as well as metabolic and hormonal conditions, the heart has the potential to alter the pattern of substrate use, which is essential to preserve energy production. Arterial concentrations of substrate, myocardial perfusion and oxygenation, and neurohumoral milieu are among the multiple factors which determine the pattern of substrate use (Bing 1965; Neely and Morgan 1974; Schelbert et al. 1986; Camici et al. 1989; Gropler et al. 1990; Lerch et al. 1992). Thus, during fasting, when plasma fatty acids are abundant, fatty acid oxidation represents the prime source of energy production for the heart. Postprandially, insulin rises in response to increased plasma glucose and lipolysis is inhibited. This alteration in plasma substrate decreases the use of fatty acids by the heart, in part because of decreased plasma fatty acid concentrations and intracellular enzyme switches. During myocardial ischemia, when blood flow (and oxygenation) to the heart is decreased, the utilization of fatty acids is diminished and that of glucose up-regulated (Myears et al. 1987). These altered patterns of substrate use can be utilized to define the changes that occur under normal, physiological conditions (Bergmann 1990; Gropler et al. 1990; Lerch et al. 1992); to identify ischemic from infarcted myocardium (Tamaki et al. 1989; Schwaiger et al. 1990); and to define the efficacy of therapeutic interventions which tend to restore metabolic function to normal (Tamaki et al. 1989; Bergmann 1990; Rubin et

al. 1996). Cardiomyopathies of diverse etiologies also alter myocardial metabolism and generally shift metabolism to a more fetal pattern, i.e. use of glucose in preference to fatty acids (Depre et al. 1999). These alterations can also be assessed noninvasively with PET (Geltman et al. 1983).

16.3
Abnormalities in Fatty Acid Oxidation

A common feature among many cardiomyopathies is the decreased ability to metabolize fatty acids (Bergmann et al. 1995; Bergmann et al. 2001). Similar to case proposed for the link between altered contractile function due to decreased fatty acid oxidation during ischemia, altered fatty acid metabolism in a number of cardiomyopathies (i.e., due to diabetes, alcohol, or virus) can result in a build-up of cellular free fatty acids and fatty acid intermediates including long-chain acyl-carnitines and fatty acyl-CoA, which have been shown to cause arrhythmias and contractile dysfunction (Lindeneg et al. 1964; Severeid et al. 1969; Lange and Sobel 1983; Corr et al. 1989; Paulson 1998). One proposed mechanism is the inhibition of ATP-translocase which shuttles ATP from the mitochondria to the contractile proteins. An alternative, or perhaps additional, mechanism leading to dysfunction is the insertion of long-chain acyl-carnitines into the sarcolemmal membrane and altered ionic flux or redox states (Lindeneg et al. 1964; Severeid et al. 1969; Lange and Sobel 1983; Corr et al. 1989; Paulson 1998). Inborn errors of fatty acid metabolism have recently been identified as the cause of a subset of inherited cardiomyopathies (Pollitt 1989; Hale and Bennett 1992; Kelly and Strauss 1994; Largillerie et al. 1995; Boles et al. 1998; Saudubray et al. 1999; Wanders et al. 1999).

Although familial inheritance of disease has long been recognized, disorders related to fatty acid metabolism as the cause of cardiomyopathy were initially reported only in the late 1970s, and specific genetic abnormalities were not identified until the late 1980s. With advances in genetic recognition, the incidence of inherited disorders in fatty acid metabolism is increasing. Current estimates of incidence are 1 in 10,000 to 15,000 live births (Kelly and Strauss 1994), which still may be an underestimation due to lack of clinical recognition, under-reporting, and difficulty in diagnosis.

Disease heterogeneity is one of the factors which make diagnosis of these defects challenging. Fatty acid oxidation disorders can involve one or multiple organ systems, with dysfunction varying widely in severity and disease processes including cardiomyopathy, skeletal muscle myopathy, hepatic dysfunction, metabolic disturbances, and/or sudden death (Pollitt 1989; Hale and Bennett 1992; Kelly and Strauss 1994; Largillerie et al. 1995; Boles et al. 1998; Saudubray et al. 1999; Wanders et al. 1999). Even in cases with documented inheritance, clinical symptoms may be minimal due to environmental factors including the frequency of feeding, the absence of acute viral illnesses, and to other precipitants yet to be defined. On the other hand, severely affected patients frequently present in infancy when an acute episode caused by fasting at the time of an infectious illness results in hypoglycemia and its sequelae as the body is unable to turn to an alternative energy source. The most severely affected patients are sometimes diagnosed on post-mortem examination after a sudden infant death syndrome (SIDS)-like

course (Pollitt 1989; Hale and Bennett 1992; Kelly and Strauss 1994; Largillerie et al. 1995; Boles et al. 1998; Saudubray et al. 1999; Wanders et al. 1999). Mildly affected individuals (often diagnosed only after loss of a sibling in infancy) may not present until later in childhood (2 to 4 years) or even remain asymptomatic throughout life. This variability occurs, in part, because the pathway for fatty acid metabolism can be disrupted at almost any point. In fact, genetic defects have been identified involving the transport of carnitine into cells, the transport of fatty acids into the mitochondria, and the specific enzymes responsible for β-oxidation of short-, medium-, long-, and very long-chain fatty acids (Fig. 16.1; Kelly and Strauss 1994). The majority of these errors follow an autosomal recessive inheritance, and the most common of these inherited defects involves the acyl coenzyme A dehydrogenases – the enzymes catalyzing the initial reaction of mitochondrial β-oxidation.

In general, defects involving long and very long-chain fatty acid metabolism and defects in the carnitine transporters cause more severe clinical disease than defects involving metabolism of the shorter fatty

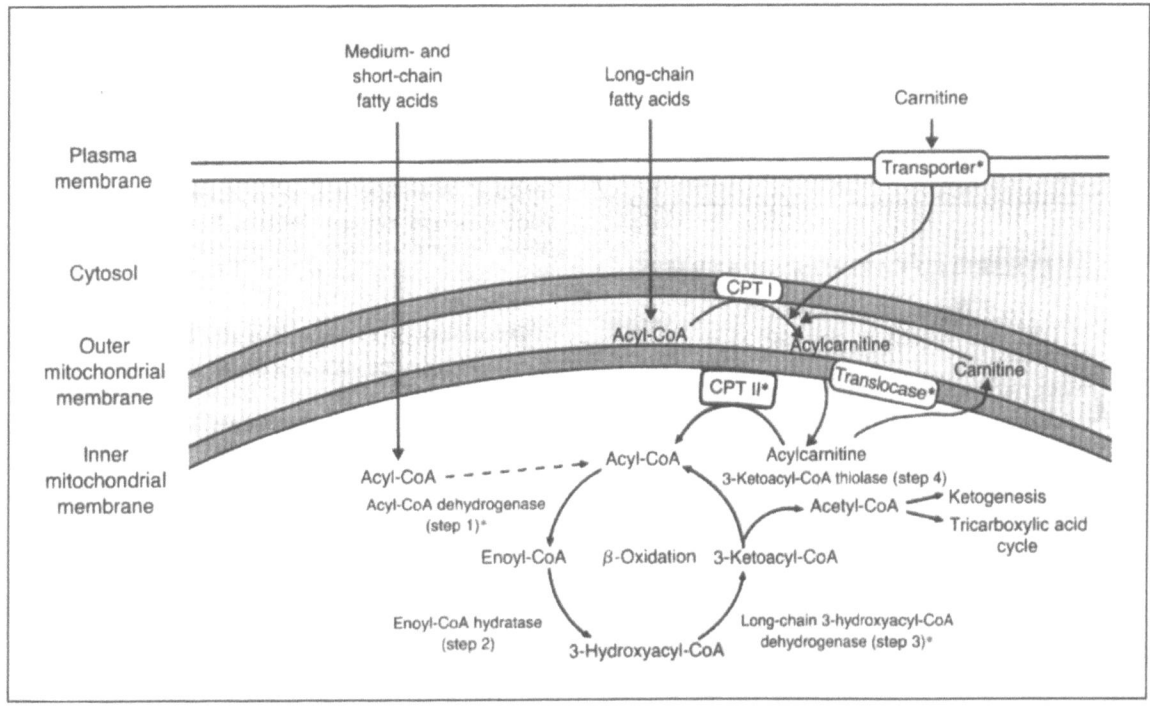

Fig. 16.1. Schematic diagram of fatty acid oxidation, including the enzymes and the reactions involved in the cellular transport and β-oxidation of fatty acids. Defects associated with inherited cardiomyopathy are marked by *asterisks*, and include defects in carnitine transport as well as deficiencies of the enzymes of mitochondrial β-oxidation. Reproduced with permission from Kelly and Strauss (1994)

acids (Kelly and Strauss 1994). Specifically, cardio-myopathy, heart failure and sudden death have been more frequently associated with defects in the carnitine transporter, carnitine-acylcarnitine translocase, carnitine-palmitoyltransferase II (CPT-II), long-chain acyl coenzyme A dehydrogenase (LCAD), and long-chain 3-hydroxyacyl-coenzyme A dehydrogenase (LCHAD; Pollitt 1989; Hale and Bennett 1992; Kelly and Strauss 1994; Largillerie et al. 1995; Boles et al. 1998; Saudubray et al. 1999; Wanders et al. 1999). Yet even in patients affected with the same genetic defect, penetrance is incomplete and phenotypic expression of disease remains highly variable.

Treatment of these inherited disorders is relatively straightforward. Therapy is aimed at decreasing plasma fatty acid levels in patients with disorders of long-chain metabolism with maneuvers such as constant feeding of carbohydrates, ingestion of short- and medium-chain fatty acids, and avoidance of fasting (Saudubray et al. 1999). In patients with defects in the carnitine transport system, oral carnitine supplementation is frequently beneficial (Kelly and Strauss 1994; Saudubray et al. 1999). In addition, it appears that enzymatic changes occur over time as the incidence of disease flares seems to abate with age. Perhaps there is induction of alternative mechanisms for metabolism such as the metabolism of fatty acids by omega-oxidation or via alternative pathways.

16.4
Assessing Cardiac Disease

Multiple approaches are available to assess cardiac function. Classically, 2-dimensional echocardiography and gated radionuclide ventriculography have been used to evaluate ventricular function. Magnetic resonance imaging (MRI) is a newer technique also used to define cardiac anatomy and myocardial contractility. Cardiac catheterization can be used to delineate cardiac and coronary anatomy and cardiac pressures.

With radioisotopes, both the function of the heart and myocardial perfusion can be assessed. Labeling of erythrocytes with technetium-99m pertechnetate permits assessment of cardiac structure and function. Using either thallium-201 or technetium-99m based perfusion agents, myocardial perfusion, wall motion, and ventricular volumes can be delineated noninvasively with either planar or, now more conventionally,

single photon emission computed tomography (SPECT). While there are some agents available for delineation of myocardial metabolism with SPECT, such as the iodinated fatty acids, PET offers significant advantages over conventional single-photon imaging. The process of positron annihilation results in the emission of 2 gamma photons with very high energy emitted approximately 180 degrees apart, yielding excellent tissue penetration and localization (Bergmann 1990, 1998). Precise correction for tissue attenuation is attainable. In addition, the positron emitting radioisotopes, oxygen-15 (^{15}O; $t_{1/2}$ = 2 min); nitrogen-13 (^{13}N; $t_{1/2}$ = 10 min); carbon-11 (^{11}C; $t_{1/2}$ = 20 min); and fluorine-18 (^{18}F; $t_{1/2}$ = 110 min) are isotopes which are of significant value as they can be incorporated into substrates of physiological interest. Their short physical half-lives permit sequential scanning with a modest radiation burden to the subjects. However, the short half-life also imposes limitations on the radiochemistry that can be performed in that chemical procedures need to be expeditious because of time-constraints for radioisotope incorporation.

16.5
Assessment of Genotypic Expression

One approach to the assessment of genetic expression is the "reporter gene" strategy. In this paradigm, a marker gene is incorporated into the therapeutic gene of interest. The reporter gene then, in some way, interacts with a radiolabeled probe. Thus, appearance of uptake of the intravenously administered probe signals incorporation of the gene and production of the product. In a mouse model transfected with a tumor containing the herpes simplex virus thymidine kinase (HSV1-tk) reporter gene, investigators have confirmed expression of the transfected gene using FIAU labeled with iodine-124 (Tjuvajew et al. 1997; Blasberg and Tjuvajev 1999). Others have used a similar reporter gene system and constructed an adenoviral vector which targeted the liver. In this case, the reporter probe was ^{18}F-ganciclovir, which is phosphorylated by the reporter gene (Gambhir et al. 1998). Infected mice showed localization of radioactivity in the liver (Fig. 16.2).

While this approach is somewhat cumbersome in that the reporter gene must be incorporated into the target or therapeutic gene and administered, it dem-

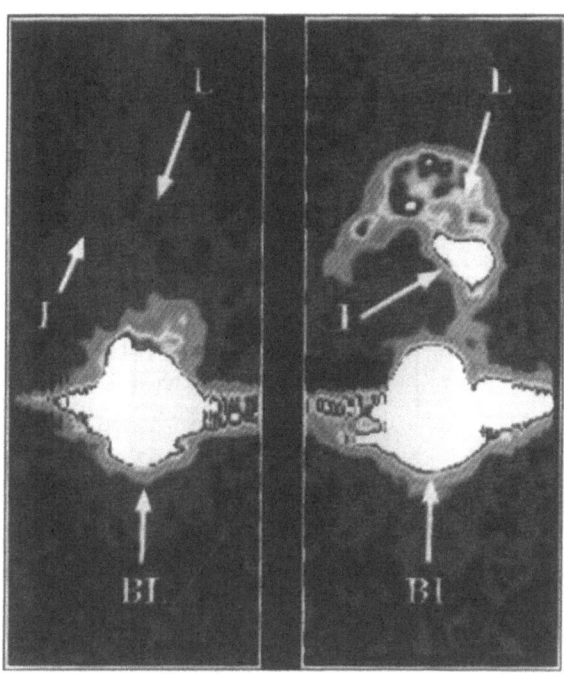

Fig. 16.2. Whole body PET scan of mice injected with control adenovirus (*left*) and virus with the associated HSV1-tk PET reporter gene (*right*). Adenovirus was administered intravenously 48 h prior to the PET study. The PET reporter probe, ^{18}F-ganciclovir, was then administered intravenously. It localizes in the liver in the mouse with the reporter gene. L, liver; I, intestine; BL, bladder. This study demonstrated the concept that the expression of administered genes could be determined noninvasively with PET. Reproduced with permission from Gambhir et al. (1998)

onstrates the ability to use noninvasive PET imaging for sequential assessment of genetic incorporation and expression. Other approaches using labeling of DNA and RNA directly are discussed elsewhere in this volume.

16.6
Assessment of Phenotypic Expression

An alternative, as well as complementary approach, is to demonstrate that a gene or gene defect performs its biochemical function by direct assessment of phenotypic expression. A number of metabolic tracers have been used for the assessment of myocardial metabolism. These include 1-^{11}C-palmitate for the assessment of long-chain fatty acid metabolism (Ter-Pogossian et al. 1980; Schelbert et al. 1986; Lerch et al. 1992; Bergmann et al. 1996); 1-^{11}C-acetate for assess-

ment of mitochondrial function (since acetate is oxidized predominantly in the mitochondria and its metabolism does not involve a carrier system; Brown et al. 1987, 1989; Herrero et al. 1996); and ^{11}C-glucose and its long-lived fluorinated analogue, ^{18}F-fluorodeoxyglucose (FDG) for the assessment of glucose metabolism and uptake (Ratib et al. 1982; Tamaki et al. 1989; Bergmann 1990, 1998; Gropler et al. 1990). Mathematical models have been developed to assess metabolism quantitatively (i.e., nmol/g/min) with each of these tracers (Fig. 16.3; Bergmann 1990, 1998). The use of the fluorinated analogue, ^{18}F-FDG, presents some problems in this respect because the uptake of FDG is not equivalent to the uptake of native glucose and a number of factors mediate this difference and must be incorporated into mathematical models to correct for these differences. In addition, for the particular case of the fluorinated analogue ^{18}F-FDG, this tracer only models the uptake of glucose from the vasculature and not subsequent metabolism since phosphorylated FDG is not a substrate for down-stream metabolism (Ratib et al. 1982; Bergmann 1990, 1998).

PET has been used extensively to characterize metabolism of the normal heart as well as under pathophysiological conditions such as ischemia and diabetes (Ter-Pogossian et al. 1980; Ratib et al. 1982; Geltman et al. 1983; Schelbert et al. 1986; Brown et al. 1987, 1989; Bergmann et al. 1989, 1996; Tamaki et al. 1989; Bergmann 1990, 1998; Gropler et al. 1990; Schwaiger et al. 1990; Lerch et al. 1992; Herrero et al. 1996; Tjuvajev et al. 1997). It has provided unique information regarding the metabolism of the heart under normal and pathophysiological conditions as well as for evaluating the effects of treatment on myocardial perfusion and metabolism (Bergmann 1990, 1998). More recently, PET has also been used to characterize metabolism of inherited defects (Kelly et al. 1993; Bergmann et al. 2001).

16.7
Clinical Application to Inherited Disease

As many as 20% of patients with cardiomyopathy have a first-degree relative with myocardial disease (Kelly and Strauss 1994), suggesting an unidentified genetic or inherited basis in many of these cases. Yet the etiology of the cardiomyopathy in most children remains

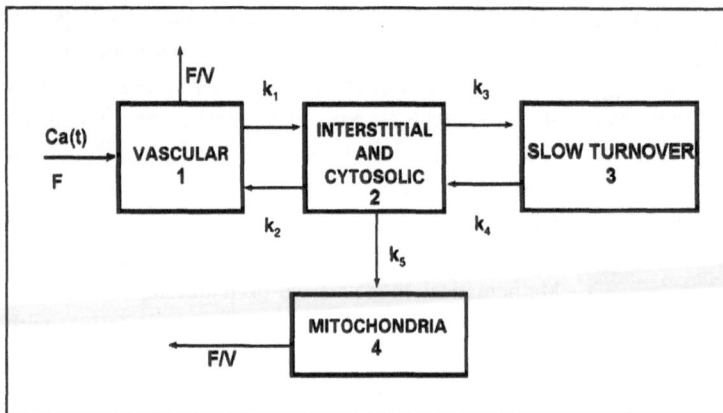

Fig. 16.3. Schematic representation of the three-compartment model used to analyze 1-^{11}C-palmitate metabolism. Compartment *1* represents the vascular space; *2*, the interstitial and cytosolic space; *3*, the slow turnover pool, including neutral lipids, amino acids and other processes; and *4*, the mitochondria. The rate constants (k_n) represent tracer turnover between compartments, and k_5 represents the irreversible import of tracer into the mitochondrial β-oxidation pathway. Reproduced with permission from Bergmann et al. (1996)

unknown, and even today the prognosis remains grim. Recently a subset of children with cardiomyopathy has been recognized, whose disease is associated with genetic defects in fatty acid metabolism (Pollitt 1989; Wanders et al. 1999). As the diagnosis of these defects remains difficult and under-recognized, these cases are likely under-reported.

Traditional testing for the diagnosis of these metabolic defects includes urine for organic acids and biopsy of skin fibroblasts and muscle for enzyme assays (Pollitt 1989; Bonnefont et al. 1990; Rhead 1990; Hale and Bennett 1992; Brivet et al. 1995; Largillerie et al. 1995; Boles et al. 1998; Saudubray et al. 1999; Wanders et al. 1999). However, the results are not always definitive, as urine can be negative for organic acids if the patient is not in an acutely ill episode, genetic testing can only be helpful for identified mutations, and enzyme assays are fraught with error. Furthermore, results of these tests, even when diagnostic, do not correlate with the severity of disease or the prognosis. Positron emission tomography offers an approach to assess the myocardial disease of this group of children. Furthermore, PET is the only noninvasive approach to evaluate myocardial metabolism at the cellular level and thus to study these metabolic defects in vivo.

In a series of studies performed by us, patients with one of two types of inherited fatty acid deficiencies were evaluated (Kelly et al. 1993; Bergmann et al. 2001; Epstein et al. 1999). These patients had either long- or very-long-chain acyl-CoA dehydrogenase deficiency. The acyl-CoA dehydrogenase deficiency prevents the β-oxidation of long-chain fatty acids in the mitochondria. The family of acyl-CoA dehydrogenase enzymes are chain-specific (i.e., there are specific enzymes for short-, medium-, long-, and very long-chain fatty acids), but the specificity for a particular

chain length is not absolute. Deficiencies in carnitine transport inhibit the shuttling of thio-esterified fatty acids from the cytosol into the mitochondria where β-oxidation occurs. Both types of deficiency result in diminished ability to utilize fatty acids (Hale and Bennett 1992; Pollitt 1989; Kelly and Strauss 1994; Largillerie et al. 1995; Boles et al. 1998; Saudubray et al. 1999; Wanders et al. 1999).

Patients with these defects were studied using PET after an overnight fast. The overnight fast is necessary to try to standardize plasma substrate levels. It should be recognized, however, that the plasma fatty acids will vary from individual to individual as will the concentrations of other substrates, hormones, and factors such as the albumin:fatty acid ratio, all of which can effect myocardial fatty acid utilization. In addition, although oral carnitine supplementation was withheld from patients with carnitine deficiencies, it is not recommended to withhold carnitine for prolonged periods of time since decreases in carnitine can precipitate an acute crisis. All subjects were studied in a whole-body tomograph that permits the simultaneous acquisition of multiple transverse slices of myocardium. Subjects were studied with a sequence of tracers designed to measure myocardial perfusion using $H_2{}^{15}O$ (Bergmann et al. 1989); long-chain fatty acid metabolism using 1-^{11}C-palmitate (Bergmann et al. 1996); and mitochondrial turnover and myocardial oxygen consumption using 1-^{11}C-acetate (Brown et al. 1987, 1989; Herrero et al. 1996). Quantitative assessments were made using validated mathematical models.

Although the metabolism of palmitate is complex, for imaging purposes the metabolism can be simplified into a compartment model (Fig. 16.3) for quantitative analysis (Bergmann et al. 1996). The model assumes that metabolism is in a metabolic steady state;

that all regions of the heart are homogeneous (a reasonable assumption in nonischemic cardiomyopathy); and that substrate supply and myocardial perfusion is equal throughout the heart, which is verified by an assessment of perfusion. The model permits assessment of the rate of entry of fatty acid into the mitochondrial compartment, assumed to be unidirectional and directly related to β-oxidation, and the import of fatty acid into a "slow-turnover" pool. This slow-turnover pool represents the incorporation of fatty acids into alternative metabolic pathways such as into triglycerides, amino acids, or membrane phospholipids. With the knowledge of myocardial perfusion and substrate supply, actual utilization of fatty acids can be derived. In addition, the concentration of fatty acid in each compartment can be calculated. Thus, even using a whole-body imaging device, intracellular concentrations and flux of fatty acid can be obtained with PET.

Using the PET assessments described above, myocardial perfusion, oxygen consumption, and fatty acid metabolism were evaluated in patients with known defects and results compared with their healthy siblings. Although healthy siblings are either homozygous-normal or heterozygous for the inherited defect, it was felt that this was a relevant "control" population. In addition, we did not feel that it was ethical to study completely healthy, non-related children, because of issues related to radiation exposure. The severity of the cardiac symptomatology was directly related to the severity of the defect in fatty acid metabolism despite analogous levels of myocardial blood flow and plasma substrate availability (Fig. 16.4; Kelly et al. 1993; Bergmann et al. 2001; Epstein et al. 1999). Myocardial perfusion and plasma levels of fatty acid were similar among patients with acyl-CoA dehydrogenase deficiency, carnitine deficiency, and healthy siblings. Palmitate utilization and oxidation, normalized to myocardial oxygen consumption (assessed with ^{11}C-acetate), was diminished in patients with β-oxidation deficiency (Table 16.1). Patients with deficiencies in the acyl-CoA dehydrogenase enzyme had more severe impairments than patients with carnitine deficiency, perhaps because the latter group was partially treated (Kelly et al. 1993; Bergmann et al. 2001; Epstein et al. 1999).

This set of studies demonstrated the ability of PET to characterize diminished fatty acid utilization in subjects with known deficiencies in inheritance of specific metabolic elements and provide a mechanism for diagnosis of these disorders. In addition, the approach should ultimately allow us to follow the progression or regression of disease, and to evaluate the efficacy of therapeutic interventions. For example, it has been shown that short-chain acyl-CoA dehydrogenase deficiency can be corrected by genetic manipulation in experimental animals (Kelly et al. 1997). If patients with a specific enzyme defect could be di-

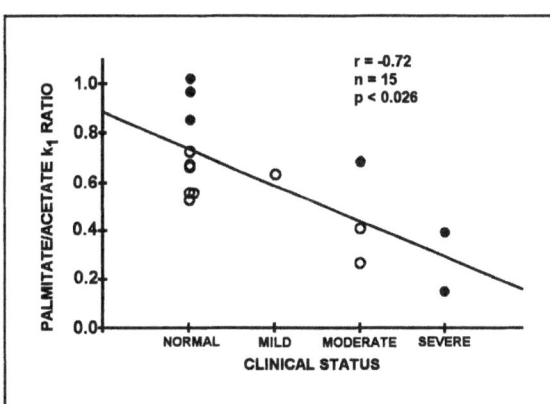

Fig. 16.4. Correlation of the ratio of palmitate normalized to acetate (which provides an assessment of overall myocardial oxygen consumption) with clinical severity of patients with long-chain acyl-CoA dehydrogenase deficiency. The severity of the clinical defects correlated with the severity of the defect in fatty acid metabolism. *Open circles* represent adults while *closed circles* represent children (who are more dependent on fatty acid metabolism). Reproduced with permission from Kelly et al. (1993)

Table 16.1. Table of results from quantitative analysis of myocardial fatty acid metabolism in healthy siblings, in patients with defects in the acyl-CoA dehydrogenase enzymes, and in patients with defects in carnitine transport. Results represent quantitative analysis of PET data for palmitate oxidation (obtained with 1-^{11}C-palmitate); myocardial oxygen consumption (MVO$_2$) based on the use of 1-^{11}C-acetate; and the percentage of MVO$_2$ accounted for by palmitate oxidation. Patients with defects in the acyl-CoA dehydrogenase enzyme system had more profound defects than patients with carnitine deficiency. The asterisk (*) represents $P<0.05$ compared with healthy siblings

	Healthy siblings	Patients with β-oxidation defects	Patients with carnitine deficiency
Palmitate oxidation (nmol/g/min)	13±6	3±3*	8±5
MVO$_2$ (μmol O$_2$/ g/min)	3.4±0.4	3.8±1.2	2.9±0.6
MVO$_2$ accounted for by palmitate oxidation (%)	9±5	2±3*	7±4

agnostically identified and a genetic therapy administered, the use of PET should allow the objective assessment of the efficacy of such a therapy. Accordingly, noninvasive approaches should be increasingly useful in our understanding of these diseases and the efficacy of therapeutic interventions. Another example would be the use of angiogenic factors to enhance tissue blood flow. This approach is being used in patients with ischemic coronary artery disease who are unable or unwilling to undergo conventional interventional strategies such as balloon angioplasty or coronary artery bypass grafting. Approaches with PET would readily be able to identify the distribution of administered gene products using a "reporter gene" strategy and determine changes in myocardial perfusion as well.

16.8
Future Directions

Molecular genetics and gene therapy have already established themselves as approaches for diagnosing and correcting inherited disorders. Experimental studies have demonstrated the great potential for correcting metabolic disorders including those involving fatty acid oxidation. As gene therapy progresses, the need for assessment of both expression of administered gene products as well as assessment of the phenotypic end points of these products will be necessary. To confirm successful DNA integration, RNA transcription and enzyme activity after gene therapy currently requires invasive tissue samples. Thus an approach that noninvasively confirms in vivo gene transcription and protein production after transfection with the therapeutic gene would be ideal. PET has already been demonstrated to have this potential. In addition, the approach offers the potential for delineation of tissue distribution and longevity with different administration and dosing protocols.

An important adjunct to the expression of genes is an objective measure of the biochemical activity of these gene products. PET can be used to noninvasively and sequentially assess myocardial metabolism in humans. Thus, it is likely that noninvasive approaches such as PET will become useful in the diagnosis of patients with inherited diseases, and for following the disease progression. In addition, as genetic therapies are developed, noninvasive approaches such as PET

should be useful in demonstrating the ability of administered gene therapy to replace defective genes and for delineation of their phenotypic correction.

■ **Acknowledgements.** The authors thank Kristine Kulage for preparation of the typescript.

16.9
References

Bergmann SR (1990) Positron emission tomography of the heart. In: Gerson M (ed) Cardiac nuclear medicine, 3rd edn. McGraw-Hill, New York, pp 299–335

Bergmann SR (1998) Cardiac positron emission tomography. Semin Nucl Med 28:320–340

Bergmann SR, Herrero P, Markham J et al (1989) Noninvasive quantification of myocardial blood flow in human subjects with oxygen-15-labeled water and positron emission tomography. J Am Coll Cardiol 14:639–652

Bergmann SR, Rubin PJ, Hartman JJ et al (1995) Detection of abnormalities in fatty acid metabolism in patients with cardiomyopathy using PET (abstract). J Nucl Med 36:142P

Bergmann SR, Weinheimer CJ, Markham J et al (1996) Quantitation of myocardial fatty acid metabolism using PET. J Nucl Med 37:1723–1730

Bergmann SR, Herrero P, Sciacca RR et al (2001) Characterization of altered myocardial fatty acid metabolism in patients with inherited cardiomyopathy. J Inherit Metabol Dis 24:657–674

Bing RJ (1965) Cardiac metabolism. Physiol Rev 45:171–213

Blasberg RG, Tjuvajev J (1999) Herpes simplex virus thymidine kinase as a marker/reporter gene for PET imaging of gene therapy. Q J Nucl Med 43:163–169

Boles RG, Buck EA, Blitzer MG et al (1998) Retrospective biochemical screening of fatty acid oxidation disorders in postmortem livers of 418 cases of sudden death in the first year of life. J Pediatr 132:924–933

Bonnefont JP, Specola NB, Vassault A et al (1990) The fasting test in paediatrics: applications of the diagnosis of pathological hypo- and hyperketotic states. Eur J Pediatr 150:80–85

Brivet M, Slama A, Saudubray JM et al (1995) Rapid diagnosis of long chain and medium chain fatty acid oxidation disorders using lymphocytes. Ann Clin Biochem 32:154–159

Brown MA, Marshall DR, Sobel BE et al (1987) Delineation of myocardial oxygen utilization with carbon-11 labeled acetate. Circulation 76:687–696

Brown MA, Myears DW, Bergmann SR (1989) Validity of estimates of myocardial oxidative metabolism with carbon-11 acetate and positron emission tomography despite altered patterns of substrate utilization. J Nucl Med 30:187–193

Camici P, Ferrannini E, Opie LH (1989) Myocardial metabolism in ischemic heart disease: basic principles and application to imaging by positron emission tomography. Prog Cardiovasc Dis 32:217

Corr PB, Creer MH, Yamada KA et al (1989) Prophylaxis of early ventricular fibrillation by inhibition of acylcarnitine accumulation. J Clin Invest 83:927–936

Depre C, Davies PJ, Taegtmeyer H (1999) Transcriptional adaptation of the heart to mechanical unloading. Am J Cardiol 83:58H–63H

Epstein S, Sciacca R, Chou RL et al (1999) Delineation of altered myocardial fatty acid metabolism in children with inherited defects of β-oxidation. J Nucl Med 40:203P

Gambhir SS, Barrio JR, Wu L et al (1998) Imaging of adenoviral-directed herpes simplex virus type 1 thymidine kinase reporter gene expression in mice with radiolabeled ganciclovir. J Nucl Med 39:2003–2011

Geltman EM, Smith JL, Beecher D et al (1983) Altered regional myocardial metabolism in congestive cardiomyopathy detected by positron tomography. Am J Med 74: 773–785

Gropler RJ, Siegel BA, Lee KJ et al (1990) Nonuniformity in myocardial accumulation of fluorine-18-fluorodeoxyglucose in normal fasted humans. J Nucl Med 31:1749–1756

Hale DE, Bennett MJ (1992) Fatty acid oxidation disorders: a new class of metabolic diseases. J Pediatr 121:1–11

Herrero P, Hartman JJ, Gropler RJ et al (1996) Quantification of myocardial perfusion with PET using carbon-11 acetate and a compartmental model in human subjects. J Nucl Med 37:83P

Ho KKL, Anderson KM, Kannel WB et al (1993) Survival after the onset of congestive heart failure in Framingham heart study subjects. Circulation 88:107–115

Kelly CL, Rhead WJ, Kutschke WK et al (1997) Functional correction of short-chain acyl-CoA dehydrogenase deficiency in transgenic mice: implications for gene therapy of human mitochondrial enzyme deficiencies. Human Mol Genet 6:1451–1455

Kelly DP, Strauss AW (1994) Inherited cardiomyopathies. N Engl J Med 330:913–919

Kelly DP, Mendelson NJ, Sobel BE et al (1993) Detection and assessment by positron emission tomography of a genetically determined defect in myocardial fatty acid utilization (long-chain acyl-CoA dehydrogenase deficiency). Am J Cardiol 71:738–744

Lange LG, Sobel BE (1983) Mitochondrial dysfunction induced by fatty acid ethyl esters, myocardial metabolites of ethanol. J Clin Invest 72:724

Largillerie C, Vianay-Saban C, Fontaine M et al (1995) Mitochondrial very long chain acyl-CoA dehydrogenases deficiency – a new disorder of fatty acid oxidation. Arch Dis Child Fetal Neonatal Ed 73:F103–F105

Lerch RA, Bergmann SR, Sobel BE (1992) Delineation of myocardial fatty acid metabolism with positron emission tomography. In: Bergmann SR, Sobel BE (eds) Positron emission tomography of the heart. Futura Publishing, Mount Kisco, NY, USA, pp 129–153

Lindeneg O, Mellemgaard K, Fabricus J et al (1964) Myocardial utilization of acetate, lactate and free fatty acids after injection of ethanol. Clin Sci 27:427–435

Myears DW, Sobel BE, Bergmann SR (1987) Substrate use in ischemic and reperfused canine myocardium: quantitative considerations. Am J Physiol Heart Circ Physiol 253:H107–H114

Neely JR, Morgan HE (1974) Relationship between carbohydrate and lipid metabolism and the energy balance of heart muscle. Annu Rev Physiol 36:423–459

Paulson DJ (1998) Carnitine deficiency-induced cardiomyopathy. Mol Cell Biochem 180:33–41

Podrid PJ, Fogel RI, Fuchs TT (1992) Ventricular arrhythmia in congestive heart failure. Am J Cardiol 69:82G–96G

Pollitt RJ (1989) Disorders of mitochondrial beta-oxidation: prenatal and early post-natal diagnosis and their relevance to Reye's syndrome and sudden infant death. J Inher Metab Dis 12 [Suppl 1]:215–230

Ratib O, Phelps ME, Huang SC et al (1982) Positron tomography with deoxyglucose for estimating local myocardial glucose metabolism. J Nucl Med 23:577–586

Rhead WJ (1990) Screening for inborn errors of fatty acid oxidation in cultured fibroblasts: an overview. In: Tanaka K, Coates PM (eds) Fatty acid oxidation: clinical, biochemical and molecular aspects. Wiley-Liss, New York, pp 365–382

Rubin PJ, Lee DS, Davilla-Roman VG et al (1996) Superiority of C-11 acetate compared with F-18 deoxyglucose in predicting myocardial functional recovery by positron emission tomography in patients with acute myocardial infarction. Am J Cardiol 78:1230–1236

Saudubray JM, Martin D, Lonlay P de et al (1999) Recognition and management of fatty acid oxidation defects: a series of 107 patients. J Inher Metab Dis 22:488–502

Schelbert HR, Henze E, Sochor H et al (1986) Effects of substrate availability on myocardial C-11 palmitate kinetics by positron emission tomography in normal subjects and patients with ventricular dysfunction. Am Heart J 111:1055–1064

Schwaiger M, Beunken R, Grover-McKay F et al (1990) Regional myocardial metabolism in patients with acute myocardial infarction assessed by positron emission tomography. J Am Coll Cardiol 15:1021–1031

Severeid L, Connor WE, Long JP (1969) The depressant effect of fatty acids on the isolated rabbit heart. Proc Soc Exp Biol Med 131:1239–1243

Tamaki N, Yonekura Y, Yamashita K et al (1989) Positron emission tomography using F-18-deoxyglucose in evaluation of coronary artery bypass grafting. Am J Cardiol 64:860–865

Ter-Pogossian NM, Klein MS, Markham J et al (1980) Regional assessment of myocardial metabolic integrity in vivo by positron-emission tomography with 11-C-palmitate. Circulation 61:242–255

Tjuvajev J, Avrill N, Safer M et al (1997) Quantitative PET imaging of HSV1-TK gene expression with [I-124]FIAU (abstract). J Nucl Med 38:239P

Wanders RJA, Vreken P, Boer MEJ den et al (1999) Disorders of mitochondrial fatty acyl-CoA beta-oxidation defects. J Inher Metab Dis 22:442–487

Molecular Background of ^{18}F-2-deoxy-D-glucose (FDG) Uptake in the Ischemic Heart

17

Silvia Egert, Markus Schwaiger

Contents

In industrialized countries coronary artery disease (CAD) remains the most common cause of death. However, the mortality from ischemic heart disease has decreased in recent years. A better understanding of risk factors combined with improvements in therapy and the development of better diagnostic techniques has significantly contributed to this process.

Preventive measures such as modification of risk factor profiles have been shown to be effective in reducing cardiovascular mortality. Therapy of advanced stages of disease has also improved substantially. Clinical progress is in part due to the increasing clinical awareness that left ventricular dysfunction in patients with CAD may be a reversible condition and often reflects underperfused but viable tissue (Rees et al. 1971). Several studies indicate that patients with poor left ventricular function and multivessel disease show an improvement of cardiac function after surgical revascularization (Mickleborough et al. 1995; Kaul et al. 1996; Miller et al. 1996; Passamani et al. 1996). Meanwhile, it is well-accepted that a dysfunctioning myocardium may develop different survival strategies to adapt to ischemic conditions and that ischemia does not necessarily lead to cell death and scarred tissue (Rahimtoola 1989; Wijns et al. 1996). Therefore, distinguishing viable from nonviable tissue is of significant clinical importance. However, the myocardial adaptations to reduced coronary flow are not completely understood. Various terms have been used to describe adaptive processes to hypoxia. Most commonly used are: "hibernation, preconditioning, and stunning." In addition, other pathophysiological processes such as apoptosis, dedifferentiation, and inflammatory response may be involved in postischemic myocardial dysfunction. Moreover, one has to be aware that acute ischemia, hibernation, and stunning as well as apoptosis and necrosis may dynamically interrelate in the development of left ventricular dysfunction.

With respect to establishing a relationship between clinical cardiac nuclear medicine and the molecular background of heart metabolism, this article will first briefly outline the myocardial substrate supply for energy production in the normal and the ischemic heart. In view of its pivotal role in ischemia, glucose metabolism will be discussed in detail. Finally, the role of apoptosis in myocardial ischemia will be described in a short paragraph.

The topics discussed in detail are as follows:
1. Myocardial substrate selection
2. The assessment of glucose uptake via FDG-PET
   - Tissue viability
   - The clinical value of FDG
3. Understanding of the FDG signal
4. Glucose metabolism in the ischemic heart
   - Experimental and pathophysiological models of acute and chronic ischemia
5. The cellular and molecular regulatory steps of glucose transport
   - Myocardial glucose transporter proteins (GLUT)
   - Regulation on the level of their kinetic activity, of translocation and of gene expression
6. Regulation steps after glucose transport
7. Ischemia and apoptosis.

17.1
Myocardial Substrate Selection

A fundamental principle of myocardial substrate utilization is the ability to gain energy from all substrates available, although a hierarchy of preferred fuels has been demonstrated. Over 30 years ago, fatty acids were identified as the major energy source in the aerobically perfused myocardium (reviewed in Neely et al. 1974). Basic publications have recorded measurements of cellular levels of intermediates and characterized elementary relationships between activities of regulatory enzymes participating in glucose and lipid metabolism (for reviews see Neely et al. 1974; Taegtmeyer et al. 1998). Fatty acids and triglycerides delivered to the myocyte are degraded to acetyl-CoA which is further metabolized to citrate in the citric acid cycle. Oxidation of fatty acids suppresses glucose utilization by inhibition of several distinct enzymes in the glycolytic pathway. The use of exogenous triglycerides for energy production depends on the availability of exogenous fatty acids, the presence of hormones such as insulin, and on the level of cardiac work (Neely et al. 1974).

Stimulated by increase in ventricular pressure (Crass et al. 1969) glucose uptake and oxidation becomes the main pathway for energy production in the postprandial state, when glucose and insulin have increased and the amount of fatty acids in blood circulation is low (Taegtmeyer 1994). The significance of glucose as an energy substrate in the heart also increases in situations of oxygen depletion such as ischemia and hypoxia (Egert et al. 1997; Wheeler 1998). Glucose inhibits fatty acid oxidation most probably by blocking the uptake of fatty acids in the mitochondria (Saddik et al. 1994). Although fatty acid and glucose metabolism in myocardium have been studied extensively, the complex interplay and regulation of these energy-generating pathways is still not fully understood.

Lactate and ketone bodies as substrates for the TCA-cycle also contribute to energy production. Lactate may be the main substrate after exercise (Gertz et al. 1988) and the concentration of ketone bodies in the blood increases with starvation, during pregnancy, and in diabetic ketoacidosis (Taegtmeyer 1983). Glycogen as an endogenous substrate is rapidly metabolized when glycogen phosphorylase is stimulated by ischemia, epinephrine, glucagon, or during an increase in cardiac workload (Goodwin et al. 1996; Henning et al. 1996). Glycogen phosphorylase is the key enzyme in regulation of glycogenolysis and its activity is increased by the cyclo-AMP-dependent protein kinase or by the calcium-activated phosphorylase kinase (Morgan et al. 1964).

The importance of understanding the complex relationship between myocardial fuel supply, utilization and contractile function has been underscored by numerous studies demonstrating that alterations in energy metabolism exist in many chronic cardiac diseases, especially ischemic heart disease. In recent years, it was shown that such chronic alterations may induce changes in myocardial gene expression (Van Bilsen et al. 1998), mainly in encoding for contractile proteins, ion pumps, and metabolic enzymes (Nadal-Ginard et al. 1989; Sadoshima et al. 1997). These genetic adaptations often represent similarities to the fetal gene expression pattern (Depre et al. 1998a).

17.2
The Assessment of Glucose Uptake via FDG-PET

As already mentioned, glucose plays a central role as an energy substrate especially in the ischemic heart. Glucose uptake consists of glucose transport and phosphorylation. It is characterized by changes in the glucose extraction rate and transport across the membrane. Glucose uptake can be directly measured by specifically labeled tracers (2-hydrogen-3-glucose)

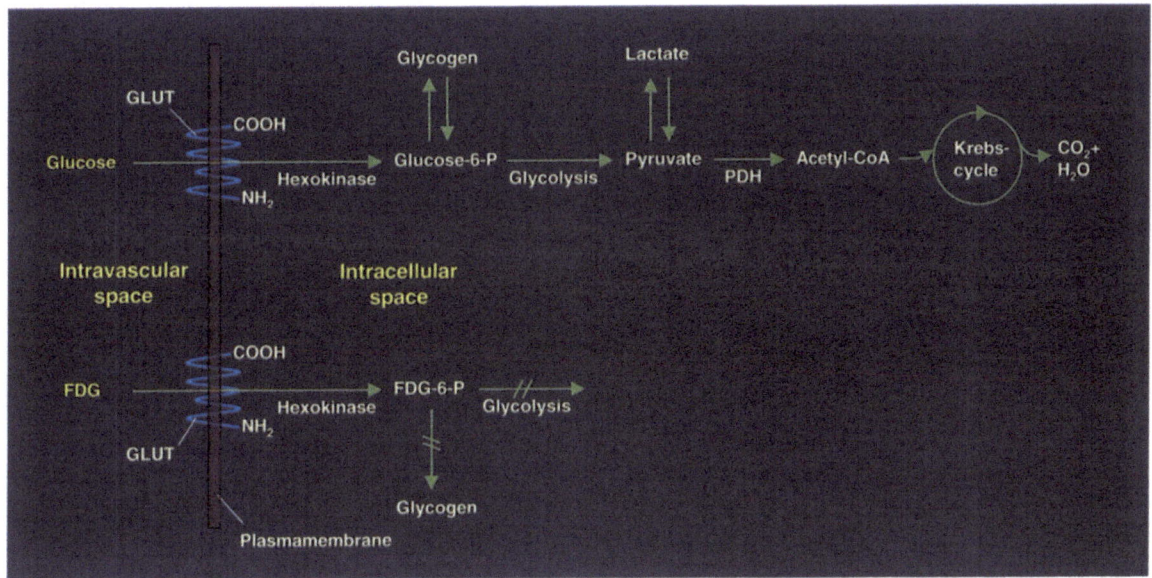

Fig. 17.1. Metabolism of glucose and FDG. Glucose and 2-[^{18}F]-fluoro-2-deoxy-D-glucose (FDG) are transported into the myocyte via glucose transporters (GLUT) and phosphorylated by hexokinase to glucose-6-phosphate (glucose-6-P) and FDG-6-phosphate (FDG-6-P). During aerobic glycolysis, glucose-6-P is converted to pyruvate. Pyruvate is metabolized to acetyl-CoA and further oxidized in the Krebs cycle. Under ischemic conditions, the pyruvate dehydrogenase complex (PDH) is inhibited leading to an increased production of lactate. FDG-6-P does not undergo further glycolytic or glycogen synthesis steps and thus is trapped in the cytoplasm

or glucose analogues (2-deoxyglucose, FDG) (Tillisch et al. 1986; Russell et al. 1997). Myocardial FDG uptake has been extensively studied in experimental and clinical approaches. Presently, FDG is a widely used tracer for the detection of viability in dysfunctional myocardium of patients by the use of positron emission tomography (PET). Like glucose, FDG is taken up into the myocyte via passive diffusion facilitated by glucose transporter proteins (Wheeler 1998). The tracer is subsequently subjected to phosphorylation by hexokinase. FDG-P is then trapped in the cytoplasm, since the very specific phosphohexose isomerase does not accept the glucose analogue as a substrate (Fig. 17.1). On account of these kinetic properties, FDG became one of the most important tracers for assessment of glucose utilization.

FDG uptake can be quantitated mainly by two different approaches: the first is based on a three-compartment model, which was established by Sokoloff et al. (1977). The first compartment is the blood plasma pool containing deoxyglucose. It interacts with a second compartment, the cytosol, which also contains deoxyglucose. The third compartment is that portion of the cytosol containing deoxyglucose-P. This model assumes that the glucose metabolism is in a steady state, that the tissue extraction fraction for glucose and deoxyglucose is small, that the blood flow is homogenous in the tissue, and that deoxyglucose diffusion between the three compartments follows first-order kinetics (for more details, see Sokoloff et al. 1977). To quantify the regional rate of glucose utilization according to this model, a correction factor has to be experimentally evaluated which converts FDG uptake to glucose uptake. This factor is called the "lumped constant" and describes the relation between a tracer analogue (FDG) and glucose in the myocardium. It has been shown that the lumped constant may change depending on the physiological or experimental conditions (Hariharan et al. 1995). The second approach for quantitating FDG uptake was developed by Patlak et al. (1983). Measurement of glucose utilization in this model is independent of the number of compartments and does not require a constant concentration of FDG in the blood plasma. Analysis of FDG kinetics with this graphical method assumes a unidirectional system with a complete and irreversible tracer trapping in a final compartment (for more mathematical details, see Patlak et al. 1983), and it involves estimations of the size and shape of the myocardium. The graphical quantification method of Patlak et al. (1983) is widely accepted for determining tissue FDG uptake with PET data.

17.2.1
Tissue Viability

Chronically dysfunctional myocardium may contain regions of viable and of nonviable tissue. To characterize tissue viability, a myocardial blood flow tracer such as ^{13}N-ammonia or ^{15}O water has to be applied in combination with FDG for two reasons. One is that a normally perfused myocardial region must be defined as a reference region for the normalization of FDG uptake (Bonow et al. 1991). Without flow measurements, areas consisting of a mixture of necrotic or scarred tissue and a significant amount of hibernating regions may not be correctly interpreted. The second reason is the finding that FDG utilization can vary throughout the normal myocardium (Gropler et al. 1991). Theoretically, three different patterns of blood flow and glucose utilization describe myocardial dysfunctioning tissue: First, normal blood flow is combined with normal or only slightly stimulated glucose uptake. This situation reflects healthy regions of the myocardium, but also regions characterized as "stunned" (Vanoverschelde et al. 1993). A recent study found that in 26 patients with chronic occlusion of a main coronary artery, the blood flow determined at rest with nitrogen-13-ammonia was almost normal in collateral-dependent left ventricular wall segments both with and without regional contractile function abnormalities (Vanoverschelde et al. 1993). Second, a reduced blood flow associated with normal or enhanced glucose uptake is often described as a "mismatch" and seems consistent with the term "hibernating myocardium" (Rahimtoola 1996). Third, a reduced blood flow accompanied by a significant reduced glucose uptake in the same areas reflects the "match" situation which is found in necrotic and scarred tissue.

The possibility of differentiating the above-mentioned metabolic situations of tissue viability in patients with CAD underlines the important role of PET in cardiology.

17.2.2
The Clinical Value of FDG

Many studies have evaluated the clinical value of FDG-PET measurements to predict tissue viability. It was shown that FDG-PET data predicted regional systolic functional recovery after coronary bypass sur-

gery with a specificity of 74–92% and a sensitivity of 73–95% (Lucignani et al. 1992; Gropler et al. 1993; Tamaki et al. 1993; Knuuti et al. 1994; Tillisch et al. 1986). These data have established FDG-PET imaging in the clinical management of CAD. Moreover, several studies investigated whether the detection of the metabolism-perfusion mismatch pattern is important for the further functional recovery of the myocardium and the survival time of patients. It was found that the extent of the perfusion-metabolism mismatch detected by ^{13}N-ammonia and FDG-PET correlated linearly with the percentage of functional improvement after coronary bypass surgery in patients with CAD and diminished left ventricular function (DiCarli et al. 1995). A recent study investigated the disease development in patients with regional wall-motion abnormalities and a mismatch pattern demonstrated by FDG-PET and ^{99m}Tc-sestamibi-SPECT over two years. They observed no adverse myocardial events in cases of successful revascularization compared with a 22% cardiac event rate in patients who received medication only (Vom Dahl et al. 1997). In combination with blood flow measurements, FDG-PET has been established as a powerful tool with demonstrated clinical value for characterizing variable patterns of blood flow and metabolism in dysfunctioning myocardium.

17.3
Understanding of the FDG Signal

The underlying regulatory mechanisms on the level of signal transduction and gene expression with respect to either the FDG signal or glucose metabolism in general are yet not completely understood. Due to the fact that dysfunctional hearts in patients may present unique mixtures of normal, adapted, and maladapted tissue regions, it is obvious that numerous molecular mechanisms communicating by a complex interplay are responsible for the individuality of disease progression. To provide an integrated understanding of the cardiac FDG signal in response to energy stress, the following pages will briefly review the current knowledge of cellular and molecular regulatory steps of myocardial glucose metabolism. The first paragraph will describe glucose metabolism in the ischemic heart with respect to various experimental and pathophysiological models of acute and chronic ischemia. Subsequently, glucose transport mediated by facilitative glucose transporter proteins (GLUT) will be discussed in detail. Regulatory steps that follow glucose transport will then be outlined. Finally, the role of apoptosis in the ischemic heart will be briefly addressed.

17.4
Glucose Metabolism in the Ischemic Heart

As already discussed, the cardiac energy metabolism depends on the oxidation of several substrates. Hormones, substrate availability, and energy demand act in a complex interrelationship to influence the participation of a substrate in the heart's energy production. Under ischemic and hypoxic conditions, glucose becomes the main source for anaerobic and aerobic energy metabolism (Liedke 1981). Several findings in various experimental models have shown that utilization of glucose and the tracer analogue FDG is stimulated during and following ischemia (Liedke 1981; Kalff et al. 1992). In dysfunctional human myocardium an increased extraction of FDG was shown in cases of acute ischemia as well as in response to chronic states of ischemia (Camici et al. 1986; Tillisch et al. 1986; Vom Dahl et al. 1994). However, although both the acute and the chronic conditions of underperfusion induce an increase in glucose utilization as measured by stimulated FDG uptake, the signaling mechanism by which these conditions mediate the increase is different. The following paragraphs describes experimental and clinical models of acute and chronic ischemia.

17.4.1
Acute Ischemia

"Acute ischemia" or "acute hypoxia" is mainly produced under experimental conditions when oxygen supplied in the perfusate is totally or partially removed for a short period of time (10–30 min). Previous studies have shown that within a few minutes after a sudden oxygen deprivation the FDG uptake increased significantly as a result of fast intracellular signaling mechanisms. The mechanism has been found to involve the translocation of glucose transporter proteins, which promotes a prompt rise in glucose uptake capacity (Kalff et al. 1992; Wheeler

1998). This short period of experimentally induced ischemia (15 min) also leads to ultrastructural changes in the cell. The first signs of injury are seen in the mitochondria. The matrix granules, which normally are present, disappear and the matrix becomes more electron-lucent (Fig. 17.2).

Clinically, acute ischemia follows immediately after a sudden arterial occlusion induced by a thrombotic event (myocardial infarction) and rapidly leads to a regional increase in myocardial FDG uptake (Vom Dahl et al. 1994). The current knowledge of the intracellular signaling mechanism in response to acute ischemia will be discussed later in this chapter (see Sect. 17.5 on glucose transport).

17.4.2
Chronic Ischemia

The term "chronic ischemia" is a collective conception and the general result of intermittent reductions of coronary flow. Depending on the individual history of the disease, several distinct strategies exist for the heart to adapt to the repetitive underperfusion of cardiac tissue. In the following, the most commonly discussed adaptive concepts are reviewed.

17.4.3
Hibernating Myocardium

In 1978, the concept of "hibernating myocardium" was defined by Rahimtoola (1989). The term means "a state of persistently impaired myocardial function at rest due to reduced coronary blood flow that can be partially or completely restored to normal if the net balance between myocardial oxygen supply and demand is favorably altered." Clinically, this concept was underlined by several studies demonstrating patients with severe left ventricular dysfunction, who progressively recovered cardiac function after coronary artery bypass grafting. The myocardial strategy of hibernation, however, does not only describe reversibility of dysfunction but also a reduced perfusion at rest in presence of severe CAD (Rahimtoola 1989; Heusch 1998). By uncoupling cardiac contractile function and blood flow, the reduction of contractile work is achieved and results in a decrease of myocardial energy demand. The reduced energy demand is thought to be the basis for the increased tolerance of myocytes to ischemia.

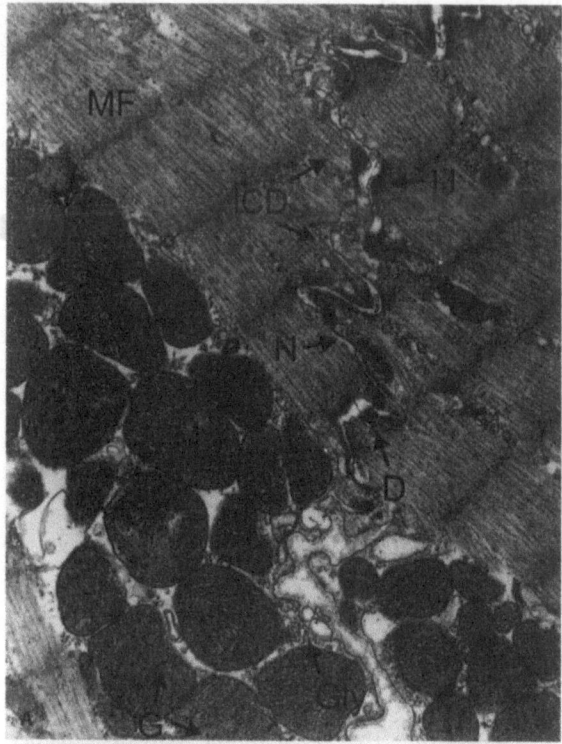

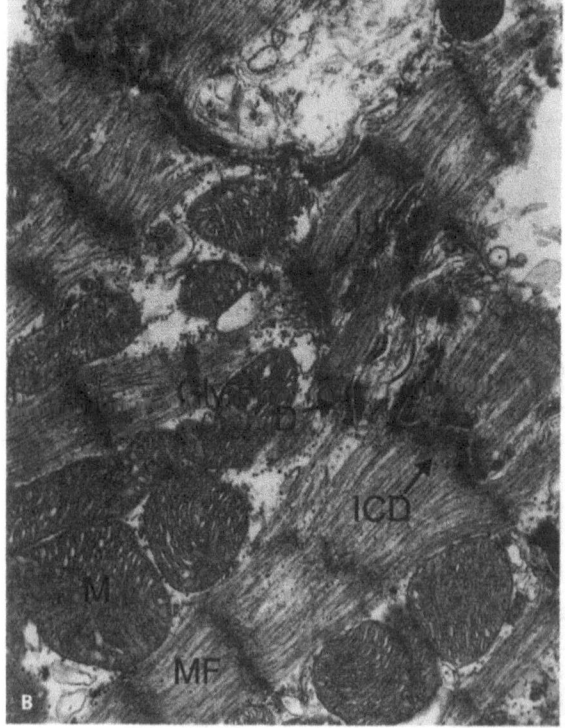

17.4.3.1
Morphologic and Metabolic Alterations in Hibernating Myocardium

The erstwhile common opinion that hibernating myocytes are intact has been found to be untrue. Transmural biopsies from hibernating myocardium obtained during surgery have revealed that many cells are characterized by dedifferentiation, degeneration, and fibrosis. Such myocytes show a perinuclear loss of contractile proteins, numerous small mitochondria, an alteration of all structural proteins resulting in a disorganization of the cytoskeleton or abnormally shaped and sized nuclei that demonstrate a heterogeneous distribution of chromatin (Borgers et al. 1993; Vanoverschelde et al. 1993; Ausma et al. 1996; Schwarz et al. 1996; Shivalkar et al. 1996). The expression of α-actinin, cardiotin, titin, and desmin is altered. Apoptosis, programmed cell death, also occurs in hibernating myocardium. Chen et al. (1997) studied pigs with experimental stenosis in the left anterior descending coronary artery (LAD), which resulted in severe systolic dysfunction. Apoptosis was detected by the in situ end labeling method and DNA laddering in the hibernating region supplied by the LAD. No apoptotic cells were found in the normal perfused control regions. The percentage of apoptotic cells correlated significantly with the severity of myocardial dysfunction. In their study, it was proposed that increasing apoptosis (and necrosis) may in part

◄───────────────────────

Fig. 17.2. Ultrastructural changes in rat myocardial tissue induced by ischemia. A Control; B ischemia. Electron microphotographs show cell ultrastructure in normally perfused rat cardiac muscle and structural alterations after 15 min of no-flow ischemia. A normal heart tissue with intact ultrastructure is seen. The plasmalemma contains junctional complexes [desmosomes (D), intermediate junction (IJ), nexus (N)] that are combined in intercalated disks (ICD). They form a tight connection between the myofibrils (MF) of two cardiac cells. Mitochondria (M) in healthy cells are composed of a dark matrix with granules (G) and intact cristae. B Signs of a beginning ischemic injury are apparent. They are first manifest in the mitochondria, the organelles most sensitive to ischemia. The dark matrix granules as seen in the normal cell have disappeared, a loss that took place in the first 10 min after the onset of ischemia. Accordingly, the matrix has become more disperse and electron-lucent. However, cristae appear to still be intact after 15 min of ischemia since no cristae fragmentation, a sign of more severe damage, is visible. Gly, glycogen granules. Magnification of both images: 27,300×

be responsible for myocyte loss and increased fibrosis in long-term hibernation. Most striking, large areas in hibernating myocytes are filled by glycogen deposits whose role still remains ill defined (Elsässer et al. 1997).

Hibernating myocardium is detected with positron emission tomography by a decrease in perfusion (as detected by decreased ^{13}N-ammonia) (Rahimtoola 1996) and an increased retention of FDG (Maki et al. 1996), suggesting enhanced glucose uptake and glycogen accumulation in the same regions (Depre et al. 1997). It can be assumed that most of the FDG taken up by the myocyte is further metabolized to the glycogen store (Doenst et al. 1999). Moreover, results in a pig model of hibernation over three months also indicated stimulated FDG utilization due to chronic ischemia (Bialik et al. 1997). This is supported by data which show a positive correlation between FDG uptake and glycogen accumulation in hibernating myocardium from humans (Depre et al. 1997) and by an increase in cellular glucose transporter protein (GLUT) 1 expression in human hibernating myocardium (Brosius et al. 1997a). Similar to other dedifferentiation processes ongoing in the hibernating myocardium, the accumulation of glycogen is a characteristic of the fetal heart (Depre et al. 1998b).

17.4.4
Preconditioning

The general concept of preconditioning derived from the observation that repeated short episodes of myocardial ischemia and reperfusion reduce the ischemic damage after a subsequent lag episode of coronary artery occlusion in the dog (Murry et al. 1986). Thus, preconditioning can be defined as the tolerance to prolonged severe ischemia achieved by one or more preceding brief ischemic episodes. Although a large number of studies were initiated in the field of intracellular signal transduction, the exact mechanism of this process is not yet defined. One important finding was that a 5-min adenosine infusion in rabbit hearts had the same protective effect as 5 min of ischemia (Doenst et al. 1999). However, since the adenosine action could not be reproduced in rat hearts, its protective role seems to be species-dependent (Kloner et al. 1998). Additional studies demonstrated that during ischemic preconditioning endogenous catecholamines activate protein kinase C by binding to the extracellu-

lar a_1-adrenoceptor (Hu et al. 1995; Mitchell et al. 1995). Further downstream signals which lead to the protecting effect are still unknown. Preconditioning has also been shown to decrease glycogen breakdown (Murry et al. 1990) and it has been suggested that increasing glycogen levels before ischemia may also have protective effects similar to preconditioning. In isolated rat hearts, FDG uptake and 2-deoxyglucose uptake are increased following several episodes of five-minute ischemia and reperfusion. This observation was accompanied by translocation of the main cardiac glucose transporter, GLUT4 (Nguyen et al. 1995; Tong et al. 2000). The mechanism by which preconditioning enhances glucose uptake and GLUT4 translocation was suggested to be mediated by p38MAP kinase since SB 202190, an inhibitor of p38MAP kinase, attenuated the increased 2-deoxyglucose-6 P accumulation in preconditioned hearts, but had no effect in non-preconditioned hearts. The same study excluded a phosphatidylinositol 3-kinase (PI3-kinase)-dependent signaling mechanism of preconditioning because the inhibition of PI3-kinase (which plays an important part in insulin-dependent signaling leading to GLUT translocation) with wortmannin (a fungal metabolite) did not block the preconditioning-induced increase in 2 deoxyglucose-6P production (Tong et al. 2000). Finally, an anti-apoptotic effect during reperfusion was also demonstrated after ischemic preconditioning, which was shown to alter the ratio between BcLX$_L$ and Bax in favor of the anti-apoptotic component (Baghelai et al. 1999).

17.4.5
Myocardial Stunning

Myocardial stunning was first described by Heyndrickx et al. (1975). They observed "prolonged but reversible myocardial dysfunction following restoration of myocardial blood flow." Therefore, the major difference between stunning and hibernating myocardium is found in myocardial perfusion, which is normal in the stunned dysfunctional heart following transient periods of ischemia (Heyndrickx et al. 1975) but chronically reduced in hibernating myocardium. Stunned myocardium is further characterized by the absence of irreversible cell damage; cardiac function completely recovers after a certain time period. Stunning has been observed experimentally and clinically (Homans et al. 1986). Experimentally, postischemic utilization of FDG was demonstrated to increase after 24 h of perfusion in a canine-stunning heart model with 3 h of occlusion, then to decrease step by step with concomitant functional recovery (Schwaiger et al. 1985). The observed postischemic increase in FDG uptake was accompanied by a myocardial lactate release. Thus, it seems reasonable to assume that the increase in FDG uptake derives from stimulation of anaerobic glycolysis and is not utilized for restoration of the glycogen pool (Schwaiger et al. 1989). Clinically, stunning has been seen after hypothermic ischemic arrest for cardiac surgery, after periods of unstable angina, and after exercise-induced ischemia (Kloner et al. 1998; Bolli 1992). An important factor for the development of stunning seems to be the ischemia-induced accumulation of free intracellular calcium, inducing a decreased myofilament responsiveness in calcium-dependent proteases (Gao et al. 1996). Another hypothesis to explain postischemic stunning discusses the generation of oxygen-derived free radicals during reperfusion following ischemia, which agents affect cellular metabolism at numerous sites (Kim et al. 1987). Myocardial oxygen consumption is normal or slightly reduced in stunned hearts.

Although not fully understood, the mechanisms of hibernation, preconditioning, and stunning all mediate specific intracellular signals which allow adaptation to chronic ischemia and reperfusion. The hibernating heart has adapted to a diminished oxygen supply without switching to anaerobic metabolism. With preconditioning, the heart gets more resistant to irreversible ischemic damage and in the stunned myocardium, repair processes occur during reduced contraction and normal oxygen supply. Therefore, chronic ischemia is a complex and sensitive interplay of numerous factors and cannot be simply explained as an imbalance between oxygen supply and demand.

17.5
Regulation of Myocardial Glucose Transport

17.5.1
Myocardial Glucose Transport Proteins (GLUT)

Except during fasting conditions, exercise, and hyperinsulinemia, the normal heart utilizes only small amounts of glucose (Barrett et al. 1984; Gertz et al. 1988; Ferrannini et al. 1993). Under pathophysiological conditions like moderate ischemia, the glucose uptake into the myocyte increases immediately (Wheeler 1998) and perhaps on a long-term basis, as reflected by left ventricular dysfunction and/or cardiac hypertrophy (Allard et al. 1994; Brosius et al. 1997b). The cellular uptake of glucose (or its tracer analogue FDG) into the myocyte is mediated by a family of glucose transport proteins (GLUT) which facilitate the diffusion process along an inward concentration gradient. Meanwhile, it is well accepted that under basal and insulin-stimulated conditions, glucose transport is the rate-limiting step in the heart (Kashiwaya et al. 1994; Charron et al. 1999). Currently, nine distinct isoforms, including one pseudogene construct, have been reported (Mueckler 1994; Doege et al. 2000a; Ibberson et al. 2000). Of these GLUT1, 3, 4, and most recently, GLUT8 have been described in heart tissue (Charron et al. 1989; Mueckler et al. 1997; Grover-McKay et al. 1999; Doege et al. 2000b). The transporters have an average molecular weight of 50 kD and contain 12 membrane-spanning domains. The largest extracellular loop contains the binding site for glucose (Mueckler 1994). In myocardium, the main GLUT isoform is GLUT4, the "insulin-responsive" transporter. Under basal conditions, GLUT4 resides to a large extent within intracellular storage vesicles (Sun et al. 1994). Induced by various stimuli including insulin, exercise, contraction, and energy stress such as ischemia, hypoxia, or anoxia, cardiac GLUT4 is rapidly recruited (within a few minutes) from its intracellular storage pool to the sarcolemma and t tubules associated with the sarcolemma (Douen et al. 1990; Cartee et al. 1991; Sun et al. 1994; Fischer et al. 1997). Immunohistochemical studies visualized by electron microscopy illustrate GLUT4 translocation in response to ischemia (Fig. 17.3). The underlying mechanism of translocation is under intensive investigation and will be discussed later in this section.

GLUT1 is an isoform found in most cell types including fetal and tumor tissue. This isoform is mainly inserted in the plasma membrane and is thought to be responsible for basal glucose transport (Mueckler 1994). It has been found to be expressed in relatively high amounts in cardiac myocytes, in contrast to skeletal muscle (Kraegen et al. 1993). This suggests that GLUT1 plays a more important functional role in the myocardium than in other insulin-responsive tissues. The high level of GLUT1 in heart also implies that it may be essential for maintaining glucose transport independently of physiological and pathophysiological stimuli. GLUT1 is present in heart tissue within the cardiomyocyte, whereas in skeletal muscle GLUT1 is located outside the myocyte. Although inserted mainly on the sarcolemma, GLUT1 may also be translocated from intracellular storage vesicles to the cell surface by stimuli such as insulin and ischemia (Egert et al. 1999a). However, compared to GLUT4, GLUT1 translocation is much less pronounced.

Previous studies have demonstrated that myocardial GLUT1 expression is upregulated by chronic ischemia in primary cultures of persistently hypoxic rat neonatal cardiac myocytes (Brosius et al. 1997a) and in vivo in rat myocardium by chronic administration of insulin in a dose-dependent manner (Lay-

butt et al. 1997). In addition, immunohistochemical detection of GLUT1 in biopsies of a normal and a hibernating region of human myocardium has found more immunostaining of this isoform in the hibernating than in the normal region of the same heart (Fig. 17.4).

The role of GLUT3 and GLUT8 in cardiac cells is still unknown. In contrast to GLUT4 and GLUT1, very small amounts of both these isoforms appear to exist in the heart (Grover-McKay et al. 1999; Doege et al. 2000b). The gene sequence and protein coding for GLUT8 was published very recently. It was suggested that the transporter may play a role in genetically manipulated mice which lack the gene for GLUT4 (Doege et al. 2000b; Ibberson et al. 2000). Since GLUT8 also was found in an intracellular compartment in parallel to its location on the plasma membrane, it was suggested that it might translocate (Carayannopoulos et al. 2000). However, the contribution of GLUT3 and GLUT8 to basal and insulin/ischemia-stimulated glucose transport appears to be of minor importance compared to GLUT4 and GLUT1.

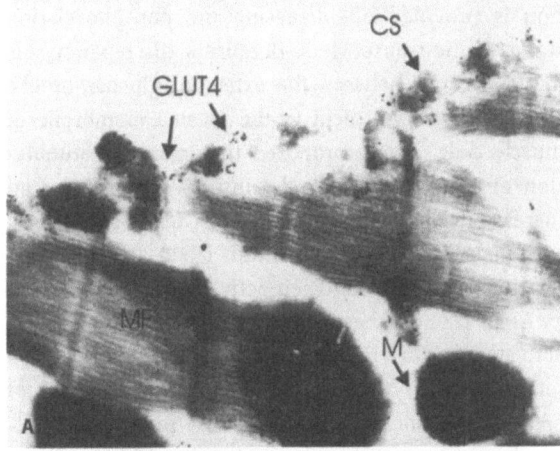

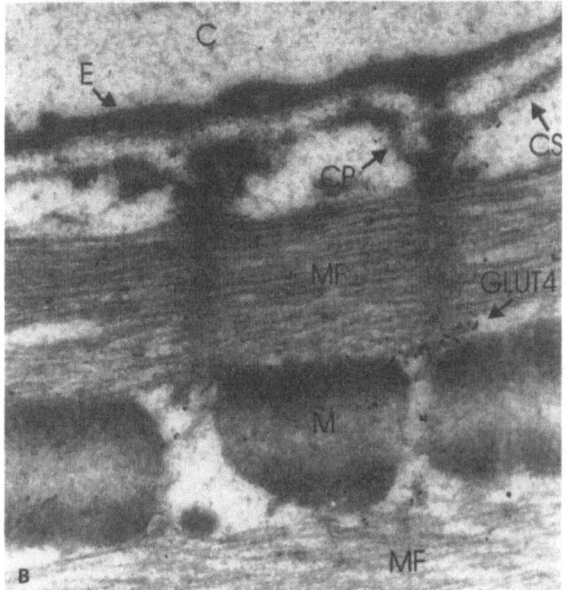

Fig. 17.3. Electron microscopic immunolocalization of GLUT4 protein in rat heart muscle. Immunoelectron microscopy was used to visualize intracellular distribution of GLUT4 protein in isolated, normal-perfused, and ischemic (15 min no-flow ischemia) rat hearts. Cryosections of glutaraldehyde/paraformaldehyde-fixed tissue were incubated with GLUT4 antibody and subsequent localized by IgG, labeled with ultra-small gold and silver enhancement. A Electron micrograph of a normal heart cryosection. Specific GLUT4 labeling is localized mostly in groups between the myofibrils (*MF*), and the mitochondria (*M*) on vesicular structures which in part may belong to the t-tubules. Labeling is also detected in small groups or as isolated gold particles on vesicular structures near the cell surface (*CS*), which latter forms coated pits (*CP*) at two positions. No noticeable gold labeling is visible on the plasma membrane of an endothelial cell (*E*) or in the capillary space (*C*). B Electron micrograph of an ischemic heart cryosection. Immunogold-labeled GLUT4 particles are predominantly found in groups, but singularly at the cell surface (*CS*) or very close to the cell surface. The GLUT4 staining at the cell surface of ischemic hearts appears to be more intense in comparison to the GLUT4 surface staining of normal-perfused rat hearts. Yet, some GLUT4 labeling is detected between the myofibrils (*MF*). However, the amount of labeled elements appears to be less than the amount of gold particles found in the same intracellular regions of control hearts. Magnification of both images: 34,700×

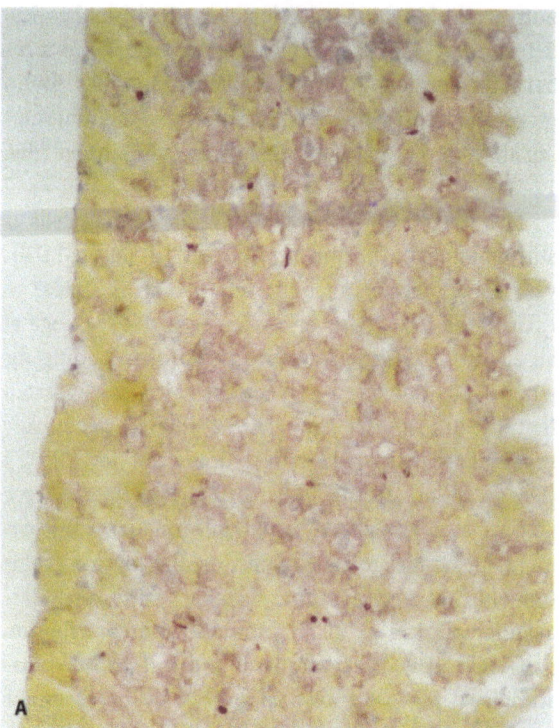

17.5.2
Regulation of GLUT Activity

The extent of glucose transport is dependent not only on the absolute amount of transporters but also on their intrinsic kinetic activity, their subcellular distribution, and on the glucose concentration gradient. The absolute number of cellular transporters is mediated by the rate of gene transcription, mRNA stability, the rate of protein translation, and by enzyme stability (Charron et al. 1999). The most important of these regulative mechanisms seems to be the rate of transcription (Zorzano et al. 1997: Santalucia et al. 1999). However, changes in the level of gene transcription occur slowly and thus may reflect the longer-term regulation that may be found during prolonged ischemia, diabetes, or hypertrophy (Brosius et al. 1997 b; Charron et al. 1999).

Another mechanism of glucose transport activation is provided by increasing the phosphorylation status of the transporters. Because a discrepancy was found to exist between the extent of glucose uptake and GLUT4 recruitment to the plasma membrane of muscle cells, it was proposed that maximal stimulation of glucose transport by insulin may require an increase in the intrinsic activity of GLUT4 subsequent to translocation (Goodyear et al. 1991). Recent investigations with p38 mitogen-activated protein kinases (MAPK) α and β have shown that these enzymes are activated by insulin and by contraction of isolated muscle cells. It was proposed, because inhibition of p38 MAPK concomitantly decreased insulin-stimulated glucose transport without affecting GLUT translocation, that the enzymes may function by modulation of the catalytic activity of glucose transporters (Sweeney et al. 1999).

Fig. 17.4. A Hibernating region; B normal region. Immunostaining of GLUT1 in human myocardial biopsies. Human myocardial biopsies were gained in one heart from regions defined before by PET scanning as "hibernating" (reduced flow and increased FDG uptake) and as "normal" (normal perfusion and FDG uptake). GLUT1 was detected in paraffin-embedded sections by immunohistochemical staining with a polyclonal rabbit anti-human GLUT1 antibody. In comparison, the staining for GLUT1 protein appears to be much more intense in sections of the hibernating region than in sections of the normal heart region. Magnification of both images: 370×

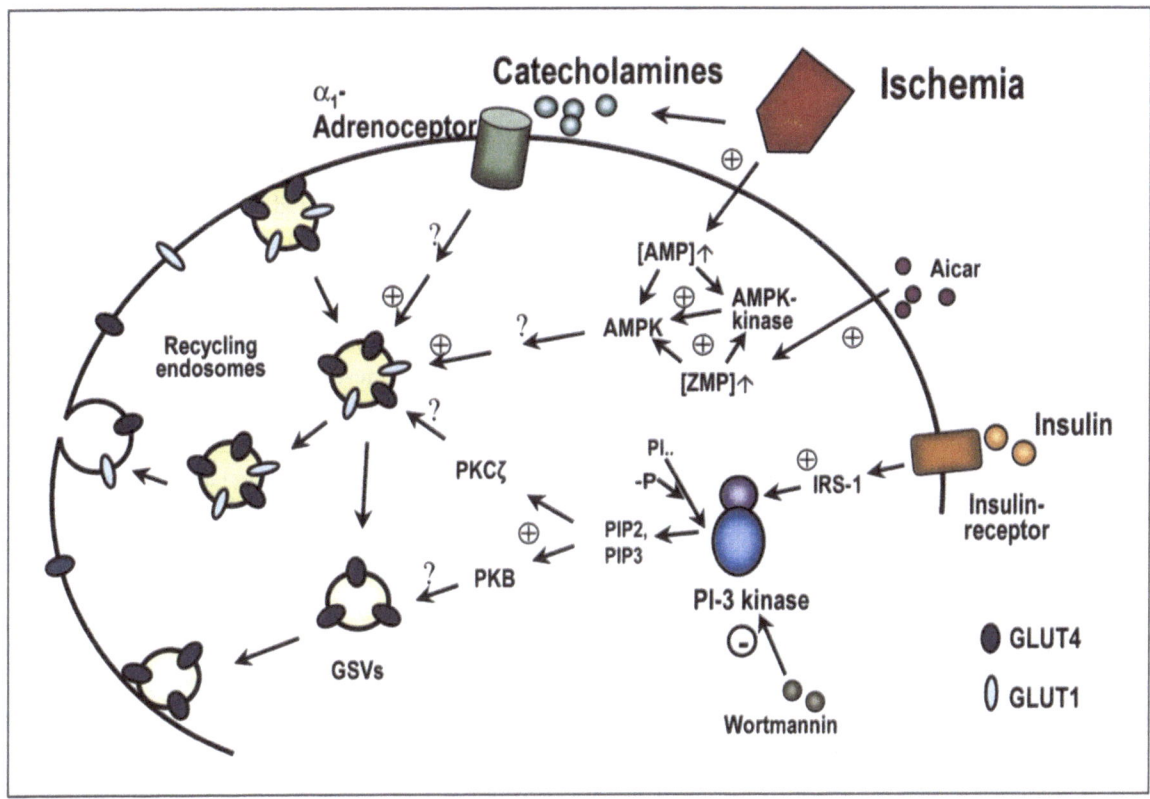

Fig. 17.5. Signaling mechanisms probably involved in glucose transporter translocation. Potential mechanisms of ischemia-induced stimulation of glucose transporter GLUT4 and GLUT1 translocation to the sarcolemma via the a_1-adrenoceptor activation and the adenosine monophosphate kinase (*AMPK*) stimulation. AMPK is also activated by the pharmacologic component 5-aminoimidazole-4-carboxamide-1-β-D-ribofuranoside (*AICAR*). The first steps of putative insulin-responsive signaling pathways are demonstrated in parallel. Phosphatidylinositol-3 kinase (*PI-3 kinase*) activity is inhibited by the fungal metabolite *wortmannin*. Insulin exerts its effects on a pool of GLUT4 storage vesicles (*GSVs*) via a constitutively active protein kinase B (*PKB*) and on recycling endosomes via a novel protein kinase C isoform (*PKCζ*), where GLUT4 is colocalized with GLUT1 and transferrin receptors. *ZMP*, AICAR monophosphate; *IRS-1*, insulin receptor 1; *PI*, phosphatidylinositol; *-P*, phosphate group

17.5.3
Regulation on the Level of Translocation

The main short-term mechanism to mediate glucose transport is the translocation of intracellularly stored GLUT to the cell surface. In myocardium, this was first demonstrated in an isolated, perfused heart model. Three conditions were examined: 15 min of no-flow ischemia, hypoxia, and insulin perfusion. All three stimuli induced a two- to four-fold increase in GLUT4 recruitment to the plasma membrane with a concomitant decrease in intracellular GLUT4 (Sun et al. 1994). In another study, a significant correlation was found in the basal state and with insulin and ischemia stimulation between the FDG uptake and the extent of GLUT4 on the plasma membrane (Wheeler 1998). These results were confirmed in an in vivo canine heart model of moderate regional ischemia (Young et al. 1997). In this model a moderate ischemia resulted in a two- to three-fold increase in glucose uptake which was accompanied by a two-fold increase in sarcolemmal GLUT4 and a 40% increase in GLUT1 on the sarcolemma. Hyperinsulinemia had similar effects on glucose uptake and GLUT translocation (Russell et al. 1998).

During the basal state, translocation occurs very slowly. With stimulation, the exocytotic process is markedly increased and is probably the main mechanism of translocation, decrease in glucose transporter internalization seeming to play a minor role (Sato et al. 1993).

17.5.3.1
Molecular Mechanism

The molecular components of the signaling pathways mediating glucose transporter translocation in response to different stimuli are under investigation. The most important stimuli in heart are insulin, ischemia, hypoxia, exercise, and catecholamines. Recent research has led to a detection of several enzymes involved in the signaling. The best pathway understood is that mediated by insulin. After insulin binds to its receptor, the activated receptor tyrosine kinase leads to subsequent binding and activation of the intracellular substrate insulin-receptor-substrate-1 (IRS-1) which binds to phosphatidylinositol-3 kinase (PI3-kinase). The association of IRS-1 with the PI3-kinase induces the phosphorylation of phosphatidylinositol (PI) and its derivates (Lam et al. 1994; Okada et al. 1994). The two potential downstream pathways activated by the phosphorylated PI products may be protein kinase B (PKB) and the novel isoforms ζ and λ of protein kinase C (Kohn et al. 1996; Bandyopadhyay et al. 1997; Fig. 17.5). The exact mechanism and how these pathways interact with the intracellular glucose transporter pool to initiate translocation are under current investigation. Preliminary results suggest that PKB causes translocation of GLUT4-containing vesicles to the plasma membrane via a route which is totally dependent on specialized vesicular accessory proteins, namely, VAMP2, syntaxin-4, and SNAP-23, which are found in GLUT4 containing intracellular membrane vesicles. They are responsible for determining trafficking specificity and formation of the core complex required for membrane fusion (Fletcher et al. 1999).

In contrast to the insulin-signaling mechanism, less information is available about the signaling pathways induced by ischemia. With respect to its important function in insulin signaling, the participation of PI3-kinase in ischemia-induced GLUT translocation was examined in isolated perfused rat hearts. Inhibition of the enzyme by the fungal metabolite wortmannin had no influence on GLUT4 and GLUT1 translocation triggered by a short-term no-flow ischemia, in contrast to its inhibitory effect on the insulin-mediated translocation of both transporters (Wheeler 1998; Egert et al. 1999a; Fig. 17.5). Therefore, ischemia does not appear to exert its effect on GLUT translocation through PI3-kinase. Other potential mediators that cause GLUT translocation in response to cellular stress have been suggested, including adenosine-monophosphate-activated protein kinase (AMPK; Russell et al. 1999), nitric oxide (NO; Balon et al. 1997), catecholamines (Egert et al. 1999b), calcium (Henriksen et al. 1989), MAP kinase activation (Denton et al. 1995), and protein kinase C (Hansen et al. 1997). AMPK, NO, and catecholamines will be discussed in more detail. AMPK activity increases in the setting of enhanced intracellular AMP and decreased creatine phosphate and is thought to act as a metabolic stress protein. The enzyme is activated during myocardial ischemia, increased contractile activity in skeletal muscle (Reibel et al. 1978; Winder et al. 1996), and pharmacologically by the compound 5-aminoimidazole-4-carboxamide-1-β-D-ribofuranoside (AICAR), which in a phosphorylated state may act as an AMP analogue (Corton et al. 1995). In vivo infusion of AICAR induced AMPK activation and GLUT4 translocation in rat myocardium. Since wortmannin did not affect the AICAR-stimulated increase in glucose uptake and GLUT4 translocation, it was concluded that AMPK-activation increases cardiac muscle glucose uptake through translocation of GLUT4 via a pathway that is independent of PI3-kinase and may be important in ischemia-induced signaling in the heart (Russell et al. 1999; Fig. 17.5).

As mentioned, another potential mediator involved in GLUT translocation may be nitric oxide. A recent investigation using transgenic mice examined the effect of their lack of endothelial nitric oxide synthase (ecNOS) on myocardial glucose transport. It was found that in hearts of ecNOS (–/–) mice glucose transport was significantly increased versus normal mice, and that 8-bromoguanosine 3',5'-cyclic monophosphate (8 Br-cGMP) (a cGMP analogue) and S-nitroso-N-acetylpenicillamine (a NO donor) stopped glucose uptake (Tada et al. 2000). These results suggest that cardiac NO production regulates myocardial glucose uptake via a cGMP-dependent mechanism, and they strongly support a pivotal role for ecNOS in this regulation. The findings may be important in the understanding of the pathogenesis of ischemic heart disease, in which NO synthesis is known to be altered.

Finally, it is also known that the release of catecholamines is increased during ischemia. This leads to an activation of α- and β-G-protein-coupled adrenoceptors and results in the regulation of various effector systems (Heusch 1990; Rockman et al. 1997). Adrenoceptor stimulation regulates FDG uptake and GLUT4 translocation since the α-agonist phenylephrine and the β-agonist isoproterenol were each shown to increase both parameters in a perfused rat heart model (Egert et al. 1999b). The same model was used to investigate whether a- and β-adrenoceptor stimulation is involved in no-flow ischemia-mediated FDG uptake and GLUT translocation. The results indicated that only the a-antagonist phentolamine effectively inhibited translocation of GLUT4 and GLUT1 in parallel to the FDG uptake, whereas blockage of β-adrenoceptors with propranolol had no effect on either ischemia-triggered GLUT translocation or on the concomitant increase in FDG. These data suggest that an a-adrenergic component is involved in the ischemia-mediated signaling pathway leading to glucose transporter translocation (Fig. 17.5). Furthermore, preliminary results indicate the predominant role of the a_1-adrenoceptor subform in ischemia-induced signaling, but not of the a_2-adrenoceptor since prazosin, a specific inhibitor of a_1-adrenoceptors, effectively blocked ischemia-stimulated GLUT4 translocation and stimulated FDG uptake in rat myocardium. In contrast, controls treated with yohimbine, a selective a_2-adrenoceptor inhibitor, were not affected upon ischemia-mediated GLUT4 translocation and increased FDG uptake (unpublished, preliminary data). However, the downstream signals connecting to the movement of GLUT-containing vesicles to the plasma membrane remain to be determined.

17.5.3.2
Regulation of the Movement of GLUT Vesicles to the Sarcolemma

The process of exocytosis and endocytosis in and out of the sarcolemma is referred to as membrane trafficking. GLUT4 molecules stored within cells not only belong to endosomes but appear to reside in several distinct but interrelated vesicles and tubulovesicular structures, with relatively smaller amounts found in the trans-Golgi network, clathrin-coated vesicles, and endosomes (Fletcher et al. 1999; Pessin et al. 1999). At the moment, the theory of at least two distinct intracellular pools of GLUT4, both of which translocate in response to insulin, is favored: (1) the unique GLUT4 storage vesicles (GSV), which characteristically contain GLUT4; and (2) proteins which are similar to proteins involved in synaptic vesicle fusion. One of the enzymes found in the GSV is VAMP2, which interacts with its fusion counterparts on the plasma membrane syntaxin 4 and SNAP-23. The VAMP2/GLUT4 vesicle population can be mobilized via PKB activation, a downstream target in insulin-responsive signaling. Another vesicle pool reflects characteristics of "recycling endosomes." In this pool GLUT4 molecules are associated with GLUT1 and the transferrin receptor and undergo continuous recycling between the plasma membrane and the cell interior independent of synaptic fusion proteins and PKB. However, the important signaling target PI3-kinase is central to the mobilization of both GLUT4-containing pools.

To further complicate this issue, it is likely that other vesicle pools exist, since GLUT4 translocation induced by ischemia or contraction occurs independently of PI3-kinase activation. Very little information is available about GLUT4 endocytosis. Insulin increases exocytosis 10- to 20-fold, whereas endocytosis is only decreased two- to three-fold and does not appear to be affected by PI3-kinase inhibitors. The GTPase protein dynamin was found to be the primary molecule implicated in endocytotic clathrin-coated vesicle formation. Insulin has been reported to induce the tyrosine phosphorylation of dynamin and the association of dynamin-GRB-2 complex with the adapter protein Shc and IRS-1, which in turn modulates the rate of dynamin GTPase activity (Pessin et al. 1999). Together with other observations it was suggested that insulin mediates GLUT4 endocytosis through a dynamin-dependent mechanism. Limiting GLUT4 endocytosis may also be a mechanism induced by ischemia to increase glucose uptake. No data are yet available to confirm this hypothesis.

17.5.4
Regulation on the Level of GLUT Expression

As already mentioned, GLUT expression is mainly regulated on the level of gene transcription. The activity of gene transcription is dependent upon factors including the age of the tissue or organ, the type of tissue, and several pathophysiological conditions. Gene transcription is affected by specific promoter regions which itself underlay activation/inactivation (Charron et al. 1999). The GLUT4 gene contains an upstream 2.4 kilobase promoter element that has been suggested to be critical for transcription. Part of the promoter reflects a myocyte enhancer factor-2-binding domain that plays an important role in GLUT4 gene transcription (Thai et al. 1998). In embryonic and neonatal myocardium the primary glucose transporter expressed is GLUT1. After birth, the substrate metabolism shifts from predominant nonoxidative glucose oxidation to fatty acid oxidation. This shift is associated with a shift in the expression of several regulatory proteins involved in the regulation of glucose and fatty acid metabolism, including GLUT proteins. The expression of GLUT1 gradually decreases, while expression of GLUT4 increases after birth until it becomes the predominant isoform in the adult heart (Santalucia et al. 1992). Generally, many GLUT isoforms have a tissue-specific expression and most isoforms have been found in various other tissues, although usually in very low amounts.

Pathophysiological changes of GLUT gene expression may be of importance. Myocardial hypertrophy, a pathological condition triggered by excess workload or cellular stress, exhibits a pattern of substrate metabolism similar to that of the fetal heart (Montessuit et al. 1999). Experiments have shown that this pattern is accompanied by an increased expression of GLUT1 mRNA. It was suggested that the signaling pathway mediating the increase in GLUT1 expression works via ras activation, which in turn induces the GLUT1 promoter (Montessuit et al. 1999). Furthermore, a decreased ratio of GLUT4 to GLUT1 expression has been detected in patients with left ventricular hypertrophy, paralleled by a decrease in insulin-stimulated glucose uptake (Paternostro et al. 1999).

Genetic manipulation of mice has provided further insights into the role of GLUT proteins in hypertrophy. A unique model of hypertrophy was presented by the engineering of mice completely lacking in GLUT4mice (Katz et al. 1995). In these animals, the hearts are considerably hypertrophied and show an increased GLUT1 level. However, neither the increase in GLUT1 expression nor of any other known GLUT isoform seems to match the absence of GLUT4 (Katz et al. 1996), suggesting the expression of a novel GLUT isoform, probably GLUT8 (also called GLUTx1). These animals have altered glucose metabolism but do not develop diabetes. The hypertrophy of GLUT4-null mice may be a consequence of energy depletion because they have diminished plasma lactate, free fatty acids, and ketone bodies. Another possibility is that hyperinsulinemia may induce their hypertrophy by changing the turnover time of proteins.

In states of relative insulin deficiency (such as streptozotocin-induced diabetes) the expression of GLUT4 and GLUT1 is markedly decreased in the heart (Kraegen et al. 1993; Charron et al. 1999). The molecular basis for regulation of in vivo GLUT4 gene expression in relative insulin deficiency has been very

difficult to answer, since insulin deficiency in vivo is complicated by the fact that compensatory counter-regulatory hormones are increased.

Prolonged ischemic and hypoxic conditions also lead to a different expression pattern of glucose transporters. In an in vivo dog heart model the LAD was occluded for 6 h to produce persistent ischemia. It was found that expression of GLUT1 was induced approximately three-fold, whereas GLUT4 expression did not change (Brosius et al. 1997b). These results suggest that GLUT1 may act as a "stress-induced protein" whose expression is elevated to help protect the myocardium from ischemic damage. GLUT1 expression is regulated by the hypoxia-inducible transcriptional factor 1α (Malhotra et al. 1999). An increased expression of GLUT1 was recently shown to protect hypoxic smooth vascular muscle cells from hypoxia-induced apoptosis; GLUT1 induced signaling through the stress-activated protein kinase pathway and diminished the mitogen-activated protein kinase-1 and c-Jun NH$_2$-terminal kinase activities that appeared to account for at least part of the reduced apoptosis (Malhotra et al. 1999).

Regulation of GLUT expression finishes the glucose transport regulation segment. However, myocardial glucose metabolism may also be controlled by its downstream steps, which now will be briefly discussed.

17.6
Regulation Steps Downstream of Glucose Transport

17.6.1
Hexokinase

After glucose is transported into the myocyte by glucose transporters, it is phosphorylated by hexokinase. Two isoforms of hexokinase exist in the heart. Hexokinase I predominates in the fetal and newborn heart, whereas hexokinase II is the main isoform in the adult myocardium (Printz et al. 1993). Intracellularly, the enzyme resides in cytosolic and mitochondrion-bound fractions and may translocate between these two compartments. The capacity to phosphorylate glucose in the adult myocardium is determined by the amount of hexokinase II, the compartmentalization of this enzyme within the cell and, in vitro, the concentration of the hexokinase inhibitor glucose-6-

phosphate. However, in vivo, increased rates of glucose transport and phosphorylation are found in spite of the rise in concentrations of Glc-6-P that occurs after insulin exposure, lactate perfusion, and during ischemia. The k_m for glucose of hexokinase II is 0.1 mM (Russell et al. 1997); the activity or gene expression of hexokinase may be altered experimentally.

An experimental hypertrophy induced by pulmonary artery clipping for 30–45 days resulted in a 26% increase in hexokinase activity in the perfused ferret heart (Do et al. 1997). Furthermore, it was demonstrated that after a 40 min period of ischemia, induced by a 60% reduction in LAD perfusion of pig hearts, the expression of hexokinase was slightly but significantly increased. The same period of ischemia but with reperfusion did not induce changes in the level of hexokinase mRNA (Feldhaus et al. 1998). These results were supported by findings of Doenst et al. (1998) who showed in perfused rat hearts that acute ischemia with reperfusion affected neither the subcellular distribution of hexokinase nor the k_M and V_{max} of hexokinase for glucose and deoxyglucose. Prolonged exercise (a 19 week treadmill-running program) likewise had no influence on the activity of hexokinase in the canine heart (Stuewe et al. 2000).

More general details of the activities and the gene expression of hexokinases in response to insulin and exercise-dependent regulation are reviewed by Vestergaard (1999), Cardenas et al. (1998), and Wasserman et al. (1998).

17.6.2
Glycogen Metabolism

Glucose-6-P may also be a substrate for glycogen synthesis. In fetal and newborn myocardium, glycogen stores represent 30% of the cell volume, whereas in the adult heart only about 2% of the cell volume is so occupied (Shelley 1961). Cardiac glycogen stores are increased during fasting, increased insulin or lactate levels and, most interestingly, in the state of hibernation (Laughlin et al. 1994; Depre et al. 1997; Moule et al. 1997). It is likely that the increase in glycogen vesicles in the hibernating heart are part of a mechanism that protects myocardial cells from ischemic damage. Glycogen breakdown is regulated by glycogen phosphorylase, an enzyme activated mainly by adrenaline or ischemia (Neely et al. 1973). The mechanism involves stimulation of cyclic AMP-dependent protein kinase (the "stress-inducible" kinase) or an increase in Ca^{2+} concentrations (Morgan et al. 1964). During acute ischemia, stimulation of glycogenolysis probably derives from the deprivation of energy phosphates. Yet, the metabolic consequences of glycogenolysis occurring during ischemia/reperfusion have been addressed from differing viewpoints. Some have reported damaging effects of glycogen breakdown with respect to cellular membrane integrity (Neely et al. 1984; Allard 1994), others have outlined the beneficial role of glycogenolysis (McEllroy et al. 1989). The contradictory opinions may reflect differences in the extent of glycogen depletion during ischemia. Complete cardiac glycogen depletion depends on the amount of stored glycogen before ischemia and on the severity of ischemia. With the beginning of ischemic contracture, glycogen stores are often depleted, and irreversible damage of the cardiomyocyte begins (Steenbergen et al. 1990).

17.6.3
Glycolysis

The first irreversible and rate-limiting step of glycolysis is the phosphorylation of Glu-6-P to Fru-1-6-P_2 catalyzed by the enzyme 6-phosphofructo-1-kinase. This enzyme is allosterically controlled by various effectors. Negative effectors are ATP, citrate, and protons (Garland et al. 1963) while positive effectors, which increase enzyme activity, are mainly AMP and Fru-2-6-P_2 (Hue et al. 1987). In the normoxic heart, phosphofructokinase (PFK-1) is mainly activated by Fru-2-6-P_2 (Depre et al. 1993). Further downstream regulatory sites in glycolysis are glyceraldehyde-3-phosphate dehydrogenase, which is inhibited by high concentrations of NADPH (Peukhurinen et al. 1982), and pyruvate kinase, which in a feed-forward mechanism is stimulated by Fru-1-6-P_2 (Kiffmeyer et al. 1991). The end product of glycolysis, pyruvate, may be metabolized in different pathways: under aerobic conditions, most of the pyruvate is oxidized by pyruvate dehydrogenase (PDH) to acetyl-CoA and then subjected to the citric acid cycle, where it is completely oxidized and $NADPH_2$ produced. The control of PDH activity is a basic control step in general glucose metabolism. Its activity depends on phosphorylation/dephosphorylation (activation) and upon the stimulation of effectors such as acetyl-CoA, NADH, adrenaline, and cAMP, whose concentrations rise during fatty acid oxidation. Pyruvate produced from glucose and lactate is an inhibitor of PDH. Under anaerobic conditions, pyruvate is reduced to lactate, and lactic acid is thus the end product of anaerobic glycolysis and released in the blood by monocarboxylate/proton symporters, which accordingly have an important function in maintaining intracellular pH (Halestrap et al. 1997).

17.7
Ischemia and Apoptosis

When protective mechanisms as induced by ischemia can no longer preserve cellular integrity, the disturbance of the relationship between energy metabolism and contractile function may induce programmed cell death (apoptosis). This cell suicide has been found under various in vivo and in vitro ischemic conditions including myocardial infarction, hibernation, preconditioned myocardium, as well as during global ischemia (cardiac surgery) (Elsässer et al. 2000). The point of no return from adaptation to maladaptation is not clearly defined. Apoptotic cells are characterized by preservation of mitochondrial and sarcolemmal structures. They develop condensation of nuclear chromatin and show cell shrinkage and budding. After programmed cell death, the remains are removed by phagocytosis (Elsässer et al. 2000). Ischemia-mediated apoptosis may be induced by different "stress-responsive" enzymes. Such proteins are for instance Jun-k, p38 MAPK, and the stress-activated protein kinase (SAPK) (Turner et al. 1988; Wang et al. 1998; Yue et al. 1998). The molecular signaling pathways by which these enzymes induce apoptosis are largely unknown. Recent studies have implicated reactive oxygen species and the resulting oxidative stress (Kannan et al. 2000), the tumor necrosis factor a (TNFa) (Krown et al. 1996), growth factor withdrawal, and ceramide signaling (Bialik et al. 1999) as playing pivotal roles in apoptosis. Programmed cell death in many cases starts with the decompartmentalization of mitochondria, accompanied by a loss of membrane potential. The decreased cardiac ATP production leads to an insertion of the pro-apoptotic protein Bax in the mitochondrial membrane and induces the development of pores through which cytochrome C, pro-caspases, apoptosis-inducing factor, and apoptotic protease-activating factor-1

(APAF-1) are released into the cytosol (Saikumar et al. 1998). A multimeric complex is formed by cytochrome C, APAF-1, and caspase 9 that activates downstream caspases leading to cell death (Kannan et al. 2000). Apoptosis being an energy-dependent process, the switch in the decision between apoptosis and necrosis therefore depends on the cellular ATP status: high ATP depletion most likely induces cell necrosis and not apoptosis (Taimor et al. 1999). This has led to the conclusion that myocardial apoptosis may be primarily found in relatively mild but long-term ischemic situations in which protective mechanisms such as hibernation and preconditioning have lost their function and pro-apoptotic pathways are stimulated. However, there are some strategies which may protect against ischemia-mediated apoptosis. Several recent studies point to an important role of glucose uptake and metabolism in preventing apoptotic damage. It was found that glucose treatment in vitro renders cardiac myocytes more resistant to hypoxia-induced apoptosis (Bialik et al. 1999; Schaffer et al. 2000). The anti-apoptotic effect of glucose transport stimulation might act via upregulation of sarcolemmal GLUT1 either by translocation of GLUT1 to the sarcolemma (Baldwin et al. 1995), or by an increased expression of GLUT1, which has been shown to inhibit a stress-activated protein kinase pathway activation (Lin et al. 2000). Last but not least, pro-apoptotic pathways are also found to be induced in the failing heart. In heart failure, GLUT1 and GLUT4 expression is decreased in parallel with numerous proteins regulating cardiac energy metabolism and contractile function (Nakao et al. 1997; Depre et al. 1998 c). Moreover, the metabolic imbalance in the failing heart makes it more vulnerable to conditions such as ischemia, overload, infarction, or the production of TNFa and reactive oxygen species, which promote apoptosis.

17.8
Conclusions

The dysfunctioning ischemic myocardium develops different strategies to adapt to its new metabolic situation. Profound alterations occur on the level of glucose metabolism, which plays a pivotal role during ischemia. The increased FDG signal represents a final result of multiple complex intracellular signaling processes induced by oxygen deficiency in the heart. The underlying molecular mechanisms are under intensive investigation. Depending on the particular ischemic situation, regulation may be mediated by the protein level (translocation of GLUT4, etc.) and/or by the level of their gene expression (upregulation of GLUT1, etc.). The various strategies of metabolic adaptation have in common the goal of protecting the heart from irreversible damage for a defined period of time. Understanding the mechanisms will help to develop specific approaches of individual therapy.

17.9
References

Allard M, Schonekess B, Henning S et al (1994) Contribution of oxidative metabolism and glycolysis to ATP production in hypertrophied hearts. Am J Physiol 267: H742–H750

Ausma J, Schaart G, Thone F et al (1995) Chronic ischemic viable myocardium in man: aspects of dedifferentiation. Cardiovasc Pathol 4:29–37

Baghelai K, Graham L, Wechsler A et al (1999) Delayed myocardial preconditioning by alpha 1 adrenoceptors involves inhibition of apoptosis. J Thorac Cardiovasc Surg 117:980–986

Baldwin SA, Barros LF, Griffith M (1995) Trafficking of glucose transporters – signals and mechanisms. Biosci Rep 15:419–426

Balon TW, Nadler JL (1997) Evidence that nitric oxide increases glucose transport in skeletal muscle. J Appl Physiol 82:359–363

Bandyopadhyay G, Standaert ML, Zhao L et al (1997) Activation of protein kinase C (alpha, beta, and zeta) by insulin in 3T3 L1 cells. Transfection studies suggest a role for PKC zeta in glucose transport. J Biol Chem 272: 2551–2558

Barrett E, Schwartz R, Francis C et al (1984) Regulation by insulin of myocardial glucose and fatty acid metabolism in the conscious dog. J Clin Invest 74:1073–1079

Bialik S, Geenen DL, Sasson IE et al (1997) Myocardial apoptosis during acute myocardial infarction in the mouse localizes to hypoxic regions but occurs independently of p53. J Clin Invest 100:1363–1372

Bialik S, Cryns VL, Drincic A et al (1999) The mitochondrial apoptotic pathway is activated by serum and glucose deprivation in cardiac myocytes. Circ Res 85:403–414

Bolli R (1992) Myocardial stunning in man. Circulation 86:1671–1691

Bonow RO, Dilsizian V, Cuocolo A et al (1991) Identification of viable myocardium in patients with chronic coronary artery disease and left ventricular dysfunction. Comparison of thallium scintigraphy with reinjection and PET imaging with ^{18}F-fluorodeoxyglucose. Circulation 83:26–37

Borgers M, Thone F, Wouters L et al (1993) Structural correlates of regional myocardial dysfunction in patients with critical coronary artery stenosis: chronic hibernation? Cardiovasc Pathol 2:237–245

Brosius FC, Nguyen N, Egert S et al (1997a) Increased sarcolemmal glucose transporter abundance in myocardial ischemia. Am J Cardiol 80:77a–84a

Brosius FC III, Liu Y, Nguyen N et al (1997b) Persistent myocardial ischemia induces GLUT1 glucose transporter gene expression in both ischemic and non-ischemic heart regions. J Mol Cell Cardiol 29:1675–1685

Camici P, Araujo L, Spinks T et al (1986) Increased uptake of 18F-fluoro-deoxyglucose in postischemic myocardium of patients with exercise-induced angina. Circulation 74:81–88

Carayannopoulos MO, Chi M, Cui Y et al (2000) GLUT8 is a glucose transporter responsible for insulin-stimulated glucose uptake in the blastocyst. Proc Natl Acad Sci USA 97:7313–7318

Cardenas ML, Cornish-Bowden A, Ureta T (1998) Evolution and regulatory role of the hexokinases. Biochim Biophys Acta 1401:242–264

Cartee G, Douen A, Ramlal T et al (1991) Stimulation of glucose transport in skeletal muscle by hypoxia. J Appl Physiol 70:1593–1600

Charron MJ, Brosius FC III, Alper SL et al (1989) A glucose transporter protein expressed predominately in insulin-responsive tissues. Proc Natl Acad Sci USA 86:2535–2539

Charron MJ, Katz EB, Olson AL (1999) GLUT4 gene regulation and manipulation. J Biol Chem 274:3253–3256

Chen C, Lijie MA, Linfert DR et al (1997) Myocardial cell death and apoptosis in hibernating myocardium. J Am Coll Cardiol 30:1407–1412

Corton JM, Gillespie JG, Hawley SA et al (1995) 5-Aminoimidazole-4-carboxamide ribonucleoside. A specific method for activating AMP-activated protein kinase in intact cells? Eur J Biochem 229:558–565

Crass MF III, McCaskill ES, Shipp JC (1969) Effects of pressure development on glucose and palmitate metabolism in perfused heart. Am J Physiol 216:1569–1576

Denton RM, Tavare JM (1995) Does mitogen-activated protein kinase have a role in insulin action? The cases for and against. Eur J Biochem 227:597–611

Depre C, Shipley G, Chen W et al (1998a) Unloaded heart in vivo replicates fetal gene expression of cardiac hypertrophy. Nat Med 4:1269–1275

Depre C, Havaux X, Dion R et al (1998b) Morphologic alterations of myocardium under left ventricular assistance. J Thorac Cardiovasc Surg 115:478–478

Depre C, Shipley G, Davies P et al (1998c) Glucose transporter isoform expression in the failing human heart (abstract). Circulation 98:I-627

Depre C, Rider MH, Veitch MK et al (1993) Role of fructose 2,6-bisphosphate in the control of heart glycolysis. J Biol Chem 268:13274–13279

Depre C, Vanoverschelde JL, Gerber B et al (1997) Correlation of functional recovery with myocardial blood flow, glucose uptake and morphologic features in patients with chronic left ventricular ischemic dysfunction undergoing coronary artery bypass grafting. J Thorac Cardiovasc Surg 113:82–87

Di Carli MF, Asgarzadie F, Schelbert HR et al (1995) Quantitative relation between myocardial viability and improvement in heart failure symptoms after revascularization in patients with ischemic cardiomyopathy. Circulation 92:3436–3444

Do E, Baudet S, Verdys M et al (1997) Energy metabolism in normal and hypertrophied right ventricle of the ferret heart. J Mol Cell Cardiol 29:1903–1913

Doege H, Bocianski A, Joost HJ et al (2000a) Activity and genomic organization of human glucose transporter 9 (GLUT9), a novel member of the family of sugar-transport facilitators predominantly expressed in brain and leukocytes. Biochem J 350:771–776

Doege H, Schürmann A, Bahrenberg G et al (2000b) GLUT8, a novel member of the sugar transport facilitator family with glucose transport activity. J Biol Chem 275:16275–16280

Doenst T, Taegtmeyer H (1998) Profound underestimation of glucose uptake by [^{18}F]2-deoxy-2-fluoroglucose in reperfused rat heart muscle. Circulation 97:2454–2462

Doenst T, Taegtmeyer H (1999) Ischemia-stimulated glucose uptake does not require catecholamines in rat heart. J Mol Cell Cardiol 31:435–443

Douen A, Ramlal T, Rastogi S et al (1990) Exercise induces recruitment of the "insulin-responsive glucose transporter". J Biol Chem 265:13427–13430

Egert S, Nguyen N, Brosius FC III et al (1997) Effects of wortmannin on insulin- and ischemia-induced stimulation of GLUT4 translocation and FDG uptake in perfused rat hearts. Cardiovasc Res 35:283–293

Egert S, Nguyen N, Schwaiger M (1999a) Myocardial glucose transporter GLUT1: Translocation induced by insulin and ischemia. J Mol Cell Cardiol 31:1337–1344

Egert S, Nguyen N, Schwaiger M (1999b) Contribution of alpha-adrenergic and beta-adrenergic stimulation to ischemia-induced glucose transporter (GLUT) 4 and GLUT1 translocation in the isolated perfused rat heart. Circ Res 84:1407–1415

Elsässer A, Schlepper M, Klövekorn VP et al (1997) Hibernating myocardium: an incomplete adaptation to ischemia. Circulation 96:2920–2931

Elsässer A, Suzuki K, Schaper J (2000) Unresolved issues regarding the role of apoptosis in the pathogenesis of ischemic injury and heart failure. J Mol Cell Cardiol 32:711–724

Feldhaus L, Liedke AJ (1998) mRNA expression of glycolytic enzymes and glucose transporter proteins in ischemic myocardium with and without reperfusion. J Mol Cell Cardiol 30:2475–2485

Ferrannini E, Santoro D, Bonadonna R et al (1993) Metabolic and haemodynamic effects of insulin on human hearts. Am J Physiol 264:E308–E315

Fischer Y, Thomas J, Sevilla L et al (1997) Insulin-induced recruitment of glucose transporter 4 (GLUT4) and GLUT1 in isolated rat cardiac myocytes. J Biol Chem 272:7085–7092

Fletcher LM, Tavare JM (1999) Divergent signaling mechanisms involved in insulin-stimulated GLUT4 vesicle trafficking to the plasma membrane. Biochem Soc Trans 27:677–683

Gao WD, Liu Y, Mellgren R et al (1996) Intrinsic myofilament alterations underlying the decreased contractility of stunned myocardium. A consequence of Ca^{2+}-dependent proteolysis? Circ Res 78:455–465

Garland PB, Randle PJ, Newsholme EA (1963) Citrate as an intermediary in the inhibition of phosphofructokinase in rat heart muscle by fatty acids, ketone bodies, pyruvate, diabetes and starvation. Nature 200:169–170

Gertz EW, Wisneski JA, Stanley WC et al (1988) Myocardial substrate utilization during exercise in humans. J Clin Invest 82:2017–2025

Goodwin G, Ahmad F, Taegtmeyer H (1996) Preferential oxidation of glycogen in isolated working rat heart. J Clin Invest 97:1409–1416

Goodyear LJ, Hirshman MF, Smith RJ et al (1991) Glucose transporter number, activity, and isoform content in plasma membranes of red and white skeletal muscle. Am J Physiol 261:E556–E561

Gropler RJ, Siegel BA, Lee KJ et al (1991) Nonuniformity in myocardial accumulation of fluorine-18-fluorodeoxyglucose in normal fasted humans. J Nucl Med 31:1749–1756

Gropler RJ, Geltman EM, Sampathkumaran K et al (1993) Comparison of carbon-11-acetate with fluorine-18-fluorodeoxyglucose for delineating viable myocardium by positron emission tomography. J Am Coll Cardiol 22:1587–1597

Grover-McKay M, Walsh SA, Thompson SA (1999) Glucose transporter 3 (GLUT3) protein is present in human myocardium. Biochim Biophys Acta 1416:145–154

Halestrap AP, Wang X, Poole RC et al (1997) Transport in heart in relation to myocardial ischemia. Am J Cardiol 80:17A–25A

Hansen PA, Corbett JA, Holloszy JO (1997) Phorbol esters stimulate muscle glucose transport by a mechanism distinct from the insulin and hypoxia pathways. Am J Physiol 273:E28–E36

Hariharan R, Bray M, Ganim R et al (1995) Fundamental limitations of [^{18}F]2-deoxy-2-fluoro-D-glucose for assessing myocardial glucose uptake. Circulation 91:2435–2444

Henning SL, Wambold RB, Schönekess BO et al (1996) Contribution of glycogen to aerobic myocardial glucose utilization. Circulation 93:1549–1555

Henriksen E, Rodnick K, Holloszy J (1989) Activation of glucose transport in skeletal muscle by phospholipase C and phorbol ester. Evaluation of the regulatory roles of protein kinase C and calcium. J Biol Chem 264:21536–21543

Heusch G (1990) a-Adrenergic mechanisms in myocardial ischemia. Circulation 81:1–13

Heusch G (1998) Hibernating myocardium. Physiol Rev 78:1055–1085

Heyndrickx GR, Millard RW, McRitchie RJ et al (1975) Regional myocardial function and electrophysiological alterations after brief coronary artery occlusion in conscious dogs. J Clin Invest 56:978–985

Homans DC, Sublett E, Dai XZ et al (1986) Persistence of regional left ventricular dysfunction after exercise-induced myocardial ischemia. J Clin Invest 77:66–73

Hu K, Nattel S (1995) Mechanisms of ischemic preconditioning in rat hearts: involvement of a_{1b}-adrenoceptors, pertussis-toxin sensitive G proteins, and protein kinase C. Circulation 92:2259–2265

Hue L, Rider MH (1987) Role of fructose 2,6-bisphosphate in the control of glycolysis in mammalian tissue. Biochem J 245:313–324

Ibberson M, Uldry M, Thorens B (2000) GLUTX1, a novel mammalian glucose transporter expressed in the central nervous system and insulin-sensitive tissues. J Biol Chem 275:4607–4612

Kalff V, Schwaiger M, Nguyen N et al (1992) The relationship between myocardial blood flow and glucose uptake in ischemic canine myocardium determined with F-18 deoxyglucose. J Nucl Med 33:1346–1353

Kannan K, Jain SK (2000) Oxidative stress and apoptosis. Pathophys 7:153–163

Kashiwaya Y, Sato K, Tsuchiya N et al (1994) Control of glucose utilization in working perfused rat heart. J Biol Chem 269:25502–25514

Katz EB, Stenbit AE, Hatton H et al (1995) Cardiac and adipose tissue abnormalities but not diabetes in mice deficient in GLUT4. Nature 377:151–155

Katz EB, Burcelin R, Tsao TS et al (1996) The metabolic consequences of altered glucose transporter expression in transgenic mice. J Mol Med 74:639–652

Kaul TK, Angihotri AK, Fields BL et al (1996) Coronary artery bypass grafting in patients with an ejection fraction of twenty percent or less. J Thorac Cardiovasc Surg 111:1001–1012

Kiffmeyer WR, Farrar WW (1991) Purification and properties of pig heart pyruvate kinase. J Protein Chem 10:585–591

Kim MS, Akera T (1987) O_2 free radicals: cause of ischemia-reperfusion injury to cardiac Na^+-K^+-ATPase. Am J Physiol 252:H252–H257

Kloner RA, Bolli R, Marban E et al (1998) Medical and cellular implications of stunning, hibernation, and preconditioning: an NHLBI Workshop. Circulation 97:1848–1867

Knuuti MJ, Saraste M, Nuutila P et al (1994) Myocardial viability: Fluorine-18-deoxyglucose positron emission tomography in prediction of wall motion recovery after revascularization. Am Heart J 127:785–796

Kohn H, Summer S, Birnbaum M et al (1996) Expression of a constitutively active Akt ser/thr kinase in 3T3 L-1 adipocytes stimulates glucose uptake and glucose transporter 4 translocation. J Biol Chem 271:31372–31378

Kraegen EW, Sowden JA, Halstead MB et al (1993) Glucose transporters and in vivo glucose uptake in skeletal and cardiac muscle: fasting, insulin-stimulation and immunoisolation studies of GLUT1 and GLUT4. Biochem J 295:287–293

Krown KA, Page MT, Nguyen C et al (1996) Tumor necrosis factor alpha-induced apoptosis in cardiac myocytes. Involvement of the sphingolipid signaling cascade in cardiac cell death. J Clin Invest 98:2854–2865

Lam K, Carpenter CL, Ruderman NB et al (1994) The phosphatidylinositol 3-kinase serine kinase phosphorylates IRS-1. J Biol Chem 269:20648–20652

Laughlin MR, Taylor J, Chesnick AS et al (1994) Non-glucose substrates increase glycogen synthesis in vivo in dog heart. Am J Physiol 267:H219–H223

Laybutt DR, Thompson AL, Cooney GJ et al (1997) Selective chronic regulation of GLUT1 and GLUT4 content by insulin, glucose, and lipid in rat cardiac muscle in vivo. Am J Physiol 273:H1309–H1316

Liedke AJ (1981) Alterations of carbohydrate and lipid metabolism in the acutely ischemic heart. Prog Cardiovasc Dis 23:321–336

Lin Z, Weinberg JM, Malhotra R et al (2000) GLUT-1 reduces hypoxia-induced apoptosis and JNK pathway activation. Am J Physiol 278:E958–E966

Lucignani G, Paoloni G, Landoni C et al (1992) Presurgical identification of hibernating myocardium by combined use of technetium-99m-hexakis-2-methoxyisobutylisonitrile single photon emission tomography and fluorine-18 fluoro-2-deoxy-D-glucose positron emission tomography in patients with coronary artery disease. Eur J Nucl Med 19:874–881

Maki M, Luotolathi M, Nuutila P et al (1996) Glucose uptake in the chronically dysfunctional but viable myocardium. Circulation 93:1658–1666

Malhotra R, Brosius FC III (1999) Glucose uptake and glycolysis reduce hypoxia-induced apoptosis in cultured neonatal rat cardiac myocytes. J Biol Chem 274:12567–12575

McElroy DD, Walker WE, Taegtmeyer H (1989) Glycogen loading improves left ventricular function of the rabbit heart after hypothermic ischemic arrest. J Appl Cardiol 4:455–465

Mickleborough LL, Maruyma H, Takagi Y et al (1995) Results of revascularization in patients with severe left ventricular dysfunction. Circulation 92 [9 Suppl]:II73–79

Miller DC, Stinson EB, Alderman EL (1996) Surgical treatment of ischemic cardiomyopathy: is it ever too late? Am J Surg 141:688–693

Mitchell MB, Meng X, Ao L et al (1995) Preconditioning of isolated rat heart is mediated by protein kinase C. Circ Res 76:73–81

Montessuit C, Thorburn A (1999) Transcriptional activation of the glucose transporter GLUT1 in ventricular cardiac myocytes by hypertrophic agonists. J Biol Chem 274: 9006–9012

Morgan HE, Parmeggiani A (1964) Regulation of glycogenolysis in muscle. II. Control of glycogen phosphorylase reaction in isolated perfused rat heart. J Biol Chem 239:2435–2439

Moule SK, Denton RM (1997) Multiple pathways involved in the metabolic effects of insulin. Am J Cardiol 80:41A–49A

Mueckler M (1994) Facilitative glucose transporters. Eur J Biochem 219:713–725

Mueckler M, Hresko RC, Sato M (1997) Structure, function and biosynthesis of GLUT1. Biochem Soc Trans 25:951–954

Murry CE, Jennings RB, Reimer KA (1986) Preconditioning with ischemia: a delay of lethal cell injury in ischemic myocardium. Circulation 75:1124–1136

Murry CE, Richard VJ, Reimer KA et al (1990) Ischemic preconditioning slows energy metabolism and delays ultrastructural damage during a sustained ischemic period. Circ Res 66:913–931

Nadal-Ginard B, Mahdavi V (1989) Molecular basis of cardiac performance: plasticity of the myocardium generated through protein isoform switches. J Clin Invest 84:1693–1700

Nakao K, Minobe W, Roden R et al (1997) Myosin heavy chain gene expression in human heart failure. J Clin Invest 100:2362–2370

Neely JR, Grotyohann LW (1984) Role of glycolytic products in damage to ischaemic myocardium. Dissociation of adenosine triphosphate levels and recovery of function of reperfused ischaemic hearts. Circ Res 55:816–824

Neely JR, Morgan HE (1974) Relationship between carbohydrate and lipid metabolism and the energy balance of heart muscle. Annu Rev Physiol 36:413–459

Neely JR, Rovetto MJ, Whitmer JT et al (1973) Effects of ischaemia on function and metabolism of the isolated working rat heart. Am J Physiol 225:651–658

Nguyen N, Liu Y, Sun DQ et al (1995) Enhanced glucose utilization following preconditioning in the isolated rat heart. J Nucl Med 36(5):47P

Okada T, Kawano Y, Sakakibara T et al (1994) Essential role of phosphatidylinositol 3-kinase in insulin-induced glucose transport and antilipolysis in rat adipocytes. J Biol Chem 269:3568–3573

Passamani E, Davis KB, Gillespie MJ et al (1996) A randomized trial of coronary artery bypass surgery. Survival of patients with a low ejection fraction. N Engl J Med 312:1665–1671

Paternostro G, Pagano D, Gnecchi-Ruscone T et al (1999) Insulin resistance in patients with cardiac hypertrophy. Cardiovasc Res 42:246–253

Patlak CS, Blasberg RG, Fenstermacher JD (1983) Graphical evaluation of blood to brain transfer constants from multiple-time uptake data. J Cereb Blood Flow Metab 3:1–7

Pessin J, Thurmond D, Elmendorff J et al (1999) Molecular basis of insulin-stimulated GLUT4 vesicle trafficking. J Biol Chem 274:2593–2596

Peukhurinen KJ, Nuutinen EM, Pietilainen EP et al (1982) Role of pyruvate carboxylation in the energy-linked regulation of pool sizes of TCA-cycle intermediates in the myocardium. Biochem J 208:577–581

Printz RL, Koch S, Potter LR et al (1993) Hexokinase II mRNA and gene structure, regulation by insulin, and evolution. J Biol Chem 268:5209–5219

Rahimtoola SH (1989) The hibernating myocardium. Am Heart J 117:211–221

Rahimtoola SH (1996) Hibernating myocardium has reduced blood flow at rest that increases with low-dose dobutamine. Circulation 94:3055–3061

Rees G, Bristow JD, Kremkau EL et al (1971) Influence of aortocoronary bypass on left ventricular performance. N Engl J Med 284:1116–1120

Reibel D, Rovetto M (1978) Myocardial ATP synthesis and mechanical function following oxygen deficiency. Am J Physiol 234:H620–H624

Rockman HA, Koch WJ, Lefkowitz RJ (1997) Cardiac function in genetically engineered mice with altered adrenergic receptor signaling. Am J Physio 272:H1553–H1559

Russell RR, Cline GW, Guthrie PH et al (1997) Regulation of exogenous and endogenous glucose metabolism by insulin and acetoacetate in the isolated working rat heart. J Clin Invest 100:2892–2899

Russell RR, Yin R, Caplan MJ et al (1998) Additive effects of hyperinsulinemia and ischemia on myocardial GLUT1

and GLUT4 translocation in vivo. Circulation 98:2180–2186

Russell RR, Bergeron R, Shulman GI et al (1999) Translocation of myocardial GLUT-4 and increased glucose uptake through activation of AMPK by AICAR. Am J Physiol 277:H643–H649

Saddik M, Gamble J, Witters LA et al (1993) Acetyl-CoA carboxylase regulation of fatty acid oxidation in the heart. J Biol Chem 268:25836–25845

Sadoshima S, Izumo S (1997) The cellular and molecular response of cardiac myocytes to mechanical stress. Annu Rev Physiol 59:551–571

Saikumar P, Dong Z, Weinberg J et al (1998) Mechanisms of cell death in hypoxia/reoxygenation injury. Oncogen 17:3341–3349

Santalucia T, Camps M, Castello A et al (1992) Developmental regulation of GLUT-1 (erythroid/HepG2) and GLUT-4 (muscle/fat) glucose transporter expression in rat heart, skeletal muscle, and brown adipose tissue. Endocrinology 130:837–846

Santalucia T, Boheler KR, Brand NJ et al (1999) Factors involved in GLUT-1 glucose transporter gene transcription in cardiac muscle. J Biol Chem 274:17626–174637

Sato S, Nishimura H, Clark AE et al (1993) Use of bismannose photolabel to elucidate insulin-regulated GLUT4 subcellular trafficking kinetics in rat adipose cells. Evidence that exocytosis is a critical site of hormone action. J Biol Chem 268:17820–17829

Schaffer SW, Croft CB, Solodushko V (2000) Cardioprotective effect of chronic hyperglycemia: effect on hypoxia-induced apoptosis and necrosis. Am J Physiol 278: H1948–H1954

Schwaiger M, Schelbert HR, Ellison D et al (1985) Sustained regional abnormalities in cardiac metabolism after transient ischemia in the chronic dog model. J Am Coll Cardiol 6:336–347

Schwaiger M, Neese RA, Araujo L et al (1989) Sustained nonoxidative glucose utilization and depletion of glycogen in reperfused canine myocardium. J Am Coll Cardiol 13:745–754

Schwarz ER, Schaper J, Dahl J vom et al (1996) Myocyte degeneration and cell death in hibernating human myocardium. J Am Coll Cardiol 26:1577–1585

Shelley HJ (1961) Cardiac glycogen in different species before and after birth. Br Med Bull 17:137–156

Shivalkar B, Maes A, Borgers M et al (1996) Only hibernating myocardium invariably shows early recovery after coronary revascularization. Circulation 94:308–315

Sokoloff L, Reivich M, Kennedy C et al (1977) The C-14 deoxyglucose measurement of local cerebral glucose utilization: theory, procedure, and normal values in the conscious and anesthetized albino rat. J Neurochem 28:897–916

Steenbergen C, Murphy E, Watts JA et al (1990) Correlation between free cytosolic calcium, contracture, ATP, and irreversible ischemic injury in perfused rat heart. Circ Res 66:136–146

Stuewe SR, Gwirtz PA, Agarwal N et al (2000) Exercise training enhances glycolytic and oxidative enzymes in canine ventricular myocardium. J Mol Cell Cardiol 32:903–913

Sun DQ, Nguyen N, DeGrado TR et al (1994) Ischemia induces translocation of the insulin-responsive glucose transporter GLUT4 to the plasma membrane of cardiac myocytes. Circulation 89:793–798

Sweeney G, Somwar R, Ramlal T et al (1999) An inhibitor of p38 mitogen-activated protein kinase prevents insulin-stimulated glucose transport but not glucose transporter translocation in 3T3-L1 adipocytes and L6-myotubes. J Biol Chem 274:10071–10078

Tada H, Thompson CI, Recchia FA et al (2000) Myocardial glucose uptake is regulated by nitric oxide via endothelial nitric oxide synthase in Langendorff mouse heart. Circ Res 86:270–274

Taegtmeyer H (1983) On the inability of ketone bodies to serve as the only energy providing substrate for rat heart at physiological work load. Basic Res Cardiol 78:435–450

Taegtmeyer H (1994) Energy metabolism in the heart: from basic concepts to clinical applications. Curr Prob Cardiol 19:57–116

Taegtmeyer H, King LM, Jones BE (1998) Energy substrate metabolism, myocardial ischemia, and targets for pharmacotherapy. Am J Cardiol 82:54K–60K

Taimor G, Lorenz H, Hofstaetter B et al (1999) Induction of necrosis but not apoptosis after anoxia and reoxygenation in isolated adult cardiomyocytes of rat. Cardiovasc Res 41:147–156

Tamaki N, Kawamoto M, Takahashi N et al (1993) Prognostic value of an increase in fluorine-18-deoxyglucose uptake in patients with myocardial infarction: comparison with stress-thallium imaging. J Am Coll Cardiol 22:1621–1627

Thai MV, Guruswamy S, Cao KT et al (1998) Myocyte enhancer factor 2 (MEF2)-binding site is required for GLUT4 gene expression in transgenic mice. Regulation of MEF2 DNA binding activity in insulin-deficient diabetes. J Biol Chem 273:14285–14292

Tillisch J, Brunken R, Marshall R et al (1986) Reversibility of cardiac wall-motion abnormalities predicted by positron emission tomography. New Engl J Med 314:884–888

Tong H, Chen W, London RE et al (2000) Preconditioning enhanced glucose uptake is mediated by p38 MAP kinase not by phosphatidylinositol 3-kinase. J Biol Chem 275:11981–11986

Turner N, Xia F, Azhar G et al (1988) Oxidative stress induces DNA fragmentation and caspase activation via the c-Jun NH2-terminal kinase pathway in H9c2 cardiac muscle cells. J Mol Cell Cardiol 30:1789–1801

Van Bilsen M, Van der Vusse G, Reneman R (1998) Transcriptional regulation of metabolic processes. Implications for cardiac metabolism. Pflügers Arch 437:2–14

Vanoverschelde JL, Wijns W, Depre C et al (1993) Mechanism of chronic regional postischemic dysfunction in humans. New insights from the study of noninfarcted collateral-dependent myocardium. Circulation 87:1513–1523

Vestergaard H (1999) Studies of gene expression and activity of hexokinase, phosphofructokinase and glycogen synthase in human skeletal muscle in states of altered insulin-stimulated glucose metabolism. Dan Med Bull 46:13–34

Vom Dahl J, Eitzman DT, Al-Aouar ZR et al (1994) Relation of regional function, perfusion, and metabolism in patients with advanced coronary artery disease undergoing surgical revascularization. Circulation 90:2356–2366

Vom Dahl J, Altehoefer C, Sheehan FH et al (1997) Effect of myocardial viability assessed by technetium-99m-sestamibi SPECT and fluorine-18-FDG PET on clinical outcome in coronary artery disease. J Nucl Med 38:742–748

Wang Y, Huang S, Sah V et al (1998) Cardiac muscle cell hypertrophy and apoptosis induced by distinct members of the p38 mitogen-activated protein kinase family. J Biol Chem 273:2161–2168

Wasserman DH, Halseth AE (1998) An overview of muscle glucose uptake during exercise. Sites of regulation. Adv Exp Med Biol 441:1–16

Wheeler TJ (1998) Translocation of glucose transporters in response to anoxia in heart. J Biol Chem 263:19447–19454

Wijns W, Vatner SF, Camici PG (1996) Hibernating myocardium (review). New Engl J Med 339:173–181

Winder WW, Hardie DG (1996) Inactivation of acetyl-CoA carboxylase and activation of AMP-activated protein kinase in muscle during exercise. Am J Physiol 270:E299–E304

Young LH, Renfu Y, Russell RR et al (1997) Low flow ischemia leads to translocation of canine GLUT-4 and GLUT-1 glucose transporters to the sarcolemma in vivo. Circulation 95:415–422

Yue TL, Ma XL, Wang X (1998) Possible involvement of stress-activated protein kinase signaling pathway and Fas-receptor expression in prevention ischemia/reperfusion-induced cardiomyocyte apoptosis by carvedilol. Circ Res 82:166–174

Zorzano A, Sevilla L, Camps M et al (1997) Regulation of glucose transport, and glucose transporters expression and trafficking in the heart: studies in cardiac myocytes. Am J Cardiol 80:65A–76A

CHAPTER 18

Imaging Studies in Substance Abuse 18

Nora D. Volkow, Joanna S. Fowler, Gene-Jack Wang, Walton W. Shreeve

Contents

18.1
Introduction

The field of imaging has seen enormous technological advances over the past 15 years, recently culminating in the application of MRI for functional imaging (fMRI). During this time technologies such as positron emission tomography (PET), single photon emission computed tomography (SPECT), and MRI have enabled linkage of the gap between basic neurosciences and clinical research. In the area of substance abuse research, these technologies have provided powerful tools that allow investigation of the mechanisms of action of drugs of abuse as well as the brain metabolic and neurochemical changes associated with addiction.

PET and SPECT are nuclear medicine instruments that detect and measure the spatial distribution and movement of radioisotopes in tissues of living subjects (Mullani and Volkow 1992; Rogers and Ackermann 1992). PET measures compounds labeled with positron (β^+) emitting radioisotopes and SPECT with single photon emitting radioisotopes. An advantage of the positron emitters is that some of those used are isotopes for the natural elements of life (Table 18.1), which enables labeling of compounds without affecting their pharmacological properties. Though labeling an organic compound with a single photon emitter such as iodine-123 (Table 18.1) results in a compound different from the parent compound, many iodine-substituted radiotracers have been developed with high biological selectivity and affinity for specific molecular targets (reviewed by Kung 1993). Positron emitters used for imaging have shorter half lives than single photon emitters. Both of these types of isotopes can be used to label ligands for specific receptor, transporter, and/or enzymatic systems to be used with PET or SPECT in order to quantify these parameters in the living human brains. In addition, PET tracers such as fluorine-18-labeled 2-deoxy-2-fluoro-D-glucose (FDG) or carbon-11-labeled 2-deoxy-D-glucose (CDG) and oxygen-15-labeled water can be used to measure regional brain glucose metabolism (rGLm) and cerebral blood flow (CBF), and SPECT tracers such as 99mTc HMPAO can also be used to measure CBF. Because rGLm and CBF under physiological conditions are tightly coupled with brain activity, these tracers can be used to assess regional brain function. However, brain metabolism can only be measured with PET and FDG or CDG because there is no SPECT counterpart for a glucose analogue. Though until recently MRI had been predominantly utilized for high spatial resolution anatomical imaging, recent technological advances have allowed its use for functional brain imaging (fMRI; Andrew 1994). It is generally believed that the activation signal generated from fMRI is due to differences in magnetic properties of oxygenated versus deoxygenated blood. During activation there are regional changes in the amount of paramagnetic deoxy-hemoglobin in brain blood vessels that are de-

Table 18.1. Typical radioisotopes for PET and SPECT. With the positron emitters the two annihilation photons (0.511 MeV), emitted about 180° apart, are detected

Radioisotope	Half-life	Decay
PET		
Carbon-11	20.4 min	β^+ (0.960 MeV)
Fluorine-18	110 min	β^+ (0.635 MeV)
Oxygen-15	2 min	β^+ (1.73 MeV)
Nitrogen-13	10 min	β^+ (1.19 MeV)
SPECT		
Iodine-123	13.3 h	γ (0.159 MeV)
Technetium-99m	6.0 h	γ (0.140 MeV)

tected with fMRI. It is hypothesized that during activation there is an excess of arterial blood delivered into the area (decrease in the ratio of deoxy-hemoglobin to oxyhemoglobin) which then gives rise to the activation signal. This imaging technique enables one to measure activation patterns with a higher spatial and temporal resolution than that achievable with PET or SPECT. However, fMRI quantification of activation is expressed relative to changes in a nonactivated condition, whereas PET is able to quantify absolute measures of rGLm and CBF during activation.

Imaging technologies can be used to investigate the neurobiology of drug abuse from three investigational perspectives: (1) the pharmacological properties of the drugs themselves, (2) the consequences of these drugs in brain function and neurochemistry, and (3) the etiology of drug abuse in regard to genetic vs. environmental factors. Also, because imaging studies can be performed in the addicted subjects themselves, they can be used to assess the patterns of brain activation associated with specific drug-related states, such as intoxication, drug craving, and drug withdrawal.. The following summarizes some of the applications of PET and SPECT imaging technologies to substance abuse research. Because of space limitations we cannot provide a complete review of these studies; instead, we provide examples to illustrates specific experimental strategies.

netics in the human brain. Additionally, the labeled drug and whole-body PET can be used to determine the target organs for the drug and its labeled metabolites, and thus to provide information on potential toxic effects as well as tissue half-lives. This strategy also allows the evaluation of the relation between the kinetics of an abused drug in the brain and the temporal relation to its behavioral effects. This is illustrated in Fig. 18.1, which shows the regional distribution for ^{11}C-cocaine and for ^{11}C-methylphenidate in the human brain. Both drugs were found to show a very similar pattern of distribution in the human brain, where they bound predominantly to dopamine (DA) transporters in the basal ganglia but they differed in their pharmacokinetics (Volkow et al. 1995). They both entered the brain rapidly (4–10 min) but the rate of clearance was significantly slower for methylphenidate than for cocaine (half life 90 and 20 min), respectively. For both drugs their initial uptake in basal ganglia paralleled the temporal changes for the perception of the "high." In the case of cocaine the reduction in the "high" followed the fast clearance of the drug in brain; whereas for methylphenidate the "high" declined rapidly despite the fact that there was still significant binding of the drug in brain (Fig. 18.2). Because the initial drug uptake was associated with the "high" and not necessarily the presence of the drug in brain, it was postulated that the rate of clearance may affect the propensity of a drug to promote the desire for frequent repeated administration. The fast uptake and clearance from brain of cocaine is compatible with repeated frequent administration as occurs classically with a cocaine binge; methylphenidate's relatively slow clearance from brain may interfere with such a pattern of administration. This in turn could account for the much lesser abuse of methylphenidate than of cocaine despite their otherwise similar pharmacological properties. These types of imaging studies enable investigation of the action of drugs in dynamic systems and in this way assess the interaction between binding characteristics, pharmacokinetics, and behavioral effects.

18.2
Imaging Pharmacokinetics of Drug Abuse

The availability of the short-lived positron emitter carbon-11 has made it possible to label drugs of abuse so that PET can be used to measure their pharmacoki-

18.3
Effects of Drug Abuse on Brain Function and Neurochemistry

Different labeled tracers can be used to evaluate the effects of drugs on brain function and neurochemistry and in this way provide information related to drug

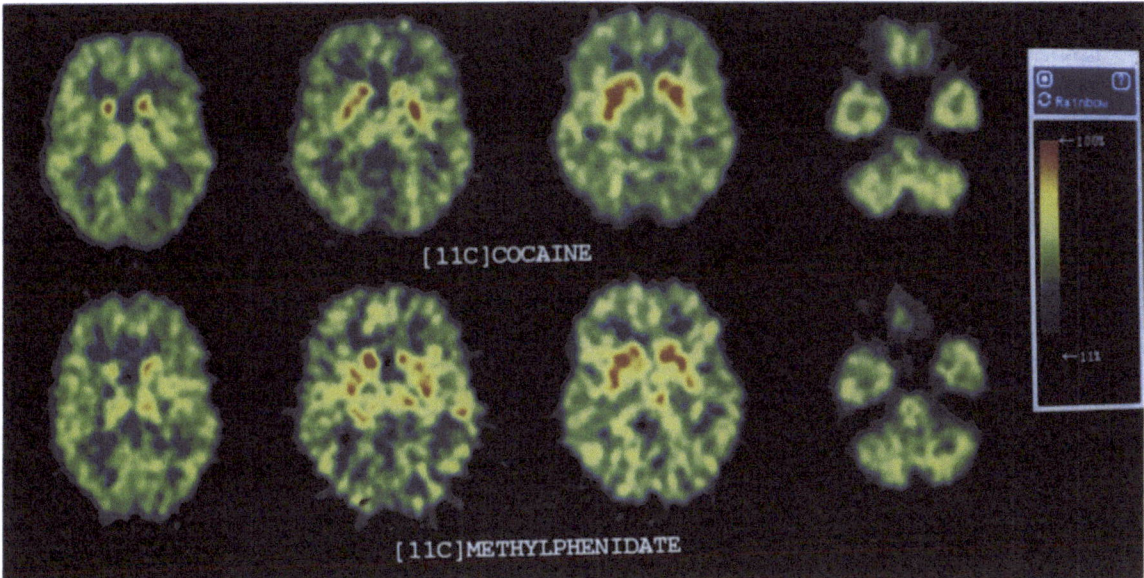

Fig. 18.1. Brain images obtained with ^{11}C-cocaine and the normal control ^{11}C-methylphenidate, scanned with both tracers. Notice the similar distribution of the two drugs in human brain, where they predominantly bind to the dopamine transporters in the basal ganglia (Volkow et al. 1995)

pharmacodynamics in the human brain. For example the effects of a drug on metabolism and CBF, on neurotransmitter activity, on transporter or receptor occupancy, and on enzyme activity can be probed. The most widely utilized approach has been to assess the effects of acute drug administration on brain glucose metabolism and on CBF. This allows analysis of the brain regions that are most sensitive to the drug, and because the studies are done in awake human subjects, it enables assessment of the relation between regional changes in metabolism or flow and the behavioral effects of the drug on addicted and non-addicted subjects. This strategy has been used to investigate the effects on brain glucose metabolism and/or CBF for most of the drugs of abuse. Cocaine, amphetamine, benzodiazepines, alcohol, and heroin have been found to decrease regional brain glucose metabolism (reviewed by London and Morgan 1993). This regional response in turn correlates with some of the drug-induced behavioral effects. Changes in CBF after drug administration have also been investigated with PET and SPECT, and currently also with fMRI (reviewed by Mathew and Wilson 1991). However, because many of the drugs of abuse have direct vasoactive properties, changes in CBF do not necessarily match the metabolic responses. Investigation of the magnitude of drug-induced changes in CBF is also relevant for understanding their toxicity as it relates to cerebrovascular pathology.

The effects of drugs of abuse on neurotransmitter activity was recently investigated with PET and SPECT for dopamine (DA) and psychostimulant drugs (Volkow et al. 1994; Laruelle et al. 1995). This response was measured using DA receptor ligands which are sensitive to its endogenous concentration. Images are obtained at baseline and after administration of a stimulant drug known to increase DA concentration. The changes in binding of the ligand with the stimulant drug are an indication of the relative changes in synaptic concentration of DA. Studies showed that the response between subjects was quite variable, that it decreased with age and that the intensity of the behavioral effects was closely correlated with the magnitude of the changes in DA concentration.

Imaging technologies can also be used to investigate the relation between behavioral effects and degree of receptor or transporter occupancy. These studies use as receptor radioligand one which is the target of the drug actions, and the studies are performed at baseline and after the drug of interest is given. For example, SPECT was used to evaluate receptor occupancy by the benzodiazepine agonist lorazepam and showed that only a very small fraction of the receptors are occupied at pharmacological doses (Sybirska et al. 1993), findings which corroborate in humans the existence of receptor reserves for benzodiazepine receptors.

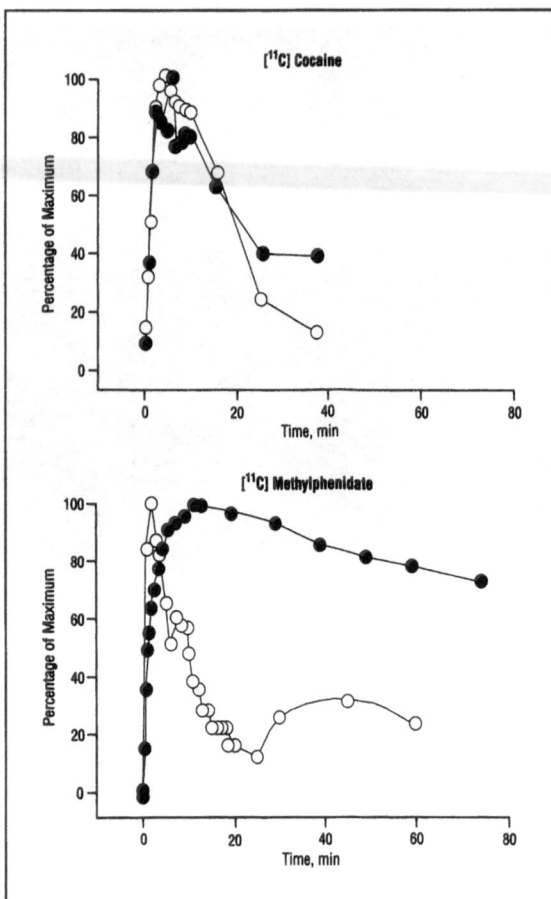

Fig. 18.2. Time-activity curves for ^{11}C-cocaine (*top*) and ^{11}C-methylphenidate (*bottom*) in the basal ganglia (*uptake*), plotted with the corresponding temporal patterns for the subjective experience of a "high" after pharmacological doses of cocaine or methylphenidate. Notice the parallelism between the "high" after cocaine and the kinetics of ^{11}C-cocaine in the basal ganglia. Notice the dissociation between the "high" from methylphenidate which falls rapidly after peaking and the slow clearance of ^{11}C-methylphenidate from the basal ganglia. *Solid circles* indicate uptake; *open circles*, subjective experience of a "high" (Volkow et al. 1995)

18.4
Imaging and the Evaluation of the Addicted Subject

PET and SPECT have been particularly valuable in studies of the addicted individual, allowing the assessment of neurochemical and functional changes in the brain. Changes in CBF or in rGLm have been documented in abusers of cocaine, marihuana, heroin, alcohol, and polysubstances. Studies with CBF have been particularly useful in documenting cerebrovascular toxicity associated with substance abuse, as was initially documented with PET and then repli-

cated with SPECT studies documenting widespread decrements in CBF in cocaine abusers (Volkow et al. 1988). Though a few isolated clinical reports of cerebrovascular damage had been reported in cocaine abusers, there was no knowledge that this was a medical complication associated with cocaine use. The imaging studies served to corroborate the clinical reports of cerebral strokes and hemorrhages in cocaine abusers and to highlight this toxic action of cocaine. Repeated studies can be done on the same subjects at different times for evaluation of therapeutic intervention. For example, imaging studies recently documented that the decrements in CBF in cocaine abusers improved with administration of buprenorphine (Levin et al. 1995). This exemplifies the value of imaging as an objective scientific tool for assessing therapeutic efficacy.

The measurement of brain glucose metabolism with FDG provides an index of brain activity not confounded by CBF changes; hence it is useful in the assessment of changes in brain function that may occur during withdrawal. For example, studies in cocaine abusers done at different times after cocaine discontinuation have shown that rGLm changes as a function of the withdrawal phase at which the studies are performed. Cocaine abusers (as well as polysubstance abusers) tested within one week of last cocaine use showed significantly higher metabolic activity in frontal brain regions and in basal ganglia than normal controls (Volkow et al. 1991; Stapleton et al. 1995). In contrast, cocaine abusers tested 1–4 months after cocaine discontinuation showed marked reduction in frontal metabolism (Fig. 18.3; Volkow et al. 1993).

Specific receptor radioligands are useful to assess the extent to which a particular neurotransmitter system is affected in addicted subjects. For example, in cocaine addiction where a dysfunction in brain DA activity has been postulated to underlie addiction, imaging studies have documented decrements in DA D_2 receptor ligands during early as well as protracted cocaine withdrawal (Fig. 18.4; Volkow et al. 1993) as well as decrements in DA metabolism (Baxter et al. 1988). Multiple tracer studies that measure glucose metabolism and/or CBF in conjunction with specific DA tracers for receptors and/or transporters enable one to assess the functional significance of changes in these DA elements. Such studies have been done to investigate the relation between brain glucose metabolism and DA D_2 receptors in cocaine abusers. A significant correlation was reported between DA D_2

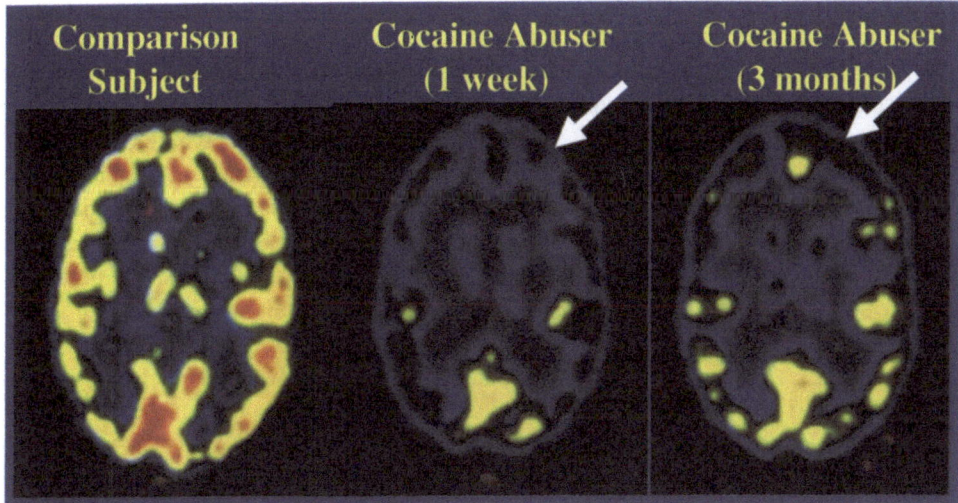

Fig. 18.3. Metabolic images obtained with FDG in a normal control and in a cocaine abuser tested 1 week and 3 months after cocaine discontinuation. Notice the reductions in metabolism in frontal brain regions when compared with the control (Volkow et al. 1993)

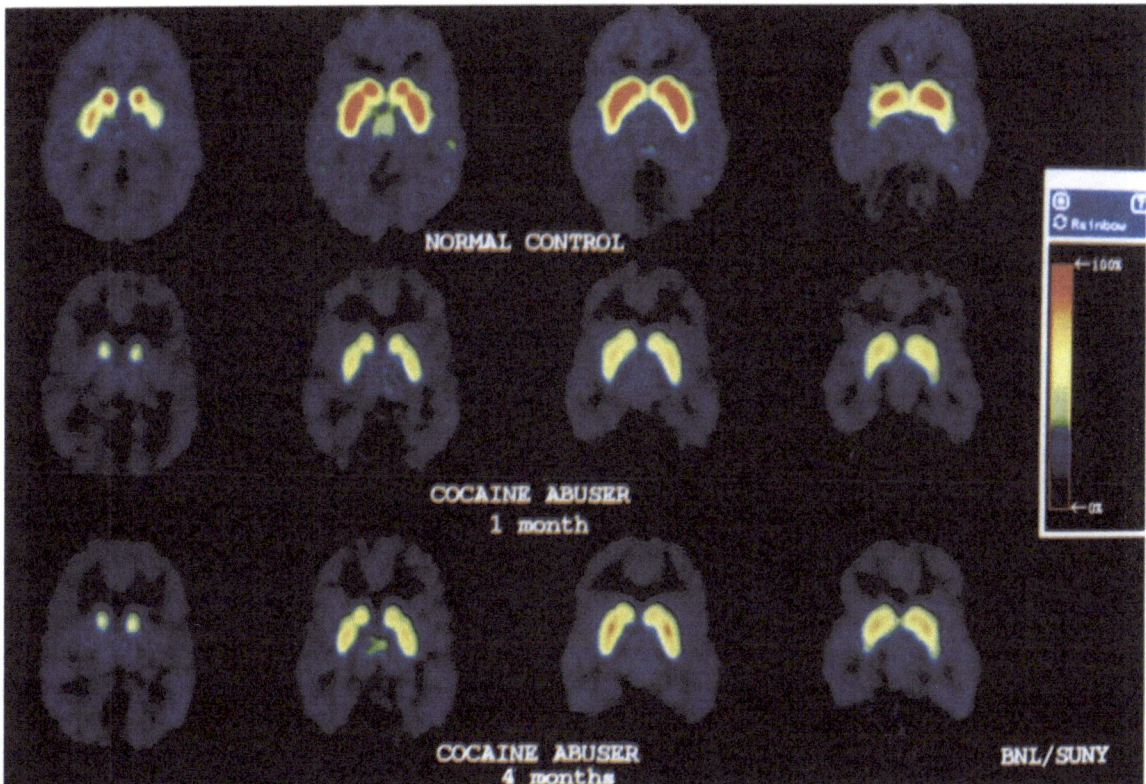

Fig. 18.4. Fluorine-18-N-methylspiroperidol images in a normal control and in a cocaine abuser tested 1 month and 4 months after last cocaine use. The images correspond to the four sequential planes where the basal ganglia are located. Notice the lower uptake of the tracer in the cocaine abuser when compared with the normal control. Notice the persistence of the decreased uptake even after 4 months of cocaine discontinuation (Volkow et al. 1993)

receptors and glucose metabolism in the orbitofrontal cortex, cingulate gyrus, and superior frontal cortex (Volkow et al. 1993). Lower values for D_2 receptor concentration were associated with lower metabolism in these brain regions.

18.5
Imaging Drug-Related States

The effects of acute pharmacological drug administration on regional brain metabolism and CBF can be used as described above to investigate the areas of the brain that have functional changes during drug intoxication. For example heroin-induced changes in scores related to euphoria were associated with heroin-induced changes in metabolic activity in the lateral occipital and primary visual cortex (London and Morgan 1993). Functional imaging strategies can also be used to assess the pattern of brain activation during drug related states triggered by behavioral interventions such as those to induce drug craving. This area of research has particularly benefited from fMRI where multiple repeated studies can be performed in the same subjects without the limitation of the radiation exposure of PET and SPECT.

18.6
Genomic Factors and Correlates

While environmental factors are undoubtedly crucial in determining the initiation and repetitive use of addictive drugs, there is mounting evidence of genetic characteristics also predisposing to their use. About 40–60% of the risk for alcohol, cocaine, nicotine, or opiate addiction is now said to be genetic (Noble 2000; Nestler 2001). Such claims are based partly on incidence of drug utilization by relatives of drug addicts, by monozygotic vs. dizygotic twins and by orphans in various family circumstances. Focusing on genomic typing and especially candidate genes, there are multiple individual studies and analyses of combined studies (Duaux et al. 2000; McKinney et al. 2000; Noble 2000), which indicate correlations between human addictive behavior and polymorphisms of genes concerned with synthesis or disposal of, or activity of receptors for, monoamines that have to do with reward and pleasure. There is particularly good evidence for association between drug disorders of various types and polymorphisms of dopaminergic neuroreceptor genes. Such studies have associated the presence of the A_1 allele of the dopamine receptor D_2 (DRD_2) Taq1 polymorphism with a variety of drug addictions and with obesity (Noble 2000). Parallel to human studies are many studies, both behavioral and genetic, in animals. Also, in animals there is now the possibility of obtaining evidence from manipulation of genes. Mice specifically lacking functional dopamine receptors ("knockout" mice) show a lack of alcohol- or opiate-reward behavior (Phillips et al. 1998; Noble 2000). On the other hand, transfection of the DRD_2 gene into the nucleus accumbens of rats previously trained to self-administer alcohol showed 43% reduction in alcohol preference and 64% decrease in alcohol intake when there was 52% increase in the expression of the DRD2 gene (Thanos et al. 2001).

Demonstration of changes in receptor activity by study with labeled ligands for receptors can provide objective evidence for phenotypic differences which are more specific and direct than behavioral analysis and more accurately correlated with genomic analysis. Autoradiography of postmortem brains of alcoholics and non-alcoholics using either 3H-spiperone or 3H-raclopride showed reduced density of D_2 dopamine receptors in the striatal regions in subjects from either group who had the A_1 allele of D_2 in either heterozygous or homozygous form (Noble 2000). An in vivo study with ^{11}C-raclopride and PET in healthy volunteers has also shown a decrease in DRD_2 availability in the striatum of those persons with the A_1/A_2 genotype compared with A_2/A_2 (Pohjalainen et al. 1998). In a group of healthy men given the psychostimulant methylphenidate (blocking dopamine transporters such as cocaine), various pleasurable or unpleasurable responses were noted and compared with DRD_2 density by a ^{11}C-raclopride/PET study on a different day (Volkow et al. 1999). The DRD_2 availability (B_{max}/K_d = density/affinity) was significantly lower in those subjects who reported pleasant feelings than in those with unpleasant ones (Fig. 18.5). More detailed studies of the phenotypes (kind, extent, and kinetics of ligand binding, related behavior, etc.) with correlation to genotypes (DR type, dopamine vs. serotonin disorder, etc.) may in the future aid diagnostic understanding and also possibly guide the variety and/or timing of therapy.

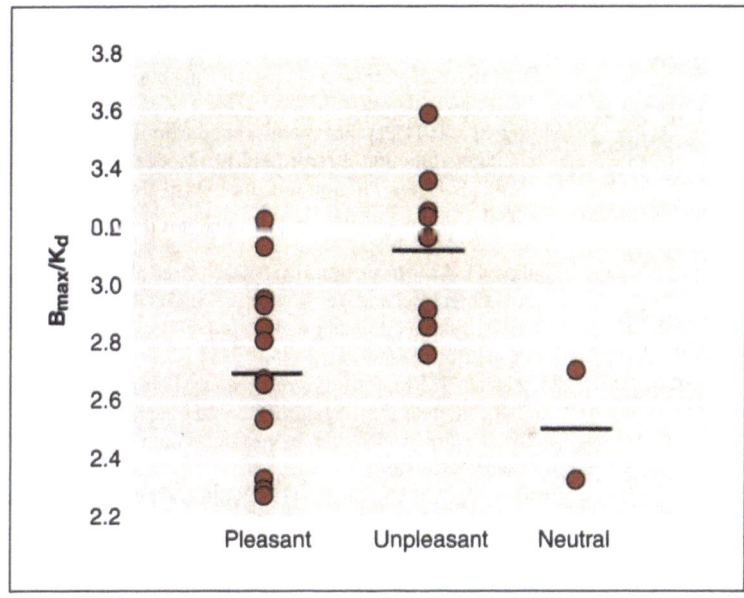

Fig. 18.5. Receptor levels (B_{max}/K_d) in 23 healthy male subjects who reported the effects of methylphenidate as pleasant, unpleasant, or neutral (Volkow et al. 1999)

18.7
Potential of Brain Imaging and Recommendations for Future Research

The potential of imaging techniques is highlighted by their ability to be used to assess the behavior of drugs of abuse in the human brain and by their ability to directly study the addicted individual. Because imaging studies are done in awake human subjects investigation of the relation between behavior and regional brain effects is allowed as well as between drug pharmacokinetics in the brain and the temporal course of pharmacological effects. Imaging technologies also enable an assessment of the effects of substances of abuse from various neurochemical and functional perspectives. While the studies described above reveal the power of imaging, the field has not yet reached its full potential. For example, with fMRI it will be possible to assess the effects of drug intoxication on the ability of the brain to respond to stimulation and/or cognitive demands. Such studies may allow the initiation of investigations into the mechanism(s) by which certain drugs facilitate the emergence of aggressive behaviors. Similarly it will allow a better understanding of the mechanism by which drug intoxication (e.g., alcohol, THC) is associated with impaired motor and cognitive performance. Another area is the use of functional imaging to determine the duration of action of a given drug on a specific functional process. Imaging may also prove to

be valuable in the evaluation of therapeutic agents for drug addiction. For example, agents of therapeutic interest could be labeled with positron emitters to assess their pharmacokinetics and bioavailability and to characterize their binding in human brain. Similarly, imaging could be used with the appropriate radiotracers to determine the relation between dose of the agent and receptor occupancy. The ability of therapeutic agents to reverse specific functional or neurochemical abnormalities in substance abusers could be evaluated with imaging strategies targeting the parameter of interest. Imaging can also be used to assess drug combinations and their potential toxicity. This is particularly relevant in cases where the drug abuser relapses, indicating the possibility of potentiation of effects of the abused substance by the therapeutic agent. Other areas where imaging has begun to be utilized are those that involve the investigation of the mechanisms by which environmental factors contribute to substance abuse. For example, the functional or neurochemical effects of a substance of abuse on the brain could be investigated under conditions where a given stress is systematically varied. Imaging may also prove of value in delineating mechanisms by which genetic factors may predispose to substance abuse by comparing functional or neurochemical responses in relation to family history and family members that are concordant or discordant for substance abuse.

18.8
Conclusions

Though still a relatively new field, imaging studies have documented specific neurochemical and functional changes in the brains of addicted subjects. They have also enabled direct investigation of the effects of drugs of abuse on the human brain. Their continued use to investigate substance abuse may help in the further delineating the actions of these drugs in the human brain and may be useful in the development of therapeutic agents for the treatment of addiction.

18.9
References

Andrew RF (1994) Introduction to nuclear magnetic resonance. In: Gillies RJ (ed) NMR in physiology and biomedicine. Academic, San Diego, pp 1–24

Baxter LR Jr, Schwartz JM, Phelps ME et al (1988) Localization of neurochemical effects of cocaine and other stimulants in the human brain. J Clin Psychiatry 49:23–26

Duaux E, Krebs MO, Loo H et al (2000) Genetic vulnerability to drug abuse. Eur Psychiatry 15:109–114

Kung HF (1993) SPECT and PET Ligands for CNS Imaging. Neurotransmissions IX (no 4). Research Biochemicals International, Massachusetts

Laruelle M, Abi-Dargham A, van Dick C et al (1995) SPECT imaging of striatal dopamine release after amphetamine challenge. J Nucl Med 36:1182

Levin JM, Mendelson JH, Holman LB et al (1995) Improved regional cerebral blood flow in chronic cocaine polydrug users treated with buprenorphine. J Nucl Med 36:1211–1215

London ED, Morgan MJ (1993) Positron emission tomographic studies on the acute effects of psychoactive drugs on brain metabolism and mood. In: London ED (ed) Imaging drug action in the brain. CRC Press, Boca Raton, pp 265–280

Mathew RJ, Wilson WH (1991) Substance abuse and cerebral blood flow. Am J Psychiatry 148:292–305

McKinney EF, Walton RT, Yudkin P et al (2000) Association between polymorphisms in dopamine metabolic enzymes and tobacco consumption in smokers. Pharmacogenetics 10:483–491

Mullani NA, Volkow ND (1992) Positron emission tomography instrumentation: a review and update. Am J Physiol Imaging 7:121–135

Nestler EJ (2001) Psychogenomics: opportunities for understanding addiction. J Neurosci 21:8324–8327

Noble EP (2000) Addiction and its reward process through polymorphisms of the D2 dopamine receptor gene: a review. Eur Psychiatry 15:79–89

Phillips TJ, Brown KJ, Burkhart-Kasch S et al (1998) Alcohol preference and sensitivity are markedly reduced in mice lacking dopamine D2 receptors. Nat Neurosci 1:610–615

Pohjalainen T, Rinne JO, Nagren K et al (1998) The A1 allele of the human D2 dopamine receptor gene predicts low D2 receptor availability in healthy volunteers. Mol Psychiatry 3:256–260

Rogers LW, Ackermann RJ (1992) SPECT instrumentation. Am J Physiol Imaging 7:105–120

Stapleton JM, Morgan MJ, Phillips RL et al (1995) Cerebral glucose utilization in polysubstance abuse. Neuropsychopharmacology 13:21–32

Sybirska E, Seibyl JP, Bremner JD et al (1993) [^{123}I]Iomazenil SPECT imaging demonstrates significant benzodiazepine receptor reserve in human and non human primate brain. Neuropharmacology 32:671–680

Thanos PK, Volkow ND, Freimuth P et al (2001) Overexpression of dopamine D2 receptors reduces alcohol self-administration. J. Neurochemistry 78:1094–1103

Volkow ND, Mullani N, Gould L et al (1988) Cerebral blood flow in chronic cocaine users. Br J Psychiatry 152:641–648

Volkow ND, Fowler JS, Wolf AP et al (1991) Changes in brain glucose metabolism in cocaine dependence and withdrawal. Am J Psychiatry 148:621–626

Volkow ND, Fowler JS, Wang G-J et al (1993) Decreased dopamine D2 receptor availability is associated with reduced frontal metabolism in cocaine abusers. Synapse 14:169–177

Volkow ND, Wang G-J, Fowler JS et al (1994) Imaging endogenous dopamine competition with [11C]raclopride in the human brain. Synapse 16:255–262

Volkow ND, Ding Y-S, Fowler JS et al (1995) Is methylphenidate like cocaine? Studies on their pharmacokinetics and distribution in human brain. Arch Gen Psychiatry 52:456–463

Volkow ND, Wang G-J, Fowler JS et al (1999) Prediction of reinforcing responses to psychostimulants in humans by brain dopamine D2 receptor levels. Am J Psychiatry 156:1440–1443

CHAPTER 19

Imaging of Cerebral Metabolism in Huntington's Disease   **19**

Torsten Kuwert

Contents

19.1
Introduction

Huntington's disease (HD) is a hereditary autosomally-dominant transmitted neurodegenerative disease. Clinically, HD is characterized by a hyperkinetic movement disorder, the so-called chorea, and progressive cognitive deterioration, i.e., dementia. Furthermore, HD may also lead to psychiatric symptoms such as depression, delinquency, alcoholism, or schizophreniform psychosis. These manifestations can precede the onset of chorea by several years and represent a considerable burden to the patient and his family. Woody Guthrie, the famous American musician, was affected by HD; the fatal consequences of this disease on his life and that of his family have been described in his biographies (Klein 1980).

The treatment of HD aims mainly at alleviating its symptoms, a causative therapy is at present not available, although several approaches are currently being tested (for a review, see Beal and Hantraye 2001). The relentless course of the disease over up to 30 years poses great psychological strain on patients and their relatives who require intensive care and counseling (for further information, see also www.hdac.org or www.hdfoundation.org). Furthermore, great strain rests on the so-called at-risk individuals (for a personal account of this experience, see Wexler 1996), i.e., subjects whose parents have been afflicted by the disease, since every gene carrier will develop HD.

George Huntington, a family physician in Huntington County (Long Island, NY) and later in Ohio, was the first to recognize that HD is hereditary (Huntington 1872). He distinguished between the so-called "chorea major" and "chorea minor," i.e., Sydenham's chorea, which is a sequelae to rheumatic fever caused by bacterial infection and was much more frequent in his time.

HD is a monogenetically transmitted disorder. The gene defect consists in an abnormally high repetition of CAG triplets in the IT-15 gene locus on chromosome 4 (The Huntington's Disease Collaborative Research Group 1993). This type of gene defect is also found in other neurodegenerative diseases such as spinocerebellar ataxias-1, -2, -3, -7 or dentatorubrol-pallidoluysian atrophy (for a review, see Paulson and Fishbeck 1996). In these diseases, the number of triplets often increases when the gene is paternally transmitted. As the number of triplets correlates with age of onset, children that have received the pathological gene from their fathers will usually become ill earlier than their father, an effect termed anticipation.

The HD gene has a 100% penetrance. The detection of more than 38 CAG repeats in a patient with typical symptoms establishes the diagnosis of HD. Although, as already mentioned above, the number of CAG triplets correlates significantly with age of onset (Trottier et al. 1994), the exact prediction of the onset of clear-cut neurological symptoms in an asymptomatic individual is still difficult.

As in other neurodegenerative disorders, HD is caused by progressive neuronal loss. Manifestations outside the brain are largely absent. The brunt of the pathology is carried by the corpora striata, two structures of gray matter located between thalami and cortex (Lange 1976; Wexler 1996). Furthermore, parts of the cortex are also regularly involved, e.g., the occipital, the prefrontal, and the parietal cortex.

Macroscopic counterparts of neurodegeneration in HD are cortical, caudate, and putaminal atrophy. These volume losses can be detected using structural imaging techniques such as X-ray, CT, or MRI (Aylward et al. 1994). Some authors have demonstrated that a reduction in putaminal volume precedes the onset of chorea by several years (Beal and Hantraye 2001); but since putaminal volumetry is difficult to perform, this technique has not found broad acceptance as a clinical tool.

19.2
PET and SPECT in Huntington's Disease

Using PET and SPECT, decreases of rCBF and regional cerebral glucose consumption (rCMRGlc) have been demonstrated in the corpora striata of HD patients (Kuhl et al. 1982; Kuwert et al. 1990; Fig. 19.1). A reduction of cortical and thalamic rCBF and rCMRGlc has also been reported. The hypometabolism involves in particular the prefrontal cortex. As expected, the severity of reduction of these variables correlates with the severity of various symptoms of HD (Kuwert et al. 1990).

Neural transmission in HD has also been the target of PET research: whereas striatal decreases of D_1 and D_2 dopaminergic receptor density usually are striking and are also found presymptomatically, dopaminergic innervation seems better preserved (Leenders et al. 1986). This can be explained by a selective death of striatal neurons, which carry the receptor, and by still-intact neurons in the substantia nigra where the dopaminergic axons originate.

Clinically, the imbalance between striatal presynaptic dopamine concentration and dopamine receptor density leads to hyperkinesis, e.g., chorea, which can be effectively treated by administration of D_2 receptor antagonists such as sulpiride. In simplification, chorea in HD would seem to be in direct contrast with the transmitter disturbances found in Parkinson's disease where there is a reduction of dopamine and an upregulation of dopamine D_2 receptors.

Recently, PET imaging studies have been conducted in a rat model of HD using an animal scanner (Araujo et al. 2000). In this model, a progressive deterioration of striatal rCMRGlc consequent to an excitotoxic lesion has been demonstrated. As in humans, presynaptic measures of dopaminergic function remained largely intact. Dopaminergic D_2 receptor density was upregulated in the first two weeks after the experimental lesion, but this density decreased below normal in the weeks thereafter. The observed upregulation is in contrast to results from previous PET studies in HD patients, although some data on a mismatch between dopamine receptor density and glucose consumption have been published in patients with Wilson's disease, another basal ganglia disorder (Schlaug et al. 1996).

Since the advent of the gene test, imaging plays only a minor role in diagnosing HD in symptomatic patients. However, several studies have shown that striatal rCMRGlc and dopamine D2 density can be reduced in asymptomatic subjects at risk of HD and in gene carriers (Leenders et al. 1986; Hayden et al. 1986; Mazziotta et al. 1987; Grafton et al. 1990; Kuwert et al. 1993; Andrews et al. 1999). This may also be the case when the striata do not show signs of atrophy on X-ray, CT, or MRI (Fig. 19.1). Molecular imaging, and FDG-PET in particular, therefore allows a very early detection of the onset of neurodegeneration. The prognosis of asymptomatic individuals exhibiting metabolic disturbances is currently a subject of research; the data available, meanwhile, show that the majority of gene-positive individuals with reduced striatal rCMRGlc develop chorea within five years (Grafton et al. 1990).

19.3
Relationship Between Results of PET and SPECT Imaging and Neurodegeneration in Huntington's Disease

It is obvious that the decreases of rCBF, rCMRGlc, and regional receptor densities observed in HD patients reflect the neurodegenerative process of HD. The relationship between these parameters of molecular imaging and this process must be discussed on three levels (Fig. 19.2):

(a) A reduction of rCMRGlc, rCBF, and regional receptor densities may reflect regional atrophy brought about by degeneration. This is a consequence of the so-called partial volume effect, dependent on the limited spatial resolution of PET and SPECT with regard to the size of the structures studied.

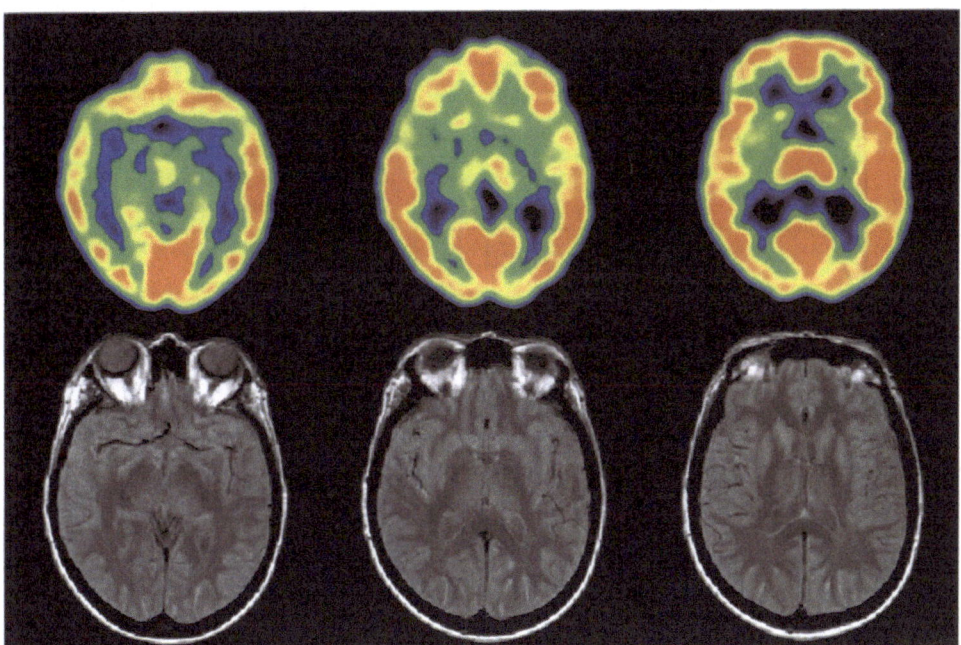

Fig. 19.1. Images of regional cerebral glucose consumption measured by FDG-PET (*upper row*) and matched T1-weighted MRIs (*lower row*) in a patient with incipient Huntington's disease. Striatal glucose consumption, which should be equal to or even higher than that in cortex, is decreased, although no atrophy of the striatum is detectable on MRI

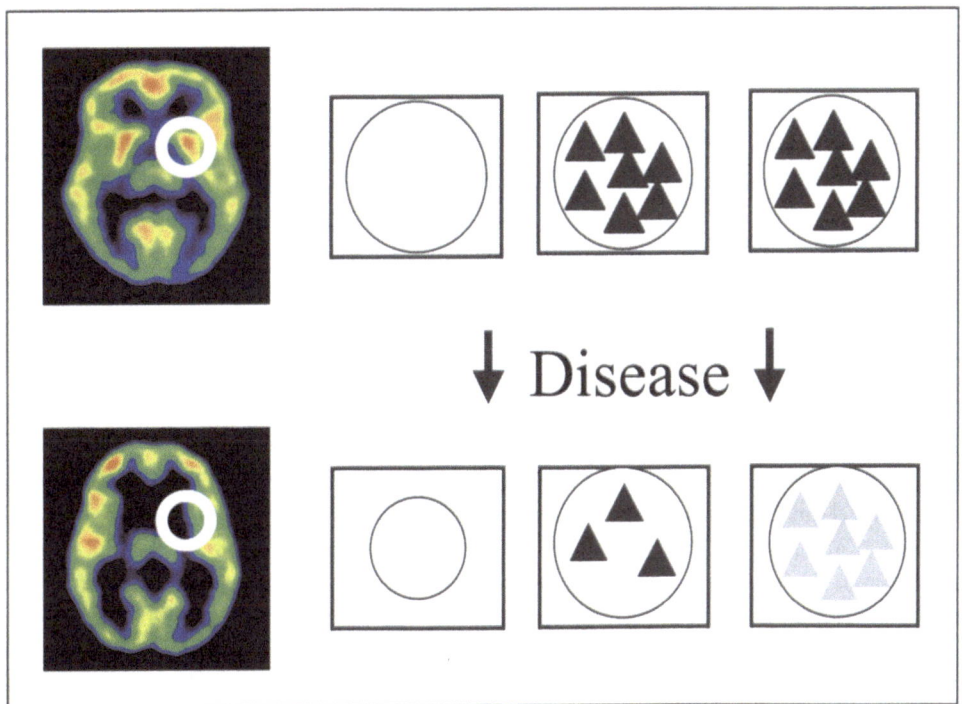

Fig. 19.2. A reduction of a PET or SPECT variable (in this case striatal glucose consumption) may be due to atrophy of the structure examined, to a reduction in the density of target cells (in this case striatal neurons), or to decreased metabolic activity of the target cells

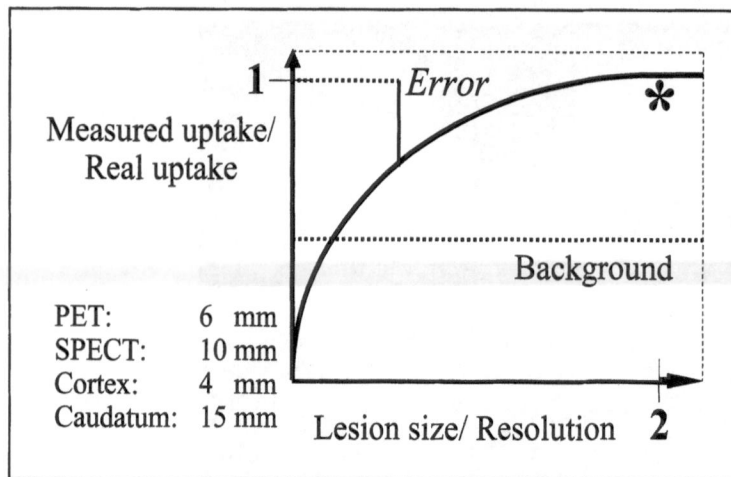

Fig. 19.3. The graph plots the ratio between the uptake of a radiopharmaceutical measured by PET or SPECT and the real uptake in the structure versus the ratio between the size of the structure and image resolution. For structures smaller than twice the image resolution measurement errors cannot be avoided

(b) A reduction of rCMRGlc, rCBF, and regional receptor densities may reflect microscopic changes in the tissue and in particular a reduction of neuronal density typical of neurodegeneration.

(c) A reduction of rCMRGlc and rCBF may reflect metabolic disturbances occurring intracellularly in the neurons affected by the disease. Analogously, decreased receptor densities may represent up- or downregulation of receptors within still-viable cells.

The partial volume effect leads to errors of measurement when the size of a structure imaged is smaller than twice the spatial resolution of the imaging system used (Fig. 19.3). With contemporary PET, image resolution is approximately 6 mm, with SPECT about 10 mm. The diameter of the caudate nucleus and the putamen is usually between 1 and 2 cm. If these structures develop atrophy such as in HD, their radioactivity concentration becomes considerably underestimated. This is not the case in normal individuals who provide the control values, so that the reduction of the PET or SPECT variable under study found when comparing patients' and control values could, at least in principle, solely be explained by the shrinkage of the degeneratively affected ganglia. Mechanism (a) has indeed been proven in many studies correlating regional volumes determined by MRI with rCMRGlc. It may overrule the effect of microscopic and biochemical alterations in later stages of the disease. However, since structures of gray matter with unremarkable morphology may also exhibit decreases of rCMRGlc or rCBF, mechanisms (b) and (c) must also be considered.

As yet, no studies directly comparing parameters of molecular imaging with regional neuronal densities are available in neurodegenerative diseases. In

HD, however, it has been shown that decreases in postmortem caudate neuronal density may be encountered in during life choreatic individuals in spite of normal caudate volume (Vonsattel et al. 1985). It therefore seems plausible to consider a reduction of neuronal density to be the correlate of reductions in striatal rCMRGlc or rCBF encountered in subjects with unremarkable MRIs.

It would be ideal to have an imaging tool to measure regional neuronal density that would correct metabolic data for the effect of cell loss in the structure studied. In principle, the magnetic resonance spectroscopic (MRS) measurement of the distribution of N-acetyl-D-aspartate (NAA) represents such a facility, since NAA is obligatorily confined to neurons, at least in the brain. Decreases of the NAA signal have been demonstrated in several disorders leading to cell loss, such as Alzheimer's disease, and also in epileptogenic foci. As yet, only few studies have correlated MRS with PET or SPECT, and for HD no such data are available.

The neurodegenerative process in HD affects primarily the spiny medium-sized neurons in the striatum. Dopamine D_2 receptor density is much more reduced than GABA receptor density (Kunig et al. 2000). This finding, obtained by PET with ^{11}C-flumazenil and ^{11}C-raclopride, suggests a compensatory upregulation of the latter receptor in a subpopulation of striatal neurons, illustrating that mechanism (c) may play a role when considering changes in radioligand binding quantified by emission tomography.

Both rCMRGlc and rCBF correlate with regional electrical neuronal activity. Thus, any loss of excitatory input into a brain region or even a single neuron is accompanied by decreases in rCMRGlc and rCBF, an effect loosely termed "diaschisis." In HD, diaschi-

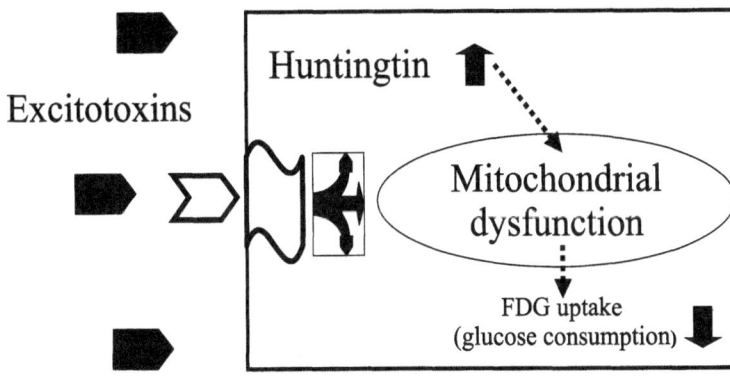

Fig. 19.4. Speculative mechanism causing neuronal death in Huntington's disease: an accumulation of the abnormal gene product in Huntington's disease, the so-called huntingtin, causes mitochondrial dysfunction leading to impaired energy metabolism and, possibly, to reduced FDG uptake. Moreover, impaired energy metabolism makes the cells vulnerable to excitotoxins, ultimately triggering cell death

sis may play a role in the decreased cortical (in particular prefrontal) rCMRGlc, although these regions also display direct pathology.

Every neuron relies on the oxidative breakdown of glucose to generate energy. This process involves the enzymes of glycolysis, of the citrate cycle, and of the respiratory chain; any damage to one of these steps may lead to a decrease in cellular glucose uptake and phosphorylation, and thus also of rCMRGlc when a high percentage of neurons in a given region of interest is affected. Given the early occurrence of rCMRGlc decreases in HD, it is tempting to speculate that a disturbance in cellular energy metabolism is the ultimate cause of neurodegeneration in HD. Some evidence indeed exists hinting in this direction, namely, that there is a decrease in the activity of respiratory-chain enzymes the brains of HD patients when analyzed postmortem (Tabrizi et al. 1999). Furthermore, a rat model mimicking the disease may be induced by using 3-nitropropionic acid, an inhibitor of the citrate cycle (Ludolph et al. 1992). Some evidence for this theory can also be found in humans: using ^{1}H-MRS, it has been repeatedly shown that the cortical concentration of lactate is increased in HD patients (Koroshetz et al. 1997). Lactate is well known to accumulate as a product of glycolysis when cellular oxidation is impaired.

What makes the hypothesis of disturbances of neuronal energy metabolism even more intriguing is the observation that the product of the HD gene, the so-called huntingtin, binds via its polyglutamine tract to glyceraldehye-3-phosphate dehydrogenase (GAPDH), an important enzyme in glycolysis (Burke et al. 1996). Integrating these lines of evidence with that of the popular mechanism of excitatory cell death, one currently pursued hypothesis on the link between geno- and phenotype of HD is that defects of energy metabolism caused by an accumulation of huntingtin make the HD neuron vulnerable to excitotoxins, thus leading to

neurodegeneration (Beal 2000; Zeron et al. 2001; Fig. 19.4).

However, ultimate proof for mitochondrial dysfunction as a cause of cell death in HD is still lacking; other mechanisms are also under study at present. Using transgenic mice as well as cells transfected with the HD gene, it has, for example, been shown that huntingtin molecules may form aggregates, possibly analogous to the amyloid deposits found in Alzheimer's disease, that inhibit not only energy metabolism but a large number of cellular functions (McGowan et al. 2000). Furthermore, work is being directed at proving that huntingtin induces apoptosis (Tobin and Signer 2000).

Using genetically manipulated animals, a more firmly founded pathogenetic theory of HD seems to be just around the corner. PET or SPECT in experimental animals carrying the HD gene offer distinct advantages for monitoring the process of neurodegeneration (Phelps 2000). Analogous to efforts aiming at imaging the brain concentration of amyloid in vitro (Dezutter et al. 1999), the development of a radiopharmaceutical that traces the cerebral accumulation of huntingtin would represent an important step in this direction.

19.4
Conclusions

Whatever the exact mechanism of pathogenesis in HD, molecular imaging with specifically designed radiopharmaceuticals may in the long term be the best way to prove the validity of a given hypothesis in the human disease. As detailed above, results from molecular imaging should be corrected wherever possible for the effects of alterations in the micro- and macrostructure brought about by the disease.

Last but maybe not least, PET and SPECT promise to be the most sensitive imaging tools for diagnosing early striatal dysfunction (Kuwert et al. 1998) and can be expected to have utmost importance in the development of effective therapeutic regimens.

■ **Acknowledgements.** The author gratefully acknowledges the support of Dr. Herwig Lange who by his longstanding devotion to the cause of the people afflicted by HD has over several years greatly stimulated the author's research work. The author wants to express his gratitude to Dr. Peter Matheja and Professor Otmar Schober from the Department of Nuclear Medicine of the Westfälische-Wilhelms-Universität Münster for providing him with an excellent FDG-PET image (Fig. 19.1). Furthermore, he thanks Dr. Alexander Schwarz from the Clinic of Nuclear Medicine of the Friedrich-Alexander-Universität Erlangen-Nürnberg for his help with the references and figures.

19.5
References

Andrews TC, Weeks RA, Turjanski N et al (1999) Huntington's disease progression. PET and clinical observations. Brain 122:2353–2363

Araujo DM, Cheney SR, Tatsukawa KJ et al (2000) Deficits in striatal dopamine D-2 receptors and energy metabolism detected by in vivo MicroPET imaging in a rat model of Huntington's disease. Exp Neurol 166:287–297

Aylward EH, Brandt J, Codori AM et al (1994) Reduced basal ganglia volume associated with the gene of Huntington's disease in asymptomatic at-risk persons. Neurology 44:823–828

Beal MF (2000) Energetics in the pathogenesis of neurodegenerative diseases. Trends Neurosci 23:298–304

Beal MF, Hantraye P (2001) Novel therapies in the search for a cure for Huntington's disease. Proc Natl Acad Sci USA 98:3–4

Burke JR, Enghild JJ, Martin ME et al (1996) Huntingtin and DRPLA proteins selectively interact with the enzyme GAPDH. Nat Med 2:347–350

Dezutter NA, Dom RJ, de Groot TJ et al (1999) 99mTc-MAMA-chrysamine G, a probe for beta-amyloid protein of Alzheimer's disease. Eur J Nucl Med 26:1392–1399

Grafton ST, Maziotta JC, Pahl JJ et al (1990) A comparison of neurological, metabolic, structural, and genetic evaluation in persons at risk of Huntington's disease. Ann Neurol 28:614–621

Hayden MR, Martin WR, Stoessl AJ et al (1986) Positron emission tomography in the early diagnosis of Huntington's disease. Neurology 36:888–894

Huntington G (1872) On chorea. Med Surg Rep 26:317–321

Klein J (1980) Woody Guthrie: a life. Ballantine Books, New York

Koroshetz WJ, Jenkins BG, Rosen BR et al (1997) Energy metabolism defects in Huntington's disease and effects of coenzyme Q_{10}. Ann Neurol 41:160–165

Kuhl DE, Phelps ME, Markham CH et al (1982) Cerebral metabolism and atrophy in Huntington's disease determined by [18]FDG and computed tomographic scan. Ann Neurol 12:425–434

Kunig G, Leenders KL, Sanchez-Pernaute R et al (2000) Benzodiazepine receptor binding in Huntington's disease: [11C]flumazenil uptake measured using positron emission tomography. Ann Neurol 47:644–648

Kuwert T, Lange HW, Langen KJ et al (1990) Cortical and subcortical glucose consumption measured by PET in patients with Huntington's disease. Brain 113:1405–1423

Kuwert T, Lange HW, Boecker H et al (1993) Striatal glucose consumption in chorea-free subjects at risk of Huntington's disease. J Neurol 241:31–36

Kuwert T, Bartenstein P, Grünwald F et al (1998) Clinical significance of positron emission tomography in neuromedicine. A position paper on the results of an interdisciplinary consensus conference. Nervenarzt 69:1045–1060

Lange HW (1976) Morphometric studies of the neuropathological changes in choreatic diseases. J Neurol Sci 28:401–425

Leenders KL, Frackowiack RS, Quinn N et al (1986) Brain energy metabolism and dopaminergic function in Huntington's disease measured in vivo using positron emission tomography. Mov Dis 1:69–77

Ludolph AC, Seelig M, Ludolph AG et al (1992) ATP deficits and neuronal degeneration induced by 3-Nitropropionic acid. Ann NY Acad Sci 11:300–302

Mazziotta JC, Phelps ME, Pahl, JJ et al (1987) Reduced cerebral glucose metabolism in asymptomatic subjects at risk for Huntington's disease. N Engl J Med 316:357–362

McGowan DP, van Roon-Mom W, Holloway H et al (2000) Amyloid-like inclusions in Huntington's disease. Neuroscience 100:677–680

Paulson HL, Fishbeck KH (1996) Trinucleotide repeats in neurogenetic disorders. Annu Rev Neurosci 19:79–107

Phelps ME (2000) PET: the merging of biology and imaging into molecular imaging. J Nucl Med 41:661–681

Schlaug A, Hefter H, Engelbrecht V et al (1996) Neurological impairment and recovery in Wilson's disease: evidence from PET and MRI. J Neurol Sci 136:129–139

Tabrizi SJ, Cleeter MW, Xuereb J et al (1999) Biochemical abnormalities and excitotoxicity in Huntington's disease brain. Ann Neurol 45:25–32

The Huntington's Disease Collaborative Research Group (1993) A novel gene containing a trinucleotide repeat that is expanded and unstable on Huntington's disease chromosomes. Cell 72:971–983

Tobin AJ, Signer ER (2000) Huntington's disease: the challenge for cell biologists. Trends Cell Biol 10:531–536

Trottier Y, Bianclana V, Mandel JL et al (1994) Instability of CAG repeats in Huntington's disease: relation to parental transmission and age of onset. J Med Genet 31:377–382

Vonsattel JP, Myers RH, Stevens TJ et al (1985) Neuropathological classification of Huntington's disease. J Neuropathol Exp Neurol 44:559–577

Wexler A (1996) Mapping fate – a memoir of family, risk, and genetic research. Random House, Times Books, New York

Zeron MM, Chen N, Moshaver A et al (2001) Mutant huntingtin enhances excitotoxic cell death. Mol Cell Neurosci 17:41–53

CHAPTER 20

Monoamine Oxidase: Radiotracer Development and Human Studies*   20

JOANNA S. FOWLER, JEAN LOGAN, NORA D. VOLKOW, GENE-JACK WANG, ROBERT R. MACGREGOR, YU-SHIN DING

Contents

20.1
Introduction

20.1.1
Discovery

In 1928, Mary Hare isolated a new enzyme which catalyzed the oxidative deamination of tyramine (Hare 1928). She called it tyramine oxidase and speculated that it "may be protective and present for the purpose of rapid detoxification of excessive amounts of tyramine absorbed from the intestine." Later Blashko and coworkers showed that this same enzyme also oxidized catecholamines (Blaschko et al. 1937). To reflect this more general reactivity, Zeller (1938) proposed the general name monoamine oxidase (MAO). In the years that followed its discovery, MAO was further characterized along with its role in the regulation of chemical neurotransmitters and as a target for therapeutic drugs and toxic substances. More recently its genetics have been studied. This chapter will focus on general aspects of MAO, on the development of radiotracers for imaging MAO A and MAO B, and on PET studies of MAO in the human brain.

20.1.2
General Features of MAO

Monoamine oxidase [MAO; amine: oxygen oxidoreductase (deaminating; flavin containing); E.C. 1.4.3.4] is an integral protein of outer mitochondrial membranes and occurs in neuronal and non-neuronal cells in the brain and peripheral organs. It oxidizes amines from both endogenous and exogenous sources, thereby influencing the concentration of neurotransmitter amines as well as many xenobiotics (eq. 20.1; Singer 1995; Richards et al. 1998). It occurs as two subtypes, MAO A and MAO B which have different inhibitor and substrate specificities (Fig. 20.1). MAO A preferentially oxidizes norepinephrine and serotonin and is selectively inhibited by clorgyline (Johnston 1968) while MAO B preferentially breaks down the trace amine phenylethylamine and is selectively inhibited by L-deprenyl, also called selegiline (Knoll and Magyar 1972). Both forms oxidize dopamine, tyramine,

* Reprinted from *Methods*, vol. 27, Fowler JS, Logan J, Volkow ND, Wang GJ, MacGregor RR, Ding YS, Monoamine oxidase: radiotracer development and human studies, pp. 263–277, 2002, with permission of Elsevier Science

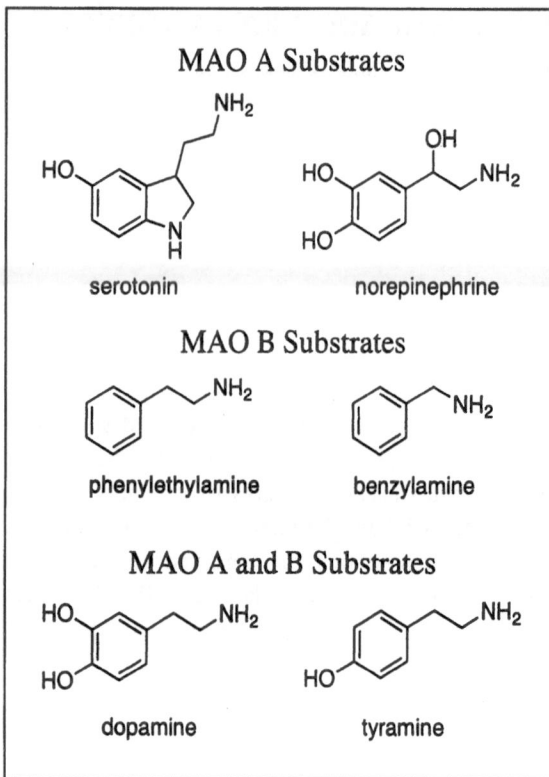

Fig. 20.1. Structures of MAO A and MAO B and MAO A and B substrates

and octopamine (Youdim and Riederer 1993). Oxidation is accompanied stoichiometrically by the reduction of oxygen to hydrogen peroxide. The relative ratios of MAO A and B are organ and species specific (see Saura et al. 1992, 1994, 1996). For example in the human brain, MAO B predominates whereas in the rat brain, MAO A is the predominant subtype. The two subtypes are also compartmentalized in different cell types in the brain with MAO B occurring predominately in glial cells and in serotonergic neurons while MAO A occurs in catecholaminergic neurons as well as in glia cells. It has been speculated that the compartmentalization of a specific MAO subtype within the neurons prevents the non-specific neuronal accumulation of neurotransmitters.

$$RCH_2NH_2 + O_2 + H_2O \xrightarrow{MAO} RCHO + NH_3 + H_2O_2$$
$$(20.1)$$

20.1.3
Genetics

MAO A and B are encoded by separate genes that are closely linked on the X chromosome and share 70% similarity in amino acid sequence (Bach et al. 1988). The loss of both MAO A and MAO B genes has been implicated in the severe mental retardation of some patients with Norrie's disease (Collins et al. 1992). Recently a family has been described in which a point mutation in the gene encoding MAO A abolished MAO A activity and is associated with a recognizable behavioral phenotype, which includes disturbed regulation of impulsive aggression (Brunner et al. 1993). With the development of molecular genetic techniques for the production of knockout animals, mice missing MAO A or MAO B have been produced and studied. MAO A knockout mice have high circulating levels of serotonin and male animals exhibit a distinct behavioral syndrome characterized by enhanced aggression (Cases et al. 1995). MAO B knockout animals have high levels of phenylethylamine, a specific substrate for MAO B, and they are resistant to 1-methyl-4-phenyl-1,2,3,5-tetrahydropyridine (MPTP) neurotoxicity (Grimsby et al. 1997). Studies in MAO B knockout mice also suggest that MAO B may regulate normal blood flow distribution (Scremin et al. 1999). Both MAO A and B knockouts show enhanced reactivity to stress. Transgenic mice overexpressing human MAO B protein have also been described. They express a 4- to 6-fold higher brain MAO B and a higher rate of dopamine metabolism, whereas liver MAO B is equal to that of control littermates (Richards et al. 1998). Transgenic animals have been valuable models for investigating the role of monoamines in psychoses and neurodegeneration and stress-related disorders (Shih et al. 1999).

20.1.4
Medical Importance

Medical interest in MAO was stimulated in the early 1950s when it was discovered that iproniazide, a drug which was being used to treat tuberculosis, elevated mood in some patients (Selikoff et al. 1952; Crane 1956). This observation suggested the possibility of treating depression pharmacologically. It was soon learned that iproniazide inhibited MAO (Zeller et al. 1955). This revelation, in part, contributed to the hypothesis that monoamine regulation may be related to mood and led to the development and application of MAO inhibitor drugs in the treatment of depression (Schildkraut 1965).

Though the MAO inhibitors have been effective antidepressants, their use has become limited by serious and sometimes lethal side effect which have come to be known as the "cheese effect" (Anderson et al. 1993). This refers to the development of hypertensive crises in individuals taking nonselective, irreversible MAO inhibitor drugs and who also ingested foods (aged cheeses, pickled meats and fish, and red wine) that contain large quantities of the vasoactive amine, tyramine. The breakdown of tyramine requires the presence of MAO in the digestive organs. A number of deaths occurred in the early years of MAO-inhibitor therapy before these serious drug-diet interactions were understood and controlled. For this reason, MAO inhibitor drugs have been largely replaced by antidepressant drugs with a more acceptable side-effect profile. Because of this initial experience with irreversible MAO inhibitor drugs, MAO's role as a vast and complex mechanism for regulating circulating catecholamines and other endogenous amines (plus dietary amines and drugs) as well as its role in regulating blood pressure has come to be appreciated and respected (Mannelli et al. 1990; Kopin 1993).

Following the initial experience with nonselective, irreversible MAO inhibitors in the treatment of depression, the selective irreversible MAO B inhibitor, L-deprenyl was developed (Knoll and Magyar 1972). It was used in combination with L-DOPA therapy in Parkinson's disease to inhibit the MAO-catalyzed oxidation of dopamine. This combination therapy was reported to have prolonged therapeutic efficacy in patients when compared to L-DOPA alone. Because L-deprenyl has a high selectivity for MAO B, leaving MAO A intact, it does not have the side effects of the nonselective irreversible MAO inhibitors (Birkmayer and Riederer 1984).

In the early 1980s another chapter in the therapeutic use of L-deprenyl unfolded when it was reported that L-deprenyl could prevent the development of Parkinson's disease in animals treated with MPTP (Langston et al. 1983, 1984; Heikkila et al. 1984). MPTP was found to be an impurity in a street drug which when ingested caused an initially puzzling outbreak of Parkinson's disease in a number of young people. In a remarkable series of studies, it was discovered that MAO B inhibition prevented MPTP-induced neurotoxicity by inhibiting the conversion of MPTP to 1-

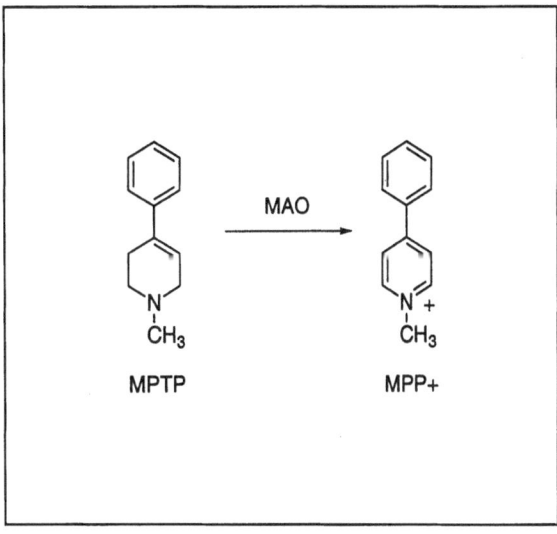

Fig. 20.2. MAO-catalyzed conversion of MPTP to MPP+

methyl-4-phenylpyridinium (MPP+) which is toxic to dopamine neurons (Fig. 20.2). This led to speculation that MAO inhibitors may have a neuroprotective effect in Parkinson's disease and to clinical trials showing that L-deprenyl significantly retarded the progression of the disease and the requirement for L-DOPA therapy (Tetrud and Langston 1989; The Parkinson's Study Group 1989). The successful use of L-deprenyl monotherapy stimulated the development of other MAO B inhibitors with enhanced neuroprotective properties and also stimulated discussions as to whether the decreased progression of the disease represented protective or symptomatic effects (Fowler et al. 1996c) – and even whether the mechanism involves MAO B inhibition (Ansari et al. 1993).

The structures of some MAO inhibitor drugs are shown in Fig. 20.3.

20.2
Radiotracer Development

Because the regional and cellular compartmentalization of MAO and its subtypes determines to a large extent the access of specific substrates to each subtype, a knowledge of the distribution of MAO A and B in the brain and the peripheral organs is a crucial element in understanding neurotransmitter regulation and in understanding the MAO inhibitory properties of drugs. Studies in humans are of special value because species variability in MAO subtype distribution

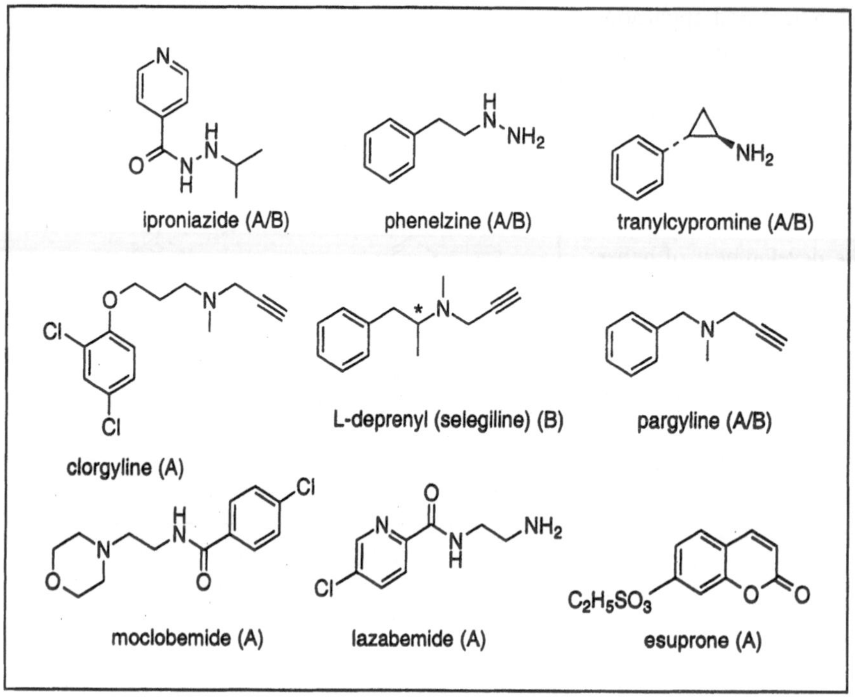

Fig. 20.3. Structures of some MAO inhibitor drugs. The *letter or letters in parenthesis* indicate subtype specificity

limits the relevance of animal measurements. In this section, we describe a number of different approaches that have been used to selectively image MAO subtypes. These include the use of (1) reversible, subtype selective radiotracers; (2) compounds which are oxidized by MAO to produce a charged, labeled product that is trapped in MAO-rich tissues (metabolically trapped radiotracers); and (3) compounds that are oxidized by MAO to produce a reactive intermediate that labels MAO by irreversible covalent attachment of the radiotracer to the enzyme (suicide inactivator radiotracers). The functional activity of MAO has also been visualized in vivo in the baboon and human heart by the kinetic behavior and deuterium isotope effects of labeled substrates, 6-^{18}F-fluorodopamine (Ding et al. 1995) and ^{11}C-phenylephrine (Raffel et al. 1996, 1999; Fig. 20.4). Human PET studies with ^{11}C-phenylephrine (which is a tracer for cardiac vesicular storage sites) showed that carbon-11 clearance from cardiac storage vesicles in vivo is sensitive to MAO but that the major determinant of clearance of carbon-11 is due primarily to leakage from vesicles. In the case of 6-^{18}F-fluorodopamine, deuterium substitution in the α (next to the amino group) and β positions, respectively, was used to determine whether MAO (which would cleave the α carbon-hydrogen bond) or dopamine-β-hydroxylase (which would

cleave the β carbon-hydrogen bond) contributed to the clearance of fluorine-18 from the heart. Both of these studies illustrate the value of the deuterium isotope effect as a mechanistic tool for essentially isolating the MAO reaction in vivo. Deuterium substitution in 6-^{18}F-fluorodopamine results in a radiotracer that closely parallels the behavior of (−)-norepinephrine. More specifically, it is likely that deuterium substitution in the alpha position protects the molecule from MAO-catalyzed oxidation in the cytosol, allowing it to be transported into the vesicle where it is probably converted to (−)-6-^{18}F-fluoronorepinephrine.

20.2.1
Reversible, Subtype-Selective Radiotracers

A number of different classes of selective, reversible inhibitors of MAO A and B have been labeled with PET or SPECT radioisotopes and evaluated for their specificity as MAO tracers in vivo. Some of these are shown in Fig. 20.5. Derivatives of the harmine alkaloids have been labeled with carbon-11 and evaluated in the monkey brain for the assessment for MAO A. Carbon-11-harmine, ^{11}C-methylharmine, ^{11}C-harmaline, and ^{11}C-brofaromine were compared in rhesus monkey, and both ^{11}C-harmine and ^{11}C-methylhar-

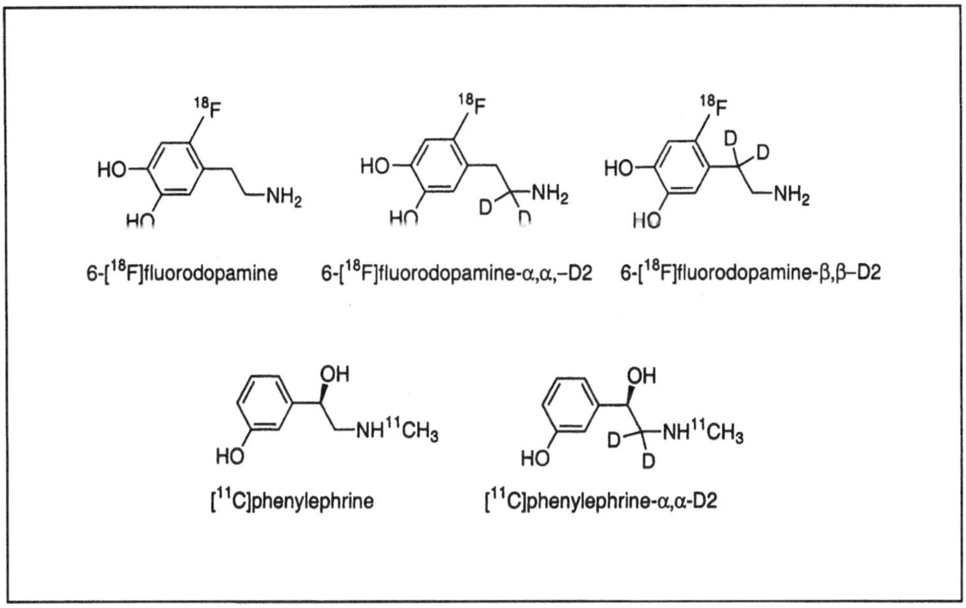

Fig. 20.4. Structures of 6-^{18}F-fluorodopamine and ^{11}C-phenylephrine and deuterium-substituted derivatives

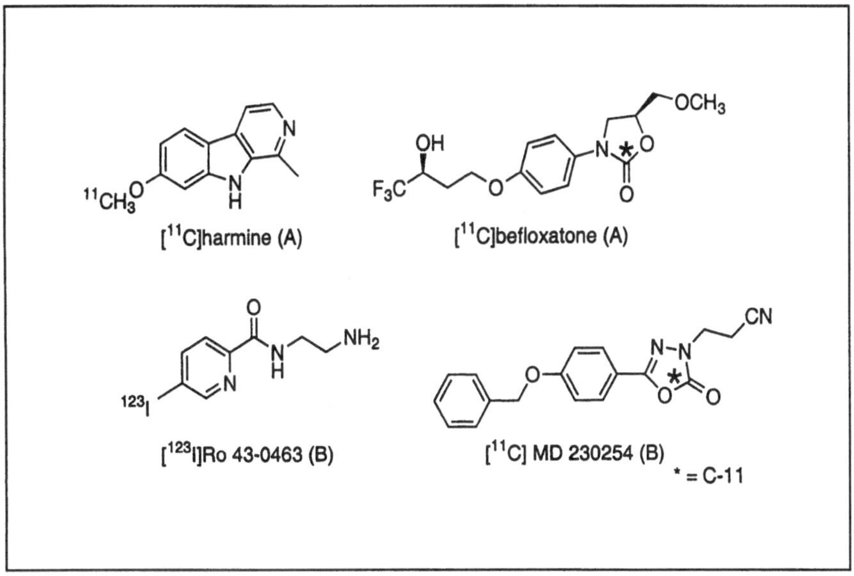

Fig. 20.5. Structures of reversibly binding radiotracers for MAO A and B for PET and SPECT studies

mine had favorable binding properties for imaging MAO A (Bergstrom et al. 1997a,b). Carbon-11-brofaromine, ^{11}C-harmaline, and ^{11}C-clorgyline did not show specific binding in this species (Ametamey et al. 1996; Bergstrom et al. 1997b) though [^{11}C]clorgyline does map MAO in humans (Fowler et al. 2001b). PET studies of the potency of MAO A inhibitor drugs have been carried out with ^{11}C-harmine (Bergstrom et al. 1997c; see description below). In addition, frozen section autoradiography with ^{11}C-harmine has

identified high MAO A enzyme binding in bladder cancer in vitro (Goller et al. 1995).

Befloxatone is an oxazolidinone derivative belonging to a new class of reversible MAO A inhibitors. It is a potent, reversible MAO A inhibitor with activity in animal models of depression (Curet et al. 1996). Interestingly, the structure does not contain a basic amine function. Carbon-11-befloxatone was synthesized via a cyclization reaction with ^{11}C-phosgene. Studies in the baboon have shown that it rapidly penetrates the brain and shows good characteristics for imaging brain MAO A (Dolle et al. 1999).

A radioiodinated derivative of moclobemide has been developed as a specific radiotracer for MAO A for SPECT. It was shown to have preferential MAO A activity in the brain and mixed MAO A and B activity in peripheral organs, possibly due to the production of labeled metabolites with MAO B activity (Rafi et al. 1996).

For SPECT and PET studies of brain MAO B, the pyridine carboxamide Ro 43-0463 has been labeled with iodine-123 and with fluorine-18 (Beer et al. 1995 a,b; Blauenstein et al. 1998). Iodine-123-Ro 43-0463 has appropriate properties for SPECT studies of MAO B in the human brain (see discussion below) whereas the ^{18}F-substituted derivative has brain uptake too low for imaging. It was speculated that the low lipophilicity of the fluorine-18 compound limited brain uptake. An oxadiazolone derivative (5-[4-(benzyloxy)phenyl]-3-(2-cyanoethyl)-1,3,4-oxadiazol-2(^{3}H)- one; MD 230254) has been labeled with carbon-11 and evaluated for MAO B imaging in the rat and baboon brain (Bernard et al. 1996). Like befloxatone, the structure does not contain a basic amine function and, like ^{11}C-befloxatone, it was synthesized from ^{11}C-phosgene. PET studies in the baboon have shown rapid binding which is reduced by L-deprenyl treatment, consistent with binding to MAO B.

20.2.2
Metabolically Trapped Radiotracers

Using a different approach, MAO has been imaged using a labeled substrate that is metabolized to produce a charged, labeled product which is intracellularly trapped. This is exemplified by ^{11}C-N,N-di-methylphenethylamine (DMPEA), a good substrate for MAO B (Inoue et al. 1985). Images reflect the intracellular trapping of MAO B-generated ^{11}C-labeled dimethylamine (Fig. 20.6). Mechanistic studies including the demonstration of a deuterium isotope effect, and PET studies in the monkey with DMPEA labeled in different positions validated its use as a MAO tracer (Hashimoto et al. 1986; Halldin et al. 1989). PET studies showed that ^{11}C was trapped in MAO-rich regions in the human brain (Shinotoh et al. 1987).

The same concept has been applied in the heart. Nitrogen-13-labeled phenylethylamine was used to demonstrate the presence of MAO B in the rat heart through the trapping of MAO-generated ^{13}N-ammonium. Deuterium-substituted phenylethylamine confirmed that ^{13}N distribution in the heart represented MAO B activity (Tominaga et al. 1987).

20.2.3
Suicide Inactivator Radiotracers

The suicide inactivator approach is another form of metabolic trapping. It involves the irreversible covalent binding of the radiotracer to the flavin cofactor of MAO, a bond that results from a highly reactive intermediate produced during MAO-catalyzed oxidation (Abeles and Maycock 1976). This approach had a precedent in the biochemical assay of MAO in which suicide inactivators were used to titrate the active centers of MAO in tissue samples (Fowler et al. 1980). The selective irreversible MAO inhibitors clorgyline (Johnston 1968) and L-deprenyl (Knoll and Magyar 1972) served as model structures to test this approach for PET imaging of MAO A and MAO B, respectively. These compounds were labeled with car-

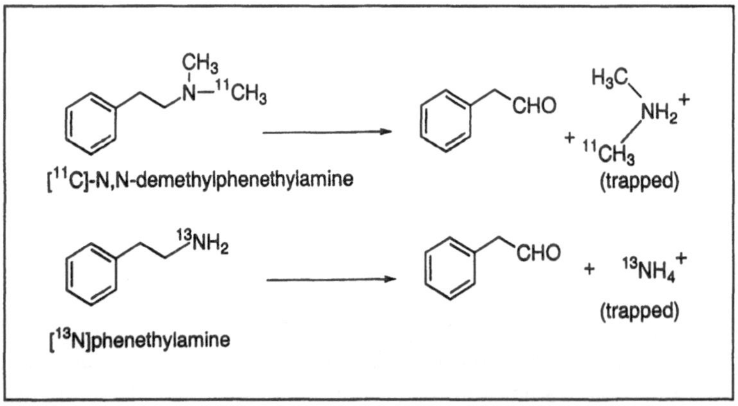

Fig. 20.6. Carbon-11- and nitrogen-13-labeled substrates for MAO B which produce labeled metabolites intracellularly trapped as a result of MAO B-catalyzed oxidation

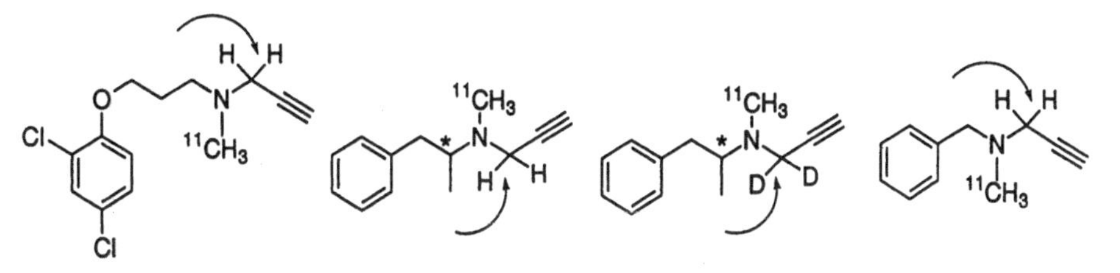

Fig. 20.7. Carbon-11-labeled suicide inactivators of MAO A (^{11}C-clorgyline) and MAO B (^{11}C-L-deprenyl and ^{11}C-L-deprenyl D$_2$), and the nonselective MAO A and B inhibitor ^{11}C-pargyline. The *arrows* indicate the bonds cleaved by MAO in the rate-limiting step of catalysis

bon-11 for early mechanistic studies in animals (Fowler et al. 1988; MacGregor et al. 1988). Another nonselective suicide inactivator, pargyline, was also labeled with carbon-11 and evaluated in animals (Ishiwata et al. 1985). The structures of these labeled compounds are shown in Fig. 20.7. Derivatives of clorgyline, deprenyl, and pargyline have also been labeled with iodine-123 and fluorine-18 and evaluated for their potential for PET and SPECT studies (Plenevaux et al. 1990; Hirata et al. 1995; Lena et al. 1995; Mukherjee and Yang 1999).

20.2.3.1
Mechanistic Studies

The kinetic scheme for suicide inactivation involves a branched pathway (Walsh 1982). The suicide inactivator (S) has a latent reactive functional group. In the case of L-deprenyl or clorgyline, the latent reactive functional group is the propargyl group, and this is unmasked within the enzyme-substrate complex (E-S) during the catalytic step (Maycock et al. 1976). The catalytically activated substrate (E-S)* then forms a covalent bond to the flavin cofactor of the enzyme causing either turnover to regenerate enzyme and form product (P; pathway 1; Scheme 20.1) or irreversible deactivation (E$_{inact}$; pathway 2). This latter process is commonly referred to as suicide inactivation (or mechanism-based inactivation) because the enzyme catalyzes its own destruction (Abeles and Maycock 1976).

The covalent attachment of the *labeled* suicide inactivator to the enzyme is at the heart of the use of the labeled suicide inactivator approach in PET imaging. Because the interaction of the functionally active enzyme and the labeled suicide inactivator results in

$$E + S \rightleftharpoons E\text{-}S \longrightarrow [E\text{-}S]^* \longrightarrow \begin{cases} E + P \text{ (pathway 1)} \\ E_{inact} \text{ (pathway 2)} \end{cases}$$

Scheme 20.1.

a covalent bond (see Fig. 20.8 for the structure of the ^{11}C-L-deprenyl-MAO B adduct), the image of radioactivity distribution after the initial distribution phase has the potential of representing functional enzyme activity. For this approach to be successful, the partition ratio must be small (i.e., the rate of enzyme inactivation (pathway 2) must be greater than the rate of turnover (pathway 1)).

Studies in mice demonstrated the feasibility of applying this technique in vivo by showing that carbon-11-labeled clorgyline and L-deprenyl are selective for brain MAO A and MAO B, respectively (MacGregor et al. 1985). PET studies with ^{11}C-L-deprenyl in baboons and in humans showed appropriate stereoselectivity (Fowler et al. 1987) and regional distribution, and blockade by pharmacological doses of L-deprenyl (Arnett et al. 1987; Fowler et al. 1994).

In addition, the deuterium isotope effect was used to probe the mechanism by which ^{11}C accumulates in the brain during a PET study with ^{11}C-L-deprenyl. The deuterium isotope effect is based on the fact that a C-D bond is more difficult to cleave than a C-H bond. The substitution of hydrogen by deuterium in a carbon-hydrogen bond can thus be used to determine whether the rate-limiting step in a reaction involves the cleavage of this particular bond. The rate-limiting step for MAO oxidation is cleavage of the C-H bond alpha to the amino group (Belleau and Moran 1963). When ^{11}C-L-deprenyl and deuterium substituted ^{11}C-L-deprenyl (where deuterium is incorpo-

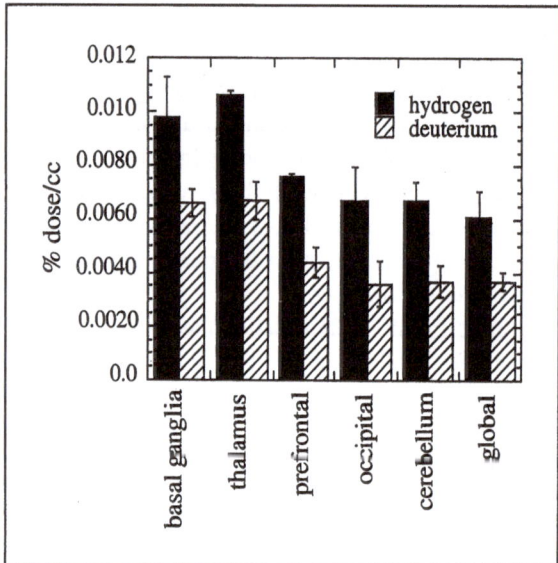

ribityl moiety

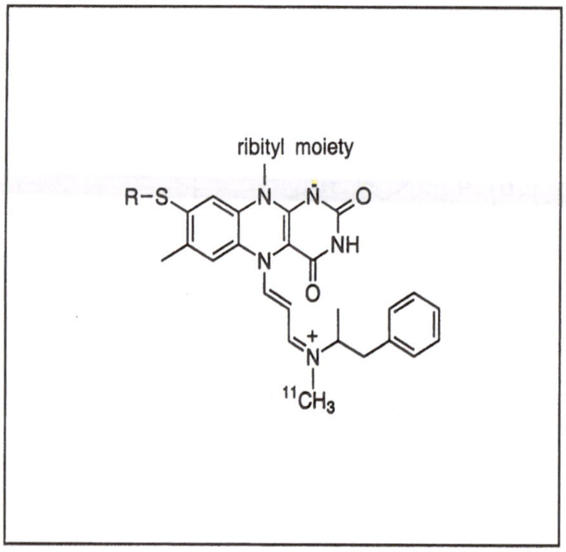

Fig. 20.8. Possible structure of the adduct between ^{11}C-L-deprenyl and MAO B. (Based on Maycock et al. 1976)

Fig. 20.9. Comparison of the uptake (% injected dose/cc; mean $\pm$ SDM) of ^{11}C-L-deprenyl (*squares*) and ^{11}C-L-deprenyl D_2 (*circles*) in different regions of the human brain ($n = 5$) (Fowler et al. 1995)

carbon-11 in brain. This observation forms the basis for the current use of deuterium substituted ^{11}C-L-deprenyl (^{11}C-L-deprenyl D_2) in human studies to increase tracer sensitivity in regions of high MAO B concentration (Fowler et al. 1995).

Deuterium substitution has also been used to demonstrate that the binding of ^{11}C-clorgyline in the human brain represents MAO A activity (Fowler et al. 2000b). In addition, pretreatment with a low dose of tranylcypromine (a nonselective irreversible MAO inhibitor; 10 mg/day for 3 days) reduced ^{11}C-clorgyline binding by an average of 58% (Fig. 20.11) demonstrating that ^{11}C-clorgyline binds to MAO A in the human brain (Fowler et al. 1996b). In contrast, treatment of human subjects with L-deprenyl (10 mg/day) did not alter ^{11}C-clorgyline binding, indicating the specificity of ^{11}C-clorgyline for MAO A (Fowler et al. 2001a). Note that this contrasts with lack of specific binding of ^{11}C-clorgyline in the rhesus monkey (Bergstrom et al. 1997b), which illustrates an unusual case where the binding of a specific radiotracer differs markedly between that species and humans (Fowler et al. 2001b).

20.2.3.2
Quantitation of Functional MAO Activity in the Human Brain

A three-compartment kinetic model has been developed and applied to the quantitation of the binding of the ^{11}C-suicide inactivators of MAO A and B in baboon and human brains. The model requires the measurement of the time course of radioactivity in the brain and in the arterial plasma (Alexoff et al. 1995). Its application allows the calculation of the model term k_1, the plasma-to-brain transfer constant related to blood flow, and λk_3, which is proportional to the concentration of catalytically active MAO molecules.

The kinetic scheme that incorporates the reversible transfer of labeled substrate between blood and brain and all of the major steps in suicide inhibition, including normal enzyme turnover as well as suicide inactivation, is given below (Fowler et al. 1988; Walsh 1982). In this model, S_p is the concentration of labeled L-deprenyl in arterial plasma, S_b the concentration of labeled tracer in brain that has not reacted with the enzyme, E the concentration of free (unbound) enzyme, E-S the enzyme-substrate complex, E_{inact} the inactivated enzyme, [E-S]* the catalytically activated intermediate, P is the product of enzyme re-

rated in the methylene group of the propargyl group; see Fig. 20.7 for the structure) are compared either in the baboon or in the human brain, there is an isotope effect (Fowler et al. 1988, 1995) manifested as a reduction in the uptake (Fig. 20.9) and the rate of binding in the brain (Fig. 20.10). This established that MAO-catalyzed cleavage of the α carbon-hydrogen bond on the propargyl group is the rate-limiting (or a major rate-contributing) step in the retention of

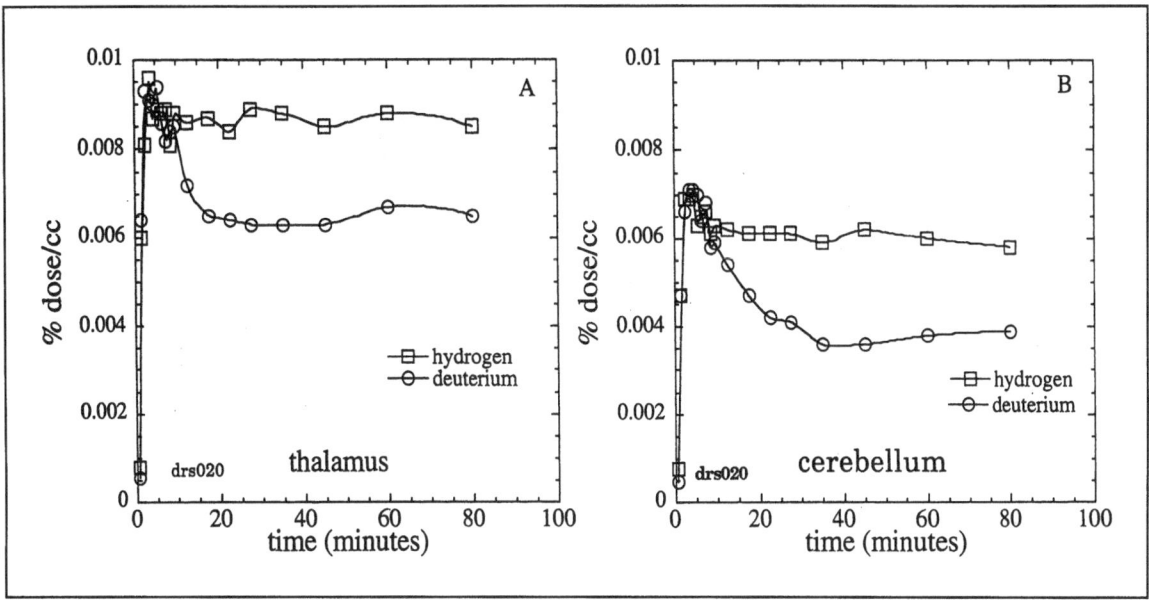

Fig. 20.10 A, B. Comparison of the time-activity curves for one subject for ^{11}C-L-deprenyl (*squares*) and ^{11}C-L-deprenyl D$_2$ (*circles*) in the human thalamus (**A**) and cerebellum (**B**) (Fowler et al. 1995)

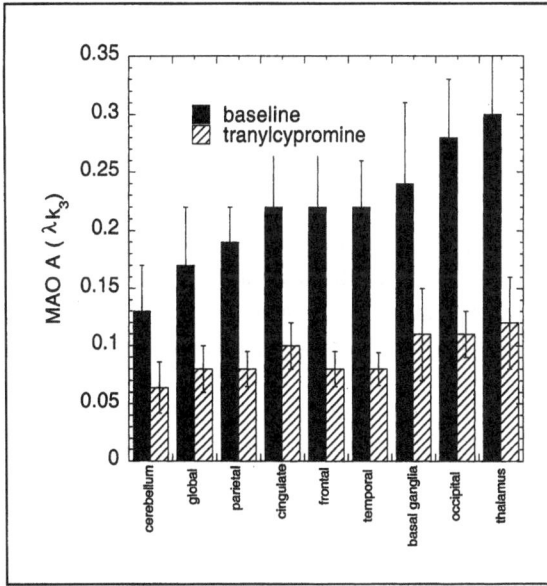

Fig. 20.11. Comparison of MAO A (as represented by the model term λk_3) for four healthy volunteers at baseline and after the non-selective MAO inhibitor drug tranylcypromine (10 mg/day) for 3 days (Fowler et al. 1996b)

$$S_p \underset{k_2}{\overset{K_1}{\rightleftarrows}} S_b + E \underset{k_4}{\overset{k_3}{\rightleftarrows}} E\text{-}S \overset{k_5}{\longrightarrow} [E\text{-}S]^* \overset{k_6}{\underset{k_7}{}} \begin{array}{l} E + P \\ E inact \end{array}$$

PET Region of Interest

Scheme 20.2.

of the enzyme substrate complex (min^{-1}), k_5 the first order constant for the formation of intermediate [E-S]* (min^{-1}), k_6 the constant for formation of the enzyme and product (min^{-1}), and k_7 the formation of inactivated enzyme (min^{-1}) (Scheme 20.2).

Since PET measures the total radioactivity in a given region of interest and cannot measure the concentrations of the intermediate species, the species in the box-labeled *PET Region of Interest* are indistinguishable to the tomograph. As a result, it is not possible to uniquely determine all constants and therefore a simpler model is used (shown below; Scheme 20.3), where S_{tr} is the concentration of labeled L-deprenyl bound to enzyme and k_3' is a kinetic term related to the processes involved in the trapping of carbon-11 in tissue.

The differential equations corresponding to this model are given by

action, k_1 the transfer constant describing the transport of substrate from plasma to brain, k_2 the transfer constant describing the transport of substrate from brain to plasma, k_3 the bimolecular rate constant (concentration^{-1} × min^{-1} of the enzyme-substrate complex), k_4 the first order constant for dissociation

$$dS_b/dt = K_1 S_p(t) - k_2 S_b - (k_3' E)S_b$$
$$dS_{tr}/dt = (k_3 E)S_b \qquad (20.2)$$

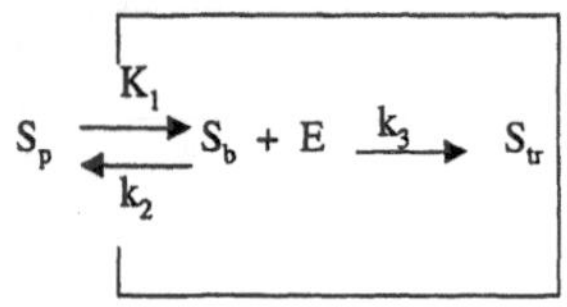

$$S_p \underset{k_2}{\overset{K_1}{\rightleftarrows}} S_b + E \xrightarrow{k_3} S_{tr}$$

PET Region of Interest

Scheme 20.3.

The model equations are solved using the arterial plasma radioactivity (corrected for the fraction of ^{11}C-radiotracer at different time points) for the input function; model constants were optimized best fit the data. The following assumptions are made:

1. The specific activity of labeled radiotracer is sufficiently high that the free enzyme concentration does not change during the course of the experiment and that the enzyme concentration is included in $k_3'E$. In the case of ^{11}C-L-deprenyl used in studies of drug efficacy (Fowler et al. 1993), the maximum injected dose was 44 μm. Using the uptake of 0.0076%/g in the basal ganglia and a MAO B concentration of 2 pmol/mg protein (200 nM) (Oreland 1991), the concentration of ^{11}C-L-deprenyl (bound to enzyme) would be 18 nM, well below both the MAO B concentration and the K_m for L-deprenyl (325 nM) (Robinson 1985). Under these conditions $k_3'E$ is constant and proportional to the free enzyme concentration. Depending on which step in the process of inactivation is rate-limiting, $k_3'E$ in the three-compartment model can be related to different rate constants in the multi-step process but should always be proportional to the free enzyme concentration (Fowler et al. 1988). On this basis, the comparison of values of $k_3'E$ for an individual at baseline to $k_3'E$ for the same individual during drug treatment and washout should permit the assessment of relative changes in enzyme concentration.

2. The kinetic term k_4, which represents loss of radioactivity from the ^{11}C-L-deprenyl-MAO B complex, is set to zero, since it is known that the turnover time of MAO B after inactivation by L-deprenyl is on the order of weeks in baboons (Arnett et al. 1987) and humans (Fowler et al. 1994).

3. Although the three model parameters (K_1, k_2, and $k_3'E$) of Scheme 20.3 can be determined, reproducibility is improved if the term compared is $\lambda k_3'E$ (where $\lambda = K_1/k_2$) rather than $k_3'E$. For simplicity we will use λk_3 rather than $\lambda k_3'E$ to refer to Scheme 20.3. This is due to the fact that k_2 and $k_3'E$ are highly correlated, a fact especially evident in regions of high enzyme concentration that may

have optimum values of both terms that are lower than expected. This last can be overcome by either fixing λ, thus reducing the number of model parameters to be optimized, or by determining all three and using the composite parameter λk_3 as the measure of enzyme concentration. Even though the transport constants K_1 and k_2 depend upon blood flow, their ratio does not (Logan et al. 1991).

A problem encountered with irreversible tracers is that if the trapping rate is very high compared to the rate of efflux (k_2), the observed uptake is related to blood flow alone and contains no information about enzyme concentration. This is the "flow-limited" situation, and is a difficulty encountered with ^{11}C-L-deprenyl when MAO B concentration is high and blood flow is low, as would be encountered in neurodegenerative disorders and aging. Deuterium substitution in ^{11}C-L-deprenyl decreases the trapping rate (λk_3) but not the delivery rate (k_1) and therefore increases the sensitivity of the uptake to MAO B (Fowler et al. 1995). Under these conditions, the three-compartment model provides a better measure of MAO concentration (Fowler et al. 1988, 1993; Lammertsma et al. 1991).

Human PET studies have shown that repeated measures of ^{11}C-L-deprenyl D_2 binding in the human brain are reproducible based on comparison of $\lambda k_3'E$ values (Logan et al. 2000). The term $\lambda k_3'E$ can also be derived from a linear form of the model equations, making use of the graphical analysis for irreversible systems (Patlak et al. 1983) which quantitates the uptake of L-deprenyl in terms of the influx constant, k_i. However, K_i is a function of blood flow and therefore it is necessary to also determine K_1; λk_3 can then be derived from K_i and K_1. The linear method is rapid but can produce biased estimates due to concomitant noise. However, using ROI analysis with low noise, results from both the linear and nonlinear methods were shown to be in good agreement (Logan et al. 2000). The graphic analysis for irreversible systems (Patlak et al. 1983) is applicable to L-deprenyl, but the influx constant K_i is a function of blood flow, whereas $\lambda k_3'E$ is not.

20.3
Human Studies

The development of radiotracers for visualizing MAO quantitatively in the human body has been driven by many factors including: (1) the possible pathophysiological role that MAO may play in aging and neurodegeneration; (2) its role as a target for therapeutic drugs and neurotoxic substances; (3) its role in the detoxification of xenobiotic amines; and (4) its proposed role as a biological marker in certain diseases and behaviors. Some of the studies concerning these are described below.

20.3.1
MAO B in the Normal Human Brain: Effects of Age

Many neuronal cells and their associated neurotransmitters and enzymes show age-related losses (Palmer and DeKosky 1993). However, MAO B is an exception. Studies in the postmortem human brain report that MAO B increases with age (Fowler et al. 1980; Galva et al. 1995; Robinson et al. 1971) and in neurodegenerative disease (Saura et al. 1994a,b). This is consistent with the compartmentalization of MAO B within glial cells (Westlund et al. 1988) and with reports that in the normal human brain the number of glial cells increases with age (Terry et al. 1987) and in neurodegenerative disease (Strolin-Benedetti and Dostert 1989; Saura et al. 1994a,b). It has been proposed that increases in brain MAO B with aging increases oxidative stress and that this may play a role in the vulnerability of the brain dopamine system to age-related degeneration (Cohen and Kesler 1999).

Brain MAO B has been measured in a group of normal healthy subjects ($n=21$; age range 23–86; 9 females and 12 males; non-smokers) using ^{11}C-L-deprenyl D_2 (Fowler et al. 1995). MAO B concentration is estimated using the model term λk_3 (Fowler et al. 1997). The regional distribution of MAO B was highest in the basal ganglia and the thalamus, with intermediate levels in the frontal cortex and the cingulate gyrus and lowest levels in the parietal and temporal cortices and the cerebellum. MAO B increased significantly with age ($P<0.004$) in all brain regions examined except the cingulate gyrus. The same patterns remained when the correlation analysis was performed separately for men and for women. The whole brain and the cortical regions and the basal ganglia, thalamus, pons, and cerebellum showed an average increase of $7.1\pm1.3\%$/decade over the age range 23–86 years. The frontal cortex showed an average rate of increase of 5.7%/decade. Regarding the correlation coefficient between age and λk_3, the model parameter for MAO B for the frontal cortex corresponded to 0.66, a value similar to those reported in postmortem studies (range 0.45–0.71). This indicates

Table 1. Comparison of λk_3 and K_1 for different brain regions for ^{11}C-L-deprenyl D_2

Brain region	MAO B (λk_3)		Blood flow (K_1)	
	r	p	r	p
Global	+0.74	0.0001	−0.62	0.003
Cingulate gyrus	0.09	0.71	−0.78	0.0001
Pons	+0.64	0.003	−0.3	0.2
Basal ganglia	+0.67	0.0008	−0.57	0.006
Thalamus	+0.71	0.0003	−0.53	0.01
Frontal cortex	+0.66	0.001	−0.76	0.0001
Cerebellum	+0.65	0.002	0.14	0.02
Parietal cortex	+0.48	0.03	−0.68	0.0006
Temporal cortex	+0.60	0.004	−0.72	0.0003

that while age is a factor contributing to the variability among subjects, it is not the only one.

Carbon-11-L-deprenyl D_2 has tracer characteristics which allow the calculation of a plasma-to-brain transfer constant (K_1), a model term related to brain blood flow. In contrast to λk_3 (MAO B) which increased with age, K_1 decreased with age in all regions except for the pons and cerebellum. The highest correlations were in the frontal, temporal, and parietal cortices. The correlation coefficients for λk_3 and for K_1 are shown in Table 20.1.

The use of the deuterium-substituted tracer was essential in this study because the rapid rate of binding of ^{11}C-L-deprenyl leads to an underestimation of MAO B levels when enzyme concentration is high, as is the case in aging. In this case, the rate of delivery of the tracer (K_1) limits the uptake and thus blood flow cannot be well resolved from MAO activity. This uncertainty is exacerbated in studies of aging where brain blood flow is expected to decrease with age while MAO B activity increases. With ^{11}C-L-deprenyl-D_2, the reduced rate of cleavage of the C-D bond results in improved sensitivity (Fowler et al. 1995).

20.3.2
MAO B Imaging in Gliosis

The known elevation of MAO B in neurodegenerative diseases and brain injury provides an opportunity to explore the use of MAO B tracers as *positive* markers for brain injury and degeneration. Neurodegenerative processes and brain injury are frequently accompanied by gliosis. Because MAO B is located in glial cells, uptake would be high where the concentration of glial cells is high. In principle this could provide a positive complement for tracers like 2-deoxy-2-^{18}F-fluoro-D-glucose (FDG) whose uptake is normally decreased in degenerative processes. For example, it is known from PET imaging studies with ^{11}C-L-deprenyl D$_2$ that MAO B is elevated in the hypometabolic regions in patients with temporal lobe epilepsy where analysis of resected tissue has confirmed gliosis (Kumlien et al. 1992, 1995). A similar observation (i.e., increased radiotracer uptake in the ipsilateral mesial temporal lobe) was made with SPECT using ^{123}I-Ro 43-0463 (Buck et al. 1998). A recent study of head trauma patients did not find an inverse relationship between MAO B and glucose metabolism, indicating that prospective studies are needed to determine the pathophysiology of hypometabolic lesions in head trauma (Fowler et al. 1999a).

Reports that MAO B is elevated in the postmortem Alzheimer's brain (Adolfsson et al. 1980; Reinikainen et al. 1988; Saura et al. 1994a,b), an observation that MAO B is expressed in astrocytes of senile plaques (Nakamura et al. 1990), and an abstract describing increased levels of ^{11}C-L-deprenyl binding in the brains of Alzheimer's patients (Bench et al. 1993) suggest a role for MAO B imaging in studies of the development and progression of Alzheimer's disease.

20.3.3
MAO B Inhibitor Drugs

The use of MAO inhibitors in the treatment of disease is grounded both in their ability to increase the bioavailability of neurotransmitters and to reduce oxidative stress. Major developments of new MAO inhibitor drugs are in the treatment of neurodegenerative disorders as well as of depression. The newer reversible MAO A inhibitors are of particular interest because of reduced harmful side-effects vis-à-vis drug-diet or drug-drug interactions (Caldecott-Hazard and Schneider 1992). Imaging studies in humans have focused on determining the efficacy and minimum effective doses of MAO inhibitor drugs and duration of drug action.

20.3.3.1
Lazabemide (Ro 19-6327)

Reports that L-deprenyl reduces the rate of progression of Parkinson's disease stimulated the development of other MAO B inhibitor drugs in order to enhance the neuroprotective effects. One of these drugs

was lazabemide (Ro 19-6327), an irreversible and highly selective MAO B inhibitor (Da Prada et al. 1988). The development of ^{11}C-L-deprenyl provided the opportunity to determine the efficacy of lazabemide to inhibit MAO B directly in the human brain. Studies were designed to determine minimum effective doses to inhibit >90% of brain MAO B for clinical trials and to determine duration of action (Bench et al. 1991; Fowler et al. 1993). One of the studies was carried out in a group of six unmedicated patients with early Parkinson's disease. Each patient received a baseline PET scan with ^{11}C-L-deprenyl and then received either 25 mg, 50 mg or 100 mg of lazabemide twice a day for 1 week. Twelve h after the last dose of lazabemide, a second PET scan was performed. Comparison of the second scan with the baseline scan showed that the 50 mg dose was sufficient to block >90% of the enzyme whereas the 25 mg dose was inadequate. A third PET scan performed 36 h after the last dose of lazabemide showed that the inhibition had completely vanished after this short drug-free interval. This study helped to establish the dose and frequency with which lazabemide would be given in clinical trials. The suggestion thus far is that the pattern of benefits of lazabemide is similar to that of L-deprenyl in patients with Parkinson's disease (Parkinson's Study Group 1996).

20.3.3.2
L-Deprenyl (Selegiline)

L-Deprenyl has an intriguing combination of catecholaminergic and neuroprotective effects and a relatively benign side-effect profile (Koller and Giron 1990). These qualities form the basis for its use in the treatment of Parkinson's disease and for the investigation of its use in treating other neurological and psychiatric disorders, including Alzheimer's disease (Sano et al. 1997), schizophrenia (Bodkin et al. 1996), and addictions to cocaine (Bartzokis et al. 1999) and smoking (Brauer et al. 2000). PET has been used to study L-deprenyl pharmacodynamics, including then duration of MAO B inhibition and its specificity for MAO B vs. MAO A.

In contrast to the rather rapidly reversible MAO B inhibitor lazabemide, PET studies have shown that MAO B inhibition persists long after the last dose of L-deprenyl. In one study, four elderly normal subjects and four unmedicated patients with a diagnosis of idiopathic Parkinson's disease received a baseline PET scan with ^{11}C-L-deprenyl and were the treated with a

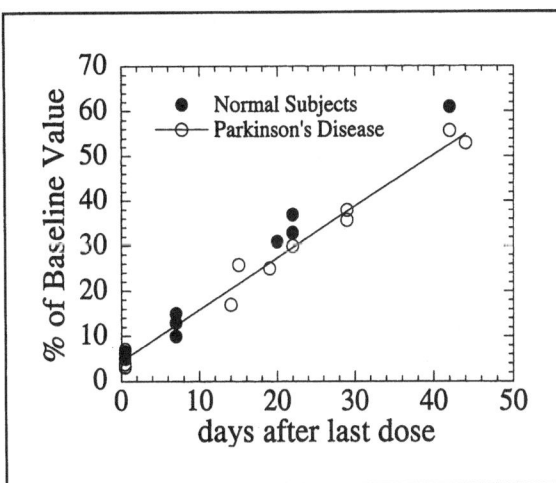

Fig. 20.12. Recovery of brain MAO B activity in elderly normal subjects and in patients with Parkinson's disease after withdrawal from L-deprenyl, which had been given for 1 week (10 mg/day). PET measurements of MAO B using ^{11}C-L-deprenyl at various times after the last dose of L-deprenyl allowed the calculation of the half-time for MAO B synthesis in the brain (about 40 days) (Fowler et al. 1994)

therapeutic dose of L-deprenyl (10 mg/day) for 1 week. A total of four PET scans (including the baseline) were performed on each subject over a six-week period following the last dose of L-deprenyl. Timing for the four scans was as follows: the first (baseline) was carried out before L-deprenyl was given; the second at 12 h after the last dose; the third at 1–2 weeks, and the fourth at 3–6 weeks after the last dose of L-deprenyl. Model equations were solved using time-activity data from different brain regions, and the input function from the arterial plasma was corrected for the presence of labeled metabolites. The half-time for recovery of the enzyme was 40 days after drug withdrawal (Fig. 20.12), demonstrating that MAO B inhibition can be maintained at a far lower dose of L-deprenyl than is currently used (Fowler et al. 1994). In addition to providing information on the duration of MAO B inhibition after L-deprenyl is withdrawn, this study also demonstrates the feasibility of measuring the rate of enzyme protein turnover. Since MAO B is an integral protein of the outer mitochondrial membrane, its recovery after irreversible inactivation requires the removal of inactivated MAO B from the membrane, the synthesis of MAO B protein in cytosolic ribosomes (a process encoded on the nuclear gene), and the insertion of the protein into the outer mitochondrial membrane (Zhuang et al. 1988).

One key issue in characterizing the molecular mechanisms contributing to the therapeutic effects of L-deprenyl is to distinguish MAO B inhibition from other mechanisms. The slow recovery of brain MAO B clearly opens up the possibility that the symptomatic effects of L-deprenyl could persist long after the last dose of the drug. Thus with a half-time of 40 days for the recovery of MAO B, a drug-free interval of several months would be required for brain MAO B to reach >90% of baseline values. This is an important issue in L-deprenyl therapy. The assessment of symptoms after a long-term drug-free interval would assure the absence of long term symptomatic effects. This would reduce ambiguity in distinguishing symptomatic from neuroprotective effects (Fowler et al. 1996 c) though ethical considerations may preclude such a lengthy drug-free interval.

Though L-deprenyl is a selective MAO B inhibitor (Knoll and Magyar 1972), its pharmacology is complex and a number of other mechanisms including partial MAO A inhibition have been proposed to account for its therapeutic effects (Riederer and Youdim 1986; Gerlach et al. 1992; Lamensdorf et al. 1996). The effect of L-deprenyl treatment on brain MAO A was investigated in six normal volunteers who received L-deprenyl (10 mg/day) for 1 week. Each subject had two PET scans with the MAO A radiotracer ^{11}C-clorgyline, one at baseline and one following L-deprenyl therapy. A three-compartment model was used to compare the plasma-to-brain transfer constant K_1, as well as λk_3 before and after treatment. L-Deprenyl treatment did not affect either brain MAO A activity or K_1 for any of the 12 brain regions examined (Fowler et al. 2001a; Fig. 20.13). This confirms that L-deprenyl is selective for MAO B, though it is possible that selectivity may not be maintained with longer administration or higher doses.

20.3.3.3
MAO A Inhibitors

There is evidence that antidepressant properties of the MAO inhibitors reside in the inhibition of MAO A and not MAO B. For this reason reversible inhibitors of MAO A which would maintain the therapeutic properties and reduce the potential for interactions with tyramine (Caldecott-Hazard and Schneider 1992) have been an attractive target in drug development.

Esuprone (Fig. 20.3) is a potent subtype-selective reversible inhibitor of MAO A (IC50: 8.4 nM). PET and the MAO A radiotracer ^{11}C-harmine were used to determine whether or not the new reversible MAO A inhibitor drug esuprone binds substantially to MAO

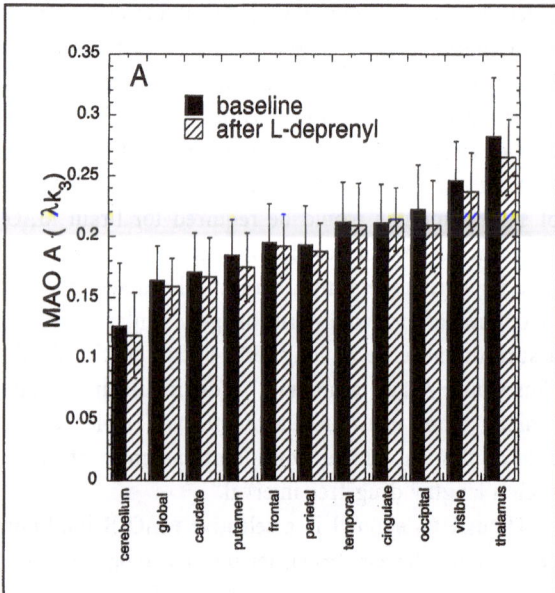

Fig. 20.13. Comparison of MAO A (as measured by ^{11}C-clorgyline and as represented by the model term λk_3; mean ±SDM) in different brain regions for six healthy volunteers at baseline and after treatment with the MAO B-selective drug L-deprenyl (10 mg/day) for 3 days. There were no significant differences indicating that selectivity for MAO B is maintained for a 1 week treatment period (Fowler et al. 2001 a)

A in the human brain (Bergstrom et al. 1997a). Volunteers were given daily doses of esuprone (800 mg) or moclobemide (300 mg twice a day) or placebo tablets for a week. PET studies with ^{11}C-harmine were performed at baseline (before drug/placebo) and on day 7 after the previous days treatment for esuprone and 11 h after the previous day's treatment with moclobemide. Both esuprone and moclobemide reduced ^{11}C-harmine binding to a similar extent relative to baseline. In the placebo group no change was observed. This study showed that esuprone had similar efficacy to inhibit MAO A as moclobemide. This study and similar studies of other drugs illustrate the possibility of obtaining information on drug efficacy directly in the brain rather than to rely on traditional pharmacokinetic measurements.

20.3.4
MAO and *Ginkgo biloba*

Extracts of *Ginkgo biloba* have been reported to reduce the symptoms of mental decline. This property is generally attributed to the extract's principal active chemical components, flavonoids, and to its terpenoids and ginkgolides and bilobalide (Kleijnen and Knipschild 1992; Curtis-Prior et al. 1999 for review). Numerous investigations have been stimulated on the mechanisms which may contribute to this protective property in aging brains. One hypothesis, based on studies in rat brain in vitro (White et al. 1996), is that extracts of *Ginkgo biloba* inhibit MAO A and B. In order to investigate whether such extracts do in fact inhibit MAO A and B in the human brain, ten normal healthy volunteers were treated for 1 month with 120 mg/day of the *Ginkgo biloba* extract EGb 761 (Fowler et al. 2000a). Carbon-11-clorgyline and ^{11}C-L-deprenyl D_2 were used to measure MAO A and B, respectively, at baseline (before Ginkgo) and after the 1-month treatment period. A three-compartment model was used to calculate the plasma to brain transfer constant K_1 which is related to blood flow and λk_3, which is proportional to the concentration of catalytically active MAO molecules. *Ginkgo biloba* administration did not produce significant changes in brain MAO A or MAO B. This study suggests that mechanisms other than MAO inhibition need to be considered as mediating its CNS effects. The lack of MAO inhibitory potency of *Ginkgo biloba* has been recently demonstrated in other systems (Porsolt et al. 2000).

20.3.5
MAO and Tobacco Smoke

It has been known for many years that platelet MAO is significantly lower in smokers than in non-smokers (Oreland et al. 1981). However, MAO levels increase in smokers who quit indicating that low MAO B is a pharmacological effects of the smoke rather than a biological characteristic of smokers (Norman et al. 1987). Similar to the findings of low platelet MAO in smokers, PET studies of normal volunteers revealed that cigarette smokers had very low brain MAO B while former smokers have normal levels (Fowler et al. 1996a). Furthermore, PET studies measuring MAO A with ^{11}C-clorgyline showed that smokers also have reduced MAO A (Fowler et al. 1996b). Inhibition is partial, with average reductions of 30% and 40% being observed for MAO A and B respectively (Fig. 20.14). This observation raises intriguing questions as to whether MAO inhibition by smoke may contribute to some of the behavioral and epidemiological features of smoking, including a decreased

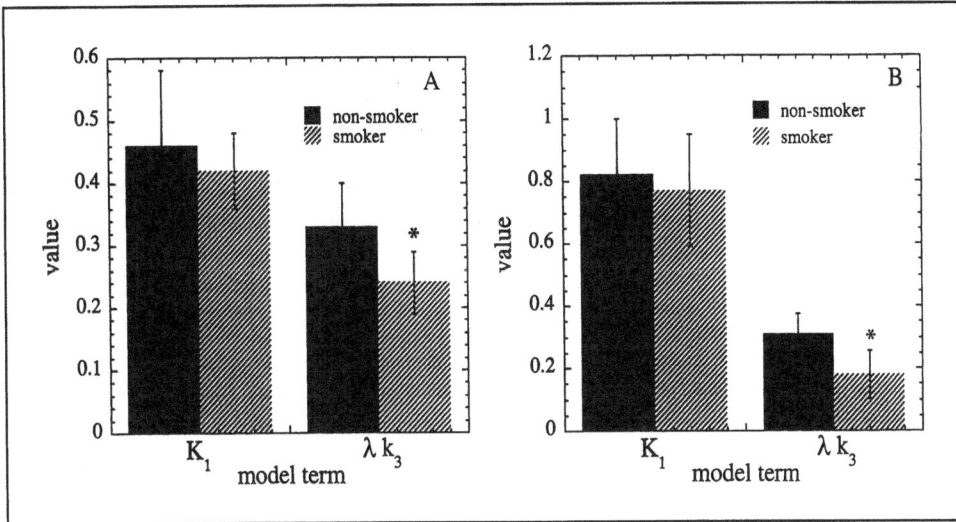

Fig. 20.14A,B. Bar graphs comparing K_1 (plasma to brain transfer constant) and λk_3 (MAO A) (**A**) and MAO B (**B**) in non-smokers and in smokers. Note that there were no sig- nificant differences in K_1, while MAO A and B were significantly reduced in the smokers' brains (Fowler et al. 1996a,b)

risk of Parkinson's (Morens et al. 1995; Gorell et al. 1999) and Alzheimer's diseases (Brenner et al. 1993) in smokers, an increased rate of smoking in depression (Glassman et al. 1990) and addictions to other substances (Henningfield et al. 1990), and lastly to a general prevalence of smoking in psychiatric illnesses (Hughes et al. 1986). Reductions in MAO A and B, in principle, could spare neurotransmitters from oxidation and reduce the production of hydrogen peroxide, a byproduct of MAO-catalyzed oxidation. MAO inhibition may act synergistically with the dopamine-releasing properties of drugs of abuse by protecting dopamine from metabolism.

Interestingly, nicotine does not inhibit platelet MAO when it is present in the concentrations normally achieved during smoking (Oreland et al. 1981) nor does it inhibit MAO B in the living baboon when administered intravenously (Fowler et al. 1998). Recently the fractionation of extracts from flue-cured tobacco leaves led to the isolation of a competitive inhibitor of human MAO-A [$k(i)=3$ µM] and MAO-B [$k(i)=6$ µM], the structure of which could be assigned as 2,3,6-trimethyl-benzoquinone by classical spectroscopic analysis and confirmed by synthesis (Khalil et al. 2000). This information may help to provide insights into some aspects of the pharmacology and toxicology of tobacco products.

While tobacco smokers have an average of 40% lower values of brain MAO B than non-smokers and former smokers, the degree of MAO B inhibition is quite variable between subjects, ranging between 17 and 67%. The variability in the level of inhibition between the smokers was not accounted for by the smoking duration (average 24 ± 13.5 years) or the frequency (average 1 ± 0.27 packs/day). Because the time interval between the PET MAO B measurements and the last cigarette varied between subjects (range 1.7– 12 h), it was of interest to assess if this time interval contributed to the variability in MAO B. A study was undertaken to determine whether MAO B activity recovered measurably after an overnight smoke free interval (Fowler et al. 2000c). Brain MAO B was measured using PET and ^{11}C-L-deprenyl D_2 in six smokers who were scanned twice once at 11.3 h (baseline) after last cigarette and once at 10 min after smoking. Brain MAO B levels as measured by the model term λk_3 levels did not differ between baseline and after smoking.

Another aspect of the pharmacodynamic relationship between tobacco smoke exposure and MAO inhibition relates to whether MAO inhibition can be detected after a *single* cigarette. For this purpose brain MAO B was measured in a group of eight non-smokers at baseline and immediately after smoking a single cigarette using ^{11}C-L-deprenyl D_2 and PET. Eight normal healthy non-smokers (35 ± 11 years) received two PET studies 2 h apart with ^{11}C-L-deprenyl D_2, one at baseline and the second 5–10 min after the subject has smoked a single cigarette (Fowler et al. 1999b). Plasma nicotine and expired carbon monoxide (CO) were measured prior to smoking and 10 min after smoking completion as an index of tobacco smoke exposure. A three-compartment model was used to calculate λk_3, a model term proportional to MAO B and K_1, the plasma-to-brain transfer constant related to brain blood flow. The average λk_3 and K_1 for 11 different brain regions did not differ significantly between baseline and smoking. These results indicate that the reduction in MAO B in smokers occurs gradually and requires chronic tobacco smoke exposure.

The observation that smokers have reduced brain MAO A and B raises the need to investigate whether MAO B inhibition may account for some of the behavioral and epidemiological features of smoking (Hughes et al. 1986). It also reinforces the importance of reporting smoking status in clinical studies and the need to reevaluate reports that low platelet MAO B is a biological marker in clinical populations where the rate of smoking is high, such as in schizophrenics (Lidberg et al. 1985). In fact, normal platelet MAO was recently reported in non-smoking patients with schizophrenia (Simpson et al. 1999).

Smoking remains a major public health problem. Yet advances in treating smoking addiction hinge on characterizing both the neuropharmacological effects of tobacco smoke and factors accounting for individual variability in smoking toxicity. Along this line recent studies reporting the use of the reversible MAO A inhibitor moclobemide (Berlin et al. 1995a,b) the combination of nicotine and L-deprenyl (Brauer et al. 2000) and the use of L-deprenyl alone (George et al. 2003) as smoking cessation treatments is an important step based on the knowledge that the effects of tobacco smoke go beyond the effects of nicotine.

20.4
Conclusions and Outlook

PET is uniquely capable of providing information on biochemical transformations in the living human body. Although most of the studies of MAO have focused on measurements in the brain, the role of peripheral MAO as a phase 1 enzyme for the metabolism of drugs and xenobiotics is gaining attention (Castagnoli et al. 1997; Strolin-Benedetti and Tipton 1998; Inoue et al. 1999). MAO is well suited for this role because its concentration in digestive organs, kidneys, and liver is high, sometimes exceeding that in the brain. Knowledge of the distribution of the MAO subtypes within different organs and different cells is important in de-

termining which substrates (and which drugs and xenobiotics) have access to which MAO subtypes. The highly variable subtype distribution with different species makes human studies critically important. In addition, the deleterious side effects of combining MAO inhibitors with other drugs and foodstuffs makes it important to know the MAO inhibitory potency of different drugs both in the brain and in peripheral organs (Ulus et al. 2000). Recently, [^{11}C]L-deprenyl was used to image MAO B in peripheral organs in humans (Fowler et al. 2002). Clearly PET can play a role in answering these questions, both in drug research and development and in discovering some of the factors which contribute to the high variability of MAO levels in different individuals.

■ **Acknowledgements.** This research was carried out at Brookhaven National Laboratory under contract DE-AC02-98CH10886 and with the US Department of Energy and supported by its Office of Biological and Environmental Research and also by the National Institutes of Health (National Institute on Drug Abuse and National Institutes of Neurological Diseases and Stroke). The authors are grateful to Melissa Lemaire for her help in preparing this manuscript.

20.5
References

Abeles RH, Maycock AL (1976) Suicide enzyme inactivators. Accounts Chem Res 9:313–319

Adolfsson R, Gottfries C-G, Oreland L et al (1980) Increased activity of brain and platelet monoamine oxidase in dementia of Alzheimer type. Life Sci 27:1029–1034

Alexoff DL, Shea C, Fowler JS et al (1995) Plasma input function determination for PET using a commercial laboratory robot. Nucl Med Biol 22:893–904

Ametany SM, Beer HF, Guenther I et al (1996) Radiosynthesis of [^{11}C]brofaromine, a potential tracer for imaging monoamine oxidase A. Nucl Med Biol 23:229–234

Anderson MC, Hasan F, McCrodden JM et al (1993) Monoamine oxidase inhibitors and the cheese effect. Neurochem Res 18:1145–1149

Ansari KS, Yu PH, Kruck PA et al (1993) Rescue of axotomized immature rat facial motor neurons by R(–)-deprenyl: stereospecificity and independence from monoamine oxidase inhibition. J Neurosci 13:4042–4053

Arnett CD, Fowler JS, MacGregor RR et al (1987) Turnover of brain monoamine oxidase measured in vivo by positron emission tomography using L-[^{11}C]deprenyl. J Neurochem 49:522–527

Bach AWJ, Lan NC, Johnson DL (1988) CDNA cloning of human liver monoamine oxidase A and B: molecular basis of differences in enzymatic properties. Proc Natl Acad Sci USA 85:4934–4938

Bartzokis G, Beckson M, Newton T et al (1999) Selegiline effects on cocaine-induced changes in medial temporal lobe metabolism and subjective ratings of euphoria. Neuropsychopharmacology 20:582–590

Beer HF, Frey LD, Haberli M et al (1995a) [^{123}I/^{18}F]N-(2-aminoethyl)-5-halogeno-2-pyridinecarbox-amides, site specific tracers for MAO B mapping with SPECT and PET. Nucl Med Biol 22:999–1004

Beer HF, Rossetti I, Frey LD et al (1995b) ^{123}I-Labeling and evaluation of Ro 43-0463, a SPET tracer for MAO B imaging. Nucl Med Biol 22:929–936

Belleau B, Moran J (1963) Deuterium isotope effects in relation to the chemical mechanism of monoamine oxidase. Ann NY Acad Sci 107:822–839

Bench CJ, Price GW, Lammertsma AA (1991) Measurement of human cerebral monoamine oxidase type B (MAO B) activity with positron emission tomography (PET): a dose ranging study with the reversible MAO B inhibitor Ro 19-6327. Eur J Clin Pharmacol 40:169–173

Bench CJ, Lammertsma AA, Dolan RJ et al (1993) Cerebral monoamine oxidase (MAO B) activity in normal subjects, Alzheimer's disease and Parkinson's disease. J Cereb Blood Flow Metab 13:S246

Bergstrom M, Westerberg G, Langstrom B (1997a) ^{11}C-Harmine as a tracer for monoamine oxidase A (MAO A): in vitro and in vivo studies. Nucl Med Biol 24:287–293

Bergstrom M, Westerberg G, Kihberg T et al (1997b) Synthesis of some ^{11}C-labeled MAO A inhibitors and their in vivo uptake kinetics. Nucl Med Biol 24:381–388

Bergstrom M, Westerberg G, Nemeth G et al (1997c) MAO A inhibition in brain after dosing with esuprone, moclobemide and placebo in healthy volunteers: in vivo studies with positron emission tomography. Eur J Clin Pharmacol 52:121–128

Berlin I, Said S, Spreux-Varocuax et al (1995a) Monoamine oxidase A and B in heavy smokers. Biol Psychiatry 33:756–761

Berlin I, Said S, Spreux-Varocuax et al (1995b) A reversible monoamine oxidase A inhibitor (moclobemide) facilitates smoking cessation and abstinence in heavy, dependent smokers. Clin Pharmacol Ther 58:444–452

Bernard S, Fuseau C, Schmid L et al (1996) Synthesis and in vivo studies of a specific monoamine oxidase B inhibitor 5-4-benzyloxy)phenyl-3-(2-cyanoethyl)-1,3,4-oxadiazo-[^{11}C]-2(3H)-one. Eur J Nucl Med 23:150–156

Birkmayer W, Riederer P (1984) Deprenyl prolongs the therapeutic efficacy of combined L-DOPA in Parkinson's disease. Adv Neurol 40:475–481

Blaschko H, Richter D, Schlossmann H (1937) The inactivation of adrenaline. J Physiol (Lond) 90:1–17

Blauenstein P, Remy N, Buck A et al (1998) In vivo properties of N-(2-aminoethyl)-5-halogeno-2-carboxamide 18F- and 123I-labeled inhibitors of monoamine oxidase B. Nucl Med Biol 25:47–52

Bodkin JA, Cohen BM, Salomon MS et al (1996) Treatment of negative symptoms in schizophrenia and schizoaffective disorder by selegiline augmentation of antipsychotic medication. A pilot study examining the role of dopamine. J Nerv Ment Dis 184:295–301

Brauer LH, Paxton DA, Rose JE (2000) Selegiline and transdermal nicotine for smoking cessation. Presented at the 6th annual meeting of the Society for Research on Nicotine and Tobacco, 18–20 Febr 2000, Arlington, Va

Brenner DE, Kukull WA, Van Belle G et al (1993) Relationship between cigarette smoking and Alzheimer's disease in a population-based case-control study. Neurology 43:293–300

Brunner HG, Nelen M, Breakefield XO et al (1993) Abnormal behavior associated with a point mutation in the structural gene for monoamine oxidase A. Science 262:578–580

Buck A, Frey LD, Blusenstein P et al (1998) Monoamine oxidase B single photon emission tomography with [^{123}I]Ro 43 0463: imaging in volunteers and patients with temporal lobe epilepsy. Eur J Nucl Med 25:464–470

Caldecott-Hazard S, Schneider LS (1992) Clinical and biochemical aspects of depressive disorders: III. Treatment and controversies. Synapse 10:141–168

Cases O, Seif I, Grimsby J et al (1995) Aggressive behavior and altered amounts of serotonin and norepinephrine in mice lacking MAO A. Science 268:1763–1766

Castagnoli N Jr, Rimoldi JM, Bloomquist J et al (1997) Potential metabolic bioactivation pathways involving cyclic tertiary amines and azarenes. Chem Res Toxicol 10:924–940

Cohen G, Kesler N (1999) Monoamine oxidase and mitochondrial respiration. J Neurochem 73:2310–2315

Collins FA, Murphey DL, Reiss AL et al (1992) Clinical, biochemical and neuropsychiatric evaluation of a patient with a contiguous gene syndrome due to microdeletion Xp11.3 including the Norrie disease locus and monoamine oxidase (MAO A and MAO B) genes. Am J Med Genet 42:127–134

Crane GE (1956) Psychiatric side effects of iproniazide. Am J Psychiatry 112:494–497

Curet O, Damoiseau G, Aubin N et al (1996) Befloxatone, a new reversible and selective monoamine oxidase-A inhibitor. I. Biochemical profile. J Pharmacol Exp Ther 277:253–264

Curtis-Prior P, Vere D, Fray P (1999) Therapeutic value of Gingko biloba in reducing symptoms of decline in mental function. J Pharm Pharmacol 51:535–541

Da Prada M, Kettler R, Keller HH et al (1988) WP, Ro 19-6327, a reversible, highly selective inhibitor of type B monoamine oxidase completely devoid of tyramine potentiating effects: comparison with selegiline. In: Dahlstrom A (ed) Progress in catecholamine research, part B. Central aspects. Liss, New York, pp 359–363

Ding Y-S, Fowler JS, Gatley SJ et al (1995) Mechanistic PET studies of 6-[^{18}F]fluorodopamine in living baboon heart: selective imaging and control of radiotracer metabolism using the deuterium isotope effect. J Neurochem 65:682–690

Dolle F, Bramoulle Y, Bottlaender M et al (1999) [^{11}C]Befloxatone, a novel highly potent radioligand for in vivo imaging monoamine oxidase A. J Labeled Compounds Radiopharm 42 [Suppl 1]:S608–S609

Fowler CJ, Wiberg A, Oreland L et al (1980) The effect of age on the activity and molecular properties of human brain monoamine oxidase. J Neural Transm 49:1–20

Fowler JS, MacGregor RR, Wolf AP et al (1987) Mapping human brain monoamine oxidase A and B with ^{11}C-suicide inactivators and positron emission tomography. Science 235:481–485

Fowler JS, Wolf AP, MacGregor RR et al (1988) Mechanistic positron emission tomography studies. Demonstration of a deuterium isotope effect in the MAO catalyzed binding of [¹¹C]L-deprenyl in living baboon brain. J Neurochem 51:1524–1534

Fowler JS, Volkow ND, Logan J et al (1993) Monoamine oxidase B (MAO B) inhibitor therapy in Parkinson's disease: the degree and reversibility of human brain MAO B inhibition by Ro 19 6327. Neurology 43:1984–1992

Fowler JS, Volkow ND, Logan J et al (1994) Slow recovery of human brain MAO B after L-deprenyl withdrawal. Synapse 18:86–93

Fowler JS, Wang G-J, Logan J et al (1995) Selective reduction of radiotracer trapping by deuterium substitution: comparison of [¹¹C]L-deprenyl and [¹¹C]L-deprenyl-D2 for MAO B mapping. J Nucl Med 36:1255–1262

Fowler JS, Wang G-J, Volkow ND et al (1996a) Inhibition of monoamine oxidase B in the brains of smokers. Nature 379:733–736

Fowler JS, Volkow ND, Wang G-J et al (1996b) Brain MAO A inhibition in cigarette smokers. Proc Natl Acad Sci USA 93:14065–14069

Fowler JS, Fazzini E, Volkow ND (1996c) Deprenyl and levodopa and Parkinson's disease progression. Ann Neurol 40:267–268

Fowler JS, Volkow ND, Wang G-J et al (1997) Age-related increases in brain MAO B in healthy human subjects. Neurobiol Aging 18:431–435

Fowler JS, Volkow ND, Logan J et al (1998) An acute dose of nicotine does not inhibit MAO B in baboon brain in vivo. Life Sci 63PL:19–23

Fowler JS, Volkow ND, Cilento R et al (1999a) Comparison of brain glucose metabolism and monoamine oxidase B (MAO B) in traumatic brain injury. Clin Positron Imaging 2:71–79

Fowler JS, Wang G-J, Volkow ND et al (1999b) Smoking a single cigarette does not produce a measurable reduction in brain MAO B in non-smokers. Nicotine Tob Res 1:325–329

Fowler JS, Wang G-J, Volkow ND et al (2000a) Evidence that Ginkgo biloba extract does not inhibit MAO A and B in the human brain. Life Sci 66 PL:141–146

Fowler JS, Logan J, Gimi R et al (2000b) Non-MAO A binding of clorgyline in white matter in human brain. J Neurochem 79:1039–1046

Fowler JS, Wang G-J, Volkow ND et al (2000c) Maintenance of brain monoamine oxidase B inhibition in smokers after overnight cigarette abstinence. Am J Psychiatry 157:1864–1866

Fowler JS, Volkow ND, Logan J et al (2001a) Evidence that L-deprenyl treatment for one week does not inhibit MAO A or the dopamine transporter in the human brain. Life Sci 68:2759–2768

Fowler JS, Ding Y-S, Logan J et al (2001b) Species differences in [¹¹C]clorgyline binding in brain. Nucl Med Biol 28:779–785

Fowler JS, Logan J, Wand GJ et al (2002) PET imaging of monoamine oxidase B in peripheral organs in humans. J Nucl Med 43:1331–1338

Galva MD, Bondiolotti GP, Olasma M et al (1995) Effect of aging on lazabemide binding, monoamine oxidase activity and monoamine metabolites in human frontal cortex. J Neural Transm [Gen Sect] 101:83–94

George TP, Vessicchio JC, Termine A et al (2003) A preliminary placebo-controlled trial of selegiline hydrochloride for smoking cessation. Biol Psychiatry 53:136–143

Gerlach M, Riederer P, Youdim MB (1992) The molecular pharmacology of L-deprenyl. Eur J Pharmacol 226:97–108

Glassman AH, Helzer JE, Covey LS et al (1990) Smoking, smoking cessation, and major depression. JAMA 264:1546–1549

Goller L, Bergstrom M, Nilsson S et al (1995) MAO-A enzyme binding in bladder cancer characterized with [¹¹C]harmine in frozen section autoradiography. Oncol Rep 2:717–721

Gorell JM, Rybicki BA, Johnson CC et al (1999) Smoking and Parkinson's disease – a dose-response relationship. Neurology 52:115–119

Grimsby J, Toth M, Karoum F et al (1997) Increased stress response and -phenylethylamine in MAO B-deficient mice. Nat Genet 17:206–210

Halldin C, Burling P, Stalnacke C-G et al (1989) ¹¹C-Labeling of dimethylphenethylamine in two different positions and biodistribution studies. Appl Radiat Isot 40:557–560

Hare MLC (1928) Tyramine oxidase. I. A new enzyme system in liver. Biochem J 22:968–979

Hashimoto K, Inoue O, Suzuki K et al (1986) Deuterium isotope effect and [¹¹C]N,N-dimethylphenethyl-amine-a,a-d2: reduction in metabolic trapping rate in brain. Nucl Med Biol 13:79–80

Heikkila RE, Manzino L, Cabbat FS et al (1984) Protection against the dopaminergic neurotoxicity of 1-methyl-4-phenyl-1,2,5,6-tetrahydropyridine by monoamine oxidase inhibitors. Nature:467–469

Henningfield JE, Clayton R, Pollen W (1990) Involvement of tobacco in alcoholism and illicit drug use. Br J Addict 85:279–292

Hirata M, Magata Y, Ohmomo Y et al (1995) Evaluation of radioiodinated iodoclorgyline as a SPECT radiopharmaceutical for MAO-A in the brain. Nucl Med Biol 22:175–180

Hughes JR, Hatsukami DK, Mitchell JE et al (1986) Prevalence of smoking among psychiatric outpatients. Am J Psychiatry 143:993–997

Inoue O, Tominaga T, Yamasaki T (1985) Radioactive N,N-dimethylphenethylamine: a selective radiotracer for the in vivo measurement of monoamine oxidase B activity in the brain. J Neurochem 44:210–216

Inoue H, Castagnoli K, Van Der Schyf C et al (1999) Species-dependent differences in monoamine oxidase A and B-catalyzed oxidation of various C4 substituted 1-methyl-4-phenyl-1,2,3,6-tetrahydropyridinyl derivatives. J Pharm Exp Ther 291:856–864

Ishiwata K, Ido T, Yanai K et al (1985) Biodistribution of a positron emitting suicide inactivator of monoamine oxidase, carbon-11 pargyline, in mice and a rabbit. J Nucl Med 26:630–636

Johnston JP (1968) Some observations upon a new inhibitor of monoamine oxidase in brain tissue. Biochem Pharmacol 17:1285–1297

Khalil AA, Steyn S, Castagnoli N (2000) Isolation and characterization of a monoamine oxidase inhibitor from tobacco leaves. Chemical research. Toxicology 13:31–35

Kleijnen J, Knipschild P (1992) Ginkgo biloba. Lancet 340:1136–1139

Knoll J, Magyar K (1972) Some puzzling effects of monoamine oxidase inhibitors. Adv Biochem Psychopharmacol 5:393–408

Koller WC, Giron LT (1990) Selegiline HCl: selective MAO-type B inhibitor. Neurology 40 [Suppl 3]:58–60

Kopin I (1993) Monoamine oxidase (MAO). Relationship to foods, poisons, and medicines. Biogenic Amines 9:355–365

Kumlien E, Hilton-Brown P, Spannare B, Gillberg P-G et al (1992) In vitro quantitative autoradiography of [³H]L-Deprenyl and [³H]-Pk 11195 binding sites in human epileptic hippocampus. Epilepsia 33:610–617

Kumlien E, Bergstrom M, Lilja A et al (1995) Positron emission tomography with [¹¹C]deuterium-deprenyl in temporal lobe epilepsy. Epilepsia 36:712–721

Lamensdorf I, Youdim MBH, Finberg JPM (1996) Effect of long-term treatment with selective monoamine oxidase A and B inhibitors on dopamine release from rat striatum in vivo. J Neurochem 67:1532–1539

Lammertsma AA, Bench CJ, Price GW et al (1991) Measurement of cerebral monoamine oxidase B activity using L-[¹¹C]deprenyl and dynamic positron emission tomography. J Cereb Blood Flow Metab 11:545–556

Langston JW, Ballard JW, Tetrud JW et al (1983) Chronic parkinsonism in humans due to a product of meperidine analog synthesis. Science 219:979–980

Langston JW, Irwin I, Langston EB et al (1984) Pargyline prevents MPTP-induced parkinsonism in primates. Science 225:1480–1483

Lena I, Ombetta J-I, Chalon S et al (1995) Iododerivative of pargyline: a potential tracer for the exploration of monoamine oxidase sites by SPECT. Nucl Med Biol 22:727–736

Lidberg L, Modin I, Oreland L et al (1985) Platelet monoamine oxidase activity and psychopathy. Psychiatry Res 4:339–343

Logan J, Dewey SL, Wolf AP et al (1991) Effects of endogenous dopamine on measures of [¹⁸F]N-methylspiroperidol binding in the basal ganglia: comparison of simulations and experimental results from PET studies in baboons. Synapse 9:195–207

Logan J, Fowler JS, Volkow ND et al (2000) Reproducibility of repeated measures of deuterium substituted [¹¹C]L-deprenyl ([¹¹C]L-deprenyl-D2) binding in the human brain. Nucl Med Biol 27:43–49

MacGregor RR, Halldin C, Fowler JS et al (1985) Selective, irreversible in vivo binding of [¹¹C]clorgyline and [¹¹C]L-deprenyl in mice: potential for the measurement of monoamine oxidase activity in brain using positron emission tomography. Biochem Pharmacol 34:3207–3210

MacGregor RR, Fowler JS, Wolf AP (1988) Synthesis of suicide inhibitors of monoamine oxidase: carbon-11 labeled clorgyline, L-deprenyl and D-deprenyl. J Labeled Compounds Radiopharm 25:1–9

Mannelli M, Pupilli C, Lanzillotti R et al (1990) Catecholamines and blood pressure regulation. Horm Res 34:156–160

Maycock AL, Abeles RH, Salach JI et al (1976) The structure of the covalent adduct formed by the interaction of 3-dimethylamine-1-propyne and the flavine of mitochondrial amine oxidase. Biochemistry 15:114–125

Morens DM, Grandinetti A, Reed D et al (1995) Cigarette smoking and protection from Parkinson's disease: false association or etiological clue. Neurology 45:1041–1051

Mukherjee J, Yang Z-Y (1999) Development of N-[3-(2′,4′-dichlorophenoxy)-2-¹⁸F-fluoropropyl]-N-methylpropargylamine (¹⁸F-fluoroclorgyline) as a potential PET radiotracer for monoamine oxidase A. Nucl Med Biol 26:619–625

Nakamura S, Kawamata T, Akiguchi I et al (1990) Expression of monoamine oxidase B inactivity in astrocytes of senile plaques. Acta Neuropathol (Berl) 80:419–425

Norman TR, Chamberlain KG, French MA (1987) Platelet monoamine oxidase: low activity in cigarette smokers. Psychiatry Res 20:199–205

Oreland L (1991) Monoamine oxidase dopamine and Parkinson's disease. Acta Neurol Scand 84 [Suppl 136]:60–65

Oreland L, Fowler CJ, Schalling D (1981) Low platelet monoamine oxidase activity in cigarette smokers. Life Sci 29:2511–2518

Palmer AM, DeKosky ST (1993) Monoamine neurons in aging and Alzheimer's disease. J Neural Transm [Gen Sect] 91:135–159

Parkinson's Study Group (1989) Effect of deprenyl on the progression of disability in early Parkinson's disease. N Engl J Med 321:1364–1371

Parkinson's Study Group (1996) Effect of lazabemide on the progression of disability in early Parkinson's disease. Ann Neurol 40:99–107

Patlak C, Fenstermacher JD, Blasberg RG (1983) Graphical evaluation of blood-to-brain transfer constants from multiple time-activity data. J Cereb Blood Flow Metab 3:1–7

Plenevaux A, Dewey SL, Fowler JS et al (1990) Synthesis of (R)-(–)- and (S)-(+)-4-fluorodeprenyl and R-(–)- and (S)-(+)-[N-¹¹C-methyl]-4-fluorodeprenyl and positron emission tomography studies in baboon brain. J Med Chem 33:2015–2019

Porsolt RD, Roux S, Drieu K (2000) Evaluation of Ginkgo biloba extract (Egb 761) in functional tests for monoamine oxidase inhibition. Arzneimittelforschung/Drug Res 50:232–235

Raffel DM, Corbett JR, del Rosario RB et al (1999) Sensitivity of [¹¹C]phenylephrine kinetics to monoamine oxidase activity in normal human heart. J Nucl Med 40:232–238

Rafi H, Chalon S, Ombetta JE et al (1996) An iodinated derivative of moclobemide as potential radioligand for brain MAO A exploration. Life Sci 58:1159–1169

Reinikainen KJ, Paljarvi L, Halonen T et al (1988) Dopaminergic system and monoamine oxidase B activity in Alzheimer's disease. Neurobiol Aging 9:245–252

Richards JG, Saura J, Luque JM et al (1998) Monoamine oxidases: from brain maps to physiology and transgenics to pathophysiology. J Neural Transm [Suppl] 52:173–187

Riederer P, Youdim MBH (1986) Monoamine oxidase activity and monoamine metabolism in brains of parkinsonian patients treated with L-deprenyl. J Neurochem 46:1359–1365

Robinson DS, Davis JM, Nies A et al (1971) Relation of sex and aging to monoamine oxidase activity of human brain, plasma and platelets. Arch Gen Psychiatry 24:536–539

Robinson JB (1985) Stereoselectivity and isoenzyme selectivity of monoamine oxidase inhibitors. Enantiomers of amphetamine, N-methylamphetamine and deprenyl. Biochem Pharmacol 34:4105–4108

Sano M, Ernesto C, Thomas RG et al (1997) A controlled trial of selegiline, alpha tocopherol or both as the treatment of Alzheimer's disease. N Engl J Med 336:1216–1222

Saura J, Kettler R, Da Prada M et al (1992) Quantitative enzyme radioautography with ^{3}H-Ro 411049 and ^{3}H-Ro 196327 in vitro: localization and abundance of MAO-A and MAO-B in rat CNS, peripheral organs, and human brain. J Neurosci 12:1977–1999

Saura J, Richards JG, Mahy N (1994a) Differential age-related changes of MAO-A and MAO-B in mouse brain and peripheral organs. Neurobiol Aging 15:399–408

Saura J, Luque AM, Cesura M et al (1994b) Increased monoamine oxidase B activity in plaque-associated astrocytes of Alzheimer brains revealed by quantitative enzyme radioautography. Neuroscience 62:15–30

Saura J, Nadal E, van den Berg B et al (1996) Localization of monoamine oxidases in human peripheral tissues. Life Sci 59:1341–1349

Schildkraut JJ (1965) The catecholamine hypothesis of affective disorders: a review of supporting evidence. Am J Psychiatry 122:509–522

Scremin OU, Holschneider DP, Chem K et al (1999) Cerebral cortical blood flow maps are reorganized in MAOB-deficient mice. Brain Res 824:36–44

Selikoff I, Robitzek E, Ornstein G (1952) Treatment of pulmonary tuberculosis with hydrazide derivatives of isonicotinic acid. JAMA 150:973–980

Shih JC, Chen K, Ridd MJ (1999) Monoamine oxidase: from genes to behavior. Ann Rev Neurosci 22:197–217

Shinotoh H, Inoue O, Suzuki K et al (1987) Kinetics of [^{11}C]N,N-dimethylphenethylamine in mice and humans: potential for measurement of brain MAO B activity. J Nucl Med 28:1006–1011

Simpson GM, Shih JC, Chen K et al (1999) Schizophrenia, monoamine oxidase and cigarette smoking. Neuropsychopharmacology 20:392–394

Singer T (1995) Monoamine oxidases: old friends hold many surprises. FASEB J 9:605–610

Strolin-Benedetti M, Dostert P (1989) Monoamine oxidase, brain ageing and degenerative diseases. Biochem Pharmacol 38:555–561

Strolin-Benedetti MS, Tipton KF (1998) Monoamine oxidases and related amine oxidases as phase I enzymes in the metabolism of xenobiotics. J Neural Transm [Suppl] 52:149–171

Terry RD, DeTeresa R, Hansen LA (1987) Neocortical cell counts in normal human adult aging. Ann Neurol 21:530–539

Tetrud JW, Langston JW (1989) The effect of deprenyl (selegiline) on the natural history of Parkinson's disease. Science 245:519–522

Tominaga T, Inoue O, Suzuki K et al (1987) [^{13}N]β-phenethylamine ([^{13}N]PEA): a prototype tracer for measurement of MAO B activity in heart. Biochem Pharmacol 36:3671–3675

Ulus IH, Maher TJ, Wurtman RJ (2000) Characterization of phentermine and related compounds as monoamine oxidase (MAO) inhibitors. Biochem Pharmacol 59:1611–1621

Walsh C (1982) Suicide substrates: mechanism based enzyme inactivators. Tetrahedron 38:871–909

Westlund KN, Denney RM, Rose RM et al (1988) Localization of distinct monoamine oxidase A and monoamine oxidase B cell populations in human brainstem. Neuroscience 25:439–456

White HL, Scates PW, Cooper BR (1996) Extracts of Ginkgo biloba leaves inhibit monoamine oxidase. Life Sci 58:1315–1321

Youdim MDH, Riederer P (1993) Dopamine metabolism and neurotransmission in primate brain in relationship to monoamine oxidase A and B inhibition. J Neural Transm 91:181–195

Zeller EA (1938) Enzymatic degradation of histamine and diamines. Helv Chim Acta 21:880–890

Zeller EA, Barsky J, Berman ER (1955) Amine oxidases. XI. Inhibition of monoamine oxidase by 1-isonicotinyl-2-isopropylhydrazine. J Biol Chem 214:267–274

Zhuang Z, Hogan M, McCauley R (1988) The in vitro insertion of monoamine oxidase B into mitochondrial outer membranes. FEBS Lett 238:185–190

CHAPTER 21

Amino Acid Transport Studies in Brain Tumors

KARL-JOSEF LANGEN

21

Contents

21.1
Introduction

Amino acids are important biological substrates that play crucial roles in virtually all biological processes. These ionic nutrients serve not only as basic modules of proteins and hormones but also as neurotransmitters, synaptic modulators, and neurotransmitter precursors. Transfer of amino acids across the hydrophobic domain of the plasma membrane is mediated by proteins that recognize, bind, and transport them from the extracellular medium into the cell, or vice versa. In the early 1960s different substrate-specific transport systems for amino acids in mammalian cells were identified (Christensen 1990). General properties of mammalian amino acid transporters were revealed, such as stereospecificity and broad substrate specificity (i.e., several amino acids share the same transport system). Functional criteria such as the type of amino acid (e.g., basic, acidic) or thermodynamic properties (energy dependence of transport) were used to classify amino acid transporters. This functional classification has been retained to date, since structural information on amino acid transporters is incomplete. The molecular identification of amino acid transporters or related proteins

started in the early 1990s, and studies on the structure-function relationship and the molecular genetics of the pathology associated with these transporters has gained considerable interest.

The use of radiolabeled amino acids for functional studies in nuclear medicine began in the early 1960s using ^{75}Se-L-methionine for pancreas scintigraphy (Blau et al. 1962). In 1979 Kloss and Leven first considered the artificial amino acid 3-^{123}I-iodo-α-methyl-L-tyrosine (IMT) as a potential tracer for nuclear medicine, but initial clinical studies for the detection of melanomas and imaging of the pancreas by planar imaging with gamma cameras were not convincing (Tisljar et al. 1979; Kloster and Bockslaff 1982).

With the advent of PET nearly all amino acids have been radiolabeled, since the replacement of a carbon atom by ^{11}C does not chemically change the molecule (Hübner et al. 1982; Vaalburg et al. 1992). These radiolabeled amino acids differ from one another with regard to the ease of synthesis, biodistribution, and formation of radiolabeled metabolites in vivo.

The most frequently used radiolabeled amino acid for PET is L-methyl-^{11}C-methionine (MET). The main reason for its wide acceptance is the convenient radiochemical production, which allows rapid synthesis with high radiochemical yield without the need for complex purification steps (Langstrom et al. 1987).

A milestone in the diagnostic use of radiolabeled amino acids was a study in a patient with a cerebral glioma with PET using MET and CT (Bergström et al. 1983). It was observed that the extent of the tumor in MET PET was much larger than the area of contrast enhancement in the CT scan. Since this patient died a few days after these investigations, the anatomical brain slice of this patient could be directly compared to PET and CT. MET PET exactly matched the tumor size while CT did not show its true extent. These results were confirmed by further studies of this group comparing MRI and CT imaging to MET PET with stereotactic serial biopsies as a reference (Mosskin et al. 1989). In contrast, PET with 2-^{18}F-fluorodeoxyglucose is useful in estimating tumor grade and prognosis of gliomas, while their delineation is difficult because of high glucose metabolism in the normal cortex (for rev. see Coleman et al. 1991).

Thus, PET studies using radiolabeled amino acids offer a great potential for neurooncology. In spite of progresses in neurosurgery, radiotherapy, and chemotherapy, the clinical outcome of patients with cerebral gliomas is still unsatisfactory. Less than 30% with high grade gliomas survive 2 years. Although cranial computer-assisted tomography (CT) and magnetic resonance imaging (MRI) are unsurpassed diagnostic modalities for the detection of cerebral space-occupying lesions, the differentiation of tumor tissue from edematous, necrotic, and fibrotic tissue is difficult with these morphological methods (Byrne 1994; Leeds and Jackson 1994; Jansen et al. 2000), which limits the efficiency of local therapy such as surgery and radiotherapy. Nevertheless, due to the short-lived positron emitter [11]C (20 min half-life), the clinical application of MET remained limited to a few PET centres that had an on-site cyclotron and radiochemistry laboratory.

Stimulated by the promising results in studies of brain tumors using PET with MET, a resumption of research on the amino acid analogue IMT started in the late 1980s (Biersack et al. 1989; Langen et al. 1990). IMT can be used with SPECT and has offered a widespread application of amino acid imaging of brain tumors. Although IMT is not incorporated into proteins it has been shown that the results of IMT SPECT are similar to those of MET PET (Langen et al. 1997, 1998). A number of studies have demonstrated the potential of SPECT with IMT for grading, diagnosis of recurrence and therapeutic response of cerebral gliomas (Kuwert et al. 1996, 1998; Guth-Tougelides et al. 1995; Weber et al. 1997, 2001; Bader et al. 1998, 1999; Weckesser et al. 2002).

Also, a number of attempts were undertaken to label amino acids with the longer lived fluorine-18(110 min half-life). The yield of the radiosynthesis of these ^{18}F-labeled amino acids, however, was rather ineffective (Coenen et al. 1989; Ogawa et al. 1996; Inoue et al. 1998; Shoup et al. 1999) so that delivery to other nuclear medicine departments in a satellite concept could not be realized. Only recently, due to nucleophilic substitution, ^{18}F-labeled amino acids have been produced

that can be synthesized with high radiochemical yields, thus offering practical advantages for routine clinical practice (Hamacher 1999; Hamacher and Coenen 2002; Wester et al. 1999).

Studies on the uptake mechanisms, characterization of transport system, and links to molecular genetics of radiolabeled amino acids have been undertaken but are still in the beginning stages. Radiolabeled amino acids assigned to specific subtypes of amino acid transporters may offer a tool to evaluate specific pathological processes.

21.2
Studies of Amino Acid Uptake: Protein Synthesis or Transport Phenomenon?

Although in clinical use for a long time, the biological meaning of increased uptake of radiolabeled amino acids is still a matter of controversy. It was long assumed that increased uptake of amino acids, e.g., MET, in brain tumors reflects an increased protein synthesis rate. Experiments in mice, however, demonstrated that inhibition of protein synthesis did not influence the uptake of MET in tumors and brain (Ishiwata et al. 1993) suggesting that alterations of amino acid transport rather than increased protein synthesis caused increased uptake in tumors. The accumulation rate of radiolabeled methionine, both in normal human brain tissue and gliomas (that did not disrupt the blood-brain barrier) decreased by 35% after infusion of branched chain amino acids (Bergström et al. 1987 a). In addition, radiolabeled L-methionine accumulated 2.4 times as much as D-methionine (Bergström et al. 1987 b). These findings indicate specific carrier-mediated uptake as an important factor governing MET uptake.

A PET study in glioma patients with the ^{18}F-labeled amino acid L-2-^{18}F-fluoro-tyrosine, which also exhibits incorporation into proteins (Coenen et al. 1989), demonstrated that the difference of uptake between gliomas and normal brain was due to an increase of the rate constant k_1 of tracer transport. In that study the rate constant k_3 describing binding to the metabolic compartment was not altered or even decreased in gliomas (Wienhard et al. 1991). Concordantly for IMT, which is not significantly incorporated into any metabolic pathway (for review see Langen et al. 2002 b) the uptake appeared to be specific and related to amino acid transport. In vivo studies in mice and rats, as well as in vitro studies, revealed that IMT transport across the blood-brain barrier is similar to its parent L-tyrosine, i.e., saturable, carrier-dependent, temperature-dependent, and cross-inhibitable by L-tyrosine (Kawai et al. 1991). SPECT studies in ten patients with brain tumors before and after infusion of natural L-amino acids demonstrated a decrease of IMT uptake by about 50% in the normal brain and in gliomas (Langen et al. 1991; Oldendorf 1991). The competitive effects were similar in gliomas and normal brain tissue so that tumor/brain ratios were independent of plasma amino acid levels. In two meningiomas and a metastasis, however, no major change of uptake during amino acid infusion was observed. In experimental rat tumors competition with macroscopic amounts of L-phenylalanine showed significant reduction of IMT uptake in the brain, a relatively small change in its uptake in tumor, but no significant reduction of IMT uptake in the muscles (Deehan et al. 1993).

The lack of competitive effects in some tissues, however, does not prove nonspecific uptake of radiolabeled amino acids since the transport of neutral amino acids into the brain differs significantly from transport into other tissues of the body. The Michaelis-Menten constant for neutral amino acid transport in tissues other than the brain is far above physiologic levels, i.e., tissues other than the brain are independent of competitive effects in vivo (Pardridge et al. 1986). In vitro studies using levels of amino acids above the physiologic range have shown competition with IMT uptake in all cell lines investigated.

Summarizing, it appears that for the radiolabeled amino acids presently used in nuclear medicine an alteration of amino acid transport is the dominating process for increased uptake.

21.3
Amino Acid Transporters and Molecular Genomics

The expression of amino acid transport systems during proliferation and malignant transformation is a subject of major interest. Amino acid transport system A is one of the few identified transport systems which is more strongly expressed in transformed and malignant cells than in normal cells and appears to be a target of protooncogene and oncogene action (Saier et al. 1988). Less data are available concerning the importance of amino acid transport system L in

this context (Kilberg et al. 1993). System L is the major amino acid transport system of the blood-brain barrier and of glial cells (Pardridge and Oldendorf 1977; Brookes 1988). A characterization of transporters has mainly been undertaken for IMT, since this tracer is widely available and has attracted considerable clinical interest.

In astrocytes of neonatal rats IMT was found to be exclusively transported via the sodium-independent system L (Kopka et al. 2001). This same system was responsible for more than 70% of IMT transport in C6 rat glioma cells (Riemann et al. 1999), in human GOS3 glioma cells (Riemann et al. 2001), in human T98 glioma cells, and in human ^{86}Hg-39 glioma cells (Langen et al. 2000). In the exponential growth phase of ^{86}Hg-39 human glioma cells, IMT transport was significantly higher than in resting cells, the increased transport being due to increased activity of system L. In addition to system L, sodium-dependent system $B^{0,+}$ was observed to contribute up to 15% of the transport in C6 rat and human GOS3 glioma cells.

In cell lines originating from extracranial tumors, a similar role played by this system L has been demonstrated for IMT uptake (for review see Langen et al. 2002b). Summarizing, IMT transport was found to be specific and dominated by sodium-independent amino acid transport system L. Non-saturable or nonspecific components of IMT transport could not be identified. A small contribution of other transport systems depends on the specific cell line and is variable, similar to the observations with natural amino acids.

A relationship between increased IMT transport and molecular expression of transporter molecules has recently been demonstrated by studies on the 4F2 antigen (Fig. 21.1; Langen et al. 2001). It was shown that system L-like IMT transport in human glioma cells is correlated to the expression of the 4F2 antigen, and that both are dependent on cell proliferation to a similar extent. The surface antigen 4F2 is thought to be involved in the trafficking and regulation of system L-like neutral amino acid transport in mammalian cells (Broer et al. 1995, 1997). The 4F2 antigen was originally described as a marker for tumor cells and activated lymphocytes (Haynes et al. 1981). The antigen belongs to a class of activation markers that appears later in the G1 phase, when quiescent cells are activated and start to proliferate (Teixeira and Kuhn 1991). It has been shown by immunoprecipitation experiments that a heavy chain

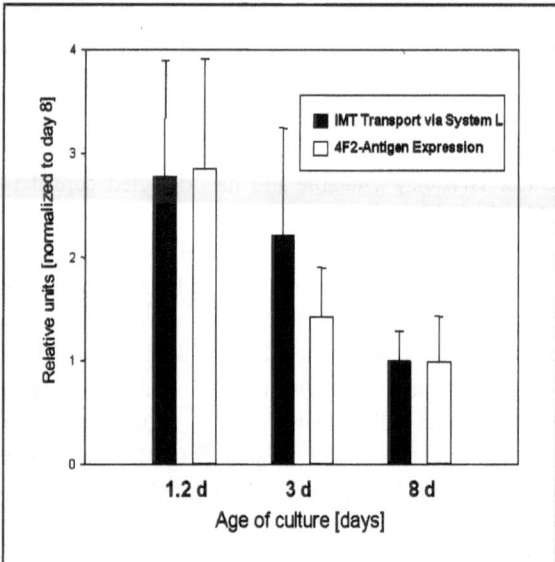

Fig. 21.1. System L-like transport of IMT and 4F2 antigen expression in [86]Hg39 human glioma cells in lag phase (*left*, 1.2 days after seeding), in exponential growth phase (*middle*, 3 days after seeding), and in plateau phase cells (*right*, 8 days after seeding). Values are given as mean±SD and normalized to plateau phase. System L-mediated IMT transport and 4F2 antigen expression was highest shortly after seeding and decreased in parallel in exponential-growth-phase and plateau-phase cells. The 4F2 antigen is part of the transporter molecule. (Reprinted from Langen et al. 2001, with permission)

(4F2hc) and a light one exist for the 4F2 surface antigen, the two subunits of the protein being linked by a disulfide bridge (Haynes et al. 1981; Hemler and Strominger 1982). All monoclonal antibodies mobilized against the 4F2 antigen are directed against the heavy chain. Both the surface antigen 4F2 and amino acid transport system L have been found to be ubiquitously expressed (Broer et al. 1995). In addition, a number of complementary DNAs have been cloned exhibiting transport patterns corresponding to system L when coexpressed with 4F2hc (Kanai et al. 1998; Mastroberardino et al. 1998; Torrents et al. 1998; Pineda et al. 1999). Mastroberardino et al. detected the permease-related protein E16 as the first light chain of 4F2hc, which latter had originally been cloned because it is induced in activated lymphocytes. Kanai et al. isolated a cDNA from rat C6 glioma cells encoding a Na[+]-independent neutral amino acid transporter designated LAT1. LAT1 was shown to be a non-glycosylated membrane protein consistent with the properties of an 4F2 light chain. By data base search, partial or incomplete sequences of LAT1 were identified (E16, TA1, ASUR4b) which had been reported as tumor-associated sequences. It was speculated that LAT1 expression is upregulated in rapidly dividing tumor cells and established cell lines to supply cells with more essential amino acids to support the continuous growth and proliferation. The 4F2 heavy chain regulates the trafficking of a family of amino acid transporters which includes the system L-associated isoforms LAT1, LAT2, the Na[+]-independent transporter for small neutral amino acids ascl, the system y[+]L-associated isoforms y[+]LAT1 and y[+]LAT2, and the glutamate/cystine exchanger xCT (Torrents et al. 1998; Pfeiffer et al. 1999; Pineda et al. 1999; Sato et al. 1999; Segawa et al. 1999; Fukasawa et al. 2000). The corresponding proteins were consistent with the properties of 4F2 light chain, suggesting that they present some of the proteins formerly referred to as 4F2 light chain.

The rapidly developing knowledge about the molecular genetics of amino acid transport systems and the capability of nuclear medicine techniques to measure such processes offer a great new potential for improved diagnostics of numerous diseases.

21.4
Clinical Imaging

The potential of radiolabeled amino acids for the diagnostic evaluation of brain tumors has been demonstrated in numerous studies (for review see Langen and Weckesser 1999; Langen et al. 2002b; del Sole et al. 2001; Jager et al. 2001). As outlined above, one of the main advantages of using amino acids has been found to be the visualization of the degree of intracerebral infiltration by gliomas (Derlon et al. 1989; Ogawa et al. 1993; Grosu et al. 2000; Weckesser et al. 2000). Although nonspecific uptake of amino acids may not be excluded (Cook et al. 1999), the method appears to be sensitive enough to differentiate high grade gliomas from non-tumorous lesions (Kuwert et al. 1996; Herholz et al. 1998), recurrences from scarring (Ogawa et al. 1991; Guth-Tougelidis et al. 1995; Kuwert et al. 1998; Bader et al. 1999), and to monitor therapeutic effects (Würker et al. 1996; Floeth et al. 2001).

In addition, it has been demonstrated that uptake of MET within the boundaries of the tumor is related to the local level of malignancy as estimated histologically on stereotactic biopsies (Goldman et al. 1997). This is of importance considering the high level of regional heterogeneity of gliomas and the potential benefit of ensuring the resection or destruction of most malignant components of the tumor when complete resection cannot be performed without unacceptable morbidity. Such a metabolically guided resection has proved feasible, using stereotactic PET MET information to direct resection of brain tumors (Levivier et al. 1999).

The role of radiolabeled amino acids for the determination of tumor grading of cerebral gliomas is controversial. Initial studies reported higher MET uptake in high-grade gliomas than in low-grade gliomas (Derlon et al. 1989). In a series of 196 consecutive patients a difference in uptake between high-grade and low-grade gliomas was found, but the overlap of the two groups was too great to allow precise grading in individual tumors (Herholz et al. 1998). For IMT some authors observed a high accuracy of IMT SPECT for the separation of high- from low-grade gliomas (Kuwert et al. 1996; Woesler et al. 1997; Weckesser et al. 2002) but others could not identify such a difference and observed a wide overlap of IMT uptake in different tumor grades (Weber et al. 1997; Bader et al. 1999; Schmidt et al. 2001). Similarly, L-1-^{11}C-tyrosine PET revealed no correlation between the protein synthesis rate and histological grading or proliferative activity of human brain tumors (Pruim et al. 1995; de Wolde et al. 1997).

For MET PET significantly longer survival times for glioma patients with low MET uptake than for such patients with high MET uptake have been reported (Kaschten et al. 1998); especially in low-grade gliomas, MET uptake has been found to be a good prognostic indicator (Ribom et al. 2001).

IMT uptake appears not to be an indicator of survival time in patients with gliomas (Schmidt et al. 2001), but the extent of the lesion on IMT SPECT was an independent predictor of prognosis (Weckesser et al. 2002). Also, after resection of primary brain tumors significantly longer survival times were reported in patients without IMT uptake than in those with residual IMT uptake (Weber et al. 2001). However, these data most likely reflect the known influence of total tumor resection on the prognosis of patients with cerebral gliomas, rather than support the predictive value of IMT uptake for individual prognosis. All in all, these studies confirm the potential of amino acids for identifying residual tumor tissue.

21.5
Perspectives

As mentioned above, ^{18}F-labeled amino acids have been presented recently which can be synthesized with high radiochemical yields due to nucleophilic substitution (Hamacher 1999; Hamacher and Coenen 2002; Wester et al. 1999). The establishment of an efficient synthesis for O-(2-^{18}F-fluoroethyl)-L-tyrosine (FET: Hamacher and Coenen 2002) represents a milestone for the introduction of amino acid imaging into routine clinical practice. It has already been demonstrated that FET yields similar results compared with MET PET in patients with cerebral gliomas (Weber et al. 2000) and a better delineation of tumor extent compared to MRI (Fig. 21.2; Pauleit et al. 2002).

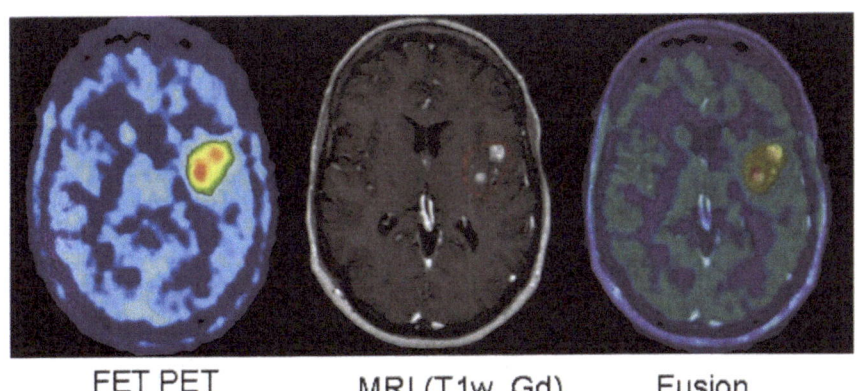

FET PET   MRI (T1w, Gd)   Fusion

Fig. 21.2. Histologically verified glioblastoma multiforme. The extent of FET accumulation in the PET scan exceeds the area of contrast enhancement in the T1-weighted MRI scan after injection of Gd-DTPA

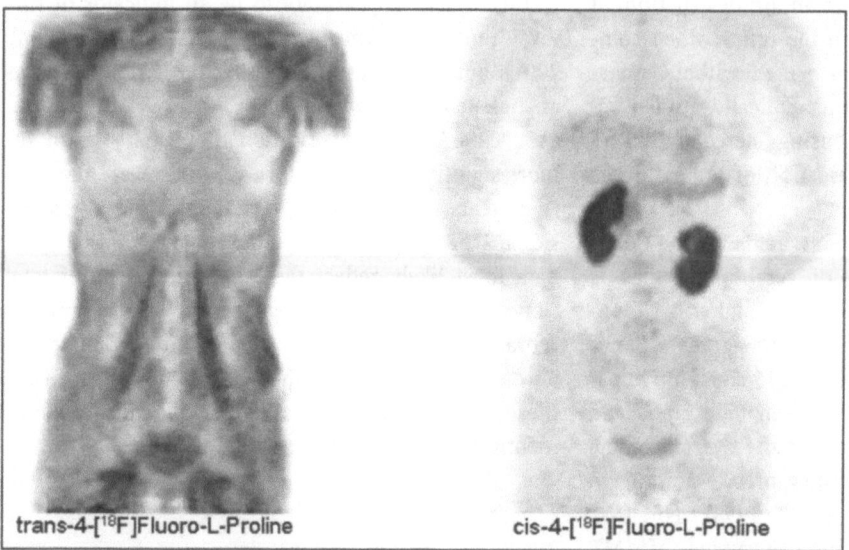

trans-4-[¹⁸F]Fluoro-L-Proline   cis-4-[¹⁸F]Fluoro-L-Proline

Fig. 21.3. Whole body PET scans of cis-FPro and trans-FPro 3 h after tracer injection. Cis-FPro in contrast to trans-FPro is incorporated into protein and is taken up by amino acid transport system A, which leads to a different physiological acceptance. (Reprinted from Langen et al. 2002a, with permission from Elsevier)

Furthermore it has been shown that FET, in contrast to 2-¹⁸F-fluoro-deoxyglucose, exhibits low uptake in non-neoplastic inflammatory cells, which finding promises a higher specificity for the detection of tumor cells than is achieved with glucose analogues (Kaim et al. 2002).

The transport characteristics of different radiolabeled amino acids may be allocated to specific molecular subtypes of amino acid transporters. For the diastereoisomeres of cis-4-¹⁸F-fluoro-L-proline (cis-FPro) and trans-4-¹⁸F-fluoro-L-proline (trans-FPro), for example, minor differences in transport characteristics have been identified, leading to considerable differences in physiological behavior (Fig. 21.3; Langen et al. 2002).

It appears that the diagnostic potential of radiolabeled amino acids will be amplified by enlarging the knowledge on the molecular mechanisms involved in the transport processes. The application of radiolabeled amino acids with selective transport characteristics will offer a new tool for a more specific evaluation of pathological processes.

21.6
References

Bader JB, Samnick S, Schaefer A et al (1998) Contribution of nuclear medicine to the diagnosis of recurrent brain tumors and cerebral radionecrosis. Radiologe 38:924–929

Bader JB, Samnick S, Moringlane JR et al (1999) Evaluation of l-3-[¹²³I]iodo-alpha-methyltyrosine SPECT and [¹⁸F]fluorodeoxyglucose PET in the detection and grading of recurrences in patients pretreated for gliomas at follow-up: a comparative study with stereotactic biopsy. Eur J Nucl Med 26:144–151

Bergström M, Collins VP, Ehrin E et al (1983) Discrepancies in brain tumor extent as shown by computed tomography and positron emission tomography using [⁶⁸Ga]EDTA, [¹¹C]glucose, and [¹¹C]methionine. J Comput Assist Tomogr 7:1062–1066

Bergström M, Ericson K, Hagenfeldt L et al (1987a) PET study of methionine accumulation in glioma and normal brain tissue: competition with branched chain amino acids. J Comput Assist Tomogr 11:208–213

Bergström M, Lundqvist H, Ericson K et al (1987b) Comparison of the accumulation kinetics of L-(methyl-¹¹C)-methionine and D-(methyl-¹¹C)-methionine in brain tumors studied with positron emission tomography. Acta Radiol 28:225–229

Biersack HJ, Coenen HH, Stöcklin G et al (1989) Imaging of brain tumors with L-3-[¹²³I]Iodo-α-methyl tyrosine and SPECT. J Nucl Med 30:110–112

Blau M, Manske RS, Bender MA (1962) Clinical experience with Selen-75 selenomethionine for pancreas visualization. J Nucl Med 3:202

Broer S, Broer A, Hamprecht B (1995) The 4F2hc surface antigen is necessary for expression of system L-like neutral amino acid-transport activity in C6-BU-1 rat glioma cells: evidence from expression studies in *Xenopus laevis* oocytes. Biochem J 312:863–870

Broer S, Broer A, Hamprecht B (1997) Expression of the surface antigen 4F2hc affects system-L-like neutral-amino-acid-transport activity in mammalian cells. Biochem J 324:535–541

Brookes N (1988) Neutral amino acid transport in astrocytes: characterization of Na+-dependent and Na+-independent components of alpha-aminoisobutyric acid uptake. J Neurochem 51:1913–1918

Byrne TN (1994) Imaging of gliomas. Semin Oncol 21:162–171

Christensen HN (1990) Role of amino acid transport and countertransport in nutrition and metabolism. Phys Rev 70:43–77

Coenen HH, Kling P, Stöcklin G (1989) Cerebral metabolism of L-[2-^{18}F]fluoro-tyrosine, a new PET tracer of protein synthesis. J Nucl Med 30:1367–1372

Coleman RE, Hoffman JM, Hanson MW et al (1991) Clinical application of PET for the evaluation of brain tumors. J Nucl Med 32:616–622

Cook GJR, Maisey MN, Fogelman I (1999) Normal variants, artefacts and interpretative pitfalls in PET imaging with 18-fluoro-2-deoxyglucose and carbon-11-methionine. Eur J Nucl Med 26:1363–1378

Deehan B, Carnochan P, Trivedi M et al (1993) Uptake and distribution of L-3-[I-125] iodo-alpha-methyl tyrosine in experimental rat tumors: comparison with blood flow and growth rate. Eur J Nucl Med 20:101–106

Del Sole A, Falini A, Ravasi L et al (2001) Anatomical and biochemical investigation of primary brain tumors. Eur J Nucl Med 28:1851–1872

Derlon JM, Bourdet C, Bustany P et al (1989) [^{11}C]L-methionine uptake in gliomas. Neurosurgery 25:720–728

De Wolde H, Pruim J, Mastik MF et al (1997) Proliferative activity in human brain tumors: comparison of histopathology and L-[1-(11)C]tyrosine PET. J Nucl Med 38:1369–1374

Floeth FW, Aulich A, Langen KJ et al (2001) MR imaging and single-photon emission CT findings after gene therapy for human glioblastoma. Am J Neuroradiol 22:1517–1527

Fukasawa Y, Segawa H, Kim JY et al (2000) Identification and characterization of a Na(+)-independent neutral amino acid transporter that associates with the 4F2 heavy chain and exhibits substrate selectivity for small neutral D- and L-amino acids. J Biol Chem 275:9690–9698

Goldman S, Levivier M, Pirotte B et al. (1997) Regional methionine and glucose uptake in high-grade gliomas: a comparative study on PET-guided stereotactic biopsy. J Nucl Med 38:1459-1462

Grosu AL, Weber W, Feldmann HJ et al (2000) First experience with I-123-alpha-methyl-tyrosine SPECT in the 3-D radiation treatment planning of brain gliomas. Int J Radiat Oncol Biol Phys 47:517–526

Guth-Tougelidis B, Muller S, Mehdorn MM et al (1995) Uptake of DL-3-123I-iodo-alpha-methyltyrosine in recurrent brain tumors. Nuklearmedizin 34:71–75

Hamacher K (1999) Synthesis of n.c.a. cis- and trans-4-[^{18}F]fluoro-L-proline, a radiotracer for PET-investigation of disordered matrix protein synthesis. J Labelled Compound Radiopharm 42:1135–1142

Hamacher K, Coenen HH (2002) Efficient routine production of the ^{18}F-labelled amino acid O-(2-[^{18}F]fluoroethyl)-L-tyrosine. Appl Radiat Isotop 57:853–856

Haynes BF, Hemler ME, Mann DL et al (1981) Characterization of a monoclonal antibody (4F2) that binds to human monocytes and to a subset of activated lymphocytes. J Immunol 126:1409–1414

Hemler ME, Strominger JL (1982) Characterization of antigen recognized by the monoclonal antibody (4F2): different molecular forms on human T and B lymphoblastoid cell lines. J Immunol 129:623–628

Herholz K, Hölzer T, Bauer B et al (1998) ^{11}C-methionine PET for differential diagnosis of low-grade-gliomas. Neurology 50:1316–1322

Hubner KF, Purvis JT, Mahaley SM Jr et al (1982) Brain tumor imaging by positron emission computed tomography using 11C-labeled amino acids. Comput Assist Tomogr 6:544–550

Inoue T, Tomiyoshi K, Higuchi T et al (1998) Biodistribution studies on L-3-[fluorine-18]fluoro-alpha-methyl tyrosine: a potential tumor-detecting agent. J Nucl Med 39:663–667

Ishiwata K, Kubota K, Murakami M et al (1993) Re-evaluation of amino acid PET studies: can the protein synthesis rates in brain and tumor tissues be measured in vivo? J Nucl Med 34:1936–1943

Jager PL, Vaalburg W, Pruim J et al (2001) Radiolabeled amino acids: basic aspects and clinical applications in oncology. J Nucl Med 42:432–445

Jansen EP, Dewit LG, van Herk M et al (2000) Target volumes in radiotherapy for high-grade malignant glioma of the brain. Radiother Oncol 56:151–156

Kaim AH, Weber B, Kurrer MO et al (2002) Autoradiographic quantification of 18F-FDG uptake in experimental soft-tissue abscesses in rats. Radiology 223:446–451

Kanai Y, Segawa H, Miyamoto K et al (1998) Expression cloning and characterization of a transporter for large neutral amino acids activated by the heavy chain of 4F2 antigen (CD98). J Biol Chem 273:23629–23632

Kaschten B, Stevenaert A, Sadzot B et al (1998) Preoperative evaluation of 54 gliomas by PET with fluorine-18-fluorodeoxyglucose and/or carbon-11-methionine. J Nucl Med 39:778–785

Kawai K, Fujibayashi Y, Saji H et al (1991) A strategy for the study of cerebral amino acid transport using iodine-123-labeled amino acid radiopharmaceutical: 3-iodo-alpha-methyl-L-tyrosine. J Nucl Med 32:819–824

Kilberg MS, Stevens BR, Novak DA (1993) Recent advances in mammalian amino acid transport. Annu Rev Nutr 13:137–165

Kloss G, Leven M (1979) Accumulation of radioiodinated tyrosine derivatives in the adrenal medulla and in melanomas. Eur J Nucl Med 4:179–186

Kloster G, Bockslaff H (1982) L-3-^{123}I-α-methyltyrosine for melanoma detection: a comparative evaluation. Int J Nucl Med Biol 9:259–269

Kopka K, Riemann B, Friedrich M et al (2001) Characterization of 3-[(123)I]iodo-L-alpha-methyl tyrosine transport in astrocytes of neonatal rats. J Neurochem 76:97–104

Kuwert T, Morgenroth C, Woesler B et al (1996) Uptake of iodine-123-α-methyltyrosine by gliomas and non-neoplastic brain lesions. Eur J Nucl Med 23:1345–1353

Kuwert T, Woesler B, Morgenroth C et al (1998) Diagnosis of recurrent glioma with SPECT and iodine-123-alpha-methyl tyrosine. J Nucl Med 39:23–27

Langen K-J, Weckesser M (1999) Recent advances of PET in the diagnosis of brain tumors. Frontiers in radiation therapy and oncology front. Radiat Ther Onkol 33:9–22

Langen KJ, Coenen HH, Roosen N et al (1990) SPECT studies of brain tumors with L-3-[^{123}I]iodo-α-methyl tyrosine: comparison with PET, ^{124}IMT and first clinical results. J Nucl Med 31:281–286

Langen KJ, Roosen N, Coenen HH et al (1991) Brain and brain tumor uptake of L-3-[^{123}I]iodo-α-methyl tyrosine: competition with natural L-amino acids. J Nucl Med 32:1225–1228

Langen KJ, Ziemons K, Kiwit JCW et al (1997) [^{123}I]iodo-α-methyltyrosine SPECT and [^{11}C]-L-methionine uptake in cerebral gliomas: a comparative Study using SPECT and PET. J Nucl Med 38:517–522

Langen KJ, Clauss RP, Holschbach M et al (1998) Comparison of iodotyrosines and methionine uptake in a rat glioma model. J Nucl Med 39:1596–1599

Langen KJ, Mühlensiepen H, Holschbach M et al (2000) Transport mechanisms of 3-[^{123}I]iodo-α-methyl-L-tyrosine in a human glioma cell line: comparison with [methyl-^{3}H]-L-methionine. J Nucl Med 41:1250–1255

Langen KJ, Bonnie R, Mühlensiepen H et al (2001) 3-[^{123}I]iodo-alpha-methyl-L-tyrosine transport and 4F2 antigen expression in human glioma cells. Nucl Med Biol 28:5–11

Langen KJ, Mühlensiepen H, Schmieder S et al (2002a) Transport of cis- and trans-4-[^{18}F]Fluoro-L-proline in F98 glioma cells. Nucl Med Biol 29:685–692

Langen KJ, Pauleit D, Heinz H et al (2002b) [^{123}I]Iodo-α-methyl-L-tyrosine: uptake mechanisms and clinical applications. Nucl Med Biol 29:625–631

Langstrom B, Antoni G, Gullberg P et al (1987) Synthesis of L- and D-[methyl-^{11}C]methionine. J Nucl Med 28:1037–1040

Leeds NE, Jackson EF (1994) Current imaging techniques for the evaluation of brain neoplasms. Curr Opin Oncol 6:254–261

Levivier M, Wikler D, Goldman S et al (1999) Positron emission tomography in stereotactic conditions as a functional imaging technique for neurosurgical guidance. In: Alexander EI, Maciunas RJ (eds) Advanced neurosurgical navigation. Thieme, New York, pp 85–99

Mastroberardino L, Spindler B, Pfeiffer R et al (1998) Amino-acid transport by heterodimers of 4F2hc/CD98 and members of a permease family. Nature 395:288–291

Mosskin M, Ericson K, Hindmarsh T et al (1989) Positron emission tomography compared with magnetic resonance imaging and computed tomography in supratentorial gliomas using multiple stereotactic biopsies as reference. Acta Radiol 30:225–232

Ogawa T, Kanno I, Shishido F et al (1991) Clinical values of PET with ^{18}F-fluorodeoxyglucose and L-methyl-^{11}C-methionine for diagnosis of recurrent brain tumor and radiation injury. Acta Radiol 31:197–202

Ogawa T, Shishido F, Kanno I et al (1993) Cerebral glioma: evaluation with methionine PET. Radiology 186:45–53

Ogawa T, Miura S, Murakami M et al (1996) Quantitative evaluation of neutral amino acid transport in cerebral gliomas using positron emission tomography and fluorine-18 fluorophenylalanine. Eur J Nucl Med 23:889–895

Oldendorf WH (1991) Saturation of amino acid uptake by human brain tumor demonstrated by SPECT. J Nucl Med 32:1229–1230

Pardridge WM, Oldendorf WH (1977) Transport of metabolic substrates through the blood-brain barrier. J Neurochem 28:5–12

Pardridge WM, Oldendorf WH, Cancilla P et al (1986) Blood-brain barrier: interface between internal medicine and the brain. Ann Intern Med 105:82–95

Pauleit D, Langen KJ, Floeth F et al (2002) Improved delineation of the tumor extension using F18-FET PET compared with MRI in cerebral gliomas? (Abstract SNM meeting) J Nucl Med 43:112P

Pfeiffer R, Rossier G, Spindler B et al (1999) Amino acid transport of y$^+$L-type by heterodimers of 4F2hc/CD98 and members of the glycoprotein-associated amino acid transporter family. EMBO J 18:49–57

Pineda M, Fernandez E, Torrents D et al (1999) Identification of a membrane protein, LAT-2, that co-expresses with 4F2 heavy chain, an L-type amino acid transport activity with broad specificity for small and large zwitterionic amino acids. J Biol Chem 274:19738–19744

Pruim J, Willemsen AT, Molenaar WM et al (1995) Brain tumors: L-[1-C-11]tyrosine PET for visualization and quantification of protein synthesis rate. Radiology 197:221–226

Ribom D, Eriksson A, Hartman M et al (2001) Positron emission tomography (11)C-methionine and survival in patients with low-grade gliomas. Cancer 92:1541–1549

Riemann B, Stogbauer F, Kopka K et al (1999) Kinetics of 3-[(123)I]iodo-L-alpha-methyltyrosine transport in rat C6 glioma cells. Eur J Nucl Med 26:1274–1278

Riemann B, Kopka K, Stogbauer F et al (2001) Kinetic parameters of 3-[(123)I]iodo-L-alpha-methyl tyrosine ([(123)I]IMT) transport in human GOS3 glioma cells. Nucl Med Biol 28:293–297

Saier MH Jr, Daniels GA, Boerner P et al (1988) Neutral amino acid transport systems in animal cells: potential targets of oncogene action and regulators of cellular growth. J Membr Biol 104:1–20

Sato H, Tamba M, Ishii T et al (1999) Cloning and expression of a plasma membrane cystine/glutamate exchange transporter composed of two distinct proteins. J Biol Chem 274:11455–11458

Schmidt D, Gottwald U, Langen KJ et al (2001) 3-[^{123}I]Iodo-α-methyl-L-tyrosine uptake in cerebral gliomas: relationship to histopathological grading and prognosis. Eur J Nucl Med 28:855–861

Segawa H, Fukusawa Y, Miyamoto K et al (1999) Identification and functional characterization of a Na$^+$-independent neutral amino acid transporter with broad substrate selectivity. J Biol Chem 274:19745–19751

Shoup TM, Olson J, Hoffman JM et al (1999) Synthesis and evaluation of [^{18}F]1-amino-3-fluorocyclobutane-1-carboxylic acid to image brain tumors. J Nucl Med 40:331–338

Teixeira S, Kuhn LC (1991) Post-transcriptional regulation of the transferrin receptor and 4F2 antigen heavy chain mRNA during growth activation of spleen cells. Eur J Biochem 202:819–826

Tisljar U, Kloster G, Ritzl F et al (1979) Accumulation of radioiodinated alpha -methyltyrosine in pancreas of mice: concise communication. J Nucl Med 20:973–976

Torrents D, Estevez R, Pineda M et al (1998) Identification and characterization of a membrane protein (y+L amino acid transporter-1) that associates with 4F2hc to encode the amino acid transport activity y+L. A candidate gene for lysinuric protein intolerance. J Biol Chem 273:32437–32445

Vaalburg W, Coenen HH, Crouzel C et al (1992) Amino acids for the measurement of protein synthesis in vivo by PET. Int J Rad Appl Instrum B 19:227–237

Weber W, Bartenstein P, Gross MW et al (1997) Fluorine-18-FDG PET and iodine-123-IMT SPECT in the evaluation of brain tumors. J Nucl Med 38:802–808

Weber WA, Dick S, Reidl G et al (2000) Correlation between postoperative 3-[(123)I]iodo-L-alpha-methyltyrosine uptake and survival in patients with gliomas. J Nucl Med 42:1144–1150

Weber WA, Wester HJ, Grosu AL et al (2000) O-(2-[^{18}F] Fluoroethyl)-L-tyrosine and L-[methyl-^{11}C]methionine uptake in brain tumours: initial results of a comparative study. Eur J Nucl Med 27:542–549

Weckesser M, Matheja P, Rickert CH et al (2000) Evaluation of the extension of cerebral gliomas by scintigraphy. Strahlenther Onkol 176:180–185

Weckesser M, Matheja P, Schwarzrock A et al (2002) Prognostic significance of amino acid transport imaging in patients with brain tumors. Neurosurgery 50:958–964

Wester HJ, Herz M, Weber W, Heiss P, Senekowitsch-Schmidtke R, Schwaiger M, Stöcklin G (1999) Synthesis and radiopharmacology of O-(2-[^{18}F]fluoroethyl)-L-tyrosine for tumor imaging. J Nucl Med 40:205–212

Wienhard K, Herholz K, Coenen HH et al (1991) Increased amino acid transport into brain tumors measured by PET of L-[2-^{18}F]fluoro-tyrosine. J Nucl Med 32:1338–1346

Woesler B, Kuwert T, Morgenroth C et al (1997) Non-invasive grading of primary brain tumors: results of a comparative study between SPET with 123I-alpha-methyl tyrosine and PET with 18F-deoxyglucose. Eur J Nucl Med 24:428–434

Würker M, Herholz K, Voges J et al (1996) Glucose consumption and methionine uptake in low-grade gliomas after iodine-125 brachytherapy. Eur J Nucl Med 23:583–586

CHAPTER 22

Lung

22

TOYOHARU ISAWA

Contents

22.1
Pulmonary Nuclear Medicine

The lung functions as a gas exchanger of oxygen and carbon dioxide, which we call respiration. Respiration consists of perfusion, ventilation, and diffusion (Fig. 22.1). Oxygen delivered to the peripheral tissues and cells, including the cells in the lungs, is mainly utilized through mitochondrial cytochrome oxidase to generate ATP (Fisher and Forman 1985). In addition to its respiratory function, the lung is responsible for such nonrespiratory functions as mucociliary clearance (an important body defense), maintenance of pulmonary epithelial permeability, uptake of various substances, cellular upkeep metabolism, and so on (Fig. 22.2).

This chapter will describe nuclear medicine techniques, nuclear imaging in particular, in the elucidation of lung function and its diagnostic applications.

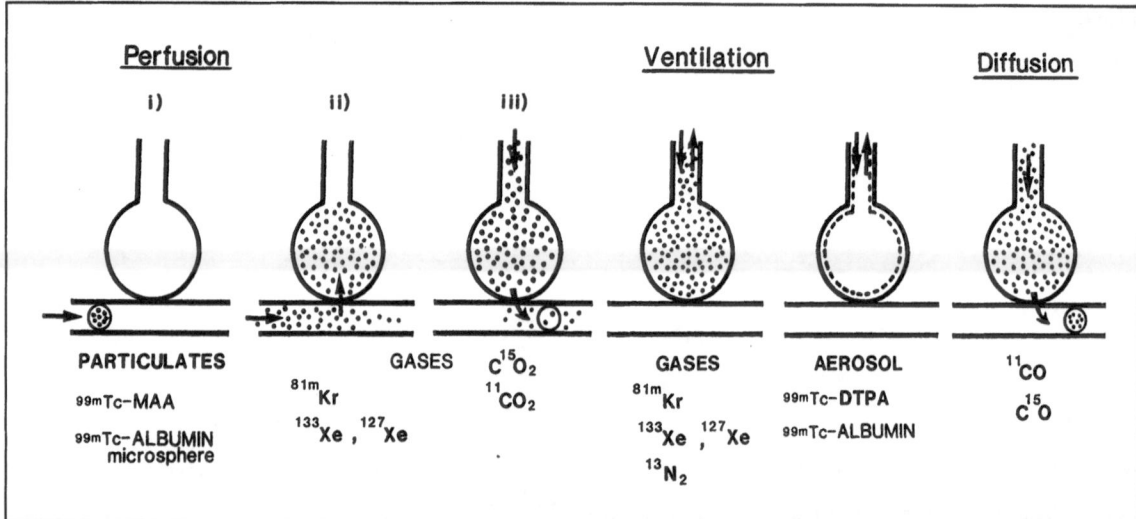

Fig. 22.1. Respiratory lung function. Perfusion in the lung can be studied with particulates such as ^{99m}Tc-MAA, or with radioactive gases such as dissolved ^{133}Xe or ^{81m}Kr, or with positron emitters. Ventilation can be investigated with radioactive gases or radioaerosols such as ^{99m}Tc-human serum albumin or ^{99m}Tc-DTPA, and diffusion by carbon monoxide labeled with positron emitters

22.2
Lung Function in Respiration

22.2.1
Ventilation

22.2.1.1
Airways

The airways are passages for air ventilation for gas exchange. Gas exchange takes place in the terminal respiratory units, the alveoli. The major function of gas exchange is to oxygenize the mixed venous blood. A volume of oxygen equal to that used by the body must be supplied continuously to the alveoli and (and thence to the pulmonary capillary beds) by ventilation; a volume of carbon dioxide nearly equal to that of the entire lung must be continuously removed by ventilation.

22.2.1.1.1
Anatomic Dead Space

The extrapulmonary airways in the thorax are the trachea, carina, and proximal part of the main bronchi. Following the dichotomous (less frequently, trichotomous) branching of the intrapulmonary airways, the airways become narrower, shorter, and more numerous as they penetrate deeply into the lung unit until they become terminal bronchioles. Characteristically, the airways proximal to the terminal bronchioles are ciliated and reinforced by the cartilage. According to Weibel, branching is repeated an average of 16 times before the airway becomes a terminal bronchiole (Weibel 1963). The airways lead the inspired gas to the gas-exchanging units of the lung distal to the terminal bronchiole. Since no gas exchange takes place in airways proximal to the terminal bronchioles, these conduits represent anatomic dead space (Weibel 1963).

22.2.1.1.2
Acinus or Respiratory Zone

The terminal bronchiole divides into respiratory bronchioles that have alveoli on their walls. After the respiratory bronchioles branch several times, the alveolar ducts are reached, which are surrounded by the alveolar sacs, consisting of the alveoli. The acinus, that is, the transitional or respiratory zone, consists of all of the lung structure distal to and subtended from the terminal bronchiole, namely, the respiratory bronchioles, alveolar ducts, alveolar sacs, and alveoli (Groskin 1993).

The acinus can be as large as 8.5 mm; the distance from the terminal bronchiole to the distal acinar edge is approximately 5–10 mm (Groskin 1993). Roentgenologically the acinus is considered as a functional or respiratory unit, in contrast to the conductive airways proximal to the terminal bronchiole.

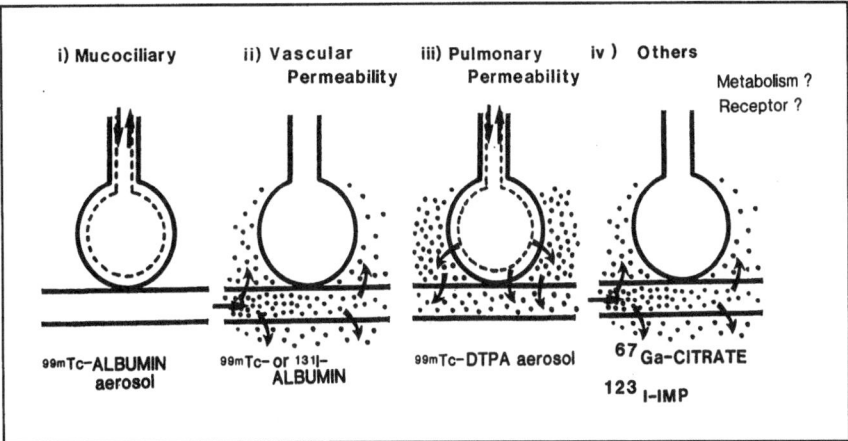

Fig. 22.2. Nonrespiratory lung function. Mucociliary clearance can be evaluated by inhaled radioaerosols such as ^{99m}Tc-human serum albumin, vascular endothelial permeability by injecting ^{99m}Tc-human serum albumin, pulmonary epithelial permeability by inhaling ^{99m}Tc-DTPA aerosols, and other nonrespiratory functions by injected or inhaled radioactive substances like ^{123}I-IMP

Although the distance from the terminal bronchiole to the most distal alveolus is very short, the respiratory zone makes up most of the lung volume (total lung volume minus the anatomic dead space). When the cross-sectional area of the airways of each branching (generation) is calculated, there is relatively little change found before the terminal bronchioles (Weibel 1963). As this level is reached the cross-sectional area increases very rapidly, West et al. (1964) compare the shape as the airway becomes acinus to that of a trumpet, or even a thumbtack. Because of this rapid change of cross-sectional area, the convective or bulk flow in the conductive zone of the airways becomes diffusive in the neighborhood of the terminal bronchioles.

This means that inhaled aerosols reaching the terminal bronchioles in convective flow tend to become deposited at or near the terminal bronchioles. Non-diffusible suspended or particulate matter fails to reach the alveolar walls and is expelled in the subsequent expiration (Altshuler et al. 1959). This is why it is difficult to deposit radioactive particulates or aerosols on the alveolar walls, and why large inspired volumes and breath holding are important for obtaining efficient alveolar deposition of inhaled radioaerosols, especially of aerosols smaller than 1 micrometer (μm) or micron (μ).

22.2.1.1.3
Cilia

Ciliated epithelial cells occur at all levels in the trachea, bronchi, and bronchioles, all the way down to the respiratory bronchioles, but are not found in the alveoli. In the smaller bronchioles ciliated cells become less frequent and their cilia shorter (Greenwood and Holland 1972). Mucus-secreting goblet cells are abundant in the epithelium of the trachea and larger bronchial airways but also become less frequent in the smaller airways, disappearing at the level of or distal to the terminal bronchioles in normal subjects. The respiratory bronchioles lack mucus-secreting cells but have other secretory cells that are believed to produce a serous secretion; this watery fluid was shown by Gil and Weibel in electron microscopic studies (1971) to be coated at the liquid-air interface with an osmophilic film that they believe to be of the same nature as the surfactant film coating the alveolar surface, but thicker.

22.2.1.2
Dead Spaces

From functional standpoint, there is a portion of lung that is wasted, does not participate in gas exchange. This is called dead space, and may be divided into the anatomic and the physiological.

Anatomic dead space is simply the volume of the conducting airways, as described above, and it increases with the size of the lung or the functional residual capacity (FRC). The portion of inspired air that traverses the airways to enter the alveoli is called alveolar ventilation, measured in volume per minute. Not all of the alveolar ventilation is equally effective in oxygenizing mixed venous blood.

Inspired air that enters alveoli lacking a pulmonary capillary blood supply is also wasted and is called "alveolar dead space." When the inspired air enters alveoli with reduced blood supply, a corresponding portion of the inspired gas is likewise wasted, and is again considered "alveolar dead space." While the anatomic dead space is determined by the anatomy of the lung, determining physiological dead space requires a functional measurement of the lung's efficiency in eliminating carbon dioxide. This can be calculated by the following equation (rewritten Bohr's equation):

$$V_D/V_T = (Pa_{CO_2} - PE_{CO_2})/Pa_{CO_2}$$

V_D refers to dead space volume, Pa_{CO_2}, carbon dioxide tension of the arterial blood, and PE_{CO_2}, carbon dioxide tension in the expired gas. Here we assume that Pa_{CO_2} is equal to the P_{CO_2} of alveolar gas (Forster et al. 1986).

When the lung is normal, physiological dead space is virtually the equivalent of anatomic dead space. Physiological dead space is increased in the presence of ventilation-perfusion inequality. The concept of physiological dead space is important in understanding mismatching of ventilation and perfusion in the lung.

22.2.1.3
Inequality of Ventilation

Measurements of ventilation distribution in the upright position show more ventilation near the lung bases than above, while those in the supine position show less difference between the lung apex and the base (West 1965). In the lateral decubitus position the lower lung has more ventilation than the upper counterpart (Isawa et al. 1969).

Why does this inequality occur? Because of the lung's own weight the intrapleural pressure becomes more negative toward the lung apex in the upright position, and the transpulmonary pressure, or the pressure to expand the lung, increases toward the lung apex and decreases near the lung base. An analogy can be found in a coiling spring. When the spring is hung from the ceiling, the distance between the coils becomes greater in the upper part of the spring than the lower part due to the weight of the spring itself. When an additional weight is applied at the lower end of the spring, the distance between the coils in the lower part of the spring also becomes larger. This type of inequality of ventilation is thus due to a gravitational effect (Milic-Emili et al. 1966). Further reasons for inequality in lung ventilation include uneven time constants (Otis et al. 1956), asymmetry of the structure of the small lung units (Engel 1978), series inequality (Cumming et al. 1966), and closing volume (Sutherland et al. 1968).

22.2.1.4
Effect of Perfusion Blockade on Ventilation

It has been believed that once the pulmonary circulation is blocked, ventilation is decreased in the lung region distal to the vascular blockade due to hypoventilation induced by a shift of ventilation to the

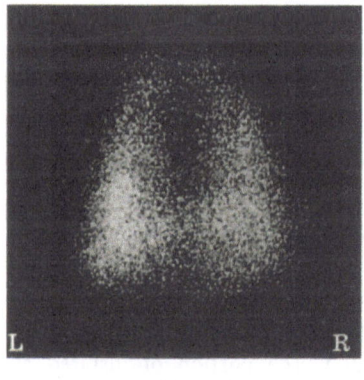

Perfusion scan
(control)

Dog 1571

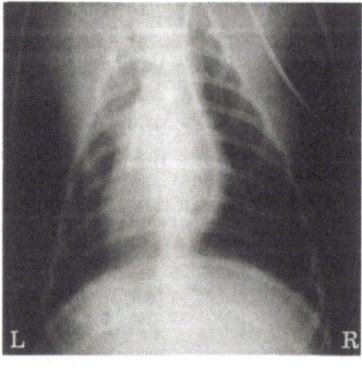

Chest X-ray

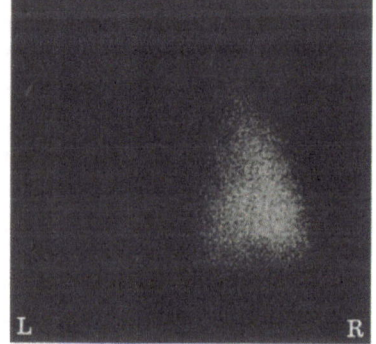

Perfusion scan
(after left p.a. occlusion)

Fig. 22.3. *Left*: perfusion lung image (posterior) of a dog (#1571) in the prone position. *Middle*: chest x-ray of same dog with a balloon catheter in the left main pulmonary artery. *Right*: perfusion lung image (posterior) immediately after left pulmonary artery occlusion by inflating the balloon

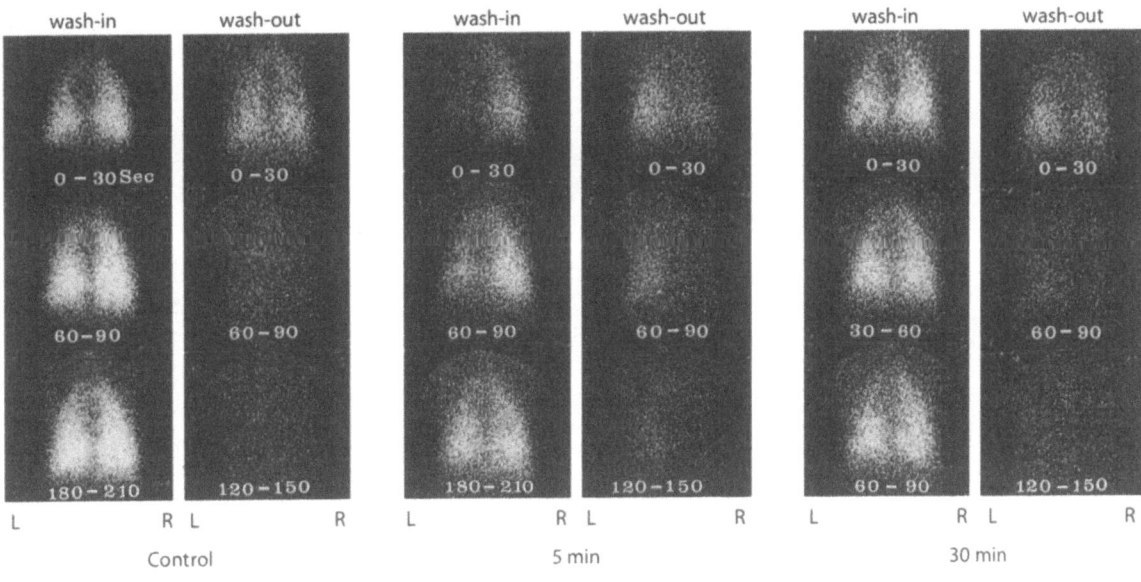

| wash-in | wash-out | | wash-in | wash-out | | wash-in | wash-out |

Dog 1571   (8% CO₂ in O₂)

Fig. 22.4. *Left panel*: ^{133}Xe gas inhalation studies (wash-in and washout phases) of the same dog (#1571) during control period. *Middle panel*: ^{133}Xe gas inhalation studies (wash-in and washout phases) of the same dog (#1571) at 5 min following occlusion of the left pulmonary artery. Note hypoventilation of the left lung during wash-in phase and retention of ^{133}Xe gas in the left lung during washout phase with air, indicating the presence of bronchoconstriction of the left lung. *Right panel*: when 8% carbon dioxide in oxygen is inhaled with ^{133}Xe gas, hypoventilation of the left lung is not manifest because of hyperventilation due to carbon dioxide inhalation, but washout of inhaled ^{133}Xe gas is still delayed in the left lung as compared with that in the right lung, indicating bronchoconstriction is not completely recovered

healthy lung regions (Venrath et al. 1952; Severinghaus et al. 1961; Swenson et al. 1961; Thomas et al. 1964). When a unilateral pulmonary artery of a dog is mechanically blocked by a balloon catheter, ^{133}Xe gas inhalation studies show that ventilation of the perfusion-blocked lung is reduced and that washout of inhaled ^{133}Xe gas from the perfusion-blocked lung is delayed, indicating that bronchoconstriction occurs there (Figs. 22.3, 22.4). Severinghaus and others have described hypoventilation and bronchoconstriction in animal and human studies (Severinghaus et al. 1961; Swenson et al. 1961). They explain that these phenomena occur because the alveolar tension of the bronchodilator carbon dioxide is decreased. When 8% CO₂ in pure oxygen was inhaled, hypoventilation was somewhat ameliorated, probably due to stimulation of the respiratory center by hyperventilation-inducing carbon dioxide. Bronchoconstriction persisted, although to a lesser degree, under the condition of 8%-CO₂-in-O₂ inhalation in our study (Fig. 22.4; Isawa et al. 1972).

When the blockade lasts longer than 6 h in dogs, however, both hypoventilation and bronchoconstriction disappear. Normal ventilation and normal washout of inhaled ^{133}Xe gas are observed in the perfusion-blocked lung as shown in Figs. 22.5, 22.6, and 22.7. As shown in Figs. 22.6 and 22.7, no evidence either of hypoventilation or bronchoconstriction exists in the left lung when the perfusion blockade lasts for longer than 6 h. Bronchoconstriction following perfusion blockade thus seems to be a transient, time-dependent phenomenon (Isawa et al. 1972).

22.2.2
Perfusion (Pulmonary Arterial Blood Flow)

Exchange of oxygen and carbon dioxide occurs in the lung as ventilation and perfusion (pulmonary arterial blood flow) meet. The perfusion distribution in the lung can be easily assessed using radioactive particulates such as ^{99m}Tc-labeled macroaggregated human serum albumin (MAA) or dissolved radioactive gases such as ^{133}Xe or ^{81m}Kr.

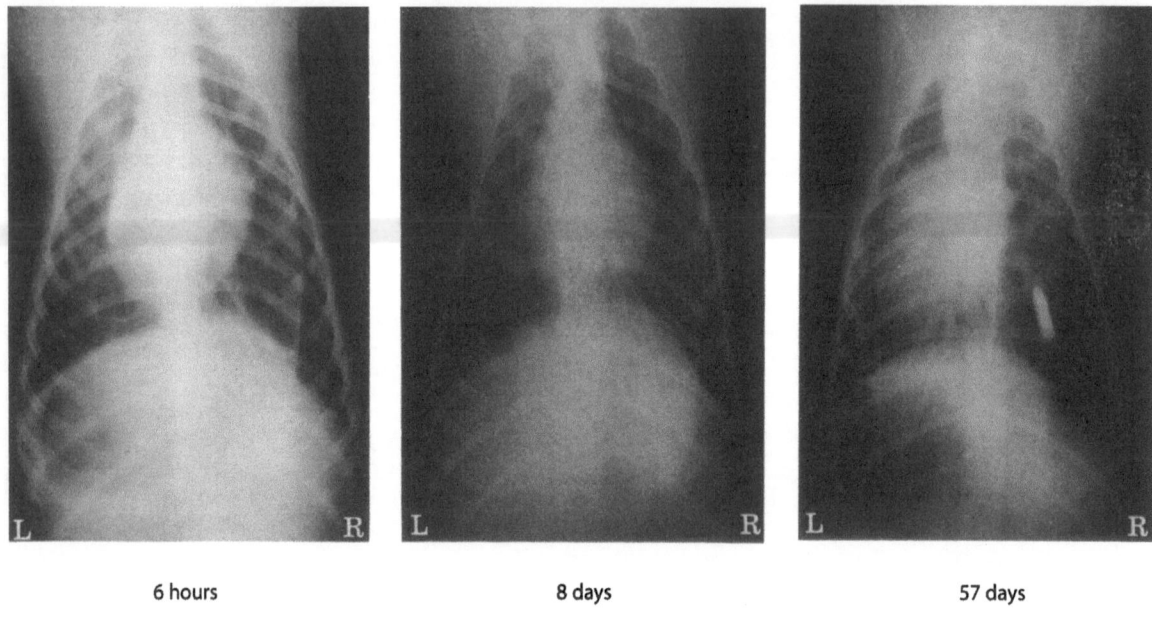

6 hours   8 days   57 days

Dog 658

Fig. 22.5. Chest x-rays of a normal dog (#658). *Left*: at 6 h following perfusion blockade of the left lung by inserting a latex balloon filled with contrast medium into the left main pulmonary artery. *Middle*: when the balloon was pulled up from the left pulmonary artery to the jugular vein on the 8th day. *Right*: for some reason the string attached to the balloon was severed and the balloon strayed into the right lower lobe pulmonary artery on the 19th day. The chest x-ray was taken on the 57th day just before sacrifice

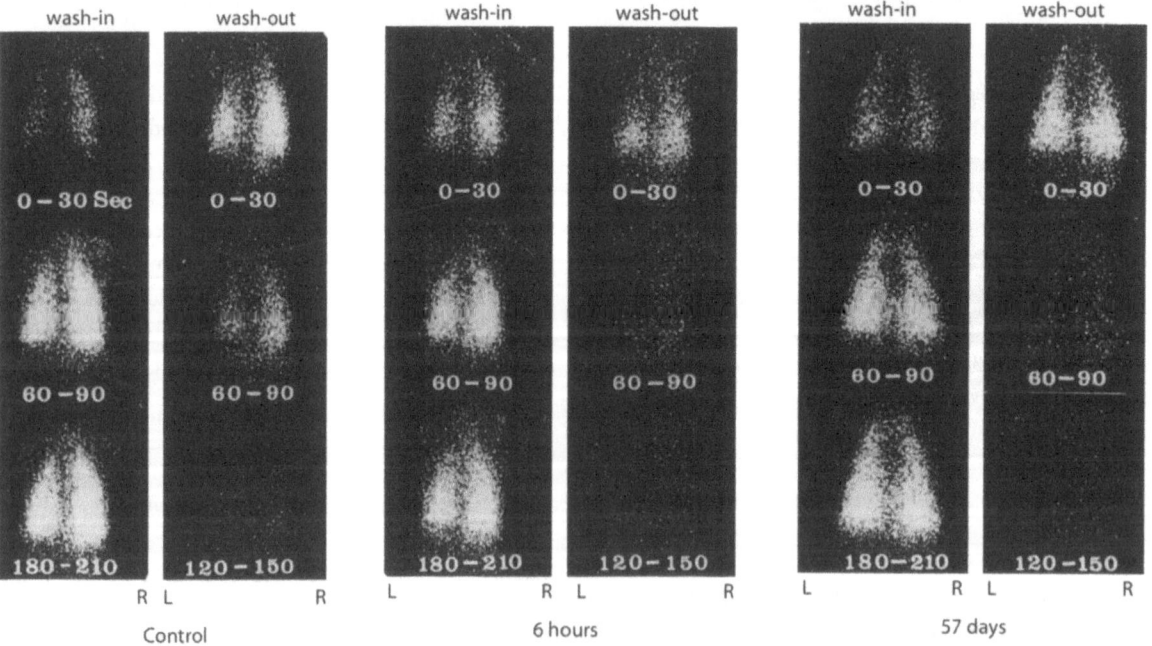

Control   6 hours   57 days

Dog 658

Fig. 22.6. Xenon-133 gas inhalation studies of the same dog (#658). *Left panel*: control period. Wash-in and washout of ^{133}Xe gas in the right and left lungs. *Middle panel*: at 6 h after occlusion of the left pulmonary artery by inserting the contrast-medium filled balloon. There is no evidence either of hypoventilation or bronchoconstriction in the left lung as compared with the right lung. *Right panel*: on the 57th day after occlusion of the left pulmonary artery by inserting the contrast medium-filled balloon. There is no evidence either of hypoventilation or bronchoconstriction in the left lung as compared with the right lung

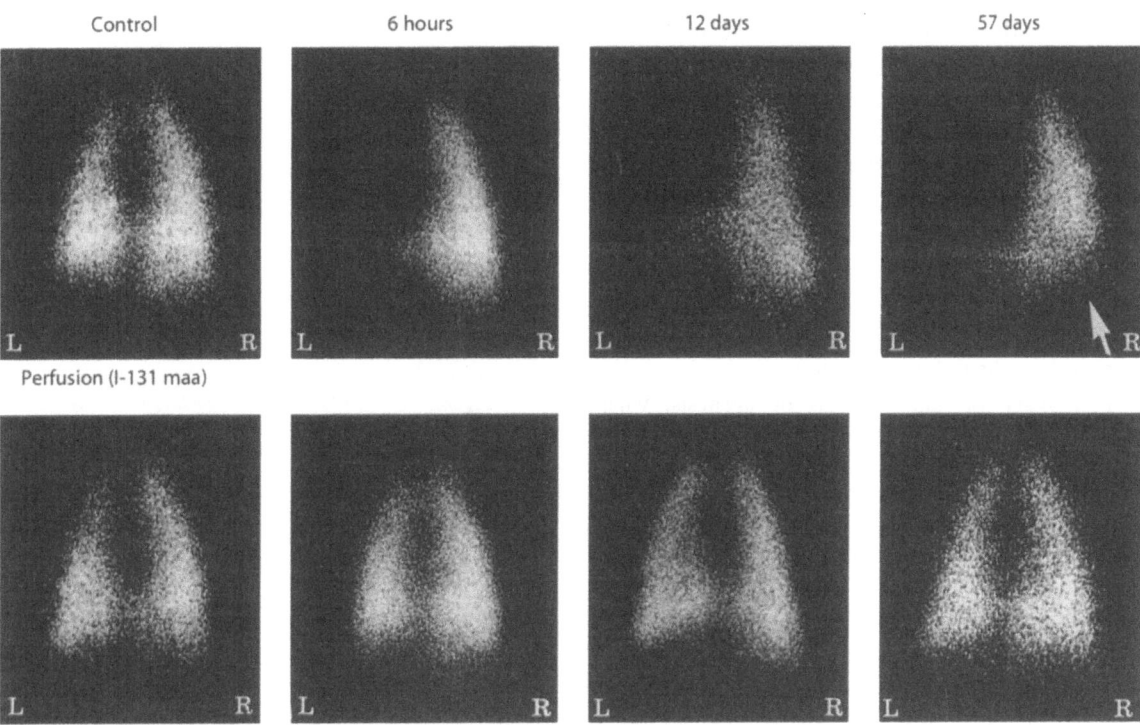

Control   6 hours   12 days   57 days

L   R   L   R   L   R   L   R

Perfusion (I-131 maa)

L   R   L   R   L   R   L   R

Ventilation (Tc-99m Albumin Aerosol)

Dog 658

Fig. 22.7. *Upper panel*: perfusion lung images before, at 6 h, and on the 12th and 57th day after occlusion of the left pulmonary artery by inserting the contrast medium-filled balloon. Perfusion in the left lung is absent at 6 h after occlusion. Perfusion in the left lung is also absent on the 12th day even though the balloon was removed from the left pulmonary artery on the 8th day. On the 57th day perfusion in the right lower lobe was also absent *Lower panel*: Corresponding ventilation in the left lung is not affected at 6 hr or on 12 and 57 days by occlusion of the left pulmonary artery

22.2.2.1
Gravitational Effect

Because the pulmonary circulation is a low-pressure system, with systolic pressure normally being 25 mmHg, diastolic pressure 8 mmHg, and mean pressure 15 mmHg (Criley and Ross 1971), the distribution of the pulmonary arterial blood flow is affected by body posture. The adult lung is approximately 30 cm long, and a hydrostatic pressure between the lung apex and the base is 30 cm blood, or 23 mmHg.

Because the lung resembles a manometer, when a tracer is injected with the subject in the upright position, the distribution of radioactivity represents the perfusion distribution in the lung at the same upright position. When the tracer is injected in the supine position, the distribution of radioactivity represents the perfusion distribution in the lung in that position. Due to the hydrostatic pressure difference when upright, the lower lung shows greater perfusion per unit lung volume than does the upper lung, while in the supine position there is only but difference in the perfusion distribution. This is the "gravitational effect" on perfusion in the lung.

Topographical relationships between alveolar pressure (P_A), pulmonary arterial pressure (P_a), and pulmonary venous pressure (P_v) in the upright lung are well explained by West's zone theory (West 1965). In the lowermost zone (zone 4) interstitial pressure (P_I) plays an important role in the regulation of perfusion (Hughes et al. 1967). The reversal of perfusion in mitral valvular diseases is considered to be due to the increase in the interstitial pressure.

22.2.2.2
Pathologic Factors Affecting Perfusion Distribution in the Lung

Pathological factors that affect perfusion distribution in the lungs are:

- Vascular obstruction such as pulmonary embolism where perfusion distal to the vascular obstruction is either absent or extremely diminished
- Parenchymal lung diseases such as pneumonia, pulmonary tuberculosis, lung abscess, pulmonary interstitial fibrosis, and cyst and bulla formation where local perfusion is decreased
- Vascular compression and/or invasion by a tumor such as bronchogenic carcinoma, especially a hilar type of carcinoma which can reduce the blood flow to an entire lung by compressing a main branch of the pulmonary artery
- Pulmonary vascular stenosis or agenesis, where perfusion distally is diminished or absent.

22.2.2.3
Effect of Alveolar Hypoxia on Regional Perfusion

In addition to the above pathological factors, the physiological factor of alveolar hypoxia (low oxygen tension in the alveoli), mostly associated with ventilatory disturbance, plays a crucial role:

- In obstructive airway diseases such as pulmonary emphysema, bronchitis, bronchial asthma, panbronchiolitis, and bronchiectasis where regional perfusion is decreased
- In bronchial obstruction due to intraluminal tumors or foreign substances such as a swallowed foreign body, leaving perfusion distal to the obstruction diminished or absent
- Experimentally when one lung (the left) is given a hypoxic gas the oxygen concentration of which is lower than that of the air, while the other is given 100% oxygen; it is observed that the more hypoxic the inspired gas, the less the perfusion in the entire left lung (Fig. 22.8), while 8% CO_2 in air given to that lung did not induce any difference in perfusion distribution (Isawa et al. 1967)

To study the effect of regional oxygen concentration on regional perfusion distribution, the right upper lobe (RUL) of a dog was isolated in vivo by a balloon catheter, as shown in Fig. 22.9, and artificially ventilated with nitrogen 10% O_2 in N_2, air, 40% O_2 in N_2, 60% O_2 in N_2, 100% O_2, and 10% CO_2 in air, while the rest of the right lung and the left lung maintained spontaneous breathing of ambient air. The regional perfusion ratio in the RUL significantly decreased, when it was artificially ventilated with hypoxic gas mixtures like N_2 and 10% O_2 in N_2. Perfusion ratios increased with increasing O_2 concentration in the gas

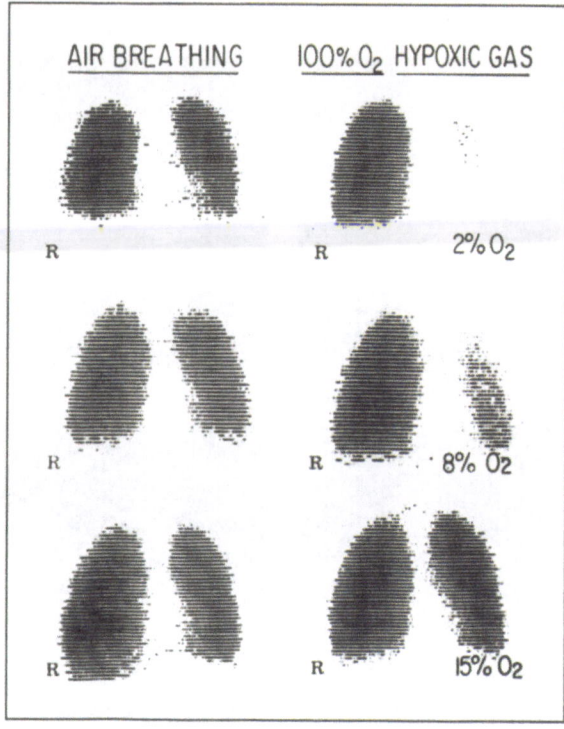

Fig. 22.8. *Upper panel: Left*: perfusion lung image when the right and left lungs are breathing air. *Right*: the right lung is breathing 100% oxygen, while the left lung is breathing 2% oxygen in nitrogen. *Middle panel: Left*: perfusion lung image when the right and left lungs are breathing air. *Right*: the right lung is breathing 100% oxygen, while the left lung is breathing 8% oxygen in nitrogen. *Lower panel: Left*: perfusion lung image when the right and left lungs are breathing air. *Right*: the right lung is breathing 100% oxygen, while the left lung is breathing 15% oxygen in nitrogen

mixtures but, interestingly, perfusion ratios were similar to those seen when the RUL spontaneously breathed air without using the balloon catheter. For comparison, Fig. 22.10 illustrates the regional perfusion ratios in the RUL of the normal and reimplanted, denervated lungs (the right lung was surgically removed once and, denervated, reimplanted.) Patterns of regional perfusion change in the RUL were similar in the normal and the reimplanted, denervated lung. Hypoxic gas mixtures induced regionally diminished perfusion, while hyperoxic gas induced increased regional perfusion distribution. The perfusion partition to the right lung as a whole, however, showed few, if any, changes as O_2 concentration varied in the exchange gas mixture for the RUL, indicating shifting of perfusion from the RUL to the remainder of the right lung, and vice versa. These results indicate that regional hypoxia induces regional

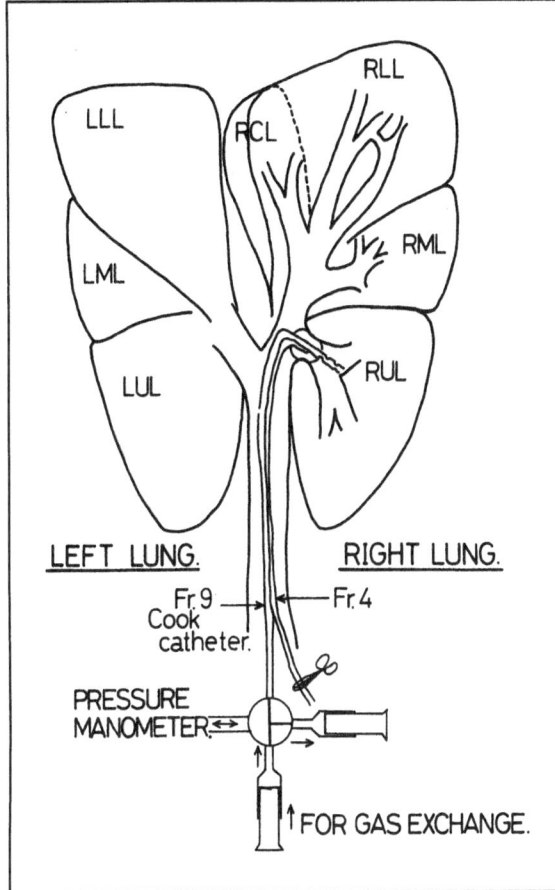

Fig. 22.9. The RUL bronchus of a dog is artificially ventilated through a balloon catheter with a gas of different oxygen concentration, while the remainder of the lungs is breathing oxygen

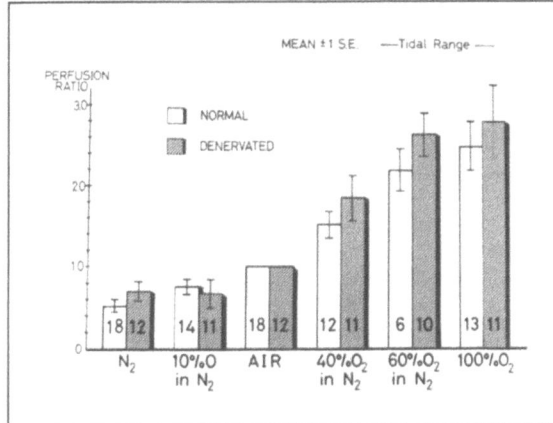

Fig. 22.10. Relative perfusion ratios in the RUL of the normal right lung (*not shaded*) and the reimplanted (denervated) right lung (*shaded*) are compared when the RUL is artificially ventilated with gas mixtures as indicated on the *x*-axis. The perfusion ratios are not significantly different between the reimplanted, denervated lungs and the normal lungs

pulmonary vasoconstriction, and that regional hyperoxia encourages recruitment of the regional pulmonary vascular beds. The O_2 sensitivity of the pulmonary vascular beds was found to decrease as the O_2 concentration in the exchange gas exceeds 60%, or the alveolar oxygen tension exceeds 250 mmHg. Pulmonary vascular responses to alveolar hypoxia and hyperoxia seem to be purely local phenomena, and nervous integrity is not essential (Isawa et al. 1978a, b).

What causes hypoxic pulmonary vasoconstriction? Theories have been proposed: (1) hypoxia acts directly to constrict the smooth muscle of the vascular beds (Staub 1963; Bergofsky and Holtzman 1967); (2) contractile interstitial cells (Kapanci et al. 1974) or neuroendocrine cells (Pack and Weddicombe 1984) act upon sensing alveolar hypoxia to induce vasoconstriction. Pulmonary vascular endothelium secretes various substances including vasoconstricting or vasodilating substances in response to various stimuli. The vasodilating substance has been called endothelium-derived relaxing factor, EDRF, identified as nitric oxide (NO)(Luscher and Vanhoutte 1988), capable of reversing hypoxic pulmonary vasoconstriction (Sprague et al. 1992; Frostell et al. 1993). Endothelium-derived endothelin 1 is a potent vasoconstricting substance (Hemsen et al. 1990). NO is considered to restrict the production and release of endothelin from the endothelium (Boulanger and Luscher 1990).

It is reported that mast cells are involved in pulmonary vasoconstriction by secreting histamine and serotonin (Haas and Bergofsky 1972). Involvement of leukotriene has also been suggested (Morgenroth et al. 1984). Hypoxic pulmonary vasoconstriction has been found to occur not only in the pulmonary arterial side but also in the pulmonary veins (Westbrook et al. 1969).

22.2.2.4
Effect of Alveolar Hypercapnia (High CO_2 Tension) on Regional Perfusion

Alveolar hypercapnia reduced regional perfusion only in the presence of extreme alveolar hyperoxia in the normal dog. In the reimplanted denervated lung regional perfusion increases even in the presence of hypercapnia under alveolar hyperoxia (Isawa et al. 1980b). The reason is not clear.

22.2.2.5
Effect of Ventilation Blockade by Bronchial Occlusion on Regional Perfusion

When the dog's right and left lungs were divided by a tracheal divider and breathed air, one side was occluded at the end of normal inspiration to maintain as much lung volume on that side as physiologically possible. After a given period of time, a suspension of either ^{99m}Tc- or ^{131}I-MAA was injected intravenously. Three minutes after injection, bronchial occlusion was released, and the complete separation of the right and left lungs confirmed.

Two of the three dogs subjected to 30 s right bronchial occlusion showed manifestly decreased perfusion in the occluded lung as compared to control values, indicating that perfusion adjustment to bronchial occlusion begins as soon as airflow stops and/or diminution in alveolar O_2 tension takes place (true in the reimplanted, denervated lung as well as in the intact dog lung). In all dogs whose unilateral bronchus was occluded for one minute or longer there was a marked decrease in perfusion on the occluded side. The gradual decrease in perfusion to the occluded right lung is clearly visualized by the serial images shown in Fig. 22.11. All dogs with reimplantation showed decreased perfusion in the reimplanted lobes or lungs and further decrease followed bronchial oc-

clusion. When the normal right lung bronchus of a dog was occluded, there was a slight decrease in perfusion and the dog became slightly hyperpneic and cyanotic. All these results seem to suggest that hypoxic pulmonary vasoconstriction occurs regionally in the lung tissue without requiring nervous system integrity (Isawa et al. 1971a).

The clinical implications derived from these studies are that obstructive airway disorders associated with localized ventilatory impairment can easily cause regional pulmonary ischemia. Localized ventilatory impairment can be induced by many conditions, e.g., conditions of increased bronchial secretions such as bronchitis, bronchial spasm as in bronchial asthma, mucosal edema as in acute and chronic respiratory infections, bronchial collapse as in pulmonary emphysema, etc. It appears that under these conditions reduced regional perfusion can occur promptly.

22.2.2.6
Effect of Alveolar Pressure on Regional Perfusion

When the RUL of a dog is ventilated through a balloon catheter under increased alveolar pressure, the regional perfusion decreases as the alveolar pressure increases. Regional perfusion was least at a maximum alveolar pressure of 14 to 19 cmH$_2$O, while it was the greatest at the tidal maximal alveolar pressure of 1 to

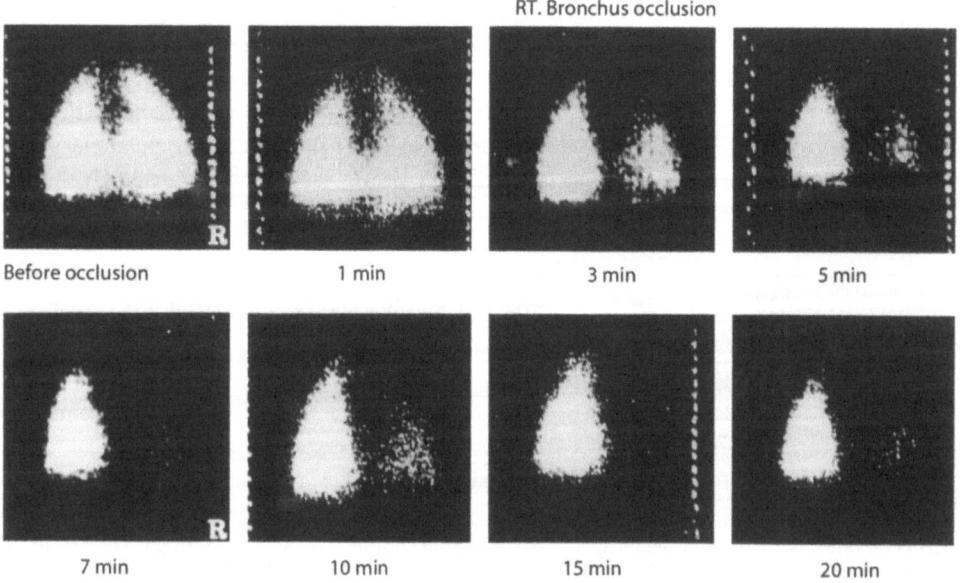

RT. Bronchus occlusion

Before occlusion   1 min   3 min   5 min

7 min   10 min   15 min   20 min

Fig. 22.11. When ventilation of the right lung of a normal dog (no. 754, posterior view) is blocked while the left lung is spontaneously breathing air, perfusion in the ventilation-blocked lung is diminished. The *indicated time* below each image is the duration of ventilation blockade

−1 cmH$_2$O when the RUL was artificially ventilated either with air or N$_2$ (Isawa et al. 1978a). Regional pulmonary vasculature in low-pressure system thus appears to be compressed by increased alveolar pressure (Whittenberger et al. 1960).

22.2.2.7
Pharmacological Effects on Regional Perfusion

22.2.2.7.1
Aminophylline

When the RUL of the normal dog lung was artificially ventilated with N$_2$, air, 60% O$_2$ in N$_2$, and 60% O$_2$ in 20% CO$_2$ in N$_2$ (while the rest of the lungs maintained spontaneous breathing of ambient air) aminophylline (0.24 mg/kg/min) did not show a vasodilating action under alveolar hypoxia [alveolar oxygen tension (PAO$_2$) ca. 40 mmHg]. On the contrary, it seemed to potentiate pulmonary vasoconstriction under severe alveolar hypoxia. When the regional PAO$_2$ became less hypoxic (PAO$_2$ ca. 70 mmHg) or higher than that in the rest of the lungs (still spontaneously breathing ambient air), aminophylline did show a definite vasodilating action. This agent also showed a vasodilating action in alveolar hypercapnia in the presence of alveolar hyperoxia (Fig. 22.12; Isawa et al. 1979). It has been reported elsewhere that aminophylline does not inhibit canine hypoxic pulmonary vasoconstriction (Benumof and Trousdale 1982).

22.2.2.7.2
Isoproterenol and Propranolol

Using the same animal model, when the RUL was artificially ventilated with N$_2$ regional pulmonary perfusion increased in the hypoxic RUL following administration of isoproterenol (0.2 μg/kg/min), but the increase was not observed when isoproterenol was administered following pretreatment with propranolol. When the RUL was artificially ventilated either with air or 60% O$_2$ in N$_2$, no change in regional perfusion occurred by adding isoproterenol. Pretreatment with propranolol before isoproterenol administration did nothing to perfusion in the RUL under the latter circumstances.

Isoproterenol did reverse regional hypoxic pulmonary vasoconstriction, but its action was blocked by pretreatment with propranolol. Propranolol per se showed no effect on pulmonary vascular responses to different alveolar oxygen tensions (Isawa et al. 1981a).

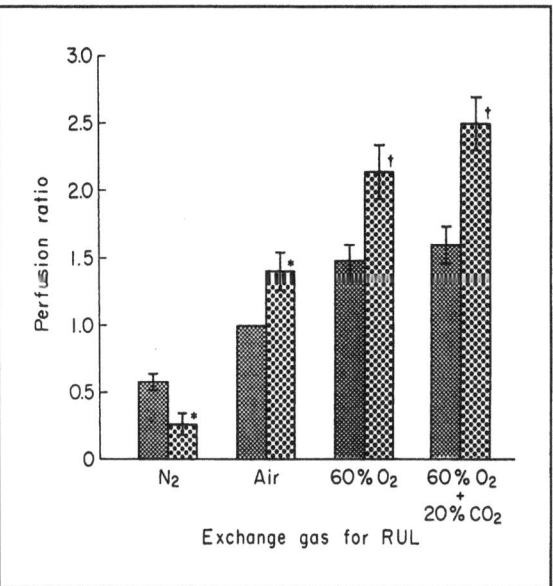

Fig. 22.12. Comparison of regional perfusion before (*left bar, small dots*) and after (*right bar, large dots*) aminophylline injection. When the RUL of a normal dog is artificially ventilated with the gas mixtures indicated on the *x-axis*, aminophylline injection induces significantly more reduced perfusion in the RUL after aminophylline injection when RUL is ventilated with nitrogen. Aminophylline increases regional perfusion when the region is ventilated with air or gas mixtures of higher oxygen concentration. This is clinically important, suggesting that aminophylline works better when regional ventilation is well maintained

In the reimplanted, denervated dog lung, however, pretreatment with propranolol did not block the action of isoproterenol to increase regional perfusion distribution in the hypoxic RUL (Isawa et al. 1982).

22.2.2.7.3
Dopamine

While the RUL was artificially ventilated with N$_2$, air, and 60% O$_2$ in N$_2$, dopamine was administered at the rate of 20 μg/kg/min. There was no change in the regional vascular responses to the different alveolar oxygen tensions in the RUL before and after dopamine administration (Isawa et al. 1981b). Furman and coworkers have reported that isoproterenol and dobutamine inhibited hypoxic pulmonary vasoconstriction but dopamine had no effect (Furman et al. 1982).

22.2.3
Diffusion

The exchange of oxygen and carbon dioxide takes place by diffusion between the blood in the pulmonary capillary beds and the air inhaled into the alveoli. The lung functions most efficiently when ventilation and blood flow are matched. Mismatching of ventilation and perfusion is the most common cause of hypoxemia.

22.2.4
Ventilation-Perfusion Relationships

The introduction by West (1965) of radioisotopes to elucidate the topographical distribution of ventilation and perfusion marked a substantial innovation in pulmonary physiology.

22.2.4.1
Normal Relationships

It has been found that both ventilation and perfusion per unit lung volume decrease from the lung base to the apex of the upright lung (Fig. 22.13 a), although the gravity effect is less for ventilation than for perfusion (Fig. 22.13 b). The upper lung zone (zone 1) is overventilated and underperfused, while the lung base (zone 3) is overperfused and underventilated, as shown in Fig. 22.13 a, b.

The use of ^{133}Xe gas following the procedures illustrated in Fig. 22.14 can clarify the ventilation distrib-

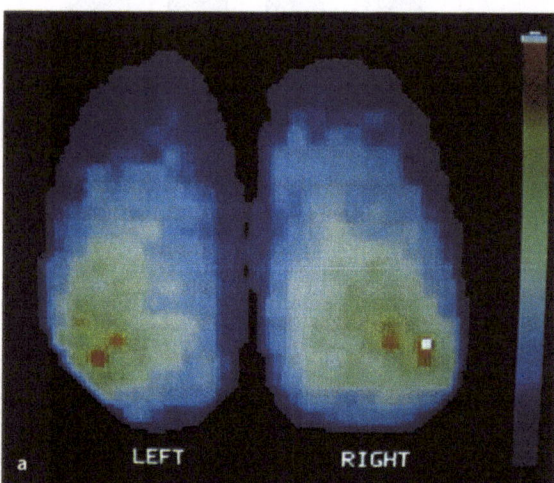
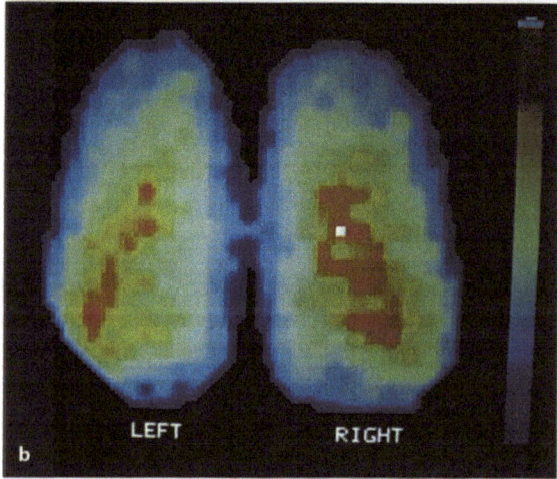

Fig. 22.13. a Perfusion distribution. Xenon-133 dissolved in saline was injected into a healthy normal 53-year-old man at TLC (total lung capacity) level with breath held in the sitting position. The distribution of radioactivity indicates perfusion at TLC. Note more perfusion distribution at the lung base than at the apex. **b** Ventilation distribution. Xenon-133 gas was inhaled from RV (residual volume) to TLC and breath was held. The distribution of radioactivity indicates ventilation distribution at TLC. The distribution of radioactivity is more uniform than in **a**, but still indicates more radioactivity in the lung base than at the apex

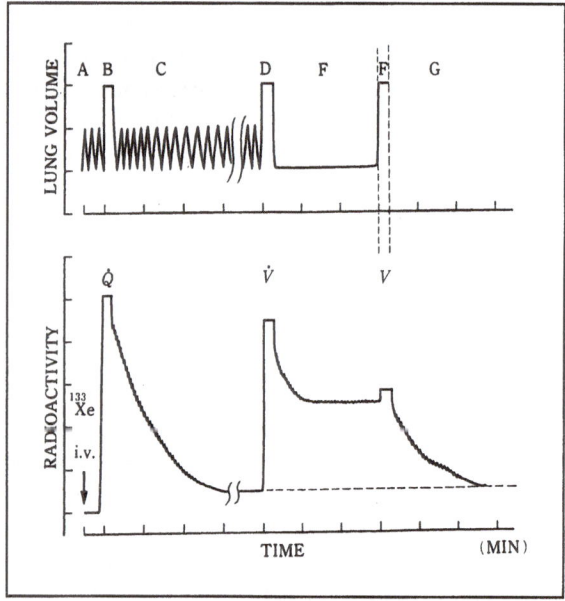

Fig. 22.14. Respiratory maneuver to obtain the perfusion ($\dot{Q}$) image, ventilation (V) image, and lung volume image ($\dot{V}$) at the same lung volume, here at TLC. The inhalation of ^{133}Xe was provided via a closed-circuit container. Dissolved ^{133}Xe was injected intravenously while the subject holds breath at TLC ($B, \dot{Q}$). Injected ^{133}Xe is washed out from the lungs by breathing air. The subject inhaled ^{133}Xe from a closed circuit from residual volume to TLC and held breath (D, V). The subject continues breathing from the closed circuit, and when radioactivity in the lungs and the closed circuit reached equilibrium, the subject breathed from the closed circuit to TLC level and held breath (F, V)

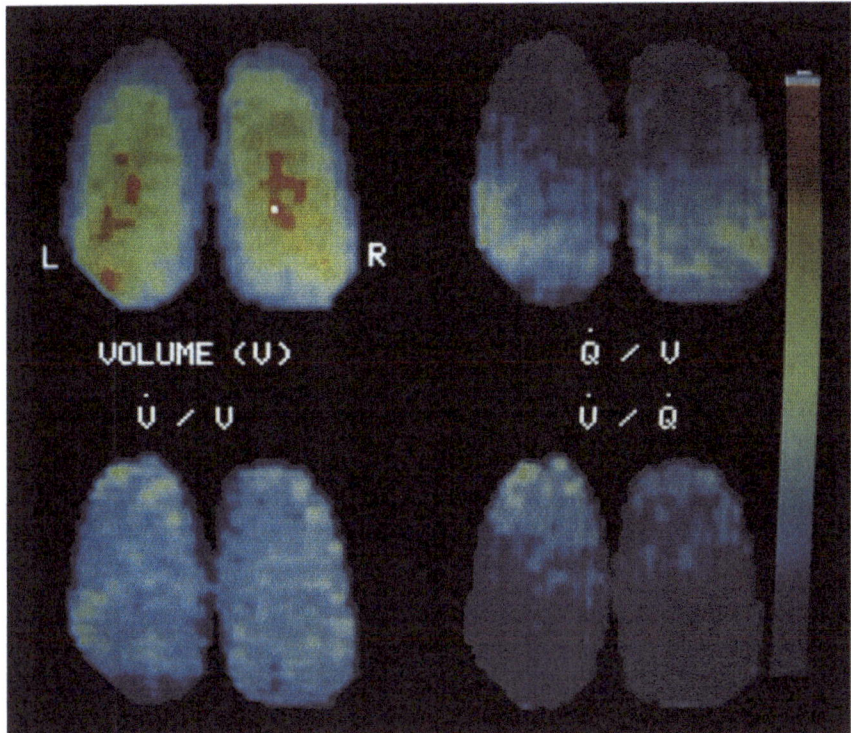

Fig. 22.15. *Upper left*: the distribution of ^{133}Xe in the lungs after equilibrium is reached between the radioactivity in the lungs and ^{133}Xe in the closed circuit, indicating the lung volume (*V*) at TLC: *Upper right*: perfusion versus lung volume ($\dot{Q}/V$); *Lower left*: ventilation versus lung volume $\dot{V}/V$). *Lower right*: ventilation versus perfusion ($\dot{V}/\dot{Q}$) or topographical ventilation-perfusion relationships in the lungs, indicating a larger $\dot{V}/\dot{Q}$ ratio at the lung apex than the lung base

ution per lung volume, perfusion distribution per lung volume, and ventilation/perfusion relationships (Fig. 22.15). Perfusion distribution can be studied by injecting ^{99m}Tc-MAA instead of dissolved ^{133}Xe gas. Ventilation distribution can be studied by inhaling ^{133}Xe gas, ^{81m}Kr gas, or Technegas (Vita Medical Ltd., Lucas Heights, NSW, Australia; Burch et al. 1986).

22.2.4.2
Pulmonary Embolism

22.2.4.2.1
Natural History and Diagnostic Trials

Pulmonary embolism is a relatively common disorder that is potentially fatal without adequate treatment. Because of the nonspecific nature of clinical, laboratory, and radiographic findings, the accurate diagnosis of pulmonary embolism is difficult without perfusion and ventilation lung imaging. The mortality in patients with pulmonary embolism who are not treated is reported to be as high as 30% (Dalen and Alpert 1975), while the mortality becomes between 2.5% and 8% if correctly diagnosed and treated (Alpert et al. 1976; Carson et al. 1992). Since anticoagulant therapy is not without risks (Porter and Jick 1977; Mant et al. 1977), accurate diagnosis is very important. Several reports about the occurrence of deep vein thrombosis (DVT) and pulmonary embolism have led to the term "economy class syndrome" (Homans 1954; Symington and Stack 1977), a reference to commercial flight in which passengers are cramped with their legs dependent and all but motionless for long stretches of time.

22.2.4.2.2
Mismatch between Perfusion and Ventilation Lung Images

Mismatch between perfusion and ventilation, or absent perfusion in the normally ventilated lung tissue, is the diagnostic hallmark of pulmonary vascular disease or pulmonary embolism when clinical signs and symptoms are compatible with the diagnosis (Fig. 22.16).

Ventilation/perfusion (V/Q) lung imaging has been shown to be a safe noninvasive technique to study re-

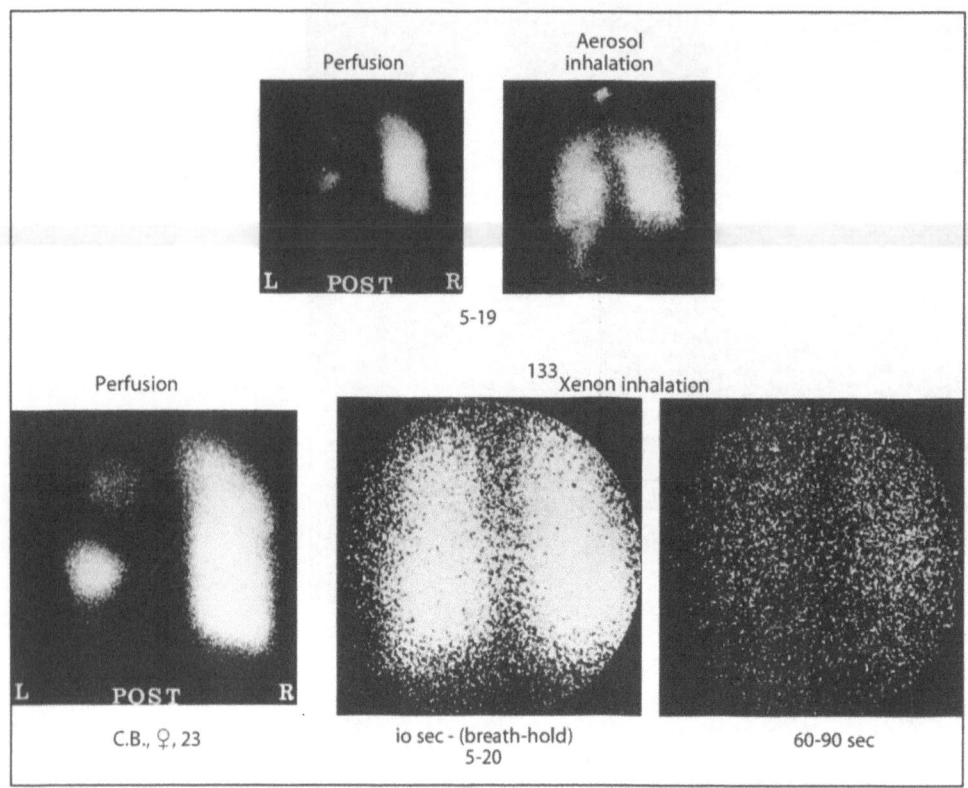

Fig. 22.16. Lung images (posterior) of a 23-year-old woman complaining of chest pain and dyspnea whose chest x-rays were within normal limits. *Upper panel: Left*: perfusion lung image. Perfusion in the left lung was virtually absent. *Right*: aerosol inhalation lung image immediately following perfusion imaging. Note perfusion and ventilation mismatch in the left lung. *Lower panel*: one day after the upper studies were done. *Left*: perfusion lung image. *Middle*: 10 s breath-holding image following a single breath of ^{133}Xe gas inhalation. *Right*: washout phase (60–90 s) with air. Note again perfusion and ventilation mismatch in the left lung and there no evidence of ^{133}Xe gas in the left lung

gional perfusion and ventilation. As has been de-scribed, virtually all parenchymal and obstructive lung diseases can cause not only decreased perfusion but also abnormal ventilation within the affected por-tions of lung. In pulmonary embolism, regional per-fusion is either absent or decreased while ventilation is preserved.

The first large scale study, the Urokinase Pulmo-nary Embolism Trial (UPET Investigators 1973), established perfusion lung scanning as an effective technique for the diagnosis of pulmonary embolism. Recovery of perfusion following pulmonary embolism was the additional subject of a study based on V/Q lung imaging and pulmonary angiography on 755 pa-tients suspected of pulmonary embolism [Prospective Investigation of Pulmonary Embolism Diagnosis (PIOPED) Investigators 1990]. This study and subse-quent comprehensive analysis established the sensitiv-ity and specificity of V/Q lung imaging for high, in-termediate, and low probability and emphasized the

importance of incorporating the clinical assessment (PIOPED Investigators 1990; Worsley and Alavi 1995). Trials to improve the diagnostic accuracy of PIOPED have subsequently been undertaken (Freitas et al. 1995; Goldberg et al. 1996; Gottschalk et al. 1996).

In ventilation-perfusion scintigraphy a "stripe sign" is proposed as a useful diagnostic sign to rule out pul-monary embolism (Sostman and Gottschalk 1982; Sostman and Gottschalk 1992).

Great care must be taken in interpreting ventila-tion and perfusion lung images in patients with con-genital heart disease (Cook and Fogelman 1996). Right to left shunts through pulmonary arteriovenous anastomoses and/or through dilated pulmonary capil-laries is reported (Panoutsopoulos et al. 2000).

A phenomenon of "reverse ventilation-perfusion mismatch" is described in which some perfusion is present in a lung region with no or disturbed ventila-tion. It has been observed in patients with mucus plugging of a unilateral lung (Shih and Bognar 1999), individuals on positive pressure ventilatory support (Slavin et al. 1985; Kim and Heyman 1989), patients with intrapulmonary functional shunting due to lung cancer (Wartski et al. 1998), and in transplanted lungs (Kuni et al. 1993).

22.2.4.2.3
Bronchoconstriction in the Early Phase of Pulmonary Embolism

Retention of inhaled ^{133}Xe gas has provided evidence of bronchoconstriction in the early phase of pulmonary embolism. Bronchoconstriction in fact was seen to appear immediately after unilateral pulmonary artery occlusion. This phenomenon, however, disappears shortly and is documented to disappear in the dog in 6 h, indicating a transient phenomenon (Isawa et al. 1972).

22.2.4.2.4
Regional Ischemia of Functional Origin or Spurious Scintigraphic Recurrence

Perfusion defects due to pulmonary embolism are resolved through thrombolysis and fragmentation of thrombi by anticoagulant therapy during the recovery course, but interestingly one often sees new perfusion abnormalities lung regions where pulmonary embolism was not originally suspected or unlikely to have affected. This phenomenon has been called "regional pulmonary ischemia of functional origin" (Isawa et al. 1970b) and "spurious scintigraphic recurrence" of pulmonary emboli (Moser et al. 1973).

22.2.4.2.5
Chronic Perfusion Defects Mimicking Acute Pulmonary Embolism

Another important fact in pulmonary embolism is that a significant chronic perfusion defect tends to remain after acute pulmonary embolism. Such patients can develop serious complications, e.g., intracerebral bleeding, from unnecessary anticoagulant therapy. The need for a control perfusion lung scans at completion of anticoagulant therapy for acute pulmonary embolism, even in patients with full resolution of symptoms, is stressed. Anyone with a V/Q abnormality should receive a copy of the scan either on a computer diskette or as an analogue image to facilitate future care (Gottschalk 2000; Wartski and Collignon 2000).

22.2.4.3
Deep Vein Thrombosis

Detection of deep vein thrombosis (DVT) is important because it often precedes pulmonary embolism. Recently ^{99m}Tc-apcitide (^{99m}Tc-P280), a small synthetic peptide binding to the GPIIb/IIIa, a receptor expressed on activated platelets, was reported to have a high enough affinity to effectively compete with endogenous fibrinogen (Taillefer et al. 1999). This agent would seem to have the potential of becoming a new gold standard for biochemical detection of DVT.

A prospective study of 788 patients with venous thrombosis, using controls with similar symptoms but in whom the disease had been excluded, has documented no increased risk of DVT among travelers (the "economy class syndrome") (Kraaijenhagen et al. 2000) and the evidence for the risk of venous thromboembolism has been seen to be only circumstantial (Geroulakos 2001).

22.2.4.4
Lung Transplantation and Lung Reduction Surgery

Nuclear medicine procedures have special indications in the follow-up of lung-transplanted patients. Ventilation and perfusion lung imaging before and after transplantation affords invaluable information regarding not only perfusion and ventilation partitions to the transplanted lung(s) or V/Q relationships but also sequential changes in V/Q relationships to warn of early rejection signs. Both ventilation and perfusion of the transplanted lungs were greater when replacing those with pulmonary fibrosis and chronic airway disease, but relative ventilation of the transplanted lung was significantly greater in patients treated for airway disease than in those with pulmonary fibrosis (Chacon et al. 1998). Most of the ventilation and perfusion was found to go to the new lung graft (Levine et al. 1994). For instance, at rest 75±13% of perfusion was directed to the transplanted lung and 67±14% of the ventilation; no significant change in fractional perfusion and ventilation was appreciated during exercise (Ross et al. 1993), the mild V/Q mismatch being retained. Patients reported improvement in quality of life. Hypoxic pulmonary vasoconstriction has been reported to be a significant problem in most unilateral lung transplants (Kuni et al. 1993). Using ^{133}Xe radiospirometry, Ikonen et al. (1996, 1997) demonstrated clinically and experimentally that the relative perfusion (Qtx) decreased and V/Qtx increased during acute rejection.

Because ventilation and perfusion lung imaging is noninvasive, V/Q imaging is also noted to be useful in the functional evaluation after surgical lung reduction procedures in patients with severe emphysema (Cooper et al. 1996).

22.2.5
Airway Diseases

22.2.5.1
Matching of Perfusion and Ventilation Lung Images

Matching of ventilation and perfusion can be seen in airway disease as shown in Fig. 22.17 a, b. In parenchymal lung disease such as pneumonia, too, matching of ventilation and perfusion is observed in lung regions corresponding to abnormal densities on chest x-rays. Thus ventilation and perfusion relationships can be called a hallmark in studying either airways disease or parenchymal lung disease.

22.2.5.2
Radioactive Gases versus Radioaerosols

Since Knipping's Isotopen-Thorakographie using ^{133}Xe gas (Knipping et al. 1955), radioactive gases such as ^{133}Xe, ^{81m}Kr, cyclotron-produced ^{15}O, ^{11}C and ^{13}N have been used to assess regional lung function. The advantages of these gases are manifold, but their utility has mostly been limited due to high cost and limited availability. An alternative to the use of radioactive gases to study regional ventilation is the use of inhaled radioactive particulates or aerosol (radioaerosol) originally developed by Taplin and Poe (1965) and Pircher et al. (1965) who used aerosols made of either ^{198}Au colloid or ^{131}I-HSA.

Regarding the comparison of radioactive gases with radioaerosols: (1) Radioaerosols are less expensive to produce; (2) radioaerosols are produced by generators from easily available radiopharmaceuticals; (3) inhaled radioaerosols stay in the lungs for a longer time (radioactive gases disappear without breath-holding), making a study of the lungs from multiple projections possible; (4) direct comparison of aerosol inhalation lung images with corresponding perfusion images is possible with good statistics; (5) dynamic aspects of lung function, such as locating "slow spaces" or determining if a bullous region communicates with an airway, can be studied only with radioactive gases; (6) repeated studies can be easily done using a short half-life gas such as ^{81m}Kr gas (or a cyclotron-produced positron gas if facilities permit); again (7) breath-holding is necessary with gases, not with aerosols; thus only one projection can be studied at a time with gases; (8) due to characteristic deposition patterns, obstructive airway diseases can be recognized with aerosol inhalation lung imaging; and (9) by using aerosols generated from non-absorbable radiopharmaceuticals, mucociliary clearance can be studied as well as ventilatory aspects (Isawa 1995).

22.2.5.3
Aerosol Generation

Radioaerosols can be generated by a jet nebulizer, an ultrasonic nebulizer, and other inhalation instruments. Agents generally used for aerosol generation are ^{99m}Tc-labeled human serum albumin (HSA), ^{99m}Tc-phytate, ^{99m}Tc-sulfur colloid, and ^{99m}Tc-diethylene triamine pentaacetate (DTPA; Isawa 1995). The size of radioaerosols usable in ventilation studies is 1–3 μ or less in "activity median aerodynamic diameter" (AMAD).

Recently Technegas, which allegedly has dual characteristics of both radioactive gas and aerosol, has been introduced for clinical use (Burch et al. 1986). Technegas is generated under 100% argon (Ar) gas from simmered pertechnetate (^{99m}TcO$_4^-$) in a carbon crucible at 2500 °C , is composed of technetium crystals covered with carbon, and each particle is 5–60 nanometer (nm) in diameter (Figs. 22.18, 22.19; Isawa et al. 1996; Senden et al. 1997). Because of its small size, it is advisable to inhale Technegas deeply followed by a few seconds of breath-holding.

22.2.5.4
Aerosol Deposition Patterns

Deposition of inhaled aerosol in the nonsmoking normal lungs is homogeneous and almost indistinguishable from the perfusion counterpart. In normal subjects nearly 40 to 50% of inhaled ^{99m}HSA aerosols whose AMADs range from 1 to 3 μ (with respective geometric standard deviation [GSD (σg)] close to 1.7) deposits in the nonciliated distal lung spaces including the alveoli during tidal resting breathing through the mouth (Miki et al. 1992).

In chronic obstructive pulmonary diseases (COPD) characteristic deposition patterns are recognized; central deposition patterns in typical emphysema patients and peripheral patchy patterns in typical

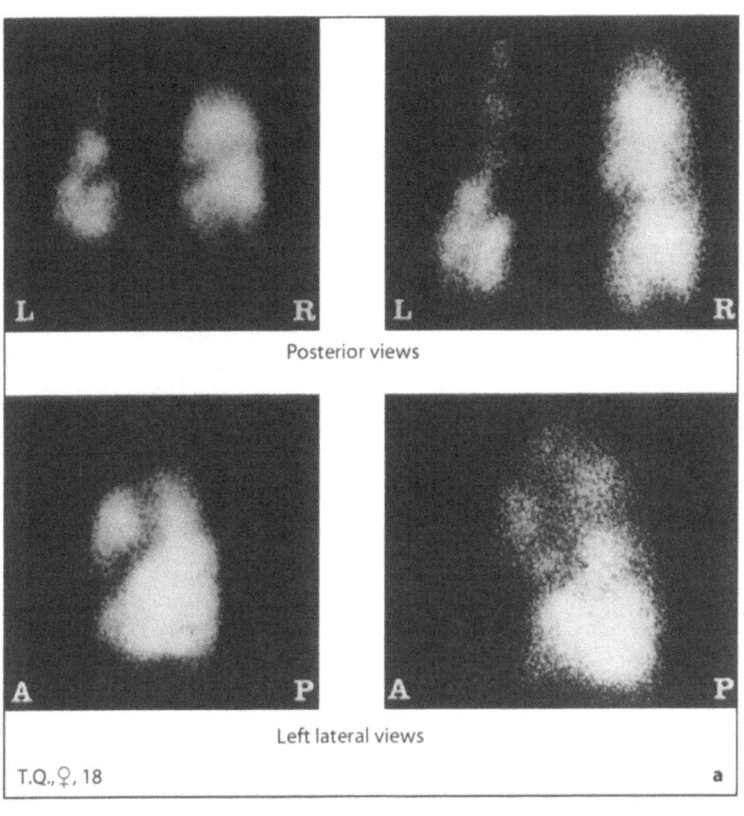

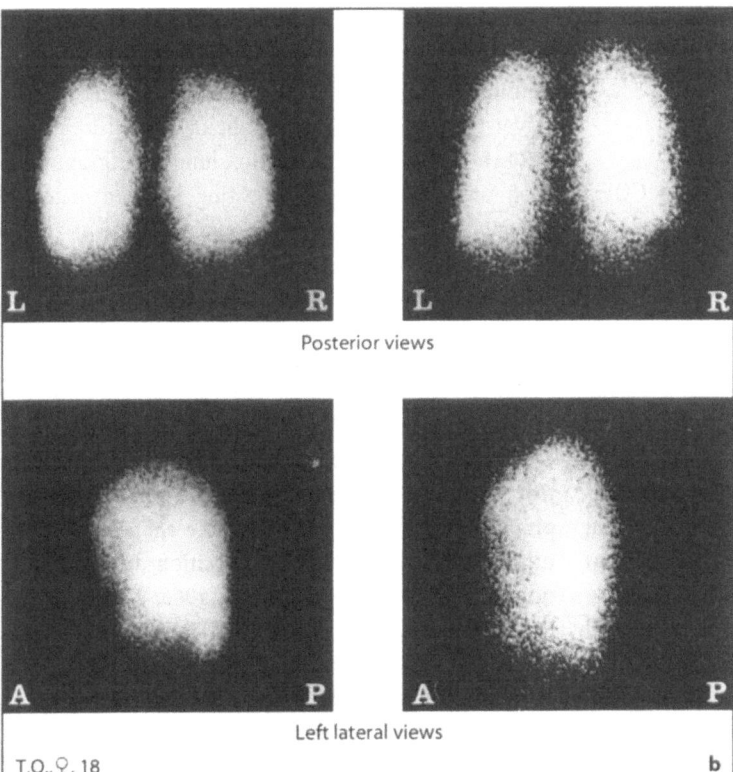

Fig. 22.17. a Perfusion (*left*) and aerosol inhalation (*right*) lung images from an 18-year-old woman complaining of chest pain, dyspnea, and wheezing who had been treated for pulmonary embolism at another hospital. Note the matched perfusion and ventilation images indicating the airway disease. *Above*, posterior views; *below*, left lateral views. **b** Perfusion (*left*) and aerosol inhalation (*right*) lung images from the same patient after 2 weeks' treatment with a bronchodilator and antibiotics under the diagnosis of asthmatic bronchitis. Perfusion and inhalation lung images returned normal. *Above*, posterior views; *below*, left lateral views

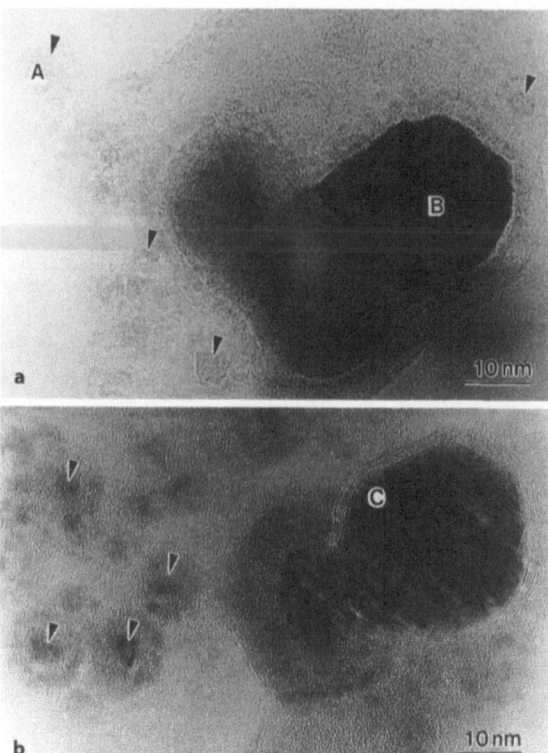

a

b

Fig. 22.18 a, b. High-resolution electron micrographs of Technegas. The large dark regions in **a** and **b** are technetium crystals covered with carbon, and the small dark regions embedded in the carbon contrast, some of which are indicated by *arrowheads*, are also small technetium crystals. (for *A*, *B*, and *C* see Fig. 22.19)

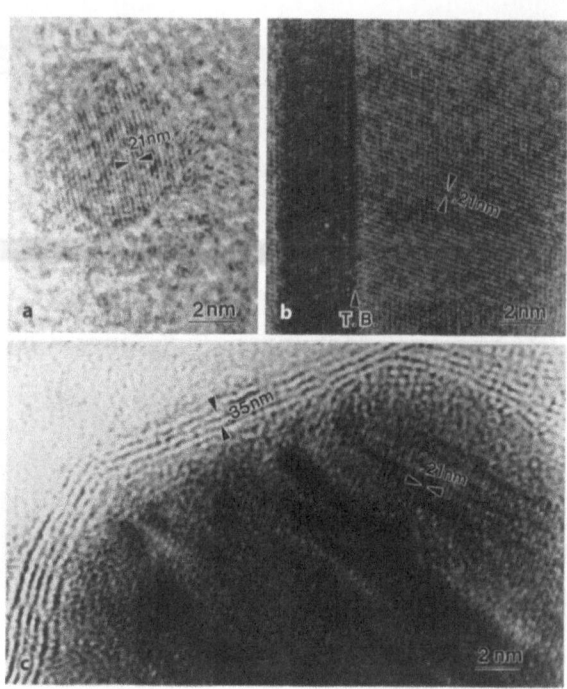

Fig. 22.19 a–c. Enlarged micrographs of regions *A*, *B*, and *C* in Fig. 22.18. Lattice fringes 0.21 nm in diameter can be seen in the technetium crystals. In b), **twin boundaries** (*T.B.*) are observed. The technetium particle in c is covered by a thin carbon layer with lattice fringes 0.35 nm in diameter

bronchitic patients, but in most patients with COPD central and peripheral patterns coexist (Fig. 22.20; Isawa et al. 1970a).

In pulmonary emphysema, an aerosol tends to deposit on the central large airways (Fig. 22.21) near the hilum, probably because the large airways collapse on deep expiration due to decreased elastic recoil of the airways, as was observed with cine-bronchography by Fraser (1961). In chronic bronchitis, edema and/or anatomical remodeling of the bronchial walls, hypersecretion and retention of bronchial mucus, and so on, may contribute to the formation of multiple hot spots, while the difference in aerosol size does not markedly affect the lung images in a normal subject (Miki et al. 1992). Generally speaking, the larger the size of aerosol, the more aerosol tends to deposit in the central larger airways than in the smaller ones in the lung parenchyma (Miki et al. 1992). In bronchiectasis both ventilation and perfusion are diminished in bronchiectatic lung regions, forming

multiple hot spots by aerosol inhalation lung scintigraphy (Bass et al. 1968; Windheim 1970; Isawa et al. 1990a). Radioaerosol inhalation lung scintigraphy has been used not only for detecting V/Q mismatch in patients with suspected pulmonary embolism (Isawa et al. 1971; Biello et al. 1979), but also for studying aerosol deposition patterns or mucociliary clearance mechanisms in the lungs of patients with various chest diseases (Isawa et al. 1984b, 1986a, 1988).

There are various factors which determine the site of aerosol deposition in the lungs including the particle size of inhaled aerosol, the inspiratory and expiratory flow rate, the composition and physicochemical properties of the carrier gases, the shape and structure of the airways, the body position during inhalation, and so on (Morrow et al. 1966; Morrow 1974; Foord et al. 1978; Agnew 1984). Because different aerosol deposition patterns can be obtained, especially in subjects suffering from various forms of lung disease, by altering inhaled aerosol size, selection of inhaled aerosol particle size is important according to the purpose of the study (Miki et al. 1992).

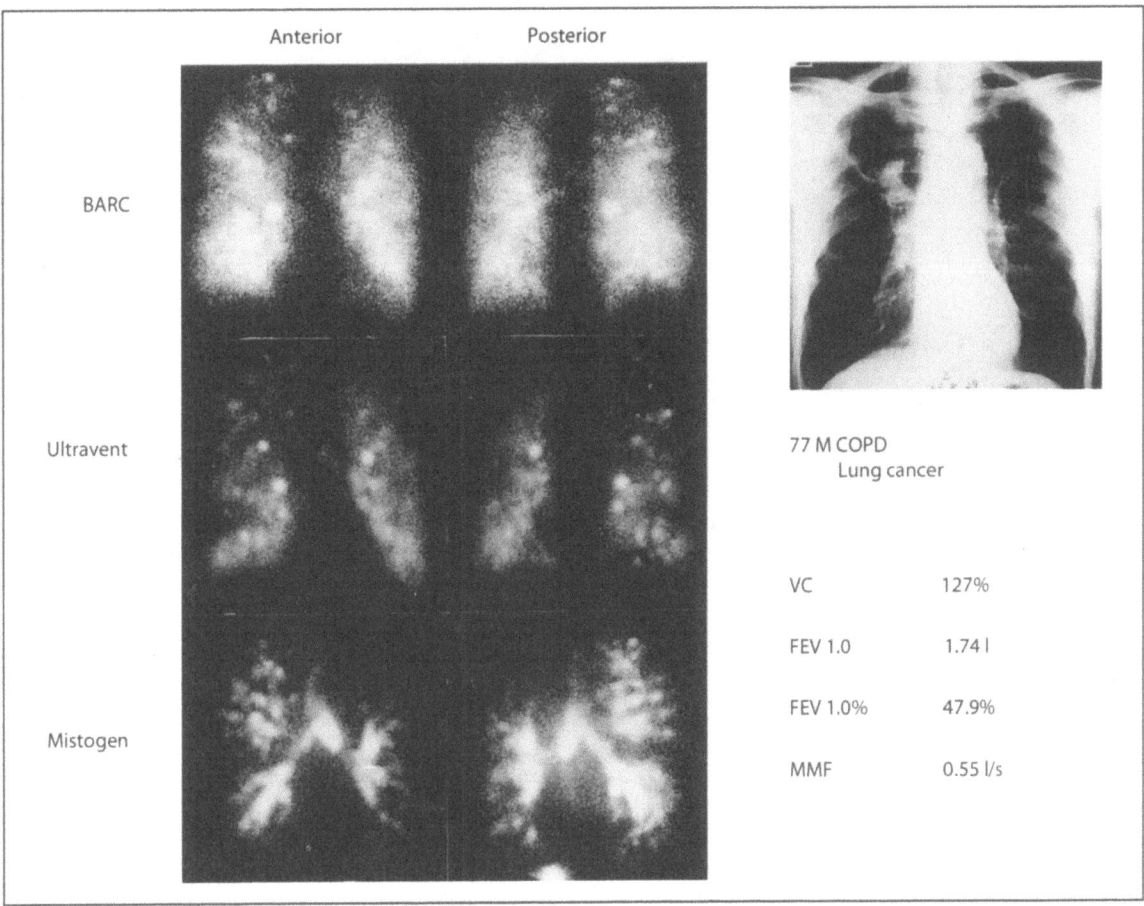

Anterior   Posterior

BARC

Ultravent

Mistogen

77 M COPD
Lung cancer

VC	127%
FEV 1.0	1.74 l
FEV 1.0%	47.9%
MMF	0.55 l/s

Fig. 22.20. Aerosol inhalation lung images of a 77-year-old patient with squamous cell carcinoma of the RUL complicated with chronic obstructive pulmonary disease. His pulmonary function tests indicated a moderate airway obstruction with vital capacity of 127%, forced expiratory volume in 1 s ($FEV_{1.0\%}$) of 47.9% and maximal mid-expiratory flow (MMF) of 0.55 l/s. In addition to inhomogeneous distribution of inhaled aerosol, multiple hot spots were seen throughout the lung for all three nebulizers. BARC (AMAD 0.84 μ, σg 1.73; Bhabha Atomic Research Centre, Bombay, India) showed the best penetration of inhaled aerosol to the lung periphery, followed by Ultra Vent (AMAD 1.04 μ, σg 1.71; Mallinckrodt Inc., St. Louis, Mo., USA). Mistogen (AMAD 1.93 μ, σg 1.52; Mistogen Equipment, Oakland, Calif., USA) produced almost bronchogram-like images, indicating poor penetration of inhaled aerosol to the lung periphery. Inhaled aerosols deposited in the central large airways as well as in the peripheral lung fields. Central deposition patterns are most prominent with Mistogen

22.2.5.5
Difference between Technegas and Genuine Gas

Comparing the distribution of inhaled ^{81m}Kr gas and Technegas in the lungs at total lung capacity, there was little difference in the distribution ratios in the right and left lungs. However, there was a statistically significant difference in the lung regions of both normal subjects and patients with pulmonary disease. Technegas tended to deposit more in the lung bases than did ^{81m}Kr gas. Despite these statistical differences, they were visually or qualitatively very similar. Thus Technegas is clinically useful as an inhalation agent, unless strictly quantitative analyses are required (Isawa et al. 1994).

Susskind et al. compared pulmonary distribution of inhaled ^{99m}Tc-DTPA aerosol and ^{81m}Kr gas in coal miners with nonembolic pulmonary disease. They found that the ^{81m}Kr gas tended to be preferentially distributed in the apical regions of the lungs, whereas the ^{99m}Tc-DTPA aerosol tended to be deposited in the lung bases, an effect attributable to gravity since the particles of ^{99m}Tc-DTPA are heavier than ^{81m}Kr gas molecules (Susskind et al. 1986). Technegas, far

Anterior　　　　　　　　　Posterior

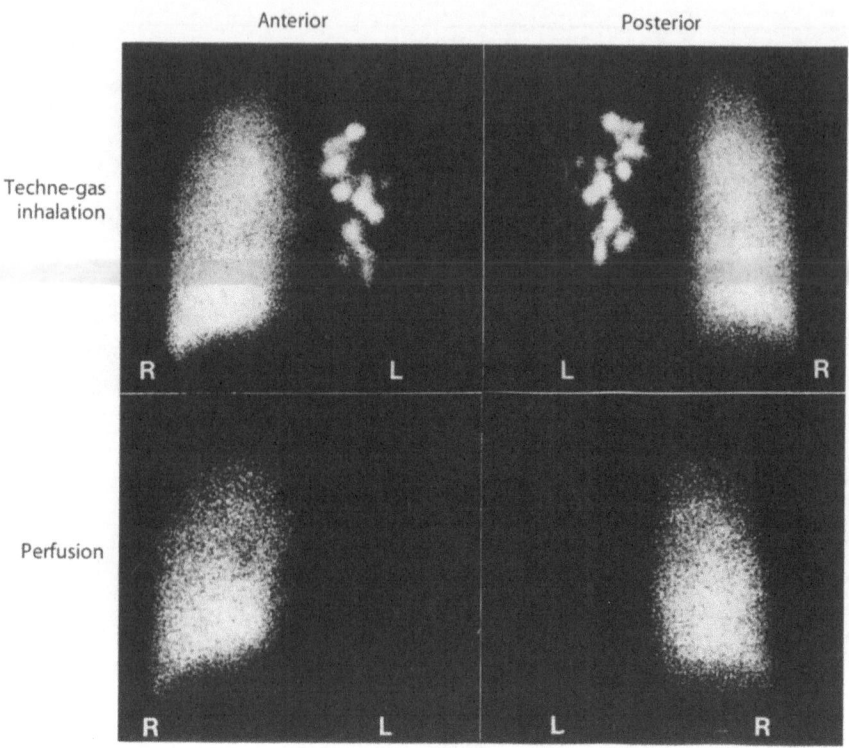

Techne-gas
inhalation

Perfusion

Fig. 22.21. Aerosol inhalation and perfusion lung images of an 18-year-old patient with Swyer-James syndrome (unilateral emphysema). The emphysematous left lung shows aerosol deposition in the central large airways and no perfusion, while the normal right lung shows normal ventilation and perfusion. *Upper panel*: aerosol inhalation lung images, anterior and posterior. Note: only the central large airways are depicted in the left lung. *Lower panel*: perfusion lung images, anterior and posterior

smaller in size than ^{99m}Tc-DTPA aerosol, also showed a significantly larger distribution in the lower third of the left lung than ^{81m}Kr gas.

Thus Technegas, though a nanometer-sized radio-aerosol, retains the characteristics of a particulate material and can never substitute for a radioactive gas. It cannot be used as a ventilation agent to study the dynamic aspects of ventilation as can ^{133}Xe gas.

22.2.5.6
Significance of Hot Spots or Excessive Radioactive Deposition

Inhaled aerosol deposits both in the conductive airways and functioning gas-exchanging space of the lung by impaction, sedimentation, and diffusion. In the functioning lung, or alveolar, space, diffusion of inhaled aerosols takes place. For relatively low inhalation flow rates, such as those in tidal breathing, the distribution of inhaled air normally depends in roughly equal measure on airway resistance and lung compliance (Milic-Emili 1977). Here airway resistance

is defined, when airflow is laminar, as the pressure difference between the mouth pressure and the alveolar pressure divided by the airflow rate. When the airflow is disturbed to become turbulent, airway resistance further increases and chances of aerosol deposition on the airways increase. Reynolds number (Re) is useful index in this regard. $Re = 2RF\rho/\eta$, where R is diameter of the airway, F is mean flow rate, and ρ and η are density and viscosity of air, respectively. Roughly speaking, when Re exceeds 2000, the flow becomes turbulent. In other words, inhaled aerosol then has more chance of impaction and sedimentation. Compliance is a measure, roughly, of the lung's tendency to conform to the shape of the bony chest, the reverse of elasticity. When the lungs lose elastic recoil as in pulmonary emphysema, they become more compliant or more difficult to return to their original resting level of the lung volume and the functional residual capacity (FRC) increases. Therefore, when the lungs become more compliant with increased airway resistance, the distribution of inhaled aerosol in the lungs is also more disturbed.

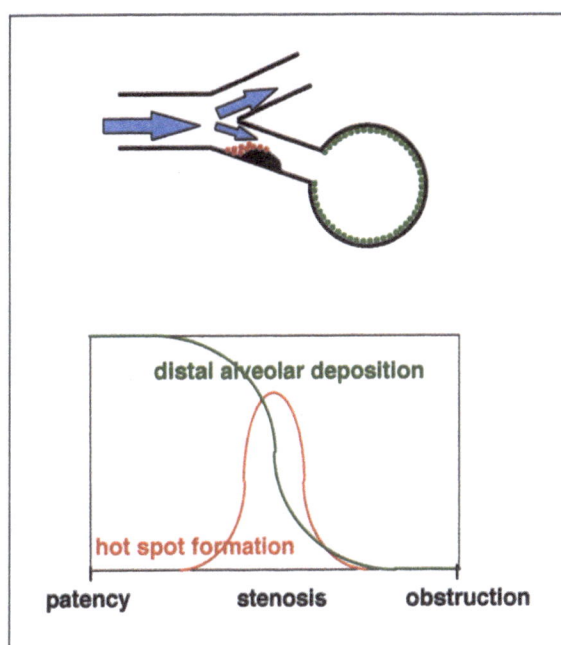

Fig. 22.22. Conceptual diagram of hot spot formation in aerosol inhalation lung imaging

Thus when the airway is patent and ventilation is normal, inhaled aerosol reaches the alveoli without depositing in the airway. When an airway obstruction occurs, the distal region shows no deposition of inhaled aerosol. When the airway is partially stenotic, a hot spot is generated due to inertial impaction of the inhaled aerosol. The more stenotic the airway becomes and the faster the airflow, the greater are the chances of impaction resulting in an airway hot spot and less alveolar deposition of inhaled aerosol. Chung et al. have described prestenotic bronchial deposition of inhaled aerosol as a new ventilation scan sign of bronchial obstruction (Chung et al. 1997). Hot spots thus indicate that while regional ventilation still exists, it is disturbed. When stenosis become excessive, the airflow at the stenotic site diminishes and a hot spot is no longer formed. Figure 22.22 conceptualizes the relationship between aerosol deposition or hot spot formation and the status of the bronchial lumen (Horikoshi et al. 2000).

Proteolysis of the alveolar walls due to impaired neutrophil elastase (NE)-anti NE balance seems to be a major cause of emphysema. Neutrophils may increase in number as a result of smoking (Hunninghake and Crystal 1983). NE is balanced by an excess of antiproteases, principally a_1-antitrypsin (AT), providing anti-NE protection to the lung parenchyma. Oxidants associated with smoking also play havoc with the lung. Molecular basis of a_1-AT has been clarified (Bradley et al. 1988).

22.3
Nuclear Medicine for Nonrespiratory Lung Function

Nonrespiratory lung functions are those not directly related to gas exchange function of the lung (Fig. 22.2). Pulmonary epithelial permeability and mucociliary clearance will be considered here.

22.3.1
Pulmonary Epithelial Permeability

The aerosol ^{99m}Tc-DTPA was first used as an inhalation agent in 1968 for studying regional diffusing capacity of the lung in patients with pulmonary sarcoidosis and alveolar proteinosis. It was expected that in these patients with decreased carbon monoxide diffusing capacity, clearance of inhaled ^{99m}Tc-DTPA aerosols from the lung would be slower; however, the result was opposite. Clearance from the lung seemed very rapid, although no quantification was feasible (Taplin and Isawa 1968).

Using computers for data analysis, Taplin et al. later found that the clearance of inhaled ^{99m}Tc-DTPA aerosol from the lung was accelerated in patients with interstitial lung diseases (Rinderknecht et al. 1980). Minty et al. found that the pulmonary clearance of inhaled ^{99m}Tc-DTPA aerosols from the lung is also accelerated in smokers (Jones et al. 1980, 1983), but it rapidly normalizes after cessation of smoking (Minty et al. 1981). The ^{99m}Tc-DTPA-aerosol inhalation method is now widely accepted as a useful test for evaluating pulmonary epithelial permeability.

A "modified Technegas" or "Pertechnegas", generated under an atmosphere of 3% O_2 and 97% Ar gas instead of 100% Ar was once claimed to be usable in place of ^{99m}Tc-DTPA aerosol (Monaghan et al. 1990, 1991). Clearance of inhaled Pertechnegas differed, however, from that of ^{99m}Tc-DTPA aerosol and was similar to that of inhaled ^{99m}Tc-pertechnetate (^{99m}TcO$_4^-$) aerosol (Isawa et al. 1995). High-resolution electron microscopic observation revealed that Pertechnegas, unlike Technegas, has no carbon coating surrounding the technetium crystals (Figs. 22.18, 22.19; Isawa et al. 1996).

22.3.1.1
Pathophysiological Basis

The alveolar-capillary barrier consists of the alveolar airway barrier in series with an endothelial barrier and in parallel with the interstitial lymphatic pathways (Staub 1983). Pulmonary epithelium forms an extremely tight barrier that is one-tenth as permeable as capillary endothelium for hydrophilic molecules and prevents the alveolar lumen from being flooded (Gorin and Stewart 1979). Permeability is likely related to the intercellular junctions, which are the main sites where hydrophilic molecules cross the membranes. The pore radius of the alveolar epithelial tight junctions is reported to be 0.4–1.0 nm whereas that of the capillary endothelium is 0.4–8.0 nm (Taylor and Gaar 1970).

After ^{99m}Tc-DTPA aerosol is inhaled, its rate of clearance depends upon integrity of the lung. When the alveolar epithelium is intact, it produces resistance to ^{99m}Tc-DTPA particles passing through the intercellular epithelial junctions; however, when the intercellular epithelial junctions are widened due to some pathological conditions, ^{99m}Tc-DTPA is cleared more rapidly. Positive end-respiratory pressure is reported to help clear inhaled ^{99m}Tc-DTPA more rapidly (Chopra et al. 1979; Marks et al. 1985; Nolop et al. 1986; Peterson et al. 1988).

22.3.1.2
Clinical Evaluation

22.3.1.2.1
Normal Subjects

In normal nonsmokers ($n = 23$) clearance half-time ($t_{1/2}$) is 78.7±18.5 min when they inhale ^{99m}Tc-DTPA aerosol with AMAD of 0.86 μ and σg of 1.75. The corresponding value with ^{99m}Tc-DTPA aerosol of 1.00 μ AMAD and 1.78 σg is 58.5±23.0 min ($n = 10$; Anazawa et al. 1991). Smaller aerosol particles tend to prolong $t_{1/2}$, although the difference is not statistically significant.

22.3.1.2.2
Smokers

Clearance of inhaled ^{99m}Tc-DTPA aerosols has been found to be faster in smokers (Jones et al. 1980, 1983). Our study showed that when seven smokers (average cigarette consumption 16 pack-years or Brinkman Index of 160) inhaled ^{99m}Tc-DTPA aerosol

with 0.86 μ AMAD and 1.75 σg, the $t_{1/2}$ was 25.7±15.8 min. The difference in $t_{1/2}$ between non-smokers and smokers was statistically significant; it is not known why smoking induces this more rapid clearance. Horseradish peroxidase labeled with transferrin and instilled into the alveoli of dogs was observed to increase in the intercellular junctions after smoking, indicating a widening of these junctions (Boucher et al. 1980; Hogg 1983). Within 3 weeks after cessation of smoking clearance acceleration has been found to normalize (Minty et al. 1981).

22.3.1.2.3
Interstitial Lung Diseases

Patients with biopsy-proven idiopathic interstitial pneumonia show significantly faster clearance of inhaled ^{99m}Tc-DTPA aerosol than do nonsmoking normal subjects, but there is no correlation between clearance of ^{99m}Tc-DTPA aerosol and diffusing capacity measured by carbon monoxide (DL$_{CO}$; Anazawa et al. 1991). Clearance is also accelerated in interstitial pneumonia due to sarcoidosis (Rinderknecht et al. 1980), systemic sclerosis (Chopra et al. 1979), radiation pneumonitis, and radiation fibrosis (Anazawa et al. 1992a).

It has been speculated that increased retractile forces due to fibrosis (Chopra et al. 1979) widen the intercellular junctions, and that the increase precedes alveolitis or a residue of subsiding active disease in sarcoidosis (Jacob et al. 1965). According to our study in which experimental fibrosis was induced in rat by bleomycin, clearance of ^{99m}Tc-DTPA aerosols definitely increased 2 weeks after instillation of bleomycin. Electron microscopic examination indicated that there was thinning, detachment, or denudation of the alveolar epithelium, and that the basement membrane was directly exposed to the alveolar surface. The widening of the intercellular junctions was, however, not confirmed. We believe that the epithelial damage and loss of epithelial covering of the basement membrane contribute to some extent to the faster clearance of inhaled ^{99m}Tc-DTPA-aerosol (Anazawa et al. 1992b).

22.3.1.2.4
Other Pathological Conditions

Increase in pulmonary clearance of inhaled ^{99m}Tc-DTPA aerosol is consistently reported with noncardiogenic pulmonary edema such as in adult respiratory distress syndrome (ARDS) and infantile respiratory

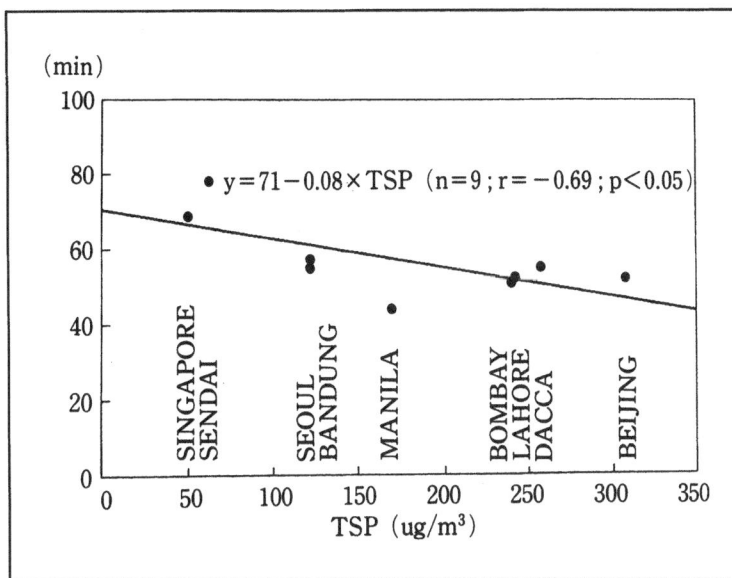

Fig. 22.23. Clearance half-time ($t_{1/2}$) of ^{99m}Tc-DTPA aerosol (AMAD 0.84 µ, σg 1.73) in normal nonsmokers is accelerated as the level of total suspended particulates (TSP) increases. The names of the cities indicate the places where the measurements in the normal subjects were done. Singapore (Singapore), Sendai (Japan), Seoul (Korea), Bandung (Indonesia), Manila (Philippines), Bombay (India), Lahore (Pakistan), Dacca (Bangladesh), Beijing (China)

distress syndrome (IRDS) (Braude et al. 1983, 1985; Mason et al. 1985; Royston et al. 1985; Coates and O'Brodovich 1986), but the increase is not a uniform finding in cardiogenic pulmonary edema. Depletion of surfactant also seems responsible for increased permeability in ARDS (Wollmer et al. 1986). The respiratory bronchioli are suggested as being the sites where permeability is accelerated (Mason et al. 1985).

In radiation pneumonitis clearance is accelerated not only in the pneumonic lesions but also in the contralateral lung where no radiological abnormalities are observed (Anazawa et al. 1992a). Radiation pneumonitis is thought to be caused by alveolar damage and alveolar edema (Gross 1977). As radiation pneumonitis resolves with steroid therapy, clearance of inhaled ^{99m}Tc-DTPA aerosol becomes normalized.

Accelerated clearance of inhaled ^{99m}Tc-DTPA aerosol from the lung has also been reported in hyaline membrane disease (Jefferies et al. 1984), long-term free-base cocaine use (Susskind et al. 1991), glue sniffing (Sundram 1995), and pneumocystis carinii pneumonia complicated with AIDS or HIV-positive hemophilia in nonsmokers (Mason et al. 1987; Meignan et al. 1990; O'Doherty et al. 1990; Van der Wall et al. 1991; Rosso et al. 1992). Especially in pneumocystis carinii pneumonia, clearance of inhaled ^{99m}Tc-DTPA aerosol returns toward normal with response to therapy. A clearance study of this aerosol is reported to demonstrate its superior diagnostic usefulness to 67-gallium (^{67}Ga) chest scans in detecting pneumocystis carinii pneumonia when chest x-rays and/or PaO$_2$ are within normal limits.

22.3.1.2.5
Urban Pollution and Pulmonary Epithelial Permeability

The $t_{1/2}$ following ^{99m}Tc-DTPA aerosol inhalation also differs in nonsmoking normal subjects living in different locations. An international cooperative study has shown that $t_{1/2}$ in healthy nonsmokers depends on the severity of air pollution, especially on the amount of suspended particulates in the air. The study was carried out under auspices of the International Atomic Energy Agency (IAEA) in ten cities in eastern, southern, and southeastern countries of Asia using the same aerosol generator (Bhabha Atomic Research Centre, Bombay, BARC; Kotrappa et al. 1977) and the same agent ^{99m}Tc-DTPA for aerosol generation (Nair et al. not yet published). Highly significant correlations were noted between $t_{1/2}$ and total suspended particulates (TSP; Fig. 22.23) and between $t_{1/2}$ and total pollutants.

22.3.2
Mucociliary Clearance

22.3.2.1
Airway Mucus and Cilia

Mucociliary clearance is the first line of defense in the respiratory system. The mucus overlying the epithelial surface of the respiratory tract is propelled upward toward the larynx by coordinated beating of the cilia of the airway epithelium. The cilia line the epithelium of the respiratory mucosa from the nasal passages to the terminal bronchioles, and are absent or poorly developed distal thereto (Sleigh 1977).

The cilia of the bronchial tree are dipped in surrounding periciliary fluid, above which there is a mucus layer. A coordinated ciliary motion at the rate of about 1000/min normally transports the mucus layer toward the oropharynx, a salient feature in the orchestration of interaction between mucus and cilia in the accomplishment of mucociliary clearance (Lucas and Douglas 1934; Wanner 1977). Ciliary action at the molecular level has been described in detail by Baum et al. (1998).

22.3.2.2
Principles of Mucociliary Imaging

22.3.2.2.1
Agents used for Mucociliary Clearance Studies

22.3.2.2.1.1
Radioactive Droplet

Any agent that stays on the airway mucus layer following placement through bronchoscopy can be used for mucociliary imaging, including $^{99m}TcO_4^-$, ^{99m}Tc-HSA, ^{99m}Tc-millimicrosphere, ^{99m}Tc-albumin microsphere, and ^{99m}Tc-MAA. Any one of these with a particle size on the order of 0.025–0.05 ml and placed through a tube inserted into the bronchoscope, is practical and convenient (Hirano 1988).

Anesthesia of the airway mucosa and oropharynx is required for inserting a bronchoscope and placing a radioactive droplet on the tracheal or bronchial mucosa. This is tedious and cumbersome and inflicts so much discomfort on the patient that the actual clinical application of this method is limited.

22.3.2.2.1.2
Radioaerosol

Radioaerosol generated either by a jet nebulizer, an ultrasonic nebulizer, a spinning-disc atomizer under special circumstances, or by other such devices, is inhaled through a mouthpiece with the nose clipped. An agent that is not absorbed or dispersed through the airway mucosa is ideal. Either ^{99m}Tc-sulfur colloid, ^{99m}Tc-phytate, or ^{99m}Tc-HSA can be used to study mucociliary clearance mechanisms. We generally use ^{99m}Tc-HSA aerosol generated by a jet nebuliz-

er or an ultrasonic nebulizer. Aerosol with σg of 1.1 or less are termed monodisperse aerosols, while those with σg greater than 1.1 are called polydisperse aerosols. The size of inhaled radioaerosol, hygroscopy, and breathing pattern determine the sites of deposition in the lungs as described elsewhere (Isawa et al. 1970a; Miki et al. 1992).

We adopt normal tidal breathing for inhaling radioaerosol. In studying mucociliary clearance aerosols of less than 5–6 μ, but greater than 1 μ in AMAD, is most suitable, as has been shown by Morrow and Yu (1993). We have personally used ^{99m}Tc-HSA aerosol of AMAD of 1.93 μ with σg of 1.52 inhaled by the normal tidal breathing for studying mucociliary clearance.

22.3.2.3
How to Study Mucociliary Clearance

22.3.2.3.1
Placement of Radioactive Droplet or Radiopaque Material

Bronchoscopic and radiographic methods have been used to study mucociliary clearance mechanisms, but these are generally invasive and require anesthesia of the airways and insertion of a bronchoscope, thus making these methods not feasible for clinical purposes (Hilding 1957; Berke and Roslinski 1971; Sackner et al. 1973; Friedman et al. 1977). For external measurement of mucociliary clearance, radioactive materials are placed on the mucosal surface of the airways and sequential imaging is performed (Baetjer 1967; Sakakura and Proctor 1972; Isawa et al. 1980a). Transport of the radioactivity placed on the mucus layer is equivalent to that of the mucus itself.

Acute toxicity of cigarette smoke to mucociliary clearance has been evaluated in dogs by this method. The transport or migration of the droplet on the mucus in the trachea is disturbed by forced smoking in a dose-response manner in the dog. Removing particulates in the cigarette smoke by glass fiber did not prevent the mucus transport from being disturbed (Figs. 22.24, 22.25; Isawa et al. 1980a). A large field-of-view gamma camera is used for this imaging.

Radioactivity is introduced onto the mucosal surface either in a radioactive droplet or via inhaled radioaerosols. As a clinical method to study mucociliary clearance the former is invasive and not practical, whereas the latter is noninvasive and readily applicable.

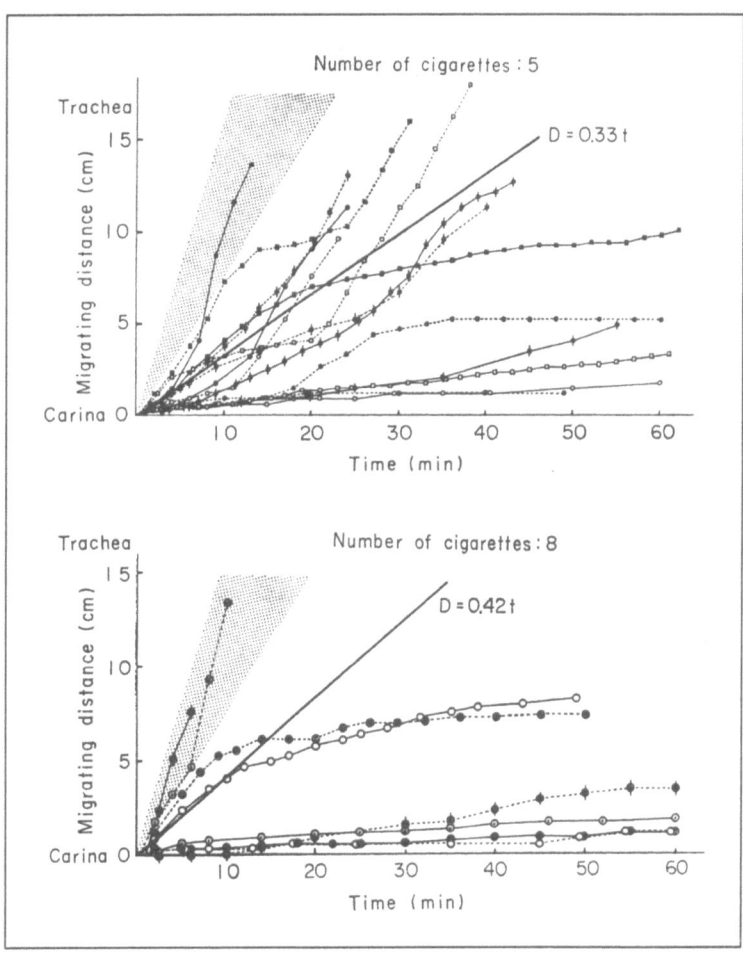

Fig. 22.24. Migrating (transport) distance of a tracer material on the dog trachea. The *x*-axis denotes the time after smoking is completed and the *y*-axis, the distance the tracer material is transported. The *shaded area* indicates one SD of the linear regression in the non-smoking normal control dogs ($n=17$). The gradient of a line means mucociliary transport velocity of a particular dog. *Upper panel:* when five non-filtered cigarettes were (forcibly) smoked. *Lower panel:* when eight non-filtered cigarettes were smoked. As the number of cigarettes smoked increases, the migrating (transport) distance of the tracer material decreases

22.3.2.3.2
Radioaerosol Inhalation Studies

22.3.2.3.2.1
Delayed or Follow-up Imaging

When delayed imaging is repeated after radioaerosol inhalation in patients with bronchogenic carcinoma, some show a hot spot which either remains constant in size or grows larger than that on an image acquired immediately after inhalation (Fig. 22.26), while in others a hot spot observed on an immediate image disappears on delayed or follow-up images. An inevitable disadvantage of this delayed, follow-up imaging method, however, is that it is not always known how the radioactivity is cleared from the airways with time because there is no guarantee that the radioactivity under consideration is the same as that on previous images, and whether a hot spot on subsequent images is the same as that on previous images is difficult to ascertain. Figure 22.27 illustrates such a puzzling case.

Immediately after aerosol inhalation there was no radioactivity on the left main bronchus. Delayed imaging at 1 h showed a radioactive glob on the left main bronchus, which disappeared within 2 h following inhalation. If mucus is always transported cephalad, it is difficult to interpret the presence of radioactivity in the left main bronchus. This finding suggests that there must be some inherent defect in this delayed imaging method for studying mucociliary clearance. Continuous measurement of radioactivity following radioaerosol inhalation instead of spotty follow-up imaging is mandatory to determine precisely what happens to the inhaled aerosols in terms of mucociliary clearance.

22.3.2.3.2.2
Radioaerosol Inhalation Lung Cine-Scintigraphy

To avoid the above drawback, radioactivity of the entire thorax including the trachea is measured continuously from immediately following radioaerosol

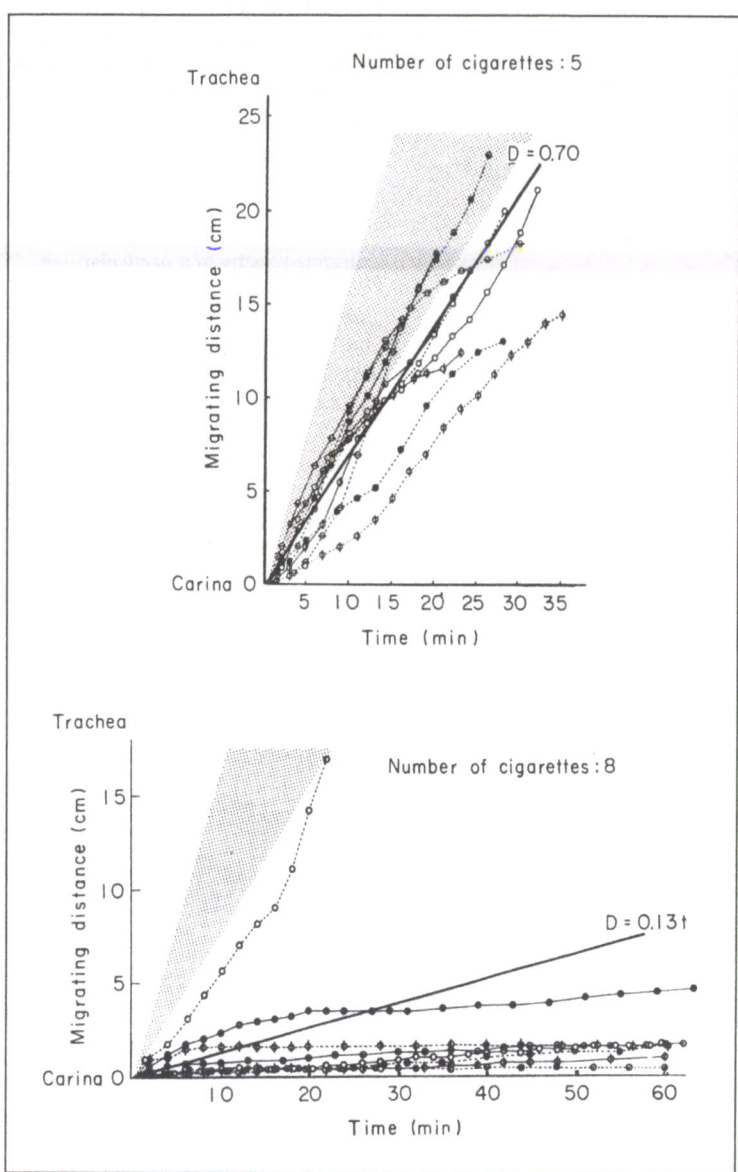

Fig. 22.25. *Upper panel:* when dogs smoke five cigarettes through a glass fiber filter which takes up more than 99.5% of particulates matter, the gradients tend to become steeper than when five cigarettes were smoked directly without using the filter. *Lower panel:* when cigarettes smoked through the glass filter are eight in number, a lesser effect of the filter becomes evident

inhalation for 80–120 min in sequential 10 s frame mode in 64×64 matrix. Each sequential 10 s data is recorded and stored in a computer. The raw data are either compiled for cinematographic display, referred to as "radioaerosol inhalation lung cine-scintigraphy," or used for quantitative analysis, or both (Isawa et al. 1981 c, 1984 a, b). "Radioaerosol inhalation lung cine-scintigraphy" is displayed in cine-mode on a cathode ray tube screen at the rate of 18 frames/s, and the cinematographic display is recorded by a movie camera or on videotape. This cinematographic display enables visualization of the features of actual mucus transport on the airways in vivo (Isawa et al. 1981 c, 1984 a, b; Teshima et al. 1989).

22.3.2.4
Quantitative Evaluation of Mucociliary Clearance

22.3.2.4.1
Extrapulmonary Airways

When a radioactive droplet is placed on the tracheal mucosal surface, transport velocity of a radioactive droplet (equivalent to the tracheal mucociliary clearance velocity) can be assessed by timing the passage of the peak radioactivity of the droplet between two measuring points, then performing the appropriate arithmetic. This method is valid only when transport is straight, upward, and constant in velocity. But, as

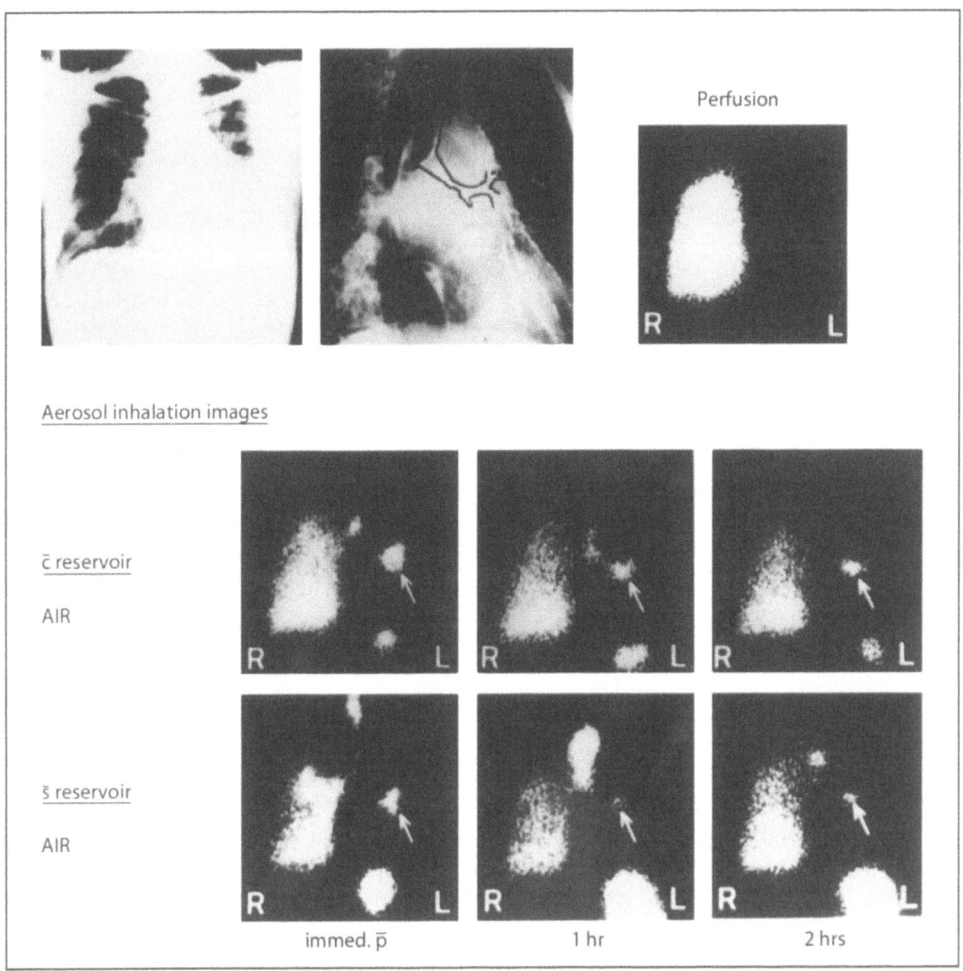

Fig. 22.26. A hot spot at the site of cancer invasion on the left main bronchus persists for 2 h or longer, indicating dis turbed mucociliary clearance. A 55-year-old man with large cell carcinoma

before, bronchoscopy is required and the method is not clinically feasible, nor is transport always straight and upward in direction, thus the development of radioaerosol inhalation lung cine-scintigraphy. Various patterns of mucus transport have been observed by the latter (Isawa et al. 1980a, 1981c, 1984a, b).

22.3.2.4.1.1
Condensed Image Mode

Quantitative analysis followed the method of Teshima et al. (1989): The tracheal portion of each frame data is selected and arranged sequentially on a computer (Fig. 22.28) and displayed on the cathode ray tube with time on the x-axis and distance on the y-axis. When mucus transport is cephalad and straight in direction and constant in velocity, the transport is seen

as a diagonal line (Fig. 22.29a). When mucus transport is greatly disturbed, and velocity is at virtual standstill, the line becomes horizontal (Fig. 22.29b). With radioaerosol inhalation, wide bands of trajectories are depicted.

22.3.2.4.1.2
Trajectory Mode

To analyze the transport of a mucus glob more accurately and quantitatively, the following method was devised and termed "trajectory mode" (Teshima et al. 1989). First, a point of interest (POI) is set at a "hot spot" on a frame data magnified from the original 64×64 matrix data by interpolation, and its location is determined mathematically and recorded in a computer. On the next frame, also similarly magnified by

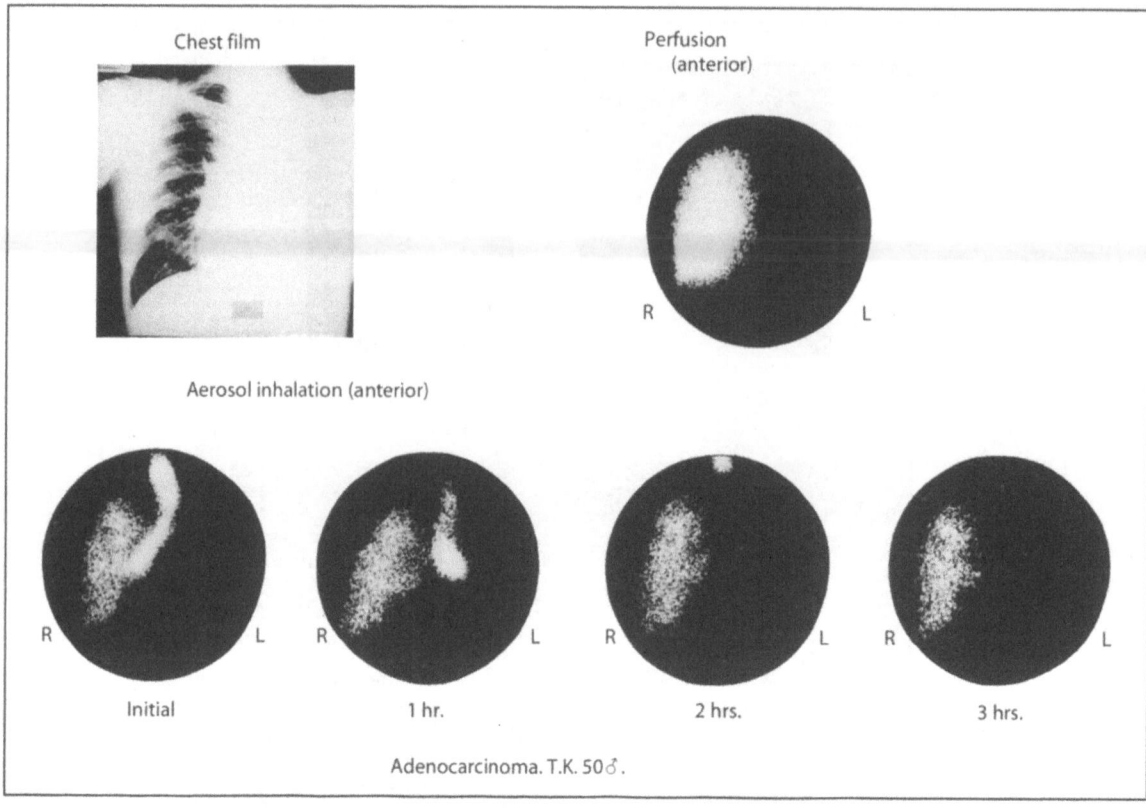

Chest film

Perfusion (anterior)

Aerosol inhalation (anterior)

| Initial | 1 hr. | 2 hrs. | 3 hrs. |

Adenocarcinoma. T.K. 50♂.

Fig. 22.27. Immediately after radioaerosol inhalation no radioactivity deposits in the left main bronchus, indicating no ventilation in the left lung. However, by 1 h a hot spot has appeared in the left main bronchus. In 3 h, the hot spot has completely disappeared. A 50-year-old man with adenocarcinoma. The left lung shows a blackout on chest x-ray and absent perfusion on perfusion lung imaging

interpolation, the change in hot spot location is measured mathematically and recorded. This procedure is repeated frame by frame, and each mathematical location is connected using B spline function, which makes a trajectory of the transport of the hot spot (Fig. 22.30). This trajectory enables a detailed analysis of mucociliary transport in the trachea (Teshima et al. 1989).

Vector analysis becomes possible by using this trajectory, as illustrated in Fig. 22.31, and various parameters can be calculated as indicated below:

- Effective transport velocity (V_{eff}) = S/T
- Apparent transport velocity (V_{app}) = s/T
- Traveling pathway index (a) = s/S
- Stasis ratio (SR) = $\sum$[number of times ($\Delta s_i = 0$)/N]

Figure 22.32 compares tracheal mucus transport velocity (effective velocity and apparent velocity) in normal nonsmokers, smokers, and patients with chronic obstructive pulmonary diseases (COPD).

Fractions of forward transport, backward transport and stasis in mucus transport can also be easily calculated. Stasis ratio (SR) between healthy smokers, non-

smokers, and patients with COPD are shown in Fig. 22.33.

22.3.2.4.1.3
Iso-Count Display

Iso-count display of a hot spot in sequential frames is also possible by delineating iso-count lines in each frame. The direction of transport can be assessed, but this method, too, is reasonable only when a radioactive droplet is placed on the trachea. Thus this method is not practically applicable to aerosol inhalation studies (Teshima et al. 1989).

22.3.2.4.2
Ciliated Airways in Lung Parenchyma

When radioaerosol is inhaled, radioactivity deposits in (A) extrapulmonary ciliated airways, (B) intrapulmonary ciliated airways, (C) nonciliated small airways in-

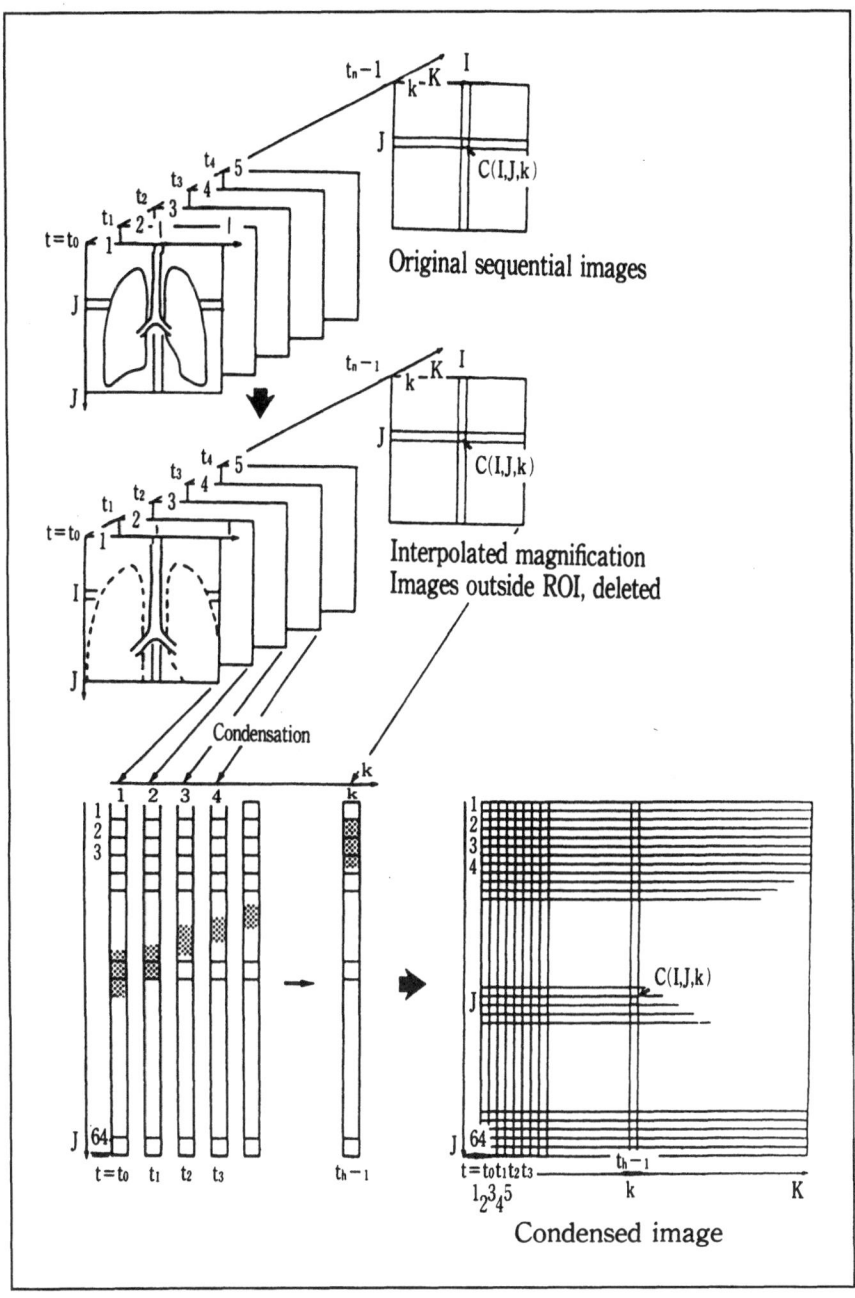

Fig. 22.28. Diagram of condensed image mode. Only the tracheal portion is arranged sequentially and a hot spot is imaged

cluding the alveolar space, and (D) esophagus and gastrointestinal tract (Fig. 22.34; Isawa et al. 1984a, b).

Disregarding the radioactivity in the gastrointestinal tract, the radioactivity at time zero, immediately after aerosol inhalation, can be written as follows:

$$A_0 + B_0 + C_0 = T_0 \tag{22.1}$$

where A, B, C represent the compartments in Fig. 22.34 and T the total radioactivity in all three

compartments. At time t, radioactivity in the compartments would be:

$$A_t + B_t + C_t = T_t \tag{22.2}$$

If radioactivity is corrected for physical decay, Eq. 22.2 becomes:

$$A_{tc} + B_{tc} + C_{tc} = T_{tc} \tag{22.3}$$

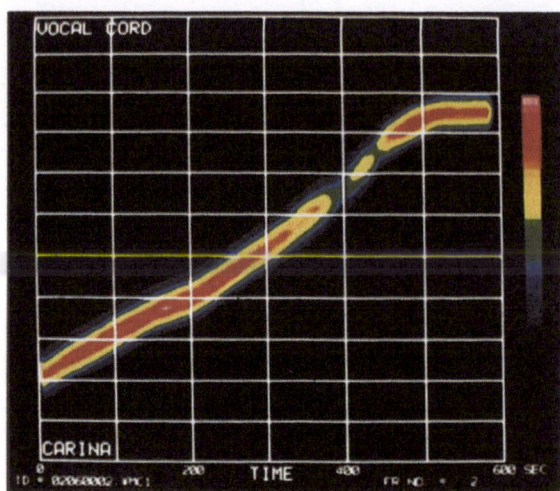

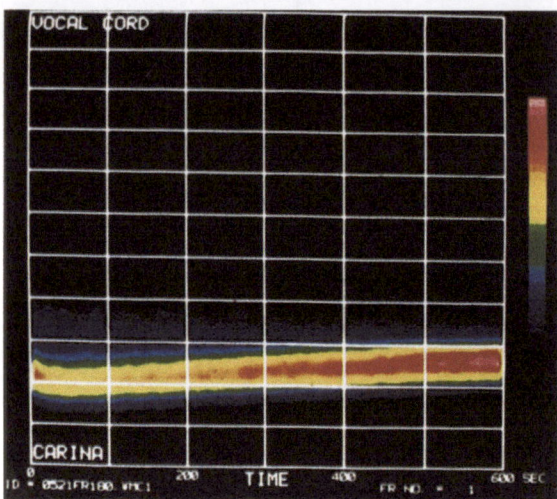

Fig. 22.29. a Transport of a radioactive droplet with mucus of a normal dog from the carina to the vocal cord by condensed image mode. **b** Transport of a radioactive droplet with mucus of a normal dog forced to smoke five cigarettes. Mucus transport is greatly disturbed

If we define the radioactivity remaining in the lung at 24 h later as the amount of radioactivity deposited in the nonciliated space of the lung, C_0, corrected for physical decay should be the same as C_{tc} ($C_0 = C_{tc}$).

Practically speaking, it is extremely difficult to measure A_t without including the contamination by radioactivity (being swallowed) in the esophagus behind the trachea. In evaluating the clearance of radioactivity over time from the lung parenchyma, which is simply mucociliary clearance in the intrapulmonary ciliated airways, only radioactivity in the extrapulmonary airways should be taken into consideration. Thus the above formulae should be rewritten as follows:

$$B_0 + C_0 = T_0 \tag{22.1'}$$

$$B_t + C_t = T_t \tag{22.2'}$$

$$B_{tc} + C_{tc} = T_{tc} \tag{22.3'}$$

Sequential measurements of radioactivity in the lung parenchyma (or in the thorax excluding the extrapulmonary mediastinal region) over 24 h are all that is required to calculate the indices that follow.

■ **Lung Retention Ratio (LRR).** LRR expresses the amount of radioactivity remaining in the lungs at time t relative to the total radioactivity initially deposited:

$$LRR(\%) = T_{tc}/T_0 \times 100$$

■ **Airway Deposition Ratio (ADR).** ADR indicates the amount of radioactivity throughout the ciliated airways relative to the total radioactivity initially deposited in the lungs:

$$ADR(\%) = B_{tc}/T_0 \times 100 = (T_{tc} - C_0)/T_0 \times 100$$

■ **Airway Retention Ratio (ARR).** ARR measures the proportion of radioactivity initially deposited in the ciliated airways that still remains there at time t:

$$ARR(\%) = B_{tc}/B_0 \times 100 = \{(T_{tc} - C_0)/(T_0 - C_0)\} \times 100$$

■ **Airway Clearance Efficiency (ACE).** ACE is the proportion of radioactivity initially deposited on the ciliated airways that has been cleared by time t:

$$ACE(\%) = (B_0 - B_{tc})/B_0 \times 100$$
$$= \{(T_0 - T_{tc})/(T_0 - C_0)\} \times 100$$

■ **Alveolar Deposition Ratio (ALDR).** ALDR indicates the proportion of total initial radioactivity remaining in the lung parenchyma at 24 h, or the proportion of radioactivity deposited in the nonciliated space of the lungs including the alveolar space at the completion of aerosol inhalation. This compartment C lacks mucociliary clearance:

$$ALDR(\%) = C_0/T_0 \times 100$$

Normal values using ^{99m}Tc-HSA aerosol with AMAD of 1.9 μ and σg of 1.7 are detailed elsewhere (Isawa et al. 1989). The normal ALDR with this aerosol is about 40%, but ALDR values may differ with the size of the

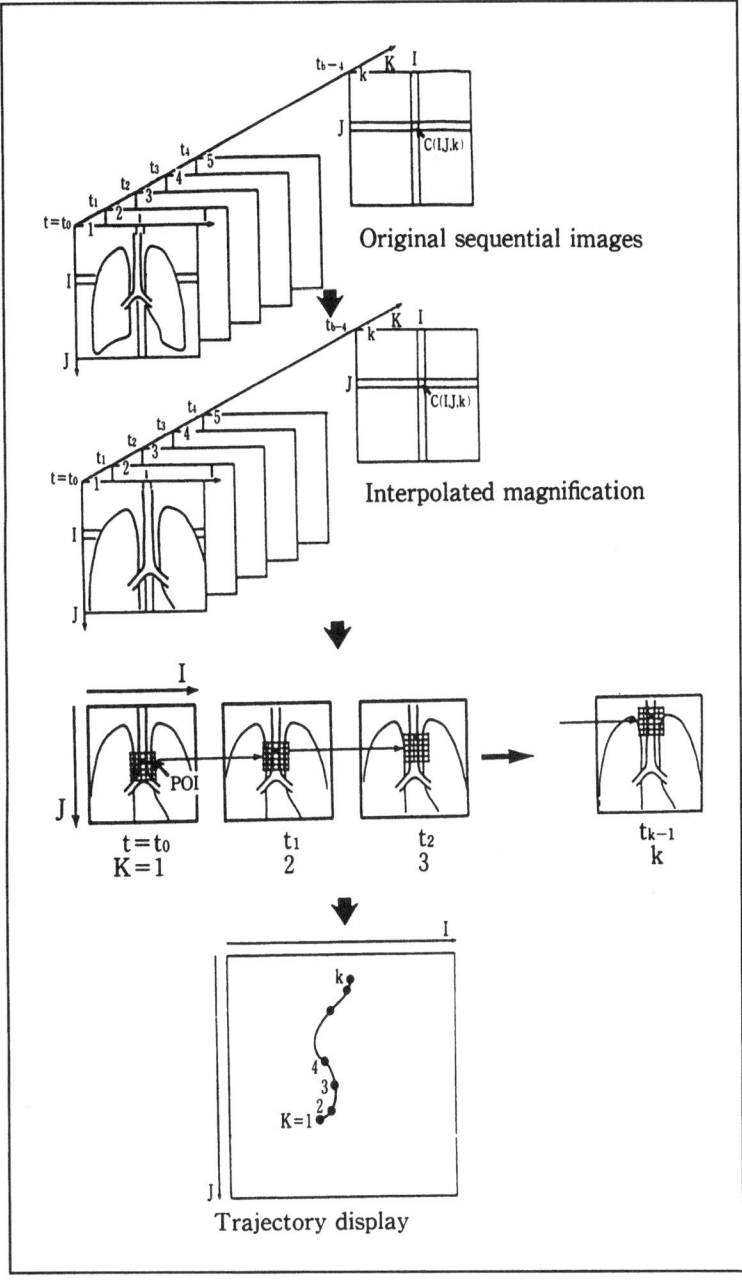

Fig. 22.30. Diagram showing the principle of the trajectory mode. The location of the point of interest is sequentially determined frame by frame, and each mathematical location is connected using B spline function

aerosol inhaled such that the smaller the aerosol size, the higher the ALDR (Miki et al. 1992). It is ideal to establish the normal ranges of each parameter at each laboratory according to the aerosol size used.

22.3.2.5
Mucociliary Clearance in Health and Disease

22.3.2.5.1
Large Airways: Trachea and Major Bronchi

22.3.2.5.1.1
Normal Subjects

According to observations by radioaerosol inhalation lung cine-scintigraphy, the transport of inhaled radioaerosol deposited in the airways in nonsmoking nor-

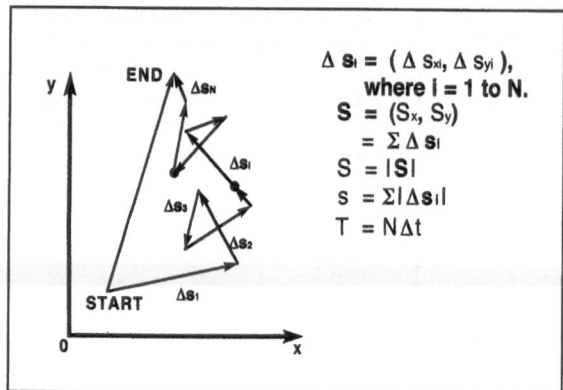

Fig. 22.31. Vector analysis of a trajectory. From *START* to *END* the actual pathway is comprised of s_1, s_2,c s_N, although macroscopically it resembles a straight line between *START* and *END*. Each s_1, s_2,c s_N is a vector and mathematical calculation is possible

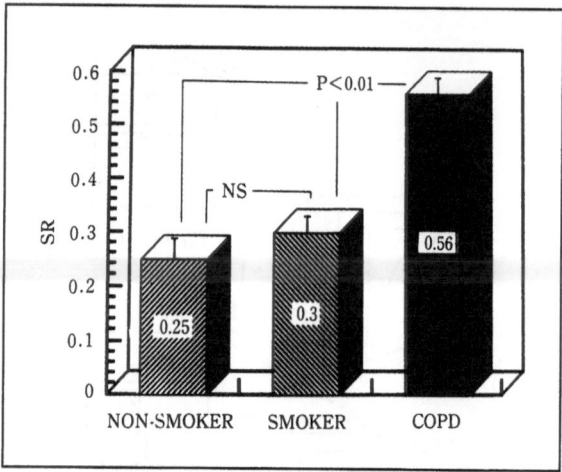

Fig. 22.33. Stasis ratio (SR) significantly differs between normal (smokers and nonsmokers) and patients with COPD

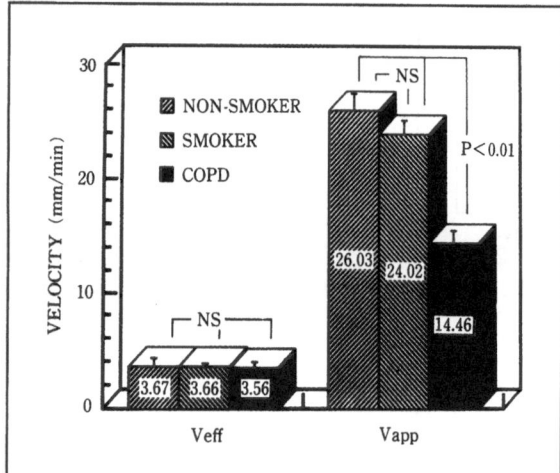

Fig. 22.32. Although there is no statistically significant difference in effective velocity (V_{eff}) between nonsmokers, smokers, and patients with chronic obstructive pulmonary disease (COPD), apparent velocity (V_{app}) significantly differs between normal subjects (smokers and nonsmokers) and patients with COPD

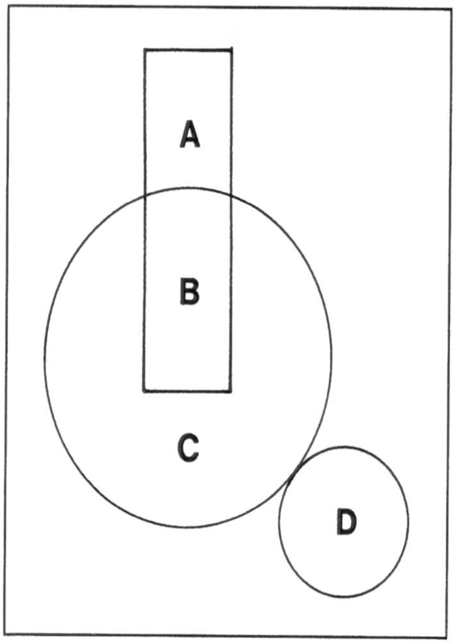

Fig. 22.34. Diagram of deposition sites of inhaled aerosol such as the trachea and the proximal portion of the major bronchi (*A*), the intrapulmonary ciliated airways (*B*), the non-ciliated airways including the alveolar space (*C*), and the digestive tract such as mouth, esophagus, stomach, etc. (*D*)

mal subjects is always axial and cephalad in direction, steady and constant in its transport velocity, and without stagnation of radioactivity in the trachea or bronchi. In smokers and some former smokers without manifest symptoms, a temporary collection of radioactive mucus is seen over the bronchi near or over the carina, although radioactive transport is still cephalad in direction with transport velocity nearly constant (Isawa et al. 1984a). Such stasis, if present, never persists long. Visible in patients with obstructive airways disease are retrograde transport (retreat),

stasis, or stagnation of mucus globs, frequent up-and-down motions of radioactive globs in the trachea or bronchi, and/or migration into the other regions of the same lung or into the bronchus of the opposite lung (Isawa et al. 1984b).

In trajectory mode, the trajectory in a normal subject is seen to be tortuous and complex, indicating that microscopically mixed forward and retrograde transports are combined with stasis, the overall transport direction, however, remaining oropharyngeal (Isawa et al. 1984b).

22.3.2.5.1.2
Obstructive Airway Diseases

In those with obstructive airway disease, mucus transport over the trachea and the major bronchi is extremely protean in its direction and transport pattern. Of 21 patients with obstructive airways disease studied by radioaerosol inhalation lung cine-scintigraphy, 14 showed temporary but frequent stopping and starting of radioactivity in the airways in the course of lung clearance. Even after mucus begins to migrate up the trachea, it tends to stop on the way. This stopping and renewed migration were repeated many times in the course of mucus transport. Migration is often accelerated by coughing and/or clearing the throat, the radioactive mucus finally being swallowed into the stomach or expectorated. Sometimes radioactivity remains at the same spot without migration until cleared by coughing. In this sense coughing appears to be the only means of upward propulsion of the mucus. In 10 patients there was reversal of mucus flow, in 5 migration or straying or radioactivity from one bronchus to that of the opposite lung, bypassing the trachea, followed by shuttling between the right and left main bronchus, finally coughed up. In 4 there was spiral or zigzag transport of radioactivity as shown in Fig. 22.35.

In other series of patients with obstructive airways disease, radioaerosol inhalation lung cine-scintigraphy frequently showed not only the shuttling of mucus between the right and left main bronchus but also migrations of mucus from one region of the lung into the different regions of the same lung (Isawa et al. 1984b).

By trajectory mode, analysis in patients with COPD reveals a simpler shape than that in normal subjects because the stasis ratio becomes higher as shown in Fig. 22.33 (Teshima et al. 1989).

22.3.2.5.1.3
Bronchiectasis

In bronchiectatic lung regions, the deposition of inhaled radioaerosols is diminished and inhomogeneous. Radioaerosol inhalation lung cine-scintigraphy has revealed that transport of inhaled radioactivity from bronchiectatic regions is greatly deranged. Regional stasis was observed in 12 of the 20 patients studied, regurgitation or reversed transport in 14, straying in 8, and spiral or zigzag motion in 1. The transport patterns were more or less a combination of these four basic abnormal patterns. When coughs occur, regurgitation and straying become more marked in the bronchiectatic regions. Only coughs can force radioactive mucus from inside the bronchiectatic regions to outside, mucociliary clearance from these areas being very inefficient without the help of coughing. These regional abnormalities in mucociliary transport seem to be responsible for the development of infections and hemoptysis in the bronchiectatic regions (Isawa et al. 1990a).

Genetic disorders, autosomal recessive, such as primary ciliary dysfunction including Kartagener syndrome (Eliasson et al. 1977), Young's syndrome (Greenstone et al. 1988), and cystic fibrosis (CF; Zielenski and Tsui 1995) culminate in bronchiectasis due to mucociliary dysfunction. The gene for CF has been identified on chromosome 7. The encoded gene product, named cystic fibrosis transmembrane conductance regulator (CFTR), corresponds to a cAMP-regulated chloride channel found almost exclusively in the

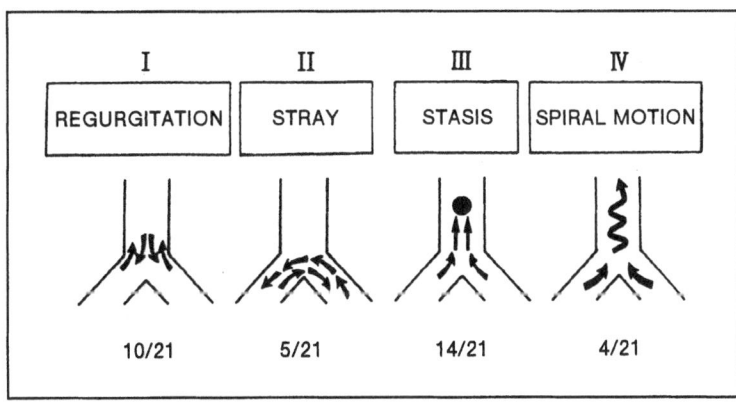

Fig. 22.35. Four abnormal mucus transport patterns in the trachea (*I–IV*). Although steady transport cephalad in direction is seen in normal nonsmokers, the four abnormal mucus transport patterns are observed in combination in pathological states. In straying, shuttling transport from one region to a different portion inside the same lung is also seen. The *numbers below* indicate frequency observed in 20 patients with COPD

secretory epithelial cells (Zielenski and Tsui 1995). Attempts have been made to transfer the normal human CFTR cDNA to the epithelium of patients with CF (Harvey et al. 1999).

22.3.2.5.1.4
Bronchogenic Carcinoma

Abnormal mucociliary transport patterns such as regurgitation, straying, stasis, and spiral or zigzag motions are seen in bronchogenic carcinoma, especially in patients with complicating COPD. Abnormal mucus transport patterns have nothing to do with the histological diagnosis of bronchogenic carcinoma but with the degrees of functional and anatomical airways obstruction.

Bronchial invasions or protrusions of cancer into large airways are often recognized as "hot spots"; these persist or disappear over time depending on the degree of mucosal damage. When a tumor is covered with intact ciliated mucosa, the hot spots eventually disappear, but they persist when the bronchial mucus is denuded by the tumor (Isawa et al. 1986c).

22.3.2.5.1.5
Idiopathic Interstitial Fibrosis and Pulmonary Vascular Disease

Transport in the large airways does not differ in these conditions from that in normal subjects unless complicated with COPD, in which case similar transport patterns are seen to those in patients with COPD (Isawa et al. 1986a).

22.3.2.5.2
Ciliated Airways in Lung Parenchyma

22.3.2.5.2.1
Normal Subjects

Using ^{99m}Tc-HSA aerosol whose AMAD is 1.93 μ with σg of 1.52, the following findings have been obtained:
- LRR is 85–90%, 80–85%, and 75–80% at 30, 60, and 90 min, respectively.
- ALDR is equivalent to the LRR at 24 h and amounts to about 40%.
- ADR immediately after inhalation is about 60% and decreases with time to 50%, 45%, and 40% at 30, 60, and 90 min, respectively.
- ARR is 80–85%, 65–70%, 60–65% at 30, 60, and 90 min, respectively.

- ACE is 15–20%, 30–35%, 35–40% at 30, 60, and 90 min, respectively.
- Otherwise normal smokers show a slightly faster clearance than nonsmokers. ALDR values are significantly larger in normal nonsmokers than in normal smokers (Isawa et al. 1989).

22.3.2.5.2.2
Obstructive Airway Diseases

As shown in Fig. 22.36, LRR itself is not distinguishable from the normal range, but ALDR is significantly less than normal range. Both ADR and ARR are higher and ACE lower than the normal range, indicating that a larger proportion of inhaled aerosol deposits in ciliated airways and that mucociliary clearance is less efficient (Isawa et al. 1984b).

22.3.2.5.2.3
Idiopathic Interstitial Fibrosis and Pulmonary Vascular Disease

All the parameters remain in the normal range in these conditions (Isawa et al. 1986a, 1988).

22.3.2.6
Pharmacological Effects

The clinical effects of mucolytic agents and bronchodilators have mostly been evaluated thus far rather subjectively on the basis of patients' sense of improvement. Quantitative and objective evaluation is possible with the present methods of radioaerosol inhalation and subsequent sequential imaging, with measurement of radioactivity of the lungs. Reports thus far regarding drug effects are rather conflicting (Sackner 1978; Matthys and Koehler 1980; Wanner 1981; Pavia et al. 1983).

22.3.2.6.1
Bromhexine

Bromhexine is claimed to liquefy mucus by breaking the mucopolysaccharide chains in the mucus, thus facilitating its removal. ACE was evaluated by radioaerosol inhalation lung scintigraphy in ten patients with various chest diseases before and after 7 days of 8 mg oral bromhexine administered three times per day.

ACE was barely significantly improved by χ^2 test ($P = 0.05$). Pulmonary function tests revealed little

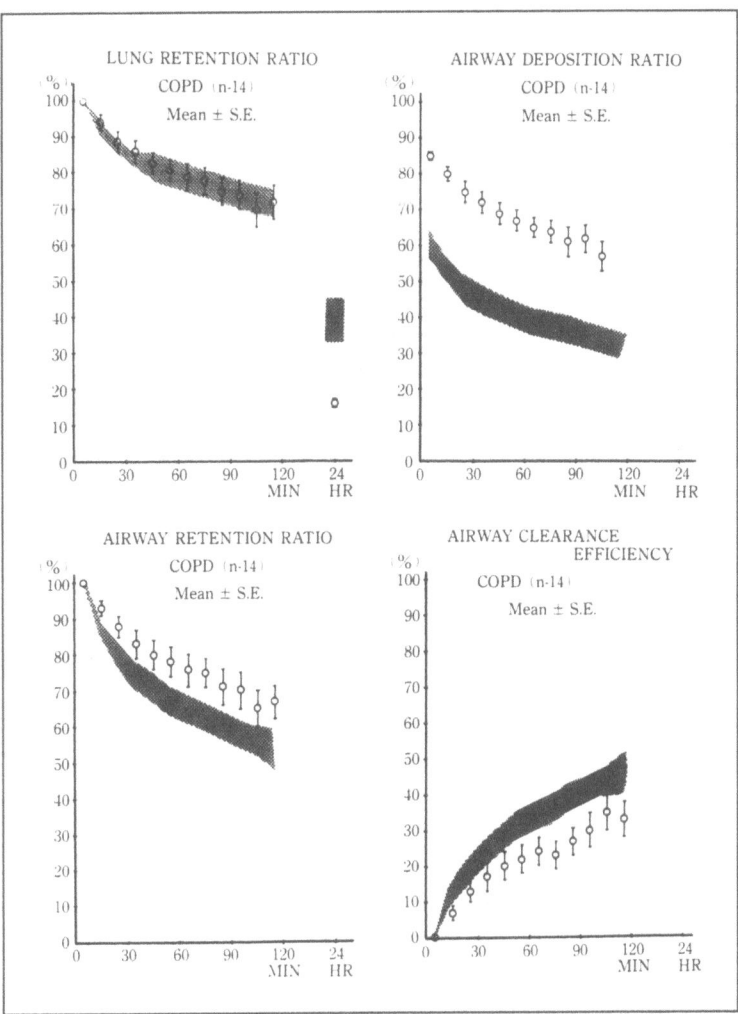

Fig. 22.36. Five parameters in patients with COPD. ALDR equivalent to LRR at 24 h and ACE are significantly decreased, and ADR and ARR increased. LRR remains within normal limits. *Bands*, means and 95% confidence intervals

change before and after the administration except for a slight increase in maximum mid-expiratory flow rate and a slight decrease in the ratio of residual volume to total lung capacity (Isawa et al. 1984 c).

22.3.2.6.2
β_2-Stimulators

It is known that ciliary beat frequency in vitro is increased by the administration of salbutamol (Van As 1974). Oral administration of salbutamol 8 mg three times per day for 7 days did not change either radioaerosol inhalation lung images or quantitative parameters compared with baseline, although pulmonary function tests showed significant bronchodilation after 7 days' administration (Isawa et al. 1986 b).

In the midst of radioactive measurement following inhalation of ^{99m}Tc-HAS aerosol, procaterol, a β_2-stim-

ulator was inhaled to determine whether change appeared in the time-activity curve. If the curve over the lungs becomes steeper, we judge that the mucociliary clearance is accelerated by the medication. However, we found neither a significant change in the slope of time-activity curves nor in any quantitative parameters in patients with bronchial asthma in remission, although spirometry indicated significant bronchodilation (Isawa et al. 1990 b).

22.3.2.6.3
Aminophylline and β_2-Stimulators

Radioaerosol inhalation lung cine-scintigraphy and pulmonary function tests were performed on ten patients with bronchial asthma in remission before and after 250 mg aminophylline infusion followed by inhalation of salbutamol. The bronchodilating effect of the combined treatment was significant; inhaled aerosol deposited more homogeneously and less centrally in the lungs. The penetration index increased from 31%±3% to 49%±7% and the ALDR from 29%±2% to 39%±1%. The ADR decreased from 72%±2% to 61%±1% immediately after the treatment. Spirometry revealed significant bronchodilation. However, there was little quantitative improvement in mucociliary clearance after the treatment (Isawa et al. 1987).

22.4
References

Agnew JE (1984) Physical properties and mechanism of deposition of aerosols. In: Clarke SW, Pavia D (eds) Aerosol and the lung, chap 3. Butterworths, London, pp 49–70

Alpert JS, Smith R, Carlson J et al (1976) Mortality in patients treated for pulmonary embolism. JAMA 236:1477–1480

Altshuler B, Palmes ED, Yarmus L et al (1959) Intrapulmonary mixing of gases studied with aerosols. J Appl Physiol 14:321–327

Anazawa Y, Isawa T, Teshima T et al (1991) Pulmonary epithelial permeability in normal subjects and patients with idiopathic interstitial pneumonia. J Jpn Soc Chest Dis 29:1439–1443

Anazawa Y, Isawa T, Teshima T et al (1992a) Changes in pulmonary epithelial permeability due to thoracic irradiation. J Jpn Soc Chest Dis 30:862–867

Anazawa Y, Isawa T, Teshima T et al (1992b) Pulmonary epithelial permeability in rats with bleomycin-induced pneumonitis. J Jpn Soc Chest Dis 30:1222–1228

Baetjer AM (1967) Effect of ambient temperature and vapor pressure on cilia-mucus clearance rate. J Appl Physiol 23:498–504

Bass H, Henderson JAM, Heckscher T et al (1968) Regional structure and function in bronchiectasis. A correlative study using bronchography and ^{133}Xe. Am Rev Respir Dis 97:598–609

Baum G, Priel Z, Roth Y et al (eds) (1998) Cilia, mucus, and mucociliary interactions. Dekker, New York, pp 1–584

Benumof JL, Trousdale FR (1982) Aminophylline does not inhibit canine hypoxic pulmonary vasoconstriction. Am Rev Respir Dis 126:1017–1019

Bergofsky EH, Holtzman S (1967) A study of the mechanisms involved in the pulmonary arterial response to hypoxia. Circ Res 20:506–519

Berke HI, Roslinski LM (1971) The roentgenographic determination of tracheal mucociliary transport rate in the rat. Am Ind Hyg Assoc J 32:174–178

Biello DR, Mattar AG, McKnight RC et al (1979) Ventilation-perfusion studies in suspected pulmonary embolism. AJR 133:1033–1037

Boucher RC, Johnson J, Inoue S et al (1980) The effect of cigarette smoke on the permeability of guinea pig airways. Lab Invest 43:94–100

Boulanger C, Luscher TF (1990) Release of endothelin from the porcine aorta: inhibition by endothelium-derived nitric oxide. J Clin Invest 85:587–590

Bradley M, Nukiwa T, Crystal RG (1988) Molecular basis of alpha-1-antitrypsin deficiency. Am J Med 84:13–31

Braude S, Nolop KB, Hughes JMB et al (1983) Comparison of lung vascular and epithelial permeability indices in the adult respiratory distress syndrome. Am Rev Respir Dis 133:1002–1005

Braude S, Apperley J, Krausz T et al (1985) Adult respiratory distress syndrome after allogenic bone marrow transplantation: evidence for a neutrophil-independent mechanism. Lancet 1:1239–1242

Burch WM, Sullivan PJ, McLaren CJ (1986) Technegas – a new ventilation agent for lung scanning. Nucl Med Commun 7:865–871

Carson JL, Kelly MA, Duff A et al (1992) The clinical course of pulmonary embolism. N Engl J Med 326:1240–1245

Chacon RA, Corris PA, Dark JH et al (1998) Comparison of the functional results of single lung transplantation for pulmonary fibrosis and chronic airway disease. Thorax 53:43–49

Chopra KS, Taplin GV, Tashkin DP et al (1979) Lung clearance of soluble aerosols of different molecular weights in systemic sclerosis. Thorax 34:63–67

Chung S-K, Kim H-H, Bahk Y-W (1997) Prestenotic bronchial radioaerosol deposition: A new ventilation scan sign of bronchial obstruction. J Nucl Med 38:71–74

Coates G, O'Brodovich H (1986) Measurement of pulmonary epithelial permeability with 99mTc-DTPA aerosol. Semin Nucl Med 16:275–284

Cook GJR, Fogelman I (1996) Pseudo pulmonary embolism in complex heart disease. J Nucl Med 37:1359–1361

Cooper JD, Patterson GA, Sundaresan RS et al (1996) Results of 150 consecutive bilateral lung volume reduction procedures in patients with severe emphysema. J Thorac Cardiovasc Surg 112:1319–1329

Criley JM, Ross RS (1971) Introduction to methods. In: Cardiopulmonary physiology. Oldsman, Tampa Tracing, pp 1–34

Cumming G, Crank J, Horsfield K et al (1966) Gaseous diffusion in the airways of the human lung. Respir Physiol 1:58–78

Dalen JE, Alpert JS (1975) Natural history of pulmonary embolism. Prog Cardiovasc Dis 17:257–270

Eliasson R, Mossberg B, Camner P et al (1977) The immotile-cilia syndrome. A congenital ciliary abnormality as an etiologic factor in chronic airway infections and male sterility. N Engl J Med 297:1–6

Engel L (1978) Gas mixing within the acinus of the lung. J Appl Physiol 54:416–423

Fisher AB, Forman HJ (1985) Oxygen utilization and toxicity in the lungs. In: Fishman AP, Fisher AB (eds) Handbook of physiology, chap 5, sect 3. The respiratory system. American Physiological Society, Bethesda, pp 231–254

Foord N, Black A, Walsh M (1978) Regional deposition of 2.5–7.5 μm diameter inhaled particles in healthy male nonsmokers. J Aerosol Sci 9:343–357

Forster RE II, DuBois AB, Briscoe WA, Fisher AB (1986) Physiologic dead space. Appendix C. In: The lung. Physiologic basis of pulmonary function tests, 3rd edn. Year Book Medical, Chicago, pp 266–267

Fraser RG (1961) Measurement of the calibre of human bronchi in three phases of respiration by cine bronchography. J Can Assoc Radiol 12:102–112

Freitas JE, Sarosi MG, Nagle CC et al (1995) Modified PIOPED criteria used in clinical practice. J Nucl Med 36:1573–1578

Friedman M, Scott FD, Poole DO et al (1977) A new roentgenologic method for estimating mucous velocity in airways. Am Rev Respir Dis 115:67–72

Frostell CG, Bromqvist H, Hedenstierna G et al (1993) Inhaled nitric oxide selectivity reverses human hypoxic pulmonary vasoconstriction without causing systemic vasoconstriction. Anesthesiology 78:427–435

Furman WR, Summer WR, Kennedy TP et al (1982) Comparison of the effects of dobutamine, dopamine, and isoproterenol on hypoxic pulmonary vasoconstriction in the pig. Crit Care Med 10:371–374

Geroulakos G (2001) The risk of venous thromboembolism. The evidence is only circumstantial. BMJ 322:188

Gil J, Weibel ER (1971) Extracellular lining of bronchioles after perfusion-fixing of rat lungs for electron microscopy. Anat Rec 169:185–200

Goldberg SN, Richardson DD, Palmer EL et al (1996) Pleural effusion and ventilation/perfusion scan interpretation for acute pulmonary embolus. J Nucl Med 37:1310–1313

Gorin AB, Stewart PA (1979) Differential permeability of endothelial and epithelial barriers to albumin flux. J Appl Physiol 47:1315–1324

Gottschalk A (2000) The chronic perfusion defect: our knowledge is still hazy, but the message is clear. J Nucl Med 41:1049–1050

Gottschalk A, Stein PD, Henry JW et al (1996) Matched ventilation, perfusion and chest radiographic abnormalities in acute pulmonary embolism. J Nucl Med 37:1636–1638

Greenstone MA, Rutman A, Hendry WF et al (1988) Ciliary function in Young's syndrome. Thorax 43:153–154

Greenwood MF, Holland P (1972) The mammalian respiratory tract surface. A scanning electron microscopic study. Lab Invest 27:296–304

Groskin SA (1993) Subsegmental anatomy of the lung. In: Groskin SA (ed) Heitzman's The lung, 3rd edn, chap 4. Mosby Year Book, St Louis, pp 43–69

Gross NJ (1977) Pulmonary effects of radiation therapy. Ann Intern Med 86:81–92

Haas F, Bergofsky EH (1972) Role of the mast cell in the pulmonary pressor response to hypoxia. J Clin Invest 51:3154–3162

Harvey BG, Leopold PL, Hackett NR et al (1999) Airway epithelial CFTR mRNA expression in cystic fibrosis patients after repetitive administration of a recombinant adenovirus. J Clin Invest 104:1245–1255

Hemsen A, Franco-Cereceda A, Matran R et al (1990) Occurrence, specific binding sites and functional effects of endothelin in human cardiopulmonary tissue. Eur J Pharmacol 191:319–328

Hilding AC (1957) Ciliary streaming in the bronchial tree and the time element in carcinogenesis. N Engl J Med 256:634–640

Hirano T (1988) Mucociliary clearance mechanisms. I. Canine tracheal mucous velocity and different particles sizes. Kohkenshi 40:107–113

Hogg JC (1983) The effect of smoking on airway permeability (editorial). Chest 83:1–2

Homans J (1954) Thrombosis of the deep leg veins due to prolonged sitting. N Engl J Med 250:148–149

Horikoshi M, Teshima T, Yanagimachi T et al (2000) ^{99m}Tc technegas imaging in lung cancer. Jpn J Clin Radiol 45:187–198

Hughes JMB, Glazier JB, Malorey JE et al (1967) Effect of interstitial pressure on pulmonary blood flow. Lancet 1:192–193

Hunninghake GW, Crystal RG (1983) Cigarette smoking and lung destruction: accumulation of neutrophils in the lungs of cigarette smokers. Am Rev Respir Dis 128:833–838

Ikonen T, Harjula AL, Kinnula V et al (1996) Selective assessment of single-lung graft function with 133Xe radiospirometry in acute rejection and infection. Chest 109:879–884

Ikonen T, Sovijarvi AR, Taskinen E et al (1997) Detection of acute rejection by 133Xe radiospirometry after single lung transplantation in an experimental porcine model. Eur Surg Res 29:12–19

Isawa T (1995) Radioaerosol lung imaging – history and pharmaceuticals. In: Bahk YW, Isawa T (eds) Radioaerosol imaging of the lung. IAEA, Vienna, pp 16–23

Isawa T, Shiraishi K, Yasuda T et al (1967) Effect of oxygen concentration in inspired gas upon pulmonary arterial blood flow. Am Rev Respir Dis 96:1199–1208

Isawa T, Okubo K, Oka S (1969) Postural effect on regional ventilation. Tohoku J Exp Med 97:101–112

Isawa T, Wasserman K, Taplin GV (1970a) Lung scintigraphy and pulmonary function studies in obstructive airway disease. Am Rev Respir Dis 102:161–172

Isawa T, Wasserman K, Taplin GV (1970b) Variability of lung scans following pulmonary embolization. A concept of regional pulmonary ischemia of functional origin. Am Rev Respir Dis 101:207–217

Isawa T, Benfield JR, Johnson DE et al (1971a) Pulmonary perfusion changes after experimental unilateral bronchial occlusion and their clinical implications. Radiology 99:355–360

Isawa T, Hayes M, Taplin GV (1971b) Radioaerosol inhalation lung scanning: its role in suspected pulmonary embolism. J Nucl Med 12:606–609

Isawa T, Taplin GV, Beazell J et al (1972) Experimental unilateral pulmonary artery occlusion. Acute and chronic effect on relative inhalation and perfusion. Radiology 102:101–109

Isawa T, Teshima T, Hirano T et al (1978a) Regulation of regional perfusion distribution in the lungs. Experimental models and effect of alveolar pressure. Tohoku J Exp Med 124:33–46

Isawa T, Teshima T, Hirano T et al (1978b) Regulation of regional perfusion distribution in the lungs. Effect of regional oxygen concentration. Am Rev Respir Dis 118:55–63

Isawa T, Teshima T, Hirano T et al (1979) Effect of aminophylline on regional perfusion distribution in the lungs. Tohoku J Exp Med 128:345–353

Isawa T, Hirano T, Teshima T et al (1980a) Effect of nonfiltered and filtered cigarette smoke on mucociliary clearance mechanism. Tohoku J Exp Med 130:189–197

Isawa T, Teshima T, Hirano T et al (1980b) Effect of alveolar hypercapnia on regional pulmonary perfusion. Tohoku J Exp Med 132:187–197

Isawa T, Teshima T, Hirano T et al (1981a) Effect of isoproterenol on regional pulmonary perfusion and its blockade by propranolol. Tohoku J Exp Med 134:59–69

Isawa T, Teshima T, Hirano T et al (1981b) Effect of dopamine on regional pulmonary perfusion. Tohoku J Exp Med 134:71–77

Isawa T, Teshima T, Hirano T et al (1981c) Radioaerosol inhalation lung cine-scintigraphy: a preliminary report. Tohoku J Exp Med 134:245–255

Isawa T, Teshima T, Hirano T et al (1982) Effect of isoproterenol on regional pulmonary perfusion in the reimplanted lung of the dog. Tohoku J Exp Med 136:163–168

Isawa T, Teshima T, Hirano T et al (1984a) Mucociliary clearance mechanisms in smoking and nonsmoking normal subjects. J Nucl Med 25:352–359

Isawa T, Teshima T, Hirano T et al (1984b) Lung clearance mechanisms in obstructive airways disease. J Nucl Med 25:447–454

Isawa T, Teshima T, Hirano T et al (1984c) Evaluation of mucociliary clearance mechanisms by radioaerosol inhalation lung scintigraphy - effect of oral bromhexine. J Jpn Soc Chest Dis 22:899–909

Isawa T, Teshima T, Hirano T et al (1986a) Mucociliary clearance mechanism in interstitial lung disease. Tohoku J Exp Med 148:169–178

Isawa T, Teshima T, Hirano T et al (1986b) Effect of oral salbutamol on mucociliary clearance mechanisms in the lungs. Tohoku J Exp Med 150:51–61

Isawa T, Teshima T, Hirano T et al (1986c) Mucociliary clearance mechanisms in lung cancer. Kohkenshi 38:147–157

Isawa T, Teshima T, Hirano T et al (1987) Effect of bronchodilation on the deposition and clearance of radioaerosol in bronchial asthma in remission. J Nucl Med 28:1901–1906

Isawa T, Teshima T, Hirano T et al (1988) Mucociliary clearance in pulmonary vascular disease. Ann Nucl Med 2:41–47

Isawa T, Teshima T, Hirano T et al (1989) Normal values for quantitative parameters for evaluation of mucociliary clearance in lungs. Tohoku J Exp Med 158:119–131

Isawa T, Teshima T, Hirano T et al (1990a) Mucociliary clearance and transport in bronchiectasis: global and regional assessment. J Nucl Med 31:543–548

Isawa T, Teshima T, Hirano T et al (1990b) Does a beta 2-stimulator really facilitate mucociliary transport in the human lungs in vivo? Am Rev Respir Dis 141:715–720

Isawa T, Teshima T, Anazawa Y et al (1994) Technegas versus krypton-81m as an inhalation agent. Comparison of pulmonary distribution at total lung capacity. Clin Nucl Med 19:1085–1090

Isawa T, Teshima T, Anazawa Y et al (1995) Inhalation of pertechnegas: similar clearance from the lungs to that of inhaled pertechnetate aerosol. Nucl Med Commun 16:741–746

Isawa T, Lee B-T, Hiraga K (1996) High-resolution electron microscopy of technegas and pertechnegas. Nucl Med Commun 17:147–152

Jacob MP, Baughman RP, Hughes J et al (1965) Radioaerosol lung clearance in patients with active pulmonary sarcoidosis. Am Rev Respir Dis 131:687–689

Jefferies AL, Coates G, O'Brodovich H (1984) Pulmonary epithelial permeability in hyaline-membrane disease. N Engl J Med 311:1075–1080

Jones JG, Lawler P, Crawley JCW et al (1980) Increased alveolar permeability in cigarette smokers. Lancet 1:66–68

Jones JG, Minty BD, Royston D et al (1983) Carbohaemoglobin and pulmonary epithelial permeability in man. Thorax 38:129–133

Kapanci T, Assimacopoulos A, Irle C et al (1974) Contractile interstitial cells. In pulmonary septa: a possible regulator of ventilation/perfusion ratio? Ultrastructural, immunofluorescence, and in vitro studies. J Cell Biol 60:375–392

Kim CK, Heyman S (1989) Ventilation/perfusion mismatch caused by positive pressure ventilatory support. J Nucl Med 30:1268–1270

Knipping HW, Bolt W, Venrath LM et al (1955) Eine neue Methode zur Prüfung der Herz- und Lungenfunktion. Die Funktionsanalyse in der Lungen- und Herzklinik mit Hilfe des radioaktiven Edelgas Xenon 133 (Isotopen-Thorakographie). Dtsch Med Wochenschr 80:1146–1147

Kotrappa P, Raghunath B, Subramanyam PSS et al (1977) Scintigraphy of lungs with dry aerosol generation and delivery system: concise communication. J Nucl Med 18:1082–1085

Kraaijenhagen RA, Haverkamp D, Koopman MM et al (2000) Travel and risk of venous thrombosis. Lancet 356:1492–1493

Kuni CC, Ducret RP, Nakhleh RE et al (1993) Reverse mismatch between perfusion and aerosol ventilation in transplanted lungs. Clin Nucl Med 18:313–317

Levine SM, Anzueto A, Peters JI et al (1994) Medium term functional results of single-lung transplantation for end-stage obstructive lung disease. Am J Respir Crit Care Med 150:398–402

Lucas AM, Douglas LC (1934) Principles of underlying ciliary activity in the respiratory tract. II. A comparison of nasal clearance in man, monkey and other mammals. Acta Otolaryngol 20:518–541

Luscher TF, Vanhoutte PM (1988) Endothelium-dependent responses in human blood vessels. Trends Pharmacol Sci 9:181–184

Mant MJ, O'Brien BD, Thong KL et al (1977) Haemorrhagic complications of heparin therapy. Lancet 1:1133–1135

Marks JD, Luce JM, Lazar NM et al (1985) Effect of increases in lung volume on clearance of aerosolized solute from human lungs. J Appl Physiol 59:1242–1248

Mason GR, Effros RM, Uszler JM et al (1985) Small solute clearance from the lungs of patients with cardiogenic and noncardiogenic pulmonary edema. Chest 88:327–334

Mason GR, Duane GB, Mena I et al (1987) Accelerated solute clearance in pneumocystis carinii pneumonia. Am Rev Respir Dis 135:864–868

Matthys H, Koehler D (1980) Effect of theophylline on mucociliary transport in man. Eur J Respir Dis 109 [Suppl]:98–102

Meignan M, Guillon JM, Denis M et al (1990) Increased lung permeability in HIV-infected patients with isolated cytotoxic T-lymphocytic alveolitis. Am Rev Respir Dis 141:1241–1248

Miki M, Isawa T, Teshima T et al (1992) Difference in inhaled aerosol deposition patterns in the lungs due to three different sized aerosols. Nucl Med Commun 13:553–562

Milic-Emili J (1977) Ventilation. In: West JB (ed) Regional differences in the lung. Academic, New York, pp 167–199

Milic-Emili J, Henderson JAM, Dolovich MB et al (1966) Regional distribution of inspired gas in the lung. J Appl Physiol 21:749–759

Minty BD, Jordan C, Jones JG (1981) Rapid improvement in abnormal pulmonary epithelial permeability after stopping cigarettes. BMJ 282:1183–1186

Monaghan P, van der Wall H, Mackey DWJ et al (1990) Detection of altered pulmonary epithelial permeability by a modified technegas technique. Eur J Nucl Med 16 [Suppl]:S149

Monaghan P, Murray IPC, Mackey DWJ et al (1991) An improved radionuclide technique for the detection of altered pulmonary permeability. J Nucl Med 32:1945–1949

Morgenroth ML, Reeves JT, Murphy RC et al (1984) Leukotriene synthesis and receptor blockade hypoxic pulmonary vasoconstriction. J Appl Physiol 56:1340–1346

Morrow PE (1974) Aerosol characterization and deposition. Am Rev Respir Dis 110:88–99

Morrow PE, Yu CP (1993) Models of aerosol behavior in airways and alveoli. In: Moren F, Dolovic MB, Newhouse MT, Newman SP (eds) Aerosols in medicine. Principles, diagnosis and therapy. Elsevier Science, Amsterdam, pp 157–193

Morrow PE, Bates DV, Fish BR et al (1966) Deposition and retention models for internal dosimetry of the human respiratory tract. Health Phys 12:173–207

Moser KM, Longo AM, Ashburn WL et al (1973) Spurious scintigraphic recurrence of pulmonary emboli. Am J Med 55:434–443

Nair G, Samuel AM, Nambi KSV et al Urban air pollution and altered lung function. An east south and south-eastern Asian regional co-ordinated study (not yet published)

Nolop KB, Maxwell DL, Royston D et al (1986) The effect of raised thoracic pressure and volume of 99mTc-DTPA clearance in humans. J Appl Physiol 60:1493–1497

Otis AB, McKerrow CB, Bartlett A et al (1956) Mechanical factors in distribution of pulmonary ventilation. J Appl Physiol 8:427–443

O'Doherty MJ, Page CL, Harrington C et al (1990) Hemophilia, AIDS and lung epithelial permeability. Eur J Haematol 44:252–256

Pack RJ, Weddicombe JB (1984) Amine containing cells of the lung. Eur J Respir Dis 65:559–578

Panoutsopoulos G, Ilias I, Christakopoulou I (2000) Transient right-to-left in massive pulmonary embolism. Ann Nucl Med 3:217–221

Pavia D, Sutton PP, Lopez-Vidriero MT et al (1983) Drug effects on mucociliary function. Eur J Respir Dis 64 [Suppl 128]:304–317

Peterson BT, James HL, McLarty JW (1988) Effects of lung volume on clearance of solutes from the air spaces of the lungs. J Appl Physiol 64:1068–1075

PIOPED Investigators (1990) Value of ventilation/perfusion scan in acute pulmonary embolism. JAMA 263:2753–2759

Pircher HJ, Temple JR, Kirsch WJ et al (1965) Distribution of pulmonary ventilation determined by radioisotope scanning. AJR 94:807–814

Porter J, Jick H (1977) Drug-related deaths among medical inpatients. JAMA 237:879–881

Rinderknecht J, Shapiro L, Krauthammer M et al (1980) Accelerated clearance of small solutes from the lungs in interstitial lung disease. Am Rev Respir Dis 121:105–117

Ross DJ, Waters PF, Waxman AD et al (1993) Regional distribution of lung perfusion and ventilation at rest and during steady-state exercise after unilateral lung transplantation. Chest 104:130–134

Rosso J, Guillon JM, Parrot A et al (1992) Technetium-99m-DTPA aerosol and gallium-67 scanning in pulmonary complications of human immunodeficiency virus infection. J Nucl Med 33:91–87

Royston D, Minty BD, Higenbottam TW et al (1985) The effect of surgery with cardiovascular bypass on alveolar-capillary barrier function in human being. Ann Thorac Surg 40:139–142

Sackner MA (1978) Effects of respiratory drugs on mucociliary clearance. Chest 73 [Suppl]:958–964

Sackner MA, Rosen MJ, Wanner A (1973) Estimation of tracheal mucous velocity by bronchofiberscopy. J Appl Physiol 34:495–499

Sakakura Y, Proctor DF (1972) The effect of various conditions on tracheal mucociliary transport in dogs. Proc Soc Exp Biol Med 140:870–879

Senden TJ, Moock KH, Gerald JF et al (1997) The physical and chemical nature of technegas. J Nucl Med 38:1327–1333

Severinghaus JW, Swenson EW, Finley TN et al (1961) Unilateral hypoventilation produced in dogs by occluding one pulmonary artery. J Appl Physiol 16:53–60

Shih WJ, Bognar B (1999) Reverse mismatched ventilation-perfusion pulmonary imaging with accumulation of technetium-99m-DTPA in a mucus plug in a main bronchus: a case report. J Nucl Med Technol 27:303–305

Slavin JD Jr, Mathews J, Spencer RP (1985) Pulmonary ventilation/perfusion and reverse mismatches in an infant. Clin Nucl Med 10:708–709

Sleigh MA (1977) The nature and action of respiratory tract cilia. In: Brain JD, Proctor DF, Reid LM (eds) Respiratory defense mechanisms, part 1. Dekker, New York, pp 247–288

Sostman HD, Gottschalk A (1982) The stripe sign: a new diagnostic sign for diagnosis of nonembolic defects on pulmonary perfusion scintigraphy. Radiology 142:737–741

Sostman HD, Gottschalk A (1992) Prospective validations of the stripe sign in ventilation-perfusion scintigraphy. Radiology 184:455–459

Sprague RS, Thiemermann C, Vane JR (1992) Endogenous endothelium-derived relaxing factor opposes hypoxic pulmonary vasoconstriction and supports blood flow to hypoxic alveoli in anesthetized rabbits. Proc Natl Acad Sci USA 89:8711–8715

Staub NC (1963) Site of action of hypoxia on the pulmonary vasculature. Fed Proc 22:453

Staub NC (1983) Alveolar flooding and clearance. Am Rev Respir Dis 127:S44–S61

Sundram FX (1995) Clinical studies of alveolar-capillary permeability using technetium-99m DTPA aerosol. Ann Nucl Med 9:171–178

Susskind H, Brill AB, Harold WH (1986) Quantitative comparison of regional distributions of inhaled Tc-99m DTPA aerosol and Kr-81m gas in coal miners' lungs. Am J Physiol Imaging 1:67–76

Susskind H, Weber DA, Volkow ND et al (1991) Increased lung permeability following long-term use of free-base cocaine (crack). Chest 100:903–909

Sutherland PW, Katsura T, Milic-Emili J (1968) Previous volume history of the lung and regional distribution of gas. J Appl Physiol 25:566–574

Swenson EW, Finley TN, Guzman SV (1961) Unilateral hypoventilation in man during temporary occlusion of one pulmonary artery. J Clin Invest 40:828–835

Symington IS, Stack BHR (1977) Pulmonary thromboembolism after air travel. Br J Chest 17:138–140

Taillefer R, Therasse E, Turpin S et al (1999) Comparison of early and delayed scintigraphy with 99m-apcitide and correlation with contrast-enhanced venography in detection of acute deep vein thrombosis. J Nucl Med 40:2029–2035

Taplin GV, Isawa T (1968) Regional alveolar liquid diffusibility by lung scintigraphy. California University, Los Angeles, report UCLA 26-686, UC-48, biology and medicine, pp 4–48

Taplin GV, Poe ND (1965) A dual lung scanning technique for evaluation of pulmonary function. Radiology 85:365–368

Taylor AE, Gaar K (1970) Estimation of equivalent pore radii of pulmonary capillary and alveolar membranes. Am J Physiol 218:1133–1140

Teshima T, Isawa T, Hirano T et al (1989) Image processing for mucociliary transport system. Respiration 8:828–835

Thomas D, Stein M, Tanabe G et al (1964) Mechanisms of bronchoconstriction produced by thromboemboli in dogs. Am J Physiol 206:1207–1212

UPET Investigators (1973) The urokinase pulmonary embolism trial. A national cooperative study. Circulation 47 [Suppl 2]:46–50

Van As A (1974) The role of selective beta 2-adrenoreceptor stimulants in the control of ciliary activity. Respiration 31:146–151

Van der Wall H, Murray IPC, Jones PD et al (1991) Optimising technetium 99m diethylene triamine pentaacetate lung clearance in patients with the acquired immunodeficiency syndrome. Eur J Nucl Med 18:235–240

Venrath H, Rothoff S, Valentin H et al (1952) Bronchospirometrische Untersuchungen bei Durchblutungsstoerungen in kleinen Kreislauf. Beitr Klinik Tuberk 107:291–294

Wanner A (1977) Clinical aspects of mucociliary transport. Am Rev Respir Dis 116:73–125

Wanner A (1981) Alteration of tracheal mucociliary transport in airway disease. Effect of pharmacologic agents. Chest 80 [Suppl]:867–870

Wartski M, Zerbib E, Regnard JF et al (1998) Reverse ventilation-perfusion mismatch in lung cancer suggests intrapulmonary functional shunting. J Nucl Med 39:1986–1989

Wartski M, Collignon MA for the THESE Study group (2000) Incomplete recovery of lung perfusion after 3 months in patients with acute pulmonary embolism treated with antiembolic agents. J Nucl Med 41:1043–1048

Weibel ER (1963) Morphometry of the lung. Academic, New York, pp 1–151

West JB (1965) Inequality of blood flow and ventilation in the normal lung. In: Ventilation/blood flow and gas exchange. Blackwell Scientific, Oxford, pp 17–33

West JB, Dollery CT, Naimark A (1964) Distribution of blood flow in isolated lung; relation to vascular and alveolar pressure. J Appl Physiol 19:713–724

Westbrook KC, Williams GD Jr, Wise WS et al (1969) Effects of acute hypoxia in dogs with an antegrade versus retrograde pulmonary blood flow. Surg Forum 20:217

Whittenberger JL, McGregor M, Berglund E et al (1960) Influence of state of inflation of the lung on pulmonary vasculature. J Appl Physiol 15:878–882

Windheim KV (1970) Perfusionsszintigraphische Untersuchungen der Lunge bei Bronchiektasien. Pneumonologie 143:193–196

Wollmer P, Evander E, Jonson B et al (1986) Pulmonary clearance of inhaled 99mTc-DTPA: effect of surfactant depletion in rabbits. Clin Physiol 6:85–89

Worsley DF, Alavi A (1995) Comprehensive analysis of the results of the PIOPED diagnosis. J Nucl Med 36:2380–2387

Zielenski J, Tsui LC (1995) Cystic fibrosis: genotypic and phenotypic variations. Annu Rev Genet 29:777–807

CHAPTER 23

Functional Genomics and the Liver

23

Ramchandra D. Lele, Vikram R. Lele

Contents

23.1

Introduction

The liver provides immense opportunities to study the functional consequences of gene expression in health and disease. Out of the estimated 50,000–100,000 genes in the human genome (current estimates down to 30,000 genes), 15,000 genes are expressed in the liver and the kidneys. A large amount of information is available now on the specific gene products, on their metabolic role in living human body, on the homeostatic circuits to which they belong, and on the way these respond to changes in environment. For instance in response to stress the liver produces many acute phase proteins, such as c-reactive protein, a_1 antitrypsin and haptoglobin, a_1 acid glycoprotein, and fibrinogen. Hepatic enzyme induction by drugs such as phenobarbitone is another example of altered gene expression in response to environment.

The liver plays a crucial role in the maintenance of metabolic homeostasis. Enzyme regulation plays a key role in homeostasis, the day-to-day maintenance of balanced and productive metabolic activity, as well as in determining the timing and nature of long-term processes such as cell division and cell differentiation that affect an organism's growth and development. The important metabolic pathways of protein, carbohydrate, and fat in the liver are depicted in Fig. 23.1. In the transition from fed to fasted state, the liver converts itself from being an organ of glucose uptake, and glycogen and fatty acid synthesis, to one of glucose production, fatty acid oxidation, and ketogen-

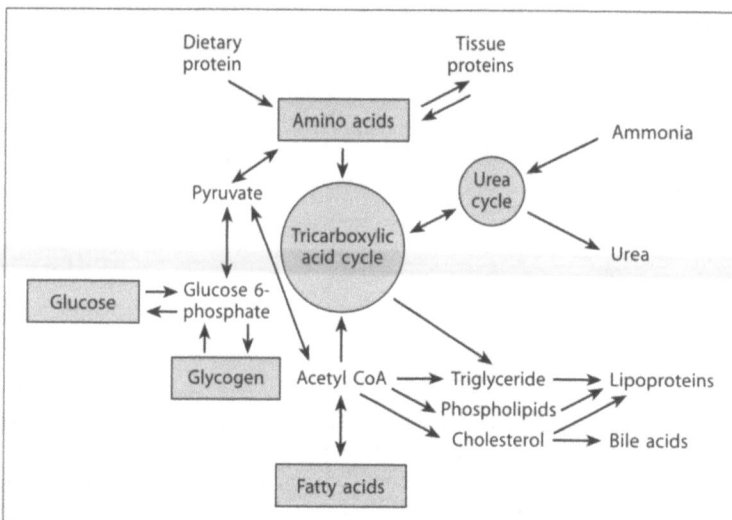

Fig. 23.1. Important metabolic pathways of protein, carbohydrate and fat in the liver

esis. Maintenance of normal blood sugar is a major homeostatic function of the liver. Hypoglycemia is a feature of diseases with hepatocellular dysfunction. On the other hand excessive hepatic glucose production is a feature of diabetes mellitus.

Many liver disorders and their manifestations can only be appreciated in the context of the functional complexity of the liver at the cellular and molecular levels. The liver processes and synthesizes multiple substances that are transported to and from other areas of the body. While in health all physiological processes occur in an orderly, regulated manner and homeostasis is maintained, severe liver injury can profoundly impair the ability of the hepatocytes to form the enzymes that catalyze a key metabolic process such as urea synthesis. The resultant inability to convert toxic ammonia to nontoxic urea is then followed by ammonia intoxication and ultimately hepatic coma.

Similarly the inability of the diseased liver to synthesise prothrombin, fibrinogen, accelerator globulin, and factor VII all necessary for blood coagulation, may lead to blood coagulation abnormalities.

The hepatocyte plays an important role in complex and diverse general metabolic processes that may be deranged as a consequence of liver disease. In addition, this metabolic diversity leads to hepatic involvement in many inborn errors of metabolism, including a wide variety of storage diseases and disorders of iron metabolism (hemochromatosis and hemosiderosis) and copper homeostasis (Wilson's disease).

All biochemical processes are resolvable into series of individual stepwise reactions. Each one of these reactions is ultimately under control of a single gene.

Understanding of dynamic physiology and biochemistry expanded greatly with the use of radioactive and stable isotopes of carbon, hydrogen, oxygen, nitrogen, phosphorus, and sulphur. The period between 1950 and 1965 saw the elucidation of the major steps in the enzymatic pathways for the biosynthesis and degradation of proteins and amino acids, purines and pyrimidines, complex carbohydrates, fatty acids, and lipids. Labeled tracers especially of carbon and hydrogen made a major contribution to these studies (Kamen 1957). The availability of PET tracers offers new opportunities for studying in vivo biochemistry. The use of PET for liver studies has been mostly in the detection of hepatomas and hepatic metastases with FDG. Studies of liver metabolism have been very limited despite the key role of the liver in the regulation of metabolism and the unique possibility of studying in vivo metabolism by PET. A sample list of such tracers is given in Table 23.1. As an illustrative example carbon-1 atom is primarily involved in oxidation of glucose via the HMP (hexose monophosphate) pathway thereby providing NADPH needed for fatty acid synthesis. When both labeled C_1 and C_6 glucose are given, a larger proportion of C_1 appears in expired air of obese subjects than normal subjects, indicating greater use of HMP pathway by the obese. Recently PET studies with Iodine-124 insulin have been initiated, which have a potential for application to evaluate the insulin resistance syndrome.

The rat hepatocyte is one of the most extensively studied of all cells from a biochemical and functional viewpoint, partly because of its availability in relatively large amounts, suitability for fractionation

Table 23.1. A sample list of radiolabeled compounds for the PET study of metabolism in the liver

	Labeled compound	Applications
Carbon-11	Glucose U	Glucose pool, turnover, oxidation by analysis of carbon-11-CO_2 in expired air
	Glucose 1	Study of Embden-Meyerhof
	Glucose 6	Pentose pathways
	Ribose 1	Study of pentose pathway
	Acetate 1	Study of cholesterol, fatty acid, triglyceride, and ketone synthesis and oxidative phosphorylation rates
	Pyruvate 2	Study of gluconeogenesis
	Lactate 2	
	Mevalonic acid 2	Study of cholesterol synthesis and turnover
	Palmitic acid 1	Turnover rate and oxidation of free fatty acids
	Linoleic acid 4	
	Amino acids	Study of protein synthesis
Fluorine-18	Thymidine	Rate of DNA synthesis
	Uridine	
	Oratic acid	
	Amino acids	Rate of protein synthesis
	Deoxyglucose	Metabolic mapping of glucose uptake and utilization
Iodine-124	Insulin	Insulin resistance syndrome

studies, and diversity of functions. Physical techniques to disrupt cells and to isolate intracellular molecules and subcellular organelles (e.g., extraction, homogenization, and differential centrifugation) and the use of the electron microscope and immuno-histochemical techniques for measurement of suitable marker enzymes and chemical components (e.g., DNA, RNA) have provided the means to develop knowledge about the function of mitochondria, microsomes (smooth and rough endoplasmic reticulum and free ribosomes), peroxisomes, lysosomes, the Golgi apparatus, the plasma membrane, and certain cytoskeletal elements. This information, summarized in Table 23.2, represents one of the major achievements of biochemical research.

The spatial arrangement and compartmentalization of enzymes, substrates, and co-factors within the cell are of cardinal significance. For example, in the hepatocyte the enzymes for glycolysis are located in the cytoplasm while the enzymes of the citric acid cycle are in the mitochondria. Histoenzymology provides a graphic and relatively physiologic picture of the pattern of enzyme distribution. Thin (2–10 µm) frozen sections of tissue are treated with a substrate for a particular enzyme. Where the enzyme is present, the product of enzyme-catalysed reaction is formed. If the product is colored and insoluble, it remains at the site of formation and thus localizes the enzyme.

Hepatocytes in culture have been useful for studying the biosynthesis and post-translational processing of proteins. Incorporation of radiolabeled amino acids into newly synthesized proteins can be followed by immunoprecipitation and polyacrylamide gel electrophoresis. Such studies have been very valuable for analysis of lysosomal storage disease (e.g., Tay-Sachs'), mitochondrial protein disorders (e.g., ornithine transcarbamylase deficiency), and disorders of cell membrane proteins (e.g., familial hypercholesterolemia, leukocyte adhesion deficiency). Use of cell cultures (human fibroblasts and lymphoblasts) have helped in the elucidation of the biochemical defects in several metabolic disorders, e.g., mucopolysaccharidosis. Studies by Neufeld and Fratantoni on "mutual or cross-correction" using fibroblasts from a patient with Hurler's syndrome and a patient with Hunter's syndrome showed that Hurler's cells excreted iduronate sulfatase (lacking in Hunter's cells) while Hunter's cells excreted a_1-iduronidase (lacking in Hurler's cells) (Fratantoni et al. 1968 a, b, 1969 a, b). These experiments provided the first indication that lysosomal enzymes could be exchanged between cells by receptor-mediated endocytosis. The practical application of this observation is seen in patients with Hurler's syndrome given bone marrow transplantation. With the newly available source of the enzyme a_1 iduronidase, the lysosomal storage material disappears from the liver in the majority of patients after bone marrow transplantation (Whitley et al. 1993).

It is not clear whether specific receptors or other hepatocyte membrane components play a role in the relative or absolute tropism of infectious agents (particularly the hepatitis viruses) that account for a large proportion of both acute and chronic liver disease. At the same time the hepatocyte may be infected by less strictly organotropic agents including other viruses (e.g., Epstein-Barr virus) and a wide variety of bacterial and parasitic organisms. The vascular supply of the liver leads to its frequent involvement in disseminated infections, including tuberculosis.

The hepatocyte has potential for regeneration and proliferation. The capacity for hepatocellular regeneration is evident in the complete recovery that usually occurs following fulminant hepatitis (due either to toxic or viral agents). A complex balance of various peptide

Table 23.2. Major intracellular organelles in the hepatocyte, their functions, and their alterations in disease

Organelle	Marker	Major function	Alterations in disease
Nucleus	DNA	DNA transcription – (DNA-directed RNA synthesis) DNA replication	Mutations causing decreased or absent proteins, enzymes transporters, receptors, etc. many genetic diseases
Mitochondrion	Glutamic dehydrogenase	Citric acid cycle, β oxidation of fatty acids, heme synthesis, oxidative phosphorylation, ATP generation	Mutations in mitochondrial DNA affecting structure of proteins or enzyme mito-chondrial cytopathies
Cytosol	Lactic dehydrogenase	Enzymes of glycolysis and fatty acid synthesis	Diminished synthesis of transport proteins to export TGs and phospholipids; fatty liver
Ribosome	High RNA content	Site of translation of mRNA into protein	
Endoplasmic reticulum	Glucose-6 phosphatase	Site of synthesis of proteins, enzymes, lipoproteins, TG	Decrease or blockage in synthesis
Rough endoplasmic reticulum		Cytochrome P450	Cytochrome P450 may acti-vate various chemicals to po-tentially toxic species, e.g., CC14
Smooth endoplasmic reticulum		Drug detoxification, bilirubin conjugation, cholesterol and bile acid synthesis	
Lysosome	Acid phosphatase	Site of many hydrolases' degradative reactions	Decreased activity of hydro-lases, accumulation of vari-ous biomolecules; lysosomal enzymes misdirected and se-creted by affected cells, e.g., I cell disease
Golgi apparatus	Galactosyl transferase	Intracellular storage of proteins; glycosylation and sulfation reactions	Decreased in GlcNAc phos-photransferase
Peroxisome	Catalase; uric acid oxidase	Degradation of long chain fatty acids; deamination of amino acids; gluconeogenesis	Decreased activity of peroxi-somal enzymes, e.g., Zellwe-ger syndrome
Cytoskeleton	No specific enzyme; markers recognized by EM or protein electro-phoresis	Microfilaments, microtubules, intermediate filaments	Depolymerized actin fila-ments; loss of tone; cholesta-sis
Plasma membrane	Na$^+$K$^+$ ATPase; 5' nucleotidase	Transport of molecules in and out of cells; intercellular adhesion and communication	Changes in oligosaccharides of glycoproteins important in permitting cancer metastasis

growth factors plays an important role in controlling hepatic cell proliferation after cell loss. Following par-tial hepatectomy or carbon tetrachloride damage, sur-viving hepatocytes express mRNA of c Myc and c fos as well as hepatocyte growth factor (HGF). HGF trig-gers liver regeneration after liver injury; it is the most potent stimulator of DNA synthesis in the mature hepa-tocyte. Epidermal growth factor (EGF) is formed in regenerating hepatocytes. TGFa (30–40% sequence homology with EGF) can bind to EGF receptors present in high density on the hepatocyte membrane and initi-

ate replication. On the other hand TGFβ is a major in-hibitor of proliferation. TGFβ_1, inhibits amino acid up-take in cultured hepatocytes.

Architecturally distorted regeneration in concert with fibrosis is an essential factor in the development of cirrhosis and leads to both disruption of blood flow through the hepatic parenchyma and to uneven hepatocellular function due to distortion of normal lobular structure.

Hepatic stellate cell (HSC or lipocyte) activation due to injury initiates repair and fibrosis. The molec-

Table 23.3. Classification of hepatic drug reactions

Type	Features	Examples
Mitochondrial cytopathies	Reye's like syndrome affects children; cirrhosis	Valproate
Acute hepatitis	Bridging necrosis; short term acute	Methyldopa, isoniazid
	Long term chronic active hepatitis	Halothane, ketoconazol
Zone 3 necrosis	Dose-dependent multi-organ failure	Carbon tetrachloride, paracetamol (acetaminophen), halothane
Steato hepatitis	Long half life; cirrhosis	Perhexiline, amiodarone
General hypersensitivity	Often with granulomas	Sulfonamides, quinidine, allopurinol
Fibrosis	Portal hypertension, Cirrhosis	Methotrexate, vinyl chloride, vitamin A
Cholestasis	Dose dependent; reversible	Sex hormones, chlorpromazine
Canalicular	Reversible obstructive jaundice	Erythromycin, nitrofurantoin
Hepatocanalicular		Azathioprine
Ductular	Age-related renal failure	Benoxyprofen
Venoocclusive disease	Dose dependent	Irradiation; cytotoxic drugs
Sinusoidal dilatation and peliosis		Azathioprine; sex hormones
Hepatic vein obstruction	Thrombotic effect	Sex hormones
Portal vein obstruction	Thrombotic effect	Sex hormones
Sclerosing cholangitis	Cholestasis	Hepatic arterial floxuridine (FUDR)
Adenoma	May rupture and bleed	Sex hormones; danazol
Hepatocellular carcinoma	Very rare	Sex and anabolic hormones

ular events leading to increased formation and reduced degradation of extracellular matrix resulting in the formation of fibrous scar are being better understood. Greater understanding will allow trials of new therapies to prevent or reverse hepatic fibrosis. Early fibrosis is reversible, but not in cirrhosis with crosslinked collagen and regenerating nodules.

Monitoring hepatic fibrosis non-invasively (without serial liver biopsies) is being attempted. Activated HSCs express PDGF receptors which can be used for imaging. By complexing cyclic peptide (C. SRNLIDC) HSA to Indium-111 via bifunctional chelators, high uptake was found 10 minutes post-injection in the cirrhotic liver and retained at 72 hours in an experimental rat model (Zhang et al. 2001).

The possibility of directly blocking the synthesis of matrix proteins by somatic gene therapy remains for the future.

The liver is a common site for both primary and metastatic malignant disease. The liver is the most frequent site of blood-borne metastases, irrespective of whether the primary is drained by systemic or portal veins. It is involved in about a third of all cancers, including half of those of stomach, colon, lung, and breast. Other frequent primary sites include esophagus, pancreas, and malignant melanoma. Prostate and ovarian metastases are exceedingly rare in the liver. While ultrasonography, CT, and MRI provide accurate anatomic information, scintigraphy based on

molecular recognition provides additional unique information. The molecular imaging approach is complied in Table 23.6. Even more than at diagnosis, future radionuclide approaches will aim at molecular surgery through delivery of Auger electrons selectively to the DNA of cancer cells. The critical requirements are high specific binding and sufficient concentration and residence time in the lesion.

The human body is exposed to drugs, chemicals (including alcohol), and various compounds in the environment. Most of there compounds undergo chemical alterations in the body catalyzed by at least 50 different enzymes, the liver being the main organ involved, containing the most of enzymes (followed by the small intestine).

Drugs can cause toxic effects able to mimic almost every naturally occurring liver disease in humans (Table 23.3). About 2% of all cases of jaundice in hospitalized patients are drug-induced; about one quarter of cases of fulminant hepatic failure are probably drug-related. In any patient with liver disease it is essential to know all drugs that have been taken over the previous three months. A 50% fall in serum transaminases within 8 days of stopping a drug is acceptable evidence of drug-induced hepatotoxicity.

The Human Genome Project has underscored the uniqueness of each individual. At least 28% of all gene loci harbor polymorphic alleles that vary among individuals. This is exemplified by the variations in handling the same drug in different individuals. The human genome encodes at least 14 families of mono-oxygenases (P450s) responsible for hydroxylation of drugs. Human tissues are estimated to contain distinct cytochrome 450s ranging from 35 to 60, with a large number of isoforms (~ 150).

Individuals who react abnormally to the anti-arrhythmia drug debrisoquine are shown to have an abnormal expression of P450 II D6. The new field of pharmacogenomics will hopefully provide information regarding key enzymes involved in drug metabolism thereby permitting development of more individualized therapies which take into account genetic variations. Selective inhibitors or inducers may be used to alter the P450 profile.

The molecular basis of several genetic disorders of the liver is depicted in Tables 23.7–23.9. Gene therapy has now become feasible for many diseases involving the liver. Because each liver cell type contains specialized uptake mechanisms for macromolecules and drugs, it is possible to deliver drugs, enzymes, or even genes specifically to the liver cells by linking these agents covalently by biodegradable bonds to carrier molecules (Pozansky et al. 1984; Merjer and van de Slnijs 1989). In gene therapy it is important to be able to monitor the location of the transferred gene and its activity at the site of transfer, as well as the duration of gene expression in the living recipient body. Radionuclide PET images can help in this assessment by the use of reporter genes whose activity can be monitored by external probes.

Liver transplantation has now become feasible. Radionuclide imaging helps in pre and post transplant evaluation of the function of the transplant and in monitoring the complications. In the treatment of heritable disorders, liver transplant may be combined with heart transplant (as in familial homozygous hypercholesterolemia with severe coronary artery atherosclerosis); with kidney transplantation (as in tyrosinemia); or with lung transplantation (as in a_1 antitrypsin deficiency). Deficiency of a_1 antitrypsin is the second most common childhood disorder for which liver transplantation is performed (Sharp 1995). Following transplantation the recipient's phenotype rapidly changes from the original (NN,ZZ with zero to <7 µmol/L serum values of a_1 antitrypsin) to that of the donor (e.g., MM with serum values 20–53 µmol/L) (Van Furth et al. 1986).

Organ (including bone marrow) transplant is to be considered as a sort of gene therapy – supplying a normal tissue which provides the missing gene product.

23.2
Functional Morphology of the Liver

The liver, the largest organ in the body, weighs 1,500 g ($\sim 3\%$ of body weight). The body's entire vascular endothelium weighs 1,800 g, of which the hepatic microcirculation forms an important component, constituting 3% of the total liver volume. The hepatic sinusoids are unique vessels in humans, lined by endothelial cells, Kupffer cells, hepatic stellate cells, and Pit cells. Unlike other vascular linings, (1) no tight junctions exist between endothelial cells and/or Kupffer cells, (2) no anatomically defined basement membrane are found under the sinusoidal lining cells, and (3) holes (fenestrae) are frequent in the endothelial lining. These three attributes allow direct access of plasma constituents (albumin, vitamins, and other essential nutrients) from the sinusoidal lumen to the surface of the hepatocytes and allows free bilateral uptake and exchange of material between the hepatocytes and plasma. The microvillous surface of the hepatocytes aids in their absorptive function.

In liver disease, particularly in the alcoholic, collagenization of the space of Disse, formation of a basal lamina beneath the endothelium, and modification of the endothelial fenestrations occurs. These changes contribute to deprivation of nutrients intended for the hepatocytes and to the development of portal hypertension.

Focal dilation of sinusoids may complicate therapy with contraceptive or anabolic steroids.

Endotheliitis, sinusoidal dilation, peliosis, and veno-occlusive disease can complicate azathioprine therapy given to renal and hepatic transplant patients.

The liver contains four types of cells:

(1) The hepatocytes (polygonal cells 25 µm size; range 13–30 µm) form 78% of the total liver volume. The adult human liver contains 250 billion hepatocytes.

(2) Kupffer cells (highly specialized macrophages) constitute 2% of total liver volume and are distributed irregularly within the sinusoidal space. There is no direct connection between two adjacent Kupffer cells. They are highly mobile scavenger cells often found within the space of Disse, or they may lie free within the sinusoidal space unattached to the endothelial cell. The cytoplasm of Kupffer cells is rich in lysosome, Golgi apparatus, and rough endoplasmic reticulum. The Kupffer

cells proliferate locally to maintain their population but in times of greater need they are reinforced from the bone marrow.

(3) Hepatic stellate cells (lipocytes or Ito cells; 1.4% of total liver volume) are resting fibroblasts in the perisinusoidal space of Disse. They are fat-storing cells, rich in vitamin A and stored retinoids. They become activated during liver injury and play a dominant role in repair and in fibrosis and collagen deposition in the subendothelial space of Disse. Their work thus leads to reduction in the number of fenestrae and loss of microvilli on the hepatocyte, both of which result in a decrease in the delivery of organic anions [including hepatic iminodiacetic acid (HIDA)/mebrofenin] into the perisinusoidal space of Disse.

(4) Pit cells are highly mobile killer T cells found in the sinusoidal space. They contain organelles necessary for removal of tumor cells and virus-infected hepatocytes (Bauwens and Wisse 1992).

The hepatic stellate cell (HSC) is the principle cell involved in repair and fibrogenesis. Following hepatic injury, HSC activation is initiated by Kupffer-cell-derived TGFβ as well as initiating factors from hepatocytes, platelets (PDGF), and lymphocytes (1L-1, TNFa). Activated stellate cells (myofibroblasts) show expression of desmin (a filamentous protein seen in muscle). They show contractile features of smooth muscle cell, synthesize endothelin-1, and may have a role in blood flow regulation.

Activated stellate cells and Kupffer cells produce metalloproteinases (collagenases, gelatinases or type IV collagenases, and stromalysins). Activated stellate cells also secrete tissue inhibitors of metalloproteinases (TIMP-1) and thus regulate both degradation of matrix as well as production of fibrous tissue. TIMP-1 is increased in the serum of pre-cirrhotic and cirrhotic alcoholic patients (Liu et al. 1994).

The hepatocyte contains a nucleus and a nucleolus, both rich in DNA. The sinusoidal membrane of the hepatocyte has multiple transport systems with partially overlapping substrate specificities. Organic anion transport protein (OATP) carries several molecules including bile acid, sulfobromophthalein and bilirubin, sodium/bile acid cotransporting protein (NTPC) carries bile acids conjugated with taurine or glycine. A Na^+ HCO_3^- cotransporter and Na^+-H^+ exchanger are involved in controlling intracellular pH. The smooth endoplasmic reticulum (SER) is composed of tubular structures containing microsomes that carry out bilirubin conjugation and detoxification of drugs and other organic anions. The SER is steroid sensitive and participates in enzyme induction when phenobarbitone is administered. The rough endoplasmic reticulum (RER) contains ribosomes and is responsible for protein synthesis including albumin and several glycoproteins listed in Table 23.2.

Lysosomes are cytoplasmic particles close to bile canaliculi and contain hydrolytic enzymes including acid phosphatase. Lysosomes perform a scavenger function and remove from blood excess material including ferritin, bile pigments, and metals such as copper.

The Golgi apparatus likewise lies close to bile canaliculi and consists of particles and vesicles. The lysosomes and Golgi apparatus together perform the task of storage entrapment and final excretion into the bile of various non-essential body constituents.

The mitochondria are scattered throughout the hepatocyte and participate in oxidative phosphorylation, heme synthesis, and the citric acid cycle.

Peroxisomes are organelles characterized by a fine granular matrix surrounded by a single membrane. They are most abundant in liver and kidney where they are 2–3 times larger (0.5 µm, range 0.2 to 1 µm) than in other tissues (0.1–0.25 µm). The major function of peroxisomes is β oxidation of long chain fatty acids with the production of H_2O_2 (instead of ATP as in mitochondria), which is degraded by catalase, abundant in peroxisomes, and the energy is lost as heat. Peroxisomal β oxidation serves gluconeogenesis via the glyoxalate cycle to succinate.

Peroxisome-proliferator-activated receptors (PPARs) were discovered in 1990. The differentiation of many cell types (hepatocyte, adipocyte, fibroblast, myocyte, keratinocyte, and monocyte/macrophage) involves PPARs. Activation of this receptor was found to induce a battery of genes involved in peroxisome proliferation in mouse liver. In a knock-out model, peroxisome proliferation did not occur. Dominant negative mutations in human PPARr is associated with severe insulin resistance, diabetes mellitus, and hypertension (Barroso et al. 1999).

The ultramicroscopic structures of the liver are shown in Fig. 23.2, pathways of endocytosis in Fig. 23.3, and the various transporters on hepatocytes are depicted in Fig. 23.4.

The bile canaliculi are a special part of the hepatocyte plasma membrane which contains transporters (mainly ATPase dependent) responsible for carrying

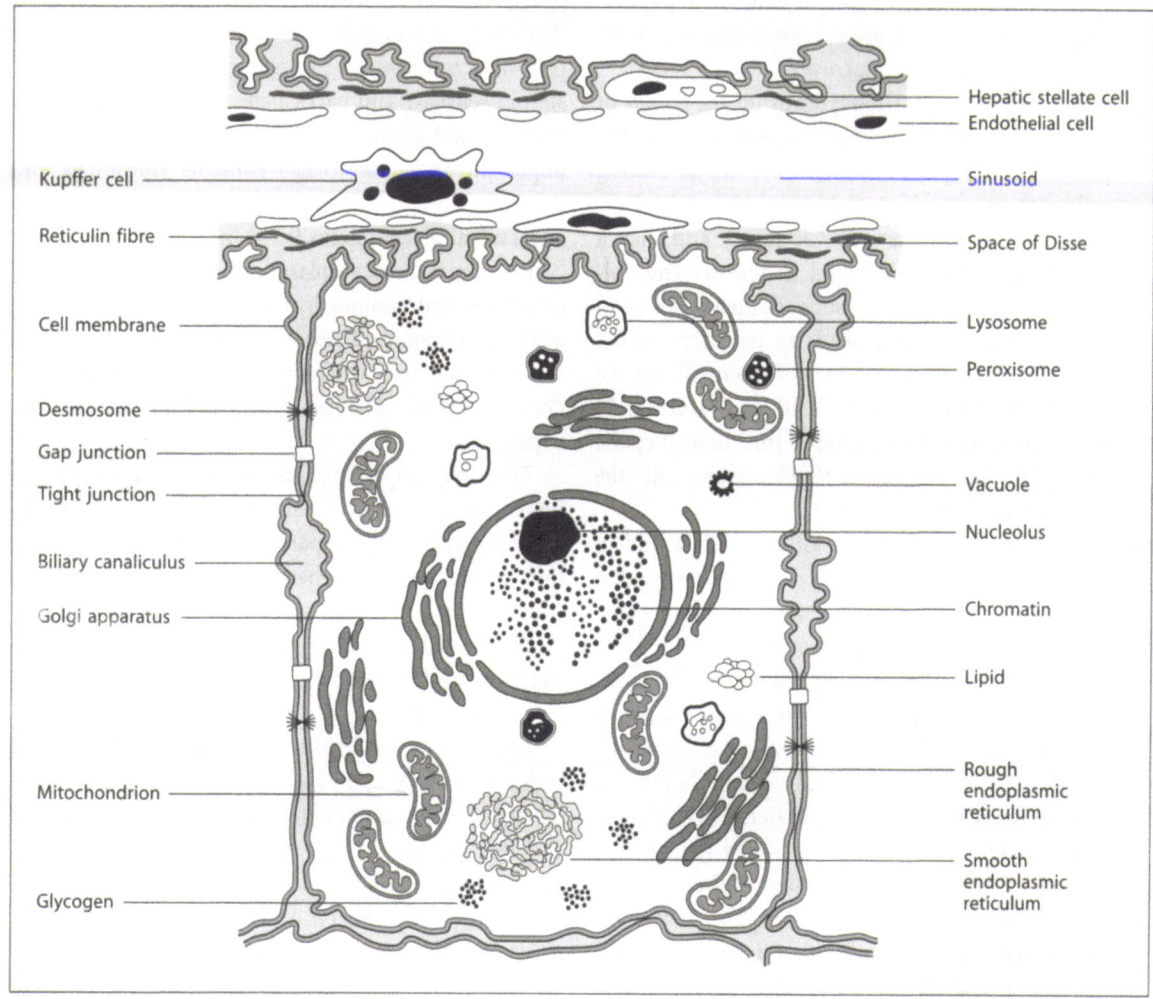

Fig. 23.2. Functional morphology of the liver. The vascular space is divided into two compartments by the endothelial at cells (*top*) sinusoidal space and (*bottom*) perisinusoidal space of Disse. Kupffer cells are located in the space of Disse. The basolateral border of the hepatocyte faces the space of Disse. Canaliculi are invaginations of the lateral wall of two adjacent hepatocytes and they join to form the canals of Hering, which in turn unite to form the interlobular ducts. The portal triad consists of a branch of the bile duct, hepatic artery, and portal vein. The canalicular surface is sealed from the rest of the intercellular surface by junctional complexes including tight junctions, gap junctions, and desmosomes

molecules into bile against steep concentration gradients. These also contain enzymes such as alkaline phosphatase and gamma glutamyl transpeptidase. The canalicular multispecific organic anion transporter (cMOAT) carries glucuronide and glutathione-s-conjugates, e.g., bilirubin diglucuronide. The canalicular bile acid transporter (cBAT) carries bile acids; cyclosporin A inhibits the cBAT. Bile acid-independent flow probably depends upon glutathione transport as well as a Cl^-/HCO_3^- exchanger for canalicular secretion of bicarbonate.

Two members of the phosphoglycoprotein family are important in canalicular transport, both ATP-dependent. MDR-1 is a transporter of organic cations and derives its name from being responsible for transporting cytotoxic drugs out of cancer cells causing multi-drug resistance. MDR$_3$ is a phospholipid translocator that acts as a flippase for phosphatidyl choline. The importance of MDR$_3$ is that without phospholipid in bile, bile acids damage the biliary epithelium and cause ductular inflammation and periductal fibrosis.

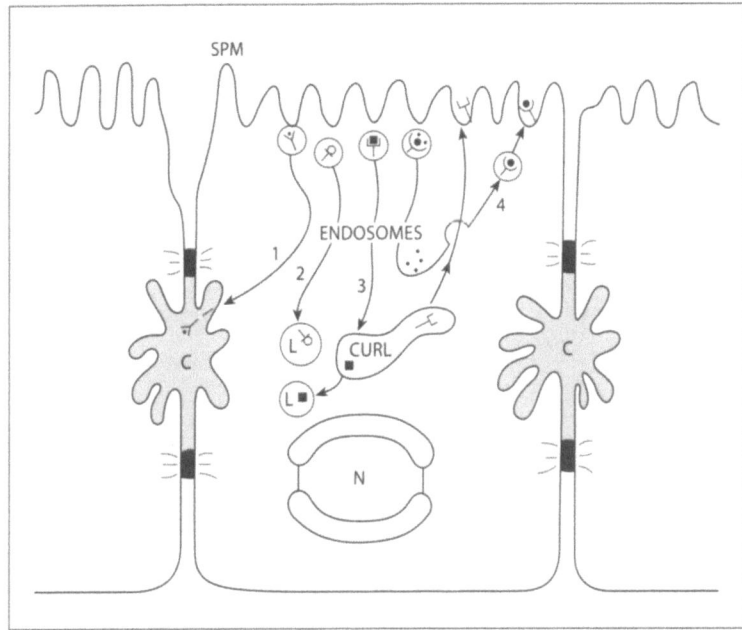

Fig. 23.3. Pathways of endocytosis from the sinusoidal plasma membrane (*SPM* sinusoidal plasma membrane, *C* bile canaliculus, *L* lysosome, *N* nucleus, *CURL* compartment of uncoupling of receptor and ligand). Receptors bound to ligand group together in coated pit. There is endocytosis resulting in a coated vesicle which then loses its clathrin coat and fuses with other vesicles to form early endosomes (the site of sorting). Subsequent pathways include: (*1*) vesicular transport to bile canaliculus where ligand and receptor are released (transcytosis; e.g., polymeric IgA), (*2*) transfer of ligand and receptor to lysosome where they are degraded, (*3*) receptor and ligand are transferred to CURL receptor and ligand separate. The receptor returns to sinusoidal plasma membrane. Ligand enters lysosome and is degraded (e.g., LDL, insulin, asialoglycoprotein). (*4*) Ligand and receptor return to plasma membrane (e.g., transferrin and its receptor after release of iron)

Normal canalicular bile secretion is accompanied by canalicular contraction and mobility involving actin-containing microfilaments influenced by many hormones (e.g., secretin) protein kinase C. An increase in cytosolic calcium inhibits bile secretion.

Cytochalasin or norethandrolone depolymerize actin filaments, resulting in loss of tone, canalicular distension, and failure to contract, with consequent canalicular ileus and cholestasis. Chlorpromazine also affects polymerization of actin leading to cholestasis.

Changes in membrane fluidity and Na^+/K^+ ATPase activity may cause cholestasis. *E. coli* endotoxin, which decreases Na^+/K^+ ATPase activity may act in a similar way.

Integrity of the canalicular membrane may be altered by disruption of either the microfilaments responsible for canalicular tone and contraction, or tight junctions. Disruption of tight junctions (caused by estrogens and phalloidin) leads to loss of the normal restrictive barrier between hepatocytes with subsequent passage of larger molecules directly into the canaliculus from the blood. There can also be regurgitation of solutes from bile to blood.

In Byler's disease, due to a defect in microfilament function or in biliary canalicular membrane, conjugated bile acids cannot be excreted leading to cholestasis. Liver transplantation has given good results.

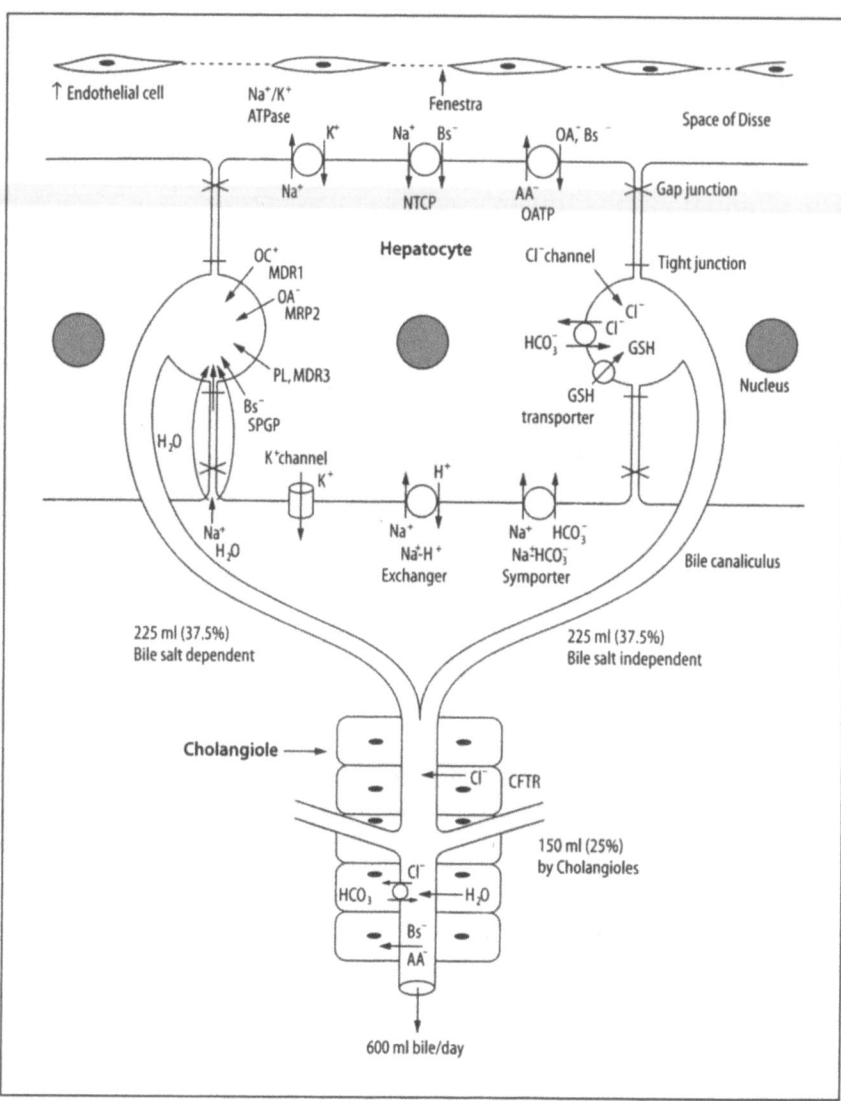

Fig. 23.4. Pathways of uptake of solutes from the space of Disse into the hepatocyte, and excretion into bile. OATP = organic anions transport-pump: organic anions (OA⁻) and bile salts (BS⁻) are exchanged for amino acids (AA⁻). NTAP = Sodium taurocholate pump for sodium-dependent bile salt-uptake. Na⁺/K⁺ ATPase = sodium potassium exchange. Canalicular transporters: organic anions through MRO₂ (multispecific organic ion transporter); organic cations through MDR-1; phospholipids through MDR3. Chloride channels, glutathione (GSH) transporter. CT/HCO₃⁻ exchanger and canalicular bile salt exporter pumps (BSEP or SPGP) also participate. Cholangioles absorb BS⁻, AA⁻ and glucose; exchange Cl⁻ for HCO₃⁻ and secrete chloride into the lumen through cystic fibrosis transmembrane regulator (CFTR). Water enters the canalicular bile through trans-cellular and paracellular routes and a combination of the two

23.3
Functions of the Liver

23.3.1
The Various Functions of the Liver as Listed According to Various Cell Types

Hepatocytes	Bile secretion / conjugation
	Phosphorylation
	Heme synthesis
	Protein synthesis – albumin; $a1$ anti-trypsin; alpha feto protein; $a2$ macro-globulin; ceruloplasmin; complement component C3, C6, and C1; fibrinogen; haptoglobin; hemopoetin; prothrombin; transferrin
	Beta oxidation
	Glycogen synthesis and gluconeogenesis
	Lipoprotein synthesis
	VLDL, LDL, HDL
	apolipoproteins B100, CII, E
	Storage of ferritin, vit B12
	Drug detoxification
Endothelial cells	Plasma filtration
	Endocytosis
	Removal of collagen, Fc fragment of IgG, C3b complement
Kupffer cells	Phagocytosis of colloids, bacteria, tumor cells, endotoxin receptors for Fc fraction, C3b complement, peroxisome PPAR proliferators' secretion of TNFa, collagenase, interleukins, and arachidonic acid. Erythrolelastosis (endogenous pyrogens) storage of iron, ferritin, hemosiderin, immune complexes
Stellate cells	
Lipocytes (Ito cells)	Storage of fats, retinoids, vit A, vit D, vit E, vit K collagen secretion – (role in repair and fibrosis)
Pit cells	Highly mobile killer T cells
Canalicular cells	Bile transit
	Water secretion
	Electrolyte transfer

23.3.2   Quantitative Assessment of Liver Function

Chronic liver diseases pass through a long period of minimum nonspecific symptoms (compensated) until the final stage of ascites, jaundice, encephalopathy and pre-coma (decompensated). Serum albumin and prothrombin (some indication of synthesis function) are usually maintained till late disease. Serial estimates of quantitative liver function in the early stages would help both in prognosis and monitoring treatment. Available tests are listed in Table 23.4. Technetium-99m mebrofenin is ideally suited for this pur-

Table 23.4. Quantitative hepatic function tests

Site	Substrate	Function
Cytosol	Galactose	Galactokinase (phosphorylation)
Microsome	Aminopyrine, caffeine	N-demethylation
	Antipyrine, lignocaine	N-deethylation
Plasma membrane	Galactose-terminated glycoprotein	Asialoglycoprotein receptor (indicator of functioning hepatic mass)
Polyribosomes	Amino acids	Synthesis of albumin, prothrombin, hepatoplastin
Multiple sites	Iminodiacetic acid	Receptor-mediated endocytosis of organic ions including bilirubin; transport to bile canaliculus and excretion into bile

pose. The hepatic extraction fraction (HEF) of this tracer in normal persons is approximately 100%, and the excretion half time is 16 min (range: 11–32 min). HEF values remain normal in early biliary disease, but decrease in hepatocellular disease. When biliary disease is severe and progressive, the HEF value begins to decrease. Excretion half time, on the other hand, increases in both hepatocellular and biliary disease, starting from the very beginning. Both parameters provide a measure of severity of disease irrespective of etiology (Krishnamurthy et al. 1983).

This direct quantitative measurement differs from and is more informative than other indirect measurements such as alkaline phosphatase, γ glutamyl transpeptidase, bilirubin, and so on. The quantification of hepatobiliary function should be an integral part of imaging with Tc-99m mebrofenin in health and disease (Doe et al. 1991).

23.4
Measurement of Hepatic Arterial and Portal Venous Blood Flow

The liver has a unique dual blood supply through the hepatic artery and portal vein. The hepatic artery supplies 25% (about 400 ml arterial blood/minute at 100–120 mmHg systolic pressure) and the portal vein 75% (about 1200 ml/min at 7–10 mm Hg pressure). The liver receives 25% of the resting cardiac output (about 1 5 liters per minute/70 kg body weight). It receives 95 ml

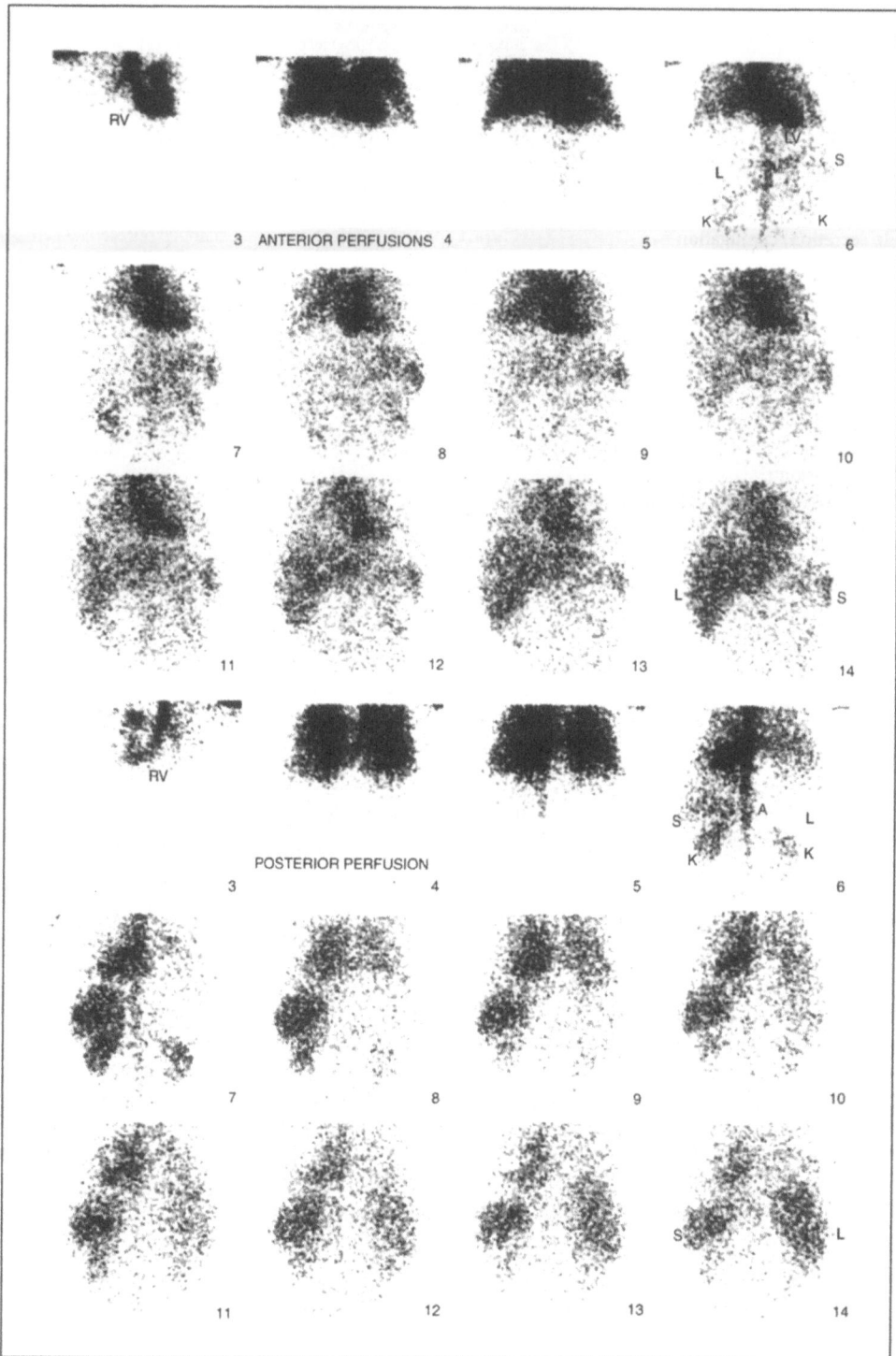

Fig. 23.5. Liver perfusion study with a dual head camera simultaneous anterior (*top*) and posterior (*bottom*) views showing passage of the radiotracer bolus serially through right ventricle (*RV*), left ventricle (*LV*), abdominal aorta (*A*), spleen (*S*), kidneys (*K*), and liver (*L*). Left lobe liver perfusion is seen in the anterior view and the perfusion of the spleen and kidneys is seen better in the posterior views

per minute per 100 mg, which is less per gram than the kidneys (360), adrenals (300), and thyroid (160), though more than the heart (70) and brain (50). The large size of the organ allows a relatively long residence time of substrates brought via its large blood flow.

Information on relative hepatic arterial and portal venous components of liver perfusion can be obtained by analyzing the first pass dynamics of both colloidal and non-diffusible tracers. Mena (1959) first noticed the difference in first pass arrival time via the two pathways following IV injection of ^{131}I-HSA. A variety of analytical techniques have been developed for the clinical application of hepatic blood flow estimation (Fleming 1981). Technetium-99m-pertechnetate, -DTPA, and -MDP merely pass through the liver, not retained by it. Tc-sulphur colloid or phytate and ^{99m}Tc-HIDA/mebrofenin also pass through the liver but are partly retained, allowing additional liver imaging. The quantity of the radiotracer reaching the liver is directly proportionate to the volume of blood supplied to the liver. The differential blood flow via hepatic artery vs. that via portal vein is calculated by two methods: (1) based on the analysis of the slope of the uptake and wash out curve and (2) calculation of area under the curve by deconvolution analysis (O'Connor et al. 1988; MacMathuna et al. 1992).

■ **Data Collection.** A fasting patient is placed supine under a large field-of-view gamma camera fitted with a low-energy, all-purpose parallel hole collimator. A dual-head camera allows simultaneous anterior and posterior perfusion imaging covering the lower part of the lungs, and the entire liver, spleen, and kidneys. About 10 mCi of the ^{99m}Tc tracer is injected as a rapid IV bolus followed by a 30 ml saline flush. Data is collected on 64×64 matrix at 2 frames per second for 100 s (Fig. 23.5). The first 30 frames are summed to form a composite image on which four regions of interest are drawn: (1) mid-part of right lobe excluding the right kidney + aorta, (2) right lower lung region, (3) spleen or left kidney (in splenectomized patients), (4) cross-talk region between right lung and liver.

Time activity curves are generated on all four regions (Fig. 23.6). The cross-talk curve from the region between the right lung and liver is scaled to the same height as the early part of the liver curve (before the arrival of hepatic arterial phase, usually between 0 and 2 s) and subtracted from the liver or spleen curves to generate the correlated liver and spleen curves. Since the hepatic and splenic artery have a common origin from the celiac artery, the time of peak activity on

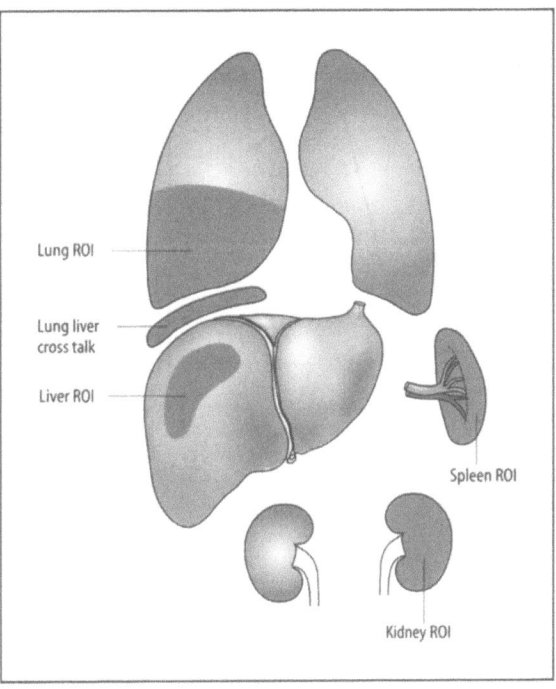

Fig. 23.6. Regions of interest (*ROI*) are drawn over the middle of the liver, right lower lung, spleen, left kidney, and cross-talk region between liver and right lung. ROIs should exclude liver margins and aorta

the splenic curve (SP) corresponds to the time TP on the liver curve and denotes the end of the hepatic arterial phases and the beginning of the portal venous phase on the liver curve. Normally there is a 7±2 second delay between the hepatic arterial and portal venous blood supply to the liver. The slope of the arterial phase (La) is measured from Ta to Ta + 7 s. The slope of the portal venous phase (LP) is measured from Tp to Tp + 7 s. Total counts are integrated and percent arterial flow is expressed using the following formula from a linear fit to the curve.

Percent hepatic arterial flow = La/La + Lp × 100
where
La = slope of hepatic arterial phase; Lp = slope of portal venous phase (Fig. 23.7)

The area method uses deconvolution analysis. Liver and spleen curves are deconvoluted with the lung curve using the modified spleen curves are integrated to give the areas under the curve AL and As respectively (Fig. 23.8).

$$\text{Arterial liver blood flow (\% Ha)} = \frac{AS}{AL} \times 100$$

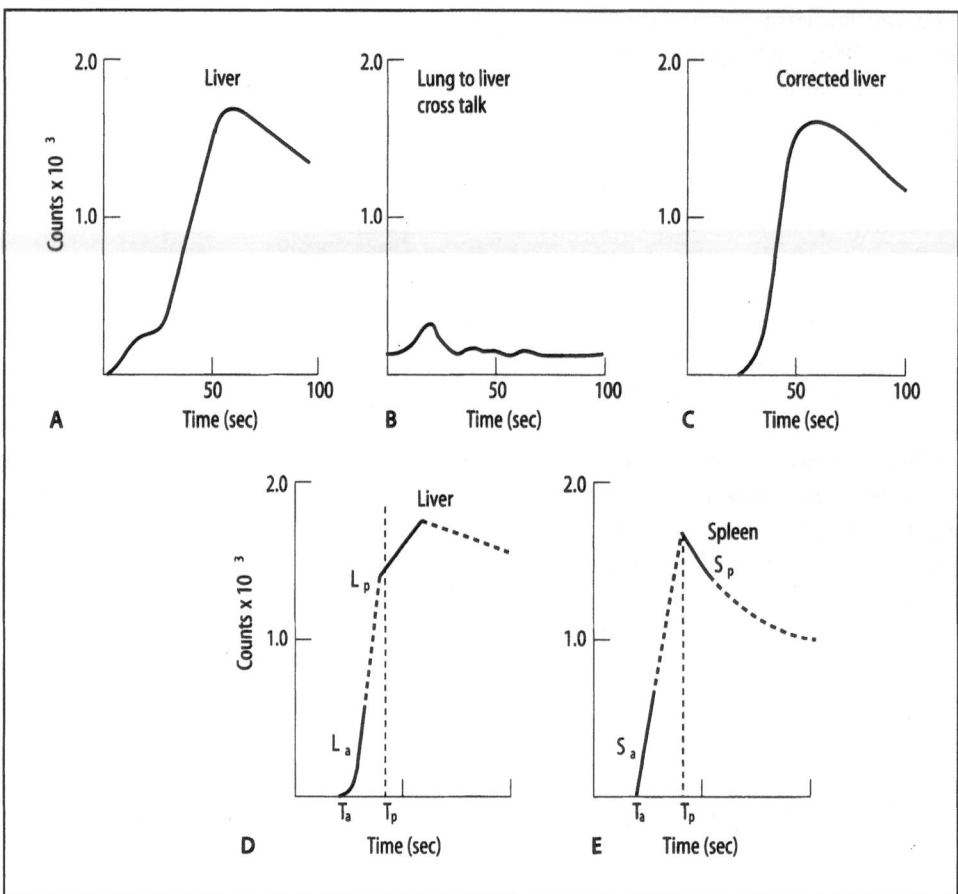

Fig. 23.7 A–E. Slope method to assess blood flow

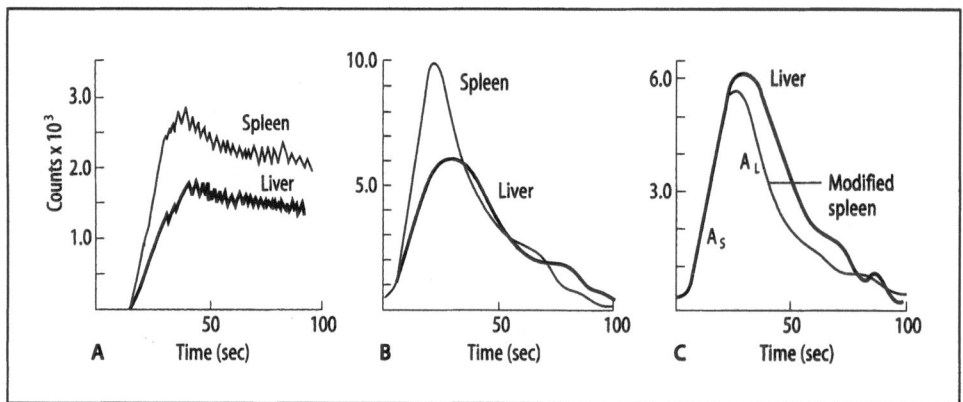

Fig. 23.8. Hepatic arterial vs portal venous blood flow by the area method. Corrected spleen and liver curves are obtained first (**A**) after subtracting the background counts from the lung to liver cross-talk region and then subjected to deconvolutional analysis (**B**). Magnitude of the spleen curve is modified so that its upslope matches with that of the liver (**C**), A_L and A_L represent the area under the modified spleen and liver curves, respectively

Portal venous flow (% PV) = 100 – %Ha.

Both the slope and area methods are found clinically useful. Arterial-to-venous ratio increases as the severity of liver disease increases from Child's class A to C (Koranda et al. 1999).

Normal median hepatic arterial flow is 22% and median portal venous flow 78%. The mean portal venous flow decrease to 68% in mild liver disease and remains below 49% in severe liver disease. In portal vein thrombosis the medial portal venous flow falls below 4%. Portal venous blood flow increases as portal venous flow and pressure tends to normalize after a successful transjugular intrahepatic portosystemic shunt (TIPS). Portal venous flow of 29% before TIPS was found to increase to 38%, indicating its therapeutic benefits (Menzel et al. 1997; Ganger et al. 1999).

23.4.1
Hepatic Lymph Flow

In cirrhosis or portal vein thrombosis, the blood flow through the liver is greatly impeded. Rise in vascular resistance in the liver can lead to a rise in capillary pressure throughout the splanchnic circulation, causing significant fluid loss from the capillaries of the intestinal tract, leading to ascites.

The rate of lymph flow from the liver is normally very high. The pores in the hepatic sinusoids are very permeable readily allowing the passage of both proteins and fluid into the lymphatic system. The lymph from the liver has protein concentration of about 6 mg/dl – only slightly less than plasma protein concentration. About half of all the lymph formed in the body under normal conditions arises from the liver.

A rise in hepatic venous pressure (due to cardiac cirrhosis or congestive heart failure) causes a corresponding rise in liver lymph flow. A rise in inferior vena cava pressure from normal (zero) to 15 mmHg can increase liver lymph flow as much as 20 times the normal flow rate. Under certain pathological conditions the excess amount of lymph formed can begin to transude through the outer surface of the liver directly into the abdominal cavity resulting in ascites.

23.5
Radionuclide Imaging of the Liver

Table 23.5 lists the various radionuclide agents used for imaging the liver, spleen, and hepatobiliary tract.

23.5.1
Imaging and Quantification of Asialoglycoprotein Receptors

The hepatocyte plasma membrane is rich in asialoglycoprotein (ASGP) receptors, not found on any other cell type in the body. The receptors are located along the basolateral and lateral domain, but not along the canalicular domain. Technetium-99m-GSA (galactosyl human serum albumin) binds to these receptors and the amount bound varies inversely with the severity of the hepatocyte dysfunction Tc-GSA is not taken up by the spleen (unlike radiocolloids) and not secreted into the bile (unlike HIDA/mebrofenin). ASGP receptor density reflects hepatocyte functional integrity accurately, much like the serum level of albumin, hepatoplastin, indocyanine green, and cholinesterase.

Almost all the intravenously injected radiotracer clears rapidly from the blood to be taken up exclusively by the liver, normally within 15 minutes. The simplest parameter is the extraction index at 15 minutes (cumulative counts in the liver for 1 min between 15 and 16 minutes to the total injected dose).

Tc-GSA uptake decreases in patients with various liver disorders including fulminant hepatic failure, chronic hepatitis and cirrhosis. Cholangiocarcinoma, hepatocellular carcinoma, benign liver growth such as adenoma, and metastatic liver lesions do not take up Tc-GSA.

Data collection and analysis begins after 4–6 h of fasting. With the patient under the gamma camera in a supine position, a bolus injection of 5 mCi (185 MBq) Tc-GSA is followed by sequential planar images (128×128 matrix) obtained at the rate of 2 frames per minutes for 20 minutes. Immediately after planar imaging is completed, SPECT images are acquired on 128×128 matrix at 10 s per step for 64 steps at 5.6 s intervals. Whole and residual liver counts are obtained by measuring organ volume from SPECT images. Whole organ volume is determined by detection of the edge for each slice and then adding all the slices. The volume of each lobe is obtained by

Table 23.5. Radionuclide tracers for imaging the liver, spleen, and hepatobiliary tract

Agent	Uptake by	Mechanism	Comments
1. Tc-GSA	Hepatocyte	ASGP receptor on cell membrane	Quantitation of functioning hepatocytes; negative in hepatoma and metastases
2. Tc-HIDA	Hepatocyte	Receptor-mediated endocytosis; secretion into biliary canaliculi along with bile	Visualization of biliary tract and gall bladder; measurement of GB and sphincter of Oddi function; prolonged retention in focal nodular hyperplasia
3. Tc-sulfur colloid	Kupffer cells	Phagocytosis of opsonin-coated radiocolloids	Also visualizes the spleen; evaluation of portal hypertension in cirrhosis; cold areas in abscess, cyst, tumor
4. Tc-rbc	Hemangioma	Blood pool	Early cold and late hot images characteristic of hemangioma
5. Tc-octreotide	Somatostatin receptors on many types of cells	Receptor-mediated endocytosis	SSRs overexpressed in many malignant tumors, especially neuro-endocrine with hepatic metastases; hot lesions
6. Gallium-67	Hepatocytes	Transferrin receptors endocytosis to lysosome	Infection and malignancy both give positive images
7. Tc-labelled leucocytes/HMPAO; ciprofloxacin	Infective focus in liver	Chemotaxis	Liver abscess; hot lesions
8. Tc-labelled monoclonal antibodies	Cancer cells expressing tumor-associated markers	Antigen-antibody interactions: alpha fetoprotein (AFP), CEA, TAG-72, etc.	Diagnosis of hepatoma (AFP) or liver metastasis from various cancers (carcinoembryonic antigen, CEA); hot lesions
9. Congo red	Amyloid	Binding to Ab protein	useful for imaging hepatic amyloid
10. Xenon-133	Fatty liver		Useful for imaging fatty liver
11. Fluorine-18 FDG	Hepatocytes	GLUT glucose transporter, hexokinase	Excessive uptake by cancer cells compared to normal cells
12. Id Urd	Hepatocytes		Increased uptake is an indicator of cellular proliferation

selecting the gall bladder fossa-inferior vena cava plane, which divides the liver into physiological right and left lobes, or by referring to the CT or MRI references. After obtaining the volume, each lobe is divided into its physiological segments, the right lobe into anterior and posterior segments, the left lobe into medial and lateral segments. The ratio of the counts in each lobe to the whole liver counts provide the liver uptake at 15 minutes (LU15) for the lobes, thus:

$$LU15 = \frac{^{16}Sc(t)dt \times 100\%}{\text{Total injected dose}}$$

Residual count ratio (RCR)

$$RCR = \frac{Rc}{Wc}$$

where Rc = residual counts for SPECT images counts from the region of the liver not to be resected.
Wc = whole liver count calculated for SPECT images.

Assessment of hepatic functional reserve is an important issue in hepatic resection especially for hepatocellular carcinoma in which a majority of patients have concomitant liver disease such as cirrhosis. Apart from estimating global hepatic function, Tc-GSA-dynamic SPECT can estimate regional hepatic functional reserve with the use of a Patlak plot for functional mapping (Hwang et al. 1999). Tc-GSA provides more accurate information about hepatic functional reserve than Indocyanine green (ICG) R 15. Clinical usefulness of scintigraphy with ^{99m}Tc-GSA for prognosis of liver cirrhosis has been shown (Sasaki et al. 1999). Estimation of predicted postoperative residual function after resection of the liver for hepatocellular carcinoma (HCC), cholangio carcinoma, and metastatic liver tumor shows good correlation with postoperative liver function. Such scintigraphy may also be ideal for predicting end-stage liver disease and timing of liver transplantation. Most of the clinical studies reported are from Japan and the radiotracer is not available in most other countries. Functional imaging and quantification need standardization to keep the variables to a minimum. Fasting for 4–6 h is essential since hepatic uptake may be variable due to post-prandial hyperperfusion of the GI tract.

In 1990 a multicenter clinical trial included 417 patients with a variety of liver disease at 23 institutions in Japan (Torizuka et al. 1992). In chronic liver disease the Tc-GSA results correlated well with serum tests and reflected the progression of the disease.

Planar images of the liver at 5 and 30 min helped to characterize mild, moderate, and severe liver dysfunction. The blood clearance and hepatic accumulation were estimated.

Tracer studies of patients with obstructive jaundice were particularly helpful in assessing the status of the hepatocyte when serum bilirubin levels were very high.

In patients with alcoholic hepatitis, liver images with Tc-phytate and Tc-GSA showed differences in the intrahepatic distribution of the two tracers. Tc-GSA revealed decreased hepatocyte function more clearly than Tc-phytate.

23.5.2
Functional Imaging of the Hepatobiliary Tract

^{99m}TC HIDA enables non-invasive and quantitative evaluation of most of the hepatobiliary diseases related to bile formation and flow. Intrahepatic cholestasis is reliably differentiated from extra-hepatic cholestasis. Cholecystokinin (CCK)-induced gall bladder ejection fraction measurement and hepatic bile clearance parameters enable detection of functional changes in the gall bladder, common bile duct, and the sphincter of Oddi, before irreversible morphologic changes take place.

Functional imaging is unique in being not only cell specific but also function-phase specific, i.e., able to delineate the phase of the cell function at fault in disease. Functional abnormalities precede morphological changes by days, weeks or months – hence early detection and treatment permit complete recovery be-

fore permanent morphologic changes take place. Additionally, with the use of computers most of the functional parameters are readily quantifiable as an integral part of imaging.

23.5.2.1
Technetium-99m Hepatic Iminodiacetic Acid

This is a generic term for a group of agents that are lidocaine analogues labeled with ^{99m}Tc. Labeling is achieved through the use of a bifunctional chelate, iminodiacetic acid (IDA). Each Tc-HIDA complex consists of two molecules each of lidocaine (ligand) an IDA (chelate) and one atom of ^{99m}Tc in the middle. The compounds of desired biokinetic characters are created simply by making chemical substitution in the benzene ring of lidocaine.

After an IV injection, the first 0–6 minute images show liver morphology much like a radiocolloid scan. The 6–10 minute images reflect canalicular bile transit. Images over the next 10–30 min provide information related to the bile transit through major intra- and extrahepatic bile duct while those from 30–45 min reflect the control of bile flow through the sphincter of Oddi, as well as the amount of bile that enters the duodenum versus the gall bladder. The residual activity left in the liver between 45–60 min indicates the degree of intra- or extrahepatic cholestasis. A study conducted between 60–90 min using fatty meal or CCK injection allows the testing of gallbladder contraction and bile emptying. It is this temporal separation of uptake and several steps in elimination that enables the differential diagnosis of various hepatobiliary diseases.

A methyl substitution at 2, 4, 5 and a bromine at 5 position (mebrofenin) confers the strongest affinity enabling HIDA to compete effectively with serum bilirubin for hepatocyte uptake. In addition to the bilirubin pathway it utilizes free fatty acid and conjugated bile acid pathways as additional routes for hepatocyte uptake, hence high serum bilirubin levels cannot completely block the hepatic uptake (unlike the first generation IDA agents).

Following IV injection, ^{99m}Tc-HIDA bound to serum albumin is carried to the liver. In the space of Disse it dissociates from the albumin receptor on the hepatocyte and undergoes receptor-mediated endocytosis, a mechanism shared by other organic anions including bilirubin. Normally 98% of the injected dose of mebrofenin is taken up by the liver and the rest is excreted by the kidney.

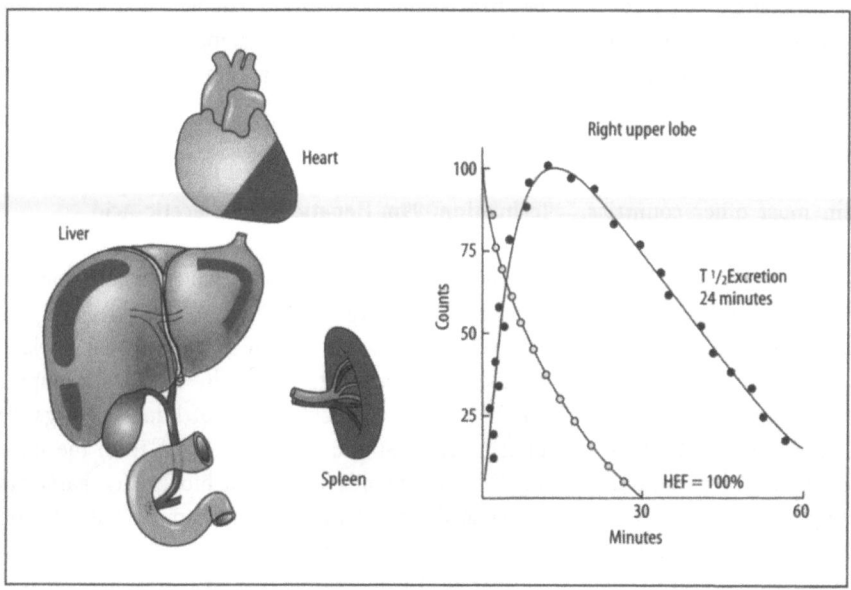

Fig. 23.9. Measurement of hepatic extraction fraction (HEF) and excretion from liver. HEF is calculated by deconvolutional analysis using heart for the input function and liver for the output function. The excretion half time ($T_{1/2}$) is measured by non-linear least-squares fit using spleen as the background

Because of the high hepatic uptake, 2.5 mCi of Tc-HIDA is enough to carry out an imaging study in patients with serum bilirubin less than 2 mg/dl, resulting in very low radiation dose. The critical organ-gallbladder receives 0.9 Rad/mCi (243 uSv for MBq) administered dose.

After hepatic uptake HIDA is transported from hepatocyte into bile canaliculi rapidly (mean excretion $T_{1/2}$ 16 minutes), and follows the path taken by the hepatic bile. This direct measurement differs from other indirect measurements such as alkaline phosphatase, gamma glutamyl transpeptidase, bilirubin, etc., as seen in Fig. 23.9.

The liver continuously secretes about 600 ml bile per day (0.4 ml/min), with part entering the gall bladder and part arriving at the duodenum depending upon the state of sphincter of Oddi.

The gall bladder is filled to its capacity of 50 ml after an overnight fast, but its wall absorbs water from the lumen during fasting, making room for fresh hepatic bile and Tc-HIDA in the lumen. Further the closed sphincter of Oddi (due to very low concentration of serum CCK during fasting) facilitates maximal increase in the basal pressure forcing bile entry into the gall bladder. Tc-HIDA is not absorbed through the GB wall. By drawing one region of interest (ROI) over the GB and another over the rest of the abdomen (excluding liver and GB), the differen-

tial bile flow between GB and intestine is measured. After an overnight fast, nearly 70% of hepatic bile enters the GB and remaining 30% directly enters the duodenum. The effects of hormones (CCK), spasmodic drugs (opiates) or spasmolytic drugs (nitrates, calcium channel blockers) on the sphincter of Oddi can be measured totally under basal conditions.

23.5.2.2
Acute Acalculous Cholecystitis

Obstruction of the cystic duct due to edema, inflammation and neutrophil, infiltration, or by stone(s) in the gall bladder neck, or irritation by bile salts or microorganisms, leads to non-visualization of the GB.

In a recent meta-analysis of 2,446 patients NM had a sensitivity of 97% and specificity of 90% (as opposed to ultrasound's 94% and 78%, respectively) for the diagnosis of acute acalculous cholecystitis.

Apart from non-visualization of the GB, a "rim" sign around the photopenic gall bladder fossa indicates acute cholecystitis with complications including gangrene and perforation. This sign is due to slow clearance of HIDA from hepatocytes adjoining the GB. The presence of a rim sign alone is not sufficient to confirm acute cholecystitis and IV morphine/pentazocin is still needed to increase the positive predictive value of the test.

Occasional rupture of the inflamed GB is shown by bile leak on the HIDA scan, a mandate for emergency surgery.

Confirmation of acute cholecystitis may take as long as 90 min, but exclusion of the diagnosis can be made in most cases within 10–30 min or as soon as the GB is visualized – 50% of normal GBs are seen within 20 min, 90% within 30 min, and 100% by 60 min.

Another approach to diagnosis of acute cholecystitis is radiolabeled leukocyte imaging. This may be preferred in patients on long-term parenteral nutrition or prolonged fasting where sludge may prevent bile entry into the GB despite a patent cystic duct.

23.5.2.3
Biliary Dyskinesia

■ **Sphincter Oddi Spasm.** A normal sphincter of Oddi is a 4–8 mm long segment that generates 5–6 phasic waves per minute, 60% of which encourage antegrade progress, 14% retrograde, and the remaining 26% are of both varieties simultaneously. CCK normally inhibits or reduces phasic wave activity and lowers the basal pressure without changing the basic sequence of wave pattern, allowing smooth passage of bile through the dilated sphincter.

The salient scintigraphic features of sphincter of Oddi spasm include a delayed hepatic peak, prolonged excretion half time, and preferential bile flow into the GB during fasting. Bile duct morphology is normal, unlike bile duct stenosis which shows tracer pooling proximal to the lesions, with a smooth tapering or an abrupt cut off of the common bile duct (CBD). In both cases the GB shows either a low or normal ejection fraction (EF) in response to CCK infusion. The bile emptied from the GB may reflux into the intrahepatic ducts and re-enter the GB soon after cessation of CCK, mostly during the GB ejection period. Unlike CT/US, scintigraphy does not require the presence of ductal dilatation for detection of obstruction.

Sphincter of Oddi spasm may be differentiated from bile duct stenosis by administration of nitrates which dilate the sphincter by relieving spasm, but have no effect on CBD stenosis. Treatment is sphincterotomy or antispasmodics.

■ **Cystic Duct Syndrome.** In this condition the GB fails to empty normally with a low EF on CCK stimulation, probably due to a decrease in the number of CCK receptors. A typical patient is a woman in her 30s or 40s with intermittent, moderate-to-severe post-prandial right upper quadrant pain, with a normal liver function test and ultrasound of the gall bladder.

Out of 283 patients with cystic duct syndrome who underwent cholecystectomy solely on the basis of a low GBEF, 90% had relief of pain and 80% had a histopathological abnormality in the GB wall or cystic duct. In another study involving 26 patients, pain was relieved in 69% but not in 31% with low GBEF. In contrast, of 35 patients with normal GBEF, 80% got pain relief without cholecystectomy.

23.5.2.4
Acute Common Bile Duct Obstruction

Following complete ligation of CBD in a dog, it is noticed that –

up to 3 h – normal hepatic uptake and secretion of bile enabling visualization of the GB and bile duct proximal to obstruction;

by 24 h – normal hepatic uptake but no biliary excretion (appearance similar to radiocolloid scan of the liver without the spleen).

In humans, for 24–48 h hepatocytes maintain good uptake and excretion of HIDA, enabling visualization of GB and CBD proximal to the level of obstruction. For 48 to 120 h after obstruction, hepatocytes continue to maintain relatively good uptake (normal hepatic extraction fraction) but excretion of HIDA into bile canaliculi stops due to a rise in CBD pressure (normal 10–12 cm water) exceeding normal hepatocyte excretory pressure (30–35 cm water).

In a patient with sudden onset of abdominal pain, elevation of serum alkaline phosphatase, and a rise in serum bilirubin of less than 10 mg/dl, non-visualization of CBD, GB, and small intestine with HIDA is diagnostic for total CBD obstruction. The cause is usually a gall stone or acute edematous pancreatitis (high serum amylase level) when the obstruction continues beyond 5 days, or serum bilirubin rises above 10 mg/dl. The hepatic extraction fraction decreases and image quality of HIDA is found to have deteriorated. At such high bilirubin levels CBD dilatation is seen on US (Juni and Reichle 1990).

Usefulness of hepatobiliary scintigraphy in the diagnosis of biliary complications after adult-to-adult living donor liver transplantation has been shown, although it has limitations as a means of differential diagnosis of non-biliary complications (Kim et al. 2002). Some patients with rejection and severely elevated serum bilirubin level showed scintigraphic findings of total biliary obstruction.

23.5.3 Radiocolloid Imaging of Liver and Spleen

The concept of nuclear hepatology was born in the late 1940s with the introduction of radiocolloids whose rate of clearance from the circulation was used as an indicator of liver function (Sheppard et al. 1947). Nuclear hepatic imaging began in 1954 with the successful imaging of the liver morphology with gold-198 colloid using an automated rectilinear scanner developed by Cassen. Imaging liver morphology became popular with the introduction of technetium-99m-labeled sulphur colloid in 1965, and the availability of gamma camera. Radiocolloids are small negatively charged particles of varying size, removed from circulation by the reticuloendothelial cells of the liver, spleen and bone marrow. The smaller the size of the particles (e.g., nanocolloid), the greater the likelihood of phagocytosis in the bone marrow. Medium-sized particles are taken up by the liver and the largest particles are likely to be phagocytosed by the spleen. Sulphur colloid particles on an average are 300 nm in size while Tc^{99m}-phytate particles are 10–20 nm size. Phytate forms microcolloids in combination with calcium.

Upon IV injection, radiocolloid particles are coated by plasma opsonins, making them susceptible to phagocytosis. In the liver the complex is encircled by pseudopods of the Kupffer cells and completely engulfed, forming a phagosome within the cytosol, which is eventually digested and its contents released back into circulation.

Normally about 85% of the injected dose of the ^{99m}Tc sulphur colloid is taken up by the liver, 7% by the spleen, 5% by the bone marrow, and the remaining 3% by other organs such as lungs and stomach. Static imaging is started 15–30 minutes after the IV injection. Static views in the anterior, right lateral, posterior, and left lateral projections give information about the functional morphology of the liver and spleen, and document hepatomegaly and splenomegaly.

There are wide variations in the normal liver shape. A 5 or 10 cm long lead marker is placed along the right costal margin to facilitate measurement of liver size and position. The same marker is used for measurement of spleen size using the posterior image.

SPECT data collection is done in a 64×64 or 128×128 matrix, clockwise for 360° with a 6° angle at each stop. This facilitates detection of lesions in depth which can be missed by planar imaging (e.g., multiple metastases).

The liver moves 1–3 cm up and down with each breath. It is useful to note this pliability which may

be lost in a diseased liver, e.g., alcoholic hepatitis, cirrhosis.

23.5.3.1
Imaging in Acute Viral or Non-Viral Hepatitis

The findings are quite nonspecific and reflect the degree of severity of hepatocellular dysfunction. A mild, nonuniform generalized decrease in uptake in early disease may progress to almost total nonvisualization of the liver in fulminant hepatic failure, with colloid shift to the spleen and bone marrow. Currently, radiocolloid liver imaging is most useful in the prognostic evaluation of chronic hepatitis and cirrhosis with portal hypertension. Sequential changes in liver size and shape, splenic size and colloid shift to the spleen, and changes in the hepatic artery/portal vein perfusion give prognostic information and also provide documentation of favorable response to treatment (e.g., interferon).

23.5.3.2
Sequential Changes in Radiocolloid Scintigraphy

1. Hepatomegaly with mild irregularity of colloid uptake: normal R:L lobe ratio, slight increased splenic uptake with splenomegaly, normal hepatic perfusion, normal pliability
2. Hepatomegaly with decreased irregular colloid uptake: mild delay in Kupffer cell extraction of colloid, moderate increase in splenic activity with marked splenomegaly or increased bone marrow activity, R:L lobe activity normal or slightly altered, mild increase in arterial component of perfusion, pliability normal or slightly impaired
3. Prominent decrease and irregularity of colloid uptake: size may be smaller right lobe with increased left lobe (well established cirrhosis), colloid shift to spleen and marrow, splenomegaly may not be marked, decrease in portal perfusion (relatively increase in hepatic artery component), decreased pliability
4. End stage cirrhosis: small liver, delay in Kupffer cell extraction of colloid and patchy uptake, marked splenomegaly, increased splenic and marrow uptake, discrete photopenic areas in liver, decrease in portal venous component of perfusion, absent pliability

In alcoholic liver disease hepatic reticuloendothelial failure may be identified on the basis of poor uptake

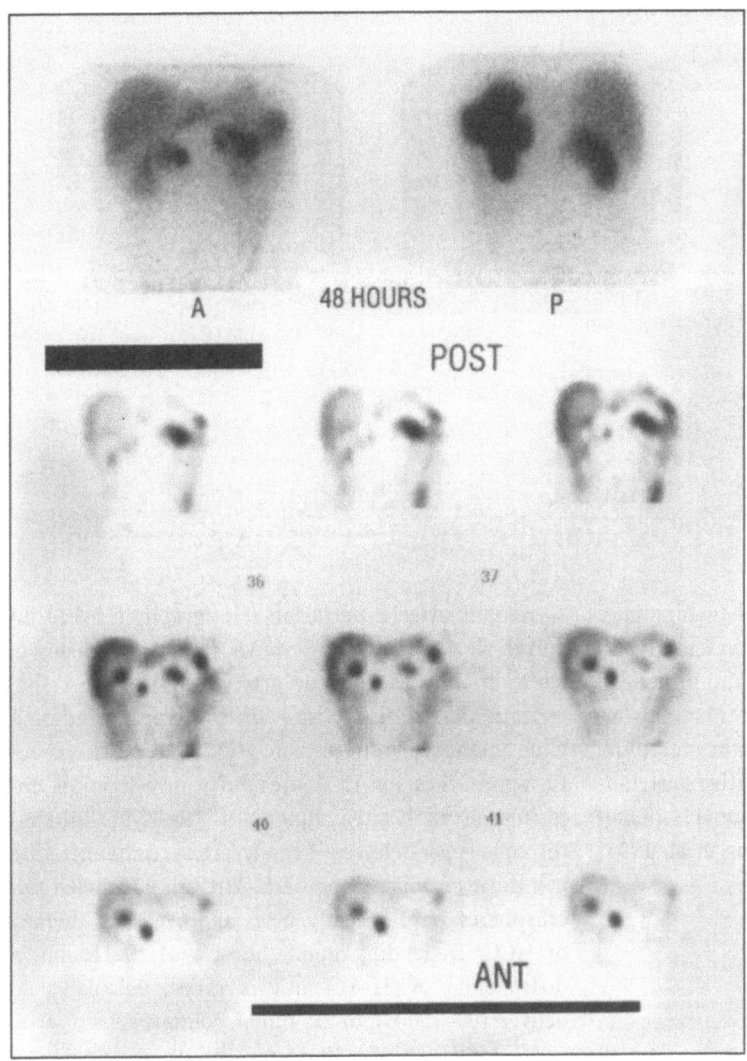

Fig. 23.10. Imaging somatostatin receptor-positive tumors and their hepatic metastases with indium-111-pentetreotide, e.g., metastatic gastrinoma: planar images (*top*) in anterior and posterior views show lesions in peripancreatic and perigastric lymph nodes. Normal activity is seen in kidneys and spleen. Coronal SPECT views (*bottom*) show multiple intrahepatic metastases, not seen in planar views. *POST* posterior, *ANT* anterior

of radiocolloid into the liver. This can be detected before severe liver dysfunction develops (Shiomi et al. 1996).

Loss of splenic reticuloendothelial function in children with sickle cell disease (SCD) can be well documented by radiocolloid spleen imaging. No splenic uptake indicates functional asplenia even with an enlarged spleen. Long-term treatment with hydroxyurea increases fetal hemoglobin (HbF) and may improve splenic function in 50% of children with SCD (Santos et al. 2002). Functional asplenia is the main cause of high morbidity and mortality due to *Streptococcus pneumoniae* infection.

In fatty infiltration or hepatitis, initial hepatomegaly may be due to hypertrophy of the left lobe (epigastric mass) and splenomegaly may not be a feature at this stage.

23.5.4
Imaging Mass Lesions in the Liver

In recent years ultrasonography (US), CT, and MRI have supplanted radionuclide liver imaging for the diagnosis of mass lesions in the liver. Ultrasonography by virtue of being a totally non-invasive, versatile, and portable modality is the preferred technique, but radionuclide imaging can supplement the information in situations where ultrasonographic criteria may be ambiguous. US combined with dynamic and static scintigraphy provides a cost-effective preferred alternative to CT or MRI.

The approach to a definite diagnosis of a mass lesion without the need for a biopsy is given in Table 23.5. Focal "cold" areas are seen in the radiocolloid

Table 23.6. Molecular basis for cancer imaging of the liver

	Primary hepatocellular cancer (HCC)	Metastases
1 Absence of Kupffer cells	Cold area on radiocolloid images	Cold area on radiocolloid images
2 Absence of asialoglycoprotein receptor	Cold lesion on GSA	Cold lesion
3 Ability to extract HIDA	Hot in 70% of cases on HDA	Cold lesion
4 Increased hexokinases and GLUT	Hot lesion on FDG	Hot lesion
5 Increased uptake of gallium 67	Hot lesion on ^{67}Ga	Hot lesion
6 Increased uptake of citrate	Hot lesion on citrate	Hot lesion
7 Increased vascularity and Na/K ATPase	Hot thallium-201	Hot ^{201}Tl
8 Increased mitochondrial activity	Hot Tc-MIBI	Hot Tc-MIBI
9 Expression of tumor-associated marker AFP	Hot on mAbAFP	Cold
10 Expression of CEA on cell surface	Hot on mAbCEA	Hot
11 Overexpression of somatostatin receptors	Hot lesion on octreotide	Hot lesion
12 Overexpression of VIP receptors	Hot lesion	Hot lesion
13 Overexpression of mRNA	Hot on RASON	Hot lesion
14 Microcalcification	Hot on MDP	Cold
15 Increased DNA synthesis	Hot on Iudr	Hot

scan in abscesses, tumors, cysts, and hemangiomas. Focal nodular hyperplasia shows hypervascularity in the early dynamic phase of HIDA scan and prolonged retention ("hot" lesion) on delayed scan. Many malignant tumours overexpress somatostatin receptors (SSR). An example of imaging SSR-positive gastrinoma with hepatic metastases with Indium-111-pentetriotide is shown in Fig. 23.10 (Krenning et al. 1995) (see also Table 23.6).

23.5.4.1
Imaging Liver Cancer

Hepatocellular cancer (HCC) occurs frequently in Africa and Asia, although it is also seen in other parts of the world. Previous hepatitis B or C infection or exposure to alfatoxin increase the likelihood of HCC. In 50–80% of patients HCC is associated with cirrhosis; 5% of cirrhosis patients eventually develop HCC, often multifocal.

Hepatoblastoma is common in children under 3 years of age and may have a genetic cause.

The diagnosis, made by US/CT/MRI, can be further reinforced or verified by the unique molecular imaging approach found in Table 23.6.

In addition to static imaging for cancer, dynamic imaging adds important diagnostic information. Primary and metastatic liver cancer derive their blood supply from the hepatic artery and not the portal vein (in contrast to the latter's 75% normal contribution). An increased hepatic perfusion index is suggestive of occult metastases when imaging results are negative. A low hepatic perfusion index decreases the likelihood.

Hepatic arterial perfusion scintigraphy (HAPS) involves slow delivery of Tc-MAA (macroaggregates of albumin) into the hepatic artery via selective catheterization. This procedure helps in planning radionuclide therapy of unresectable HCC. Several therapeutic approaches for HCC are under investigation. Injection of radioactive lipiodol (^{188}Re-N$_2$S$_2$ lipiodol) through super-selective hepatic artery catheterization is a most promising approach. Yttrium-90-labeled microspheres have recently been approved for therapy of HCC. Depending upon the size of the lesion, a dose range of 31–164 mCi is given, calculated to deliver 100–150 Gy to the tumor volume.

^{131}-AntiHBxMoAb 10–15 mCi by transhepatic arterial catheterization has been tried to convert large non-resectable tumors into small resectable ones.

23.5.5
Imaging Apoptosis and Necrosis

Cell death by apoptosis is an important mechanism in many liver diseases; cell death by necrosis can also occur. Apoptosis and necrosis are phenomena which are often observed simultaneously in tissues or cell cultures exposed to the same stimulus. For example, a high dose of radiation or chemotherapy causes necrosis while a low dose causes apoptosis. Intracellular ATP levels determine cell death fate by apoptosis or necrosis.

Apoptosis, first described by Kerr et al. in 1972, is characterized by morphological changes such as cell shrinkage, chromatin condensation, nuclear and cytoplasmic membrane blebbing, and formation of membrane-bound apoptotic bodies. In contrast, necrosis is characterized by an early loss of membrane permeability, cell swelling, dilatation of cytoplasmic vesicles, and random fragmentation of DNA. In contrast to necrosis, DNA degradation occurs in apoptosis before loss of membrane integrity.

Radford et al. proposed that early apoptosis is induced by DNA strand breaks while delayed apoptosis is due to chromosome damage from unrepaired or mismatch repair breaks.

Apoptosis pathways can be ^{53}P dependent or ^{53}P independent.

In response to stress, two common programs of gene expression are activated: the acute phase response and the heat shock response. In the acute phase response, acute phase reactants, synthesized in the liver, accumulate in the plasma to maintain systemic homeostasis. In the heat-shock response, a generic cellular response to maintain intracellular homeostasis, proteins are synthesized within the cells that cannot be measured in the blood. Some heat-shock proteins are "molecular chaperons" which have a role in folding, stabilizing and translocating newly synthesized proteins. Other heat-shock proteins are involved in apoptosis. Accumulated injuries within a cell can reach a threshold beyond which apoptosis occurs.

In acute viral hepatitis, due to various viral agents, the cell death of hepatocytes is probably by apoptosis (for instance the Councilman bodies first described in yellow fever). Dead cells are quickly removed so that little or no dead tissue is seen under the microscope (contrast ischemic infarct – death by necrosis). The presence of activated phagocytic Kupffer cells is shown by periodic acid-Schiff staining.

Apoptosis may represent a host defence mechanism against viral infection, wherein an infected cell can prevent viral replication and spread by dying.

Apoptosis is a common and important mechanism of cell death in immunologically mediated cytotoxicity, e.g., autoimmune hepatitis, primary biliary cirrhosis, primary sclerosing cholangitis, and liver transplant rejection.

Cells undergoing apoptosis express on their cell membrane phosphatidyl serine (PS), which is not normally so expressed. Annexin V (mol. wt. 36 KDa) is a physiological human protein with a high degree of affinity (Ca^{2+} dependent) for the PS on the cell membrane. It has been used to detect early stages of apoptosis in experimental models. Two preparations of radiolabeled annexin V are now available for imaging apoptosis in vivo. Technetium-99m-labeled phen-

thiolate-annexin V is commercially available as Apomate Kit (Thesus Imaging Inc.), 300–500 mBq, as is ^{99m}Tc-labeled iminothiolate-annexin V (Mallinckrodt), 600 MBq (Blankenberg et al. 1999).

In patients with myocardial infarction undergoing reperfusion therapy, delayed images (15–20 h post IV injection) of the radiotracer on scintigraphy clearly showed an intense accumulation of the tracer at the site of infarct, indicating presence of externalized PS on the cardiomyocyte membrane.

The problem in imaging apoptosis in the kidneys and liver is that 21±6% of the injected tracer accumulates in the normal kidney and 12.8±2.7% in the normal liver. Additionally, it has a long residence time in the liver and a long biological half life. It remains to be seen if late images can identify hepatocytes undergoing apoptosis. Delayed kidney images have shown positive results in experimental kidney transplant rejection.

Ability to image apoptosis allows evaluation of possible therapies to prevent or reverse apoptosis. For example, enhanced glucose transport early after cellular stress protects the cells by preventing the induction of apoptosis. The practice of glucose-insulin potassium drip following acute myocardial infarction has a rational basis providing growth factor (via insulin which is similar to IGF-1 and enhances glucose uptake), protecting the cardiac muscle from apoptosis.

In vivo quantification of "hot-spot" annexin V uptake within an area of acute myocardial infarction is possible. Compared to Tc-sestamibi, annexin V uptake can be seen in a smaller defect. This approach has detected allograft rejection in rat liver transplantation (Ogura et al. 2000).

23.5.5.1
Imaging Hepatic Necrosis

A new radiopharmaceutical Tc-glucarate has become available for imaging necrosis. The tracer accumulates in the nuclear histones of injured cell nuclei. Experimental evidence has shown that it does not accumulate in normal and ischemic cells, but only in infarcted cells. Clinical studies in acute myocardial infarction and cerebral infarcts have been successful. It remains to be seen if focal necrosis in acute viral hepatitis or fulminant hepatic necrosis shows a positive glucarate image.

Acute hepatic infarction (ischemic hepatitis) is seen in patients in coronary care units in the setting of low cardiac output, decreased hepatic blood flow, and passive venous congestion. Zone 3 necrosis without inflammation results. Serum transaminases and lactic dehydrogenase values rise rapidly and strikingly, and serum bilirubin and alkaline phosphatase increase slightly. Mortality is high. It remains to be explored if annexin V or glucarate imaging can visualize there changes.

23.5.6
Pleiotropic Glycoprotein Expression

Most cells in the body have transmembrane channels consisting of a class of phosphorylated glycoproteins that transport toxic agents of various kinds entering the cell back into circulation (efflux). The liver, kidney, and alimentary tract express these channels which are called pleiotropic glycoproteins (P_{gp}; 170 kDa 1280 amino acids, 12 Trans-membrane regions, 2 ATP binding folds).

Hepatocellular carcinoma (HCC) as well as a wide variety of cancers, including solid tumors and hematologic malignancies, have been shown to overexpress P_{gp}. This efflux pump expels a variety of drugs from the cell and thus is a mediator of multiple drug resistance (MDR) encountered in cancer chemotherapy. P_{gp} can be shown to be increased in resistant cancer cells in vitro as well as in vivo. Increased P_{gp} expression in cancer cells is correlated with poor prognosis to chemotherapy. MDR to cancer chemotherapy can be secondary (treatment failure occurring after several months of chemotherapy) or primary (chemotherapy never effective in the first place).

MDR can be induced by transfecting P_{gp} into cells which did not possess the P_{gp}.

In all, some hundred different types of phosphoglycoproteins have been identified and nearly all cells in the human body carry one or some of them. The function of the transporter proteins may be inhibited by substances like verapamil, quinidine, and cyclosporin A. Newer compounds have also been developed that bind irreversibly and specifically with high affinity to these proteins, and thus block the transporter channels. Monoclonal antibodies against the specific proteins are also available.

For testing nature and function of the MDR transport proteins, metal complexes of isonitriles now play a major role. Being lipophilic, with a cationic charge, they can enter the cell and tend to concentrate in the mitochondria. In cells with functioning MDR transporters, the technetium-99m-labeled isonitrile derivative called sestamibi is excreted, and the rate of loss

from the cells grades the function of the transporter. This has been well tested by introducing into the cellular genome the wild-type gene as the transporter protein. A series of tests, also with specific blocking of the transporter, has ascertained the reliability of the MDR transporter assay with sestamibi.

Sequential imaging with Tc-sestamibi in humans showed the liver to have two functional responses. The individual rates of tracer loss from the liver can be compared with the heart (which has no P_{gp} expression). This will help to determine the MDR phenotype in the healthy human. Treatment of hepatocytes with antioxidants (e.g., 1 mm ascorbic acid and 2% dimethyl sulfoxide) for 3 days markedly suppresses MDR-1 mRNA and P_{gp} expression. This may be exploited therapeutically.

The effect of rifampicin (anti-TB drug) on Pgp expression in the liver was prospectively studied in 80 patients, indications of such expression being determined before and one month after starting drug therapy. Initial Tc-sestamibi washout at one hour was 50–55% in 17 of 20 patients. The washout was increased significantly after 1 month of rifampicin therapy, thereby indicating that increased expression of P_{gp} by the liver cells is induced by rifampicin.

The immediate challenge is to evaluate the predictive value of sestamibi studies for detecting the MDR phenotype before starting cancer chemotherapy, and to appropriately correlate these results with biochemical and molecular biological findings obtained from tissue specimens.

The long-term task is to link specific gene expression to a well and easily observable function of membrane bound proteins on diverse cells in the intact body.

23.6
Diabetes and Liver

23.6.1
Insulin, Glucagon and Diabetes Mellitus

The pancreatic acini secrete digestive enzyme into the gut (exocrine function), while the one to two million islet cells of Langerhans of four distinct cell types have an endocrine function.

Alpha cells	~25%	glucagon
Beta cells	~60%	insulin
Delta cells		somstostatin
PP cells		pancreatic polypeptide

Both insulin and glucagon are synthesized as large pre-prohormones. In the Golgi apparatus the prohormones are packaged in granules and then largely cleaved into free hormone plus peptide fragments. In the beta cells insulin and connecting (C) peptide, which connects the two peptide chains of insulin, are released into the circulating blood in equimolar amounts. Proinsulin, insulin, and C peptide levels can be measured by radioimmunoassay (RIA). Proinsulin, which has a much lower biological activity than insulin, can now be separately measured by special RIA kits. A higher proinsulin/insulin (PRO/IRI) ratio is a predictor for developing future insulin-dependent diabetes mellitus (IDDM) (Rodriguez-Villar 1997) as well as for future diabetes of the non-dependent variety (NIDDM) (Haffner 1995). Some progress has been made in relating abnormal patterns of proinsulin processing and insulin production to specific mutations (Steiner 1995).

Approximately 50% of the insulin and glucagon in the portal vein is metabolized on first pass in the liver. Most of the remaining hormone is metabolized by the kidneys.

Presently known mutations in the insulin receptor gene appear to decrease insulin-induced tyrosine phosphorylation and post-receptor signal transduction (Taylor 1990).

23.6.2
Metabolic Effects of Insulin

Insulin has rapid, intermediate and delayed actions.

Seconds = rapid action – increased glucose and amino acid uptake into cells
Minutes = intermediate – stimulation of protein synthesis, inhibition of protein degradation, activation and inactivation of enzymes
Hours   = delayed action – increased transcription

23.6.2.1
Insulin Effects on Carbohydrate Metabolism

In skeletal muscle insulin facilitates glucose diffusion down its concentration gradient from the blood into cells by increasing the number of glucose transporters (GLUT-4), recruited from a cytoplasmic pool of vesicles, in the cell membrane. Insulin-dependent glucose uptake is restricted to the postprandial period when insulin is secreted. During exercise glucose transport to muscle cells is non-insulin dependent, via transporter GLUT-1.

In the liver, insulin:

(1) promotes glucose uptake by inducing glucokinase which increases the phosphorylation of glucose to glucose-6 phosphate
(2) increases glycogen synthesis by activating glycogen synthase as well as by increasing glucose uptake
(3) directs the flow of glucose through glycolysis by increasing the activity of key glycolytic enzymes, e.g., phosphofructokinase and pyruvatekinase, thus: glucose $\Rightarrow$ pyruvate $\Rightarrow$ acetyl co A $\Rightarrow$ malonyl Co A
(4) decreases the hepatic output of glucose
   a inhibits glycogen phosphorylase
   b inhibits glucose-6 phosphatase
   c decreases amino acid uptake into liver
   d decreases activity or levels of key gluconeogenic enzymes phosphoenol pyruvate carboxykinase (PEPCK) and fructose-1-6 diphosphatase (Fig. 23.11)
(5) enhances synthesis of fatty acids by stimulating acetyl Co A carboxylase which converts acetyl Co A to malonyl Co A (rate-limiting step)

In the liver, insulin inhibits the oxidation of fatty acids (because of increased conversion of acetyl Co A to malonyl Co A which inhibits carnitine acyl transferase, the enzyme responsible for shuttling fatty acids from the cytoplasm into the mitochondria for beta oxidation and conversion to ketoacids. Insulin is thus antiketogenic.

In adipose tissue:

(1) insulin promotes glucose uptake into adipose cells by increasing glucose transporters in the cell membrane. Subsequently the metabolism of glucose to a-glycerol phosphate provides the glycerol that is needed for esterification of fatty acids for storage as triglycerides
(2) insulin inhibits hormone-sensitive lipase. This decreases the rate of lipolysis of triglycerides and the release of stored fatty acids into circulation
(3) insulin induces lipoprotein lipase present in the capillary wall and splits circulating triglycerides into fatty acids, necessary for their transport into fat cells

23.6.2.2
Insulin Effect on Protein Metabolism

Insulin is an anabolic hormone, increasing uptake of several amino acids from the blood into cells. This

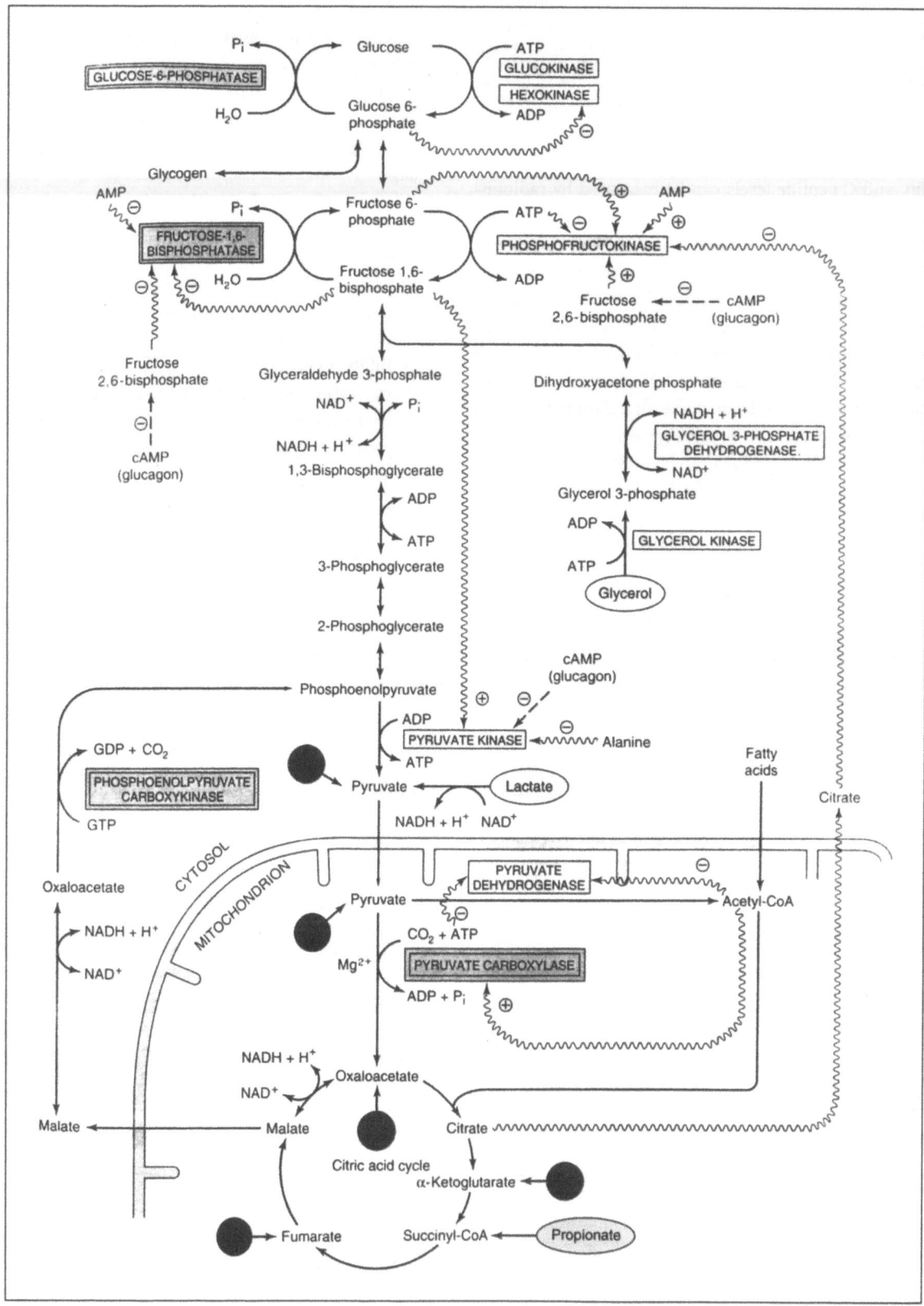

Fig. 23.11. Major pathways of regulation of gluconeogenesis and glycolysis in the liver, with the sites of action of insulin and glucagon

limits the rise in the plasma levels of certain amino acids after a protein meal. Hepatotrophic insulin increases protein synthesis by stimulating both gene transcription and translation of mRNA. It inhibits catabolism of proteins and therefore decreases the release of amino acids from muscle. Insulin, like growth hormone, is essential for growth.

23.6.2.3
Insulin/Glucagon Ratio in the Blood

The usual molar ratio of insulin to glucagon is approximately 2.0. Under conditions of fasting, the ratio decreases to 0.5 or less because of glucagon release, while after a normal meal the ratio may increase to >10. The high insulin-glucagon ratio minimizes the magnitude and duration of postprandial hyperglycemia. Insulin stimulates cyclic GMP while glucagon stimulates cyclic AMP. At the molecular level, the ratio of cGMP and cAMP is a crucial determinant of the metabolic pathway taken.

23.6.2.4
Tracer Studies in Diabetes

Carbon-14 studies in carbohydrate metabolism and oxidation of glucose in diabetic human subjects were performed by Shreeve (Shreeve and Hennes 1957), as were the effects of insulin on the turnover of plasma carbohydrates and lipids (Shreeve 1966). The influence of insulin and tolbutamide on plasma glucose turnover in humans was studied with ^{14}C-glucose (Searle et al. 1959). Impaired oxidation of ^{14}C-labeled galactose by the alcoholic or diabetic liver was demonstrated (Shreeve 1987).

■ **Somatostatin.** Prosomatostatin is a dimer formed by a union of two short 14AA peptides. Somatostatin (SS) secretion is stimulated by factors related to the ingestion of food, including increased blood levels of glucose, amino acids, and fatty acids, and to a number of gastrointestinal hormones. SS inhibits gastrointestinal motility, secretion and absorption, and is a potent inhibitor of insulin and glucagon secretion. It delays the assimilation of nutrients from the gastrointestinal tract and the utilization of absorbed nutrients by the liver and peripheral cells.

23.6.3
Hyperinsulinemia and Insulin Resistance

The development of the technique of radioimmunoassay (RIA, by Yalow and Berson 1959) opened the door for the measurement of peptide hormones with high degree of sensitivity, specificity, and reliability, and for the study of in vivo hormone regulation. It was discovered that in many maturity-onset diabetes patients, the levels of immuno-reactive insulin (contrary to expectation) were higher than normal in the presence of hyperglycemia. Further, mild type II diabetics showed some capability of increasing their insulin secretion in response to increased caloric intake. Thus the paradigm changed from insulin deficiency to insulin resistance.

DeFronzo et al. (1979) described the glucose clamp technique as a method of quantifying insulin secretion and insulin resistance. The clamp technique revealed that subjects with a normal oral glucose tolerance test can be insulin resistant (Hollenbeck and Reaven 1987).

Using ^{31}P NMR spectroscopy, high energy phosphate intermediaries such as ATP, phosphocreatine, and glucose-6 phosphate can be assessed in muscle and liver. By employing a hyperglycemic hyperinsulinemic clamp, the rate at which IV-infused I-carbon-13 glucose is incorporated into muscle can be measured using ^{31}P NMRS. Compared to normal subjects the increment in glucose-6 phosphate was significantly blunted in type II diabetes suggesting a defect in GLUT4 and/or hexokinase activity.

Carbon-13 NMRS was used to measure the rate of hepatic glycogenolysis. The increased rate of glucose production in type II DM could be entirely attributed to increased rate of gluconeogenesis (Magnussen et al.). Metformin decreases hepatic gluconeogenesis.

Human diabetes is multigenic in origin but clues from a single gene mutation (db/db mutant mice) show that the basic defect in the liver cell is non-responsiveness to the suppressive signal of insulin to the rate-limiting enzyme for gluconeogenesis, namely, phosphoenolpyruvate carboxykinase (PEPCK). Normally PEPCK is effectively reduced by a small physiological rise in circulating insulin. In the db/db mutant, in spite of hyperinsulinemia, there is marked elevation of PEPCK. Carbon-14/carbon-11 lactate and alanine are converted to glucose at a higher rate than normal, resulting in overproduction of glucose by the liver. There is also an increased level of glucagon, indicating the insensitivity of islet α cells to glucose suppression. Administration of antiglucagon antibodies reduces the hepatic glucose output (Shafrir 1988). In the single-gene mutant mice, hepatic insulin resistance is selective. It does not extend to enzymes of lipogenesis which remain responsive to insulin, hence the obesity.

The ability of the islet β cells to compensate for the insulin resistance by secreting more insulin is a crucial determinant of the glucose tolerance. When the β islet cells are unable to sustain the hypersecretion of insulin to maintain normoglycemia, blood glucose starts to rise. Higher than normal plasma free fatty acids (FFAs) in the face of hyperinsulinemia is an indicator of insulin resistance.

23.6.4
Insulin Resistance Syndrome (IRS)

Reaven (1988) discussed the role of insulin resistance in a variety of clinically important conditions such as visceral obesity (waste-hip ratio >0.9) hypertension, diabetes, and atherosclerosis. It has been suggested that endothelial dysfunction may be an initial or primary event (Steinberg et al. 1996). A cardiovascular dysmetabolic syndrome (CDS) is proposed as a culmination of insulin resistance and endothelial dysfunction (Deedwania 1998).

The endothelium is a dynamic tissue that maintains the integrity of the vasculature. Endothelium-derived nitric oxide (NO) is crucial in maintaining the vascular tone in a normally vasodilated state; it inhibits platelet and leukocyte adhesion and migration, and also inhibits vascular smooth muscle migration and growth. Defects in NO production or excessive inactivation of NO lead to unopposed vasoconstriction (through locally produced endothelin and angiotensin). NO induces and stabilizes the NF-κB inhibitor IκBα, thus attenuating the activation of NF-κB, an important mediator of inflammation.

Insulin promotes vasodilation by stimulating NO production from endothelial cells. Inhibition of NO synthase with L-N monomethyl arginine completely inhibits insulin-induced blood flow.

The cause for the common defect in insulin-mediated glucose uptake and insulin-mediated NO production may be the fact that both actions of insulin involve the same signaling pathway that via phosphatidyl inositol kinase (PI3K). In two epidemiological studies, the serum concentrations of von Wille-

brand factor (VWF), a marker of endothelial dysfunction, were correlated with those of insulin, a surrogate marker for insulin resistance. Thus IRS is a marker for peripheral endothelial dysfunction.

23.6.5
Strategies to Decrease Insulin Resistance

Regular physical exercise, cessation of tobacco use, L-arginine supplement (8 g/d) in diet, antioxidants to decrease lipid peroxidation and oxidative stress, estrogen therapy in post-menopausal women, ACE inhibitors, all help to reduce resistance to insulin (IR) and improve endothelial function. Metformin, as stated, decreases hepatic glucose production. Pioglitazone and other thiazolidinediones (rosiglitazone) act on IR peroxisome proliferator-activated receptor (PPARr), which is expressed at high levels in adipose tissue. It increases peripheral glucose utilization by increasing GLUT-4 and insulin-sensitizing action in muscle and adipose tissue (Kramel et al. 2001).

23.7
Therapy of Genetic Liver Diseases

Mutations can occur in genes coding for structural proteins (e.g., albumin) or enzymes, receptors, transporters, etc. Many proteins are targeted to mitochondria, lysosomes, other intracellular compartments, or the extracellular space, and they may be mutated in ways that affect the targeting mechanisms. Thus a mutation in the locus for a lysosomal enzyme may lead to no protein formation, to the production of a protein which never reaches the lysosome, or to production of a protein that reaches the lysosome but is unstable or catalytically impaired. Numerous examples of all types of mutations mentioned above are found in genetic liver diseases.

Table 23.7 lists the genetic disorders due to mutant genes in the liver causing a decrease in the production of various proteins.

A therapeutic strategy for this type of defect is simply to administer the missing protein (e.g., in analbuminemia or agammaglobulinemia). In the future recombinant DNA technology and genetic engineering should make many such proteins available, e.g., factor VIII (antihemophilic globulin).

Table 23.7. Genetic disorders due to mutant genes in the liver causing decrease in production of various proteins

Disease	Chromosomal location
Analbuminemia	Ch 4 q 1.13
Agammaglobulinemia	Ch X-q 2.21
Abetalipoproteinemia	Ch 4 q 2.24
Aceruloplasminemia	Ch 3 q 2.24
Thyroxin-binding globulin deficiency	
Sex-hormone binding globulin deficiency	
Transcobalamine II deficiency	Ch 22 q 1.11
Ferritin deficiency	
Fibronectin deficiency	
Fatty acid binding protein deficiency	
Anti-thrombin III deficiency	Ch 1 q 1.23
Protein C deficiency	Ch 2 p 1.13
Protein S deficiency	Ch 3 q 1.11
Von Willebrand's disease	Ch 12 p 1.12
Factor V deficiency	Ch 1. q 44
Factor VIII deficiency hemophilia	Ch x q 2.21
Factor IX deficiency Christmas disease	
Factor X deficiency	Ch 13 q 3.32
Dysfibrinogenemia a type	Ch 4 q 2.26
Dysplasminogenemia	Ch 6 q 2.26
Protein C inhibitor deficiency	Ch 14 q 3.31

Table 23.8 lists genetic disorders due to mutant hepatic enzyme. Therapeutic strategies for enzyme defects are:

(1) to replace the missing enzyme if possible. For example in Gaucher's disease, the lysosomal β glucocerebrosidase enzyme is deficient. This could be replaced by IV infusion of placental glucocerebrosidase (deglycosylated)

(2) to increase activity of mutant enzyme by supplying large amounts of co-factors, e.g., to inject vitamin B12 for methyl malonic aciduria

(3) to increase activity of mutant enzymes by induction, e.g., of phenobarbitone in Crigler-Najjar syndrome, type 1

(4) to limit substrate for the deficient enzyme by way of dietary restrictions, e.g., low in phenylalanine in phenyl ketonuria, diet free of lactose in galactosemia, diet free of fructose and sucrose in fructose intolerance, corn starch diet for glycogen storage disease type III

(5) to remove excess of stored material, e.g. by removal of iron by repeated venesection in hemochromatosis, of copper by penicillamine in Wilson's disease, of excess cholesterol by cholestyramine in familial hypercholesterolemia

(6) to use antioxidants like Vit E for reducing mitochondrial oxidative injury in Wilson's disease or glucose loading and cimetidine in acute porphyria

(7) to transplant organs (bone marrow, liver, kidney)

(8) to inaugurate gene therapy.

DNA technology and site-directed mutagenesis may be used to modify enzyme specificity or catalytic efficiency. These techniques ultimately will facilitate the design and introduction into human subjects of enzymes with specific desired properties.

Table 23.9 lists mutant genes leading to a defective transport system in the liver.

23.7.1
Gene Therapy for Liver Diseases

In theory gene therapy can involve gene replacement, gene correction, or gene augmentation in which foreign genetic material is introduced into a somatic cell to compensate for the defective product of a mutant gene. Augmentation is the sole type of gene therapy available at present. The gene of interest is usually introduced via a viral vector or plasmid-liposome complex.

Gene therapy using a "transplant" of autologous hepatocytes, genetically corrected with recombinant LDL receptors via retroviral vectors, was successful in patients with familial homozygous hypercholesterolemia with a follow-up of 18 months (Grossman et al. 1995).

23.8
Hepatic Transplantation

23.8.1
Indications for Liver Transplantation

Liver transplant may be called for in:

1. All patients with end-stage cirrhosis – serum albumin < 3 mg % and prothrombin time > 5 s over control, intractable ascites, and failure of medical and sclerotherapy to control variceal bleeding.

2. Cholestatic liver disease: primary biliary cirrhosis, biliary atresia, primary sclerosing cholangitis.

3. Fulminant liver disease in previously healthy persons.

4. Malignant liver disease: primary hepatocellular carcinoma, hepatoblastoma, hemangioendothelioma.

Table 23.8. Genetic disorders due to mutant hepatic enzymes

Disease	Mutant enzyme defect	Chromosome location	Clinical consequences
Alkaptonuria	Hepatic homogentisic acid hydroxylase		Black urine, "India ink" ochronosis, osteoarthritis
Phenylketonuria	Phenylalanine hydroxylase	Ch 12 q 2.24	Mental deficiency
Galactosemia	Galactose 1 phosphate; uridyl transferase; inability to convert to glucose-I-phosphate		Hepatomegaly; accumulation of galactose-1 phosphate
Hereditary tyrosinemia	Fumaryl acetoacetate hydrolase		Progressive liver and renal tubular damage
Hereditary fructose intolerance	Aldose β-impaired cleavage of fructose-1 phosphate	Ch 9	Hepatomegaly
a_1 Antitrypsin deficiency	Accumulation of polymerized protein units prevents export into circulation	Ch 14	Low serum level of aIAT; emphysema; cirrhosis; HCC
Wilson's disease	Copper-transporting ATPase; defective transfer of copper to apoceruloplasmin; no copper excretion in bile	Ch 13 q 1.14	Low serum ceruloplasmin; copper toxicity; cirrhosis; basal ganglia dysfunction
Hemochromatosis	CYS 282 tyr mutation	Ch 6 p$_I$ 21	Cirrhosis; HCC; arthropathy
Cholesteryl ester	Cholesteryl	Ch 10 q 2.23	Hepatomegaly
Storage disorders	Ester hydrolase		Benign condition
Gaucher's disease	Lysosomal β glucocerebrosidase	Ch 1 q 1.23	Hepatosplenomegaly; cirrhosis
Niemann-Pick disease	Lysosomal sphingomyelinase		Cirrhosis, portal hypertension
Gilbert's disease	bilirubin glucuronidase UGT 1*1	Ch	Life long asymptomatic; unconjugated hyperbilirubinemia
Crigler-Najjar syndrome	bilirubin glucuronide transferase	Ch 2 q 3.37	Hyperbilirubinemia, kernicterus; death within 1 y of birth
Acute intermittent porphyria	Porphobilinogen deaminase; secondary induction of alpha aminolevulinic acid (ALA); peripheral neuropathy	Ch 1 p 31	Overproduction of porphobilinogen and ALA; abdominal pain; psychosis
Porphyria cutanea tarda	Uroporphyrinogen decarboxylase		Cutaneous photosensitivity, blistering, scarring, hypertrichosis, pigmentation
Hereditary coproporphyria	Coproporphyrinogen oxidase	Ch 3 q 1.11	Neurocutaneous increase in urinary coporphyrin and protoporphyrin
Variegate porphyria		Ch 14 q 3.31	Neuropsychiatric symptoms
Glycogen storage disease			
Type 0	Glycogen synthetase		Hypoglycemia
Type Ia (von Gierke's disease)	Glucose-6-phosphatase		Mental retardation
Type Ib	G-6-P-translocase		Hepatomegaly; hypoglycaemia
Type II Pompe's disease	Acid a glucosidase	Ch 17 q 2.24	Cardiomegaly; muscular hypotonia
Type III (Cori's disease)	Amylo-1,6 glucosidase		Hepatomegaly
Type IV (Andersen's)	Branching enzyme		Hepatosplenomegaly
Type V (McArdle's syndrome)	Muscle phosphorylase (liver normal)		Muscle cramps, myoglobinuria
Type VI (Hers' disease)	Liver phosphorylase	Ch 14 q 2.21	growth retardation; hepatomegaly
Type VIII	Phosphorylase activation in liver		
Type IXa IXb	Phosphorylase kinase		Short stature; hepatomegaly

Table 23.8 (continued)

Disease	Mutant enzyme defect	Chromosome location	Clinical consequences
Mucopolysaccharidoses			
MPs I (Hurler's disease)	a_1 iduronidase	Ch 4 p 1.16	Hepatomegaly
MPs II (Hunter's disease)	Iduronate sulfatase		Hepatomegaly
MPs III (Sanfillippo's syndrome A)	Heparan N-sulfatase		Hepatomegaly
MPs III (Sanfillippo's syndrome B)	a-N-acetylglucosaminidase		
MPs IV (Morquio's syndrome)	Galactose-6 sulfatase	Ch 16 q 2.24	
MPs VIMar	N-acetyl galactosamine 4 sulfatase		Hepatosplenomegaly
MPs VII (Sly's syndrome)	β-glucuronidase		Hepatosplenomegaly

Table 23.9. Mutant genes leading to defective transport systems

Disease	Genetic defect	Chromosome location	Clinical consequences
Familial hypercholesterolemia	Mutant LDL receptor		Xanthomata; death from coronary atherosclerosis before age 30
Dubin-Johnson syndrome	Defective ATP-dependent canalicular transport		Benign, intermittent; jaundice of conjugated bilirubin
Rotor's syndrome	Defective hepatic uptake of bilirubin		Nonvisualization of liver, biliary tract, and GB on HIDA scan
AVED (ataxia due to Vit E deficiency)	a-tocopherol transfer protein deficiency, inability to transfer Vit E to lipoprotein		Ataxia, progressive peripheral neuropathy
Byler's disease	Defect in micro-filament function in biliary canalicular membrane	Ch 18	Conjugated bile cannot be excreted, cholestasis; death before 8 years of age
Cystic fibrosis	Mutant in CFTR	Ch 7 q 1.11	Abnormal sodium chloride and calcium content in secretions, increased viscosity
Gilbert's disease hyperbilirubinemia	Hepatic transport abnormality in bile duct epithelium		Lifelong asymptomatic; unconjugated bilirubin

5. Metabolic liver disorders: end stage disease or premalignant change.

Table 23.10 lists the candidates for liver transplant according to age (children or adults).

The first liver transplantation was performed in 1955 by Welch in a dog (Welch 1955). The first human cadaver liver transplantation was performed in 1963 by Starzl et al. (Starzl 1963). Early clinical results were disappointing. The advent of modern immunosuppressive drugs such as cyclosporine has improved survival rates in both children and adults suffering from previously fatal end stage liver disease. In children survival rates are 82% at 1 year, 80% at 3 years, 78% at 5 years, and 76% at 10 years.

23.8.2
Types of Liver Transplant

There are essentially three types of transplants – (1) cadaver liver, (2) living donor liver, and (3) auxiliary liver. Figure 23.12 depicts the various types: during cadaver transplant the entire recipient liver is removed and replaced by the cadaver liver (orthotopic). In the case of living donor, either the lateral segment or the entire left lobe of the donor liver replaces the entire native liver of the recipient.

An auxiliary liver transplant is done for patients with fulminant hepatic failure, where in the donor liver acts as a transient support until the native liver recovers; the donor liver is placed below the native liver or replaces the left lobe of the native liver.

Table 23.10. Candidates for liver transplantation according to age

Children	Adults
α_1-antitrypsin deficiency	Alcoholic cirrhosis
Alagille's syndrome	Chronic hepatitis B or C with cirrhosis
Biliary atresia	Cryptogenic cirrhosis
Byler's disease (familial intrahepatic cholestasis)	Primary biliary cirrhosis
Crigler-Najjar syndrome type 1	Secondary biliary cirrhosis
Congenital hepatic fibrosis	Primary sclerosing cholangitis
Cystic fibrosis	Caroli's disease
Fabry's disease	Severe methotrexate hepatotoxicity
Familial hypercholesterolemia	Fulminant hepatitis
Galactosemia	Hepatic vein thrombosis
Glycogen storage disease	Hepatocellular carcinoma
Lysosomal storage disease	Hepatic adenomas
Mitochondrial respiratory chain defects	
Neonatal hemochromatosis	
Neonatal hepatitis	
Protoporphyria	
Primary coagulation disorders (deficiency of F VIII, IX, protein C)	
Primary oxaluria type I	
Urea cycle defects	
Wilson's disease	
Zellweger's syndrome	

Cholecystectomy is performed as an integral part of liver transplantation to prevent future complications due to gall stones. The transplantation procedure involves: (1) hepatectomy of the native liver, the blood flow through the vena cava, portal vein, and hepatic artery being interrupted; (2) donor liver revascularization in the recipient; and (3) a biliary reconstruction stage in which a choledochostomy is performed in adults, or a choledochojejunostomy is performed in children weighing less than 16 kg.

23.8.2.1
Living Donor Liver Transplant

Although the liver is one solid organ, its distinct and separate segmental and lobar anatomy and physiology allow resection into smaller portions for transplantation. Resection of the donor liver is performed along the physiological planes. The lateral segment of the left lobe (areas 2 and 3) or the entire left lobe (areas 2, 3, 4A, and 4B) is transplanted as shown in Fig. 23.13 (which depicts segmental anatomy and physiology). An orthotopic liver transplant with end to end anastomosis is the most common type. In auxiliary transplant the donor liver is placed either in the orthotopic (middle) or heterotopic position (bottom). The first living donor liver transplantation was performed in Brazil in 1989 by Raia et al. (1989). The left lobe of the liver from a mother was transplanted successfully into her son in Australia (Strong et al. 1990). After addressing the ethical issues, numerous other centers round the world have started performing living donor transplantation. The major advantages of living donor vs. cadaver donor is (1) reduction in waiting time, (2) high-quality of donor liver, and (3) immunological similarity of the donor.

Postoperative complications are a major concern both in donor and recipient. Thirteen out of 100 donors had complications, including biliary leak, infection, and injury to bile duct and spleen. Donor survival is 100% and recipient survival ranges from 80–94% in experienced centers.

Early complications after orthotopic liver transplantation have been reviewed (Mazariegos et al. 1999).

Complications of liver transplantation are frequent, both in the immediate postoperative period such as bile leak, vascular thrombosis, infection, and hemorrhage, and at various times thereafter such as rejection. Their early recognition is essential for proper management. Rejection and infection are the two most common complications, followed by impaired blood flow, bile duct stenosis, biliary leakage, and hemorrhage. Rejection and infection reduces hepatocyte function, which markedly reduces the uptake and clearance of hepatobiliary tracers by the liver. Percutaneous liver biopsy is needed to differentiate the two conditions.

Other complications identifiable by hepatobiliary scintigraphy include compromized arterial or portal venous flow, characterized by delayed or diminished uptake of radiotracer into the grafted liver. In patients with biliary stenosis or obstruction, good extraction of the radiotracer by the transplanted liver is seen, but excretion is delayed.

Six percent of organ transplant recipients will develop lymphoproliferative malignancy, skin cancer, Kaposi's sarcoma (many related to immunosuppression), usually within five years of transplantation. Yearly cancer surveillance (probably with FDG PET) is essential for all patients post-transplant.

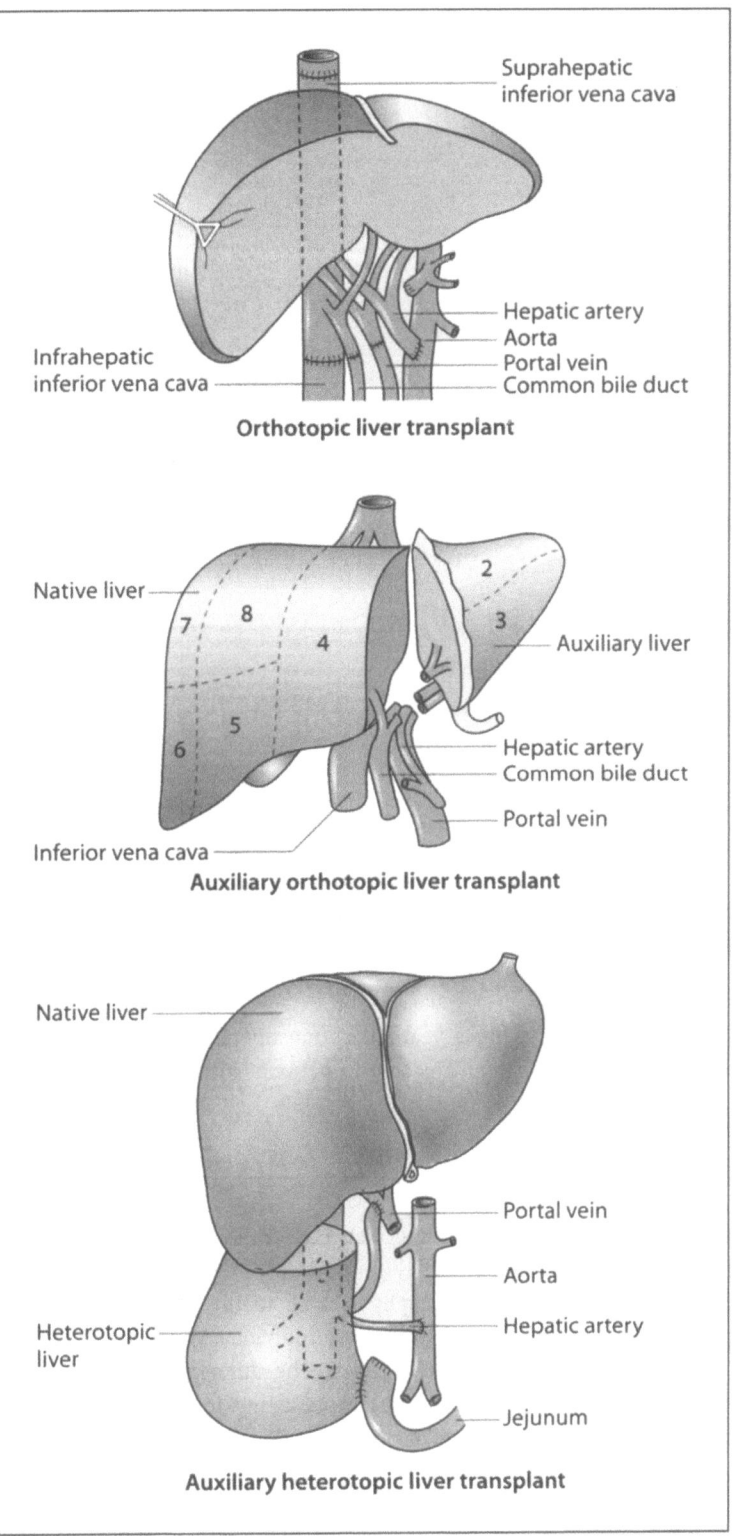

Suprahepatic
inferior vena cava

Infrahepatic
inferior vena cava

Hepatic artery
Aorta
Portal vein
Common bile duct

Orthotopic liver transplant

Native liver

2
3
Auxiliary liver

Hepatic artery
Common bile duct

Portal vein

Inferior vena cava

Auxiliary orthotopic liver transplant

Native liver

Portal vein

Aorta

Hepatic artery

Heterotopic
liver

Jejunum

Auxiliary heterotopic liver transplant

Fig. 23.12. Liver transplantation: (*top*) orthotopic liver transplant with end to end anastomosis is the most common type. In auxiliary transplant the donor liver is placed either in the orthotopic (*middle*) or heterotopic position (*bottom*)

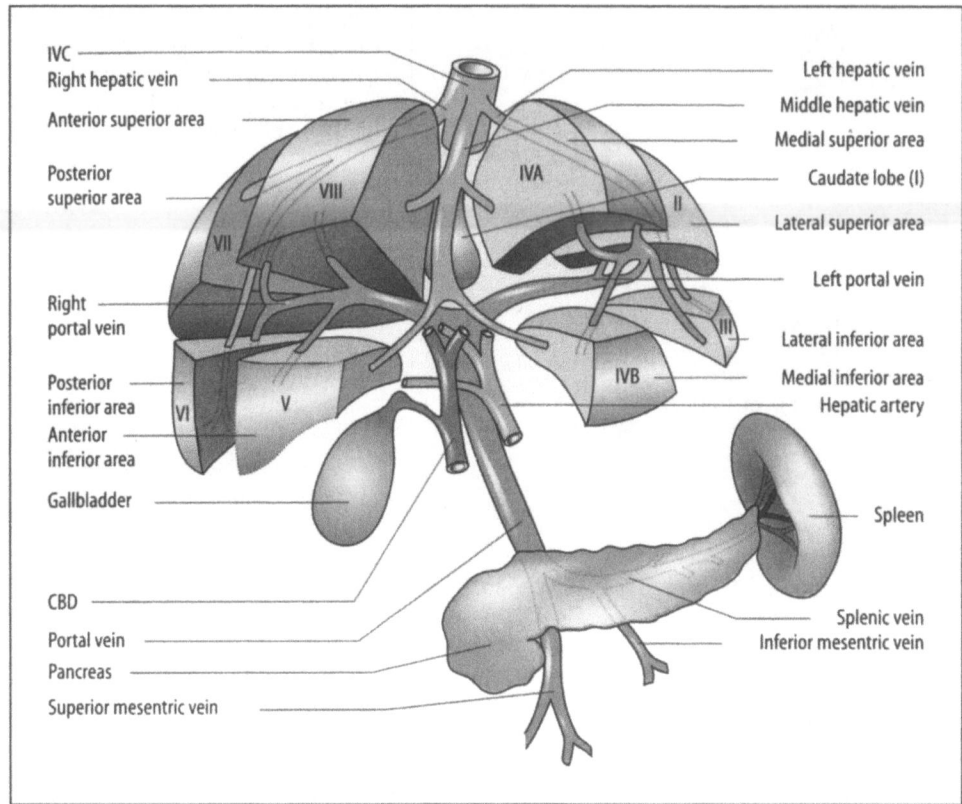

Fig. 23.13. Hepatic lobes and the hepatic and portal venous system. The segmental anatomy and physiology permit resection of the donor liver along physiological planes for transplantation, removing the lateral segment of the left lobe (*areas II, III*) or the entire left lobe (*areas II, III, IVA and IVB*)

The differences between primary non-function and rejection require attention.

Clinically, primary non-function (days 1–2) presents as liver failure, with encephalopathy, persistent acidosis, and coagulopathy, and requires treatment different from that of acute or chronic rejection.

Acute rejection occurs after 5–30 days but within 3 months of transplantation, rejection thereafter considered chronic. Unlike the kidney, the liver does not manifest a hyperacute rejection. Rejection is a microvascular phenomenon wherein fibrin deposits, and inflammatory and immune cells cluster within and around the capillaries, manifesting an obliterative angiopathy and chronic ischemia. Chronic ischemia leads to obliteration of bile ducts (vanishing bile duct syndrome) and intrahepatic cholestasis. Rejection and intrahepatic cholestasis due to other causes manifest similar scintigraphic findings, hence cannot be distinguished except by liver biopsy. Rejection is managed by a combination of several immunosuppressive agents.

After 5 years, immunosuppression can often safely be stopped except in patients with autoimmune hepatitis.

23.8.3
Differentiation of Native versus Donor Liver Function

Cholescintigraphy is able to measure the total liver function as well as the individual functions of the donor and native livers. The differential liver function information helps in adjusting the dose of immunosuppressive agents.

Auxiliary liver transplant is mostly restricted to patients with fulminant hepatic failure, which heals slowly over several months to attain full functional recovery. The immunosuppressive dose needs to be increased when there is rejection of the donor liver. An improvement in the native liver function calls for a reduction in the immunosuppressive dose or total discontinuation. The donor liver undergoes spontaneous regression and atrophy when the native liver recovers its full function, and steroids are safely withdrawn at this time. The donor liver is surgically removed if it does not undergo spontaneous atrophy and interferes with the function of the native liver.

23.9
References

Barroso L et al (1999) Dominant negative mutation in human PPARγ associated with severe insulin resistance, hypertension and diabetes mellitus. Nature 402:880–883

Bauwens L, Wisse E (1992) Pit cells in the liver. Liver 12:3

Blankenberg FG et al (1999) Imaging apoptosis (programmed cell death) with Tc-99m Annexin V. J Nucl Med 40:184–191

DeFronzo RA, Tobin JD, Andres R (1979) Glucose clamp technique: a method for quantifying insulin secretion and resistance. Am J Physiol 237:E214–E223

Deedwania PC (1998) The deadly quartet revisited. Amer J Med 105:15–35

Doe E, Krishnamurthy GT, Eklem MJ, Gilbert S, Brown PH (1991) Quantification of hepatobiliary function as an integral part of imaging with Tc-99m mebrofenin in health and disease. J Nucl Med 32:48–57

Flemming (1981) Hepatic blood flow estimation. J Nucl Med 22:18

Fratantoni JC, Hall CW, Neufeld EF (1968a) Hurler and Hunter syndromes: mutual correction of the defect in cultured fibroblasts. Science 162:570–572

Fratantoni JC, Hall CW, Neufeld EF (1968b) The defect in Hurler's and Hunter's syndromes: faulty degradation of mucopolysaccharide. Proc Natl Acad Sci USA 60:699–706

Fratantoni JC, Hall CW, Neufeld EF (1969a) The defect in Hurler and Hunter syndromes. 2. Deficiency of specific factors involved in mucopolysaccharide degradation. Proc Natl Acad Sci USA 64:360–366

Fratantoni JC, Neufeld EF, Uhlendorf BW, Jacobson CB (1969b) Intrauterine diagnosis of the hurler and hunter syndromes. N Engl J Med 280:686–688

Ganger DR, Klapman JB, McDonald V, Matalon TA, Kaur S, Rosenblate H, Kane R, Saker M, Jensen DM (1999) Transjugular intrahepatic portosystemic shunt (TIPS) for Budd-Chiari syndrome or portal vein thrombosis: review of indications and problems. Am J Gastroenterol 94:603–608

Grossman M et al (1995) A pilot study of ex vivo gene therapy for homozygous familial hypercholesterolemia. Nature Medicine 1(11):1148–1154

Haffner SM, Stern MP, Miettinen H, Gingerich R, Bowsher RR (1995) Higher pro-insulin and specific insulin are both associated with a parental history of diabetes in non-diabetic Mexican-American subjects. Diabetes 44:1156–1160

Hollenbeck C, Reaven GM (1987) Variations in insulin-stimulated glucose uptake in healthy individuals with normal glucose tolerance. J Clin Endocrinol Metab 64:1169–1173

Hwang EH, Taki J, Shuke N, Nakajima K, Kinuya S, Konishi S, Michigishi T, Aburano T, Tonami N (1999) Preoperative assessment of residual hepatic functional reserve using 99mTc-DTPA-galactosyl-human serum albumin dynamic SPECT. J Nucl Med 40:1644–1651

Juni JE, Reichle R (1990) Measurement of hepatocellular function with deconvolutional analysis: application in the differential diagnosis of acute jaundice. Radiology 177:171–175

Kament MD (1957) Isotopic Tracers in Biology, 3rd ed. Academic Press, New York

Kerr JF, Wyllie AH, Currie AR (1972) Apoptosis: a basic biological phenomenon with wide-ranging implications in tissue kinetics. Br J Cancer 26:239–257

Kim JS, Moon DH, Lee SG, Lee YJ, Park KM, Hwang S, Lee HK (2002) The usefulness of hepatobiliary scintigraphy in the diagnosis of complications after adult-to-adult living donor liver transplantation. Eur J Nucl Med Mol Imaging 29:473–479

Koranda P et al (1999) Hepatic Perfusion changes in patients with cirrhosis; indices of hepatic arterial flow. Clin Nucl Med 24:507–510

Kramel D et al (2001) Insulin sensitizing effect of rosiglitazone by regulation of glucose transporters in muscle and fat of Zuker rats. Metabolism 50(11):294–300

Krenning EP, Kwekkeboom DJ, Pauwels S et al (1995) Somatostatin receptor scintigraphy. In: Freeman LM (ed) Nuclear medicine annual. Raven, New York, pp 1–50

Krishnamurthy GT, Bobba VR, McConnell D, Turner F, Mesgarzadeh M, Kingston E (1983) Quantitative biliary dynamics: introduction of a new noninvasive scintigraphic technique. J Nucl Med 24:217–223

Krishnamurthy S, Krishnamurthy GT (1997) Biliary dyskinesia: role of the sphincter of Oddi, gallbladder and cholecystokinin. J Nucl Med 38:1824–1830

Liu J, Rosman AS, Leo MA, Nagai Y, Lieber CS (1994) Tissue inhibitor of metalloproteinase is increased in the serum of precirrhotic and cirrhotic alcoholic patients and can serve as a marker of fibrosis. Hepatology 19:1418–1423

Mena (1959) Difference in first pass arrival time after i.v. inj HSA. Amer J Diges Dis 4:19

MacMathuna P et al (1992) Non-invasive diagnosis of portal vein occlusion by radionuclide angiography. Gut 33:1671–1674

Magnussen I et al (1992) Increased rate of gluconeogenesis in type 2 diabetes mellitus: a C-13 NMR study. J Clin Invest 10(4):1323–1327

Mazariegos GV et al (1999) Early complications after orthotopic liver transplantation. Surg Clin N Amer 79:109–129

Menzel J, Schober O, Reimer P, Domschke W (1997) Scintigraphic evaluation of hepatic blood flow after intrahepatic portosystemic shunt (TIPS). Eur J Nucl Med 24:635–641

O'Conner MK et al (1988) Hepatic arterial and portal venous components of liver blood flow from computerized radionuclide angiography. J Nucl Med 29:466–472

Ogura Y et al (2000) Tc-99m Annexin V imaging: diagnosis of allograft rejection in an experimental rodent model of liver transplantation. Radiology 214:795–800

Raia S, Nery JR, Mies S (1989) Liver transplantation from liver donors. Lancet 2:497

Reaven GM (1988) Role of insulin resistance in human disease. Diabetes 37:1595–1607

Rodrigues Viller (1997) High pro-insulin level in late pre-IDDM stage. Diab Res Clin Pract 37(2):145–148

Santos A, Pinheiro V, Anjos C, Brandalise S, Fahel F, Lima M, Etchebehere E, Ramos C, Camargo EE (2002) Scintigraphic follow-up of the effects of therapy with hydroxyurea on splenic function in patients with sickle cell disease. Eur J Nucl Med Mol Imaging 29:536–541

Sasaki N, Shiomi S, Iwata Y, Nishiguchi S, Kuroki T, Kawabe J, Ochi H (1999) Clinical usefulness of scintigraphy with 99mTc-galactosyl-human serum albumin for prognosis of cirrhosis of the liver. J Nucl Med 40:1652–1656

Searle GK et al (1959) Effect of insulin and tolbutamide on plasma glucose turnover in humans-C-14 glucose

Shafrir E (1988) Non-recognition of insulin as a gluconeogenesis suppressant: a manifestation of selective hepatic insulin resistance in several animal species with type 2 diabetes. In: Shafrir E, Renold AE (eds) Lessons from animal diabetes. Libbey, London, p 304

Sharp HL (1995) Wherefore art thou liver disease associated with alpha-1 antitrypsin deficiency? Hepatology 22:666

Sheppard CW, Wells EB, Habu PE, Goodel JBP (1947) Studies of the distribution of intravenously administered radiocolloid clearance as an indicator of liver function. J Lab & Clin Med 32:274–277

Shiomi S, Kuroki T, Ueda T, Takeda T, Nishiguchi S, Nakajima S, Kobayashi K, Ochi H (1996) Diagnosis by routine scintigraphy of hepatic reticuloendothelial failure before severe liver dysfunction. Am J Gastroenterol 91:140–142

Shreeve WW, Hennes AR (1957) Effect of adrenal steroid hormones on the metabolic fate of C^{14} labeled acetate in human subjects. Quarterly Progress Report (Brookhaven National Laboratory, Medical Department)

Shreeve WW (1966) Effects of insulin on the turnover of plasma carbohydrates and lipids. Am J Med 1966 40:724–734

Shreeve WW (1987) Impaired oxidation of carbon-labeled galactose by alcoholic or diabetic liver in vivo. Nuklearmedizin 26:159–166

Starzl TS, Marchioro TL, Von Kaulla KN et al (1963) Homotransplantation of the liver in humans. Surg Gynecol Obst 117:655–676

Steinberg HO, Chaker H, Leaming R et al (1996) Obesity/insulin resistance is associated with endothelial dysfunction: implication for the syndrome of insulin resistance. J Clin Invest 97:2601–2610

Steiner (1995) Abnormal pattern of proinsulin processing and insulin production due to specific mutations

Strong RW, Lynch SV, Ong TH et al (1990) Successful liver transplant from a living donor to her son. N Engl J Med 322:1505–1507

Taylor SI, Kodowsaki T (1990) Mutation in insulin receptor gene in insulin-resistant patients. Diabetes Care 13:565–575

Torizuka K, Ha-Kawa SK, Kudo M, Kubota Y, Yamamoto K, Itoh K, Nagao K, Uchiyama G, Koizumi K, Sasaki Y, et al (1992) Phase III multi-center clinical study on 99mTc-GSA, a new agent for functional imaging of the liver (in Japanese). Kaku Igaku 29:159–181

Van Furth R, Kramps JA, Van der Putten AB et al (1986) Change in a-1 antitrypsin phenotype after orthotopic liver transplant. Clin Exp Immunol 66:669

Welch CS (1955) A note on transplantation of whole liver in dogs. Transpl Bull 2:54

Whitley CB et al (1993) Long term outcome of Hurler Syndrome following bone marrow transplantation. Amer J Med Genetics 46(2):209–218

Yalow, Berson (1959) Assay of plasma insulin in human subjects by immunological methods. Nature 184:1648–1649

Zhang et al (2001) Imaging hepatic fibrosis with Indium-11 C-SRNLIDC

Recommended Reading

1. Krishnamurthy GT, Krishnamurthy S, Nuclear hepatology. Springer, Berlin Heidelberg New York

2. Oxford Textbook of Hepatology

3. Sherlock S, Dooley J (1997) Diseases of the liver and biliary system, 10th edn. Blackwell, Oxford

4. Zakim D, Boyer TD (1996) Hepatology: a textbook of liver diseases, 3rd edn. Saunders, Philadelphia

CHAPTER 24

Molecular Gastrointestinal Scintigraphy

24

JEAN-LUC C. URBAIN, MARIE-CHRISTIANE M. VEKEMANS, LEON S. MALMUD

Contents

24.1
Introduction

Over the past 30 years, the description of the scintigraphic procedures to perform gastrointestinal imaging studies and their findings in multiple diseases have been described in journal articles and textbook chapters (Urbain et al. 2000, 2002). The aim of this contribution is to provide the nuclear medicine scientist with the basic genetic knowledge of various diseases as it relates to the gastrointestinal tract and may affect the result of gastrointestinal scintigraphic studies. First, the basic, molecular and genetic abnormalities of diseases affecting the motility of the gastrointestinal tract are explained. The molecular onco-

genesis of gastrointestinal tumors is then described and illustrated with positron emission tomography (PET) examples. The neuroendocrine system of the gastrointestinal tract is then reviewed with a "molecular scope". Finally, the most promising avenue in molecular gastrointestinal scintigraphy is evoked.

24.2
Genetic Diseases and GI Transit Abnormality

24.2.1
Salivary Glands

The diagnosis of Sjögren syndrome requires the documentation of decreased lacrimal function, dryness in the mouth and evidence of autoimmunity dysfunction. It is the second most frequent autoimmune disorder after rheumatoid arthritis. The scintigraphic salivary gland challenge test with pertechnetate has proven to be very useful as part of the diagnostic workup for this syndrome.

Autoimmune disorders are typically classified in two groups: the organ-specific diseases and the systemic diseases. In organ-specific autoimmunity (e.g. Hashimoto's thyroiditis, Grave's disease, multiple sclerosis, etc.), antigens peculiar to a cell type are the target of an aggressive immune response. The non-organ-specific disorders include diseases such as systemic sclerosis, rheumatoid arthritis and Sjögren syndrome.

These diseases of systemic autoimmunity occur in individuals genetically predisposed. In these individuals, one or multiple environmental influences trigger T-cell dysregulations and the pathogenic production of autoantibodies reacting to self antigens. The genes that regulate normal immune function and play a key role in autoimmunity encode the major histocompatibility complex (MHC) or human lymphocyte antigen (HLA) on chromosome 6, the T-cell receptors (TCR), immunoglobulins, the complement and the antigen processing and presentation molecules. Since women are affected twice as often as men, the sex-related genes are also involved.

The MHC gene products are grouped in two categories. The class I molecules are present on the surface of most nucleated cells. They bind short peptides arising from the cytoplasm of the cell and present them to cytotoxic T cells. MHC class II molecules, also designated DR and DQ, are present on monocytes, B cells and other antigen-presenting cells. They bind larger peptides derived from extracellular proteins and interact with T-cell receptors on CD4+ lymphocytes that regulate the immunoglobulin production.

The HLA DR3 antigen has been associated with Sjögren syndrome. Antinuclear antibodies are present in approximately 90% of patients with SS. Antibodies to SS-A (anti-Ro) and to SS-B (anti-La) are also found in patients with SS.

24.2.2
Esophagus

The esophagus is a long muscular tube which extends from the cricoid cartilage to the stomach. The proximal third of the esophagus consists of striated muscle, the distal third is composed of smooth muscle, and the middle portion is a transitional mixture of the two. The upper esophageal sphincter, the body, and the lower esophageal sphincter are the distinguishable anatomic structure. The coordinated function of these structures conveys swallowed liquid and food from the mouth to the stomach and clears residual substances. Esophageal motility follows a precisely coordinated pattern characterized by primary, secondary and tertiary contractions.

Esophageal motility disturbances are classically categorized into primary or secondary disorders depending on the pathophysiologic mechanisms. Achalasia, diffuse esophageal spasm, nutcracker esophagus and non-specific motor disorders account for the so-called primary esophageal motility disorders. Neuromuscular and connective tissue disorders encompass diseases such as scleroderma, systemic lupus erythematosus, Raynaud's disease and dermatopolymyositis. Vertebral cervicopathy and diabetes also may exhibit esophageal transit abnormality. Anatomic alterations due to tumor, diverticula, surgery, trauma, infection may also lead to secondary motor disorders of the esophagus.

Systemic sclerosis (SS) is a generalized vascular and fibrotic autoimmune disease that affects multiple organs including the gastrointestinal tract. Diminished smooth muscle motor activity of the distal two-thirds of the esophagus is the most frequent finding and results in dysphagia and esophageal reflux. Gastric emptying might be delayed; small bowel involvement results in malabsorption and bacterial overgrowth. Colonic involvement results in stasis, constipation and diverticula. As for all autoimmune diseases, the pathogenesis is felt to involve interacting genetic and environmental factors. The molecular pathophysiology of systemic sclerosis is incompletely understood and the primary etiologies remain a mystery. Hypotheses include: an altered ability of T lymphocytes to distinguish between self and non-self antigens, the clonal expansion of self-reactive T cells involving the major histocompatibility complex (MHC) or the heteromultimeric T-cell receptor (TCR). Immunologic features of SS include the presence of specific autoantibodies against centromeres, topoisomerase I, RNA polymerase I, II, III, fibrillarin, a ribonucleoprotein (U3RNP), U1RNP. Cytokines such as interleukins IL-1, IL-6, IL-8, TGF-μ, PDGF, fibronectin, endothelin and VEGF have also been implicated in the pathogenesis of the disease. The anti-topoisomerase I autoantibody is associated with DR5.

Polymyositis (PM) and dermatomyositis (DM) are autoimmune inflammatory diseases of the striated muscles that may present distinctively or with other autoimmune disorders such as rheumatoid arthritis (RA), systemic lupus erythematosus (SLE) or systemic sclerosis. The peak onset of this disease is in the fifth decade. Most autoimmune diseases have been associated with genes in the class II of the MHC. PM and DM have been associated with the DR3 antigen of the MHC.

24.2.3
Stomach

The human stomach consists of two functionally integrated, but electro-mechanically distinct, portions. The proximal stomach, which encompasses the fundus and the proximal corpus, functions as a reservoir for solid and liquid food and controls the emptying of liquids. The distal stomach, which includes the mid and distal corpus and the antrum, is characterized by peristaltic contractions that break down solid food into small particles and mix them with gastric secretions to form a semiliquid juice that is then emptied into the duodenum.

Antral hypomotility and impaired fasting motor activity are likely factors of delayed gastric emptying in connective tissue disorders. Delayed gastric emptying is present in 50% of patients with systemic sclerosis. In SLE, PM and DM, delayed emptying may be asymptomatic. The clinical spectrum of SLE ranges from mild symptoms such as rash and arthritis to acute life-threatening involvement of the central nervous system and/or the kidneys. The immunologic features of SLE include a positive nuclear antibody test (ANA) and more specific antibodies against double-stranded DNA. The first recognized genetic markers for SLE include the MHC class I allele B8 and class II DR2 and DR3. Patients with homozygous deficiency of C1q, C1r-C1s, C3, C4 or C2 have a high frequency of lupus-like illness.

Delayed gastric emptying in symptomatic and asymptomatic diabetics is mainly accounted for by a prolonged lag phase, an impairment of the proximal stomach function and a reduction in the antral motor activity. A markedly prolonged linear gastric emptying of solids and a delayed emptying of liquids characterize advanced diabetic gastroparesis. In patients with type II diabetes of recent onset, an accelerated gastric emptying for solids was recently described. The delay in gastric emptying in type II diabetes could be related to higher postprandial levels of glucose.

Type I diabetes is characterized by destruction of pancreatic β cells and insulitis. Autoantibodies present are directed against a variety of β cell antigens, including insulin, glutamic acid decarboxylase (GAD65 and 67), membrane proteins that are homologous to tyrosine phosphatases and islet neuroendocrine gangliosides. At least 13 susceptibility loci present on various chromosomes (2, 6, 7, 11, 14 and 15) have been identified as type I diabetes mellitus markers. Type II diabetes is polygenic and develops in response to both genetic and environmental factors. Mitochondrial diabetes mellitus refers to the association of type II diabetes with or without deafness and mitochondrial DNA mutations. Maturity onset diabetes of the young (MODY) is characterized by the onset of diabetes at an early age and by an autosomal dominant inheritance. There are currently at least 4 subtypes of MODY identified. In MODY 1, patients have a mutation in the hepatocyte nuclear factor (HNF)-4 alpha gene on chromosome 20q. Mutations of the glucokinase gene on chromosome 7q are present in MODY 2. Patients with MODY 3 have muta-

tions in their HNF-4 alpha gene on chromosome 12q. Inactivating mutations in the insulin promoter factoral (IPF-1) gene are responsible for MODY 4. At least six types of mutations in the insulin gene have been described. Patients with these mutations are heterozygotes and display hyperproinsulinemia. More than 40 mutations of the insulin receptor gene are known to affect receptor synthesis, processing, insulin binding and tyrosine kinase activity. However, the role of these mutations in type II diabetes remains to be established. Genome scanning and positional cloning have been used to define the multiple genes that could be involved in late-onset type II diabetes. The polygenicity of this disorder is reflected by the multidiabetogenes described so far and the absence of clear identification of susceptibility loci.

In chronic gastritis, emptying of liquids is usually normal, while emptying of solids is delayed. Stasis of both liquids and solids has been described in atrophic gastritis, while in pernicious anemia delayed gastric emptying for solids only is observed. Parietal cell antibodies (PCA) are present in 75–90% of patients with pernicious anemia. The major parietal cell antigen in pernicious anemia is the H+/K+-ATPase alpha and beta subunits acid pump. In patients with atrophic gastritis, the prevalence of PCA is 59%. It seems that there might be an association between PCA and the HLA DR5 genotype.

24.2.4
Small Bowel and Colon

The clinical applications of scintigraphic studies of small bowel and colon transit are limited and not widely used in the diagnostic workup of gastrointestinal motor function disturbances. Several methods have been proposed to investigate the entire gastrointestinal tract in a single test. The most practical and physiological test uses ^{111}In-DTPA-labeled water with an egg sandwich solid test meal. The egg sandwich is labeled with ^{99m}Tc to simultaneously study gastric emptying of solids.

In normal subjects, water empties rapidly from the stomach (90% within 2 h) and a geometric mean of abdominal counts (GMC) is obtained at 2–3 h to determine 100% of the administered ^{111}In-DTPA activity. The terminal ileum filling rate is determined and the input bolus into the colon is used to measure the progression of the food through the colon using the geometric mean center (GMC) method.

In normal subjects, the ascending colon empties in a linear manner after an initial lag phase suggesting its storage role. A linear progression through the colon is also demonstrated when using the GMC analysis.

Patients with diarrhea have an accelerated small bowel transit time and/or colonic transit. The most common clinical application of colon transit scintigraphy is the evaluation of patients with idiopathic constipation. In patients with idiopathic constipation, the GMC test enables to differentiate colonic inertia, that displays a slow transit throughout the entire length of the colon, from pelvic obstruction of defecation that is associated with an abnormal retention in the rectosigmoid.

Over the past decade, many enterocyte-specific genes regulating the digestion and absorption of luminal nutrients have been cloned and sequenced. Although not directly involved in bowel transit regulation, these genes are important to recognize as they are involved in intestinal disease that affects the gastrointestinal transit.

Crohn's disease and ulcerative colitis are characterized by diarrhea and abdominal pain. The molecular pathophysiology of these inflammatory bowel disorders remains largely unknown. However, transgenic mice with disruption of their cytokines, T-cell receptor or the N cadherin gene are promising research avenues.

Malabsorption and steatorrhea are part of the clinical symptoms of patients with celiac sprue. The alcohol-soluble component of gluten, gliadin, triggers the inflammation/atrophy of the small bowel villi. In this disease there is a very strong link with HLA class II haplotypes.

Adult-onset hypolactasia is responsible for lactose intolerance and its cohort of symptoms: abdominal pain, diarrhea, bloating and excessive flatulence. Low levels of NF-LPH-1, an intestine-specific transactivator that binds to a DNA sequence close to the TATA promoter box have been described.

24.3
Molecular Carcinogenesis and the GI Tract

Although most of the details describing the signaling pathways remain to be identified, the genetic alterations of most gastrointestinal cancers are fairly well characterized. The sequence of genomic insults occurring during the lifespan of colorectal tumors for example is one of the best understood models of molecular oncogenesis. Gastrointestinal cancers account for a large proportion of human tumors. They are most often incurable when metastases have occurred. Esophageal, gastric and colorectal cancers have etiologic relationship with environmental factors.

24.3.1
Esophagus

The high prevalence of esophageal cancer in patients with smoking and alcohol ingestion habits is well documented. Chronic esophagitis is associated with a premalignant lesion, the so-called Barrett's esophagus. Microsatellite instability and p53 mutations are early genetic abnormalities in this disorder. P53 mutation, loss of heterozygosity at the APC locus, overexpression of epidermal growth factor receptor (EGFR) and HER2/neu, amplification of the cyclin D1, gene mutation of the p16 cyclin-dependent kinase inhibitor are found in significant proportion in esophageal adenocarcinomas and squamous cell cancers.

24.3.2
Stomach

Gastric cancer is the second most common malignancy in the world. At the present time, there is no clear road map of the genomic insults characterizing progression and other aspects of tumor biology for benign and malignant gastric tumors. P53 and APC abnormalities, microsatellite instability and overexpression of HER2/neu all occur at high frequency.

24.3.3
Pancreas

Pancreatic carcinoma is one of the most difficult tumors to visualize and treat. P53 mutations, microsatellite instability, K-ras mutations, overexpression of EGFR, inactivation of the tumor suppressor gene DPC4 located on gene 18q21 are common in pancreatic cancers.

24.3.4
Colon and Rectum

It is now well established that a large number of colorectal cancers have a sequential tumorigenetic development pattern going from normal to hyperproliferative mucosa, small polyps, large polyps with dysplasia, carcinoma in situ, invasive tumors and finally metastatic cancers.

Hereditary nonpolyposis colon cancers (HNPCC) represent about 5–10% of colorectal cancers. The molecular oncogenesis basis of these syndromes is related to mutations in the DNA mismatched repair system leading to microsatellite dinucleotides and trinucleotide repeats. These alterations are secondary to replication errors in the microsatellites.

Familial adenomatous polyposis (FAP) is an autosomal dominant disorder that is characterized by the development of multiple polyps in the colon. The risk of colorectal cancers in these patients is close to 100%. The adenomatous polyposis coli (APC) gene at 3q21 is mutated in the germline of most patients with AFP.

Deletion in the colon carcinoma gene (DCC) (18p) occurs in 70–80% of colorectal carcinomas. It usually precedes the deletion or mutation of the 17p gene that maps p53. Activating mutations in the ras protooncogenes (H-ras, K-ras, N-ras) are also frequently mutated in colorectal carcinoma. Mutations in ras, APC and p53 are found both in HNPCC and non-HNPCC.

24.3.5
Molecular Imaging of Gastrointestinal Cancers

The imaging of gene expression (Imagene) in tumors using peptidomimetic and nucleic acid molecules is being actively pursued by multiple academic research laboratories, pharmaceutical and radiopharmaceutical companies.

Malignant cells exhibit increased glycolytic metabolism and in many instances increased glucose transporter gene expression. Seven distinct glucose transporters have been sequenced thus far. The erythrocyte-brain glucose transporter or GLUT1 was the first facilitative transporter to be cloned. Because of its ubiquity and overexpression in most cancers, GLUT1 has been used as a tool for PET imaging with fluorodeoxyglucose (FDG).

Over the past decade, ^{18}F-FDG has become the best non-invasive functional tool for the diagnosis,

staging and restaging of a variety of cancers expressing the GLUT1 transporter. Fluorodeoxyglucose is recognized by GLUT 1 as glucose, enters the cell and then the glycolytic pathway. It is phosphorylated in the 6 position by the enzyme hexokinase. Phosphorylated 2-deoxy-fluoro-D-glucose is a very poor substrate for phosphohexose isomerase or glucose-6-phosphate-dehydrogenase and is trapped within the cell, enabling imaging of the tissue and organs that have accumulated the tracer.

Immunostaining studies have demonstrated a high level of expression of the GLUT1 receptor in Barrett's metaplasia (BM), human gastric carcinoma and colorectal cancer. The imaging of the GLUT1 gene expression (Imagene) is illustrated in Figs. 24.1–24.3. The expression of GLUT1 in Barrett's metaplasia is a late event during the neoplastic progression and correlates with aggressive biologic behavior (Younes et al. 2000). Advanced BM can be nicely documented with ^{18}F-FDG scintigraphy (Fig. 24.1). Gastric carcinoma phenotype also displays an overexpression of the GLUT1 transporter (Fig. 24.2). It is associated with depth of invasion, lymphatic permeation, venous invasion, lymph node and distant metastases (Kawamura et al. 2001). The GLUT2, 3 and 4 expression has also been reported in gastric cancers. In colorectal carcinomas (Fig. 24.3), GLUT1 expression is a good predictor of the grade of

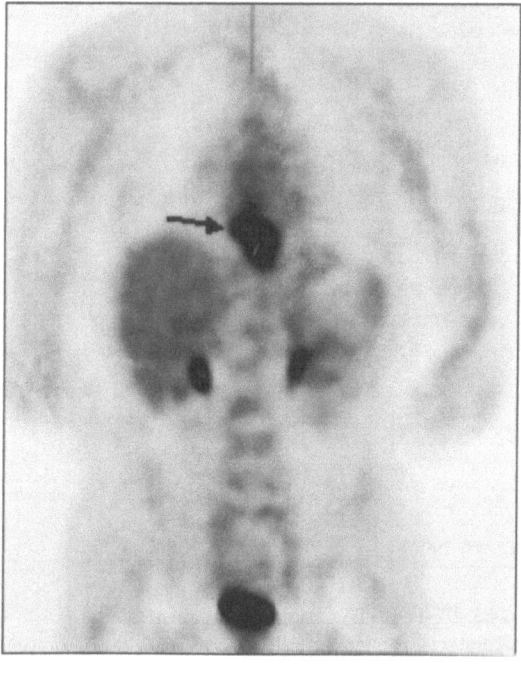

Fig. 24.1. ^{18}F-FDG/PET imaging in a patient with Barrett esophagus

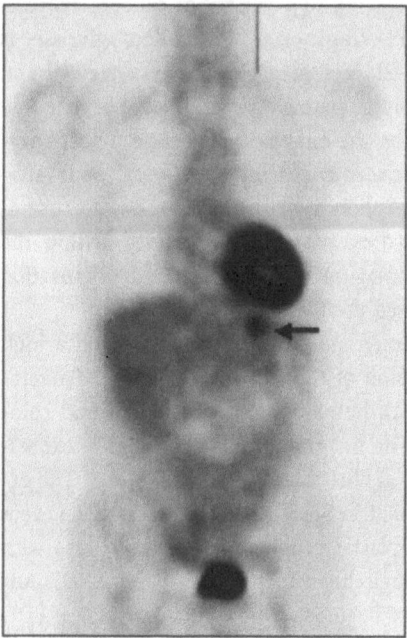

Fig. 24.2. ^{18}F-FDG/PET imaging in a patient with gastric carcinoma

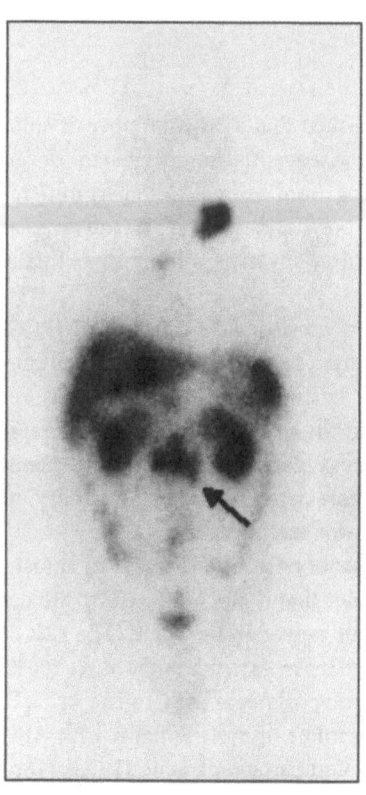

Fig. 24.4. ^{111}In-DTPA-octreotide study in a patient with a GI carcinoid syndrome

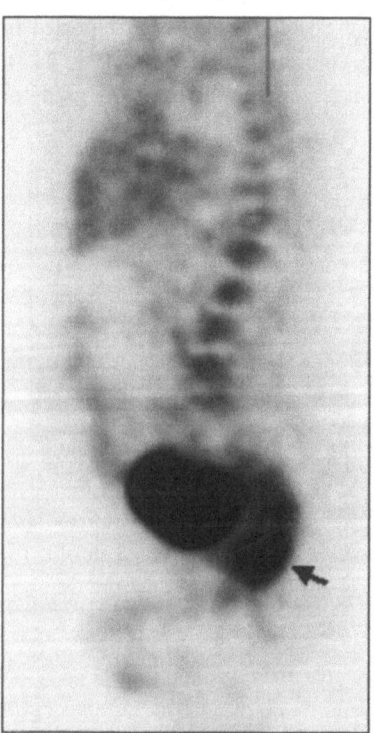

Fig. 24.3. ^{18}F-FDG/PET imaging in a patient with colorectal carcinoma

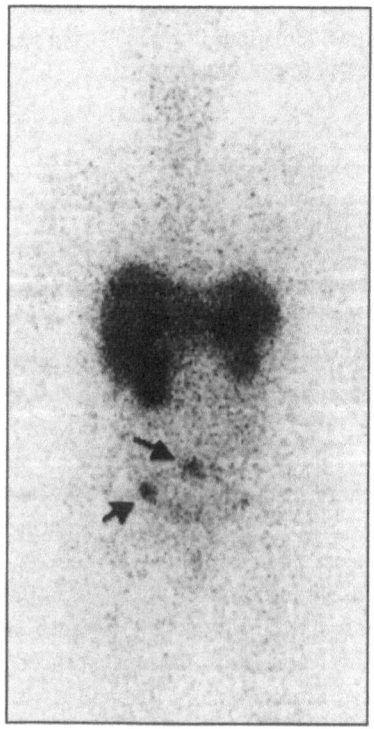

Fig. 24.5. ^{111}In-DTPA-octreotide study in a patient with a neuroendocrine GI tumor

malignancy and prognosis; it does also correlate well with the Duke's stage (Sakashita et al. 2001).

24.4
Molecular Biology/Oncology of the Gastrointestinal Neuroendocrine System

24.4.1
The Endocrine Enteric System

The gut is the oldest and largest endocrine organ of the human body. It consists of endocrine cells scattered among the other cell types of the gastrointestinal tract. Most of the secretory products of the GI tract endocrine system or GI peptides were isolated in the 1970s and 80s. They are typically divided in three groups depending on their function. Gut peptides that function mainly as hormones include for example the following hormones: insulin, gastrin, motilin, secretin, glucagon and related gene products such as GLP-1, glicentin and oxyntomodulin. Gastrointestinal peptides that act principally as neuropeptides encompass: gastrin releasing peptide (GRP), vasoactive intestinal peptide (VIP), calcitonin gene-related peptide (CGRP), galanin, neuropeptide Y, and thyrotropin releasing hormone (TRH). Gut peptides that function as hormones, neuropeptides or paracrine agents include: cholecystokinin (CCK), corticotropin releasing factor, endothelin, neurotensin, and somatostatin (SST). Advances in molecular biology have made it possible not only to characterize the genes producing these peptides and hormones but also to clone and sequence most of their receptors.

Besides the neuronal and hormonal GI peptides, small non-peptide chemical messengers such as the biogenic amines, prostanoids and nitric oxide, growth factors, cytokines and immune mediators modulate the gastrointestinal functions.

24.4.2
Somatostatins and Their Receptors

Somatostatin (SST), like other peptide hormones, is synthesized as a large precursor protein (proSST) that is processed to generate a 28 (SST-28) and 14 (SST-14) amino acids peptide. The human somatostatin gene is located on the long arm of chromosome 3 and encodes for both forms. Somatostatin-secreting cells are found in high concentration in the central and peripheral nervous system, the endocrine pancreas (D cells) and gut. There are two types of somatostatin-producing cells in the gut: D cells in the mucosa and neurons that are intrinsic to the submucous and myenteric plexuses.

There are five structurally related subtypes of SST cognate receptor (SSTR) genes. They encode for seven transmembrane domains, G protein-coupled receptor proteins that bind natural and synthetic SST peptides. SSTR1-4 bind both SST-14 and SST-28. SSTR5 has two subtypes and displays relative selectivity for SST-28. SSTR2 is the most widely and abundant SST expressed receptor.

Naturally, the 2-3 min short half lives of SST peptides has limited their pharmacologic use in therapy and imaging. Synthesis of structural analogs with enhanced metabolic stability and labeling properties have revolutionized the diagnosis and treatment of neuroendocrine tumors.

24.4.3
Endocrine Tumors of the Gastrointestinal Tract

Gastroenteropancreatic neuroendocrine tumors are named according to the scientist who discovered the related clinical symptoms and/or to the main peptide secreted by the tumor. Gastrinoma, VIPoma, glucagonoma, somatostatinoma, Pipoma, neurotensinoma, enteroglucagonoma and other tumors secreting specific GI peptides have been documented.

The Zollinger-Ellison syndrome has been unequivocally associated with rapid gastric emptying of both solids and liquids. A defective inhibitory mechanism of gastric emptying due to extensive inflammation of gastric mucosa may be responsible. Oncofetal expression of the gastrin gene in the pancreas leads to gastrinoma in the adult pancreas. In one-third of patients, gastrinoma is associated with the Multiple Endocrine Neoplasia type 1 (MEN1: pituitary, pancreatic and parathyroid tumors). At the time of diagnosis, 50% of patients with gastrinoma have liver metastases.

The Verner Morrison syndrome was first described in 1958 in patients with pancreatic cancers and severe secretory diarrhea and electrolytes losses. Oversecretion of VIP is responsible for the syndrome. Ninety percent of VIPoma originate in the pancreas and are malignant. They can also be part of the MEN1 syndrome that has been mapped to gene 11, locus 13 by linkage analysis.

The pathophysiology of the carcinoid syndrome is related to the hypersecretion of 5-hydroxytryptamine and other substances that cause flushing, diarrhea, heart disease and bronchospasm. Seventy five percent of carcinoid tumors arise in the enterochromaffin cells that are scattered in the submucosa of the intestine. Irrespective of its association with MEN1 carcinoid tumors display a loss of heterozygosity on locus 13 of chromosome 11 (11q13).

24.4.4
SSTR Type 2 Imaging

Most neuroendocrine tumors, particularly of the gastrointestinal tract, overexpress the SSTR2 receptor at their surface. This phenotype has been exploited advantageously to treat and image these tumors with somatostatin analog. Octreotide is a long acting cyclic octapeptide that has a high affinity for the SSTR type 2 and 5. Labeled with ^{111}In-DTPA, it has been used over the past decade to diagnose and localize enterogastric neuroendocrine tumors (Figs. 24.4 and 24.5). Treatment of these tumors, particularly carcinoid tumors, with high dose of octreotide-based ligand coupled to α- or β-emitting isotopes is currently under clinical investigation.

24.5
The Brain-Gut Axis

There is now a strong body of data supporting the existence of a bidirectional brain-gut axis. The richly innervated nerve plexuses and neuroendocrine association of the enteric nervous system and its connection with the spinal, autonomic and central nervous system are the backbone network of this axis.

The modulation of feeding activity, postprandial satiety, motility, and painful visceral perception occur via several neurotransmitters and neuropeptides found in the CNS and the gastrointestinal tract. The projection of visceral afferent fibers to somatotropic, emotional and cognitive areas of the CNS produces a variety of interpretations based on prior learning, cognitive and emotional state.

Brain neurotransmission depends on the biochemical interactions of neuronal and neuroendocrine ligands and receptors. These processes occur at low concentrations, typically in the nanomolar to the femtomolar range. Because of their exquisite sensitivity to radiolabeled probes, PET and SPECT techniques are well suited for the characterization and analysis of brain receptor activation by various methods of gut stimulation.

The widespread availability of ^{18}F-FDG and other tracers such as the dopaminergic and opiate receptor ligands will give molecular imaging clinicians and researchers a vast array of opportunities to study brain activation in healthy subjects and in patients with various GI ailments.

24.6
References

Kawamura T, Kusakabe T, Sugino T et al (2001) Expression of glucose transporter-1 in human gastric carcinoma: association with tumor aggressiveness, metastasis and patient survival. Cancer 92:634–641

Sakashita M, Aoyama N, Minami R et al (2001) Glut 1 expression in T1 and T2 stage colorectal carcinomas: its relationship to clinicopathological features. Eur J Cancer 37:204–209

Urbain JL, Johnson CL, Vekemans MC (2000) Gastrointestinal nuclear medicine. In: Baert AL, Heuck FHW, Youker JE, Brady LW, Heilmann HP (eds) Diagnostic nuclear medicine. Springer, Berlin Heidelberg New York, pp 123–134

Urbain JL, Vekemans MC, Malmud LS (2002) In: Sandler MP, Coleman RE, Patton JA, Wackers FJ, Gottschalk A (eds) Diagnostic nuclear medicine, 4th edn. Williams and Wilkins, Philadelphia

Younes M, Lechago J, Chakraborty S et al (2000) Relationship between dysplasia, p53 protein accumulation, DNA ploidy and Glut1 overexpression in Barrett metaplasia. Scand J Gastroenterol 35:131–137

Recommended Reading

Jameson JL (ed) (1998) Principles of molecular medicine. Humana Press

Feldman M, Friedman LS, Sleisenger MH (eds) (2002) Sleisenger and Fordtran's gastrointestinal and liver disease: pathophysiology, diagnosis, management, 7th ed, 2 vols. WB Saunders, Philadelphia

Becker KL (ed) (2001) Principles and practice of endocrinology and metabolism. Lippincott Williams & Wilkins

CHAPTER 25

The Kidney and Genital System

25

KEITH E. BRITTON

Contents

25.1
Some Renal Tubular Transport Disorders

25.1.1
Introduction

The proximal renal tubule has many properties, one of which is the transport of substances from the capillary lumen to the lumen of the tubule and vice versa. It is a two-stage process. There is transport across the capillary side of the tubular cell, intracellular movement of the agent and then transport across the luminal side into the filtered fluid. The high affinity of a tubular organic acid transporter protein has been utilized since the 1940s for the measurement of renal clearance, using para amino hippurate (PAH) (Smith 1951). The introduction of its radioactive analogue radio-iodinated ortho-iodo-hippurate (OIH, Hippuran) led to the development of probe renography, reviewed by Britton and Brown (1971), and gamma camera radionuclide studies. Probe renography was mainly undertaken with ^{131}I-OIH and the camera work with ^{123}I-OIH. The expense and reduced availability of ^{123}I led to the testing of ^{99m}Tc-labeled substitutes. ^{99m}Tc-MAG3 mercapto acetyl triglycine, is commercially available (Mallinkrodt Corporation) and is widely used (Al-Nahhas et al. 1988). ^{99m}Tc-EC, ethylene dicysteine, has slightly better tubular transport than MAG3 (Ozker et al. 1994; Gupta et al. 1995; Kostadinova and Simeonova 1995) (Table 25.1). It does not require a boiling step but it is only available in Eastern Europe (Izinta, Hungary).

PAH clearance measurements require continuous infusion of tens of grams of PAH and a tubular maximum can be demonstrated. When radioiodinated OIH is compared with PAH simultaneously, the clearance of OIH is less, which partly depends on competition from PAH since the radioactive tracer is present in only milligram amounts. The kinetics of uptake follow those of Michaelis/Menten as for enzyme-substrate interaction, thus neither PAH nor OIH clearance can be complete. Taking PAH as the standard, the following table shows their relative tubular excretion and protein binding.

This organic acid transporter is also that for penicillin and its analogues, probenecid and some non-steroidal anti-inflammatory drugs which may interfere with the uptake of these radiopharmaceuticals. In renal conditions with tubular damage uptake of I*OIH, MAG3 and EC may continue but the luminal tubular transport fails so that there is intracellular retention of these agents. This gives rise to what has been called "parenchymal stasis", when the activity time curve rises to a plateau in the absence of pelvic dilatation or resistance

Table 25.1. Tubular excretion of radiopharmaceuticals

	Tubular excretion	Protein binding
PAH	100%	35%
I*OIH	83%	70%
^{99m}Tc MAG3	65%	92%
^{99m}-EC	75%	35%

Table 25.2. Amino acid disorders

Transport	Amino acid	Transport disorder
Dibasic amino acids	Lysine arginine cysteine	Cystinuria type I
Cysteine transport	Cysteine	Cystinuria
Neutral amino acids	Glycine etc.	Hartnup disorder
Dicarboxylic acids	Glutamate aspartate	Recessive disorder

to outflow. Such may be seen in the nephrotic syndrome, chronic glomerulonephritis, incipient acute tubular necrosis and Fanconi's syndrome.

Some compounds of medical importance are not secreted by the organic acid transport system but by the organic base transporter. These include penicillamine, thiamine, and cyclosporine. It is well known that conventional clearance measurements may not demonstrate impairment of renal function when measured serum levels of cyclosporine are normal and yet renal impairment appears later. In the author's laboratory, a new radiopharmaceutical which is basic, ^{99m}Tc-DACH, diaminocyclohexane, was developed and tested. In volunteer studies oral thiamine depressed the clearance of ^{99m}Tc-DACH but not that of I*OIH (Padhy et al. 1993). Subsequently it was shown that renal impairment could be detected when cyclosporine was given to patients for psoriasis using ^{99m}Tc-DACH as compared to ^{99m}Tc-DTPA (Sonmezoglu et al. 1998). This was also true for cyclosporine treatment of renal transplants (Datseris et al. 1995a). The complexity of the microvascular and tubular events in renal transplant rejection versus cyclosporine toxicity made it unlikely that this agent will be taken up for clinical use in this context.

25.1.2
Carbohydrate Transport

Carbohydrate reabsorption is through a low affinity sodium/glucose counter transport system (2 glucose for 1 sodium ion) and a high affinity low capacity sodium/glucose linked transporter system (SGLT1) similar to that in the intestine. Its gene is on chromosome 22$_q$ 11.2 (Hediger et al. 1987). All filtered glucose is reabsorbed up to a threshold level at which time it appears in the urine as in diabetes mellitus. The threshold may be lower than normal causing renal glycosuria with a normal blood sugar level. Defects in the sodium/glucose link transporter genes have been identified. Pento-

suria and fructosuria are rare recessive disorders. Hereditary fructose intolerance is one cause of the Fanconi syndrome where there is excessive urinary excretion of glucose, amino acids, bicarbonate and phosphate leading to hypokalaemic metabolic acidosis.

25.1.3
Amino Acid Transport

Over 95% of all amino acids are filtered, reabsorbed in the proximal tubules and metabolized. The transport systems are described in more detail in the following sections (see also Table 25.2).

25.1.4
Cystinuria Type I

This is associated with excessive excretion of dibasic amino acids, cysteine, lysine, arginine and ornithine. There is also an impaired intestinal absorption of these compounds. The tubular brush border appears defective (Wright 1994) with impaired dibasic amino acid transport protein, rBAT. The gene for this is on chromosome 2p21. The presentation is usually that of renal colic, due to renal stones in patients in their 20s or 30s. Normal cysteine excretion is under 200 mg per day. Stones form at over 300 mg per day due to the insolubility of cysteine at this concentration. Particular crystals are seen on microscopy. The urine turns pink with the cyanide-prusside test. The treatment is a high fluid intake, D-penicillamine or thiola, both of which have many side effects, or captopril to bind and solubilize the cysteine.

The stones may be large and even a stag horn calculus may form. They are sufficiently opaque to be seen radiologically. They require surgery or lithotripsy and they recur. The patients with these stones (and the much more common calcium stones) may be monitored by serial plain X-ray to show stone formation and by renal radionuclide studies to show the development of obstructive nephropathy and/or obstructing uropathy, both prior to treatment and subsequently. The obstruction may be acute or chronic. Acute obstruction is assessed by plain X-ray, intravenous urography and ultrasound. Chronic obstruction is evaluated by ultrasound and renal radionuclide studies.

25.1.5
Chronic Obstructive Uropathy

The presence of a chronic obstructing uropathy, which is the change in the outflow tract due to an increased resistance to flow above normal, may be indicated radiologically through intravenous urography or ultrasound evidence of a dilated renal pelvis. The dilatation does not mean obstruction because it may be due to a large baggy unobstructed pelvis and lack of dilatation e.g. due to a stiff outflow tract or due to oliguria does not exclude obstruction. Obstructive nephropathy is the effect of an increased resistance to outflow on renal function. Its physical basis should be appreciated. Force (from cardiac output) acts on a resistance (the outflow tract) and this gives rise to a pressure. Pressure is a consequence of a force acting on a resistance. A pressure difference between two points gives a pressure gradient and if there is fluid in this system, there will be flow from the higher to the lower pressure. If there is flow between the two points then tracer will take a certain time to move from one to the other, the transit time. In chronic outflow obstruction the force of the heart is unchanged causing filtration but the resistance to outflow is increased. The pressure gradient changes from the source of the force, the glomerulus, to the site of the resistance. Thus there is a pressure change throughout the whole kidney from the site of the glomerular filtration to the stone causing the resistance. Pressure is equal and opposite in all directions so that there is no such thing as 'back pressure'. If the pressure increases in the tubules marginally over that in the peritubular capillaries, then there will be increased salt and water reabsorption by that kidney in the proximal tubules. Therefore the flow in the lumens of the nephrons will be reduced and the time that the tracer takes to go from the glomerulus to the site of the resistance will be prolonged. Non-reabsorbable solutes such as ^{99m}Tc-MAG3 will take longer to transit along the nephrons. The resistance of flow is thus transduced to a prolongation of the renal transit time.

The transit time is measured by deconvolution analysis of the activity time curve from the left ventricle and that from part of the renal parenchym which excludes the renal pelvis. The result is a mean parenchymal transit time, MPTT, which is made up of a minimum transit time common to all nephrons and dependent on urine flow and the difference between this and the MPTT which is called the parenchymal transit time index, PTTI. This relates to salt and water reabsorption in the proximal tubule. This reabsorption is increased with the increase in the pressure gradient due to an increased resistance to outflow and is a sensitive indicator of obstructive nephropathy (Britton et al. 1979, 1987; Britton and Maisey 1998). The normal value of PTTI is less than 156 s.

An alternative approach is to intervene usually 18–20 min into the study by giving an intravenous injection of furosemide, also called frusemide or Lasix (40 mg adult, 0.5 mg/kg in a child to a maximum of 20 mg). The diuresis induced helps to clear a dilated baggy unobstructed pelvis of active urine and causes a rapid fall in the activity time curve (Kletter and Nurnberger 1989; O'Reilly et al. 1996). Increased resistance to outflow is associated with no response to frusemide or inappropriately poor response. Appropriateness of response is judged by comparing the rate of fall of the activity time curve post-frusemide with the previous rate of rise of the activity time curve during the first few minutes before there is loss of activity from the kidney. If only a little activity goes into the kidney then it is appropriate that only a little comes out. This can be measured as an outflow efficiency, which equates the cumulative output from the kidney with its previous input (Chaiwatanarat et al. 1993). Normally in 30 min, 82–98% or more of the activity taken up by the kidney is excreted by the kidney in response to frusemide. The lower limit of normal is taken as 78%. These measurements are applicable to all situations of suspected or actual chronic resistance to outflow before consideration of surgery, lithotripsy or other therapy.

Serial studies may be used to monitor the patient if no treatment is considered necessary and to monitor the response to treatment. The percentage contribution of each kidney to total renal function is measured directly during the first 1.5–2.5 min of the activity time curve. In the context of chronic resistance to outflow a restorative operation or approach is recommended if the one kidney contributes 15% or more of total function, whereas a contribution of less

Table 25.3. Some genetic renal disorders

Disorder	Inheritance	Chromosome	Gene	Product	References
Polycystic kidney type I	Dom	16p 3.3	PKD1	Polycystin	Gabow (1993)
Cystinuria type I	Rec	2p16	SLC3A1	Dibasic amino acid transporter	Pras et al (1994)
Nephrogenic diabetes insipidus	X	Xq28	AVPR2	Vasopressin 2 receptor	Rosenthal et al (1992)
	Rec	12q 13	AQPR2	Aquaporin 2	Deen et al (1994)
Von Hippel Lindau	Dom	3p 25–26	VHL	G7 protein	Latif et al (1993)
Alport syndrome	X	Xq 22	COL4 A5	A5 (iv) collagen	Barker et al (1990)
Tuberous sclerosis type 2	Dom	16p 13.3	TSc2	Tuberin	European Consortium (1993)
Wilm's tumor	Dom	11p13	WT1	Zinc finger protein	Call et al (1990)

Dom, dominant inheritance; Rec, recessive; X, X linked.

than 5% in an adult is unlikely to show recoverable function in response to treatment.

25.1.6
Nephrogenic Diabetes Insipidus

Diabetes insipidus is characterized by an unremitting diuresis. It is typically due to loss of posterior pituitary function through tumor, trauma, surgery or infection, so there is loss of the neurohypophyseal hormone, arginine vasopressin, AVP. This is now treated with a nasal spray of AVP. Nephrogenic diabetes insipidus is an example of a genetically based tubular receptor disorder (Table 25.3), whereby there are defects in the AVP- dependent receptors in the renal collecting ducts. This is not treatable with AVP spray and is only treatable by matching the fluid loss with a high fluid intake. This disorder is congenital and may be due either to mutations of the gene coding for the vasopressin V2 receptor (AVPR2) on the X chromosome Xq 28 or one of the genes responsible for the vasopressin-regulated water channel, aquaporin (aqp2) of chromosome 12q13 (Oksche and Rosenthal 1998). The gene for the V2 receptor is on the X chromosome with evident gender-related familial inheritance. The gene for aquaporin is autosomal recessive. Currently 70 different mutations of the V2 receptor gene and 13 of the aquaporin gene have been reported. When the defect is severe in the obvious clinical case the diagnosis and management with a high water intake are straightforward and the need for measurement of body fluid compartments is unlikely.

The loss of the AVP receptor was demonstrated in vivo by Britton et al. (1977) who compared the renal handling of ^{131}I-IOH and ^{131}I-vasopressin in patients with pituitary diabetes insipidus, nephrogenic diabetes insipidus and a normal control patient. In nephrogenic diabetes insipidus, but not in pituitary diabetes insipidus, there was no uptake of the ^{131}I-AVP by the kidney as demonstrated by an activity time curve that was flat, whereas in the same patient an ^{131}I-IOH curve was quite normal (Fig. 25.1).

In the few cases where mutations have been associated with a mild phenotype (Bichet et al. 1998) and possibly in heterozygotes for the more severe defects in aquaporin function, there could be a subclinical degree of vulnerability to dehydration in stressful situations. Such individuals could benefit from recognition of genetic content and possible need for special monitoring of body water components. Moreover there could be as yet undiscovered mutations in the responsible genes and a wider occurrence of mild nephrogenic diabetes insipidus than presently supposed. ^{123}I-AVP could be, but has not as yet been, used to demonstrate whether this defect occurs widely.

25.2
Hypertension

25.2.1
Introduction

George was a fastidious man nearing his 50th birthday. He had become increasingly irritated over the past five months by having to get up around 2 a.m. each night to pass his water. Then it started, a pounding headache at the back of the head waking

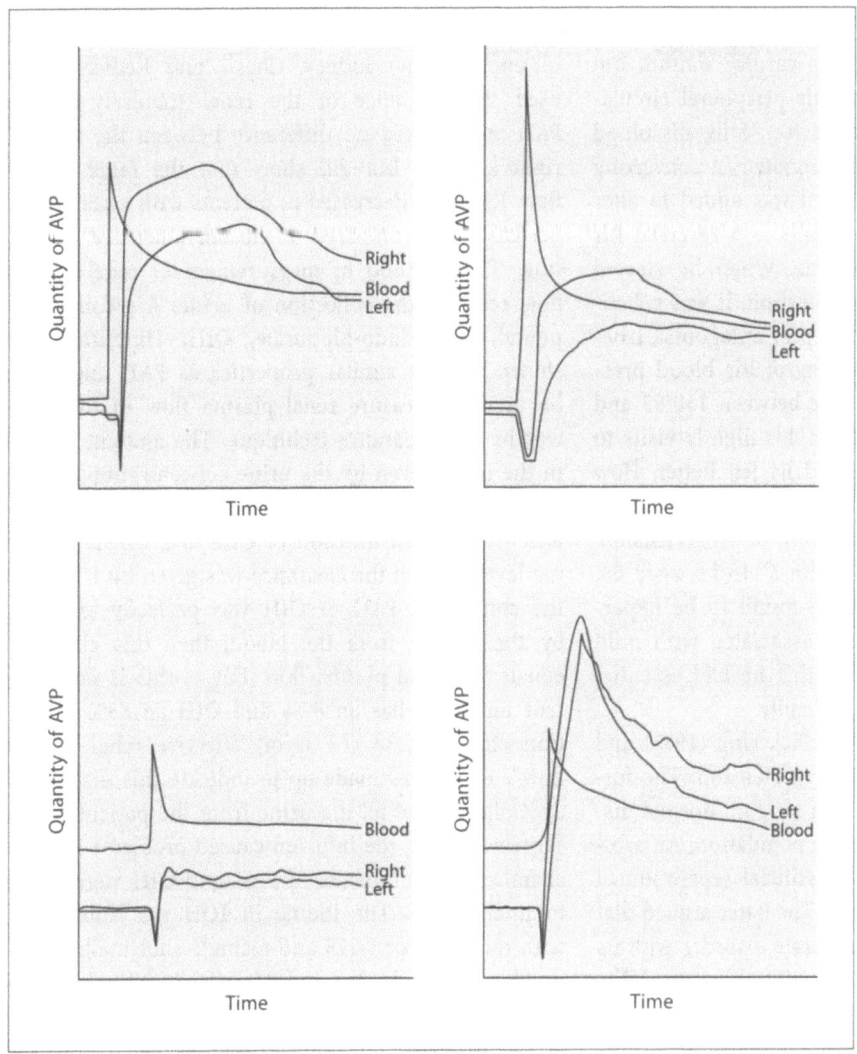

Fig. 25.1. Renographic studies with ^{131}I arginine vasopressin, ^{131}I AVP in nephrogenic diabetes insipidus. *Top left*, normal person ^{131}I AVP activity time curves; *top right*, patient with pituitary diabetes insipidus, normal ^{131}I AVP activity time curve; *bottom right*, normal [131]I Hippuran renogram in nephrogenic diabetes insipidus whose ^{131}I AVP activity time curve; *bottom left*, is completely flat indicating lack of receptor uptake, yet normal renal function

him at 7 a.m. A persistent throbbing radiating from his neck over his ears and scalp happening every morning. The headache persisted until lunchtime and was partly relieved by paracetamol which he found necessary to take. The appointment with his doctor revealed his high blood pressure 180/110 lying, 175/115 standing. His pulse was regular and his heart had a prominent aortic second sound. Changes were found in his eyes on ophthalmoscopy with relative narrowing of the arteries and dilatation of the veins. There was loss of the lucency of the arterial wall causing apparent "nipping" of the underlying vein as

it passed over it. The urine showed a trace of protein but no sugar. His blood tests of electrolytes, urea, uric acid, and serum creatinine showed no abnormality. His cholesterol was a little raised and his blood sugar was normal. Urine collection for catecholamines showed no excess. His ECG showed some left axis deviation. His renal radionuclide study was normal (see table below describing "Renal radionuclide study" technique). He was advised to lose weight and to reduce his salt intake and his saturated fat intake. He was started on a diuretic, bendrofluazide, to increase the loss of sodium chloride from his body with a potential side effect of increasing his urinary excretion of potassium ions. Then a beta blocker, atenalol, was started to reduce his cardiac output, but with a side effect of affecting his peripheral circulation giving him cold hands and feet. Still, his blood pressure remained high so an angiotensin converting enzyme (ACE) inhibitor enalapril was added to alter his angiotensin II production. This also reversed his tendency to lose potassium ions. When he started coughing as a result of this medication, it was substituted with an angiotensin II receptor antagonist, irbesartan, on which 24-h monitoring of his blood pressure showed an acceptable range between 150/95 and 130/80. His headaches were gone, his nightly visits to the toilet became infrequent and he felt better. How had all this come about?

He had a history in the family of hypertension. His mother had died recently with a stroke, aged 69. His elder brother had also been found to be hypertensive. His father had angina, associated with mild hypertension. It was concluded that he had essential hypertension which runs in the family.

Two Great British physicians, Pickering (1965) and Platt (1948) had argued about hypertension. The former felt it was just a top end of a random normal distribution of blood pressure in the population. An arbitrary cut off of 160/95 was an artificial separation of what was a graded phenomenon. The latter argued that essential hypertension was a separate disorder with its own Gaussian distribution overlapping the normal distribution, but with recognizable genetic and phenotypic differences. Platt lost the argument at that time but the quest for the differences has continued.

25.2.2 Renal Clearance

Goldblatt et al. (1934) had shown that if you narrow the renal artery in a rabbit, the blood pressure goes up. If this narrowing was removed before the other kidney was damaged by the high blood pressure then the hypertension was reversible. This phenomenon was shown to be due to the secretion of renin which cleaves a liver-produced protein, angiotensinogen, to produce angiotensin I. A converting enzyme acts on angiotensin I to produce angiotensin II. This is a potent vasoconstrictor with salt-retaining properties leading to swelling of arterioles whose narrowing increases the peripheral resistance to flow. Angiotensin II also was later shown to stimulate adrenal aldosterone production causing salt retention through the

kidney. The question then was whether essential hypertension was due to narrowing of the renal arteries in one or other kidney. Chasis and Redich (1941) used the clearance of the renal tubularly-secreted PAH and showed no difference between the left and right kidneys, but did show that the renal plasma flow, RPF, was decreased in patients with essential hypertension as compared to the normotensive population. This method of measurement of renal plasma flow requires the collection of urine. A second compound, ortho-iodo-hippurate, OIH, Hippuran, was shown to have similar properties as PAH and could be used to measure renal plasma flow in the same way by this clearance technique. The amount of OIH in the urine given by the urine concentration, U, and the volume of urine passed in a certain time, V, collected during an infusion of OIH to a constant plasma level, P, then the clearance was given by $U \times V/P$. If the compound PAH or OIH was perfectly extracted by the kidney from the blood, then this clearance equals the renal plasma flow. But as this is not quite true and PAH has an 87% and OIH an 83% extraction efficiency, so the term "effective renal plasma flow", ERPF, was made up to indicate this discrepancy. Collection of all the urine from the patient over a set time during the infusion caused problems and the chemical measurements of PAH and OIH were prone to interference. The iodine in IOH was substituted with radio ^{131}I or I-125 and a single shot method was developed by Blaufox and Merrill (1966) where the clearance was in effect redefined as the loss of a compound from the blood to the kidney and not to the urine. The rate of loss from the blood equals the rate of uptake by the kidney.

This requires drugs such as OIH and later mercapto acetyl triglycine, MAG3, and ethylene dicysteine, EC, that have their only exit from the body through the kidney. The rate of loss from the blood was measured by collection of serial blood samples, P_1, P_2, P_3 etc., typically at 5, 15, 30, 45, 60 and 120 min and measuring the radioactivity in each sample. Knowing the activity and the volume of the dose of I*OIH injected, D, then the clearance, that is the ERPF, is equal to the D*Lambda/P_0.

Lambda is the time determined from the washout rate and has units of per second. It equals 0.693/$T_{1/2}$, where $T_{1/2}$ is the time for half of the OIH to have left the blood. P_0 is obtained by extrapolation of the values of P_1, P_2, P_3 etc. back to the zero time axis to give the imagined plasma concentration as if all the OIH had been initially distributed evenly through the whole of the plasma injected. D is the activity per ml injected and P is the activity per ml in plasma, so if there is a point of equilibrium, then ERPF $= V \times$ Lambda.

V is the volume of distribution when all the dose is uniformly mixed in the body fluids. At this time the rate of input to the extra-cellular fluid, ECF, from the plasma and the rate of input from the ECF to the plasma are equal. This occurs at about 44 min after injection so the collection of a plasma sample at that time will show a concentration of activity in it that is directly proportional to the ERPF. This is the basis of the single plasma sample measurement technique for assessing ERPF. It requires steady state conditions and reasonable renal function to be accurate. It requires the construction of a standard curve where serial measurements in patients with different levels of renal function have had the single sample plasma activity measured against a full clearance technique.

The physiological basis of this technique depends on creating a physiological model that represents as far as possible the physiological reality. The plasma is considered to be a compartment into which the activity is injected. This is exchanging activity with another compartment, the extra-cellular fluid. It is exchanging activity with the red blood cells in the plasma and it has an exit from plasma through the kidneys alone. This compartmental model however has assumptions for it to function. The first is that steady state conditions are necessary during the measurement and the second is that the rate of mixing in each compartment should be much faster than the rate of exchange between compartments. Neither of these assumptions are quite true in the real clinical situation. Hence there is a 'biological error' on the estimate of effective renal plasma flow as well as a measurement error. The presence of biological variability as well as measurement error is a constant problem in nuclear medicine.

25.2.3
Cortical and Juxtamedullary Nephrons

Nephrons are of two distinct types, the cortical and the juxtamedullary (JM) nephrons, although there is some overlap between them. Cortical nephrons make up about 85% and their glomeruli lie in the outer two-thirds of the cortex. Each cortical nephron has a juxtaglomerular apparatus (JGA), which is formed as a syncytium between the muscular afferent arteriole and the early part of the distal tubule, the macula densa, of the same glomerulus, together with some interstitial mesangial cells and the almost muscle-free efferent arteriole. The muscle fibres of the afferent arterioles are altered and bear granules containing the enzyme, renin. The substrate for renin is the angiotensinogen present in plasma and the JGA. Enzymic breakage of its leucyl-leucine bond by renin produces a decapeptide, angiotensin I, which in turn forms the substrate for a plasma and tissue converting enzyme, which converts it into angiotensin II, an octapeptide. This is a potent constrictor of arterioles including the afferent arterioles of cortical nephrons and the efferent arterioles of JM nephrons and a stimulus to aldosterone release by the adrenals. All the enzymes necessary to produce angiotensin II are apparently bound to larger molecules in the isolated JGA.

The resistance to the flow of blood to cortical nephrons depends on the tone of the muscular wall of the afferent arteriole. Paradoxically this tone depends on the renal perfusion pressure itself in the following way: a fall in perfusion pressure reduces the peri-tubular capillary pressure around the proximal tubule and this enhances the passive component of salt and water reabsorption. Less salt and water is delivered to the JGA, which senses this and in turn reduces the local release of renin and thence angiotensin, so that the afferent arteriole tends to relax. This reduces the afferent arteriolar resistance thereby maintaining flow at a lower renal perfusion. This ability to maintain flow in spite of variation in perfusion pressure between physiological limits is called *autoregulation* and depends on the control loop outlined above (Britton 1968). The autoregulatory response sets off a further control loop. The lack of requirement for renin locally leads to its systemic release and the production in the plasma of angiotensin II, which constricts peripheral vessels. Its release of aldosterone aids sodium reabsorption in the distal tubule and, with the increase in vascular tone, tends to reverse the drop in perfusion pressure.

Alternatively afferent vasoconstriction occurs in response to a rise in perfusion pressure through increased delivery of salt to the JGA stimulating the local renin-angiotensin system (Thurau et al. 1972). Increased tone can be initiated additionally by a rise in sympathetic activity, which can override the autoregulatory response. Certain systemic disorders, such as renovascular hypertension, occur when pathology affects these renal self-regulatory mechanisms, and certain renal disorders, such as acute renal failure in shock result from an overriding of the local auto-regulatory system by changes in the systemic control systems.

In the normal resting state afferent arterioles are relaxed and the resistance to flow through cortical nephrons is less than the resistance to flow through the JM nephrons. This, as well as their greater numbers, accounts for the higher blood flow through cortical nephrons. Reduction in flow by 20% can occur in normal daily life, for example, from sudden noise, change in posture or mild stress. A relaxed atmosphere is essential for reliable renal studies.

The afferent arterioles to the JM nephrons are poorly muscular and the main control site is at their efferent arterioles. These respond to AVP, angiotensin II and sympathetic activity by vasoconstriction and relax in response to prostaglandins and nitrates. Flow to JM nephrons is difficult to characterize since for a given efferent arteriolar resistance increased afferent arteriolar flow just leads to increased JM glomerular filtration. Thus, JM efferent arteriolar flow is not easily accessible to measurement. JM nephrons also lack a JGA and thus have no afferent autoregulation. The ratio of cortical nephrons to JM nephrons in anatomical terms is about 6:1. In physiological terms, the ratio of cortical nephron function to JM function is much more variable. The intralobular artery supplies afferent arterioles to the glomeruli of JM nephrons before those to cortical nephrons, so flow through the glomeruli of JM nephrons precedes that through glomeruli of cortical nephrons.

25.2.4
Intrarenal Blood Flow

Intrarenal distribution of blood flow has some importance in renal physiology. The conservation of water depends on medullary blood flow, which is controlled by the JM efferent arterioles. It is a general assumption that various drugs, particularly diuretics, act similarly on all nephrons, yet consideration of their differing effects on renal physiology, for example free-water clearance, may be explicable in terms of separate actions on cortical and JM nephrons. Certain diseases appear to affect the two populations differently. Examples mainly affecting JM nephrons are analgesic nephropathy and the atrophy of outflow obstruction.

A non-invasive approach for measuring the ratio of cortical to JM nephron flows in man using probe renography and ^{131}I-OIH was put forward by Britton and Brown (1971), who demonstrated changes of flow distribution with alteration of salt intake. Plasma renin activity was shown to increase, as the 'cortical ne-

phron' component of flow decreased (Wilkinson et al. 1977) and the technique has been validated against microsphere distribution in animals (Wilkinson et al. 1978; Britton et al. 1980).

The measurement of intrarenal blood flow distributions between the cortical nephrons and the JM nephrons is based on the fact that the long loops of Henle of the JM nephrons have a longer mean transit time for non-reabsorbable solutes such as ^{123}I-OIH than the short loops of Henle of the cortical nephrons. Therefore the OIH transit time distribution obtained from a kidney should be bimodal, the earlier larger mode representing the fraction of cortical and the later smaller mode the JM nephron flow. Since the quantity of ^{123}I-OIH administered is well below the tubular maximum, the uptake of ^{123}I-OIH by each population of nephrons is proportional to flow, each population being identified by its nephron transit time. The technique has now been adapted to the use of ^{123}I-OIH with the gamma camera (Gruenewald et al. 1981; Nimmon et al. 1982, 2000; Britton 1985; Britton et al. 1986).

To make these measurements the renal parenchyma is first identified separate from the pelvis. To outline the pelvicalyceal area the computer is programmed to construct a mean time picture which is a functional image displaying the mean tracer time for each pixel of the 64×64 matrix, so the pixels with long mean times in the calyces and pelvis have high intensities. Further regions of interest are then constructed to exclude the high intensity, easily identified renal pelvis and calyces from the parenchyma.

To emphasize the dominant components in each zone of the cortex to aid subsequent mathematical analysis, the parenchymal region is further divided into an outer zone which will contain a sample of glomeruli mainly of cortical nephrons but few glomeruli of JM nephrons, and a middle zone containing both cortical and JM nephron glomeruli. Using the activity-time curve from the heart region of interest or the probe over the chest as the renal input and curves from the outer and middle zones, separate deconvolution analysis by the inverse matrix algorithm method are performed to give the retention of activity in each zone. These impulse retention functions of each zone would result from a theoretical spike injection to the renal artery with recirculation. The earlier vascular component is excluded and differentiations of these corrected retention functions give the distribution of transit times through the outer and middle zones, respectively. After the data have under-

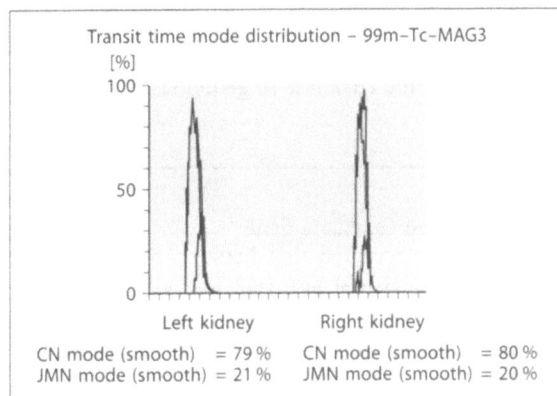

Transit time mode distribution – 99m-Tc-MAG3

Left kidney
CN mode (smooth) = 79 %
JMN mode (smooth) = 21 %

Right kidney
CN mode (smooth) = 80 %
JMN mode (smooth) = 20 %

Fig. 25.2. Intrarenal blood flow. The transit time distribution through the cortical nephrons, *tall peak*, and the juxtamedullary nephrons, *small peak*, are shown for both kidneys. The percentage contribution of the cortical nephrons is 80% and the juxtamedullary nephrons 20%, similar in each kidney in this example. The method of obtaining this data is described in the text

gone a noise reduction process using a cross-correlation technique, the leading edge of the outer zone transit time-distribution curve is fitted to the leading edge of the middle zone transit time-distribution curve. Subtraction of the two curves results in a new transit time-distribution curve which represents the transit of tracer through the JM nephrons. By comparing the heights of these cortical and JM nephron transit time-distribution curves, the flow to the cortical nephrons is obtained as a percentage of the total flow to the nephrons (Fig. 25.2).

In summary, the fraction of flow that takes the shorter mean transit time is the fraction of flow to the cortical nephrons, and the fraction of flow that takes the longer mean transit time is the fraction of flow to the JM nephrons. These fractions are applied to the flows to each kidney to give the flows to each nephron population in each kidney in milliliters per minute. Simulation studies have shown that the standard deviation for these results is 6.8% when ^{123}I-OIH is given in the dose used for this study. The technique requires high count rate data to reduce the statistical noise in the analysis and therefore is only applicable to kidneys with good function using tubularly secreted radiopharmaceuticals. When the mean transit times of the two populations are similar the differentiation of cortical from JM nephrons cannot be made. A 60-s difference is required for a 95% probability of separating the two components (Nimmon et al. 1982).

25.2.5
A Genetic Basis for Essential Hypertension

In essential hypertension, the renal images are of similar size, the renograms are symmetrical, relative functions and peak times are also similar. The mean parenchymal transit times are normal. Renal blood flow is reduced equally in each kidney. Britton (1981) asked the question 'As the renal plasma flow is reduced to each kidney symmetrically in essential hypertension, is the reduction of renal blood flow only to the cortical nephrons or only to the juxtamedullary nephrons or to both equally?' He showed that in 12 normotensive subjects, mean blood pressure 121/77, the cortical nephron flow as a percentage of total renal plasma flow was 83.9±0.7% (SEM). However in 21 patients with essential hypertension with a mean blood pressure of 154/101, the cortical nephron flow as a percentage of total was 74.6±1.5%. This difference was significant at $P < 0.002$. This finding has been confirmed by Miles (1991) and Miles et al. (1994) using dynamic X-ray CT. In other words, in essential hypertension, there is an apparent reduction of cortical nephron flow with normal or increased JM nephron flow. This could come about through overconstriction of afferent arterioles to cortical nephrons such as could occur with a resetting of the autoregulatory cortical nephron control system (Britton 1981).

Consider a genetically determined disorder of this control system. A physiological rise in blood pressure in the course of everyday activities might give an over-compensatory increase in cortical nephron afferent arteriolar tone reducing cortical nephron plasma flow and peritubular capillary pressure. This in turn would enhance the reabsorption of proximal tubular water and electrolytes, notably salt and calcium (Earley and Friedler 1966). Plasma and extra-cellular volume would tend to increase but it would be compensated for by the continuing flow to the JM nephrons. As the JM nephrons do not autoregulate, a rise in blood pressure causes their plasma flow to increase, which tends to washout the osmolar concentration gradient in the medulla, thereby reducing the concentrating ability of the kidney. This would account for the nocturia that is found early in many patients such as George with essential hypertension. Because of the reduced concentrating ability, the extra-cellular and plasma volumes may not increase despite the retention of salt by the kidneys. Thus at homeostasis in essential hypertension, there would be relative salt retention due to the lower cortical afferent arterial flow and relatively reduced peritubular tubular capillary pressure at the cortical nephrons. A relative reduction in plasma volume would occur due to the increased blood pressure and increased post glomerular flow of the juxtamedullary nephrons. The retained salt is distributed through all parts of the body, including the afferent arterioles supplying muscle (Tobian and Binion 1952). The tendency for tissues to retain salt may be enhanced by circulating sodium transport inhibitor (Poston et al. 1981). The diuretic response to a saline load in essential hypertension is excessive and may be attributed to the failure of the relatively high JM nephron flow to maintain adequate urine concentrating ability. Before intrarenal flow distribution was measured, it was apparently assumed that the reduction of renal plasma flow in essential hypertension meant that the JM nephron flow was reduced. Such an explanation as above was therefore discounted.

In conclusion it may be postulated that there is an overcontraction of cortical nephron afferent arterioles in response to a physiological rise in blood pressure in the genetically hypertensive and that the insidious effects of this overcontraction lead to changes in salt homeostasis and to essential hypertension. A minimal salt intake of 60 mmol/day may be needed (de Wardener and MacGregor 1980) for the expression of this postulated disorder of cortical nephron autoregulation. Dependence on salt intake would help to create the phenotypic response consistent with the gradation of normotension to hypertension found in most populations.

What is the mechanism of this increased afferent arteriolar tone?

Britton (1981) demonstrated that angiotensin converting enzyme inhibitors, ACEI, were able to increase the cortical nephron flow in patients with essential hypertension. Using captopril the cortical nephron flow as a percentage of total renal blood flow increased to 91.8±1.6%. Al-Nahhas et al. (1990) showed in a study of ramipril versus placebo that the cortical nephron flow increased from 207±7 ml/min to 257±21 ml/min with ramapril, while it caused a reduction of blood pressure from an average of 172/107 on placebo to 148/90. If, as it appears, that ACEI are able to increase cortical nephron flow, this implies that the intra-renal renin-angiotensin system subserving autoregulation is involved, hence a component of this system could be one genotype for essential hypertension.

The search for phenotypic expression of genotype markers for essential hypertension has increased in intensity with the advances in genomics.

25.2.6
Angiotensinogen Candidate Gene

It is generally accepted that between 30–50% of the prevalence of essential hypertension is due to genetic susceptibility (Ward 1990). The genetic basis of essential hypertension has been shown to be complex. Krushkal et al. (1999) have found four chromosomal regions, 2p, 5q, 6q, 15q, linked to systolic blood pressure in family studies of essential hypertension. A few rare autosomal monogenetic causes have been identified (Table 25.4).

The renin–angiotensin system maintains blood pressure (MacGregor et al. 1981) and subserves renal autoregulation (Britton 1968). It is therefore thought to be involved in the aetiology of essential hypertension through a number of mechanisms, one of which is the reduction of cortical nephron flow (Britton 1981). This may be an important and measurable phenotypic expression of the genetic basis of essential hypertension, many of whose candidate genes concern the renin-angiotensin system.

Williams and Hollenberg (1991) have divided essential hypertension patients into the modulators who handle a salt load normally with a direct relationship between salt loading and salt excretion and the non-modulators who have impaired handling of a salt load. This non-modulating group are a phenotype seen in 40% of essential hypertension. They interpreted this altered responsiveness to an increased intrarenal angiotensin-II production (Williams et al. 1992). A correlation has also been found between the plasma angiotensinogen (AGT) concentration and the level of blood pressure in essential hypertension (Watt et al. 1992). Thus an increased plasma AGT diffusing into the juxtaglomerular apparatus (JGA) of the afferent arteriole of the cortical nephrons would lead to an increased local production of angiotensin-II causing a rise in their arteriolar tone and a reduction in cortical nephron flow.

The AGT gene has been intensively studied as a 'candidate' gene for essential hypertension (Jeunemaitre et al. 1992). Many variants have been found of which the most cited is the $M^{235}T$ variant, where threonine is encoded instead of methionine, with an odds ratio of 1.2 on meta-analysis (Kunz et al. 1997). This variant is associated with an increased level of

Table 25.4. Hypertension due to rare autosomal monogenetic disorders

Name	Inheritance	Expression	Genetic cause	Chromosome	References
Liddle's syndrome	Dominant	Hyperkalaemia	Altered a and b	16p	Liddle et al (1963); Shimkets et al (1994); Hansson et al (1995); Baker et al (1998)
		Suppressed renin	Subunits of epithelial sodium channel		
		Suppressed aldosterone	Up-regulated		
		Salt-sensitive hypertension	Amiloride sensitive		
			Epithelial sodium channel		
Gordon's syndrome	Dominant	Hyperkalaemia	Thiazide sensitive	1,17	Mansfield et al (1997); Gordon et al (1970)
			Gene not known		
Glucocorticoid reme-diable aldosteronism	Dominant	Mild hypokalaemia	Chimera between genes for aldosterone synthetase and 11β hydroxylase	8p	Lifton et al (1992); Lifton (1995)
		ACTH stimulates aldosterone secretion			
Hypertension and brachydactyly	Dominant	Posterior inferior cerebellar artery anomaly	Gene not known	12p	Schuster et al (1996)
Apparent mineral corticoid excess	Recessive	Short stature	Defective	16q	Mune and White (1996)
		Early hypertension Hypokalaemic alkalosis	11β Hydroxysteroid dehydrogenase 2		
		Suppressed renin/ aldosterone Cortisol excess	Spironolactone-sensitive hypertension		

circulating AGT. However several of these studies showed no significant association including that of Caulfield et al. (1994). They used a marker dinucleotide repeat sequence flanking the AGT gene on chromosome 1q42-43 and found this strongly linked to essential hypertension in 63 white European families where two or more members had essential hypertension. Similar linkage was found in an Afro-Caribbean population (Caulfield et al. 1995). However a multicentre European study was not supportive (Brand et al. 1998). Corvol et al. (1999) discuss the difficulties inherent in candidate gene studies.

An $A^{20}C$, cytosine for adenine, promoter polymorphism in the AGT gene was shown to be linked to essential hypertension (Zhao et al. 1999) and associated with raised plasma AGT levels (Ishigami et al. 1999). This was also noted for an $A^{30}P$, proline for adenine (Nakajami et al. 1999), and for $C^{532}T$ (Paillard et al. 1999). These candidate genes all use essential hypertension as the phenotype. If the hypothesis of over-autoregulation of cortical nephron flow is an important contributor to the aetiology and if only 40% of essential hypertension patients have a genetic susceptibility, then the combination of essential hypertension and reduced cortical nephron flow (plus non-modulator status if a link between that status and reduced cortical nephron flow can be confirmed) should sharpen up this subgroup of essential hypertension patients for genetic analysis in the evaluation and validation of a candidate gene.

The contribution of the measurements of renal plasma flow, relative renal function and intrarenal flow distribution by nuclear medicine techniques would be important in separating this large ACEI sensitive group from the rare monogenetic thiazide, amiloride, glucocorticoid and spironolactone sensitive causes of essential hypertension set out in Table 25.4. The tailoring of a particular treatment to a particular genetically characterized individual with essential hypertension is in prospect (O'Byrne and Caulfield 1998).

25.2.7
Renovascular Disorder

Renovascular disorder is a pathophysiological term given to the demonstration that hypertension is due to disease of the large or small blood vessels supplying the kidney. Large vessel disease includes renal artery stenosis [as in the Goldblatt et al. (1934) studies] and fibromuscular hyperplasia, where the renal artery appear to take on a support function for an unusually mobile kidney. The common small vessel diseases are glomerulonephritis, which is usually symmetrical, or pyelonephritis, which is usually asymmetrical following recurrent urinary tract infection. Other causes include diabetes mellitus, autoimmune diseases and the stretching of small renal vessels by the cysts of polycystic disease (see Table 25.3). These all have in common an increase in proximal tubular salt and water reabsorption due to the marginally reduced peritubular capillary pressure as compared with the intraluminal tubular pressure and the stimulation of the re-

lease of renin, at least initially, and the activation of the renin-angiotensin system.

The increased reabsorption of salt and water from the proximal tubules reduces the flow therein and thus prolongs the transit time of nonreabsorbable solutes such as ^{99m}Tc MAG3. The prolongation of the peak time and the reduction of the second phase of the renogram on the side of the renovascular disorder have been used as indicators of renovascular disorder for decades (Britton and Brown 1971). The measurement of the mean parenchymal transit time (MPTT) is a more recent and reliable indicator (Al-Nahhas et al. 1989).

Smith (1956) showed that in hypertensive individuals with a small kidney on one side, in 25%-50% of occasions removal of the small kidney relieved the hypertension. His challenge was how to choose the small kidney with renovascular disorder from those without, the removal of which would not benefit the hypertension (Luke et al. 1968). The above findings from renography go a long way to solve this problem, but the addition of prior captopril administration as a 'stress' test for renovascular disorder has made the prediction much more reliable (Nally et al. 1995; Taylor and Nally 1995, 1996) as shown in a multicentre study (Fommei et al. 1993) (Fig. 25.3).

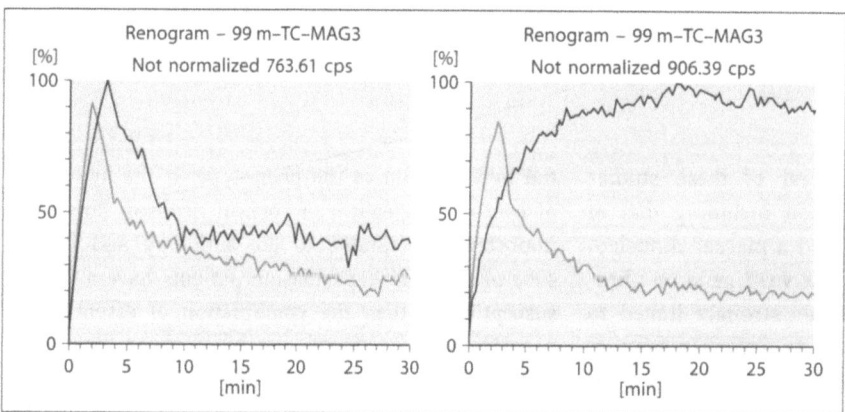

Fig. 25.3. Results of renal radionuclide studies in renovascular disorder. ^{99m}Tc MAG3, mercapto acetyl triglycine, renal radionuclide study in a patient with right renal artery stenosis. *Left,* baseline renograms. *Right,* post captopril 25 mg orally. *Baseline,* the left kidney shows a normal renogram with a sharp peak. The right kidney shows a peak delayed by a minute as compared to the left, and a slightly impaired second phase and borderline raised MPTT indicating probable renovascular disorder. *Post-Captopril,* the left kidney's renogram remains normal. The right kidney's renogram now shows further impaired renal function, 37% of total, a reduced and prolonged second phase with a delayed peak, and very abnormal MPTT of 564 s.

	Baseline Relative function	MPTT	Captopril Relative function	MPTT
Left	55%	116	63%	155
Right	45%	248	37%	564

Normal mean parenchymal transit time, MPTT, is <240 s. Captopril has enhanced the abnormality of the right renogram in a fashion typical of renovascular disorder of which renal artery stenosis is one cause

With the addition of the MPTT it is helpful in assessing the hypertensive diabetic (Datseris et al. 1995 b) and in the hypertensive patient with renal failure (Datseris et al. 1994). The method is outlined in the "Renal radionuclide study" table below. Once the presence of a renovascular disorder is shown and small vessel causes are excluded on clinical and investigative grounds, then angiography is able to show the large vessel disease leading to angioplasty or renal artery bypass surgery. Relief of hypertension and the return of the renographic findings towards normal indicate the success of this intervention. Renal artery stenosis is common in the atherosclerotic elderly as is essential hypertension. Only the application of renography can distinguish the functionally significant stenosis from the narrowed artery that is not causing hypertension.

For the patient, the taking of treatment for hypertension is essentially like paying the premium of an insurance policy, in that long-term reduction of blood pressure reduces the incidence of the complications of heart failure, stroke and renal impairment. The incidence of myocardial infarction also falls if other risk factors are controlled at the same time.

Renal radionuclide study
Renography (^{99m}Tc-MAG 3)

Radiopharmaceutical:	^{99m}Tc-MAG3, mercapto acetyl triglycine
Energy:	140 KeV activity: 100–50 MBq (3–4 mCi)
Collimator:	Low energy general purpose or high resolution
Procedure:	Patient should be normally hydrated. If doubt give 400 ml water 30 min before the start. Patient should empty bladder before the start
	Set up two-way tap infusion set for injection and saline flush
	Position patient reclining against the camera face which is tilted back 15° so kidneys fall back against the muscles. Ensure left ventricle and kidneys are in the posterior field of view
Dynamic:	Injection given (good bolus or flush system) Start computer: 180×10-s frames, 128×128 resolution (10-s frames are optimal for deconvolution analysis for transit times) For frusemide (furosemide, Lasix), inject 40 mg (0.5 mg/kg in child to maximum 20 mg) after 100–110 frames on computer (at 17–18 min). Warn the patient of the diuretic effects. All patients should empty the bladder after the study. A postmicturition standing posterior image of kidneys and bladder is often helpful

If captopril renogram is requested, monitor blood pressure before captopril 25 mg is given orally, then at 5-min intervals. If the diastolic pressure falls by 10 mmHg, start renography as it indicates captopril is already absorbed, if not start at 60 min. Record for 30 min

If a micturition study is requested after renography, start computer as micturition starts: 180×1 s frames, 128×128 resolution

---

25.3
Prostate Cancer

25.3.1
Introduction

Cancer is a genetic disease due to serial mutations in somatic cells. Cancer is also a clonal disease, which means that it arises from a somatic mutation in one particular cell. Following the example of colorectal cancer, it appears that a series of five or six mutations in the same cell are needed to turn it from a normal cell to a malignant cell with metastatic potential. Mutations are chance events. It may be the chance of cancer is increased by environmental factors such as cigarette smoking or by the pre-existing alteration of a gene to an oncogene in one of the stages required for cancer development. A gene is a series of consecutive base pairs on a chromosome. An oncogene is a gene that has been demonstrated to be involved in the transformation of a normal cell to a cancer cell. Most oncogenes are tumor suppressors and it is their deletion that releases an inhibitory control for cellular growth. Some oncogenes are tumorigenic by producing a stimulatory oncoprotein.

The demonstration by Watson and Crick (1953) of the helical nature of DNA and the fact that the base pairs were related – adenine to thymine and cytosine to guanine – provided the structure for genetic development. The carrying of genes through to the offspring and their combination from those provided by each parent transmit the information necessary for the individual to develop from the fertilized cell. This recognition gave the impression that the DNA molecule is a stable, robust, unchanging entity, both in the somatic cell as well as in the oocyte or sperm. The facts are somewhat different. The primary structure of DNA is not stable. DNA is in a state of flux, for between 1000 and 10,000 DNA molecules are altered each day in each cell and these alterations are repaired by a whole series of normal repair mechanisms (Lindahl 1993). DNA damage occurs through oxidation, hydrolysis, depurination or methylation, through misinsertion in pairing, or through slippage, where an extra base is slipped in without proper pairing. Since this "mutation" is so frequent, the wonder is that every cell is not a cancer cell. The contribution of natural radiation to perhaps four mutations per day in a cell is totally lost in the normal 'mutation' rate inherent in the labile DNA structure (see Chap. 33). If alterations to DNA were to become stable, then they may represent 1 of the 200 oncogenes which have been identified as leading to cancer. The fact that this does not happen is because of DNA repair, the first line of protection against mutagenesis. There are mechanisms for base excision and repair of the defect and for cutting out a series of bases which may be misaligned as well as the repair of single strand DNA breaks. Double strand breaks are more difficult to repair. Successful DNA repair is part of the complex adaptive response to noxious stimuli. If there is damage to the repair mechanism this loss allows mutagenesis and cancer induction. The genetic form of colorectal cancer unrelated to polyposis has been shown to be due to damaged DNA repair (Lynch and Lynch 1995) as has the skin cancer which occurs with minimal exposure to sunlight in the syndrome of ataxia telangectasia (Cox et al. 1986).

Cells have three possible fates: differentiate, proliferate or die by apoptosis. Apoptosis is a form of programmed cell death. If damaged DNA escapes the DNA repair system then it is recognized as such by the P53 system and a series of steps are initiated whereby the cells shrink through internal digestion by enzymes called caspases and die. The cancer cell has to find a way of avoiding this mechanism of self-destruction. It may do this in several ways of which the commonest is through a mutation in the P53 gene so it fails to recognize damaged DNA and the oncogene persists. In the treatment of cancer different chemotherapies are aimed at inducing apoptosis in the cancer cell as does moderate irradiation of cancer. High levels of chemotherapy and radiation cause necrosis where cancer cells leak, so their contents are spread out. Necrosis can induce an immunological response.

Just as the DNA of normal genes induces messenger RNA which in turn encodes particular amino acid combinations to form a particular protein by the ribosomal assembly system, so oncogenes encode altered messenger RNA to produce altered proteins, oncoproteins. These oncoproteins may be a basis for detection of cancer themselves, but it is usually their effects in up-regulating antigens or receptors, transport proteins or growth factors that help to make the cancer cell succeed in a population of normal cells.

25.3.2
Imaging Prostate Cancer

Whereas radiological techniques require the cancer to form a mass in tissue, displacing tissue or infiltrating tissue so that the cancer can be detected by contrast, nuclear medicine does not require a mass but aims to detect the cancer cell through its subtle differences from the normal cell by a process of identification. Radiology is limited by size, for example, a lymph node of 1 cm diameter is called normal, yet in order to enlarge a lymph node to a size greater than normal, cancer cells must have been present dividing in the normal sized node. The identification process by nuclear medicine is not dependent on size. Indeed a radioactive pinhead is detectable if it has enough radioactivity on it, although because of the poor resolution of nuclear medicine equipment, it may appear much larger than it really is. Detection of cancer therefore depends on detecting activity bound to or in the cancer cell. For the use of ^{18}F de-oxyglucose, FDG, by positron emission tomography, PET, the glucose transporter protein-1 is up-regulated by a factor of 5–10 and the hexokinase enzyme which phosphorylates FDG is up-regulated by a factor of 2–3, hence cancer cells are determined to be different from normal cells by their increased utilization of FDG. Antigens on a cancer cell may be up-regulated so that there are 1000–50,000 binding sites per cell for an appropriately chosen antibody. Receptors may be up-regulated on a cancer cell from 500–5000 times as an appropriate target for a radiolabelled peptide. It is these amplifications that make nuclear medicine such a powerful technique for detecting cancer through these three differing approaches.

FDG with PET detection can be considered a 'catch all' approach as any tumor, inflammation, granuloma or infection which increases the utilization of sugar may be detected. Leukocytes in particular are avid consumers of glucose. Nevertheless there is a relationship between the rate of glucose utilization and the malignancy of the tissue. FDG PET is thus very sensitive in detecting a lesion but its specificity is context dependent. A solitary pulmonary nodule with high uptake of FDG is likely to be due to lung cancer.

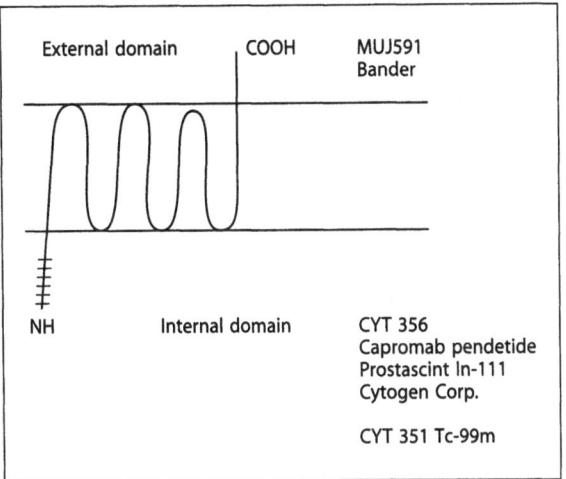

Fig. 25.4. Prostate-specific membrane antigen. A diagram of the cell wall membrane of the prostate cancer cell showing diagrammatically the external domain to which the MUJ 591 antibody binds and the internal domain to which the capromab penteotide, Prostascint antibody binds, CYT 356, as well as the technetium-99m version, CYT 351

In a patient with lung cancer uptake of FDG in a node in the mediastinum is likely to be due to lung cancer in this context. However for prostate cancer the rate of glucose utilization is generally low, particularly in the early stages. A prostate cancer may be present for many years before it becomes clinically evident through prostatic enlargement. Its metastases tend also to have weak uptake of FDG so PET is not a good method for detecting prostate cancer (Effert et al. 1996; Shreve et al. 1996; Hoh et al. 1998). Prostate cancer-related receptors and antigens hold more promise.

25.3.3
Prostate Cancer Phenotypes

Prostate cancer is becoming the commonest cause of death due to cancer in man. 350,000 new cases with 45,000 deaths occur in the USA and 10,000 deaths in the UK annually. The aetiology is still unclear. Benign prostate hypertrophy is not a risk factor. High vitamin A and vitamin C intake and cadmium are thought to be dietary factors. Having a male relative with prostate cancer increases the risk twofold and with two male relatives the risk increases ninefold. However, this family history is seen in less than 5% of patients presenting with prostate cancer. Nevertheless, just as for colorectal and breast cancer, the study

of the few patients with hereditary cancer has helped to illuminate the common sporadic occurrence of the cancer. Bratt (2000) has reviewed hereditary prostate cancer and described prostate cancer susceptibility genes. HPC-1 hereditary prostate cancer gene-1 (Cooney et al. 1997) has a locus on chromosome 1q. A further susceptibility gene was located on the long arm of the X-chromosome, Xq 27–28 (Xu et al. 1998). Perinchery et al. (2000) have demonstrated deletion of Y chromosome-specific genes, particularly in advanced stages and grades of human prostate cancer. Alfonso et al. (1999) showed loss of heterozygosity on chromosome 13 in advanced stage prostate cancer and Strup et al. (1999) have found allelic loss on chromosome 16. The breast cancer genes Brca1 and Brca2 have been implicated, since breast cancer shares the features of hormone dependency, a tendency to bone metastases and some familial associations with prostate cancer (Sigurdsson et al. 1997). These studies show that there is no simple genetic basis for prostate cancer and thus from the imaging and therapy point of view the hope that a specific oncoprotein for targeting with a radiolabelled antibody could be identified in relation to these oncogenes is unlikely to be realized.

The alternative approach is to seek constituents of the prostate cancer cell and particularly its cell surface that are demonstrably different in structure and or expression from the normal prostate cell. These factors may be identified inside the cell or on the cell surface. They include receptors for hormones or autocrine growth factors that stimulate prostate cancer cell growth and antigens that are highly expressed in prostate cancer cells as compared to normal. Guate et al. (1999) looked at neurotrophin receptors in normal and human prostate cancer tissue and found increased expression of a transmembrane glycoprotein and a tyrosine kinase receptor protein Trk A, but a decrease in the expression of Trk C. Nelson and Carducci (2000) investigated the role of endothelin-1 and endothelin receptor antagonists in prostate cancer. Oh et al. (2000) have investigated whether there were genotype changes in the tumor necrosis factor-a promoter regions associated with human prostate cancer and found a relationship. The Bcl-2 proto-oncogene produces an oncoprotein known to inhibit apoptosis and this protein was found to be expressed both in low and high grade prostatic intra-epithelial neoplastic tissue (Baltaci et al. 2000). Watson and Fitzpatrick (2000) have looked for target sites for manipulating apoptosis in prostate cancer. It can thus be seen that, although this is an area undergoing active research, no particular receptor has yet been identified to characterize prostate cancer cells as different from normal prostate cells.

The situation with antigen expression is somewhat more encouraging. Monoclonal antibodies were first created against enzymes liberated by prostate cancer: prostatic acid phosphatase, PAP (Vihko et al. 1989), and prostate specific antigen (PSA). These antibodies were radiolabelled for imaging but were not successful partly because high levels of circulating PAP or PSA bound the monoclonal antibody before it reached its target. The identification of a prostate specific membrane antigen (PSMA) (Lopes et al. 1990), also called prostate membrane specific antigen (PMSA), and the development of a monoclonal antibody against this antigen has changed the situation.

Wright et al. (1995) showed up-regulation of this antigen in primary and metastatic prostate cancer. Wynant et al. (1991) first showed immunoscintigraphy of prostate cancer using the indium-111 monoclonal antibody then called 7E11-C5.3 (CYT 356). This antibody is now commercially available as Prostascint, also known as [111]In Capromab pendetide (Cytogen Corporation) and over 22,000 patients have been studied using this monoclonal antibody labelled with ^{111}In for imaging prostate cancer in the USA. Minor side effects have been reported in only four percent of patients. The absorbed radiation dose is at the high end of the diagnostic range (Mardirossian et al. 1996).

Investigation of PSMA has shown that it is a transmembrane glycoprotein (100,000 Dalton) which has intracellular and extracellular domains (Troyer et al. 1995) (Fig. 25.4). The PSMA gene has been cloned and sequenced (Israeli et al. 1993) and mapped to chromosome 11q14 (Rinker-Schaeffer et al. 1995). PSMA appears to be a novel folate hydrolase (Pinto et al. 1996) with neuropeptidase features (Carter et al. 1996). It has been shown that the CYT-356 monoclonal antibody binds primarily with the intracellular domain and therefore presumably requires increased permeability of the prostate cancer cell or internalization of the antibody to demonstrate its successful binding. It has also been labelled with [99m]Tc (Stalteri et al. 1997) and shown to be clinically useful (Feneley et al. 1996; Chengazi et al. 1997).

Liu et al. (1997, 1998) have developed a new monoclonal antibody, J591, against the extracellular domain of PSMA with the intention of better targeting and imaging properties. This antibody may inter-

nalize the antigen and also reacts with neoangiogenic endothelium. It has been labelled with [131]I (Goldsmith et al. 2000) and with ^{99m}Tc in the author's laboratory where it has been used successfully for imaging in 15 patients. [90]Y humanized J591 is being evaluated experimentally for radioimmunotherapy (Smith-Jones et al. 2000).

25.3.4
Clinical Applications

Prostate cancer is diagnosed by digital rectal examination and/or a rising PSA. Then a transrectal ultrasound is used to guide prostate biopsy, and 6–8 specimens are obtained. The number of positive biopsies is used to make a Gleason score. The higher the score the more likely the spread to local lymph nodes. Patients with localized disease thought to be confined to the prostate have a choice of treatment: radical prostate surgery, implantation of radioactive ^{125}I seeds (Ragde et al. 1998), cryoablation or external beam radiotherapy. Current tests make the diagnosis of suitability for these treatments by exclusion. A bone scan is performed to exclude bone metastases. CT and MRI may be performed to try and identify whether local lymph nodes are enlarged or there is involvement of seminal vesicles as well as information about the prostate itself. However, as indicated in the introduction, a normal sized node does not exclude the presence of cancer. Nodal involvement alters prognosis in that if a single node demonstrated at surgery should contain prostate cancer, this reduces the 5-year survival from 80% to 35%. Patients may also undergo a pelvic lymph node dissection as a sampling procedure usually by laparoscopy before proceeding to radical surgery, but in about a quarter of cases the sampling procedure fails to identify an involved node subsequently found at radical surgery.

Prostate radioimmunoscintigraphy is in principle able to show that the cancer is confined to the prostate (Fig. 25.5); or that there is evidence of extension through the capsule either posteriorly or towards the seminal vesicles; or that the obturator node is involved on one or both sides; or that there is evidence of lymph node involvement in the iliac chain; or related to the aorta; or in the presacral region (Maguire et al. 1993; Babaian et al. 1994; Kahn et al. 1994; Murphy et al. 1997; Hinkle et al. 1998 a, b; Manyak 1998). In practice it can only detect relatively macroscopic rather than microscopic histology. Clear capsular in-

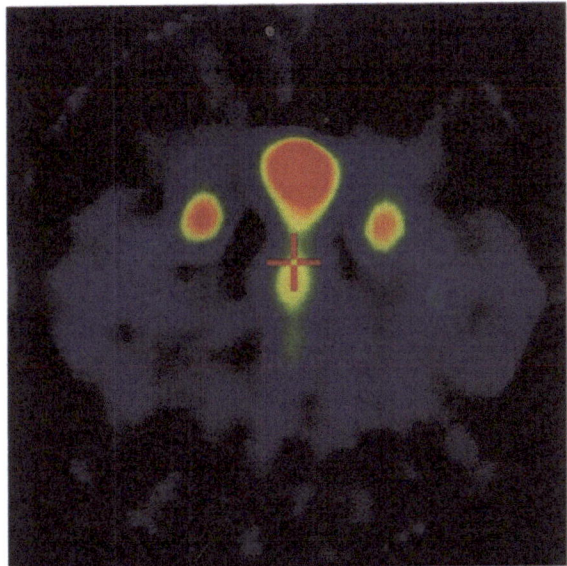

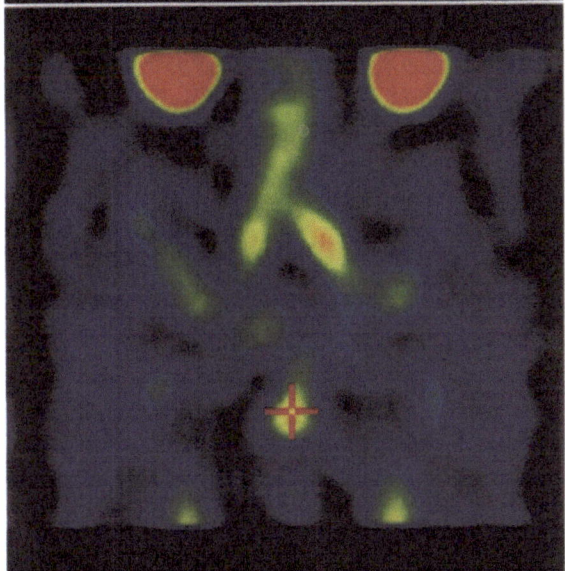

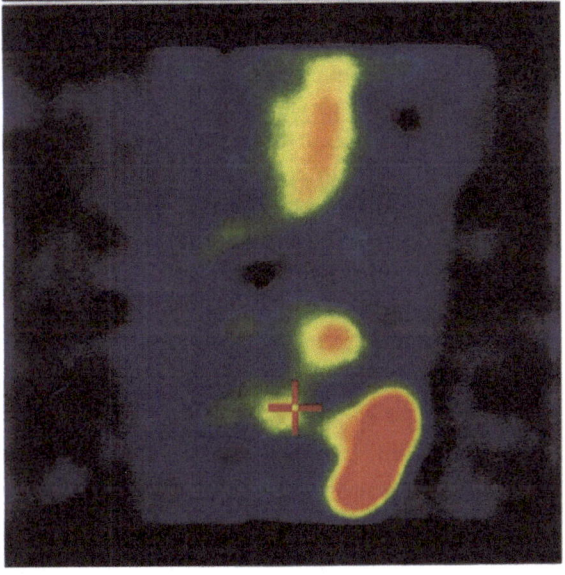

vasion with extension posteriorly of the prostate may be demonstrated but histopathological evidence of micropathological invasion cannot. Obturator node involvement down to 2 mm disease can be demonstrated but micrometastases cannot. However, the technique needs very careful attention to detail: top quality well-maintained double-headed gamma camera system, preferably with a thick crystal; slow rotation speeds, accurate repositioning of the patient at each study time; and obtaining the same information content on the late images as the early ones. Interpretation may be difficult and requires a careful comparison of the early image which often shows tortuous vascular structures and a late image where the activity in the vessels is reduced. Nodes that occur in relation to blood vessels may be demonstrated. An alternative approach is to use a double radionuclide setting of the gamma camera for ^{99m}Tc-radiolabelled blood cells on the third day of the Indium-111 Prostatscint imaging. Comparison of the two images with co-location protocols allows the demonstration that a site of uptake is in a lymph node and not a tortuous blood vessel. Britton et al. (2000) have reviewed the requirements for interpreting prostate radioimmunoscintigraphy and the technique is outlined in the Appendix.

After primary surgery or radiotherapy, the serum PSA is monitored regularly. It should be unrecordable immediately after surgery and stay so until recurrence occurs when the PSA rise is a reliable indicator of the current disease. However a problem exists after surgery when the PSA is detectable and fluctuates within the normal range. Demonstration of bone metastases using the bone scan and/or MRI is relatively straightforward. Many recurrences occur in soft tissue either in the prostate bed, pre-sacrally or along the lymph node chain. Serial prostatic radioimmunoscintigraphy is being investigated in this situation. Kahn et al. (1997,

Fig. 25.5. Prostate cancer imaged with ^{99m}Tc MUJ 591 at 24 h. Top transverse section, middle coronal section, bottom sagittal section. The site of the prostate cancer confined to the prostate is marked with a *red cross*. On the transverse section anteriorly the bulb of the penis and the two external iliac arteries are seen in *red*. In the coronal section at the top of the image the two kidneys are seen in *red* and in *yellow* the aorta in its bifurcation. In the sagittal section inferiorly the bulb of the penis is seen behind which is the prostate cancer *cross* and above which is the small bladder in *red*. Uptake in the bone marrow of the lumbar sacral spine is seen superiorly. Conclusion: prostate cancer confined to the prostate. No evidence of extra prostatic spread of the cancer is seen. This means that it is operable by radical prostatectomy

1998) and Levesque et al. (1998) showed that the response to salvage radiotherapy after failed radical prostatectomy could be predicted using Indium-111 Prostascint. A negative prostate immunoscintigraphy would be a reason to delay radiotherapy which is sometimes given just in case. A patient with a rising PSA and a negative bone scan and radiology should therefore be imaged to determine the presence of lymph node involvement (Berry et al. 1998; Polascik et al. 1999). Liver, spleen (Naseem et al. 1998) and even brain metastases (Garcia-Morales et al. 2000) may be demonstrated. While there may be still no effective therapy, nevertheless the identification of the source of a rising PSA level helps the surgeon and the patient to explain the phenomenon which may be a major source of anxiety. Radioimmunotherapy with these antibodies labeled with a beta-emitting radionuclide is under evaluation (Smith-Jones et al. 2000).

Bone metastases are generally only positive on radioimmunoscintigraphy during the active invasion stage when the PSA level is rising. Local radiotherapy or hormone ablation treatment impairs antibody uptake. Bone scan changes in patients who receive a variety of therapies and whose PSA is typically falling may be positive due to changes related to healing and scarring, but negative on radioimmunoscintigraphy.

In conclusion, nuclear medicine techniques have an increasing role in the detection of cancer and the management of patients with cancer, of which prostate radioimmunoscintigraphy is one example.

Appendix

Prostate Immune Study: ^{111}In-Prostascint

- Radiopharmaceutical: ^{111}In-CYT356 monoclonal antibody.
- Energy: 171 and 245 KeV activity: 200–400 MBq.
- Collimator: Medium energy.
- Post injection time: 10–60 min, 24, 48 or 72 h.
- The 24-h image may be omitted.

The Planar and SPET images are best done with a double head 'thick' crystal camera system (the squat view with a single head camera).

■ Note. Patients need a full explanation of the procedure and are required to sign a consent form. Patients with allergy to foreign protein, i.e. reaction to inoculation, are excluded. Vital signs may be monitored pre- and post-injection.

Procedure

1. Inject slowly in an antecubital vein 200–400 MBq (5–10 mCi) ^{111}In-CYT356.
2. Patient empties bladder, then at 10–60 min post injection, take the following planar views in order: anterior and posterior pelvis, anterior and posterior upper abdomen (128 resolution, 800 Kc), ensuring the camera is horizontal.
3. Patient again empties bladder. Position for squat views having the single head camera as low as possible, tilted backwards angled at 45°. The patient sits on a stool, and leans forward, resting on a pillow on a lower stool. Check position so that the kidneys and liver are not in the field of view. Take one squat view, 128 resolution, without markers for 400 Kc.
4. Then do SPET of pelvis, 360° time per angle 20 s.
5. Repeat (2–4) at 24 h and 48 or 72 h, but in order 3, 2, 4: SPET time per angle 60 s.

Improved Procedure

1a. Prior to the injection of [111]In Prostascint, undertake a ^{99m}Tc-MDP, methylene diphosphonate or equivalent bone scan. Immediately after completion of the 3-h views inject the [111]In Prostascint and set up SPET/CT 'Hawkeye' for dual radionuclide capture of ^{99m}Tc and [111]In.
5a. Prior to the 48-h or 72-h SPET/CT, do a ^{99m}Tc blood pool scan – inject a pyrophosphate/stannous kit intravenously, then 20 min later draw back blood into a large syringe containing 370 MBq (10 mCi) ^{99m}Tc pertechnetate. Allow to mix and reinject for blood pool imaging. Set up SPET/CT for ^{99m}Tc and ^{111}In dual radionuclide capture.
6. Compare bone scan SPET with early [111]In Prostascint for anatomical registration.
   Compare early [111]In Prostascint SPET with late [111]In Prostascint SPET.
   Compare early [111]In Prostascint SPET with ^{99m}Tc blood pool SPET.
   Compare late [111]In Prostascint SPET/CT with ^{99m}Tc blood pool SPET.

The comparison is best made by co-registration of the images with transfer of a region of interest around an abnormality or suspected abnormality on the late ^{111}In Prostascint SPET to the ^{99m}Tc blood pool SPET to the bone scan and to the early ^{111}In Prostascint SPET. Anatomical co-registration of the Prostascint SPET of the pelvis to the X-ray CT, followed by coregistration of the other images, will aid radiotherapy planning.

Prostate Immune (^{99m}Tc)

Patient needs to sign consent form, after explanation and checking for allergy to foreign protein.

Isotope:	^{99m}Tc
Energy:	140 KeV
Form:	CYT 351 or MUJ 591
Activity:	Approx. 800 MBq
Collimator:	Millenium (VPC 35) and Varicam (VPC 45)
Postinjection time:	10 min to approx. 1 h and 24 h
Static scan:	On single head camera: Squat view 400 Kc (max. 600 s), 128×128 resolution: Have camera as low as possible, tilted backwards (angled at 45°). The patient sits on a pillow on a swivel chair and leans forward elbows resting on a pillow on a lower stool. On double head camera 800 Kc, 128×128 (max. 600 s per view) Post and Ant pelvis, Post and Ant abdomen Post and Ant chest, up to neck (in some patients this may already be covered)
SPET:	64×64 Resolution, zoom 1.28, 6° steps. At 1 h p.i. use 45 s/slice (total 22 min) – this shows vascular areas At 24 h p.i. use 90 s/slice (total 45 min)

25.4
References

Afonso A, Emmert-Buck MR, Duray PH et al (1999) Loss of heterozygosity on chromosome 13 is associated with advanced stage prostate cancer. J Urol 162:922–926

Al-Nahhas A, Jafri RA, Britton KE et al (1988) Clinical experience with 99mTc-MAG3, mercaptoacetyltriglycine and a comparison with 99mTc DTPA. Eur J Nucl Med 14:453–462

Al-Nahhas A, Marcus AJ, Bomanji J et al (1989) Validity of the mean parenchymal transit time as a screening test for the detection of functional renal artery stenosis in hypertensive patients. Nucl Med Commun 10:807–815

Al-Nahhas A, Nimmon CC, Britton KE et al (1990) The effect of ramipril, a new angiotensin converting enzyme inhibitor on cortical nephron flow and effective renal plasma flow in patients with essential hypertension. Nephron 54:47–52

Babaian RJ, Sayer J, Podoloff DA et al (1994) Radioimmunoscintigraphy of pelvic lymph nodes with 111-Indium labelled monoclonal antibody CYT 356. J Urol 152:1952–1955

Baker EH, Dong YB, Sagnella GA et al (1998) Association of hypertension with T594M mutation in a beta subunit of epithelial sodium channels in black people resident in London. Lancet 351:1388–1392

Baltaci S, Orhan D, Ozer G et al (2000) Bcl-2 Proto-oncogene expression in low and high grade prostatic intraepithelial neoplasia. Br J Urol Int 85:158–159

Blaufox MD, Merrill JP (1966) Simplified hippuran clearance. Nephron 3:274–281

Barker DF, Hoskikka SL, Zhou J et al (1990) Identification of mutations in the COL4A5 collagen gene in Alport syndrome. Science 248:1224–1227

Berry MG, Feneley MR, Domizio P et al (1998) Evaluating the return of prostate adenocarcinoma. J R Soc Med 91:641–643

Bichet DG, Turner M, Morin D (1998) Vasopressin receptor mutations causing nephrogenic diabetes insipidus. Proc Assoc Am Physicians 110:387–394

Brand E, Chatelain N, Keavney B et al (1998) Evaluation of the angiotensinogen locus in human essential hypertension: a European study. Hypertension 31:725–729

Bratt O (2000) Hereditary prostate cancer. Br J Urol Int 85:588–589

Britton KE (1968) Renin and renal autoregulation. Lancet 2:329–333

Britton KE (1981) Essential hypertension: a disorder of cortical nephron control? Lancet ii:900–902

Britton KE (1985) Radioisotope renal function studies in essential hypertension. Cardiology 72 [Suppl 1]:22–29

Britton KE, Brown NJG (1971) Clinical renography. Lloyd Luke, London

Britton KE, Maisey MN (1998) Kidney and urinary tract. In: Maisey MN, Britton KE, Collier BD (eds) Clinical nuclear medicine, 3rd edn. Chapman and Hall Medical, London, pp 389–424

Britton KE, Khokar AM, Brown NJG et al (1977) A non-invasive test for receptor binding applied to nephrogenic diabetes insipidus. Postgrad Med J 53:374–377

Britton KE, Nimmon CC, Whitfield HN et al (1979) Obstructive nephropathy: successful evaluation with radionuclides. Lancet i:905–907

Britton KE, Bernardi M, Wilkinson SP et al (1980) Validation of a non-invasive method of estimating intrarenal plasma flow distribution. In: Hollenberg NK, Lange S (eds) Radionuclides in nephrology. Thieme, Stuttgart, pp 204–208

Britton KE, Nawaz MK, Nimmon CC et al (1986) Total and intrarenal flow distribution in healthy subjects: technique, acute effects of Ibopamine and of Indoramin. Nephron 43:265–273

Britton KE, Nawaz MK, Whitfield HN et al (1987) Obstructive nephropathy: comparison between parenchymal transit time index and Frusemide diuresis. Br J Urol 59:127–132

Britton KE, Feneley MR, Jan H et al (2000) Prostate cancer: the contribution of nuclear medicine. Br J Urol Int 86 [Suppl 1]:135–142

Call KM, Buckler AJ, Rose EA et al (1990) Isolation and characterisation of a zinc finger polypeptide gene at the human chromosome 11 Wilms' tumor locus. Cell 60:509–520

Carter RE, Feldman AR, Coyle JT (1996) Prostate-specific membrane antigen is a hydrolase with substrate and pharmacological characteristics of a neuropeptidase. Proc Natl Acad Sci USA 93:749–753

Caulfield M, Lavender P, Farrall M et al (1994) Linkage of the angiotensinogen gene to essential hypertension. N Engl J Med 330:1629–1633

Caulfield M, Lavender P, Newell-Price J et al (1995) Linkage of the angiotensinogen gene locus to human essential hypertension in African Caribbeans. J Clin Invest 96:687–692

Chaiwatanarat T, Padhy AK, Bomanji JB et al (1993) Validation of renal output efficiency as an objective quantitative parameter in the evaluation of upper urinary tract obstruction. J Nucl Med 334:845–848

Chasis H, Redish J (1941) Effective renal plasma flow in the separate kidneys of subjects with essential hypertension. J Clin Invest 20:655–661

Chengazi VU, Feneley MR, Ellison D et al (1997) Imaging prostate cancer with Technetium-99m-7E11-C 5.3 (CYT-351). J Nucl Med 38:675–682

Cooney K, McCarthy J, Lange E et al (1997) Prostate cancer susceptibility locus on chromosome 1q: a confirmatory study. J Natl Cancer Inst 89:955–999

Corvol P, Persu A, Gimenez-Roqueplo A-P et al (1999) Seven lessons from two candidate genes in human essential hypertension. Hypertension 33:1324–1331

Cox R, Debenham PG, Masson WK et al (1986) Ataxia telangectasia: a human mutation giving high frequency misrepair of DNA double strand scissions. Mol Biol Med 3:219–224

Datseris IE, Bomanji JB, Brown EA et al (1994) Captopril renal scintigraphy in patients with hypertension and chronic renal failure. J Nucl Med 35:251–254

Datseris IE, Boletis J, Papadakis E et al (1995 a) 99mTc Diamino Cyclohexane (DACH) a renal tubular agent with cationic transport excretion in renal transplant patients. Eur J Nucl Med 22:835

Datseris IE, Sonmezoglu K, Siraj QH et al (1995 b) Predictive value of captopril transit renography in essential hypertension and diabetic nephropathy. Nucl Med Commun 16:4–9

De Wardener HE, MacGregor GA (1980) Dahl's hypothesis that a saluretic substance may be responsible for a sustained rise in arterial pressure: its possible role in essential hypertension. Kidney Int 18:1–9

Deen PMT, Verdijk KMA, Nine VAM et al (1994) Requirement of human renal water channel aquaporin-2 for vasopressin-dependent concentration of urine. Science 264:92–95

Earley LF, Friedler RM (1966) Effects of combined renal vasodilatation and pressor agents on renal hemodynamics and tubular reabsorption of sodium. J Clin Invest 45:542–551

Effert PJ, Bares R, Handt S et al (1996) Metabolic imaging of untreated prostate cancer by Positron Emission Tomography with 18-Fluorine labelled deoxyglucose. J Urol 155:994–998

European Chromosome 16 Tuberous Sclerosis Consortium (1993) Identification and characterization of the tuberous sclerosis gene on chromosome 16. Cell 75:1305–1315

Feneley MR, Chengazi VU, Kirby RS et al (1996) Prostatic radioimmunoscintigraphy: preliminary results using technetium-labelled monoclonal antibody CYT 351. Br J Urol 77:373–381

Fommei E, Ghione S, Hilson AJW et al (1993) Captopril radionuclide test in renovascular hypertension: a European multicentre study. Eur J Nucl Med 20:635–644

Gabow PA (1993) Autosomal dominant polycystic kidney disease. N Engl J Med 329:332–342

Garcia-Morales F, Chengazi VU, O'Mara RE (2000) Detection of brain metastasis with Indium-111 capromab pendetide (Prostascint) due to prostate carcinoma. Urology 55:286xiii–286xv

Goldblatt H, Lynch J, Hanzal RF (1934) Studies on experimental hypertension. I. The production of persistent elevation of systolic blood pressure by means of renal ischaemia. J Exp Med 59:47–379

Goldsmith SJ, Kostakoglu L, Vallabhajosula S et al (2000) Evaluation of anti-PSMA antibody, [131]I J591 in the treatment of prostate cancer. J Nucl Med 41 [Suppl]:80P

Gordon RD, Geddes RA, Pawsey CG et al (1970) Hypertension and severe hyperkalaemia associated with suppression of renin and aldosterone and completely reversed by dietary sodium restriction. Austr Ann Med 19:287–294

Gruenewald SM, Nimmon CC, Nawaz MK et al (1981) A non-invasive gamma camera technique for the measurement of intrarenal flow distribution in man. Clin Sci 61:385–389

Guate JL, Fernandez N, Lanzas JM et al (1999) Expression of p75[LNGFR] and Trk neurotrophin receptors in normal and neoplastic human prostate. Br J Urol Int 84:495–502

Gupta NK, Bomanji JB, Waddington W et al (1995) Technetium-99m-L-L. Ethylene dicysteine scintigraphy with patients with renal disorders. Eur J Nucl Med 22:617–624

Hansson J, Nelson-Williams C, Suzuki H et al (1995) Hypertension caused by a truncated epithelial sodium channel subunit: genetic heterogeneity of Liddle syndrome. Nat Genet 11:76–82

Hediger MA, Coady MJ, Ikaeda TS et al (1987) Expression cloning and cDNA sequencing of the Na+/glucose cotransporter. Nature 330:379–381

Hinkle GH, Burgers JK, Neal CE et al (1998 a) Multicentre radioimmunoscintigraphic evaluation of patients with prostate carcinoma using Indium-111 Capromab Pendetide. Cancer 83:739–747

Hinkle GH, Burgers JK, Olsen JO et al (1998 b) Prostate cancer abdominal metastases detected with Indium-111 capromab pendetide. J Nucl Med 39:650–652

Hoh CK, Seltzer MA, Franklin J et al (1998) Positron emission tomography in urological oncology. J Urol 159:347–356

Ishigami T, Tamura K, Fujita T et al (1999) Angiotensinogen gene polymorphism near transcription site and blood pressure: role of a T-to-C transition in intron 1. Hypertension 34:430–434

Israeli RS, Powell CT, Fair WR et al (1993) Molecular cloning of a complementary DNA encoding a prostate-specific membrane antigen. Cancer Res 53:227–230

Jeunemaitre X, Soubrier F, Kotelevtsev YV et al (1992) Molecular basis of human hypertension. Role of angiotensinogen. Cell 71:169–180

Kahn D, Williams RD, Seldin DW et al (1994) Radioimmunoscintigraphy with 111-Indium labelled CYT-356 for the detection of occult prostate cancer recurrence. J Urol 152:1490–1495

Kahn D, Haseman MK, Libertino J et al (1997) Indium-111 capromab pendetide (Prostascint) imaging of patients with rising PSA post-prostatectomy. J Urol 158:157–204

Kahn D, Williams RD, Haseman MK et al (1998) Radioimmunoscintigraphy with ^{111}In labelled capromab pendetide predicts prostate cancer response to salvage radiotherapy after failed radical prostatectomy. J Clin Oncol 16:284–289

Kletter K, Nurnberger N (1989) Diagnostic potential of diuresis renography: limitations by the severity of hydronephrosis and by impairment of renal function. Nucl Med Commun 10:51–61

Kostadinova I, Simeonova A (1995) The use of 99mTc-EC Captopril test in patients with hypertension. Nucl Med Commun 116:128–131

Krushkal J, Ferrell R, Mockrin SC et al (1999) Genome wide linkage analysis of systolic blood pressure using highly discordant siblings. Circulation 99:1407–1410

Kunz R, Kreutz R, Beige J et al (1997) Association between the angiotensinogen 235T-variant and essential hypertension in whites: a systematic review and methodological appraisal. J Hypertens 30:1331–1337

Latif F, Tory K, Gnara J et al (1993) Identification of the von Hippel-Lindau disease tumor suppressor gene. Science 260:1317–1320

Levesque PE, Nieh PT, Zinman LN et al (1998) Radiolabelled monoclonal antibody Indium-111-labelled CYT-356 localises extraprostatic recurrent carcinoma after prostatectomy. Urology 51:978–984

Liddle GW, Bledsoe T, Coppage WS (1963) A familial renal disorder simulating primary aldosteronism but with negligible aldosterone secretion. Trans Assoc Am Physicians 76:199–213

Lifton RP (1995) Genetic determinants of human hypertension. Proc Natl Acad Sci USA 92:8545–8551

Lifton RP, Dluly RG, Powers M et al (1992) A chimeric 11 beta-hydroxylase/aldosterone synthetase gene causes glucocorticoid-remediable aldosteronism and human hypertension. Nature 355:262–265

Lindahl T (1993) Instability and decay of the primary structure of DNA. Nature 362:709–715

Liu H, Moy P, Kim S et al (1997) Monoclonal antibodies to the extracellular domain of Prostate-specific membrane antigen also react with tumor vascular epithelium. Cancer Res 57:3629–3634

Liu H, Rajasekaran AK, Moy P et al (1998) Constitutive and antibody induced internalisation of Prostate-specific membrane antigen. Cancer Res 58:4055–4060

Lopes AD, Davis WL, Rosenstraus MJ et al (1990) Immunohistochemical and pharmacokinetic characterisation of the site specific immunoconjugate CYT 356 derived from antiprostate monoclonal antibody 7E11C5. Cancer Res 50:6423–6429

Luke RG, Kennedy AC, Briggs JD et al (1968) Result of nephrectomy in hypertension associated with unilateral renal disease. Br Med J 3:764–768

Lynch HT, Lynch JF (1995) Clinical implications of advances in molecular genetics of colorectal cancer. Tumori 81 [Suppl]:19–29

MacGregor GA, Markandu ND, Roulston JE (1981) Maintenance of blood pressure by the renin-angiotensin system in normal man. Nature 291:329–331

Maguire RT, Pascucci VL, Maroli AN et al (1993) Immunoscintigraphy in patients with colorectal, ovarian and prostate cancer: results with site-specific immunoconjugates. Cancer 72:3453–3462

Mansfield TA, Simon TB, Farfel Z et al (1997) Multilocus linkage of familial hyperkalaemia and hypertension, pseudohypoaldosteronism type II, to chromosomes 1q31–42 and 17p11-q21. Nature Genet 16:202–205

Manyak MJ (1998) Clinical applications of radio-immunoscintigraphy with prostate specific antibodies for prostate cancer. Cancer Control 5:493–499

Mardirossian G, Brill AB, Dwyer KM et al (1996) Radiation absorbed dose from Indium-111-CYT 356. J Nucl Med 37:1583–1588

Miles KA (1991) Measurement of tissue perfusion by dynamic computed tomography. Br J Radiol 64:409–412

Miles KA, Hayball M, Dixon AK (1994) Functional imaging of changes in human intrarenal perfusion using quantitative dynamic computed tomography. Invest Radiol 29:911–914

Mune T, White PC (1996) Apparent mineralocorticoid excess: genotype is correlated with biochemical phenotype. Hypertension 27:1193–1199

Murphy GP, Maguire RT, Rogers B et al (1997) Comparison of serum PSMA, PSA levels with results of Cytogen-356 Prostascint scanning in prostatic cancer patients. Prostate 32:281–285

Nakajimi T, Cheng T, Rohrwasser A et al (1999) Functional analysis of a mutation occurring between two in-frame AUG codons of human angiotensinogen. J Biol Chem 274:35749–35755

Nally JW, Chen C, Fine E et al (1995) Diagnostic criteria of renovascular hypertension with Captopril renography. Am J Hypertension 4:S749–S752

Naseem MS, January HA, Britton KE et al (1998) Splenic metastases from adenocarcinoma of the prostate. Br J Urol Int 82:131–138

Nelson JB, Carducci MA (2000) The role of Endothelin-1 and Endothelin receptor antagonists in Prostate cancer. Br J Urol Int 85 [Suppl 2]:45–48

Nimmon CC, Britton KE, Gruenewald S et al (1982) Intrarenal distribution of blood flow: development of a gamma camera technique using ^{123}I orthoiodohippurate, OIH, and its validation in man. In: Joekes AM, Constable AR, Brown NJG et al (eds) Radionuclides in Nephrology. Academic, London, pp 55–63

Nimmon CC, Samal M, Backfrieder W et al (2000) An improved method for the determination of the intrarenal transit time distribution. Eur J Nucl Med 27:994 (abstract)

O'Byrne S, Caulfield M (1998) Genetics of hypertension: therapeutic implications. Drugs 56:203–214

Oh BR, Sasaki M, Perinchery G et al (2000) Frequent genotype changes at –308 and 488 regions of the Tumour Necrosis Factor-a (TNF-a) gene in patients with prostate cancer. J Urol 163:1584–1587

Oksche A, Rosenthal W (1998) The molecular basis of nephrogenic diabetes. J Mol Med 76:326–337

O'Reilly P, Aurell M, Britton KE et al (1996) Consensus in diuresis renography. J Nucl Med 37:1872–1876

Ozker R, Onsel C, Kabasakal L et al (1994) Technetium 99m-N, N Ethylene dicysteine: a comparative study of renal scintigraphy with Technetium 99m MAG3 and iodine 131 OIH in patients with obstructive renal disease. J Nucl Med 35:840–845

Padhy AK, Solanki KK, Bomanji JB et al (1993) Clinical evaluation of 99mTc Diaminocyclohexane, a renal agent with cationic transport: results in healthy normal volunteers. Nephron 65:294–298

Paillard F, Chansel D, Brand E et al (1999) Genotype-phenotype relationships for the renin-angiotensin-aldosterone system in a normal population. Hypertension 34:423–429

Perinchery G, Sasaki M, Angan A et al (2000) Deletion of Y-Chromosome specific genes in human prostate cancer. J Urol 163:1339–1342

Pickering G (1965) Hyperpiesis: high blood pressure without evident cause: essential hypertension. Br Med J 2:1021–1026

Pinto JT, Suffoletto BP, Berzin TM et al (1996) Prostate-specific membrane antigen: a novel folate hydrolase in human prostatic carcinoma cells. Clin Cancer Res 2:1445–1451

Platt R (1948) Severe hypertension in young persons: a study of 50 cases. Quart J Med 17:83–93

Polascik TJ, Manyak MJ, Haseman MK et al (1999) Comparison of clinical staging algorithms and 111-indium-capromab pendetide immunoscintigraphy in the prediction of lymph node involvement in high risk prostate carcinoma patients. Cancer 85:1586–1592

Poston L, Sewell RB, Wilkinson SP et al (1981) Evidence for a circulating sodium transport inhibitor in essential hypertension. Br Med J ii:847–849

Pras E, Aber NAK, Sentijevich I et al (1994) Localization of a gene causing cystinuria to chromosome 2p. Nat Genet 6:415–419

Ragde H, Elgamal A-AA, Snow PB et al (1998) Ten year disease free survival after transperineal sonography-guided I-125 brachytherapy with or without 45-Gray external beam irradiation in the treatment of patients with clinically localised, low to high Gleason grade prostate carcinoma. Cancer 83:989–1001

Rinker-Schaeffer CW, Hawkins AL, Su SL et al (1995) Localization and physical mapping of the prostate-specific membrane antigen (PSM) gene to human chromosome 11. Genomics 30:105–108

Rosenthal W, Seibold A, Antaramian A et al (1992) Molecular identification of the gene responsible for congenital nephrogenic diabetes insipidus. Nature 359:233–235

Schuster H, Wienker TE, Bahring S et al (1996) Severe autosomal dominant hypertension and brachydactyly in a unique Turkish kindred maps to human chromosome 12. Nature Genet 13:98–100

Shimkets RA, Warnock DG, Bositis CM et al (1994) Liddle's syndrome: Heritable human hypertension caused by mutations in the B subunit of the epithelial sodium channel. Cell 79:407–414

Shreve PD, Grossman HB, Gross MD et al (1996) Metastatic prostate cancer; initial findings of PET with 2-deoxy-2 (F-18) fluoro-d-glucose. Radiology 199:751–756

Sigurdsson S, Thorlacius S, Tomasson J et al (1997) BRCA2 mutation in Icelandic prostate cancer patients. J Mol Med 75:758–761

Smith HW (1951) The kidney: structure and function in health and disease. Oxford University Press, Oxford

Smith HW (1956) Unilateral nephrectomy in hypertensive disease. J Urol 76:685–701

Smith-Jones PM, Vallabhajosula S, Navarro V et al (2000) Y-90 huJ591 Mab specific to PSMA: Radioimmunotherapy RIT studies in nude mice with prostate cancer LN CaP tumor. Eur J Nucl Med 27:951 (abstract)

Sonmezoglu K, Erdil TY, Demir M et al (1998) Evaluation of renal function in low dose Cyclosporine treated patients using Technetium-99m diamino cyclohexane: a cationic tubular excretion agent. Eur J Nucl Med 25:1630–1636

Stalteri MA, Mather SJ, Belinka BA et al (1997) Site specific conjugation and labelling of prostate antibody 7E11-C5.3 (CYT-351) with Technetium-99m. Eur J Nucl Med 24:651–654

Strup SE, Pozzatti RO, Florence CD et al (1999) Chromosome 16 Allelallelic loss analysis of a large set of micro dissected prostate carcinomas. J Urol 162:590–594

Taylor A, Nally JV (1995) Clinical applications of renal scintigraphy. Am J Radiol 164:31–41

Taylor A, Nally JV (1996) Consensus report on ACE inhibitor renography for the detection of renovascular hypertension. J Nucl Med 37:1876–1882

Thurau K, Dahlheim H, Gruner A et al (1972) Activation of renin in the single juxta medullary apparatus by sodium chloride in the tubular fluid at the macula densa. Circ Res 31 [Suppl II]:182–186

Tobain L, Binion JT (1952) Tissue cations and water in arterial hypertension. Circulation 5:754–758

Troyer JK, Feng Q, Beckett ML et al (1995) Biochemical characterisation and mapping of the 7E11C5 epitope of the prostate specific membrane antigen. Urol Oncol 1:29–37

Vihko P, Heikkila J, Konturri M et al (1984) Radioimaging of the prostate and metastases of prostatic carcinoma with 99m-Tc labelled prostatic acid phosphatase specific antibodies and their Fab fragments. Ann Clin Res 16:51–52

Ward R (1990) Familial aggregation and genetic epidemiology of blood pressure. In: Laragh JH, Brenner BM (eds) Hypertension: pathophysiology, diagnosis and management, vol 1. Raven, New York, pp 81–100

Watson JD, Crick DHC (1953) Molecular structure of nucleic acids: a structure for deoxyribonucleic acid. Nature 171:737–738

Watson RWG, Fitzpatrick JM (2000) Target sites for manipulating apoptosis in prostate cancer. Br J Urol Int 85 [Suppl 2]:38–44

Watt GCM, Harrap SB, Foy CJW et al (1992) Abnormalities of glucocorticoid metabolism and the renin-angiotensin system: a four-corners approach to the identification of genetic determinants of blood pressure. J Hypertens 10:473–482

Wilkinson SP, Smith IK, Clarke M et al (1977) Intrarenal distribution of plasma flow in cirrhosis as measured by transit renography: relationship with plasma renin activity and sodium and water excretion. Clin Sci Mol Med 52:469–475

Wilkinson SP, Bernardi M, Pearce PC et al (1978) Validation of 'transit renography' for the determination of the intrarenal distribution of plasma flow: comparison with the microsphere method in the anaesthetised rabbit and pig. Clin Sci Mol Med 55:277–283

Williams GH, Hollenberg NK (1991) Non-modulating hypertension: a subset of sodium sensitive hypertension. Hypertension 17 [Suppl I]:I-81–I-85

Williams GH, Dluhy RG, Lifton RP et al (1992) Non-modulation as an intermediate phenotype in essential hypertension. Hypertension 20:788–796

Wright EM (1994) Cystinuria defect expresses itself. Nature Genet 6:328–329

Wright GL, Haley C, Beckett ML et al (1995) Expression of prostate specific membrane antigen in normal benign and malignant prostate tissues. Urol Oncol 1:18–29

Wynant GE, Murphy GP, Horoszewicz JS et al (1991) Immunoscintigraphy of prostate cancer: preliminary results with ^{111}In labelled monoclonal antibody 7E11-C5.3 (CYT-356). Prostate 18:229–241

Xu J, Meyers D, Freije D et al (1998) Evidence for prostate cancer susceptibility locus on the X chromosome. Nature Genet 20:175–179

Zhao YY, Zhou J, Narayanan CS et al (1999) A Role of C/A polymorphism at –20 on the expression of human angiotensinogen gene. Hypertension 33:108–115

CHAPTER 26

Biomolecular Magnification Imaging of Musculoskeletal Diseases 26

Yong-Whee Bahk, Soo-Kyo Chung, June-Key Chung

Contents

26.1
Introduction

Yong-Whee Bahk, Soo-Kyo Chung

The human musculoskeletal system has four functions: (1) body structuring with weight bearing and protection against extraneous forces, (2) locomotion, (3) calcium storage and liberation, and (4) hematopoiesis. Of these, the structure or anatomy and the molecular profile or metabolism of bone calcium can be imaged using ^{99m}Tc methyl diphosphonate (MDP) or hydroxydiphosphonate (HDP), calcium salt analogues, and hematopoiesis can be graphically assessed using ^{52}Fe, ^{59}Fe, ^{99m}Tc nanocolloid, and ^{99m}Tc-labeled anti-NCA95 antibody. Bones are hardened with calcium salts and, under the influence of calcitonin, bone calcium is mobilized into general circulation to maintain homeostasis. Being closely integrated with other tissues in general calcium metabolism, bone serves as the largest reservoir of calcium in the human body (97%) (Williams et al. 1989). The mobilization of calcium from bone results in decalcification that occurs in various conditions such as immobilization, inflammation, arthritis, osteoporosis, renal osteodystrophy, and reflex sympathetic dystrophy. Actually, live bones are ceaselessly engaged with the deposition and removal of calcium salts in the form of bone production and resorption mediated through the activities of osteoblasts and osteoclasts. Altered calcium metabolism, either local or systemic, can be assessed using radiography, CT, MRI, scintigraphy, and neutron activation analysis. However, bone scintigraphy can uniquely image both anatomy and molecular or metabolic profile at the same time (Holms 1978; Smith 1986; Bahk 2000; Etchebehere et al. 2001). In addition, denatured muscle can also be imaged by ^{99m}Tc-MDP or -HDP scan. Bone marrow is the largest hematopoietic organ in the human producing erythropoietic precursor cells, granulocytes, and reticuloendothelial cells. Each of these cells can be separately imaged using appropriate radiopharmaceuticals. Bone marrow scan will be discussed in detail under a separate section. From the view point of molecular nuclear medicine and for the sake of a categorical description, it seems warranted to classify skeletal disorders into two major groups. The first group consists of disorders that are associated with genetic imbalance or heredity and the second group consists of disorders that are mere histopathological entities in nature with no known association with genome problems. Disorders in the first group, clinically by far less common in occurrence than the second group, result from autosomal or sex chromosomal imbalance or mutations. Well known autosomal and chromosomal disorders include Turner syndrome, Klinefelter syndrome, and trisomy defects. Mucopolysaccharidoses and osteochondrodysplasias are other major groups of genetic disorders. The former disorders, caused by genetically determined deficiencies of lysosomal enzymes that degrade mucopolysaccharides (McAlister and Herman 1995), include Hurler's disease, Hunter's syndrome, and Morquio's disease and the latter Marfan's syndrome, osteopetrosis, osteogenesis imperfecta (Goldman 1995), multiple cartilaginous exostoses, and others. Certain skeletal disorders are known to be the result of the action of several different genes and hence referred to as polygenic disorders. Rheumatoid arthritis, ankylosing spondylitis, and Reiter's syndrome belong to this category. These disorders can be assessed by antigen test, the HLA-B27 antigen in particular (Morris et al. 1974; Kahn 1988), and constitute excellent indications for bone scintigraphy (Kim et al. 1999; Bahk 2000).

With widespread use of antigen testing and gene typing, an increasing number of skeletal disorders have become recognized as having a genetic background. Some examples are reactive arthritis (Leirisalo et al. 1982), diffuse idiopathic skeletal hyperostosis (Shapiro et al. 1976), slipped capital femoral epiphysis (Mullaji et al. 1993), and hydroxyapatite crystal deposition disease (Pinals and Short 1965; McCarty and Gatter 1966; Amor et al. 1977). The second category of skeletal disorders consists of simple histopathologic entities that have no known cause-and-effect relationship with genetic abnormality. Disorders in this category include infection, nonspecific inflammation, osteochondroses, avascular bone necrosis, trauma, recreational injuries, periarticular soft-tissue rheumatism, muscular rheumatism syndromes, metabolic-nutritional disorders, tumors, and tumorous conditions. Clinically, these disorders constitute major indications for bone scintigraphy and hence make up the main body of current presentation.

26.2
Essential Anatomy of the Musculoskeletal System

The musculoskeletal system consists of bone, joint, and muscle with the musculotendinous unit. Bone is engaged in complex biomechanical, metabolic, and endocrine activities like the brain, heart, and tumors. Bone tissues are produced, maintained, and eliminated by the ceaseless osteoblastic and osteoclastic activity. Principal roles played by these cells are to maintain bone integrity and body calcium homeostasis in general by balancing between the ratios of production and resorption of collagenous matrices and governing mineralization processes. Collagen production is common to various connective tissues but mineralization is unique to bone cells. Skeletal muscles are rich in actin and myosin whose interactions effectuate contraction (Williams et al. 1989). It is composed of a large number of muscle fibers (cells). Muscle fibers, individually invested by the endomysium, are grouped in fascicles enveloped in successive connective tissue sheaths. Variable numbers of fascicles compose a skeletal muscle that is ensheathed by the epimysium. Tendon is a specialized connective tissue that unites to muscle bellies forming the musculotendinous unit on one side and attaches to the periosteum, fibrous capsule of joint, or directly to bone on the other side.

26.3
Imaging Modalities for the Musculoskeletal System

As mentioned above, four imaging modalities are currently in use for the diagnosis, treatment, monitoring, and clinical research of musculoskeletal diseases. They are radiography, CT, MRI, and scintigraphy. The first three modalities are excellent means of defining the anatomy. Radiography is valued for its high spatial resolution (5–10 lp/mm), CT for the transaxial sectional image of bone and joint, and MRI for the three-dimensional image of the bone marrow, articular cartilage, muscle, tendon, and ligament. On the other hand, ^{99m}Tc-MDP or -HDP bone scintigraphy can portray not only the anatomy but uniquely the metabolic or molecular profile in a number of musculoskeletal disorders (Kim et al. 1999; Bahk 2000). Using the pinhole magnification technique, the spatial resolution of bone scintigraphy is raised up to those of CT and MRI (2 lp/cm) significantly refining the image quality. In addition, the three-phase bone scintigraphy provides the information about vascularity and single photon emission computed topography (SPECT) can separate the plane of interest from overlapping structures enhancing the contrast. Recently introduced pinhole-SPECT, a hybrid of the magnification technique and SPECT, enhances both spatial resolution and contrast at the same time (Bahk et al. 1998 a).

26.4
Advantages of Nuclear Bone Scintigraphy

As already mentioned, bone scintigraphy has two distinct advantages. The one is its capability to portray molecular or metabolic alterations in musculoskeletal diseases and the other is exquisite diagnostic sensitivity. For example, bone scintigraphy can often detect acute infections of bone and joint days before, and malignant metastasis weeks before, radiographic signs and clinical symptoms appear. Covert fracture and contusion, especially in porotic and anatomically complex bone, shin splints and avascular osteonecrosis are other important indications for bone scan. It is an excellent diagnostic tool for rhabdomyolysis, myositis, and musculotendinous unit strain. Pinhole scan greatly enhances diagnostic acumen facilitating differential diagnosis, for example, of various spinal diseases (Bahk et al. 1987) and showing diagnostic signs in a number of skeletal diseases. Some typical instances include a "hotter spot within hot area" appearance of osteoid osteoma (Kim et al. 1992), the "C" or "inverted C" sign of Tiezte's disease (Yang et al. 1994), and enthesopathies in Reiter's syndrome (Kim et al. 1999). Its usefulness in pediatric bone diseases has been amply documented (Treves et al. 1995; Connolly et al. 1998; Etchebehere et al. 2001).

On the other hand, scintigraphy using ^{67}Ga citrate or ^{111}In- or ^{99m}Tc-labelled granulocytes is a valuable adjunct to the diagnosis of infective diseases of the musculoskeletal system. ^{67}Ga and ^{201}Tl can be used for the imaging of osteosarcoma and Ewing's sarcoma (Nadel 1993) and ^{67}Ga for the imaging of soft tissue tumors such as rhabdomyosarcoma (Kaufman et al. 1977). In addition, the quantitation of scintigraphic findings has been proposed by Pitt and Sharp (1985). This method is based on the calculation of the bone-to-soft tissue, bone-to-bone, and bone-to-lesion ratios of radioactivity. The clearance of ^{99m}Tc-MDP or -HDP, photon absorptiometry, and quantitative bone scintigraphy can be applied to the study of osteoporosis and osteomalacia.

26.5
Bone Adsorption of ^{99m}Tc Phosphates and Phosphonates

The mechanism with which ^{99m}Tc phosphate and phosphonate are deposited in bone is not fully understood. It is known, however, that the deposition is strongly influenced by the metabolic activity, vascularity, surface bone area available to extracellular fluid, and calcium content of bone. Metabolically active and highly vascularized metaphysis retains 1.6 times more ^{99m}Tc than the less active diaphysis of long bone (Silverstein et al. 1975). Thus, as emphasized, bone scintigraphy is well suited to the imaging of vascularity- and metabolism-related alterations in normal growing bone as well as morbid bones. Typical clinical situations are bone metastasis, fracture, infection, rickets, Paget's disease of bone and renal osteodystrophy. Another influential factor is the nature of the calcium phosphate existing in bone, which is indicated by the Ca/P molar ratio. Francis et al. (1980) experimentally showed that diphosphonates were more avidly adsorbed to immature amorphous calcium phosphate (Ca/P = 1.35) than to mature hydroxyapatite crystal (Ca/P = 1.66). The low Ca/P salt exists in rapidly calcifying front of the osteoid matrix in the physis of growing long bone, whereas crystalline hydroxyapatite exists in the cortex. Various theories were proposed regarding the site of phosphate deposition. Jones et al. (1976) suggested that a small amount of phosphate chemisorbs at kink and dislocation sites on the surface of hydroxyapatite crystal, whereas Rosenthall and Kaye (1975) pointed at the organic matrix to be a calcium salt deposition site. According to Francis et al. (1981) diphosphonates were laid down almost exclusively on the surface of inorganic calcium phosphate. Evidence in favor of this finding was provided by an autoradiographic study (Guillermart et al. 1980).

26.6
Radiopharmaceuticals for Musculoskeletal Scintigraphy

The clinically useful properties of technetium-99m (^{99m}Tc) were fully documented by Richards (1960) and Harper et al. (1965) and stannous triphosphate complex of ^{99m}Tc was first introduced by Subramanian and McAfee (1971) as a potent bone scan agent. Within a short period of time, ^{99m}Tc polyphosphate, pyrophosphate, and diphosphonate were developed in series. Chemically, the phosphate compounds contain multiple phosphate residues [P-O-P] with the simplest form being pyrophosphate that has two residues. Phosphonate is a compound with P-C-P bonds instead of P-O-P. Currently, diphosphonates are widely used as ^{99m}Tc methylene diphosphonate (MDP) and hydroxymethylene diphosphonate (HDP).

^{99m}Tc efficiently labels diphosphonates yielding ^{99m}Tc-MDP or -HDP that has a strong avidity for hydroxyapatite crystals in mineral phase, especially at the sites where new bone is actively laid down as in the physis of growing bone and fracture. Diphosphonates are rapidly distributed after intravenous injection in the extracellular fluid space and about half are taken up by bone and the remaining unfixed are excreted by glomerular filtration (Alazraki 1988). They are adsorbed onto calcium of hydroxyapatite (Fogelman and Carr 1980). The amount of radiopharmaceuticals accumulated in bone at 1 h after injection is 58% with MDP, 48% with HDP, and 47% with pyrophosphate (Davis and Jones 1976). Blood levels of HDP are 10% at 1 h after intravenous injection and fall to 6%, 4%, and 3% at 2 h, 3 h, and 4 h after the injection, respectively (Mallinckrodt Medical 1996).

An advantage of this preparation is that an optimum blood level can be reached already at 1 h after injection so that scanning can be started earlier without increasing the dosage of tracer.

The usual dose of tracer for adults is 740–1110 MBq (20–30 mCi). For obese adults, it may be increased to 11–13 MBq (0.3–0.35 mCi)/kg. For children, the optimal dose is 9–11 MBq (0.25–3.0 mCi)/kg, with a minimum of 40–90 MBq (1.1–2.4 mCi) (Donohoe et al. 1998). The maximum dose for children should not exceed the dose for an adult. In children, the administered dose can be scaled on the basis of body surface area (Pediatric Task Group ENMA 1990). The scan agents are subject to oxidation, hence care should be exercised to avoid introducing air into multidose vials. Quality control is a must before the administration of radiopharmaceutical.

Radiopharmaceuticals used for the imaging of skeletal muscle include ^{67}Ga citrate and ^{201}Tl chloride. ^{67}Ga citrate is used for the imaging of soft-tissue sarcoma (Kaufman et al. 1977) and ^{201}Tl for pediatric tumors (Nadel 1993). Bone-seeking radiopharmaceuticals such as ^{99m}Tc-MDP and -HDP are also avidly concentrated by acutely necrotized muscle (Brill 1983) and used for the diagnosis of myolysis (Suzuki et al. 1974; Bahk 2000). The mechanism with which bone seekers accumulate in necrotic muscle is explained by the influx of calcium ions and subsequent deposition as in myocardial necrosis (Werner et al. 1977).

26.7
Methodological Considerations

Scintigraphic methods utilized for the imaging of the musculoskeletal system include planar whole-body scintigraphy, planar spot scintigraphy, planar pinhole scintigraphy, planar SPECT, and pinhole SPECT. In addition, ^{111}In- and ^{99m}Tc-leukocyte and ^{67}Ga citrate scintigraphy are used for the diagnosis of skeletal infections.

26.7.1
Planar Whole-Body Bone Scintigraphy

Whole-body bone scintigraphy is unique in that it can visualize the skeleton in its entirety on a single pair of the anterior and posterior views (Fig. 26.1). It

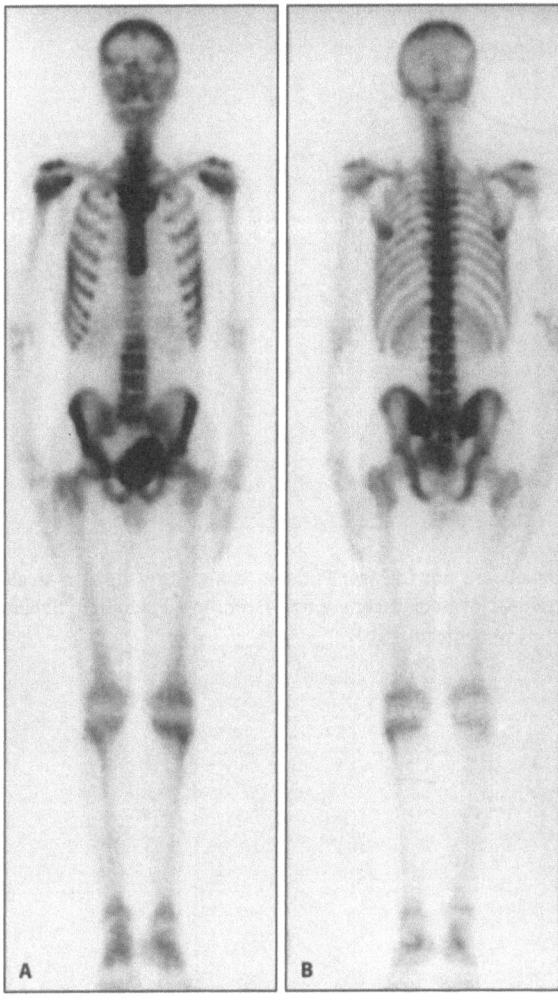

Fig. 26.1 A, B. Normal anterior (A) and posterior (B) whole-body bone scans show the entire skeleton from vertex to toes. Note dynamic metabolic or molecular representation of the skeleton with accentuated ^{99m}Tc diphosphonate uptake in the axial and periarticular bones that are under constant stress and strain

is thus extremely useful for the assessment of multifocal or systemic skeletal diseases. The imaging can be started as early as 1.5 h after the intravenous administration of tracer when ^{99m}Tc-HDP is used. Such an early commencement of scanning yields more photons per injected unit dose of tracer improving the system sensitivity.

26.7.2
Planar Spot Bone Scintigraphy

Planar spot scintigraphy covers a portion of the skeleton, usually the region of interest (Fig. 26.2 A). Baseline scan is the anterior or posterior view, and special view(s) can be added so that the pathology in question is visualized most advantageously. The lateral, oblique, tangential, and tilted views are commonly used as additions and modifications. Generally, the resolution of the planar view is not high enough to portray fine anatomy.

26.7.3
Three-Phase Bone Scintigraphy

Three-phase bone scintigraphy is a noninvasive means of semiquantitative assessment of vascularity in musculoskeletal disorders, inflammatory diseases in particular. It can efficiently distinguish bone infection from cellulitis. Methodologically, it consists of imaging of arterial blood flow, blood pool, and bone uptake. The blood flow images are taken in rapid sequence after the bolus injection of tracer. A recommended imaging scheme is immediate post-bolus-injection angiography (16 serial frames of 2- to 4-s images), blood-pool scan within 10 min post-injection, and delayed static bone imaging at 1.5–4 h or 24 h post-injection (Fig. 26.3 A–C).

26.7.4
Planar Pinhole Bone Scintigraphy

Pinhole scintigraphy is either planar or tomographic in mode. The planar pinhole bone scan can be performed using any single-head gamma camera system by simply replacing multihole collimator with a pinhole collimator. It enhances the resolution through optical magnification (Treves et al. 1995; Bahk 2000). It can generate truly magnified images of any selected bone and/or joint in any desired projections (Fig. 26.2 B). The dual-head gamma camera can also be used for the same purpose (Bahk et al. 1998b). Dual-head pinhole scanning simultaneously produces a pair of highly magnified images in the anterior and posterior projections or the medial and lateral projections. This mode can eliminate the blind zone that inevitably exists in the background of a single-head

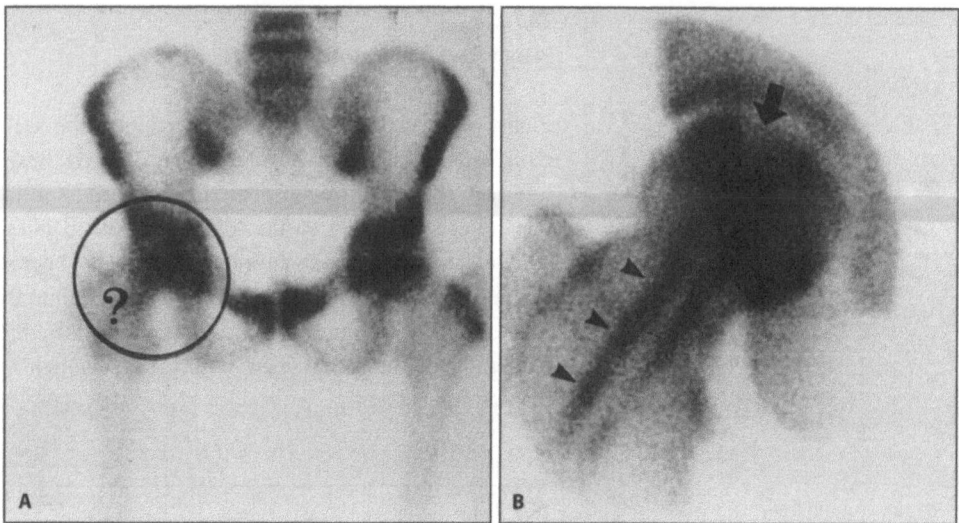

Fig. 26.2. Anterior planar bone scan of avascular necrosis in the right femoral head treated with vascularized fibular graft shows irregularity (?) of the right hip joint (A). Details are not imaged. Pinhole scan clearly shows a small residual avascular zone (*arrow*) and well adopted fibular graft (*arrowheads*) (B)

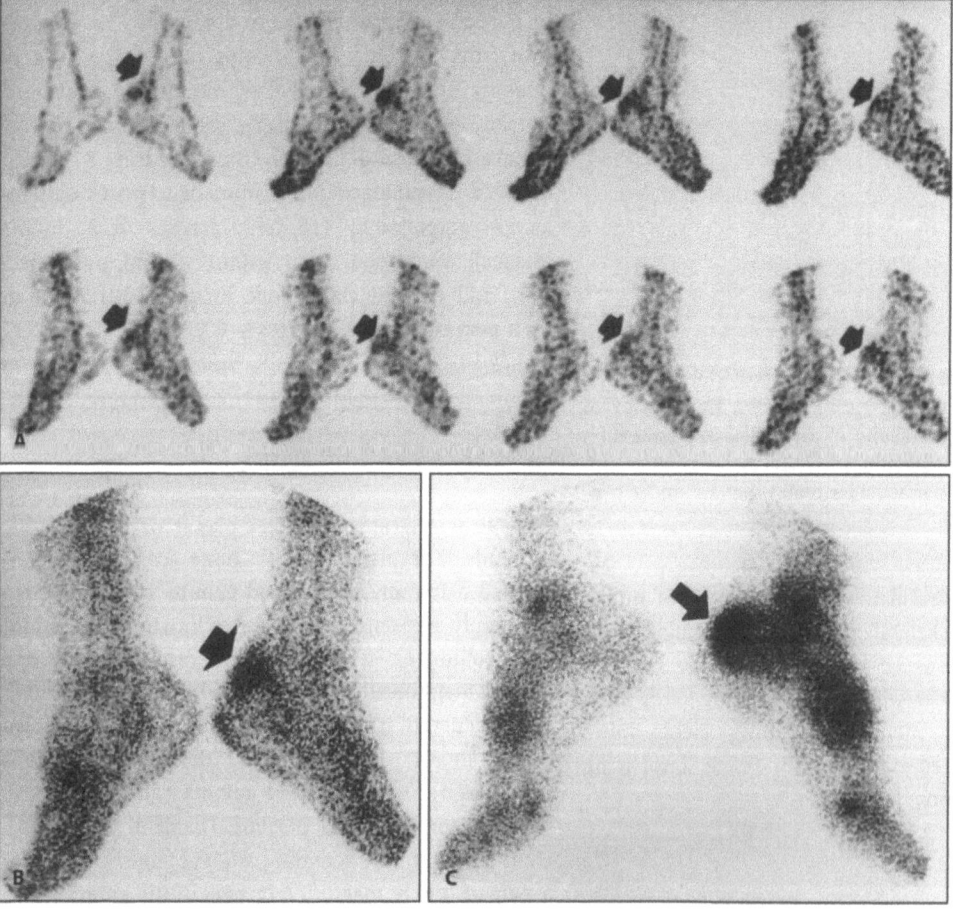

Fig. 26.3 A–C. Three-phase bone scintigraphy of acute osteomyelitis of the calcaneus. Serial blood flow images show increased arterial flow in the posterior aspect of the left ankle (A). Blood pool (B) and bone uptake (C) are also markedly increased in retrocalcaneal infection

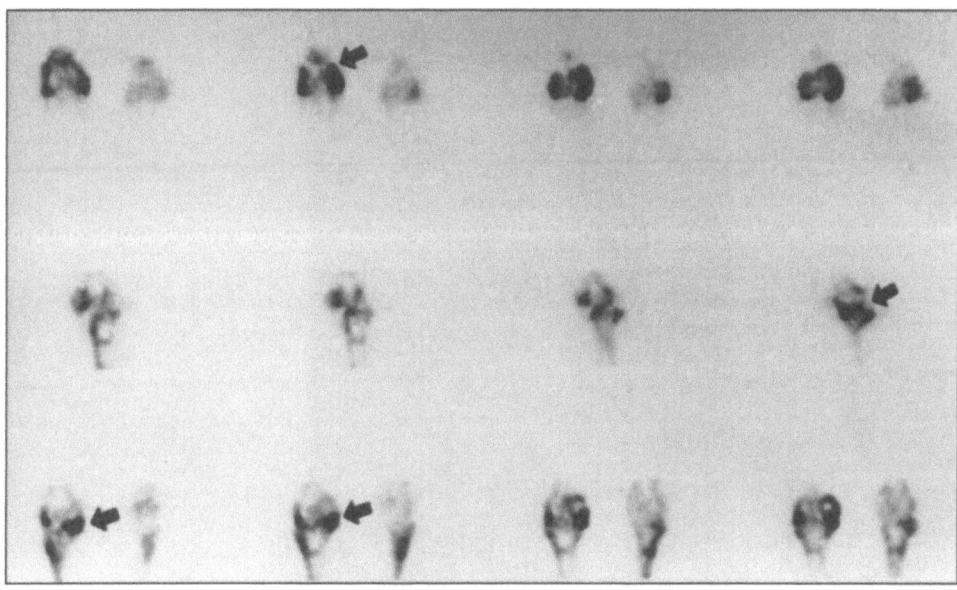

Fig. 26.4. SPECT of osteoarthritis of the right knee shows irregular small spotty hot areas with narrowed articular space (*arrows*). Transaxial (*top panel*), sagittal (*middle panel*), and coronal view (*bottom panel*). Scan contrast is improved but spatial resolution remains low

pinhole scan. The time required for pinhole scanning is 15–20 min when a pinhole collimator having a 4-mm aperture is used. The value of the spatial resolution of a pinhole scan is 2 line pairs/cm, that is two times greater than the 1 line pair/cm of plain planar scintigraphy. It is closer to those of CT scan and MR imaging (Huda and Slone 1995).

26.7.5
Planar Single Photon Emission Computed Tomography (SPECT) of Bone

Planar SPECT produces sectional images of the skeleton. Sectioning separates a selected plane or small volume of an organ from others that overlie or underlie it, enhancing the image contrast up to six-fold (Jaszczak et al. 1977). SPECT can also be used for the quantification of radioactivity distribution. Images of SPECT can be threedimensionally reconstructed. It is, however, to be noted that SPECT does not improve the spatial resolution. The detector of SPECT may be single, dual, or triple. SPECT is utilized for the diagnosis of complex bone structures such as the spine, pelvis, and hip. Typical applications include low back pain and hip and knee joint diseases (Fig. 26.4). Generally, it is not suitable for the imaging of small, flat, or thin bones.

26.7.6
Pinhole Single Photon Emission Computed Tomography (SPECT) of Bone

Pinhole SPECT, a hybrid of pinhole scan and SPECT, can generate sectional images that are optically magnified. This hybrid technique enhances both spatial resolution and image contrast, greatly improving diagnostic feasibility (Bahk et al. 1998a). It can be achieved using a single-head SPECT gamma camera system that is provided with the filtered backprojection algorithm and a Butterworth filter. The resolution of pinhole SPECT is comparable to that of CT (Fig. 26.5). Because of the limited range of the detector gyration of the gamma camera system available, this new technique can be used only for the imaging of the distal appendicular skeleton such as the ankle and wrist.

26.7.7
^{111}In- and ^{99m}Tc-Leukocyte and ^{67}Ga Citrate Scintigraphy in Skeletal Infections

As mentioned, acute uncomplicated infection of bone and joint can be detected by ^{99m}Tc-MDP or -HDP scintigraphy reinforced with three-phase bone scan

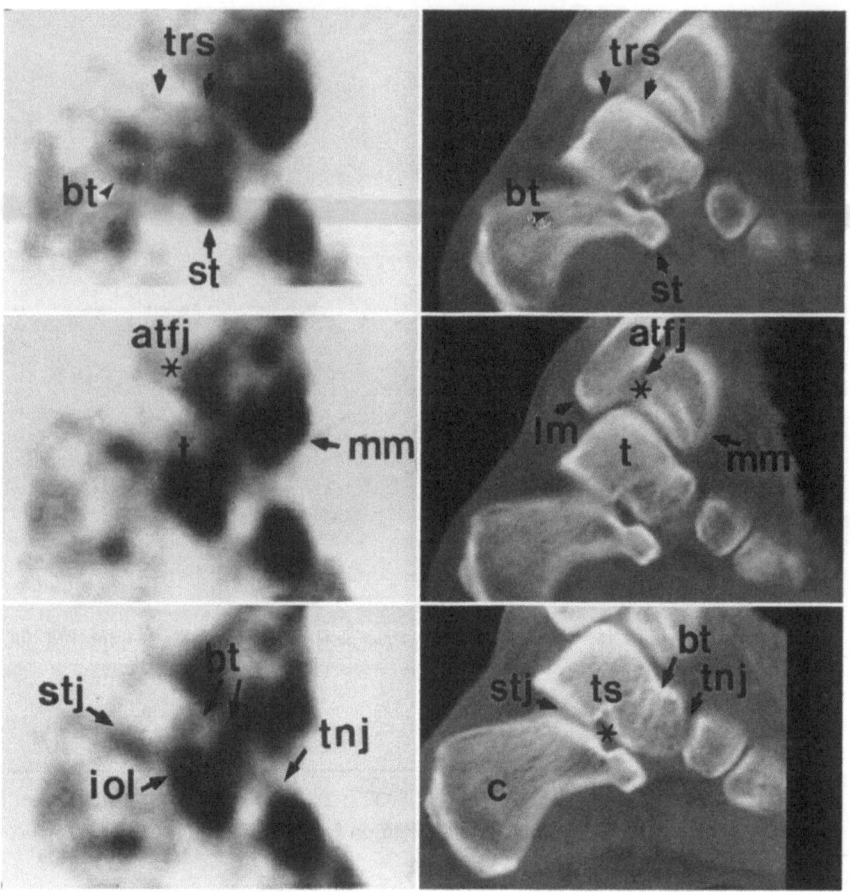

Fig. 26.5. Sagittal pinhole SPECT (*left column*) of a normal ankle shows markedly enhanced resolution which is well comparable to that of CT (*right column*). *trs, bt, st, atfj, mm, t, iol, tnj, c,* and *ts* represent trochlear surface of the t alus, bone trabeculae, sustentaculum tali, anterior tibiofibular joint, medial malleolus, talus, interosseous ligament, talonavicular joint, calcaneus and tarsal sinus, respectively

and pinhole technique. However, white blood cells labeled with ^{111}In or ^{99m}Tc are used for the specific diagnosis of bone infection that is superimposed on fracture, operative wound, or prosthesis. One latest clinical study has confirmed the accuracy of ^{111}In-leukocyte scintigraphy in the diagnosis of post-traumatic and postoperative cranial and spinal infections (Medina et al. 2000). In order to replace expensive ^{111}In with readily available ^{99m}Tc, leukocytes are labeled with hexamethyl-propylene amine oxime (HMPAO). The image quality of ^{99m}Tc-HMPAO leukocyte scan is acceptable (Fig. 26.6) and has been shown to be superior to that of ^{111}In-leukocyte scan at least in intestinal imaging (Peters 1994). Unfortunately, however, ^{99m}Tc-HMPAO leukocytes avidly accumulate in the axial skeleton even in normal state, detracting from its value. A ^{99m}Tc-HMPAO leukocyte scan is generally more suited for rapid screening of acute infection, whereas ^{111}In-leukocyte scan is more suited for chronic infection. [67]Ga citrate indiscriminately accumulates both in infection and tumor. Some radiolabeled antibodies such as ^{99m}Tc-antigranulocyte antibodies were tested earlier only to find out that they were neither reliable nor advantageous (Hotze et al. 1992; McAfee et al. 1991).

26.8
Clinical Applications of Bone Scintigraphy

As mentioned earlier, the spectrum of clinical applications of bone scintigraphy has become remarkably broadened and versatile. Bone scintigraphy is now clinically used for the diagnosis of not only bone infections, traumatic injuries, metabolic disorders, tu-

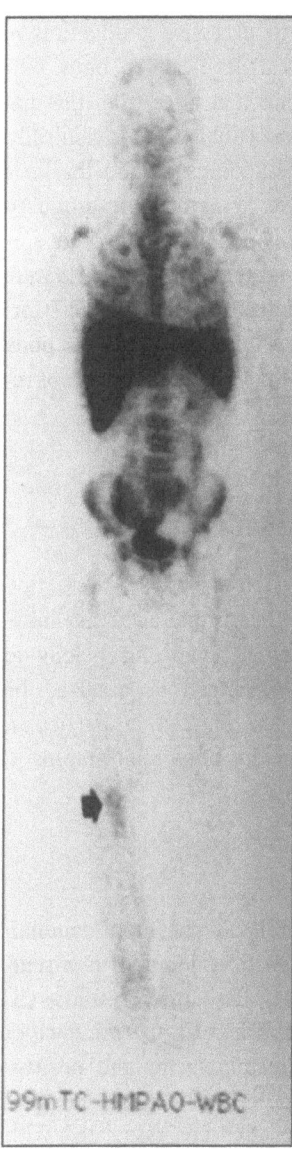

Fig. 26.6. Anterior whole-body ^{99m}Tc-HMPAO-labelled white-blood-cell scan shows selective leukocyte uptake in an infective focus in the right proximal tibia (*arrow*)

mors, and arthritides but for the diagnosis and documentation of a number of autosomal and polygenic disorders of the skeleton.

26.8.1
Autosomal and Polygenic Skeletal Disorders

The widespread use of antigen typing and the development of genome techniques identifies an increasing number of skeletal disorders that are heritable or as-

sociated with the imbalance of multiple genes. This section describes an attempt made at the scintigraphic representation of some relatively common autosomal and polygenic skeletal disorders within space limitation.

26.8.1.1
Autosomal Skeletal Disorders

Representative autosomal skeletal disorders are osteochondrodysplasias and dysostoses. Of a long list of these disorders, we have so far personally performed pinhole scan study only in Albers-Schönberg's disease and multiple cartilaginous exostoses.

26.8.1.1.1
Albers-Schönberg's Disease

Albers-Schönberg's disease is benign autosomal dominant osteopetrosis. There is also renal tubular acidosis and usually some mental retardation associated with cerebral calcification. The cause of the disease is traced to a deficiency of the isozyme, carbonic anhydrase (CA) II. Four different mutations have been identified in the CA II gene, which maps to chromosome 8q22 (Sly and Hu 1995). The disease is also called chalk bones or marble bone disease, and is characterized by the persistent existence of primary spongiosa that has failed to be eliminated by absorption. Pathological features include systemic osteosclerosis with thickened cortex and obliterated marrow, defective tubulation, and metaphyseal clubbing. The condition is usually asymptomatic and often detected accidentally by pathologic fracture. Bone scintigraphy is useful for panoramic portrayal of systemic involvement with preferential high tracer uptake in the metaphyses of tubular bones (Fig. 26.7 A). Scintigraphy is also valuable in detecting fracture and superimposed bone infection (Park and Lambertus 1977). Pinhole scan is characterized by the clubbing with intense tracer uptake of tubular bone ends and the flaring of vertebral endplates with an anvil appearance (Fig. 26.7 B).

26.8.1.1.2
Multiple Cartilaginous Exostoses

Multiple cartilaginous exostoses, also called hereditary multiple exostoses and diaphyseal aclasia, is a fairly common benign autosomal dominant disease of scintigraphic interest. There is genomic mapping of different

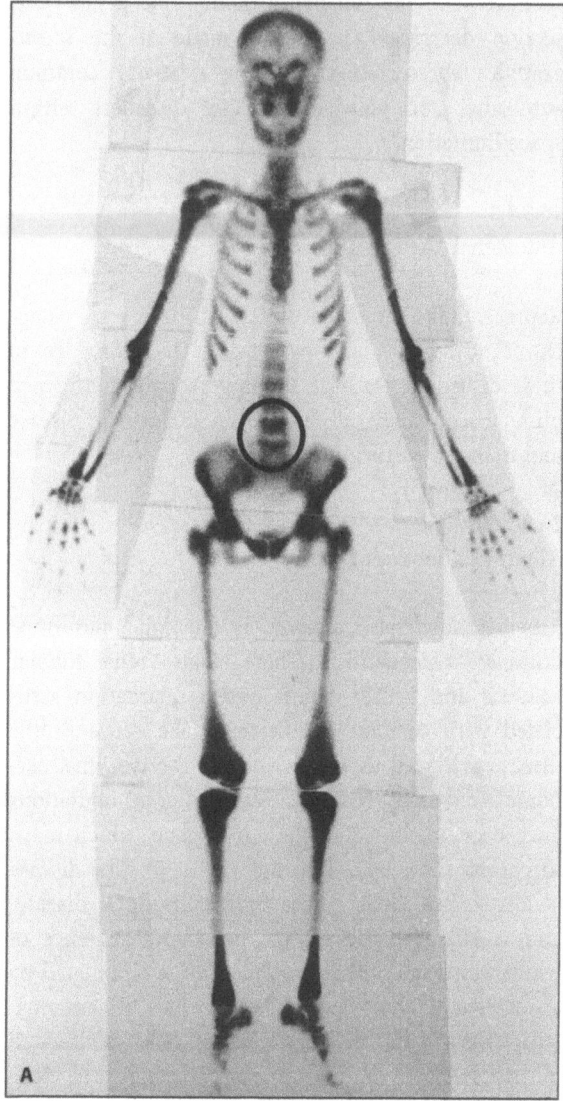

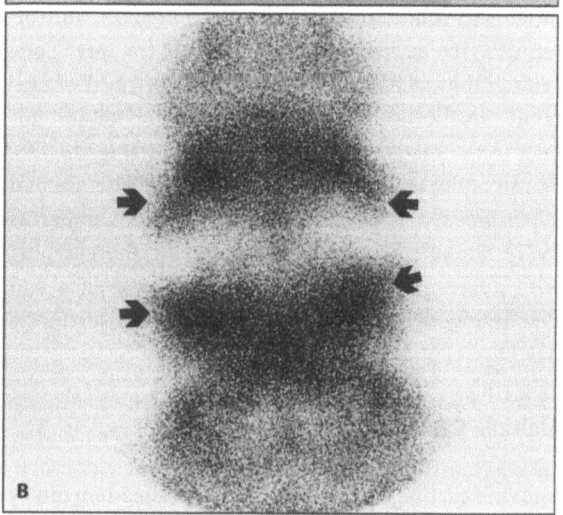

types to chromosomes 8, 11, or 19 (Sly and Hu 1995). This disease is ascribed to excessive production of spongiosa with resultant cartilage-capped bony outgrowths or osteochondromas that arise from the diaphyseal aspect of the physis (Rubin 1964). Clinically, exostoses present as painless lumps around the knee and elbow typically in boys. Symptoms are produced when exostoses mechanically compress or irritate surrounding vessels, nerves, and tendons. About 3% were estimated to undergo malignant transformation (Gordon et al. 1981). Bone scan is useful for the whole-body mapping and assessment of the metabolic profile of individual lesions (Fig. 26.8).

26.8.1.2
Polygenic Skeletal Disorders

Polygenic skeletal disorders are caused by the action of several different genes. Rheumatoid arthritis, ankylosing spondylitis, and Reiter's syndrome belong to this category. These disorders can be assessed by antigen typing, the HLA-B27 antigen in particular, and are excellent indications for bone scintigraphy.

26.8.1.2.1
Rheumatoid Arthritis

Rheumatoid arthritis is probably the most common inflammatory disorder of synovial joints. It is a scintigraphically interesting articular disorder whose diagnosis can firmly be established by the presence of a high-titer positive rheumatoid factor and positive histocompatibility HLA-B27 or HLA-DR4 antigen. Different subtypes of the latter are associated with different racial groups. A concordance rate of 30% for monozygotic twins and 9% for dizygotic twins suggests a polygenic inheritance (Fugger et al. 1995). Women are affected two to three times more often than men. It may affect any synovial joints with a strong proclivity to small joints in the hands, wrists, and feet. Involvement is characteristically polyarticular and symmetrical on both sides of the body. Early

Fig. 26.7. Anterior whole-body bone scan in a 36-year-old woman with osteopetrosis shows increased tracer uptake in both axial and appendicular bones producing a superscan. Kidneys are not visible. Note that, unlike in metastasis, superscan in osteopetrosis uniformly involves both axial and appendicular bones (A). Pinhole scan of L4 and 5 portrays characteristic anvil appearance (B)

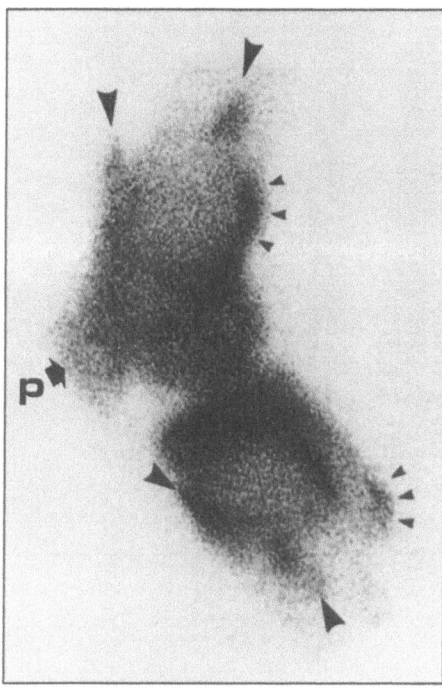

Fig. 26.8. Lateral pinhole scan of the right distal femur and proximal tibia in a 22-year-old woman with multiple exostoses shows bone outgrowths with increased tracer uptake in the tips

pathological alterations are synovial congestion, edema, and exudation. Hyperemia and disuse cause marked osteoporosis in periarticular bones. Granulation and pannus formation follow destroying cartilages and bones with resultant articular narrowing and deformity. Whole-body scan is an ideal means of portraying symmetrical and polyarticular involvement of this disease (Fig. 26.9 A). Pinhole bone scintigraphy is useful for the simultaneous assessment of anatomical and molecular alterations in great detail (Fig. 26.9 B) and three-phase scan provides the information regarding the vascularity that reflects disease activity. In the acute stage, vascularity becomes increased and in remission reverts to normal. Pinhole SPECT can show characteristic synovial alteration in rheumatoid arthritis (Bahk et al. 1998 a; Bahk 2000).

26.8.1.2.2
Ankylosing Spondylitis

Ankylosing spondylitis and Reiter's syndrome are seronegative spondyloarthropathies of genetic and scintigraphic interest. Both disorders are remarkably as-

sociated with HLA-B27 antigen. More than 90% of patients with ankylosing spondylitis possess HLA-B27 or HLA-DR4 antigen has also been shown to be associated with characteristic clinical features of this disorder (Kahn 1988; Fugger et al. 1995). Ankylosing spondylitis is a nonspecific inflammatory disease that involves the sacroiliac joints and spine. The disease starts from the synovial sacroiliac joint usually symmetrically on both sides and gradually spreads to the discovertebral, apophyseal, costovertebral, and neurocentral articulations of the thoracolumbar, lumbar, lumbosacral, thoracic, and cervical spine in ascending order. In the late stage, the annulus fibrosus, the anterior longitudinal ligaments, and the interspinous ligaments become ossified giving rise to a bamboo spine appearance. Whole-body bone scintigraphy shows increased tracer accumulation in the sacroiliac joints and syndesmophytes and ossified ligaments of the lower thoracic and lumbar spine diffusely obliterating the vertebral contour, intervertebral disc spaces, and facet joints giving rise to a bamboo appearance (Fig. 26.10). Bone scan can reveal the metabolic status of ankylosing disease. Thus, tracer uptake is increased in the early and intermediate stage when the disease is engaged with florid inflammation and reduced in the late ossifying stage when the disease is metabolically quiescent (Bahk 2000).

26.8.1.2.3
Reiter's Syndrome

Reiter's syndrome is another seronegative spondyloarthropathy of genetic and scintigraphic interest. As in ankylosing spondylitis, markedly increased frequency of antigen HLA-B27 is well recognized in this disorder. A frequency of 96% has been reported (Morris et al. 1974). Presence of HLA-B27 antigen predisposes to Reiter's syndrome after exposure to an infectious agent and clinical expression may be markedly influenced by genetic factors (Fugger et al. 1995). Clinically, this syndrome consists of the triad of urethritis, arthritis, and conjunctivitis. The disease mechanism is not yet established but possible interaction between several different infective organisms and a specific genetic background is being seriously considered. Enthesopathy appears to be a prominent feature (Groshar et al. 1997; Kim et al. 1999) (Fig. 26.11). Clinical manifestations include tendinitis, fasciitis, sausage digit, or paravertebral enthesitis alone or in combination. Spur and osteophytosis are common findings in the late stage. The skeletal in-

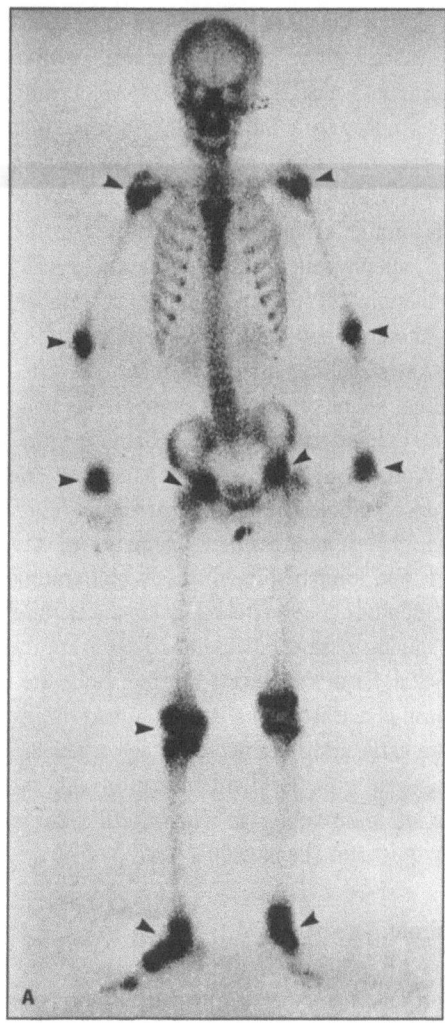

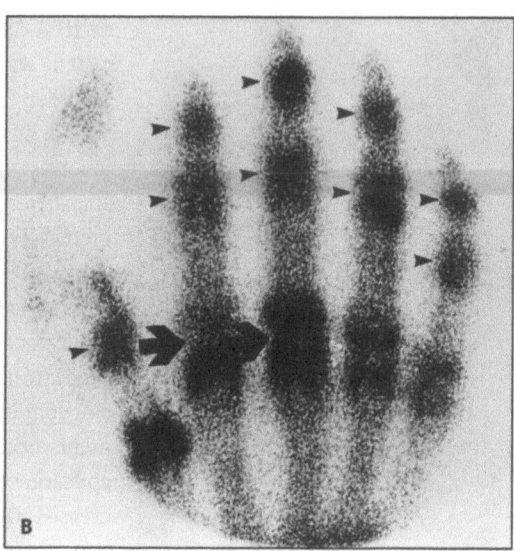

Fig. 26.9. Anterior whole-body scan of patient with rheumatoid arthritis shows panoramic display of symmetrical polyarthritis (A). Dorsal pinhole scan of the right hand portrays characteristic intense tracer uptake in para-articular bones that are the sites of inflammation and porosis (B)

volvement of Reiter's syndrome is typically asymmetrical and pauciarticular. Bone scintigraphy has been shown to be a sensitive detector of enthesopathy in the absence of radiographic alterations in 14.1% of cases (Kim et al. 1999).

26.8.1.3
Other HLA-Associated Skeletal Disorders

Other HLA-associated skeletal disorders of scintigraphic interest include reactive arthritis, diffuse idiopathic skeletal hyperostosis (Forestier's disease), slipped capital femoral epiphysis, and calcific periarthritis (hydroxyapatite crystal deposition disease).

26.8.1.3.1
Reactive Arthritis

Reactive arthritis, a well-known rheumatic disorder, is initiated or triggered by a variety of infective agents without direct articular contamination. Sacroiliitis in ankylosing spondylitis and Reiter's syndrome is a typical example of reactive arthritis. Leirisalo et al. (1982) reported that the sacroiliitis in Reiter's syndrome is more common in patients with positive HLA-B27 antigen. Bone scan shows tracer to intensely accumulate in the sacroiliac joints, characteristically in the synovial compartment.

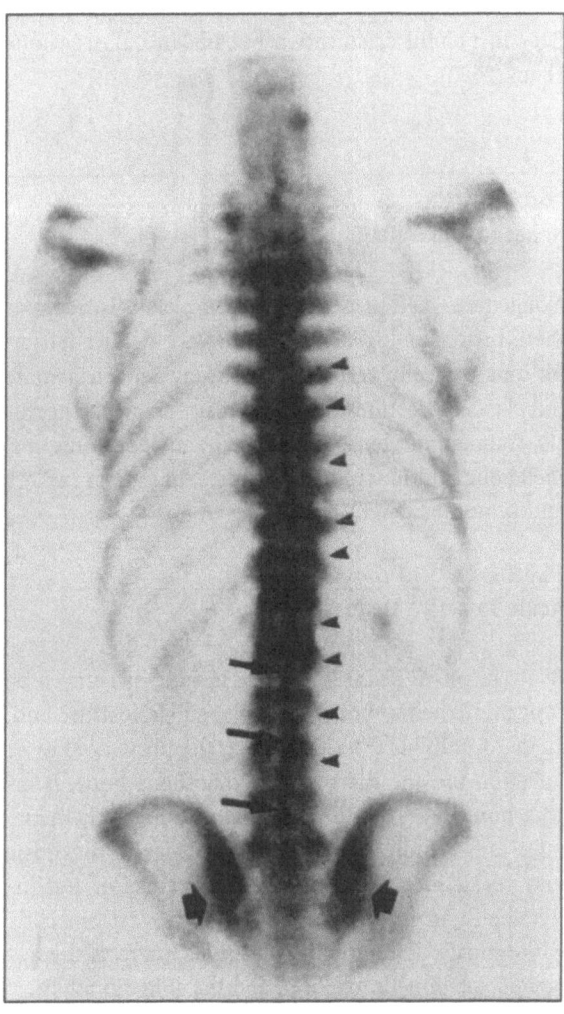

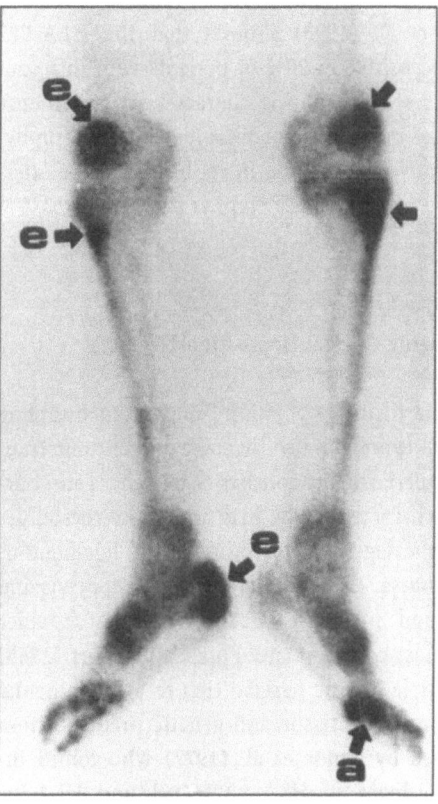

Fig. 26.11. Anterior bone scan of both knees and legs of patient with Reiter's syndrome shows asymmetric patchy hot lesions in patellae, proximal tibias, right calcaneus, and left metatarsals. *e* and *a* denote enthesopathy and arthropathy, respectively

Fig. 26.10. Posterior scan of axial skeleton of patient with ankylosing spondylitis with "bamboo spine" shows diffuse obliteration of spinal contour and disc spaces (*arrowheads*), increased tracer uptake in the interspinous ligaments (*long arrows*) and sacroiliac joints (*short arrows*)

26.8.1.3.2
Diffuse Idiopathic Skeletal Hyperostosis (Forestier's Disease)

Diffuse idiopathic skeletal hyperostosis is a disorder characterized by the ossification of the anterior longitudinal ligament with pointed hyperostosis that is localized to the anterolateral aspects of the vertebral body-intervertebral disc junctions in the thoracolumbar and cervical spine. In 34% of patients, the HLA-B27 antigen was positive (Shapiro et al. 1976). Bone scan is featured by increased tracer accumulation in the anterolateral aspects of several continuous affected vertebrae. Intervertebral disc spaces are obliter-

ated with pointed or bumpy tracer uptake in the lateral aspects. On occasion, tracer uptake is also increased in the costovertebral-apophyseal joints and spinous processes where interspinous ligaments insert. This scan finding is at variance with the classic description that apophyseal joints are usually not affected. Much like in ankylosing spondylitis, tracer uptake is prominent in the early stage of Forestier's disease, whereas it is reduced in the late stage when hyperostosis becomes ossified and mature.

26.8.1.3.3
Slipped Capital Femoral Epiphysis

Slipped capital femoral epiphysis is the medioposterior spillage of the capital epiphysis of the femur in children and adolescents. Slippage occurs in the zone of hypertrophic chondrocytes of actively growing physeal plate. Trauma, excessive physical activity, growth spurt, and obesity are suggested as contributing factors. This condition has been associated with high incidence of genetic markers. Indeed, one study by Mullaji et al. (1993) showed that the HLA-B27 antigen was positive in 20% of patients with this condition. Bone scintigraphy is characterized by intense tracer uptake in the flattened capital femoral epiphysis with relative photopenia in its base and the lateroanterior buckling of the metaphysis.

26.8.1.3.4
Calcific Periarthritis
(Hydroxyapatite Crystal Deposition Disease)

Calcific periarthritis is a painful and often disabling disorder that is precipitated by acute and chronic trauma. Calcific periarthritis consists of tendinitis and bursitis and is characterized by hydroxyapatite crystal deposition in the degenerated tendinous and ligamentous tissues and bursa. The shoulder, hand, wrist, pelvis and hip, knee, and ankle are affected with the shoulder being the most common site. Pinals and Short (1965) related multiple calcific periarthritis to a fundamental defect in connective tissue and genetic predisposition was proposed by Amor et al. (1977) who found increased prevalence of HLA-A2 (66%) and HLA-Bw (34%) in patients with this disorder. Radiography is useful to show calcific deposition around the joint. Ordinary planar bone scan is not so informative but pinhole scan can detect calcification in this disorder and occasionally diagnose tendinitis and bursitis even in the absence of radiographic calcification (Fig. 26.12). A unique feature of bone scintigraphy is that it portrays the metabolic profile of such calcification. Indeed, it has been shown that the tracer positively accumulates in painful calcification but not in quiescent one (Bahk 2000).

26.8.2
Nongenetic Skeletal Disorders

Nongenetic skeletal disorders are skeletal disorders of mere histopathological nature that are neither heritable nor basically associated with any known genetic imbalance. Infections, osteoarthrosis, avascular necrosis, traumatic lesions, recreational and sports injuries, metabolic and nutritional disorders, and most tumors and tumorous conditions belong to this group.

26.8.2.1
Acute Infective Diseases of Bone

Acute infective diseases of bone include osteomyelitis, pyogenic osteitis, bone abscess, and periostitis, and, as these individual terms denote, the primary sites of infection are the marrow and cancellous bone, compact bone, cortex, and periosteum, respectively. Acute infections rapidly progress and the delay in diagnosis may result in a serious irreversible sequel or not infrequently an intractable chronic form. Osteomyelitis is the most common bone infection. It affects the appendicular bone in children and the spine in adults.

Bone scan is a well established, sensitive diagnostic tool of acute bone infections (Duszynski et al. 1975; Gilday et al. 1975). It can detect infections in the early stage often in the absence of radiographic alteration (Capitanio and Kirkpatrick 1970). Reinforced with three-phase and pinhole scan, the sensitivity and accuracy can be greatly enhanced. Pinhole scan can point to the specific sites of individual infections. Thus, active osteomyelitis of long bone shows intense tracer accumulation in the marrow space of the metaphysis where endarteries are distributed favoring bacterial embolization (Fig. 26.13) and in the compact bone, cortex, and periosteum in infective osteitis, cortical abscess, and infective periostitis (Fig. 26.14), respectively. It is to be noted that rarely acute osteomyelitis is indicated not by hot area but by photon defect (Fig. 26.15) (Russin and Staab 1976; Lee et al. 2000). The spine is a common site of infection in adults. It is either hematogenous or directly introduced through traumatic or surgical wound. Infective spondylitis creates a characteristic

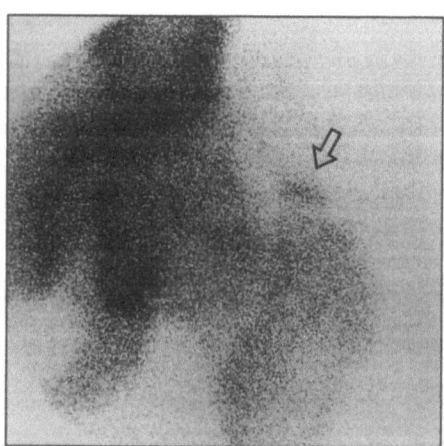

Fig. 26.12. Anterior pinhole scan of the left hip with painful calcific bursitis shows a small area of increased tracer uptake in the supratrochanteric region (*arrow*). [From Bahk (2000), reprinted with permission]

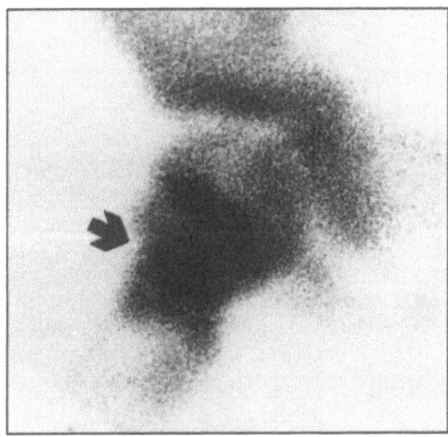

Fig. 26.13. Anterior pinhole scan of the right proximal femoral metaphysis in a 12-year-old boy with acute hematogenous osteomyelitis shows intense tracer accumulation (*arrow*). Note that the lesion is well confined to metaphysis and delimited by physeal line

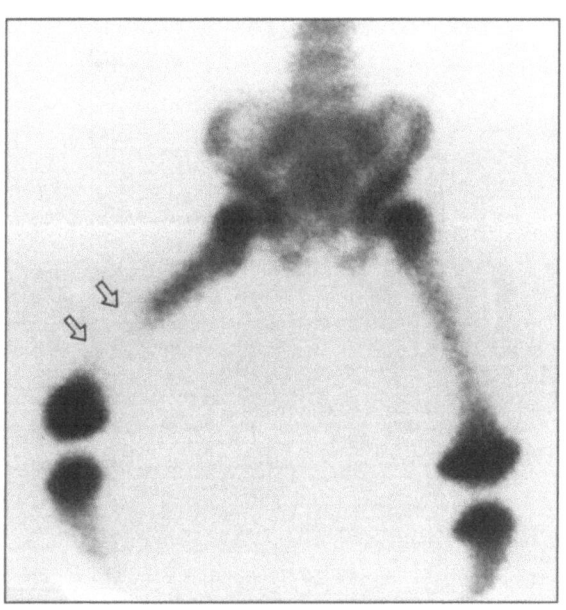

Fig. 26.15. Anterior bone scan of the right femur in a 1-year-old boy with acute osteomyelitis shows a large photon defect (*arrows*). The lesion was a staphylococcal infection. [Courtesy of Drs. S.M. Lee, S.K. Bae, and M.R. Cho; Lee et al. (2000)]

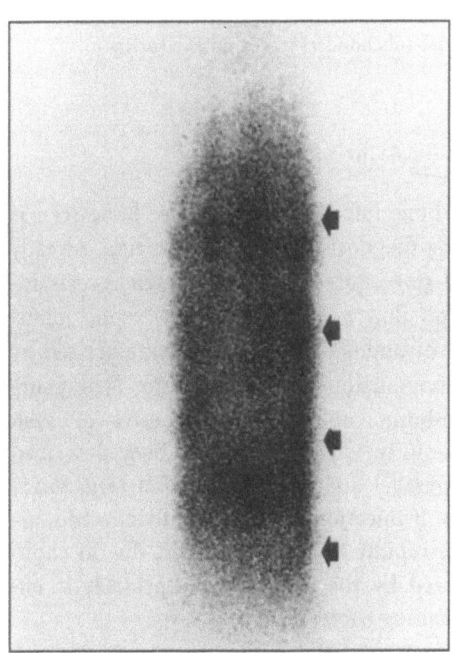

Fig. 26.14. Anterior pinhole scan of the right tibia with acute periostitis shows characteristic location of pathological uptake in the periosteum (*arrows*). (Adapted from Bahk 2000)

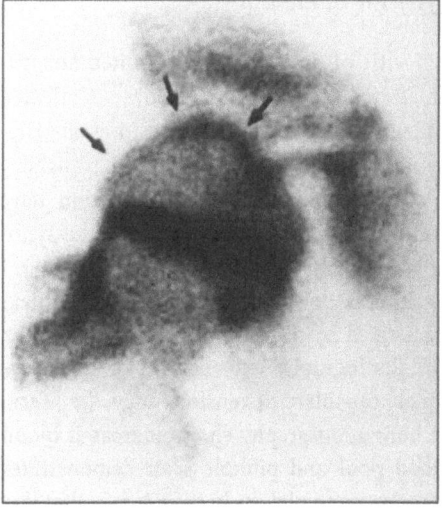

Fig. 26.16. Anterior pinhole scan of the right hip in an 8-year-old girl with transient synovitis shows cap-like increased tracer uptake in the synovial-subchondral bone complex of the femoral head (*arrows*)

sandwich sign (Bahk et al. 1987). Sclerosing osteomyelitis of Garré is a rare variant of osteomyelitis that produces marked sclerosis in the mandible and long bone diaphysis. It is attended by little suppuration (Waldfogel and Vasey 1980). Pinhole scan can distinguish this condition from periostitis and osteitis by localizing

tracer uptake to diaphyseal bone marrow when it involves a long bone. Three-phase bone scan is useful in differentiating between bone and soft tissue infection. The arterial blood flow and blood pool may become increased in both conditions but increased bone uptake is seen only in bone infection (Fig. 26.3 A–C).

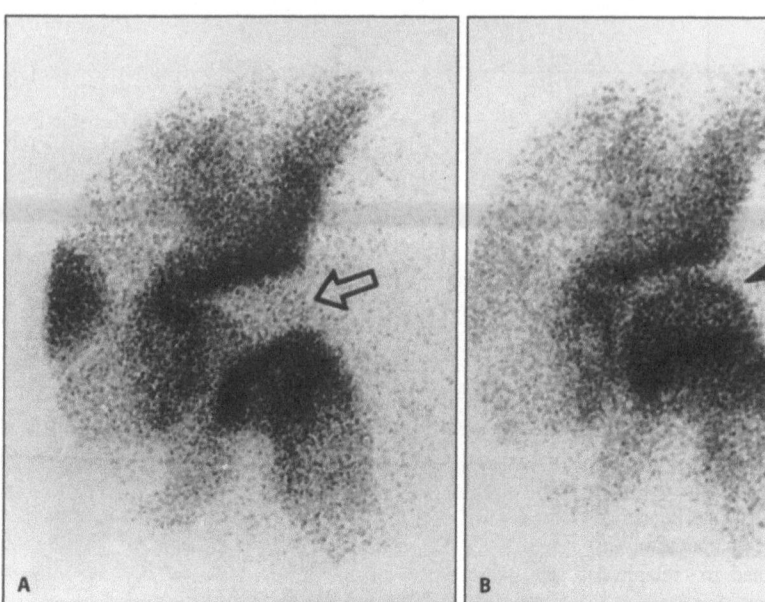

Fig. 26.17. Anterior pinhole scan of the left hip in a 10-year-old boy with transient synovitis with effusion before (A) and after aspiration (B). Before aspiration, the capital femoral epiphysis was totally photopenic due to increased intra-articular pressure and vascular compromise (*open arrow*). After aspiration, vascularity returned to normal with typical synovial-subchondral tracer uptake (*arrow*)

26.8.2.2
Transient Synovitis of the Hip

Transient synovitis of the hip is a self-limited nonspecific inflammatory disease of infancy. Boys are affected more often than girls. It is also referred to as the irritable hip syndrome, observation hip, and transitory coxitis. Infection and hypersensitivity reaction have been held to be the likelier causes. In 1–7%, the synovitis is related to Perthes disease although the cause-and-effect relation is not established. Pathological feature is characterized by diffuse synovial inflammation with effusion. Radiography is usually not informative but occasional capsular distension may be seen. Three-phase bone scintigraphy shows increased blood flow and blood pool and pinhole scan demonstrates curvilinear tracer accumulation in the synovia that line the femoral head and acetabular fossa (Fig. 26.16). The finding reflects increased blood flow in inflamed synovia through the anastomotic vascular channels (Rosenthall 1987). The follow-up scan in an uncomplicated case reveals rapid normal return with bed rest. On occasion, when synovial effusion is large in amount the capital femoral epiphysis may become totally photopenic as a result of transient vascular compromise caused by increased intra-articular pressure. Following the aspiration of effusion, increased tracer deposition quickly returns to normal (Fig. 26.17).

26.8.2.3
Acute Infective Arthritis

As in acute bone infections, three-phase bone scan is helpful in the diagnosis of infective arthritis. Already in the early stage, three-phase bone scan reveals increased blood flow and blood pool in septic joints. The finding of pinhole bone scan is characterized by a bizarre accumulation of tracer in the synovium-subchondral-bone complex. The intensity of such tracer uptake in infective arthritis has been described to roughly parallel the grade of infection (Fig. 26.18) (Bahk 2000). If infection is not promptly checked, articular space rapidly becomes narrowed due to chondrolysis caused by the action of the proteolytic enzyme of offending bacteria.

26.8.2.4
Osteoarthritis

Osteoarthritis is a ubiquitous joint disorder which is mainly associated with age and environmental stress, although . there is some genetic tendency (see Chap. 13). Pathologically, it is characterized by slow progressive degeneration of joint leading to cartilage destruction and cystic change of subchondral bone. Unlike in rheumatoid arthritis, synovial inflammation is not a conspicuous feature in osteoarthritis. Involve-

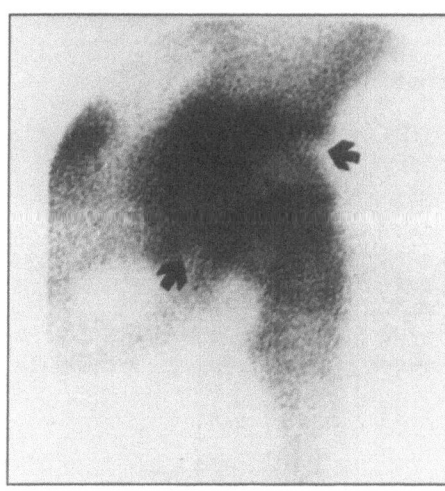

Fig. 26.18. Anterior pinhole scan of the left hip in a 13-year-old boy with acute pyogenic arthritis shows intense tracer uptake in the femoral head and acetabulum (*arrows*). Joint space appears narrowed suggesting cartilaginous destruction. Note that tracer uptake in pyarthrosis is much more intense than in simple transient synovitis (see Fig. 26.16)

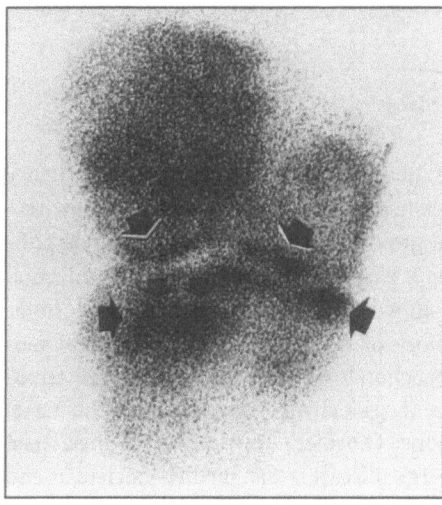

Fig. 26.19. Anterior pinhole scan of the right knee with degenerative osteoarthritis shows multiple mottled and segmental areas of increased tracer uptake in the para-articular subchondral bones (*arrows*) with the narrowing of the medial femorotibial compartment (*right arrow*). [Adapted from Bahk (2000)]

ment is random, oligoarticular, and asymmetric. The knee, hip, and spine are the most commonly affected sites. Bone scan reveals spotty, patchy, or segmental tracer uptake that is single or multiple in number (Fig. 26.19). It is not infrequent to see positive scan changes in radiographically normal joints. Spur and

hyperostosis, the common sequels of osteoarthritis, concentrate various amounts of tracer according to the evolutional stage or the metabolic activity. An early spur that is radiographically small and inconspicuous concentrates tracer rather intensely, whereas fully mature and prominent spur tends to concentrate tracer only minimally (Bahk et al. 1990a, 2000).

26.8.2.5
Reflex Sympathetic Dystrophy

Reflex sympathetic dystrophy (RSD), also referred to as Sudeck's atrophy, posttraumatic osteoporosis, and traumatic angiospasm, is a rheumatic disorder of much scintigraphic interest. Symptoms and signs include pain, tenderness, swelling, vasomotor and sensory disturbance, hypertrichosis, hyperhidrosis, and skin atrophy. Trauma, brain diseases, myocardial infarction, infection, and tumor have been related to this condition. Pathogenetically, the internuncial pool theory is widely held as a likelier cause. Recent identification of sympathetic vasoactive intestinal peptide (VIP)-containing nerve fibers innervated at the cortex and bone-periosteal junction has provided a biochemical basis for this theory (Hohmann et al. 1986). The VIP released from such sympathetic nerve fibers induces hyperemia and dramatic bone resorption as observed in RSD. Three-phase bone scintigraphy is useful for the dynamic assessment of increased vascularity in RSD (Kozin et al. 1981). Pinhole bone scan demonstrates peculiar small spotty tracer uptake in the peripheries of small bones of the ankle with RSD. Most recently, pinhole SPECT has shown such tracer uptake to coincide in location with cortex and probably bone-periosteal junction at enthesis, the site of the insertion of ligament and tendon (Fig. 26.20) (Bahk et al. 1998a).

26.8.2.6
Avascular Necrosis of Bone

Avascular necrosis of bone or osteonecrosis is the consequence of the deprivation of blood flow. Common causes include trauma, embolism, thrombosis, elevated bone marrow pressure, irradiation, and vasculitis. Basically, avascular bones cannot accumulate tracer. However, the ordinary planar scan often shows tracer to be spuriously increased in avascular bone, especially when ischemic bone is small in volume. The apparent increase in tracer uptake is presumably due to watershed phenomenon or reactive bone

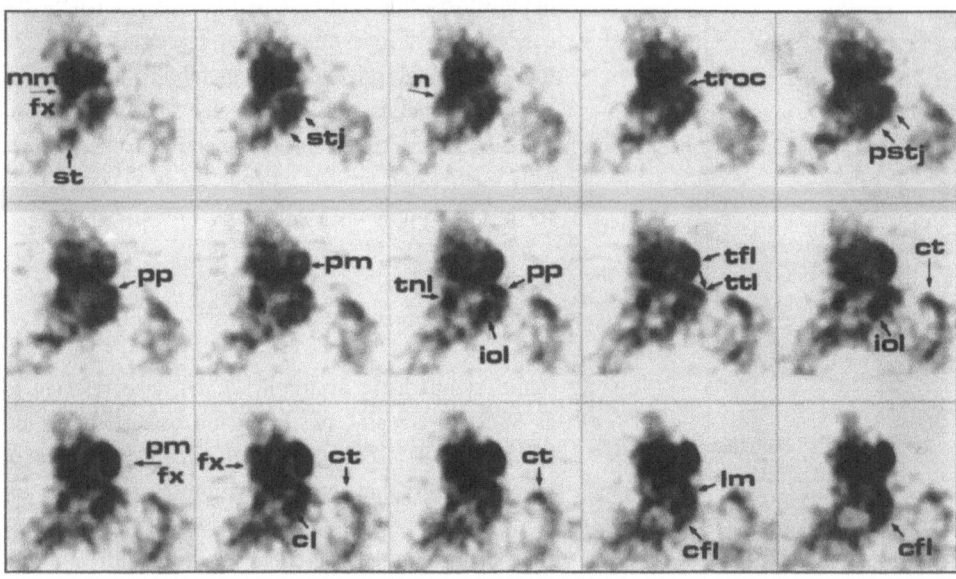

Fig. 26.20. High-resolution sagittal pinhole-SPECT scans of right ankle with reflex sympathetic dystrophy shows spotty hot areas in the peripheries of tarsal bones and ligamentous insertions. *mm, stj, n, troc, pp, pm, tnl, iol, tfl, ttl, ct,* and *cfl* denote medial malleolus, subtalar joint, talar neck, trochlea, posterior process, posterior malleolus, talonavicular ligament, interosseous ligament, tibiofibular ligament, tibiotalar ligament, calcaneal tip, and calcaneofibular ligament, respectively. Observe the imaging of subtle molecular alterations in the dystrophy

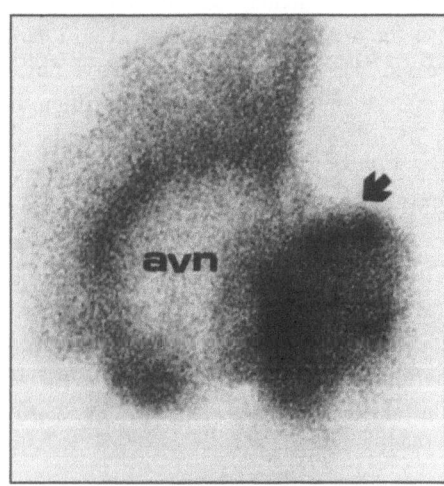

Fig. 26.21. Anterior pinhole scan of the left hip shows photon defect in avascular necrosis of the femoral head (*avn*) and intense tracer uptake in neck fracture (*arrow*)

change obliterating small photopenic area of ischemia. Thus, if an avascular area is large enough or magnified using pinhole scan it can be portrayed as cold area as such. Well-known avascular necroses include Legg-Calvé-Perthes disease, steroid induced osteonecrosis, Chandler's disease, and acute femoral head necrosis incidental to the fracture of the femoral neck (Fig. 26.21).

26.8.2.7
Osteochondroses

Contrary to the initial unitary concept of primary avascular osteonecrosis, osteochondroses are now recognized as a group of heterogeneous pathologic entities. Common clinical features include predilection for actively growing bone, habitual exposure to or frequent history of trauma, and local pain and tenderness. Osteochondrosis occurs in the capital femoral epiphysis (Legg-Calvé-Perthes disease), the tarsal navicular bone (Koehler's disease), the metatarsal head (Freiberg's disease), the medial clavicular end (Friedrich's disease), the secondary ossification center of the vertebra (Scheuermann's disease), and the tibial tubercle (Osgood-Schlatter's disease). Scintigraphic findings depend on underlying pathology. A large avascular osteonecrosis is indicated by cold area, whereas the associated microfracture or infraction causes tracer to intensely accumulate presenting as a hot area. Pinhole bone scintigraphy can provide the morphological information regarding the size, shape, texture, and location of osteochondrosis and assess the degree of initial avascularity and subsequent revascularization with new bone formation (Fig. 26.22) (Conway 1995; Bahk 2000).

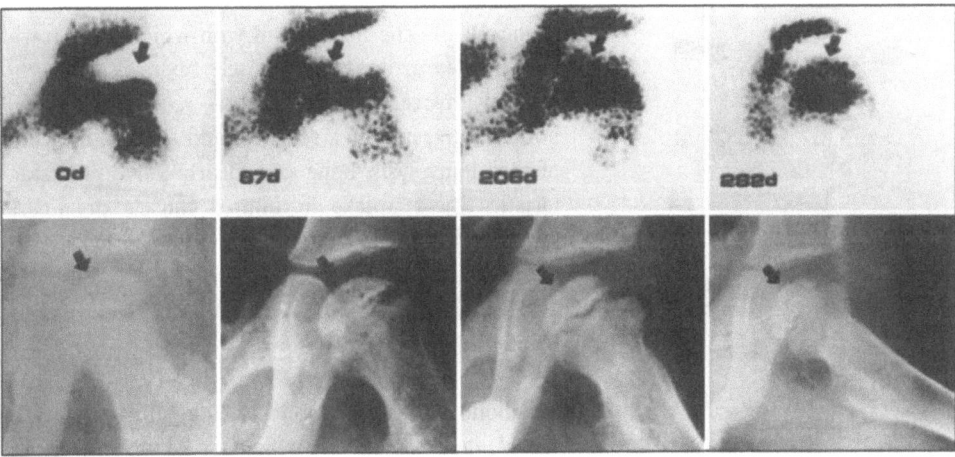

Fig. 26.22. Serial pinhole scans (*top panel*) and radiograms (*bottom panel*) of the left hip with Legg-Calvé-Perthes disease show gradual recovery of photon defect of avascular area in the capital femoral epiphysis over a period of 282 days. Note unimpressive radiogram (*arrows*)

26.8.2.8
Traumatic and Sports Injuries of Bone

With unprecedented mechanization of everyday life and limitless spread of various leisure sports, traumatic and sports injuries of bone have become one of the most important medical conditions. Radiography, CT, MRI, and bone scintigraphy are used for skeletal traumatology. Of these, bone scintigraphy is particularly useful for the diagnosis and management of traumatic and sports injuries (Matin 1983; Bahk 2000). It is ideal for the initial screening of patients with whole-body trauma, for the diagnosis of fractures and bruise in porotic ribs and spine and complex bones such as the pelvis and sacrum (Roger 1992) (Fig. 26.23), and for the differential diagnosis of shin splints and stress fracture (Brill 1983; Zwas et al. 1987). Bone scintigraphy reinforced with pinhole scan can portray even a very subtle focal alteration in traumatized bone that can be detected by no other means (Bahk 2000).

26.8.2.9
Periarticular Soft-Tissue and Muscular Rheumatism Syndromes

Periarticular soft-tissue rheumatism syndromes include bursitis, tenosynovitis, and enthesitis. Enthesitis is defined as nonspecific inflammation of the insertions of tendons, ligaments, and articular capsule to bone (Resnick and Niwayama 1983). These disorders are tender, painful, and often disabling. Bone scan portrays characteristic tracer accumulation at the

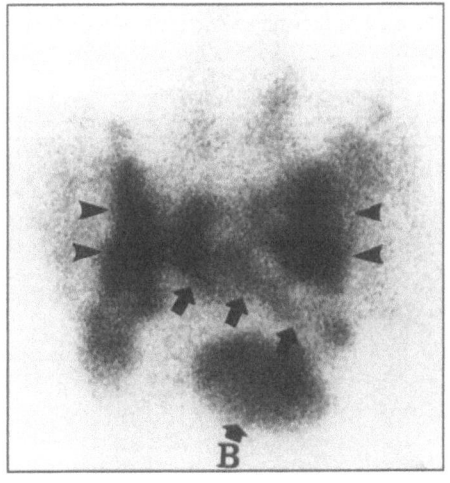

Fig. 26.23. Posterior pinhole scan of the sacrum in an 80-year-old woman with insufficiency fractures shows transverse tracer uptake in the mid-sacrum (*arrows*) and both iliac ala (*arrowheads*) giving a butterfly appearance. *B* denotes the urinary bladder

sites specific to individual disorders. For example, tracer accumulates in the greater tubercle in subdeltoid bursitis, in the acromion in supra-acromial bursitis (Fig. 26.24), in the greater trochanteric region in trochanteric bursitis (Fig. 26.12), and in the upper retrocalcaneal surface in Achilles tendinitis.

The muscular and musculotendinous rheumatism syndromes of scintigraphic interest include myositis ossificans, rhabdomyolysis, and musculotendinous unit injuries. Bone scintigraphy is useful for the demonstration of the tracer accumulated in denatured or

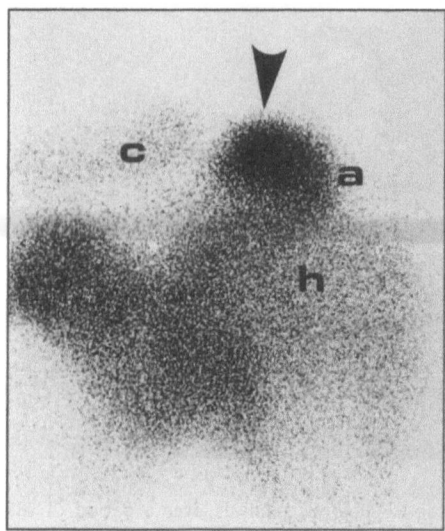

Fig. 26.24. Anterior pinhole scan of the left shoulder shows intense tracer uptake in the tip of the acromion (*a*) due to supra-acromial bursitis (*arrow head*). *c* and *h* denote the clavicle and the head of the humerus, respectively. [Adapted from Bahk (2000)]

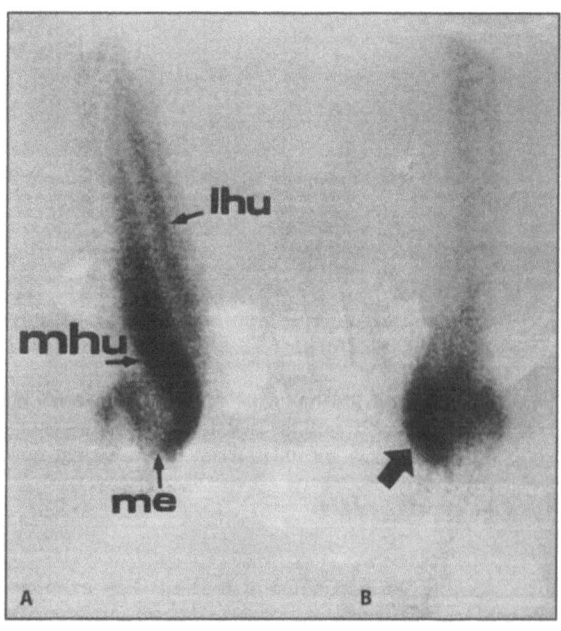

Fig. 26.25 A, B. Lateral ^{99m}Tc-HDP bone scans of the right (**A**) and left (**B**) upper arms in a dumbbell user respectively show peculiar bundle-like tracer uptake in the medial and lateral head units (*mhu* and *lhu*) of the triceps denoting musculotendinous unit injury and intense tracer uptake in the left medial epicondyle denoting epicondylitis (*arrow*). [Adapted from Bahk (2000)]

calcified muscle fibers and musculotendinous unit (Bahk 2000). The mechanism with which bone tracer accumulates in denatured muscle fibers is similar to that of myocardial necrosis. ^{99m}Tc-MDP or -HDP bone scintigraphy is a sensitive and specific indicator of rhabdomyolysis. Bone scan clearly portrays characteristic tracer uptake in injured muscle or muscle group and musculotendinous unit (Fig. 26.25).

26.8.2.10
Metabolic Diseases of Bone

Metabolic diseases of bone are due to a variety of causes such as vitamin deficiency or excess, undernourishment, endocrine disorders, renal failure, and disturbed calcium and phosphorus metabolism. Clinical entities are diverse and manifestations are complex. This chapter describes involutional osteoporosis, osteomalacia, rickets, and renal osteodystrophy which can be diagnosed by bone scintigraphy.

26.8.2.10.1
Involutional Osteoporosis

Involutional osteoporosis includes postmenopausal porosis and senile porosis. Osteoporosis is a condition in which bone mass is significantly reduced. Porotic bones are brittle, fragile, and hence liable to fracture. Postmenopausal osteoporosis has been related to low estrogen level, reduced physical activity, and inadequate nutritional state. Postmenopausal osteoporosis is characterized by the disproportionate reduction of trabecular bone mass compared to cortical bone mass, whereas the senile osteoporosis is characterized by proportionate loss of cortical and trabecular bone mass. The important complication of osteoporosis is fracture that commonly involves the spine and distal radius. Other fracture sites include the femoral neck, proximal humerus, tibia, and pelvis. Women suffer twice as frequently as men. Whole-body bone scan may reveal generalized decrease in skeletal tracer uptake and pinhole scan reveals the thinning of cortices of long bones and sparse endplates of vertebrae. Not infrequently, the endplates of porotic vertebrae may become depressed concentrating tracer intensely giving rise to "fish vertebrae" appearance (Fig. 26.26). (See Chap. 13, Sect. 13.5.3 for further discussion of osteoporosis.)

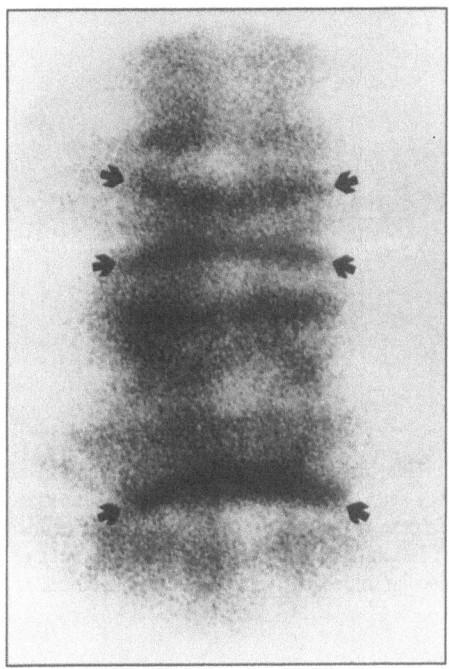

Fig. 26.26. Anterior pinhole scan of the lower lumbar spine in a patient with fish-vertebra deformity due to advanced osteoporosis shows typical concave depression of the endplate (*arrows*)

26.8.2.10.2
Hyperparathyroidism

Hyperparathyroidism may be primary, secondary, or tertiary. Primary hyperparathyroidism is caused by increased parathormone production with excessive mobilization of bone calcium. Bone scintigraphy may show diffuse increased tracer accumulation in the calvarium, mandible, sternum, and shoulder bones. Pinhole scintigraphy can portray linear tracer uptake in the subperiosteal bone absorption in the phalangeal shafts and acrolysis of the tufts. In the calvarium, pinhole scintigraphy may show a salt-and-pepper appearance (Bahk 2000).

26.8.2.10.3
Renal Osteodystrophy

Renal osteodystrophy is a condition that develops in hyperparathyroidism secondary to chronic renal insufficiency. Bone scintigraphy has been shown to be more sensitive than radiography in the diagnosis of osteodystrophy and osteomalacia (Fogelman and Carr 1980). The scintigraphic findings of renal osteodystrophy include increased tracer uptake in the sternum

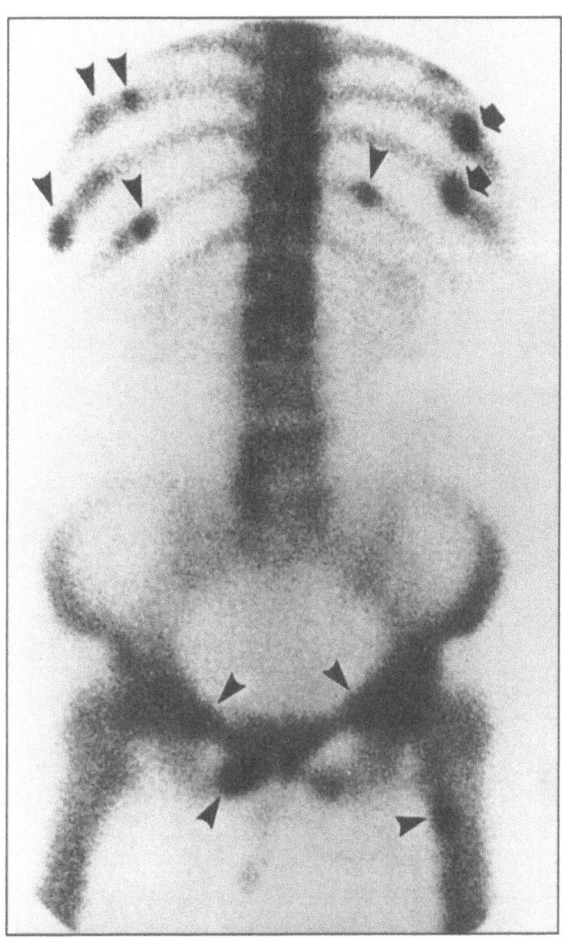

Fig. 26.27. Composite spot scans of the thoracolumbar and lumbopelvic skeleton in a patient with hyperparathyroidism with multiple infractions in the ribs, ischiopubic rings, and left proximal femur (*arrowheads*). [From Bahk (2000), reprinted with permission]

and costochondral junctions that are beaded. "Hot" patella may be seen in metabolic bone disease but it can also be seen in other diseases such as chondromalacia, metastasis, disuse osteoporosis, and even in normal condition. Pinhole bone scintigraphy is useful for the diagnosis of stimulated focal bone turnover in Looser's infraction and diffuse change in malacic skeleton at large (Fig. 26.27). It can be also utilized for the assessment of subperiosteal bone resorption, cystic change, and osteosclerosis.

26.8.2.10.4
Rickets and Osteomalacia

Both rickets and osteomalacia are characterized by deficient mineralization of osteoid. The basic differ-

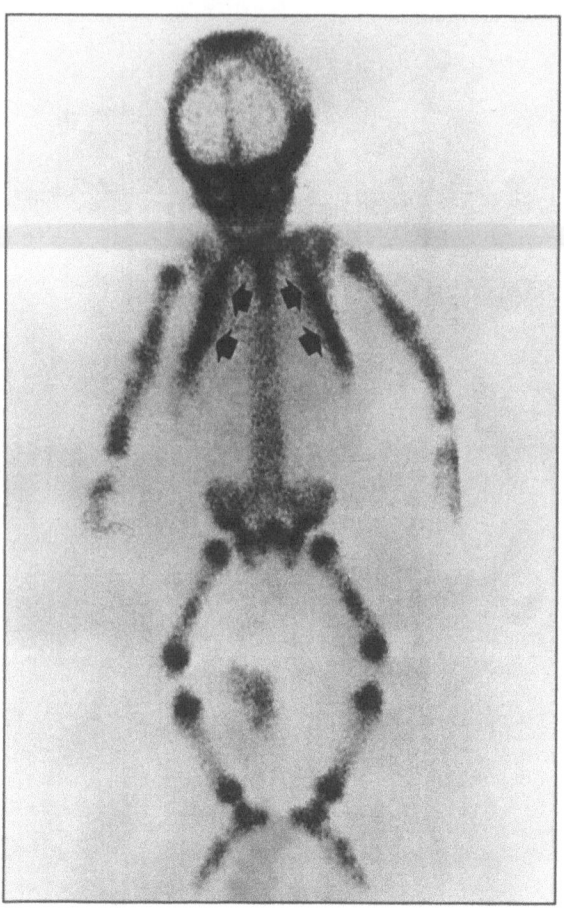

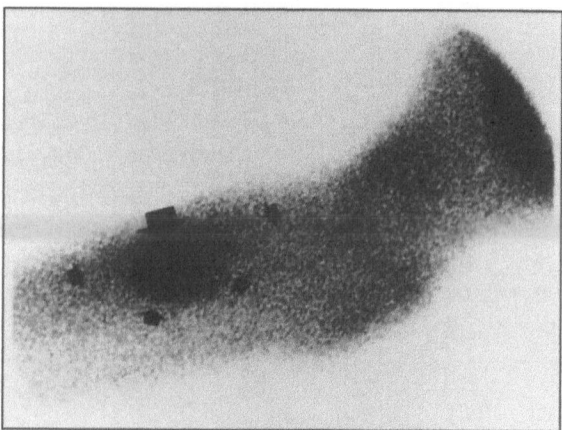

Fig. 26.29. Oblique pinhole scan of the right femur in an 11-year-old boy with osteoid osteoma shows a bean-sized area of very intense tracer uptake (*large arrow*) surrounded by an area of less intense tracer uptake (*small arrows*) representing vascular nidus and reactive bone alteration, respectively

infractions. In rickets, bone scan may show increased tracer uptake in the flared long bone metaphyses and beaded costochondral junctions that give rise to a rosary appearance. Joint spaces appear spuriously widened due to small dystrophic ossification centers and bulky uncalcified cartilaginous zone.

26.8.2.11
Bone Tumors

26.8.2.11.1
Primary Bone Tumors

The diagnostic yield of planar bone scintigraphy of primary bone tumor was generally much lower than that of radiography and its diagnostic specificity was seriously doubted. However, the systematic application of the pinhole magnification technique to the diagnosis of bone tumors has indicated that bone scintigraphy can make accurate diagnosis of most benign bone tumors such as unicameral bone cyst, osteoma, osteochondroma, and osteoid osteoma (Bahk 1996) (Fig. 26.29). In addition and importantly, this imaging technique can assess the metabolic status and evolutional stage of such tumors (Bahk 2000). Malignant tumors such as osteosarcoma and Ewing's sarcoma can be imaged along with the information about soft tissue invasion and distant metastasis (Fig. 26.30).

Fig. 26.28. Anterior whole-body scan of the skeleton in a 1-year-old boy with rickets shows general increase in tracer uptake producing a superscan. Prominent tracer uptake can be seen in clubbed long bone metaphyses and beaded costochondral junctions (*arrows*). [From Bahk (2000), reprinted with permission]

ence between two disorders is that the former disease occurs in actively growing bone and the latter in mature adult bone. Causes include deficiency of vitamin D and its active form [1,25-dihydroxyvitamin D3] and disturbed calcium-phosphorus metabolism. Bone scan has been shown to be more sensitive than radiography (Fogelman and Carr 1980). The scintigraphic manifestations of rickets and osteomalacia can be divided into general and local. Whole-body scan shows general increase in tracer uptake in the skeleton creating a superscan. The kidneys show little or no tracer uptake either due to renal dysfunction or the drain of available bone-seeking tracer by avaricious malacic bone (Fig. 26.28). Superscan is seen more typically in renal osteodystrophy. Small spotty hot areas in cortical bones with superscan represent bone

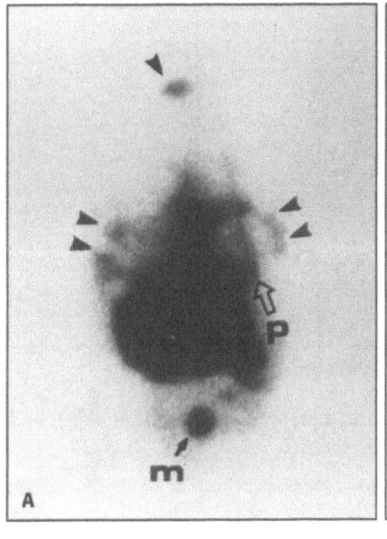
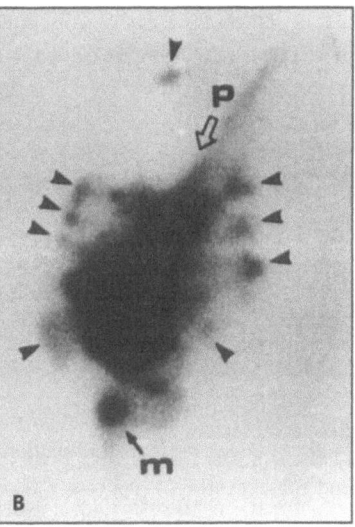

Fig. 26.30 A, B. Anterior (**A**) and lateral (**B**) planar scans of the right distal femur in a 21-year-old man with osteosarcoma shows a large hot area in the main tumor with sunburst-like tracer uptake denoting muscular invasion (*arrowheads*). *p* and *m* denote periosteal reaction and metastasis to tibia, respectively

26.8.2.11.2
Metastatic Bone Tumor

Metastatic bone tumor is one of the most rewarding indications of bone scintigraphy. Bone scintigraphy is highly sensitive and useful for the assessment of dissemination and therapeutic effect and for prognostication. Nevertheless, there are still unanswered questions about the appropriateness of the bone scintigraphy in patients with early cancers (McNeil 1978a; Perez et al. 1983; Yeh et al. 1995; Becker 1998; Jacobson and Fogelman 1998). Consensus would appear that bone scan is better reserved for patients with a malignancy of higher stage.

Bone scintigraphy can detect metastases weeks ahead of radiography and occasionally reveal both primary and metastasis on one and the same scan. It has been reported that 10–40% of metastases detected by bone scan have negative radiography, whereas less than 5% of patients with radiographic abnormality have negative scintigraphy (Denardo et al. 1972; Alazraki 1988). According to Pistenma et al. (1975), the incidence of false-negative bone scan is less than 3%. About 80% of metastases occur in the axial bone and 10% each in the skull and long bones (McNeil 1978b). In the absolute majority, bone metastases are multiple and only about 7% present with a solitary lesion. Breast and prostatic cancers tend to spread to the spine via vertebral veins, whereas lung cancer spreads hematogenously to randomized sites. Metastases from renal and thyroid cancer frequently present as photopenic lesions. The incidence of photopenic metastasis was reported to be 2% which is an underestimation.

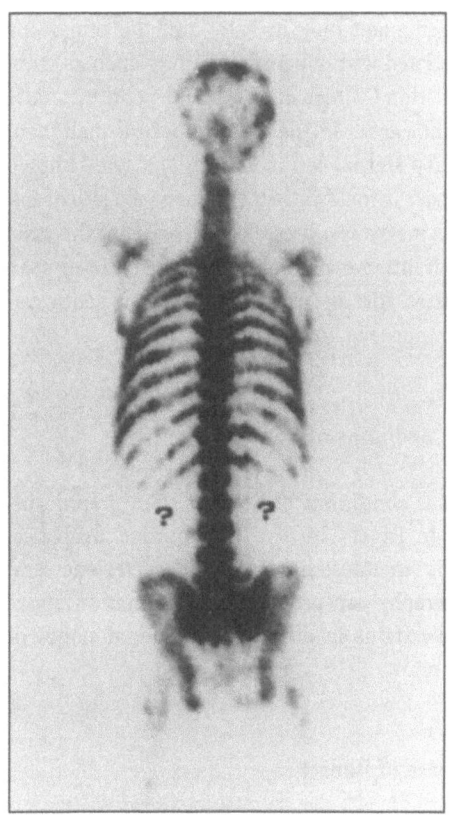

Fig. 26.31. Posterior scan of the skeleton with wide-spread cancer metastases shows intense tracer accumulation in the axial skeleton giving rise to a superscan appearance. Note the absence of tracer accumulation in the appendicular bones and kidneys (*?*)

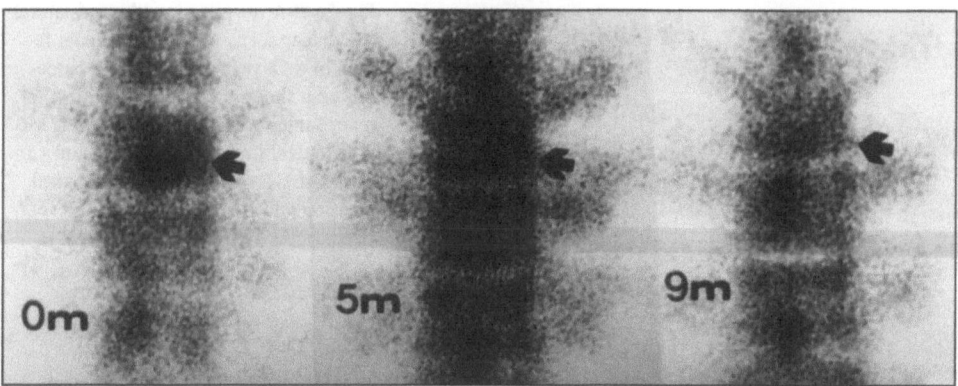

Fig. 26.32. Post-chemotherapy follow-up of metastatic breast carcinoma. Initial posterior pinhole scan of T7 shows a patchy area of intense tracer uptake (*0 month*) and subsequent scans show post-chemotherapy flare up (*5 months*) and eventual resolution (*9 months*)

If one uses pinhole scan the incidence will be significantly increased (Bahk 2000). A solitary hot area in the rib is malignant only in 9.8% and the remaining 90.2% are related with a benign etiology such as trauma or irradiation (Tumeh et al. 1985). In contrast, 68% of solitary hot areas in the axial bone are malignant and a solitary sternal lesion in patients with known primary cancer is metastasis if trauma is excluded. Extensive bone metastases create "superscan" of the axial skeleton with little renal uptake (Fig. 26.31). Bone scan can be utilized for the evaluation of therapeutic response of tumor (Fig. 26.32).

26.8.2.12
Tumorous Conditions of Bone

The tumorous conditions of bone of scintigraphic interest include Paget's disease and fibrous dysplasia. Scintigraphic manifestations are characteristic and bone scintigraphy can provide the information about the metabolic status in different evolutional stages of these diseases.

26.8.2.12.1
Paget's Disease of Bone

Paget's disease of bone is common affecting as many as some 3% of the European population. It was once considered to be extremely rare in Asia but appears on the slow increase. The main clinical symptoms are pain, tenderness, and deformity. Regarding the etiology, virus has been held as a promising and plausible agent. Virus-like inclusions were found in the osteoclasts of pagetic bone in 100% (Mills and Singer

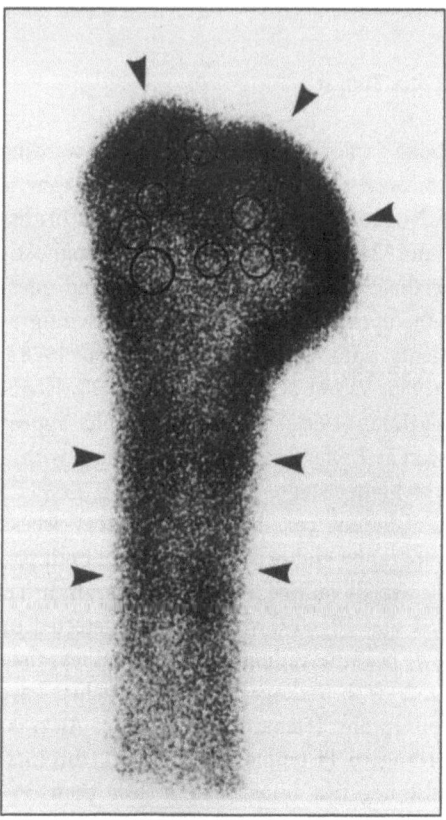

Fig. 26.33. Anterior pinhole scan of the right humerus with florid pagetic alteration shows intense tracer accumulation in thickened cortices (*arrowheads*) with constricted marrow and spotty photopenia due to pumiceous bones (*circles*). [Adapted from Bahk (2000)]

1976). The pelvic bones, spine, long bones, skull, scapula, and clavicle are commonly affected. Pathologically, Paget's disease is characterized by bone resorption and blastic production with cortical thickening, widening, and pumiceous alteration, and frequent fracture. Whole-body scan is ideal for the mapping of disease that is polyostotic in 50–65% of cases. Tracer accumulation is extremely intense. Such a scintigraphic finding along with peculiar distribution pattern has been described to be diagnostic of Paget's disease (Serafini 1976). Pinhole scan, however, reveals the tracer to specifically accumulate in cortices and peripheries of affected bones. Occasionally, there may be seen intermingled photopenic areas representing pumiceous alteration (Fig. 26.33). This is pathognomonic (Bahk et al. 1995). Bone scan can be used for semiquantitative assessment of the metabolic status of pagetic bone and therapeutic response to medical treatment.

26.8.2.12.2
Fibrous Dysplasia

Fibrous dysplasia is a nonhereditary, benign, hamartomatous disease of bone. This disease is characterized by metaplastic production of fibrous tissue stroma and curled spicules of woven bone formed therefrom. It is either monostotic or polyostotic with the monostotic type being radiographically 20 times as frequent as the polyostotic. Common sites of monostotic involvement include the femur (36%), tibia (19%), skull (17%), and ribs (10%) (Schajowicz 1981). Dysplastic bone lesions, sexual precocity, and cutaneous pigmentation are known as Albright's triad. Pinhole scan is useful for the mapping of polyostotic lesions and detection of pathological fracture that is a common complication. The metabolic status of fibrous dysplasia in different stages of disease can be assessed using pinhole scan (Baek et al. 1997; Lee et al. 1998). Osseous lesion of fibrous dysplasia that is opaque on radiogram concentrates tracer more avidly than the fibrous lesion that is radiolucent (Fig. 26.34).

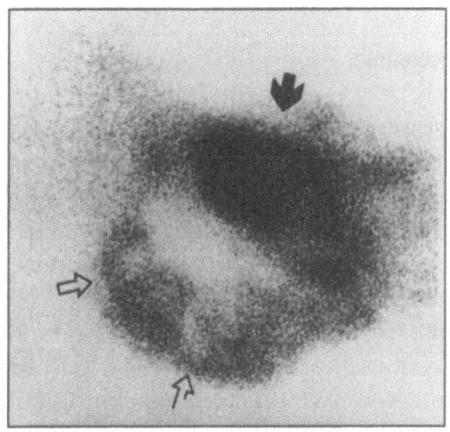

Fig. 26.34. Anterior pinhole scan of the right ischiopubic ring with fibro-osseous dysplasia shows increased tracer uptake in osseous lesion (*arrows*) and relatively low tracer uptake in fibrous lesion (*arrowheads*). Note how well bone scintigraphy portrays metabolic status of different pathologies in fibro-osseous dysplasia. [From Bahk (2000), reprinted with permission]

26.9
Bone Marrow Scintigraphy

June-Key Chung, Yong-Whee Bahk

26.9.1
Introduction

Bone marrow is one of the largest organs in the human body weighing 1,600–3,600 mg in adults. One-half of it is hematopoietic red marrow that occupies the axial skeleton and proximal one-thirds of femora and humeri. In young children, bone marrow extends further out into the mid-shafts of those bones. Bone marrow is the site of various diseases including blood dyscrasias, infections, metabolic disorders, trauma, and tumors. With the recent advent of refined and potent radiopharmaceuticals such as ^{99m}Tc nanocolloid and ^{99m}Tc-labeled anti-nonspecific cross-reacting antigen (NCA) 95 antibody, bone marrow scan is drawing acute attention as one of the typical molecular imaging methods.

26.9.2
Radiopharmaceuticals

The radiopharmaceuticals are capable of imaging three different types of marrow cells: (a) erythropoietic precursor cells, (b) reticuloendothelial cells (REC) and (c) granulopoietic cells. Using phagocytosis, ^{99m}Tc colloids can image REC. However, the accumulation in the red marrow of currently available ^{99m}Tc colloids is not large enough to produce a marrow image of sufficient quality. Also, disparity may exist between the locations of REC and hematopoietic cells in different hematological disorders. Theoretically, ^{52}Fe and ^{59}Fe can be used for the imaging of erythropoietic bone marrow but their improper physical characteristics prevent practical utilization. ^{111}In-chloride as an iron substitute has been tested but proven to be not satisfactory (Lilien et al. 1973). ^{111}In-chloride is expensive. ^{99m}Tc nanocolloid and ^{99m}Tc-labeled anti-NCA95 antibody are two representative agents for bone marrow scintigraphy.

26.9.2.1
^{99m}Tc Nanocolloids

Nanocolloids, less than 80 nm in diameter, are produced from antimony sulfide colloids or microaggregated albumin. Compared to conventional colloidal compounds, they accumulate more avidly in bone marrow than in the liver and spleen. For example, only 5% of the injected activities of ^{99m}Tc sulfur colloids accumulate in the bone marrow and 95% in the liver and spleen, whereas 15–50% of the injected ^{99m}Tc nanocolloids accumulate in the bone marrow.

26.9.2.2
^{99m}Tc Labeled-Anti-NCA95 Antibody

Nonspecific cross-reacting antigens (NCAs) are epitopes of carcinoembryonic antigen (Kuroki et al. 1984). Among several NCAs, NCA-95, having a molecular weight of 95 kDa, is present in human granulocytes and granulopoietic cells (Buchegger et al. 1984). Among monoclonal antibodies (MoAbs) specific for NCA-95 antigen, BW 250/183 and CEA-79.4 MoAbs are known to have an affinity constant of 2–9×10^9 L/mol and 0.5–2×10^5 epitopes per granulocyte, respectively (Steinstresser et al. 1988; Choi et al. 1995). These antibodies can be directly labeled with ^{99m}Tc. The exposure of an antibody to a reducing agent changes some of the antibody's disulfide bonds to free thiol groups, which then react with reduced ^{99m}Tc (Rhodes et al. 1986).

Granulopoietic cells are present in great abundance in bone marrow, 50–100 times more than in peripheral blood. Naturally, when intravenously injected, most ^{99m}Tc-labeled anti-NCA95 antibodies bind to bone marrow cells (Choi et al. 1995). Actually, the uptake in bone marrow is estimated to be 2.5–4.5 times greater than that in the spleen or liver (Fig. 26.35). As to the efficacy of uptake, hematopoietic marrow uptake is 2–4 times higher with ^{99m}Tc-labeled anti-NCA95 antibodies than ^{99m}Tc nanocolloids. Conversely, the hepatic and splenic uptake of the former agent is much less. It is, however, to be noted that significant increase in hepatic and splenic uptake has been shown to occur after repeated administrations of murine monoclonal antibodies.

26.9.3
Methods

26.9.3.1 ^{99m}Tc Nanocolloids

Bone marrow can be imaged after the intravenous injection of 185–740 MBq of ^{99m}Tc nanocolloids. Images may be obtainable as early as 20–30 min after injection. However, it is desirable to allow 45–60 min to maximize the target-to-background ratio and to improve the image quality. The shielding of the liver and spleen may be helpful for the evaluation of the thoracic and lumbar spines.

26.9.3.2
^{99m}Tc Anti-NCA95 Antibody Immunoscintigraphy

Bone marrow can also be imaged 4–5 h after the intravenous injection of 370–555 MBq of ^{99m}Tc anti-NCA95 antibodies. The amount is 0.2–0.5 mg. No adverse pharmacological reaction or significant change in vital signs has been observed in normal subjects. Unlike with ^{99m}Tc nanocolloids, there is no need of shielding the liver and spleen.

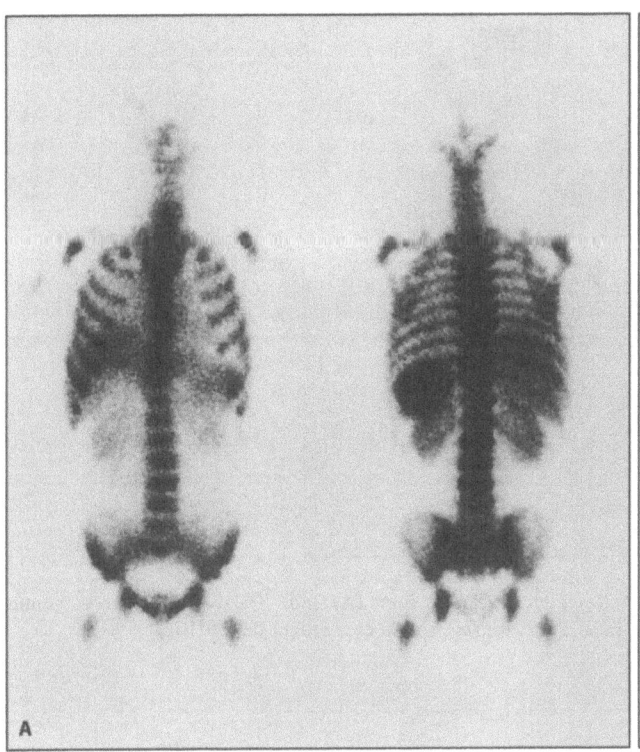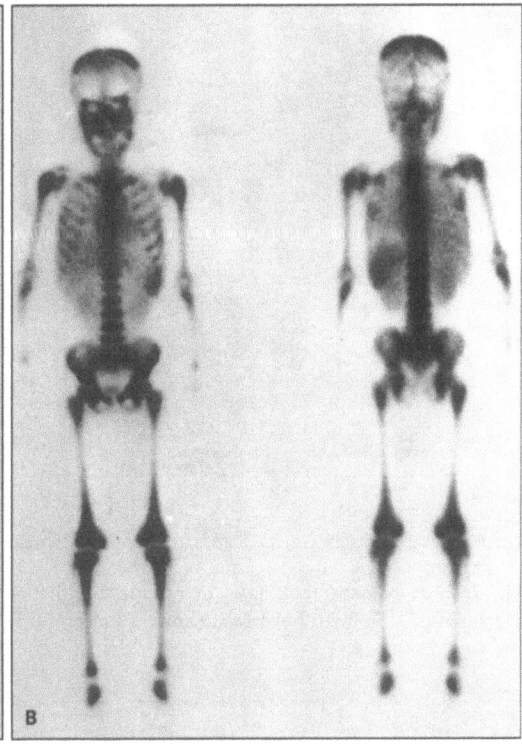

Fig. 26.35 A, B. Anterior and posterior ^{99m}Tc-NCA95 marrow immunoscintigrams in normal adult (**A**) and child (**B**). Marrow uptake is excellent with low hepatosplenic uptake. Note the difference in abundance of red marrow between adult and child with rich peripheral distribution in child

26.9.4
Clinical Applications

26.9.4.1
Bone and Bone Marrow Metastasis

As described earlier, ^{99m}Tc methylene diphosphonate (MDP) or hydroxydiphosphonate (HDP) bone scan is the standard method of investigating suspected skeletal metastasis in cancer patients. However, the spatial resolution and specificity of the ordinary planar bone scan is low making it difficult to differentiate between benign disease and metastasis (Bahk 2000). Kamby et al. (1987) reported the sensitivity of bone scan in the detection of skeletal metastasis in patients with breast cancer to be as high as 96% but specificity to be only 66%. Michel et al. (1991) found the specificity in patients with non-small-cell lung cancer was as low as 24%.

Correlation with radiography may reduce false scan interpretations, but the problem cannot adequately be resolved by planar bone scan alone. Radio-

graphically, a loss of some 50% of the mineral content of bone is required before a lesion can be detected (Edelstyn et al. 1967). Thus, radiography is generally insensitive during the early stage of metastasis. CT and MRI are useful to evaluate radiographically unexplained bone scan lesions, but limited availability, high cost, and difficulties in whole-body assessment limit their routine application. Moreover, CT is rather insensitive for marrow metastases and MRI may occasionally show false positive results in certain situations (Flickinger and Sanal 1994).

Recently, interest has been shown in the use of bone marrow scan in bone metastases that are imaged as *photon defects* (Figs. 26.36 and 26.37). Approximately 90% of metastases occur in bone marrow through hematogenous spread. Cancer cells are carried to the bone marrow by blood stream to be anchored and implanted in the medullary sinusoidal arteries. Here, tumor cells are facilitated to extravasate into marrow space because blood flow is sluggish, the endothelium is loose, and the basement membrane is absent. Bone marrow metastases can occur without cortical involvement, for example, in breast cancer

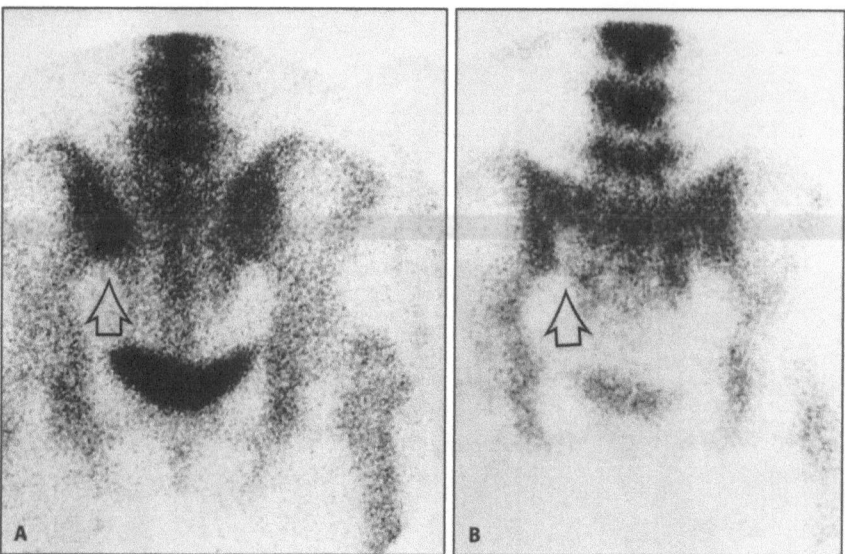

Fig. 26.36 A, B. Bone metastasis in a patient with prostate carcinoma. ^{99m}Tc-MDP bone scan shows a hot lesion in the right sacroiliac joint (A) and ^{99m}Tc-NCA95 marrow immunoscintigram shows concordant defect (B)

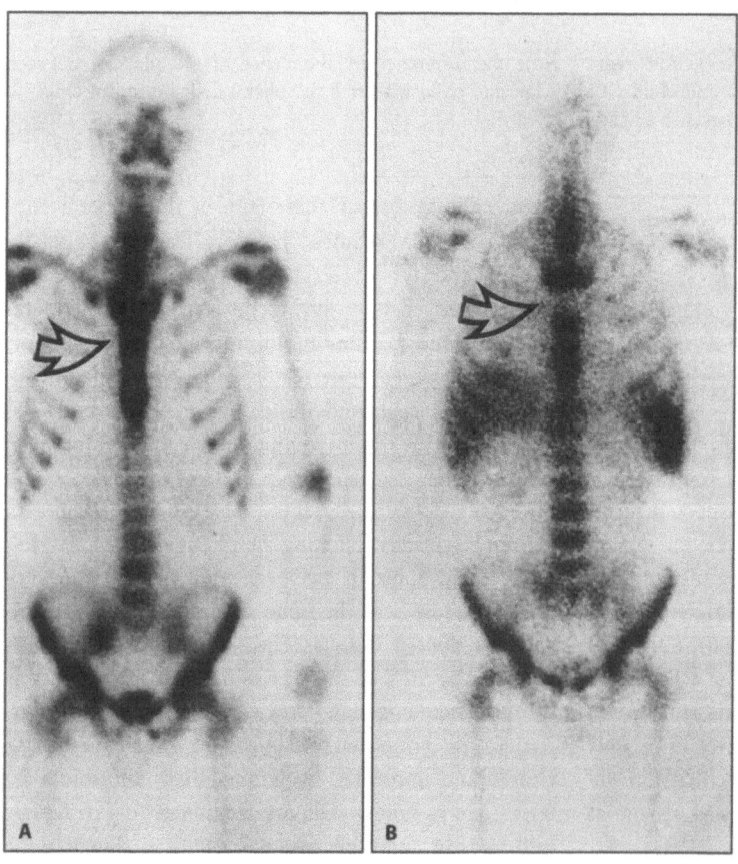

Fig. 26.37 A, B. Sternal metastasis in a patient with breast carcinoma. ^{99m}Tc-MDP bone scan shows an equivocal lesion in the sternum (A), whereas ^{99m}Tc-NCA95 bone marrow immunoscintigram reveals a prominent defect (B)

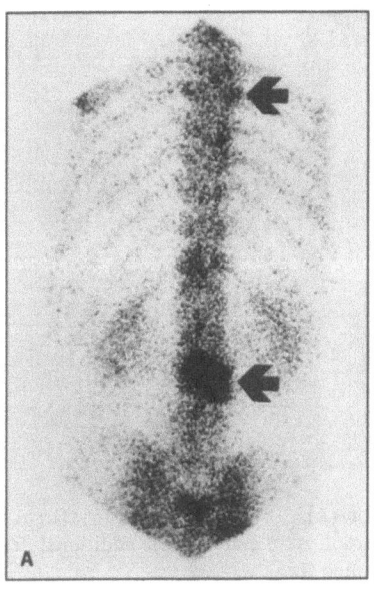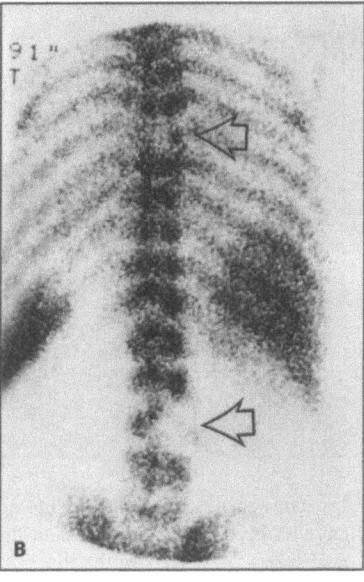

Fig. 26.38 A, B. Bone metastases in a patient with gastric carcinoma. ^{99m}Tc-MDP bone scan shows a prominent uptake in L3 and equivocal focus in T7 (*arrows*). Other areas of irregular uptake are due to degenerative spondylosis (**A**). ^{99m}Tc-NCA95 marrow immunoscintigram shows concordant defect in L4 and an additional defect in T7 (**B**) (*arrows*)

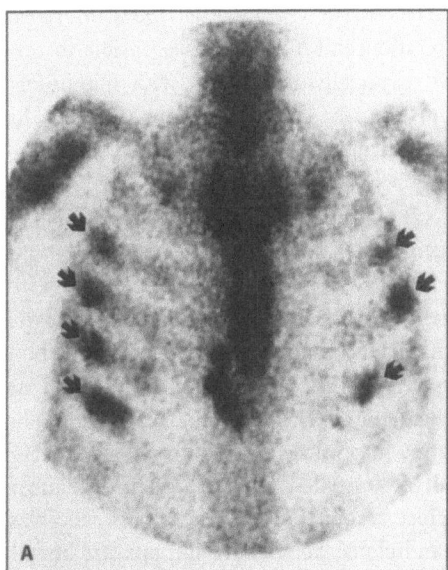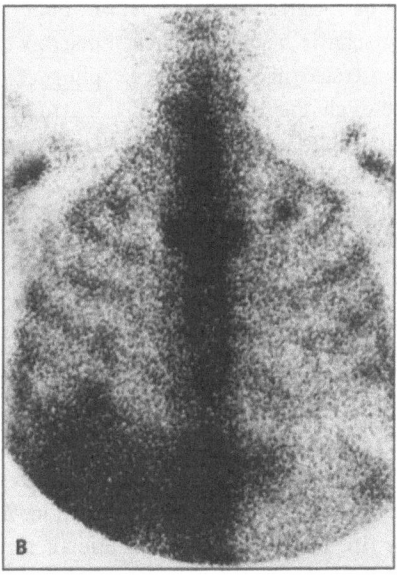

Fig. 26.39 A, B. Mismatch between bone scan and marrow scan in rib fractures. ^{99m}Tc-MDP bone scan shows multiple hot areas in fractured ribs (**A**) but ^{99m}Tc-NCA95 marrow scan fails to show them (**B**)

(Kamby et al. 1987). The scintigraphic detection of marrow invasion before bone invasion may lead to early institution of treatment increasing life expectancy (Fig. 26.38).

Immunoscintigraphy has been shown to be a sensitive approach to define the presence and extent of malignant bone marrow infiltration (Reske et al. 1989; Duncker et al. 1990). We assessed the capability of immunoscintigraphy to distinguish malignant lesions from benign ones in patients with metastatic bone tumors with equivocal ^{99m}Tc-MDP bone scan (Lee et al. 1995). The sensitivity and specificity of immunoscintigraphy in skeletal metastases in cancer patients with equivocal bone scan were 100% and 79%, respectively (Fig. 26.39). There were false positives and they were due to degenerative arthritis in the spine with fat replacement of bone marrow. Our study suggested that, in tumor patients with equivo-

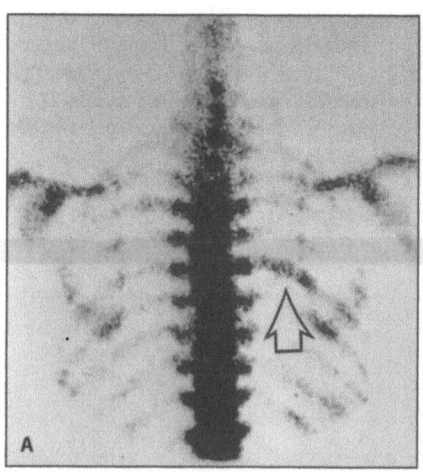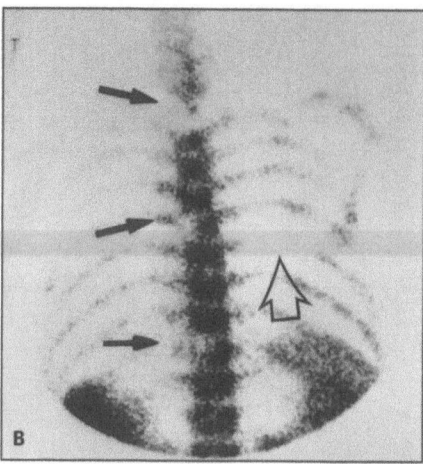

Fig. 26.40 A, B. Concordance of bone scan and marrow scan in metastases. ^{99m}Tc-MDP bone scan in a patient with hepatocellular carcinoma metastases shows segmental increased tracer uptake in right ribs 6 and 7 and no discernible abnormality in the spine (A). ^{99m}Tc-NCA95 marrow scan reveals concordant defect in rib 6 and multiple additional defects in the thoracic spine (B)

cal bone scan, the likelihood of skeletal metastasis is very low when bone marrow scan is negative. When marrow scan defect is demonstrated in concordance with positive bone scan, although the likelihood of metastasis may increase, radiography is a mandate to exclude benign processes such as the involution of vertebral red marrow (Cooper et al. 1992), focal bone necrosis (Haubold-Reuter et al. 1993), Paget's disease, and bone infarction (Yuasa et al. 1991).

Dunker et al. (1990) assessed the usefulness of immunoscintigraphy in patients with breast cancer suspected of having bone metastasis, and found marrow defects in 25 of 32 patients and bone invasion was subsequently confirmed in 23 (92%) patients. ^{99m}Tc-MDP bone scan could detect metastases in 17 of 32 patients, whereas anti-NCA95 antibody scan could detect more metastatic sites in 12 of 17 patients. Other authors also reported the superiority of bone marrow scan to bone scan in the diagnosis of metastases in patients with carcinomas of the lung, kidney, bladder, and prostate (Widding et al. 1990; Bourgeois et al. 1992) (Fig. 26.40).

26.9.4.2
Hematological Disorders

Bone marrow disorders are the primary cause of many hematological diseases. Patients are usually evaluated by bone marrow aspiration and biopsy, but these techniques are invasive and sometimes inadequate, not accurately representing all changes present in bone marrow tissue (Padhy et al. 1987). Moreover, marrow aspiration and biopsy are susceptible to sampling errors, especially in patients with nonuniform marrow. The scintigraphy of hematopoietic cells has an important place in this situation.

26.9.4.2.1
Aplastic Anemia

Aplastic anemia is a life-threatening hematological disorder characterized by pancytopenia and bone marrow depletion. Most cases are acquired, and decrease in hematopoiesis is secondary to offending agents on bone marrow cells. This disease also occurs as the result of inherited disorders such as Fanconi anemia. A close association also exists between aplastic anemia and clonal hemopathy and aplastic anemia and paroxysmal nocturnal hemoglobinuria. Even in patients with acquired aplastic anemia, a genetic predisposition is suggested by the finding of high incidence of HLA class II antigen DR2 and Dpw3. Bone marrow scan can be used for detecting aplastic anemia and evaluating its functional status. Untreated, the great majority of patients show homogeneously decreased uptake of either anti-NCA-95 antibody or ^{99m}Tc colloid in hematopoietic marrow (Najean et al. 1980; Chung et al. 1996) (Fig. 26.41).

Bone marrow scan can be also used for the assessment of response to treatment. After an effective medication, photopenia may become reverted to normal or increased uptake (Fig. 26.42). Occasionally, heteroge-

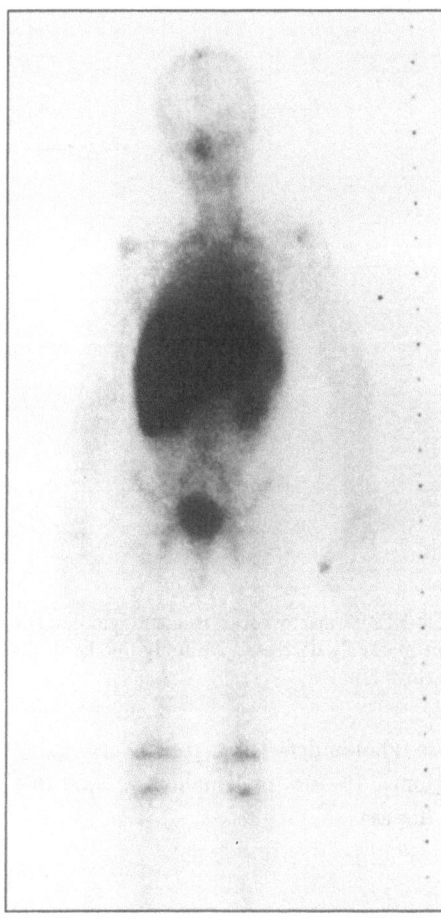

Fig. 26.41. Bone marrow immunoscintigram in aplastic anemia. ^{99m}Tc-NCA95 marrow scan shows generalized decrease in antibody uptake in the entire bone marrow

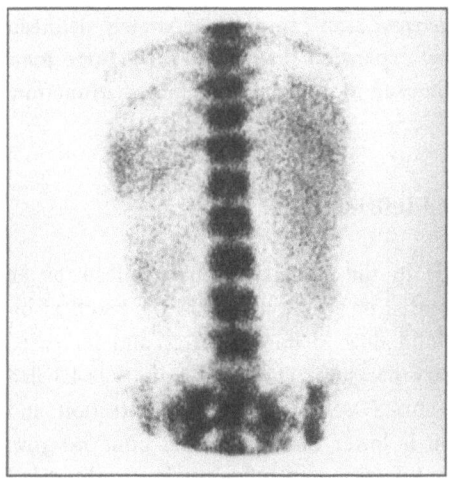

Fig. 26.42. ^{99m}Tc-NCA95 marrow scan in aplastic anemia after oxymetholone therapy shows increased marrow uptake in the axial skeleton. Compare with the desolated antibody uptake in untreated patient shown in Fig. 26.41 (different case)

neous tracer uptake may be seen during or after treatment. The damage to microenvironments of bone marrow seems to play an important role in its pathophysiology (Hotta et al. 1990). The peripheral blood analysis was not different between the groups with decreased and irregular uptake (Chung et al. 1996). Padhy et al. (1987) found post-therapeutic improvement of bone marrow scan findings and tracer uptake index to take place much earlier than appreciable changes in bone marrow cellularity or peripheral blood analyses. In some of our patients with normal or increased marrow uptake, hemoglobin levels improved earlier than leukocyte and platelet counts. This finding is in accord with the finding that erythroid recovery preceded the recovery of granulocytes and platelets in patients with aplastic anemia treated with androgen (Najean et al. 1980).

26.9.4.2.2
Myelodysplastic Syndrome

Myelodysplastic syndrome, a term used to describe an abnormal marrow-proliferation state that merges into acute myelogenous leukemia, is characterized by ineffective hematopoiesis or increased intramedullary destruction of mature hematopoietic cells. It is likely that this syndrome will be shown to involve the effect of retroviruses and/or the mutation of cellular proto-oncogenesis, such as RAS, retinoblastoma gene and growth-factor-related genes. The antibody scan in this syndrome may show heterogeneous uptake with areas of decreased and increased uptake (Chung et al. 1996) (Fig. 26.43). The accumulation of blasts that do not bind to antibody is considered to cause these small focal defects observed on the immunoscintigraphic scan of bone marrow.

26.9.4.2.3
Myeloproliferative Diseases

Myeloproliferative diseases are a group of disorders with variable phenotypes not only between patients but in the same patient in the course of his or her illness. The group includes polycythemia vera, essential thrombocythemia, myeloid metaplasia, and myelofibrosis. Bone marrow scan is usually normal in polycythemia vera in the early stage, but the splenic uptake becomes gradually increased and, as myelofibrosis and myeloid metaplasia are established, the peripheral marrow becomes expanded (Munz 1984). Bone marrow scan is also unremarkable in the early stage of myelofibrosis but, in the late stage, the tracer

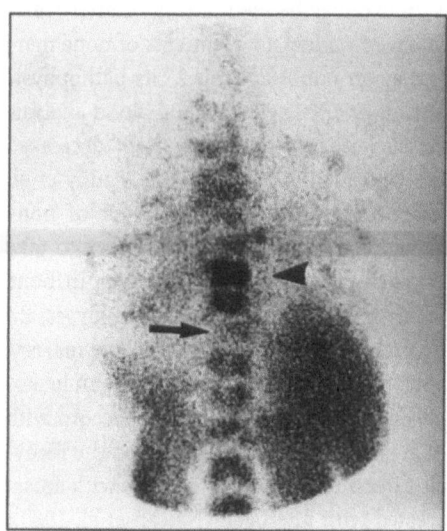

Fig. 26.43. ^{99m}Tc-NCA95 marrow immunoscintigram in a patient with myelodysplastic syndrome shows increased uptake (*arrowhead*) and decreased uptake (*arrow*) of antibody in the spine

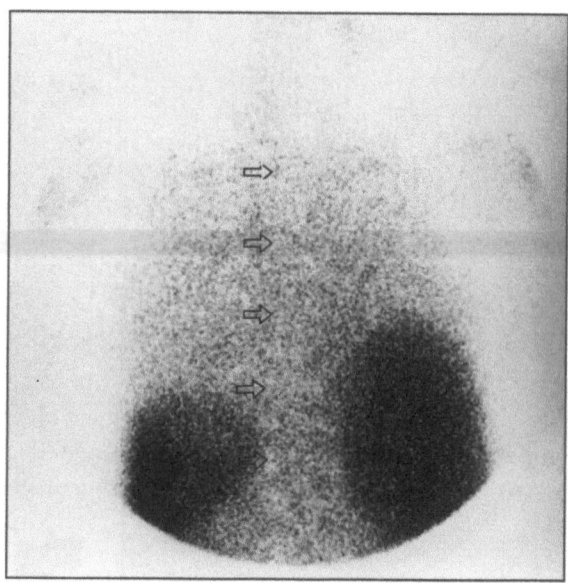

Fig. 26.44. ^{99m}Tc-NCA95 marrow scan in acute lymphocytic leukemia shows markedly decreased antibody uptake in the entire marrow (*arrows*)

uptake becomes decreased in the central marrow and conversely increased in the spleen.

26.9.4.2.4
Hemolytic Disorders

Bone marrow scan is unremarkable in the early stage of hemolytic disorders but peripheral marrow expansion may be seen in the late stage. In sickle cell disease, marrow scan may show extensive peripheral expansion with irregular photon defects (Datz and Taylor 1985).

26.9.4.2.5
Leukemia

In myelogenous leukemia, myelocytes express NCA-95 to variable degrees, and the anti-NCA-95 antibody uptake may range from normal to marked in degree. In acute lymphocytic leukemia, tracer uptake becomes markedly decreased due to the lack of lymphoid cell binding sites (Chung et al. 1996) (Fig. 26.44).

26.9.4.2.6
Multiple Myeloma and Lymphomas

In multiple myeloma and lymphomas, bone marrow scan demonstrates photon defects and the expansion of the peripheral bone marrow. The incidence of photon defect greatly varies and depends upon the type of disease. Photon defects are seen in 40–50% of multiple myeloma, 10–70% of lymphomas, and 10% of Hodgkin's disease.

26.9.4.2.7
Other Disorders

In patients with iron deficiency anemia, pure red cell aplasia, thalassemia minor, and Evan's syndrome, bone marrow scan reveals marrow expansion. It is to be noted that bone marrow scan can more accurately delineate bone marrow expansion than ^{99m}Tc-MDP bone scan, and it can be used to evaluate bone marrow function.

26.9.4.3
Infection and Inflammation

Osteomyelitis in the axial skeleton manifests as an area of decreased or absent uptake as the result of interrupted blood flow to bone marrow and increased intramedullary pressure. The uptake of ^{99m}Tc-labeled colloids or anti-NCA95 antibodies in infection and inflammation is lower than in normal bone marrow. However, the infection and inflammation in the distal long bone, where there is no red marrow, manifest as an area of increased uptake of anti-NCA95 antibodies because they bind circulating granulocytes and accumulate in an infective focus (Fig. 26.45).

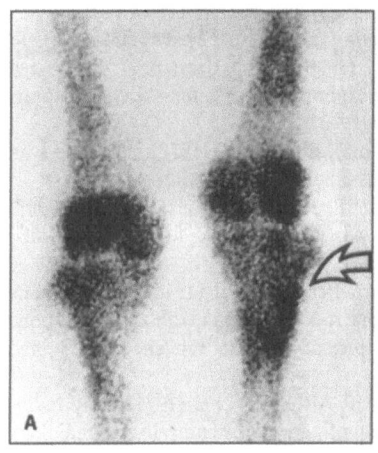
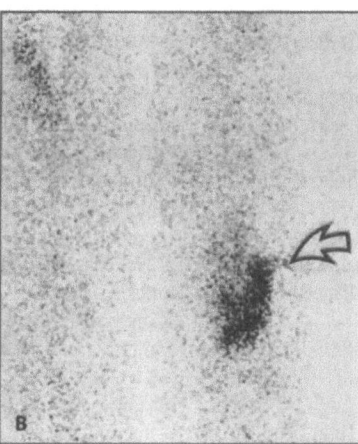

Fig. 26.45 A, B. ⁹⁹ᵐTc-MDP bone scan and ⁹⁹ᵐTc-NCA95 marrow scan in acute osteomyelitis. Bone scan shows increased tracer uptake in the proximal left tibia (A) and marrow scan shows concordant antibody uptake (B)

26.9.5
Future Direction

26.9.5.1
Immunoscintigraphy

When murine monoclonal antibodies are injected into patients, they are recognized by the body as foreign proteins, and, hence, capable of initiating an immune response to form human anti-mouse antibody (HAMA). The resulting immune response may be severe when the patient is hypersensitive. If human monoclonal antibodies could be used in the place of their murine counterparts, many problems of immunogenicity would be overcome. Chimeric antibodies, in which a large portion of mouse immunoglobulin is replaced by human immunoglobulin, have been developed for anti-NCA95 antibody using molecular biological techniques (Higuchi et al. 1998). Furthermore, it is now possible to introduce combining sites of mouse antibody into human immunoglobulin to produce even more human-like antibody, termed a 'humanized' antibody.

Using genetic engineering techniques, small fragment-like components of antibodies, grown in bacteria rather than in hybridoma cells, have been recently produced. Binding of an antigen requires an interaction between combining sites on both the heavy and light chains. Constructs consisting of these two variable combining site regions, known as Fv fragments, is engineered with the same antigen-binding characteristics (Perkins and Pimm 1991). A small peptide is used as a link between the two variable regions and a suitable radionuclide can be labeled with this moiety, which has been shown to be less immunogenic than intact immunoglobulin. Other advantages are enhanced ability to penetrate target tissues due to reduced size and even target tissue distribution.

26.9.5.2
Peptides

Peptides are compounds comprised of about 80 amino acids and a molecular weight of less than 10,000 Da. Since the discovery of peptide receptors and the synthesis of small biologically active peptides, radiolabeled peptide-based radiopharmaceuticals have opened a new era in nuclear medicine (Blok et al. 1999). Due to their small size, peptide molecules exhibit favorable pharmacokinetic characteristics such as rapid blood clearance and uptake by target tissues. Molecular engineering techniques now permit the synthesis of a wide range of biologically active peptides that carry chelating groups in their structures which leave their receptor binding properties unaffected (Okarvi 1999). New peptides for bone marrow scans are now being investigated in several world research centers.

26.10
References

Alazraki N (1988) Radiopharmaceuticals. In: Resnick D, Niwayama G (eds) Diagnosis of bone and joint disorders, 2nd edn. Saunders, Philadelphia, pp 463–465

Amor B, Cherot A, Delbarre et al (1977) Hydroxyapatite rheumatism and HLA markers. J Rheumatol Suppl 3:101–104

Baek JH, Lee SY, Kim SH et al (1997) Pinhole bone scintigraphic manifestation of fibrous dysplasia. Korean J Nucl Med 31:452–458

Bahk YW (1996) Pinhole scanning in tumors and tumorous conditions of bone: a new imaging approach to skeletal oncology. J Orthop Sci 1:70–89

Bahk YW (2000) Combined scintigraphic and radiographic diagnosis of bone and joint diseases. 2nd edn. Springer, Berlin Heidelberg New York

Bahk YW, Kim OH, Chung SK (1987) Pinhole collimator scintigraphy in differential diagnosis of metastasis, fracture, and infections of the spine. J Nucl Med 28:447–451

Bahk YW, Park YH, Chung SK et al (1995) Bone pathologic correlation of multimodality imaging in Paget's disease. J Nucl Med 36:421–426

Bahk YW, Chung SK, Park YH et al (1998a) Pinhole SPECT imaging in normal and morbid ankles. J Nucl Med 39:130–139

Bahk YW, Kim SH, Chung SK et al (1998b) Dual-head pinhole bone scintigraphy. J Nucl Med 39:1444–1448

Bahk YW, Kim SH, Chung SK et al (2000) Pinhole bone scan and pinhole bone SPECT findings of reflex sympathetic dystrophy syndrome. J Nucl Med Proceedings of 47th Annual Meeting No. 1426

Becker W (1998) A changing role for bone scintigraphy in oncology: the road from routine imaging screening to patient-based screening. Eur J Nucl Med 25:359–361

Block D, Feitsma REIJ, Vermeij P et al (1999) Peptide radiopharmaceuticals in nuclear medicine. Eur J Nucl Med 26:1511–1519

Brill DR (1983) Sports nuclear medicine. Bone imaging for lower extremity pain in athletes. Clin Nucl Med 8:101–106

Bourgeois P, Thimpson J, Feremans W et al (1992) Bone marrow scintigraphy in lung carcinomas using nano-sized colloids: when is it useful and how useful is it? Nucl Med Commun 13:421–428

Buchegger F, Schreyer M, Carrel S et al (1984) Monoclonal antibodies identify a CEA cross-reacting antigen of 95 kDa (NCA-95) distinct in antigenicity and tissue distribution from the previously described NCA of 55 kDa. Int J Cancer 33:839–845

Capitanio MA, Kirkpatrick JA (1970) Early roentgen observations in acute osteomyelitis. AJR 108:488–496

Choi CW, Chung J-K, Lee DS, Lee MC et al (1995) Development of bone marrow immunoscintigraphy using a ^{99m}Tc-labeled anti-NCA-95 monoclonal antibody. Nucl Med Biol 22:117–123

Chung J-K, Yeo JS, Lee DS et al (1996) Bone marrow scintigraphy using Technetium 99m-labeled antigranulocyte antibody in hematologic disorders. J Nucl Med 37:978–982

Connolly LP, Treves ST, Connolly SA et al (1998) Pediatric skeletal scintigraphy: applications of pinhole magnification. RadioGraphics 18:341–351

Conway JJ (1995) Radionuclide evaluation of Legg-Calvé-Perthes disease. In: Treves ST (ed) Pediatric nuclear medicine, 2nd edn. Springer, Berlin Heidelberg New York

Cooper M, Miles KA, Wraight EP, Dixon AK (1992) Degenerative disc disease in the lumbar spine: another cause for focally reduced activity on bone marrow scintigraphy. Skel Radiol 21:247–249

Datz FL, Taylor A Jr (1985) The clinical use of radionuclide bone marrow imaging. Semin Nucl Med 15:239–259

Davis MA, Jones AG (1976) Comparison of ^{99m}Tc labeled phosphate and phosphonate agents for skeletal imaging. Semin Nucl Med 6:19–31

Denardo GL, Jacobson SJ, Raventos A (1972) ^{85}Sr bone scan in neoplastic disease. Semin Nucl Med 2:18–30

Donohoe KJ, Henkin RE, Rotal HD et al (1998) Procedure guideline for bone scintigraphy: 1.0. J Nucl Med 37:1903–1906

Duncker CM, Carrio I, Berna L et al (1990) Radioimmune imaging of bone marrow in patients with suspected bone metastases from primary breast cancer. J Nucl Med 31:1450–1455

Duszynski DO, Kuhn JP, Afshani E et al (1975) Early radionuclide diagnosis of acute osteomyelitis. Radiology 117:337–340

Edelstyn GA, Gillespie PJ, Grebbell FS (1967) The radiological demonstration of osseous metastasis. Experimental observations. Clin Radiol 18:158–162

Etchebehere ECSC, Caron M, Pereira JA et al (2001) Activation of the growth plate on three-phase bone scintigraphy: the explanation for the overgrowth of fractured femurs. Eur J Nucl Med 28:72–80

Flickinger FW, Sanal SM (1994) Bone marrow MRI: techniques and accuracy for detecting breast cancer metastasis. Magn Reson Imaging 12:829–835

Fogelman I, Carr D (1980) A comparison of bone scanning and radiology in the evaluation of patients with metabolic bone disease. Clin Radiol 31:321–326

Francis MD, Ferguson DL, Tofe AJ et al (1980) Comparative evaluation of three diphosphonates: in vivo adsorption (C-14 labelled) and in vivo osteogenic uptake (Tc-99m complexed). J Nucl Med 21:1185–1189

Francis MD, Horn PA, Tofe AJ (1981) Controversial mechanism of technetium-99m deposition on bone. J Nucl Med (abstract) 22:72

Fugger L, Tisch R, Liblau R et al (1995) The role of human major histocompatibility complex (HLA) genes in disease. In: Scriver CR, Beaudet AL, Sly WS, Valle D (eds) Metabolic and molecular bases of inherited disease I, 7th edn, chap 9. McGraw-Hill, New York, pp 555–585

Gilday DL, Eng B, Paul DJ et al (1975) Diagnosis of osteomyelitis in children by combined blood pool and bone imaging. Radiology 177:331–335

Goldman AB (1995) Heritable diseases of connective tissue, epiphyseal dysplasia, and related conditions. In: Resnick D (ed) Diagnosis of bone and joint disorders, 3rd edn. Saunders, Philadelphia

Gordon SL, Buchanan J, Lodda RL (1981) Hereditary multiple exostoses: report of a kindred. J Med Genet 18:428–430

Groshar D, Rosenbaum M, Rosner I (1997) Enthesopathies, inflammatory spondyloenthesopathies and bone scintigraphy. J Nucl Med 38:2003–2005

Guillermart A, Le Page A, Galy YG et al (1980) Bone kinetics of calcium-45 and pyrophosphate labelled with technetium 96. An autoradiographic evaluation. J Nucl Med 21:466–470

Harper PV, Lathrop KA, Jiminez F et al (1965) Technetium 99m as a scan agent. Radiology 85:1–9

Haubold-Reuter BG, Duewell S, Schilcher BR et al (1993) The value of bone scintigraphy, bone marrow scintigraphy and fast spinecho magnetic resonance imaging in staging of patients with malignant solid tumors: a prospective study. Eur J Nucl Med 20:1063–1069

Higuchi T, Inoue T, Sarwar M et al (1998) Tc-99m-labelled chimeric human/mouse antigranulocyte antibody bone marrow scintigraphy: a preliminary clinical study. Nucl Med Commun 19:463–474

Hohmann EL, Elde RP, Rysavy JA et al (1986) Innervation of periosteum and bone by sympathetic vasoactive intestinal polypeptide-containing nerve fibers. Science 232:868–871

Holms RA (1978) Quantification of skeletal Tc-99m labelled phosphates to detect metabolic bone disease. J Nucl Med 19:330–331

Hotta T, Murate R, Inoue C et al (1990) Patchy haemopoiesis in long-term remission of idiopathic aplastic anemia. Eur J Haematol 45:73–77

Hotze AL, Griele B, Ooverbeck B et al (1992) Technetium-99m-labeled anti-granulocyte antibodies in suspected bone infections. J Nucl Med 33:526–531

Huda W, Slone R (1995) Screen/film radiography. In: Huda W, Slone R (eds) Review of radiologic physics. Williams and Wilkins, Baltimore

Jacobson AF, Fogelman I (1998) Bone scanning in clinical oncology: does it have a future? Eur J Nucl Med 25:1219–1223

Jaszczak RJ, Murphy PH, Huard D (1977) Radionuclide emission computed tomography of the head with ^{99m}Tc and scintillation camera. J Nucl Med 18:373–380

Jones AG, Francis MD, Davis MA (1976) Bone scanning: radionuclide reaction mechanisms. Semin Nucl Med 6:3–18

Kahn MA (1988) Ankylosing spondylitis and heterogeneity of HLA-B27. Semin Arthritis Rheum 18:134–141

Kamby C, Vejborg I, Daugaard S et al (1987) Clinical and radiological characteristics of bone metastases in breast cancer. Cancer 60:2254–2261

Kaufman JH, Cedermark BJ, Parthasarathy KL et al (1977) The value of Ga-67 scintigraphy in soft-tissue sarcoma and chondrosarcoma. Radiology 123:131–134

Kim JY, Chung SK, Park et al (1992) Pinhole bone scan appearance of osteoid osteoma. Korean J Nucl Med 26:160–163

Kim SH, Chung SK, Bahk YW et al (1999) Whole-body and pinhole bone scintigraphic manifestations of Reiter's syndrome: distribution patterns and early and characteristic signs. Eur J Nucl Med 26:163–170

Kozin F, Soin JS, Ryan LM et al (1981) Bone scintigraphy in the reflex sympathetic dystrophy syndrome. Radiology 138:37–43

Kuroki M, Koga Y, Masuoka Y (1984) Monoclonal antibodies to carcinoembryonic antigen: a systematic analysis of antibody specificities by using related normal antigens and evidence for allotypic determinants on carcinoembryonic antigen. J Immunol 133:2090–2097

Lee KH, Chung J-K, Choi CW et al (1995) Technetium-99m-labeled antigranulocyte antibody bone marrow scintigraphy. J Nucl Med 36:1800–1805

Lee SM, Bae SK, Cho MR (2000) Acute osteomyelitis shown as a cold lesion on bone scan. Korean J Nucl Med 34:516–520

Lee SY, Baek JH, Kim SH et al (1998) Metabolic profile of fibro-osseous dysplasia using pinhole bone scan. Eur J Nucl Med (abstract)

Leirisalo M, Skylv G, Kousa M et al (1982) Followup study on patients with Reiter's disease and reactive arthritis, with special reference to HLA-B27. Arthritis Rheum 25:249–259

Lilien DL, Berger HG, Anderson DP et al (1973) Indium-111-chloride: a new agent for bone marrow imaging. J Nucl Med 14:184–186

Mallinckrodt Medical Inc (1996) Technescan HDP kit for the preparation of technetium Tc-99m oxidronate (revised 8/1996). Mallinckrodt Medical Inc, St Louis, Mo

Matin P (1983) Bone scintigraphy in the diagnosis and management of traumatic injury. Semin Nucl Med 13:104–122

McAfee JG, Gagne G, Subramanian G et al (1991) The localization of indium-111-leukocytes, gallium-67-polyclonal IgG and other radioactive agents in acute focal inflammatory lesions. J Nucl Med 32:2126–2131

McAlister WH, Herman TE (1995) Osteochondrodysplasias, dysostoses, chromosomal aberrations, mucopolysaccharidoses, and mucolipidoses. In: Resnick D, Niwayama G (eds) Diagnosis of bone and joint disorders, 3rd edn. Saunders, Philadelphia

McCarty DJ, Gatter RA (1966) Recurrent acute inflammation associated with focal apatite crystal deposition. Arthritis Rheum 9:804–819

McNeil BJ, Pace PD, Gray EB et al (1978a) Preoperative and follow-up bone scans in patients with primary carcinoma of the breast. Surg Gynec Obst 147:745–748

McNail BJ (1978b) Rationale for the use of bone scans in selected metastatic and primary bone tumors. Semin Nucl Med 8:336–345

Medina M, Viglietti AL, Gozzeli L et al (2000) Indium-111 labelled white blood cell scintigraphy in cranial and spinal septic lesions. Eur J Nucl Med 27:1473–1480

Michel F, Soler M, Inhof E, Perruchound AP (1991) Initial staging of non-small cell lung cancer: value of routine radioisotope bone scanning. Thorax 46:469–473

Mills BG, Singer FR (1976) Nuclear inclusions in Paget's disease of bone. Science 194:201–202

Morris R, Metzger AL, Bluestone R et al (1974) HL-A W27 – a clue to the diagnosis and pathogenesis of Reiter's syndrome. N Engl J Med 290:554–556

Mullaji AB, Emery RJH, Joysey VC et al (1993) HLA and slipped capital femoral epiphysis. J Orthop Rheumatol 6:167–169

Munz DL (1984) Bone marrow imaging: basic concepts and clinical results. Nuklearmedizin 4:251–268

Nadel HR (1993) Thallium-201 for oncological imaging in children. Semin Nucl Med 23:243–54

Najean Y, Le Danvic M, Le Mercier N et al (1980) Significance of bone marrow scintigraphy in aplastic anemia: concise communication. J Nucl Med 21:213–218

Okarvi SM (1999) Recent developments in Tc-99m-labelled peptide-based radiopharmaceuticals: an overview. Nucl Med Commun 20:1093–1112

Padhy AK, Garg A, Kochupilai V et al (1987) Marrow uptake index: a quantitative scintigraphic study of bone marrow in aplastic anemia. Thymus 10:137–146

Park HM, Lambertus J (1977) Skeletal and reticuloendothelial imaging in osteopetrosis: case report. J Nucl Med 18:1091–1095

Pediatric Task Group EANM Member (1990) A radiopharmaceutical schedule for imaging in pediatrics. Eur J Nucl Med 17:127–129

Perez DJ, Powles TJ, Milan J et al (1983) Detection of breast carcinoma metastases in bone: relative merits of X-rays and skeletal scintigraphy. Lancet 10:613–657

Perkins AC, Pimm MV (1991) Immunoscintigraphy: practical aspects and clinical applications. Wiley-Liss, New York

Peters AM (1994) The utility of [99mTc] HMPAO-leukocytes for imaging infection. Semin Nucl Med 14:110–127

Pinals RS, Short CL (1965) Calcific periarthritis involving multiple sites. Arthritis Rheum 8:462

Pistenma DA, McDougall IR, Kriss JP (1975) Screening for bone metastases. JAMA 255:46–50

Pitt WR, Sharp PF (1985) Comparison of quantitative and visual detection of new focal bone lesions. J Nucl Med 26:230–236

Reske SN, Karstens JH, Gloeckner MW et al (1989) Radioimmunoimaging of bone marrow. Results in patients with breast cancer and skeletal metastases and patients with malignant melanoma. Lancet 2:299–301

Resnick D, Niwayama G (1983) Enthesis and enthesopathy. Radiology 146:1–9

Rhodes BA, Zamora PO, Newall KD et al (1986) Technetium-99m labeling of murine monoclonal antibody fragments. J Nucl Med 27:685–693

Richards P (1960) A survey of the production at Brookhaven National Laboratory of Radioisotopes for medical research. Comitato Nazionale Ricerche Nucleari, Rome (Congresso nucleare, vol 2)

Roger LF (1992) Radiology of skeletal trauma, 3rd edn. Churchill Livingstone, New York

Rosenthall L (1987) Synovitis. In: Fogelman I (ed) Bone scanning in clinical practice. Springer, Berlin Heidelberg New York

Rosenthall L, Kaye M (1975) Technetium-99m-pyrophosphate kinetics and imaging in metabolic bone disease. J Nucl Med 16:33–39

Rubin P (1964) Dynamic classification of bone dysplasias. Year Book Med Publishers, Chicago

Russin LD, Staab EV (1976) Unusual bone-scan findings in acute osteomyelitis: case report. J Nucl Med 17:617–619

Schajowicz F (1981) Tumors and tumor-like lesions of bone and joints. Springer, Berlin Heidelberg New York

Serafini AN (1976) Paget's disease of the bone. Semin Nucl Med 6:47–58

Shapiro RF, Utsinger PD, Wiesner KB et al (1976) The association of HL-A B27 with Forestier's disease (vertebral ankylosing hyperostosis). J Rheumatol 3:4–8

Sly WS, Hu PY (1995) The carbonic anhydrase II deficiency syndrome: osteopetrosis with renal tubular acidosis and cerebral calcification. In: Scriver CR, Beaudet AL, Sly WS, Valle D (eds) Metabolic and molecular bases of inherited disease III, 7th edn, chap 137. McGraw-Hill, New York, pp 4113–4124

Smith ML (1986) Quantitative 99mTc diphosphonate uptake measurements. In: Fogelman I (ed) Bone scanning in clinical practice. Springer, Berlin Heidelberg New York

Silverstein EB, Francis MD, Tofe AJ et al (1975) Distribution of ^{99m}Tc-Sn-diphosphonate and free ^{99m}Tc-pertechnetate in selected soft and hard tissues. J Nucl Med 16:58–61

Steinstraesser A, Schorlemmer HU, Schwarz A et al (1988) A novel 99mTc-labeled antibody for in vivo targeting of granulocyte. J Nucl Med 29:925

Subramanian G, McAfee JG (1971) A new complex of ^{99m}Tc for skeletal imaging. Radiology 99:92–96

Suzuki Y, Hisada K, Takeda M (1974) Demonstration of myositis ossificans by ^{99m}Tc pyrophosphate bone scanning. Radiology 111:663–664

Treves ST, Connolly LP, Kirkpatrick AB et al (1995) Bone. In: Treves ST (ed) Pediatric nuclear medicine, 2nd edn. Springer, New York Berlin Heidelberg

Tumeh SS, Beadle G, Kaplan WD (1985) Clinical significance of solitary rib lesions in patients with extraskeletal malignancy. J Nucl Med 26:140–143

Waldfogel FA, Vasey H (1980) Osteomyelitis: the past decade. N Engl J Med 303:360–370

Werner JA, Botvinick EH, Shames DM et al (1977) Clinical application of technetium-99m stannous pyrophosphate infarct scintigraphy. West J Med 127:464–478

Widding A, Stilbo I, Hansen SW et al (1990) Scintigraphy with nanocolloid Tc-99m in patients with small cell lung cancer, with special reference to bone marrow and hepatic metastasis. Eur J Nucl Med 16:717–719

Williams PL, Warwick R, Dyson M et al (1989) Gray's anatomy, 37th edn. Churchill Livingstone, Edinburgh

Yang WJ, Bahk YW, Chung SK et al (1994) Pinhole skeletal scintigraphic manifestations of Tietze's disease. Eur J Nucl Med 21:947–952

Yeh KA, Fortunato L, Ridge JA et al (1995) Routine bone scanning in patients with T1 and T2 breast cancer: a waste of money. Ann Surg Oncol 2:319–324

Yuasa K, Sugimura K, Okizuka AH et al (1991) Bone infarction and fat island appearing as local defects in radionuclide bone marrow imaging. Kaku Igaku 28:91–96

Zwas ST, Elkanovitch R, Frank G (1987) Interpretation and classification of bone scintigraphic findings in stress fractures. J Nucl Med 82:452–457

Part IV Tumors

CHAPTER 27

Molecular Imaging in Oncology

27

DAVID A. PIWNICA-WORMS, GARY D. LUKER, CAROLYN ANDERSON, RICHARD L. WAHL

Contents

through noninvasive in vivo investigation of cellular and molecular events involved in normal and pathologic processes.

To begin to investigate molecular signals in oncology in vivo, researchers already have developed and characterized methods for imaging endogenous proteins, such as receptors, transporters or enzymes and their respective functions. More recently, methods also have been validated to analyze the specific expression of a gene of interest, typically by quantifying the activity of an exogenous reporter gene. In the following sections, we will review several principles of selected applications of radiotracer technologies for molecular imaging in oncology with PET and SPECT.

27.1
Introduction

A first draft of the human genome has been completed (Pennisi 2000) and, while much work remains to complete a refined map, this accomplishment is expected to lead to new medical treatments, drugs, and ultimately cures previously thought impossible. In this emerging "post-genomic era", wherein functionality will be added to this vast array of genetic information, opportunity exists for imaging to play a significant role in both basic and translational research as related to functional genomics. Herein, the overall objective is focused on adding function through molecular imaging. Molecular imaging is broadly defined as the characterization and measurement of biological processes in living animals, model systems and humans at the cellular and molecular level using remote imaging detectors. The goal is to advance our understanding of biology and medicine

27.2
Molecular Imaging of Endogenous Protein Function in Cancer in vivo

27.2.1
Glucose Transporters and Hexokinase

Positron emission tomography (PET) with [18F]fluoro-2-deoxy-D-glucose (FDG) may be characterized as one of the first "molecular imaging" techniques validated in the basic sciences as well as in a clinical setting. Hexokinase catalyzes the initial, rate-limiting step in glycolysis, which is phosphorylation of glucose to glucose-6-phosphate. An analog of glucose, fluoro-2-deoxy-D-glucose, enters cells through the same pathways of facilitated diffusion as glucose, i.e., via glucose transporters (Glut1 and Glut4), and is subsequently phosphorylated by hexokinase. However, FDG-6-phosphate is not further metabolized significantly and, because it is negatively charged, remains trapped within cells. Imaging net glucose transport and hexokinase activity can be achieved by incorporating [18F] into FDG, enabling detection of the trapped, phosphorylated metabolite by PET. Pharmacological doses of FDG actually inhibit glycolysis in vivo, but imaging with this compound can be performed safely because only tracer amounts are needed for PET. Imaging of glycolysis with 2-[^{18}F]FDG is a widely used method for detecting a specific molecular target in vivo, and this radiotracer has an established clinical utility for detection of normal and pathologic function in brain and heart. Additionally, imaging with PET and 2-[^{18}F]FDG is becoming increasingly important in the diagnosis and management of patients with cancer, particularly for detection of metastatic disease (reviewed in Hoekstra et al. 2000). In the context of molecular imaging in oncology, imaging with 2-[^{18}F]FDG fulfills all criteria for selective imaging of a specific enzymatic activity, but is limited per se in that the target enzyme, hexokinase, is not specific for cancer. Thus, the success of the approach does not reside in a high degree of selectivity in terms of the molecular target, but rather, the clinical utility is manifest in the enhanced functional activity of hexokinase (and transport) displayed in malignancy. This represents an example of the interesting interplay between imaging a protein target versus protein function and will be further reviewed in detail below.

While FDG was developed as an agent to image the high glycolytic rate of the brain, glucose is utilized to supply energy to many tissues in the body. The tracer FDG is used more commonly in most PET centers to image the high glycolytic rates of many cancers than the metabolic activity of the brain. Warburg recognized the increased rates of glycolysis present in tumors over 70 years ago (Warburg 1931). Several groups have shown the feasibility of FDG for targeting animal tumors, human tumor xenografts of a variety of types, and there is now an extensive clinical literature showing efficacy of PET with FDG in a variety of human tumors (Larson et al. 1980; Wahl et al. 1991, 1993 b). Thus, FDG PET could be considered the first routine molecular imaging procedure performed using dedicated quantitative PET devices.

The rate of glucose utilization is closely regulated. While hexokinase is believed to be the rate limiting enzyme in the use of glucose in the brain, and thus the target enzyme activity imaged by FDG, a number of factors affect FDG uptake in the body beyond hexokinase levels (Marshall et al. 1979), depending on the tissue. As outlined above, FDG is phosphorylated to FDG-6-phosphate by hexokinase, and retained in the cell as FDG-6-phosphate. However, FDG must first reach the target cell. Thus, like all radiotracers, delivery to the target tissue is necessary. Since cancers typically have low rates of blood flow versus normal tissues, delivery can be a limiting factor. Thus, preliminary data from various investigators has shown that delivery of tracer (flow) and FDG uptake are reasonably well correlated (Zasadny et al. 2000). This is not unique to FDG. Rather, all molecular imaging tracers given intravenously must reach tumor targets by flow. Thus, if flow is very low, net uptake must, by definition, be low. Vascular permeability in tumors is rarely an issue, so passage through vascular endothelium by FDG is not a major barrier. However, if there is high interstitial pressure in the tumor, it may be more difficult for FDG to reach the center of the tumor.

The transport kinetics of FDG across tumor cell membranes potentially can also affect the final imaging signal. It has been shown by several groups that Glut1 protein levels are increased on the cell surface of many cancer cells (Brown and Wahl 1993; Younes et al. 1996). Glut1 levels also are increased by placing cells in an hypoxic environment (Clavo et al. 1995). In hypoxia, accumulation of FDG and accumulation of FDG-6-phosphate increase as glucose transporter levels increase. While hexokinase activity may also change, an adaptive response of hypoxic tumor cells is increased expression of glucose transporters on the surface. This phenomenon can preserve, in all likelihood, glucose metabolism at levels needed for survival. Several groups have shown that the tumor cells with highest levels of Glut1 expression are located in what are apparently hypoxic areas of tumors, near necrotic regions (Kubota et al. 1993; Brown et al. 1996).

The importance of transport of FDG has been well shown in the myocardium and skeletal muscle. These tissues have predominantly Glut4 glucose transporters, which are insulin sensitive. Soon after insulin is given or myocardial cells become ischemic, Glut4 moves from the cytoplasm to the cell surface, increasing transport of glucose quite rapidly (Brosius et al. 1997). This phenomenon is generally regarded as one in which transport, and not hexokinase levels, are governing the rate of FDG uptake. Similarly, humans with partial Glut1 deficiency are prone to retarded mental development and seizures, which respond to glucose-rich diets and which are worsened by stresses reducing glucose transport (Klepper et al. 1999). In addition, transgenic mice with high levels of Glut4 or Glut1 on skeletal muscle are typically hypoglycemic or at least, less prone to diabetes, than animals with lower levels of glucose transporter expression (Marshall et al. 1999). Thus, glucose transporter levels can affect tissue uptake of FDG.

Experimental hyperglycemia or glucose clamp studies have shown high glucose levels reduce tumor FDG uptake (Wahl et al. 1992). Somewhat paradoxically, low glucose levels induced by insulin do not increase tumor FDG uptake (Torizuka et al. 1998). This is because in non-diabetic animal models insulin redirects FDG to Glut4-positive muscle cells and the myocardium, reducing blood bioavailability of the tracer and also reducing the available FDG to tumor tissue. There are some circumstances in diabetics where insulin can increase FDG uptake levels in tumors, but these must be precisely timed so that glucose levels are low as well as insulin levels, so that FDG is not redirected to normal tissues. The success of FDG for tumor imaging is substantially dependent on targeting glucose utilization in the fasted state. Only then is myocardial glucose metabolism switched to a fatty acid or ketone body mode, and muscle glycolytic needs are quite modest. With a low background level of muscle uptake, tumors can be effectively detected with PET.

Studies performed in vitro and in animals with breast cancer have shown that FDG uptake in tumors is positively correlated with Glut1 levels and with the number of viable cancer cells (Brown et al. 1996) (Fig. 27.1). To date, correlations of glucose metabolism with Glut1 have been better than with hexokinase levels, but neither is precisely correlated (Aloj et al. 1999). It should be realized that just because hexokinase message is present does not mean there are high levels of protein. Similarly, immunohistochemical staining for protein does not necessarily indicate that hexokinase is active, as ATP is required, and,

furthermore, hexokinase (often HK2) must be in close association with the mitochondria to be effective (Wilson 1995; Golshani-Hebroni et al. 1997; Mathupala et al. 1997). Another clear result of animal studies is that FDG uptake is not unique to cancer cells. Since FDG broadly traces glucose metabolism rather than tumor-specific glucose metabolism, FDG uptake can occur in other glucose-utilizing cells, such as macrophages and leukocytes (Mochizuki et al. 2001). This uptake can be confusing in clinical interpretation of images, but it illustrates that the biology is being properly mapped. The pitfall is that the tracer is not specific to only tumors. Indeed, detection of infections and inflammation with FDG is a very promising area of research. Thus, a desire exists for more tumor-specific agents, which may be more restricted in their spectrum of activity.

Increased glucose utilization is usually present at baseline in cancer cells and other normal tissues. However, glucose utilization rises when cells are stimulated. Thus, glucose utilization is a potential downstream readout of a variety of signaling pathways, which stimulate or repress cancer cells. For example, estrogen receptor (ER) binding with agonists rapidly increases FDG uptake in ER-positive normal rodent uterus. Indeed, it has been shown that increased FDG uptake can occur in breast cancers following receptor binding with an ER-binding drug, tamoxifen (Dehdashti et al. 1999). While tamoxifen is primarily an estrogen antagonist, the drug also possesses a partial agonist effect causing the increased FDG uptake. The tracer signal from scans before and after treatment

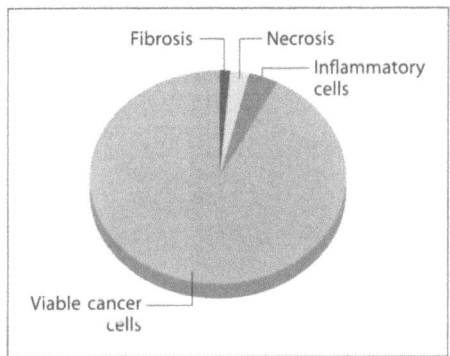

Fig. 27.1. Distribution of [^{18}F]fluoro-2-deoxy-D-glucose in breast cancer

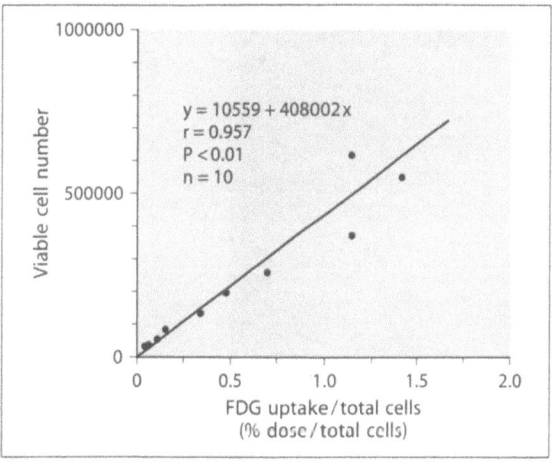

Fig. 27.2. [^{18}F]Fluoro-2-deoxy-D-glucose uptake in total cells versus viable cells. (Reproduced with permission)

suggests that an increase in the FDG uptake after tamoxifen predicts a response. Again, this general tracer can be useful for the downstream readout of more complex processes such as receptor ligand binding and signal transduction.

Glucose utilization is rather well correlated with viable cell number in vitro and in vivo (Higashi et al. 1993) (Fig. 27.2). This property makes FDG an attractive agent for monitoring cytotoxic cancer therapy response. In a wide number of studies, effective chemotherapy reduces FDG uptake substantially and often completely (Wahl et al. 1993; Smith 1998). The molecular/metabolic changes due to effective therapy appear to precede changes in tumor anatomy, such as in breast cancer treatment response. Rapid declines in tracer uptake are generally indicative of a good response and this has been shown in several studies. A rise in FDG uptake with cytotoxic therapy is generally indicative of a failure to respond.

The ability to quantify FDG uptake in vivo with PET makes it a very attractive tool. However, users of the technology must be aware that our detection of FDG with PET is not at a microscopic level. Thus, current imaging devices are able to detect very small tumors, smaller than those seen by CT in many cases, but often fail to find tumors <5 mm in size, particularly in a setting of higher background activity about the tumors. This limitation will remain, although PET scanners will almost certainly increase in their sensitivity and resolution.

At present, in many common cancers, PET can detect lesions which are not clearly identified by CT or other methods. This includes common cancers like lung cancer, colon cancer, melanoma and lymphoma. Identifying the abnormality on PET is useful, but determining where it is precisely located is challenging. The clear identification of variable normal structures and their uptake of FDG is needed. Software approaches have been employed in this regard, but are challenging if there is not similar positioning of patients between studies (Wahl et al. 1993). Combined PET/CT devices to determine both the presence and precise anatomic location of abnormal foci of FDG uptake are emerging as highly desirable methods to combine information from differing imaging modalities (Beyer et al. 2000) (Fig. 27.3).

A complete description of the clinical uses of FDG PET imaging in cancer is beyond the scope of this chapter. However, the major areas fall into lesion characterization, cancer metastasis detection, and

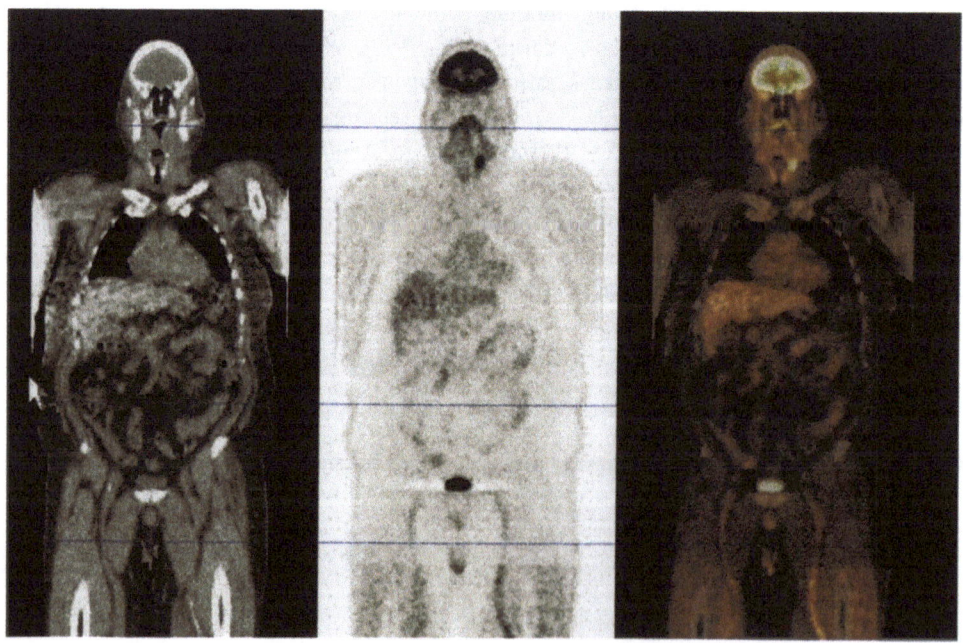

Fig. 27.3. Whole body fusion computed tomography (CT) and [¹⁸F]fluoro-2-deoxy-D-glucose positron emission tomography (FDG PET) images. The *left image* represents a mid section coronal plane reconstructed from a whole body transaxial CT exam, while the *middle image* represents a matched coronal tomogram from an FDG PET exam of the same patient. *On the right* is a false-color fusion image of the two exams

treatment response assessment. Characterization of solitary pulmonary lesions as most likely malignant or most likely benign can be achieved using FDG PET imaging. Sensitivity values of FDG PET for detecting cancer in solitary pulmonary nodules has been reported to be about 95% for lesions >7 mm in size, with a specificity of about 80%. False negative results can occur in hypocellular tumors like bronchioloalveolar cancers and in diabetics. False positive scans are reported in some inflammatory processes such as tuberculosis (Maisey et al. 1999).

For tumor staging, FDG PET is more accurate than CT in several common cancers such as lung cancer, colorectal cancer, melanoma, lymphoma and others. FDG PET is not as accurate in the detection of brain metastases as in other locations due to the intense normal FDG uptake in the brain. For treatment response, several papers show that rapid declines in FDG uptake after treatment are indicative of a likely longer-term response of cancer to treatment.

Some cancers, such as prostate cancer, are not as FDG avid as, for example, lung cancer. Thus, while FDG is an excellent tracer of glucose metabolism, it is a less than perfect tracer of cancer. In addition, FDG is not identical to glucose in its biodistribution patterns. For example, FDG is substantially excreted in the urine unchanged, but glucose in a non-diabetic is typically reabsorbed nearly totally by the kidneys. Thus, while similar to glucose, FDG is not identical.

In sum, FDG represents the first PET molecular imaging agent that has achieved widespread clinical application in cancer imaging. While it traces glucose metabolism and may substantially depend on hexokinase activity levels for accumulation, overall accumulation in tumors is dependent on a variety of factors including delivery. All molecular tracers will face similar delivery issues. The FDG experience is a clear success story in the field of oncologic molecular imaging.

27.2.2
Somatostatin Receptors-Type 2

The earliest reported peptide-based tumor receptor imaging agents that were developed and investigated in cancer patients were radiolabeled analogs of the hormone somatostatin. Somatostatin is a 14-amino-acid peptide involved in the regulation and release of a number of hormones, including growth hormone, thyroid-stimulating hormone and prolactin. Somatostatin receptors (SSTR) are found on the cell surface and occur in a number of different normal organ systems including the central nervous system, the gastrointestinal tract, and the exocrine and endocrine pancreas (Guillemin 1978; Reichlin 1983a, b). A large number of human tumors express somatostatin receptors (Reubi et al. 1990), particularly somatostatin receptor-subtype 2 (SSTR2). Somatostatin has a very short biological half-life, but analogs such as octreotide have much longer residence times (Bauer et al. 1982). Fig. 27.4 shows structures of several of the SSTR2 ligands that will be discussed.

Octreotide, an 8-amino acid somatostatin analog, has been labeled with ^{111}In using DTPA (Bakker et al. 1991a) and is approved for human use in the United States and Europe as a diagnostic imaging agent for neuroendocrine tumors (Krenning et al. 1992, 1993). This agent shows low nanomolar affinities for somatostatin receptors in neuroendocrine tumors (Bakker et al. 1991), and >80% of ^{111}In-DTPA-octreotide clears rapidly through the kidneys as the intact complex (Krenning et al. 1992). The combination of sensitivity, specificity and favorable clearance characteristics has made this radiopharmaceutical the prototype for a new area of peptide-based receptor ligands for cancer imaging. Figure 27.5 shows a whole-body image taken 24 h post-injection of ^{111}In-DTPA-octreotide in a woman with Zollinger-Ellison syndrome and an elevated gastrin level. There was no evidence for gastrinoma; however, the patient had a history of a long-standing, stable meningioma adjacent to the left cavernous sinus, which corresponded with increased activity of ^{111}In-DTPA-octreotide.

Somatostatin analogs, including octreotide, RC-160 and the peptide P829 developed by Diatide, Inc., also have been labeled with a variety of other radionuclides for diagnostic imaging by gamma scintigraphy or PET. These include ^{99m}Tc (Pearson et al. 1996; Vallabhajosula et al. 1996; Bangard et al. 2000; Decristoforo et al. 2000), ^{64}Cu (Anderson et al. 1995, 2001), ^{68}Ga (Smith-Jones et al. 1994; Henze et al. 2001), ^{18}F (Guhlke et al. 1994) and ^{86}Y (Brockmann et al. 1995; Wester et al. 1997). In the following paragraphs, agents that have been evaluated in clinical studies are discussed.

^{68}Ga and ^{67}Ga were first labeled to octreotide using the bifunctional chelator (BFC) desferrioxamine-B (DFO) (Smith-Jones et al. 1994; Stolz et al. 1994) to evaluate the potential of ^{68}Ga-DFO-octreotide as a PET agent for detection of neuroendocrine cancer.

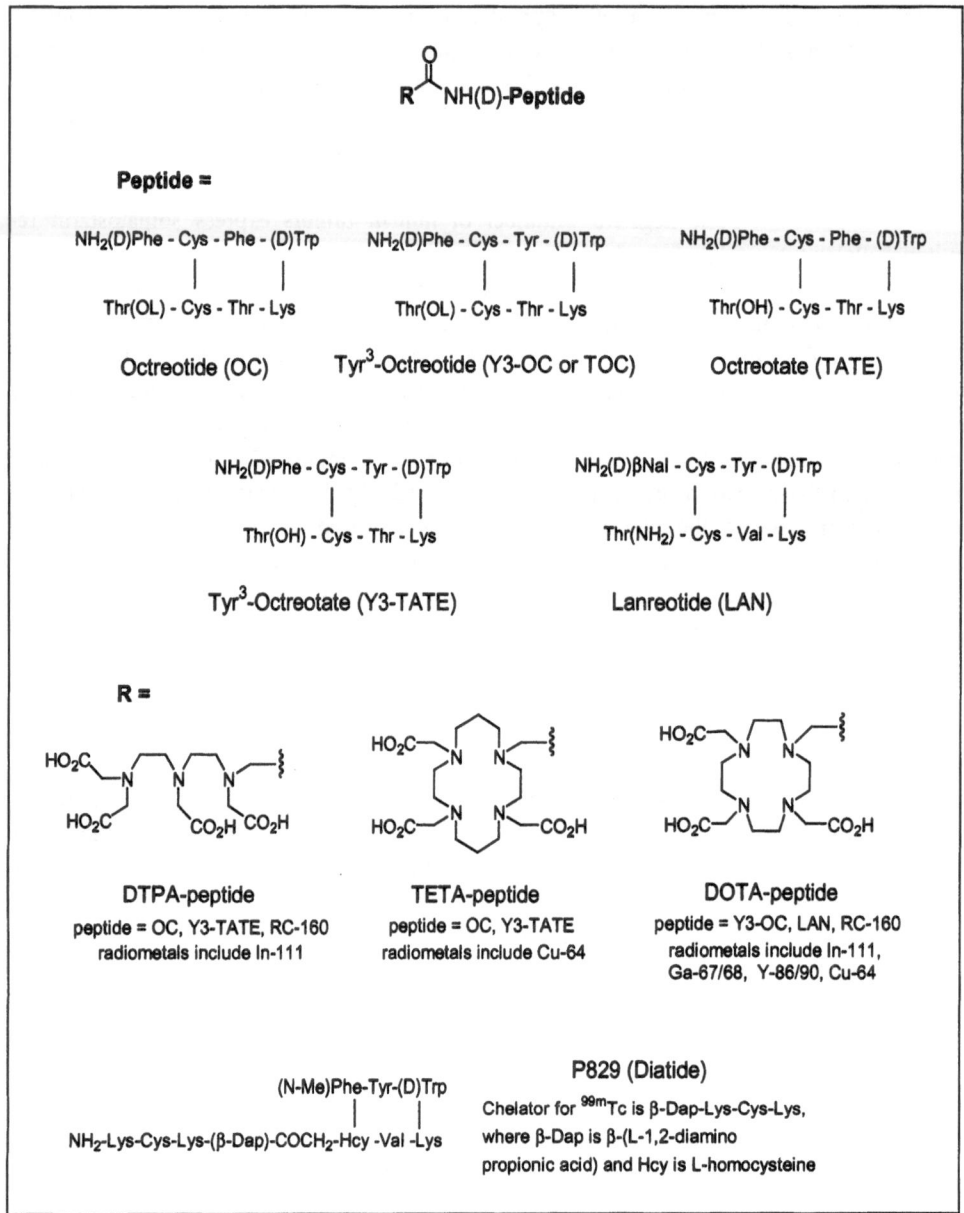

Fig. 27.4. Somatostatin analogs used for diagnostic imaging and therapy of somatostatin receptor-subtype 2-positive tumors

Although animal studies of this agent were promising, to date no human studies have been reported. More recently, ^{68}Ga-DOTA-Tyr^3-octreotide (^{68}Ga-DOTATOC) was evaluated as a PET-imaging agent in patients with meningiomas (Henze et al. 2001). This agent shows considerable promise, since even small meningiomas (7-mm diameter) showed high tracer uptake that was clearly delineated from normal tissue. The results with ^{68}Ga-DOTATOC were significantly better than those reported with ^{111}In-DTPA-OC and SPECT, where there were difficulties in detecting meningiomas with diam-

eters < 2.7 cm (Bohuslavizki et al. 1996). The increase in resolutions for smaller tumors using ^{68}Ga-DOTA-TOC is likely a result of the combination of PET, which in general has higher resolution than SPECT, and the marked increase in SSTR2 affinity of ^{68}Ga-DOTATOC (IC$_{50}$ = 2.5 nM) compared to ^{111}In-DTPAOC (IC$_{50}$ = 22 nM).

Octreotide has been conjugated to the BFC TETA, for labeling with ^{64}Cu (Anderson et al. 1995). ^{64}Cu-TETA-OC showed high affinity for the SSR both in vitro and in vivo and cleared primarily through the

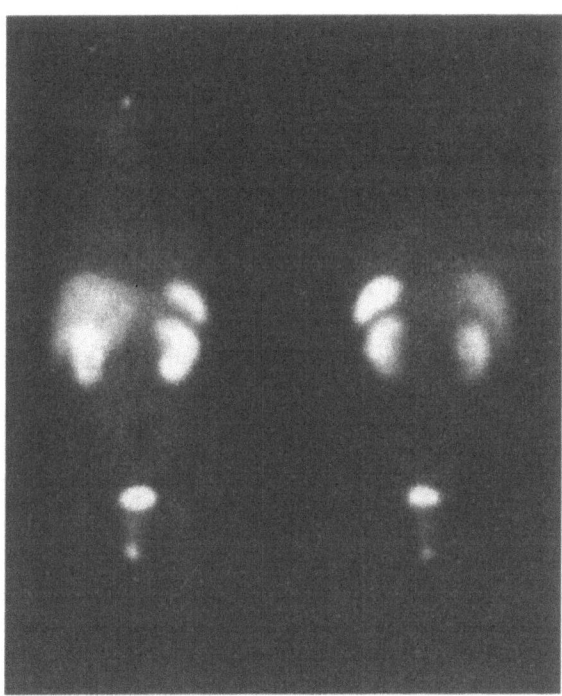

Fig. 27.5. Whole body gamma scintigraphy image of a 64-year-old woman who was injected with ^{111}In-DTPA-octreotide and imaged 24 h post-injection. The patient was under evaluation for elevated gastrin levels and recurrent peptide ulcers. Anterior (*left*) and posterior (*right*) images showed no evidence of gastrinoma; however, the intense increased activity near the base of the skull was consistent with the patient's history of a long-standing stable meningioma adjacent to the left cavernous sinus. The activity observed in the liver, kidneys, spleen and bladder is the typical normal organ clearance pattern for ^{111}In-DTPA-octreotide. (Used with permission, MIR Nuclear Medicine Web-based Teaching File, copyright Washington University School of Medicine)

kidneys, with very low liver accumulation. ^{64}Cu-TETA-OC was evaluated as a PET-imaging agent for neuroendocrine tumors in eight patients (Anderson et al. 2001). Preliminary results showed that in two patients, ^{64}Cu-TETA-OC and PET detected more SSTR-positive lesions than the currently used agent, ^{111}In-DTPA-OC, and gamma scintigraphy/SPECT. In one patient, ^{111}In-DTPA-OC and gamma scintigraphy showed low uptake in a lung lesion that was not observed with ^{64}Cu-TETA-OC and PET. In addition to PET imaging, ^{64}Cu-labeled somatostatin analogs have also been evaluated for targeted radiotherapy in tumor-bearing animal models, with promising results (Anderson et al. 1998; Lewis et al. 1999).

Technetium and rhenium have been complexed to somatostatin analogs via chelates formed from peptides containing the amino acids glycine, cysteine and lysine (Pearson et al. 1996). Using small peptides as chelators for peptide receptor ligands greatly simplifies the synthesis of these chelate-receptor ligand bioconjugates. Two of the peptides evaluated in vitro by Pearson et al., ^{99m}Tc-labeled P587 and P829 (Pearson et al. 1996), were studied further in animal models (Vallabhajosula et al. 1996). These ^{99m}Tc-labeled peptides were both found to have high binding affinity to the SSR in tumor-bearing rats and had clearance characteristics favorable for further clinical investigations (Palestro et al. 1997). The ^{99m}Tc-labeled peptide, P829 (NeoTect®), was approved by the FDA in August, 1999 for imaging lung cancer.

Other ^{99m}Tc-labeled somatostatin analogs have also been investigated as alternatives for ^{111}In-DTPA-OC in the imaging of somatostatin receptor-positive tumors. The peptide hydrazinonicotinyl-TOC (HYNIC-TOC) has been labeled with ^{99m}Tc in the presence of the co-ligands tricine and ethylenediamine diacetic acid (EDDA) (Decristoforo et al. 2000). In this study, EDDA was found to be the most optimal co-ligand with respect to having the best tumor:non-tumor ratios in a tumor-bearing mouse model. In another study, ^{99m}Tc-tricine-HYNIC-TOC was evaluated in patients with SSTR-positive tumors and compared directly with ^{111}In-DTPA-OC (Bangard et al. 2000). The plasma clearance of ^{99m}Tc-tricine-HYNIC-TOC was less than optimal, since 10% of the activity remained in the plasma after 3 h, and activity cleared very slowly thereafter; however, uptake in the liver, spleen, kidney and lung was comparable between ^{99m}Tc-tricine-HYNIC-TOC and ^{111}In-DTPA-OC. A lesion-by-lesion comparison between ^{99m}Tc-tricine-HYNIC-TOC and ^{111}In-DTPA-OC showed comparable results, with ^{99m}Tc-tricine-HYNIC-TOC being somewhat more sensitive for imaging abdominal lesions than ^{111}In-DTPA-OC while ^{111}In-DTPA-OC had advantages for imaging hepatic lesions.

An exciting development of the diagnostic imaging of SSTR-positive tumors is the application of radiolabeled somatostatin analogs for targeted radiotherapy of cancer (for a review of the patent literature, see Anderson and Lewis 2000). The first reported use of a radiolabeled somatostatin analog for therapy was the administration of high doses of ^{111}In-DTPA-OC, showed antiproliferation effects of SSTR-positive tumor in one patient (Krenning et al. 1994). Although ^{111}In is generally used for gamma scintigraphy due to its high abundance of gamma emissions, ^{111}In also decays by electron capture, which results in the emission of several low energy Auger and conversion electrons per decay. Since this report, there have been several other clinical studies with ^{111}In-DTPA-OC for therapy reported in the literature, with generally favorable results. In a report where 21 patients were treated with high multiple radiotherapeutic doses of ^{111}In-DTPA-OC at the University of Rotterdam and in Brussels, no major side effects were observed (Krenning et al. 1999). Of these 21 patients who received cumulative treatments of at least 20 GBq (540 mCi), eight patients resulted in stable disease, while six patients showed tumor shrinkage. In a study in the US, 27 patients completed treatments with high doses of ^{111}In-DTPA-OC at LSU Medical Center while another 13 received treatment at Yale University (McCarthy et al. 2000). Of these patients, 17/27 and 9/13, respectively, had objective responses which included both biochemical (improvement in hormone levels) and radiographic (tumor shrinkage) responses.

Another agent in clinical therapy trials for patients with SSTR-positive tumors is ^{90}Y-DOTATOC. Novartis has sponsored a Phase I clinical trial and has established the safety and tolerability of the dose of ^{90}Y-DOTATOC (or ^{90}Y-SMT 487) selected for further study (Smith et al. 2000). In another clinical trial carried out between research groups in Milan and Basel, the dosage, safety and therapeutic efficacy of ^{90}Y-DOTATOC was evaluated in 30 patients (Paganelli et al. 2001). No major acute undesirable reactions were observed up to 2.59 GBq (70 mCi) per cycle, although five patients had moderate gastrointestinal toxicity, four had grade 2 nausea and one had grade 1 vomiting. Three patients developed delayed grade 1 toxicity at a dose of 6.66–7.77 GBq (180–210 mCi). The response rate was impressive, with 63.3% of patients having stable disease and 23.3% showing tumor regression (partial or complete responses).

Lutetium-177-DOTA-Tyr[3]-octreotate has been evaluated pre-clinically (de Jong et al. 2001; Lewis et al. 2002), and preliminary clinical studies have recently been presented (Kwekkeboom et al. 2001). Twenty-six patients had cumulative doses of at least 300 mCi of ^{177}Lu-DOTA-Tyr[3]-octreotate, and side effects included nausea (31%), vomiting (9%) and mild abdominal discomfort (11%). Mild blood toxicities were also observed. Tumor shrinkage was observed in eight patients, with one patient having a partial remission.

A novel approach of combining gene therapy with targeted radiotherapy using radiolabeled somatostatin analogs is being investigated at the University of Alabama-Birmingham (Rogers et al. 2000). The adeno-

virus encoding the gene for SSTR2, under the control of the cytomegalovirus promoter (AdCMVSSTr2), was administered i.p. to mice bearing i.p. SK-OV-3.ip1 tumors to induce tumor expression of SSTR2. This resulted in tumor localization of i.p. injected ^{111}In-DTPA-OC (Rogers et al. 1999) as well as ^{64}Cu-TETA-OC (Buchsbaum et al. 1999). Therapy experiments were performed in the SK-OV-3.ip1 model with two i.p. injections of AdCMVSSTr2 and two corresponding i.p. injections of ^{64}Cu-TETA-OC 2 days after each viral injection. This resulted in increased survival time of the tumor-bearing mice (Rogers et al. 2000).

27.2.3
Sodium/Iodide Symporter (NIS)

NIS is a membrane glycoprotein that couples transport of iodine and sodium into cells, using the sodium gradient maintained by the sodium/potassium ATPase. Secondary active transport of iodine occurs in the thyroid gland, lactating breast epithelium, salivary glands, and gastric mucosa (Carrasco 1993). Radioactive isotopes of iodine (such as ^{123}I and ^{131}I) or ^{99m}Tc-pertechnetate have a long and established record in the practice of nuclear medicine for diagnosis and targeted therapy of thyroid gland pathology, exploiting the capacity of functional NIS to concentrate these tracers within target cells (Tazebay et al. 2000). NIS transport is maintained in thyroid malignancies, albeit at decreased activity relative to normal thyroid tissue, enabling detection of recurrent and/or metastatic thyroid cancers as sites of enhanced accumulation of radiotracer ("hot spots"). These imaging techniques are commonly performed in most nuclear medicine departments, and they are an integral part of management and therapy of patients with thyroid cancer. However, recent studies indicate that functional imaging of NIS may have significance for detection and therapy of cancers other than thyroid.

Boland and co-workers used an adenoviral vector to transfer the NIS gene to a variety of different tumor cell lines (Boland et al. 2000). NIS functioned normally in these cells, as evidenced by greatly enhanced accumulation of ^{123}I and sensitivity to killing by [131]I. Adenoviral-mediated delivery of NIS to tumors implanted in mice also conferred specific accumulation of ^{123}I as detected by imaging and analysis of tumor specimens. Thus, identifying NIS function in vivo may become important in the future for monitoring efficacy of this gene therapy approach to targeted radiotherapy of cancers.

Recent work also suggests possible diagnostic and therapeutic applications of imaging NIS in breast cancer. In mouse models of breast adenocarcinoma, some tumors expressed functional NIS, as evidenced by specific accumulation of ^{99m}Tc-pertechnetate on both imaging studies and direct analysis of tumor specimens (Tazebay et al. 2000). Tumors lacking NIS and normal breast tissue did not accumulate the radiotracer. Although patients with breast cancer were not imaged in this study, expression of NIS was detected by immunohistochemistry in more than 80% of breast cancer specimens. Further studies are needed to validate these initial observations, but the data imply that molecular imaging of NIS function could become an important tool in detection and follow-up of primary and metastatic breast cancer.

27.2.4
Multidrug Resistance P-Glycoprotein

Emergence of multidrug resistance (MDR) is a major obstacle to successful chemotherapy of cancer. Several of the first characterized mechanisms of MDR include transporter-mediated resistance conferred by increased expression of the M_r 170,000 transmembrane glycoprotein, P-glycoprotein (Pgp), the product of the *MDR1* gene (Juliano and Ling 1976; Gros et al. 1986; Shen et al. 1986; Gottesman and Pastan 1993; Bosch and Croop 1996) and a related M_r 190,000 membrane glycoprotein, the multidrug resistance-associated protein (MRP1) (Cole et al. 1992; Flens et al. 1996). Pgp and MRP1 are members of the ATP-binding cassette (ABC) superfamily of membrane transport proteins (Gottesman and Pastan 1993) and confer resistance to an overlapping array of structurally and functionally unrelated toxic xenobiotics, natural product drugs and, in the case of MRP1, conjugated compounds (Lautier et al. 1996; Keppler et al. 1997). Several other ABC transporters have been reported to be associated with MDR, including the lung resistance protein (LRP) (Scheper et al. 1993), the breast cancer resistance protein (BCRP/MXR/ABCP; a recently cloned ABC "half-transporter") (Allikmets et al. 1998; Doyle et al. 1998; Miyake et al. 1999), and MRP3 (Zeng et al. 1999). Cells in culture exhibiting MDR by selection in cytotoxic drugs or transfection with these recombinant transporters generally show reduced net drug accumulation and altered intracellular drug distribution.

MDR1 Pgp and related ABC transporters have been targets for cancer therapy on two fronts. First, reversal of multidrug resistance in tumor cells by nontoxic agents that block the transport activity of these ABC proteins has been an important target of pharmaceutical development (Ford and Hait 1990). When co-administered with a cytotoxic agent, these antagonists, known as MDR modulators, enhance net accumulation of cytotoxic compounds within the tumor cells. Several high potency modulators discovered by targeted synthesis or combinatorial chemistry in combination with high-throughput screening are now in clinical trials (Gaveriaux et al. 1991; Hyafil et al. 1993; Dantzig et al. 1996; Germann et al. 1997; Rabindran et al. 1998; Newman et al. 2000). Second, transgenic expression of the *MDR1* gene has been explored for hematopoietic cell protection in the context of cancer chemotherapy (Podda et al. 1992; Sorrentino et al. 1992; Hanania et al. 1995), wherein Pgp could protect hematopoietic progenitor cells from chemotherapy-induced myelotoxicity. Hematopoietic cells transduced via retroviral-mediated transfer of the *MDR1* gene have shown preferential survival after treatment of the animal with MDR drugs (Hanania et al. 1995) and recent pilot clinical data support the approach (Moscow et al. 1999). For proper use of MDR modulators as well as monitoring MDR gene therapy in chemotherapeutic protocols, identification of transporter-mediated resistance could guide the choice of agents and provide important prognostic information for cancer patients. Thus, non-invasive molecular imaging with a transport substrate serving as a surrogate marker of chemotherapeutic agents may identify those tumors and tissues in which ABC transporter proteins are expressed and active.

To achieve this goal, several gamma-emitting compounds have been synthesized, validated and characterized as transport substrates for *MDR1* Pgp (Piwnica-Worms 2000). One of the best characterized is *hexakis*(2-methoxyisobutyl isonitrile) ^{99m}Tc(I) (^{99m}Tc-Sestamibi), a widely available radiopharmaceutical which may enable scintigraphic analysis of MDR (Piwnica-Worms et al. 1993, 1995; Ballinger et al. 1995a, b; Cordobes et al. 1996). The targeted synthesis and validation of other single-photon radiopharmaceuticals (Herman et al. 1995; Ballinger et al. 1996, 1997; Luker et al. 1997a; Crankshaw et al. 1998; Sharma and Piwnica-Worms 1999; Chen et al. 2000) and positron emission tomography (PET) agents (Hendrikse et al. 1999; Sharma et al. 2000) for imaging MDR have been reported.

Most of these radiopharmaceuticals are cationic and modestly hydrophobic, which is similar to many chemotherapeutic drugs in the MDR phenotype. For example, [^{99m}Tc]Sestamibi accumulates within cells in response to the physiologically negative mitochondrial and plasma membrane potentials (Piwnica-Worms et al. 1990, 1995). However, functional *MDR1* Pgp mediates net outward transport of [^{99m}Tc]Sestamibi from cells, reducing cellular accumulation of the radiopharmaceutical. Thus, cellular accumulation of [^{99m}Tc]Sestamibi into drug-sensitive tumor cells is high and translates into a "hot spot" on scintigraphic images or a slow washout rate from a tumor focus. Conversely, expression of *MDR1* Pgp transports the tracer out of cells, thereby resulting in reduced net accumulation. This will be detected either as a "cold" tumor or as a rapid washout rate from a tumor focus or Pgp-expressing tissue. Similar results are obtained with another widely available radiopharmaceutical, [1,2-*bis*{*bis*(2-ethoxyethyl)phosphino}ethane]$_2$O$_2$^{99m}Tc (V) (Tc-Tetrofosmin) (Chen et al. 2000), as well as with selected agents in pre-clinical development (Crankshaw et al. 1998; Sharma et al. 2000).

Several clinical studies have shown the feasibility of interrogating Pgp transport activity in vivo with imaging cameras commonly available in nuclear medicine facilities (Kostakoglu et al. 1994, 1998; Bom et al. 1997; Chen et al. 1997; Del Vecchio et al. 1997a,b; Luker et al. 1997; Barbarics et al. 1998; Ciarmiello et al. 1998). In general, ^{99m}Tc-Sestamibi pharmacokinetic data are extracted from tumor images over time and correlated with immunohistochemical assessment of Pgp expression in the tumor specimen. For example, Del Vecchio et al. determined rates of efflux of ^{99m}Tc-Sestamibi in 30 patients with untreated breast cancer (Del Vecchio et al. 1997b). Dynamic imaging of the primary tumor was performed for 4 h following injection of ^{99m}Tc-Sestamibi and tumor specimens were obtained for quantitative autoradiography of Pgp. Rates of efflux of ^{99m}Tc-Sestamibi were 2.7-fold greater in tumors overexpressing *MDR1* Pgp compared with tumors that expressed Pgp at a level comparable to benign breast lesions. Estimates of sensitivity and specificity for in vivo detection of *MDR1* Pgp using ^{99m}Tc-Sestamibi were 80% and 95%, respectively. From these data, the authors concluded that efflux rate constants of ^{99m}Tc-Sestamibi may be used for non-invasive identification of *MDR1* Pgp in breast cancer.

In another prospective study of 46 lung cancer patients, Kostakoglu and colleagues (Kostakoglu et al. 1998) calculated washout rates and the degree of ^{99m}Tc-Sestamibi accumulation in the lung tumors by analysis of early (30 min) and delayed (180 min) SPECT and planar images of the thorax. These were correlated with Pgp expression as determined by immunohistochemical analysis of tumor sections with the mAb JSB-1. A strong inverse correlation was again observed between tumor-to-background ratios at 45 min and levels of Pgp expression. However, no appreciable correlation between tumor washout rates of ^{99m}Tc-Sestamibi and Pgp expression in lung cancer was found, although the initial SPECT imaging was not completed until approximately 45–60 min post-injection which may have lead to underestimation of early washout events. The investigators correctly pointed out that the MRP1 gene may be more relevant than *MDR1* Pgp in mechanistically determining the MDR phenotype of lung cancer (Giaccone et al. 1996), thus accounting for the low numbers of strongly positive Pgp-expressing tumors in their study. Given that MRP1 may also transport ^{99m}Tc-Sestamibi (Crankshaw et al. 1996; Hendrikse et al. 1998; Moretti et al. 1998; Chen et al. 2000), this also could potentially contribute to the lack of correlation of washout rates in lung cancer with Pgp expression alone.

In another study of 30 patients with untreated lung cancer (Fukumoto et al. 1999), dual isotope SPECT scans of the thorax were performed 10 min and 2 h after intravenous co-injection of ^{99m}Tc-Tetrofosmin and ^{210}Tl. Patients were then treated either sequentially or concurrently with radiation and chemotherapy (cisplatin plus etoposide). The relation between therapeutic response and tumor retention of tracer was analyzed. Regardless of whether therapy was sequential or concurrent, 14 of 18 tumors with high ^{99m}Tc-Tetrofosmin retention exhibited a favorable response to chemotherapy, whereas all 12 tumors with low ^{99m}Tc-Tetrofosmin retention did not respond. The investigators conclude that ^{99m}Tc-Tetrofosmin SPECT is a useful tool in vivo for prediction of radioresistance and MDR in lung cancer.

In a study of 34 patients with brain tumors, early and delayed SPECT images were obtained at 10 min and 3 h after intravenous co-injection of ^{99m}Tc-Sestamibi and ^{210}Tl (Nagamachi et al. 1999). The investigators found that tumor/normal tissue uptake ratios (both early uptake ratios and delayed uptake ratios) were higher with ^{99m}Tc-Sestamibi than ^{210}Tl. Furthermore, while the numbers were small, delayed uptake ratios of ^{99m}Tc-Sestamibi were inversely correlated with the presence of Pgp in brain metastases as determined by immunohistochemistry.

Furthermore, in addition to the in vitro laboratory studies above, blockade of Pgp-mediated extrusion of [^{99m}Tc]-Sestamibi in tissues and tumors in vivo has been observed by treatment with Pgp modulators such as PSC 833 (valspodar) or dipyridamole (Chen et al. 1997; Luker et al. 1997; Lim et al. 2000). A two-step protocol for imaging MDR reversal in patients is under evaluation which may provide a non-invasive method to determine the effectiveness of MDR modulation. Herein, after baseline imaging, administration of a potent modulating agent and reinjection of the tracer is performed. After administration of the modulator, those tissues or tumors showing higher accumulation of the tracer and/or reduced washout rates would indicate specific blockade of *MDR1* Pgp. Thus, these radiopharmaceuticals enable non-invasive mapping of specific molecular events in vivo and provide a basis for further exploration of molecular imaging of Pgp in cancer and gene therapy.

The prospective value of high tumor clearance rates of ^{99m}Tc-Sestamibi to predict poor therapeutic outcomes also was evaluated in locally advanced breast cancer. In a study by Ciarmiello and colleagues (Ciarmiello et al. 1998), 39 patients with stage III disease were enrolled in a prospective clinical trial to receive pre-treatment mammoscintigraphy with ^{99m}Tc-Sestamibi before neoadjuvant chemotherapy. Breast tumor clearance of ^{99m}Tc-Sestamibi was determined out to 4 h post-injection and half-times calculated from the image sets. Patients were subsequently treated with standard chemotherapy (epirubicin) for 6 weeks and then underwent surgery within 3 weeks of completion of chemotherapy. Tumor burden was assessed by pathologic examination of the mastectomy specimens. Of the 39 patients, 17 showed rapidly effluxing tumors, and 15 of these 17 showed a highly cellular residual tumor indicating lack of response to neoadjuvant chemotherapy. Conversely, 22 patients showed prolonged tumor clearance (half time of >204 min) and, of these, only eight showed highly cellular residual tumor at the time of surgery. Interestingly, for patients with a half time of <164 min, which was approximately two standard deviations faster than the mean half time of high Pgp-expressing breast tumors (Del Vecchio et al. 1997b), none showed evidence of pathologic response to neoadjuvant chemotherapy. Thus, pre-treatment scintigraphy

of advanced breast tumors was highly predictive of subsequent response to neoadjuvant chemotherapy.

27.2.5
Cell Membrane Permeant Peptides

Peptide-based imaging agents are desired for molecular imaging applications because of their potential for specific detection of a target and the availability of well-characterized moieties for stable chelation of isotopes such as ^{99m}Tc. Fragments of antibodies (single chain fragments, diabodies, or minibodies) (Wu et al. 2000) and small peptides (Krenning et al. 1993; Liu and Edwards 1999) have been used successfully to detect cell surface proteins such as carcinoembryonic antigen or SSTR2 receptors. However, peptides and small proteins typically have poor permeability through the plasma membrane of cells, thus limiting molecular targets to proteins that contain extracellular domains. Imaging receptors on the cell surface also potentially limits detection of the tracer, because signal amplification through transport into cells or enzymatic activity does not occur.

Investigators have identified several small proteins and peptides that permeate plasma membranes of intact cells. Included among these peptides are the third helix of the homeodomain of Antennapedia (Derossi et al. 1996) and viral proteins such as Herpes simplex virus VP22 (Elliot and O'Hare 1997), HIV-1 Rev protein (Kubota et al. 1989), and the basic domain of HIV-1 Tat protein (Frankel and Pabo 1988; Fawell et al. 1994). The mechanism through which these peptides enter cells remains uncertain, although it does not appear to depend upon known mechanisms of receptor-mediated endocytosis. Not only do the peptides themselves transduce into cells, but they also mediate entry of other peptides and full-length proteins (Fawell et al. 1994; Elliot and O'Hare 1997; Nagahara et al. 1998), fluorophores (Vives et al. 1997), and derivatized superparamagnetic iron oxide particles (Josephson et al. 1999).

We have shown that peptides based on the basic domain of Tat and an attached chelator for metals can be used to deliver ^{99m}Tc into cells in vitro (Polyakov et al. 1999, 2000) (Fig. 27.6). These peptides quickly (<2 min) accumulate and concentrate within cells in culture while maintaining stable chelation of the radioisotope. In addition, conjugation of the peptide with fluorescein can directly reveal the intracellular localization of the peptide when cells are viewed

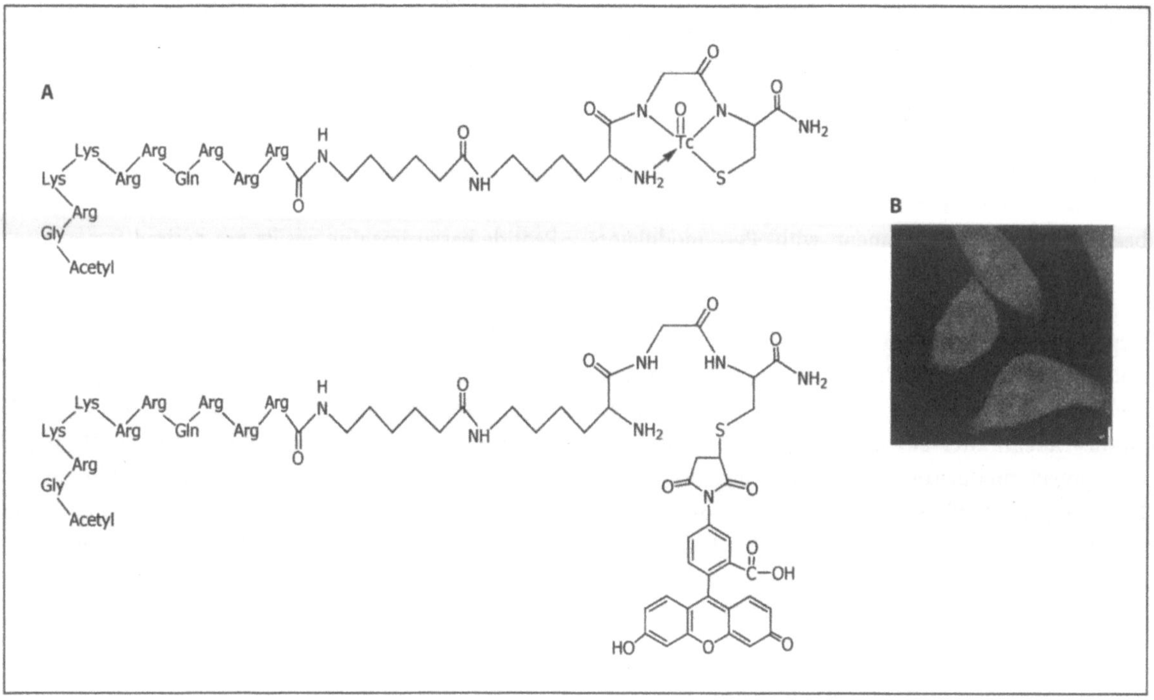

Fig. 27.6. A Structure of a [^{99m}Tc]Tat-peptide complex (*top*) and a fluorescein-5-maleimide conjugate (*bottom*). **B** Cellular accumulation of the fluorescein Tat-peptide conjugate in human KB-3-1 tumor cells. Cells were incubated with peptide conjugate for 10 min at 37 °C and then fixed. Note fluorescence from labeled peptide residing in both cytosolic and nuclear (nucleolar) compartments. Bar = 5 μm

under a fluorescence microscope (Fig. 27.6B). Washout of Tat peptides from cells also demonstrates similar rapid kinetics. When injected into mice, radiolabeled Tat peptides distribute throughout the body, reaching peak accumulation in organs and tissues within minutes. The peptide subsequently clears primarily through the kidneys during the first hour after injection. Thus, the pharmacokinetic data suggest that Tat peptides may be useful agents for imaging.

To make these peptides useful for molecular imaging, a peptide sequence specifically recognized by an intracellular enzyme could be inserted between the transduction domain and the peptide chelator for the radioisotope. Using this approach, intracellular trapping of radiotracer would occur only in cells that have the desired enzymatic activity, similar to accumulation of phosphorylated FDG in cells that overexpress hexokinase. This strategy for selective targeting with Tat peptides has been exploited previously by incorporating the recognition sequence for HIV protease between the Tat domain and a precursor form of caspase 3, an enzyme that is activated in apoptotic cell death (Vocero-Akbani et al. 1999). Only cells infected with HIV cleave the Tat peptide and release functional caspase 3, resulting in specific killing of these cells. Tat peptides also have been shown to transduce a functional enzyme (β-galactosidase) into cells in living mice (Schwarze et al. 1999), implying that other reporter genes for oncologic molecular imaging could be transduced in vivo.

27.3
Imaging Exogenous Gene Expression in Cancer: Reporter Gene Concept

Reporter genes are commonly used in molecular biology to monitor expression and/or repression of a gene of interest. Typical reporter genes consist of a chimeric gene linking an endogenous or exogenous enhancer or promoter to an enzyme (luciferase or β-galactosidase) or fluorophore [green fluorescent protein (GFP)]. In all cases, the reporter must be introduced into the target cell or tissue, using a variety of methods including transfection of DNA, transduction with viral vectors, or incorporation into the genome of transgenic animals. The reporter gene can then be used to detect activation of the promoter and/or enhancer of interest, which is inferred to duplicate expression of the endogenous gene(s) controlled by the same promoter elements. Ideally, the magnitude and time course of reporter gene activity should parallel the strength and duration of expression of the endogenous target gene.

Reporter genes used for studies in cultured cells and tissue specimens have limited utility for detecting and quantifying gene expression in intact animals. Hence, researchers have had to adapt and characterize different reporter systems for use in molecular imaging. For detection of gene expression with PET, most studies have used either heterologous enzymes such as Herpes simplex virus-1 thymidine kinase (HSVTK) or receptors such as the dopamine-2 (Herschman et al. 2000) or SSR-2 receptors (Zinn et al. 2000). As compared with direct imaging of small quantities of mRNA, these systems have the potential to provide signal amplification at two biological levels and thereby increase sensitivity for detecting and localizing gene expression. First, a single mRNA transcript is translated into many copies of a receptor or enzyme. Second, each molecule of an enzyme can catalyze intracellular trapping of a large number of reporter probe molecules, thus theoretically providing greater signal compared to a receptor system which binds a single reporter ligand at a time. However, enzyme systems as well as receptors have been used successfully for molecular imaging of gene expression (Herschman et al. 2000). Another fundamental advantage of reporter gene/reporter probe systems is that once validated, the reporter gene can theoretically be cloned into an appropriate vector and any gene of interest can be interrogated with the same validated reporter probe. For PET, this provides a significant efficiency compared to the time and constraints inherent to traditional routes of synthesizing, labeling and validating a new and different radioligand for every new receptor or protein of interest.

27.3.1
Dopamine-2 Receptor

The dopamine-2 receptor (D2R) is expressed in the plasma membrane of cells, primarily in the striatum of the brain and the pituitary gland. Several different ligands for PET or SPECT imaging of D2R have been validated, including 3-(2'-^{18}F-fluoroethyl)spiperone (^{18}F-FESP) and ^{123}I-iodobenzamine. Investigators

have shown that expression of D2R can be detected in living mice, using either a replication-incompetent adenovirus that targets the reporter gene to liver or an implanted tumor that stably expresses D2R (Maclaren et al. 1999). In both models, binding of ^{18}F-FESP to D2R as measured by PET in vivo has a high correlation with in vitro assays of D2R mRNA expression, amount of protein, and ligand-receptor binding. By PET imaging, significantly more accumulation of ^{18}F-FESP was measured in tumors transfected with D2R than tumors without this receptor. More recently, D2R has been used in conjunction with HSVTK to quantify relative gene expression from the first and second genes in bicistronic transcripts, showing the feasibility of using PET to monitor expression of a therapeutic gene in vivo (Yu et al. 2000).

27.3.2
Herpes Simplex Virus-1 Thymidine Kinase and Cytosine Deaminase

Viral or bacterial enzymes that convert a prodrug into a toxic metabolite originally were investigated for use as suicide genes in therapy of cancer. Two of the most frequently used enzymes for therapeutic applications, cytosine deaminase (CD) from Escherichia coli and thymidine kinase from Herpes simplex virus-1 (HSVTK), also have been used for molecular imaging of gene expression. CD, an enzyme that is absent in mammalian cells, converts cytosine to uracil; thus, only cells that express the transgene convert 5-fluorocytosine to the cytotoxic 5-fluorouracil. However, imaging of CD with radiotracers has been limited by slow uptake of 5-fluorocytosine and rapid diffusion of 5-fluorouracil out of cells (Haberkorn et al. 1996). Conversely, HSVTK has been used extensively as a PET reporter for gene expression in vivo and will be reviewed more fully below.

The underlying principle for use of HSVTK in both therapeutic and imaging applications is manifest in the lower specificity of HSVTK kinase activity, enabling phosphorylation of nucleoside analogs, such as ganciclovir, compared with endogenous thymidine kinases normally present in mammalian cells. Following active transport of a nucleoside analog across the cell membrane, the agent is selectively phosphorylated by viral thymidine kinase. The monophosphorylated nucleoside is trapped within cells and subsequently converted to a nucleoside triphosphate. Approximately 50% of cell-associated nucleoside is converted to the triphosphate form and incorporated into DNA within 1 h (Haubner et al. 2000), where it acts to terminate DNA synthesis. These metabolic steps result in trapping of nucleoside analogs like ganciclovir selectively within cells that express HSVTK. In the absence of the viral thymidine analogs, nucleoside analogs are not phosphorylated by native enzymes and do not accumulate within cells. In pharmacological doses, incorporation of nucleoside analogs into DNA causes cell death. However, tracer amounts of radiolabeled nucleoside analogs can be trapped inside cells without causing toxicity, thus enabling repetitive imaging of HSVTK activity with SPECT or PET.

Use of radiolabeled nucleoside analogs to localize expression of HSVTK in living animals first was reported almost two decades ago by Saito et al. (1982, 1984). These authors administered ^{14}C-2'-fluoro-5-methyl-1-β-D-arabinosyluracil (FMAU) to rats infected with HSV-1 and then used selective trapping of the drug to quantify and map sites of viral infection by autoradiography of brain sections. Accumulation of FMAU was correlated highly with sites of focal infection, as determined by immunoperoxidase staining for viral antigens. The authors also proposed that appropriately labeled nucleoside analogs could be used for clinical imaging of thymidine kinase activity.

Since these initial reports, several different technical innovations and improvements have been made to enable imaging of HSVTK in vivo. Radiolabeled nucleoside analogs appropriate for in vivo detection of HSVTK by SPECT or PET have been developed and characterized, including uracil nucleoside derivatives (such as 5-iodo-2'-fluoro-2'-deoxy-1-β-D-arabinofuranosyl-5-iodouracil [FIAU] labeled with ^{124}I or ^{131}I) and derivatives of acycloguanosine [such as 8-[^{18}F]fluoro-9-[(2-hydroxy-1-(hydroxymethyl)ethoxy)-methyl]guanine (8-[^{18}F]-fluoroganciclovir; [^{18}F]FGCV) and 9-[(3-[^{18}F]fluoro-1-hydroxy-2-propoxy)methyl]-guanine ([^{18}F]FHPG)] (recently reviewed in Blasberg and Tjuvajev 1999; Gambhir et al. 1999a). To enable imaging of gene expression and other biologic processes in small animals such as mice, the development of microPET has been a significant advance in technology. The first-generation microPET scanner has a volumetric resolution of approximately 2 mm^3 (Charziioannou et al. 1999), allowing imaging of reporter gene activity at the level of an organ or tissue (Gambhir et al. 1999). Scanners that perform both PET and MR imaging have been developed for humans (Shao et al. 1997), allowing co-registration of functional PET images with anatomic localization of MRI. A combined

scanner for microPET and CT or MR will provide similar benefits for molecular imaging in small animals.

To introduce HSVTK into animals for imaging, investigators have either transfected tumor cells with HSVTK in vitro and subsequently implanted these cells into mice to form tumors or delivered the reporter gene into liver via adenoviral vectors (Tjuvajev et al. 1999; Gambhir et al. 2000; Haubner et al. 2000; Hospers et al. 2000). For almost all published reports of imaging HSVTK as a reporter gene, expression of the enzyme has been driven by heterologous viral promoters, such as promoters of early genes in cytomegalovirus, SV40, or Rous sarcoma virus. These promoters provide constitutive, high-level amounts of enzyme that likely exceed expression driven by most endogenous promoters. Investigators currently are working to further develop this technology for monitoring activity of endogenous mammalian promoters. Transgenic mice that use HSVTK as a marker for expression or deletion of a specific gene also have been proposed. Progress in these areas of research will increase greatly the number and types of biological hypotheses that can be addressed by molecular imaging of oncologic gene expression in vivo.

27.4
Conclusions

Conventional radiology is focused largely on defining the anatomic basis of disease. As the human genome project is refined and data are translated into medical practice, one can speculate that an increasing proportion of cancers will be diagnosed and treated before there is manifest evidence of anatomic change. Thus, molecular imaging in oncology will likely interface with future trends in the science and clinical practice of molecular oncology, diagnosis, and therapy.

■ **Acknowledgements.** This educational project and work from the Washington University Molecular Imaging Center were supported by the U.S. National Institutes of Health grant P50 CA94056 and Department of Energy contract DE FG02 94ER61885.

27.5 References

Allikmets R, Schrimi L, Hutchinson A et al (1998) A human placenta-specific ATP-binding cassette gene (ABCP) on chromosome 4q22 that is involved in multidrug resistance. Cancer Res 58:5337–5339

Aloj L, Caraco C, Jagoda E et al (1999) Glut-1 and hexokinase expression: relationship with 2-fluoro-2-deoxy-D-glucose uptake in A431 and T47D cells in culture. Cancer Res 59:4709–4714

Anderson CJ, Lewis JS (2000) Radiopharmaceuticals for targeted radiotherapy of cancer. Exp Opin Ther Patents 10:1057–1069

Anderson CJ, Pajeau T, Edwards W et al (1995) In vitro and in vivo evaluation of copper-64-octreotide conjugates. J Nucl Med 36:2315–2325

Anderson CJ, Jones LA, Bass LA et al (1998) Radiotherapy, toxicity and dosimetry of copper-64-TETA-Octreotide in tumor-bearing rats. J Nucl Med 39:1944–1951

Anderson CJ, Dehdashti F, Cutler PD et al (2001) Copper-64-TETA-octreotide as a PET imaging agent for patients with neuroendocrine tumors. J Nucl Med 42:213–221

Bakker WH, Albert R, Bruns C et al (1991a) [111In-DTPA-D-Phe]-octreotide, a potential radiopharmaceutical for imaging of somatostatin receptor-positive tumors: synthesis, radiolabeling and in vitro validation. Life Sci 49:1583–1591

Bakker WH, Krenning EP, Reubi JC et al (1991b) In vivo application of [In-111-DTPA-D-Phe]-octreotide for detection of somatostatin receptor-positive tumors in rats. Life Sci 49:1593–1601

Ballinger J, Hua H, Berry B et al (1995a) ^{99m}Tc-Sestamibi as an agent for imaging P-glycoprotein-mediated multidrug resistance: in vitro and in vivo studies in a rat breast tumour cell line and its doxorubicin-resistant variant. Nucl Med Commun 16:253–257

Ballinger JR, Sheldon KM, Boxen I et al (1995b) Differences between accumulation of ^{99m}Tc-MIBI and Tl-201-thallous chloride in tumor cells: role of P-glycoprotein. Q J Nucl Med 39:122–128

Ballinger JR, Bannerman J, Boxen I et al (1996) Technetium-99m-tetrofosmin as a substrate for P-glycoprotein: in vitro studies in multidrug-resistant breast tumor cells. J Nucl Med 37:1578–1582

Ballinger JR, Muzzammil T, Moore M (1997) Technetium-99m-Furifosmin as an agent for functional imaging of multidrug resistance in tumors. J Nucl Med 38:1915–1919

Bangard M, Behe M, Guhlke S et al (2000) Detection of somatostatin receptor-positive tumours using the new ^{99m}Tc-tricine-HYNIC-D-Phe^1-Tyr^3-octreotide: first results in patients and comparison with ^{111}In-DTPA-D-Phe^1-octreotide. Eur J Nucl Med 27:628–637

Barbarics E, Kronauge J, Cohen D et al (1998) Characterization of P-glycoprotein transport and inhibition in vivo. Cancer Res 58:276–282

Bauer W, Briner U, Doepfner W et al (1982) SMS 201-995. Life Sci 31:1133–1140

Beyer T, Townsend D, Brun T et al (2000) Combined PET/CT scanner for clinical oncology. J Nucl Med 41:1369–1379

Blasberg R, Tjuvajev J (1999) Herpes simplex virus thymidine kinase as a marker/reporter gene for PET imaging of gene therapy. Q J Nucl Med 43:163–169

Bohuslavizki KH, Brenner W, Braunsdorf WEK et al (1996) Somatostatin receptor scintigraphy in the differential diagnosis of meningioma. Nucl Med Commun 17:302–310

Boland A, Ricard M, Opolon P et al (2000) Adenovirus-mediated transfer of the thyroid sodium/iodide symporter gene into tumors for a targeted radiotherapy. Cancer Res 60:3484–3492

Bom H, Kim Y, Lim S et al (1997) Dipyridamole modulated Tc-99m-Tc-sestamibi (mibi) scintigraphy: a predictor of response to chemotherapy in patients with small cell lung cancer. J Nucl Med 38:240P

Bosch I, Croop J (1996) P-glycoprotein multidrug resistance and cancer. Biochim Biophys Acta 1288:F37–F54

Brockmann J, Rosch F, Herzog H et al (1995) In vivo uptake kinetics and dosimetry calculations of 86Y-DTPA-octreotide with PET as a model for potential endotherapeutic octreotides labelled with 90Y. J Labelled Compounds Radiopharm 37:519–521

Brosius FR, Nguyen N, Egert S et al (1997) Increased sarcolemmal glucose transporter abundance in myocardial ischemia. Am J Cardiol 80:77A–84A

Brown R, Wahl R (1993) Overexpression of Glut-1 glucose transporter in human breast cancer. An immunohistochemical study. Cancer 72:2979–2985

Brown R, Leung J, Fisher S (1996) Intratumoral distribution of tritiated fluorodeoxyglucose in breast carcinoma: correlation between Glut-1 expression and FDG uptake. J Nucl Med 37:1042–1047

Buchsbaum DJ, Rogers BE, Khazaeli MB et al (1999) Targeting strategies for cancer radiotherapy. Clin Cancer Res 5 [Suppl]:3048s–3055s

Carrasco N (1993) Iodide transport in the thyroid gland. Biochim Biophys Acta 1154:65–82

Charziioannou A, Cherry S, Shao Y et al (1999) Performance evaluation of microPET: a high-resolution lutetium oxyorthosilicate PET scanner for animal imaging. J Nucl Med 40:1164–1175

Chen C, Meadows B, Regis J et al (1997) Detection of in vivo P-glycoprotein inhibition by PSC 833 using Tc-99m-Tc-Sestamibi. Clin Cancer Res 3:545–552

Chen W, Luker K, Dahlheimer J et al (2000) Effects of MDR1 and MDR3 P-glycoproteins, MRP1 and BCRP/MXR/ABCP on transport of Tc-99m-Tc-tetrofosmin. Biochem Pharmacol 60:413–426

Ciarmiello A, Del Vecchio S, Silvestro P et al (1998) Tumor clearance of technetium-99m-Sestamibi as a predictor of response to neoadjuvant chemotherapy for locally advanced breast cancer. J Clin Oncol 16:1677–1683

Clavo A, Brown R, Wahl R (1995) 2-Fluoro-2-deoxy-D-glucose (FDG) uptake into human cancer cell lines is increased by hypoxia. J Nucl Med 36:1625–1632

Cole SPC, Bhardwaj G, Gerlach JH et al (1992) Overexpression of a transporter gene in a multidrug-resistant human lung cancer cell line. Science 258:1650–1654

Cordobes M, Starzec A, Delmon-Moingeon L et al (1996) Technetium-99m-Sestamibi uptake by human benign and malignant breast tumor cells: correlation with mdr gene expression. J Nucl Med 37:286–289

Crankshaw C, Piwnica-Worms D (1996) Tc-99m-Tc-Sestamibi may be a transport substrate of the human multidrug resistance-associated protein (MRP). J Nucl Med 37:247P

Crankshaw C, Marmion M, Luker G et al (1998) Novel Tc(III)-Q-complexes for functional imaging of the multidrug resistance (MDR1) P-glycoprotein. J Nucl Med 39:77–86

Dantzig A, Shepard R, Cao J et al (1996) Reversal of P-glycoprotein-mediated multidrug resistance by a potent cyclopropyldibenzosuberane modulator, LY335979. Cancer Res 56:4171–4179

Decristoforo C, Melendez-Alafort L, Sosabowski JK et al (2000) ^{99m}Tc-HYNIC-Tyr^3-Octreotide for imaging somatostatin-receptor-positive tumors: preclinical evaluation and comparison with ^{111}In-octreotide. J Nucl Med 41: 1114–1119

Dehdashti F, Flanagan F, Mortimer J et al (1999) Positron emission tomographic assessment of "metabolic flare" to predict response of metastatic breast cancer to antiestrogen therapy. Eur J Nucl Med 26:51–56

De Jong M, Breeman WAP, Bernard BF et al (2001) [^{177}Lu-DOTA^0-Tyr^3]Octreotate for somatostatin receptor-targeted radionuclide therapy. Int J Cancer 92:628–633

Del Vecchio S, Ciarmiello A, Pace L et al (1997a) Fractional retention of technetium-99m-sestamibi as an index of P-glycoprotein expression in untreated breast cancer patients. J Nucl Med 38:1348–1351

Del Vecchio S, Ciarmiello A, Potena MI et al (1997b) In vivo detection of multidrug resistance (MDR1) phenotype by technetium-99m-sestamibi scan in untreated breast cancer patients. Eur J Nucl Med 24:150–159

Derossi D, Calvet S, Trembleau A et al (1996) Cell internalization of the third helix of the Antennapedia homeodomain is receptor-independent. J Biol Chem 271:18188–18193

Doyle L, Yang W, Abruzzo L et al (1998) A multidrug resistance transporter from human MCF-7 breast cancer cells. Proc Natl Acad Sci USA 95:15665–15670

Elliot G, O'Hare P (1997) Intercellular trafficking and protein delivery by a Herpes virus structural protein. Cell 88:223–233

Fawell S, Seery J, Daikh Y et al (1994) Tat-mediated delivery of heterologous proteins into cells. Proc Natl Acad Sci USA 91:664–668

Flens MJ, Zaman GJR, Valk P van der et al (1996) Tissue distribution of multidrug resistance protein. Am J Pathol 148:1237–1247

Ford JM, Hait WN (1990) Pharmacology of drugs that alter multidrug resistance in cancer. Pharmacol Rev 42:155–199

Frankel A, Pabo C (1988) Cellular uptake of the tat protein from human immunodeficiency virus. Cell 55:1189–1193

Fukumoto M, Yoshida D, Hayase N et al (1999) Scintigraphic prediction of resistance to radiation and chemotherapy in patients with lung carcinoma: technetium 99m-tetrofosmin and thallium-201 dual single photon emission computed tomography study. Cancer 86:1470–1479

Gambhir S, Barrio J, Herschman H et al (1999a) Assays for noninvasive imaging of reporter gene expression. Nucl Med Biol 26:481–490

Gambhir S, Barrio J, Phelps M et al (1999b) Imaging adenoviral-directed reporter gene expression in living animals with positron emission tomography. Proc Natl Acad Sci USA 96:2333–2338

Gambhir S, Bauer E, Black M et al (2000) A mutant herpes simplex virus type 1 thymidine kinase reporter gene shows improved sensitivity for imaging reporter gene expression with positron emission tomography. Proc Natl Acad Sci USA 97:2785–2790

Gaveriaux C, Boesch D, Jachez B (1991) PSC 833 a non-immunosuppressive cyclosporin analog, is a very potent multidrug-resistance modifier. J Cell Pharmacol 2:225–234

Germann U, Ford P, Schlakhter D et al (1997) Chemosensitization and drug accumulation effects of VX-710, verapamil, cyclosporin A, MS-209, and GF120918 in multidrug resistant HL60/ADR cells expressing the multidrug resistance-associated protein MRP. Anticancer Drugs 8:141–155

Giaccone G, van Ark-Otte J, Rubio G et al (1996) MRP is frequently expressed in human lung cancer cell lines, in non-small cell lung cancer and in normal lungs. Int J Cancer 66:760–767

Golshani-Hebroni S, Bessman S (1997) Hexokinase binding to mitochondria: a basis for proliferative energy metabolism. J Bioenerget Biomembr 29:331–338

Gottesman M, Pastan MI (1993) Biochemistry of multidrug resistance mediated by the multidrug transporter. Annu Rev Biochem 62:385–427

Gros P, Ben Neriah Y, Croop JM et al (1986) Isolation and expression of a complementary DNA that confers multidrug resistance. Nature 323:728–731

Guhlke S, Wester H-J, Bruns C et al (1994) (2-[^{18}F]Fluoropropionyl-D(phe^1)-octreotide, a potential radiopharmaceutical for quantitative somatostatin receptor imaging with PET: synthesis, radiolabeling, in vitro validation and biodistribution in mice. Nucl Med Biol 21:819–825

Guillemin R (1978) Peptides in the brain: the new endocrinology of the neuron. Science 202:390–402

Haberkorn U, Oberdorfer F, Gebert J et al (1996) Monitoring gene therapy with cytosine deaminase: in vitro studies using tritiated-5-fluorocytosine. J Nucl Med 37:87–94

Hanania E, Fu S, Roninson I et al (1995) Resistance to taxol chemotherapy produced in mouse marrow cells by safety-modified retroviruses containing a human MDR-1 transcription unit. Gene Ther 2:279–284

Haubner R, Avril N, Hantzopoulos P et al (2000) In vivo imaging of herpes simplex virus type 1 thymidine kinase gene expression: early kinetics of radiolabelled FIAU. Eur J Nucl Med 27:283–291

Hendrikse N, Franssen E, van der Graaf W et al (1998) ^{99m}Tc-sestamibi is a substrate for P-glycoprotein and the multidrug resistance-associated protein. Br J Cancer 77:353–358

Hendrikse N, Franssen E, van der Graaf W et al (1999) Visualization of multidrug resistance in vivo. Eur J Nucl Med 26:283–293

Henze M, Schuhmacher J, Hipp P et al (2001) PET imaging of somatostatin receptors using [^{68}Ga]DOTA-D-Phe^1-Tyr^3-octreotide: first results in patients with meningiomas. J Nucl Med 42:1053–1056

Herman LW, Sharma V, Kronauge JF et al (1995) Novel hexakis(areneisonitrile)technetium(I) complexes as radioligands targeted to the multidrug resistance P-glycoprotein. J Med Chem 38:2955–2963

Herschman H, Barrio J, Satyamurthy N et al (2000) Progress toward in vivo imaging of reporter gene expression using positron emission tomography. Am Soc Clin Oncol Educ Book Spring 169–177

Higashi K, Clavo A, Wahl R (1993) Does FDG uptake measure the proliferative activity of human cancer cells? In vitro comparison with DNA flow cytometry and ³H-thymidine uptake. J Nucl Med 34:414–419

Hoekstra C, Paglianiti I, Hoekstra O et al (2000) Monitoring response to therapy in cancer using [18F]-2-fluoro-2-deoxy-D-glucose and positron emission tomography: an overview of different analytical methods. Eur J Nucl Med 27:731–743

Hospers G, Calogero A, van Waarde A et al (2000) Monitoring of herpes simplex virus thymidine kinase enzyme activity using positron emission tomography. Cancer Res 60:1488–1491

Hyafil F, Vergely C, Du Vignaud P et al (1993) In vitro and in vivo reversal of multidrug resistance by GF120918, an acridonecarboxamide derivative. Cancer Res 53:4595–4602

Josephson L, Tung C-H, Moore A et al (1999) High-efficiency intracellular magnetic labeling with novel superparamagnetic-Tat peptide conjugates. Bioconjug Chem 10:186–191

Juliano RL, Ling V (1976) A surface glycoprotein modulating drug permeability in Chinese hamster ovary cell mutants. Biochim Biophys Acta 455:152–162

Keppler D, Leier I, Jedlitschky G (1997) Transport of glutathione conjugates and glucuronides by the multidrug resistance proteins MRP1 and MRP2. Biol Chem 378:787–791

Klepper J, Fischbarg J, Vera J et al (1999) Glut1-deficiency: barbiturates potentiate haploinsufficiency in vitro. Pediatr Res 46:677–683

Kostakoglu L, Elahi N, Kirarli P et al (1994) Clinical validation of the influence of P-glycoprotein on technetium-99m-Sestamibi uptake in malignant tumors. J Nucl Med 38:1003–1008

Kostakoglu L, Kirath P, Ruacan S et al (1998) Association of tumor washout rates and accumulation of technetium-99m-MIBI with expression of P-glycoprotein in lung cancer. J Nucl Med 39:228–234

Krenning EP, Bakker WH, Kooij PPM et al (1992) Somatostatin receptor scintigraphy with Indium-111-DTPA-D-Phe-1-Octreotide in man: metabolism, dosimetry and comparison with Iodine-123-Tyr-3-Octreotide. J Nucl Med 33:652–658

Krenning EP, Kwekkeboom D, Bakker W et al (1993) Somatostatin receptor scintigraphy with [In-111-DTPA-D-Phe] and [I123-Tyr]octreotide: the Rotterdam experience with more than 1000 patients. Eur J Nucl Med 20:716–731

Krenning EP, Kooij PPM, Bakker WH et al (1994) Radiotherapy with a radiolabeled somatostatin analogue, 111In-DTPA-Phe1-Octreotide: a case history. Ann NY Acad Sci 733:496–506

Krenning EP, de Jong M, Kooij PPM et al (1999) Radiolabelled somatostatin analogue(s) for peptide receptor scintigraphy and radionuclide therapy. Ann Oncol 10 [Suppl 2]:S23–S29

Kubota S, Siomi H, Satoh T et al (1989) Functional similarity of HIV-1 rev and HTLV-1 rex proteins: identification of a new nucleolar-targeting signal in rev protein. Biochem Biophys Res Commun 162:963–970

Kubota K, Kubota R, Yamada S (1993) FDG accumulation in tumor tissue. J Nucl Med 34:419–421

Kwekkeboom DJ, Kam BL, Bakker WH et al (2001) Treatment with 177Lu-DOTA0-Tyr3-octreotate in patients with somatostatin receptor positive tumors: preliminary results (abstract). J Nucl Med 42:37P

Larson S, Grunbaum Z, Rasey J (1980) Positron imaging feasibility studies: selective tumor concentration of 3H-thymidine, 3H-uridine, and 14C-2-deoxyglucose. Radiology 134:771–773

Lautier D, Canitrot Y, Deeley R et al (1996) Multidrug resistance mediated by the multidrug resistance protein (MRP) gene. Biochem Pharmacol 52:967–977

Lewis JS, Lewis MR, Cutler PD et al (1999) Radiotherapy and dosimetry of 64Cu-TETA-Tyr3-Octreotate in a somatostatin receptor-positive tumor-bearing rat model. Clin Cancer Res 5:3608–3616

Lewis JS, Wang M, Laforest R et al (2002) Toxicity and dosimetry of ¹⁷⁷Lu-DOTA-Y3-octreotate in a rat model. Int J Cancer (in press)

Lim S-C, Park K-O, Kim Y-C et al (2000) Comparison of Tc-99m Sestamibi, serum neuron-specific enolase and lactate dehydrogenase as predictors of response to chemotherapy in small cell lung cancer. Cancer Biother Radiopharm 15:381–386

Liu S, Edwards D (1999) ⁹⁹ᵐTc-labeled small peptides as diagnostic radiopharmaceuticals. Chem Rev 99:2235–2268

Luker G, Rao V, Crankshaw C et al (1997a) Characterization of phosphine complexes of technetium (III) as transport substrates of the multidrug resistance (MDR1) P-glycoprotein and functional markers of P-glycoprotein at the blood-brain barrier. Biochemistry 36:14218–14227

Luker GD, Fracasso PM, Dobkin J et al (1997b) Modulation of the multidrug resistance P-glycoprotein: detection with Tc-99m-Tc-Sestamibi in vivo. J Nucl Med 38:369–372

Maclaren D, Gambhir S, Satyamurthy N et al (1999) Repetitive, non-invasive imaging of the dopamine D2 receptor as a reporter gene in living animals. Gene Ther 6:785–791

Maisey M, Wahl R, Barrington S (1999) Atlas of clinical positron emission tomography. Arnold, Oxford University Press, London, ISBN 0-340-740981

Marshall B, Hansen P, Ensor N et al (1999) Glut-1 or Glut-4 transgenes in obese mice improve glucose tolerance but do not prevent insulin resistance. Am J Physiol 276: E390–E400

Marshall M, Neal F, Goldberg DM (1979) Isoenzymes of hexokinase, 6-phosphogluconate dehydrogenase, phosphoglucomutase and lactate dehydrogenase in uterine cancer. Br J Cancer 40:380–390

Mathupala S, Rempel A, Pedersen P (1997) Aberrant glycolytic metabolism of cancer cells: a remarkable coordination of genetic, transcriptional, post-translational, and mutational events that lead to a critical role for type II hexokinase. J Bioenerget Biomembr 29:339–343

McCarthy KE, Woltering EA, Anthony LB (2000) In situ radiotherapy with ^{111}In-pentetreotide. Q J Nucl Med 44:88–95

Miyake K, Mickley L, Litman T et al (1999) Molecular cloning of cDNAs which are highly overexpressed in mitoxantrone-resistant cells: demonstration of homology to ABC transport genes. Cancer Res 59:8–13

Mochizuki T, Tsukamoto E, Kuge Y et al (2001) FDG uptake and glucose transporter subtype expressions in experimental tumor and inflammation models. J Nucl Med 42:1551–1555

Moretti J-L, Cordobes M, Starzec A et al (1998) Involvement of glutathione in loss of technetium-99m-MIBI accumulation related to membrane MDR protein expression in tumor cells. J Nucl Med 39:1214–1218

Moscow J, Huang H, Carter C et al (1999) Engraftment of MDR1 and NeoR gene-transduced hematopoietic cells after breast cancer chemotherapy. Blood 94:52–61

Nagahara H, Vocero-Akbani A, Synder E et al (1998) Transduction of full-length TAT fusion proteins directly into mammalian cells: TAT-p27^{Kip1} induces cell migration. Nat Med 4:1449–1452

Nagamachi S, Jinnouchi S, Ohnishi T et al (1999) The usefulness of Tc-99m-Tc-MIBI for evaluating brain tumors: comparative study with Tl-201 and relation with P-glycoprotein. Clin Nucl Med 24:765–772

Newman M, Rodarte J, Benbatoul K et al (2000) Discovery and characterization of OC144-093, a novel inhibitor of P-glycoprotein-mediated multidrug resistance. Cancer Res 60:2964–2972

Paganelli G, Zoboli S, Cremonesi M et al (2001) Receptor-mediated radiotherapy with ^{90}Y-DOTA-D-Phe^1-Tyr^3-octreotide. Eur J Nucl Med 28:426–434

Palestro CJ, Bitton R, Tomas MB et al (1997) P829: a technetium-labeled peptide for imaging tumors possessing somatostatin receptors. J Nucl Med 38:236P

Pearson DA, Lister-James J, McBride WJ et al (1996) Somatostatin receptor-binding peptides labeled with technetium-99m: chemistry and initial biological studies. J Med Chem 39:1361–1371

Pennisi E (2000) Human genome: finally, the book of life and instructions for navigating it. Science 288:2304–2307

Piwnica-Worms D (2000) Functional identification of multidrug resistance gene expression in vivo. Am Soc Clin Oncol Educ Book Spring 178–184

Piwnica-Worms D, Kronauge J, Chiu M (1990) Uptake and retention of hexakis (2-methoxy isobutyl isonitrile) technetium(I) in cultured chick myocardial cells: mitochondrial and plasma membrane potential dependence. Circulation 82:1826–1838

Piwnica-Worms D, Chiu M, Budding M et al (1993) Functional imaging of multidrug-resistant P-glycoprotein with an organotechnetium complex. Cancer Res 53:977–984

Piwnica-Worms D, Rao V, Kronauge J et al (1995) Characterization of multidrug-resistance P-glycoprotein transport function with an organotechnetium cation. Biochemistry 34:12210–12220

Podda S, Ward M, Himelstein A et al (1992) Transfer and expression of the human multiple drug resistance gene into live mice. Proc Natl Acad Sci USA 89:9676–9680

Polyakov V, Sharma V, Dahlheimer J et al (1999) Synthesis and characterization in vitro of membrane permeant peptide conjugates for imaging and radiotherapy. J Labelled Compounds Radiopharm 42:S4–S6

Polyakov V, Sharma V, Dahlheimer J et al (2000) Novel Tat-peptide chelates for direct transduction of technetium-99m and rhenium into human cells for imaging and radiotherapy. Bioconjugate Chem 11:762–771

Rabindran S, He H, Singh M et al (1998) Reversal of a novel multidrug resistance mechanism in human colon carcinoma cells by fumitremorgin C. Cancer Res 58:5850–5858

Reichlin S (1983a) Somatostatin, part 1. N Engl J Med 309:1495–1501

Reichlin S (1983b) Somatostatin (part 2). New Engl J Med 309:1556–1563

Reubi JC, Kvols LK, Krenning EP et al (1990) Distribution of somatostatin receptors in normal and tumor tissue. Metabolism [Suppl 2] 39:78–81

Rogers BE, McClean SF, Kirkman RL et al (1999) In vivo localization of [111In]-DTPA-D-Phe1-octreotide to human ovarian tumor xenografts induced to express the somatostatin receptor subtype 2 using an adenoviral vector. Clin Cancer Res 5:383–393

Rogers BE, Zinn KR, Buchsbaum DJ (2000) Gene transfer strategies for improving radiolabeled peptide imaging and therapy. Q J Nucl Med 44:208–223

Saito Y, Price R, Rottenberg D et al (1982) Quantitative autoradiographic mapping of herpes simplex virus encephalitis with a radiolabeled antiviral drug. Science 217:1151–1153

Saito Y, Rubenstein R, Price R et al (1984) Diagnostic imaging of herpes simplex virus encephalitis using a radiolabeled antiviral drug: autoradiographic assessment in an animal model. Ann Neurol 15:548–558

Scheper R, Broxterman H, Scheffer G et al (1993) Overexpression of a M_r 110,000 vesicular protein in non-P-glycoprotein-mediated multidrug resistance. Cancer Res 53:1475–1479

Schwarze S, Ho A, Vocero-Akbani A et al (1999) In vivo protein transduction: delivery of a biologically active protein into the mouse. Science 285:1569–1572

Shao Y, Cherry S, Farahani K et al (1997) Simultaneous PET and MR imaging. Phys Med Biol 42:1965–1970

Sharma V, Piwnica-Worms D (1999) Metal complexes for therapy and diagnosis of drug resistance. Chem Rev 99:2545–2560

Sharma V, Beatty A, Wey S-P et al (2000) Novel gallium(III) complexes transported by MDR1 P-glycoprotein: potential PET imaging agents for probing P-glycoprotein-mediated transport activity in vivo. Chem Biol 7:335–343

Shen DW, Fojo A, Chin JE et al (1986) Human multidrug-resistant cell lines: increased mdr1 expression can precede gene amplification. Science 232:643–645

Smith CM, Liu J, Chen T et al (2000) OctreoTher: ongoing early clinical development of a somatostatin-receptor-targeted radionuclide antineoplastic therapy. Digestion 62 [Suppl 1]:69–72

Smith T (1998) FDG uptake, tumour characteristics and response to therapy: a review. Nucl Med Commun 19:97–105

Smith-Jones PM, Stolz B, Bruns C et al (1994) Gallium-67/Gallium-68-[DFO]-Octreotide – a potential radiopharmaceutical for PET imaging of somatostatin receptor-positive tumors: synthesis and radiolabeling in vitro and preliminary in vivo studies. J Nucl Med 35:317–325

Sorrentino B, Brandt S, Bodine D et al (1992) Selection of drug-resistant bone marrow cells in vivo after retroviral transfer of human MDR1. Science 257:99–103

Stolz B, Smith-Jones PM, Albert R et al (1994) Biological characterization of [67 Ga] or [68 Ga] labelled DFO-octreotide (SDZ-216-927) for PET studies of somatostatin receptor positive tumors. Horm Metab Res 26:452–459

Tazebay U, Wapnir I, Levy O et al (2000) The mammary gland iodide transporter is expressed during lactation and in breast cancer. Nat Med 6:871–878

Tjuvajev J, Chen S, Joshi A et al (1999) Imaging adenoviral-mediated herpes virus thymidine kinase gene transfer and expression in vivo. Cancer Res 59:5186–5193

Torizuka T, Fisher S, Brown R et al (1998) Effect of insulin on uptake of FDG into experimental mammary carcinoma in diabetic rats. Radiology 208:499–504

Vallabhajosula S, Moyer BR, Lister-James J et al (1996) Preclinical evaluation of technetium-99m-labeled somatostatin receptor-binding peptides. J Nucl Med 37:1016–1022

Vives E, Brodin P, Lebleu B (1997) A truncated HIV-1 Tat protein basic domain rapidly translocates through the plasma membrane and accumulates in the cell nucleus. J Biol Chem 272:16010–16017

Vocero-Akbani A, Heyden N, Lissy N et al (1999) Killing HIV-infected cells by transduction with an HIV protease-activated caspase-3 protein. Nature Med 5:29–33

Wahl R, Hutchins G, Buchsbaum D et al (1991) 18F-2-deoxy-2-fluoro-d-glucose (FDG) uptake into human tumor xenografts: feasibility studies for cancer imaging with PET. Cancer 67:1544–1550

Wahl R, Henry C, Eithier S (1992) Serum glucose effects on tumor and normal tissue accumulation of 18F-fluoro-2-deoxy-D-glucose (FDG) in rodents with mammary carcinoma. Radiology 183:643–647

Wahl R, Quint L, Cieslak R et al (1993 a) Anatometabolic tumor imaging: fusion of FDG PET with CT or MRI to localize foci of increased activity. J Nucl Med 34:1190–1197

Wahl R, Zasadny K, Helvie M et al (1993 b) Metabolic monitoring of breast cancer chemohormonotherapy using positron emission tomography (PET): initial evaluation. J Clin Oncol 11:2101–2111

Warburg O (1931) The metabolism of tumors. Smith, New York, pp 1–129

Wester H-J, Brockmann J, Rosch F et al (1997) PET-Pharmacokinetics of ^{18}F-octreotide: a comparison with ^{67}Ga-DFO- and ^{86}Y-DTPA-octreotide. Nucl Med Biol 24:275–286

Wilson J (1995) Hexokinases. Rev Physiol Biochem Pharmacol 126:65–198

Wu A, Yazaki P, Tsai S et al (2000) High-resolution micro-PET imaging of carcinoembryonic antigen-positive xenografts by using a copper-64-labeled engineered antibody fragment. Proc Natl Acad Sci USA 97:8495–8500

Younes M, Lechago L, Somoano J et al (1996) Wide expression of the human erythrocyte glucose transporter Glut 1 in human cancers. Cancer Res 56:1164–1167

Yu Y, Annala A, Barrio J et al (2000) Quantification of target gene expression by imaging reporter gene expression in living animals. Nat Med 6:933–937

Zasadny K, Helvie M, Hutchins G et al (2000) Monitoring tumor blood flow and volume of distribution during chemohormonotherapy of breast cancer with dedicated PET and 15-0 water. Radiology 217 [Suppl]:219

Zeng H, Bain L, Belinsky M et al (1999) Expression of multidrug resistance protein-3 (multispecific organic anion transporter-D) in human embryonic kidney 293 cells confers resistance to anticancer drugs. Cancer Res 59:5964–5967

Zinn K, Buchsbaum D, Chaudhuri T et al (2000) Noninvasive monitoring of gene transfer using a reporter receptor imaged with a high-affinity peptide radiolabeled with Tc-99m or Re-188. J Nucl Med 41:887–895

Chapter 28

Nuclear (PET/SPECT) Imaging of Gene Expression: Methods and Applications

28

Ronald G. Blasberg, Juri G. Gelovani Tjuvajev

Contents

28.1
Introduction and Background

The past decade has witnessed a remarkable increase in knowledge and understanding of the genetics and molecular biology of human disease, particularly in human cancer (Culver 1994). Significant progress in the understanding of the molecular-genetic mechanisms of many diseases have been achieved with the advent of the modern molecular-biological assays (e.g., Western and Northern blots, PCR, RT-PCR, ELISA, etc.). The development of transgenic animal models of human disease, where the molecular basis of that disease can be studied in a living organism, have provided new insights into disease development, progression and treatment. In parallel with these new developments in genetics and molecular biology, the imaging sciences have also made remarkable advances in technology for visualizing tissue structure and function. They include new instruments (e.g., microPET, microCT, high-field strength magnets and novel optical and ultrasound technology) for imaging small animals. There have been corresponding improvements in clinical imaging instrumentation as described in other chapters of this book. The disciplines of molecular/cellular biology and non-invasive imaging rarely interacted until the 1990s. In the mid-1990s, several investigators and the NCI began to promote "cellular and molecular imaging" as a hybrid or convergence of the developments cited above. This convergence and interaction between the "molecular" and "imaging" sciences has been strongly supported by the NCI and was designated as one of six "Extraordinary Scientific Opportunities for Investment". More recently, other NIH institutes and the Department of Energy (DOE) in the US have also developed "molecular imaging" program support.

The convergence of molecular biology and non-invasive imaging was timely and addresses several experimental objectives. For example, ex vivo molecular assays in animal models of disease and in transgenic animals require invasive sampling procedures. Tissue sampling may not always adequately represent the biochemical or pathological process under investigation due to tissue heterogeneity, which is especially characteristic of cancer. Furthermore, temporal studies that employ molecular-biological assays require large numbers of animals that are sacrificed at specific time points in order to achieve a statistically significant temporal profile. The development of versatile and sensitive assays to monitor molecular-genetic and cellular processes in vivo would be of considerable value in human subjects, as well as in experimental animal models of different diseases. Non-invasive imaging of molecular-genetic and cellular processes would compliment existing ex vivo molecular-biological assays, and can provide a spatial as well as a temporal dimension to our understanding of various diseases. It will also accelerate the development of new and more effective therapeutic strategies.

Recent progress in our understanding of the molecular-genetic mechanisms of many diseases has already led to the application of new biologically-based approaches in therapy. However, a limitation of many biological-based therapies has been our inability to achieve controlled and effective delivery of biologically active molecules to target tissues (e.g., tumor cells or their surrounding matrix). This condition is particularly limiting for biologically active compounds that have very short tissue half-lives and exhibit site-specific therapeutic and toxic effects. The new gene-based therapies can provide control over the level, timing and duration of action of many biologically active products by including specific promoter/activator elements in the genetic material transferred. Methods are actively being developed for controlled gene delivery to various somatic tissues and tumors using novel gene constructs, and for controlling gene expression using cell specific, replication-activated, and drug-controlled expression systems (Jolly 1994). A non-invasive, clinically applicable method for quantitatively imaging the expression of transduced genes in target tissue or specific organs would be of considerable value. The capability to image gene expression will facilitate the monitoring and evaluation of gene therapy in human subjects by defining the *location(s)*, *magnitude* and *persistence* of gene expression over time (Tjuvajev et al. 1995; Blasberg and Tjuvajev 1997).

The readers of this chapter are urged to consult other reviews of imaging transgene expression and molecular-biological reporter gene imaging that have been written (Jones 1996; Blasberg and Tjuvajev 1997; Gambhir et al. 2000a; Phelps 2000; Tavitian 2000; Berger and Gambhir 2001). In addition, there are two new journals in the process of coming to press – *Molecular Imaging* (MIT Press; the official journal of the Society of Molecular Imaging, www.molecularimaging.com) and *Molecular Imaging and Biology* (Elsevier, the official journal of the Academy of Molecular Imaging (http://149.142.143.206/index.html). The *European Journal of Nuclear Medicine* now includes "Molecular Imaging" in both its name and in publication topics

28.2
Molecular Imaging Strategies

Two imaging strategies – "direct" and "indirect" – will be described, and examples of each will be discussed. "Direct imaging" can be defined in terms of a probe-target interaction, whereby the resultant image of probe localization and magnitude (image intensity) is directly related to its interaction with the target epitope or enzyme. Examples of direct imaging are common in nuclear medicine and include monoclonal antibody targeting of a particular cell membrane epitope, or imaging the activity of a particular enzyme (e.g., hexokinase) or transporter using an enzyme- or transporter-specific probe (e.g., deoxyglucose). Indirect molecular imaging is a little more complex in that it may involve multiple components. One example of indirect imaging that is now being widely used is "reporter imaging" which usually includes a "marker/reporter gene" and a "marker/reporter probe". The "reporter gene" product can be an enzyme that converts a "reporter probe" to a metabolite that is selectively trapped within transduced cells. The enzymatic amplification of the probe-signal facilitates imaging the magnitude and location of reporter gene expression. This reporter paradigm is essentially an in vivo radiotracer enzyme assay that reflects reporter gene expression. Indirect imaging paradigms are currently more widely used in molecular imaging and will be discussed in greater detail below. Indirect molecular imaging will be the focus of this chapter.

28.2.1
Direct Imaging

Direct imaging strategies are based on imaging the target directly, usually with use of a target-specific probe. One strategy involves the "binding" of a radiolabeled ligand or probe directly to the target (e.g., receptor, antigen, epitope, etc.). Examples of the direct imaging paradigm are common in nuclear medicine. PET imaging of receptor density/occupancy using small radiolabeled molecular probes are widely used, particularly in neuroscience research. Imaging cell surface-specific antigens or epitopes with radiolabeled antibodies is another example of direct molecular imaging that has developed over the past 30 years. These studies represent some of the first "molecular imaging" applications and are used in both animal and clinical nuclear medicine research.

28.2.1.1
Ligands and Neuroreceptors

In neuroscience research, different imaging probes (tracers) have been synthesized to study several neurotransmitter systems. The dopaminergic and serotonergic transmitter systems have been a major focus of efforts to selectively trace synthesis, storage, release, post-synaptic binding and reuptake, of neurotransmitters. Dopamine-derived tracers are used in diseases of the basal ganglia, whereas serotonin-, benzodiazepine-, and opiate-derived tracers are used in lesions of the cerebral cortex. PET has demonstrated the progressive loss of dopaminergic terminals and dopamine synthetic capacity in the evolution of Parkinson's disease, MPTP intoxication and Lesch-Nyhan's syndrome, as well as the release of dopamine after administration of cocaine and amphetamine (Gjedde 2001). Increased synaptic dopamine levels and release of dopamine following motor activity and cognition has been shown (Schneider et al. 1995). Furthermore, neuroleptic occupancy of dopamine receptors in the striatum of patients with schizophrenia (Kasper et al. 1999), and loss of muscarinic and nico-

tinergic receptors in Alzheimer's disease (Zubieta et al. 2001), and loss of benzodiazepine and opiate receptors in stroke, epilepsy, and Huntington's chorea have been demonstrated (Hatazawa and Shimosegawa 1998; Duncan 1999; Willoch et al. 1999; Koepp and Duncan 2000; Heiss 2001). Altered opiate receptor occupancy has been demonstrated in chronic pain and drug abuse (Jones et al. 1999; Kling et al. 2000), and release of opiates in analgesia (Casey et al. 2000).

Other radiolabeled probes and PET imaging have been used to investigate the neurochemical basis of epilepsy. For example, ^{11}C-flumazenil binds to the central benzodiazepine receptor (cBZR)-gamma-aminobutyric acid (GABA) (Sanabria-Bohorquez et al. 2000). The complex of opiate receptors have been studied in vivo by PET imaging with ^{11}C-diprenorphine, ^{18}F-cyclofoxy and ^{11}C-carfentanil (Koepp and Duncan 2000), and ^{11}C-deprenyl has been used to study monoamine oxidase B (Duncan 1999). Further understanding of the processes that underlie the epilepsy (and other diseases) will come from the development of new radiolabeled ligands that are specific for excitatory amino acid receptors, the subtypes of the opioid receptors, and the GABA-B receptor.

28.2.1.2
Antibodies and Peptides

Imaging cell-specific surface antigens or epitopes with radiolabeled antibodies was one of the first examples of "molecular imaging". Tumor (target) cell imaging with radiolabeled proteins has developed over the past 30 years. Initial studies involved the direct radiolabeling of native antibodies (e.g., IgG, MW 150 kDa) (Brumley and Kuhn 1995). Antibody-targeted imaging has evolved considerably with the introduction of genetic engineering, which provides a powerful approach for redesigning antibodies for use in oncologic applications in vivo. Recombinant fragments have been produced that retain high affinity for target antigens, and display a combination of rapid, high-level tumor targeting with concomitant clearance from normal tissues and the circulation in animal models. An important first step was cloning and engineering of antibody heavy and light chain variable domains into singlechain recombinant fragments (scFvs; molecular weight, 25–27 kDa), in which the variable regions are joined via a synthetic linker peptide sequence. Although scFvs themselves showed limited tumor uptake in preclinical and clinical studies, they provide a useful building block for intermediate-sized recombinant fragments.

Covalently linked dimers or non-covalent dimers of scFvs (also known as diabodies) show improved targeting and clearance properties due to their higher molecular weight (55 kDa) and increased avidity. Further gains can be made by using larger recombinant fragments, such as the scFv-CH3 fusion protein that self-assembles into a bivalent dimer of 80 kDa (a minibody). Ease of engineering and expression, combined with novel specificities that will arise from advances in genomic and combinatorial approaches to target discovery, starts a new era of recombinant antibodies for biological imaging. The reader is referred to several recent reviews that cover this topic in greater detail (Divgi 1996; Cerqueira 1999; Mariani 1999; Moffat et al. 1999; Wu and Yazaki 2000; Boerman et al. 2001; Kalofonos et al. 2001).

More recently, the visualization of various tumor cell surface receptors using radiolabeled regulatory peptides is a novel strategy that is being pursued in oncology research. These regulatory peptides are small, readily diffusible, and potent natural substances with a wide spectrum of receptor-mediated actions. High affinity receptors for these radiolabeled peptides are over-expressed in many cancers and represent molecular targets for cancer diagnosis and therapy. The somatostatin analogs were the first class of receptor-binding peptides that reached clinical application. ^{111}In-DTPA-[D-Phe1]-octreotide is the first and only radiolabeled peptide which has obtained regulatory approval in Europe and the United States to date. Extensive clinical studies involving several thousands of patients have shown that the major clinical application of somatostatin receptor scintigraphy is the detection and staging of gastrointestinal and pancreatic neuroendocrine tumors (carcinoids). In these tumors, octreotide scintigraphy is superior to any other staging method (Bombardieri et al. 2001). However, octreotide sensitivity and accuracy in other neoplasms is more limited.

Other radiolabeled peptides have been studied. For example, vasoactive intestinal peptide (VIP) has been shown to visualize the majority of gastrointestinal adenocarcinomas, as well as some neuroendocrine tumors, including insulinomas (Virgolini et al. 1994). The CCK-B receptor is widely expressed on human small-cell lung cancer and ^{111}In-labeled DTPA derivatives of gastrin have shown excellent targeting of CCK-B receptor expressing tissues in animals and patients (Reubi et al. 1998). Many other peptide-based radioligands are being developed, including the gastrin-releasing peptide bombesin, neurotensin, substance-P, pan-somatostatin (somatostatin derivatives which bind to all five receptor subtypes) and glucagon-like peptide-1 (glp-1) analogs (the latter for the specific detection of insulinomas) (Kaltsas et al. 2001). Imaging results using radioiodinated or ^{18}F-labeled RGD-containing glycopeptide suggested that this compound may be suitable for non-invasive determination of the alpha(V)beta3 integrin in tumors and therapy monitoring (Haubner et al. 2001 a, b). Future work is likely to define a much larger number of novel and potentially useful peptide-based radioligands for clinical application. A recent review summarizes peptide-based radiopharmaceuticals which are presently commercially available or are in advanced stages of clinical testing (Behr et al. 2001).

28.2.1.3
Antisense and Aptomer Probes

Oligonucleotide probes have been developed to specifically hybridize to target mRNA or proteins in vivo, and this work has recently been reviewed (Stalteri and Mather 2001). The use of antisense probes is a direct approach to imaging endogenous gene expression at a transcriptional level, and is an alternative to the "indirect" reporter-imaging approach described below. Antisense strategies involve the use of radiolabeled oligonucleotides (RASONs). RASONs are small oligonucleotides that are complimentary to a small segment of target mRNA or DNA, and could potentially target any specific mRNA or DNA sequence. In this context, imaging specific mRNAs with RASONs is a "direct" approach for imaging molecular genetic events.

Some efficacy for imaging endogenous gene expression has been reported using the ^{111}In-RASONs and a gamma camera. Rapid localization of the c-myc oncogene-targeted ^{111}In-RASONs to a tumor expressing c-myc oncogene was demonstrated in mice at 2 h after tracer injection (Dewanjee et al. 1994). The biodistribution of ^{125}I-RASONs to TGFα was reported in athymic mice with a human mammary tumor xenograft following intratumoral injection (Cammilleri et al. 1996). Pharmacokinetics, metabolism, and elimination of 20-mer RASONs after intravenous and subcutaneous administration has been assessed with a whole body autoradiography in mice (Phillips et al. 1997). Recently, the biodistribution and pharmacokinetics of three different types of nonspecific ^{18}F-RASONs (phosphodiester, phosphorothioate, and 2'-O-methylated) was reported in baboons with PET (Tavitian et al. 1998).

Similarly, radiolabeled aptamer probes, RASONs, can be used to target specific proteins as well as complimentary nuclide sequences of endogenous mRNA and DNA. Nuclease-resistant aptamers identified from randomized nucleic acid libraries represent a novel class of drug candidates. Aptamers are synthesized chemically and therefore can be readily modified with functional groups that modulate their properties. Recently, the synthesis and radioiodination of a stannyl oligodeoxyribonucleotide (ODN) was undertaken to evaluate a gamma ray-emitting ODN ligand for thrombus imaging in vivo. Synthesis of the ODN was based on modified beta-cyanoethyl phosphoramidite automated chemistry with an organotin nucleoside (dU*) coupled to a thrombin-binding aptamer sequence. A high yield, high specific activity product was obtained with high target specificity and affinity (Dougan et al. 1997). Others describe preparation, initial characterization, and functional properties of a nuclease-resistant vascular endothelial growth factor (VEGF) aptamer anchored in liposome bilayers through a lipid group on the aptamer. While the high-affinity binding to VEGF is maintained, the plasma residence time of the liposome-anchored aptamer is considerably improved compared with that of the free aptamer. The lipid group attachment and/or liposome anchoring leads to a dramatic improvement in inhibitory activity of the aptamer toward VEGF-induced endothelial cell proliferation in vitro and vascular permeability increase and angiogenesis in vivo (Willis et al. 1998). Current research in radiolabeled aptamer imaging is focused on prolonging the clearance of intact radiolabeled aptamers from blood to improve targeting by using various derivatized oligonucleotides (Dougan et al. 2000). Nevertheless, the efficacy and utility of radiolabeled aptamers for non-invasive imaging has yet to be established.

Several serious limitations of RASON and aptamer imaging remain, including: (a) low specificity of localization (low number of target mRNA molecules per cell, poor stability of binding, degradation by H-RNAse); (b) high background activity (high nonspecific extra- and intracellular distribution); (c) limited tracer delivery (poor vascular and cell membrane permeability; can not penetrate blood-brain barrier); and (d) slow clearance (slow washout of non-bound oligonucleotides from intracellular space); (e) radiolabeled metabolites (large amounts of nonspecific, non-RASON, radioactivity due to radiolabeled metabolites which confound interpretation of the images).

28.2.1.4
Imaging Enzyme Activity

The molecular imaging paradigm that is most widely used clinically is an assay of enzymatic activity; namely, imaging hexokinase activity (glucose utilization) with 2'-deoxyglucose [DG] or 2'-fluoro-2'-deoxyglucose [FDG] (Gambhir et al. 2001; Reske and Kotzerke 2001). This imaging paradigm is based on an enzyme-specific radiolabeled probe and an enzyme-metabolic trapping mechanism. The enzyme-specific trapping of a radiolabeled probe provides an amplification of the signal (radioactivity) in target tissue. Such enzymatic signal amplification strategies can offer substantial enhancements for direct molecular imaging. Currently, imaging tumors with FDG is the most widely used PET-imaging modality in oncology, although DG was originally developed to study brain function in awake animals (Sokoloff et al. 1977) and FDG imaging was developed for similar studies in human subjects (Reivich et al. 1977). Nevertheless, whole body FDG imaging has become a mainstay in staging the extent of disease for many cancers and has become a prerequisite imaging study (along with CT and MR imaging) prior to embarking on "curative" surgery (Delbeke and Martin 2001), and is increasingly being used to evaluate treatment response (Young et al. 1999; Hoekstra et al. 2000).

Another example of an imaging strategy that uses enzyme-specific metabolic trapping of a radiolabeled probe is imaging DNA synthesis and cell proliferation. Thymidine kinase (TK) is an enzyme that is expressed during S-phase of the cell cycle and the level of TK expression reflects the proliferative activity of the tissue or tumor. TK phosphorylates thymidine and closely related analogues; the phosphorylated probe is effectively trapped in the cell (similar to the trapping of deoxyglucose phosphate). In addition, some phosphorylated probes are further metabolized and incorporated into the DNA of dividing cells. The largest number of clinical studies imaging cell-proliferative activity have used [11C]thymidine ([11C]TdR) (Mankoff et al. 1998; Shields et al. 1990, 1996a). The advantages of using [11C]TdR are that it is a naturally occurring nucleoside, it is rapidly incorporated into DNA, and can be imaged with PET. However, imaging cell proliferation with [11C]TdR has several disadvantages: (1) a large fraction of radiolabeled metabolites rapidly appear in blood and tissue; (2) the short (20 min) half-life of [11C] that precludes sufficient time for tissue washout and body clearance of these metabolites; and (3) complex modeling is required to obtain reliable measures and parametric images of proliferative activity. Although the amounts of radiolabeled metabolites is lower when using [2-^{11}C]TdR as compared to [methyl-^{11}C]TdR, a substantial fraction of [2-^{11}C]TdR metabolites persists at the time of imaging (Shields et al. 1992; Van der Borght et al. 1994); at least 25% of blood radioactivity is reflected in background tissue activity due to CO_2 alone.

Radiolabeled thymidine analogues, such as 5-iodo-2'-deoxy-uridine (IUdR) and 5-bromo-2'-deoxy-uridine (BrUdR), have also been used in a few studies to image tumor-proliferative activity. An advantage of IUdR and BrUdR is the substantially longer physical half-life of the radionuclide used for labeling (e.g., ^{123}I, 13 h; ^{124}I, 4 days; ^{131}I, 8 days; ^{76}Br, 16 h). Both IUdR and BrUdR share the disadvantage of imaging substantial background of radiolabeled metabolites (*I$^-$ and ^{76}Br$^-$, respectively). The long physical half-life of these radiolabels provides the opportunity to implement a "washout" (late imaging) strategy to reduce background radioactivity due to radioactive iodine and bromine – the major radiolabeled metabolites of IUdR and BrUdR, respectively. This late imaging strategy increases image specificity and simplifies the analysis of the images (Tjuvajev et al. 1994; Blasberg et al. 2000). Both IUdR and BrUdR have similar rates of dehalogenation; 80% of the administered dose is dehalogenated during the first 5–10 min after i.v. administration (Hamptom and Eidinoff 1961; Kriss et al. 1963). Radiolabeled iodide from IUdR is cleared fairly rapidly by the kidney; about 79–80% is excreted in the urine during the first 24 h (Prusoff et al. 1960; Welch et al. 1960; Calabresi et al. 1961). Liberated bromine from BrUdR is cleared more slowly by the kidney and remains distributed in the extracellular fluid; only 20% of bromide is excreted in the urine during the 24 h and a substantial fraction is incorporated into bone (Kriss et al. 1963). This results in a higher background of bromide radioactivity compared to that of iodide, and makes BrUdR less desirable than IUdR for in vivo imaging of DNA synthesis and cell proliferation. In one study with [^{76}Br]BrUdR and PET (Bergstrom et al. 1998), the level of background radioactivity was too high to obtain useful images even after a 24-h period of washout and diuresis.

The use of "non-degradable" radiolabeled pyrimidine nucleoside analogues has recently been proposed (Conti et al. 1995; Shields et al. 1996b, 1998). Most of the published studies report the use of 2'-FMAU, 2'-FMRU, 2'-FFAU, 3'-FLT (3'-FMRU), and 2'-FBrAU. An important characteristic of the 2-fluoro-arabino ("up") and 2-fluoro-ribo ("down") substitution on the sugar moiety is that it provides a protective group against enzymatic cleavage of the N-glucosidic bond by nucleoside phosphorylases. The increased stability of N-glucosidic bond results in a significant prolongation of plasma half-life of these compounds (they are excreted largely unchanged in the urine) and a substantially higher input function (blood concentration-time integral) in comparison to TdR and IUdR. The 5-fluoro-substituted pyrimidine analogue, [5-^{18}F]FUdR, is less avidly incorporated into DNA and are incorporated into RNA as well (Ishiwata et al. 1984; Tsurumi et al. 1990) because FUdR is an analogue of 2'-deoxy-uridine (not thymidine). In blood and tissue (predominantly liver), [5-^{18}F]FUdR is rapidly hydrolyzed to [5-^{18}F]-fluoro-uracil ([5-^{18}F]FU) which is catabolized into [^{18}F]FBAL and avidly accumulated by liver and small intestine whereas the radiolabeled metabolite [^{18}F]fluoride accumulates in bone (despite fairly rapid excretion of fluoride by kidneys). More studies are needed to fully assess the feasibility of these nucleoside analogues in clinical oncology practice.

28.2.1.5
Transporters

The application of *radioiodide* accumulation, as detected by thyroid scintigraphy, has been used for over 50 years in the diagnosis and treatment of thyroid disease (Mazaferri 2000). Historically, thyroid scintigraphy is the oldest method of molecular imaging. Active iodide transport in the thyroid, salivary glands, stomach and lactating mammary gland is mediated by the Na$^+$/I$^-$ symporter (NIS), an intrinsic plasma membrane glycoprotein (Dai et al. 1996; Spitzweg et al. 1998; Tazebay et al. 2000). A cDNA clone encoding rat NIS (rNIS) has been isolated and encodes a protein of 618 amino acids that is highly homologous to the subsequently cloned human NIS (hNIS, 643 amino acids) (Smanik et al. 1996). In all tissues where it is functionally expressed, NIS mediates inward 'uphill' translocation of I$^-$ against a concentration gradient by coupling to the inward 'downhill' translocation of Na$^+$. Therefore, NIS is a symporter as it translocates both Na$^+$ and I$^-$ simultaneously and in the same direction. The driving force for NIS-mediated I$^-$ transport is generated by Na$^+$-K$^+$ ATPase (Bagchi and Fawcett 1973; Weiss et al. 1984a; Carrasco 1993; De la Vieja et al. 2000). In the thyroid, I$^-$ accumulation is stimulated by thyroid stimulating hormone (TSH)

(Weiss et al. 1984b; Vassart and Dumont 1992). In addition, as part of the thyroid hormone biosynthetic pathway, accumulated I^- is incorporated into tyrosyl residues on the large thyroglobulin (Tg) molecule by a process known as I^- organification (Taurog 2000). Iodinated Tg gives rise to T_3 and T_4 (for a detailed description of thyroid hormone biosynthesis see Taurog 2000). By contrast, I^- accumulation in extra-thyroidal tissues is not regulated by TSH (De la Vieja et al. 2000) and NIS-mediated I^- accumulation is blocked by the specific competitive inhibitors, including thiocyanate and perchlorate (Carrasco 1993; Nagataki and Yokohyama 1996; Eskandari et al. 1997).

NIS activity plays a major diagnostic and therapeutic role in the management of differentiated thyroid carcinoma. On scintigraphy, most thyroid cancers exhibit decreased I^- uptake relative to the surrounding normal thyroid tissue. Nevertheless, sufficient I^- transport activity in differentiated thyroid cancer cells is retained for ^{131}I radioablation therapy to be effective. This has proven to be true against most well differentiated malignant cells in the thyroid gland itself or in distant metastases after thyroidectomy (Mazaferri 2000). Radioiodide ablation destroys occult microscopic carcinomas and also any remaining normal thyroid tissue. Pre- and post-ablative $^{131}I^-$ or $^{124}I^-$ total body scintigraphic or PET imaging permits the search for distant metastases or persistent carcinoma. Given the lower accumulation of I^- observed in most thyroid cancers compared to that in normal thyroid, it had long been expected that NIS expression was decreased in thyroid cancer cells. Several researchers have explored ways to induce NIS transcription in thyroid cancer, thus seeking to improve the I^- transport in thyroid cancer cells and thereby the effectiveness of radioiodide therapy (Venkataraman et al. 1999; Schmutzler and Kohrle 2000).

The transport of *amino acid*s across tumor cells and their supporting vasculature is increased in malignant transformation (Busch et al. 1959; Isselbacher 1972). This increase in transport may be associated with specific cell surface changes in transformed cells. For example, amino acid transport system A is one of the few identified transport systems that is more highly expressed in transformed and malignant cells. This up-regulation appears to be a target of proto-oncogene and oncogene action (Saier et al. 1988). In general, however, the process of malignant transformation requires that cells acquire and use nutrients efficiently for energy production, protein synthesis, and cell division. Increased transport of amino acids

most likely reflects a specific malignant cell-based demand for increased amino acid availability. With respect to imaging tumors with amino acids, the available data suggests that the increased transport rate of amino acids may be more important than protein synthesis. Several processes contribute to amino acid demand and utilization in addition to protein synthesis, including amino acid transamination and transmethylation, metabolism to nonprotein precursors, and their use as energy substrates (Souba 1993; Stryer 1995).

Many amino acids have been radiolabeled by replacement of a carbon atom with ^{11}C. This radiolabeling does not chemically change the molecule and provides the opportunity to image with PET $[^{11}C]$-radiolabeled amino acids. The radiolabeling of naturally occurring amino acids differ with respect to ease of synthesis and the formation of radiolabeled metabolites in vivo. For these and other reasons, $[^{11}C$-methyl]-methionine has been most extensively studied and used clinically. More recently, several amino acid analogues such as $[^{123}I]$iodomethyltyrosine (IMT) (Matheja and Schober 2001), L-3-$[^{18}F]$fluoro-a-methyltyrosine (FMT) (Inoue et al. 1998), O-2-$[^{18}F]$fluoroethyl-L-tyrosine (FET) (Wester et al. 1999a), $[^{18}F]$fluoro-L-phenylalanine (Kubota et al. 1996), $[^{18}F]$-1-amino-3-fluorocyclobutane-1-carboxylic acid (FACBC) (Shoup et al. 1999), $[^{18}F]$fluoro-L-proline (Wester et al. 1999b), and $[^{11}C$-methyl]-aaminoisobutyric acid (Orlefors et al. 1998) have been studied. The tryptophan metabolite ^{11}C-labeled 5-hydroxytryptophan has been used to study carcinoid tumors (Bading et al. 1996). Uptake of several amino acids in neuroendocrine tumors appears to be irreversible and specific.

In human subjects, the concept of amino acid incorporation into brain and tumor protein is *not* well substantiated (*no* tissue sampling, protein extraction and radioactivity assay data). Only kinetic analyses of time-activity PET data have been reported, and amino acid transport may be *more important* for tumor imaging in patients than amino acid incorporation into protein. This conclusion is based on the results of several dynamic PET studies using methionine (Roelcke et al. 1996), fluorotyrosine (Wienhard et al. 1991), and fluorophenylalanine (Miura et al. 1992; Ito et al. 1995). These results suggest that the incorporation of radiolabeled amino acids (such as methionine) into brain and brain tumor protein is much less likely to effect the images in patient studies compared to that observed in animals bearing tumors.

28.2.2
Indirect Imaging

"Indirect" imaging is currently a widely used strategy for radionuclide-based molecular imaging (PET, SPECT, gamma camera and autoradiography). Indirect imaging strategies are also widely used in optical- (Hooper et al. 1990; Mayerhofer et al. 1995; Bennett et al. 1997; Welsh and Kay 1997; Rehemtulla et al. 2000) and MR-based (Weissleder et al. 1990, 1991, 1997; Kayyem et al. 1995; Louie et al. 2000) molecular imaging techniques (*these two imaging technologies will not be discussed in this chapter*).

Most "indirect" molecular imaging paradigms involve the use of *reporter-transgene* technology and specific probes that can be used to produce an image. The paradigm involves several steps, including the initiation of transcription (that can be controlled by specific promoter/enhancer elements), the process of DNA transcription and stabilization of mRNA, and subsequent translation of mRNA into the gene product (a protein). In this manner, a reporter expression cassette can be designed to provide information about endogenous gene regulation, mRNA stabilization and specific protein-protein interactions.

A "foreign" (nonhost) reporter transgene (e.g., Herpes simplex virus (type 1) thymidine kinase HSV1-*tk*) can be used and placed under the control of upstream promoter/enhancer elements. These promoter/enhancer elements can be "always turned on" with constitutive promoters (e.g., LTR, RSV, CMV). Constitutively expressing reporter genes can be used to monitor and track the distribution and expression of vectors in gene therapy protocols. They have been described as "beacon" systems for defining the location, magnitude and persistence of gene expression over time. Alternatively, reporter genes can be placed under the control of promoter/enhancer elements that are "sensors" of endogenous molecules and transcription factors. In this case the promoter/enhancer elements act as a "switch". The switch is sensitive to specific molecules (signals) in the cell and will initiate reporter gene transcription only in the presence of these specific endogenous signals. The promoter can be cell specific, restricting expression of the transgene to certain cells or to specific transcription factors in multiple cell types. Several non-invasive imaging paradigms have been described and it has recently been shown that transcriptional regulation of endogenous (host tissue) gene expression can be imaged using PET (Dou-

brovin et al. 2001a). Several reporter gene, indirect imaging paradigms will be discussed in greater detail below and will be the focus of this chapter.

The general paradigm for quantitative reporter gene imaging using nuclear imaging technology will be described first. This will be followed by a discussion of several reporter transgenes with a focus on the Herpes simplex virus type 1 thymidine kinase gene (HSV1-*tk*) and a comparison of different radiolabeled probes available for imaging HSV1-*tk* expression. Other reporter transgenes will also be briefly described. The latter part of this chapter will focus on specific examples of imaging the expression of endogenous host genes using nuclear imaging technology. This is a very rapidly moving field and new PET and gamma camera reporter imaging paradigms are being reported frequently.

28.3
Reporter Gene Imaging

28.3.1
Paradigm for Indirect Nuclear Imaging of Gene Expression

A paradigm for imaging transgene expression was initially described in 1995 (Tjuvajev et al. 1995) and is shown in Fig. 28.1. This paradigm requires the appropriate *combination* of a reporter transgene and a reporter probe. The transgene encodes for an enzyme (e.g., HSV1-*tk*) that leads to the trapping of a radiolabeled probe. Alternatively, a reporter gene can encode for an intracellular and/or extracellular receptor (e.g., D2R) that would "irreversibly" bind a radiolabeled probe (see below). In either case, the level of probe accumulation must be shown to be proportional to the level of gene expression; namely, the level of gene-product being expressed. It is also important to note that imaging transgene expression is independent of the vector used to transfect/transduce target tissue; namely, any of several currently available vectors can be used (e.g., retrovirus, adenovirus, adeno-associated virus, lentivirus, plasmids, liposomes, etc.).

The common feature for all vectors is the cDNA expression cassette containing the reporter transgene(s) of interest (e.g., HSV1-*tk*), but it is important to note that the arrangement of the expression cassette can be varied. The reporter transgene(s) can be driven by any promoter of choice. The promoter can

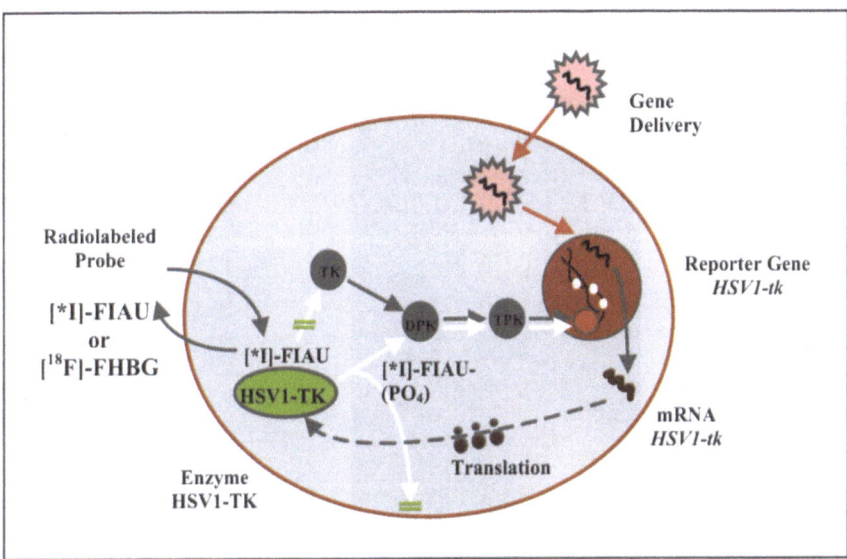

Fig. 28.1. Schematic for imaging HSV1-*tk* reporter gene expression with reporter probes FIAU and FHBG. The HSV1-*tk* gene complex is transfected into target cells by a vector. Inside the transfected cell, the HSV1-*tk* gene is transcribed to HSV1-*tk* mRNA and then translated on the ribosomes to a protein (enzyme), HSV1-*tk*. After administration of a radiolabeled probe and its transport into the cell, the probe is phosphorylated by HSV1-*tk* (gene product). The phosphorylated radiolabeled probe does not readily cross the cell membrane and is "trapped" within the cell. Thus, the magnitude of probe accumulation in the cell (level of radioactivity) reflects the level of HSV1-*tk* enzyme activity and level of HSV1-*tk* gene expression

be "constitutive" (leading to continuous transcription), or it can be "inducible" (leading to controlled expression or sensitive to the level of endogenous transcription factors). The promoter can also be cell specific, allowing expression of the transgene to be restricted to certain cells.

The imaging paradigm involves administering a radiolabeled probe which is selectively bound or metabolized (e.g., phosphorylated) and trapped by interaction with the gene product (e.g., an enzyme) in the transduced cell. In this manner the bound or metabolized (phosphorylated) probe accumulates selectively in transduced tissue, and the level of probe accumulation is proportional to the level of gene product being expressed. This paradigm for reporter gene imaging is considered "indirect" because the images depend on the level of gene product (enzyme) expression, and not on the level of reporter gene transcription per se. These issues will be discussed in greater detail below, along with the importance of "marker/reporter" probe selection. For now, it may be useful to consider the reporter imaging paradigm as an enzymatic radiotracer assay that reflects reporter gene expression. Viewed from this perspective, reporter gene imaging is similar to imaging hexokinase activity with fluorodeoxyglucose (FDG).

The first demonstration of imaging of HSV1-*tk* gene expression in vivo using PET is shown in Fig. 28.2 (Blasberg and Tjuvajev 1997; Tjuvajev et al. 1998). This example also shows the first PET images of in vivo transduction of an established wild-type tumor with a retrovirus containing the HSV1-*tk* gene.

28.3.2
Characteristics of a Good Reporter Gene – Good Reporter Probe Combination

The following characteristics are considered ideal, if not essential for in vivo transgene imaging using non-invasive nuclear imaging techniques: (1) A marker/reporter transgene is usually a "foreign gene" that is not present (or not normally expressed) in the host tissue, and the gene product is an enzyme or receptor that can be expressed in transduced or transfected host cells. The enzyme or receptor must be non-toxic to host cells and lead to accumulation of the reporter probe within transduced or transfected cells. The enzyme or receptor should also not significantly perturb normal cell function. (2) A marker/reporter probe is chosen to match the marker/reporter transgene; it is a compound that is not metabolized (or

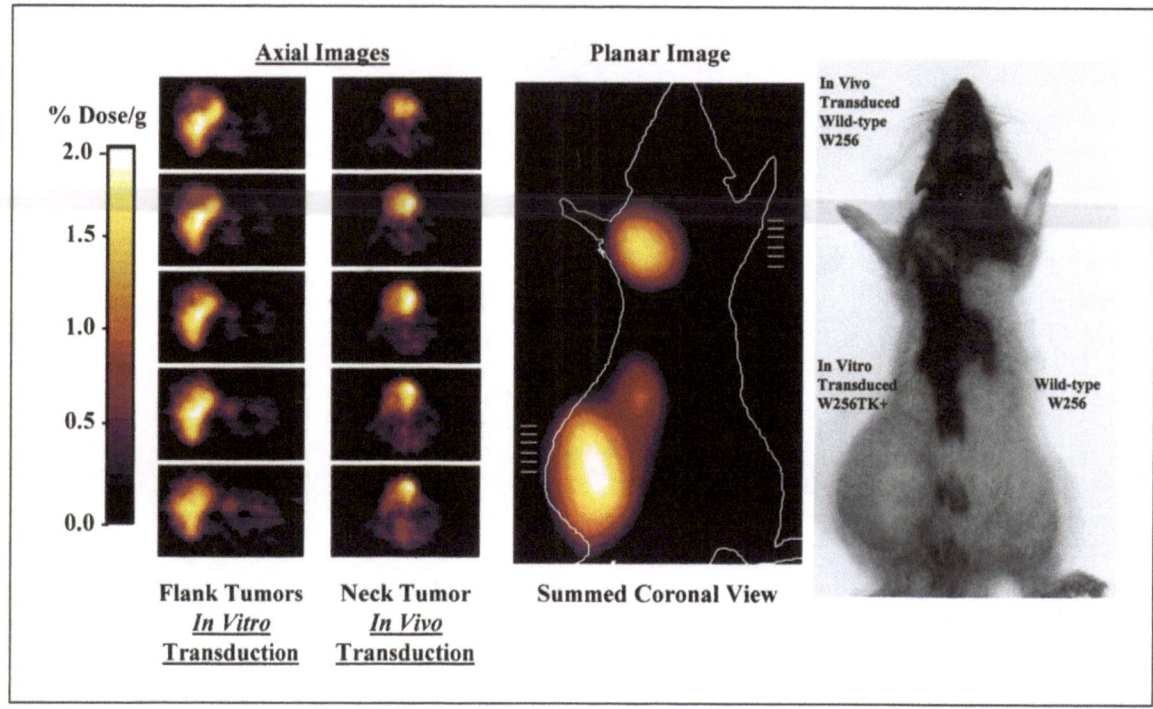

Fig. 28.2. [^{124}I]FIAU PET imaging of HSV1-*tk* gene expression. Three tumors were produced in rnu/rnu rats. A W256TK+ tumor (positive control) was produced from stably transduced W256TK+ cells in the left flank, and wild-type W256 tumors were produced in the dorsum of the neck and in the right flank (negative control). The neck tumor (wild-type) was inoculated with 10^6 gp-STK-A2 vector-producer cells (retroviral titer: 10^6–10^7 cfu/ml) to induce in vivo transduction of the tumor. No carrier-added [^{124}I]-FIAU (25 µCi) was injected i.v. 14 days after gp-STK-A2 cell inoculation and PET imaging was performed at 30 h later. Localization of radioactivity is clearly seen in left flank tumor (positive control) and in vivo transduced neck tumor (test), but only low background levels of radioactivity were observed in the right flank tumor (negative control)

only slowly metabolized) by the host, and does not accumulate in host tissues that do not express the marker/reporter gene. The probe can be radiolabeled with appropriate isotopes for clinical imaging using gamma camera, SPECT, or PET techniques. (3) The marker/reporter probe must cross cell membranes readily and be freely exchangeable for interaction with an "external" (cell membrane) receptor or epitope, or an "internal" (intracellular) enzyme or receptor; it must be rapidly metabolized or bound by the marker/reporter gene product and be effectively trapped by or within transduced or transfected cells throughout the period of imaging, and it must accumulate to levels that are measurable by existing clinical imaging techniques. (4) The accumulation of the "probe" in transduced or transfected cells must reflect the activity of the maker/reporter transgene product and, thereby, the expression of the marker/reporter transgene in transduced or transfected tissue. Further details for an ideal marker/reporter gene imaging assay are described elsewhere (Tjuvajev et al. 1995; Blasberg and Tjuvajev 1997).

28.4
HSV1-*tk* as a "Reporter Gene"

HSV1-*tk* was selected as an example of a "reporter gene" for several reasons. HSV1-*tk* has been extensively studied; it is essentially non-toxic in humans and is currently being used in clinical gene therapy protocols as a "susceptibility" gene for treatment of cancer (in combination with ganciclovir) (Moolten 1986; Borrelli et al. 1988). The gene product of HSV1-*tk* is an enzyme, HSV1 thymidine kinase (HSV1-*tk*). We and others have shown that HSV1-*tk* can be used as a "reporter gene" as well as a "therapeutic gene" (Tjuvajev et al. 1995, 1996, 1998; Blasberg and Tjuvajev 1997; Gambhir et al. 1998, 1999 a, b). This represents an ideal situation where the "therapeutic" and "reporter" genes are the same. In HSV1-*tk* gene therapy protocols, identifying the location and magnitude of HSV1-*tk* enzyme activity by non-invasive imaging would provide a highly desirable measure of gene expression (following successful gene transduction) and provide a basis from which the timing of ganciclovir treatment can be optimized.

28.4.1
Reporter Gene Enhancements

Genetic alteration of reporter genes to increase the imaging capability of existing reporter/marker probes has been proposed (Gambhir et al. 2000b). Genetic engineering is used to produce small alterations in the structure of the gene product (e.g., an enzyme) which will result in increased enzymatic activity and "trapping" of a particular probe. This approach focuses on optimizing the active (enzymatic) site of the gene product to enhance sensitivity and specificity of reporter gene imaging with a particular probe or class of probes. The more traditional approach in nuclear medicine has focused on optimizing the chemical structure of the probe to the wild-type enzyme (reporter gene product) in order to enhance sensitivity and specificity of reporter gene imaging. For example, the use of site-mutated variants of wild-type HSV1-*tk* as a "super reporter gene" has been proposed to enhance the sensitivity of radiolabeled acycloguanosine probes for imaging HSV1-*tk* expression (Gambhir et al. 2000b). This proposal developed from earlier work that demonstrated enhanced viral cytotoxicity to ganciclovir and acyclovir, and to enhanced drug phosphorylation with site-mutated HSV1-*tk* variants compared to wild-type HSV1-*tk* (Black et al. 1996, 2001). This work is discussed further below, after a discussion of two classes of reporter probes for imaging the expression of wild-type HSV1-*tk*.

28.4.2
"Reporter Probes" for HSV1-*tk*

The HSV1-*tk* gene product is a protein (an enzyme), HSV1 thymidine kinase (HSV1-TK). HSV1-TK phosphorylates thymidine, like mammalian TKs. Unlike mammalian TKs, HSV1-TK has less substrate specificity than human thymidine kinase 1 (hTK1) and phosphorylates a wider range of compounds, including acycloguanosines (e.g., acyclovir, ACV; ganciclovir, GCV; penciclovir, PCV) and 2'-fluoro-nucleoside analogues of thymidine (e.g., 5-iodo-2'-fluoro-2'-deoxy-1-β-D-arabino-furanosyl-uracil (FIAU)) (see

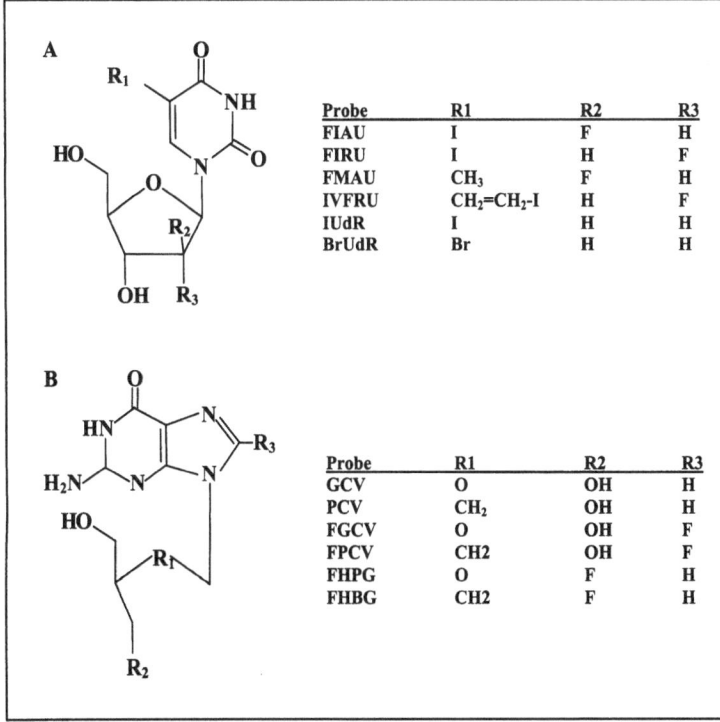

Probe	R1	R2	R3
FIAU	I	F	H
FIRU	I	H	F
FMAU	CH₃	F	H
IVFRU	CH₂=CH₂-I	H	F
IUdR	I	H	H
BrUdR	Br	H	H

Probe	R1	R2	R3
GCV	O	OH	H
PCV	CH₂	OH	H
FGCV	O	OH	F
FPCV	CH2	OH	F
FHPG	O	F	H
FHBG	CH2	F	H

Fig. 28.3 A, B. Chemical structures of pyrimidine (A) and acycloguanosine (B) nucleoside probes for imaging HSV1-*tk* gene expression. Derivatives of thymidine as probes for HSV1-*tk* (*top panel*), and derivatives of acycloguanosine as probes for HSV1-*tk* (*bottom panel*). See Table 28.1 for chemical structure of the abbreviations

Fig. 28.3). This difference in enzyme specificity permits the development and use of radiolabeled probes that are relatively specific for HSV1-*tk* in comparison to mammalian TK. The initial phosphorylation of the acycloguanosines and thymidine analogues is a rate determining step. Cellular enzymes then convert acycloguanosine and thymidine analogue monophosphates to di- and triphosphates that are effectively trapped within transduced cells. The 2'-fluoro-nucleoside analogues of thymidine (e.g., FIAU) are incorporated into cellular DNA, whereas the acycloguanosine analogues act as DNA chain terminators.

The acycloguanosines and the 2'-fluoro-1-β-D-arabinofuranosyl and 2'-fluoro-1-β-D-ribofuranosyl analogues of thymidine were selected for study because they were developed as potential anti-viral agents. Some have been radiolabeled and studied as potential imaging agents of viral infections in the past (Saito et al. 1982, 1984; Price et al. 1983; Tovell et al. 1987; Iwashina et al. 1988). These compounds are phosphorylated by HSV1 thymidine kinase (HSV-*tk*), but are poorly or not at all phosphorylated by mammalian thymidine kinase; namely, they accumulate in virally infected (transduced) cells and minimally in noninfected (nontransduced) dividing cells. They can be radiolabeled with several different radionuclides, including ^{11}C and ^{18}F, or carrier-free ^{131}I, ^{123}I and ^{124}I, appropriate for clinical imaging with PET (or SPECT) in many nuclear medicine departments. The longer half-life of the iodine radionuclides provides an advantage with respect to production and distribution of the radiopharmaceutical, and with respect to the clearance of radiolabeled metabolites prior to imaging. The shorter half-life of [^{11}C]- and [^{18}F]-labeled probes provides an advantage for repeated imaging over short time intervals in a single subject.

28.4.3
2'-Fluoro Analogues of Thymidine

An important characteristic of the 2'-fluoro substitution on the arabinose or ribose moiety is that it provides a protective group against enzymatic cleavage of the N-glycosidic bond by nucleoside phosphorylases. The increased stability of the N-glycosidic bond results in a significant prolongation of plasma half-life of these compounds and they are excreted largely unchanged in the urine (Feinberg et al. 1985; Tovell et al. 1988; Machida et al. 1995). As a result, a greater amount of non-degraded radiolabeled drug is delivered to the target tis-

sues and confounding problems associated with imaging radiolabeled metabolites in both target and surrounding tissue are reduced. The 2'-fluoro analogues of thymidine include 2'-fluoro-2'-deoxy-1-β-D-arabino-furanosyl-5-iodouracil [FIAU], 2'-fluoro-2'-deoxy-5-iodo-1-β-D-ribofuranosyl-uracil [FIRU], 2'-fluoro-2'-deoxy-5-methyl-1-β-D-arabino-furanosyl-uracil [FMAU], and 2'-fluoro-2'-deoxy-5-iodovinyl-1-β-D-ribofurano-syl-uracil [IVFRU] (see Fig. 28.3).

The advantages of "later imaging", after a period of time to allow for "washout" and body clearance of the non-phosphorylated analogue (and any radiolabeled metabolites that may have formed) have been discussed previously (Tjuvajev et al. 1995, 1996, 1998; Blasberg and Tjuvajev 1997). Late imaging, 12–48 h after i.v. administration and an extended period of "washout", *cannot* be performed with either [^{11}C]- and [^{18}F]-labeled 2'-fluoro nucleosides. Recent studies involving [^{124}I]FIAU and PET imaging of HSV1-*tk* transduced and non-transduced xenografts demonstrated that "early" FIAU imaging can also be performed (Gelovani-Tjuvajev et al. 2002). For the 10–20 min post-[^{124}I]FIAU image acquisition, a 0.70% dose/g difference and a 2.5-fold ratio between the transduced and non-transduced xenografts were demonstrated. Corresponding values at 60 and 120 min were a 0.94% and 1.00% dose/g difference, and a 3.7- and 4.2-fold ratio, respectively, between the transduced and non-transduced tumors. Good quality images can be obtained within the first 30 min to 2 h following FIAU administration (Gelovani-Tjuvajev et al. 2002). Thus, the concern we previously raised with respect to a potential limitation of [^{11}C]- or [^{18}F]-labeled 2'-fluoro analogues may not be valid.

28.4.4
Acycloguanosine Analogues

Guanosine analogues are also effectively phosphorylated by the HSV1-*tk* enzyme, but not by mammalian thymidine kinase. The acycloguanosine analogues that have been radiolabeled and studied as potential HSV1-*tk*- imaging probes include: acyclovir [ACV], ganciclovir [GCV], fluoroganciclovir [FGCV], penciclovir [PCV], fluoropenciclovir [FPCV], 9-[3-fluoro-1-hydroxy-2-propoxymethyl] guanine [FHPG] and 9-[4-fluoro-3-(hydroxymethyl)butyl] guanine [FHBG] (see Fig. 28.3). The acycloguanosines are potentially desirable probes because they can be radiolabeled with fluorine-18 (half-life 110 min), which provides the opportunity for repetitive imaging of HSV1-*tk* expression (e.g., every 6–8 h if needed). Because 2′-fluoro position of the 2′-fluoro analogues of thymidine (e.g., FIAU) is not easily labeled with fluorine-18, the emphasis of several chemistry groups has been directed towards the development of new HSV1-*tk* probes by radiolabeling acycloguanosine derivatives with [18]F. Radiolabeled GCV and other acycloguanosine analogs [^{18}F]FGCV, [^{18}F]FPCV, and [^{18}F]FHPG have been shown to be resistant to metabolic degradation and metabolically stable in blood and liver (Faulds and Heel 1990; Gambhir et al. 1999a, b, 2000b; Iyer et al. 2001a). In contrast to many of the 2′-fluoro analogues of thymidine, the acycloguanosines are not incorporated into replicating DNA, but act as DNA chain terminators because of the absence of a 3′ hydroxyl (Faulds and Heel 1990).

A comparison of acycloguanosine analogs as potential HSV1-*tk* imaging probes has been performed and is based on paired probe accumulation studies in HSV1-*tk* transduced cells in vitro (Gambhir et al. 1999b, 2000b; Namavari et al. 2000; Iyer et al. 2001a). A comparison of values normalized to that of GCV (set at 1.0) yields the following nucleoside/GCV values: FGCV ~0.5 (Gambhir et al. 1998, 1999b; Namavari et al. 2000), PCV ~2.0 (Gambhir et al. 2000b; Iyer et al. 2001), FPCV ~0.2 (Gambhir et al. 2000b), FPCV ~1.0 (Iyer et al. 2001a), FIAU ~23 (Tjuvajev et al. 1995). A direct comparison of FHBG and FHPG with other acycloguanosine analogs has not yet appeared (as of 2/02).

Table 28.1. Radiolabeled compounds (probes) that have been used to study and image HSV1 thymidine kinase

Radiolabeled probe	Chemical name	References
Thymidine derivatives		
FIAU (^{123}I, [124]I, ^{125}I, ^{131}I)	2′-Fluoro-2′-deoxy-1-β-D-arabino-furanosyl-5-iodouracil	Reichman et al (1975); Tovell et al (1988); Tjuvajev et al (1995, 1996, 1998, 1999a, b); Wiebe et al (1999); Bengel et al (2000); Haubner et al (2000); Bennett et al (2001); Brust et al (2001); Doubrovin et al (2001a); Jacobs et al (2001a, b); Ponomarev et al (2001b); Gelovani-Tjuvajev et al (2002); Qiao et al (2002)
FIRU (^{125}I)	2′-Fluoro-2′-deoxy-5-iodo-1-β-D-ribofuranosyl-uracil	Tovell et al (1988); Wiebe et al (1999)
FMAU (^{11}C)	2′-Fluoro-2′-deoxy-5-methyl-1-β-D-arabinofuranosyl-uracil	Conti et al (1995); Bading et al (2000); de Vries et al (2000)
IVFRU (^{123}I, ^{125}I, ^{131}I)	2′-Fluoro-2′-deoxy-5-iodovinyl-1-β-D-ribofuranosyl-uracil	Wiebe et al (1999); Morin et al (1997a, b, 2000)
IUdR (^{3}H)	2′-Deoxy-5-iodo-1-β-D-ribofuranosyl-uracil	Tjuvajev et al (1995)
Acycloguanosine derivatives		
ACV (8-^{14}C, 8-^{3}H)	9-[(2-Hydroxy-1-ethoxy)methyl]guanine	Gambhir et al (1998); Brust et al (2001)
GCV (8-^{14}C, 8-^{3}H)	9-[(2-Hydroxy-1-(hydroxymethyl)ethoxy)methyl]-guanine	Tjuvajev et al (1995); Gambhir et al (1998); Brust et al (2001)
PCV (8-^{3}H)	9-[4-hydroxy-3-(hydroxymethyl)butyl]guanine	Gambhir et al (2000b); Iyer et al (2001a)
FGCV (8-^{18}F)	8-Fluoro-9-[(2-hydroxy-1-(hydroxymethyl)ethoxy)methyl]guanine	Gambhir et al (1999b); Namavari et al (2000)
FPCV (8-^{18}F)	8-Fluoro-9-[4-hydroxy-3-(hydroxymethyl)butyl]guanine	Gambhir et al (2000b); Iyer et al (2001a)
FHPG (3-^{18}F)	9-[3-Fluoro-1-hydroxy-2-propoxymethyl]guanine	Alauddin et al (1996, 1999); Bading et al (1997); de Vries et al (2000); Hospers et al (2000); Brust et al (2001); Hustinx et al (2001); Shiue et al (2001); Gelovani-Tjuvajev et al (2002)
FHBG (4-^{18}F)	9-[4-Fluoro-3-(hydroxymethyl)butyl]guanine	Alauddin and Conti (1998); Alauddin et al (2001); Shiue et al (2001); Yaghoubi et al (2001a, b); Gelovani-Tjuvajev et al (2002)

IUdR, iododeoxyuridine; ACV, acyclovir; GCV, ganciclovir; PCV, penciclovir; FGCV, fluoroganciclovir; FPCV, fluoropenciclovir

28.4.5
Comparison of 2'-Fluoro Nucleoside and Acycloguanosine Probes for Imaging *HSV-tk* Expression

A major problem with comparing different radiolabeled uracil- and guanosine-based probes is that no independent measure of HSV1-*tk* enzyme activity in the transduced cell line or tumor xenograft is presented in most studies reported to date. Usually, only a comparison between transduced and wild-type cells or xenografts is presented, and this comparison is highly dependent on the level of HSV1-*tk* enzyme activity in the transduced cell line or tissue being compared in each study. Since the level of HSV1-*tk* enzyme expression is likely to be different in the different transduced cells and tissue reported in the literature (Table 28.1), it is not possible to perform a rigorous comparison.

To establish a stable transduced cell line for comparing different radiolabeled probes, HSV1-*tk* was transduced into several different tumor cell lines using retroviral vectors (Tjuvajev et al. 1995). Several RG2TK clones, as well as wild-type RG2 and bulk culture RG2TK+ cells, were obtained and compared with respect to FIAU accumulation and two independent measures of HSV-*tk* expression: (a) sensitivity to the anti-viral drug ganciclovir, a functional measure of HSV1-*tk* expression (ED_{50}), and (b) normalized IOD values of HSV1-*tk* mRNA identified on Northern blot analysis. Accumulation of FIAU, expressed as the FIAU/TdR uptake ratio, was proportional to the levels of HSV1-*tk* mRNA in corresponding cell lines, and, as expected, a highly significant inverse relationship was also observed between FIAU uptake and ganciclovir sensitivity (Tjuvajev et al. 1995). Additional cell lines, including W256 tumor cells, were transduced with the STK and STLEO retroviruses (containing the HSV1-*tk* gene). Following the transductions, clonal cell lines were derived from the bulk culture and were shown to express different levels of HSV1-*tk*. The level of FIAU accumulation in each of the cell lines was compared to their sensitivity to ganciclovir (an independent assay of HSV1-*tk* expression); a composite plot of this data is shown in Fig. 28.4. These results demonstrate a consistency between FIAU accumulation and ganciclovir sensitivity which is independent of cell line or transduction vector.

HSV1-*tk* enzyme levels are likely to be higher following HSV1-*tk* gene transduction with adenoviral vectors at high multiplicity of infection (MOI) com-

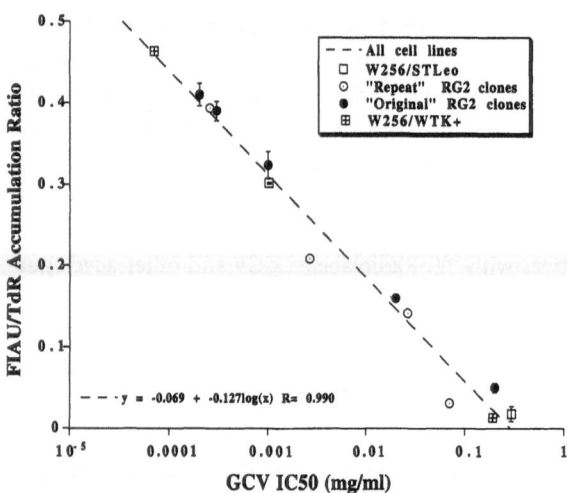

Fig. 28.4. Comparative assays of HSV1-*tk* expression. The radiotracer accumulation ratio (FIAU/TdR) which normalizes FIAU uptake for cell proliferation is compared to a functional assay of HSV1-*tk* expression; namely, sensitivity (IC_{50}) to the antiviral drug, ganciclovir. Four different sets of stably transduced cell lines (clones) expressing HSV1-*tk* were studied. A highly reproducible relationship between these two assays is demonstrated which is independent of cell line and retroviral transduction vector

pared to that with retroviral vectors or with liposome-mediated plasmid vectors (Feng et al. 1997; Robbins and Ghivizzani 1998). Most radiolabeled probe comparisons are based on results obtained in different cell cultures or in different animals and many are performed at different times. Furthermore, the radiotracer results are frequently reported in different units which further complicates a comparison of different radiolabeled probes. Thus, HSV1-*tk* probe efficacy comparisons derived from a review of the literature and a comparison of different studies are qualitative at best.

To address this issue, some studies have reported double-label and triple-label probe comparison studies, or sequential studies comparing two or more probes using the same transduced cell line or transduced xenograft. The first study to report such a comparison (Tjuvajev et al. 1995), demonstrated a >20-fold difference between FIAU and GCV accumulation in transduced RG2TK+ cells in culture. A more recent multi-label study normalized the in vitro data to acyclovir (which is accumulated less avidly in *HSV1-tk-transduced cells than GCV (Gambhir et al. 1998); the nucleoside/ACV ratio values were* ∼1–8 for FHPG and ∼40–100 for FIAU (Brust et al. 2001). The calculated FIAU/FHPG ratios in these experiments in different transduced cell lines varied between 5- and

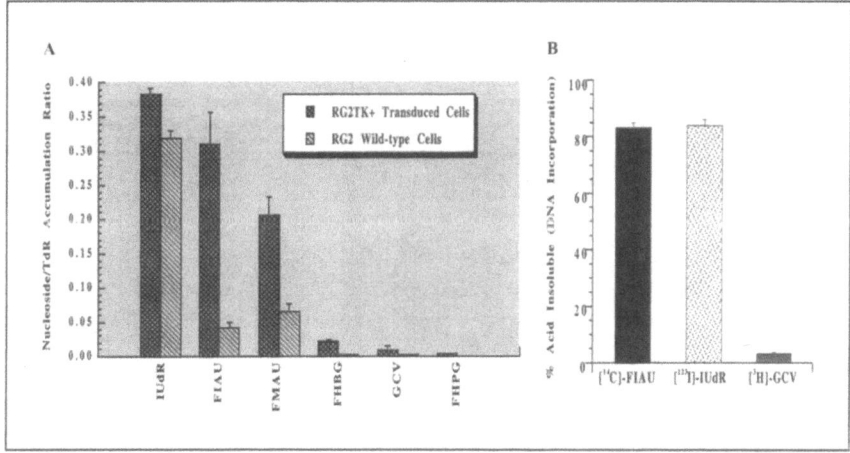

Fig. 28.5 A, B. Comparison of different radiolabeled probes for imaging HSV1-*tk* expression. Radiotracer accumulation BAR graph TK+ and wild-type. The substrate accumulation ratio, normalized to thymidine, is shown for RG2TK+ cells and RG2 cells (A). The acid-insoluble fraction of 2-[^{14}C]-FIAU is compared with that of [^{123}I]-IUdR and 8-[^{3}H]-ganciclovir in RG2TK+ cells for a 120-min, triple-label in vitro experiment (B)

100-fold. Another in vitro study involving paired comparisons between FIAU and FHPG and between FIAU and FHBG accumulation in transduced RG2TK+ cells demonstrated a 41- and 17-fold difference, respectively (Gelovani-Tjuvajev et al. 2002). A comparison of probe accumulation in the *same* stably transduced (RG2TK+) and wild-type (RG2) cell lines is shown in Fig. 28.5. Several different nucleoside analogues and acycloguanosine probes are compared (Fig. 28.5A), and the data are normalized to the rate of thymidine accumulation to account for small differences in proliferation rate of the cells in culture (Tjuvajev et al. 1995). The acid-insoluble fractions of FIAU and IUdR accumulation in the RG2TK+ cell line at 2 h were similar, whereas the acid-insoluble fraction after incubation with ganciclovir (GCV) was much lower (Fig. 28.5B). This suggests that most of the measured FIAU and IUdR derived radioactivity is in DNA, in contrast to that derived from GCV. It has also been reported (Haberkorn et al. 1998) that only a small fraction of cell-derived radioactivity is in the acid-insoluble fraction when transduced, HSV1-*tk* expressing cells are incubated with 8-[^{3}H]-ganciclovir. HPLC analysis of the soluble fraction demonstrated a time-dependent shift of GCV radioactivity to phosphorylated metabolites in transduced cells (but *not* in wild-type control cells); the phosphorylated metabolites accounted for 3.6, 14, 42 and 91% of soluble radioactivity after 2, 4, 24 and 48 h incubation, respectively, suggesting that phosphorylation may be rate-limiting (Haberkorn, personal communication).

The slow cell uptake of ganciclovir and its low affinity for HSV1-*tk* suggest that GCV is not an ideal substrate for the nucleoside transport systems or for imaging the expression of HSV1-*tk* (Haberkorn et al. 1998), and this may apply to other radiolabeled acycloguanosine analogs as well.

Morin et al. described the synthesis and in vitro uptake of 2′-substituted analogues of (E)-5-(2-[^{125}I]iodovinyl)-2′-deoxyuridine in tumor cells transduced with the HSV1-*tk* gene (Morin et al. 1997a, b, 2000; Wiebe et al. 1999). These compounds are similar to FIAU, but have the iodovinyl substitution in the 5-position of the uracil moiety. It was demonstrated that (E)-5-(2-[^{125}I]iodovinyl)-2′-fluoro-2′-deoxyuridine ([^{125}I]-FIVRU) accumulated in HSV1-*tk*-transduced KBALB murine sarcoma cells in vitro better than (E)-5-(2-[^{125}I]iodovinyl)-2′-fluoro-2′-arabinouridine ([^{125}I]-FIVAU) and (E)-5-(2-[^{125}I]iodovinyl)-2′-arabinouridine ([^{125}I]-IVAU), but no comparison was done with radioiodinated FIAU. The same group demonstrated successful imaging of HSV1-*tk* expression in retrovirally transduced murine tumors using a gamma camera and (E)-5-(2-[^{131}I]iodovinyl)-2′-fluoro-2′-deoxyuridine ([^{131}I]-FIVRU) (Morin et al. 1997b).

In vivo comparisons have also been performed. A recent study directly comparing two probes, FIAU and FHPG (double-label experiments), demonstrated that the FIAU/FHPG ratio in HSV1-*tk*-transduced tumors increased from 21 at 30 min to 119 at 4 h (Ponomarev et al. 2001a). It was also noted that FHPG cleared rapidly from the transduced tumor; a maxi-

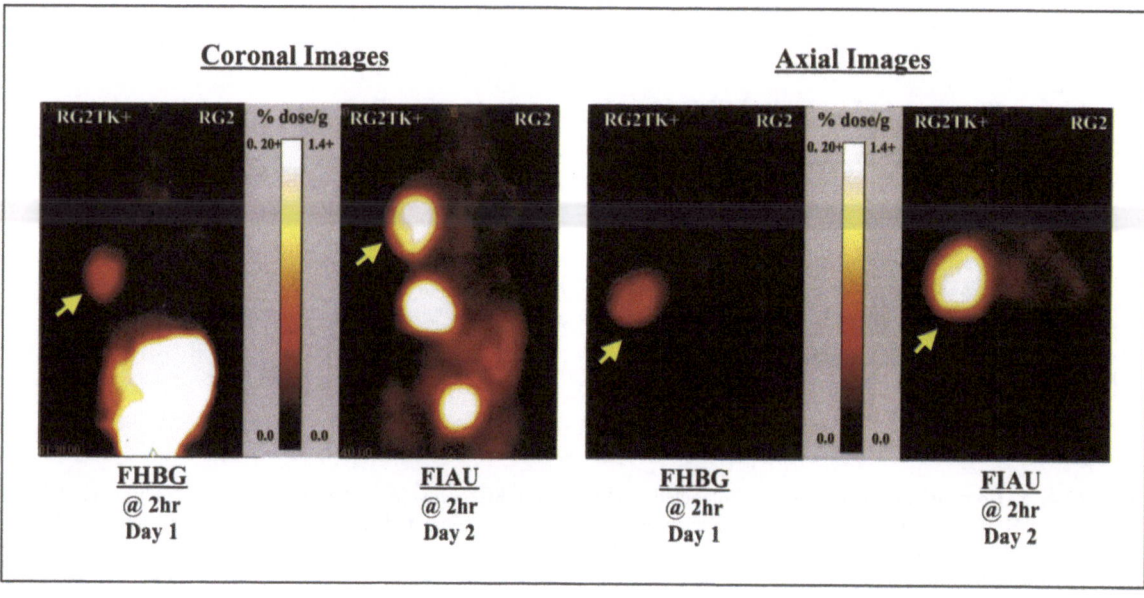

Fig. 28.6. Comparison of FIAU and FHBG as imaging probes for HSV1-*tk* gene expression. Paired PET imaging studies comparing [^{124}I]FIAU and [^{18}F]FHBG were performed within 24 h of each other. Comparable 2-h coronal and axial PET images are shown for the same rat; 1.35 mCi [^{18}F]FHBG was injected on day 1 and 0.19 mCi [^{124}I]FIAU was injected on day 2. The location of the RG2TK+ xenograft is indicated by the *arrow*; the wild-type (non-transduced) control xenograft is located in the contralateral shoulder. A linear scaling of image intensity was performed to achieve similar background intensity in all the images; background radioactivity in the thorax was 0.024 and 0.18% dose/g for FHBG and FIAU, respectively. No FHBG or FIAU radioactivity is visualized above background in the control RG2 xenograft. Several tissues had substantially higher values than indicated on the intensity scale (FHBG: intestine and bladder; FIAU: stomach and bladder)

mum value of 2.2±0.2% dose/g was observed at 1 h that fell to 0.2±0.03% dose/g at 4 h post-injection. In contrast, FIAU remained essentially constant during this interval (12.7±2.4 and 10.5±3.9% dose/g, respectively). A comparison involving sequential [^{18}F]FHPG and [^{124}I]FIAU PET imaging in the same set of animals demonstrated similar findings (Fig. 28.6).

In vivo studies involving double label tissue sampling experiments and sequential imaging experiments have been performed with FIAU and FHPG as well as with FIAU and FHBG. These studies demonstrated similar differences between the three probes (Gelovani-Tjuvajev et al. 2002). At 2 h post-injection of the two radiolabeled probes there was a 59- and 21-fold difference in radioactivity concentration (% dose/g) for FIAU/FHPG and FIAU/FHBG, respectively, in transduced RG2TK+ xenografts (Gelovani-Tjuvajev et al. 2002). The time course of [^{124}I]FIAU, [^{18}F]FHPG and [^{18}F]FHBG radioactivity (% dose/g) in transduced RG2TK+ and wild-type RG2 xenografts further illustrates significant differences between these probes (Fig. 28.7).

FIAU rapidly accumulates post-injection and reaches a plateau in transduced RG2TK+ xenografts during the 2-h period of imaging, whereas FIAU is being cleared from the wild-type RG2 xenograft. In contrast, FHBG and FHPG (data not shown) are being cleared from both TK-transduced and wild-type xenografts over the two imaging period. FHBG clearance from RG2TK+ xenografts appears to approach a plateau by 2 h, whereas FHBG clearance from the RG2 xenografts fits a single exponential.

28.4.6
Reporter Gene Enhancements for Acycloguanosine Analogues

As noted above, enhanced viral cytotoxicity to ganciclovir and acyclovir as well as enhanced drug phosphorylation was obtained with site-mutated HSV1-*tk* variants compared to wild-type HSV1-*tk* (Black et al. 1996, 2001). Initial results obtained with the HSV1-

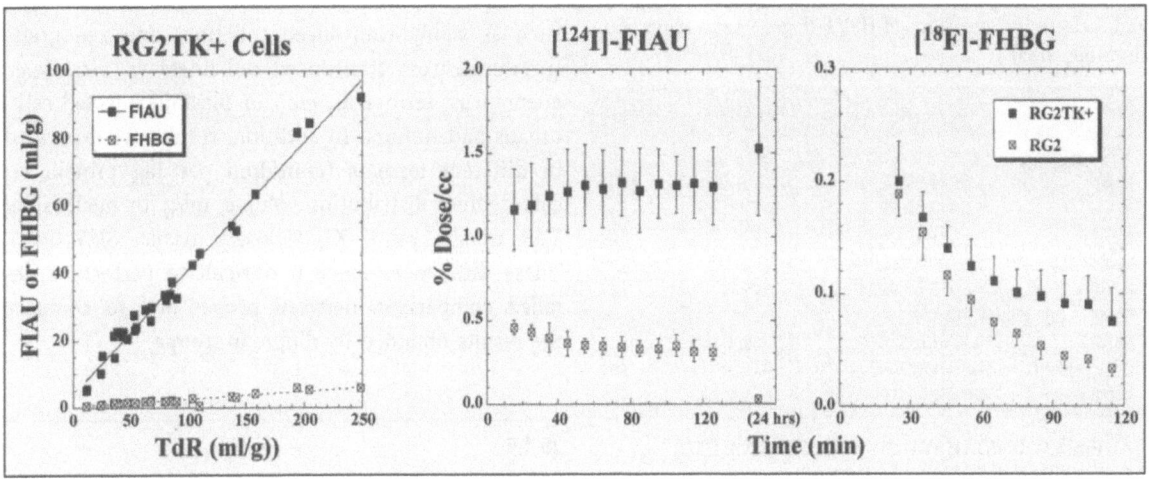

Fig. 28.7. FIAU and FHBG uptake and washout profiles in vitro and in vivo. Paired nucleoside accumulation studies in RG2TK+ transduced cells in vitro (*left graph*) and in RG2TK+ and RG2 xenografts in vivo (*right graph*). [^{14}C]FIAU (*solid square*) and [^{18}F]FHBG (*open square*) accumulation data (ml medium/g cells), compared to that of [^{3}H] TdR, are shown for RG2TK+ cells (*left graph*). A plot of [^{124}I]FIAU and [^{18}F]FHBG accumulation in RG2TK+ and RG2 xenografts are shown for 10-min imaging frames; mean values ± SD are shown; *n=6 (right graph)*

sr39tk mutant demonstrated the potential for enhanced imaging sensitivity for PCV and FPCV, but this enhancement was modest ($\sim$2-fold) (Gambhir et al. 2000b). The $\sim$2-fold enhancement in probe accumulation contrasts with the $\sim$300-fold difference in GCV IC$_{50}$ observed between HSV1-*sr39tk* and wildtype HSV1-*tk* C6 transfectants (Black et al. 2001). This disparity in probe accumulation and GCV IC$_{50}$ has not been explained. One consideration is the effect of "slow" transport (flux) of acycloguanosine analogues across cell membranes. A question has been raised as to whether cell membrane transport of PCV and FPCV was limiting in the radiolabeled probe accumulation studies run over a 15 to 240-min period (Gambhir et al. 2000b). In contrast, the GCV IC$_{50}$ studies were run over a 3-day period, providing a longer period for GCV to enter the transduced cells and exert a cytotoxic effect (Black et al. 1996, 2001).

The use of mutant enzymes provides a new approach to enhance sensitivity and specificity of reporter gene imaging. Genetic site mutation technology may also minimize competition from endogenous compounds, such as thymidine for the HSV1-*sr39tk* enzyme, which may be important for in vivo applications where intracellular thymidine levels can not be controlled. The approach of selecting mutant enzymes with enhanced affinity for a given marker/reporter probe should be applicable to other marker/reporter gene assays and should be a powerful general method to complement the search for alternate marker/reporter probes.

28.4.7
Data Analysis

In order to evaluate the proficiency of "marker probes" for imaging of HSV1-*tk* expression, two parameters were developed: (1) "sensitivity", defined as the change in probe uptake (normalized to TdR) divided by the change in HSV1-*tk* expression, and (2) "selectivity", defined as "sensitivity" divided by probe uptake due to endogenous (mammalian) TK (Tjuvajev et al. 1995). These two related measures were considered useful for imaging considerations because both intensity (e.g., high "sensitivity") and contrast (the ability to differentiate between HSV1-*tk* and endogenous TK; e.g., high "selectivity") in the images are required. It is important to note that the "selectivity" measure should *not* be interpreted as a relative sensitivity of the probe to HSV1-*tk* vs. endogenous TK. Also, neither of the measures represents a fraction of some ideal value (i.e. they do not have a maximum of one); they are intended only for the direct comparison of different probes.

A summary comparison of the "sensitivity" and "selectivity" indices is shown in Table 28.2. Note that the level of HSV1-*tk* expression was taken to be the difference (i.e., change, Δ) in the probe/TdR ratio between transduced and wild-type cells. This value was normalized (divided by) an independent measure of HSV1-*tk* expression (the difference in the HSV1-*tk*

Table 28.2. Comparison of different probes for "sensitivity" and "selectivity" measures of HSV1-*tk* gene expression (data from Fig. 28.5)

Probe	Sensitivity[a]	Selectivity[b]	(n)
Thymidine analogues			
FIAU	0.0083±0.0013	0.204±0.042	(16)
FIRU[c]	0.0017±0.0004	0.404±0.254	(3)
FMAU[c]	0.0043±0.0010[d]	0.069±0.027[d]	(3)
FEAU[c]	0.00021±0.00015[d]	0.091±0.071	(6)
IUdR	0.0019±0.0024[d]	0.006±0.001[d]	(3)
Guanosine analogues			
FHPG[c]	0.00011±0.00002[d]	0.300±0.108	(2)
Ganciclovir	0.00021±0.00015[d]	0.091±0.071	(6)
FHBG[c]	0.00065±0.00021[d]	0.666±0.277	(2)

[a] Δ (Probe/TdR)/D (HSV1-*tk* mRNA/S28 RNA IOD)
[b] Sensitivity[a]/(probe/TdR) for RG2 cells
[c] New preliminary data, not previously presented
[d] Significantly different from FIAU values ($p < 0.05$)
(n), Number of concurrent RG2TK+ and RG2 experiments

mRNA/S28 RNA IOD ratio between transduced and wild-type cells) in the calculation of the "sensitivity index". From these results it can readily be seen that (a) the accumulation of ganciclovir, FHPG and FHBG by RG2TK+ cells was small compared to the other probes and may not provide a sufficient signal for imaging (poor sensitivity), and (b) IUdR and FMAU do not distinguish well between transduced RG2TK+ and non-transduced RG2 cells (poor selectivity).

28.4.8
Assessments Based on the In Vitro and In Vivo Data

A detailed comparison of all the radiolabeled probes studied for imaging HSV1-*tk* marker/reporter gene expression has yet to be reported in the literature. Several key problems exist with respect to comparing results obtained from studies performed by different groups. The first relates to variable and largely unknown levels of HSV1-*tk* gene expression in the test systems that are used, and to the absence of a commonly used method to assay for HSV1-*tk* enzyme activity. Three independent assays have been reported: (1) sensitivity (IC-50) of transduced cells in culture to an antiviral drug (e.g., ganciclovir, acyclovir), a functional assay; (2) quantitation of HSV-*tk* mRNA levels in transduced tissue or cells in culture; (3) an in vitro radiotracer enzyme assay (phosphorylation of [3H]-ganciclovir) of transduced tissue or cells in culture, a functional assay. Other factors relate to differ-

ent cell lines and in vivo models used in the assays, such as stably transduced cell lines and xenografts (produced from transduced cell lines) vs. viral (e.g., adenovirus, retrovirus, etc.) or plasmid infected cells, organs and tumors. In addition, results are expressed in different formats (cpm/dpm per mg protein vs. cell/medium distribution volume, ml/g, or nucleoside/TdR uptake ratio, Ki, %dose/g tissue, SUV, etc.). These differences make it difficult to perform a detailed comparison between probes and to compare the results obtained by different groups.

28.4.9
Comment

The sensitivity and dynamic range of a PET reporter probe for imaging the expression of a particular reporter gene is an important consideration that will impact on both animal and patient studies. As of 2/02, the most widely used reporter genes for nuclear (PET) imaging are wild-type HSV1-*tk* and HSV1-*sr39tk*, and the two most widely used radiolabeled reporter probes are FIAU and FHBG. The results of the 2-h imaging studies shown in Figs. 28.6 and 28.7 clearly demonstrate the sensitivity and dynamic range advantages of FIAU over FHBG for imaging wild-type HSV1-*tk* expression. A comparison of wild-type HSV1-*tk* FIAU imaging vs. HSV1-*sr39tk* FHBG reporter gene imaging, normalized to the level of gene product – TK enzyme, has not been reported.

The ability to image "late", 24 or more hours after FIAU administration, provides an additional advantage; namely, the opportunity to achieve high image specificity due to physiological washout of background radioactivity and the retention of the radioactivity in HSV1-*tk*-transduced tissue. Although FIAU is usually labeled with [124]I for PET studies (or with ^{123}I or ^{131}I for gamma camera scintigraphy) and "late" imaging is usually performed, the feasibility of "early" FIAU imaging has been demonstrated. Good quality images can be obtained within the first 30 min to 2 h following FIAU administration. Thus, radiolabeling FIAU with ^{18}F or ^{11}C is an alternative strategy to using FHBG; it would provide additional positron-emitting FIAU probes with high HSV1-*tk* sensitivity and dynamic range. In addition, the relatively short half-life of these radionuclides would facilitate repeated or sequential PET imaging of HSV1-*tk* expression in the same patient or subject.

Greater sensitivity and dynamic range provide the ability to image lower levels of reporter gene expression over a wider range of expression levels. This advantage translates into several practical benefits: (1) requires lower quantities of vector for successful transduction of target tissue with the reporter system; (2) provides a wider range of image intensity for more accurate quantitation of reporter expression; (3) provides a more sensitive measure of weakly activated reporter systems. For example, the higher sensitivity and wider dynamic range of PET imaging with FIAU allows for more accurate monitoring of small changes in the activity of various signal transduction pathways, when expression of the HSV1-*tk* reporter gene is controlled by a signal transduction-specific *cis*-acting promoter/enhancer element.

The development of new reporter probes for imaging HSV1-*tk* (and other reporter gene-reporter probe combinations) and the application of genetic engineering of reporter genes to enhance the sensitivity and dynamic range of a particular PET reporter probe or class of reporter probes is in its very early stages of implementation. This new and exciting genetic approach to reporter gene enhancement compliments the more traditional approach of modifying the chemical structure of the reporter probe. Clearly, both approaches should be pursued and this is likely to lead to other reporter transgene-reporter probe combinations in the future.

28.5
Other Reporter Systems

28.5.1
Enzyme-based Reporters

Other novel enzyme-based reporter gene-reporter probe combinations are being developed. One reporter system includes the *E. coli* xanthine phosphoribosyl transferase (XPRT) gene with [8-^{11}C]- or [8-^{18}F]-labeled xanthine (Ponomarev et al. 2001a). This system is expected to be very useful in the brain since the radiolabeled xanthine probes will readily cross the blood-brain barrier. Two other reporter systems include the *H. influenza* uracil phosphoribosyl transferase (UPRT) gene and [5-^{18}F]fluorouracil as the reporter probe (Koutcher et al. 2000), and the *S. cerevisiae* cytosine deaminase gene (yCD) whose expression can be quantitatively monitored by NMR spectroscopy of 5-fluorocytosine (5FC) metabolism in yCD-transduced tumors (Stegman et al. 1999).

Nuclear imaging of CD-transduced tumors with [5-^{18}F]fluorocytosine has been largely unsuccessful (Haberkorn et al. 1996) because the deaminated metabolite, [5-^{18}F]fluorouracil (5FU), rapidly exits from the transduced cells. To address the problem of rapid egress of 5FU from transduced cells, current investigations are assessing the efficacy of a yCD/UPRT fusion gene using cold 5FC and NMR spectroscopy, as well as PET imaging using [5-^{18}F]5FC and [5-^{18}F]5FU. In this paradigm, the yCD component of the fusion gene converts 5FC to 5FU and the UPRT component of the fusion gene converts 5FU to the nucleotide monophosphate (5-fluorouridine 5'-monophosphate, 5-FUMP) which is trapped within transduced cells and further metabolized to 5-FdUMP. The yCD/UPRT fusion gene can function as both a therapeutic gene to selectively treat transduced tumors with 5-FC (Tiraby et al. 1998; Adachi et al. 2000; Erbs et al. 2000) and as a potential reporter gene for magnetic resonance spectroscopy (and possible MR imaging), as well as a reporter gene for PET imaging using [5-^{18}F]5FC or [5-^{18}F]5FU. This combined therapeutic (pro-drug activation and enhancement) and reporter function is similar to that described above for the HSV1-*tk* gene, and may provide the opportunity for both MR and PET monitoring levels of gene expression as well as the levels of activated drug in transduced tumors.

28.5.2
Cell Membrane Receptors

A reporter gene can also encode for an extracellular-facing or intracellular receptor (e.g., hD2R, or hSSTR2) that binds or transports a radiolabeled or paramagnetic probe. The human dopamine 2 receptor (hD2R) (MacLaren et al. 1999; Liang et al. 2001) and the human somatostatin receptor subtype-2 (hSSTR2) (Rogers et al. 1999, 2000; Hemminki et al. 2001) genes have been suggested as potential reporter genes for human studies. An example of hD2R imaging is shown in Fig. 28.8. Both human genes have limited expression in the body; hD2R expression is limited to the striatal-nigral system of the brain and high hSSTR2 expression is largely limited to carcinoid tumors. This approach is a very clever strategy because there are established complimentary radiolabeled probes for each of these reporter genes; 3-(2'-[^{18}F]fluoroethyl)spiperone (FESP) for hD2R imaging (Barrio et al. 1989), and [^{111}In]DTPA-octreotide (a

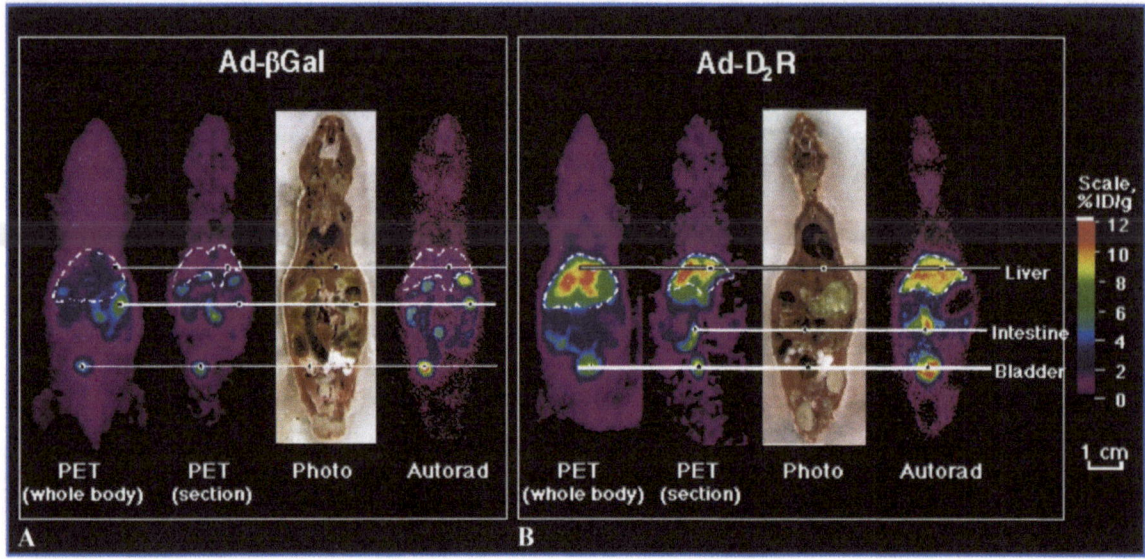

Fig. 28.8A, B. PET and autoradiographic images of fluoro-ethylspiperone following Ad-D₂R and Ad-Bgal virus administration. Nude mice were injected via the tail vein with: A 9×10^9 p.f.u. of Ad-Bgal virus, or B 9×109 p.f.u. Ad₂R virus. At 2 days after virus administration both mice were injected via the tail vein with ^{18}FESP (200 µCi, 200 µl). At 3 h after the ^{18}FESP injection the animals were imaged with micro-PET. A whole body coronal projection image of the ^{18}F activity distribution is displayed on the *left*. The liver outline is shown in *white*. The second images from the *left* are coronal sections, approximately 2 mm thick, from the micro-PET. After their PET scans, the mice were killed, frozen, sectioned and photographed (third image from *left*); the images on the *right* are autoradiographs of the tissue sections. The *color scale* represents the per cent injected dose per gram of tissue (%ID/g). Images are displayed on the same quantitative color scale to allow signal intensity comparisons among the panels (from MacLaren et al. 1999)

complimentary radiolabeled somatostatin analogue) for hSSTR2 imaging (Ur et al. 1993). Furthermore, both probes are approved for human administration. Both of these reporter systems have distinct benefits with respect to initiating molecular/reporter imaging in human subjects. However, receptor expression on the external surface of cells or on specific intracellular organelles is a complex process. Receptor expression involves intracellular trafficking and cell membrane incorporation that is likely to be altered under different conditions and different disease states. Furthermore, cell surface receptor expression may account for only a small part of total receptor protein produced by transcriptional activation of the reporter gene. It remains to be shown whether imaging receptor-based reporter systems (e.g., the *hD2R* and *hSSTR2* reporter gene systems) will provide a consistent and reliable measure of reporter gene expression under variable stress or altered conditions. In any case, the level of probe accumulation (level of radioactivity) must be shown to be proportional to the level of gene expression.

28.5.3
Transporter-based Reporters

The Na-I symporter has been proposed as a reporter gene (see Sect. 28.2.1.5). Several in vitro experiments concerning NIS-based gene therapy for both diagnostic and therapeutic purposes have been reported, in which NIS-mediated radioiodide uptake was used to visualize and destroy malignant tumor cells. rNIS-transduced tumor cells (melanoma, and ovarian, liver and colon carcinoma) exhibited I⁻ uptake activity (Mandell et al. 1999). In vitro experiments showed that these transduced cells could be destroyed by accumulation of ^{131}I. Tissue-specific, androgen-dependent I⁻ uptake activity has also been induced in prostate cancer cells in vitro by PSA (prostate-specific antigen) promoter-directed NIS expression (Spitzweg et al. 1999). In a different study, xenografts from a NIS-expressing human prostate cancer cell line established in nude mice were reported to actively accumulate as much as 25–30% of administered I⁻ in vivo (Spitzweg et al. 2000). Strikingly, the size of the xenograft tumors in these mice was significantly reduced after a single intraperitoneal injection of a therapeutic dose (3 mCi) of ^{131}I (Spitzweg et al. 2000).

28.6
Reporter Gene Imaging (Examples)

28.6.1
Imaging the Expression of Therapeutic Transgenes

A non-invasive, clinically applicable method for imaging the expression of successful gene transduction in target tissue or specific organs of the body would be of considerable value. It would facilitate the monitoring and evaluation of gene therapy in human subjects by defining the *location*, *magnitude* and *persistence* of gene expression over time. Targeting gene therapy to particular tissue (e.g., tumor) or specific organs is an increasingly active area of research with 288 related articles published in 1991, 1534 articles in 1996, and 4687 articles in 2001 based on a PubMed search.

Several issues that are important for clinical optimization of gene therapy remain unresolved in many current clinical protocols: (1) Has gene transduction or transfection been successful? (2) Is the distribution of the transduced or transfected gene localized to the target organ or to target tissue, and is the distribution in the target optimal? (3) Is the level of transgene expression in the target organ or tissue sufficient to result in a therapeutic effect? (4) Does the transduced or transfected gene localize to any organ or tissue at sufficient levels to induce unwanted toxicity? (5) In the case of combined pro-drug-gene therapy protocols, when is transgene expression maximum (optimal) and when is the optimal time to initiate treatment with the pro-drug? (6) How long does transgene expression persist in the target and other tissues?

We and others have proposed that non-invasive imaging techniques (including gamma camera, SPECT or PET) using selected marker/reporter gene-marker/reporter probe combinations will provide a practical and clinically useful way to identify successful gene transduction and expression in patients undergoing gene therapy, and that radiotracer imaging of transgene expression will be able to address the questions posed above. Although one could argue that biopsies of target tissue could be performed and that imaging is not critical, imaging provides some clear advantages. These include: (a) the ability to re-

peatedly assess gene expression which is very difficult using a biopsy approach, (b) the absence of any perturbation of the underlying tissue which occurs with biopsy procedures, and (c) a map of the entire body which provides an assessment not only of the target organs (tissue), but other potential sites of gene expression as well, which could be important for toxicity issues.

HSV1-*tk* has the advantage of being both a "therapeutic gene" (combined with ganciclovir treatment) and a "reporter gene" (using an appropriate radiolabeled probe, such as FIAU or FHBG). Experimental validation of this approach has been demonstrated in animal models of colorectal metastases to the liver treated with adenoviral-mediated HSV1-*tk* gene transfer and ganciclovir ("suicide" gene therapy) (Tjuvajev et al. 1999a; Qiao et al. 2002), or treatment with conditionally replicating, oncolytic herpes viruses that constitutively express the HSV1-*tk* gene (Bennett et al. 2001; Jacobs et al. 2001a). However, most therapeutic transgenes do *not* lend themselves to direct imaging of their transgene product. Most therapeutic transgene products lack appropriate ligands or probes that can be radiolabeled and used to generate images that define the magnitude of transgene expression. The time and cost involved in developing "new" ligands and probes for each therapeutic transgene and the effort required to validate each new imaging paradigm is a significant impediment to generalize this approach. It is therefore reasonable to consider alternative strategies for "indirect" imaging of therapeutic gene expression that use established marker/reporter gene constructs. The sections below discuss several different indirect approaches that can be used to image a wide range of different therapeutic genes with a small number of established reporter gene-reporter probe combinations, and achieve the objective of monitoring therapeutic gene expression.

Several strategies have been discussed (Fig. 28.9): one strategy uses a fusion gene containing cDNA from both the marker/reporter and therapeutic genes (Fig. 28.9A); a second strategy uses a bicistronic IRES-linked therapeutic and reporter gene (Fig. 28.9B); a third strategy uses a double vector carrying two separate genes. All three strategies are based on demonstrating a proportional and a constant relationship in the co-expression of two or more transgenes over a wide range of expression levels (Fig. 28.9C).

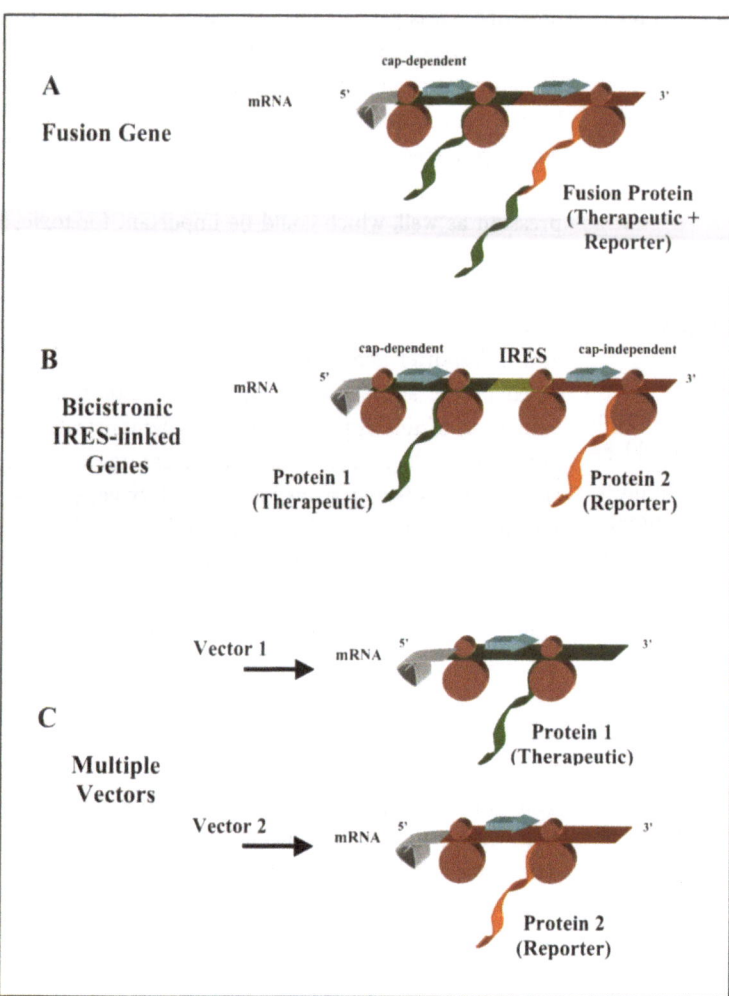

Fig. 28.9 A–C. Strategies for imaging and monitoring therapeutic gene expression. Use of a fusion gene containing cDNA from both the marker/reporter gene and the therapeutic gene (**A**); use of a bicistronic IRES-linked therapeutic and reporter gene (**B**); use of multiple vectors carrying two or more separate genes (**C**)

28.6.1.1
Fusion Genes

Strict co-expression of two proteins in equimolar amounts can only be achieved by a fusion gene encoding both gene products. This approach is based on existing fusion-gene technology, where two genes are "linked" in a fixed and definable manner. One of the two genes is a "reporter" gene and is used to provide information about the second gene – e.g., a "therapeutic" gene. Transcription of the fusion gene to mRNA occurs (under the control of upstream promoter/enhancer elements), and translation of the mRNA proceeds to yield a single "fusion protein" (the gene product, Fig. 28.9 A). The fusion protein is a hybrid of the two individual proteins (therapeutic and reporter). An important aspect of fusion genes and their gene products (fusion proteins) is that the

expression of the linked genes is "fixed" in a known 1:1 ratio. In this way, one component of the fusion gene-fusion protein can be a "reporter" and the other component can have "therapeutic" characteristics. Thus, information obtained by imaging the "reporter" component will provide corresponding information about the "therapeutic" component.

The principle of fusion gene technology in treatment protocols was demonstrated some time ago (Hammer et al. 1984; Selden et al. 1987). More recently the application of double pro-drug activation gene therapy using the *E. coli cytosine deaminase* (CD)-*herpes simplex virus 1 thymidine kinase* (HSV1-*tk*) fusion gene (CD/TK) with 5FC, GCV, and radiotherapy was developed and is currently under evaluation for treatment of different tumors (Rogulski et al. 1997, 2000). The efficacy of non-invasive imaging for monitoring CD/TK fusion gene expression with [124I]FIAU

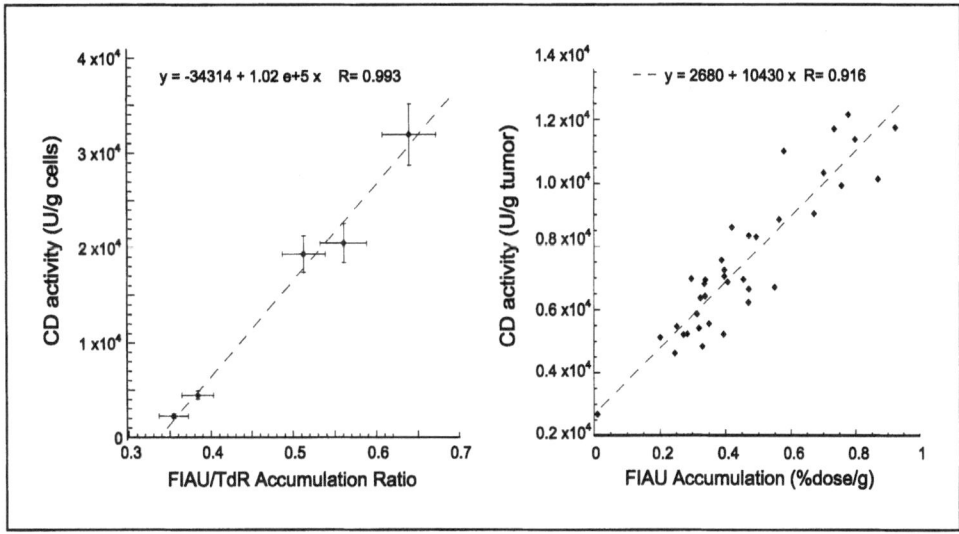

Fig. 28.10. CD/TK fusion gene assays in vitro and in vivo. Functional co-expression of the CD and HSV1-*tk* subunits of the CDglyTK fusion gene product (fusion protein) in different single cell-derived clones of W256CDTK+ cells (in vitro experiments, *left panel*). Functional co-expression of the CD and HSV1-*tk* subunits of CDglyTK fusion gene product (fusion protein) in different s.c. W256CDTK+ tumors (in vivo experiments, *right panel*). CD enzyme activity is plotted against FIAU accumulation (radiotracer assay for HSV1-*tk* enzyme activity)

and PET was performed to confirm this treatment-imaging paradigm (Hackman et al. 2002). Walker-256 tumor cells were transduced with a retroviral vector bearing the *CD/TK* fusion gene. The activity of HSV1-*tk* and CD subunits of the CD/TK gene product were assessed in different single cell-derived clones of W256CD/TK cells using the radiotracer FIAU accumulation assay and a CD enzyme assay in cell homogenates, respectively. A linear relationship was observed over a wide range of CD/TK expression levels between the levels of CD and HSV1-*tk* subunit expression in corresponding clones in vitro (Fig. 28.10 A). Multiple s.c. xenografts produced from corresponding CD/TK clones expressing different levels of CD/TK activity were produced in rats. PET imaging of HSV.1-*tk* subunit activity with [124I]FIAU and postmortem analysis of CD enzyme activity in corresponding tumors showed a significant correlation (Fig. 28.10 B). These results demonstrated a highly consistent relationship between the activities of the two subunits of the CDglyTK gene product. Knowing this relationship, parametric images of CD subunit activity of the CD/TK fusion protein were generated from the [124I]FIAU PET images (Hackman et al. 2002). Thus, PET imaging of CDglyTK expression could aid in the development, assessment, and optimization of clinical tumor gene therapy protocols by non-invasively defining the location, magnitude, and persistence of the therapeutic gene expression in target tissues.

"Proof of principle" for the application of fusion gene imaging was initially demonstrated using the HSV1-*tk*/eGFP (enhanced green fluorescent protein) fusion gene (Loimas et al. 1998; Jacobs et al. 1999). This dual-reporter fusion gene reporter encodes for the corresponding HSV1-*tk*/eGFP fusion protein (Fig. 28.11). This dual reporter construct provides the opportunity to perform whole body optical fluorescence imaging (OFI) of HSV1-*tk*/GFP expression, as well as non-invasive nuclear imaging (gamma camera, SPECT, or PET) with [*I]FIAU (or other suitable probe).

There are a number of potential disadvantages of the fusion gene approach. First, the fusion construct may not result in a functional gene product. Namely, the resultant fusion protein may be inactive or subactive. This could be due to a change in the conformational structure of the native protein and result in an alteration in the subcellular localization (intracellular trafficking) of the fusion protein, in the binding or enzymatic activity of its "therapeutic component", or to a loss of activity of its "reporter component", or to both and result in a completely inactive gene product. Second, transcription modulation of the fusion mRNA, or post-translation differences such as a change in the clearance (breakdown) of the fusion protein may be significantly different compared to the two native proteins (gene products of the two native genes). Any one or a combination of these different

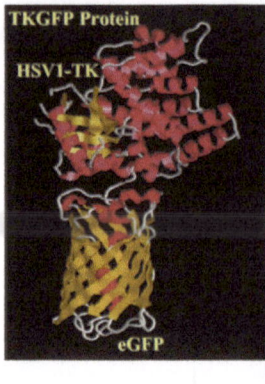
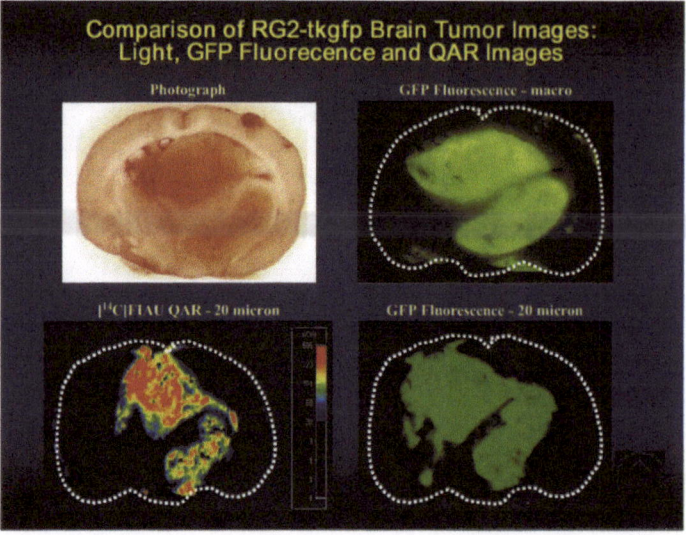

Fig. 28.11. Imaging the HSV1-TKeGFP fusion gene: composite QAR and fluorescence images. Multi-modality HSV1-TKeGFP reporter gene imaging. The HSV1-TKeGFP fusion protein is shown in the *upper left panel*, and corresponding images are seen in the quadrants of the *right panel*. This dual reporter construct provides the ability to monitor gene expression both in vitro (or in situ) by fluorescence microscopy and FACS analysis, and in vivo by whole body optical fluorescence imaging or by nuclear imaging. The spatial correspondence between histology and in situ fluorescence in a rat brain xenograft model (RG2-TK/GFP) is shown in the *upper panels*. Corresponding quantitative autoradiographic ([^{14}C]FIAU – HSV1-*tk*) and in situ fluorescence (GFP) images are shown in the *lower panels*

effects could have a significant impact on the level of the fusion gene product, and, thereby, on the level of its biological activity. Third, fusion proteins are larger than the two corresponding native proteins and more likely to generate an immunological response and limit the clinical application of the fusion gene construct. Thus, fusion gene technology cannot be generalized and it is not likely to be widely applicable in clinical imaging of therapeutic gene expression. However, when the fusion gene product is functional and non-immunogenic, it is a very useful approach, as demonstrated above for dual reporter constructs utilizing both nuclear as well as optical reporter genes.

28.6.1.2
Bicistronic IRES-linked Genes

Another approach that was recently described and validated involves the proportional expression of two cis-linked genes, using an internal ribosomal entry site (IRES) element within a single bicistronic transcription cassette (Pelletier 1988; Ghattas 1991; Jackson 1995; Sachs et al. 1997). The IRES element enables translation initiation within the bicistronic mRNA, thus permitting gene co-expression by cap-dependent translation of the first cistron and cap-independent, IRES-mediated translation of the second cistron (Fig. 28.9 B).

In the first report describing the use of indirect imaging strategies, a type II IRES element derived from the 5' untranslated region of encephalomyocarditis virus (EMCV) was used because of its apparent lack of tissue-specificity relative to other IRES types (Tjuvajev et al. 1997 a, b). The expression of HSV1-*tk* was used to monitor the coordinated expression of LNGFR (low affinity nerve growth factor receptor) cis-linked by the EMCV IRES with the HSV1-*tk* gene, following transduction with the SFG-NIT retroviral vector. There was a linear relationship between the level of LNGFR expression on the cell surface and the level of HSV1-*tk* gene expression (as measured by FIAU accumulation and sensitivity to GCV) in different clones of SFG-NIT-transduced Jurkat cells, and this proportionality was maintained in all single-cell clones tested and over a wide range of gene expression levels. Furthermore, the level of HSV1-*tk* gene expression in SFG-NIT-transduced Jurkat cell clones was within the range that was adequate for in vivo imaging.

It was also shown that monitoring the level of HSV1-*tk* expression using radiolabeled FIAU and gamma camera imaging can be used to monitor the co-expression of another, second gene (*Lac Z* gene) in cells and xenografts (Tjuvajev et al. 1999b). In this paradigm, a MoMVL based retroviral vector STLEO bearing the HSV1-*tk* gene and *LacZ*/NeoR fusion gene linked by an IRES sequence was used for transduction. The level of HSV1-*tk* and *Lac Z* expression was measured in different RG2STLEO and W256-STLEO clones (and in tumors produced from the clones). A comparison of enzyme activities demonstrated that the expression of the two genes is proportional and constant over a wide range of expression levels in both the in vitro and in vivo assays and in the two different cell lines (Fig. 28.12). The observed correlation between the levels of *Lac Z* and HSV1-*tk* gene expression, both in vitro and in vivo (Fig. 28.12 B–D), demonstrates the potential for monitoring therapeutic gene transfer and expression (*Lac Z* in the above example) by non-invasive imaging of

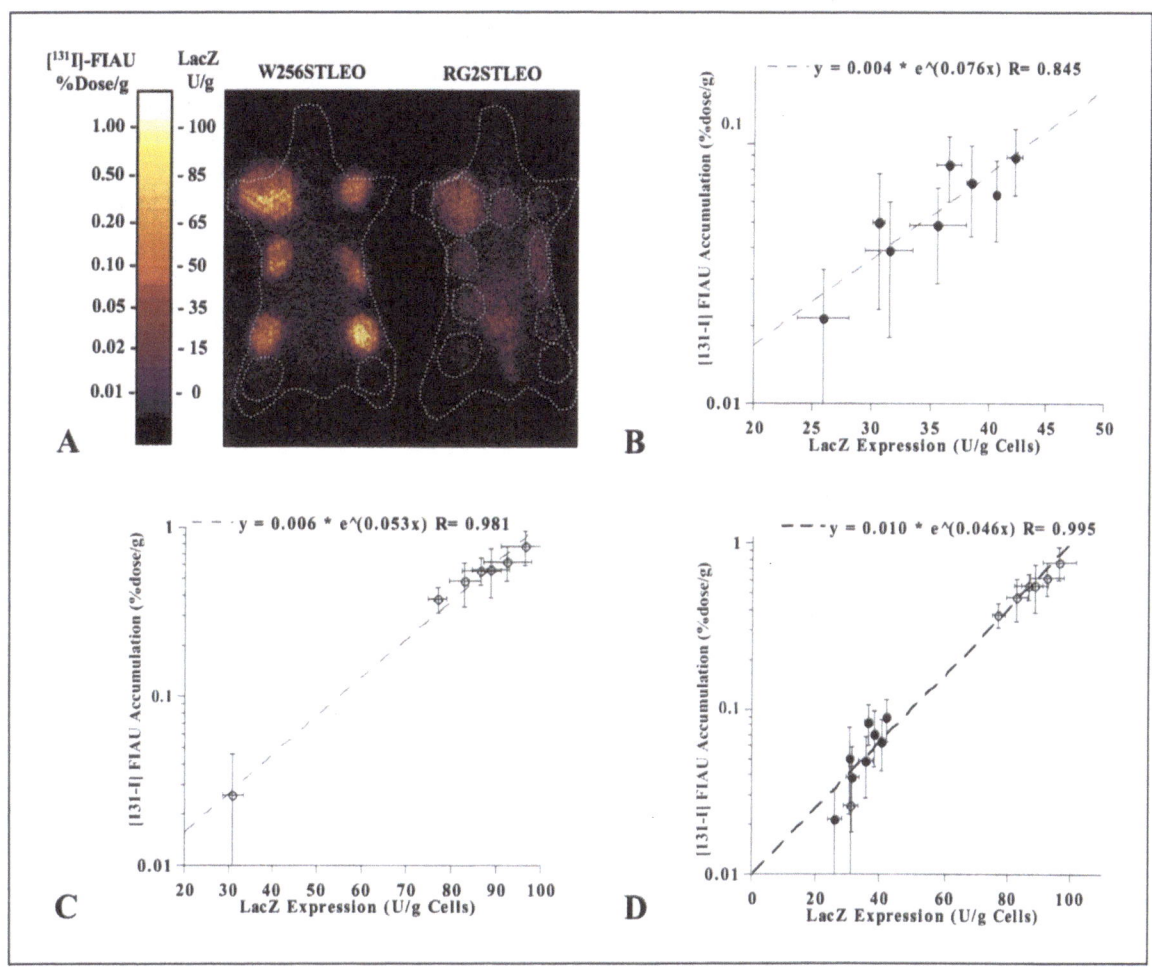

Fig. 28.12 A–D. Imaging *cis*-linked HSV1-*tk* and *LacZ* gene co-expression. Gamma camera images of [^{131}I]-FIAU accumulation in W256SLEO and RG2STLEO subcutaneous tumors derived from clonal cell lines, 24 h after tracer administration reflect HSV1-*tk* expression (**A**). The *white dashed line* outlines the contours of the animal and tumors with relatively low activity and wild-type control tumors (lowest tumor on the right thigh). Some residual bladder activity is still present in RG2STLEO tumor-bearing rats. Note the difference in [^{131}I]-FIAU accumulation between different tumors. The [^{131}I]-FIAU images of HSV1-tk expression were also converted into parametric images of *lacZ* gene expression based on the relationship defined in **D**, and this is indicated by the color-coded intensity bar showing units of β-galactosidase activity (U/g) in **A**. The levels of HSV1-tk gene expression measured in different tumors (%dose/g [^{131}I]-FIAU) were plotted against measures of *lacZ* gene expression (U/g) obtained in corresponding tumor samples for different W256SLEO (**B**) and RG2STLEO (**C**) clonal tumors. The relationship between the two measures was defined by regression analysis. The same relationship was observed when the RG2STLEO and W256STLEO data were combined (**D**)

the HSV1-*tk* marker/reporter gene. Knowing this relationship, parametric images of *LacZ* gene expression in vivo could be generated from the HSV1-*tk* images (Sachs et al. 1997) as shown in Fig. 28.12 A.

Several other constructs including pCMV-RL-IRES-HSV1-*sr39tk*, PCMV-D2R-IRES-HSV1-*sr39tk* and others have been studied to demonstrate correlated expression of the genes placed proximal and distal to a type II ECMV IRES (Yu et al. 2000). In both cell culture and tumor-bearing mice, a correlation between expression of a "target gene" [renilla luciferase (RL) or the dopamine D2 receptor] and a PET marker/reporter gene (mutant herpes simplex virus type 1 thymidine kinase, HSV1-*sr39tk*) was observed. A high degree of correlation was obtained between the expression of target and marker/reporter genes both in stably transfected cells in culture using biochemical assays ($r^2 > 0.97$) and in tumor-bearing mice ($r^2 > 0.95$) using microPET analysis with FPCV (for HSV1-*sr39tk* expression) and FESP (for D2R expression) to image the expression of these reporter genes (Yu et al. 2000).

The imaging paradigms described above are not confined to one type of vector; the imaging paradigm can be extended to liposome-encapsulated DNA, naked plasmid vectors, and other viral vectors, because IRES-mediated co-expression is determined at the translational level. Eventually, it will be important to assess if the IRES-based vectors are reliable indicators of transgene co-expression in different tissues, taking into account the half-life of the encoded proteins. These considerations will be important when using non-invasive imaging to assess organ (tissue) specificity, as well as the level and duration of transgene expression.

28.6.1.3
Multiple Vector Strategy

Another approach to achieve proportional co-expression of one or more therapeutic genes and a reporter gene involves the administration of multiple vectors bearing these genes in various combinations (Fig. 28.9 C). This approach has been tested by co-administering two distinct but otherwise identical adenoviruses, one expressing a therapeutic transgene and the other expressing the reporter transgene, to track the therapeutic gene's expression. Each vector has identical envelope characteristics and identical promoter/enhancer elements driving the expression of both therapeutic and reporter transgenes. There

are three key requirements for successful implementation of this approach are: (1) the multiple vectors transfect and transduce target organs and tissue proportionate to their ratio in the administered cocktail; (2) the therapeutic and reporter genes are co-expressed proportionally in all target organs and tissue; and (3) the proportionality of co-expression is constant over a wide range of expression levels.

Several recent reports have provided encouraging results in support of using multiple vectors (adenovirus) to deliver reporter and therapeutic genes to target organs. One study demonstrated the efficacy of monitoring the adenoviral mediated IL2 and GM-CSF expression (gene therapy) by imaging of HSV1-*tk* co-expression (Doubrovin et al. 2001 b). Similar results have been shown using two PET reporter genes, each inserted into separate adenoviral vectors (Ad) (Yaghoubi et al. 2001 a). One gene was a mutant herpes simplex virus type 1 thymidine kinase (HSV1-*sr39tk*) and the other was a dopamine-2 receptor (D2R), each regulated by the same cytomegalovirus (CMV) promoter. This study demonstrated that cells co-infected with equivalent titers of Ad-CMV-HSV1-*sr39tk* and Ad-CMV-D(2)R expressed both reporter genes with good correlation ($r^2 = 0.93$). Similarly, a high correlation ($r^2 = 0.97$) was observed between the expression of both reporter genes in the livers of mice co-infected via tail-vein injection with equivalent titers of these two adenoviruses (Fig. 28.13). MicroPET imaging of HSV1-*sr39tk* and D(2)R expression with 9-(4-[^{18}F]fluoro-3-hydroxy-methylbutyl) guanine ([^{18}F]FHBG) and 3-(2-[^{18}F]fluoroethyl)spiperone ([^{18}F]FESP), utilizing several adenovirus-mediated delivery routes, demonstrated the feasibility of evaluating relative levels of transgene expression in living animals, using this approach.

The imaging studies in animals described above indicate that proportional co-expression of therapeutic and reporter genes is obtained with multiple adenovirus vectors, and that this may be a reliable way to assess therapeutic gene expression. A key observation will be to demonstrate that this proportionality is maintained in different tissues and over extended periods of time. This is necessary when non-invasive imaging is used to assess organ (tissue) specificity, as well as the level and duration of therapeutic transgene expression. Non-invasive imaging of therapeutic gene expression would be of considerable value in many ongoing and future clinical gene therapy trials by defining the location, specificity, magnitude, and persistence of gene expression over time.

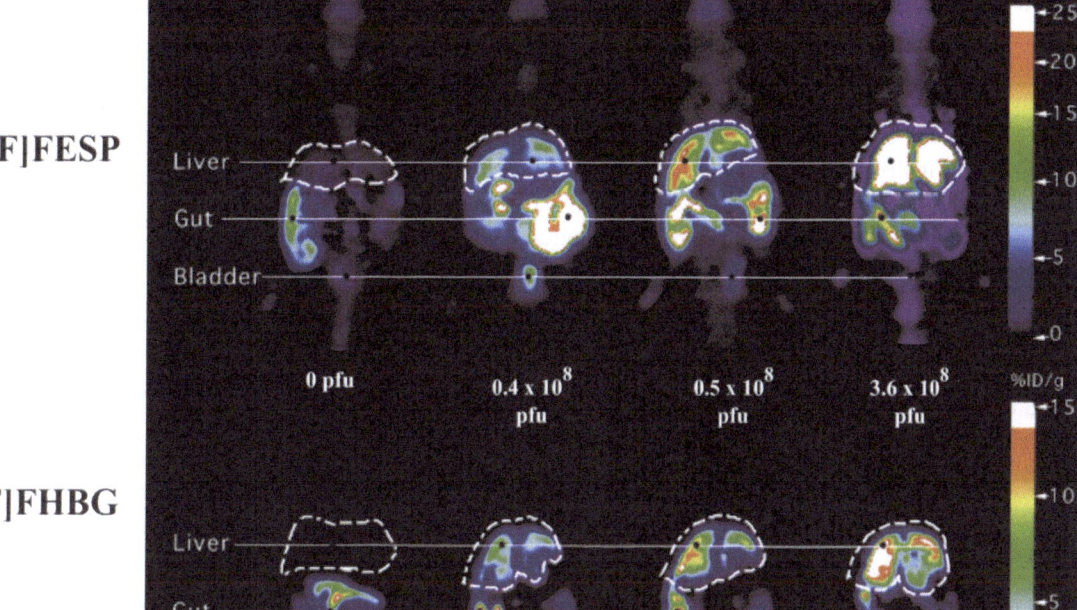

Fig. 28.13. MicroPET whole body [^{18}F]FESP and [^{18}F]FHBG images of Ad-CMV-D$_2$R and Ad-CMV-HSV1-*sr39tk* localization. Equal viral titers of Ad-CMV-D$_2$R and Ad-CMV-HSV1-*sr39tk* were injected into the tail vein of nude mice. After 3 days, the mice were injected with [^{18}F]FHBG and imaged by MicroPET. The next day, the same mice were injected with [^{18}F]FESP and re-imaged with microPET. The images illustrate the biodistribution and degree of [^{18}F]FHBG and [^{18}F]FESP accumulation in individual nude mice co-injected with different, but equivalent titers of Ad-CMV-HSV1-*sr39tk* and Ad-CMV-D2R. From *left to right*, the mice received 0 p.f.u., 0.4×10^8 p.f.u., 0.5×10^8 p.f.u. and 3.6×10^8 p.f.u. The *upper panel* illustrates biodistribution of [^{18}F]FESP and the *lower panel* that of [^{18}F]FHBG. Images are the average of several whole body slices containing the liver. The images are scaled based on the injected dose (from Yaghoubi et al. 2001)

28.6.1.4
Conclusions

An advantage of fusion gene, IRES-linked genes, and multiple vector approaches is that only a small number of "reporter gene"–"reporter probe" combinations are necessary to provide information about a very large number of therapeutic transgenes. However, each of the three imaging strategies require validation of proportional and constant co-expression of both the "reporter" and the "therapeutic" transgene over a wide range of expression levels and over time post-administration. These *indirect* imaging strategies provide the opportunity for a much wider application of transgene imaging in clinical practice for several reasons. First, most therapeutic transgenes do *not* lend themselves to direct imaging of their transgene product, and most therapeutic transgene products lack appropriate ligands or probes that can be radiolabeled and used to generate images that define the magnitude of transgene expression. Second, the cost and time involved in developing and validating "new" ligands and probes for imaging therapeutic transgenes can be substantial. Therefore, it is reasonable to consider alternative "indirect" imaging strategies (fusion gene, IRES-linked or multiple vector strategies) to image therapeutic gene expression; strategies that use established marker/reporter gene constructs.

28.6.2
Reporter Gene-based Molecular-Biological Assays

In principle, the established (existing) methods for non-invasive imaging reporter gene expression described above can be introduced into existing reporter gene based molecular biological assay systems. In such assay systems, reporter gene expression is linked to an endogenous molecular genetic process of interest. These molecular genetic processes of interest include regulation of endogenous gene expression at the transcriptional and post-transcriptional levels, and the activation (or inactivation) of specific signal transduction pathways and different protein-protein interactions.

Transcription and gene expression levels are typically monitored by Northern and Western blotting. In the radiotracer paradigm, the activation of the promoter of the endogenous gene (and hence the activity of the upstream signaling events) is also the final step in activation of the reporter gene. The reporter system is constructed so the reporter transgene (HSV1-tk) is driven by the same enhancer/promoter elements as the endogenous gene. The reporter's level of activity represents a direct measure of endogenous promoter activity, and thus an indirect measure of the pathway leading to the transcription factor(s) which binds to the enhancer/promoter elements of the endogenous gene. The reporter gene usually encodes for an enzyme (CAT, β-galactosidase, luciferase, HSV1-tk, etc.) that can be measured easily, inexpensively and quantitatively. Reporter enzyme assays are replacing tedious blotting procedures, and are widely used due to their convenience, sensitivity and ability to provide high spatial resolution and to generate reliable in vivo data.

Two reporter systems, *cis* and *trans*, will be described and examples of these reporter systems will be presented. There is a significant difference between the *cis*- and *trans*-reporter systems, and this will be described below. The initial results using *cis*- and *trans*-reporter systems show that in vivo imaging of endogenous gene expression is feasible. In addition, the advantages of combined in vitro, *in situ*, and in vivo imaging of endogenous gene expression using the TKeGFP dual reporter/fusion gene approach will be illustrated.

28.6.2.1
Cis-Reporter System

In *cis*-reporter systems, the expression of the reporter gene is mediated by endogenous activated transcription factors or by a complex formed by an activated transcription factor that binds to the *cis*-acting enhancer element of the reporter gene cDNA. Extracellular signals may trigger the sequential activation of a series of intracellular signaling molecules, such as protein kinases and phosphatases. The final step in the activation of many pathways is the binding of an activated transcription factor to specific enhancer elements found in the promoters of various cellular genes. This binding to enhancer/promoter elements modulates the transcription of these genes. In this way the transcription level of cellular genes reflects the activation status of the involved signaling events.

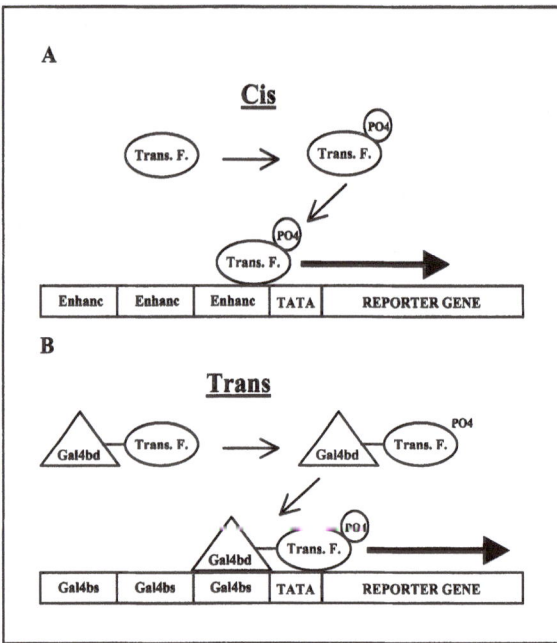

Fig. 28.14 A, B. Combined paradigm of the *cis*-reporter system and the *trans*-reporter system. For the *cis*-reporter system, a transcription factor (*Trans.F.*) is activated (phosphorylated) during the activation of a particular signal transduction pathway (**A**). The activated transcription factor binds to the specific *cis*-acting DNA enhancer element (*Enhanc.*), and initiates transcription of the reporter gene. For the *trans*-reporter system, a GAL4 DNA-binding domain – transcription factor fusion protein (GAL4bd_Trans.F.) is activated (e.g., phosphorylated) during the activation of a particular signal transduction pathway (**B**). The activated GAL4bd_Trans.F. fusion protein binds to the GAL4 DNA-binding site (GAL4bs) of the reporter construct, and initiates transcription of a reporter gene

To determine the activation of the promoter, and hence signaling events, different vectors have been designed that use a reporter gene in place of the endogenous cellular gene. *Cis*-reporting systems typically feature a *cis*-reporting plasmid, retroviral, or adenoviral vector that contains a reporter gene driven by a basic promoter element (TATA box) joined to repeats of *cis*-acting DNA elements, enhancers (Fig. 28.14A). Since most enhancer elements can be regulated by more than one transcription factor, each enhancer element can be used to monitor more than one signal transduction pathway. The *cis*-reporting systems can be used for studying the effects of a new gene, growth factor or drug. An example of a reporter construct that uses a specific artificial enhancer that is sensitive to and binds the activated p53 transcription factor is shown in Fig. 28.15A (Doubrovin et al. 2001a).

After a vector expressing the *cis*-reporter system of interest is transfected into mammalian cells, various exogenous manipulations (such as a change in the cellular milieu or environment, exposure to irradiation, treatment with a candidate drug or specific metabolic inhibitor) can be used to induce transcription factor production and activation, which in turn will bind to the enhancer elements of the reporter construct and initiate or enhance transcription of the reporter gene. Increased reporter enzyme activity indicates increased transcriptional activation of the reporter gene and involvement of the signaling pathways converging at these *cis*-acting enhancer elements in cellular responses to exogenous stimuli or drug action. *Cis*-reporter systems have been widely used to study the regulation of expression of different endogenous genes and signal transduction pathways [e.g., TGF*a* (Liu et al. 1996; Kretzschmar et al. 1999; Wotton et al. 1999), SRF (Belaguli et al. 1999), AP1 (Glinghammar et al. 1999), p53 (Tang et al. 1998; Tan et al. 1999), HIF1*a* (Gerber et al. 1997; Varma and Cohen 1997), etc.].

28.6.2.2
Examples of *Cis*-Reporter System Imaging

Monitoring activity of p53 pathway-regulated genes by PET imaging: Several important biological processes are affected by endogenous expression and stabilization of p53, including G1/G2 arrest and apoptosis. A recent paper assessed the efficacy of the *cis*-p53/TKeGFP reporter system in vivo (Doubrovin et al. 2001a) and showed that p53-dependent gene ex-

pression can be imaged with [^{124}I]FIAU and PET. A retroviral vector (*Cis*-p53/TKeGFP) was constructed by placing the herpes simplex virus type 1 thymidine kinase (*tk*)-enhanced green fluorescent protein (e*GFP*) fusion gene (TKeGFP, a dual-reporter gene) under control of a p53-specific artificial response element (Fig. 28.15A). DNA damage-induced up-regulation, stabilization and activation of endogenous p53 caused significant activation of p53 transcriptional activity in transduced U87 cells and xenografts. This up-regulation was monitored with the *cis*-p53/TKeGFP reporter construct, and the level of reporter expression (level of radioactivity in the transduced cells and xenografts) correlated with the expression of p53-dependent genes (including p21). This was the first demonstration that a *cis*-reporter system (*cis*-p53/TKeGFP) was sufficiently sensitive to image endogenous gene expression using non-invasive PET imaging (Fig. 28.15 D).

Fluorescent microscopic analysis was also performed on xenograft tissue sections taken from the animals that were imaged with PET. A low incidence of fluorescent positive U87-*cis*-p53/TKeGFP cells was observed in untreated (non-induced) U87-*cis*-p53/TKeGFP xenografts (Fig. 28.16A). In contrast, U87-*cis*-p53/TKeGFP xenografts from animals treated with BCNU had a high density of fluorescent cells that was comparable to that in BCNU-treated RG2-TKeGFP xenografts (positive control). RT-PCR analysis of tumor tissue samples demonstrated increased HSV1-*tk*/eGFP and p21 mRNA levels induced by BCNU treatment in U87-*cis*-p53/TKeGFP xenografts compared to non-treated (non-induced) U87-*cis*-p53/TKeGFP xenografts (Fig. 28.16C) (Doubrovin et al. 2001a).

This study showed that an established PET reporter gene (TKeGFP) can be sufficiently driven (expressed) by the presence and binding of an endogenous transcription factor (activated p53) to specific enhancer elements of the reporter construct. The specific enhancer elements are used as a "sensory" component to detect the presence of a particular protein, transcription factor or intracellular process (e.g., level of endogenous p53 transcriptional activity). The enhancer elements also function as a "switch" when placed up-stream of a minimal basic promoter. In this case, the binding of a transcription factor (activated p53) to the "sensor/switch" induces transcription of the reporter gene (*TKeGFP*). The amount of reporter gene product reflects the endogenous level of the transcription factor, and this will determine the level of radioactivity in the PET image following adminis-

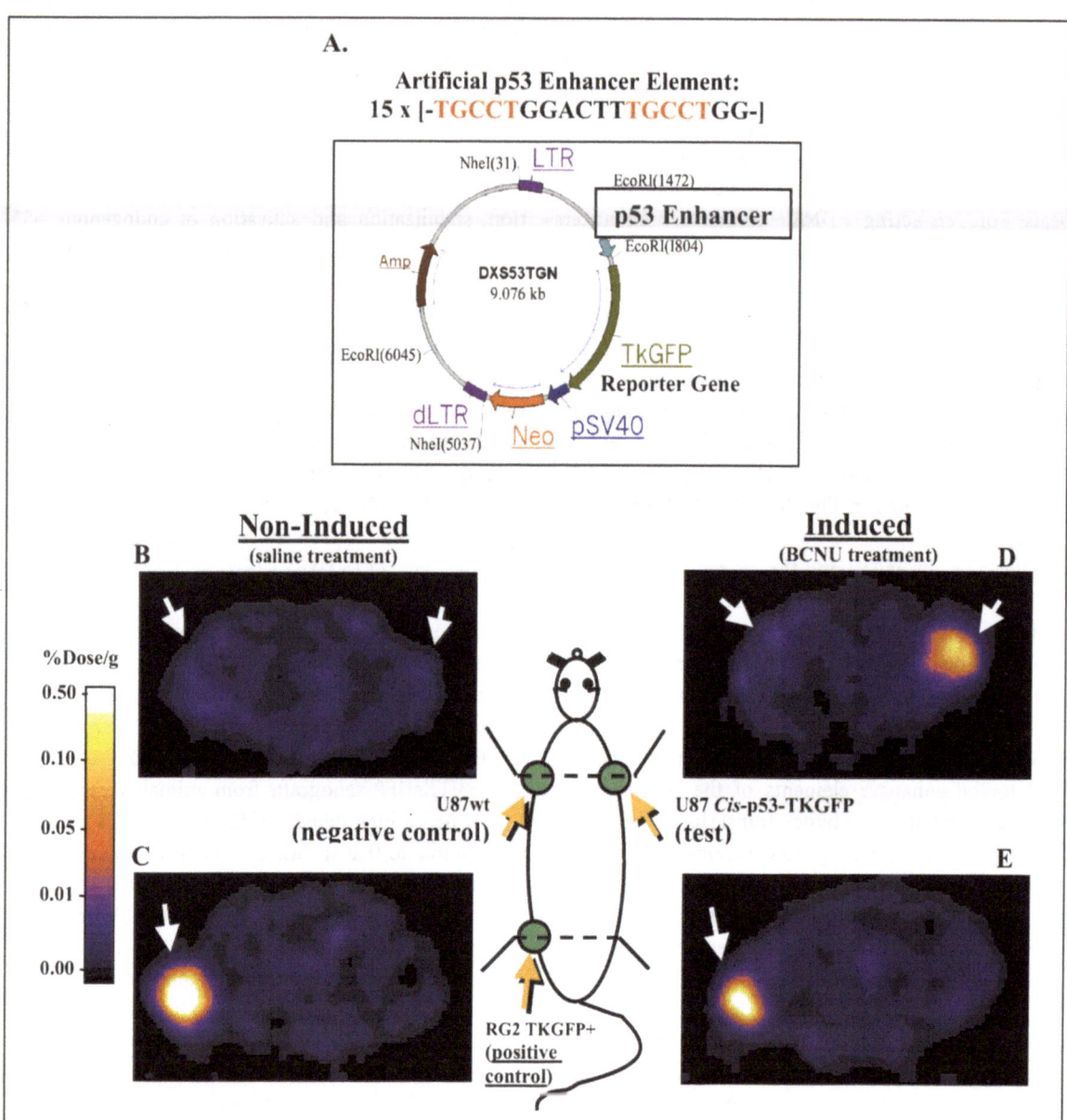

Fig. 28.15 A–E. *Cis*-p53/TKeGFP vector construct and PET imaging of activated endogenous p53 expression. Structure of the DXS53TGN retroviral vector bearing the *cis*-p53/TKeGFP reporter system (**A**). Transaxial PET images (GE Advance tomograph) through the shoulder (**B, D**) and pelvis (**C, E**) of two rats are shown; the images are color-coded to the same radioactivity scale (% dose/g). An untreated animal is shown on the *left*, and a BCNU-treated animal is shown on the *right*. Both animals had three s.c. tumor xenografts: U87p53TKeGFP (test) in the right shoulder, U87 wild-type (negative control) in the left shoulder, and RG2TKeGFP (positive control) in the left thigh. The nontreated animal on the *left* shows localization of radioactivity only in the positive control tumor (RG2TKeGFP); the test (U87p53TKeGFP) and negative control (U87wt) tumors are at background levels. The BCNU-treated animal on the *right* shows significant radioactivity localization in the test tumor (*right shoulder*) and in the positive control (*left thigh*), but no radioactivity above background in the negative control (*left shoulder*)

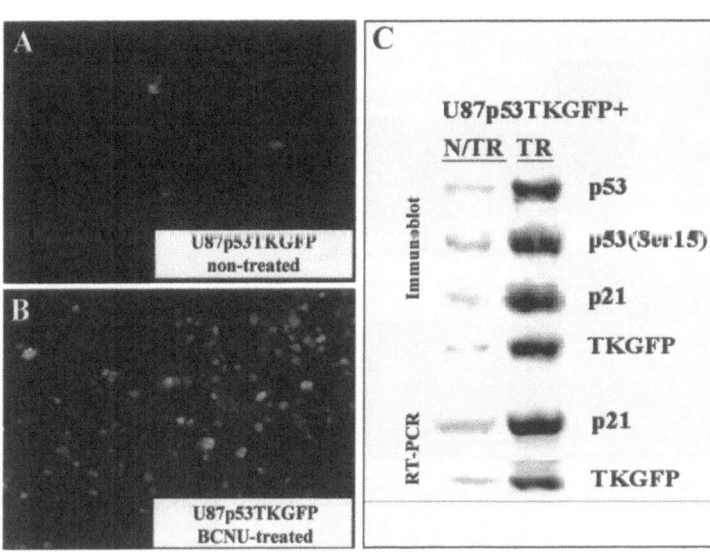

Fig. 28.16 A–C. Assessment of *cis*-p53/TKeGFP reporter system in vivo. Fluorescence microscopy images of U87p53/TKeGFP s.c. xenograft samples obtained from non-treated rats (**A**) and rats treated with 40 mg/kg BCNU i.p. (**B**). The same U87p53/TKeGFP s.c. tumor samples obtained from non-treated (*N/TR*) and BCNU-treated (*TR*) animals were assessed for the levels of activated p53 (Ser15 phosphorylated), total p53 protein, p21 and TKeGFP proteins by immunoblot analysis and for the levels of p21 and TKeGFP mRNAs by RT-PCR (**C**)

tration of [^{124}I]FIAU. Similarly, the number and intensity of in situ fluorescent cells will depend on the level of TKeGFP reporter gene expression.

This is an example of an indirect imaging motif, where the reporter component (e.g., a reporter transgene) is placed under the control of a sensory component (e.g., specific enhancer/promoter elements upstream from the reporter transgene), and the resultant PET and fluorescent images reflect the activity of the intracellular process detected by the sensory element(s) of the reporter construct. The PET images in Fig. 28.15 show up-regulation of genes in the p53 signal transduction pathway (p53-dependent downstream genes such as p21) in response to DNA damage induced by BCNU chemotherapy. PET imaging of p53 transcriptional activity in tumors using the *cis*-p53TKeGFP reporter system could be used to assess the response to radiation therapy, new drugs or novel therapeutic approaches, including gene therapy strategies based on p53 over-expression.

Monitoring T-cell activation by PET imaging: T-cell activation is an essential component of the immune response in normal and disease states. Non-invasive whole body imaging of T-cell activity would be of a considerable value; it could provide the identification of sites and the temporal dynamics of T-cell activation, and could provide a way to monitor trafficking of T cells during primary and memory responses. A recent study demonstrated that highly specific images of T-cell activation could be obtained (Ponomarev et al. 2001b). To visualize T-cell activation in vivo, a T-cell receptor (TCR)-dependent,

NFAT-sensitive TKeGFP dual-reporter system was developed. This reporter system was used to obtain whole body PET images as well as whole body optical fluorescence images (OFI). A retroviral vector (*cis*-NFAT/TKeGFP) was generated by placing the dual-reporter fusion gene (TKeGFP) under control of the response element for nuclear factor of activated lymphocytes (NFAT). The TCR-specific NFAT family of transcription factors contains five distinct members. NFATs 1–4 share a highly conserved DNA binding domain and contribute to regulating a number of target genes, including IL-2 and other cytokines (Aramburu et al. 1997; Rao et al. 1997; Crabtree 1999; Kiani et al. 2000).

A human T-cell leukemia cell line (Jurkat) that expresses a functional TCR was transduced with the *cis*-NFAT/TKeGFP reporter vector and used in these studies. Known activators of T-cells (anti-CD3 and anti-CD28 antibody) produced significantly higher levels of TKeGFP reporter gene expression (increased GFP fluorescence, increased levels of HSV1-*tk* mRNA and increased radiolabeled probe accumulation) in *cis*-NFAT/TKeGFP+ Jurkat cells compared to non-treated or non-transduced Jurkat cells. PET imaging of mice with local *cis*-NFAT/TKeGFP+ Jurkat cell infiltrates demonstrated significantly higher levels of radioactivity localized to the transduced and anti-CD3/CD28-stimulated Jurkat cell infiltrates compared to corresponding controls (Fig. 28.17 A, B).

Whole body optical fluorescence images and FACS analysis of the Jurkat/dcmNFATtgn infiltrates was performed following PET imaging (Fig. 28.17 C, D).

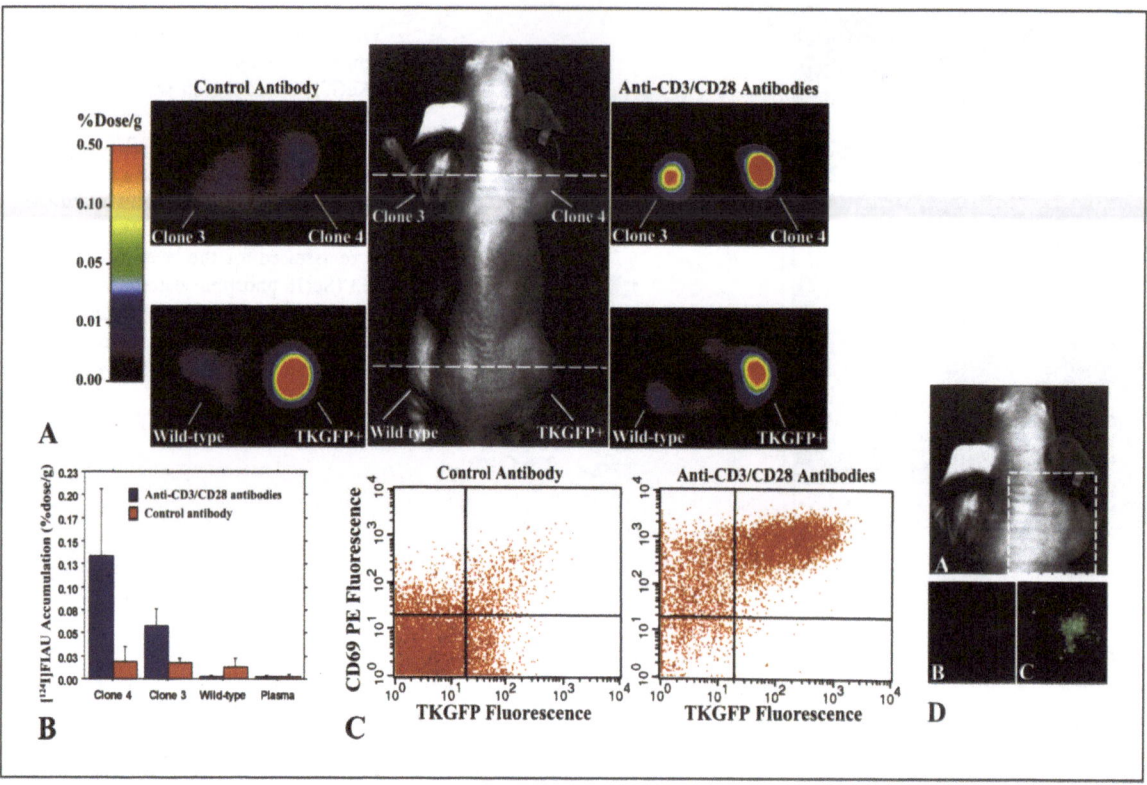

Fig. 28.17 A–D. *Cis*-NFAT/TKeGFP PET and fluorescence imaging. Imaging NFAT-TKeGFP reporter system activity with [124I]FIAU and PET (GE Advance) and subsequent tissue sample assays. Photographic image of a typical mouse bearing different s.c. infiltrates (*middle panel*); transaxial PET images of TKeGFP expression in a mouse treated with control antibody (*left panel*) and anti-CD3/CD28 antibodies (*right panel*) were obtained at the levels indicated by the *dashed lines* (**A**). [124I]FIAU accumulation (% dose/g) in tissue samples of the Jurkat/dcmNFATtgn clone 3 and 4 infiltrates, wild-type Jurkat infiltrates, and blood plasma, obtained after PET imaging (**B**). FACS profiles of TKeGFP and CD69 (marker of activated T-cells) expression in a tissue sample from the same Jurkat/dcmNFATtgn clone 4 infiltrate that was imaged with PET (**C**). TKeGFP- optical fluorescence imaging of a s.c. Jurkat/dcmNFATtgn (clone 4) infiltrate in a live mouse (**A**); before (**B**) and after treatment (**C**) with anti-CD3/CD28 antibodies that activate T-cells via the NFAT transcription factor (**D**)

The induction of TKeGFP and CD69 (a marker of activated T cells) expression in the Jurkat/dcmNFATtgn infiltrates was clearly seen in the anti-CD3/CD28 (induced) treated animals, as compared to the control (non-induced) animals (Fig. 28.17 D). The Jurkat/dcmNFATtgn infiltrates from the anti-CD3/CD28-treated animals had a very large fraction (>60%) of cells over-expressing both TKeGFP and CD69 (Fig. 28.17 C). The fluorescence-FACS analysis of the transduced Jurkat cell infiltrates yielded results that were very similar to those obtained by PET imaging and radiotracer assay.

This study demonstrated that the *cis*-NFAT-TKeGFP reporter system was sufficiently sensitive to detect transcriptional up-regulation in the TCR-dependent NFAT-mediated signal transduction pathway of Jurkat cells (a model for T lymphocytes) using [124I]FIAU and a clinical PET tomograph. A strong correlation between TKeGFP expression and up-regulation of T-cell activation markers (CD69 and IL-2 production) was shown in both the in vitro and in vivo studies. These results demonstrated that: (1) activation of the NFAT signal transduction pathway occurs after TCR stimulation, and (2) PET imaging of T-lymphocyte activation in tumors following TCR engagement is feasible using the described TKeGFP-based *cis*-reporter system. PET imaging of T-lymphocyte activation using the NFAT-reporter system could potentially be used to assess efficacy and optimize T-cell doses in immunotherapies, and to evaluate T cell responses in new immunotherapeutic approaches and could be used to assess the efficacy of novel anti-tumor vaccines.

The p53 and NFAT studies described above demonstrate that PET imaging of endogenous gene expression is feasible using a *cis*-reporter system and that PET imaging can be used to detect changes in the level of endogenous gene expression. In addition, these studies demonstrate the advantages of a dual reporter construct (e.g., TKeGFP) for combined PET and fluorescence imaging, and quantification by both radiotracer and FACS analysis.

28.6.2.3
Trans-Reporter System

There are significant differences between the *cis*- and *trans*-reporter system (Fig. 28.14). The *cis*-reporter system is based on endogenous activated transcription factors or complexes that bind to the *cis*-acting enhancer element of the reporter gene. The *cis*-acting promoter drives expression of the reporter gene by recruiting endogenous transcription factors that bind to upstream enhancer elements, which may or may not be strong enough to initiate transcriptional activity. *trans*-reporting systems have several attractive features and can address different issues. For example, the *trans*-reporting system can be used when the specific promoter of the target gene is not strong enough to induce the expression of the reporter gene to levels that are measurable by the reporter system (e.g., CAT assay, GFP fluorescence, PET imaging). Furthermore, some enhancer (promoter) elements may not be very specific and may bind several different transcription factors. The *trans*-reporter system allows one to study the activation of specific transcription factors in specific signal transduction pathways. *Trans*-reporter systems are also useful for studying the in vivo effects of new genes, growth factors, different drugs as well as exogenous/extracellular stimuli on the activation of different intracellular kinases and other signaling molecules which may lead to the activation of these kinases (e.g., protein-protein interactions) (Brown et al. 1999; Hansen et al. 1999; Hocevar et al. 1999).

Trans-reporter systems include two components (Fig. 28.14 B): (1) a unique *trans*-activator fusion construct (GAL4VP16) that is driven by a specific promoter choice (e.g., CEA) and can function as a "sensory" element, and (2) a reporter element that encodes multiple GAL4-binding sites upstream from a basic minimal promoter (TATA) and the reporter gene (e.g., HSV1-*tk*). Both components can be placed in the same expression cassette and can be delivered by a single vector (plasmid), or can be delivered separately using

two vectors (plasmids). The GAL4VP16 *trans*-activator protein consists of the yeast GAL4 DNA-binding domain (GAL4db) and the papilloma virus transcription factor VP16. In the *Trans*-system, a weak *cis*-acting promoter can drive the expression of the very potent transcription factor VP16 that is fused to a specific DNA-binding protein GAL4 (GAL4VP16 *trans*-activator protein). Even a small amount of the GAL4VP16 *trans*-activator protein, a transcription factor produced by the "first"-*cis*-acting transcription step, can be sufficient to initiate substantial transcription of the reporter gene in the "second" transcription step. This is because the GAL4VP16 transcription factor is relatively stable and can bind repeatedly to the GAL4-specific DNA sequence and repeatedly initiate transcription of the reporter gene. This cycling "amplification" mechanism results in very high levels of reporter gene expression and accounts for the high sensitivity of the *trans*-reporter system. In the *trans*-reporter construct, expression of the GAL4VP16 *trans*-activator fusion protein is induced by a specific gene promoter of choice, even a weak promoter such as the CEA or PSA promoter (see below). *Trans*-reporter systems have been widely used to study regulation of both gene expression and signal transduction pathway activity [e.g., JNK (Brown et al. 1999), TGFβ signaling (Hocevar et al. 1999), MAP kinase (Hansen et al. 1999), BRCA1 (Haile and Parvin 1999), histone acetylase (Ikeda et al. 1999), HIF-1α (Kallio et al. 1998; Ema et al. 1999), Elk1 (Chung et al. 1998; Sharif and Sharif 1999), PSA protein (Sadar 1999), etc.].

28.6.2.4
Example of *Trans*-Reporter System Imaging

Monitoring carcinoembryonic antigen (CEA) gene expression by *trans*-reporter gene imaging. A recent set of experiments compared the activity of different *Cis*-acting promoters (CEA, RSV, and CMV) driving the expression of the CAT reporter gene to that of a *trans*-CEA/CAT reporter system. Adenoviral vector constructs and transfection of MOD mouse colorectal carcinoma cells was performed in culture and the level of CAT expression was assayed in vitro (Qiao et al. 2002) (Fig. 28.18). Despite the high levels of constitutive CEA expression by the MOD cells, the level of CAT reporter gene expression driven by the CEA promoter (*cis*CEA/CAT construct) was quite low. In comparison, CAT reporter activity following transfection with the constitutive RSV and CMV viral promoter constructs was 10- and 75-fold higher than that with the CEA promoter,

Reporter Constructs

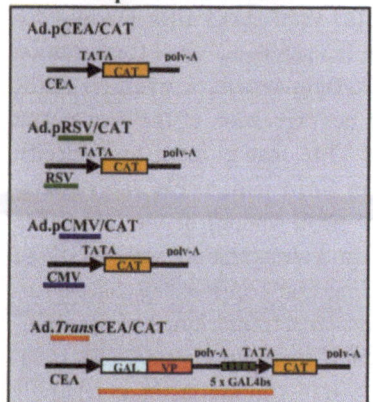

MOD murine breast carcinoma cells

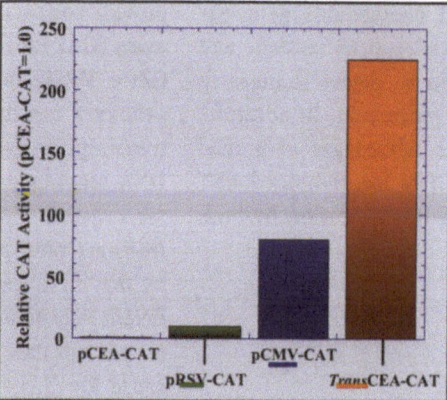

Fig. 28.18. Comparative activity of different *cis-* and *trans*-promoter constructs. Transfection of plasmids containing different promoter constructs for the CAT (chloramphenicol acetyl transferase) reporter gene into CEA-positive MOD cells was performed. The CAT assay was used to compare the relative activity of the CEA (carcinoembryonic antigen) promoter (the pCEA-CAT = 1) with constitutive viral promoters (RSV: *green bar*; CMV: *blue bar*) and the binary (*trans*) system (*red bar*) – GVP: GAL4/VP16 fusion protein. G5B: minimal TATA promoter with five GAL4-binding sites

respectively (Fig. 28.18). However, the highest level of CAT expression was induced by the *trans*CEA/CAT reporter system; the level was more than 200-fold higher than that induced by the *cis*CEA/CAT reporter system. These results indicated that the *cis*CEA/HSV1-*tk* reporter system would have poor sensitivity for detection of endogenous CEA expression in hepatic tumors and adjacent liver. Therefore, a *trans*CEA/HSV1-*tk* reporter system was constructed to provide high sensitivity for the CEA transcription factor and, correspondingly, high levels of HSV1-*tk* expression.

The feasibility for non-invasive imaging of adenoviral-mediated CEA-specific HSV1-*tk* gene expression was assessed using an animal model of metastatic colon carcinoma to the liver (Doubrovin et al. 2000; Qiao et al. 2002). The Ad*trans*CEA-*tk* adenoviral vector was constructed to deliver the *trans*-reporter system under the control of the CEA promoter-driving expression of the GAL4bd-VP16 fusion protein. The GAL4bd-VP16 fusion protein then binds to the GAL4-specific DNA-binding site (GAL4bs) upstream the HSV1-*tk* gene, and transactivates HSV1-*tk* expression. The structure of this adenoviral vector (Ad*trans*CEA/HSV1-*tk*) is similar to the structure of the adenoviral vector bearing the *trans*CEA/CAT reporter system that was previously tested in vitro (Fig. 28.18, bottom construct in left panel).

The gamma camera images demonstrated highly specific localization of [¹³¹I]FIAU-derived radioactivity to the area of Ad*trans*CEA-*tk*-injected MOD tumors in the liver, whereas the MCA-26-transfected tumors accumulated significantly lower levels of radioactivity ($p < 0.05$) (Fig. 28.19). There was no accumulation of [¹³¹I]FIAU-derived radioactivity in tumors that were injected with the control vector or buffer alone (images not shown). Histological and QAR comparisons also revealed highly specific localization of [¹³¹I]FIAU-derived radioactivity in transfected tumor nodes and remote tumor cell clusters in the liver. In Ad*trans*CEA/HSV1-*tk*-injected animals, the level of radioactivity that accumulated in the peritumoral liver tissue was significantly lower than that in AdRSV/HSV1-*tk*-injected animals. The relatively high levels of peritumoral liver radioactivity in Ad.RSV/HSV1-*tk* injected animals can be explained by the lateral spread of the virus (and maybe by re-circulation of the virus), and by the constitutive expression of HSV1-*tk* induced by the RSV promoter demonstrating the absence tumor-specificity). In contrast, Ad*trans*CEA/HSV1-*tk* vector-mediated HSV1-*tk* expression is limited to tumor tissue that expresses the CEA. Even though the Ad*Trans*CEA/HSV1-*tk* vector may infect surrounding liver tissue, little or no liver radioactivity is seen because adjacent liver does not produce CEA. Animals injected with the AdDL312 vector (non-HSV1-*tk*-bearing control) did not show any accumulation of radioactivity above background levels, including the tumor and peritumoral liver tissue. These results indicate that non-invasive imaging of promoter-specific HSV1-*tk* reporter gene expression is feasible using the *trans*-reporter system, and is sensitive enough to discriminate between different

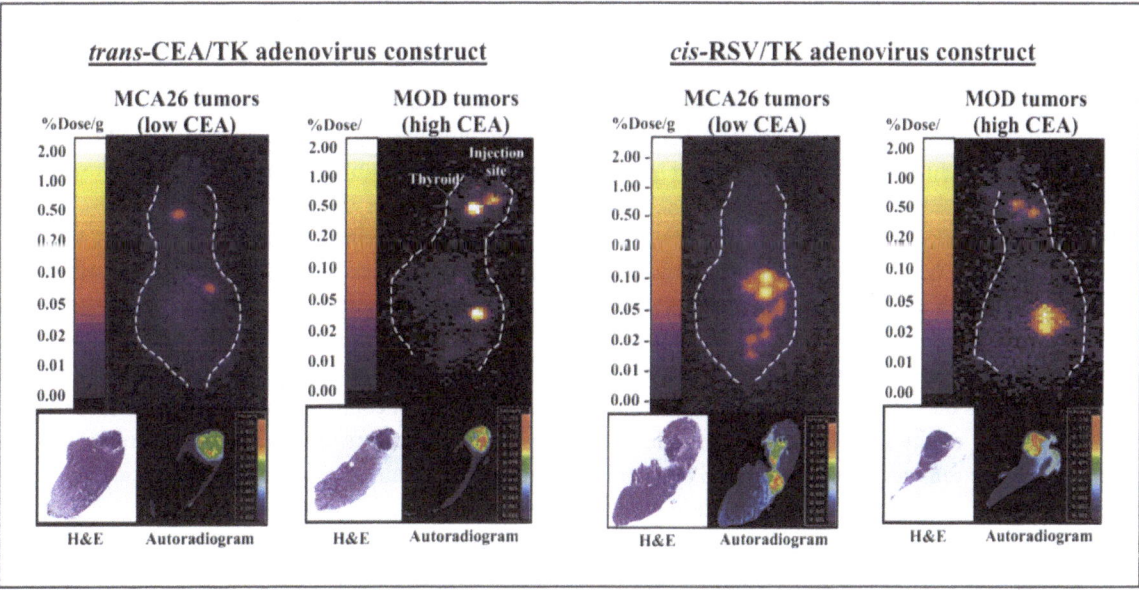

Fig. 28.19. In vivo imaging of adenovirus-mediated HSV-*tk* expression. Gamma camera and quantitative autoradiographic images of [^{131}I]FIAU accumulation in mice bearing MCA26 (low CEA expressing) and MOD (high CEA expressing) intrahepatic tumors. Two adenovirus vectors are compared: one bearing the CEA-sensitive *trans*-CEA/TK (binary) construct and another bearing the constitutively expressing RSV/TK construct. Note the highly localized accumulation of [^{131}I]FIAU radioactivity to the tumors injected with the CEA-sensitive ADV-CEA/TK construct; there is little or no radioactivity observed in adjacent liver tissue. In contrast, the ADV-RSV/TK construct shows constitutive TK expression in both the tumor and adjacent liver. Not surprisingly, considerably more hepatic toxicity was observed following ganciclovir treatment in animals injected with the ADV-RSV/TK vector compared to those injected with the ADV-RSV/TK vector. No tumor or hepatic accumulation of radioactivity was observed in animals injected with the control vector ADV/DL312 (negative control, images not shown). Note the residual radioactivity at the injection site (orbital vein) and accumulation of [^{131}I]-iodide in the thyroid (indicating some breakdown of [^{131}I]FIAU)

levels of CEA expression. The Ad*Trans*CEA/HSV1-*tk* vector allows for tumor-specific (CEA-specific) expression of the HSV1-*tk* gene, and is more suitable for clinical cancer gene therapy of colorectal carcinoma metastases to the liver.

Monitoring prostate-specific antigen promoter activity (PSE) by *trans*-reporter gene imaging: a similar example using the GAL4VP16 *trans*-activator in a "two-step transcriptional amplification" (TSTA) system was recently reported (Iyer et al. 2001 b). This system was necessary because the "one-step" *cis*-PSE reporter construct used to transiently transfect androgen-sensitive LNCaP prostate cancer cells was not sufficient to produce an adequate reporter signal following androgen stimulation. The in vitro studies demonstrated a ~50-fold increase in the firefly luciferase reporter assay and a ~12-fold increase in the HSV1-*sr39tk* reporter radiotracer assay. A very encouraging result was the demonstration that the *trans*-reporter system was androgen concentration sensitive, suggesting a continuous rather than binary reporter response. However, as observed with the CEA-*trans* reporter system above, the in vivo imaging comparison of the *trans*- and *cis*-reporter systems showed less dramatic differences than that obtained by the in vitro analyses. Only a ~5-fold difference in response to androgen stimulation was observed between PSE *trans*- and *cis*-reporter systems in the in vivo firefly luciferase optical imaging experiments.

Nevertheless, these are very exciting results and illustrate how molecular biology and non-invasive imaging can complement each other. We are at the beginning of a new era, and there are many technical and methodological issues that remain to be addressed.

28.6.3
Transgenic Animals

The use of transgenic animals carrying a marker/reporter gene offers unique possibilities for tracking a single animal repeatedly over time during experimental manipulations. Transgenic animals expressing the HSV1-*tk* gene from tissue-specific promoters have been developed to perform cell-specific ablation following administration of pharmacologic levels of pro-drugs such as GCV (Mathis et al. 2000). A transgenic mouse model in which the HSV1-*tk* reporter gene is driven by the albumin promoter has recently been studied (Green et al. 2000). Albumin promoter-HSV1-*tk* transgenic mice have been imaged on a microPET with both [^{18}F]FPCV and [^{18}F]FHBG, and clearly demonstrate accumulation of the reporter probe in the liver of transgenic mice. Localized reporter probe accumulation in the liver is the result of tissue specific transcriptional activation of the HSV-*tk* marker/reporter gene by the albumin promoter. The albumin-HSV1-*tk* mice will be very useful for comparing alternate probes in vivo and for assessing the reproducibility of assays.

Transgenic mice are also being developed in which the expression of the HSV1-*tk* reporter gene is driven by an artificial promoter that is specific to hypoxia-inducible factor type one alpha (HIF-1*a*). HIF-1*a* is an oxygen-regulated transcriptional activator and is also activated under some non-hypoxic conditions; HIF-1*a* plays essential roles in mammalian development, physiology and disease pathogenesis. The *cis*-HIF-1*a*/TKeGFP reporter system in the transgenic-reporter mouse will allow visualization of hypoxic/angiogenic responses of the host normal tissues to local injury as well as the response of peritumoral tissue and stroma as compared to that of xenografted tumors and their supporting stroma. The HIF-1*a*/HSV1-*tk* transgenic reporter mouse could be used to study the involvement of tumor and adjacent stroma in hypoxia/hyponutrition-mediated neoangiogenesis. By crossing the HIF-1*a*/HSV1-*tk* transgenic mouse with other transgenic mice that develop specific types of tumors following specific gene knockouts (tumor suppressor genes), knock-ins (tumor-promoter onco-genes) or specific gene mutations, it will be possible to study the effect of such genetic mutations on the HIF-1*a* pathway and neoangiogenesis using non-invasive imaging techniques.

28.7
Prospects for the Future

The primary applications of non-invasive in vivo marker/reporter gene imaging are likely to be: (i) quantitative monitoring of transduction efficacy in gene thera-py and animal research protocols by imaging the location and extent of transgene expression; (ii) non-invasive assessments of endogenous molecular events in transduced tissue using different reporter gene imaging technologies; (iii) monitoring cell trafficking and replication of immune-competent cells following ex vivo transduction; (iv) repetitive, quantitative evaluation of reporter gene expression over time in both patients and transgenic animals during longitudinal experiments. Those applications involving therapeutic genes will involve the use of bicistronic and fusion gene vectors that express therapeutic and reporter gene products from a common transcript, or the use of multiple transfer vectors where each vector expresses a different therapeutic or reporter gene.

Imaging reporter gene expression in living subjects is a rapidly evolving focus of molecular imaging research. Studies from several different centers have validated reporter gene imaging using positron emission tomography (PET), single photon emission computed tomography (SPECT), quantitative autoradiography (QAR), magnetic resonance imaging (MRI) and spectroscopy (MRS), as well as in vivo fluorescence imaging with wild-type and mutant fluorescent proteins and in vivo bioluminescence imaging using firefly or renilla *luciferase* reporter genes. Imaging radiolabeled probes with PET has several advantages, including high sensitivity, the capability for quantitative assessments, the ability to perform repeated measurements over time, and the ability to directly translate the imaging paradigms from animals to humans. These advantages will assure that radionuclide-based imaging will remain at the forefront of molecular and gene imaging in the future.

Non-invasive reporter gene imaging will have both pre-clinical and direct clinical applications in the study of various diseases where different molecular events need to be monitored by whole body imaging over time. These include: (1) assessment of trafficking for different cell populations (i.e., bone marrow transplantation, T-cell and NK-cell vaccines, stem cell therapy, etc.) and their activation, proliferation or differentiation; (2) assessment of the efficacy and specificity of various gene therapy vectors (i.e., non-replicating and replicating adenoviral and herpes viral vectors, AAV, lentiviral, plasmid/liposomal vectors, etc.); (3) indirect monitoring of various endogenous genes and activities of different signal transduction pathways (i.e., oncogene activity, growth factor receptor-mediated, pro-apoptotic, or differentiation-inducing signaling, etc.).

Many applications of in vivo reporter gene imaging will develop over the next several years. Three applications related to cancer therapy are likely to be introduced into the clinic in the near future. One application involves visualization of therapeutic gene expression following transduction of primary tumors and/or metastases with "suicide" genes (i.e., HSV1-*tk*, CD) which sensitize the transduced tumor to cytotoxic prodrugs (i.e., ganciclovir, 5FC). In this therapeutic/imaging protocol, PET will be performed to assess the efficacy and tumor specificity of vector delivery and therapeutic gene expression; optimization of viral vector dose, mode and timing of administration could be performed and correlated with treatment response.

Another application involves intratumoral administration of different types of conditionally replicating oncolytic viruses that constitutively over-express the HSV1-*tk* gene. In this paradigm, PET imaging of HSV1-*tk* expression would be performed to assess the efficacy and tumor specificity of oncolytic virus delivery, replication, and lateral spread through the tumor. PET imaging of HSV1-*tk* reporter gene expression could potentially visualize the wave of the oncolytic virus spreading through the tumor and identify the presence of virus in adjacent or remote normal tissues.

A third application could involve the administration of tumor-specific cytotoxic T-cells following ex vivo transduction with a reporter gene under control of a NFAT-promoter. PET imaging of reporter gene expression could be used to visualize the sites of T-cell activation upon recognizing tumor-specific antigens. This could provide for improvements in anti-tumor sensitization procedures and optimization of T-cell dose and schedule of administration. A similar approach could be used for imaging the distribution and engraftment of stem cells in the treatment of various diseases, including cancer.

Future research directions should also be aimed at the development of novel non-immunogenic reporter genes for human applications; development of novel reporter gene-reporter substrate combinations that would be suitable for CNS imaging (with probes that cross the BBB easily); the development of reporter systems and for imaging key oncogenic signal transduction pathways that are involved in tumorigenesis, angiogenesis, tumor progression, invasion, organ/tissue-specific metastasis, and tumor maintenance. Development of transgenic reporter mice that allow for imaging the endogenous expression/activity of vari-

ous genes and signal transduction pathways would significantly further our understanding of many incurable diseases including cancer. The ability to noninvasively image transgenic animals will also foster the development of novel and more effective drugs to treat these diseases.

28.8
References

Adachi Y, Tamiya T, Ichikawa T et al (2000) Experimental gene therapy for brain tumors using adenovirus-mediated transfer of cytosine deaminase gene and uracil phosphoribosyltransferase gene with 5-fluorocytosine. Hum Gene Ther 11:77–89

Alauddin MM, Conti PS (1998) Synthesis and preliminary evaluation of 9-(4-[18F]-fluoro-3-hydroxymethylbutyl)-guanine ([18F]FHBG): a new potential imaging agent for viral infection and gene therapy using PET. Nucl Med Biol 25:175–180

Alauddin MM, Conti PS, Mazza SM et al (1996) 9-[(3-[18F]-fluoro-1-hydroxy-2-propoxy)methyl]guanine ([18F]-FHPG): a potential imaging agent of viral infection and gene therapy using PET. Nucl Med Biol 23:787–792

Alauddin MM, Shahinian A, Kundu RK et al (1999) Evaluation of 9-[(3-18F-fluoro-1-hydroxy-2-propoxy)methyl]-guanine ([18F]-FHPG) in vitro and in vivo as a probe for PET imaging of gene incorporation and expression in tumors. Nucl Med Biol 26:371–376

Alauddin MM, Shahinian A, Gordon EM et al (2001) Preclinical evaluation of the penciclovir analog 9-(4-[(18)F]fluoro-3-hydroxymethylbutyl)guanine for in vivo measurement of suicide gene expression with PET. J Nucl Med 42:1682–1690

Aramburu J, Rao A, Klee CB (1997) Calcineurin: from structure to function. Curr Top Cell Regul 36:237–295

Bading JR, Kan-Mitchell J, Conti PS (1996) System A amino acid transport in cultured human tumor cells: implications for tumor imaging with PET. Nucl Med Biol 23:779–786

Bading JR, Alauddin MM, Fissekis JH et al (1997) Pharmacokinetics of F-18 fluorohydroxy-propoxymethylguanine (FHPG) in primates. J Nucl Med 38:43P

Bading JR, Shahinian AH, Bathija P, Conti PS (2000) Pharmacokinetics of the thymidine analog 2'-fluoro-5-[(14)C]-methyl-1-beta-D-arabinofuranosyluracil ([(14)C]FMAU) in rat prostate tumor cells. Nucl Med Biol 27:361–368

Bagchi N, Fawcett DM (1973) Role of sodium ion in active transport of iodide by cultured thyroid cells. Biochim Biophys Acta 318:235–251

Barrio JB, Satyamurthy N, Huang SC et al (1989) 3-(2'-[18F]fluoroethyl)spiperone: in vivo biochemical and kinetic characterization in rodents, nonhuman primates, and humans. J Cereb Blood Flow Metab 9:830–839

Behr TM, Gotthardt M, Barth A, Behe M (2001) Imaging tumors with peptide-based radioligands. Q J Nucl Med 45:189–200

Belaguli NS, Zhou W, Trinh TH et al (1999) Dominant negative murine serum response factor: alternative splicing within the activation domain inhibits transactivation of serum response factor binding targets. Mol Cell Biol 19:4582–4591

Bengel FM, Anton M, Avril N et al (2000) Uptake of radiolabeled 2'-fluoro-2'-deoxy-5-iodo-1-beta-D-arabinofuranosyluracil in cardiac cells after adenoviral transfer of the herpes virus thymidine kinase gene: the cellular basis for cardiac gene imaging. Circulation 102:948–650

Bennett J, Duan D, Engelhardt JF, Maguire AM (1997) Real-time, noninvasive in vivo assessment of adeno-associated virus-mediated retinal transduction. Invest Ophthalmol Vis Sci 38:2857–2863

Bennett JJ, Tjuvajev JG, Johnson P et al (2001) Positron emission tomography imaging for herpes virus infection: implications for oncolytic viral treatments of cancer. Nat Med 7:859–863

Berger F, Gambhir SS (2001) Recent advances in imaging endogenous or transferred gene expression utilizing radionuclide technologies in living subjects: applications to breast cancer. Breast Cancer Res 3:28–35

Bergstrom M, Lu L, Fasth KJ et al (1998) In vitro and animal validation of bromine-76-bromodeoxyuridine as a proliferation marker. J Nucl Med 39:1273–1279

Black ME, Newcomb TG, Wilson HM, Loeb LA (1996) Creation of drug-specific herpes simplex virus type 1 thymidine kinase mutants for gene therapy. Proc Natl Acad Sci USA 93:3525–3529

Black ME, Kokoris MS, Sabo P (2001) Herpes simplex virus-1 thymidine kinase mutants created by semi-random sequence mutagenesis improve prodrug-mediated tumor cell killing. Cancer Res 61:3022–3026

Blasberg R, Tjuvajev J (1997) Berlin-Heidelberg in vivo monitoring of gene therapy by radiotracer imaging. Impact of molecular biology and new technical developments on diagnostic imaging. Springer, Berlin Heidelberg New York, pp 161–189 (Ernst Shering Research Foundation Workshop 22)

Blasberg RG, Roelcke U, Weinreich R et al (2000) Imaging brain tumor proliferative activity with [124I]iododeoxyuridine. Cancer Res 60:624–635

Boerman OC, Dams ET, Oyen WJ et al (2001) Radiopharmaceuticals for scintigraphic imaging of infection and inflammation. Inflamm Res 50:55–64

Bombardieri E, Maccauro M, De Deckere E et al (2001) Nuclear medicine imaging of neuroendocrine tumours. Ann Oncol 12 [Suppl 2]:51P–61P

Borrelli E, Heyman R, Hsi M, Evans RM (1988) Targeting of an inducible toxic phenotype in animal cells. Proc Natl Acad Sci USA 85:7572–7576

Brown JD, DiChiara MR, Anderson KR et al (1999) MEKK-1, a component of the stress (stress-activated protein kinase/c-Jun N-terminal kinase) pathway, can selectively activate Smad2-mediated transcriptional activation in endothelial cells. J Biol Chem 274:8797–8805

Brumley CL, Kuhn JA (1995) Radiolabeled monoclonal antibodies. AORN J 62:343–350; 353–355; 356–358; 361–362

Brust P, Haubner R, Friedrich A et al (2001) Comparison of [18F]FHPG and [124/125I]FIAU for imaging herpes simplex virus type 1 thymidine kinase gene expression. Eur J Nucl Med 28:721–729

Busch H, Davis JR, Honig GR et al (1959) The uptake of a variety of amino acids into nuclear proteins of tumors and other tissues. Cancer Res 19:1030–1039

Calabresi P, Cardoso S, Finch S et al (1961) Initial clinical studies with 5-iodo-2'-deoxyuridine. Cancer Res 21:550–559

Cammilleri S, Sangrajrang S, Perdereau B et al (1996) Biodistribution of iodine-125 tyramine transforming growth factor alpha antisense oligonucleotide in athymic mice with a human mammary tumour xenograft following intratumoral injection. Eur J Nucl Med 23:448–452

Carrasco N (1993) Iodide transport in the thyroid gland. Biochim Biophys Acta 1154:65–82

Casey KL, Svensson P, Morrow TJ et al (2000) Selective opiate modulation of nociceptive processing in the human brain. J Neurophysiol 84:525–533

Cerqueira MD (1999) Current status of radionuclide tracer imaging of thrombi and atheroma. Semin Nucl Med 29:339–351

Chung KC, Gomes I, Wang D et al (1998) Raf and fibroblast growth factor phosphorylate Elk1 and activate the serum response element of the immediate early gene pip92 by mitogen-activated protein kinase-independent as well as -dependent signaling pathways. Mol Cell Biol 18:2272–2281

Conti PS, Alauddin MM, Fissekis JR, Schmall B, Watanabe KA (1995) Synthesis of 2'-fluoro-5-[11C]-methyl-1-beta-D-arabinofuranosyluracil ([11C]-FMAU): a potential nucleoside analog for in vivo study of cellular proliferation with PET. Nucl Med Biol 22:783–789

Crabtree GR (1999) Generic signals and specific outcomes: signaling through Ca2+, calcineurin, and NF-AT. Cell 96:611–614

Culver KW, Blaese RM (1994) Gene therapy for cancer. Trends Genet 10:174–178

Dai G, Levy O, Carrasco N (1996) Cloning and characterization of the thyroid iodide transporter. Nature 379:458–460

De la Vieja A, Dohan O, Levy O, Carrasco N (2000) Molecular analysis of the sodium/iodide symporter: impact on thyroid and extrathyroid pathophysiology. Physiol Rev 80:1083–1105

Delbeke D, Martin WH (2001) Positron emission tomography imaging in oncology. Radiol Clin North Am 39:883–917

De Vries EF, van Waarde A, Harmsen MC et al (2000) [(11)C]FMAU and [(18)F]FHPG as PET tracers for herpes simplex virus thymidine kinase enzyme activity and human cytomegalovirus infections. Nucl Med Biol 27:113–119

Dewanjee MK, Ghafouripour AK, Kapadvanjwala M et al (1994) Noninvasive imaging of c-myc oncogene messenger RNA with indium-111-antisense probes in a mammary tumor-bearing mouse model. J Nucl Med 35:1054–1063

Divgi CR (1996) Status of radiolabeled monoclonal antibodies for diagnosis and therapy of cancer. Oncology (Huntington) 10:939–953

Doubrovin M, Chen SH, Ponomarev V et al (2000) Noninvasive imaging of CEA-specific expression of adenoviral-mediated HSV1-*tk* gene transfer in tumors using a trans-reporter system (abstract). SNM Abstract – TRANS, 3rd Annual Meeting of the American Society of Gene Therapy, Denver, 31 May–3 June 2000

Doubrovin M, Ponomarev V, Beresten T et al (2001a) Imaging transcriptional regulation of p53 dependent genes with positron emission tomography in vivo. Proc Natl Acad Sci USA 98:9300–9305

Doubrovin M, Lin A, Beresten T et al (2001b) In vivo radionuclide imaging of adenoviral mediated co-expression of IL2, GM-CSF, and HSV1-tk genes in tumor xenografts in mice [Abstract]. 4th American Society of Gene Therapy annual meeting, Seattle, 30 May–4 June 2001, p 75

Dougan H, Hobbs JB, Weitz JI, Lyster DM (1997) Synthesis and radioiodination of a stannyl oligodeoxyribonucleotide. Nucleic Acids Res 25:2897–2901

Dougan H, Lyster DM, Vo CV et al (2000) Extending the lifetime of anticoagulant oligodeoxynucleotide aptamers in blood. Nucl Med Biol 27:289–297

Duncan JS (1999) Positron emission tomography receptor studies in epilepsy. Rev Neurol (Paris) 155:482–488

Ema M, Hirota K, Mimura J et al (1999) Molecular mechanisms of transcription activation by HLF and HIF1alpha in response to hypoxia: their stabilization and redox signal-induced interaction with CBP/p300. EMBO J 18:1905–1914

Erbs P, Regulier E, Kintz J et al (2000) In vivo cancer gene therapy by adenovirus-mediated transfer of a bifunctional yeast cytosine deaminase/uracil phosphoribosyltransferase fusion gene. Cancer Res 60:3813–3822

Eskandari S, Loo DD, Dai G, Levy O, Wright EM, Carrasco N (1997) Thyroid Na$^+$/I$^-$ symporter. Mechanism, stoichiometry, and specificity. J Biol Chem 272:27230–27238

Faulds D, Heel RC (1990) Ganciclovir A review of its antiviral activity, pharmacokinetic properties and therapeutic efficacy in cytomegalovirus infections. Drugs 39:597–638

Feinberg A, Leyland-Jones B, Fanucchi MP et al (1985) Biotransformation and elimination of [2-14C]-1-(2-deoxy-2'-fluoro-beta-D-arabinofuranosyl)-5-iodocytosine in immuno-suppressed patients with herpesvirus infections. Antimicrob Agents Chemother 27:733–738

Feng M, Jackson WH Jr, Goldman CK et al (1997) Stable in vivo gene transduction via a novel adenoviral/retroviral chimeric vector. Nat Biotechnol 15:866–870

Gambhir SS, Barrio J, Wu L et al (1998) Imaging of adenoviral directed herpes simplex virus type 1 thymidine kinase gene expression in mice with ganciclovir. J Nucl Med 39:2003–2011

Gambhir SS, Barrio JR, Herschman HR, Phelps ME (1999a) Assays for noninvasive imaging of reporter gene expression. Nucl Med Biol 26:481–490

Gambhir SS, Barrio JR, Phelps ME et al (1999b) Imaging adenoviral-directed reporter gene expression in living animals with positron emission tomography. Proc Natl Acad Sci USA 96:2333–2338

Gambhir SS, Herschman HR, Cherry SR et al (2000a) Imaging transgene expression with radionuclide imaging technologies. Neoplasia 2:118–138

Gambhir SS, Bauer E, Black ME et al (2000b) A mutant herpes simplex virus type 1 thymidine kinase reporter gene shows improved sensitivity for imaging reporter gene expression with positron emission tomography. Proc Natl Acad Sci USA 97:2785–2790

Gambhir SS, Czernin J, Schwimmer J et al (2001) A tabulated summary of the FDG PET literature. J Nucl Med 42[Suppl]:1P–93P

Gelovani-Tjuvajev J, Doubrovin M, Akhurst T et al (2002) Comparison of radiolabeled nucleoside probes (FIAU, FHBG and FHPG) for PET imaging of HSV1-tk gene expression. J Nucl Med 43:1072–1083

Gerber HP, Condorelli F, Park J, Ferrara N (1997) Differential transcriptional regulation of the two vascular endothelial growth factor receptor genes. Flt-1, but not Flk-1/KDR, is up-regulated by hypoxia. J Biol Chem 272:23659–23667

Ghattas SM (1991) The encephalomyocarditis virus internal ribosomal entry site allows efficient coexpression of two genes from a recombinant provirus in cultured cells and in embryos. Mol Cell Biol 11:5848

Gjedde A (2001) Receptor mapping in living human beings by means of positron emission tomography. Ugeskr Laeger 163:5199–5205

Glinghammar B, Holmberg K, Rafter J (1999) Effects of colonic lumenal components on AP-1-dependent gene transcription in cultured human colon carcinoma cells. Carcinogenesis 20:969–976

Green LA, Nguyen K, Bauer E et al (2000) Indirect monitoring of endogenous gene expression by imaging PET reporter gene expression in transgenic mice (abstract). 47th Meeting of the Society of Nuclear Medicine, St Louis, 3–7 June 2000. JNM [Suppl], p 81

Haberkorn U, Oberdorfer F, Gebert J et al (1996) Monitoring gene therapy with cytosine deaminase: in vitro studies using tritiated-5-fluorocytosine. J Nucl Med 37:87–94

Haberkorn U, Khazaie K, Morr I et al (1998) Ganciclovir uptake in human mammary carcinoma cells expressing herpes simplex thymidine kinase. Nucl Med Biol 25:367–373

Hackman T, Doubrovin M, Balatoni J, Beresten T, Ponomarev V, Beattie B, Finn R, Blasberg R, Tjuvajev JG (2002) Imaging expression of cytosine deaminase – herpes virus thymidine kinase fusion gene (CD/TK) expression with [^{124}I]FIAU and PET. Mol Imaging 1:36–42

Haile DT, Parvin JD (1999) Activation of transcription in vitro by the BRCA1 carboxyl-terminal domain. J Biol Chem 274:2113–2117

Hammer RE, Palmiter RD, Brinster RL (1984) Partial correction of murine hereditary growth disorder by germ-line incorporation of a new gene. Nature 31:65–67

Hamptom E, Eidinoff M (1961) Administration of 5-iododeoxyuridine-I 131 in the mouse and rat. Cancer Res 21:345–352

Hansen TV, Rehfeld JF, Nielsen FC (1999) Mitogen-activated protein kinase and protein kinase A signaling pathways stimulate cholecystokinin transcription via activation of cyclic adenosine 3',5'-monophosphate response element-binding protein. Mol Endocrinol 13:466–475

Hatazawa J, Shimosegawa E (1998) Imaging neurochemistry of cerebrovascular disease with PET and SPECT. Q J Nucl Med 42:193–198

Haubner R, Avril N, Hantzopoulos PA et al (2000) In vivo imaging of herpes simplex virus type 1 thymidine kinase gene expression: early kinetics of radiolabelled FIAU. Eur J Nucl Med 27:283–291

Haubner R, Wester HJ, Burkhart F et al (2001a) Glycosylated RGD-containing peptides: tracer for tumor targeting and angiogenesis imaging with improved biokinetics. J Nucl Med 42:326–336

Haubner R, Wester HJ, Weber WA et al (2001b) Noninvasive imaging of alpha(v)beta3 integrin expression using 18F-labeled RGD-containing glycopeptide and positron emission tomography. Cancer Res 61:1781–1785

Heiss WD (2001) Imaging the ischemic penumbra and treatment effects by PET. Keio J Med 50:249–256

Hemminki A, Belousova N, Zinn KR et al (2001) An adenovirus with enhanced infectivity mediates molecular chemotherapy of ovarian cancer cells and allows imaging of gene expression. Mol Ther 4:223–231

Hocevar BA, Brown TL, Howe PH (1999) TGF-beta induces fibronectin synthesis through a c-Jun N-terminal kinase-dependent, Smad4-independent pathway. EMBO J 18:1345–1356

Hoekstra CJ, Paglianiti I, Hoekstra OS et al (2000) Monitoring response to therapy in cancer using [18F]-2-fluoro-2-deoxy-D-glucose and positron emission tomography: an overview of different analytical methods. Eur J Nucl Med 27:731–743

Hooper CE, Ansorge RE, Browne HM, Tomkins P (1990) CCD imaging of luciferase gene expression in single mammalian cells. J Biolumin Chemilumin 5:123–130

Hospers GA, Calogero A, Waarde A van et al (2000) Monitoring of herpes simplex virus thymidine kinase enzyme activity using positron emission tomography. Cancer Res 60:1488–1491

Hustinx R, Shiue CY, Alavi A et al (2001) Imaging in vivo herpes simplex virus thymidine kinase gene transfer to tumour-bearing rodents using positron emission tomography. Eur J Nucl Med 1:5–12

Ikeda K, Steger DJ, Eberharter A, Workman JL (1999) Activation domain-specific and general transcription stimulation by native histone acetyltransferase complexes. Mol Cell Biol 19:855–863

Inoue T, Tomiyoshi K, Higuichi T et al (1998) Biodistribution studies on L-3-[fluorine-18]fluoro-alpha-methyl tyrosine: a potential tumor-detecting agent. J Nucl Med 39:663–667

Ishiwata K, Ido T, Kawashima K et al (1984) Studies on 18F-labeled pyrimidines II. Metabolic investigation of 18F-5-fluorouracil, 18F-5-fluoro-2'-deoxyuridine and 18F-5-fluorouridine in rats. Eur J Nucl Med 9:185–189

Isselbacher KJ (1972) Sugar and amino acid transport by cells in culture: differences between normal and malignant cells. N Engl J Med 286:929–933

Ito H, Hatazawa J, Murakami M et al (1995) Aging effect on neurtral amino acid transport at the blood-brain barrier measured with L-[2-18F]-fluorophenylalanine and PET. J Nucl Med 36:1232–1237

Iwashina T, Tovell DR, Xu L et al (1988) Synthesis and antiviral activity of IVFRU, a potential probe for the non-invasive diagnosis of herpes simplex encephalitis. Drug Des Deliv 3:309–321

Iyer M, Barrio JR, Namavari M et al (2001a) 8-[18F]Fluoropenciclovir: an improved reporter probe for imaging HSV1-tk reporter gene expression in vivo using PET. J Nucl Med 42:96–105

Iyer M, Wu L, Carey M et al (2001b) Two-step transcriptional amplification as a method for imaging reporter gene expression using weak promoters. Proc Natl Acad Sci USA 98:14595–14600

Jackson RJ (1995) Internal initiation of translation in eukaryotes: the picornavirus paradigm and beyond (review). RNA 1:985–1000

Jacobs A, Dubrovin M, Hewett J et al (1999) Functional co-expression of HSV-1 thymidine kinase and green fluorescent protein: implications for noninvasive imaging of transgene expression. Neoplasia 1:154–161

Jacobs A, Tjuvajev JG, Dubrovin M et al (2001a) Positron emission tomography-based imaging of transgene expression mediated by replication-conditional, oncolytic herpes simplex virus type 1 mutant vectors in vivo. Cancer Res 61:2983–2995

Jacobs A, Bräunlich I, Graf R et al (2001b) Quantitative kinetics of [124I]FIAU in Cat and Man. J Nucl Med 42:467–475

Jolly D (1994) Viral vector systems for gene delivery. Cancer Gene Ther 1:51–64

Jones AK, Kitchen ND, Watabe H et al (1999) Measurement of changes in opioid receptor binding in vivo during trigeminal neuralgic pain using [11C] diprenorphine and positron emission tomography. J Cereb Blood Flow Metab 19:803–808

Jones T (1996) The imaging science of positron emission tomography. Eur J Nucl Med 23:807–813

Kallio PJ, Okamoto K, O'Brien S et al (1998) Signal transduction in hypoxic cells: inducible nuclear translocation and recruitment of the CBP/p300 coactivator by the hypoxia-inducible factor-1alpha. EMBO J 17:6573–6586

Kalofonos HP, Karamouzis MV, Epenetos AA (2001) Radioimmunoscintigraphy in patients with ovarian cancer. Acta Oncol 40:549 557

Kaltsas G, Korbonits M, Heintz E et al (2001) Comparison of somatostatin analog and meta-iodobenzylguanidine radionuclides in the diagnosis and localization of advanced neuroendocrine tumors. J Clin Endocrinol Metab 86:895–902

Kasper S, Tauscher J, Kufferle B et al (1999) Dopamine- and serotonin-receptors in schizophrenia: results of imaging-studies and implications for pharmacotherapy in schizophrenia. Eur Arch Psychiatry Clin Neurosci 249 [Suppl 4]:83–89

Kayyem JF, Kumar RM, Fraser SE, Meade TJ (1995) Receptor-targeted co-transport of DNA and magnetic resonance contrast agents. Chem Biol 2:615–620

Kiani A, Rao A, Aramburu J (2000) Manipulating immune responses with immunosuppressive agents that target NFAT. Immunity 12:359–372

Kling MA, Carson RE, Borg L et al (2000) Opioid receptor imaging with positron emission tomography and

[(18)F]cyclofoxy in long-term, methadone-treated former heroin addicts. J Pharmacol Exp Ther 295:1070–1076

Koepp MJ, Duncan JS (2000) PET: opiate neuroreceptor mapping. Adv Neurol 83:145–156

Koutcher JA, Matei C, Doubrovin M, Zakian K, Sadelain M, Tjuvajev JG (2000) In vivo 19F nuclear magnetic resonance measurements of enhanced conversion of 5FU to fluoronucleotides after UPRT gene transduction (abstract). 91th Annual Meeting, American Association for Cancer Research, San Francisco, 1–5 April 2000, p 379

Kretzschmar M, Doody J, Timokhina I, Massague J (1999) A mechanism of repression of TGFbeta/Smad signaling by oncogenic Ras. Genes Dev 13:804–816

Kriss J, Maruyama Y, Tung L, Bond S, Revesz L (1963) The fate of 5-bromodeoxyuridine, 5-bromodeoxycytidine, and 5-iododeoxycytidine in man. Cancer Res 23:260–268

Kubota K, Ishiwata K, Kubota R et al (1996) Feasibility of fluorine-18-fluorophenylalanine for tumor imaging compared with carbon-11-L-methionine. J Nucl Med 37:320–325

Liang Q, Satyamurthy N, Barrio JR et al (2001) Noninvasive, quantitative imaging in living animals of a mutant dopamine D2 receptor reporter gene in which ligand binding is uncoupled from signal transduction. Gene Ther 8:1490–1498

Liu F, Hata A, Baker JC et al (1996) A human protein acting as a BMP-regulated transcriptional activator. Nature 381:622–623

Loimas S, Wahlfors J, Janne J (1998) Herpes simplex virus thymidine kinase-green fluorescent protein fusion gene: new tool for gene transfer studies and gene therapy. Biotechniques 24:614–618

Louie AY, Huber MM, Ahrens ET et al (2000) In vivo visualization of gene expression using magnetic resonance imaging. Nat Biotechnol 18:321–325

Machida H, Watanabe Y, Kano F et al (1995) Deglycosylation of antiherpesviral 5-substituted arabinosyluracil derivatives by rat liver extract and enterobacteria cells. Biochem Pharmacol 49:763–766

MacLaren DC, Gambhir SS, Satyamurthy N et al (1999) Repetitive, non-invasive imaging of the dopamine D2 receptor as a reporter gene in living animals. Gene Ther 6:785–791

Mandell RB, Mandell LZ, Link CJ Jr (1999) Radioisotope concentrator gene therapy using the sodium/iodide symporter gene. Cancer Res 59:661–668

Mankoff DA, Shields AF, Graham MM et al (1998) Kinetic analysis of 2-[carbon-11]thymidine PET imaging studies: compartmental model and mathematical analysis. J Nucl Med 39:1043–1055

Mariani G (1999) Imaging of pancreatic adenocarcinoma with radiolabeled monoclonal antibodies. Ann Oncol 10 [Suppl 4]:37P–40P

Matheja P, Schober O (2001) 123I-IMT SPECT: introducing another research tool into clinical neuro-oncology? Eur J Nucl Med 28:1–4

Mathis C, Hindelang C, LeMeur M, Borrelli E (2000) A transgenic mouse model for inducible and reversible dysmyelination. J Neurosci 20:7698–7705

Mayerhofer R, Araki K, Szalay AA (1995) Monitoring of spatial expression of firefly luciferase in transformed zebrafish. J Biolumin Chemilumin 10:271–275

Mazaferri EL (2000) Carcinoma of the follicular epithelium. In: Braverman LE, Utiger RD (eds) The thyroid: a fundamental and clinical text. Lippincott-Raven, Philadelphia, pp 904–930

Miura S, Murakami M, Kanno I et al (1992) Phenylalanine transport in the living human brain by a dynamic PET of L-[2-18F]-fluorophenylalanine. Nipon Rinsho 50:1457–1460

Moffat FL Jr, Gulec SA, Serafini AN et al (1999) A thousand points of light or just dim bulbs? Radiolabeled antibodies and colorectal cancer imaging. Cancer Invest 17:322–334

Moolten FL (1986) Tumor chemosensitivity conferred by inserted herpes thymidine kinase gene: paradigm for a prospective cancer control strategy. Cancer Res 46:5276–5281

Morin KW, Atrazheva ED, Knaus EE, Wiebe LI (1997a) Synthesis and cellular uptake of 2′-substituted analogues of (E)-5-(2-[125I]iodovinyl)-2′-deoxyuridine in tumor cells transduced with the herpes simplex type-1 thymidine kinase gene. Evaluation as probes for monitoring gene therapy. J Med Chem 40:2184–2190

Morin KW, Knaus EE, Wiebe LI (1997b) Non-invasive scintigraphic monitoring of gene expression in a HSV-1 thymidine kinase gene therapy model. Nucl Med Commun 18:599–605

Morin KW, Knaus EE, Wiebe LI et al (2000) Reporter gene imaging: effects of ganciclovir treatment on nucleoside uptake, hypoxia and perfusion in a murine gene therapy tumour model that expresses herpes simplex type-1 thymidine kinase. Nucl Med Commun 21:129–137

Nagataki S, Yokohyama N (1996) Other factors regulating thyroid function. Autoregulation: effects of iodide. In: Braverman LE, Utiger RD (eds) The thyroid. Lippincott-Raven, Philadelphia, pp 241–247

Namavari M, Barrio JR, Toyokuni T et al (2000) Synthesis of 8-[18F]fluoroguanine derivatives: in vivo probes for imaging gene expression with positron emission tomography. Nucl Med Biol 27:157–162

Orlefors H, Sundin A, Ählstrom H et al (1998) Positron emission tomography with 5-hydroxytryptophan in neuroendocrine tumors. J Clin Oncol 16:2534–2541

Pelletier S (1988) Internal initiation of translation of eukaryotic mRNA directed by a sequence derived from poliovirus RNA. Nature 334:320

Phelps ME (2000) PET: the merging of biology and imaging into molecular imaging. J Nucl Med 41:661–681

Phillips JA, Craig SJ, Bayley D et al (1997) Pharmacokinetics, metabolism, and elimination of a 20-mer phosphorothioate oligodeoxynucleotide (CGP 69846A) after intravenous and subcutaneous administration. Biochem Pharmacol 54:657–68

Ponomarev V, Serganova I, Ageyeva L et al (2001a) A new reporter gene for multi-modality imaging: xanthine phosphoribosyl transferase and red fluorescent protein fusion [Abstract]. 48th Annual Meeting of the Society for Nuclear Medicine. Toronto, Canada, 23–27 June 2001, p 70

Ponomarev V, Doubrovin M, Lyddane C et al (2001b) Imaging TCR-dependent NFAT-mediated T-cell activation with positron emission tomography in vivo. Neoplasia 3:480–488

Price R, Cardle K, Watanabe K (1983) The use of antiviral drugs to image herpes encephalitis. Cancer Res 43:3619–3627

Prusoff W, Jaffe J, Gunther H (1960) Studies in the mouse of the pharmacology of 5-iododeoxyuridine, an analogue of thymidine. Biochem Pharmacol 3:110–121

Qiao J, Doubrovin M, Bernhard VS et al (2002) Tumor-specific transcriptional targeting of suicide gene therapy. Gene Ther 9:168–175

Rao A, Luo C, Hogan PG (1997) Transcription factors of the NFAT family: regulation and function. Annu Rev Immunol 15:707–747

Rehemtulla A, Stegman LD, Cardozo SJ et al (2000) Rapid and quantitative assessment of cancer treatment response using in vivo bioluminescence imaging. Neoplasia 2:491–495

Reichman U, Watanabe KA, Fox JJ (1975) A practical synthesis of 2′-deoxy-2′-fluoro-β-D-arabinofuranose derivatives. Carbohydr Res 42:233–240

Reivich M, Kuhl D, Wolf A et al (1977) Measurement of local cerebral glucose metabolism in man with 18F-2-fluoro-2-deoxy-d-glucose. Acta Neurol Scand 64 [Suppl]: 190P–191P

Reske SN, Kotzerke J (2001) FDG-PET for clinical use. Results of the 3rd German Interdisciplinary Consensus Conference, "Onko-PET III". Eur J Nucl Med 28:1707–1723

Reubi JC, Waser B, Schaer JC et al (1998) Unsulfated DTPA- and DOTA-CCK analogs as specific high-affinity ligands for CCK-B receptor-expressing human and rat tissues in vitro and in vivo. Eur J Nucl Med 5:481–490

Robbins PD, Ghivizzani SC (1998) Viral vectors for gene therapy. Pharmacol Ther 80:35–47

Roelcke U, Radu E, Ametamey S et al (1996) Association of rubidium and C-methionine uptake in brain tumors measured by positron emission tomography. J Neurooncol 27:163–171

Rogers BE, McLean SF, Kirkman RL et al (1999) In vivo localization of [(111)In]-DTPA-D-Phe1-octreotide to human ovarian tumor xenografts induced to express the somatostatin receptor subtype 2 using an adenoviral vector. Clin Cancer Res 5:383–393

Rogers BE, Zinn KR, Buchsbaum DJ (2000) Gene transfer strategies for improving radiolabeled peptide imaging and therapy. Q J Nucl Med 44:208–223

Rogulski KR, Kim JH, Kim SH, Freytag SO (1997) Glioma cells transduced with an Escherichia coli CD/HSV-1 TK fusion gene exhibit enhanced metabolic suicide and radiosensitivity. Hum Gene Ther 8:73–85

Rogulski KR, Zhang K, Kolozsvary A, Kim JH, Freytag SO (1997) Pronounced antitumor effects and tumor radiosensitization of double suicide gene therapy. Clin Cancer Res 3:2081–2088

Rogulski KR, Wing MS, Paielli DL, Gilbert JD, Kim JH, Freytag SO (2000) Double suicide gene therapy augments the antitumor activity of a replication-competent lytic adenovirus through enhanced cytotoxicity and radiosensitization. Hum Gene Ther 11:67–76

Sachs AB, Sarnow P, Hentze MW (1997) Starting at the beginning, middle, and end: translation initiation in eukaryotes. Cell 89:831–838

Sadar MD (1999) Androgen-independent induction of prostate-specific antigen gene expression via cross-talk between the androgen receptor and protein kinase A signal transduction pathways. J Biol Chem 274:7777–7783

Saier MH Jr, Daniels GA, Boerner P, Lin J (1988) Neutral amino acid transport systems in animal cells: potential targets of oncogene action and regulators of cellular growth. J Membr Biol 104:1–20

Saito Y, Price RW, Rottenberg DA et al (1982) Quantitative autoradiographic mapping of herpes simplex virus encephalitis with radiolabeled antiviral drug. Science 217:1151–1153

Saito Y, Rubenstein R, Price RW et al (1984) Diagnostic imaging of herpes simplex virus encephalitis using radiolabeled antiviral drug: autoradiographic assessment in animal model. Ann Neurol 15:548–558

Sanabria-Bohorquez SM, Labar D, Leveque P et al (2000) [11C]flumazenil metabolite measurement in plasma is not necessary for accurate brain benzodiazepine receptor quantification. Eur J Nucl Med 27:1674–1683

Schmutzler C, Kohrle J (2000) Retinoic acid redifferentiation therapy for thyroid cancer. Thyroid 10:393–406

Schneider JS, Lidsky TI, Hawks T et al (1995) Differential recovery of volitional motor function, lateralized cognitive function, dopamine agonist-induced rotation and dopaminergic parameters in monkeys made hemi-parkinsonian by intracarotid MPTP infusion. Brain Res 672:112–117

Selden RF, Skoskiewicz MJ, Howie KB, Russell PS, Goodman HM (1987) Implantation of genetically engineered fibroblasts into mice: implications for gene therapy. Science 36:714–718

Sharif TR, Sharif M (1999) A high throughput system for the evaluation of protein kinase C inhibitors based on Elk1 transcriptional activation in human astrocytoma cells. Int J Oncol 14:327–335

Shields AF, Lim K, Grierson J, Link J, Krohn KA (1990) Utilization of labeled thymidine in DNA synthesis: studies for PET. J Nucl Med 31:337–342

Shields AF, Graham MM, Kozawa SM et al (1992) Contribution of labeled carbon dioxide to PET imaging of carbon-11-labeled compounds. J Nucl Med 33:581–584

Shields AF, Mankoff D, Graham MM et al (1996a) Analysis of 2-carbon-11-thymidine blood metabolites in PET imaging. J Nucl Med 37:290–296

Shields AF, Grierson JR, Kozawa SM, Zheng M (1996b) Development of labeled thymidine analogs for imaging tumor proliferation. Nucl Med Biol 23:17–22

Shields AF, Grierson JR, Dohmen BM et al (1998) Imaging proliferation in vivo with [F-18]FLT and positron emission tomography. Nat Med 4:1334–1336

Shiue GG, Shiue CY, Lee RL et al (2001) Simplified one-pot synthesis of 9-[(3-[18F]fluoro-1-hydroxy-2-propoxy)-methyl]guanine([18F]FHPG) and 9-(4-[18F]fluoro-3-hydroxymethylbutyl)guanine ([18F]FHBG) for gene therapy. Nucl Med Biol 28:875–883

Shoup TM, Olson J, Hoffman JM et al (1999) Synthesis and evaluation of [¹⁸F]1-amino-3-fluorocyclobutane-1-carboxylic acid to image brain tumors. J Nucl Med 40:331–338

Smanik PA, Liu Q, Furminger TL, Ryu K, Xing S, Mazzaferri EL, Jhiang SM (1996) Cloning of the human sodium iodide symporter. Biochem Biophys Res Commun 226:339–345

Sokoloff L, Reivich M, Kennedy C et al (1977) The [14C]deoxyglucose method for the measurement of local cerebral glucose utilization: theory, procedure, and normal values in the conscious and anesthetized albino rat. J Neurochem 28:897–916

Souba WW (1993) Glutamine and cancer. Ann Surg 218:715–728

Spitzweg C, Joba W, Eisenmenger W, Heufelder AE (1998) Analysis of human sodium iodide symporter gene expression in extrathyroidal tissues and cloning of its complementary deoxyribonucleic acids from salivary gland, mammary gland, and gastric mucosa. J Clin Endocrinol Metab 83:1746–1751

Spitzweg C, Zhang S, Bergert ER et al (1999) Prostate-specific antigen (PSA) promoter-driven androgen-inducible expression of sodium iodide symporter in prostate cancer cell lines. Cancer Res 59:2136–2141

Spitzweg C, O'Connor MK, Bergert ER et al (2000) Treatment of prostate cancer by radioiodine therapy after tissue-specific expression of the sodium iodide symporter. Cancer Res 60:6526–6530

Stalteri MA, Mather SJ (2001) Hybridization and cell uptake studies with radiolabelled antisense oligonucleotides. Nucl Med Commun 22:1171–1179

Stegman LD, Rehemtulla A, Beattie B et al (1999) Noninvasive quantitation of cytosine deaminase transgene expression in human tumor xenografts with in vivo magnetic resonance spectroscopy. Proc Natl Acad Sci USA 96:9821–9826

Stryer L (1995) Biochemistry, 4th edn. Freeman, New York

Tan M, Li S, Swaroop M et al (1999) Transcriptional activation of the human glutathione peroxidase promoter by p53. J Biol Chem 274:12061–12066

Tang HY, Zhao K, Pizzolato JF et al (1998) Constitutive expression of the cyclin-dependent kinase inhibitor p21 is transcriptionally regulated by the tumor suppressor protein p53. J Biol Chem 273:29156–29156

Taurog AM (2000) Thyroid hormone synthesis. In: Braverman LE, Utiger RD (eds) The thyroid: a fundamental and clinical text. Lippincott-Raven, Philadelphia, pp 75–79

Tavitian B (2000) In vivo antisense imaging. Q J Nucl Med 44:236–255

Tavitian B, Terrazzino S, Kuhnast B et al (1998) In vivo imaging of oligonucleotides with positron emission tomography. Nat Med 4:467–471

Tazebay UH, Wapnir IL, Levy O et al (2000) The mammary gland iodide transporter is expressed during lactation and in breast cancer. Nat Med 6:871–878

Tiraby M, Cazaux C, Baron M, Drocourt D, Reynes JP, Tiraby G (1998) Concomitant expression of E. coli cytosine deaminase and uracil phosphoribosyltransferase improves the cytotoxicity of 5-fluorocytosine. FEMS Microbiol Lett 167:41–49

Tjuvajev JG, Macapinlac HA, Daghighian F et al (1994) Imaging of brain tumor proliferative activity with iodine-131-iododeoxyuridine. J Nucl Med 35:1407–1417

Tjuvajev JG, Stockhammer G, Desai R et al (1995) Imaging the expression of transfected genes in vivo. Cancer Res 55:6126–6132

Tjuvajev JG, Finn R, Watanabe K et al (1996) Noninvasive imaging of herpes virus thymidine kinase gene transfer and expression: a potential method for monitoring clinical gene therapy. Cancer Res 56:4087–4095

Tjuvajev JG, Gallardo H, Joshi A et al (1997a) Bi-cistronic co-expression of a cell surface marker gene and the HSV1-tk gene in jurkat cells: Implications for noninvasive imaging in vivo. Cancer Gene Ther 4 [Suppl]:44P

Tjuvajev JG, Avril N, Lindsley L et al (1997b) Monitoring the expression of therapeutic genes by noninvasive imaging of the HSV1-tk marker gene using double-gene vector constructs. Cancer Gene Ther 4 [Suppl]:43P–44P

Tjuvajev JG, Avril N, Oku T et al (1998) Imaging herpes virus thymidine kinase gene transfer and expression by positron emission tomography. Cancer Res 58:4333–4341

Tjuvajev JG, Chen SH, Joshi A et al (1999a) Imaging adenoviral-mediated herpes virus thymidine kinase gene transfer and expression in vivo. Cancer Res 59:5186–5193

Tjuvajev JG, Joshi A, Callegari J et al (1999b) A general approach to the non-invasive imaging of transgenes using cis-linked herpes simplex virus thymidine kinase. Neoplasia 1:315–320

Tovell D, Yacyshyn H, Misra H et al (1987) Effect of acyclovir on the uptake of ^{131}I-labelled 1-(2'fluoro-2'-deoxy-beta-D-arabinofuranosyl)-5-iodouracil in herpes infected cells. J Med Virol 22:183–188

Tovell DR, Samuel J, Mercer JR et al (1988) The in vitro evaluation of nucleoside analogues as probes for use in the noninvasive diagnosis of herpes simplex encephalitis. Drug Des Deliv 3:213–221

Tsurumi Y, Kameyama M, Ishiwata K et al (1990) 18F-fluoro-2'-deoxyuridine as a tracer of nucleic acid metabolism in brain tumors. J Neurosurg 72:110–113

Ur E, Bomanji J, Mather SJ et al (1993) Localization of neuroendocrine tumours and insulinomas using radiolabelled somatostatin analogues, 123I-Tyr3-octreotide and 111In-pentatreotide. Clin Endocrinol (Oxf) 38:501–506

Vander Borght T, Pauwels S, Lambotte L et al (1994) Brain tumor imaging with PET and 2-[carbon-11]thymidine. J Nucl Med 35:974–982

Varma S, Cohen HJ (1997) Co-transactivation of the 3' erythropoietin hypoxia inducible enhancer by the HIF-1 protein. Blood Cells Mol Dis 2:169–176

Vassart G, Dumont JE (1992) The thyrotropin receptor and the regulation of thyrocyte function and growth. Endocr Rev 13:596–611

Venkataraman GM, Yatin M, Marcinek R, Ain KB (1999) Restoration of iodide uptake in dedifferentiated thyroid carcinoma: relationship to human Na$^+$/I$^-$ symporter gene methylation status. J Clin Endocrinol Metab 84:2449–2457

Virgolini I, Raderer M, Kurtaran A et al (1994) Vasoactive intestinal peptide-receptor imaging for the localization of intestinal adenocarcinomas and endocrine tumors. N Engl J Med 331:1116–1121

Weiss SJ, Philp NJ, Grollman EF (1984a) Iodide transport in a continuous line of cultured cells from rat thyroid. Endocrinology 114:1090–1098

Weiss SJ, Philp NJ, Ambesi-Impiombato FS, Grollman EF (1984b) Thyrotropin-stimulated iodide transport mediated by adenosine 3′,5′-monophosphate and dependent on protein synthesis. Endocrinology 114:1099–1107

Weissleder R, Reimer P, Lee AS, Wittenberg J, Brady TJ (1990) MR receptor imaging: ultrasmall iron oxide particles targeted to asialoglycoprotein receptors. Am J Roentgenol 155:1161–1167

Weissleder R, Lee AS, Fischman AJ et al (1991) Radiology 181:245–249

Weissleder R, Simonova M, Bogdanova A et al (1997) MR imaging and scintigraphy of gene expression through melanin induction. Radiology 204:425–429

Welch A, Jaffe J, Cardoso S et al (1960) Studies on the pharmacology of 5-iododeoxyuridine in animals and man. Proc Am Assoc Cancer Res 3:161

Welsh S, Kay SA (1997) Reporter gene expression for monitoring gene transfer. Curr Opin Biotechnol 8:617–622

Wester HJ, Herz M, Weber W et al (1999a) Synthesis and radiopharmacology of O-(2-[^{18}F]fluoroethyl)-L-tyrosine for tumor imaging. J Nucl Med 40:205–212

Wester HJ, Herz M, Senekowitsch-Schmidtke R et al (1999b) Preclinical evaluation of 4-[^{18}F]fluoroprolines: diastereomeric effect on metabolism and uptake in mice. Nucl Med Biol 26:259–265

Wiebe LI, Knaus EE, Morin KW (1999) Radiolabelled pyrimidine nucleosides to monitor the expression of HSV-1 thymidine kinase in gene therapy. Nucleosides Nucleotides 18:1065–1066

Wienhard K, Herholz K, Coenen HH et al (1991) Increased amino acid transport into brain tumor measured by PET of L-(2–18F)Fluorotyrosine. J Nucl Med 32:1338–1346

Willis MC, Collins BD, Zhang T et al (1998) Liposome-anchored vascular endothelial growth factor aptamers. Bioconjug Chem 9:573–582

Willoch F, Tolle TR, Wester HJ et al (1999) Central pain after pontine infarction is associated with changes in opioid receptor binding: a PET study with 11C-diprenorphine. Am J Neuroradiol 20:686–690

Wotton D, Lo RS, Lee S, Massague J (1999) A Smad transcriptional corepressor. Cell 97:29–39

Wu AM, Yazaki PJ (2000) Designer genes: recombinant antibody fragments for biological imaging. Q J Nucl Med 44:268–283

Yaghoubi SS, Wu L, Liang Q et al (2001a) Direct correlation between positron emission tomographic images of two reporter genes delivered by two distinct adenoviral vectors. Gene Ther 8:1072–1780

Yaghoubi SS, Barrio JR, Dahlbom M et al (2001b) Human pharmacokinetic and dosimetry studies of [^{18}F]FHBG: a reporter probe for imaging herpes simplex virus type-1 thymidine kinase reporter gene expression. J Nucl Med 42:1225–1234

Young H, Baum R, Cremerius U et al (1999) Measurement of clinical and subclinical tumour response using [18F]-fluorodeoxyglucose and positron emission tomography: review and 1999 EORTC recommendations. European Organization for Research and Treatment of Cancer (EORTC) PET Study Group. Eur J Cancer 35:1773–1782

Yu Y, Annala AJ, Barrio JR et al (2000) Quantification of target gene expression by imaging reporter gene expression in living animals. Nat Med 6:933–937

Zubieta JK, Koeppe RA, Frey KA et al (2001) Assessment of muscarinic receptor concentrations in aging and Alzheimer disease with [11C]NMPB and PET. Synapse 39:275–287

Development of DNA-based Radiopharmaceuticals Carrying Auger-Electron Emitters for Anti-gene Radiotherapy

29

IGOR G. PANYUTIN, THOMAS A. WINTERS, LUDWIG E. FEINENDEGEN, RONALD D. NEUMANN

Contents

29.1
Introduction

Most gene therapy methods currently used to correct gene expression rely on either gene introduction, modification, replacement or antisense approaches. Typically, gene therapy seeks to add-back missing functionality. Antisense therapy is employed to reduce or eliminate expression of specific gene products. A more direct manipulation of gene function at the DNA level may be advantageous in circumstances where gene replacement alone is not sufficient to prevent or treat disease due to continued expression of defective gene products, or for cases in which continuous long-term antisense treatment is impractical or impossible. Consequently, exploration of approaches to change gene function directly at the DNA level may, in some cases, provide an important alternative to current methods of gene therapy. Targeting specific genes unique to cancer cells may also be a way to selectively eliminate such cells via induced cell necrosis or apoptosis; and, perhaps, even a way to genetically "correct" some aspect of neoplastic cell behavior to cause reversion to a non-neoplastic phenotype.

29.2
Background

Cancer cells can be selectively destroyed by internal radionuclide radiotherapy, as is evident from the success of radioiodine therapy for thyroid cancer. Radionuclide treatments are designed to irradiate tumor cells yet minimize the dose received by normal cells. The radionuclide must be chosen for its range of likely effect, i.e., the emitted energy should be maximally absorbed within the DNA of the targeted tumor cells. Auger-electron-emitting radionuclides, or Auger emitters (A Ettr), are ideally suited for intranuclear radionuclide therapy, provided that they can be positioned close to the genomic DNA (Feinendegen 1975). Decay of an A Ettr produces a cascade of low energy electrons named after the French physicist Pierre Auger who first discovered them in 1929. For example, the decay of ^{125}I results in an average emission of about 20 electrons of varying energy in addition to a gamma photon; producing a highly positively charged tellurium daughter atom. Most of the Auger electrons from ^{125}I have energies below 1 keV and ranges less than several nanometers. The simultaneous action of the low energy electrons produces a high-density energy disposition in the vicinity of the decay site, and in that way resembles an alpha particle effect (Fig. 29.1). Significant damage in adjacent molecular structures is the immediate effect. Decay of ^{125}I incorporated into one strand of a DNA duplex produces strand breaks located within 10 bp from the decay site with an efficiency close to one double strand break (DSB) per decay (Martin and Haseltine 1981).

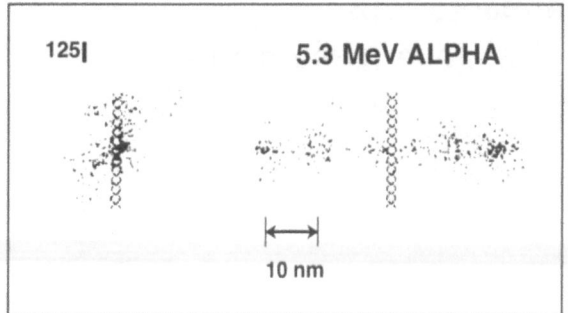

Fig. 29.1. Reactive chemical species produced by decay of ^{125}I and 5.3 MeV alpha particle. Helix represents DNA. (from Wright et al. 1990)

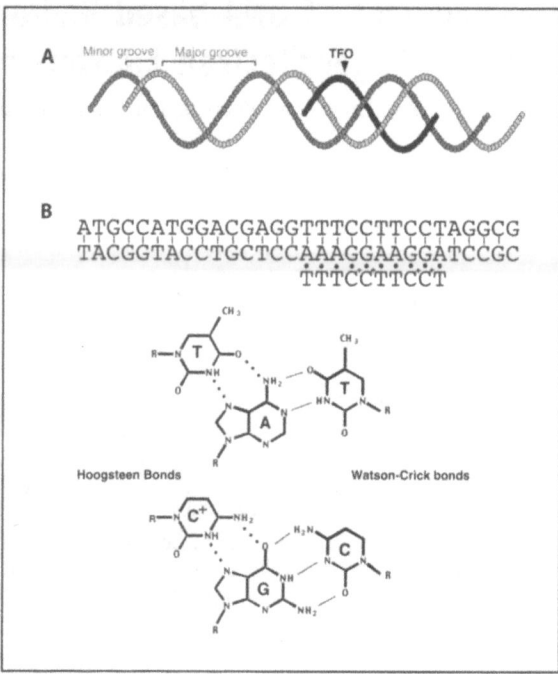

Fig. 29.2. Triplex-forming oligonucleotides in the major groove of the duplex (A), third strand sequence recognition (B) T-A·T and C-G·C$^+$ triads stabilized by Watson-Crick and Hoogsteen hydrogen bonds

This is in contrast to more typical ionizing radiation with more widely distributed direct and indirect effects mediated in part by reactive oxygen species.

Various "carrier" molecules have been used to deliver A Ettr to tumor cells. A precursor of DNA synthesis, I-UdR (Feinendegen et al. 1966) and a DNA receptor-ligand, estradiol (DeSombre et al. 1992), are currently used. But the full potential of Auger effect (AE) therapy is not realized if the A Ettr, at the moment of radiodecay, is too far from the genomic DNA; or, in other words, if the DNA is "out of reach" of the Auger effect. In 1994 we proposed a completely new approach that we call anti-gene radiotherapy (AR) to address this problem. In our design the target is not the total DNA of a cell, but the specific DNA sequence of a gene within the genome of the cell. Anti-gene radiotherapy optimally utilizes the sub-nanometer effect range of A Ettr to allow targeting of most of the radiodamage to a selected gene sequence while producing minimal damage to the rest of the genome and other cell components. Clearly, AR requires a carrier molecule that exhibits enough specificity for a selected DNA sequence to deliver the A Ettr to that specific sequence and not to other sites in the genome. As our initial carrier molecule we selected short synthetic oligonucleotides that are able to form a sequence-specific triple helix with the target sequence, so-called triplex-forming oligonucleotides (TFO; Helene 1991).

Synthetic oligodeoxyribonucleotides (ODN) have drawn much attention in recent years as tools for manipulation of gene expression. Investigators have shown that after entering a cell and binding to a DNA target (the anti-gene approach) ODNs are capable of altering expression of the targeted gene (reviewed in Helene 1991). In anti-gene experiments the

ODN is thought to form a triple helix (or triplex) with the target DNA duplex. It is well established that at normal (physiological) conditions two complementary DNA strands form a double helix of the B-form type proposed by Watson and Crick. In this form purines (A and G) and pyrimidines (T and C) are bound via the Watson-Crick hydrogen bond scheme resulting in AT and GC base pairs. However, interactions between complementary DNA strands are not limited to the B-form. Some DNA sequences under certain circumstances can form alternative structures such as left-handed Z-DNA, triple- and four-stranded structures (van Holde and Zlatanova 1994; Frank-Kamenetskii and Mirkin 1995; Soyfer and Potoman 1995; Wells et al. 1998).

In triple-stranded DNA (or triplexes) the third strand is located in the major grove of the B-DNA duplex and is stabilized by so-called Hoogsteen hydrogen bonds with purines of the duplex (Fig. 29.2) (Frank-Kamenetskii and Mirkin 1995; Soyfer and Potoman 1995). Thus, the bases of the DNA triplex form triads stabilized by Watson-Crick (duplex) and Hoogsteen (third strand) hydrogen bonds. In the center of all known triads are always the purines since the pyrimidines do not have enough donor and acceptor groups to form both Watson-Crick and Hoogsteen hydrogen bonds. Therefore, classical triplexes can be formed with a target duplex that has one strand consisting of only purines and the other, correspondingly, pyrimidines in order to avoid energetically unfavorable strand switch of the Hoogsteen bonds between strands of the duplex.

Depending upon whether the third strand consists of purines or pyrimidines triplexes can be divided into two groups: purine $(Y-R \cdot R)$ and pyrimidine $(Y-R \cdot Y)$ triplexes (the dash represents the Watson-Crick bonds and the dot represents the Hoogsteen bonds; Frank-Kamenetskii and Mirkin 1995). The canonical base triads of the pyrimidine triplex are $T-A \cdot T$ and $C-G \cdot C^+$. In order to form Hoogsteen bonds C in the third strand must be protonated (Mirkin and Frank-Kamenetskii 1994). As a result, the pyrimidine triplexes are most stable at low pH. The canonical base triads for the purine triplex are $C-G \cdot G$ and $T-A \cdot A$. Bivalent metal cations, such as Mg^{++} or Mn^{++}, are required for stabilization of the purine triplexes. Other base triads such as $T-A \cdot T$ and $C-G \cdot A^+$ are also possible in the purine triplexes. In fact, under certain conditions, the stability of the triplexes built of $C-G \cdot G$ and $C-G \cdot T$ triads is greater than that of $C-G \cdot G$ and $T-A \cdot A$ triads. Therefore, since T can be present in the third strand of the purine triplex, such structures are often referred to as purine-motif triplexes.

Triplex DNA structures were first discovered more than 40 years ago by Felsenfeld, Davies and Rich in mixtures of RNA polymers such as poly(A)/poly(U). But only recently was it realized that short synthetic homopyrimidine or purine-rich oligonucleotides can be designed to form sequence-specific triplexes with corresponding homopurine-homopyrimidine sites on duplex DNA. This immediately attracted a great deal of attention when it became clear that such triplex-forming oligonucleotides (TFO) might be universal drugs that exhibit sequence-specific recognition of duplex DNA (Giovannangeli and Helene 1997). The high specificity of TFO-DNA recognition, which is comparable to that of complementary strands pairing in Watson-Crick duplexes, has led to the development of an "antigene" strategy, the goal of which is to modulate gene activity using TFO.

For the past 10 years much work was done to improve triplex stability and the specificity of the TFO binding. Chemical modifications of both TFO sugar-phosphate backbone and bases have been developed that significantly increase TFO affinity to the target and prevent TFO self-association. New triplex-specific intercalators have been developed that provide additional stabilization of a TFO to the target. Numerous studies were also focused on the mode of action of the TFO. It was shown that a triplex formation can block transcription and replication at the target sequence. In order to enhance the antigene activity of the TFO, they were modified with chemical- and photo-activated groups (Helene 1991; Giovannangeli et al. 1997; Chan and Glazer 1997).

The requirement for a polypurine-polypyrimidine target sequence imposes a specific limitation on the genomic sites that can be effectively targeted by TFO. Though significant efforts have been directed to overcome this limitation, the problem is still unsolved in general (Chan and Glazer 1997). Fortunately, polypurine-polypyrimidine sequences are ubiquitous in eukaryotic genomes and are often found in important regulatory regions of genes. They are believed to play a role in regulation of DNA functions but the exact mechanism is yet to be established. There is also no recipe for selecting the best binding TFO for a given polypurine-polypyrimidine sequence. The general rule is that to form a stable triplex the target sequence must not only be homopurine but also G-rich. There are also several other principles for the TFO design, but researchers commonly have to screen several candidates to select the best binding TFO for a target.

TFO have recently been demonstrated to be effective vehicles for sequence-specific delivery of DNA-damaging agents such as TFO-bound psoralen and nitrogen mustards to targets within DNA (Kraemer and Seidman 1989; Giovannangeli et al. 1997; Lukhtanov et al. 1997; Belousov et al. 1998; Majumdar et al. 1998). These studies prove the ability of TFOs to mediate highly precise positioning of mutagenic compounds. However, the lesions produced by chemical mutagens are repaired efficiently, and thus result in relatively low overall mutation frequencies. Consequently, even if TFO-mediated targeting is efficient, modulation of target gene activity will only occur infrequently. These results illustrate the dependence of effective TFO-mediated anti-gene targeting on the biological response to the type of damage produced in the target DNA. Therefore, effective TFO-mediated anti-gene strategies should employ a TFO-bound DNA-damaging agent that produces complex lesions (preferably involving a loss of coding information) which invoke cellular DNA repair mechanisms with low repair fidelity.

Such lesions are exemplified by multiply damaged sites (MDS) involving DNA DSB. These DSBs are a high frequency product of high-LET radiation (Howell et al. 1994; Jones et al. 1994; Ward 1995). Therefore, the highly localized high-LET-like damage produced by the AE fulfills these requirements perfectly.

Decay of A Ettr delivered by TFO to a specific DNA sequence should produce damage mainly localized in this particular sequence, primarily in the form of DNA DSB. Attempts at repair of this damage by cellular enzymes appear to cause gene-specific mutations and result in a "knock-out" of the targeted gene (Gibbs et al. 1987; Ward 1994; Mezhevaya et al. 1999; Purmal et al. 1994). DSB produced by the AE are known to be highly cytotoxic. Therefore, cells containing target DNA sequences for triplex formation should be significantly more sensitive to damage from the A Ettr-labeled TFO (A Ettr-TFO) than cells that do not contain the target sequence. For example, if the target sequence selected is part of a viral genome integrated into mammalian genomic DNA or the target sequence appears as a result of genomic rearrangements and/or amplifications often associated with cancers, then A Ettr-TFO directed against such sequences will specifically kill or mutate these cells.

In contrast, current radiotherapies employing low-LET radiation such as electrons, β-particles, X- and γ-radiation cause cellular biologic effects largely by indirect mechanisms involving reactive oxygen species (ROS), such as OH radicals, that are formed primarily by water radiolysis. Direct effects on DNA are less frequent with low-LET radiation. Base modifications and single-strand breaks (SSB) produced by low-LET radiation are repaired more efficiently than DSB (Ward 1994; Chaudhry and Weinfeld 1995). The method of AR we are proposing is of a highly rational nature, permitting case-specific design of TFO carrier molecules based upon preselected intragenic target sequences, allowing the production of easily identifiable radiation effects within the target region. However, an A Ettr located in DNA produces per decay a most complex set of DNA lesions. Most of these lesions from A Ettr in general, and ^{125}I in particular, can resemble respective lesions from low- and high-LET radiation, yet are expected also to differ because of their particular spatial-temporal creation from an A Ettr decay (Painter et al. 1974; Feinendegen et al. 1977; Sundell-Bergman and Johanson 1980; Pomplun et al. 1996). AR's effectiveness as a therapy is strictly dependent upon the biological response to the AE. Since the immediate effect of TFO-mediated AR is AE-induced DNA damage, cellular mechanisms of DNA repair are the primary biological response by

which this method generates biological effects. Consequently, prior to designing and instituting AR in vivo trials, we require a better understanding of cellular DNA repair responses to AE-induced DNA damage.

29.3
Development of a Plasmid Model

The first step in our AR research required a demonstration that decay of A Ettr in a triplex indeed produced DNA DSB with reasonable efficiency. For this purpose we designed an experimental system based on short circular plasmid DNA that can be easily cloned and purified in sufficient amount for in vitro experiments. We inserted into the plasmid a polypurine·polypyrimidine sequence from the *nef* gene of HIV and designed a pyrimidine TFO that could specifically bind to this sequence (Panyutin and Neumann 1994). We developed a primer extension method of labeling TFO with commercially available ^{125}I-dCTP and a procedure for subsequent purification of the labeled product from the reaction mixture using magnetic beads. Iodine-125 is a classical A Ettr, both well-studied and easily available; thus, we used it in most of our in vitro experiments. We showed that after being incorporated into TFO, carried into a triplex, and, thus, located in trans relative to the double helix, ^{125}I was able to produce DSB with an efficiency of about 0.8 DSB per decay. The DSB were distributed within a very short molecular distance (±5 base pairs or ±1.7 nm) around the radionuclide decay site. This was the first demonstration of site-specific cleavage of duplex DNA by the decay of ^{125}I delivered to the target sequence by TFO, and the first direct biochemical confirmation of the "one-break-per-one-decay" dogma.

To further examine the efficiency of DNA cleavage per TFO, we next tried our approach on another plasmid model containing a polypurine·polypyrimidine stretch of the human HPRT gene and multiple-A Ettr-labeled TFO (Panyutin and Neumann 1996). A GT-motif TFO was designed to specifically bind to this HPRT sequence and was labeled with (on average) two ^{125}I-dCTP per TFO. Decay of ^{125}I in the bound TFO was shown to cause sequence-specific DSB in the target HPRT sequence cloned into the plasmid DNA. No sequence-specific breaks were observed if ^{125}I-labeled TFO were not bound to the plasmid DNA. After 60 days of decay (one ^{125}I half-life) approximately a quarter of all plasmid molecules contained sequence-specific DSB, corresponding to about 0.3 site-specific DSB per decay. Sequencing gel analysis showed that the DNA breaks were distributed within a few bases of the maxima at those DNA bases opposite to the positions of ^{125}I in the TFO.

By comparing the data on the pyrimidine triplex (HIV sequence-containing plasmid) and the purine triplex (HPRT sequence-containing plasmid) it became clear that the efficiency of breaks per decay strongly depends on the position of the radioiodine vis à vis the nucleic acid phosphodiester backbone within the triplex structure. This observation was recently confirmed theoretically by one of our collaborators. They used a molecular model of triplex DNA with a known position of ^{125}I, Monte Carlo-calculated electron tract structures and ensuing chemistry codes to simulate the distribution of breaks in both DNA strands. Based on our earlier comparison of our experimental and theoretical data we proposed a new method of studying DNA conformation that we call "radioprobing" (Panyutin and Neumann 1997; Karamychev et al. 1999).

29.4
Targeting a Single Copy Gene in Genomic DNA

The next important test of our AR approach was to demonstrate that the TFO could efficiently bind and cleave a target sequence within the context of the total genomic DNA in vitro. For this, total DNA was purified from two human cell lines and mixed in a test tube with ^{125}I-TFO designed to target the HPRT gene that is present as a single copy in the human genome. The Southern blot analysis revealed sequence-specific breaks in the HPRT target sequence demonstrating that TFO were able to target a unique sequence out of the total DNA in human cells (Panyutin and Neumann 1996). The method we developed for the detection of ^{125}I-TFO-produced breaks in the genomic DNA was used later for cell culture experiments.

29.5
Delivery of TFO into Cells

An important step in developing AR is the efficient delivery of ^{125}I-TFO into cells, and specifically to the target DNA sequence within a cell nucleus. We tried a wide variety of intracellular delivery methods for nucleic acids developed by others for gene insertion, replacement and antisense therapies. These methods include: permeabilization of the cell membrane, electroporation, and liposome-mediated delivery. All these methods were optimized for the extremely low concentrations of ^{125}I-TFO that we used. To measure cellular uptake we usually used radioiodinated TFO, while for the studies of intracellular distribution, we used fluorescently labeled TFO (Delporte et al. 1997; Sedelnikova et al. 1998, 1999). Several experiments showed that the best cellular uptake, i.e., up to 40% of added TFO, could be obtained with a specific formulation of cationic liposomes. We also characterized the liposome formulations and conditions of treatment so that nearly all intracellular TFO were released from the endosomal vesicles and localized predominantly in cell nuclei.

In the course of these delivery studies we measured the nonspecific, i.e., non-bound TFO, radiotoxicity of ^{125}I-TFO (Sedelnikova et al. 1998). We found a dramatic difference in radiotoxicity between ^{125}I-TFO that were delivered into a nucleus, but not bound to genomic DNA, and the DNA-incorporated ^{125}I-IUdR. This observation of lesser toxicity confirms that despite intranuclear localization, ^{125}I-ODN are not highly radiotoxic if not in close proximity to the genomic DNA. Therefore, relatively high intracellular concentrations of unbound ^{125}I-TFO can be tolerated without causing significant cell death due to the overall nonspecific radiation dose to the cells.

29.6
Development of TFO for In Vivo Experiments

Using ^{125}I as our A Ettr we proved the principle of anti-gene radiotherapy in vitro, by demonstrating that ^{125}I-labeled TFO were able to produce sequence-specific breaks in purified plasmids and genomic DNA. However, the relatively long half-life of ^{125}I (60 days) impedes the use of this radionuclide for any in vivo trials of AR. For experiments in animal models and any future clinical trials, radionuclides with physical half-lives more comparable to the biological half-life of oligonucleotides in vivo (which is in the range of hours) are required.

The phosphodiester type of TFO that we used in vitro has a very short biological half-life in vivo due to their susceptibility to naturally occurring nucleases. In order to extend the biological half-life of TFO, different chemical modifications of their backbone and bases have been suggested. Certain chemical modifications of TFO were also shown to significantly stabilize triplexes (Chan and Glazer 1997).

We did several experiments to test triplex formation with different TFO formulations and to develop methods of labeling these TFO with other A Ettr (Karamychev et al. 1997, 2000b, 2000c; Reed et al. 1997). Except for ^{125}I, ^{123}I and ^{77}Br most of the available A Ettr are radiometals. Therefore, we developed a method to incorporate ^{111}In into TFO through the chelating agent DTPA that can be easily adapted for other radiometallic A Ettr. In the course of these studies we measured for the first time the yield and distribution of direct DNA breaks produced by ^{123}I and ^{111}In (Karamychev et al. 2000c). These measurements crucially help us to understand the nature of the Auger effect and the molecular mechanisms of DNA damage by A Ettr.

29.7
Targeted DNA Breaks in Cell Nuclei and Permeabilized Cells

For these in situ cell culture experiments we used as a target the multidrug-resistance gene (mdr1). This gene is about 50 times amplified in the human KB-VI carcinoma cell line (i.e. these cells have on average 50 copies of the mdr1 vs. two copies in a normal cell). Our choice of an amplified DNA target sequence was to increase the chances of assaying any specific binding. In addition, mdr1 is a very desirable target in the future in vivo trials of AR. The problem of multidrug resistance is a major problem in cancer chemotherapy. Knocking out the mdr1 gene by AR may help to overcome resistance to some of the most commonly used anticancer drugs (Fojo et al. 1985).

For the purpose of increasing TFO stability, we synthesized chemically modified TFO for these experiments. Phosphodiester pyrazolopyrimidine dG (ppG)-modified TFO complimentary to the polypurine-polypyrimidine region of the mdr1 gene were synthesized and labeled with ^{125}I-dCTP at several cytosines. Experiments with a plasmid containing the mdr1 polypurine-polypyrimidine region and with purified genomic DNA confirmed the ability of our ^{125}I-TFO to recognize and cleave the target sequence. ^{125}I-

TFO were delivered into KB-VI cells with several delivery systems. The cellular uptake and intracellular distribution were monitored by using fluorescently labeled TFO. DNA from the ^{125}I-TFO-treated cells was recovered and analyzed for the sequence-specific cleavage in the mdr1 target by Southern hybridization. We showed that ^{125}I-TFO in nanomolar concentrations can recognize and cleave the target sequence in the mdr1 gene in situ in digitonin-permeabilized cells and in isolated nuclei. This was the first demonstration that a DNA target in its normal physiological form, i.e., in a complex with nucleosomes, could be cleaved with ^{125}I-TFO. Thus, the nearly 50 times amplification of the mdr1 gene in KB-VI cells affords a very useful model for evaluation of our methods to produce sequence-specific DNA DSB for AR.

Next, using the described approach we attempted to target the mdr1 gene within intact KB-VI cells. For that purpose we used TFO conjugated with a nuclear localization signal peptide (NLS) that, as before, were delivered into cells using cationic liposomes. This was done either alone or in the presence of an excess of a "ballast" oligonucleotide with an unrelated sequence. In all cases nuclear localization of TFO and survival of the cells after treatment has been confirmed by fluorescent microscopy. Breaks in the gene target were analyzed by restriction enzyme digestion of the DNA recovered from the TFO-treated cells followed by Southern hybridization with DNA probes flanking the target sequence. We have found that TFO/NLS conjugates cleave the target in a concentration-dependent manner regardless of the presence of the "ballast" oligonucleotide. In contrast, TFO without NLS cleaved the target only in the presence of an excess of the "ballast". We hypothesize that TFO and TFO/NLS are delivered into the nucleus by different pathways; TFO/NLS probably utilize a nuclear pore transport that results in release of TFO/NLS within the nucleus allowing them to bind to the target. TFO without NLS most likely became trapped by some non-specific oligonucleotide-binding factors that eventually transport them into the nucleus but such delivery does not lead to binding to the target. The presence of an excess of the "ballast" oligonucleotides may allow some TFO to escape the trapping. Our results provide a new insight into the mechanism of intracellular transport of oligonucleotides and open new avenues for improvement of the efficacy of antigene therapies.

29.8
Radioprobing

We discovered in the course of our experiments a new method to study DNA structures in solution, and we named it "radioprobing". It is based on an analysis of DNA strand breaks produced by the decay of ^{125}I incorporated into a DNA duplex or triplex (Panyutin and Neumann 1997; Karamychev et al. 1999, 2000a; Malkov et al. 2000). It had been first shown by others, that decay of ^{125}I incorporated into a DNA duplex produced DNA strand breaks that were localized within a few base pairs from the position of the radioiodine; 90% of these breaks occurred within 10 bp from the ^{125}I site. It is generally assumed that these strand breaks are due to energy deposition in the DNA sugar-phosphate moiety since damage to the DNA bases does not immediately lead to strand scission (Nikjoo et al. 1994). The frequencies of these breaks depend on the distances between the sugar phosphate bond and the ^{125}I decay, i.e., the further from the decay, the lower the probability of a strand break. Radioprobing then is our analysis of the frequency and distribution of DNA breaks. We first applied it to the detection of fine conformational changes in a DNA duplex when it is in a DNA protein complex by using the DNA/cyclic AMP receptor protein complex as an example (Karamychev et al. 1999). Our initial data show that radioprobing is capable of detecting a change of a few Ås in internucleotide distances. Thus, radioprobing can be useful for the study of conformational changes in nucleic acids in solution in the "intermediate range" from 10 Å to 30 Å. The advantage of ^{125}I radioprobing over chemical and enzymatic tests that are widely used for footprinting assays of nucleic acid-protein complexes is that the frequency of breaks produced by the AE depends primarily on the distance to the target atom and is not affected by protein-induced alterations in the chemical reactivity of that atom. Therefore, radioprobing, like NMR, allows one to obtain information on interatomic distances and, in principle, to reconstruct the 3D-structure of nucleic acids in aqueous complexes with proteins. Radioprobing requires only a small amount of material, does not demand special equipment and can be applied to large nucleoprotein assemblages which make it a very promising new method for structural biology.

29.9
Radioprinting

Our in vitro experiments showed that triplexes formed at plasmid and TFO concentrations as low as 10^{-9} M. Once formed, the triplex was remarkably stable and could withstand even 10 min incubation at 65 °C. We could therefore deliver preformed TFO-plasmid complexes into cultured human cells (Sedelnikova et al. 1999; Shamsul Hoque et al. 1999). To monitor the TFO-plasmid complexes while inside cells we created a technique that we call "radioprinting" – a method for detecting or monitoring binding of radiolabeled oligonucleotides to DNA targets. The method is again based on our analysis of DNA breaks induced by the highly localized AE. Radioprinting the plasmids we recovered allowed us for the first time to show that triplexes preformed on covalently closed circular DNA remained stable inside cells for at least 48 h. Based on these findings we proposed using radiolabeled TFO for indirect labeling of any intact plasmid DNA. As an applications demonstration, we showed that the intracellular distribution of fluorescein-labeled TFO was different when they were liposome-delivered into cultured human cells alone or in a complex with a plasmid. In the latter case, the fluorescence was detected in nearly all the cells while detection of the plasmid by use of the conventional marker gene (β-galactosidase) revealed expression of the gene in only about half of the cells. This method can thus be applied to measure delivery of plasmid DNA (or any pre-labeled TFO-DNA complex) into cells and perhaps someday, in combination with PET, serve as a method to monitor gene-introduction therapies.

29.10
Determining the Enzymatic Requirements for SSB Repair

The specific biochemical mechanism involved in the repair of any DNA lesion is highly dependent upon the structure of the lesion itself. Different lesions (SSB, DSB, base damage, base mismatches, bulky adducts, etc.) are processed for repair by distinct repair pathways, and in many cases the enzymes involved in DNA repair are highly specific for particular chemical modifications within the DNA (Lehmann et al. 1992; Friedberg et al. 1995). Since the DNA damage produced by AE is primarily SSB and DSB, elements of the base excision repair (BER) and the nonhomologous end-joining (NHEJ) pathways would be expected to participate in their repair. As discussed below, SSB resemble an intermediate in the BER DNA repair pathway, which is responsible for the removal and replacement of damaged nucleotide bases. In contrast, DSB are repaired by either homologous recombination, or nonhomologous recombination mechanisms. In human cells 80% of DSB are repaired by rejoining the nonhomologous break ends in a multi-step repair pathway (often leading to deletions at the break site) called nonhomologous end-joining (Jeggo 1998). The roles of the BER and NHEJ pathways were thus examined for the repair of AE-induced SSB and DSB in our initial studies.

The end group chemistry of AE-induced SSB and DSB has not been systematically determined. However, AE-induced DNA strand breaks appear to share at least two characteristics with strand breaks produced by low-LET radiation: AE-induced strand breaks presumably result from the destructive loss of at least one nucleotide, and the 5′-end group of the break is phosphate (5′-P; Panyutin and Neumann 1994, 1996, 1997). No determination of 3′-end structure has yet been reported.

The decay of ^{125}I in DNA is thought to produce at least between 3 and 6 SSBs per DSB as many as 2/3 of which may be due to indirect low-LET-like ionization effects (Painter et al. 1974; Feinendegen et al. 1977; Sundell-Bergman and Johanson 1980; Pomplun et al. 1996). In the case of low-LET ionizing radiation, SSB end groups have been found to consist of two equally distributed chemical forms (Fig. 29.3). These breaks consist of at least a one nucleotide gap containing a 5′-P and either a 3′-phosphoglycolate (3′-PG) or a 3′-phosphate (3′-P; Henner et al. 1983a, b). Therefore, we have chosen this lesion to model the majority of SSB produced by decay of A Ettr.

The loss of coding information and the lack of a 3′-OH at radiation-induced SSB suggests that these lesions are processed by a repair pathway similar to that of BER or apurinic/apyrimidinic (AP) site repair. BER repair involves the removal of a damaged base from DNA via cleavage of the N-glycosidic bond between the base and sugar moieties of a nucleotide, resulting in the formation of a baseless, or abasic, site. This baseless apurinic/apyrimidinic (AP) site is then incised by a specialized AP endonuclease producing a nicked DNA strand possessing a 3′-OH end which serves as a substrate for replacement of the missing

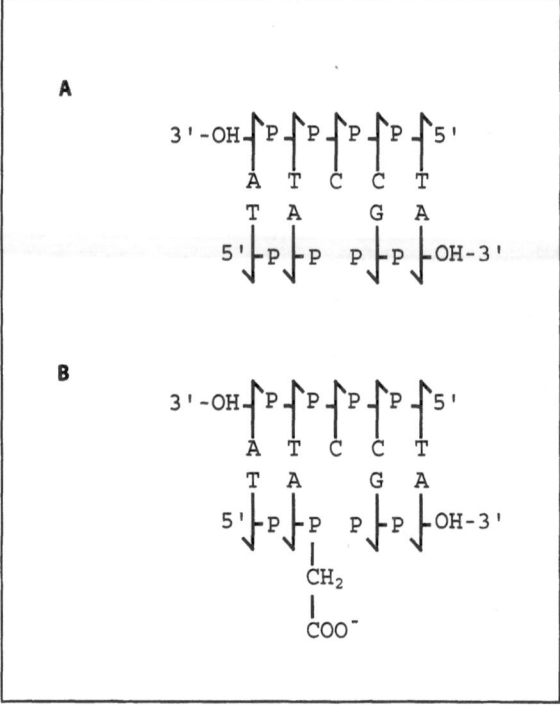

Fig. 29.3 A, B. The two forms of single-strand breaks produced (with apparent equimolar yield) by either ionizing radiation or oxidative processes. **A** The lower strand of the DNA duplex contains a nick opposite the C nucleotide. The strand break ends flanking the position of the missing nucleotide are both phosphates (5′-P and 3′-P). **B** The strand break ends flanking the missing nucleotide in (B) are 3′phosphoglycolate (3′-PG) and 5′-P

nucleotide by DNA polymerase using the complementary DNA strand as a template. Resealing the nicked DNA with DNA ligase completes the repair process. Therefore, presumably the 3′-end of an SSB lesion is converted to a 3′-end-OH, the lost coding information is restored in a template-dependent manner by DNA polymerase and repair is completed by the action of DNA ligase. In the case of BER and AP site repair, these reactions have been shown to be catalyzed by human apurinic/apyrimidinic endonuclease 1 (HAP1) and DNA polymerase β in association with DNA li-

Fig. 29.4. A Construction of the defined site-specific 3′-phosphoglycolate/5′-phosphate-containing single-strand break vector substrate. **B** Radiation-induced single-strand break vector substrate lesion detail. The positions of the 3′-PG-containing 17 mer is indicated by the right bar. The position of the downstream 5′-P-containing 17 mer is indicated by the left bar. The location of the missing nucleotide is indicated by the *arrow*. Restriction enzyme recognition sites are indicated above the sequence

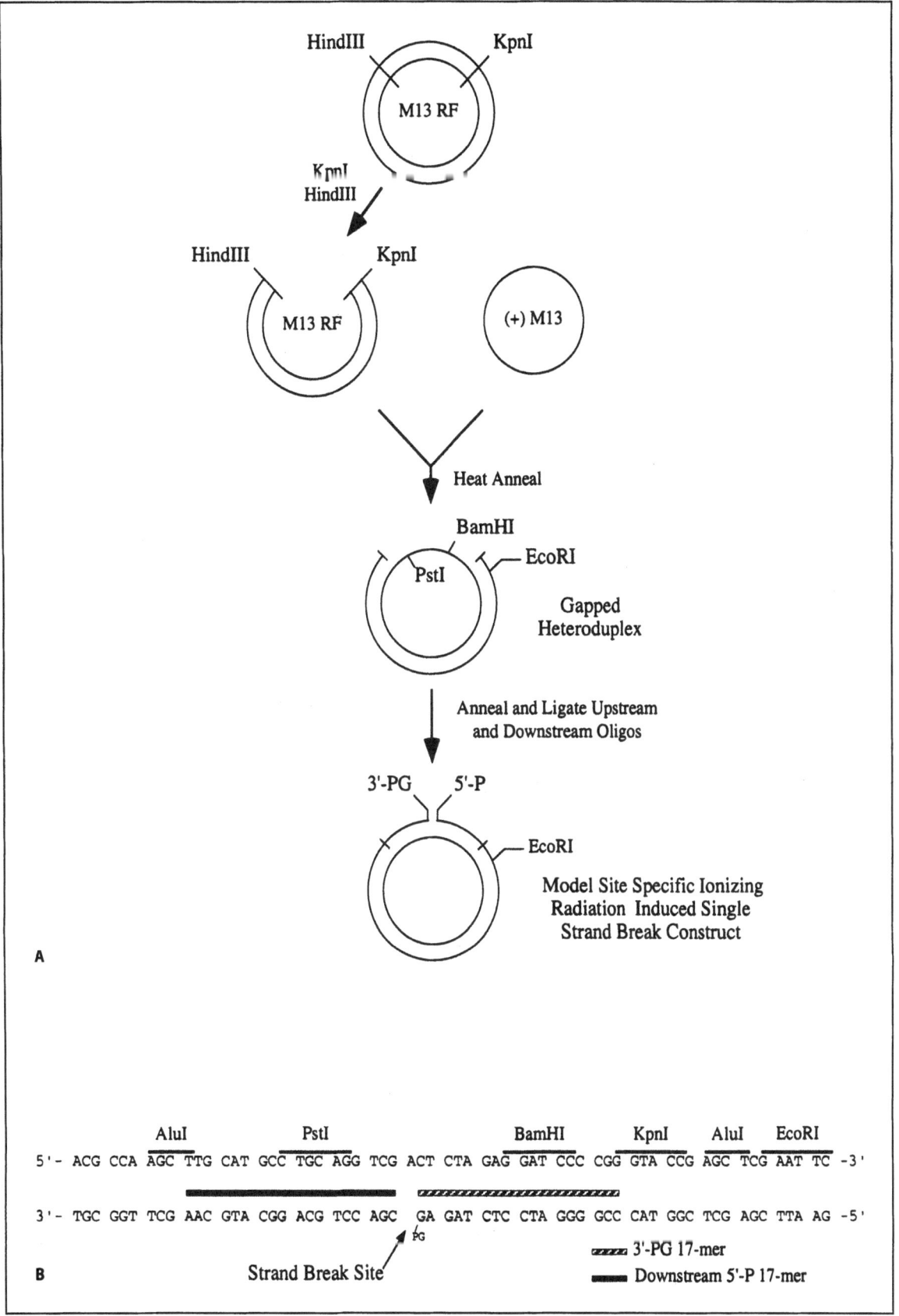

5' - ACG CCA AGC TTG CAT GCC TGC AGG TCG ACT CTA GAG GAT CCC CGG GTA CCG AGC TCG AAT TC -3'

3' - TGC GGT TCG AAC GTA CGG ACG TCC AGC GA GAT CTC CTA GGG GCC CAT GGC TCG AGC TTA AG -5'

gase I or DNA ligase III and XRCC1 (Chen et al. 1991; Kubota et al. 1996; Prasad et al. 1996; Bennett et al. 1997; Cappelli et al. 1997; Nash et al. 1997) and/ or by DNA polymerase δ/ε in association with DNA ligase I (Matsumoto et al. 1994; Frosina et al. 1996; Nealon et al. 1996; DeMott et al. 1998).

We constructed a defined site-specific SSB substrate based upon the M13 bacteriophage vector genomic DNA that consists of a one nucleotide gap flanked by a 5'-P and 3'-PG (Fig. 29.4). The vector substrate was employed in a series of SSB repair reconstitution experiments using human HeLa cells that had been extracted and chromatographically fractionated to isolate all potential repair enzyme activities (Winters et al. 1999).

DNA polymerase β was found to be responsible for up to 67% of nucleotide incorporation at the lesion site following excision of the 3'-PG blocking group. DNA polymerase δ/ε was also capable of up to $\sim 30\%$ of nucleotide incorporation at the lesion site. In addition, repair reactions catalyzed by DNA polymerase β were found to be most effective in the presence of DNA ligase III; while those catalyzed by DNA polymerase δ/ε appeared to be more effective in the presence of DNA ligase I. We also demonstrated that the repair initiating 3'-PG excision reaction was not dependent upon HAP1 activity, as judged by inhibition of HAP1 with neutralizing HAP1-specific polyclonal antibody. This antibody has been demonstrated to specifically bind HAP1 protein by western blotting and activity assays have demonstrated the antibodies can specifically eliminate, or neutralize, HAP1 enzyme activity. These results support the idea that the SSB repair-initiating reaction may be dependent upon other recently identified and partially characterized enzymes (Chen et al. 1991; Winters et al. 1994; Sarker et al. 1995). These results differentiate SSB repair from BER and AP site repair based upon the lack of a role for HAP1 in the repair reaction, which is a common feature in both BER and AP site repair. The SSB repair results also establish the complex dual pathway nature of radiation-induced SSB repair, and define the complete enzyme utilization pattern for both repair pathway branches.

The recent observation that oxidatively induced SSB are mutagenic and almost exclusively result in base substitutions following repair in human cells (Dar and Jorgensen 1995), in conjunction with our determination of the biochemical mechanism of their repair, defines a biological response to radiation-induced SSBs. This information allows correlation of the biological consequences of radiation-induced

SSBs with the specific enzymes and biochemical mechanisms responsible for these consequences. As a result, accurate statistical predictions should now be possible with respect to the ultimate fate and biological impact of an individual SSB, even under conditions which are known to affect the activity of enzymes in the repair pathway.

29.11
Determining the Biological Response for DSB Repair

DSB are thought to be the more biologically important form of AE-induced DNA damage produced both directly and indirectly by the AE. Consequently, the effectiveness of TFO-mediated AE would be expected to depend most upon cellular responses to AE-induced DSBs that result in mutagenic disruption of target gene function.

To test this hypothesis we employed a mutagenesis assay based upon the human shuttle vector pSP189 (Mezhevaya et al. 1999). This plasmid contains the E. coli SupF tyrosine suppressor tRNA gene as the mutagenesis target, as well as eukaryotic and prokaryotic replication origins that permit it to be shuttled back and forth between eukaryotic and prokaryotic cells. Damage to the SupF gene that results in a mutation after repair and replication in human cells will disrupt tRNA gene function (Kraemer and Seidman 1989). Mutants are detected by recovering repaired plasmids from the human cell extracts and transforming them into an E. coli strain that contains a suppressible amber mutation in its β-galactosidase gene. Plating the transformed bacteria onto media containing x-gal and IPTG permits identification of colonies bearing wild type and mutant plasmids by blue/white colony screening. Those bacterial colonies in which the cells contain a plasmid bearing the wild type SupF gene will suppress the β-galactosidase mutation and appear blue due to their ability to metabolize x-gal. In contrast, colonies in which the cells contain a plasmid bearing a mutated SupF gene that is incapable of suppressing the host cell β-galactosidase mutation, will not metabolize x-gal and appear as unpigmented (white) colonies.

Figure 29.5 shows the pSP189 plasmid target sequence aligned with the ^{125}I labeled TFO PSP3. Following ^{125}I decay accumulation, a complex mixture of damage which included open circular (SSB-containing), linear (DSB-containing) and covalently closed

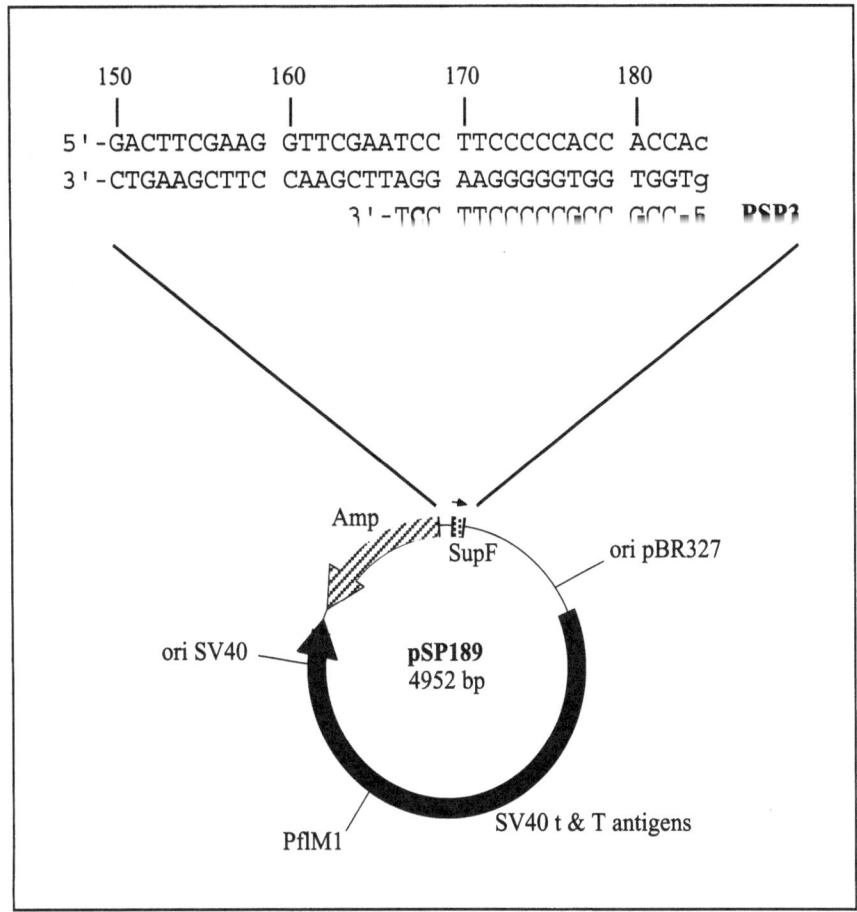

Fig. 29.5. Human shuttle vector pSP189. The plasmid contains the bacterial *SupF* suppressor tRNA gene as the mutagenesis reporter gene, a portion of which has been expanded and numbered according to tRNA convention in the upper panel. The triplex forming oligonucleotide PSP3 is shown aligned with its target sequence. The position of the [5-^{125}I] cytosine residue is indicated in bold at position 15 of the TFO

(undamaged) DNA was recovered. Generation of 2861 bp and 2091 by fragments following cleavage of the damaged DNA with PflM1 restriction endonuclease demonstrated ~70% of the ^{125}I-induced DSB to be site-specific at the decay site within the *SupF* gene. Bulk-damaged DNA and the isolated site-specific DSB-containing DNA were then separately transfected into human WI38VA13 cells and allowed to repair prior to recovery and analysis of mutants. Bulk-damaged DNA had a relatively low mutation frequency of 2.7×10^{-3}, similar to that reported for random breaks induced by the radiomimetic drug bleomycin (Sarker et al. 1995). In contrast, the isolated linear DNA-containing site-specific DSB had an unusually high mutation frequency of 7.9×10^{-1}. This is nearly 300-fold greater than that observed for the bulk-damaged DNA mixture, more than 1.5×10^4-fold greater than background, and more than 100-fold greater than random oxidatively induced DSB pro-

duced by bleomycin. The ^{125}I-TFO-induced DSB mutation frequency was also nearly 7-fold greater than that reported for the repair of a 5'-psoralen-conjugated anti-parallel TFO targeted to crosslink *SupF* positions 166 and 167; and more than 2.5-fold greater than the mutation frequency observed for the same lesion repaired in nucleotide excision repair defective XPV cells (Raha et al. 1996).

The ^{125}I-TFO-induced mutation spectra displayed a high proportion of deletion mutants concentrated at the ^{125}I position within the *SupF* gene for both bulk-damaged DNA and isolated linear DNA. Both spectra were characterized by complex mutations with mixtures of changes. However, mutations recovered from the linear site-specific DSB-containing DNA presented a much higher proportion of complex deletion mutations.

Results obtained with the *SupF* mutagenesis assay indicate that we can expect a high frequency (nearly

1:1) of target gene inactivation for each ^{125}I-TFO-positioned DSB. However, although the mutagenesis assay provides information about the products of the DSB repair reaction, it does not provide any information about the actual biochemical events that produce the rejoined products. As with SSB repair we were interested in determining the enzymatic processes involved in DSB repair. A clear understanding of the biochemistry underlying DSB repair would not only allow us to make more accurate predictions about the fate of a DSB in a given cellular background, it would also open up the possibility of manipulating DSB repair activity to help produce outcomes of our choice.

As discussed above, mammalian cells primarily repair DSB by NHEJ (Jeggo 1998). Genetic studies, using radiosensitive mammalian cell lines, and animals deficient in V(D)J recombination, have been useful in identifying several proteins (Ku 70 and Ku 86, DNA-PKcs, DNA ligase IV, XRCC4, Rad 50/Xrs-2/MreII complex) that are involved in the DSB repair process (Fig. 29.6) (Jeggo et al. 1991; Lees-Miller et al. 1995; Gu et al. 1997; Riballo et al. 1999). However, due to the lack of a simple in vitro assay, the exact biochemical mechanism of repair remains unknown.

The chemical and physical structure of the DSB end group is also an important consideration when investigating NHEJ, because it may directly affect the pathway's ability to repair a break. Several DSB repair methods based on end joining of linear plasmids have been described in the literature (North et al.

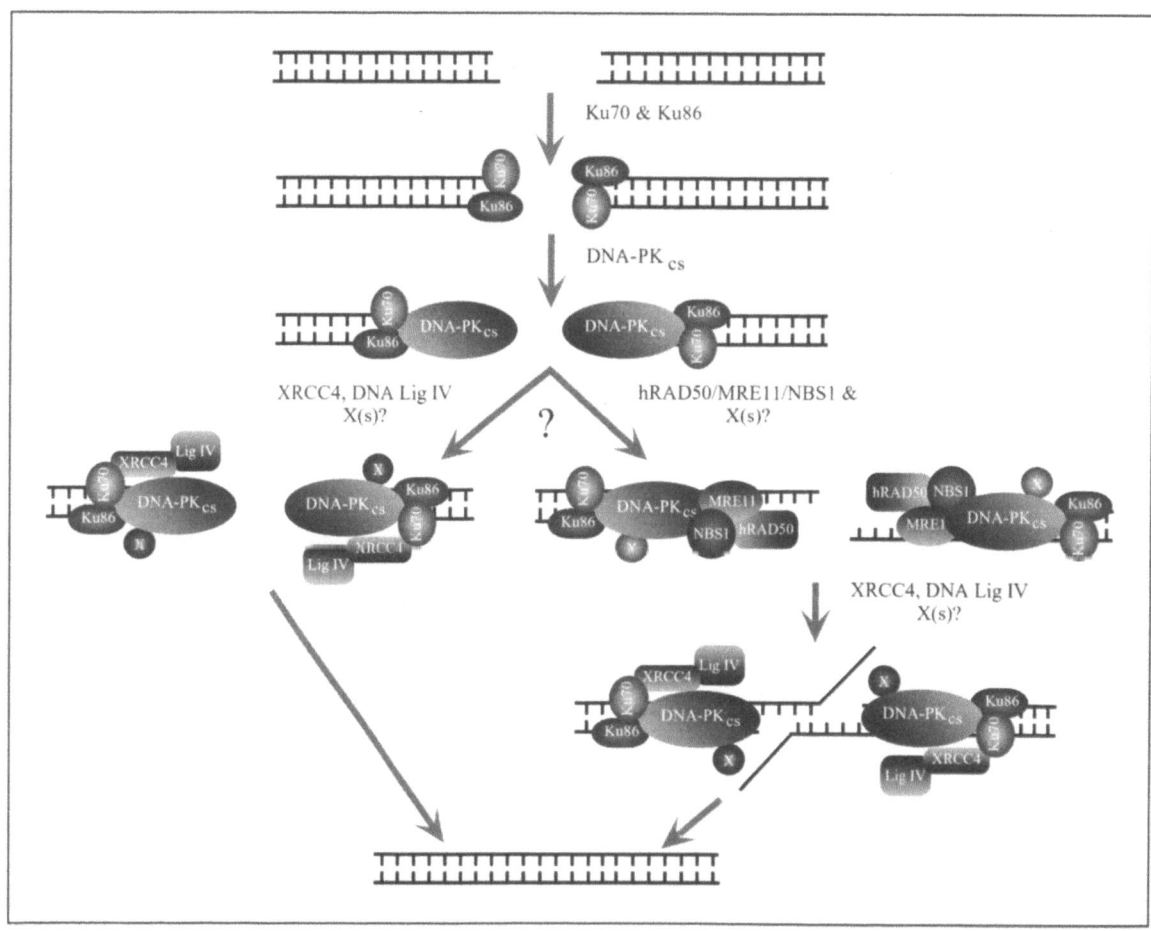

Fig. 29.6. A schematic representation of illegitimate recombination represented by the nonhomologous end joining (NHEJ) and single-strand annealing DSB repair pathways. DSB are recognized and protected by the Ku heterodimer which in turn recruits additional components of the pathway. Those factors that determine whether a DSB is processed by direct end joining (left hand branch), or requires 3'-end resection followed by single-strand annealing, are not well understood and are represented by a question mark (?) in the figure. The exact mechanisms by which complex DSB are repaired most likely require participation of proteins that are as yet unknown (X)

1990; Mason et al. 1996; Baumann and West 1998; Feldmann et al. 2000). All of these assays use restriction enzyme cut plasmid DNA as the repair substrate. This does not adequately model the structure of DSBs produced by ionizing radiation, endogenous metabolic processes, or many other DNA-damaging agents. Radiation-induced DSBs, including those produced by the AE, are formed by the destructive loss of at least one nucleotide in each DNA strand at the break site. In contrast to restriction enzyme-cut DNA, the resulting end structure is complex, consisting of nucleotide fragments that render the ends unligatable in the absence of nucleolytic processing (Chen et al. 1991; Gu et al. 1996; Suh et al. 1997). Strand break ends blocked by fragmented nucleotides is not the extent of lesion complexity produced at DSBs. In addition to blocked ends, nucleotides proximal to the break site may also become damaged in one or both strands, and at either or both sides of the discontinuity. Such lesions are referred to as clustered, or multiply damaged sites (MDS) (Ward 2000; Milligan et al. 2001; Blaisdell and Wallace 2001). Further, the yield of MDS type DSB is expected to increase with increasing radiation LET (Hoglund and Stenerlow 2001).

Therefore, to study NHEJ repair using substrates modeled on those expected in vivo, we have developed an in vitro DSB end-joining assay that employs plasmid DNA substrates linearized by a variety of agents, including restriction enzymes, the radiomimetic drug bleomycin, ^{60}Co γ, and ^{125}I-TFO (Pastwa et al. 2001a). These damaged DNA substrates represent a range in DSB complexity, from the easily ligatable restriction enzyme cut DNA to the highly complex MDS type DSB of high LET radiation. Restriction enzyme cut DNA represents the simplest form of DSB, while bleomycin-induced DSB are intermediate between restriction enzyme DSB and γ-ray-induced DSB. Similarly, ^{60}Co γ-ray-induced DSB are intermediate in complexity between bleomycin-induced DSB and high LET DSB like those expected from AE. The end group structure is only known with precision for restriction enzyme cut DNA (3′-OH and 5′-P), and bleomycin-induced DSB. Bleomycin-induced DSB are similar to the 3′-PG strand breaks described above for ionizing radiation, and are distributed almost equally between 3′-PG blocked blunt ends, and ends containing a 3′-PG and a single base 5′-overhang (Povirk et al. 1989). Low LET radiation like ^{60}Co γ-rays, are thought to produce a continuum of DSB structures that range between the relatively simple breaks depicted in Fig. 29.3, to the more complex MDS type DSB. High LET radiation, and the high LET-like AE, are thought to predominately produce DSB of the MDS type. Although as indicated above, no direct analysis of either AE-induced DSB or high LET DSB structures have been reported.

Using our assay we show that human HeLa cell extracts support end joining of complex DSB and form multimeric plasmid products from all of the substrates tested. However, end joining was end-group dependent and decreased as the cytotoxicity of the DSB-inducing agent increased. The initial repair rate for complex non-ligatable bleomycin-induced DSBs was 6-fold less than that of similarly configured (blunt ended), but less complex (ligatable) restriction enzyme-induced DSBs. Also, repair of DSBs produced by γ-rays was 15-fold less efficient than repair of restriction enzyme-induced DSBs. Repair of ^{125}I-TFO-induced DSB was at least 2-fold lower than that of γ-ray induced DSB. These results indicate that as DSB structural complexity increases, repair by NHEJ becomes more inhibited. In addition to assessing DSB repair, we have been able to employ our assay to examined several structural characteristics for site-specific ^{125}I-TFO-induced DSBs.

By using the bacterial enzymes, endonuclease IV and T4 DNA ligase, we have demonstrated the high LET-like DSBs produced by ^{125}I to be blocked by 3′-nucleotide fragments (Pastwa et al. 2001b). Endonuclease IV is a class II AP endonuclease that is capable of cleaving 3′-nucleotide fragments at DNA strand breaks to produce a 3′-OH which can serve as a substrate for ligation and polymerization (Weinfeld et al. 1997). The activity of T4 DNA ligase is limited to rejoining directly opposable strand breaks possessing 3′-OH and 5′-P ends. Up to 2% of ^{125}I-TFO-induced DSB in our experiments can be rejoined by T4 DNA ligase following endonuclease IV processing. This indicates that at least a subpopulation of these breaks must contain 3′-nucleotide fragments that are recognizable by endonuclease IV. No determination of 3′-end structure had been reported prior to this finding.

Moreover, these breaks are opposable and ligatable following 3′-end group hydrolysis, suggesting that they most likely consist of a blunt end structure. Conversely, the lack of rejoining with the remaining 98% of the ^{125}I-TFO-induced DSB suggests that these breaks consist of a more complex structure than the 2% of ends that were capable of supporting rejoining. As discussed above, the 5′-end group chemistry of AE-induced DSB have been indirectly determined by

Maxam-Gilbert sequencing (Panyutin and Neumann 1994, 1996). Since at least a small subset of AE-induced DSB are ligatable following 3'-end processing, the presence of 5'-PO$_4$ end groups at these DSB is confirmed. Furthermore, the conformation of these ends must either be directly opposable blunt ends, or complementary overhands of equal length. Identification of a complex 3'-end structure at ^{125}I-induced DSBs in conjunction with the comparatively low yield of rejoined products, observed in time course reactions, suggest that DSB produced by the AE are much more structurally complex than those produced by γ-rays and the radiomimetic drug bleomycin.

These results, taken together with our earlier data (Panyutin and Neumann 1994) showing that given appropriate TFOs and targets, DSBs can be produced at frequencies approaching one DSB/decay, demonstrate that by virtue of the complex nature of AE-induced DSBs, TFOs linked to A Ettr can be highly effective mutagens for site-specific disruption of gene function. Furthermore, our mutation data for the bulk-damaged DNA mixture in conjunction with our findings concerning the mechanisms of SSB repair indicate that non-targeted, non-DSB, AE-induced DNA damage would be efficiently repaired and relatively innocuous.

29.12
References

Baumann P, West SC (1998) DNA end-joining catalyzed by human cell-free extracts. Proc Natl Acad Sci 95:14066–14070

Belousov ES, Afonia IA, Kutyavin IV et al (1998) Triplex targeting of a native gene in permeabilized intact cells: covalent modification of the gene for the chemokine receptor ccr5. Nucleic Acids Res 26:1324–1328

Bennett RA, Wilson DM, Wong D, Demple B (1997) Interaction of human apurinic endonuclease and DNA polymerase beta in the base excision repair pathway. Proc Natl Acad Sci USA 94:7166–7169

Blaisdell JO, Wallace SS (2001) Abortive base-excision repair of radiation-induced clustered DNA lesions in Escherichia coli. Proc Natl Acad Sci USA 98:7426–7430

Cappelli E, Taylor R, Cevasco M et al (1997) Involvement of XRCC1 and DNA ligase III gene products in DNA base excision repair. J Biol Chem 272:23970–23975

Chan PP, Glazer PM (1997) Triplex DNA: fundamentals, advances, and potential applications for gene therapy. J Mol Med 75:267–282

Chaudhry MA, Weinfeld M (1995) Induction of double-strand breaks by S1 nuclease, mung bean nuclease and nuclease P1 in DNA containing abasic sites and nicks. Nucleic Acids Res 23:3805–3809

Chen DS, Herman T, Demple B (1991) Two distinct human DNA diesterases that hydrolyze 3'-blocking deoxyribose fragments from oxidized DNA. Nucleic Acids Res 19:5907–5914

Dar ME, Jorgensen TJ (1995) Deletions at short direct repeats and base substitutions are characteristic mutations for bleomycin-induced double- and single-strand breaks, respectively, in a human shuttle vector system. Nucleic Acids Res 23:3224–3230

Delporte C, Panyutin IG, Sedelnikova OA et al (1997) Triplex-forming oligonucleotides can modulate aquaporin-5 gene expression in epithelial cells. Antisense Nucleic Acid Drug Dev 7:523–529

DeMott MS, Zigman S, Bambara RA (1998) Replication protein A stimulates long patch DNA base excision repair. J Biol Chem 273:27492–27498

DeSombre ER, Shafii B, Hanson RN et al (1992) Estrogen receptor-directed radiotoxicity with Auger electrons: specificity and mean lethal dose. Cancer Res 52:5752–5758

Feinendegen LE (1975) Biological damage from the Auger effects, possible benefits. Radiat Environ Biophys 12:85–99

Feinendegen LE, Bond VP, Hughes WL (1966) 125-I-DU (5-iodo-2-deoxyuridine) in autoradiographic studies of cell proliferation. Exp Cell Res 43:107–119

Feinendegen LE, Henneberg P, Tisljar-Lentulis G (1977) DNA strand breakage and repair in human kidney cells after exposure to incorporated iodine-125 and cobalt-60 γ-rays. Curr Topics Rad Res Q 12:436–452

Feldmann E, Schmiemann V, Goedecke W et al (2000) DNA double-strand break repair in cell-free extracts from Ku80-deficient cells: implications for Ku serving as an alignment factor in non-homologous DNA end joining. Nucleic Acids Res 28:2585–2596

Fojo AT, Whang-Peng J, Gottesman MM, Pastan I (1985) Amplification of DNA sequences in human multidrug-resistant KB carcinoma cells. Proc Natl Acad Sci USA 82:7661–7665

Frank-Kamenetskii MD, Mirkin SM (1995) Triplex DNA structures. Annu Rev Biochem 64:65–95

Friedberg EC, Walker GC, Siede W (1995) DNA repair and mutagenesis. ASM Press, Washington DC

Frosina G, Fortini P, Rossi O et al (1996) Two pathways for base excision repair in mammalian cells. J Biol Chem 271:9573–9578

Gibbs RA, Camakaris J, Hodgson GS, Martin RF (1987) Molecular characterization of 125I decay and X-ray-induced HPRT mutants in CHO cells. Int J Radiat Biol Relat Stud Phys Chem Med 51:193–199

Giovannangeli C, Helene C (1997) Progress in developments of triplex-based strategies. Antisense Nucleic Acids Drug Dev 7:413–421

Giovannangeli C, Diviacco S, Labrousse V et al (1997) Accessibility of nuclear DNA to triplex-forming oligonucleotides: the integrated HIV-1 provirus as a target. Proc Natl Acad Sci USA 94:79–84

Gu XY, Bennett RA, Povirk LF (1996) End-joining of free radical-mediated DNA double-strand breaks in vitro is blocked by the kinase inhibitor wortmannin at a step preceding removal of damaged 3' termini. J Biol Chem 271:19660–19663

Gu YS, Jin SF, Gao YJ et al (1997) Ku70-deficient embryonic stem cells have increased ionizing radiosensitivity, defective DNA end-binding activity, and inability to support V(D)J recombination. Proc Natl Acad Sci USA 94:8076–8081

Helene C (1991) The anti-gene strategy: control of gene expression by triplex-forming-oligonucleotides. Anticancer Drug Des 6:569–584

Henner WD, Grunberg SM, Haseltine WA (1983a) Enzyme action at 3′ termini of ionizing radiation-induced DNA strand breaks. J Biol Chem 258:15198–15205

Henner WD, Rodriguez LO, Hecht SM, Haseltine WA (1983b) Gamma-ray induced deoxyribonucleic acid strand breaks, J Biol Chem 258:711–713

Hoglund E, Stenerlow B (2001) Induction and rejoining of DNA double-strand breaks in normal human skin fibroblasts after exposure to radiation of different linear energy transfer: possible roles of track structure and chromatic organization. Radiat Res 155:818–825

Howell RW, Azure MT, Narra VR, Rao DV (1994) Relative biological effectiveness of alpha-particle emitters in-vivo at low-doses. Radiat Res 137:352–360

Jeggo PA (1998) Identification of genes involved in repair of dna double-strand breaks in mammalian cells. Radiat Res 150:S80–S91

Jeggo PA, Tesmer J, Chen DJ (1991) Genetic-analysis of ionizing-radiation sensitive mutants of cultured mammalian-cell lines. Mutat Res 254:125–133

Jones GDD, Boswell TV, Ward JF (1994) Effects of postirradiation temperature on the yields of radiation-induced single-strand and double-strand breakage in SV40 DNA. Radiat Res 138:291–296

Karamychev VN, Panyutin IG, Reed MW, Neumann RD (1997) Effect of radionuclide linker structure on DNA cleavage by 125I-labeled oligonucleotides. Antisense Nucleic Acid Drug Dev 7:549–557

Karamychev VN, Zhurkin VB, Garges S et al (1999) Detecting the DNA kinks in a DNA-CRP complex in solution with iodine-125 radioprobing. Nat Struct Biol 6:747–750

Karamychev VN, Neumann RD, Panyutin IG, Zhurkin VB (2000a) Folding of RNA and DNA strands in transcription complex as revealed by iodine-125 radioprobing. J Biomol Struct Dyn 11:155–167

Karamychev VN, Panyutin IG, Meyong-Kon K et al (2000b) DNA cleavage by In-111 labeled oligonucleotides. J Nucl Med 41:1093–1101

Karamychev VN, Reed MW, Neumann RD, Panyutin IG (2000c) Distribution of DNA strand breaks produced by I-123 and In-111 in synthetic oligodeoxynucleotides. Acta Oncol 39:687–692

Kraemer KH, Seidman MM (1989) Use of supf, an escherichia-coli tyrosine suppressor transfer-RNA gene, as a mutagenic target in shuttle-vector plasmids. Mutat Res 220:61–72

Kubota Y, Nash RA, Klungland A et al (1996) Reconstitution of DNA base excision-repair with purified human proteins: interaction between DNA polymerase beta and the XRCC1 protein. EMBO J 15:6662–6670

Lees-Miller SP, Godbout R, Chan DW et al (1995) Absence of 0350 subunit of DNA-activated protein kinase from a radiosensitive human cell line. Science 267:1183–1185

Lehmann AR, Hoeijmakers JHJ, Vanzeeland AA et al (1992) Workshop on DNA-repair. Mutat Res 273:1–28

Lukhtanov EA, Mills AG, Kutyavin IV et al (1997) Minor groove DNA alkylation directed by major groove triplex forming oligodeoxyribonucleotides. Nucleic Acids Res 25:5077–5084

Majumdar A, Khorlin A, Dyatkina N et al (1998) Targeted gene knockout mediated by triple helix forming oligonucleotides. Nat Genet 20:212–214

Malkov L, Panyutin IG, Neumann RD et al (2000) Tracing the nucleic acids strands in large nucleoprotein complexes with Auger electrons: structure of a RecA-three stranded DNA complex. J Mol Biol 299:629–640

Martin RF, Haseltine WA (1981) Range of radiochemical damage to DNA with decay of iodine-125. Science 213:896–898

Mason RM, Thacker J, Fairman MP (1996) The joining of non-complementary DNA double-strand breaks by mammalian extracts. Nucleic Acids Res 24:4946–4953

Matsumoto Y, Kim K, Bogenhagen DF (1994) Proliferating cell nuclear antigen-dependent abasic site repair in Xenopus laevis oocytes: an alternative pathway of base excision DNA repair. Mol Cell Biol 14:6187–6197

Mezhevaya K, Winters TA, Neumann RD (1999) Gene targeted DNA double strand break induction by 125I-labeled triplex forming oligonucleotides is highly mutagenic following repair in human cells. Nucleic Acids Res 27:4282–4290

Milligan JR, Aguilera JA, Paglinawan RA et al (2001) DNA strand break yields after post-high LET irradiation incubation with endonuclease-III and evidence for hydroxyl radical clustering. Int J Radiat Biol 77:155–164

Mirkin SM, Frank-Kamenetskii MD (1994) H-DNA and related structures. Annu Rev Biophys Biomol Struct 23:541–576

Nash RA, Caldecott KW, Barnes DE, Lindahl T (1997) XRCC1 protein interacts with one of two distinct forms of DNA ligase III. Biochemistry 36:5207–5211

Nealon K, Nicholl ID, Kenny MK (1996) Characterization of the DNA polymerase requirement of human base excision repair. Nucleic Acids Res 24:3763–3770

Nikjoo H, O'Neill P, Terrissol M, Goodhead DT (1994) Modelling of radiation-induced DNA damage: the early physical events. Int J Radiat Biol 66:453–457

North P, Ganesh A, Thacker J (1990) The rejoining of double-strand breaks in DNA by human cell extracts. Nucleic Acids Res 18:6205–6210

Painter RB, Young BR, Burki HJ (1974) Non-repairable strand breaks induced by 125I incorporated into mammalian DNA. Proc Natl Acad Sci USA 71:4836–4838

Panyutin IG, Neumann RD (1994) Sequence-specific DNA double-strand breaks induced by triplex forming 125I labeled oligonucleotides. Nucleic Acids Res 22:4979–4982

Panyutin IG, Neumann RD (1996) Sequence-specific DNA breaks produced by triplex-directed decay of iodine-125. Acta Oncol 35:817–823

Panyutin IG, Neumann RD (1997) Radioprobing of DNA: distribution of DNA breaks produced by decay of 125I incorporated into a triplex-forming oligonucleotide correlates with geometry of the triplex. Nucleic Acids Res 25:883–887

Pastwa E, Neumann RD, Winters TA (2001a) In vitro repair of complex unligatable oxidatively induced DNA double-strand breaks by human cell extracts. Nucleic Acids Res 29:E78–E78

Pastwa E, Mezhevaya K, Neumann RD, Winters TA (2001b) Repair of radiation-induced DNA double-strand breaks is dependent upon radiation quality and the structural complexity of double strand breaks. Radiat Res 159:251–261

Pomplun E, Terrissol M, Demonchy M (1996) Modelling of initial events and chemical behaviour of species induced in DNA units by Auger electrons from 125I, 123I and carbon. Acta Oncol 35:857–862

Povirk LF, Han YH, Steighner RJ (1989) Structure of bleomycin-induced DNA double-strand breaks: predominance of blunt ends and single-base 5' extensions. Biochemistry 28:5808–5814

Prasad R, Singhal RK, Srivastava DK et al (1996) Specific interaction of DNA polymerase beta and DNA ligase I in a multiprotein base excision repair complex from bovine testis. J Biol Chem 271:16000–16007

Purmal AA, Lampman GW, Kow YW, Wallace SS (1994) The sequence context-mispairing of 5-hydroxycytosine and 5-hydroxyuridine in-vitro. DNA Damage 726:361–363

Raha M, Wang G, Seidman MM, Glazer PM (1996) Mutagenesis by third-strand-directed psoralen adducts in repair-deficient human cells: high frequency and altered spectrum in xeroderma pigmentosum variant. Proc Natl Acad Sci USA 93:2941–2946

Reed MW, Panyutin IG, Hamlin D et al (1997) Synthesis of 125I-labeled oligonucleotides from tributylstannylbenzamide conjugates. Bioconjug Chem 8:238–243

Riballo E, Critchlow SE, Teo SH et al (1999) Identification of a defect in DNA ligase IV in a radiosensitive leukaemia patient. Curr Biol 9:699–702

Sarker AH, Watanabe S, Seki S et al (1995) Oxygen radical-induced single-strand DNA breaks and repair of the damage in a cell-free system. Mutat Res DNA Repair 337:85–95

Sedelnikova OA, Panyutin IG, Thierry AR, Neumann RD (1998) Radiotoxicity of iodine-125-labeled oligodeoxyribonucleotides in mammalian cells. J Nucl Med 39:1412–1418

Sedelnikova OA, Panyutin IG, Luu AN, Neumann RD (1999) The stability of DNA triplexes inside cells as studied by iodine-125 radioprinting. Nucleic Acids Res 27:3844–3850

Shamsul Hoque AT, Panyutin IG, Baum BJ (1999) Use of triplex-forming oligonucleotides and adenoviral constructs for studying the regulation of gene expression. Methods 18:266–272

Soyfer VN, Potaman VN (1995) Triple-helical nucleic acids. Springer, Berlin Heidelberg New York

Suh D, Wilson DM, Povirk LF (1997) 3'-Phosphodiesterase activity of human apurinic/apyrimidinic endonuclease at DNA double-strand break ends. NAR 25:2495–2500

Sundell-Bergman S, Johanson KJ (1980) Repairable and unrepairable DNA strand breaks induced by decay of ^{3}H and ^{125}I incorporated into DNA of mammalian cells. Radiat Environ Biophys 18:239–248

Van Holde K, Zlatanova J (1994) Unusual DNA structures, chromatin and transcription. Bioessays 16:59–68

Ward JF (1994) The complexity of DNA damage: relevance to biological consequences. Int J Radiat Biol 66:427–432

Ward JF (1995) Radiation mutagenesis – the initial DNA lesions responsible. Radiat Res 143:355

Ward JF (2000) Complexity of damage produced by ionizing radiation. Cold Spring Harbor Symp Quant Biol 65:377–382

Weinfeld M, Lee J, Gu RQ et al (1997) Use of a postlabelling assay to examine the removal of radiation-induced DNA lesions by purified enzymes and human cell extracts. Mutat Res Fund Mol M 378:127–137

Wells RD, Collier DA, Hanvey JC et al (1998) The chemistry and biology of unusual DNA structures adopted by oligopurine oligopyrimidine sequences. FASEB J 2:2939–2949

Winters TA, Henner WD, Russell PS et al (1994) Removal of 3'-phosphoglycolate from DNA strand-break damage in an oligonucleotide substrate by recombinant human apurinic/apyrimidinic endonuclease 1. Nucleic Acids Res 22:1866–1873

Winters TA, Russell PS, Kohli M et al (1999) Determination of human DNA polymerase utilization for the repair of a model ionizing radiation-induced DNA strand break lesion in a defined vector substrate. Nucleic Acids Res 27:2423–2433

Wright HA, Hamm RN, Turner JE, Howell RW, Rao DV, Sastry KSR (1990) Calculations of physical and chemical reactions with DNA in aqueous solution from Auger cascades. Radiat Prot Dosimetry 31:59–62

Part V Special Topics

CHAPTER 30

The Role of Nuclear Medicine in Relation to Alternative Modalities 30

CHRISTOPH BREMER, RALPH WEISSLEDER

Contents

30.1
Optical Imaging

Optical imaging techniques such as fluorescent microscopy or bioluminescence have for a long time been used in vitro in molecular biology (Contag et al. 1998; Flotte et al. 1998; Moore et al. 1998a, b; Simonova et al. 1999). Reporter genes with an optical signature (e.g. fluorescence, bioluminescence) can be linked to genetic regulatory elements that can reveal spatial and temporal information about a variety of biological processes at the level of transcription (Contag et al. 1995, 1998). Several technical advances in the field of photon generation as well as photon detection have opened the door for taking these techniques one step further and probe for optical molecular beacons in vivo. Application of light in the near infrared range (NIR) which travels more efficiently through the tissue-compared visible light (Chance 1998), highly sensitive CCD-technology and various techniques such as phase modulation, ultra-fast imaging with early arriving photons and diffuse optical tomography are currently being explored to apply optical imaging to detection of deeper structures in vivo

(Boas et al. 1994; Wu et al. 1997; Chance 1998; Mahmood et al. 1999; Ntziachristos et al. 2000). Recently, diffuse optical tomography has successfully been applied to detect breast lesions in a clinical setting (Ntziachristos et al. 2000). Besides the macroscopic diagnostic level, at the microscopic level, 10 μm resolution is achievable using optical coherence tomography (Boppart et al. 1997; Tearney et al. 1997), a technique analogous to ultrasound. Subsurface imaging of fluorescence (Mahmood et al. 1999; Weissleder et al. 1999) can be brought to the subcellular level in mouse models with intravital microscopy (Fukumura et al. 1998; Monsky et al. 1999; Dellian et al. 2000). Bioluminescence imaging (Contag et al. 1997, 2000; Sweeney et al. 1999) can be used to follow transgene expression in the whole animal, albeit at somewhat lower resolution. Various optical imaging marker genes and optical imaging probes have been developed over the last years and will be briefly discussed in the following sections.

30.1.1
Fluorescent Proteins

Green fluorescent protein (GFP) from the jellyfish *Aequorea victoria* is a protein fluorescing in the range of the visible light (excitation 489 nm; emission 508 mm). Genes encoding for GFP has been linked to specific "genes of interest" or therapeutic vectors monitoring the expression of the latter. GFP can be expressed in a variety of living organisms as fusion proteins for tracking individual proteins and study cellular dynamics (Prasher et al. 1992; Chalfie et al. 1994; Simonova et al. 1999). GFP has been extensively used to study a variety of aspects in cell biology (Prasher et al. 1992; Chalfie et al. 1994; Cole et al. 1996; Girotti and Banting 1996; Yokoe and Meyer 1996; Presley et al. 1997).

Recently, first advances were made to use GFP as an imaging marker gene for optical imaging in vivo. A genetically engineered 9L cell line constitutively expressing GFP has been used for imaging research to directly quantitate the angiogenic tumor burden (Moore et al. 1998 b). The same GFP-expressing melanoma cells have also been employed to quantify the tumor burden in locoregional lymph nodes during cancer spread in vitro (Moore et al. 1998 a).

In a model of transgenic mice expressing GFP under the control of the promoter for vascular endothelial growth factor (VEGF) the expression pattern of VEGF in models of wound healing and tumor-induced angiogenesis could be studied (Fukumura et al. 1998). Using this model, it was demonstrated that a large amount of VEGF in tumors (here reported by GFP expression) is produced by surrounding stromal cells rather than tumor cells themselves (Fukumura et al. 1998).

GFP can be easily employed as an imaging marker gene in a variety of different therapeutic vectors and thus serve as a monitoring tool for gene therapy. One drawback of this system, however, is the poor depth penetration of light in the visible range (only 1–2 mm in tissue). This hurdle might, at least in part, be overcome since red-shifted mutants of the GFP protein are currently under development, which will ultimately allow for more efficient tissue translumination (Matz et al. 1999).

30.1.2
Bioluminescence

Luciferases are enzymes that emit light through ATP- and O_2-dependent oxidation of luciferin. Luciferase genes have been cloned from a large number of organisms, including bacteria, firefly (*Photinus pyralis*), coral (*Renilla*), jellyfish (*Aequorea*), and dinoflagellates (*Gonyaulax*). The luciferase from the North American firefly, for example, releases green light during the oxidation of luciferin. They therefore behave like light bulbs emitting photons from within the body – comparable to radiation generated by commonly used isotopes. Sensitive imaging systems have been built to detect quantitatively small numbers of cells or organism expressing luciferase as a transgene (Sweeney et al. 1999). The use of bioluminescence as reporters in living mammalian cell cultures has been described for assessment of gene expression (Thompson et al. 1995; Contag et al. 1997). Ex vivo luciferase assays have been used to assess levels of expression of the reporter in mammalian tissues and to monitor tumor burden and infection (Morrey et al. 1991; Zhang et al. 1994; Thompson et al. 1995; Hickey et al. 1996).

In vivo human cervical carcinoma cell lines (HeLa) transfected with an optical reporter gene encoding for a modified firefly luciferase could be monitored using a sensitive ICCD-camera (Edinger et al. 1999). As little as 1×10^3 transfected tumor cells in the peritoneal cavity and 1×10^4 cells in the subcutaneous location were detected in vivo using by bioluminescence (Edinger et al. 1999). Using this technique, it was also possible to monitor growth and regression of labeled human cervical carcinoma (HeLa) cells engrafted into immunodeficient mice during chemotherapy (Sweeney et al. 1999). This model thus allows for real time, quantitative, spatial analysis of neoplastic cells and gene expression in vivo. Advantages of the luciferase detection system is that it is not found in mammalian organisms and thus operates with no background noise (Contag et al. 1995, 1997, 1998; Siragusa et al. 1999; Zhang et al. 1999). However, it will be confined to research applications since luciferin is unlikely to be administered to patients.

30.1.3
Near-Infrared Fluorescent Imaging (NIRF)

NIRF imaging, as opposed to the previously described techniques, relies on fluorochromes with an excitation and emission bandwidth in the near infrared range (≥ 650 nm). This wavelength, however, guarantees maximum penetration of the photons in the tissue which offers some better depth resolution (Chance 1998). Combined with the novel techniques described above, photon detection of several centimeters depth is feasible by NIRF (Chance 1998; Ntziachristos et al. 2000).

Fluorochromes in the NIRF range can be applied to design activatable optical contrast agents. The underlying principle of these molecular beacons is that fluorochromes with an overlapping excitation and emission spectrum are mounted in close proximity on a carrier molecule (Weissleder et al. 1999). In this native state mutual energy transfer takes place amongst the fluorochromes which leads to efficient quenching of their signal (Weissleder et al. 1999). Thus in the native state of the probe almost no optical signal will be detected. Only after proteolytic cleavage of the molecule by proteases, fluorochromes are being released and a bright fluorescent signal can be detected (Weissleder et al. 1999; Tung et al. 1999, 2000).

The design of the first generation of these optical contrast agents allowed for monitoring of lysosomal proteases such as cathepsin B and H (Weissleder et al. 1999). Using this approach, LX tumor in the submillimeter range implanted in athymic nude mice could be easily visualized (Weissleder et al. 1999). Biochemical refinements of the probe, however, allowed to improve probe specificity and pinpoint activation to specific enzymes (Tung et al. 1999, 2000). Using a cathepsin D-specific probe, tumor cell lines stably transfected with human cathepsin D confirmed enzyme-specific activation, while the non-transfected cells did not generate any NIRF signal (Tung et al. 1999). In vivo, cathepsin D-transfected tumors were easily distinguishable by a bright fluorescent signal from nontransfected control tumors (Tung et al. 2000).

More recently an optical probe was designed to probe for matrix metalloprotease-2 (MMP-2), an enzyme crucially involved in tumor progression and angiogenesis (Edwards and Murphy 1998; Bremer et al. 2001; Fang et al. 2000; Nelson et al. 2000). Treatment efficacy of MMP-2 inhibitors on MMP-2-expressing fibrosarcomas could be evaluated in vivo using this approach (Bremer et al. 2001). Imaging with protease-specific optical contrast agents thus serves a potent tool for detection of transgene expression and for molecular target assessment of novel protease inhibitors in vivo (Tung et al. 1999, 2000). The techniques described can potentially be applied to a variety of other endogenous proteases such as kinases, transferases, or polymerases.

30.2
Molecular Imaging by MRI/MRS

The major advantage of applying MR for molecular diagnostics is the excellent spatial resolution, complete depth penetration and the opportunity to derive additional physiological parameters from a living organism (e.g. by MRA or functional MR; Johnson et al. 1993; Smith et al. 1994). Because of the natural insensitivity of MRI for label detection, robust cellular amplification strategies are required to confer ade-

quate sensitivity of label detection. This is usually achieved by using targeted and/or "smart" MR contrast agents coupled with biological amplification strategies. Imaging marker genes that have been employed for MR-based molecular diagnostics are various enzyme systems and receptors with internalizing functions.

30.2.1
Enzymes as Imaging Marker Genes

30.2.1.1
Tyrosinase

Melanins are in part, due to their high metal-binding capacities, responsible for the high signal intensity of some melanomas on T1-weighted MR images (Okazaki et al. 1985; Enochs et al. 1989, 1997; Isiklar et al. 1995). Cells transfected with a vector bearing a cDNA insert coding for human tyrosinase yielded an eightfold higher signal intensity compared to the nontransfected control cell line (Kwon et al. 1988; Weissleder et al. 1997). The enzyme system involved in melanogenesis and in particular tyrosinase as one of the key enzymes can thus potentially be exploited to function as an imaging marker gene for MRI (Kwon et al. 1988; Weissleder et al. 1997). Moreover, melanins exhibit some cytotoxicity so that tyrosinase may serve not only diagnostic but also therapeutic purposes in oncological gene therapy protocols. Recently, tyrosinase mutants could be generated that are characterized by higher enzyme activity, different sorting patterns in non-melanotic cells and low endogenous toxicity (Simonova et al. 2000). When compared to the wild-type tyrosinase transfectants, truncated mutant expression resulted in higher mRNA levels which paralleled higher enzyme activity of the truncated mutants. Overall, these results indicate that the developed tyrosinase mutants hold promise as prodrug activation systems for tumoral gene therapy and as imaging marker genes.

30.2.1.2
Smart MR Contrast Agents

Smart MR contrast agents exploit specific enzymes as imaging marker genes in vivo. These agents are characterized by enzyme-induced conformational changes, which significantly alter their imaging properties in vivo (i.e. shorter T1-relaxation). To date, various approaches are under investigation to develop smart MR probes capable of detecting small amounts of specific enzyme activity in vivo.

β-Galactosidase (β-Gal) is a marker enzyme frequently used in molecular biology in tissue sections or in vitro assays in which the cleavage of an indicator substrate yields an opaque blue precipitate. An MR contrast agent that can be converted by β-Gal has recently been developed (Louie et al. 2000). In this contrast agent, access of water to the centrally chelated Gd is blocked by galactopyranose, a substrate removed by cleavage through β-Gal. Cleavage of the substrate makes Gd accessible to water and thus induces R1 changes detectable by MR. This system could be employed to monitor β-Gal expression in *Xenopus* embryos (Ahrens et al. 1998; Louie et al. 2000).

Other approaches to design smart MR contrast agents exploit spontaneous polymerization of paramagnetic products by oxidoreductase-mediated catalysis (A. Bogdanov, MGH; personal communication). The resulting paramagnetic polymers exhibit significantly stronger relaxation effects than substrate monomers resulting in a manifold MR signal amplification (A. Bogdanov, MGH; personal communication). This substrate thus also allows probing for marker-enzyme activity by means of MRI.

30.2.2
Transgene Imaging

Receptor-mediated cell internalization of superparamagnetic iron oxide particles represents another efficient approach of imaging gene expression by MR in vivo (Weissleder et al. 2000). The native transferrin receptor (TfR) is overexpressed in a variety of tumor cells and has thus been extensively investigated in cancer research (Cotten et al. 1993; Thorstensen and Romslo 1993) and for imaging applications (Kayyem et al. 1995; Koretsky et al. 1996). It was shown that receptor expression and regulation can be visualized by MR imaging, when a genetically engineered transferrin receptor (ETR) is targeted with a transferrin-labeled superparamagnetic iron oxide (Weissleder et al. 2000; Moore et al. 1998c). Expression levels of TfR-transfected 9L-glioma cells were correlated with MR signal intensities on T2-weighted images (Weissleder et al. 2000). In vivo, tumors expressing this up-regulated TfR were easily distinguishable from the same tumor cell line lacking the receptor by lower SNRs

compared to the control (Weissleder et al. 2000). These studies have provided proof of principle that it is feasible, through the mechanism of receptor overexpression, to image receptor gene expression using MR (Weissleder et al. 2000).

30.2.3
Magnetic Resonance Spectroscopy (MRS)

The ability of magnetic resonance spectroscopy to quantitatively detect specific metabolites has also been employed for monitoring gene expression in vivo. An imaging marker gene suitable for ^{31}P-magnetic resonance spectroscopy (^{31}P-MRS) is arginine kinase (AK) which is the invertebrate correlate to the vertebrate enzyme creatine kinase (CK) (Walter et al. 2000). The product of AK, phosphoarginine (PArg) served as an MRS-detectable marker so that tissue transduction with an AK-bearing vector was readily visible by a characteristic PArg peak in ^{31}P-MRS (Walter et al. 2000). A similar MRS imaging marker gene has been described for ^{19}F-MRS in vivo. Cytosine deaminase from yeast (yCD) (Stegman et al. 1999) converts the non-toxic pro-drug 5-fluorocytosine (5-FC) to the antimetabolite 5-fluorouracil (5-FU) widely used in conventional systemic chemotherapy. Both metabolites (5 FC and 5-FU) are detectable by a characteristic spectrum in ^{19}F-MRS in vivo. This concept not only allows for monitoring of the expression level of the IMG (yCD), but also measures the therapeutic levels of the chemotherapeutic agent (Stegman et al. 1999).

Advantages of MRS are direct quantification of metabolites in vivo and acquisition of time-resolved quantitative data on gene expression, which may potentially allow for noninvasive determination of kinetics of gene expression.

30.4
References

Ahrens ET, Rothbacher U, Jacobs RE et al (1998) A model for MRI contrast enhancement using T1 agents. Proc Natl Acad Sci USA 95:8443–8448

Boas DA, O'Leary MA, Chance B et al (1994) Scattering of diffuse photon density waves by spherical inhomogeneities within turbid media: analytic solution and applications. Proc Natl Acad Sci USA 91:4887–4891

Boppart SA, Tearney GJ, Bouma BE et al (1997) Noninvasive assessment of the developing Xenopus cardiovascular system using optical coherence tomography. Proc Natl Acad Sci USA 94:4256–4261

Bremer C, Tung C, Weissleder R (2001) In vivo molecular target assessment of matrix metalloproteinase inhibition. Nat Med 7:655–656

Chalfie M, Tu Y, Euskirchen G et al (1994) Green fluorescent protein as a marker for gene expression. Science 263:802–805

Chance B (1998) Near-infrared images using continuous, phase-modulated, and pulsed light with quantitation of blood and blood oxygenation. Ann NY Acad Sci 838:29–45

Cole NB, Smith CL, Sciaky N et al (1996) Diffusional mobility of Golgi proteins in membranes of living cells. Science 273:797–801

Contag CH, Contag PR, Mullins JI et al (1995) Photonic detection of bacterial pathogens in living hosts. Mol Microbiol 18:593–603

Contag CH, Spilman SD, Contag PR et al (1997) Visualizing gene expression in living mammals using a bioluminescent reporter. Photochem Photobiol 66:523–531

Contag CH, Jenkins D, Contag PR et al (2000) Use of reporter genes for optical measurements of neoplastic disease in vivo (in process citation). Neoplasia 2:41–52

Contag PR, Olomu IN, Stevenson DK et al (1998) Bioluminescent indicators in living mammals. Nat Med 4:245–247

Cotten M, Wagner E, Birnstiel ML (1993) Receptor-mediated transport of DNA into eukaryotic cells. Methods Enzymol 217:618–644

Dellian M, Yuan F, Trubetskoy VS et al (2000) Vascular permeability in a human tumour xenograft: molecular charge dependence. Br J Cancer 82:1513–1518

Edinger M, Sweeney TJ, Tucker AA et al (1999) Noninvasive assessment of tumor cell proliferation in animal models. Neoplasia 1:303–310

Edwards DR, Murphy G (1998) Cancer. Proteases – invasion and more. Nature 394:527–528

Enochs WS, Hyslop WB, Bennett HF et al (1989) Sources of the increased longitudinal relaxation rates observed in melanotic melanoma. An in vitro study of synthetic melanins. Invest Radiol 24:794–804

Enochs WS, Petherick P, Bogdanova A et al (1997) Paramagnetic metal scavenging by melanin: MR imaging. Radiology 204:417–423

Fang J, Shing Y, Wiederschain D et al (2000) Matrix metalloproteinase-2 is required for the switch to the angiogenic phenotype in a tumor model. Proc Natl Acad Sci USA 97:3884–3889

Flotte TR, Beck SE, Chesnut K et al (1998) A fluorescence video-endoscopy technique for detection of gene transfer and expression. Gene Ther 5:166–173

Fukumura D, Xavier R, Sugiura T et al (1998) Tumor induction of VEGF promoter activity in stromal cells. Cell 94:715–725

Girotti M, Banting G (1996) TGN38-green fluorescent protein hybrid proteins expressed in stably transfected eukaryotic cells provide a tool for the real-time, in vivo study of membrane traffic pathways and suggest a possible role for ratTGN38. J Cell Sci 109:2915–2926

Hickey MJ, Arain TM, Shawar RM et al (1996) Luciferase in vivo expression technology: use of recombinant mycobacterial reporter strains to evaluate antimycobacterial activity in mice. Antimicrob Agents Chemother 40:400–407

Isiklar I, Leeds NE, Fuller GN et al (1995) Intracranial metastatic melanoma: correlation between MR imaging characteristics and melanin content. Am J Roentgenol 165:1503–1512

Johnson GA, Benveniste H, Black RD et al (1993) Histology by magnetic resonance microscopy. Magn Reson Q 9:1–30

Kayyem JF, Kumar RM, Fraser SE et al (1995) Receptor-targeted co-transport of DNA and magnetic resonance contrast agents. Chem Biol 2:615–620

Koretsky A, Lin Y, Schorle H et al (1996) Genetic control of MRI contrast by expression of the transferrin receptor. Int Symp Magn Reson Med, New York, p 5471

Kwon BS, Haq AK, Kim GS et al (1988) Cloning and characterization of a human tyrosinase cDNA. Prog Clin Biol Res 256:273–282

Louie AY, Huber MM, Ahrens ET et al (2000) In vivo visualization of gene expression using magnetic resonance imaging. Nat Biotechnol 18:321–325

Mahmood U, Tung CH, Bogdanov A Jr et al (1999) Near-infrared optical imaging of protease activity for tumor detection. Radiology 213:866–870

Matz MV, Fradkov AF, Labas YA et al (1999) Fluorescent proteins from nonbioluminescent Anthozoa species (see comments; published erratum appears in Nat Biotechnol 1999, 17:1227). Nat Biotechnol 17:969–973

Monsky WL, Fukumura D, Gohongi T et al (1999) Augmentation of transvascular transport of macromolecules and nanoparticles in tumors using vascular endothelial growth factor. Cancer Res 59:4129–4135

Moore A, Sergeyev N, Bredow S et al (1998a) A model system to quantitate tumor burden in locoregional lymph nodes during cancer spread. Invasion Metastasis 18:192–197

Moore A, Marecos E, Simonova M et al (1998b) Novel gliosarcoma cell line expressing green fluorescent protein: a model for quantitative assessment of angiogenesis. Microvasc Res 56:145–153

Moore A, Basilion JP, Chiocca EA et al (1998c) Measuring transferrin receptor gene expression by NMR imaging. Biochim Biophys Acta 1402:239–249

Morrey JD, Bourn SM, Bunch TD et al (1991) In vivo activation of human immunodeficiency virus type 1 long terminal repeat by UV type A (UV-A) light plus psoralen and UV-B light in the skin of transgenic mice. J Virol 65:5045–5051

Nelson AR, Fingleton B, Rothenberg ML et al (2000) Matrix metalloproteinases: biologic activity and clinical implications. J Clin Oncol 18:1135–1149

Ntziachristos V, Yodh AG, Schnall M et al (2000) Concurrent MRI and diffuse optical tomography of breast after indocyanine green enhancement. Proc Natl Acad Sci USA 97:2767–2772

Okazaki M, Kuwata K, Miki Y et al (1985) Electron spin relaxation of synthetic melanin and melanin-containing human tissues as studied by electron spin echo and electron spin resonance. Arch Biochem Biophys 242:197–205

Prasher DC, Eckenrode VK, Ward WW et al (1992) Primary structure of the Aequorea victoria green-fluorescent protein. Gene 111:229–233

Presley JF, Cole NB, Schroer TA et al (1997) ER-to-Golgi transport visualized in living cells (see comments). Nature 389:81–85

Simonova M, Weissleder R, Sergeyev N et al (1999) Targeting of green fluorescent protein expression to the cell surface. Biochem Biophys Res Commun 262:638–642

Simonova M, Wall A, Weissleder R et al (2000) Tyrosinase mutants are capable of prodrug activation in transfected non-melanotic cells. Cancer Res 60:6656–6662

Siragusa GR, Nawotka K, Spilman SD et al (1999) Real-time monitoring of Escherichia coli O157:H7 adherence to beef carcass surface tissues with a bioluminescent reporter. Appl Environ Microbiol 65:1738–1745

Smith BR, Johnson GA, Groman EV et al (1994) Magnetic resonance microscopy of mouse embryos. Proc Natl Acad Sci USA 91:3530–3533

Stegman LD, Rehemtulla A, Beattie B et al (1999) Noninvasive quantitation of cytosine deaminase transgene expression in human tumor xenografts with in vivo magnetic resonance spectroscopy. Proc Natl Acad Sci USA 96:9821–9826

Sweeney TJ, Mailander V, Tucker AA et al (1999) Visualizing the kinetics of tumor-cell clearance in living animals. Proc Natl Acad Sci USA 96:12044–12049

Tearney GJ, Brezinski ME, Bouma BE et al (1997) In vivo endoscopic optical biopsy with optical coherence tomography (see comments). Science 276:2037–2039

Thompson EM, Adenot P, Tsuji FI et al (1995) Real time imaging of transcriptional activity in live mouse preimplantation embryos using a secreted luciferase. Proc Natl Acad Sci USA 92:1317–1321

Thorstensen K, Romslo I (1993) The transferrin receptor: its diagnostic value and its potential as therapeutic target. Scand J Clin Lab Invest Suppl 215:113–120

Tung CH, Bredow S, Mahmood U et al (1999) Preparation of a cathepsin D sensitive near-infrared fluorescence probe for imaging. Bioconjug Chem 10:892–896

Tung C, Mahmood U, Bredow S et al (2000) In vivo imaging of proteolytic activity using a novel molecular reporter. Cancer Res 60:4953–4958

Walter G, Barton ER, Sweeney HL (2000) Noninvasive measurement of gene expression in skeletal muscle. Proc Natl Acad Sci USA 97:5151–5155

Weissleder R, Simonova M, Bogdanova A et al (1997) MR imaging and scintigraphy of gene expression through melanin induction. Radiology 204:425–429

Weissleder R, Tung CH, Mahmood U et al (1999) In vivo imaging of tumors with protease-activated near-infrared fluorescent probes. Nat Biotechnol 17:375–378

Weissleder R, Moore A, Mahmood U et al (2000) In vivo magnetic resonance imaging of transgene expression. Nat Med 6:351–355

Wu J, Perelman L, Dasari RR et al (1997) Fluorescence tomographic imaging in turbid media using early-arriving photons and Laplace transforms. Proc Natl Acad Sci USA 94:8783–8788

Yokoe H, Meyer T (1996) Spatial dynamics of GFP-tagged proteins investigated by local fluorescence enhancement (see comments). Nat Biotechnol 14:1252–1256

Zhang L, Hellstrom KE, Chen L (1994) Luciferase activity as a marker of tumor burden and as an indicator of tumor response to antineoplastic therapy in vivo. Clin Exp Metastasis 12:87–92

Zhang W, Contag PR, Madan A et al (1999) Bioluminescence for biological sensing in living mammals. Adv Exp Med Biol 471:775–784

CHAPTER 31

MR Contrast Agents for Molecular and Cellular Imaging

31

JEFF W.M. BULTE, L. HENRY BRYANT JR., JOSEPH A. FRANK

Contents

31.1

Introduction

Although the principles of NMR were elucidated in the 1940s, the ability to produce images of humans necessitated the construction of large bore magnets in the late 1970s. Since the installation of the first MRI unit in the US, the hardware and software has continued to evolve, yet all the images produced by MRI depend on the same basic principles. A full discussion of the basic principles of magnetic resonance imaging is beyond the scope of this chapter; a short overview is provided here. In brief, protons (positively charged hydrogen nuclei) can be induced to emit signals, and this can be subsequently processed into images. MR takes advantage of the natural abundance of hydrogen atoms in (intra- and extra-)cellular water, lipids, proteins, and other more complex molecules, all of which are readily found in most tissues. When a subject is placed in a strong magnetic field the water hydrogen nuclei (protons) in the body align with the main (z) axis of the magnet in an equilibrium state. The protons spin and precess around the main axis of the magnetic field resulting in a net magnetization vector of which the frequency of precession is proportional to the strength of the magnetic field. To perturb the hydrogen nuclei from their equilibrium position, excitation radiofrequency (RF) pulses at the appropriate precessional (i.e. Larmor) frequency are used. The RF pulse originates from the head or body coil of the MR unit which acts as antennae to transmit pulses and receive signals from the body. This excitation effectively tips the proton spins away from the direction of the main magnetic field into a perpendicular plane or xy axis. The duration and magnitude of the RF pulse determines the degree of excitation produced. A 90 degree RF pulse brings the proton spins into the xy plane; a 180 degree RF pulse produces twice the degree of rotation (either into the z or opposite xy plane). The NMR signal is then induced in the coils surrounding the subject. These signals decay exponentially with specific time constants or relaxation times known as the longitudinal relaxation time T1 and the transverse relaxation time T2. The T1 and T2 relaxation times reflect the two distinct ways the NMR signal disappears following an RF excitation pulse returning the magnetization vector to its equilibrium position. By varying the pattern and timing of the RF pulse in combination with altering the main magnetic field using additional magnetic field gradients, one can impart spatial information to the NMR signals and thus create an MR image.

The combination of specifically orchestrated RF and magnetic field gradient pulses used to create an MR image is known as a pulse sequence. Pulse sequences used for MR imaging are generally known as spin echo (SE), gradient recalled echo (GRE), and inversion recovery (IR) pulse sequences. Spin echo pulse sequences are the most versatile and commonly used sequence to image patients today. The SE pulse sequence produces images that favor or are "weighted" towards one or the other relaxation times, i.e. T1 weighted (T1 W) or T2 weighted (T2 W). The weighting is produced by altering the scanning variables known as repetition time (TR) and echo time

(TE). The TR is the time between the repetition of pulse sequences, while TE is the interval time at which 180° RF pulses are given to generate echoes.

MR image contrast is largely determined by the nuclear magnetic relaxation times of tissues. MR contrast agents shorten the relaxation times and can affect T1 as well as T2 by dipole-dipole interactions. Their net effectiveness is expressed as relaxivity (R), which represents the reciprocal of the relaxation time per unit concentration of metal, with units $mM^{-1}s^{-1}$. MR contrast agents can be classified into certain categories depending on their property of choice. One simplified classification is "T1 agents" and "T2 agents". This classification is according to their predominant effect on relaxation, but because of dephasing effects, the T2 relaxivity (R2) is always higher than R1 (even for T1 agents).

Contrast agents that primarily affect T1 are all based on paramagnetic chelates containing Gd(III), Mn(II) or Fe(III). To increase their relaxivity and/or blood half-life, they can be linked to a larger molecule or polymer backbone. Examples of such "carrier" molecules include albumin (Lauffer and Brady 1985; Schmiedl et al. 1986; Lauffer et al. 1998), poly-L-lysine (Shreve and Aisen 1986; Bogdanov et al. 1993), dextran (Gibby et al. 1989; Wang et al. 1990), liposomes (Unger et al. 1993; Storrs et al. 1995; Trubetskoy et al. 1995), and Starburst dendrimers (Dendritech, Inc., Midland, MI) (Wiener et al. 1994; Bryant et al. 1999). Contrast agents that predominantly affect T2 include Dy(III) containing polymers (Bulte et al. 1998b; Zaharchuk et al. 1998) or macromolecules (Bulte et al. 2000b) and superparamagnetic iron oxides (Stark et al. 1988; Shen et al. 1993; Bulte and Brooks 1997), where the magnetically active core is either maghemite (γFe_2O_3) or magnetite (Fe_3O_4).

In general these superparamagnetic iron oxide nanoparticles have, on a (milli)molar metal basis, a significantly higher relaxivity than the paramagnetic (gadolinium) chelates. In addition, there are usually several thousand iron atoms per particle, thus amplifying their effectiveness as a contrast agent. For example, when a particle (size range in the order of 10–20 nm) containing 5000 iron atoms is linked to a few mabs and exhibits a measured relaxivity of 100 $mM^{-1}s^{-1}$, the actual molecular relaxivity will be in the order of 500,000 $mM^{-1}s^{-1}$. But perhaps even more important is the fact that these magnetic nanoparticles can be detected with scanning techniques that are very sensitive to local differences in magnetic susceptibility and microscopic field inhomogeneities, which causes a rapid

dephasing of protons (T2 shortening), and can be detected without refocusing of 180° pulses (T2* effect). Of paramount importance here is that water protons at distant sites can be affected, leading to a "blooming effect", i.e. an amplification of signal changes. Depending on the applied field strength, chemical environment, and tissue biodistribution, contrast agents enhance the relaxation of water protons by different mechanisms. Much of our understanding has come from variable-field relaxometry, and analyzing the resulting nuclear magnetic relaxation dispersion (NMRD) profiles. In our laboratory, a custom-built instrument is available that can obtain both 1/T1 and 1/T2 NMRD profiles. Figure 31.1 shows an example of such profiles for an ensemble of superparamagnetic iron oxide nanoparticles. We consider this variable-field T1–T2 relaxometer a key instrument for the development and analysis of new contrast agents. It also serves as a routine instrument to determine the specific uptake of cellular and molecular contrast agents in vitro, analogous to the use of a liquid scintillation counter in nuclear medicine. It is important to develop and evaluate theories of the underlying magnetic properties of the contrast agents with regard to attributing the relaxation mechanisms of water protons interacting with inner- or outer-sphere effects of the paramagnetic ion or iron oxide core (Bulte et al. 1999a, b; Roch et al. 1999). By understanding these contrast mechanisms, intelligent selection of agents that will deliver more "bang for the buck" can be made and therefore allow for the synthesis of appropriate molecular or cellular contrast agents.

MR contrast agents can also be classified as targeted and/or activated agents. Targeted contrast agents are directed at a specific receptor expressed by certain cells, leading to a selective uptake and retention of the agent in the tissue. These contrast agents are then either metabolized or excreted. Activated contrast agents are designed to be chemically responsive to physiological and metabolic states within cells and organs. Targeted and activated MR contrast agents are now of great research interest because they can potentially expand the role of MR imaging to areas beyond routine morphologic/pathologic imaging, that is, into the realm of functional metabolic imaging of normal and disease states including altered gene expression. The non-invasive nature of MR imaging, its high spatial resolution (up to 20 μm isotropic, i.e. near cellular resolution), and the ability of 3D volume acquisition continuously or repeatedly at different time points, all make contrast-enhanced MR

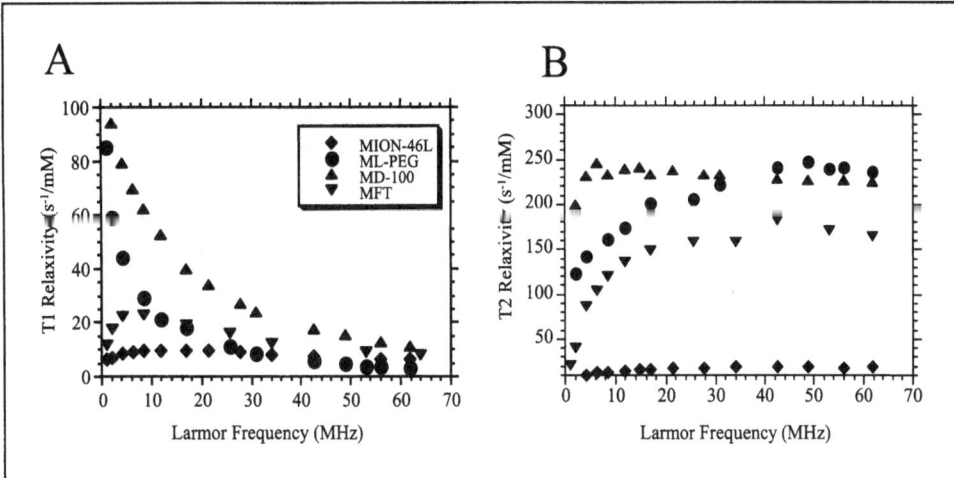

Fig. 31.1. T1 (**A**) and T2 (**B**) nuclear magnetic resonance dispersion (NMRD) profiles of various superparamagnetic iron oxide nanoparticles, obtained at 37 °C using a variable-field T1-T2 relaxometer. Shown are the profiles for MION-46L (Bulte et al. 1999), ML-PEG (PEGylated magnetoliposomes; Bulte et al. 1999), MD-100 (magnetodendrimers; Bulte et al. 2000) and MFT (magnetoferritin; Bulte et al. 1994)

imaging a unique and powerful tool to study biological processes at the molecular and cellular level. Indeed, the interest in the development of molecular and cellular MR contrast agents appears to have increased significantly over the last few years, and the present chapter attempts to give an overview of the latest developments in the field. First, however, we describe some of the paramagnetic, non-targeted macromolecular agents that have been developed recently for imaging of the blood pool, since these agents can be further derivatized for their use as molecular or cellular agents.

31.2
Blood Pool Agents

Several macromolecular-based MR contrast agents have been synthesized and studied as blood pool imaging agents. These include the albumins (Lauffer et al. 1986; Schmiedl et al. 1986; Brasch 1991), the dextrans (Wang et al. 1990; Li et al. 1992), and the polylysines (Spanoghe et al. 1992; Bogdanov et al. 1993). However, each of these have drawbacks when used in vivo. The albumins tend to be immunogenic and their flexible composition does not provide the necessary rigidity to obtain high relaxivities. The dextrans, at a molecular weight needed to increase vascular retention (MW >9400), exhibit a decrease in their relaxivities which has been attributed to limited water exchange. The synthesized polylysines have a broad distribution of molecular weights and therefore do not exist as a single chemical entity. Dendrimers overcome most of these obstacles. They are monodisperse rigid spheres with very narrow ranges of molecular weight distributions (Tomalia 1994). They are available in a variety of sizes and peripheral functional groups and their chemical derivatization protocols (such as covalent attachment of Gd chelates) are reproducible.

Dendrimer gadolinium poly-chelates are a class of MR imaging agents with large proton relaxation enhancements and high molecular relaxivities. Wiener et al. (1994) first demonstrated that the covalent attachment of gadolinium chelates to dendrimers have the potential to be blood-pool MR T1 imaging agents for use in MR angiography. The synthesis involved the covalent attachment of the acyclic Gd-DTPA chelate to a G2 and G6 dendrimer utilizing a stable thiourea linkage between the chelate and dendrimer. These dendrimer-based MR imaging agents had molar relaxivities up to six times higher than for clinically used gadolinium chelates because of the higher molecular weight of the dendrimer. The authors demonstrated the potential usefulness of these agents for vascular imaging by being able to delineate the vascular system of a rat for at least up to 1 h. Not necessarily confined to complexation of Gd for T1-weighted MRI, a dysprosium chelate has been attached to a dendrimer in a similar fashion and opens up the opportunity for T2* MR imaging agents to utilize the macromolecular characteristics that the dendrimer provides (Bulte et al. 1998b). The incorporation of Dy provides a unique T2 relaxation agent which may be important for tissue perfusion studies using MRI.

There are three main parameters that dictate achieving maximum T1 relaxivity for dendrimer-based gadolinium chelates; the exchange time of the water molecule being bound (at least transiently) to the Gd and in bulk solution; the electron relaxation time of the Gd; and the tumbling time of the Gd in solution (Lauffer 1987). The observed relaxivities for the blood pool imaging agents are derived mainly from the longer tumbling times. Since the shortest (fastest) of the three parameters modulate(s) the relaxivity, the longer tumbling time allows the other two parameters to dominate. The characteristic peak in the relaxivities at about 25 MHz has been attributed to the dispersion of the Gd electron relaxation time which is magnetic-field dependent. A limitation in achieving the full expected relaxivities, because of the slow exchange time of the interacting water molecule, was observed and verified by ^{17}O NMR studies of the G3, G4, and G5 dendrimer-based MR imaging agents (Toth et al. 1996). It was concluded that modification of the chelating ligand may result in faster water exchange and therefore higher relaxivities. The plateau for the T1 relaxivities of the G5, G7, G9 and G10 dendrimer-based Gd-DOTA complexes in Fig. 31.2 provides evidence of even more severe limitation of achieving full relaxivities as the generation

(molecular weight) of the dendrimer increases (Bryant et al. 1999). Since contrast depends on the coordinated water molecule interacting with bulk water, a long residence time at the Gd limits the relaxivity and therefore would limit the observed contrast.

The dendrimer-based MR agents were found to have similar blood pool properties as Gd-DTPA-polylysine in pigs (Adam et al. 1994). No statistical differences in relative signal intensities were observed in various organs. In rats, strong tumor rim enhancement and detailed angiographic definition of peritumoral vessels was observed (Schwickert et al. 1995). The vascular enhancement of a rat can be seen in Fig. 31.3 at 120 min after injection of a similar dendrimer-based gadolinium chelate. In MR imaging of canine breast tumors, a delayed tumor clearance was observed compared to the clinically used gadopentate dimeglumine (Adam et al. 1996). A minimum effective dose of 0.02 mmol/kg of dendrimer-based gadolinium chelate was needed for visualization of the mediastinum, abdomen and lower limbs of rabbits on 3D time of flight magnetic resonance angiography of the body (Bourne et al. 1996). The pharmacokinetic and biodistribution was found to depend on the molecular weight of the dendrimer as well as the type of terminal groups (Margerum et al. 1997). As shown for the biodistribution of G10-Gd-DOTA in the rat in Fig. 31.4, these macromolecules are rapidly taken up by the liver. Hepatic localization may be decreased and blood half-lives increased by the covalent attachment of polyethylene glycol (Demsar et al. 1998).

A kinetic theory for describing the dynamic properties of dendrimer-based gadolinium chelates has been developed (Demsar et al. 1998). The method has clinical applications based on its potential for pixel-by-pixel mapping. The first covalent attachment of tetraaza macrocycles to the terminal phosphorous group of a phosphorous-sulfur containing dendrimer has been achieved (Prévôté et al. 1999). The coordination of Gd to the coupled macrocycle still needs to be explored, but if possible, it would open up the possibility of having two nuclei, ^{157}GD and ^{31}P, which are detectable by MRI and may be useful in multinuclear MRS. In addition, the reported prototropic exchange of the bound water protons for tetraamide phosphonate macrocycles may allow for responsive MR agents (Zhang et al. 2001) (see also Sect. 31.6).

In addition to conjugation of the Gd chelate to a macromolecule, targeting of relatively low molecular weight monomeric Gd chelates to proteins already present in vivo allows the small Gd chelates to behave as

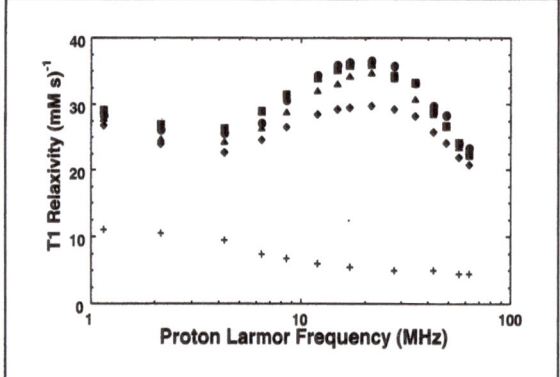

Fig. 31.2. T1 NMRD profiles for G5 (*diamonds*), G7 (*pyramids*), G9 (*circles*) and G10 (*squares*)-Gd-DOTA at 37 °C (Bryant et al. 1999). For comparison, the profile for Gd-DOTA-BzNO$_2$ (*crosses*) is shown. Notice the characteristic peak in the relaxivities for the blood pool agents compared to the extravascular agent as a result of the increased molecular weight

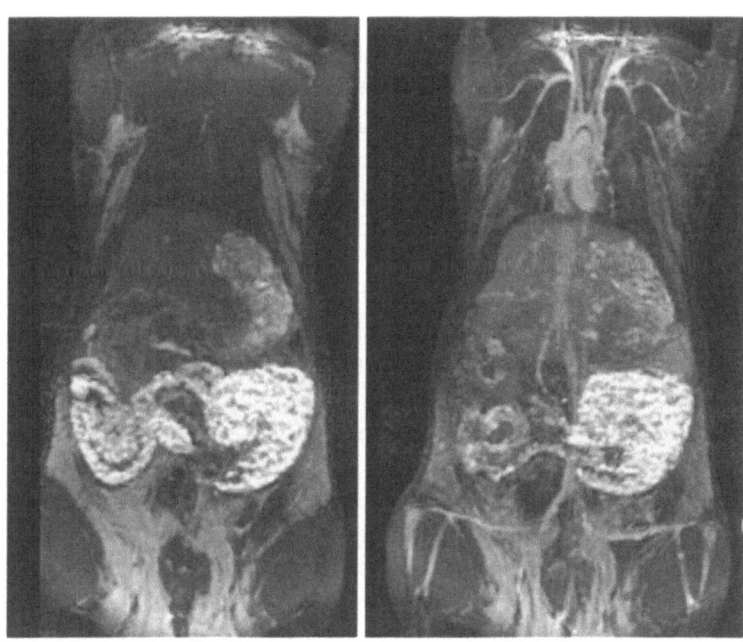

Fig. 31.3. 3D Time-of-flight MR angiogram of a rat before (*left*) and 120 min (*right*) after injection of 0.05 mmol/kg G9-Gd-DOTA (Bryant et al. 2000b)

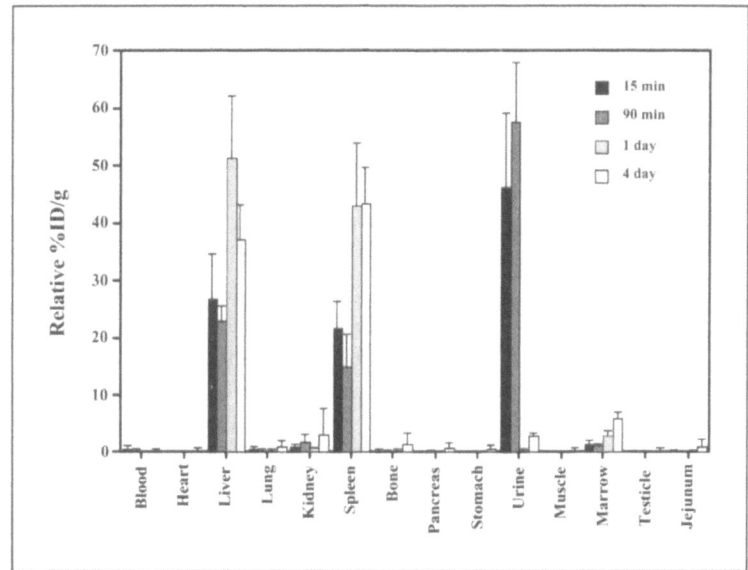

Fig. 31.4. Biodistribution of G10-Gd-DOTA after injection of 10 μCi of ^{153}Gd (1.0 μmole total Gd) (Bryant et al. 2000)

macromolecular MR contrast agents (Aime et al. 1998; Caravan et al. 1999). An in vivo equilibrium exists between the monomeric Gd chelate being "free" and bound to the macromolecular protein. When there is non-covalent interaction of the Gd chelate with the macromolecular protein, the relaxivity of the Gd ion significantly increases (by a factor of about 10). The macromolecule has long vascular retention times and can be used as an 'in-situ' blood pool agent. When the Gd chelate is "free" in vivo it behaves as other clini-

cally used Gd chelates such as gadopentate dimeglumine and undergoes renal excretion. Since the Gd chelate is in equilibrium and the monomeric Gd chelate is continuously being removed by the system, eventually all of the Gd chelate will be removed by "FDA-approved routes". A good example of such an agent is MS-3253, a derivative of Gd-DTPA, developed by Lauffer et al. (Parmelee et al. 1997) which interacts with human serum albumin (HSA) in a noncovalent interaction between the hydrophobic pockets on HSA (presumably two) and a carbon-backbone-attached diphenylcy-clohexyl phosphonooxymethyl group on the DTPA. The agent has gone through phase 3 clinical trials, and good vascular enhancement was observed in the hind-quarters of rabbits for up to at least 1 h after injection of MS-325 (Lauffer et al. 1998). Another derivatized Gd-DTPA chelate is MP-2269, which has the potential to interact with proteins (Wallace et al. 1998). MP-2269 has the methyl of one of the carboxymethyl groups derivatized with a lipophilic chain. It has been shown that the increase in the number of benzyloxy-methyl groups covalently attached to the carbon-back-bone of DTPA or DOTA increases the interaction with HSA and therefore increases the relaxivities (Aime et al. 1996; Cavagna et al. 1997). Besides targeting HSA, a sulfonamide-derivatized Gd(III)DTPA complex has been synthesized to selectively target carbonic anhydrase and the relaxometric investigations carried out (Anelli et al. 2000). No interaction with HSA was observed.

The chelates used have been either derivatized DTPA or DOTA in which a lipophilic moiety has been attached to the acyclic or macrocyclic core. Recently, a Gd(III)calixarene complex has been shown, based on relaxometric studies, to interact with HSA presumably through the tetra-phenyl rings (Bryant et al. 2000c). An approximate 10-fold increase in the T1 relaxivity was observed at clinically relevant field strengths and allows the potential to develop a new class of blood pool agents based on calixarenes.

31.3
Magnetically Labeled Molecules

In 1975, Köhler and Milstein introduced the monoclonal antibody (mab) technology (Köhler and Milstein 1975), creating the opportunity to produce antibodies in large quantities with a high degree of purity and specific for a single antigenic epitope. With the ability to conjugate radioisotopes to mabs it was shown a few years later that tumor nodules could be detected specifically and non-invasively, initially using ^{131}I-labeled anti-CEA mabs (Mach et al. 1981). This was followed by reports that improved radioimmunodetection could be achieved when Fab or F(ab′)$_2$ fragments were used instead of the intact immunoglobulin (Larson et al. 1983; Moldofsky et al. 1984).

When MR imaging was introduced in the early 1980s along with gadolinium chelates as MR contrast agents, the preparation and use of magnetically labeled antibodies seemed a natural extension of the earlier work carried out with radiolabeled antibodies. In this way the detailed anatomic information on the MR images could then potentially be specifically altered or marked in order to detect (neoplastic) disease in its earliest stages. The first reports on the use of Gd-DTPA-labeled mab appeared around 1985 and described the attachment of a few to a maximum of 15 chelates per antibody (Unger et al. 1985; Anderson-Berg et al. 1986; Curtet et al. 1986; Macrì et al. 1989). The outcome of these early studies was disappointing, in that no specific contrast enhancement could be observed. Using the exact same tumor-animal model system but [153]Gd instead of [157]Gd, Anderson-Berg et al. (1986) showed that this is an issue of sensitivity: radioisotopes can be detected as tracer molecules in nanomolar concentrations, whereas paramagnetic chelates require micro- to millimolar doses in order to be detectable. Since the amount of linkable chelates per molecule is limited to approximately 5–10 (too heavily loaded mabs lose their immunoreactivity), other strategies had to be pursued.

One strategy is the use of large carrier molecules such as poly-L-lysine (Shreve and Aisen 1986; Göhr-Rosenthal et al. 1993; Curtet et al. 1998) or crosslinked albumin-gelatin complexes (Kornguth et al. 1987), which are loaded with chelates, and covalently linked to mabs. In this approach, about 50–100 chelates can be bound per mab. A detailed review about the achieveable "molecular" relaxivity (i.e. total relaxation enhancement per mab or molecule taking into account all attached metal ions) using different strategies has been published elsewhere (Nunn et al. 1997). However, it should be kept in mind that in addition to (not always that simple) dose-signal enhancement requirements there are physiological biodistribution barriers which eventually impose their own limitations on the feasibility of immunospecific imaging using paramagnetically labeled antibodies. For instance, larger molecules or complexes may have a reduced blood half-life and/or vascular permeability, both of which can reduce the specific uptake in the target tissue. Another factor is the total dose needed for specific (tumor) detection; an administered dose of 2 mg antibody per mouse (100 mg/kg) translates to 7 g of protein for an equivalent human study.

Instead of linear polymers, such as poly-L-lysine, spherical molecules or complexes that allow multiple attachment of chelates may be used. Dendrimers are ideal molecules for this purpose: a molecular relaxiv-

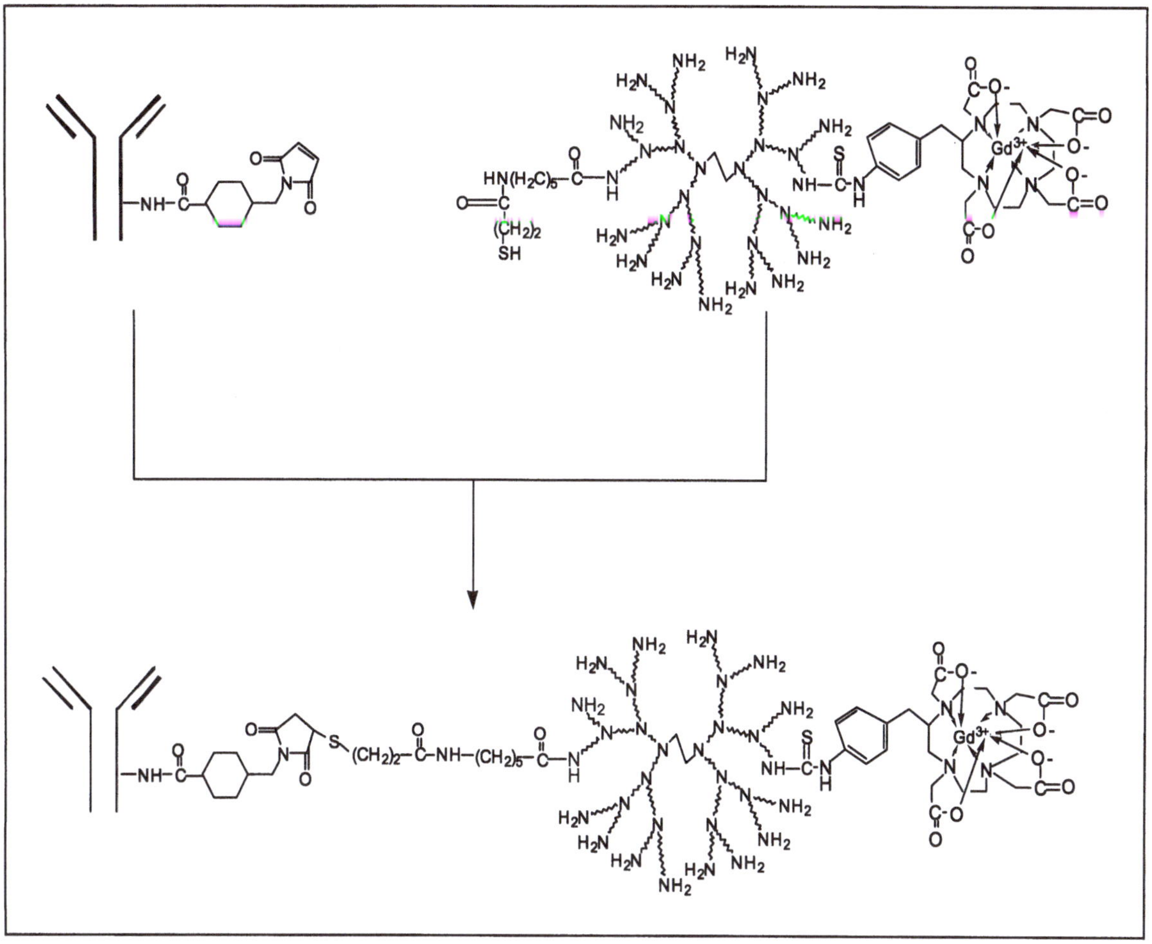

Fig. 31.5. Schematic diagram for the covalent attachment of dendrimer-Gd-DOTA to a monoclonal antibody or other biological entity

ity of 60,000–80,000 $mM^{-1}s^{-1}$ can be achieved for a generation 10 (G10) dendrimer (Bryant et al. 1999). These molecules allow the attachment of mabs (Wu et al. 1994) or other receptor-specific molecules (Wiener et al. 1997), and are currently being further explored for that purpose. As shown schematically in Fig. 31.5, the chemistry for the attachment of monoclonal antibodies or other receptor-targeted molecules should be straightforward. Alternatively, particulate emulsions or liposomes may be employed (Lanza et al. 1998; Sipkins et al. 1998, 2000; Anderson et al. 2000), but these large complexes may exhibit limited tissue penetration.

Immunoglobulins can also be covalently linked to the polysaccharide coat of the iron oxide-coated dextrans using a periodate-oxidation/borohydride-reduction method, which, through the formation of Schiff bases as intermediates, covalently links the amine (lysine) groups of the mab to the alcohol groups of the

dextran (Sanderson and Wilson 1971; Dutton et al. 1979). MION-(46L) iron oxide nanoparticles have been conjugated this way to polyclonal IgG for detection of induced inflammation (Weissleder et al. 1991), to mab fragments for the specific visualization of cardiac infarct (Weissleder et al. 1992), and to intact mabs for immunospecific detection of intracranial small cell lung carcinoma (Remsen et al. 1996), ICAM-1 gene expression (Bulte et al. 1998a), and oligodendrocyte progenitors (Bulte et al. 1999d). Alternative ways of attaching mabs to magnetic nanoparticles include glutaraldehyde crosslinking (Renshaw et al. 1986), complexing through ultrasonication (Cerdan et al. 1989; Suwa et al. 1998), using the biotin-streptavidin system (Bulte et al. 1992) and amine-sulfhydryl group linkage (Tiefenauer et al. 1993, 1996). For in vivo applications, limited success (e.g. true specific immunodetection) has been achieved thus far but this is likely to improve with the development of smaller nanoparticles that facilitate endothelial penetration and exhibit longer blood half lives.

Either paramagnetic chelates or magnetic nanoparticles can be linked to molecules other than mabs in order to confer specificity for a targetable receptor. For the group of paramagnetic agents, it has been demonstrated that "folated" gadolinium-dendrimers can be targeted in vitro to folate-receptor bearing leukemic cells (Wiener et al. 1997), and induce significant specific changes in relaxation times that is inhibitable by free, non-conjugated folate. This may be used for in vivo imaging of folate receptor-overexpressing tumors (Konda et al. 2000), but further work including the use of non-targeted polymer controls is needed. Another approach of conferring specificity to a paramagnetic label is to link it to an antisense oligonucleotide; a specific proton relaxation enhancement has been achieved for 5S rRNA as a macromolecular target and its labeled complimentary 6mer antisense sequence (Hines et al. 1999).

Studies in the 1990s demonstrated the ability to conjugate superparamagnetic iron oxide particles (SPIO) to arabinogalactan allowing for directed uptake by the asialoglycoprotein receptor on hepatocytes versus using SPIO coated with dextrans which are taken up by the reticuloendothelial system (RES) (Josephson et al. 1990; Reimer et al. 1990). Similar results were obtained when asialofetuin was used as a coating (Schaffer et al. 1993), and may be useful for improved detection of hepatocellular carcinoma. (Synthetic) peptides can also be linked to MION-46 or other very small iron oxide particles. For instance, cholecystokinin- (Reimer et al. 1994) and secretin- (Shen et al. 1996) linked particles have been employed for MR visualization of their respective pancreatic receptor and may aid in the diagnosis of pancreatic cancer. Transferrin is another example of a targetable protein, since certain tumors are known to overexpress transferrin receptors. Transferrin-iron oxide particles have been used for specific detection of gliosarcoma (Moore et al. 1998; Weissleder et al. 2000) and breast carcinoma (Kresse et al. 1998), with and without transfection of the Tfr-encoding gene, respectively.

In general, iron oxide nanoparticles require stabilization in order to prevent aggregation. Most commonly this is accomplished by a surface coating of dextran. Another approach has been to use an apoferritin coat to synthesize superparamagnetic iron oxides (Meldrum et al. 1992). The measured $1/T1$ and $1/T2$ of 8 and 175 mM^{-1} (Fe) s^{-1}, respectively, at body tempera-

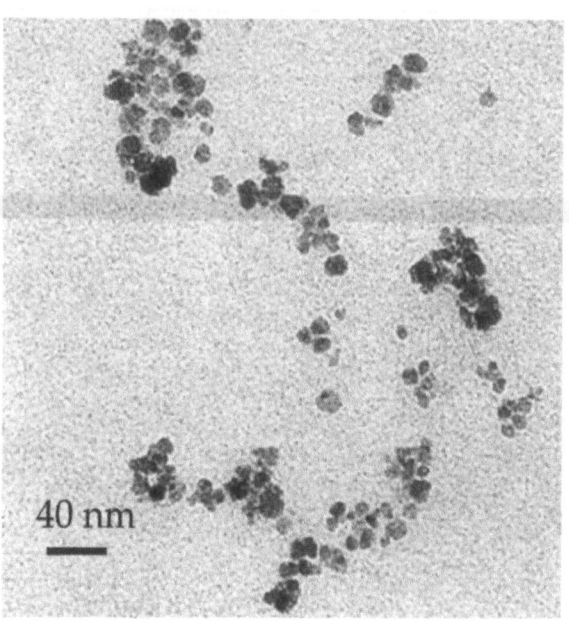

Fig. 31.6. Transmission electron microscopy results for magnetodendrimer oligomers synthesized with 100:1 Fe:dendrimer (G=4.5) stoichiometry (Bulte et al. 2000; courtesy of T. Douglas)

ture and clinical field strengths (Bulte et al. 1994) and the unusually high R2/R1 ratio of 22 is thought to arise from ideal core composition, with no evidence of crystalline paramagnetic inclusions. The nanodimensional biomimetic protein cage for the iron oxide core is highly conserved across species and its matrix is convenient for complexing molecules, which may provide the desired specificity for further development of targeted contrast agents. In vivo targeting of magnetoferritin did not result in binding to ferritin receptors as was demonstrated using apoferritin pre-saturation studies (Bulte et al. 1995).

Another approach for the synthesis and stabilization of ferromagnetic iron oxide nanoparticles is the use of carboxylated poly(amidoamine) PAMAM dendrimers (Bulte et al. 2000a). Oxidation of Fe(II) at slightly elevated pH and temperature results in the formation of highly soluble nanocomposites of iron oxides and dendrimer which are stable under a wide range of temperature and pH. These aggregates appear to have an overall size of 20–30 nm, consistent with the oligomeric nature of the composite material (Fig. 31.6). NMRD profiles of solvent (water) protons revealed unusually high T1 and T2 relaxivities, which make these materials excellent candidates as contrast agents for MR imaging (see below).

31.4
Magnetically Labeled Cells

Our understanding of the biological function of cells and their interaction with(in) tissues comes mostly from a static viewpoint obtained by light or electron microscopy, techniques which are basically unaltered since their inception about 350 and 60 years ago, respectively. In order to determine the history and fate of transplanted cells, including their migration in vivo, cells are currently labeled ex vivo using a vital dye (e.g. a fluorochrome), a thymidine analogue (e.g. BRDU), or a transfected gene (e.g. LacZ or GFP), which can be visualized using (immuno)histochem-

ical procedures following tissue removal at a particular given time point. Obviously, the use of therapeutic cells in humans will require a technique that can monitor their fate non-invasively and preferably repeatedly, in order to take a momentary "snapshot" assessment of the cellular biodistribution at a particular given time point (see Fig. 31.7). MR imaging offers the "dye and let live" approach: if cells can be labeled ex vivo with MR contrast agents, then their fate could possibly be monitored in vivo non-invasively and repeatedly, so that the dynamics of cellular movements and interaction with the surrounding tissues can be obtained. Clearly, this technique holds enormous potential and could potentially revolutionize the field of cell biology.

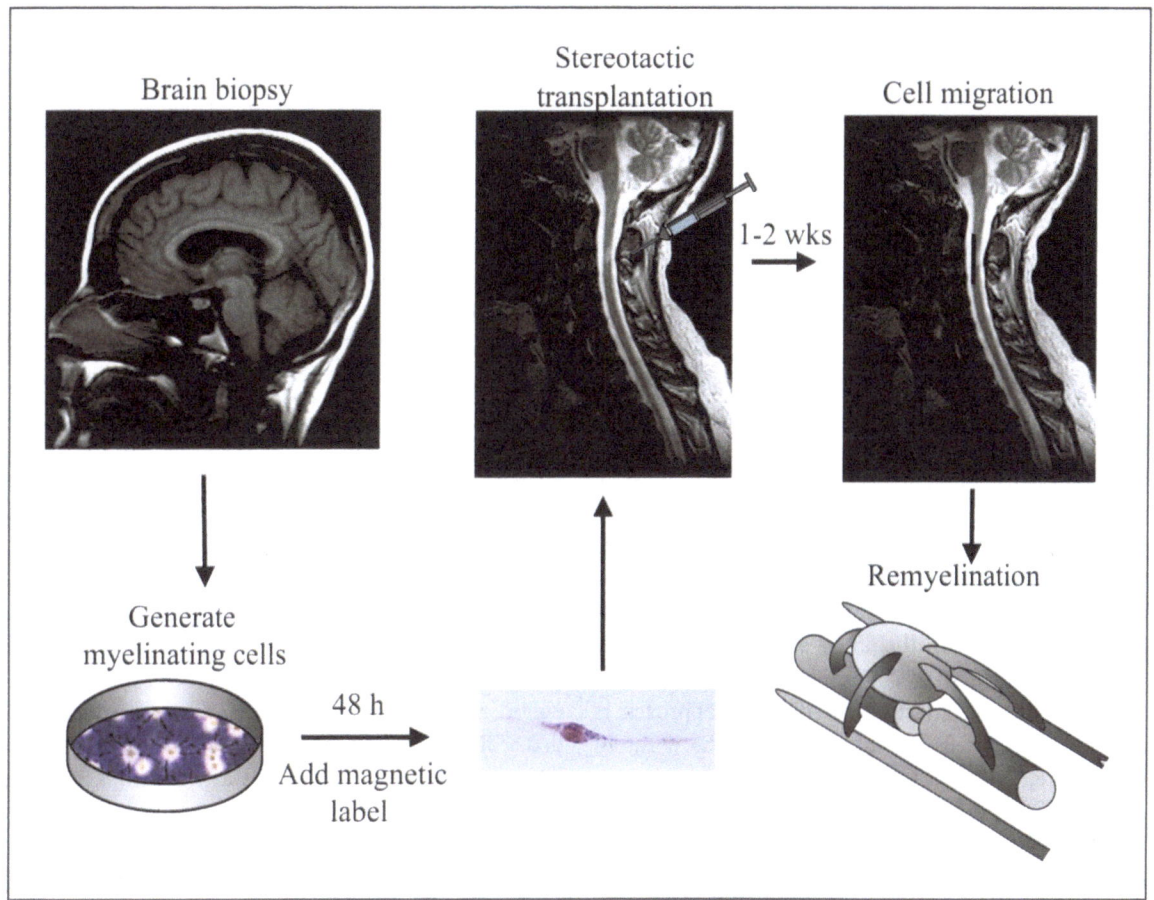

Fig. 31.7. Schematic representation of the potential role of MR imaging in monitoring cell-based therapies. Shown is an example of a patient with multiple sclerosis who receives a dose of magnetically labeled cells near an upper spinal cord lesion. As source of myelinating cells, in lieu of autologous brain cells, one could also envisage the future use of permanent neural or embryonic stem cell cultures. After a short interval of time the patient undergoes another MRI scan in order to assess the migration and biodistribution of transplanted cells. In this protocol, the scan will then determine if the transplantation was successful or, alternatively, it may suggest that a second grafting (further away from the lesion) is needed

Requirements for the "vital dye", obviously, is no alteration of the function or longevity of the labeled cell, so it needs to be internalized into the cellular cytoplasm or nucleus (membrane-bound contrast agents are likely to interfere with cell-tissue interactions and may detach easily from the cell membrane). The second requirement is that, while not overloading and potentially killing the cell, a sufficient degree of labeling needs to be achieved in order to be able to detect the cells by MR imaging. Iron oxide particles are naturally a first choice as candidates for cellular contrast agents, given their high relaxivity, T2* signal amplification effects, limited toxicity (cells naturally need and contain iron), and biodegradability.

In the late 1980s, clinical studies on the biodistribution of ex vivo ^{111}In-labeled tumor infiltrating lymphocytes (TILs) and peripheral blood lymphocytes (PBLs) were carried out on patients undergoing adoptive cellular immunotherapy (Griffith et al. 1989; Fisher et al. 1989), and this stimulated attempts to label lymphocytes with superparamagnetic iron oxides. Strategies to prepare magnetically labeled lymphocytes included incubation with liposome-encapsulated iron oxide particles (Bulte et al. 1993), incubation with non-derivatized dextran-coated iron oxide particles (Yeh et al. 1993, 1995; Weissleder et al. 1997a; Schoepf et al. 1998; Dodd et al. 1999; Sipe et al. 1999), lectin-mediated uptake (Bulte et al. 1996), and uptake mediated by the tat-peptide (Josephson et al. 1999; Lewin et al. 2000). The HIV-1 tat-peptide contains a membrane-translocating signal that efficiently shuttles MION nanoparticles into cells. In addition, granulocytes (neutrophils) have been labeled with iron oxides to image their localization in areas of infection and inflammation (Krieg et al. 1995), analogous to clinical nuclear medicine studies using ^{111}In-labeled leukocytes. In the above studies, magnetically labeled white blood cells were administered systemically (intravenous injection). A different approach is to inject tagged cells in situ (locally) in the tissue of interest, i.e. to transplant the cells.

The central nervous system has been a primary area of interest for neurotransplantation studies of iron oxide-labeled cells. Several groups have demonstrated that it is possible to depict magnetically labeled cells at the injection site (Norman et al. 1992; Hawrylak et al. 1993; Franklin et al. 1999). Our group demonstrated, for the first time, that it is possible to visualize *cell migration*, at least up to 10 mm away from the site of transplantation (Bulte et al. 1999d) (see Fig. 31.8). In this study, oligodendrocyte progenitors were first incu-

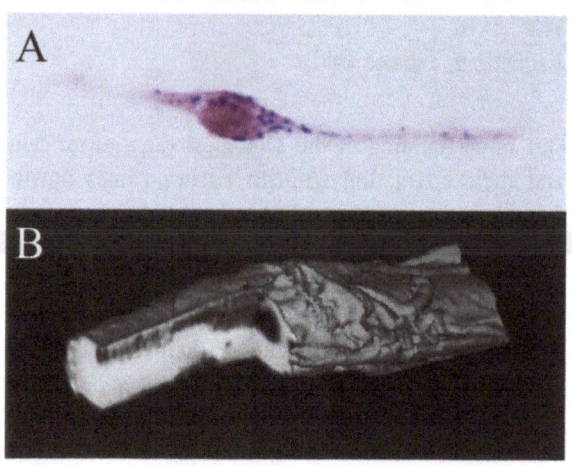

Fig. 31.8. A Prussian Blue staining of MION-46L-OX-26-labeled CG-4 oligodendrocyte progenitor cells (Bulte et al. 1999). Note the presence of numerous endocytotic vesicles that contain the internalized iron-containing magnetic probe. **B** 3D reconstruction of a 78-μm resolution ex vivo MR image dataset of rat spinal cord. The myelin-deficient (md) animal was given MION-46L-OX-26-labeled cells, and the spinal cord was removed at 10 days post transplantation. Note the migration of cells away from the injection site

bated with iron oxide particles that were covalently linked to anti-transferrin receptor (Tfr) mabs. Upon binding of the construct, the Tfr is being crosslinked by the mab which results in an internalizing signal. The receptor-mediated endocytosis then stashes away the iron oxide particles in endosomes. Following transplantation of only 5×10^4 magnetically tagged cells, the dark areas on the MR images corresponded to the cellular migration mainly within the dorsal column, and corresponded to the areas of new myelination. While the imaging in that study was performed at high resolution ex vivo, we have since demonstrated that these and similar cells can be monitored serially in vivo, even using lower resolution clinical MRI systems (Fig. 31.9) (Bulte et al. 2000c). In parallel with these new technologies, another breakthrough development occurred that will have profound implications for the use of cellular therapies, namely the isolation and successful propagation of human embryonic stem (ES) cells (Shamblott et al. 1998; Thomson et al. 1998). Using mouse ES cells, a number of different groups have shown the near unlimited potential of these cells to become differentiated, specific cell types that can be used to repair defunct or damaged tissue. The initial MR tracking studies used dextran-coated MION-46L iron oxide nanoparticles as the magnetic tag, following covalent attachment of an anti-transferrin receptor monoclonal antibody to induce receptor-mediated en-

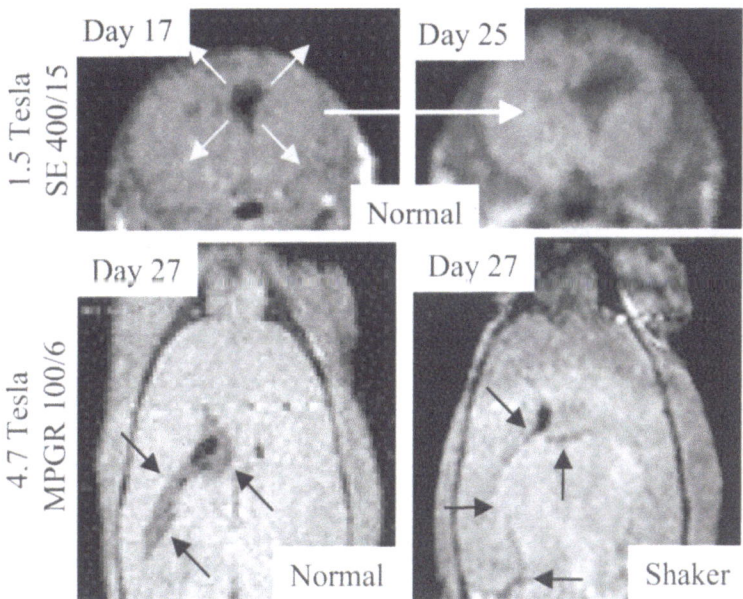

Fig. 31.9. In vivo tracking of MION-46L-OX-26-labeled oligodendroglial progenitors prepared from neural stem cells (Bulte et al. 2000). *Top row*: Axial MR images of a wild type LE rat 17 and 25 days post transplantation obtained using a clinical 1.5 Tesla imaging system. *Bottom row*: Coronal MR images of a wild type and shaker LE (*les*) rat 27 days post transplantation obtained using a 4.7-Tesla animal imaging unit

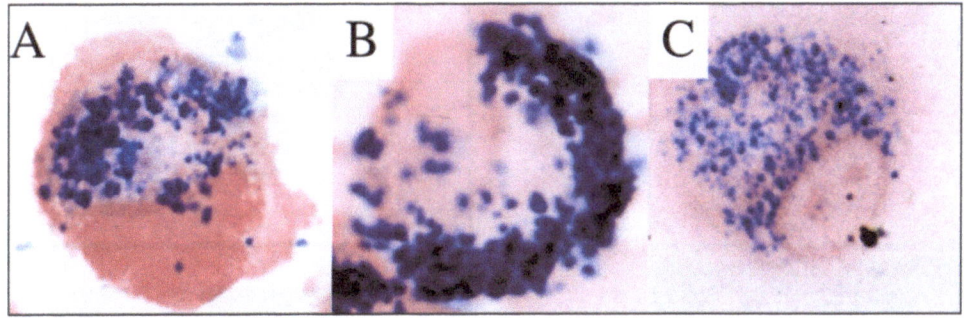

Fig. 31.10 A–C. Prussian Blue staining of magnetodendrimer-labeled cells. A CG-4 rat oligodendrocyte progenitor cells; B HeLa human cervix carcinoma cells; C human mesenchymal stem cells. Note the absence of iron-containing endosomes in the nucleus. Cells were labeled for 48 h at 25 (A, B) or 10 μg Fe/ml, trypsinized and washed before staining

docytosis. In addition, for improved cellular magnetic labeling, these particles can be derivatized with a short HIV-tat peptide (Josephson et al. 1999; Lewin et al. 2000). MION-46L is a magnetic label that has been specifically developed and optimized for blood pool imaging, lymphography, and in vivo targeting (with a long blood half life and an ultrasmall size being one of its hallmark features), but not for magnetic tagging of cells ex vivo. Its small size compromises the magnetic susceptibility and T2 relaxivity (Shen et al. 1993; Bulte et al. 1999 a, b), and custom-tailored derivatization of the dextran coat is a cumbersome process suitable only for magnetic labeling of cells that express the targeted receptor. Magnetic particles that can label cells non-specifically, regardless of tissue origin or animal species, yet show high cellular uptake ratios are highly de-

sirable to further develop the field of cellular MR imaging. In collaboration with Dr. Trevor Douglas at Temple University, we have developed magnetodendrimers as a versatile new class of magnetic tags that can efficiently label mammalian cells, including human neural stem cells (NSC) and mesenchymal stem cells (MSC), through a non-specific membrane adsorption process with subsequent intracellular localization in endosomes (Fig. 31.10) (Bulte et al. 2000 a, 2001 a–c). Magnetodendrimers induce sufficient MR cell contrast at incubated doses as low as 1 μg Fe/ml and, when containing between 17 and 27 pg Fe/cell, labeled cells exhibit a T2 relaxivity as high as 24–39 s^{-1} mM Fe (Fig. 31.11). Labeled cells are unaffected in their viability and proliferating capacity, and labeled human NSC differentiate normally into neurons and their processes

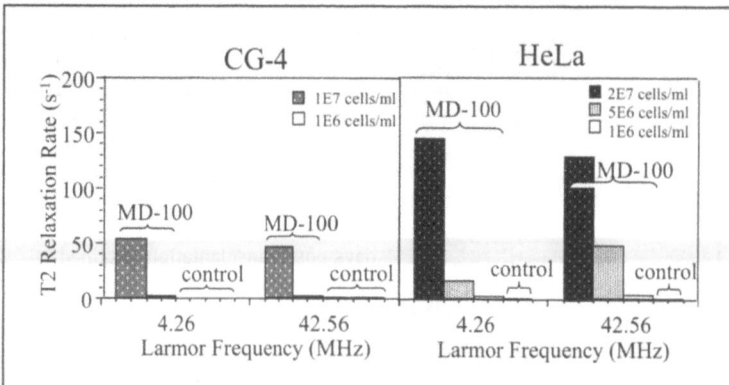

Fig. 31.11. T2 relaxation rate enhancement of magnetodendrimer-labeled CG-4 rat oligodendrocyte progenitor and HeLa human cervix carcinoma cells (Bulte et al. 2000). Cells were labeled for 48 h at 25 µg Fe/ml, washed, counted and resuspended in gelatin. Non-labeled cells (control) were included. Note the dramatic increase in 1/T2, even for the lower cell concentrations

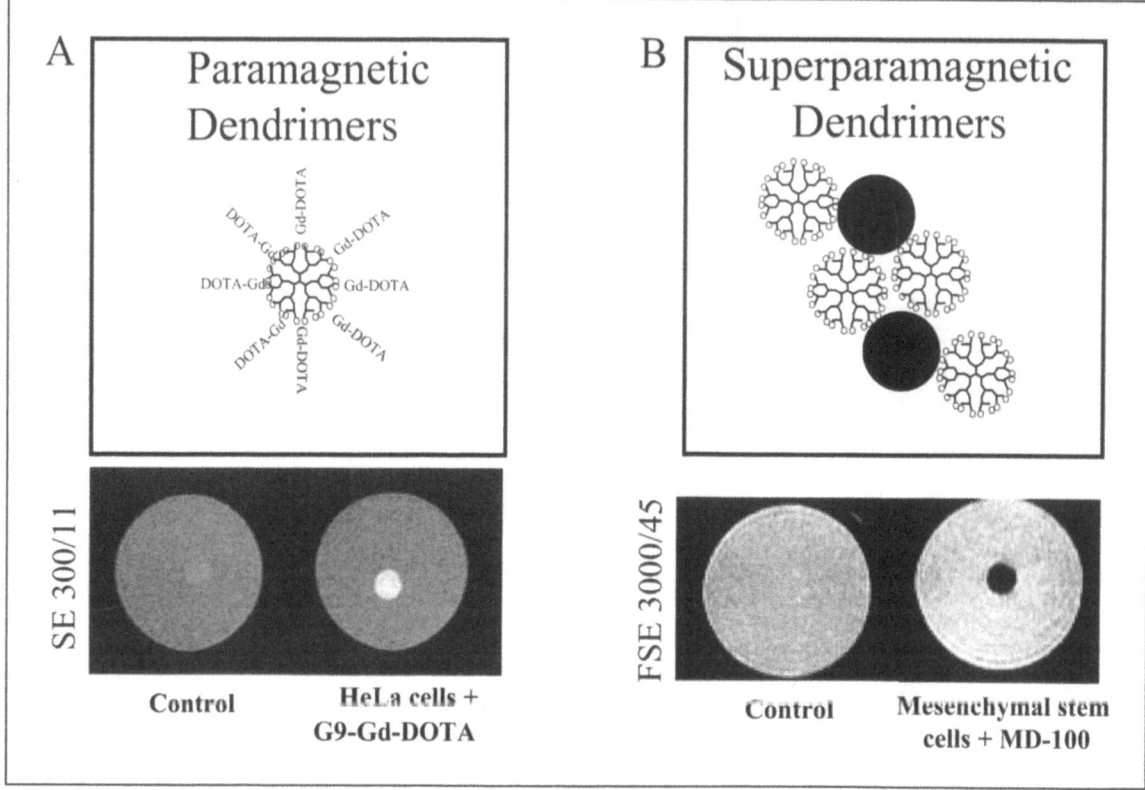

Fig. 31.12 A, B. Dendrimers as cellular imaging agents. **A** Spin echo T1-weighted MR images of human cervix carcinoma (HeLa) cells labeled with Gd-DOTA-containing paramagnetic dendrimers results in a contrast of the cell pellet that is hyperintense (Bryant et al. 2000). **B** Fast spin-echo T2-weighted MR images of human mesenchymal stem cells labeled with magnetodendrimers results in a contrast of the cell pellet that is hypointense (Bulte et al. 2001)

(Bulte et al. 2001a). Magnetically labeled oligospheres (oligodendroglial progenitors prepared from NSC) could be easily detected in vivo using a clinical 1.5 Tesla imaging system, with a good histopathological correlation with X-gal staining for the transfected LacZ gene (Bulte et al. 2001c). We have also shown that mouse ES cells can be highly labeled with magnetodendri-

mers, and that transplanted cells can be tracked down using high-resolution MR imaging in a spinal cord contusion model (Bulte et al. 2001b). Thus, the availability of magnetodendrimers opens up the possibility of MR tracking of a wide variety of cell transplants, and may help guide future stem cell-based therapies.

Paramagnetic chelates may also be used for magnetic tagging of cells. An example of the use of paramagnetic vs. (superparamagnetic) dendrimers is shown in Fig. 31.12. It was shown that by injecting Gd-DTPA-dextran in just one single cell of an early (16-cell) stage developing frog embryo, the embryonic cell lineages and movement of differentiating cells could be followed using MR microscopic imaging (Jacobs and Fraser 1994). Transferrin conjugated to poly-L-lysine has been used to co-transfer Gd chelates and DNA into cells (Kayyem et al. 1995), and using the HIV-tat-peptide described above, cells have been tagged with Gd and Dy chelates (Bhorade et al. 2000). The advantage of using Gd chelates is that the inner-sphere water coordination may be manipulated in order to create "on–off" cellular switches; our group is currently pursuing this approach using paramagnetically labeled dendrimers (Bryant et al. 2000a).

31.5
Axonal and Neuronal Tracing

Peripheral nerves are normally isointense with the surrounding tissue and are therefore difficult to detect individually by MR imaging. Following injection of wheat germ agglutinin (WGA)-conjugated dextran-coated iron oxide particles into the rabbit forearm muscle, the particles are taken up and transported by median nerves, which can then be easily detected as separate structures on the images (Filler 1994). Similarly, using a rat model of focal crush injury to the sciatic nerve, it was shown that these nerves can become traceable following injection of MION particles directly at the site of injury (Enochs et al. 1993). Binding and transport of the iron oxides were comparable to slow axonal transport with a speed of about 5 mm a day. Interestingly, although WGA is being reported to have a high specific affinity for neurons, similar results were obtained for WGA-coated and non-coated particles that were either negatively or positively charged (Enochs et al. 1993; Everdingen et al. 1994). Another approach to detect nerves and, in particular, to trace neuronal connections is to use intravitreal injection of paramagnetic manganese ions (Mn^{2+}), allowing MRI visualization of the olfactory pathway (Pautler et al. 1998). The mechanism of the neuronal cellular uptake of Mn^{2+} is not clear and possibly results from binding to voltage and ligand-gated Ca^{2+}-transporters. This pioneering

work has sprang forward from the group's earlier observations that systemic Mn^{2+} administration plus opening of the blood-brain barrier enabled MRI detection of neuronal activation, presumably through mimicking of the calcium influx necessary for release of neurotransmitters (Lin and Koretsky 1997).

31.6
Imaging of Gene Expression and Enzyme Activity – "Activated" Contrast Agents

With the advent of gene therapy, that is, therapeutic cellular delivery of DNA encoding for missing or defective genes, there is an urgent need for a non-invasive technique that can monitor the cellular uptake, host DNA integration, and functional expression. The primary strategy to accomplish this would be co-transfection with a reporter gene, i.e. a gene of which the functional expression can be visualized directly (the assumption is that both genes will be co-expressed). While other imaging techniques have shown that this is indeed feasible, e.g. positron-emission tomography (PET) tracers linked to antiviral drugs that bind to the product of the reporter gene thymidine kinase (HSV-*tk*) (Tjuvajev et al. 1995, 1996, 1998; Gambhir et al. 1999) or bioluminescent and fluorescent proteins such as firefly luciferase and green fluorescent protein in the case of optical imaging (Contag et al. 2000), none of these techniques offer the microscopic resolution and deep tissue penetrating capabilities of MR imaging. However, as pointed out earlier, the sensitivity of MRI for detectable tracers is low.

Ideally the reporter gene would encode for the cellular expression and synthesis of a superparamagnetic iron oxide. We know that eukaryotic cells are capable of biological synthesis of magnetite: magnetoreceptor-bearing cells containing iron oxide particles have recently been identified in migrating trout (Diebel et al. 2000), which use the earth's magnetic field for navigation. Similar particles can also be found in bees, salmon, pigeons and other birds. Studies on the genes involved with the preparation of magnetosomes (single domain magnetite particles coated with a lipid bilayer) have focussed on the magnetotactic bacterium *Magnetospirillum* sp. AMB-1 (Matsunaga 1997; Matsunaga et al. 1997). In order to achieve cellular expression of magnetite it is unlikely that just one single gene is needed; rather it would require a set of reporter genes of which the correct insertion, transcription and assembly may be extremely complex. Accomplishing this daunting task, however, would be of immense value and would make MRI the primary imaging technique for the in vivo detection of gene expression. In the meantime Koretsky et al., who earlier pioneered a non-invasive detection of gene expression through use of ^{31}P NMR spectroscopy-detectable tracers (creatine kinase; Koretsky et al. 1990), have inserted transferrin receptor-transfected tumor cells into the mouse, and were able to detect significant signal changes as a result of the increased uptake of iron (Koretsky et al. 1996). The use of Tfr as a potential reporter gene, however, may have some complexities to it as the transferrin-bound iron is initially paramagnetic (predominant T1-effect), and the time course of subsequent metabolization and transformation into antiferromagnetic ferritin (predominant T2-effect) may vary widely among different cell types. Another approach is to target iron oxide particles bound to transferrin to the Tfr-transfected cell lines (Moore et al. 1998; Weissleder et al. 2000), although this can be viewed as being merely a model of detecting an overexpressed receptor (e.g. other receptor-ligand systems may be used) rather than an attempt to use the intrinsic contrast-enhancing properties of a reporter gene. Another potential reporter gene for inducing "endogenous contrast", analogous to Koretsky's Tfr approach, is tyrosinase, an oxidoreductase that is an enzyme essential for the overproduction of melanin. Transfected cells incubated with high levels of iron show an increased uptake of iron and appear bright on T1-weighted images (Enochs et al. 1997; Weissleder et al. 1997b).

Outside the imaging field, one of the most commonly used reporter genes is LacZ, that encodes for the enzyme β-galactosidase, and can be visualized (invasively) histochemically by conversion of the X-gal substrate resulting in a Prussian Blue stained end product. Meade et al. have created a paramagnetic Gd chelate in which the only accessible water-binding site (for inner sphere relaxation) is blocked by a galactose group (Moats et al. 1997). In the presence of the β-galactosidase enzyme, the blocking sugar cap is removed, and the MR contrast agent "switched on." Indeed, the use of such an "activated" (or "functional," "smart," "intelligent," "responsive") contrast agent has allowed tracking of the gene expression and cellular differentiation pattern of the Xenopus laevis embryo: by injecting a galactose-capped Gd chelate in both cells of a two cell-stage embryo, but β-galactosidase DNA into only one cell, it is possible to follow the de-

velopment of transfected progenitor cells (which all exhibit contrast enhancement) by MR microscopy (Louie et al. 2000). However, the enzymatic cleavage is irreversible. Other activated MR agents have recently been developed which are reversible and report the surrounding pH in vitro. A monomeric tetra-amide-phosphonate appended Gd-DOTA derivative has been synthesized having a maximum T1 relaxivity at pH = 6 (Zhang et al. 2001). Increasing the pH to 8.5 results in about a 50% decrease in T1. A pyridine-containing Gd macrocyclic complex may show a similar T1 behavior with pH (Hall et al. 1998). It has been suggested that prototropic exchange at pH = 6 of the slowly exchanging water molecule maximizes the observed relaxivity.

A polymeric polyion complex (p-pic) involving the interaction of polymerized Gd-DTPA with poly-methylacrylate and the T1 as a function of pH has recently been reported (Mikawa et al. 1998). There was an approximately 50% decrease in T1 when going from pH 5 to pH 7 using a unimolar ratio of the methylacrylate polymer with polymerized Gd-DTPA. Further addition of base increased T1. Only about 20% of the available DTPA ligands in the polymerized DTPA have Gd ions coordinated to them. It is reasonably assumed that the free carboxylates of the excess DTPA (negative charge) are electrostatically interacting with the protonated amines of the poly-methylacrylate. The electrostatic interaction blocks the water from coordinating to the inner-sphere of the Gd ions. A change in the pH changes the degree of protonation for the amines, resulting in dissociation of the p-pic and ready access of the water to the Gd. A similar but unexpected finding has been found for the dendrimer-based MR agents (Bryant et al. 2000a). In the presence of a proprietary PLUS reagent consisting of dipeptides attached to alkyl amines, a four-fold decrease in the relaxivity of the G9-Gd-DOTA was observed in solution. As shown in Fig. 31.13, the relaxivities of the dendrimeric pic (d-pic) were found to be pH dependent and reversible. At pH 6, the d-pic was intact. Upon increasing the pH there was dissociation of the d-pic which resulted in an increase in the relaxivities. At pH = 9 there was complete dissociation of the d-pic and the relaxivities returned to those of the dendrimer-Gd-DOTA. The pH dependence of the relaxivities was reversible and additional studies revealed no dissociation of Gd or cleavage of the Gd-DOTA. It was also observed that an increase in the ionic strength dissociated the d-pic. It may be possible to monitor the pH and/or ionic strength ex-

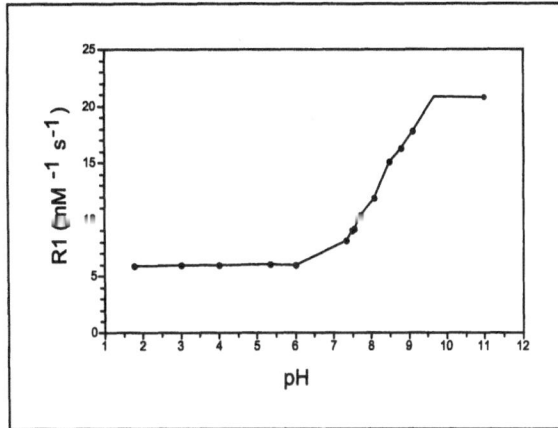

Fig. 31.13. The pH dependence of the water proton relaxivity of G9-Gd-DOTA at 61 MHz, 23 °C (Bryant et al. 2000)

tra- and/or intracellularly and we are pursuing such research.

Further examples of activated contrast agents include bioactivation of a pro-drug and induction of relaxation enhancement by the enzyme alkaline phosphatase (McMurry et al. 1998) and the development of a calcium-sensitive MR contrast agent (Li et al. 1999). Non-invasive, three-dimensional visualization of such an intracellular secondary messenger concentration would be very valuable; however, for all these "smart" approaches, we should realize that two free parameters are being introduced: (1) the concentration of the enzyme or messenger, and (2) the concentration of the administered contrast agent. Since both will be unknown and may vary in time, a correct interpretation of the obtained contrast may actually prove more difficult than expected (Robert Muller, personal communication). In addition, blocking of the contrast effect by a removable cap may not be complete but perhaps 80%, as a result of the remaining outer-sphere relaxation effect (Michael Tweedle, personal communication), complicating the interpretation possibly even further. Nevertheless, it is expected that these "smart" contrast agent technologies and variations thereof will mark the dawn of a new era and potentially revolutionize the field of molecular and cellular biology.

31.7
References

Adam G, Neuerburg J, Spuntrup E et al (1994) Gd-DTPA-Cascade-polymer: potential blood pool contrast agent for MR imaging. J Magn Reson Imaging 4:462–466

Adam G, Muhler A, Spuntrup D et al (1996) Differentiation of spontaneous canine breast tumors using dynamic magnetic resonance imaging with 24-gadolinium-DTPA-cascade-polymer, a new blood-pool agent-preliminary experience. Invest Radiol 31:267–274

Aime S, Botta M, Fasano M et al (1996) Gd(III) complexes as contrast agents for magnetic resonance imaging; a proton relaxation enhancement study of the interaction with human serum albumin. J Biol Inorg Chem 1:312–319

Aime S, Botta M, Fasano M, Terreno E (1998) Lanthanide (III) chelates for NMR biomedical applications. Chem Soc Rev 27:19–29

Anderson SA, Rader RK, Westlin WF et al (2000) Magnetic resonance contrast enhancement of neovasculature with $a_v\beta_3$-targeted nanoparticles. Magn Reson Med 44:433–439

Anderson-Berg WT, Strand M, Lempert TE et al (1986) Nuclear magnetic resonance and gamma camera tumor imaging using gadolinium-labeled monoclonal antibodies. J Nucl Med 27:829–833

Anelli PL, Bertini I, Fragai M et al (2000) Sulfonamide-functionalised gadolinium DTPA complexes as possible contrast agents for MRI: a relaxometric investigation. Eur J Inorg Chem 4:625–630

Bhorade RM, Weissleder R, Nakakoshi T et al (2000) Macrocyclic chelators with paramagnetic cations are internalized into mammalian cells via a HIV-Tat derived membrane translocation peptide. Bioconj Chem 11:301–305

Bogdanov AA, Weissleder RW, Frank HW et al (1993) A new macromolecule as a contrast agent for MR angiography – preparation, properties, and animal studies. Radiology 187:701–706

Bourne MW, Margerun L, Hylton N et al (1996) Evaluation of the effects of intravascular MR contrast media (gadolinium dendrimer) on 3D time of flight magnetic resonance angiography of the body. J Magn Reson Imaging 6:305–310

Brasch RC (1991) Rationale and applications for macromolecular Gd-based contrast agents. Magn Reson Med 22:282–287

Bryant LH, Brechbiel MW, Wu C et al (1999) Synthesis and relaxometry of high-generation (G = 5, 7, 9, and 10) PAMAM dendrimer-DOTA-gadolinium chelates. J Magn Reson Imaging 9:348–352

Bryant LH Jr, Bulte JWM, Combs CA et al (2000a) Dendrimer-based cellular MR contrast agents: Development of a molecular "off-switch" for a macromolecular contrast agent. Proceedings of the International Society for Magnetic Resonance in Medicine, 8th annual meeting, Denver, Colorado, p 377

Bryant LH Jr, Jordan EK, Bulte JWM et al (2000b) Blood clearance of high-generation dendrimer-based gadolinium chelates: the study of a potentially saturable site (mechanism) in the rat. ISMRM, 8th Scientific Meeting of the International Society for Magnetic Resonance in Medicine, Denver, April 2000

Bryant LH Jr, Yordanov AT, Linnoila JJ et al (2000c) First noncovalently bound calix[4]arene-Gd(III)-albumin complex. Angew Chem Int Ed 39:1641–1643

Bulte JWM, Brooks RA (1997) Magnetic nanoparticles as contrast agents for MR imaging. In: Häfeli U, Schütt W, Teller J, Zborowski M (eds) Scientific and clinical applications of magnetic carriers. Plenum, New York, pp 527–543

Bulte JWM, Hoekstra Y, Kamman RL et al (1992) Specific MR imaging of human lymphocytes by monoclonal antibody guided dextran-magnetite particles. Magn Reson Med 25:148–157

Bulte JWM, Ma LD, Magin RL et al (1993) Selective MR imaging of labeled human peripheral blood mononuclear cells by liposome mediated incorporation of dextran-magnetite particles. Magn Reson Med 29:32–37

Bulte JWM, Douglas T, Mann S et al (1994) Magnetoferritin: characterization of a novel superparamagnetic MR contrast agent. J Magn Reson Imaging 4:497–505

Bulte JWM, Douglas T, Mann S et al (1995) Initial assessment of magnetoferritin biokinetics and proton relaxation enhancement in rats. Acad Radiol 2:871–878

Bulte JWM, Laughlin PG, Jordan EK et al (1996) Tagging of T cells with superparamagnetic iron oxide: uptake kinetics and relaxometry. Acad Radiol 3:S301–S303

Bulte JWM, Verkuyl JM, Herynek V et al (1998a) Magnetoimmunodetection of (transfected) ICAM-1 gene expression. Proceedings of the International Society for Magnetic Resonance in Medicine, 6th annual meeting, Sydney, Australia

Bulte JWM, Wu C, Brechbiel MW et al (1998b) Dy-DOTA-PAMAM dendrimers as macromolecular T2 contrast agents: preparation and relaxometry. Invest Radiol 33:841–845

Bulte JWM, Brooks RA, Moskowitz BM et al (1999a) Relaxometry, magnetometry, and EPR evidence for three magnetic phases in the MR contrast agent MION-46L. J Magn Magn Mater 194:217–223

Bulte JWM, Brooks RA, Moskowitz BM et al (1999b) Relaxometry and magnetometry of the MR contrast agent MION-46L. Magn Reson Med 42:379–384

Bulte JWM, De Cuyper M, Despres D, Frank JA (1999c) Short- vs. long-circulating magnetoliposomes as bone marrow-seeking MR contrast agents. J Magn Reson Imaging 9:329–335

Bulte JWM, Zhang S-C, van Gelderen P et al (1999d) Neurotransplantation of magnetically labeled oligodendrocyte progenitors: MR tracking of cell migration and myelination. Proc Natl Acad Sci USA 96:15256–15261

Bulte JWM, Douglas T, Strable E et al (2000a) Magnetodendrimers as a new class of cellular contrast agents. Proceedings of the International Society for Magnetic Resonance in Medicine, 8th annual meeting, Denver, Colorado, p 2061

Bulte JWM, Greenfield MT, Caravan P (2000b) Molecular factors that determine Curie spin relaxation in Dy-chelates. Proceedings of the International Society for Magnetic Resonance in Medicine, 8th annual meeting, Denver, p 2059

Bulte JWM, Zhang S-C, van Gelderen P et al (2000c) 3D MR tracking of magnetically labeled oligosphere transplants: initial in vivo experience in the LE (Shaker) rat brain. Proceedings of the International Society for Magnetic Resonance in Medicine, 8th annual meeting, Denver, Colorado, p 383

Bulte JWM, Douglas T, van Gelderen P et al (2001a) Cellular imaging using magnetodendrimers: application to human stem cells and neoplastic cells in vivo. Proceedings of the International Society for Magnetic Resonance in Medicine, 9th annual meeting, Glasgow, Scotland

Bulte JWM, Lu J, Zywicke H et al (2001b) 3D MR tracking of magnetically labeled embryonic stem cells transplanted in the contusion injured rat spinal cord. Proceedings of the International Society for Magnetic Resonance in Medicine, 9th annual meeting, Glasgow, Scotland

Bulte JWM, Witwer B, Zhang S-C et al (2001c) Serial MR imaging of magnetodendrimer-tagged oligosphere brain transplants. Proceedings of the International Society for Magnetic Resonance in Medicine, 9th annual meeting, Glasgow, Scotland

Caravan P, Ellison JJ, McMurry TJ, Lauffer RB (1999) Gadolinium(III) chelates as MRI contrast agents: structures, dynamics, and applications. Chem Rev 99:2293–2352

Cavagna FM, Naggioni F, Castelli PM et al (1997) Gadolinium chelates with weak binding to serum proteins – a new class of high-efficiency, general purpose contrast agents for magnetic resonance imaging. Invest Radiol 32:780–796

Cerdan S, Lötscher HR, Künnecke B, Seelig J (1989) Monoclonal antibody-coated magnetite particles as contrast agents in magnetic resonance imaging of tumors. Magn Reson Med 12:151–163

Contag CH, Jenkins D, Contag PR, Negrin RS (2000) Use of reporter genes for optical measurements of neoplastic disease in vivo. Neoplasia 2:41–52

Curtet C, Tellier C, Bohy J et al (1986) Selective modification of NMR relaxation time in human colorectal carcinoma by using gadolinium-diethylenetriaminepentaacetic acid conjugated with monoclonal antibody 19-9. Proc Natl Acad Sci USA 83:4277–4281

Curtet C, Maton F, Havet T et al (1998) Polylysine-Gd-DTPA$_n$ and polylysine-Gd-DOTA$_n$ coupled to anti-CEA F(ab?)$_2$ fragments as potential immunocontrast agents – relaxometry, biodistribution, and magnetic resonance imaging in nude mice grafted with human colorectal carcinoma. Invest Radiol 33:752–761

Demsar F, Shames DM, Roberts TPL et al (1998) Kinetics of MRI contrast agents with a size ranging between Gd-DTPA and albumin-Gd-DTPA: use of cascade-Gd-DTPA-24 polymer. Electro Magnetobiol 17:283–297

Diebel CE, Proksch R, Green CR et al (2000) Magnetite defines a vertebrate magnetoreceptor. Nature 406:299–302

Dodd SJ, Williams M, Suhan JP et al (1999) Detection of single mammalian cells by high-resolution magnetic resonance imaging. Biophys J 76:103–109

Dutton AH, Tokuyasu KT, Singer SJ (1979) Iron-dextran antibody conjugates: general method for simultaneous staining of two components in high-resolution immuno-electron microscopy. Proc Natl Acad Sci USA 76:3392–3396

Enochs WS, Schaffer B, Bhide PG et al (1993) MR imaging of slow axonal transport in vivo. Exp Neurology 123: 235–242

Enochs WS, Petherick P, Bogdanova A et al (1997) Paramagnetic metal scavenging by melanin: MR imaging. Radiology 204:417–423

Filler AG (1994) Axonal transport and MR imaging: prospects for contrast agent development. J Magn Reson Imaging 4:259–267

Fisher B, Packard BS, Read EJ et al (1989) Tumor-localization of adoptively transferred In-111 labeled tumor infiltrating lymphocytes in patients with metastatic melanoma. J Clin Oncol 7:250–261

Franklin RJM, Blaschuk KL, Bearchell MC et al (1999) Magnetic resonance imaging of transplanted oligodendrocyte precursors in the rat brain. Neuroreport 10:3961–3965

Gambhir SS, Barrio JR, Phelps ME et al (1999) Imaging adenoviral-directed reporter gene expression in living animals with positron emission tomography. Proc Natl Acad Sci USA 96:2333–2338

Gibby WA, Bogdan A, Ovitt TW (1989) Cross-linked DTPA polysaccharides for magnetic-resonance imaging – synthesis and relaxation properties. Invest Radiol 24:302–309

Göhr-Rosenthal S, Schmitt-Willich H, Ebert W, Conrad J (1993) The demonstration of human tumors on nude mice using gadolinium-labelled monoclonal antibodies for magnetic resonance imaging. Invest Radiol 28:789–795

Griffith KD, Read EJ, Carrasquillo JA et al (1989) In vivo distribution of adoptively transferred indium-111-labeled tumor infiltrating lymphocytes and peripheral-blood lymphocytes in patients with metastatic melanoma. J Natl Cancer Inst 81:1709–1717

Hall J, Häner R, Aime S et al (1998) Relaxometric and luminescence behaviour of triaquahexaazamacrocyclic complexes, the gadolinium complex displaying a high relaxivity with a pronounced pH dependence. New J Chem 627–631

Hawrylak N, Ghosh P, Broadus J et al (1993) Nuclear magnetic resonance (NMR) imaging of iron oxide-labeled neural transplants. Exp Neurol 121:181–192

Hines JV, Ammar GM, Buss J, Schmalbrock P (1999) Paramagnetic oligonucleotides: contrast agents for magnetic resonance imaging with proton relaxation enhancement effects. Bioconj Chem 10:155–158

Jacobs RE, Fraser SE (1994) Magnetic resonance microscopy of embryonic cell lineages and movements. Science 263:681–684

Josephson L, Groman EV, Menz E et al (1990) A functionalized superparamagnetic iron oxide colloid as a receptor directed MR contrast agent. Magn Reson Imaging 8:637–646

Josephson L, Tung C-H, Moore A, Weissleder R (1999) High-efficiency intracellular magnetic labeling with novel superparamagnetic-tat peptide conjugates. Bioconj Chem 10:186–191

Kayyem JF, Kumar RM, Fraser SE, Meade TJ (1995) Receptor-targeted co-transport of DNA and magnetic resonance contrast agents. Chem Biol 2:615–620

Köhler G, Milstein C (1975) Continuous cultures of fused cells secreting antibody of predefined specificity. Nature 256:495–497

Konda SD, Aref M, Brechbiel M, Wiener EC (2000) Development of a tumor targeting MR contrast agent using the high affinity folate receptor – work in progress. Invest Radiol 35:50–57

Koretsky AP, Brosnan MJ, Chen L et al (1990) NMR detection of creatine-kinase expressed in liver of transgenic mice – determination of free ADP levels. Proc Natl Acad Sci USA 87:3112–3116

Koretsky AP, Lin Y-J, Schorle H, Jaenisch R (1996) Genetic control of MRI contrast by expression of the transferrin receptor. Proceedings of the International Society for Magnetic Resonance in Medicine, 4th annual meeting, New York, p 69

Kornguth SE, Turski PA, Perman WH et al (1987) Magnetic resonance imaging of gadolinium-labeled monoclonal antibody polymers directed at human T lymphocytes implanted in canine brain. J Neurosurg 66:898–906

Kresse M, Wagner S, Pfefferer D et al (1998) Targeting of ultrasmall superparamagnetic iron oxides (SPIO) to tumor cells in vivo by using transferrin-receptor pathways. Magn Reson Med 40:236–242

Krieg FM, Andres RY, Winterhalter KH (1995) Superparamagnetically labelled neutrophils as potential abscess-specific contrast agent for MRI. Magn Reson Imaging 13:393–400

Lanza GM, Lorenz CH, Fischer SE et al (1998) Enhanced detection of thrombi with a novel fibrin-targeted magnetic resonance contrast agent. Acad Radiol 5:S173–S176

Larson SM, Brown JP, Wright PW et al (1983) Imaging of melanoma with I-131-labeled monoclonal-antibodies. J Nucl Med 24:123–129

Lauffer RB (1987) Paramagnetic metal complexes as water proton relaxation agents for NMR imaging: theory and design. Chem Rev 87:901–927

Lauffer RB, Brady TJ (1985) Preparation and water relaxation properties of proteins labeled with paramagnetic metal chelates. Magn Reson Imaging 3:11–16

Lauffer RB, Brady TJ, Brown RD et al (1986) 1/T1 NMRD profiles of solutions of Mn2+ and Gd3+ protein chelate conjugates. Magn Reson Med 3:541–548

Lauffer RB, Parmelee DJ, Dunham SU et al (1998) MS-325: albumin-targeted contrast agent for MR angiography. Radiology 207:529–538

Lewin M, Carlesso N, Tung C-H et al (2000) Tat peptide-derivatized magnetic nanoparticles allow in vivo tracking and recovery of progenitor cells. Nat Biotechnol 18:410–414

Li KCP, Quisling RG, Armitage FE et al (1992) In vivo MR evaluation of Gd-DTPA conjugated to dextran in normal rabbits. Magn Reson Med 10:3439–3444

Li W-H, Fraser SE, Meade TJ (1999) A calcium-sensitive magnetic resonance imaging contrast agent. J Am Chem Soc 121:1413–1414

Lin YJ, Koretsky AP (1997) Manganese ion enhances T-1-weighted MRI during brain activation: an approach to direct imaging of brain function. Magn Reson Med 38:378–388

Louie AY, Hüber MM, Ahrens ET et al (2000) In vivo visualization of gene expression using magnetic resonance imaging. Nat Biotechnol 18:321–325

Mach JP, Buchegger F, Forni M et al (1981) Use of radiolabeled monoclonal anti-CEA antibodies for the detection of human carcinomas by external photoscanning and tomoscintigraphy. Immunol Today 2:239–249

Macrì MA, De Luca F, Maraviglia B et al (1989) Study of proton spin lattice relaxation variation induced by paramagnetic antibodies. Magn Reson Med 11:283–287

Margerum LD, Campion BK, Koo M et al (1997) Gadolinium(III) DO3A macrocycles and polyethylene glycol coupled to dendrimers-effect of molecular weight on physical and biological properties of macromolecular magnetic resonance imaging contrast agents. J Alloys Comp 249:185–190

Matsunaga T (1997) Genetic analysis of magnetic bacteria. Mater Sci Eng C 4:287–289

Matsunaga T, Kamiya S, Tsujimura N (1997) Production of a protein (enzyme, antibody, protein A)-magnetite complex by genetically engineered magnetic bacteria Magnetospirillum Sp. AMB-1. In: Häfeli U, Schütt W, Teller J, Zborowski M (eds) Scientific and clinical applications of magnetic carriers. Plenum, New York, pp 287–294

McMurry TJ, Dunham SU, Dumas SA et al (1998) Bioactivated MRI contrast agents: preliminary results in sensing alkaline phosphatase via changes in protein binding. Proceedings of the International Society for Magnetic Resonance in Medicine, 6th annual meeting, Sydney, Australia, p 636

Meldrum FC, Heywood BR, Mann S (1992) Magnetoferritin: in vitro synthesis of a novel magnetic protein. Science 257:522–523

Mikawa M, Miwa N, Bräutigam M et al (1998) A pH-sensitive contrast agent for functional magnetic resonance imaging (MRI). Chem Lett 7:693–694

Moats RA, Fraser SE, Meade TJ (1997) A "smart" magnetic resonance imaging agent that reports on specific enzymatic activity. Angew Chem Int Ed Engl 36:726–728

Moldofsky PJ, Sears HF, Mulhern CB Jr et al (1984) Detection of metastatic tumor in normal-sized retroperitoneal lymph nodes by monoclonal antibody imaging. N Engl J Med 311:106–107

Moore A, Basilion JP, Chiocca EA, Weissleder R (1998) Measuring transferrin receptor gene expression by NMR imaging. Biochim Biophys Acta 1402:239–249

Norman AB, Thomas SR, Pratt RG et al (1992) Magnetic resonance imaging of neural transplants in rat brain using a superparamagnetic contrast agent. Brain Res 594:279–283

Nunn AD, Linder KE, Tweedle MF (1997) Can receptors be imaged with MRI agents? Quart J Nucl Med 41:155–162

Parmelee DJ, Walovitch RC, Ouellet HS, Lauffer RB (1997) Preclinical evaluation of the pharmacokinetics, biodistribution, and elimination of MS-325, a blood pool agent for magnetic resonance imaging. Invest Radiol 32:741–747

Pautler RG, Silva AC, Koretsky AP (1998) In vivo neuronal tract tracing using manganese-enhanced magnetic resonance imaging. Magn Reson Med 40:740–748

Prévôté D, Donnadieu B, Moreno-Manas M et al (1999) Grafting of tetraazamacrocycles on the surface of phosphorous-containing dendrimers. Eur J Org Chem 7:1701–1708

Reimer P, Weissleder R, Lee AS et al (1990) Receptor imaging: application to MR imaging of liver cancer. Radiology 177:729–734

Reimer P, Weissleder R, Shen T et al (1994) Pancreatic receptors: initial feasibility studies with a targeted contrast agent for MR imaging. Radiology 193:527–531

Remsen LG, McCormick CI, Roman-Goldstein S et al (1996) MR of carcinoma-specific monoclonal antibody conjugated to monocrystalline iron oxide nanoparticles: the potential for noninvasive diagnosis. Am J Neuroradiol 17:411–418

Renshaw PF, Owen CS, Evans AE, Leigh JS Jr (1986) Immunospecific NMR contrast agents. Magn Reson Imaging 4:351–357

Roch A, Muller RN, Gillis P (1999) Theory of proton relaxation induced by superparamagnetic particles. J Chem Phys 110:5403–5411

Sanderson CJ, Wilson DV (1971) A simple method for coupling proteins to insoluble polysaccharides. Immunology 20:1061–1065

Schaffer BK, Linker C, Papisov M et al (1993) MION-ASF: biokinetics of an MR receptor agent. Magn Reson Imaging 11:411–417

Schmiedl U, Ogan MD, Moseley MR, Brasch RC (1986) Comparison of the contrast-enhancing properties of albumin-(Gd-DTPA) and Gd-DTPA at 2.0 T: an experimental study in rats. Am J Roentgenol 147:1263–1270

Schoepf U, Marecos E, Melder R et al (1998) Intracellular magnetic labeling of lymphocytes for in vivo trafficking studies. BioTechniques 24:642–651

Schwickert HC, Roberts TPL, Muhler A et al (1995) Angiographic properties of Gd-DTPA-24-cascade-polymer – a new macromolecular MR contrast agent. Eur J Radiol 20:144–150

Shamblott MJ, Axelman J, Wang SP et al (1998) Derivation of pluripotent stem cells horn cultured human primordial germ cells. Proc Natl Acad Sci USA 95:13726–13731

Shen T, Weissleder R, Papisov M, Bogdanov A, Brady TJ (1993) Monocrystalline iron oxide nanocompounds (MION): physicochemical properties. Magn Reson Med 29:599–604

Shen TT, Bogdanov A Jr, Bogdanova A et al (1996) Magnetically labeled secretin retains receptor affinity to pancreas acinar cells. Bioconj Chem 7:311–316

Shreve P, Aisen AM (1986) Monoclonal antibodies labeled with polymeric paramagnetic ion chelates. Magn Reson Med 3:336–340

Sipe JC, Filippi M, Martino G et al (1999) Method for intracellular magnetic labeling of human mononuclear cells using approved iron contrast agents. Magn Reson Imaging 17:1521–1523

Sipkins DA, Cheresh DA, Kazemi MR et al (1998) Detection of tumor angiogenesis in vivo by a_vb_3-targeted magnetic resonance imaging. Nat Med 4:623–626

Sipkins DA, Gijbels K, Tropper FD et al (2000) ICAM-1 expression in autoimmune encephalitis visualized using magnetic resonance imaging. J Neuroimmunol 104:1–9

Spanoghe M, Lanens D, Dommissee R et al (1992) Proton relaxation enhancement by means of serum-albumin and poly-L-lysine labeled with DTPA-Gd3+-relaxivities as a function of molecular weight and conjugation efficiency. Magn Reson Med 10:913-917

Stark DD, Weissleder R, Elizondo G et al (1988) Superparamagnetic iron oxide: clinical application as a contrast agent for MR imaging of the liver. Radiology 168:297-301

Storrs RW, Tropper FD, Li HY et al (1995) Paramagnetic polymerized liposomes as new recirculating MR contrast agents. J Magn Reson Imaging 5:719-724

Suwa T, Ozawa S, Ueda M et al (1998) Magnetic resonance imaging of esophageal squamous cell carcinoma using magnetite particles coated with ant-epidermal growth factor receptor antibody. Int J Cancer 75:626-634

Thomson JA, Itskovitz-Eldor J, Shapiro SS et al (1998) Embryonic stem cell lines derived from human blastocysts. Science 282:1145-1147

Tiefenauer LX, Kühne G, Andres RY (1993) Antibody-magnetite nanoparticles: in vitro characterization of a potential tumor-specific contrast agent for magnetic resonance imaging. Bioconj Chem 4:347-352

Tiefenauer LX, Tschirky A, Kühne G, Andres RY (1996) In vivo evaluation of magnetite nanoparticles for use as a tumor contrast agent in MRI. Magn Reson Imaging 14:391-402

Tjuvajev JG, Stockhammer G, Desai R et al (1995) Imaging the expression of transfected genes in vivo. Cancer Res 55:6126-6132

Tjujavev JG, Finn R, Watanabe K et al (1996) Noninvasive imaging of herpes virus thymidine kinase gene transfer and expression: a potential method for monitoring clinical gene therapy. Cancer Res 56:4087-4095

Tjuvajev JG, Avril N, Oku T et al (1998) Imaging herpes virus thymidine kinase gene transfer and expression by positron emission tomography. Cancer Res 58:4333-4341

Tomalia DA (1994) Starburst/cascade dendrimers: fundamental building blocks for a new nanoscopic chemistry set. Adv Mater 6:529-539

Toth E, Pubanz D, Vauthey S et al (1996) The role of water exchange in attaining maximum relaxivities for dendrimeric MRI contrast agents. Chem Eur J 2:1607-1615

Trubetskoy VS, Canillo JA, Milshtein A et al (1995) Controlled delivery of Gd-containing liposomes to lymph nodes - surface modification may enhance MRI contrast properties. Magn Reson Imaging 13:31-37

Unger EC, Totty WG, Neufeld DM et al (1985) Magnetic resonance imaging using gadolinium labeled monoclonal antibody. Invest Radiol 20:693-700

Unger EC, Shen D-K, Fritz TA (1993) Status of liposomes as MR contrast agents. J Magn Reson Imaging 3:195-198

van Everdingen KJ, Enochs WS, Bhide PG et al (1994) Determinants of in vivo MR imaging of slow axonal transport. Radiology 193:485-491

Wallace RA, Haar JP Jr, Miller DB et al (1998) Synthesis and preliminary evaluation of MP-2269: a novel, nonaromatic small-molecule blood-pool MR contrast agent. Magn Reson Med 40:733-739

Wang S-C, Wikström MS, White DL et al (1990) Evaluation of Gd-DTPA-labeled dextran as an intravascular MR contrast agent: imaging characteristics in normal rat tissues. Radiology 175:483-488

Weissleder R, Lee AS, Fischman AJ et al (1991) Polyclonal human immunoglobulin G labeled with polymeric iron oxide: antibody MR imaging. Radiology 181:245-249

Weissleder R, Lee AS, Khaw BA et al (1992) Antimyosin-labeled monocrystalline iron oxide allows detection of myocardial infarct: MR antibody imaging. Radiology 182:381-385

Weissleder R, Cheng H-C, Bogdanova A, Bogdanov A Jr (1997a) Magnetically labeled cells can be detected by MR imaging. J Magn Reson Imaging 7:258-263

Weissleder R, Simonova M, Bogdanova A et al (1997b) MR imaging and scintigraphy of gene expression through melanin induction. Radiology 204:425-429

Weissleder R, Moore A, Mahmood U et al (2000) In vivo magnetic resonance imaging of transgene expression. Nat Med 6:351-354

Wiener EC, Brechbiel MW, Brothers H et al (1994) Dendrimer-based metal chelates - a new class of magnetic resonance imaging contrast agents. Magn Reson Med 31:1-8

Wiener EC, Konda S, Shadron A et al (1997) Targeting dendrimer-chelates to tumors and tumor cells expressing the high-affinity folate receptor. Invest Radiol 32:748-754

Wu C, Brechbiel MW, Kozak RW, Gansow OA (1994) Metal-chelate-dendrimer-antibody constructs for use in radioimmunotherapy and imaging. Biorg Med Chem Lett 4:449-454

Yeh T-C, Zhang W, Ildstad ST, Ho C (1993) Intracellular labeling of T-cells with superparamagnetic contrast agents. Magn Reson Med 30:617-625

Yeh T-C, Zhang W, Ildstad ST, Ho C (1995) In vivo dynamic MRI tracking of rat T-cells labeled with superparamagnetic iron-oxide particles. Magn Reson Med 33:200-208

Zaharchuk G, Bogdanov AA, Marota JJA et al (1998) Continuous assessment of perfusion by tagging including volume and water extraction (CAPTIVE): a steady-state contrast agent technique for measuring blood flow, relative blood volume fraction, and the water extraction fraction. Magn Reson Med 40:666-678

Zhang S, Wu K, Sherry AD (2001) Gd3+ complexes with slowly exchanging bound-water molecules may offer advantages in the design of responsive MR agents. Invest Radiol 36:82-86

CHAPTER 32

Magnetic and Optical Molecular Imaging
Live Cell and Tissue Fluorescence: Imaging Techniques and Methods

CHRISTIAN A. COMBS

Contents

32.1
Fluorescence Imaging of Live Cells and Tissues

Optical imaging of living cells provides a powerful suite of techniques for non-invasive linking of phenotypic expression at the biochemical level to individual genotypes. A plethora of fluorescent markers are available for selectively marking cellular components through monoclonal antibodies, specific ligand interactions, and through covalent bonding (Tsien and Waggoner 1995; Mason 2000). Imaging of fluorescent marker distribution within living cells has enabled measurement of many active physiological processes including protein transport, membrane potential, and free ion distributions (Terasaki and Dailey 1995). These markers, in conjunction with optical sectioning techniques such as confocal or multiphoton microscopy, allow interrogation of cellular dynamics to diffraction-limited spatial resolution even in thick tissues or cells. The main purpose of the first two sections of this chapter are to introduce the reader to various techniques and methods of fluorescence imaging of live cells and thick tissues. The descriptions are aimed at the interdisciplinary scientist unfamiliar with fluorescence imaging of thick (> 5–$10\ \mu M$) specimens. The fundamentals of optical sectioning through various forms of confocal microscopy and multiphoton microscopy will be addressed first, followed by a description of new or improved fluorescence techniques for dynamic live cell imaging. These descriptions will also emphasize the limitations and practical consideration of these techniques.

32.1.1
Confocal Laser Scanning Microscopy (CLSM)

The advantage of confocal microscopes over conventional wide-field microscopes is the ability to reject out-of-focus background light. This makes it possible to optically section thick cells or tissues, and to generate three-dimensional views at high resolution (White et al. 1987; Inoue 1995). A confocal microscope optically sections by restricting the field of view of the objective lens and the excitation illumination to a single spot. An image is acquired by scanning the illumination source (usually a laser beam) over the specimen and collecting the emitted fluorescence through an exit pinhole placed before the photodetector. In this manner, light originating from out of focus regions of the specimen is excluded from detection and the image is generated point by point. The thickness of the optical section is then a function of the size of the pinhole in the intermediate image plane and the numerical aperture (NA) of the lens (Sandison et al. 1995; Wilson 1995). Thus, field of view (or temporal resolution) is sacrificed to gain axial resolution. While this technique can provide crisp clear optical sectioning capability, there are sacrifices made over conventional wide-field fluorescence microscopy.

Two main disadvantages of conventional CLSM are a slow speed of acquisition and potential photodamage (Pawley 1995). Photodamage can arise from the high illumination intensities produced from focused laser light. During 3D-image generation, although emission light is captured from one image plane at a time as the plane of focus is moved through the specimen, excitation light is not restricted. Therefore, the cone of excitation light is exciting, and potentially photobleaching, all planes of focus regardless of which plane is being imaged at a time. Thus, photodamage can be considerable during 3D-imaging. The second drawback to conventional CLSM is the byproduct of raster scanning. Raster scanning allows for the sectioning capabilities of CLSM but is much slower than wide field camera microscopy. Typically a 512×512 pixel 8 bit image will take 1–2 s to acquire compared to conventional WF microscopy were images can be acquired at video rates or higher.

32.1.2
Two-Photon Excitation Microscopy

Two-photon (2P) excitation microscopy is a complimentary optical sectioning technique to confocal microscopy. Two-photon excitation occurs when two photons are simultaneously absorbed by a fluorophore. Two photons of twice the wavelength that would normally excite a fluorophore absorbed in a single quantized event provide the same energy as single photon absorption at one-half the wavelength. For example, two red photons (~ 700 nm) absorbed at the same time would excite a fluorophore that would normally be excited by ultraviolet light (~ 350 nM). To initiate enough 2P absorption events to excite fluorophores effectively requires very high photon densities, typically much higher than are usually used for epi-fluorescent imaging (Denk et al. 1995; Denk and Svoboda 1997). In 2P excitation microscopy high photon densities are achieved at the focal plane of the specimen both by temporal and spatial means. Temporal crowding is achieved using mode-locked (pulsed) lasers. Typically the lasers are pulsed on the femto- or pico-second timescale at high powers (mW levels) and although the peak powers are high, the average powers are low due to the duration of the pulses. Spatial crowding is achieved by the optics of the microscope. The focusing of the laser beam through the optics results in crowding of the photons at the plane of focus. It is precisely this crowding that allows for the optical sectioning cap-

abilities of multi-photon excitation. Above and below the focal plane the photon density is not high enough to elicit significant two-photon absorption events (Denk et al. 1995; Denk and Svoboda 1997). In addition, no pinhole is necessary in the emission pathway due to the excitation occurring only in the focal plane. Therefore all the emission light is collected and is not limited by a pinhole in the intermediate image plane.

The main advantage of two-photon microscopy is that excitation occurs predominantly at the focal plane of the objective (Denk et al. 1995). In CLSM excitation light is wasted above and below the focal plane in a cone even though emission information is restricted to light coming from the focal plane by the pinhole placed before the detector. Thus, in two-photon microscopy there is less photobleaching and photodamage in out of focus planes of the specimen. The second main advantage of 2P excitation is that long wavelength light penetrates much deeper into tissues. This provides for 2–3 times deeper penetration than confocal microscopy and allows for information to be gathered in much thicker cells or tissues (> 200 μm) (Centonze and White 1998).

Two-photon excitation microscopy has been demonstrated in a variety of applications. Denk and colleagues have pioneered the use of 2P microscopy in calcium imaging in small structures in brain slices (Svoboda et al. 1996; Helmchen et al. 1999; Yuste et al. 1999; Cox et al. 2000). In-vivo functional imaging of the brain has also been conducted (Svoboda et al. 1997). The need to image to depths of hundreds of microns in both of these examples precluded the use of one-photon excitation or wide-field microscopy. The utility of 2P excitation has also found a niche in developmental biology. Many groups are imaging developing embryos which are very sensitive to photobleaching and photodamage (Summers et al. 1996; Squirrell et al. 1999). Other examples include imaging intrinsic NADH fluorescence to access metabolic state in the cornea and pancreatic β-cells (Bennett et al. 1996; Poston and Knobel 1999), imaging green fluorescent protein (Niswender et al. 1995; Potter et al. 1996), and for experiments involving UV uncaging of metabolites in living cells (Brown et al. 1999a; Wang et al. 2000).

The major limitations of two-photon excitation microscopy are that tuning and maintenance of the lasers (typically Ti:Sapphire or ND:YLF) are more complicated than conventional lasers used in CLSM, photobleaching in the focal plane can be more pronounced (Patterson and Piston 2000), two-photon excitation is sensitive to optical aberrations due to refractive index mismatch and spherical abberation (Denk et al. 1995), and that the two photon spectrum for a fluorophore may differ from the corresponding single-photon spectrum (Denk et al. 1995). In addition, for thin samples or cultured cells there is often no advantage to using two-photon excitation microscopy. The other potential limitation is that this technique shares the same limitation as CLSM in that it is a raster scanning application and is therefore relatively slow and may not be appropriate for imaging very fast biological events.

32.1.3
Fast Imaging

32.1.3.1
Disc-Scanning Microscopy

One means of obtaining confocal images at video rates is through the use of a disc-scanning confocal microscope. The basic difference from standard CLSM is that instead of having a single pinhole placed before the detector, a large number of pinholes are used (Petran et al. 1968, 1985). In this arrangement the illumination source is transmitted through a modified Nipkow disc containing thousands of pinholes arranged in a helical pattern (Petran et al. 1985). Spinning the disc to fill in the spaces between the pinholes forms the image. The most efficient types of disc-scanning confocal microscope employ a second disc fitted with thousands of micro lenses aligned with the pinholes on the other disc (Ichihara et al. 1996). Using this arrangement, coupled with sensitive cameras, confocal images can be collected at rates > 30 frames/s.

The main limitation for disc-scanning confocal microscopes is illumination efficiency. The first disc-scanning-type confocal microscopes transmitted around 1% of the light due to the small area that the pinholes occupied on the disc. The modern Yokagawa high-speed confocal system (with microlenses) transmits much more light (40–60% of the illuminating beam) (Ichihara et al. 1996; Inoue and Spring 1997). For weakly fluorescent specimens this still presents a problem and requires long camera integration times which defeats the purpose of this type of imaging. Another small disadvantage is that pinhole size cannot be changed during an experiment as is the case on modern CLSM systems. Therefore the optical slice is dependent on the pinhole sizes on the disc and the numerical

aperture of the objective lens. In addition, no commercial systems have been developed that can transmit UV wavelength light through the microlenses.

32.1.3.2
Deconvolution

An alternative to disc-scanning systems for fast imaging is computational deconvolution of standard widefield image series obtained by video-rate microscopy techniques. Deconvolution algorithms are employed post-acquisition to mathematically remove out of focus light from the 2-D sections collected (Castleman 1979; Agard et al. 1989). These algorithms de-blur each individual image using a detailed knowledge of the image degradation introduced by the microscope optics. The correction algorithm can be based on empirical measurements or on an idealized point-spread function of the light as it passes through the focal plane of the specimen (Holmes et al. 1995). These algorithms can also be used to improve the quality of confocal images to remove the limited out of focus haze that passes the pinhole (Shaw 1995). The main limitations of digital deconvolution are that it is computationally intensive (slow and/or requiring a fast processor), and that it is model dependent (Shaw 1995; Inoue and Spring 1997). Moreover, samples thicker than 60–100 μm usually contain too much information for deconvolution to be effective. Many commercial companies are available to provide software alone or for building a complete system for deconvolution image acquisition and processing.

32.1.4
In Vivo Imaging

True in vivo imaging requires examination of cells in their native environment within the body. A primary difficulty is bringing the light from a live sample to an imaging device. Recently, confocal endoscopes have been developed and demonstrated on rats (Sabharwal et al. 1999). In addition, endoscopic fluorescence spectral imaging has been performed in the gastrointestinal tract of humans (Zeng et al. 1999). The main limitation of these endoscopy approaches is the depth of penetration of the light. Again, 100–200 μM is the limit at which these systems can peer into the target tissue (Sabharwal et al. 1999). Thus, these spectral imaging systems are often limited to measurements of tissue surface level topology. Despite these limitations endoscopic spectral imaging has shown differences between cancerous and non-cancerous tissues in vivo (Zeng et al. 1999). Coupling these systems to two-photon microscopy may increase the usefulness of these systems for imaging cells in intact organs. In experiments where perfused whole hearts from mice were examined, two-photon microscopy showed closely packed individual cardiac myocytes distributed along capillary beds in their native environment (Ohler et al. 2000). This powerful new technology may allow for the study of the metabolic interactions of individual myocytes and their extracellular environment in the native state.

32.2
Selected Live Cell Fluorescence Imaging Methods

The correlation of physiological processes with light intensity changes of fluorescent probes in cells is complicated by many factors (photobleaching, compartment hydration state, compartment pH, dye quenching, etc.). In addition, in studies of fluorescent metabolites in cells, it is often difficult to interpret if light changes are due to changes in cellular production or consumption of the metabolite. Recent advances in fluorescent methods have reduced this ambiguity and have even provided information at resolutions greater than the diffraction limitations of the microscope optics. Dynamic processes such as enzyme activity, molecular diffusion, and protein–protein interactions can be measured using techniques such as fluorescence resonance energy transfer (FRET) and fluorescence recovery after photobleaching (FRAP).

32.2.1
Fluorescence Resonance Energy Transfer (FRET) Microscopy

This technique is most often used to examine the interactions between protein partners in close proximity. FRET relies on the non-radiative (not mediated by a photon) interaction between two fluorescent molecules (a donor and an acceptor molecule) that are separated by very small distances (10–100 Ångstroms) (Clegg 1995) because the efficiency of the energy transfer varies as the inverse of the sixth power

of the distance between the chromophores. In simple terms, as the donor fluorophore (D) is excited it transfers energy to the acceptor molecule (A) through long-range dipole-dipole couplings. These couplings can occur if the fluorescence spectrum of D and the absorbance spectra of A overlap. Given these couplings, the fluorescent behavior (photobleaching rate, lifetime, ratioed intensity) of both the donor and the acceptor can provide information on their molecular interactions.

The dependence of FRET on molecular proximity make it a powerful tool for making molecular measurements on a scale past the diffraction limit of the microscope optics. In addition, reporter constructs have been manufactured to measure many different types of cellular processes by FRET. For instance, Zaccolo et al. (2000) have tagged protein kinase A with two mutant green fluorescent proteins. The FRET between the two moieties in this protein is sensitive to cAMP levels. In another instance, Miyawaki et al. (1997) have developed a fluorescent indicator system for Ca^{2+} based on the interactions of calmodulin and green fluorescent proteins. Other applications for FRET in live cell imaging include receptor ligand interactions (Sako et al. 2000), phosphorylation state of proteins (Ng et al. 1999), organization of proteins in membranes (Kenworthy and Edidin 1998; Kenworthy et al. 2000), and other protein-protein interactions (Wouters et al. 1998).

One of the main limitations to using FRET to examine molecular interactions within cells is labeling the proteins with the appropriate fluorophores. The development of many types of fluorescent protein (GFP, YFP, CFP etc.) for protein tagging in live cells has made this process much easier. Another problem is having fluorophores that produce enough FRET fluorescence above background noise. This is particularly necessary where the FRET effect may be small or where other factors such as photobleaching complicate interpretation of the data.

32.2.2
Fluorescence Recovery after Photobleaching (FRAP) Microscopy

Photobleaching of fluorophores during imaging is usually a complicating factor in fluorescent imaging applications, however, it can also be used as a unique tool. FRAP most often involves photobleaching the fluorophore from a region of the cell and measuring the rate at which the non-bleached fluorophore diffuses or is transported back into the photobleached area (Peters et al. 1974; Axelrod et al. 1976; Edidin et al. 1976). Also, if diffusion or active transport is eliminated as a means of fluorescence recovery then FRAP can be used to measure local resynthesis of a metabolite (Combs and Balaban 2001). In practice, FRAP experiments involve first measuring the emission of a fluorophore excited by a laser in a cellular region or membrane. The fluorophore in the region is rapidly bleached by a short intense pulse of intense excitation light then the fluorescence from this region is monitored over time. The rate of fluorescence recovery can then be used to model the mechanism responsible for recovery.

In this manner FRAP has been used to measure the mobility of cell surface proteins (Storrie et al. 1994), fluorophore diffusion coefficients (Axelrod et al. 1976; Soumpasis 1983; Periasamy and Verkman 1998), protein transport and molecular dynamic events (Nehls et al. 2000), and cellular enzyme activity (Combs and Balaban 2001). In general, these studies have yielded information on a 2D level, due to the photobleaching pulse affecting areas above and below the plane of interest. Recently FRAP has been combined with multiphoton excitation to restrict photobleaching to the plane of interest to measure true 3D diffusion in solutions (Brown et al. 1999b). The combination of FRAP and multiphoton excitation microscopy may provide valuable information on 3D transport and reaction kinetics in live cells.

32.3
References

Agard DA, Hiraoka Y, Shaw P, Sedat (1989) Fluorescence microscopy in three dimensions. Methods Cell Biol 30:353–377

Axelrod D, Koppel DE, Schlessinger J et al (1976) Mobility measurement by analysis of fluorescence photobleaching recovery kinetics. Biophys J 16:1055–1069

Bennett BD, Jetton TL, Ying G et al (1996) Quantitative subcellular imaging of glucose metabolism within intact pancreatic islets. J Biol Chem 271:3647–3651

Brown EB, Shear JB, Adams SR et al (1999a) Photolysis of caged calcium in femtoliter volumes using two-photon excitation. Biophys J 76:489–499

Brown EB, Wu ES, Zipfel W, Webb WW (1999b) Measurement of molecular diffusion in solution by multiphoton fluorescence photobleaching recovery. Biophys J 77:2837–2849

Castleman KR (1979) Digital image processing. Prentice-Hall, Englewood Cliffs, NJ

Centonze VE, White JG (1998) Multiphoton excitation provides optical sections from deeper within scattering specimens than confocal imaging. Biophys J 75:2015–2024

Clegg RM (1995) Fluorescence resonance energy transfer. Curr Opin Biotechnol 6:103–110

Combs CA, Balaban RS (2001) Direct imaging of dehydrogenase activity within living cells using enzyme-dependent fluorescence recovery after photobleaching. Biophys J 80:2018–2028

Cox CL, Denk W, Tank DW, Svoboda K (2000) Action potentials reliably invade axonal arbors of rat neocortical neurons. Proc Natl Acad Sci USA 97:9724–9728

Denk W, Svoboda K (1997) Photon upmanship: why multiphoton imaging is more than a gimmick. Neuron 18:351–357

Denk W, Piston DW, Webb WW (1995) Two-photon molecular excitation in laser scanning microscopy. In: Pawley JB (ed) Handbook of biological confocal microscopy. Plenum Press, New York, pp 445–458

Edidin M, Zagyansky Y, Lardner TJ (1976) Measurement of membrane protein lateral diffusion in single cells. Science 191:466–468

Helmchen F, Svoboda K, Denk W, Tank DW (1999) In vivo dendritic calcium dynamics in deep-layer cortical pyramidal neurons. Nat Neurosci 2:989–996

Holmes TJ, Bhattacharyya S, Cooper JA et al (1995) Light microscopic images reconstructed by maximum likelihood deconvolution. In: Pawley JB (ed) Handbook of confocal microscopy. Plenum, New York, pp 389–402

Ichihara A, Tanaami T, Isozaki K et al (1996) High-speed confocal fluorescence microscopy using a Nipkow scanner with microlenses for 3-D imaging of single fluorescent molecule in real time. Bioimages 4:57–62

Inoue S (1995) Foundations of confocal scanned imaging in light microscopy. In: Pawley JB (ed) Handbook of biological confocal microscopy. Plenum, New York, pp 1–17

Inoue S, Spring KR (1997) Video microscopy: the fundamentals. Plenum, New York

Kenworthy AK, Edidin M (1998) Distribution of a glycosylphosphatidylinositol-anchored protein at the apical surface of MDCK cells examined at a resolution of < 100 A using imaging fluorescence resonance energy transfer. J Cell Biol 142:69–84

Kenworthy AK, Petranova N, Edidin M (2000) High-resolution FRET microscopy of cholera toxin B-subunit and GPI-anchored proteins in cell plasma membranes. Mol Biol Cell 11:1645–1655

Mason WT (2000) Fluorescent and luminescent probes for biological activity. Academic, New York

Miyawaki A, Llopis J, Heim R et al (1997) Fluorescent indicators for Ca^{2+} based on green fluorescent proteins and calmodulin. Nature 388:882–887

Nehls S, Snapp EL, Cole NB et al (2000) Dynamics and retention of misfolded proteins in native ER membranes. Nat Cell Biol 2:288–295

Ng T, Squire A, Hansra G et al (1999) Imaging protein kinase C-alpha activation in cells. Science 283:2085–2089

Niswender KD, Blackman SM, Rohde L et al (1995) Quantitative imaging of green fluorescent protein in cultured cells: comparison of microscopic techniques, use in fusion proteins and detection limits. J Microsc 180:109–116

Ohler A, Xu W, O'Rourke B (2000) Fluorescence imaging of perfused whole hearts with sub-cellular resolution using two-photon excitation. Biophys Soc Annu Meet 2000

Patterson GH, Piston DW (2000) Photobleaching in two-photon excitation microscopy. Biophys J 78:2159–2162

Pawley JB (1995) Fundamental limits in confocal microscopy. In: Pawley JB (ed) Handbook of biological confocal microscopy. Plenum, New York, pp 19–38

Periasamy N, Verkman AS (1998) Analysis of fluorophore diffusion by continuous distributions of diffusion coefficients: application to photobleaching measurements of multicomponent and anomalous diffusion. Biophys J 75:557–567

Peters R, Peters J, Tews KH, Bahr W (1974) A microfluorimetric study of translational diffusion in erythrocyte membranes. Biochim Biophys Acta 367:282–294

Petran M, Hadravsky M, Egger MD et al (1968) Tandem scanning reflected light microscope. J Opt Soc Am 58:661–664

Petran M, Hadravsky M, Boyde A (1985) The tandem scanning reflected light microscope. Scanning 7:97–108

Piston DW, Knobel SM (1999) Quantitative imaging of metabolism by two-photon excitation microscopy. Methods Enzymol 307:351–368

Potter SM, Wang CM, Garrity PA, Fraser SE (1996) Intravital imaging of green fluorescent protein using two-photon laser-scanning microscopy. Gene 173:25–31

Sabharwal YS, Rouse AR, Donaldson L et al (1999) Slit-scanning confocal microendoscope for high-resolution in vivo imaging. Appl Opt 38:7133–7158

Sako Y, Minoghchi S, Yanagida T (2000) Single-molecule imaging of EGFR signalling on the surface of living cells. Nat Cell Biol 2:168–172

Sandison DR, Williams RM, Wells KS et al (1995) Quantitative fluorescence confocal laser scanning microscopy (CLSM). In: Pawley JB (ed) Handbook of biological confocal microscopy. Plenum, New York, pp 39–53

Shaw PJ (1995) Comparison of wide-field/deconvolution and confocal microscopy for 3D imaging. In: Pawley JB (ed) Handbook of biological confocal microscopy. Plenum, New York, pp 373–387

Soumpasis DM (1983) Theoretical analysis of fluorescence photobleaching recovery experiments. Biophys J 41:95–97

Squirrell JM, Wokosin DL, White JG, Bavister BD (1999) Long-term two-photon fluorescence imaging of mammalian embryos without compromising viability. Nat Biotechnol 17:763–767

Storrie B, Pepperkok R, Stelzer EHK, Kreis TE (1994) The intracellular mobility of a viral membrane glycoprotein measured by confocal microscope fluorescence recovery after photobleaching. J Cell Sci 107:1309–1319

Summers RG, Piston DW, Harris KM, Morrill JB (1996) The orientation of first cleavage in the sea urchin embryo, Lytechinus variegatus, does not specify the axes of bilateral symmetry. Dev Biol 175:177–183

Svoboda K, Tank DW, Denk W (1996) Direct measurement of coupling between dendritic spines and shafts. Science 272:716–719

Svoboda K, Denk W, Kleinfeld D, Tank DW (1997) In vivo dendritic calcium dynamics in neocortical pyramidal neurons. Nature 385:161–165

Terasaki M, Dailey ME (1995) Confocal microscopy of living cells. In: Pawley JB (ed) Handbook of biological confocal microscopy, Plenum, New York, pp 327–346

Tsien RY, Waggoner A (1995) Fluorophores for confocal microscopy. In: Pawley JB (ed) Handbook of biological confocal microscopy. Plenum, New York, pp 267–277

Wang SS, Khiroug L, Augustine GJ (2000) Quantification of spread of cerebellar long-term depression with chemical two-photon uncaging of glutamate. Proc Natl Acad Sci USA 97:8635–8640

White JG, Amos WB, Fordham M (1987) An evaluation of confocal versus conventional imaging of biological structures by fluorescent light microscopy. J Cell Biol 105:41–48

Wilson T (1995) The role of the pinhole in confocal imaging systems. In: Anonymous (ed) Handbook of biological confocal microscopy. Plenum, New York, pp 167–182

Wouters FS, Bastiaens PI, Wirtz KW, Jovin TM (1998) FRET microscopy demonstrates molecular association of non-specific lipid transfer protein (nsL-TP) with fatty acid oxidation enzymes in peroxisomes. EMBO J 17:7179–7189

Yuste R, Majewska A, Cash SS, Denk W (1999) Mechanisms of calcium influx into hippocampal spines: heterogeneity among spines, coincidence detection by NMDA receptors, and optical quantal analysis. J Neurosci 19:1976–1987

Zaccolo M, De Giorgi F, Cho CY et al (2000) A genetically encoded, fluorescent indicator for cyclic AMP in living cells. Nat Cell Biol 2:25–29

Zeng H, Weiss A, MacAulay C (1999) System for fast measurements of in vivo fluorescence spectra of the gastrointestinal tract at multiple excitation wavelengths. Appl Opt 38:7157–7158

CHAPTER 33

Biological Effects of Low Doses of Ionizing Radiation: Damage versus Protection

33

LUDWIG E. FEINENDEGEN

Contents

33.1
Introduction

Medical applications of ionizing radiation frequently provoke the question of risk that may be associated with a planned radiological or nuclear medical procedure. In fact, the apprehension about a potential detriment following exposure to ionizing radiation in clinical diagnosis or therapy is often overwhelming to the patient. The consequence is not rarely a flat refusal of treatment. The reason for the relatively widely spread radiation phobia in the general public is partly rooted in the radiation-protection regulations and their interpretation by non-specialists, including the news media. The problem also often stems from the lack of knowledge by professionals who do not have the proper training in the biological effects of ionizing radiation. Patients need to be properly informed and instructed in order both to alleviate personal apprehension and fear, and to adhere to professional standards that demand the best possible treatment of the patient with minimization of the risk-benefit ratio. It thus appears fitting to treat the question of biological and health effects of low doses of ionizing radiation in the context of this book.

Biological and health effects of low doses of ionizing radiation are currently broadly debated and studied in basic research and by epidemiological means, and are being reassessed for radiation protection and for the regulation of administrative agencies. The reason for this is the fact that, first, the cancer incidence following low doses of low-LET radiation is not significantly different from that in a non-irradiated control population, and that, secondly, over the last two decades evidence has increasingly shown the incompatibility of cellular reactions per unit dose at high doses with those per unit dose at low doses.

Biological and health effects of relatively high doses are rather well known from studies on single cells, tissues, and whole organisms involving almost all forms of life from microorganisms to mammals (Bond et al. 1966; Hall 2000). In contrast, interest in the effects of low doses, especially of low-LET type radiation, has arisen only over the past two decades, mainly because of the advent of refined and sensitive analytical procedures. Responses of complex biological systems to low doses from different types of ionizing radiation were found to differ from responses expected from data obtained at high doses (Sugahara et al. 1992; UNSCEAR 1994; Academie des Sciences 1995; DOE/NIH 2000). It became increasingly clear that low doses of ionizing radiation may exert a dual response in the exposed mammalian tissues and its cells (Feinendegen et al. 1999). One response entails damage mainly to DNA and is rather well known from studies with high doses. The other response is not seen at high doses and is expressed in various forms of adaptive protection against an array of endogenous and exogenous toxins such as reactive oxygen species as well as ionizing radiation. The present chapter attempts to briefly review these recent developments and to put the dual response pattern into the frame of dosimetry at the basic level of biological organization, that of cells and comparable tissue micromasses.

There are inherent difficulties with conventional dosimetry when applied to the scenario of heterogeneously and often non-stochastically distributed energy-deposition events in tissues. This is, indeed, frequently typical for low doses and so nearly always encountered in nuclear medicine that the difficulty has been extensively discussed and put into a practically applicable frame with recommendations for nuclear medicine (ICRU 2002). The expression of risks as a function of dose under conditions of heterogeneous and often non-stochastic distribution of energy absorption at the cellular level brings options of applying dosimetry. Crucial here is the choice of the cause-effect level to be studied, be it the microscopic or macroscopic level. Since acute as well as late effects of ionizing radiation such as cancer arise at the cellular level of biological organization, absorbed doses to tissue micromasses may appear more revealing than absorbed dose to whole tissue, especially under conditions of low-dose exposure. Approaching dosimetry at the microscopic level also allows one to sum up energies absorbed per micromass over all exposed micromasses to equal the total energy absorbed in the exposed system as such. This overcomes the dilemma caused by the fact that equal absorbed doses in tissue masses of different sizes are not equal to the total energy absorbed in these different masses. A recent summarizing review for the nuclear medicine community (Feinendegen and Pollycove 2001) focused on the probabilities of various types of cell responses in complex tissues as a function of absorbed doses at the level of tissue micromasses.

Indeed, biological tissue functions through concerted contributions of the tissue-constituent cells. These are embedded in networks of signaling, which assure coordination of cell actions and responses for the sake of both coherent tissue function and maintenance of tissue integrity. The latter serves the former. It does so at a homeostatically controlled level in face of various exposures to an array of potentially toxic agents from the environment and endogenously from metabolism. Tissues are indeed complex adaptive systems (Gell-Mann 1994). Because of the immense complexity of mammalian tissues, the relationship between the amount of a potentially toxic agent in tissue and tissue response to this agent appears indeed not simple.

In order to comprehend tissue alterations such as in the development of cancer, the consequences of all cellular responses in a tissue system must be considered. This scrutiny appears especially crucial in the case of exposure to low doses and dose-rates of ionizing radiation, when different types of radiation per unit absorbed dose cause different numbers of energy-deposition events of different sizes at the level of cells and corresponding tissue micromasses. Any tissue effect including cancer is a consequence of all cell responses in the irradiated tissue (Feinendegen 1991) from irradiated cells and from unirradiated neighboring cells by way of signaling and via factors that are transferred from irradiated cells, e.g., by bystander effects, and from tissue matrix (Nagasawa and Little 1992; Mothersill and Seymour 1997; Barcellos-Hoff and Brooks 2001). This makes it desirable to express low-dose effects to tissue not as a function of the conventional term absorbed dose to whole tissue, i.e., energy concentration in whole tissue, but in the context of energies absorbed per tissue micromass summed over all exposed micromasses. To do so requires attention to:

1. The relationship between absorbed dose in tissue and in tissue micromasses
2. The spectrum of cellular responses to doses in tissue micromasses
3. The contribution of various cell responses to the generation of tissue effects, including cancer

This type of analysis is unconventional but has the advantage of remaining in concordance with the fundamental quantities and units for ionizing radiation (ICRU 1998). Moreover, it allows for flexibility in assessing biological tissue effects as a consequence of all types of responses of all cells as a function of energy deposited in micromasses in tissues exposed to ionizing radiation of different qualities.

33.2
Relationship Between Absorbed Dose to Tissue and Dose to its Micromasses

33.2.1
Absorbed Dose

The definition of absorbed dose (ICRU 1998) expresses energy deposition per unit mass and not energy absorbed in any larger amount of mass such as an organ or the whole body. The unit of absorbed dose (D) is the gray: 1 Gy (100 rad) = 1 J/kg. At a sufficiently high value of absorbed dose from external irradiation, absorbed dose in large mass is identical to the absorbed dose in a any small mass of the same exposed tissue; but the total energy absorbed in the two masses is not the same. The assessment of effects as a function of absorbed dose may thus demand adjustments regarding the definition of the mass that is critical for the effect to develop.

In this presentation, the principal gross-sensitive tissue micromass is taken to be that of an average mammalian cell. The cell is the basic unit of life, with structural and functional components interacting through signaling networks and in mutual signal exchanges with other cells and matrix throughout a given tissue. The average mammalian cell is taken to have a spherical volume of 1 ng mass (Feinendegen et al. 1994).

Penetrating ionizing radiation causes the deposition of energy from particle tracks that arise stochastically throughout the exposed mass (ICRU 1983). The mean energy deposited by a single particle track in traversing a tissue micromass of 1 ng is here expressed as a "microdose" and the event of a microdose is a "microdose-hit." Large absorbed doses D in the tissue create large numbers of microdose-hits per tissue micromass and the sum of microdoses per given micromass, here denoted cell-dose, is then nearly identical to D. As D decreases, the number of microdose-hits per micromass is reduced. When it falls far enough below a mean value of one per micromass, the dose to each micromass becomes either zero or the microdose from a single track traversing the micromass, and only a fractional number of micromasses experiences a microdose-hit.

The individual microdoses conform to a spectrum that is given by the radiation quality. The microdose may formally equate with specific energy z_1, and its fluency-derived mean is $\bar{z}_1$ (ICRU 1983). The number of microdose events, i.e., hits, N_H, per number of exposed micromasses, N_E, is then $D/\bar{z}_1$. Absorbed dose D is thus the ratio of N_H, to N_E, multiplied by $\bar{z}_1$:

$$D = \bar{z}_1(N_H/N_E) \tag{1}$$

Obviously, because $\bar{z}_1$ is constant for a given radiation quality, the only variable with D is (N_H/N_E). Therefore, when a high-LET radiation produces higher values of $\bar{z}_1$, fewer hits, i.e., lower N_H will deliver the same D as does low-LET radiation with its lower values of z_1 and larger N_H (Feinendegen et al. 1985). Thus $\bar{z}_1$ and (N_H/N_E) are inversely related to each other, and the ratios between $\bar{z}_1$ and different values of D are specific to different radiation qualities.

When on average of five or more microdose-hits from penetrating radiation occur per micromass, more than 99% of the exposed micromasses are hit and the total dose to a given micromass, i.e., cell-dose, becomes increasingly equal to tissue dose. With on average one hit per micromass, 37% of the micromasses receive one hit, 26% have more than one hit, 2% receive four hits, and 37% have zero hits (Feinendegen and Graessle 2003). When the fraction of exposed micromasses that are hit is less than or equal to 0.2, it will be unlikely that more than a single hit occurs in any individual micromass. Equation 1 indicates this to be also true when the absorbed dose D is less than 0.2 of the value $\bar{z}_1$. This level at times defines a "low" dose. Thus, an absorbed dose in tissue D of 250 kVp X-rays with $\bar{z}_1 = 0.9$ mGy would need to be below about 0.2 mGy to generate not more than about one microdose-hit per micromass. In contrast, D as high as 70 mGy would meet the one-hit criterion for 4 MeV alpha particles, where $z_1 = 350$ mGy. Irrespective of these considerations, D below about 200 mGy is generally considered as low.

The focus on energy depositions at the micromass level in tissues at low doses allows the expression of divergent cellular response probabilities for differing numbers of microdose hits in the exposed micromasses, as well as for different radiation qualities.

33.2.2
Dose Rate

The microdose approach shows that dose rate, i.e., D per unit time t, in tissue expresses repeated hits to its micromasses. Thus, using Eq. 1,

$$D/t = \bar{z}_1 (N_H/N_E) 1/t)$$

or:

$$D/t = \bar{z}_1 (t N_E/N_H)$$

and with t_x for $(t N_E/N_H)$

$$D/t = \bar{z}_1/t_x \qquad (2)$$

The denominator t_x (t N_E/N_H) in Eq. 2 is equal to $t\bar{z}_1/D$ (see Eq. 1). The denominator expresses, for a given radiation quality, the average time interval between two consecutive microdose-hits in a given micromass (ICRU 1983; Feinendegen et al. 1985).

Varying the time interval t_x may either enhance or limit the full expression of radiation-induced responses of any type in the affected cells. It may be long enough for a hit not to interfere with a response to a preceding hit. This is likely the case, for instance, at chronic exposure to 1 mGy of 100 kVp X-rays per year, which causes on average one microdose-hit of 1 mGy per micromass per year. A chronic exposure to 100 mGy of 250 kVp X-rays per year would give an average time interval of 3.24 days between two consecutive hits per micromass and each hit would result in a mean microdose of 0.9 mGy; 31 such hits would occur per 100 cells per day – an important consideration in view of potential bystander effects. A chronic exposure to 330 mGy of 250 kVp X-rays per year would bring on average one microdose-hit per micromass per day. An annual chronic exposure to 150 mGy of ^{137}Cs-gamma-radiation would cause per day on average one microdose-hit of 0.4 mGy per micromass.

The above average values of time intervals and microdoses, of course, include two different probability distributions. One is for the stochastic incidence of microdose-hits per micromass from penetrating radiation, as discussed above. The other distribution here expresses the probability of microdose value per hit according to the measured particle spectrum from a given radiation quality (Feinendegen and Graessle 2003).

33.3
The Spectrum of Cellular Responses to Microdoses

33.3.1
Endogenous Induction of Damage and Related Signaling Effects

It appears appropriate in the context of low-dose effects to first consider cell responses to potentially toxic agents from non-radiation sources (Feinendegen et al. 1995; Lindahl 1996; Pollycove and Feinendegen 2003). Such toxic agents constantly arise from oxidative metabolism in the form of reactive oxygen species (ROS) and may come from micronutrient deficiencies, as well as from various toxic compounds in the environment (Beckman and Ames 1998; DOE/NIH 2000). As presented in more detail elsewhere, ROS come from mitochondria into cytoplasm of mammalian cells in vivo at an average rate of close to 10^9 per cell per day, with minimal ROS bursts frequently occurring from various metabolic reactions (Beckman and Ames 1998). These ROS cause many oxidative reactions throughout the cell and attack renewable molecules, lipids, and proteins with concomitant temporary changes in intracellular signaling (Stadtman and Berlett 1998; Finkel and Holbrook 2000). On average about 10^6 oxidative DNA damages (DNA oxyadducts) are estimated to occur per cell per day, i.e., on average about 10 per second (Pollycove and Feinendegen 2003). Despite physiologically extremely effective repair systems, some injuries escape repair and leave permanent DNA alterations. On average, one permanent DNA alteration such as a mutation is likely to occur from endogenous sources per mammalian cell each day. Most cells with such damage accumulation leave their tissue through cell differentiation and senescence and others may be removed by apoptosis or immune responses. Yet, these DNA alterations in surviving cells are held to contribute to both aging and spontaneous carcinogenesis (Beckman and Ames 1998; Finkel and Holbrook 2000). The probability, however, of non radiation-induced, i.e., spontaneous, oncogenic transformation of a human hemopoietic stem cell causing lethal leukemia has been estimated to remain at only about 10^{-11} (Feinendegen et al. 1995).

The various types of ROS potentially cause both cellular damage and signaling changes. The different categories of cellular protection such as defense mechanisms, DNA repair, and damage removal including apoptosis, may be stimulated by suprabasal bursts of ROS as they occur in the course of normal metabolism (Ramana et al. 1998; Chandra et al. 2000; Finkel and Holbrook 2000; Sen et al. 2000). These data suggest that cellular ROS may directly or indirectly produce and suppress DNA alterations depending on ROS concentration.

33.3.2
Endogenously Caused and Radiation-induced Damage

Radiation-induced primary damage to DNA appears proportional to absorbed dose in the low dose region. This statement relies on measurements that come from isolated DNA as well as DNA extracted from cell populations in culture and in complex tissues (Hall 2000). DNA damage from low-dose irradiation must be viewed in the context of endogenously caused damage. For instance, a Compton electron track of about 6 keV from 100 kVp X-rays ranges over less than 1 μm in tissue and creates about 200 ROS with considerable clustering along its track. The ROS from low-LET radiation in hit cells are responsible for some 60% of the resulting DNA damage; about 40% of the damage comes from direct electron interactions with the DNA. In all, a 10 mGy average cell-dose from low-LET radiation causes at least 10 base changes, some 10 single strand breaks (SSBs), 0.4 double strand breaks (DSBs), and less than 5 intermolecular cross links.

A chronic exposure to about 2 mGy of low-LET radiation per year produces on average one microdose-hit per micromass at a rate of about two to six times a year, depending on radiation quality with its particular mean microdose. Thus, 2 mGy per year from ^{60}Co γ-radiation brings on average one microdose-hit of about 0.3 mGy every 2 months, and this dose rate from 100 kVp X-rays brings on average one microdose-hit of about 6 keV every 6 months. Correspondingly, on average some 70 to 200 ROS are generated per microdose-hit each time in less than a microsecond (Feinendegen 2002). The resulting persistent DNA damages per average cell per day are many orders of magnitude lower than the incidences of DNA damage from the non-radiation sources discussed above.

The preponderance of endogenously caused DNA damages also holds regarding damage severity. Endogenous ROS primarily cause single DNA oxyadducts at a relatively low rate of about 10 per second per cellular genome. On the other hand, background radiation-induced DNA damage is relatively rare as it occurs per cell at average time intervals of months. But this damage results per microdose-hit from frequently clustered ROS and from direct particle interactions. The result is that a relatively large fraction of the total DNA alterations is of the multiple-damage-site type (Ward 1988). Indeed, the probability of a DSB per primary DNA alteration from ionizing radiation is estimated to be 10^5 times higher than that per ROS attack on DNA from endogenous sources (Pollycove and Feinendegen 2003). Nevertheless, the high incidence of endogenously generated DNA oxyadducts per day alone is held to generate so many DSB-type lesions per average tissue cell per day that they outnumber the DSB incidence per day caused by background radiation by a factor up to 10^3 (Pollycove and Feinendegen 2003).

It thus appears that DNA damages from endogenous non-radiation sources far outnumber even the more severe DNA damages from radiation exposure at background levels. One may speculate that DNA damage accumulation from any source eventually conditions a cell to become susceptible to apoptosis, which may be triggered, for example, by a microdose-hit with its supra-basal burst of ROS from background radiation (Chandra et al. 2000; Feinendegen 2002).

33.3.3
Cellular Defense, Repair, and Damage Removal

The enormous vulnerability of the mammalian genome to endogenously and environmentally generated toxins other than radiation may be the evolutionary cause for most effective biochemical mechanisms that provide for cell and tissue protection (Lindahl 1996). Both damage and protection may be induced by endogenous toxins such as ROS as well as by low-dose irradiation. The cell in all likelihood does not distinguish between ROS produced endogenously and by low-dose irradiation even though their topography of ROS generation in the cell is different. Type and extent of both damage and protection apparently vary with species, cell type, cell cycle, and metabolism (Alberts et al. 1994; Hall 2000).

The physiological protective mechanisms operate by: (a) elimination of toxic agents, especially ROS, by different biophysical and biochemical scavenging systems; (b) repair of DNA damage of various kinds such as base changes, SSBs, DSBs, and intermolecular cross-links; and (c) removal of cells with certain types and degrees of damage, for example, by apoptosis, necrosis, premature differentiation, or by competent immune responses.

Following irradiation, cells apparently initiate these protective mechanisms in at least two ways. One of these leads to quick responses, for instance, to apoptosis or repair of cellular damage, such as to DNA, which may take from minutes to hours after exposure depending on the type of damage (Friedberg et al. 1995; Hall 2000). Another response specifically appears with a delay of several hours following low doses of ionizing radiation and elicits stress response-like reactions that may last up to several weeks and provide improved protection in a manner that is also called adaptive response (Wolff 1998), as further discussed below.

33.3.3.1
Immediate Repair and Damage Removal

Regarding immediate removal of cellular damage, repair of DNA base changes, SSBs, and DSBs has been studied extensively after high-dose irradiation (Alberts et al. 1994; Hanawalt 1995; Friedberg et al. 1995; Ohyama and Yamada 1998; Wallace 1998; Hall 2000; Wood et al. 2001). DNA repair begins almost immediately after the damage has occurred. Different base changes are repaired within about 10 min to 1 h (Jaruga and Dizdaroglou 1996). SSBs are usually repaired with a half time of less than 10 min, whereas the repair half times for DSBs are longer than 30 min (Frankenberg-Schwager 1990). Damage removal by signal-induced cell death (apoptosis) is readily seen within hours after irradiation (Potten 1977; Yamada and Hashimoto 1998). Misrepaired and unrepaired damage may make viable daughter cells more susceptible to oncogenic transformation, i.e., may cause "genomic instability", even after many cell divisions (Little 2000). These cells are considered to be more vulnerable to renewed toxic attacks and to become prone to apoptosis. The probability of an oncogenic transformation of a human hemopoietic stem cell in vivo with lethal consequences is very low, about 10^{-13} to 10^{-14} per microdose-hit of 1 mGy of low-LET radiation such as 100 kVp X-rays (Feinendegen et al. 1995, 2000). The corresponding risk per 1 mGy microdose per cell for a DSB is about 10^{-2} and that for chromosomal aberrations 10^{-4}. The large ratio of incidences of chromosomal aberrations and oncogenic transformation, about 10^9 to 10^{10}, suggests the existence of effective scavenging systems that prevent development of lethal tumors in humans.

33.3.3.2
Adaptive Protection Response

Regarding adaptive protection, increasing evidence in the literature indicates corresponding responses to occur in mammalian cells in vivo and in vitro after single low doses of X- or γ-radiation (Sugahara et al. 1992; UNSCEAR 1994; Academie des Sciences 1995; DOE/NIH 2000). In view of direct or indirect low dose-induced bursts of ROS from microdose-hits, consecutive signaling effects may at least in part be responsible for low dose-induced adaptive protection, which except for apoptosis is not seen at high doses. The resulting biochemical reactions develop relatively slowly over up to a few hours, may last for several weeks, and resemble physiological stress responses that protect against DNA damage from any source, whether it originates endogenously or from renewed irradiation. Adaptive protection categories after single low-dose, low-LET irradiation, are as follows:

- Stimulation of the radical detoxification system appearing to a maximum at 4 h after irradiation and slowly declining thereafter over several hours to days. In mouse bone marrow in vivo the effect lasts for about 6 h and causes an elevation of free glutathione in parallel with a delayed and temporary reduction of the activity of thymidine kinase to some 70% of control (Feinendegen et al. 1984, 1987, 1995). In other low dose-irradiated rodent tissues, increased levels of superoxide dismutase (SOD) occurred in parallel with decreased lipid peroxidation lasting for weeks (Yamaoka 1991, 1992) and an elevated level of glutathione was involved in an increase in natural killer cell activity (Kojima et al. 2002). ROS detoxification was also linked to gene activation. Thus, mRNAs for glutathione synthesis-related proteins in the mouse liver became elevated after low-dose gamma irradiation (Kojima et al. 1998). The low dose increased intracellular glutathione in RAW 264.7 cells with its maximum between 3 and 6 h after exposure was mediated by transcriptional regulation of the gamma-glutamylcysteine synthetase gene, predominantly through the AP-1 binding site in its promoter (Kawakita et al. 2003).

- Protection against chromosomal aberrations seen after high-dose irradiation such as 4 Gy, or after exposure to other DNA damaging agents (Wolff et al. 1988). This protection again appears to reach a maximum at about 4 h after low-dose irradiation and lasts up to about 3 days, as reported for various human cells in vivo as well as in culture (UN-SCEAR 1994). This adaptive response probably involves a several-fold enhancement of the DNA repair rate (Ikushima et al. 1996; Le et al. 1998). A similar adaptive response appeared regarding micronuclei formation in human fibroblasts (Azzam et al. 1994). In these cells, conditioning doses from 1 to 500 mGy were equally effective; this also indicates that at the lowest dose, when about 40% of the cells do not have a microdose-hit, a bystander effect is involved in causing the adaptive protection (Broome et al. 2002). Inhibition of DNA synthesis and cell growth in rat glial cells in culture by a high dose of X-rays was reduced following a conditioning low-dose exposure, when the cells were obtained from young rats; the adaptive response decreased with age of the donor rats. This adaptive response involved protein-kinase C (PCK), DNA-dependent protein-kinase (DNA-PK), and phosphatidylinositol 3-kinase (PI3K), as well as the activity of the ataxia-telangiectasia gene (ATM) (Miura et al. 2002).

- Damage removal in vivo, for instance, by way of an induced immune competence (James and Makinodan 1990; Anderson 1992). This is associated with an increased number of circulating cytotoxic lymphocytes, and may cause a reduction in the incidence of cancer and of metastases to less than one third of control (Hashimoto et al. 1999). This response has its maximum at about 0.2 Gy (Sakamoto et al. 1997) and can last for several weeks (Makinodan 1992).

- Damage removal in vivo, for instance, by signal-induced cell death (apoptosis). It usually occurs within hours after high-dose irradiation. Low dose-induced apoptosis of pre-damaged cells and healthy cells may be a major route of in vivo removal of oncogenically transformed cells (Potten 1977; Kondo 1988, 1993, 1999; Norimura et al. 1996; Ohyama and Yamada 1998; Yamada and Hashimoto 1998). Damaged cells also may exit the system by premature differentiation and maturation to senescence, which was observed to follow low-dose irradiation via bystander effect in microbeam experiments directed to single cells in complex tissue (Belyakov et al. 2002). The various mechanisms of protection may be directly or indirectly linked with transient changes in the activity

of the G_1 cell cycle checkpoint (Boothman et al. 1993). Another mechanism in this category of damage removal is known to occur in a number of tissue culture cell types by way of hypersensitivity to low-dose radiation, which disappears at higher doses (Joiner et al. 1996, 1999). This hypersensitivity in some cells is linked to the cell cycle (Short et al. 2003). The hypersensitivity disappeared in a number of culture cells at about 4 h, but not immediately, after a single low-dose, low-LET irradiation (Joiner 2002). Radiation-induced predisposition to genetic instability in culture cells also declined following low-dose irradiation (Suzuki et al. 1998). Both these data indicate prevention of damage removal by way of low dose-induced DNA repair. This is consistent with the observation in rat thymocytes, where the incidence of apoptosis first declined at low doses and only rose with higher doses (Shu-Zheng et al. 1996). The induction of apoptosis apparently requires a certain level of DNA damage.

Such adaptive protection also appears to generate a reduction in spontaneous oncogenesis. In fact, single low doses of low-LET radiation of tissue culture cells initiated (not immediately but with a delay of 1 day) a significant reduction in spontaneous clonogenic transformation to about one third of control (Azzam et al. 1996; Redpath and Antoniono 1998). In mice heterozygous for the Trp-53 gene a single low dose of low-LET radiation when given at the age of about 2 months significantly delayed the appearance of "spontaneous" lymphoma and also spinal osteosarcoma later in life (Mitchel et al. 2003). A review of tumor development following low-dose, low-LET radiation in rodents supports the existence of a threshold dose (Tanooka 2001).

The protective responses listed above involve changes in gene expression (Amundson et al. 1999; DOE/NIH 2000). An example for DNA repair gene activation refers to the telangiectasia gene (Miura et al. 2002). Human fibroblasts in culture showed DNA repair in the course of adaptive protection against micronucleus formation following high-dose irradiation; the repair was more effective in the gene-poor chromosome than in the gene-rich chromosome of the cells (Broome et al. 1999). Another recent presentation showed that exposure of human skin fibroblasts in culture to a single dose of 20 mGy γ-radiation caused more than 100 genes to change their expression within 2 h. This gene group in part related to stress-response genes and was different from the group of genes that in parallel cultures concomitantly responded to 500 mGy (Golder-Novoselsky et al. 2002).

Except for apoptosis, all the above protective responses to single exposure tend to be expressed maximally after less than 0.1 and not more than 0.5 Gy X- or γ-radiation and to increasingly fail with higher doses depending on type of adaptive protection, on cell type, and on species, as summarized previously (Feinendegen et al. 2000; Feinendegen 2002). In most mammalian cells so far examined, the maximal expression of adaptive protection occurred above 5 mGy and below about 200 mGy.

Adaptive protection appears as physiological expression of both: (a) cellular capabilities for maintaining integrity of tissue structure and function, and (b) cellular adaptation in order to cope with a renewed or continued exposure to intra- and extra-cellular toxins such as ROS, be they from endogenous sources or ionizing radiation. Despite the disparity of the examined systems and responses, a common pattern of data can be discerned. In fact, adaptive protection following low doses of low-LET radiation appears to be the consequence of changes in cellular signaling and to be ubiquitous.

It follows: (1) DNA damage in mammalian cells from non-radiation sources far outweighs DNA damage from background radiation exposure. (2) Induction of adaptive protection outweighs damage at doses well below 200 mGy low-LET radiation. (3) The delayed appearing and temporarily lasting protective responses to low doses appear to primarily operate against DNA damage from non-radiation sources. (4) At higher absorbed doses in tissue and the corresponding cell doses, cell and DNA damage appears increasingly to overrule, negate, or annihilate the more subtle signaling effects causing protection after low doses.

33.4
Cell Responses in the Generation of Tissue Effects

33.4.1
Tissue System with Cellular Elements

The dual cell response to low-dose irradiation, lasting DNA damage on the one hand and delayed appearance of temporary protection on the other, can be related in a model in order to understand the patho-genesis of tissue effects. The various components of the dual response, as presented above, became known mainly from measurements after single low-LET irradiation of multicellular systems in culture or living tissues rather than of single cells.

Single cells in culture systems have been separately irradiated using microbeams of different LET-types, and effects were registered throughout the cell population. Clearly, unirradiated cells being neighbors to irradiated cells showed responses as bystander effects, which proved to be a source of DNA damage (Brooks et al. 1974; Nagasawa and Little 1992; Mothersill and Seymour 1997; Azzam et al. 1998; Barcellos-Hoff and Brooks 2001; Sawant et al. 2001 a, b). There is evidence (by no means conclusive thus far) that bystander effects may also initiate protective responses in non-irradiated cells (Matsumoto et al. 2001; Sawant et al. 2001 b; Belyakov et al. 2002; Lehnert and Iyer 2002; Broome et al. 2002). Moreover, physiological intercellular signaling that also involves matrix function appears to be less disrupted by low doses than high doses; and low doses appear to modulate rather than destroy the balance between damage and repair in tissues (Barcellos-Hoff and Brooks 2001). Indeed, data that are stochastically generated and summarily measured in multicellular systems, and not single cells, include intercellular bystander effects and those from any extracellular signaling that may affect the ensuing results. It is increasingly clear that radiation effects in cell populations and tissues are always the consequence of any and all cell responses in the exposed system. In this way, the tissue as a whole should be seen as a system composed of elements with different radiation sensitivities and responses, but reacting as a whole (Feinendegen 1991; Barcellos-Hoff and Brooks 2001).

Regarding generation of data, observations of cell systems rather than single cells readily express one or more categories of adaptive protection at low doses of low-LET radiation, where damaging effects are hardly or not at all measurable. As explained above, when tissue doses of low-LET radiation result on average in more than five microdose-hits per micromass, average cell-doses become equal to tissue doses and the average cell-dose is always a multiple of microdoses. With high-LET radiation, mean microdoses per hit micromass are relatively high. Therefore, a low tissue dose here cannot be equated with an average cell dose. Yet, overall tissue effects may be related to hit number and values of microdoses.

33.4.2
From Dose-Risk Function
to Microdose-Hit-Number Effectiveness Function

A given tissue effect at low tissue doses from stochastically distributed microdose-hits from a given type of radiation is here expressed as consequence of all effect probabilities per number of mean microdose-hits in the number of micromasses exposed. Assuming for the case of cancer induction the probability of radiation-induced oncogenic transformation to be proportional to the number of microdose-hits up to a certain value in the low dose range, and taking a to be the proportionality constant, the well-known relationship between risk R and tissue dose D:

$$R = a\,D \tag{3}$$

may be transformed into microdosimetric terms.

Let R be the risk of cancer, i.e., the cancer incidence in an irradiated tissue. R gives the ratio of the number of cells from which observable radiation-induced cancer arises (N_q) to the number of exposed micromasses (N_E). Thus, substituting for R the ratio N_q/N_E and for D using Eq. 1: $D = \bar{z}_1(N_H/N_E)$, the following equation results:

$$N_q/N_E = (a\,\bar{z}_1)N_H/N_E \tag{4}$$

This equation expresses for a given radiation quality the incidence of cancer, not as a function of energy absorbed per mass, but as a function of the number of microdose-hits of a given radiation quality in all micromasses of the exposed system. The conventionally used dose-risk function has thus been transformed into a microdose-hit-number-effectiveness function or "hit-number-effectiveness function" for a given radiation quality in the exposed system (Bond et al. 1995). This function focuses on relating observed tissue effects such as cancer to the probability of cellular responses to microdose-hits in the system when exposed to a given radiation quality.

It is understood that replacing the conventional dose-risk function by the microdose-hit-number-effectiveness function for a given radiation quality entails various constraints, which may come from heterogeneous hit densities and from the characteristics of particle track structures in tissues. The term microdose in tissue, on the other hand, offers the advantage of being consistent with the conventional quantities and units of ionizing radiation (ICRU 1998).

33.4.3
Cancer Incidence as a Function of the Sum
of All Cellular Response Probabilities
per Mean Microdose-Hit over All Hits

33.4.3.1
Microdoses of Low-LET Radiation

The present approach is a first approximation and confronts enormous complexity. It nonetheless leads to a model that does not rely on specific steps in the process of oncogenic transformation of a cell and its becoming the seed of a malignant tumor, instead the probability of a cancer to appear is put into the context of balance between probabilities per microdose-hit of oncogenic transformation and of various protections against oncogenic transformation and tumor development, be they induced by non-radiation toxins or by ionizing radiation.

The probability of a radiation-induced malignant transformation in a cell causing tumor development derives from the measured incidence of a malignant tumor in the observed organism. The following assumptions are made: (1) a malignant tumor arises from a single cell; (2) cancer incidence obtained at higher doses in the linear region of the dose-risk function indicates the incidence of radiation-induced oncogenic cell transformation; (3) the probability of a radiation-induced oncogenic cell transformation per unit dose within a given range of higher doses is the same in the low dose region. On these assumptions, the probability of radiation-induced oncogenic cell transformation per mean microdose-hit of a given radiation quality is denoted p_{ind}. In case a radiation-induced enhancement of p_{ind} occurs, such as through genetic instability of the progeny of a hit cell, or through other cancer-enhancing effects on that cell, the fractional enhancement of p_{ind} per hit is here described by the probability p_{enh}. The probability of oncogenic cell transformation caused by non-radiation toxins such as endogenous ROS, i.e., of spontaneous carcinogenesis, per oncogenically transferable cell during its life time is denoted by p_{spo}.

The adaptive protection responses have been defined above. They appear delayed and operate over different lengths of time from hours to weeks after irradiation and do not develop at higher doses, i.e., with increasing numbers of microdose-hits. These responses express cell conditioning and are different from the acutely operating repair mechanisms. The cumulative probability of all operating protection against any oncogenic transformation and tumor development is here expressed by p_{prot}. Because protection depends on both dose D $[\bar{z}_1 N_H/N_E]$ and its duration after exposure, t_p, the term p_{prot} becomes p_{prot} (D, t_p).

The relation between the number of microdose-hits of a given radiation quality in the exposed micromasses and the various cellular response probabilities is here first analyzed at single low-dose exposures with given numbers of microdose-hits (Feinendegen et al. 1995, 2000). With:

p_{spo} = lifetime probability of spontaneous oncogenic cell transformation with potential cancer development, per cell at a given age

p_{ind} = probability of radiation-induced oncogenic cell transformation with potential cancer development, per average hit

p_{enh} = fractional enhancement of p_{ind} per average hit

p_{prot} (D, t_p) = cumulative probability of protective mechanisms that operate against, and would reduce, DNA damage and cancer in tissue by a certain fraction, i.e., against p_{spo}, p_{ind}, and p_{enh} per average hit and with N_q/N_E and N_H/N_E as explained above:

$$N_q/N_E = [p_{ind} + p_{ind}p_{enh} - p_{prot}(D, t_p)p_{spo}$$
$$- p_{prot}(D, t_p)p_{ind}$$
$$- p_{prot}(D, t_p)p_{ind}p_{enh}]N_H/N_E$$

or

$$N_q/N_E = [p_{ind}(1 + p_{enh})$$
$$- p_{prot}(D, t_p)(p_{spo} + p_{ind}$$
$$+ p_{ind}p_{enh})]N_H/N_E \qquad (5)$$

Combining Eqs. 4 and 5:

$$a = [p_{ind}(1 + p_{enh})$$
$$- p_{prot}(D, t_p)(p_{spo} + p_{ind}$$
$$+ p_{ind}p_{enh})]/\bar{z}_1 \qquad (6)$$

The positive and negative terms contained in a in Eq. 6 determine the shape of the dose-risk function for tissue effects. As discussed above, whereas radiation-induced oncogenic cell transformation with potential cancer development increases in proportion to the number of microdose-hits of low-LET radiation, the probability of adaptive protection mainly against non-radiation-induced cancer tends to become zero with increasing numbers of microdose-hits. It is a justifiable assumption that the radiation-induced adaptive protection at low instant numbers of microdose-hits of the low-LET type, $[p_{prot}(D, t_p)(p_{spo}+p_{ind}+p_{ind}p_{enh})N_H/N_E]$, reduces the generation and accumulation of DNA damage

that is overwhelmingly from endogenous sources. If the probability of cancer induction would be equal to the probability of protection per microdose-hit, as defined above, the value of a under these premises would become zero. It would be negative in case the probability of adaptive protection attains a larger value than the probability of cancer induction. In the latter case, a hormetic tissue effect would appear (Feinendegen et al. 1995, 1999, 2000).

As an example, neglecting any enhancement, Eq. 5 can be simplified to:

$$N_q/N_E = [p_{ind} - p_{prot}(D, t_p)(p_{spo} + p_{ind})]N_H/N_E]$$

and letting N_q/N_E be zero, then:

$$[N_H/N_E]p_{prot}(D, t_p) = [N_H/N_E]p_{ind}/(p_{spo} + p_{ind})$$

With above-named values of $p_{ind} = 10^{-13}$–10^{-14} for an N_H/N_E value of low-LET type, and of $p_{spo} = 10^{-11}$, the value of $p_{prot}(D, t_p) = 10^{-2}$–$10^{-3}$ would result. In other words, the coefficient a would be zero already at a protection probability of as little as 10^{-2}–10^{-3} per mean microdose-hit from low-LET radiation (Feinendegen et al. 1995, 2000). As indicated above, $[N_H/N_E] p_{prot}(D)$ can attain the value of 0.3 and last an average of 30 days, for instance as an expression of a low-dose-stimulated immune response. If the average time span between cellular oncogenic transformation and tumor occurrence is 5 years, i.e., 1825 days, the time-corrected value of $[N_H/N_E]p_{prot}(D)$, now denoted by $[N_H/N_E] p_{prot}(D, t_p)$, becomes 0.3 (30/1825) = $\sim 10^{-2}$–10^{-3}. With the assumptions above, this degree of protection would closely compensate for the value of $[N_H/N_E \, p_{ind}]$.

Indeed, the value of a only becomes constant at higher doses of low-LET radiation and consequently with higher numbers of microdose-hits in the number of micromasses exposed to this radiation, when the probability of adaptive protection is zero. It therefore appears likely that the value of a is not constant at the ranges of D, in which protection operates. In addition to experimental results, some epidemiological data also support this reasoning (Pollycove and Feinendegen 2001).

33.4.3.2
High-LET Microdoses

Concerning the situation with high-LET radiation, the corresponding relatively high microdose values, $\bar{z}_1$, for instance of 350 mGy for 4 MeV a-particles, may be ineffective with regard to $p_{prot}(D, t_p)$ in the hit cells. However, p_{ind}, $p_{ind}p_{enh}$ and p_{spo} in exposed tissues may be offset by $p_{prot}(D, t_p)$ if protective mechanisms are initiated in non-hit cells through intercellular signaling such as bystander effects and tissue-specific extracellular signaling. Indeed, bystander effects alone should be considered to operate between hit and non-hit cells to both induce damage and signal for protection upon single and chronic irradiation. The available data are scant and come mostly from tissue culture studies; in principle they do not contradict but also rather support the occurrence of protective effects in non-irradiated cells that are neighbors of high-LET-irradiated cells (Matsumoto et al. 2001; Sawant et al. 2001b; Belyakov et al. 2002; Lehnert and Iyer 2002). Investigation is needed to determine the degree to which bystander effects operate in either way, i.e., to cause damage or to protect, in tissues that are exposed to very low doses and dose rates of low- or high-LET radiation.

33.4.3.3
Microdoses at Low Dose Rate

With respect to low-dose rates, the time interval between consecutive microdose-hits t_x of a given radiation quality in the exposed tissue, as seen in Eq. 2, is obviously crucial in conjunction with the time of acute DNA repair and with the relatively longer time t_p over which the adaptive protection operates. If the value of t_x is large compared with t_p, Eq. 5 remains valid. If t_x becomes shorter than t_p, yet remains longer than the time of acute DNA repair, the effects of a hit may interfere with the protective responses that are elicited by the preceding hit. If t_x becomes even shorter, DNA repair may be affected. Thus, Eq. 5 demands adjustment (Feinendegen and Graessle 2003). A factor $[F(t_x)]$ may serve to correct for dose rate-dependent interferences with acute DNA repair and adaptive protection. For example, the factor can correct for a relatively short t_x in case it reduces protection and thus increases risk; or with increasing protection it may reduce risk, for instance, through repetitive protection or induction of apoptosis of predamaged cells. Increased protection must also be considered to cause increased risk when a protected cell survives to become oncogenically transformed; as stated above, the corresponding probabilities would be expressed either by p_{ind} or p_{enh}. Despite the complexities that are here involved in assessing tissue risk from low dose rate of a given radiation quality, the

present approach at the very least allows experimental verification and new experimental avenues and interpretations. Thus, in line with Eq. 5:

$$N_q/N_E = [p_{ind}(1 + p_{enh}) - p_{prot}(D, t_p)(p_{spo} + p_{ind} + p_{ind}p_{enh})\{F(t_x)\}]N_H/N_E \quad (7)$$

By introducing here microdosimetric terms for D (see Eq. 2):

$$N_q/N_E = [p_{ind}(1 + p_{enh}) - p_{prot}([\bar{z}_1 t/t_x], t_p)(p_{spo} + p_{ind} + p_{ind}p_{enh})\{F(t_x)\}]N_H/N_E \quad (8)$$

In this context, a unique set of recent experiments on life shortening and the appearance of thymic lymphoma in mice living from birth on tritium water is relevant (Yamamoto et al. 1995, 1998). Since a 1 mGy whole body dose from tritium beta particles corresponds to a mean microdose per micromass of 1 mGy, i.e., about 6 keV/ng, the listed dose rate can be easily converted, for example, to microdose-hits per micromass occurring at average time intervals of t_x. Life shortening and tumor development was only observed if the dose rate increased to beyond 1 mGy per day. This then amounts on average to repetitive microdose-hits per micromass of 1 mGy with an average t_x of less than 1 day. If the average t_x was longer than about 1 day, no tumors appeared and no life shortening was observed.

Another analysis reviewed published data on the mutagenic effects of low-LET radiation at increasing dose rates beginning with 0.01 mGy per minute. This study revealed, first, a reduction of mutagenesis to a minimum with increasing dose rates to the range of 1 to 10 mGy per minute in various mammalian cell lines in culture, and in the range of 0.1 to 1 mGy per minute in mouse spermatogonia. Mutagenesis then increased with higher dose rates (Vilenchik et al. 2000). The authors attribute the observed phenomenon of inverse dose-rate effect to an optimal induction of error-free DNA repair in a dose-rate region of minimal mutability. The diminished activation of repair at minimal dose rates was considered to reflect a low ratio of induced to spontaneous DNA damage, much of which stems from endogenous ROS. In fact, the results suggested a genetically programmed optimization of response to radiation in the minimal-mutability dose-rate region, possibly depending on cell replication. In these experiments, the average time interval between consecutive microdose-hits per micromass is in the range of minutes in the dose-rate region of minimal mutability in spermatogonia. Thus, the data likely express a protective response that remains operating throughout repetitive hits with their relatively large, i.e., suprabasal, bursts of ROS, and deserve further examination. These data also agree in principle with the observation that the induction of adaptive protection appears to depend on the dose rate of the conditioning dose in experiments where the protection is tested by repeated irradiation at higher doses (Shadley and Wienke 1989; Broome et al. 2002).

33.5
Conclusions

Low doses of low-LET radiation evoke a spectrum of biochemical and functional cell and tissue responses in various mammalian cells that express adaptive cell protection and are not observed at high doses. These responses vary, as with oncogenic transformation with potential cancer development, according to species, cell type, cell metabolism, and cell cycle. The probability of low dose-induced cancer is put into the context of balance between probabilities of oncogenic transformation and of various adaptive protections against oncogenic transformation and tumor development, whether this is induced by non-radiation toxins or by ionizing radiation.

In order to dissect the factors at a given radiation quality that contribute to low dose-induced tissue effects including cancer, it appears prudent to pay attention to: (1) the relationship between absorbed dose to tissue and energy deposition from particle tracks in defined tissue micromasses, i.e. microdoses, per number of exposed micromasses of this tissue; (2) the probabilities of various cellular responses per mean microdose-hit of a given radiation quality at increasing numbers of microdose-hits per exposed micromasses; and (3) the contribution of all cellular response probabilities over all microdose-hits in the exposed tissue to the generation of tissue effects, including cancer. The analysis for the case of penetrating ionizing radiation, i.e., stochastic distribution of microdose-hits from a given radiation quality per exposed tissue micromasses leads to the following conclusions:

1. Ionizing radiation causes DNA damage in the exposed tissue according to the number of microdose-hits, by direct and probably bystander cell effects.

2. Acute low doses of low-LET radiation, and thus instant low numbers of microdose-hits per number of exposed micromasses, can initiate various categories of protective responses. Except for apoptosis, these increasingly fail, then disappear as the number of microdose-hits per number of exposed micromasses rises to a level that is equivalent to a tissue dose of about 200 mGy. High microdoses at low tissue doses likely initiate both damage and protective responses via intercellular signaling such as bystander effects. The protective response categories involve:
   - Cellular defenses such as radical detoxification
   - DNA repair involving various pathways
   - Cell removal by immune response
   - Cell removal through intracellular signaling such as by apoptosis (which also occurs at high doses).

3. Depending on radiation quality, non-radiation-induced DNA damage far outweighs radiation-induced damage at relatively low numbers of microdose-hits per number of exposed micromasses, so that adaptive protection induced by these microdoses is expected to mainly operate against non-radiation-induced DNA damage.

4. Cancer induction appears proportional to the degree of DNA damage; thus, radiation-induced adaptive protection mainly against non-radiation-induced DNA damage may reduce the "spontaneous" cancer incidence. This supports experimental observations and some epidemiological data.

5. Regarding dose rate, the spacing in time between consecutive microdose-hits from a given radiation quality per number of exposed micromasses influences cellular responses; with higher dose rates, i.e., with decreasing time intervals between microdose-hits, the incidence of DNA damage may first decrease and then rise and accumulate to eventually cause tissue failure.

6. In view of the limited power of epidemiological data on radiation-induced cancer and the unequivocal broad evidence of low dose-induced adaptive protection, the linear no-threshold hypothesis on radiation-induced cancer appears invalid. It should be replaced for a given radiation quality by a form of dose-risk function or microdose-hit-number-effectiveness function that includes both linear and non-linear terms.

■ **Acknowledgements.** The author gratefully acknowledges the critical reading of the manuscript by Dr. C.A. Sondhaus of the University of Arizona, Tucson, AZ, USA, and expresses deep appreciation to the long and fruitful influencing collaboration and discussions on biological consequences of low-dose irradiation with his early mentors Dr. V.P. Bond and the late Dr. E.P. Cronkite, and the late Dr. K.I. Altman, and Drs. J. Booz, T.M. Fliedner, M. Frazier, R.D. Neumann, M. Pollycove, and C.A. Sondhaus.

33.6
References

Academie des Sciences, Institut de France (1995) Problems associated with the effects of low doses of ionizing radiations. Lavoisier, TecDoc, Paris (Rapport de l'Academie des Sciences, no 38)

Alberts B, Bray D, Lewis J, Raff M, Roberts K, Watson JD (eds) (1994) Molecular biology of the cell, 3rd edn. Garland, New York

Amundson SA, Do KT, Fornace AJ (1999) Induction of stress genes by low doses of gamma rays. Radiat Res 152:225–231

Anderson RE (1992) Effects of low-dose radiation on the immune response. In: Calabrese EJ (ed) Biological effects of low level exposures to chemicals and radiation. Lewis Pub Inc, Chelsea, MI, pp 95–112

Azzam EI, de Toledo SM, Raaphorst GP, Mitchel REJ (1994) Résponse adaptive au rayonnement ionisant des fibroblastes des peau humain. Augmentation de la vitesse de reparation de l'ADN et variation de l'expression des génes. J de Chimie Physique 91:931–936

Azzam EI, Toledo SM de, Raaphorst GP et al (1996) Lowdose ionizing radiation decreases the frequency of neoplastic transformation to a level below the spontaneous rate in C3H 10T1/2 cells. Radiat Res 146:369–373

Azzam EI, de Toledo SM, Gooding T et al (1998) Intercellular communication is involved in the bystander regulation of gene expression in human cells exposed to very low fluency of alpha particles. Radiat Res 150:497–504

Barcellos-Hoff MH, Brooks AL (2001) Extracellular signaling through the microenvironment: a hypothesis relating carcinogenesis; bystander effects, and genomic instability. Radiat Res 156:618–627

Beckman KD, Ames BN (1998) The free radical theory of aging matures. Physiol Rev 78:547–581

Belyakov OV, Folkard M, Mothersill C et al (2002) Bystander-induced apoptosis and premature differentiation in primary urothelial explants after charged particle microbeam irradiation. Radiat Prot Dosimetry 99:249–251

Bond VP, Fliedner TM, Archambeau JO (1966) Mammalian radiation lethality: a disturbance in cellular kinetics. Academic, New York

Bond VP, Benary V, Sondhaus CA et al (1995) The meaning of linear dose-response relations, made evident by use of absorbed dose to the cell. Health Phys 68:786–792

Boothman DA, Meyers M, Odegaard E, Wang M (1996) Altered G1 checkpoint control determines adaptive survival responses to ionizing radiation. Mutation Res. 358:143–153

Brooks AL, Retherford JC, McClellan RO (1974) Effects of $^{238}PuO_2$ particle number and size on the frequency and distribution of chromosome aberrations in the liver of the Chinese hamster. Radiat Res 59:693–709

DOE/NIH, Feinendegen LE, Neumann RD (eds) (2000) Cellular responses to low doses of ionizing radiation. Workshop of the US Department of Energy (DOE), Washington, DC, and the National Institutes of Health (NIH), Bethesda, MD, 27–30 Apr 1999. Mary Woodward Lasker Center, Cloister, NIH; DOE Report Publication SC-047

Broome EJ, Brown DL, Mitchel REJ (2002) Dose response for adaptation to low doses of ^{60}Co-gamma rays and ^3H beta particles in normal human fibroblasts. Radiat Res 158:181–186

Chandra J, Samali A, Orrenius S (2000) Triggering and modulation of apoptosis by oxidative stress. Free Radic Biol Med 29:323–333

Feinendegen LE (1991) Radiation risk of tissue late effect, a net consequence of probabilities of various cellular responses. Eur J Nucl Med 18:740–751

Feinendegen LE (1999) The role of adaptive responses following exposure to ionizing radiation. Hum Exp Toxicol 18:426–432

Feinendegen LE (2002) Reactive oxygen species in cell responses to toxic agents. Huma Exp Toxicol 21:85–90

Feinendegen LE, Pollycove M (2001) Biologic response to low doses of ionizing radiation: detriment versus hormesis, part 1. Dose responses of cells and tissues. J Nucl Med 42:17N–27N

Feinendegen LE, Graessle DH (2003) Energy deposition in tissue during chronic irradiation and the biological consequences. Br J Radiol [Suppl 26] (in press)

Feinendegen LE, Muehlensiepen H, Lindberg C et al (1984) Acute and temporary inhibition of thymidine kinase in mouse bone marrow cells after low-dose exposure. Intern J Radiat Biol 45:205–215

Feinendegen LE, Booz J, Bond VP et al (1985) Microdosimetric approach to the analysis of cell responses at low dose and low dose rate. Radiat Prot Dosimetry 13:299–306

Feinendegen LE, Muehlensiepen H, Bond VP et al (1987) Intracellular stimulation of biochemical control mechanisms by low-dose low-LET irradiation. Health Phys 52:663–669

Feinendegen LE, Bond VP, Booz J (1994) The quantification of physical events within tissue at low levels of exposure to ionizing radiation. ICRU-News 2:9–13

Feinendegen LE, Loken MK, Booz J et al (1995) Cellular mechanisms of protection and repair induced by radiation exposure and their consequences for cell system responses. Stem Cells 13 [Suppl 1]:7–20

Feinendegen LE, Bond VP, Sondhaus CA et al (1999) Cellular signal adaptation with damage control at low doses versus the predominance of DNA damage at high doses. CR Acad Sci Paris Life Sci 322:245–251

Feinendegen LE, Bond VP, Sondhaus CA (2000) The dual response to low-dose irradiation: induction vs. prevention of DNA damage. In: Yamada T, Mothersill C, Michael BD, Potten CS (eds) Biological effects of low dose radiation. Excerpta medica. International Congress Series 1211. Elsevier, Amsterdam, pp 3–17

Finkel T, Holbrook NJ (2000) Oxidants, oxidative stress and the biology of aging. Nature 408:239–247

Frankenberg-Schwager M (1990) Induction, repair and biological relevance of radiation-induced DNA lesions in eukaryotic cells. Rad Environ Biophys 29:273–292

Friedberg EC, Walker GC, Siede W (1995) DNA repair and mutagenesis. ASM Press, Washington DC

Gell-Mann M (1994) The quark and the jaguar. Freeman, New York

Golder-Novoselsky E, Ding L-H, Chen F et al (2002) Radiation response in normal HSF (human skin fibroblasts): cDNA microarray analysis. DOE Low Dose Radiation Research Program Workshop III (abstract). Office of Biological and Environmental Research, US Department of Energy, Washington DC

Hanawalt PC (1995) DNA repair comes of age. Mutat Res 336:101–113

Hall EJ (2000) Radiobiology for the radiologist, 5th edn. Lippincott, Williams and Wilkins, Philadelphia

Hashimoto S, Shirato H, Hosokawa M et al (1999) The suppression of metastases and the change in host immune response after low-dose total-body irradiation in tumor-bearing rats. Radiat Res 151:717–724

Ikushima T, Aritomi H, Morisita J (1996) Radioadaptive response: efficient repair of radiation-induced DNA damage in adapted cells. Mutat Res 358:193–198

International Commission on Radiation Units and Measurements (ICRU) (1983) Microdosimetry. ICRU-Report 36. ICRU, Bethesda MD

International Commission on Radiation Units and Measurements (ICRU) (1998) Fundamental Quantities and units for ionizing radiation. ICRU-Report 60. ICRU, Bethesda MD

International Commission on Radiation Units and Measurements (ICRU) (2002) Absorbed-dose specification in nuclear medicine. ICRU-Report 67. ICRU, Bethesda MD

James SJ, Makinodan T (1990) T-cell potentiation by low dose ionizing radiation: possible mechanisms. Health Physics 59:29–34

Jaruga P, Dizdaroglou M (1996) Repair of products of oxidative DNA base damage in human cells. Nucleic Acid Res 24:1389–1394

Joiner MC, Larnbin P, Malaise EP et al (1996) Hypersensitivity to very low single radiation doses: its relationship to the adaptive response and induced radioresistance. Mutat Res 358:171–183

Joiner MC, Lambin P, Marples B (1999) Adaptive response and induced resistance. Compt Rend Acad Sci Paris. Life Sci 322:167–175

Joiner MC (2002) Personal communication

Kawakita Y, Ikekita M, Kurozumi R, Kojima SP (2003) Increase of intracellular glutathione by low-dose gamma-ray irradiation is mediated by transcription factor AP-1 in RAW 264.7 cells. Biol Pharm Bull 26:19–23

Kojima S, Matsuki O, Nomura T et al (1998) Induction of mRNAs for glutathione synthesis-related proteins in the mouse liver by low doses of γ-rays. Biochim Biophys Acta 1381:312–318

Kojima S, Ishida H, Takahashi M et al (2002) Elevation of glutathione induced by low-dose gamma rays and its involvement in increased natural killer activity. Radiat Res 157:275–280

Kondo S (1988) Altruistic cell suicide in relation to radiation hormesis. Int J Radiat Biol 53:95–102

Kondo S (1993) Health effects of low level radiation. Kinki Univ Press, Osaka, Japan; Medical Physics, Madison WI, USA

Kondo S (1999) Evidence that there are threshold effects in risk of radiation. J Nucl Sci Technol 36:1–9

Le XC, Xing JZ, Lee J et al (1998) Inducible repair of thymine glycol detected by an ultrasensitive assay for DNA damage. Science 280:1066–1069

Lehnert BE, Iyer R (2002) Exposure to low-level chemicals and ionizing radiation: reactive oxygen species and cellular pathways. Hum Exp Toxicol 21:65–69

Lindahl T (1996) The Croonian Lecture, 1996: endogenous damage to DNA. Philos Trans R Soc Lond B Biol Sci 351:1529–1538

Little JB (2000) Radiation carcinogenesis. Carcinogenesis 21:397–404

Makinodan T (1992) Cellular and subcellular alteration in immune cells induced by chronic, intermittent exposure in vivo to very low dose of ionizing radiation (ldr) and its ameliorating effects on progression of autoimmune disease and mammary tumor growth. In: Sugahara T, Sagan LA, Aoyama T (eds) Low-dose irradiation and biological defense mechanisms. Excerpta Medica, Amsterdam, pp 233–237

Matsumoto H, Hayashi S, Hatashita M et al (2001) Induction of radioresistance by nitric oxide-mediated bystander effect. Radiat Res 155:387–396

Mitchel REJ, Jackson JS, Morrison DP, Carlisle SM (2003) Low doses of radiation increase the latency of spontaneous lymphomas and spinal osteosarcomas in cancer prone, radiation sensitive Trp53 heterozygous mice. Radiat Res 159:320–327

Miura Y, Abe K, Urano S, Furuse T, Noda Y, Tatsumi K, Suzuki S (2002) Adaptive response and influence of aging: effects of low-dose irradiation on cell growth of cultured glial cells. Int J Radiat Biol 78:911–921

Mothersill C, Seymour CB (1997) Medium from irradiated human epithelial cells but not human fibroblasts reduces the clonogenic survival of unirradiated cells. Int J Radiat Biol 71:421–427

Nagasawa H, Little JB (1992) Induction of sister chromatid exchanges by extremely low doses of alpha-particles. Cancer Res 52:6394–6396

Norimura T, Nomoto S, Katsuki M et al (1996) p 53 dependent apoptosis suppresses radiation teratogenesis. Nature Med 2:577–580

Ohyama H, Yamada T (1998) Radiation-induced apoptosis: a review. In: Yamada T, Hashimoto Y (eds) Apoptosis, its roles and mechanisms. Business Center for Academic Societies, Japan, Tokyo, pp 141–186

Pollycove M, Feinendegen LE (2001) Biologic response to low doses of ionizing radiation: detriment versus hormesis, part 2. Dose responses of organisms. J Nucl Med 42:26N–32N

Pollycove M, Feinendegen LE (2003) Radiation-induced versus endogenous DNA damage: possible effect of inducible protective responses in mitigating endogenous damage. Hum Exp Toxicol (in press)

Potten CS (1977) Extreme sensitivity of some intestinal crypt cells to X and γ irradiation. Nature 269:518–521

Ramana CV, Boldogh I, Izumi T et al (1998) Activation of apurinic/apyrimidinic endonuclease in human cells by reactive oxygen species and its correlation with their adaptive response to genotoxicity of free radicals. Proc Natl Acad Sci USA 95:5061–5066

Redpath JL, Antoniono RJ (1998) Introduction of an adaptive response against spontaneous neoplastic transformation in vitro by low-dose gamma radiation. Radiat Res 14:517–520

Sakamoto K, Myojin M, Hosoi Y et al (1997) Fundamental and clinical studies on cancer control with total or upper half body irradiation. J Jpn Soc Ther Radiol Oncol 9:161–175

Sawant SG, Randers-Pehrson G, Geard CR et al (2001a) The bystander effect in radiation oncogenesis I. Transformation in C3H T1/2 cells in vitro can be initiated in the unirradiated neighbors of irradiated cells. Radiat Res 155:397–401

Sawant SG, Randers-Pehrson G, Metting NF et al (2001b) Adaptive response and the bystander effect induced in C3H 10T1/2 cells in culture. Radiat Res 156:177–180

Sen K, Sies H, Baeurle P (eds) (2000) Redox regulation of gene expression. Academic Press, San Diego

Shadley JD, Wiencke JK (1989) Induction of the adaptive response by X-rays is dependent on radiation intensity. Int J Radiat Biol 56:107–118

Short SC, Woodcock M, Marples B, Joiner MC (2003) Effects of cell cycle phase on low-dose hyper-radiosensitivity. Int J Radiat Biol 79:99–105

Stadtman ER, Berlett BS (1998) Reactive oxygen-mediated protein oxidation in aging and disease. Drug Metab Res 30:225–243

Sugahara T, Sagan LA, Aoyama T (eds) (1992) Low-dose irradiation and biological defense mechanisms. Excerpta Medica, Amsterdam

Shu-Zheng L, Yin-Chun Z, Ying M, Xu S, Jian-Xiang L (1996) Thymocyte apoptosis in response to low-dose radiation. Mutation Res 358:185–191

Suzuki K, Kodama S, Watanabe M (1998) Suppressive effect of low-dose preirradiation on genetic instability induced by X rays in normal human embryonic cells. Radiat Res 150:656–662

Tanooka H (2001) Threshold dose-response in radiation carcinogenesis: an approach from chronic β-irradiation experiments and a review of non-tumor doses. Int J Rad Biol 77:541–551

UNSCEAR (1994) Sources and effects of ionizing radiation, Annex B, Adaptive responses to radiation in cells and organisms. United Nations, New York

Vilenchik MM, Alfred G, Knudson AG Jr (2000) Inverse radiation dose-rate effects on somatic and germ-line mutations and DNA damage rates. Proc Natl Acad Sci USA 97:5381–5386

Wallace SS (1998) Enzymatic processing of radiation-induced free radical damage in DNA. Radiat Res 150 [Suppl]:60–79

Ward JF (1988) DNA damage produced by ionizing radiation in mammalian cells: Identities, mechanisms of formation, and reparability. Prog Nucleic Acid Res Mol Biol 35:95–125

Wolff S (1998) The adaptive response in radiobiology: evolving insights and implications. Environ Health Perspect 106:277–283

Wolff S, Afzal V, Wienke JK et al (1988) Human lymphocytes exposed to low doses of ionizing radiations become refractory to high doses of radiation as well as to chemical mutagens that induce double-strand breaks in DNA. Int J Radiat Biol 53:39–49

Wood RD, Mitchell M, Sgouros J, Lindahl T (2001) Human DNA repair genes. Science 291:1284–1289

Yamada T, Hashimoto Y (eds) (1998) Apoptosis, its roles and mechanisms. Business Center for Academic Societies, Japan, Tokyo

Yamamoto O, Seyama T, Jo T et al (1995) Oral administration of tritiated water (THO) in mouse II. Tumours development. Int J Radiat Biol 68:47–54

Yamamoto O, Seyama T, Ito A et al (1998) Oral administration of tritiated water (THO) in mouse III. Low dose-rate irradiation and threshold dose-rate for radiation risk. Int J Radiat Biol 73:535–541

Yamaoka K (1991) Increased SOD activities and decreased lipid peroxide in rat organs induced by low X-irradiation. Free Radic Biol Med 11:3–7

Yamaoka K, Edamatsu R, Mori A (1992) Effects of low dose x-ray irradiation on old rats – SOD activity, lipid peroxide level, and membrane fluidity. In: Sugahara T, Sagan LA, Aoyama T (eds) Low-dose irradiation and biological defense mechanisms. Excerpta Medica, Amsterdam, London New York Tokyo, pp 419–422

Subject Index